INSTRUCTOR'S SOLUTIONS MANUAL

EMMETT M. LARSON

Brevard Community College

MATHEMATICAL IDEAS
TENTH EDITION

MATHEMATICAL IDEAS
EXPANDED TENTH EDITION

CHARLES D. MILLER

VERN E. HEEREN
American River College

JOHN HORNSBY
University of New Orleans

PEARSON

Addison Wesley

Boston   San Francisco   New York
London   Toronto   Sydney   Tokyo   Singapore   Madrid
Mexico City   Munich   Paris   Cape Town   Hong Kong   Montreal

Reproduced by Pearson Addison-Wesley from electronic files supplied by the author.

Copyright © 2004 Pearson Education, Inc.
Publishing as Pearson Addison-Wesley, 75 Arlington Street, Boston, MA 02116

All rights reserved. No part of this publication may be reproduced, stored in a retrieval system, or transmitted, in any form or by any means, electronic, mechanical, photocopying, recording, or otherwise, without the prior written permission of the publisher. Printed in the United States of America.

ISBN   0-321-17027-X

2 3 4 5 6 VHG 06 05 04

Contents

Chapter 1   The Art of Problem Solving
1.1 Solving Problems by Inductive Reasoning...1
1.2 An Application of Inductive Reasoning: Number Patterns...5
1.3 Strategies for Problem Solving...9
1.4 Calculating, Estimating, and Reading Graphs..15
Chapter 1 Test..17

Chapter 2   The Basic Concepts of Set Theory
2.1 Symbols and Terminology..19
2.2 Venn Diagrams and Subsets...21
2.3 Set Operations and Cartesian Products..23
2.4 Cardinal Numbers and Surveys..30
2.5 Infinite Sets and Their Cardinalities...37
Chapter 2 Test..39

Chapter 3   Introduction to Logic
3.1 Statements and Quantifiers...41
3.2 Truth Tables and Equivalent Statements..43
3.3 The Conditional and Circuits..48
3.4 More on the Conditional...55
3.5 Analyzing Arguments with Euler Diagrams..57
Extension: Logic Puzzles..61
3.6 Analyzing Arguments with Truth Tables...64
Chapter 3 Test..72

Chapter 4   Numeration and Mathematical Systems
4.1 Historical Numeration Systems..75
4.2 Arithmetic in the Hindu-Arabic System...78
4.3 Conversion Between Number Bases..84
4.4 Finite Mathematical Systems..90
4.5 Groups...92
Chapter 4 Test..94

Chapter 5   Number Theory
5.1 Prime and Composite Numbers..96
5.2 Selected Topics from Number Theory...100
5.3 Greatest Common Factor and Least Common Multiple..102
5.4 Clock Arithmetic and Modular Systems..107
5.5 The Fibonacci Sequence and the Golden Ratio...114
Extension: Magic Squares..118
Chapter 5 Test..123

Chapter 6   The Real Numbers and Their Representations
6.1 Real Numbers, Order, and Absolute Value..125
6.2 Operations, Properties, and Applications of Real Numbers..................................126
6.3 Rational Numbers and Decimal Representation..129
6.4 Irrational Numbers and Decimal Representation..135
6.5 Applications of Decimals and Percents...139
Extension: Complex Numbers...142
Chapter 6 Test..144

Contents

Chapter 7   The Basic Concepts of Algebra
7.1 Linear Equations..146
7.2 Applications of Linear Equations...153
7.3 Ratio, Proportion, and Variation...161
7.4 Linear Inequalities..168
7.5 Properties of Exponents and Scientific Notation..175
7.6 Polynomials and Factoring...179
7.7 Quadratic Equations and Applications...184
Chapter 7 Test..196

Chapter 8   Graphs, Functions, and Systems of Equations and Inequalities
8.1 The Rectangular Coordinate System and Circles...200
8.2 Lines and Their Slopes...205
8.3 Equations of Lines and Linear Models...216
8.4 An Introduction to Functions: Linear Functions, Applications, and Models.......................225
8.5 Quadratic Functions, Applications, and Models..229
8.6 Exponential and Logarithmic Functions, Applications, and Models...................................236
8.7 Systems of Equations and Applications...242
Extension: Using Matrix Row Operations to Solve Systems...264
8.8 Linear Inequalities, Systems, and Linear Programming..270
Chapter 8 Test..279

Chapter 9   Geometry
9.1 Points, Lines, Planes, and Angles..284
9.2 Curves, Polygons, and Circles..287
Extension: Geometric Constructions..289
9.3 Perimeter, Area, and Circumference..289
9.4 The Geometry of Triangles: Congruence, Similarity, and the Pythagorean Theorem.........296
9.5 Space Figures, Volume, and Surface Area...305
9.6 Transformational Geometry...310
9.7 Non-Euclidean Geometry, Topology, and Networks...314
9.8 Chaos and Fractal Geometry..316
Chapter 9 Test..318

Chapter 10   Trigonometry
10.1 Angles and Their Measures..321
10.2 Trigonometric Functions of Angles...324
10.3 Trigonometric Identities...327
10.4 Right Triangles and Function Values...332
10.5 Applications of Right Triangles...341
10.6 The Laws of Sines and Cosines; Area Formulas..349
10.7 The Unit Circle and Graphs..359
Chapter 10 Test..363

Chapter 11   Counting Methods
11.1 Counting by Systematic Listing...367
11.2 Using the Fundamental Counting Principle...377
11.3 Using Permutations and Combinations..381
11.4 Using Pascal's Triangle and the Binomial Theorem..387
11.5 Counting Problems Involving "Not" and "Or"..390

Contents

Chapter 11 Test......394

Chapter 12 Probability
12.1 Basic Concepts......398
12.2 Events Involving "Not" and "Or"...... 405
12.3 Events Involving "And"...... 410
12.4 Binomial Probability...... 418
12.5 Expected Value...... 424
12.6 Estimating Probabilities by Simulation...... 428
Chapter 12 Test......429

Chapter 13 Statistics
13.1 Frequency Distributions and Graphs...... 433
13.2 Measures of Central Tendency...... 437
13.3 Measures of Dispersion...... 442
13.4 Measures of Position...... 448
13.5 The Normal Distribution...... 452
Extension: How to Lie with Statistics......458
13.6 Regression and Correlation...... 459
Chapter 13 Test......464

Chapter 14 Consumer Mathematics
14.1 The Time Value of Money...... 467
Extension: Annuities......476
14.2 Consumer Credit...... 478
14.3 Truth in Lending...... 487
14.4 Purchasing a House...... 493
14.5 Investing...... 499
Chapter 14 Test......505

Chapter 15 Graph Theory
15.1 Basic Concepts...... 508
15.2 Euler Circuits...... 511
15.3 Hamilton Circuits...... 513
15.4 Trees and Minimum Spanning Trees...... 519
Chapter 15 Test......525

Chapter 16 Voting and Apportionment
16.1 The Possibilities of Voting...... 527
16.2 The Impossibilities of Voting...... 544
16.3 The Possibilities of Apportionment...... 562
16.4 The Impossibilities of Apportionment...... 570
Chapter 16 Test......587

Appendix The Metric System......595

1.1 EXERCISES

1. This is an example of a deductive argument because a specific conclusion, "you can expect it to be ready in four days" is drawn from the two given premises.

2. This is an example of a deductive argument because the specific conclusion, "you will feel better," is drawn from the two given premises.

3. This is an example of inductive reasoning because you are reasoning from a specific pattern to the conclusion that "It will also rain tomorrow".

4. This is an example of inductive reasoning because you are reasoning from the specific pattern of all boys (so far) to the conclusion that the next child will also be a boy.

5. This represents deductive reasoning since you are moving from a general rule (addition) to a specific result (the sum of 95 and 20).

6. This represents deductive reasoning since you are moving from a general rule (subtraction property of equality) to a specific result (the subtraction of 12 from both sides).

7. This is a deductive argument where you are reasoning from the two given premises. The first, "If you build it, they will come" is a general statement and the conclusion, "they will come" is specific.

8. This is deductive reasoning because (as in Exercise 7) you are reasoning from the given premises, the first being general, to a specific conclusion "Socrates is mortal."

9. This is an example of inductive reasoning because you are reasoning from a specific pattern of all previous attendees to a conclusion that the next one "I" will also be accepted into medical school.

10. This is an example of inductive reasoning because you are reasoning from a specific historical pattern where the summer flowers have alternated their colorings from summer to summer to a specific conclusion about the color of this summer's flowers.

11. This is an example of inductive reasoning because you are reasoning from a specific pattern to a generalization as to what is the next element in the sequence.

12. This is an example of inductive reasoning because you are reasoning from a specific pattern to a generalization as to what will happen with her next release.

13. Writing exercise

14. Writing exercise; answers will vary.

15. Each number in the list is obtained by adding 3 to the previous number. The most probable next term is $18 + 3 = 21$.

16. The probable next term is 38 because each term in the list (after the first) can be obtained by adding 5 to the previous term.

17. Each number in the list is obtained by multiplying the previous number by 4. The most probable next term is $4 \times 768 = 3072$.

18. The probable next term is 1 because each term following the first can be obtained by taking one half of the previous term (or dividing the preceding term by 2).

19. Each number is a multiple of 3. Beginning with the third term, each number in the sequence is the sum of the two previous terms.
$$3 = 3 \times 1$$
$$6 = 3 \times 2$$
$$9 = (3 + 6)$$
$$15 = (6 + 9)$$
$$24 = (9 + 15)$$
$$39 = (15 + 24)$$

The most probable next term is
$$63 = (24 + 39).$$

20. 11/13 is the probable next term. The apparent pattern is to add 2 to both numerator and denominator.

21. The numerators and denominators are consecutive counting numbers. The probable next term is 11/12.

22. The next term is probably 36, which is 6^2. Notice that the other numbers can be computed as $1^2, 2^2, 3^2, 4^2, 5^2$.

23. The most probable next term is $6^3 = 216$. Observe the sequence:
$$1 = 1^3$$
$$8 = 2^3$$
$$27 = 3^3$$
$$64 = 4^3$$
$$125 = 5^3.$$
This sequence is made up of the cubes of each counting number.

24. The next term is probably 56. Note that each term (after the first) may be computed by adding successively 4, 6, 8, 10, and 12 to each preceding term. Thus, it follows that a probable next term would be
$$42 + 14 = 56.$$

25. The probable next term is 52. Note that each term (after the first) may be computed by adding successively 5, 7, 9, and 11 to each preceding term. Thus, it follows that a probable next term would be
$$39 + 13 = 52.$$

26. The probable next term is -7. Observe that every other integer is a successive odd integer and negative, or a successive even and positive.

27. The probable next term is 5 since the sequence of numbers seems to add one more 5 each time the 5's precede the number 3.

28. The probable next term is 2 since the sequence of numbers seems to add one more 2 each time the 2's follow the number 8.

29. There are many possibilities. One such list is 10, 20, 30, 40, 50, ...

30. There are many possibilities. One such list is 1, −2, 3, −4, 5, −6, 7, −8, ...

31.
$$(9 \times 9) + 7 = 88$$
$$(98 \times 9) + 6 = 888$$
$$(987 \times 9) + 5 = 8888$$
$$(9876 \times 9) + 4 = 88,888$$

Observe that on the left, the pattern suggests that the digit 5 will appended to the first number. Thus, we get $(98,765 \times 9)$ which is added to 3. On the right, the pattern suggests appending another digit 8 to obtain 888,888. Therefore,

$$(98,765 \times 9) + 3 = 888,885 + 3 = 888,888.$$

By computation, the conjecture is verified.

32.
$$(1 \times 9) + 2 = 11$$
$$(12 \times 9) + 3 = 111$$
$$(123 \times 9) + 4 = 1111$$
$$(1234 \times 9) + 5 = 11,111$$

Observe that on the left, the pattern suggests that the digit 5 will be appended to the first number. Thus, $(12,345 \times 9) + 6 = 111,111$ which can be verified using a calculator to compute the left side of the equation.

33.
$$3367 \times 3 = 10,101$$
$$3367 \times 6 = 20,202$$
$$3367 \times 9 = 30,303$$
$$3367 \times 12 = 40,404$$

Observe that on the left, the pattern suggests that 3367 will be multiplied by the next multiple of 3, which is 15. On the right, the pattern suggests the result 50,505. The pattern suggests the following equation:

$$3367 \times 15 = 50,505.$$

Multiply 3367×15 to verify the conjecture.

34.
$$15,873 \times 7 = 111,111$$
$$15,873 \times 14 = 222,222$$
$$15,873 \times 21 = 333,333$$
$$15,873 \times 28 = 444,444$$

Observe that on the left, the pattern suggests that 15,873 will be multiplied by the next multiple of 7, which is 35. On the right, the pattern suggests the result 555,555. The pattern suggests the following equation:

$$15,873 \times 35 = 555,555.$$

Multiply $15,873 \times 35$ to verify the conjecture.

35.
$$34 \times 34 = 1156$$
$$334 \times 334 = 111,556$$
$$3334 \times 3334 = 11,115,556$$

The pattern suggests the following equation:

$$33,334 \times 33,334 = 1,111,155,556.$$

Multiply $33,334 \times 33,334$ to verify the conjecture.

36.
$$11 \times 11 = 121$$
$$111 \times 111 = 12,321$$
$$1111 \times 1111 = 1,234,321$$

The pattern suggests the following equation:

$$11,111 \times 11,111 = 123,454,321.$$

Multiply $11,111 \times 11,111$ to verify the conjecture.

37.
$$3 = \frac{3(2)}{2}$$
$$3 + 6 = \frac{6(3)}{2}$$
$$3 + 6 + 9 = \frac{9(4)}{2}$$
$$3 + 6 + 9 + 12 = \frac{12(5)}{2}$$

The pattern suggests the following equation:

$$3 + 6 + 9 + 12 + 15 = \frac{15(6)}{2}.$$

Since both the left and right sides equal 45, the conjecture is verified.

38.
$$2 = 4 - 2$$
$$2 + 4 = 8 - 2$$
$$2 + 4 + 8 = 16 - 2$$
$$2 + 4 + 8 + 16 = 32 - 2$$

This pattern suggests the following equation:

$$2 + 4 + 8 + 16 + 32 = 64 - 2.$$

Since both the left and right sides equal 62, the conjecture is verified.

39.
$$5(6) = 6(6-1)$$
$$5(6) + 5(36) = 6(36-1)$$
$$5(6) + 5(36) + 5(216) = 6(216-1)$$
$$5(6) + 5(36) + 5(216) + 5(1296) = 6(1296-1)$$

Observe that the last equation may be written as:
$$5(6^1) + 5(6^2) + 5(6^3) + 5(6^4) = 6(6^4 - 1).$$

Thus, the next equation would likely be:
$$5(6) + 5(36) + 5(216) + 5(1296) + 5(6^5) = 6(6^5 - 1) \text{ or,}$$
$$5(6) + 5(36) + 5(216) + 5(1296) + 5(7776) = 6(7776 - 1).$$

40.
$$3 = \frac{3(3-1)}{2}$$
$$3 + 9 = \frac{3(9-1)}{2}$$
$$3 + 9 + 27 = \frac{3(27-1)}{2}$$
$$3 + 9 + 27 + 81 = \frac{3(81-1)}{2}$$

Observe that the last equation may be written as:
$$3^1 + 3^2 + 3^3 + 3^4 = \frac{3(3^4 - 1)}{2}.$$

Thus, the next equation would likely be:
$$3^1 + 3^2 + 3^3 + 3^4 + 3^5 = \frac{3(3^5 - 1)}{2}, \text{ or}$$
$$3 + 9 + 27 + 81 + 243 = \frac{3(243-1)}{2}$$
$$363 = \frac{3(242)}{2}$$
$$= \frac{726}{2}$$
$$= 363.$$

Thus, the conjecture is verified.

41.
$$\frac{1}{2} = 1 - \frac{1}{2}$$
$$\frac{1}{2} + \frac{1}{4} = 1 - \frac{1}{4}$$
$$\frac{1}{2} + \frac{1}{4} + \frac{1}{8} = 1 - \frac{1}{8}$$
$$\frac{1}{2} + \frac{1}{4} + \frac{1}{8} + \frac{1}{16} = 1 - \frac{1}{16}$$

Observe that the last equation may be written as
$$\frac{1}{2^1} + \frac{1}{2^2} + \frac{1}{2^3} + \frac{1}{2^4} = 1 - \frac{1}{2^4}.$$

The next equation would be
$$\frac{1}{2^1} + \frac{1}{2^2} + \frac{1}{2^3} + \frac{1}{2^4} + \frac{1}{2^5} = 1 - \frac{1}{2^5}, \text{ or}$$
$$\frac{1}{2} + \frac{1}{4} + \frac{1}{8} + \frac{1}{16} + \frac{1}{32} = 1 - \frac{1}{32}.$$

Using the common denominator 32 for each fraction, the left and right side add (in each case) to 31/32. The conjecture is, therefore, verified.

42.
$$\frac{1}{1 \cdot 2} = \frac{1}{2}$$
$$\frac{1}{1 \cdot 2} + \frac{1}{2 \cdot 3} = \frac{2}{3}$$
$$\frac{1}{1 \cdot 2} + \frac{1}{2 \cdot 3} + \frac{1}{3 \cdot 4} = \frac{3}{4}$$
$$\frac{1}{1 \cdot 2} + \frac{1}{2 \cdot 3} + \frac{1}{3 \cdot 4} + \frac{1}{4 \cdot 5} = \frac{4}{5}$$

This pattern suggests the following equation:
$$\frac{1}{1 \cdot 2} + \frac{1}{2 \cdot 3} + \frac{1}{3 \cdot 4} + \frac{1}{4 \cdot 5} + \frac{1}{5 \cdot 6} = \frac{5}{6}.$$

This is verified by:
$$\frac{1}{2} + \frac{1}{6} + \frac{1}{12} + \frac{1}{20} + \frac{1}{30} =$$
$$\frac{30}{60} + \frac{10}{60} + \frac{5}{60} + \frac{3}{60} + \frac{2}{60} =$$
$$\frac{30 + 10 + 5 + 3 + 2}{60} =$$
$$\frac{50}{60} = \frac{5}{6}.$$

43. $1 + 2 + 3 \ldots + 200$
Pairing and adjoining the first term to the last term, the second term to the second to last term, etc., we have:
$$1 + 200 = 201, \ 2 + 199 = 201, \ 3 + 198 = 201, \ \ldots.$$
There are 100 of these sums. Therefore,
$$100 \times 201 = 20,100.$$

44. $1 + 2 + 3 + \ldots + 400$
Pairing and adjoining the first term to the last term, the second term to the second to last term, etc., we have:
$$1 + 400 = 401, \ 2 + 399 = 401, \ 4 + 398 = 401, \ \ldots.$$
There are 200 of these sums. Therefore,
$$200 \times 401 = 80,200.$$

45. $1 + 2 + 3 + \ldots + 800$
Pairing and adjoining the first term to the last term, the second term to the second to last term, etc., we have:
$$1 + 800 = 801, \ 2 + 799 = 801, \ 4 + 798 = 801, \ \ldots.$$
There are 400 of these sums. Therefore,
$$400 \times 801 = 320,400.$$

46. $1 + 2 + 3 + \ldots + 2000$.
    Observe that there would be 1000 pairs
    (e.g., $1 + 2000 = 2001, 2 + 1999 = 2001, \ldots$).
    Thus,
    $$1000 \times 2001 = 2,001,000.$$

47. $1 + 2 + 3 + \ldots + 175$.
    Note that there are an odd number of terms. So consider omitting, for the moment, the last term and take $1 + 174 = 175, 2 + 173 = 175, 3 + 172 = 175$, etc. There are $(174/2) = 87$ of these pairs in addition to the last term. Thus,
    $$(87 \times 175) + 175 = 15,400.$$

48. Writing exercise

49. $2 + 4 + 6 + \ldots + 100 = 2(1 + 2 + 3 + \ldots + 50)$
    $= 2[25(1 + 50)]$
    $= 2(1275)$
    $= 2550.$

50. $4 + 8 + 12 + \ldots + 200 = 4(1 + 2 + 3 + \ldots + 50)$
    $= 4[25(1 + 50)]$
    $= 4(1275)$
    $= 5100.$

51. The pattern in Figures (a) − (d) shows a clockwise shading of the middle squares. In Figures (e) and (f), the pattern of the shading of the outer square corners appears in the clockwise directions. By inductive reasoning, we conclude that the lower right-hand corner should be shaded next. The next figure in the sequence is shown below.

52. To find any term, choose the term directly above it and add to it the two preceding terms. If there are fewer than two terms, add as many as there are. Thus, the next row would be predicted to be

    0 2 10 30 60 90 102 90 60.

53. These are the number of chimes a clock rings, starting with 12 o'clock, if the clock rings the number of hours on the hour and 1 chime on the half-hour. The next most probable number is the number of chimes at 3:30, which is 1.

54. E, which represents the first letter of eleven. (One, Two, Three, and so on.)

55. (a) Here are three examples.
    $$\begin{array}{ccc} 623 & 841 & 584 \\ -\,326 & -\,148 & -\,485 \\ \hline 297 & 693 & 99 \end{array}$$
    In each result, the middle digit is always 9, and the sum of the first and third digits is always 9 (considering 0 as the first digit if the difference has only two digits).

    (b) Writing exercise

56. Try several numbers. The last result will always be half of the number added in step (b).

57. $142,857 \times 1 = 142,857$
    $142,857 \times 2 = 285,714$
    $142,857 \times 3 = 428,571$
    $142,857 \times 4 = 571,428$
    $142,857 \times 5 = 714,285$
    $142,857 \times 6 = 857,142$

    Each result consists of the same six digits, but in a different order.
    $$142,857 \times 7 = 999,999$$
    Thus, the pattern doesn't continue.

58. $12,345,679 \times 9 = 111,111,111$
    $12,345,679 \times 18 = 222,222,222$
    $12,345,679 \times 27 = 333,333,333$

    Observe that the multiples 9, 18, and 27 are in turn given by $9 = 9 \times 1, 18 = 9 \times 2$, and $27 = 9 \times 3$. The pattern indicates multiplying $12,345,679$ by 72, which is given by (9×8), obtaining the answer $888,888,888$.

59. Count the chords and record the results.

No. of points	No. of chords	No. of chords added
2	1	1
3	3	2
4	6	3
5	10	4
6	15	5

    By the pattern, 6 more chords would be added for a total of 21 chords.

60. Try several numbers. For example, suppose the age chosen is 50 years and the amount of change in your pocket is 35 cents. Then the numbers generated are
    $$4 \times 50 = 200,$$
    $$200 + 10 = 210,$$
    $$25 \times 210 = 5250$$
    $$5250 - 365 = 4885,$$
    $$4885 + 35 = 4920,$$
    $$4920 + 115 = 5035.$$

    The final number, 5035, provides the information.

61. Writing exercise

62. Writing exercise

1.2 EXERCISES

1. 1 4 11 22 37 56 <u>79</u>
      3 7 11 15 19 <u>23</u>
         4 4 4 4 (4)

    Each line represents the difference of the two numbers above it. The number 23 is found from adding the predicted difference, (4), in line three to 19 in line 2. And 79 is found by adding 23, in line two, to 56 in line one. Thus, our next term in the sequence is 79.

2. 3 14 31 54 83 118 <u>159</u>
      11 17 23 29 35 <u>41</u>
         6 6 6 6 (6)

    The number 41 is found from adding a predicted difference, (6), to 35. And 159 is found from adding 41 to 118. Thus, our next term in the sequence is 159.

3. 6 20 50 102 182 296 <u>450</u>
      14 30 52 80 114 <u>154</u>
         16 22 28 34 <u>40</u>
            6 6 6 (6)

    Thus, our next term in the sequence is $154 + 296 = 450$.

4. 1 11 35 79 149 251 <u>391</u>
      10 24 44 70 102 <u>140</u>
         14 20 26 32 <u>38</u>
            6 6 6 (6)

    Thus, our next term in the sequence is
    $$140 + 251 = 391.$$

5. 0 12 72 240 600 1260 2352 <u>4032</u>
      12 60 168 360 660 1092 <u>1680</u>
         48 108 192 300 432 <u>588</u>
            60 84 108 132 <u>156</u>
               24 24 24 (24)

    Thus, our next term in the sequence is
    $$1680 + 2352 = 4032.$$

6.   2 57 220 575 1230 2317 <u>3992</u>
        55 163 355 655 1087 <u>1675</u>
           108 192 300 432 <u>588</u>
              84 108 132 <u>156</u>
                 24 24 (24)

    Thus, our next term in the sequence is
    $$1675 + 2317 = 3992.$$

7. 5 34 243 1022 3121 7770 16799 <u>32758</u>
      29 209 779 2099 4649 9029 <u>15959</u>
         180 570 1320 2550 4380 <u>6930</u>
            390 750 1230 1830 <u>2550</u>
               360 480 600 <u>720</u>
                  120 120 (120)

    Thus, our next term in the sequence is
    $$15959 + 16799 = 32{,}758.$$

8. 3 19 165 771 2503 6483 14409 <u>28675</u>
      16 146 606 1732 3980 7926 <u>14266</u>
         130 460 1126 2248 3946 <u>6340</u>
            330 666 1122 1698 <u>2394</u>
               336 456 576 <u>696</u>
                  120 120 (120)

    Thus, our next term in the sequence is
    $$14{,}266 + 14{,}409 = 28{,}675.$$

9. 1 2 4 8 16 31 (<u>57</u>) 99
      1 2 4 8 15 <u>26</u> 42
         1 2 4 7 11 <u>16</u>
            1 2 3 4 <u>5</u>
               1 1 1 (1)

    The next term of the sequence is 57. Following this pattern, we predict that the number of regions determined by 8 points is 99. Use $n = 8$ in the formula
    $$\frac{n^4 - 6n^3 + 23n^2 - 18n + 24}{24}.$$
    $$\frac{8^4 - 6 \times 8^3 + 23 \times 8^2 - 18 \times 8 + 24}{24}$$
    $$= \frac{4096 - 3{,}072 + 1472 - 144 + 24}{24}$$
    $$= \frac{2376}{24}$$
    $$= 99.$$

    Thus, the result agrees with our prediction.

10. (a) $1^2 + 3(1) + 1 = 5,$
        $2^2 + 3(2) + 1 = 11,$
        $3^2 + 3(3) + 1 = 19,$ and
        $4^2 + 3(4) + 1 = 29.$

    (b) 5 11 19 29 <u>41</u>
          6 8 10 <u>12</u>
            2 2 (2),
    where $12 + 29 = 41.$

    (c) $5^2 + 3(5) + 1 = 25 + 15 + 1 = 41.$
    The results agree.

11. $(1 \times 9) - 1 = 8$
$(21 \times 9) - 1 = 188$
$(321 \times 9) - 1 = 2888$

By the pattern, the next equation is
$$(4321 \times 9) - 1 = 38,888.$$
To verify, calculate left side and compare,
$$38,889 - 1 = 38,888.$$

12. $(1 \times 8) + 1 = 9$
$(12 \times 8) + 2 = 98$
$(123 \times 8) + 3 = 987$

By the pattern, the next equation is
$$(1234 \times 8) + 4 = 9876.$$
To verify,
$$9872 + 4 = 9876.$$

13. $999,999 \times 2 = 1,999,998$
$999,999 \times 3 = 2,999,997$

By the pattern, the next equation is
$$999,999 \times 4 = 3,999,996.$$
To verify, multiply left side to get,
$$3,999,996 = 3,999,996.$$

14. $101 \times 101 = 10,201$
$10,101 \times 10,101 = 102,030,201.$

By the pattern, the next equation is
$$1,010,101 \times 1,010,101 = 1,020,304,030,201.$$
To verify,
$$1,020,304,030,201 = 1,020,304,030,201.$$

15. $3^2 - 1^2 = 2^3$
$6^2 - 3^2 = 3^2$
$10^2 - 6^2 = 4^3$
$15^2 - 10^2 = 5^3$

Following this pattern, we see that the next equation will start with 21^2 since $15 + 6 = 21$. This equation will be $21^2 - 15^2 = 6^3$. The left side is $441 - 225 = 216$. The right side also equals 216.

16. $1 = 1^2$
$1 + 2 + 1 = 2^2$
$1 + 2 + 3 + 2 + 1 = 3^2$
$1 + 2 + 3 + 4 + 3 + 2 + 1 = 4^2$

By the pattern, the next equation is
$$1 + 2 + 3 + 4 + 5 + 4 + 3 + 2 + 1 = 5^2.$$
To verify, $25 = 25$.

17. $2^2 - 1^2 = 2 + 1$
$3^2 - 2^2 = 3 + 2$
$4^2 - 3^2 = 4 + 3$

Following this pattern, we see that the next equation will be
$$5^2 - 4^2 = 5 + 4.$$
To verify, the left side is $25 - 16 = 9$. The right side also equals 9.

18. $1^2 + 1 = 2^2 - 2$
$2^2 + 2 = 3^2 - 3$
$3^2 + 3 = 4^2 - 4$

By the pattern, the next equation is
$$4^2 + 4 = 5^2 - 5.$$
To verify, evaluate each side of the equation to get
$$16 + 4 = 25 - 5$$
$$20 = 20.$$

19. $1 = 1 \times 1$
$1 + 5 = 2 \times 3$
$1 + 5 + 9 = 3 \times 5$

The last term on the left side is 4 more than the previous last term. The first factor on the right side is the next counting number; the second factor is the next odd number. Thus, the probable next equation is
$$1 + 5 + 9 + 13 = 4 \times 7.$$
To verify, calculate both sides to arrive at $28 = 28$.

20. $1 + 2 = 3$
$4 + 5 + 6 = 7 + 8$
$9 + 10 + 11 + 12 = 13 + 14 + 15$

By the pattern, the next equation is
$$16 + 17 + 18 + 19 + 20 = 21 + 22 + 23 + 24.$$
To verify, calculate both sides to arrive at $90 = 90$.

21. $1 + 2 + 3 + \ldots + 300$
$$S = \frac{300(300+1)}{2} = \frac{90300}{2} = 45,150$$

22. $1 + 2 + 3 + \ldots + 500$
$$S = \frac{500(500+1)}{2} = \frac{250500}{2} = 125,250$$

23. $1 + 2 + 3 + \ldots + 675$

$$S = \frac{675(675 + 1)}{2} = \frac{456300}{2} = 228{,}150$$

24. $1 + 2 + 3 + \ldots + 825$

$$S = \frac{825(825 + 1)}{2} = \frac{681450}{2} = 340{,}725$$

25. $1 + 3 + 5 + 7 + \ldots + 101$
Note that

$$n = \frac{1 + 101}{2} = 51 \text{ terms, so that}$$

$$S = 51^2 = 2601.$$

26. $1 + 3 + 5 + 7 + \ldots + 49$
Note that

$$n = \frac{1 + 49}{2} = 25 \text{ terms, so that}$$

$$S = 25^2 = 625.$$

27. $1 + 3 + 5 + \ldots + 999$
Observe that

$$n = \frac{1 + 999}{2} = 500 \text{ terms, so that}$$

$$S = 500^2 = 250{,}000.$$

28. $1 + 3 + 5 + 7 + \ldots + 301$
Observe that

$$n = \frac{1 + 301}{2} = 151 \text{ terms, so that}$$

$$S = 151^2 = 22{,}801.$$

29. Since each term in the second series is twice that of the first series, we might expect the sum to be twice as large or

$$S = 2 \times \frac{n(n+1)}{2} = n(n+1).$$

30. Writing exercise

31. Writing exercise

32. Writing exercise

33.
Figurate Number	1st	2nd	3rd	4th	5th	6th	7th	8th
Triangular	1	3	6	10	15	21	28	36
Square	1	4	9	16	25	36	49	64
Pentagonal	1	5	12	22	35	51	70	92
Hexagonal	1	6	15	28	45	66	91	120
Heptagonal	1	7	18	34	55	81	112	148
Octagonal	1	8	21	40	65	96	133	176

34. The first five hexagonal numbers are

$$1 = 1$$
$$6 = 1 + 5$$
$$15 = 1 + 5 + 9$$
$$28 = 1 + 5 + 9 + 13$$
$$45 = 1 + 5 + 9 + 13 + 17.$$

35. $8(1) + 1 = 9 = 3^2$; $8(3) + 1 = 25 = 5^2$; $8(6) + 1 = 49 = 7^2$; $8(10) + 1 = 81 = 9^2$.

36. The triangular numbers are $1, 3, 6, 10, 15, \ldots$.

$$1 \div 3 = 0, \text{ remainder } 1$$
$$3 \div 3 = 1, \text{ remainder } 0$$
$$6 \div 3 = 2, \text{ remainder } 0$$
$$10 \div 3 = 3, \text{ remainder } 1$$
$$15 \div 3 = 5, \text{ remainder } 0$$
$$21 \div 3 = 7, \text{ remainder } 0$$

The pattern of remainders is $1, 0, 0, 1, 0, 0, \ldots$.

37. The square numbers are $1, 4, 9, 25, 36, \ldots$.

$$1 \div 4 = 0, \text{ remainder } 1$$
$$4 \div 4 = 1, \text{ remainder } 0$$
$$9 \div 4 = 2, \text{ remainder } 1$$
$$16 \div 4 = 4, \text{ remainder } 0$$
$$25 \div 4 = 6, \text{ remainder } 1$$
$$36 \div 4 = 9, \text{ remainder } 0$$

The pattern of remainders is $1, 0, 1, 0, 1, 0, \ldots$.

38. Creating the first ten, or so, pentagonal numbers, we have: $1, 5, 12, 22, 35, 51, 70, 92, 117, 145, \ldots$. Dividing each by 5 (n = 5) generates the following sequence of remainders: $1, 0, 2, 2, 0, 1, 0, 2, 2, 0, \ldots$. Creating the first twelve, or so, hexagonal numbers, we have: $1, 6, 15, 28, 45, 66, 91, 120, 153, 190, 231, 276, \ldots$. Dividing each by 6 (n = 6) generates the following sequence of remainders: $1, 0, 3, 4, 3, 0, 1, 0, 3, 4, 3, 0, \ldots$. That is, in each case, we have generated a repeating sequence of numbers.

39. The square number 25 may be represented by the sum of the two triangular numbers 10 and 15. The square number 36 may be represented by the sum of the two triangular numbers 15 and 21.

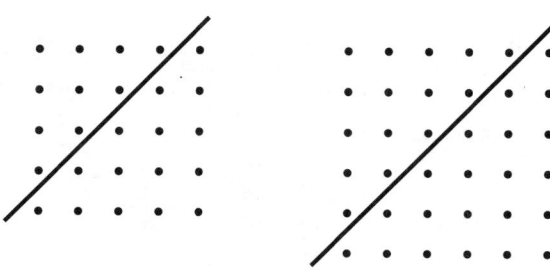

40. (a) For $n = 2$ to $n = 8$, the fractions

$$\frac{n\text{th } square \; number}{(n+1)\text{th } square \; number}$$

are: $\frac{2^2}{3^2} = \frac{4}{9}, \frac{3^2}{4^2} = \frac{9}{16},$
$\frac{4^2}{5^2} = \frac{16}{25}, \frac{5^2}{6^2} = \frac{25}{36},$
$\frac{6^2}{7^2} = \frac{36}{49}, \frac{7^2}{8^2} = \frac{49}{64},$ and
$\frac{8^2}{9^2} = \frac{64}{81}.$

(b) For $n = 2$ to $n = 8$, the fractions

$$\frac{n\text{th triangular number}}{(n+1)\text{th triangular number}}$$

are: $\frac{3}{6}, \frac{6}{10}, \frac{10}{15}, \frac{15}{21}, \frac{21}{28}, \frac{28}{36},$ and $\frac{36}{45}.$

(c) The fraction formed by two consecutive square numbers is always reduced to lowest terms, while the fraction formed by two consecutive triangular numbers is never reduced to lowest terms.

41.

n	2	3	4	5	6	7	8
A Square of n	4	9	16	25	36	49	64
B (Square of n) + n	6	12	20	30	42	56	72
C One-half of Row B entry	3	6	10	15	21	28	36
D (Row A entry) − n	2	6	12	20	30	42	56
E One-half of Row D entry	1	3	6	10	15	21	28

(a) The results are 3 (e.g., $\frac{4+2}{2}$), 6, 10, 15, 21, 28, 36; or triangular numbers.

(b) 1 (e.g., $\frac{2^2 - 2}{2}$), 3, 6, 10, 15, 21, 28, 36; or triangular numbers.

42. The smallest integer is $N = 6$. This would be the third triangular number and the second hexagonal number.

43. To find the sixteenth square number, use

$$S_n = n^2 \text{ with } n = 16.$$
$$S_n = 16^2 = 256.$$

44. To find the eleventh triangular number, use

$$T_n = \frac{n(n+1)}{2} \text{ with } n = 11.$$
$$T_{11} = \frac{11(12)}{2} = 66.$$

45. To find the ninth pentagonal number, use

$$P_n = \frac{n(3n-1)}{2} \text{ with } n = 9.$$
$$P_9 = \frac{9(26)}{2} = 117.$$

46. To find the seventh hexagonal number, use

$$H_n = \frac{n(4n-2)}{2} \text{ with } n = 7.$$
$$H_7 = \frac{7(26)}{2} = 91.$$

47. To find the tenth heptagonal number, use

$$Hp_n = \frac{n(5n-3)}{2} \text{ with } n = 10.$$
$$Hp_{10} = \frac{10(47)}{2} = 235.$$

48. To find the twelfth octagonal number, use

$$O_n = \frac{n(6n-4)}{2} \text{ with } n = 12.$$
$$O_{12} = \frac{12(68)}{2} = 408.$$

49. Since each coefficient in parentheses appears to step up by 1, we would predict:

$$N_n = \frac{n(7n-5)}{2}.$$
$$N_6 = \frac{6(37)}{2} = 111.$$

This verifies our prediction for $n = 6$.

50. Using Exercise 50 to find the tenth nonagonal number, we have

$$N_n = \frac{n(7n-5)}{2} \text{ with } n = 10.$$
$$N_{10} = \frac{10(65)}{2} = 325.$$

51. The triangular numbers are
1, 3, 6, 10, 15, 21, 28, 36, 45, ...
Adding consecutive triangular numbers, for example,
$1 + 3 = 4, 3 + 6 = 9, 6 + 10 = 16, \ldots,$
will give square numbers.

52. The triangular numbers are
1, 3, 6, 10, 15, 21, 28, 36, 45, ...
The squares of each of these numbers are
1, 9, 36, 100, 225, 441, 784, 1296, ...
Adding consecutive squares, for example,
$1 + 9 = 10, 9 + 36 = 45,$ etc.,
will give triangular numbers.

53. In each case, you get a perfect cube number. That is, if we take the 2nd and 3rd triangular numbers 3 and 6;
$$6^2 - 3^2 = 36 - 9 = 27.$$
which is the perfect cube number, 3^3.

54. $T_{n-1} = \dfrac{(n-1)(n)}{2}$ with $n = 3$.

   $T_2 = \dfrac{2(3)}{2} = 3.$

   Multiply by 3: $3 \times 3 = 9$.
   Add $n = 3$: $9 + 3 = 12$.
   The result is the nth pentagonal number.

1.3 EXERCISES

1. Choose a sock from the box labeled *red and green socks*. Since it is mislabeled, it contains only *red* socks or only *green* socks, determined by the sock you choose. If the sock is green, relabel this box *green socks*. Since the other two boxes were mislabeled, switch the remaining label to the other box and place the label that says *red and green* socks on the unlabeled box. No other choice guarantees a correct relabeling, since you can remove only one sock.

2. OHIO is another state such that each capatalized letter has vertical and horizontal symmetry.

3. The total number of dots on each die is $1+2+3+4+5+6 = 21$. Thus the top die has $(21 - \text{dots showing})$, unseen dots, or $21 - (1+2+3) = 21 - 6 = 15$. The middle die has $21 - (4+6) = 21 - 10 = 11$. The bottom die has $21 - (5+1) = 21 - 6 = 15$ dots not shown. The total is $15 + 11 + 15 = 41$ dots not shown. This is option D.

   Alternatively, since each die has 21 dots there a $21 \times 3 = 63$ total dots. Thus, there are $63 - 22 = 41$ unseen dots.

4. One strategy is to set up an equation as a model of the problem and solve. Let x = year of birth. Then $x + 2002 - (x + 10) - (x + 50) + (2002 - x) = 80$, where present age is found by $2002 - x$. Solving the equation we get:
$$\begin{aligned}
x + 2002 - (x+10) - (x+50) + (2002 - x) &= 80 \\
x + 2002 - x - 10 - x - 50 + 2002 - x &= 80 \\
-2x + 4004 - 60 &= 80 \\
-2x + 3944 &= 80 \\
-2x &= -3864 \\
x &= 1932.
\end{aligned}$$

   Thus, Mr. Green's current age is $2002 - 1932 = 70$ yr.

5. Visualize (or create unfolded box strip) with "1" on top, and folding "2", "3", and "4" around the middle. Option A satisfies this result.

6. Careful! Since you are the bus driver, the answer is *your* age.

7. One example of a solution follows.

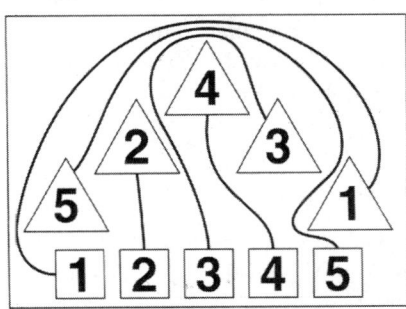

8. You must place the 0 in the bottom-left square. No other choice guarantees you a win.

9. By trial an error, the following arrangement will work:

   | 9 | 7 | 2 | 14 | 11 | 5 | 4 | 12 | 13 | 3 | 6 | 10 | 15 | 1 | 8 | (or reverse the order reading right to left).

10. One strategy is to organize a table such as the one which follows. Let x = Chris's current age.

	Current age	Past age	Elapsed no. of years
Pat	24	x	$24 - x$
Chris	x	$x - (24 - x)$	

    Since Chris's past age can be represented as
    $$x - (24 - x) = -24 + 2x = 2x - 24,$$
    and Pat's current age, 24, is twice that of Chris's past age, we have
    $$\begin{aligned}
    24 &= 2(2x - 24) \\
    24 &= 4x - 48 \\
    72 &= 4x \\
    18 &= x.
    \end{aligned}$$

    Thus, Chris's current age is 18 years.

11. Use trial and error. One possible solution is as follows.

12. Use trial and error. The following figure represents a solution where each region has a sum of 26.

    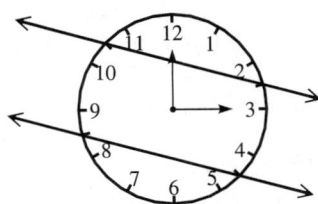

13. For units column assume that 1 is borrowed from the a digit. This suggests that $b = 9$, since $12 - 9 = 3$. To arrive at 7 in the tens column, we know that 8 must be subtracted form 15. Thus, $a = 6$ (remember that we borrowed one from that column, also). We borrowed one from the 7, as well, so that $c = 6 - 4 = 2$ in the hundreds column. Thus,

$$a + b + c = 6 + 9 + 2 = 17.$$

    This is represented by option "D."

14. Use a calculator to find the square root of each number. Only 329,476 has a square root, 524, without a decimal remainder. Thus,

$$524^2 = 329476.$$

15. This exercise can be solved algebraically. If we let

    D = the total distance of the trip, and
    x = the distance traveled while asleep,

    then

$$x + \frac{1}{2}x = \frac{1}{2}D$$
$$2x + x = D$$
$$3x = D$$
$$x = \frac{1}{3}D.$$

    Thus, the distance traveled while asleep is $\frac{1}{3}$ of the total distance traveled.

16. One strategy would be to begin counting the smallest squares (10). Then count the two-square rectangles slanted from lower left to upper right (6) followed by the number of three-square rectangles in the same direction (2). There is the same number count for the rectangles slanted lower right to upper left, so double these results Finally count the number of larger squares (5).

Number of small squares =	10
Number of two-square rectangles =	12
Number of three-square rectangles =	4
Number of larger squares =	5
Total number of rectangles =	31

17. Fill the big bucket. Pour into the small bucket. This leaves 4 gallons in the larger bucket. Empty the small bucket. Pour from the big bucket to fill up the small bucket. This leaves 1 gallon in the big bucket. Empty the small bucket. Pour 1 gallon from the big bucket to the small bucket. Fill up the big bucket. Pour into the small bucket. This leaves 5 gallons in the big bucket. Pour out the small bucket. This leaves exactly 5 gallons in the big bucket to take home. The above sequence is indicated by the following table.

Big bucket	7	4	4	1	1	0	7	5	5
Small bucket	0	3	0	3	0	1	1	3	0

18. This problem can be solved in many ways. Let's solve it in a way that shows the pattern. In order to see the pattern, we will not simplify the expressions that we generate for each day. Let n be the number of nuts on Monday morning.

Day	Number of Nuts Each Morning
M	n
T	$\frac{n}{2} + \frac{32}{2}$
W	$\frac{n}{2^2} + \frac{32}{2^2} + \frac{32}{2}$
T	$\frac{n}{2^3} + \frac{32}{2^3} + \frac{32}{2^2} + \frac{32}{2}$
F	$\frac{n}{2^4} + \frac{32}{2^4} + \frac{32}{2^3} + \frac{32}{2^2} + \frac{32}{2}$

    Simplifying this last expression, setting it equal to the remaining nuts on Saturday morning, and solving yields:

$$\frac{n}{16} + 2 + 4 + 8 + 16 = 38$$
$$\frac{n}{16} = 8$$
$$n = 128.$$

19. Count systematically.

15	1×1 rectangles
10	1×2 rectangles
5	1×3 rectangles
12	2×1 rectangles
8	2×2 rectangles
4	2×3 rectangles
9	3×1 rectangles
6	3×2 rectangles
3	3×3 rectangles
6	4×1 rectangles
4	4×2 rectangles
2	4×3 rectangles
3	5×1 rectangles
2	5×2 rectangles
1	5×3 rectangles
90	total rectangles

    This gives a total of 90 rectangles.

20. The following represents a solution found by trial and error.

		3	5	
7	1	8	2	
		4	6	

21. One strategy is to assume the car was driving near usual highway speed limits (55 − 75 mph). We begin by trying 55 mph. In two hours the car would have traveled 110 miles. Adding 110 miles to the odometer reading, 15051. We get $110 + 15051 = 16061$ miles, which is palindromic. Thus, the speed of the car was 55 mile per hour.

22. Since 12 fish = 30 coconuts, 1 fish = $\frac{30}{12} = \frac{5}{2}$ coconuts. Since 20 coconuts = 50 bananas, 1 coconut = $\frac{50}{20} = \frac{5}{2}$ bananas. Thus,

$$\begin{aligned} 1 \text{ hammock} &= 100 \text{ fish} \\ &= 100(1 \text{ fish}) \\ &= 100\left(\frac{5}{2} \text{ coconuts}\right) \\ &= 250 \text{ coconuts} \\ &= 250(1 \text{ coconut}) \\ &= 250\left(\frac{5}{2} \text{ bananas}\right) \\ &= 625 \text{ bananas}. \end{aligned}$$

23. Similar to Example 5 in the text, we might examine the units place and tens place for repetitive powers of 7 in order to explore possible patterns.

$7^1 =$	07	$7^5 =$	16,807
$7^2 =$	49	$7^6 =$	117,649
$7^3 =$	343	$7^7 =$	823,543
$7^4 =$	2401	$7^8 =$	5,764,801

   Since the final two digits cycle over four values, we might consider dividing the successive exponents by 4 and examining their remainders. (Note: We are using inductive reasoning when we assume that this pattern will continue and will apply when the exponent is 1997.) Dividing the exponent 1997 by 4, we get a remainder of 1. This is the same remainder we get when dividing the exponent 1 (on 7^1) and 5 (on 7^5). Thus, we expect that the last two digits for 7^{1997} would be 07 as well.

24. It will be darkest when the fewest digital segments are lit on the clock. This occurs at 1:11. The room will be brightest when the most digital segments are lit on the clock. This occurs at 10:08.

   1:11   10:08

25. A kilogram of $10 gold pieces is worth twice as much as a half a kilogram of $20 gold pieces. (The denomination has nothing to do with the value, only the weight does!)

26. Similar to Example 5 in the text, we might examine the units place for repetitive powers of 3 in order to explore possible patterns.

$3^1 =$	3	$3^5 =$	243	$3^9 =$	19683
$3^2 =$	9	$3^6 =$	729	$3^{10} =$	59049
$3^3 =$	27	$3^7 =$	2187	$3^{11} =$	177147
$3^4 =$	81	$3^8 =$	6561	$3^{12} =$	531441

   If we divide the exponent 324 by 4 (since the pattern of the units digit cycles after every 4th power), we get a reminder of 0. Noting that in the line of 3^4, where each exponent when divided by 4 yields a remainder of 0, we reason inductively that 3^{324} has the same units digit 1.

27. Similar to Example 5 in the text (and Exercise 26 above), we might examine the units place for repetitive powers of 7 in order to explore possible patterns.

$7^1 =$	7	$7^5 =$	16,807
$7^2 =$	49	$7^6 =$	117,649
$7^3 =$	343	$7^7 =$	823,543
$7^4 =$	2,401	$7^8 =$	5,764,801

   Since the units digit cycles over four values, we might consider dividing the successive exponents by 4 and examining their remainders. Divide the exponent 491 by 4 to get a quotient of 122 and a remainder of 3. Reasoning inductively, the units digit would be the same as that of 7^3 and 7^7, which is 3.

28. Since her final amount was $8 and this represents half of the remaining money after buying a train ticket and spending $4, we reason that she had $2 \times \$8 + \$4 = \$20$ after buying the train ticket. And the ticket cost the same or $20 for a total of $40 spent after the purchase of the book. Therefore, she started with $10 (cost of book) + $40 = $50.

29. The final number is 37, which represents the sum of 12 and the previous result, so subtract 12 from 37 to get 25. Since this result was half the previous result, multiply 25 by 2 to get 50. Now find half of 50, or 25. This is the square of the positive number, so find the positive square root of 25, or 5, which is the original number.

30. At end of 1st day, the frog has a net progression of 1 foot; day 2: 2 feet; day 3: 3 feet; ...; day 16: 16 feet (it crawls up 4 feet from 15 to 19 feet and then falls back 3 feet to 16 feet); on the 17th day it crawls up 4 feet from the 16-foot level, which takes it to the top before slipping back.

31. To find the minimum number of socks to pull out, guess and check. There are two colors of socks. If you pull out 2 socks, you could have 2 of one color or 1 of each color. You must pull out more than 2 socks. If you pull out 3 socks, you might have 3 of one color or 1 of one color and 2 of the other. In either case, you have a matching pair, so 3 is the minimum number of socks to pull out.

32. Counting directly we get 9 (1-unit squares) plus 4 (4-unit squares) plus 1 (9-unit square). This equals 14 squares in total.

33. To count the triangles, it helps to draw sketches of the figure several times. There are 5 triangles formed by two sides of the pentagon and a diagonal. There are 4 triangles formed with each side of the pentagon as a base, so there are $4 \times 5 = 20$ triangles formed in this way. Each point of the star forms a small triangle, so

there are 5 of these. Finally, there are 5 triangles formed with a diagonal as a base. In each, the other two sides are inside the pentagon. (None of these triangles has a side common to the pentagon.) Thus, the total number of triangles in the figure is
$5 + 20 + 5 + 5 = 35$.

34. Set 4 opposite 12, then 8 half way around from 4 and 12. Opposite 8 must be 16 to allow for three equally spaced numbers (children) between each of these values. Note children 1, 2, and 3 stand between 16 and 4. So there are 16 children in the circle.

35. Use trial and error to find the smallest perfect number. Try making a chart such as the following one.

Number	Divisors other than itself	Sum
1	None	
2	1	1
3	1	1
4	1, 2	3
5	1	1
6	1, 2, 3	6

Six is the smallest perfect number.

36. Becky's mother named her third child Becky, since Becky is the only child left after Penny and Nichole have been named.

37. Working backward, we see that if the lily pad doubles its size each day so that it completely overs the pond on the twentieth day, the pond was half covered on the previous (or nineteenth) day.

38. It is a *palindrome*, since it reads the same backwards as forwards.

39. Solve by drawing a sketch. The following figure satisfies the description. Only three birds are needed.

40. By trial and error, squaring the counting numbers 44, 45, and 46 then subtracting 76, we get
$$44^2 - 76 = 1860$$
(Not a possible birth date for a living author.)
$$45^2 - 76 = 1949$$
(Possible birth date for a living author.)
$$46^2 - 76 = 2040$$
(Not a possible birth date for a living author.) Other counting numbers take one further from a possible birth date; thus, the answer is 1949.

41. From condition (2), we can figure that since the author is living now, the year must be 196_, since $9 - 3 = 6$. Then, from condition (1),
$$23 - (1 + 9 + 6) = 7,$$
so the year is 1967.

42. B, since either Bill or Bob must be the tallest person.

43. By Eve's statement, Adam must have $2 more than Eve. But according to Adam, a loss of $1 from Eve to Adam gives Adam twice the amount that Eve has. By trial and error, the counting numbers 5 and 7 are the first to satisfy both conditions. Thus Eve has $5, and Adam has $7.

44. The answer is 14, since the correct problem is
$$\begin{array}{r} 435 \\ 826 \\ + \ 147 \\ \hline 1408. \end{array}$$

45. The first digit in the answer cannot be 0, 2, 3, or 5, since these digits have already been used. It cannot be more than 3, since one of the factors is a number in the 30's, making it impossible to get a product over 45,000. Thus, the first digit of the answer must be 1. To find the first digit in the 3-digit factor, use estimation. Dividing a number between 15,000 and 16,000 by a number between 30 and 40 could give a result with a first digit of 3, 4, or 5. Since 3 and 5 have already been used, this first digit must be 4. Thus, the 3-digit factor is 402. We now have the following.

$$\begin{array}{r} 4\ 0\ 2 \\ \times \quad\ \ 3 \\ \hline 1\ \ 5, \quad\ \end{array}$$

To find the units digit of the 2-digit factor, use trial and error with the digits that have not yet been used: 6, 7, 8, and 9.

$36 \times 402 = 14,472$ (Too small and reuses 2 and 4)
$37 \times 402 = 14,874$ (Too small and reuses 4)
$38 \times 402 = 15,276$ (Reuses 2)
$39 \times 402 = 15,678$ (Correct)

The correct problem is as follows.

$$\begin{array}{r} 4\ 0\ 2 \\ \times \quad 3\ 9 \\ \hline 1\ \ 5,\ 6\ 7\ 8 \end{array}$$

Notice that a combination of strategies was used to solve this problem.

46. Add the diagonal elements together to get 15. Each row, column, and other diagonal must also add to 15. This yields the following perfect square.

6	1	8
7	5	3
2	9	4

47. Notice that the first column has three given numbers. Thus,
    $$34 - (6 + 11 + 16) = 1$$
    is the first number in the second row. (Note you could use the diagonal to solve for missing number in the same manner.) Then,
    $$34 - (1 + 15 + 14) = 4$$
    is in the second row, third column. The diagonal from upper left to lower right has three given numbers. Therefore,
    $$34 - (6 + 15 + 10) = 3$$
    is in the fourth row, fourth column. Continue filling in the missing numbers until the magic square is completed.

6	12	7	9
1	15	4	14
11	5	10	8
16	2	13	3

48. Solve this problem by making a list. First, find the ways he can use pennies to make 15 cents.

    15 pennies
    10 pennies, 1 nickel
    5 pennies, 1 dime
    5 pennies, 2 nickels

    Find additional ways he can use nickels.

    3 nickels
    1 nickel, 1 dime

    There are 6 ways to make 15 cents, so there are 6 ways he can pay 15 cents for a chocolate mint.

49. 25 pitches: Game tied 0 to 0 going into the 9th inning. Each pitcher has pitched a minimum of 24 pitches (three per inning). The winning pitcher pitches 3 more (fly ball/out) pitches for a total of 27. The losing (visiting team) pitcher pitches 1 more (for a total of 25) which happens to be a home run, thus, losing the game by a score of 1–0. Note: the same result occurs if the losing pitcher gives up one homerun in any inning.

50. Solve this problem by trial and error. Write the names of the natural numbers in the English language. One is the least natural number, but its letters are not in alphabetical order. You will find that forty is the least natural number that has its letters in alphabetical order.

51. The two children row across. One stays on the opposite bank, and the other returns. One soldier rows across, and the child on the opposite bank then rows back. The two children row across. One stays, and the other returns. Now another soldier rows across. This process continues until all the soldiers are across.

52. For three weighings, first balance four against four. Of the lighter four, balance two against the other two. Finally, of the lighter two, balance them one against the other. To find the bad coin in two weighings, divide the eight coins into groups of 3, 3, 2. Weigh the groups of three against each other on the scale. If the groups weigh the same, the fake coin is in the two left out and can be found in one additional weighing. If the two groups of three do not weigh the same, pick the lighter group. Choose any two of the coins and weigh them. If one of these is lighter, it is the fake; if they weigh the same, then the third coin is the fake.

53. A sketch may be helpful in solving this problem. The person takes the goat across and returns alone. On the second trip, the person takes the wolf across and returns with the goat. On the third trip, the goat is left on the first side while the person takes the cabbage across. Then the person returns alone and brings the goat back across.

54. Use the "guess and check" strategy and deductive reasoning to solve this problem. Since Ms. Thompson always tells the truth, she is not the woman on the left. She cannot be the woman in the middle, since the woman in the middle says, "I'm Ms. Johnson." From left to right, they are Johnson, Andersen, and Thompson.

55. Draw a sketch, visualize, or cut a piece of paper to build the cube. The cube may be folded with Z on the front.

    Then, E is on top and M is on the left face. This places Q opposite the face marked Z. (D is on the bottom and X is on the right face.)

56.

57. This may be worked algebraically or in reverse as Example 2 in text. Multiplying 2 by 10 gives 20. Subtract 8 to give 12 and then square to get 144. Add 52 to get 196. This represents a number times itself. The number is 14, from the fact that $14 \times 14 = 196$ (or the square root of $196 = 14$). The quotient must be 21 since $21 - \frac{1}{3} \times 21 = 14$. Multiplying 21 by 7, we get 147, which represents 3 times the original number plus 3/4 of that same product. The original number must be 28 since $3 \times 28 = 84$ and 3/4 of 84 is 63. And $84 + 63 = 147$.

58. If each digit cost $.75 (as in buying house numbers), then the sum of the 5 numerals would be $5 \times \$.75 = \$3.75$.

59. A solution, found by trial and error, is shown here.

    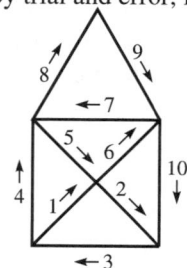

60. A solution, found by trial and error, is shown here.

    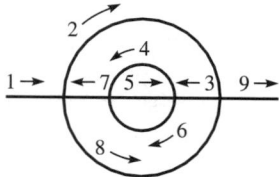

61. Solve this problem by looking for a pattern. 1/7 has a decimal representation of .142857..., where the group of 6 digits, 142857, is repeated indefinitely. When 100 is divided by 6, the remainder is 4, so the fourth digit of the repeating group, which is 8, is the 100th digit of the decimal representation.

62. None. To see this result, set two books beside each other and note the location of page 1 of the 1st book and the location of the last page of the 2nd book.

63. Common sense tells you that the CEO is a woman.

64. A strategy would be to assume the first possible age for the teenage girl, 13, and check by adding 2. This gives 15, not a perfect square. Trying the next possible age, 14, adding 2 gives the perfect square 16. Check by subtracting 10 from 14 to get 4 and note that $4^2 = 16$.

Thus, her age is 14 years, and her brother's age is $14 - 5 = 9$ years.

65. Jessica is married to James or Dan. Since Jessica is married to the oldest person in the group, she is not married to James, who is younger than Cathy. So Jessica is married to Dan, and Cathy is married to James. Since Jessica is married to the oldest person, we know that Dan is 36. Since James is older than Jessica but younger than Cathy, we conclude that Cathy is 31, James is 30, and Jessica is 29.

66. One strategy is to list all possible combinations of coins: 2 quarters; 5 dimes; 10 nickels; 1 quarter, 2 dimes, 1 nickel; 1 quarter, 1 dime, 3 nickels; 1 quarter, 5 nickels; 1 dime, 8 nickels; 2 dimes, 6 nickels; 3 dimes, 4 nickels; 4 dimes, 2 nickels; for a total of 10 ways.

67. This is a problem with a "catch." The obvious answer is that only one month, February, has 28 days. However, the problem does not specify exactly 28 days, so any month with at least 28 days qualifies. All 12 months have 28 days.

68. They must be February and March. This can occur only in a non-leap year.

69. This is a problem with a "catch." Someone reading this problem might go ahead and calculate the volume of a cube 6 feet on each side, to get the answer 216 cubic feet. However, common sense tells us that since holes are by definition empty, there is no dirt in a hole.

70. The products will always differ by 1. For example the initial terms are 1, 1, 2, 3 so that $(1)(3) = 3$ and $(1)(2) = 2$ and $3 - 2 = 1$.

71. "Madam, I'm Adam."

72. Since the number is palindromic and $e = f$, we must have $a = g$, and $b = c = e = f$. Since a, b, and c are multiples of 4 and $b = 2 \times a$, we must have $a = 4$ and $b = c = 8$. Also, $e = f = 8$. And

$$a + b + c + d + e + f + g = 47$$
$$4 + 8 + 8 + d + 8 + 8 + 4 = 47$$
$$d = 47 - 40$$
$$d = 7.$$

Thus, the phone number is $488 - 7884$.

73. The maximum number of squares is 6. The following represents one of several possible arrays.

	X	X
X		X
X	X	

74. One solution is
$1 + 2 + 3 + 4 + 5 + 6 + 7 + (8 \times 9) = 100$.

1.4 EXERCISES

Using a graphing calculator, such as the TI-83, we would enter the expressions as indicated on the left side of the equality then push [Enter] to arrive at the answer. When using scientific or other types of calculators some adjustments will have to be made. See observations related to the solutions for Exercise 8 and Exercise 13 below. It is a good idea to review your calculator handbook for related examples.

1. $39.7 + (8.2 - 4.1) = 43.8$
2. $2.8 \times (3.2 - 1.1) = 5.88$
3. $\sqrt{5.56440921} = 2.3589$
4. $\sqrt{37.38711025} = 6.1145$
5. $\sqrt[3]{418.508992} = 7.48$
6. $\sqrt[3]{700.227072} = 8.88$
7. $2.67^2 = 7.1289$
8. $3.49^3 = 42.508549$ Observe that for many calculators, the symbol "^" will be used to indicate the exponent prior to inserting the value 3.
9. $5.76^5 = 6340.338097$
10. $1.48^6 = 10.50921537$

Observe that when using a calculator, the numerator must be grouped in parenthesis as must the denominator. This will make the last operation (the indicated) division.

11. $\dfrac{(14.32 - 8.1)}{(2 \times 3.11)} = 1$
12. $\dfrac{(12.3 + 18.276)}{(3 \times 1.04)} = 9.8$
13. $\sqrt[5]{1.35} = 1.061858759$. Observe that many scientific calculators have only the $\sqrt[2]{}$ function built into the calculator. For an index larger than 2, you may want to think of the nth root of a number b as equivalent to the exponential expression $b^{\frac{1}{n}}$. For example, $\sqrt[5]{1.35} = (1.35)^{\frac{1}{5}}$. Then use your exponentiation function (button) to calculate the $5th$ root of 1.35.
14. $\sqrt[6]{3.21} = 1.214555893$
15. $\dfrac{\pi}{\sqrt{2}} = 2.221441469$
16. $\dfrac{2\pi}{\sqrt{3}} = 3.627598728$
17. $\sqrt[4]{\dfrac{2143}{22}} = 3.141592653$
18. $\dfrac{12345679 \times 72}{\sqrt[3]{27}} = 296296296$
19. Choose a five-digit number such as 73,468.
$$73468 \times 9 = 661212$$
$$6 + 6 + 1 + 2 + 1 + 2 = 18$$
$$1 + 8 = 9$$
Choose a six-digit number such as 739,216.
$$739216 \times 9 = 6652944$$
$$6 + 6 + 5 + 2 + 9 + 4 + 4 = 36$$
$$3 + 6 = 9$$
Yes, the same result holds.
20. Squaring the numbers 15, 25, 35, 45, 55, 65, 75, and 85, we get
$$15^2 = 225,$$
$$25^2 = 625,$$
$$35^2 = 1225,$$
$$45^2 = 2025,$$
$$55^2 = 3025,$$
$$65^2 = 4225,$$
$$75^2 = 5625,$$
$$85^2 = 7225.$$
It appears that to square a two-digit number ending in 5, we can multiply the number in the tens position by the next counting number and add the two digits, 25, to the end of the result. Thus, we can get 95^2 by $9 \times 10 = 90$ followed by the two digits 25 or $95^2 = 9025$.
21. $(-3) \times (-8) = 24$
$(-5) \times (-4) = 20$
$(-2.7) \times (-4.3) = 11.61$

Multiplying two negative numbers gives a <u>positive</u> product.
22. $5 \times (-4) = -20;$
$-3 \times 8 = -24;$
$2.7 \times (-4.3) = -11.61$

Multiplying a negative number and a positive number gives a <u>negative</u> product.
23. $5.6^0 = 1;\ \pi^0 = 1;\ 2^0 = 1;\ 120^0 = 1;\ .5^0 = 1$

Raising a nonzero number to the power 0 gives a result of <u>1</u>.
24. $1^2 = 1;\ 1^3 = 1;\ 1^{-3} = 1;\ 1^0 = 1;\ 1^{13} = 1.$

Raising 1 to any power gives a result of <u>1</u>.

25. $\dfrac{1}{7} \approx .1428571$

    $\dfrac{1}{(-9)} \approx -.1111111$

    $\dfrac{1}{3} \approx .3333333$

    $\dfrac{1}{(-8)} \approx -.1250000$

    The sign of the reciprocal of a number is the same as the sign of the number.

26. $(5/0) = $ ERROR; $(9/0) = $ ERROR; $(\pi/0) = $ ERROR; $(-3/0) = $ ERROR; $(0/0) = $ ERROR.

    Dividing a number by 0 gives an ERROR message on a calculator. (Division by 0 is not allowed.)

27. $(0/8) = 0; (0/2) = 0; (0/(-3)) = 0; (0/\pi) = 0$.

    Zero divided by a nonzero number gives a quotient of 0.

28. $(-3) \times (-4) \times (-5) = -60$
    $(-3) \times (-4) \times (-5) \times (-6) \times (-7) = -2520$
    $(-3) \times (-4) \times (-5) \times (-6) \times (-7) \times (-8) \times (-9)$
    $= -181440$

    Multiplying an *odd* number of negative numbers gives a negative product.

29. $(-3) \times (-4) = 12$;
    $(-3) \times (-4) \times (-5) \times (-6) = 360$;
    $(-3) \times (-4) \times (-5) \times (-6) \times (-7) \times (-8) = 20160$

    Multiplying an *even* number of negative numbers gives a positive product.

30. $\sqrt{-3}; \sqrt{-5}; \sqrt{-6}; \sqrt{-10}$
    Taking the square root of a negative number gives an error message on a calculator.

31. Writing exercise

32. The result is the three digit number you started with. Dividing by 7, then 13, and then 11 is the same as dividing by 1,001 ($7 \times 11 \times 13 = 1001$). A three-digit number, abc, multiplied by 1001 gives abcabc, so we just reversed the process.

33. The result of multiplying any digit except 0 by 429 and then multiplying the result by 259 is a six-digit number consisting of only the digit you started with. Multiplying by 429 and 259 is the same as multiplying by 111,111 ($429 \times 259 = 111111$), which gives the same digits.

34. A cycle will form that has its sixth number equal to the first number chosen. For example, choosing 5 and 6 gives

    $\dfrac{6+1}{5} = 1.4; \dfrac{1.4+1}{6} = .4; \dfrac{.4+1}{1.4} = 1; \dfrac{1+1}{.4} = 5; \dfrac{5+1}{1} = 6$

    as the first seven terms.

35. $(100 \div 20) \times 14,215,469 = \underline{71077345}$. One of the biggest petroleum companies in the world is ShELLOIL.

36. $\dfrac{10 \times 10,609}{\sqrt{4}} = \underline{53045}$ "Its got to be ShOES."

37. $60^2 - \dfrac{368}{4} = \underline{3508}$ The electronic manufacturer BOSE produces the Wave Radio.

38. $187^2 + \sqrt{1600} = \underline{35009}$. Have you ever read Mother GOOSE nursery rhymes.

39. Writing exercise

40. (a) 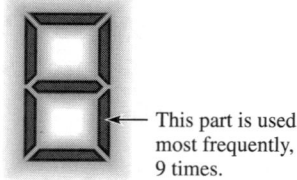 ← This part is used most frequently, 9 times.

    (b) The part used least is shown in the figure.

    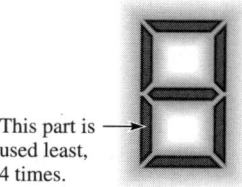 This part is used least, 4 times.

    (c) The digit 8 uses the most parts (all seven of them).

    (d) The digit 1 uses the fewest parts (two of them).

41. $431 \div 9 \approx 47.889$. Since more than 47 are needed, we require 48 pages.

42. $204 \div 18 \approx 11.333$. Since more than 11 are needed, we require 12 drawers.

43. $400 \div 30 \approx 13.333$. Since more than 13 are needed, we require 14 containers.

44. $155 \div 24 \approx 6.45833$. Since more than 6 teachers are needed, 7 are required.

45. $140,000 \div 80 \approx 160,000 \div 80 = \$2000$; option B.

46. $2,009 \div 50 \approx 2000 \div 50 = 40$ hours for one way. Thus, when traveling both ways, the time is approximated by $2 \times 40 = 80$ hours; option C.

47. $34,671 \div 1005 \approx (35,000/1000) = 35$; option A.

48. $90824.2 \div 88 \approx (90,000/90) = 1000$; option C.

49. 2008 yards $\div$ 392 attempts
    $\approx$ 2000 yards $\div$ 400 attempts
    $= 5$; option D.

50. $(40.5 \text{ meters}) \times (13.5 \text{ meters}) \approx (40 \text{ meters})$
    $\times (15 \text{ meters}) = 600$ square meters; option D.

51. By inspection of the graph, the answer is DIRECT at 45% of the market.

52. The number of home subscribers to Primestar is 16% of 12,000,000 or

$$(.16) \times 12000000 = 1,920,000.$$

53. Since 15% of the subcribers are C-Band (large dish), there are $(100 - 15)\% = 85\%$ of 12,000,000 subsribers who use small dishes. Thus, there are

$$(.85) \times 12000000 = 10200000,$$

or 10,200,000 subscribers.

54. Primestare has 1,920,000 (Exercise 52) subscribers and C-Band has $(.15) \times 12000000 = 1,800,000$ subscribers. Thus, Primestare has

$$1920000 - 1800000 = 120,000 \text{ more subscribers.}$$

Note: One can also subtract the percent values then use this result to find the number of subscribers.

55. By inspection of the graph (and using given units-millions), we see that there were 150,000,000 e-mail boxes.

56. By the graph there were just over 50,000,000 e-mail boxes in 1995 and 300,000,000 e-mail boxes in 2001. Thus, the growth was just less than $300,000,000 - 50,000,000 = 250,000,000$ e-mail boxes.

57. Finding 150 (million) on the vertical axis and reading horizontally to the right, we see that the corresponding year is 1998.

58. The number of e-mail boxes increased by 50,000,000 between 2000 and 2001. Thus, assuming the same increase in 2002, there would be approximately $300,000,000 + 50,000,000 = 35,000,000$ in 2002.

59. The average price increased only between 1993 and 1994.

60. The general trend since 1994 has been decreasing prices, since the graph falls from left to right.

61. The average price was almost $2000 in 1996 and about $1300 in 1999.

62. The average price in 1994 was about $2400 and about $1300 in 1999. So the decline in price is about $2400 - $1300 = $1100.

63. We would expect Detroit to be in the 70s.

64. A cold front appears to be moving towards Atlanta.

65. It will be hot and stormy in Miami.

66. The temperature in Augusta (Maine) is in the 60s.

67. There appears to be a trough over Texas.

68. Probably not, since rain is not predicted for Cleveland.

CHAPTER 1 TEST

1. This is an example of inductive reasoning, since you are reasoning from a specific pattern to the general conclusion that he will again exceed his annual sales goal.

2. This is a deductive argument because you are reasoning from the stated general property to the specific result, 101^2 is a natural number.

3. Observing a pattern such as $1 = 1^1, 4 = 2^2, 27 = 3^3$, leads us to the possibility that the fourth term may be $4^4 = 256$, the fifth term $5^5 = 3125$ and so on. Since the next term (6th term) is given as 46656, we have an added check for our pattern as $6^6 = 46656$. (The nth term of the sequence is n^n.)

4. The specific pattern seems to indicate that the second factor in the product is a multiple of 17 and the digits on the right side of the equation increase by 1. If this pattern is correct, then the next term in the sequence would be

$$65,359,477,124,183 \times 68 = 4,444,444,444,444,444,$$

since $4 \times 17 = 68$. This can be verified by multiplying $65,359,477,124,183 \times 68$ on your calculator.

5. 3 11 31 69 131 223 <u>351</u>
    8 20 38 62 92 <u>128</u>
      12 18 24 30 <u>36</u>
        6 6 6 (6)

6. Using the method of Gauss, we have $1 + 250 = 251$, $2 + 249 = 251 \ldots$ etc. There are $\frac{250}{2} = 125$ such pairs, so the sum can be calculated as $125 \times 251 = 31,375$.

7. The next predicted octagonal number is 65, since the next equation on the list would be $65 = 1 + 7 + 13 + 19 + 25$, where $25 = 19 + 6$.

8. Beginning with the first five octagonal numbers and applying the method of successive differences, we get
1 8 21 40 65 96 133 176
  7 13 19 25 31 37 43
    6 6 6 6 (6) (6).

Dividing each octagonal number by 4 we get the following pattern of remainders: 1, 0, 1, 0, 1, 0, 1, 0 …

9. After the first two terms (both of which are 1), we can find the next by adding the two proceeding terms. That is, to get the $3rd$ term, add $1 + 1 = 2$; the $4th, 1 + 2 = 3$; the $5th$ term, $2 + 3 = 5$; and so forth.

10. You will have 16 nights from the original 16 candles. Your can create $16/4 = 4$ more candles from the stubs. This gives 4 more nights. The remaining 4 stubs create 1 more candle for use 1 more night. Thus, there is a total of $16 + 4 + 1 = 21$ complete nights of candle light.

11. Examine the units place for repetitive powers of 9 in order to explore possible patterns.

$$9^1 = 9 \quad 9^3 = 729 \quad 9^5 = 59049$$
$$9^2 = 81 \quad 9^4 = 6561 \quad 9^6 = 531441$$

If we divide the exponent 1997 by 2 (since the pattern of the units digit cycles after every $2nd$ power), we get a reminder of 1. Noting that in the line of 9^1, where each exponent when divided by 2 yields a remainder of 1, has a units digit of 9, we reason inductively that 9^{1997} has the same units digit, 9.

12. There are 5 smaller triangles representing the extremities of the inside star. There are 5 triangles outside (between) the extremities of the star. Each (outside) line segment forms the base of (5) isosceles triangles that have their apex at each point of the star. Using the line segment connecting two points of the star as a base, two triangles can be formed; one (outside) with a point on the star as an opposite vertex; and one (inside) with opposite vertex at the intersection of any two lines forming the star. There is a total of 10 of these isosceles triangles. This gives a complete total of 35 triangles.

13. $629 + 154 = 783$ is one of several solutions.

14. We might group pages the following way
$(1,2)(3,4)(5,6)(7,8)\ldots(53,54)(55,56)(57,58)(59,60)$
Observe that pages $(1,2)$ will be printed on the same sheet as pages $(59,60)$. In the same manner, pages $(3,4)$ will be printed on the same sheet as $(57,58)$, etc. Thus, pages $(7,8)$ will be printed on the same sheet of paper as pages $(53,54)$. So, if the sheet with page 7 is missing, then 8, 53, and 54 will also be missing.

15. Observe the following patterns on successive powers of 11, 14, and 16 in order to determine the units value of each term in the sum $11^{11} + 14^{14} + 16^{16}$.

$$11^1 = 11 \quad 14^1 = 14 \quad 16^1 = 16$$
$$11^2 = 121 \quad 14^2 = 196 \quad 16^2 = 256$$
$$11^3 = 1331 \quad 14^3 = 2744 \quad 16^3 = 4096$$
$$14^4 = 38146$$

Thus, we would expect 11^{11} to have the same unit digit value of 1. Since powers of 14 have units digits which cycle between 4 and 6, we observe that division of the exponents by 2 yield remainders of 1 or 0. We might expect the same pattern to continue to 14^{14}. Division of the exponent by 2 gives a remainder of 0. We get the same remainder, 0, for all even powers on 14, and each of these numbers has a units digit of 6. The powers of 16 seem to all have the same unit value of 6. Thus, if we add the units digits $1 + 6 + 6 = 13$, we see that the units digit of this sum is 3.

16. Making the following observations

$$9 \times 1 = 9$$
$$9 \times 2 = 18 \ (1 + 8 = 9)$$
$$9 \times 3 = 27 \ (2 + 7 = 9)$$
$$9 \times 4 = 36 \ (3 + 6 = 9)$$
$$9 \times 5 = 45 \ (4 + 5 = 9)$$
$$\ldots$$

suggests that the sum of the digits in the product will always be 9.

17. $\sqrt{98.16} = 9.907572861$ But answers may vary depending upon what calculator you are using.

18. $3.25^3 = 34.328125$

19. The ratio of made shots to those attempted is approximately $200/500 = 2/5$. So in 5 attempts, we would expect him to make about 2 shots; option B.

20. (a) Since the graph slopes down from left to right, the decade where the percent of women decreased is $1990 - 2000$.

    (b) The maximum was reached in 1990.

    (c) Scanning horizontally to the right from 27% (on the vertical axis), we see that the corresponding year is about 1980 (on the horizontal axis).

2.1 EXERCISES

1. $\{2, 4, 6, 8\}$ matches C, the set of even positive integers less than 10.

2. $\{x|x$ is an even integer greater than 4 and less than 6$\}$ matches G, the empty set, since there is no even integer larger than 4 and smaller than 6.

3. $\{\ldots, -4, -3, -2, -1\}$ matches E, the set of all negative integers.

4. $\{\ldots, -6, -4, -2, 0, 2, 4, 6, \ldots\}$ matches A, the set of all even integers.

5. $\{2, 4, 8, 16, 32\}$ matches B, the set of the five least positive integer powers of 2, since each element represents a successive power of 2 beginning with 2^1.

6. $\{\ldots, -5, -3, -1, 1, 3, 5, \ldots\}$ matches D, the set of all odd integers.

7. $\{2, 4, 6, 8, 10\}$ matches H, the set of the five least positive integer multiples of 2, since this set represents the first five positive even integers. Remember that all even numbers are multiples of 2.

8. $\{1, 3, 5, 7, 9\}$ matches F, the set of odd positive integers less than 10.

9. The set of all counting numbers less than or equal to 6 can be expressed by listing as $\{1, 2, 3, 4, 5, 6\}$.

10. The set of all whole numbers greater than 8 and less than 18 can be expressed by listing as $\{9, 10, 11, 12, 13, 14, 15, 16, 17\}$.

11. The set of all whole numbers not greater than 4 can be expressed by listing as $\{0, 1, 2, 3, 4\}$.

12. The set of all counting numbers between 4 and 14 can be expressed by listing as $\{5, 6, 7, 8, 9, 10, 11, 12, 13\}$.

13. In the set $\{6, 7, 8, \ldots, 14\}$, the ellipsis (three dots) indicates a continuation of the pattern. A complete listing of this set is $\{6, 7, 8, 9, 10, 11, 12, 13, 14\}$.

14. The set $\{3, 6, 9, 12, \ldots, 30\}$ contains all multiples of 3 from 3 to 30 inclusive. A complete listing of this set is $\{3, 6, 9, 12, 15, 18, 21, 24, 27, 30\}$.

15. The set $\{-15, -13, -11, \ldots, -1\}$ contains all integers from -15 to -1 inclusive. Each member is two larger than its predecessor. A complete listing of this set is $\{-15, -13, -11, -9, -7, -5, -3, -1\}$.

16. The set $\{-4, -3, -2, \ldots, 4\}$ contains all integers from -4 to 4. Each member is one larger than its predecessor. A complete listing of this set is $\{-4, -3, -2, -1, 0, 1, 2, 3, 4\}$.

17. The set $\{2, 4, 8, \ldots, 256\}$ contains all powers of two from 2 to 256 inclusive. A complete listing of this set is $\{2, 4, 8, 16, 32, 64, 128, 256\}$.

18. In the set $\{90, 87, 84, \ldots, 69\}$, each member after the first is found by subtracting 3 from the previous member. The set contains multiples of 3 in decreasing order, from 90 to 69 inclusive. A complete listing is $\{90, 87, 84, 81, 78, 75, 72, 69\}$.

19. A complete listing of the set $\{x \mid x$ is an even whole number less than 11$\}$ is $\{0, 2, 4, 6, 8, 10\}$. Remember that 0 is the first whole number.

20. The set $\{x \mid x$ is an odd integer between -8 and 7$\}$ can be represented as the listing $\{-7, -5, -3, -1, 1, 3, 5\}$.

21. The set of all counting numbers greater than 20 is represented by the listing $\{21, 22, 23, \ldots\}$.

22. The set of all integers between -200 and 500 is represented by the listing $\{-199, -198, -197, \ldots, 499\}$.

23. The set of Great Lakes is represented by {Lake Erie, Lake Huron, Lake Michigan, Lake Ontario, Lake Superior}.

24. The set of United States presidents who served after Lyndon Johnson and before William Clinton is represented by {George Bush, Ronald Reagan, Jimmy Carter, Gerald Ford, Richard Nixon}.

25. The set $\{x \mid x$ is a positive multiple of 5$\}$ is represented by the listing $\{5, 10, 15, 20, \ldots\}$.

26. The set $\{x \mid x$ is a negative multiple of 6$\}$ is represented by the listing $\{-6, -12, -18, -24, -30, \ldots\}$.

27. The set $\{x \mid x$ is the reciprocal of a natural number$\}$ is represented by the listing $\{1, 1/2, 1/3, 1/4, 1/5, \ldots\}$.

28. The set $\{x \mid x$ is a positive integer power of 4$\}$ is represented by the listing $\{4, 16, 64, 256, 1024, \ldots\}$ since $4^1 = 4, 4^2 = 16$, and so forth.

29. The set of all rational numbers may be represented using set-builder notation as $\{x \mid x$ is a rational number$\}$.

30. The set of all even natural numbers may be represented using set-builder notation as $\{x \mid x$ is an even natural number$\}$.

31. The set $\{1, 3, 5, \ldots, 75\}$ may be represented using set-builder notation as $\{x \mid x$ is an odd natural number less than 76$\}$.

32. The set $\{35, 40, 45, \ldots, 95\}$ may be represented using set-builder notation as $\{x \mid x$ is a multiple of 5 between 30 and 100$\}$.

33. The set $\{2, 4, 6, \ldots, 32\}$ is finite since the cardinal number associated with this set is a whole number.

34. The set $\{6, 12, 18\}$ is finite since the cardinal number is a whole number.

35. The set $\{1/2, 2/3, 3/4, \ldots\}$ is infinite since there is no last element, and we would be unable to count all of the elements.

36. The set $\{-10, -8, -6, \ldots\}$ is infinite since there is no last element, and therefore its cardinal number is not a whole number.

37. The set $\{x \mid x$ is a natural number greater than 50$\}$ is infinite since there is no last element, and therefore its cardinal number is not a whole number.

38. The set $\{x \mid x$ is a natural number less than 50$\}$ is finite since there are a countable number of elements in the set.

39. The set $\{x \mid x$ is a rational number$\}$ is infinite since there is no last element, and therefore its cardinal number is not a whole number.

40. The set $\{x \mid x$ is a rational number between 0 and 1$\}$ is an infinite set. One could never finish counting the elements of this set because between every two rational numbers, we can always find another rational number.

41. For any set A, $n(A)$ represents the cardinal number of the set, that is, the number of elements in the set. The set $A = \{0, 1, 2, 3, 4, 5, 6, 7\}$ contains 8 elements. Thus, $n(A) = 8$.

42. The set $A = \{-3, -1, 1, 3, 5, 7, 9\}$ contains 7 elements. Thus, $n(A) = 7$.

43. The set $A = \{2, 4, 6, \ldots, 1000\}$ contains 500 elements. Thus, $n(A) = 500$.

44. The set $A = \{0, 1, 2, 3, \ldots, 3000\}$ contains 3001 elements. Thus, $n(A) = 3001$.

45. The set $A = \{a, b, c, \ldots, z\}$ has 26 elements (letters of the alphabet). Thus, $n(A) = 26$.

46. The set $\{x \mid x$ is a vowel in the English alphabet$\}$ has 5 members since there are 5 vowels, a, e, i, o, and u. Thus, $n(A) = 5$.

47. The set A = the set of integers between -20 and 20 has 39 members. The set can be indicated as $\{-19, -18, \ldots, 18, 19\}$, or 19 negative integers, 19 positive integers, and 0. Thus, $n(A) = 39$.

48. The set A = set of current US senators has 100 members (two from each state). Thus, $n(A) = 100$.

49. The set $A = \{1/3, 2/4, 3/5, 4/6, \ldots, 27/29, 28/30\}$ has 28 elements. Thus, $n(A) = 28$.

50. The set $A = \{1/2, -1/2, 1/3, -1/3, \ldots, 1/10, -1/10\}$ has 18 (nine negative and nine positive) elements. Thus, $n(A) = 18$.

51. Writing exercise

52. Writing exercise

53. The set $\{x \mid x$ is a real number$\}$ is well defined since we can always tell if a number is real and belongs to this set.

54. The set $\{x \mid x$ is a negative number$\}$ is well defined since we can tell whether a particular number belongs to this set by observing its sign.

55. The set $\{x \mid x$ is good athlete$\}$ is not well defined since set membership, in this case, is a value judgment, and there is no clear-cut way to determine whether a particular athlete is "good."

56. The set $\{x \mid x$ is skillful typist$\}$ is not well defined since set membership is a value judgment

57. The set $\{x \mid x$ is a difficult course$\}$ is not well defined since set membership is a value judgment, and there is no clear-cut way to determine whether a particular course is "difficult."

58. The set $\{x \mid x$ is a counting number less than 2$\}$ is well defined since we can always tell if a number satisfies the conditions and, hence, belongs to this set. Note that there is only one element, 1, in the set.

59. $5 \boxed{\in} \{2, 4, 5, 6, 7\}$ since 5 is a member of the set.

60. $8 \boxed{\in} \{3, -2, 5, 7, 8\}$ since 8 is a member of the set.

61. $-4 \boxed{\notin} \{4, 7, 8, 12\}$ since -4 is not contained in the set.

62. $-12 \boxed{\notin} \{3, 8, 12, 18\}$ because -12 is not a member of the set.

63. $0 \boxed{\in} \{-2, 0, 5, 9\}$ since 0 is a member of the set.

64. $0 \boxed{\notin} \{3, 4, 6, 8, 10\}$ since 0 in not a member of the set.

65. $\{3\} \boxed{\notin} \{2, 3, 4, 6\}$ since the elements are not sets themselves.

66. $\{6\} \boxed{\notin} \{3, 4, 6, 8, 10\}$ since the set $\{6\}$ is not a member of the second set even though the number 6 is.

67. The statement $3 \in \{2, 5, 6, 8\}$ is false since the element 3 is not in the set.

68. The statement $6 \in \{-2, 5, 8, 9\}$ is false since the element 6 is not a member of the set.

69. The statement $b \in \{h, c, d, a, b\}$ is true since b is contained in the set.

70. The statement $m \in \{l, m, n, o, p\}$ is true since m is contained in the set.

71. The statement $9 \notin \{6, 3, 4, 8\}$ is true since 9 is not a member of the set.

72. The statement $2 \notin \{7, 6, 5, 4\}$ is true since 2 is not a member of the set.

73. The statement $\{k, c, r, a\} = \{k, c, a, r\}$ is true since both sets contain exactly the same elements. 74. The statement $\{e, h, a, n\} = \{a, h, e, n\}$ is true since both sets contain exactly the same elements. 75. The statement $\{5, 8, 9\} = \{5, 8, 9, 0\}$ is false because the second set contains a different element from the first set, 0. 76. The statement $\{3, 7, 12, 14\} = \{3, 7, 12, 14, 0\}$ is false because the second set contains a different element from the first set, 0.

77. The statement $\{x \mid x$ is a natural number less than 3$\}$ $= \{1, 2\}$ is true since both represent sets with exactly the same elements.

78. The statement $\{x \mid x$ is a natural number greater than 10$\}$ $= \{11, 12, 13, \ldots\}$ is true since both represent sets with exactly the same elements.

79. The statement $4 \in A$ is true since 4 is a member of set A.

80. The statement $8 \in B$ is true since 8 is a member of set B.

81. The statement $4 \notin C$ is false since 4 is a member of the set C.

82. The statement $8 \notin B$ is false since 8 is a member of the set B.

83. Every element of C is also an element of A is true since the members, 4, 10, and 12 of set C, are also members of set A.

84. The statement, every element of C is also an element of B, is false since the element, 12, of set C is not also an element of set B.

85. Writing exercise

86. Writing exercise

87. An example of two sets that are not equivalent and not equal would be $\{3\}$ and $\{c, f\}$. Other examples are possible.

88. An example of two sets that are equal but not equivalent is impossible. If they are equal, they have the same number of elements and must be equivalent.

89. An example of two sets that are equivalent but not equal would be $\{a, b\}$ and $\{a, c\}$. Other examples are possible.

90. Two sets that are both equal and equivalent would be $\{5\}$ and $\{4 + 1\}$. Other examples are possible.

91. (a) The stocks with share volumes ≥ 188.7 million are those listed in the set {Viacom (Class B), Trans World Airlines, Harken Energy, Echo Bay Mines}.

    (b) The stocks with share volumes ≤ 188.7 million are those listed in the set {Echo Bay Mines, JTS, Nabors Industries, Hasbro, Royal Oak Mines, Grey Wolf Industries, IVAX}.

92. (a) Since there are 220 calories in the candy bars that Jamie likes, she must burn off $3 \times 220 = 660$ calories. The following sets indicate activities that will burn off the required number of calories and will take no more that two hours: $\{r\}, \{g, s\}, \{c, s\}, \{v, r\}, \{g, r\}, \{c, r\},$ and $\{s, r\}$.

    (b) The required number of calories to burn off is $5 \times 220 = 1100$ calories. The following sets indicate activities taking less than or equal to three hours and burning off the required number of calories: $\{v, g, r\}, \{v, c, r\}, \{v, s, r\}, \{g, c, r\}, \{g, s, r\},$ and $\{c, s, r\}$.

2.2 EXERCISES

1. $\{p\}, \{q\}, \{p, q\}, \emptyset$ matches F, the subsets of $\{p, q\}$.

2. $\{p\}, \{q\}, \emptyset$ matches B, the proper subsets of $\{p, q\}$. Note that the set $\{p, q\}$, itself, though a subset, is not a proper subset.

3. $\{a, b\}$ matches C, the complement of $\{c, d\}$, if $U = \{a, b, c, d\}$.

4. $\emptyset$ matches D, the complement of U.

5. U matches A, the complement of $\emptyset$.

6. $\{a\}$ matches E, the complement of $\{b\}$, if $U = \{a, b\}$.

7. $\{-2, 0, 2\}$ $\boxed{\not\subseteq}$ $\{-2, -1, 1, 2\}$

8. $\{M, W, F\}$ $\boxed{\not\subseteq}$ $\{S, M, T, W, Th\}$ since the element "F" is not a member of the second set.

9. $\{2, 5\}$ $\boxed{\subseteq}$ $\{0, 1, 5, 3, 4, 2\}$

10. $\{a, n, d\}$ $\boxed{\subseteq}$ $\{r, a, n, d, y\}$

11. $\emptyset$ $\boxed{\subseteq}$ $\{a, b, c, d, e\}$, since the empty set is considered a subset of any given set.

12. $\emptyset$ $\boxed{\subseteq}$ $\emptyset$, since the empty set is considered a subset of every set including itself.

13. $\{-7, 4, 9\}$ $\boxed{\not\subseteq}$ $\{x \mid x$ is an odd integer$\}$ since the element "4" is not an element of the second set.

14. $\{2, 1/3, 5/9\}$ $\boxed{\subseteq}$ the set of rational numbers since 2, 1/3, and 5/9 are rational numbers.

15. $\{B, C, D\}$ $\boxed{\subseteq}$ $\{B, C, D, F\}$ and $\{B, C, D\}$ $\boxed{\subseteq}$ $\{B, C, D, F\}$, i.e., both.

16. $\{\text{red, blue, yellow}\}$ $\boxed{\subseteq}$ $\{\text{yellow, blue, red}\}$.

17. $\{9, 1, 7, 3, 5\}$ $\boxed{\subseteq}$ $\{1, 3, 5, 7, 9\}$

18. $\{S, M, T, W, Th\}$ $\boxed{\not\subseteq}$ $\{M, W, Th, S\}$; therefore, neither.

19. $\emptyset$ $\boxed{\subset}$ $\{0\}$ or $\emptyset$ $\boxed{\subseteq}$ $\{0\}$, i.e., both.

20. $\emptyset$ $\boxed{\subseteq}$ $\emptyset$ only.

21. $\{-1, 0, 1, 2, 3\}$ $\boxed{\not\subseteq}$ $\{0, 1, 2, 3, 4\}$; therefore, neither. Note that if a set is not a subset of another set, it can not be a proper subset either.

22. $\{5/6, 9/8\}$ $\boxed{\not\subseteq}$ $\{6/5, 8/9\}$ since either member of the first set is not a member of the second. Thus, neither is the correct answer.

23. $A \subset U$ is true since all sets must be subsets of the Universal set by definition, and U contains at least one more element than A.

24. $C \subset U$ is true since all sets must be subsets of the Universal set by definition, and U contains at least one more element than C.

25. $D \subseteq B$ is false since the element "d" in set D is not also a member of set B.

26. $D \subseteq A$ is false since the element "d" in set D is not also a member of set A.

27. $A \subset B$ is true. All members of A are also members of B, and there are elements in set B not contained in set A.

28. $B \subseteq C$ is false since the elements "a" and "e" in set B are not also members of set C.

29. $\emptyset \subset A$ is true since the empty set, $\emptyset$, is considered a subset of all sets. It is a proper subset since there are elements in A not contained in $\emptyset$.

30. $\emptyset \subseteq D$ is true since the empty set, $\emptyset$, is considered a subset of all sets.

31. $\emptyset \subseteq \emptyset$ is true since the empty set, $\emptyset$, is considered a subset of all sets including itself. Note that all sets are subsets of themselves.

32. $D \subset B$ is false. The element "d", though a member of set D, is not also a member of set B.

33. $D \not\subseteq B$ is true. Set D is not a subset of B because the element "d", though a member of set D, is not also a member of set B.

34. $A \not\subseteq B$ is false. Set A is a subset of set B since all of the elements in set A are also in set B.

35. There are exactly 6 subsets of C is false. Since there are 3 elements in set C, there are $2^3 = 8$ subsets.

36. There are exactly 31 subsets of B is false. Since there 5 elements in set B, there are $2^5 = 32$ subsets.

37. There are exactly 3 subsets of A is false. Since there are 2 elements in set A, there are $2^2 = 4$ subsets.

38. There are exactly 4 subsets of D is true. Since there are 2 elements in set D, there are $2^2 = 4$ subsets.

39. There is exactly one subset of $\emptyset$ is true. The only subset of $\emptyset$ is $\emptyset$ itself.

40. There are exactly 127 proper subsets of U is true. Since there are 7 elements in set U, there are $2^7 = 128$ subsets of U. Since the set U, itself, is not a proper subset, there is one less or 127 proper subsets.

41. The Venn diagram does not represent the correct relationships among the sets since C is not a subset of A. Thus, the answer is false.

42. The Venn diagram shows that C is a subset of B and that B is a subset of U. This is a correct relationship since all members of set C are also members of set B, and B is a subset of U. Thus, the answer is true.

43. Since the given set has 3 elements, there are $2^3 = 8$ subsets and $2^3 - 1 = 7$ proper subsets.

44. Since the given set has 4 elements, there are $2^4 = 16$ subsets and $2^4 - 1 = 15$ proper subsets.

45. Since the given set has 6 elements, there are $2^6 = 64$ subsets and $2^6 - 1 = 63$ proper subsets.

46. Since the given set has 7 elements, which are the days of the week, there are $2^7 = 128$ subsets and $2^7 - 1 = 127$ proper subsets.

47. The set $\{x \mid x \text{ is an odd integer between } -6 \text{ and } 4\} = \{-5, -3, -1, 1, 3\}$. Since the set contains 5 elements, there are $2^5 = 32$ subsets and $2^5 - 1 = 32 - 1 = 31$ proper subsets.

48. The set $\{x \mid x \text{ is an odd whole number less than } 4\} = \{1, 3\}$. Since the set contains 2 elements, there are $2^2 = 4$ subsets and $2^2 - 1 = 3$ proper subsets.

49. The complement of $\{1, 4, 6, 8\}$ is $\{2, 3, 5, 7, 9, 10\}$, that is, all of the elements in U not also in the given set.

50. The complement of $\{2, 5, 7, 9, 10\}$ is $\{1, 3, 4, 6, 8\}$.

51. The complement of $\{1, 3, 4, 5, 6, 7, 8, 9, 10\}$ is $\{2\}$.

52. The complement of $\{1, 2, 3, 4, 6, 7, 8, 9, 10\}$ is $\{5\}$.

53. The complement of $\emptyset$, the empty set, is $\{1, 2, 3, 4, 5, 6, 7, 8, 9, 10\}$, the universal set.

54. The complement of the universal set, U, is the empty set, $\emptyset$.

55. In order to contain all of the indicated characteristics, the universal set $U = \{$Higher cost, Lower cost, Educational, More time to see the sights, Less time to see the sights, Cannot visit relatives along the way, Can visit relatives along the way$\}$.

56. Since F contains the characteristics of the flying option, $F' = \{$Lower cost, Less time to see the sights, Can visit relatives along the way$\}$.

57. Since D contains the set of characteristics of the driving option, $D' = \{$Higher cost, More time to see the sights, Cannot visit relatives along the way$\}$.

58. The set of element(s) common to set F and D is $\{$Educational$\}$.

59. The set of element(s) common to F' and D' is $\emptyset$, the empty set, since there are no common elements.

60. The set of element(s) common to F and D' is $\{$Higher cost, More time to see the sights, Cannot visit relatives along the way$\}$.

61. The only possible set is $\{A, B, C, D, E\}$. (All are present.)

62. The possible subsets of four people would include $\{A, B, C, D\}, \{A, B, C, E\}, \{A, B, D, E\}, \{A, C, D, E\}$, and $\{B, C, D, E\}$.

63. The possible subsets of three people would include $\{A, B, C\}, \{A, B, D\}, \{A, B, E\}, \{A, C, D\}, \{A, C, E\}, \{A, D, E\}, \{B, C, D\}, \{B, C, E\}, \{B, D, E\}$, and $\{C, D, E\}$.

2.3 SET OPERATIONS AND CARTESIAN PRODUCTS

64. The possible subsets of two people would include $\{A, B\}, \{A, C\}, \{A, D\}, \{A, E\}, \{B, C\}, \{B, D\}, \{B, E\}, \{C, D\}, \{C, E\}$, and $\{D, E\}$.

65. The possible subsets consisting of one person would include $\{A\}, \{B\}, \{C\}, \{D\}$ and $\{E\}$.

66. The set indicating that no people get together (no one shows up) is $\emptyset$.

67. Adding the number of subsets in Exercises 61 – 66, we have $1 + 5 + 10 + 10 + 5 + 1 = 32$ ways that the group can gather.

68. They are the same: $32 = 2^5$. The number of ways that people, from a group of five, can gather is the same as the number of subsets there are of a set of five elements.

69. (a) Consider all possible subsets of a set with four elements (the number of bills). The number of subsets would be $2^4 = 16$. Since 16 includes also the empty set (and we must choose one bill), we will subtract one from this or $16 - 1 = 15$ possible sums of money.

    (b) Removing the condition says, in effect, that we may also choose no bills. Thus, there are $2^4 = 16$ subsets or possible sums of money; It is now possible to select no bills.

70. (a) Consider all possible subsets of a set with five elements (the number of coins). The number of subsets would be $2^5 = 32$. But the 32 includes also the empty set, and we must select at least one coin. Thus, there are $32 - 1 = 31$ sums of money.

    (b) Removing the condition says, in effect, that we may also choose no coins. Thus, there are $2^5 = 32$ subsets or possible sums of coins including no coins.

71. (a) There are s subsets of B that do not contain e. These are the subsets of the original set A.

    (b) There is one subset of B for each of the original subsets of set A, which is formed by including e as the element of that subset of A. Thus, B has s subsets which do contain e.

    (c) The total number of subsets of B is the sum of the numbers of subsets containing e and of those not containing e. This number is $s + s$ or $2s$.

    (d) Adding one more element will always double the number of subsets, so we conclude that the formula 2^n is true in general.

72. Writing exercise

2.3 EXERCISES

1. The intersection of A and B, $A \cap B$, matches B, the set of elements common to both A and B.

2. The union of A and B, $A \cup B$, matches F, the set of elements that are in A or in B or in both A and B.

3. The difference of A and B, $A - B$, matches A, the set of elements in A that are not in B.

4. The complement of A, A', matches C, the set of elements in the universe that are not in A.

5. The Cartesian product of A and B, $A \times B$, matches E, the set of ordered pairs such that each first element is from A and each second element is from B, with every element of A paired with every element of B.

6. The difference of B and A, $B - A$, matches D, the set of elements of B that are not in A.

7. $X \cap Y = \{a, c\}$ since these are the elements that are common to both X and Y.

8. $X \cup Y = \{a, b, c, e, g\}$ since these are the elements that are contained in X or Y (or both).

9. $Y \cup Z = \{a, b, c, d, e, f\}$ since these are the elements that are contained in Y or Z (or both).

10. $Y \cap Z = \{b, c\}$ since these are the elements that are common to both Y and Z.

11. $X \cup U = \{a, b, c, d, e, f, g\} = U$. Observe that any set union with the universal set will give the universal set.

12. $Y \cap U = \{a, b, c\} = Y$ since these are the elements that are common to both Y and U. Note that any set intersected with the universal set will give the original set.

13. $X' = \{b, d, f\}$ since these are the only elements in U not contained in X.

14. $Y' = \{d, e, f, g\}$ since these are the only elements in U not contained in Y.

15. $X' \cap Y' = \{b, d, f\} \cap \{d, e, f, g\} = \{d, f\}$

16. $X' \cap Z = \{b, d, f\} \cap \{b, c, d, e, f\} = \{b, d, f\}$

17. $X \cup (Y \cap Z) = \{a, c, e, g\} \cup \{b, c\} = \{a, b, c, e, g\}$
    Observe that the intersection must be done first.

18. $Y \cap (X \cup Z) = \{a, b, c\} \cap \{a, b, c, d, e, f, g\} = \{a, b, c\}$ Observe that the union must be done first.

19. $(Y \cap Z') \cup X = (\{a, b, c\} \cap \{a, g\}) \cup \{a, c, e, g\} = \{a\} \cup \{a, c, e, g\} = \{a, c, e, g\} = X$

20. $(X' \cup Y') \cup Z = (\{b, d, f\} \cup \{d, e, f, g\}) \cup \{b, c, d, e, f\} = \{b, d, e, f, g\} \cup \{b, c, d, e, f\} = \{b, c, d, e, f, g\}$

21. $(Z \cup X')' \cap Y = (\{b, c, d, e, f\} \cup \{b, d, f\})' \cap \{a, b, c\} = \{b, c, d, e, f\}' \cap \{a, b, c\} = \{a, g\} \cap \{a, b, c\} = \{a\}$

22. $(Y \cap X')' \cup Z' = (\{a, b, c\} \cap \{b, d, f\})' \cup \{a, g\} = \{b\}' \cup \{a, g\} = \{a, c, d, e, f, g\} \cup \{a, g\} = \{a, c, d, e, f, g\}$

23. $X - Y = \{e, g\}$ Since these are the only two elements that belong to X and not to Y.

24. $Y - X = \{b\}$. Since this is the only element that belongs to Y and not to X.

25. $X' - Y = \{b, d, f\} - \{a, b, c\} = \{d, f\}$. Observe that we must find X' first.

26. $Y' - X = \{d, e, f, g\} - \{a, c, e, g\} = \{d, f\}$. Observe that we must find Y' first.

27. $X \cap (X - Y) = \{a, c, e, g\} \cap \{e, g\} = \{e, g\}$. Observe that we must find $X - Y$ first.

28. $Y \cup (Y - X) = \{a, b, c\} \cup \{b\} = \{a, b, c\}$

29. $A \cup (B' \cap C')$ is the set of all elements that are in A, or are not in B and not in C.

30. $(A \cap B') \cup (B \cap A')$ is the set of all elements that are in A but not in B, or in B but not in A.

31. $(C - B) \cup A$ is the set of all elements that are in C but not in B, or they are in A.

32. $B \cap (A' - C) = B \cap (A' \cap C')$ is the set of elements that are in B and are not in A and not in C.

33. $(A - C) \cup (B - C)$ is the set of all elements that are in A but not C, or are in B but not in C.

34. $(A' \cap B') \cup C'$ is the set of all elements that are not in A and not in B, or are not in C.

35. The smallest set representing the universal set U is $\{e, h, c, l, b\}$.

36. A', the complement of A, is the set of all effects in U that are not adverse effects of alcohol use: $A' = \{e, c\}$.

37. T', the complement of T, is the set of effects in U that are not adverse effects of tobacco use: $T' = \{l, b\}$.

38. $T \cap A$ is the set of adverse effects of both tobacco and alcohol use: $T \cap A = \{h\}$.

39. $T \cup A$ is the set of all adverse effects that are either tobacco related or alcohol related:
$T \cup A = \{e, h, c, l, b\} = U$.

40. $T \cap A'$ is the set of adverse tobacco related effects that are not alcohol related: $T \cap A' = \{e, c\}$.

41. $B \cup C$ is the set of all tax returns showing business income or filed in 1999.

42. $A \cap D$ is the set of all tax returns with itemized deductions and selected for audit.

43. $C - A$ is the set of all tax returns filed in 1999 without itemized deductions.

44. $D \cup A'$ is the set of all tax returns selected for audit or without itemized deductions.

45. $(A \cup B) - D$ is the set of all tax returns with itemized deductions or showing business income, but not selected for audit.

46. $(C \cap A) \cap B'$ is the set of all tax returns filed in 1999 and with itemized deductions but not showing business income.

47. $A \subseteq (A \cup B)$ is always true since $A \cup B$ will contain all of the elements of A.

48. $A \subseteq (A \cap B)$ is not always true since $A \cap B$ may not contain all the elements of A.

49. $(A \cap B) \subseteq A$ is always true since the elements of $A \cap B$ must be in A.

50. $(A \cup B) \subseteq A$ is not always true since the elements of $A \cup B$ may contain elements not in A.

51. $n(A \cup B) = n(A) + n(B)$ is not always true. If there are any common elements to A and B, they will be counted twice.

52. $n(A \cap B) = n(A) - n(B)$ is not always true. If A and B are different sized sets and have no elements in common, then $n(A \cap B) = 0$ and $n(A) - n(B) \neq 0$.

53. $n(A \cup B) = n(A) + n(B) - n(A \cap B)$ is always true, since any elements common to sets A and B which are counted twice by $n(A) + n(B)$ are returned to a single count by the subtraction of $n(A \cap B)$.

54. $n(A \cap B) = n(A) + n(B) - n(A \cup B)$ is always true since this equation is equivalent to that of Exercise 53 which we may solve for $n(A \cap B)$.

55. (a) $X \cup Y = \{1, 2, 3, 5\}$

    (b) $Y \cup X = \{1, 2, 3, 5\}$

    (c) For any sets X and Y,
    $$X \cup Y = Y \cup X.$$
    This conjecture indicates that set union is a commutative operation.

56. (a) $X \cap Y = \{1, 3\}$

    (b) $Y \cap X = \{1, 3\}$

    (c) For any sets X and Y,
    $$X \cap Y = Y \cap X.$$
    This conjecture indicates that set intersection is a commutative operation.

57. (a) $X \cup (Y \cup Z) = \{1, 3, 5\} \cup (\{1, 2, 3\} \cup \{3, 4, 5\})$
    $= \{1, 3, 5\} \cup \{1, 2, 3, 4, 5\}$
    $= \{1, 3, 5, 2, 4\}$

    (b) $(X \cup Y) \cup Z = (\{1, 3, 5\} \cup \{1, 2, 3\}) \cup \{3, 4, 5\}$
    $= \{1, 3, 5, 2\} \cup \{3, 4, 5\}$
    $= \{1, 3, 5, 2, 4\}$

(c) For any sets X, Y, and Z,
$$X \cup (Y \cup Z) = (X \cup Y) \cup Z.$$

This conjecture indicates that set union is an associative operation.

58. (a) $X \cap (Y \cap Z) = \{1, 3, 5\} \cap (\{1, 2, 3\} \cap \{3, 4, 5\})$
$= \{1, 3, 5\} \cap \{3\}$
$= \{3\}$
(b) $(X \cap Y) \cap Z = (\{1, 3, 5\} \cap \{1, 2, 3\}) \cap \{3, 4, 5\}$
$= \{1, 3\} \cap \{3, 4, 5\}$
$= \{3\}$
(c) For any sets X, Y, and Z,
$$X \cap (Y \cap Z) = (X \cap Y) \cap Z.$$

This conjecture indicates that set intersection is an associative operation.

59. (a) $(X \cup Y)' = \{1, 3, 5, 2\}' = \{4\}$
(b) $X' \cap Y' = \{2, 4\} \cap \{4, 5\} = \{4\}$
(c) For any sets X and Y,
$$(X \cup Y)' = X' \cap Y'.$$

Observe that this conjecture is one form of DeMorgan's Laws.

60. (a) $(X \cap Y)' = \{1, 3\}' = \{2, 4, 5\}$
(b) $X' \cup Y' = \{2, 4\} \cup \{4, 5\} = \{2, 4, 5\}$
(c) For any sets X and Y,
$$(X \cap Y)' = X' \cup Y'.$$

Observe that this conjecture is the other form of De Morgan's Laws.

61. (a) $X \cup \emptyset = \{1, 3, 5\} \cup \emptyset = X$
(b) For any set X,
$$X \cup \emptyset = X.$$

62. (a) $X \cap \emptyset = \{1, 3, 5\} \cap \emptyset = \emptyset$
(b) For any set X,
$$X \cap \emptyset = \emptyset.$$

63. The statement $(3, 2) = (5 - 2, 1 + 1)$ is true.

64. The statement $(10, 4) = (7 + 3, 5 - 1)$ is true.

65. The statement $(6, 3) = (3, 6)$ is false. The parentheses indicate an ordered pair (where order is important) and corresponding elements in the ordered pairs must be equal.

66. The statement $(2, 13) = (13, 2)$ is false. Corresponding elements in ordered pairs must be equal.

67. The statement $\{6, 3\} = \{3, 6\}$ is true since order is not important when listing elements in sets.

68. The statement $\{2, 13\} = \{13, 2\}$ is true since order is not important when listing elements in sets.

69. The statement $\{(1, 2), (3, 4)\} = \{(3, 4), (1, 2)\}$ is true. Each set contains the same two elements, the order of which is unimportant.

70. The statement $\{(5, 9), (4, 8), (4, 2)\} = \{(4, 8), (5, 9), (4, 2)\}$ is true. Each set contains the same three elements, the order of which is unimportant.

71. To form the Cartesian product $A \times B$, list all ordered pairs in which the first element belongs to A and the second element belongs to B:
With $A = \{2, 8, 12\}$ and $B = \{4, 9\}$,
$A \times B = \{(2, 4), (2, 9), (8, 4), (8, 9), (12, 4), (12, 9)\}$.

To form the Cartesian product $B \times A$, list all ordered pairs in which the first element belongs to B and the second element belongs to A:
$B \times A = \{(4, 2), (4, 8), (4, 12), (9, 2), (9, 8), (9, 12)\}$.

72. For $A = \{3, 6, 9, 12\}$ and $B = \{6, 8\}$,
$A \times B =$
$\{(3, 6), (3, 8), (6, 6), (6, 8), (9, 6), (9, 8), (12, 6), (12, 8)\}$;
$B \times A =$
$\{(6, 3), (6, 6), (6, 9), (6, 12), (8, 3), (8, 6), (8, 9), (8, 12)\}$.

73. For $A = \{d, o, g\}$ and $B = \{p, i, g\}$,
$A \times B = \{(d, p), (d, i), (d, g), (o, p), (o, i), (o, g), (g, p), (g, i), (g, g)\}$;
$B \times A = \{(p, d), (p, o), (p, g), (i, d), (i, o), (i, g), (g, d), (g, o), (g, g)\}$.

74. For $A = \{b, l, u, e\}$ and $B = \{r, e, d\}$,
$A \times B = \{(b, r), (b, e), (b, d), (l, r), (l, e), (l, d), (u, r), (u, e), (u, d), (e, r), (e, e), (e, d)\}$;
$B \times A = \{(r, b), (r, l), (r, u), (r, e), (e, b), (e, l), (e, u), (e, e), (d, b), (d, l), (d, u), (d, e)\}$.

75. For $A = \{2, 8, 12\}$ and $B = \{4, 9\}$,
$n(A \times B) = n(A) \times n(B) = 3 \times 2 = 6$, or by counting the generated values in Exercise 71 we also arrive at 6.
In the same manner, $n(B \times A) = 2 \times 3 = 6$.

76. For $A = \{d, o, g\}$ and $B = \{p, i, g\}$,
$n(A \times B) = n(A) \times n(B) = 3 \times 3 = 9$, or by counting the generated values in Exercise 73 we also arrive at 9.
In the same manner, $n(B \times A) = 3 \times 3 = 9$.

77. For $n(A) = 35$ and $n(B) = 6$,
$n(A \times B) = n(A) \times n(B) = 35 \times 6 = 210$.
$n(B \times A) = n(B) \times n(A) = 6 \times 35 = 210$.

78. For $n(A) = 13$ and $n(B) = 5$,
$n(A \times B) = n(A) \times n(B) = 13 \times 5 = 65$.
$n(B \times A) = n(B) \times n(A) = 5 \times 13 = 65$.

26 CHAPTER 2 THE BASIC CONCEPTS OF SET THEORY

79. To find $n(B)$ when $n(A \times B) = 36$ and $n(A) = 12$, we have:
$$n(A \times B) = n(A) \times n(B)$$
$$36 = 12 \times n(B)$$
$$3 = n(B).$$

80. To find $n(A)$ when $n(A \times B) = 100$ and $n(B) = 4$, we have:
$$n(A \times B) = n(A) \times n(B)$$
$$100 = n(A) \times 4$$
$$25 = n(A).$$

81. Let $U = \{a, b, c, d, e, f, g\}$,
$A = \{b, d, f, g\}$, and $B = \{a, b, d, e, g\}$.

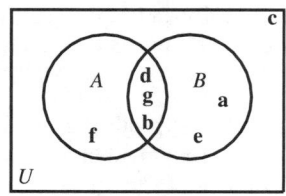

82. Let $U = \{5, 6, 7, 8, 9, 10, 11, 12, 13\}$,
$M = \{5, 8, 10, 11\}$, and $N = \{5, 6, 7, 9, 10\}$.

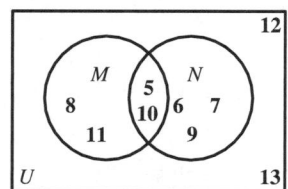

83. The set operations for $B \cap A'$ indicate those elements in B and not in A.

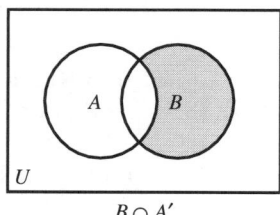

$B \cap A'$

84. The set operation for $A \cup B$ indicates those elements in A or in B.

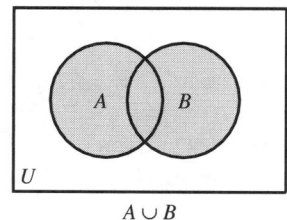

$A \cup B$

85. The set operations for $A' \cup B$ indicate those elements not in A or in B.

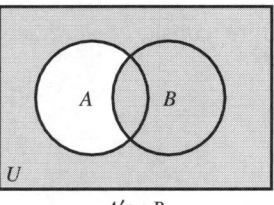

$A' \cup B$

86. The set operations for $A' \cap B'$ indicate those elements not in A and, at the same time, not in B.

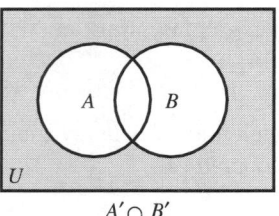

$A' \cap B'$

87. The set operations for $B' \cup A$ indicate those elements not in B or in A.

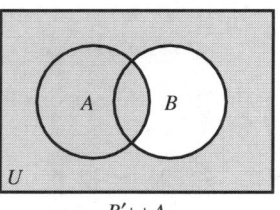

$B' \cup A$

88. The set operations for $A' \cup A$ indicate all elements not in A along with those in A, that is, the universal set, U.

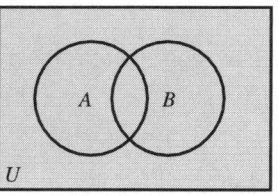

$A' \cup A = U$

89. The set operations $B' \cap B$ indicate those elements not in B and in B at the same time, and since there are no elements that can satisfy both conditions, we get the null set (empty set), $\emptyset$.

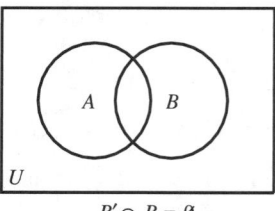

$B' \cap B = \emptyset$

90. The set operation $A \cap B'$ indicate all elements in A which are not in B.

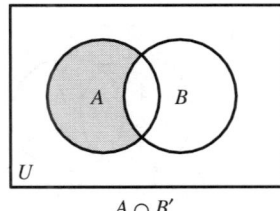

$A \cap B'$

91. The indicated set operations mean those elements not in B or those not in A as long as they are also not in B. It is a help to shade the region representing "not in A" first, then that region representing "not in B." Identify the intersection of these regions (covered by both shadings). As in algebra, the general strategy when deciding which order to do operations is to begin inside parentheses and work out.

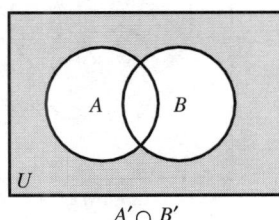

$A' \cap B'$

Finally, the region of interest will be that "not in B" along with (union of) the above intersection—$(A' \cap B')$. That is, the final region of interest is given by

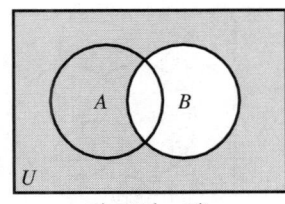

$B' \cup (A' \cap B')$

92. The set operations $(A \cap B) \cup B$ indicate all those elements in A and B at the same time or those in B. This is the same set as B itself.

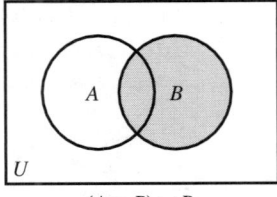

$(A \cap B) \cup B$

93. The complement of U, U', is the set of all elements not in U. But by definition, there can be no elements outside the universal set. Thus, we get the null (or empty) set, $\emptyset$, when we complement U.

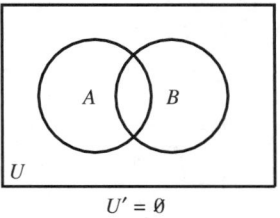

$U' = \emptyset$

94. The complement of the empty set, $\emptyset'$, is the set of all elements outside of $\emptyset$. But this is the universal set, U.

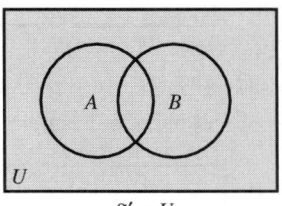

$\emptyset' = U$

95. Let $U = \{m, n, o, p, q, r, s, t, u, v, w\}$,
$A = \{m, n, p, q, r, t\}$,
$B = \{m, o, p, q, s, u\}$, and
$C = \{m, o, p, r, s, t, u, v\}$.

Placing the elements of these sets in the proper location on a Venn diagram will yield the following diagram.

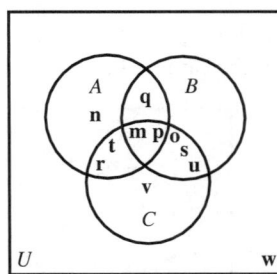

It helps to identify those elements in the intersection of A, B, and C first, then those elements not in this intersection but in each of the two set intersections (e.g., $A \cap B$, etc.), next, followed by elements that lie in only one set, etc.

96. Let $U = \{1, 2, 3, 4, 5, 6, 7, 8, 9\}$,
$A = \{1, 3, 5, 7\}$,
$B = \{1, 3, 4, 6, 8\}$, and
$C = \{1, 4, 5, 6, 7, 9\}$.

Placing the elements of these sets in the proper location on a Venn diagram will yield the following diagram.

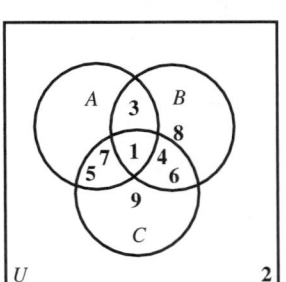

97. The set operations $(A \cap B) \cap C$ indicate those elements common to all three sets.

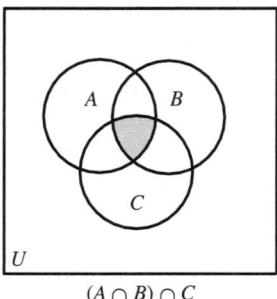

$(A \cap B) \cap C$

98. The set operations $(A \cap C') \cup B$ indicate those elements which are in A and outside C, or are found in B.

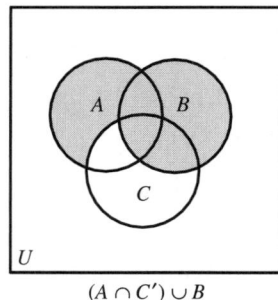

$(A \cap C') \cup B$

99. The set operations $(A \cap B) \cup C'$ indicate those elements in A and B at the same time along with those outside of C.

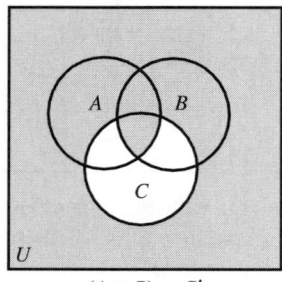

$(A \cap B) \cup C'$

100. The set operations $(A' \cap B) \cap C$ indicate those elements not in A but which are in B and also in C.

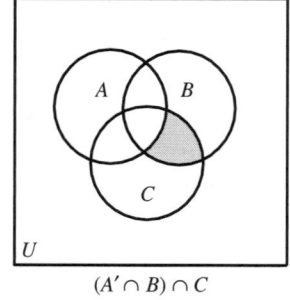

$(A' \cap B) \cap C$

101. The set operations $(A' \cap B') \cap C$ indicate those elements that are in C while simultaneously outside of both A and B.

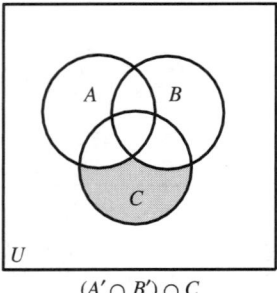

$(A' \cap B') \cap C$

102. The set operations $(A \cup B) \cup C$ indicate all elements in A or in B or in C, that is, all elements in each of the three sets.

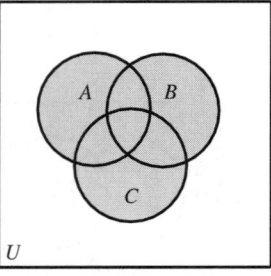

$(A \cup B) \cup C$

103. The set operations $(A \cap B') \cup C$ indicate those elements that are in A and at the same time outside of B, along with those in C.

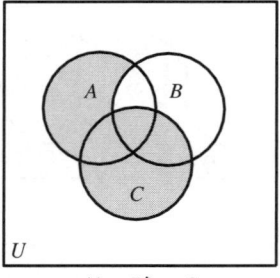

$(A \cap B') \cup C$

104. The set operations $(A \cap C') \cap B$ indicate those elements which are in A and outside of C while at the same time, inside B.

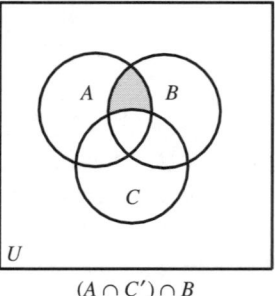

$(A \cap C') \cap B$

105. The set operations $(A \cap B') \cap C'$ indicate the region in A and outside B and at the same time outside C.

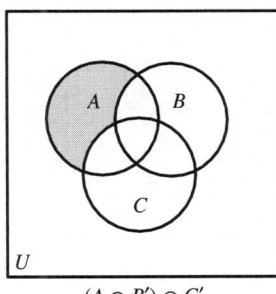

$(A \cap B') \cap C'$

106. The set operations $(A' \cap B') \cup C$ indicate those elements outside of A and outside of B at the same time, along with all elements of C.

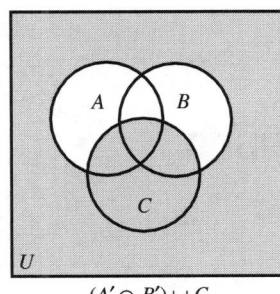

$(A' \cap B') \cup C$

107. The set operations $(A' \cap B') \cup C'$ indicate the region that is both outside A and at the same time outside B, along with the region outside C.

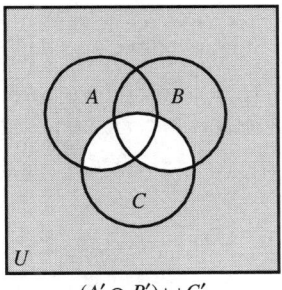

$(A' \cap B') \cup C'$

108. The set operations $(A \cap B)' \cup C$ indicate those elements not common to A and B, along with all elements of C.

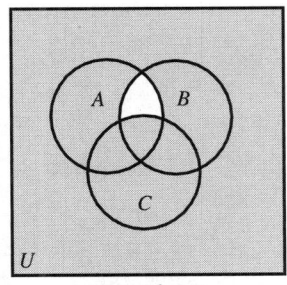

$(A \cap B)' \cup C$

109. The shaded area indicates the region $(A \cup B)'$ or $A' \cap B'$.

110. Since only $A - B$ is unshaded, we get the shaded region by $(A - B)'$. This may also be indicated by the set $(A \cap B')'$ or, using one form of De Morgan's law, as $(A' \cup B)$.

111. Since this is the region in A or in B but, at the same time, outside of A and B, we have the set $(A \cup B) \cap (A \cap B)'$ or $(A \cup B) - (A \cap B)$.

112. Since this region represents all elements in A which are outside of B, the area can be indicated by the set $A \cap B'$ or $A - B$.

113. The shaded area may be represented by the set $(A \cap B) \cup (A \cap C)$; that is, the region in the intersection of A and B along with the region in the intersection of A and C or, by the distributive property, $A \cap (B \cup C)$.

114. The shaded area is in A and, at the same time, outside the union of B with C. This region can be represented by the set $A \cap (B \cup C)'$ or $A \cap (B' \cap C')$.

115. The region is represented by the set $(A \cap B) \cap C'$, that is, the region outside of C but inside both A and B, or $(A \cap B) - C$.

116. The shaded area is the part of $B \cup C$ which is outside of A. This set is $(B \cup C) \cap A'$ or $(B \cup C) - A$.

117. If $A = A - B$, then A and B must not have any common elements, or $A \cap B = \emptyset$.

118. $A = B - A$ is true only if $A = B = \emptyset$.

119. $A = A - \emptyset$ is true for any set A.

120. $A = \emptyset - A$ is true only if A has no elements, or $A = \emptyset$.

121. $A \cup \emptyset = \emptyset$ is true only if A has no elements, or $A = \emptyset$.

122. $A \cap \emptyset = \emptyset$ is true for any set A.

123. $A \cap \emptyset = A$ is true only if A has no elements, or $A = \emptyset$.

124. $A \cup \emptyset = A$ is true for any set A.

125. $A \cup A = \emptyset$ is true only if A has no elements, or $A = \emptyset$.

126. Since $A \cap A = A$, for $A \cap A = \emptyset$, A must be the empty set, or $A = \emptyset$.

127. $A \cap A' = \emptyset$.

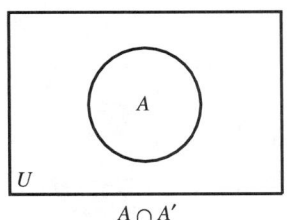 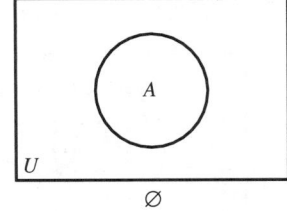

$A \cap A'$ $\emptyset$

Thus, by the Venn diagrams, the statement is always true.

128. $A \cup A' = U$.

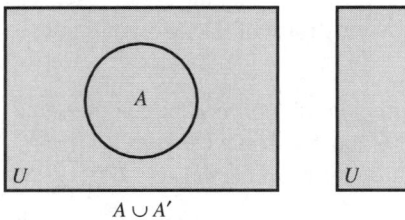

Thus, by the Venn diagrams, the statement is always true.

129. $(A \cap B) \subseteq A$.

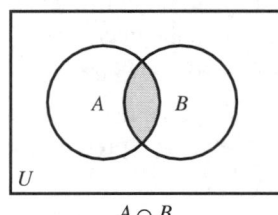 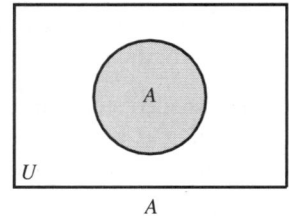

Thus, by the Venn diagrams, the shaded region is in A; therefore, the statement is always true.

130. $(A \cup B) \subseteq A$.

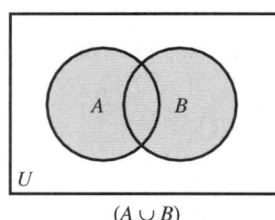 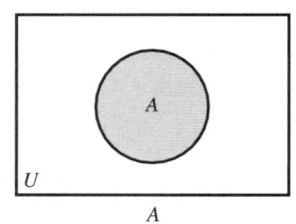

Thus, the statement is not always true.

131. If $A \subseteq B$, then $A \cup B = A$.

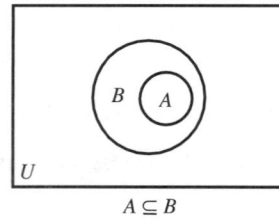 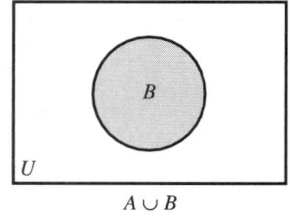

Thus, the statement is not always true.

132. If $A \subseteq B$, then $A \cap B = B$.

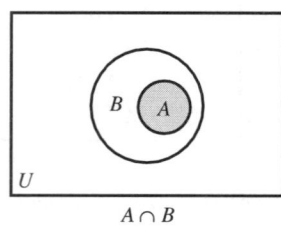 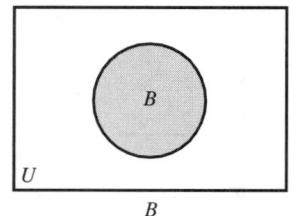

Thus, the statement is not always true.

133. $(A \cup B)' = A' \cap B'$ (De Morgan's second law).

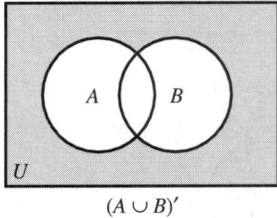 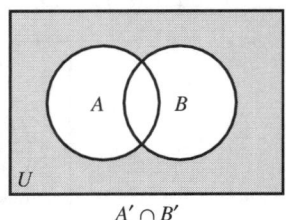

Thus, by the Venn diagrams, the statement is always true.

134. Writing exercise

135. Writing exercise

136. (a) $Q \cup H = \{x \mid x \text{ is a real number}\}$, since the real numbers are made up of all rational and all irrational numbers.

(b) $Q \cap H = \emptyset$, since there are no common elements.

2.4 EXERCISES

1. (a) $n(A \cap B) = 1$ since A and B share only one element.

   (b) $n(A \cup B) = 9$, since there are a total of 9 elements in A or in B.

   (c) $n(A \cap B') = 6$ since there are 6 elements which are in A and, at the same time, outside B.

   (d) $n(A' \cap B) = 2$ since there are 2 elements which are in B and, at the same time, outside A.

   (e) $n(A' \cap B') = 5$ since there are 5 elements which are outside of A and, at the same time, outside of B.

2. (a) $n(A \cap B) = 6$ since A and B have 6 elements in common.

   (b) $n(A \cup B) = 8$ since there are a total of 8 elements in A or in B.

   (c) $n(A \cap B') = 0$ since there are 0 elements which are in A and, at the same time, outside B.

   (d) $n(A' \cap B) = 2$ since there are 2 elements which are in B and, at the same time, outside A.

   (e) $n(A' \cap B') = 9$ since there are 9 elements which are outside of A and, at the same time, outside of B.

3. (a) $n(A \cap B \cap C) = 1$ since there is only one element shared by all three sets.

   (b) $n(A \cap B \cap C') = 2$ since there are 2 elements in A and B while, at the same time, outside of C.

   (c) $n(A \cap B' \cap C) = 6$ since there are 6 elements in A and C while, at the same time, outside of B.

(d) $n(A' \cap B \cap C) = 7$ since there are 7 elements which are outside of A while, at the same time, in B and C.

(e) $n(A' \cap B' \cap C) = 8$ since there are 8 elements outside of A and outside of B while, at the same time, inside of C.

(f) $n(A \cap B' \cap C') = 3$ since there are 3 elements in A which, at the same time, are outside of B and outside of C.

(g) $n(A' \cap B \cap C') = 4$ since there are 4 elements outside of A and, at the same time, outside of C but inside of B.

(h) $n(A' \cap B' \cap C') = 5$ since there are 5 elements which are outside all three sets at the same time.

4. (a) $n(A \cap B \cap C) = 1$ since there is only one element shared by all three sets.

(b) $n(A \cap B \cap C') = 3$ since there are 3 elements in A and B while, at the same time, outside of C.

(c) $n(A \cap B' \cap C) = 4$ since there are 4 elements in A and C while, at the same time, outside of B.

(d) $n(A' \cap B \cap C) = 0$ since there are 0 elements which are outside of A while, at the same time, in B and C.

(e) $n(A' \cap B' \cap C) = 2$ since there are 2 elements outside of A and outside of B while, at the same time, in C.

(f) $n(A \cap B' \cap C') = 10$ since there are 10 elements in A which, at the same time, are outside of B and outside of C.

(g) $n(A' \cap B \cap C') = 2$ since there are 2 elements outside of A and, at the same time, outside of C but inside of B.

(h) $n(A' \cap B' \cap C') = 5$ since there are 5 elements which are outside all three sets at the same time.

5. Using the Cardinal Number Formula,
$n(A \cup B) = n(A) + n(B) - n(A \cap B)$, we have
$n(A \cup B) = 8 + 14 - 5 = 17$.

6. Using the Cardinal Number Formula,
$n(A \cup B) = n(A) + n(B) - n(A \cap B)$, we have
$25 = 15 + 12 - n(A \cap B)$. Solving for $n(A \cap B)$, we get $n(A \cap B) = 2$.

7. Using the Cardinal Number Formula,
$n(A \cup B) = n(A) + n(B) - n(A \cap B)$, we have
$30 = n(A) + 20 - 6$. Solving for $n(A)$, we get $n(A) = 16$.

8. Using the Cardinal Number Formula,
$n(A \cup B) = n(A) + n(B) - n(A \cap B)$, we have
$55 = 35 + n(B) - 15$. Solving for $n(B)$, we get $n(B) = 35$.

9. Use deduction to complete the cardinalities of the unknown regions. For example since $n(B') = 30$, there are 13 elements in B $[n(U) - n(B') = 43 - 30]$; therefore, 8 elements are in B that are not in A $[n(B) - n(A \cap B) = 13 - 5]$. Since there is a total of 25 elements in A and 5 are accounted for in $A \cap B$, there must be 20 elements in A that are not in B. This leaves a total of 33 elements in the regions formed by A along with B. Thus, there are 10 elements left in U that are not in A or in B. Completing the cardinalities for each region, we arrive at the following Venn diagram.

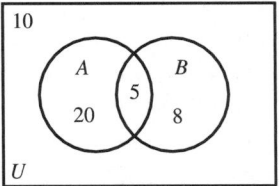

10. Using the Cardinal Number Formula, we find that
$$n(A \cup B) = n(A) + n(B) - n(A \cap B)$$
$$25 = 19 + 13 - n(A \cap B)$$
$$n(A \cap B) = 19 + 13 - 25$$
$$= 7.$$

Then $n(A \cap B') = 19 - 7 = 12$ and
$n(B \cap A') = 13 - 7 = 6$.
Since $n(A') = 11$ and
$n(B \cap A') = 6$, we have
$n(A \cup B)' = n(A' \cap B') = 11 - 6 = 5$.

Completing the cardinalities for each region, we arrive at the following Venn diagram.

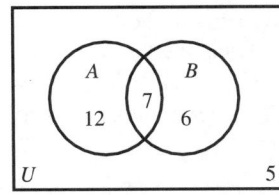

11. Use deduction to complete the cardinalities of each region outside of $A \cap B$. Since there is a total of 13 elements in A and 8 accounted for in the intersect of A and B, there must be 5 elements in A outside of B. Since there is a total of 15 elements in the union of A and B with 13 accounted for in A, there must be 2 elements in B which are not in A. The region $A' \cup B'$ is equivalent (by De Morgan's law) to $(A \cap B)'$, or the elements outside the intersection. Since this totals 11, we must have 4 $[11 - 7]$ elements outside the union but inside U. Completing the cardinalities for each region, we arrive at the following Venn diagram.

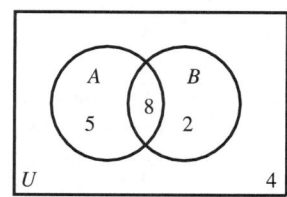

12. Since $n(B) = 28$ and $n(A \cap B) = 10$, we have
    $n(B \cap A') = 28 - 10 = 18$. Since $n(A') = 25$ and
    $n(B \cap A') = 18$, it follows that $n(A' \cap B') = 7$.
    By De Morgan's laws, $A' \cup B' = (A \cap B)'$.
    So, $n[(A \cap B)'] = 40$.
    Thus, $n(A \cap B') = 40 - (18 + 7) = 15$.
    Completing the cardinalities for each region, we arrive at the following Venn diagram.

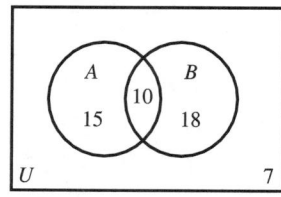

13. Fill in the cardinal numbers of the regions, beginning with the $(A \cap B \cap C)$. Since there is a total of 15 elements in $A \cap C$ of which 6 are accounted for in $A \cap B \cap C$, we conclude that there are 9 elements in $A \cap C$ but outside of B. Similarly, there must be 2 elements in $B \cap C$ but outside A and 4 elements in $A \cap B$ but outside of C. Since there are 26 elements in C of which we have accounted for 17 $[9 + 6 + 2]$, there must be 9 elements in C but outside of A or B. Similarly, there are 12 elements in B outside of A or C and 5 elements in A outside of B or C. And finally, adding the elements in the regions of A, B, and C gives a total of 47. Thus, the number of elements outside of A, B, and C is $n(U) - 47$ or $50 - 47 = 3$. Completing the cardinalities for each region, we arrive at the following Venn diagram.

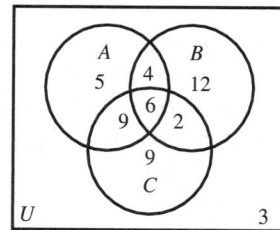

14. Fill in the cardinal numbers of the regions, beginning with $(A \cap B \cap C)$. Since $n(A \cap B \cap C) = 15$ and $n(A \cap B) = 35$, we have
    $n(A \cap B \cap C') = 35 - 15 = 20$. Since $n(A \cap C) = 21$, we have $n(A \cap C \cap B') = 21 - 15 = 6$. Since $n(B \cap C) = 25$, we have
    $n(B \cap C \cap A') = 25 - 15 = 10$. Since $n(C) = 49$, we have $n(C \cap A' \cap B') = 49 - (6 + 15 + 10) = 18$.

Since $n(A) = 57$, we have
$n(A \cap B' \cap C') = 57 - (20 + 15 + 6) = 16$. Since $n(B') = 52$, we have
$n(A \cup B \cup C)' = n(A' \cap B' \cap C') =$
$52 - (16 + 6 + 18) = 12$.
Completing the cardinalities for each region, we arrive at the following Venn diagram.

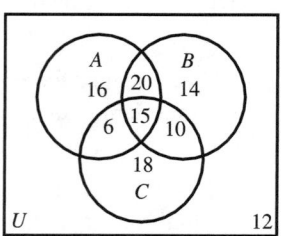

15. Fill in the cardinal numbers of the regions, beginning with $n(A \cap B \cap C) = 6$. Since there are 21 elements in A and B and 6 elements in all three sets, the number of elements in A and B but not in C is given by $n(A \cap B \cap C') = 21 - 6 = 15$. Since $n(A \cap C) = 26$, we have those elements in A and C but not in B as $n(A \cap C \cap B') = 26 - 6 = 20$. Since $n(B \cap C) = 7$, $n(B \cap C \cap A') = 7 - 6 = 1$. Since $n(A \cap C') = 20$, we have those elements in A but not in either B or C as $n(A \cap (B \cup C)') = 20 - 15 = 5$. Since $n(B \cap C') = 25$, we have those elements in B but not in A or C as $n(B \cap (A \cup C)') = 25 - 15 = 10$. Since $n(C) = 40$, we have those elements in C but not in A or B as
$n(C \cap (A \cup B)') = 40 - (20 + 6 + 1) = 13$. Observe that $n(A' \cap B' \cap C') = n(A \cup B \cup C)'$ (by De Morgan's). That is, there are 2 elements outside the union of the three sets.

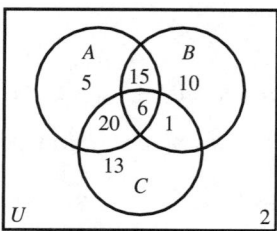

16. Fill in the cardinal numbers of the regions, beginning with the $(A \cap B \cap C)$. $n(A \cap B \cap C) = 5$ and $n(A \cap C) = 13$, so $n(A \cap C \cap B') = 13 - 5 = 8$.
    $n(B \cap C) = 8$, so $n(B \cap C \cap A') = 8 - 5 = 3$.
    $n(A \cap B') = 9$, so $n(A \cap B' \cap C') = 9 - 8 = 1$.
    $n(A) = 15$, so $n(A \cap B \cap C') = 15 - (1 + 8 + 5) = 1$.
    $n(B \cap C') = 3$, so $n(B \cap A' \cap C') = 3 - 1 = 2$.
    $n(A' \cap B' \cap C') = n(A \cup B \cup C)' = 21$.

Completing the cardinalities for each region, we arrive at the following Venn diagram.

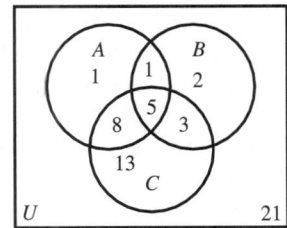

17. Complete a Venn diagram showing the cardinality for each region. Let S = the set of CDs featuring Paul Simon, G = the set of CDs featuring Art Garfunkel. Beginning with $n(S \cap G) = 5$ and $n(S) = 8$, we conclude that $n(S \cap G') = 8 - 5 = 3$. Since $n(G) = 7$, we conclude that $n(S' \cap G) = 7 - 5 = 2$. There are 12 CDs on which neither sing, so $n(S \cup G)' = 12$

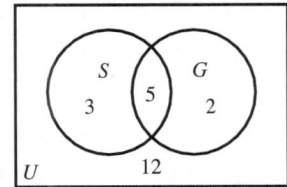

Interpreting the resulting cardinalities we see that:

(a) There are 3 CDs which feature only Paul Simon.

(b) There are 2 CDs which feature only Art Garfunkel.

(c) There are 10 CDs which feature at least one of these two artists.

Complete a Venn diagram showing the cardinality for each region. Let W = set of projects Kent LaVoie writes. Let P = set of projects Kent LaVoie produces.

Begin with $(W \cap P)$. $n(W \cap P) = 2$. Since $n(W) = 5$, $n(W \cap P') = 5 - 2 = 3$. Since $n(P) = 7$, $n(P \cap W') = 7 - 2 = 5$.

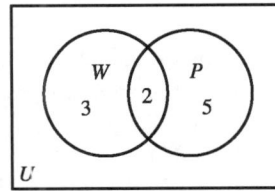

Interpreting the resulting cardinalities we see that:

(a) He wrote but did not produce $n(W \cap P') = 3$ projects.

(b) He produced but did not write $n(P \cap W') = 5$ projects.

18. Complete a Venn diagram showing the cardinality for each region. Let W = set of projects Kent LaVoie writes. Let P = set of projects Kent LaVoie produces.

Begin with $(W \cap P)$. $n(W \cap P) = 2$. Since $n(W) = 5$, $n(W \cap P') = 5 - 2 = 3$. Since $n(P) = 7$, $n(P \cap W') = 7 - 2 = 5$.

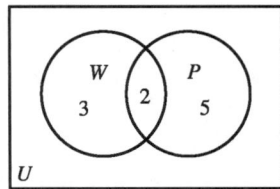

Interpreting the resulting cardinalities we see that:

(a) He wrote but did not produce, $n(W \cap P') = 3$ projects.

(b) He produced but did not write, $n(P \cap W') = 5$ projects.

19. Let U be the set of mathematics majors receiving federal aid. Since half of the 48 mathematics majors receive federal aid, $n(U) = \frac{1}{2}(48) = 24$. Let $P, W,$ and T be the sets of students receiving Pell Grants, participating in Work Study, and receiving TOPS scholarships, respectively.

Construct a Venn diagram and label the number of elements in each region. Since the 5 with Pell Grants had no other federal aid, 5 goes in set P, which does not intersect the other regions. The most manageable data remaining are the 2 who had TOPS scholarships and participated in Work Study. Place 2 in the intersection of W and T. Then since 14 altogether participated in Work Study, $14 - 2 = 12$ is the number who participated in Work Study but did not have TOPS scholarships or Pell Grants; in symbols, $n(W \cap T' \cap P')$. Also, $4 - 2 = 2$ is the number who had TOPS scholarships, but did not participate in Work Study nor had Pell Grants, $n(T \cap W' \cap P')$. Finally,

$$n(P' \cap W' \cap T') = n(P \cup W \cup T)'$$
$$= 24 - (5 + 12 + 2 + 2)$$
$$= 3.$$

is the number in the region not included in the sets $P, W,$ and T. The completed Venn diagram is as follows.

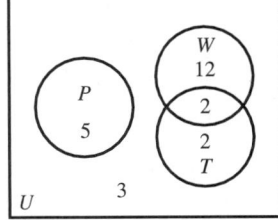

(a) Since half of the 48 mathematics majors received federal aid, the other half, 24 math majors, received no federal aid.

(b) There were only 2 students who received more than one of these three forms of aid. This is shown by the number in the region $T \cap W$.

(c) Since 24 students received federal aid, but only 21 are accounted for in the given information (the sum of the numbers in the three circles), there were 3 math majors who received other kinds of federal aid.

(d) The number of students receiving a TOPS scholarship or participating in Work Study is $12 + 2 + 2 = 16$.

20. Construct a Venn diagram and label the number of elements in each region. Choose set N to represent those who saw *The Natural*, set F to represent those who saw *Field of Dreams*, and set R to represent those who saw the *Rookie*. Begin with the region indicating the intersection of all three sets, $n(N \cap F \cap R) = 2$. Since $n(N \cap F) = 6, n(N \cap F \cap R') = 6 - 2 = 4$. Since $n(N \cap R) = 8, n(N \cap R \cap F') = 8 - 2 = 6$. Since $n(F \cap R) = 10, n(F \cap R \cap N') = 10 - 2 = 8$. Since $n(N) = 17$, the number of elements inside N and not in F or R is $17 - (6 + 2 + 4) = 5$. Since $n(F) = 17$, the number of elements inside F and not in N or R is $17 - (4 + 2 + 8) = 3$. Since $n(R) = 23$, the number of elements inside R and not in N or F is $23 - (6 + 2 + 8) = 7$. Since $n(U) = 55$, there are $55 - (5 + 4 + 3 + 6 + 2 + 8 + 7) = 55 - 35 = 20$ elements outside the three sets. That is, $n(N \cup F \cup R)' = 20$.
The completed Venn diagram is as follows.

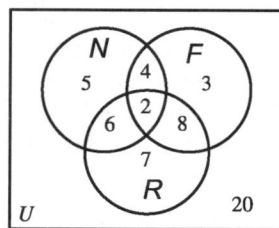

(a) There are $4 + 8 + 6 = 18$ that have seen exactly two of these movies.

(b) There are $5 + 3 + 7 = 15$ that have seen exactly one of these movies.

(c) There are $n(S \cup G \cup H)' = 20$ that have seen none of these movies.

(d) There are $n(S \cap G' \cap H') = 5$ students who have seen *The Natural* but neither of the others.

21. Let $S = \{\text{people who like Spañada}\}$,
$R = \{\text{people who like Ripple}\}$, and
$B = \{\text{people who like Boone's Farm Apple wine}\}$.

Construct a Venn diagram to represent the survey data beginning with the region representing the intersection of S, R, and B. Rather than representing each region as a combination of sets and set operations, we will label the regions a–h. Since 93 like all three, place 93 in region d. Since 96 like Spañada and Boone's with region d accounting for 93 of the 96, place $96 - 93 = 3$ in region e. Similarly, place $94 - 93 = 1$ in region g, and $95 - 93 = 2$ in region c. The 99 people who like Boone's are in regions d, e, f, and g, and regions d, e, and g account for $93 + 3 + 1 = 97$ people. Thus, region f represents 2 people. Similarly, region b represents 1 person, and region h represents 0 people. Adding the values placed in regions b–h gives 102, which is the total number of people surveyed, so region a represents 0 people.

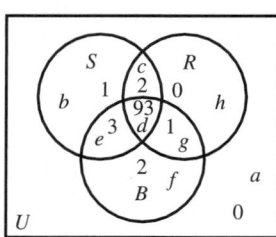

(a) Region a represents people preferring none of the three. There are 0 people in region a.

(b) $S \cap R'$ contains regions b and e, which represent a total of 4 people.

(c) B' contains regions a, b, c, and h, which account for 3 people.

(d) $R \cap S' \cap B' = R \cap (S \cup B)'$ is region h, which represents 0 people.

(e) Regions c, e, and g contain 6 people who prefer exactly two kinds of wine.

22. Let U be the set of people interviewed, and let M, E, and G represent the sets of people using microwave ovens, electric ranges, and gas ranges, respectively.

Construct a Venn diagram and label the cardinal number of each region, beginning with the region $(M \cap E \cap G)$. $n(M \cap E \cap G) = 1$. Since $n(M \cap E) = 19, n(M \cap E \cap G') = 19 - 1 = 18$. Since $n(M \cap G) = 17, n(M \cap G \cap E') = 17 - 1 = 16$. Since $n(G \cap E) = 4, n(G \cap E \cap M') = 4 - 1 = 3$. Since $n(M) = 58, n(M \cap G' \cap E') = n(M \cap (G \cup E)') = 58 - (18 + 16 + 1) = 23$. Since $n(E) = 63$, $n(E \cap M' \cap G') = 63 - (18 + 3 + 1) = 41$.

Since $n(G) = 58$,
$n(G \cap M' \cap E') = n(G \cap (M \cup E)') =$
$58 - (16 + 3 + 1) = 38$. Also, $n(M \cup E \cup G)' = 2$
(these are the people who cook with only solar energy).

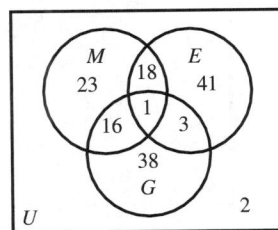

The sum of the numbers in all the regions is 142, while only 140 people were interviewed. Therefore, Robert's data is incorrect and he should be reassigned again.

23. Let L, P, and T represent the sets of songs about love, prison, and trucks, respectively.

    Construct a Venn diagram and label the cardinal number of each region.

    $n(T \cap L \cap P) = 12$.
    Since $n(P \cap L) = 13, n(P \cap L \cap T') = 13 - 12 = 1$.
    Since $n(T \cap L) = 18, n(T \cap L \cap P') = 18 - 12 = 6$.

    We have $n(T \cap P \cap L') = 3$, $n(P \cap L' \cap T') = 2$, and $n(P' \cap L' \cap T') = 8$. Since $n(L) = 28$,
    $n(L \cap P' \cap T') = 28 - (1 + 12 + 6) = 9$. Since
    $n(T \cap P') = 16, n(T \cap P' \cap L') = 16 - 6 = 10$.

    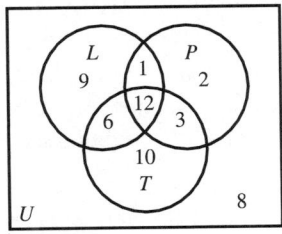

    (a) The total in all eight regions is 51, the total number of songs.

    (b) $n(T) = 6 + 12 + 3 + 10 = 31$.

    (c) $n(P) = 1 + 2 + 12 + 3 = 18$.

    (d) $n(T \cap P) = 12 + 3 = 15$.

    (e) $n(P') = n(U) - n(P) = 51 - 18 = 33$.

    (f) $n(L') = n(U) - n(L) = 51 - 28 = 23$.

24. Construct a Venn diagram as indicated in the textbook hint—a set for fat (F), for male (M), and for red (R). Observe that anything outside the F circle is thin, anything outside the M (rooster) circle is a hen, and anything outside the R (red) circle is brown.

    The associated regions and cardinalities are as given as follows:
    Fat red roosters— $n(F \cap R \cap M) = 9$,
    Fat red hens—$n(F \cap R \cap M') = 2$,
    Fat roosters—$n(F \cap M) = 26$,
    Fat chickens—$n(F) = 37$,
    Thin brown hens—$n(F' \cap R' \cap M') = 7$,
    Thin brown roosters—$n(F' \cap R' \cap M) = 18$,
    Thin red roosters—$n(F' \cap R \cap M) = 6$,
    Thin red hens—$n(F' \cap R \cap M') = 5$.

    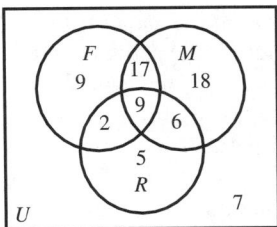

    Therefore, by observation and deduction, there are

    (a) $n(F) = 37$ fat chickens,

    (b) $n(R) = 22$ red chickens,

    (c) $n(M) = 50$ male chickens,

    (d) $n(F \cap M') = 11$ fat, but not male chickens,

    (e) $n(R' \cap F') = 25$ brown, but not fat chickens,

    (f) $n(R \cap F) = 11$ red and fat chickens.

25. Construct a Venn diagram and label the cardinal number of each region beginning with the intersection of all three sets, $n(W \cap F \cap E) = 80$. Since $n(E \cap F) = 90$,
    $n(E \cap F \cap W') = 90 - 80 = 10$.
    Since $n(W \cap F) = 95$,
    $n(W \cap F \cap E') = 95 - 80 = 15$.
    Since $n(E \cap F) = 90$,
    $n(E \cap F \cap W') = 90 - 80 = 10$.
    Since $n(F) = 140$,
    $n(F \cap W' \cap E') = 140 - (15 + 10 + 80) = 35$.
    $n(W' \cap F' \cap E') = n(W \cup F \cup E)' = 10$.
    Since $n(W \cap E)$ is not given, it is not obvious how to label the three remaining regions. We need to use the information that $n(E') = 95$. The only region not yet labeled that is outside of E is $(W \cap E' \cap F')$. Since

    $n(E') = 95, n(W \cap E' \cap F') = 95 - (10 + 35 + 15)$
    $\qquad = 35$. Since
    $n(W) = 160, n(W \cap E \cap F') = 160 - (35 + 15 + 80)$
    $\qquad = 30$. Since
    $n(E) = 130, n(E \cap W' \cap F') = 130 - (30 + 10 + 80)$
    $\qquad = 10$.

    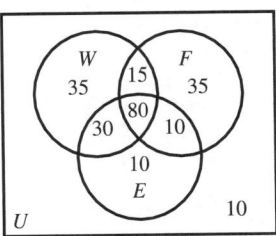

    Add the cardinal numbers of all the regions to find that the total number of students interviewed was 225.

26. Construct a Venn diagram to represent the survey data beginning with the region representing the intersection of $S, B,$ and C. Rather than representing each region as a combination of sets and set operations, we will label the regions a–h. There are 21 patients in Nadine's survey that are in the intersection of all three sets, i.e., in region d. Since there are 31 patients in $B \cap C$, we can deduce that there must be 10 patients in region c. Similarly since there are 33 patients in $B \cap S$, there must be 12 patients in region e. From the given information, there is a total of 51 patients in regions $c, d, e,$ and g. Thus, there are $51 - (10 + 21 + 12) = 8$ patients in region g. Since there are 52 patients in S, we can deduce that there are $52 - (12 + 21 + 8) = 11$ patients in region f. Similarly, there are 4 patients in region b, and 7 patients in region h. There is a total of 73 patients found in regions b–h. Thus, there must be 2 patients in region a.

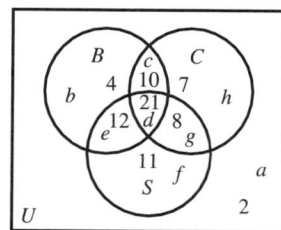

(a) The number of these patients who had either high blood pressure or high cholesterol levels, but not both is represented by regions $b, e, g,$ and h for a total of 31 patients.

(b) The number of these patients who had fewer than two of the indications listed are found in regions $a, b, f,$ and h for a total of 24 patients.

(c) The number of these patients who were smokers but had neither high blood pressure nor high cholesterol levels are found in region f, which has 11 members.

(d) The number of these patients who did not have exactly two of the indications listed would be those excluded from regions $c, e,$ and g (representing patients with exactly two of the indications). We arrive at a total of 45 patients.

27. (a) The set $A \cap B \cap C \cap D$ is region 1.

(b) The set $A \cup B \cup C \cup D$ includes the regions 1, 2, 3, 4, 5, 6, 7, 8, 9, 10, 11, 12, 13, 14, and 15.

(c) The set $(A \cap B) \cup (C \cap D)$ includes the set of regions $\{1, 3, 9, 11\} \cup \{1, 2, 4, 5\}$ or the regions 1, 2, 3, 4, 5, 9, and 11.

(d) The set $(A' \cap B') \cap (C \cup D)$ includes the set of regions $\{5, 13, 8, 16\} \cap \{1, 2, 3, 4, 5, 6, 7, 8, 9, 10, 12, 13\}$, which is represented by regions 5, 8, and 13.

28. Let $F, B, T,$ and G represent the sets of those who watch football, basketball, tennis, and golf respectively. Construct a Venn diagram, similar to the figure in Exercise 27 in the text, with the four sets. Be careful to indicate the number of elements in each region. Begin at the bottom of the list and work upwards.

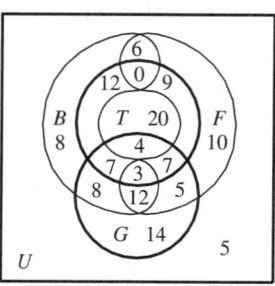

(a) The region which represents those who watch football, basketball, and tennis, but not golf, is in the north central portion of the diagram and has 0 elements. Thus, the answer is none.

(b) Adding cardinalities associated with the regions representing exactly one of these four sports, we get $8 + 10 + 14 + 20 = 52$ viewers.

(c) Adding cardinalities associated with the regions representing exactly two of these four sports, we get $6 + 9 + 5 + 8 + 12 + 4 = 44$ viewers.

29. (a) $n(W \cap O) = 6$, coming from the intersection of second row with the third column.

(b) $n(C \cup B) = 473$, coming from the union of the first row along with the first column is given by $95 + 390 - 12$. Observe that the 12 are counted twice when adding the totals for the respective row and column and hence must be subtracted (once).

(c) $n(R' \cup W') = n(R \cap W)' = 835$, which is the number of army personnel outside of the intersection of the second column (R) with the second row (W) or $840 - 5$.

(d) $n((C \cup W) \cap (B \cup R)) = 12 + 29 + 4 + 5 = 50$. This comes from adding the numbers in the first two rows (representing $C \cup W$) that are also in the first two columns (representing $B \cup R$).

(e) Using the cardinal number formula,
$$\begin{aligned} n((C \cap B) &\cup (E \cap O)) \\ &= n(C \cap B) + n(E \cap O) - n(C \cap B \cap E \cap O) \\ &= 12 + 285 - 0 = 297. \end{aligned}$$

(f) $\begin{aligned} n(B \cap (W \cup R)') &= n(B \cap (W' \cap R')) \\ &= n(B \cap W' \cap R') \\ &= n(B \cap W') \\ &= n(B) - n(B \cap W) \\ &= 390 - 4 \\ &= 386. \end{aligned}$

30. (a) $n(J \cap G) = 9$, coming from the intersection of the first row with the first column.

(b) $n(S \cap N) = 9$, coming from the intersection of second row and the third column.

(c) $n(N \cup (S \cap F)) = 20$ since there are 20 players who are in either N (total of 15) or in S and F (just 5), at the same time.

(d) $n(S' \cap (G \cup N)) = 20$ since there are $9 + 4 + 5 + 2 = 20$ players who are not in S, but are in G or in N.

(e) $n((S \cap N') \cup (C \cap G')) = 27$. There are 27 players who are in S but not in N (12 + 5), or who are in C but not in G (8 + 2).

(f) $n(N' \cap (S' \cap C')) = 15$. There are 15 (9 + 6) players who are not in N and at the same time are not in S and not in C.

31. Writing exercise

32. Writing exercise

2.5 EXERCISES

1. The set {3} has the same cardinality as B, {26}. The cardinal number is 1.

2. The cardinality of {a, b, c} is the same as D, $\{x, y, z\}$. The cardinal number is 3.

3. The set $\{x \mid x \text{ is a natural number}\}$ has the same cardinality as A, $\aleph_0$.

4. The set $\{x \mid x \text{ is a real number}\}$ has the same cardinality, c, as C.

5. The set $\{x \mid x \text{ is an integer between 0 and 1}\}$ has the same cardinality, 0, as F since there are no members in either set.

6. The set $\{x \mid x \text{ is an integer that satisfies } x^2 = 25\}$ has the same cardinality, 2, as E. Note that both sets contain the same elements $\{5, -5\}$.

7. One correspondence is:

$$\begin{array}{ccc} \{\text{I}, & \text{II}, & \text{III}\} \\ \updownarrow & \updownarrow & \updownarrow \\ \{x, & y, & z\}. \end{array}$$

Other correspondences are possible.

8. It is not possible to place the sets {a, b, c, d} and {2, 4, 6} in a one-to-one correspondence because the two sets do not have the same cardinal number. In other words, the two sets are not equivalent.

9. One correspondence is:

$$\begin{array}{cccccc} \{\text{a}, & \text{d}, & \text{i}, & \text{t}, & \text{o}, & \text{n}\} \\ \updownarrow & \updownarrow & \updownarrow & \updownarrow & \updownarrow & \updownarrow \\ \{\text{a}, & \text{n}, & \text{s}, & \text{w}, & \text{e}, & \text{r}\}. \end{array}$$

Other correspondences are possible.

10. One correspondence, pairing each president with his wife, is:

$$\begin{array}{ccc} \{\text{Reagan}, & \text{Clinton} & \text{Bush}\} \\ \updownarrow & \updownarrow & \updownarrow \\ \{\text{Nancy}, & \text{Hillary}, & \text{Laura}\}. \end{array}$$

Other correspondences are possible.

11. $n(\{a, b, c, d, \ldots, k\}) = 11$. By counting the number of letters a through k, we establish the cardinality to be 11.

12. $n(\{9, 12, 15, \ldots, 36\}) = 10$. By counting the multiples of 3 from 9 to 36, we see that the cardinality is 10.

13. $n(\emptyset) = 0$ since there are no members.

14. $n(\{0\}) = 1$ since there is 1 element.

15. $n(\{300, 400, 500, \ldots\}) = \aleph_0$ since this set can be placed in a one-to-one correspondence with the counting numbers (i.e., is a countable infinite set).

16. $n(\{-35, -28, -21, \ldots, 56\}) = 14$ because a complete listing of this set would show 14 elements where each element after the first is found by adding 7. Another way to evaluate the cardinality is to find the difference between the largest and smallest, $56 - (-35) = 91$. Divide 91 by 7 to find how many times 7 has been added to the first element: $\frac{91}{7} = 13$. Since 7 has been added 13 times to the first element to get to 91, we have 13 numbers plus the first number, or 14 in total, which is the cardinal number for the set.

17. $n(\{-1/4, -1/8, -1/12, \ldots\}) = \aleph_0$ since this set can be placed in a one-to-one correspondence with the counting numbers.

18. $n(\{x \mid x \text{ is an even integer}\}) = \aleph_0$ since this set can be placed in a one-to-one correspondence with the counting numbers.

19. $n(\{x \mid x \text{ is an odd counting number}\}) = \aleph_0$ since this set can be placed in a one-to-one correspondence with the counting numbers.

20. $n(\{b, a, 1, 1, a, d\}) = n(\{b, a, 1, d\}) = 4$ since there are only 4 distinct elements in the set.

21. $n(\{\text{Jan, Feb, Mar}, \ldots, \text{Dec}\}) = 12$ since there are twelve months indicated in the set.

22. $n(\{\text{Alabama, Alaska, Arizona}, \ldots, \text{Wisconsin, Wyoming}\}) = 50$, since there are fifty states of the United States.

23. "$\aleph_0$ bottles of beer on the wall, $\aleph_0$ bottles of beer, take one down and pass it around, $\boxed{\aleph_0}$ bottles of beer on the wall." This is true because $\aleph_0 - 1 = \aleph_0$.

24. (a) Let {a, b, c} represent the set of three actors and {d, e, f} represent the set of three roles played by the actors. As we can see, there are 6 different correspondences that can be shown.

{a, b, c}
↕ ↕ ↕
{d, e, f}

{a, b, c}
↕ ↕ ↕
{d, f, e}

{a, b, c}
↕ ↕ ↕
{e, d, f}

{a, b, c}
↕ ↕ ↕
{e, f, d}

{a, b, c}
↕ ↕ ↕
{f, e, d}

{a, b, c}
↕ ↕ ↕
{f, d, e}

(b)
{Mike Myers, Julia Roberts, William Shatner}
↕ ↕ ↕
{Austin Powers, Erin Brockovich, Captain James T. Kirk}.

25. The answer is both. Since the sets {u, v, w} and {v, u, w} are equal sets (same elements), they must then have the same number of elements and thus are equivalent.

26. The sets {48, 6} and {4, 86} are equivalent because they contain the same number of elements (same cardinality) but not the same elements.

27. The sets {X, Y, Z} and {x, y, z} are equivalent because they contain the same number of elements (same cardinality) but not the same elements.

28. The sets {lea} and {ale} are equivalent because they contain the same number of elements (one) but not the same elements.

29. The sets $\{x \mid x \text{ is a positive real number}\}$ and $\{x \mid x \text{ is a negative real number}\}$ are equivalent because they have the same cardinality, c. They are not the equal since they contain different elements.

30. The sets $\{x \mid x \text{ is a positive rational number}\}$ has a cardinality of $\aleph_0$ while the set $\{x \mid x \text{ is a negative rational number}\}$ has a cardinality of c. Since the two sets do not contain the same elements and do not have the same cardinality, they are not equal nor are they equivalent.

Note that each of the following answers shows only one possible correspondence.

31. $\{2, 4, 6, 8, 10, 12, \ldots, 2n, \ldots\}$
   ↕ ↕ ↕ ↕ ↕ ↕ ↕
   $\{1, 2, 3, 4, 5, 6, \ldots, n, \ldots\}$

32. $\{-10, -20, -30, -40, \ldots, -10n, \ldots\}$
   ↕ ↕ ↕ ↕ ↕
   $\{1, 2, 3, 4, \ldots, n, \ldots\}$

33. $\{1{,}000{,}000 \ 2{,}000{,}000 \ 3{,}000{,}000, \ldots, 1{,}000{,}000n, \ldots\}$
   ↕ ↕ ↕ ↕
   $\{1, 2, 3, \ldots, n, \ldots\}$

34. We can establish a one-to-one correspondence between the set of odd integers and the set of counting numbers as follows.
   $\{1, -1, 3, -3, 5, -5, \ldots, a, \ldots\}$
   ↕ ↕ ↕ ↕ ↕ ↕ ↕
   $\{1, 2, 3, 4, 5, 6, \ldots, n, \ldots\}$

   Where $a = n$ if n is odd and $a = 1 - n$ if n is even.

35. $\{2, 4, 8, 16, 32, \ldots, 2^n, \ldots\}$
   ↕ ↕ ↕ ↕ ↕ ↕
   $\{1, 2, 3, 4, 5, \ldots, n, \ldots\}$

36. $\{-17, -22, -27, -32, \ldots, -5n - 12, \ldots\}$
   ↕ ↕ ↕ ↕ ↕
   $\{1, 2, 3, 4, \ldots, n, \ldots\}$

37. The statement "If A and B are infinite sets, then A is equivalent to B" is not always true. For example, let $A = $ the set of counting numbers and $B = $ the set of real numbers. Each has a different cardinality.

38. The statement, "If set A is an infinite set and set B can be put in one-to-one correspondence with a proper subset of A, then B must be infinite," is not always true. For example, let $A = $ the set of counting numbers and $B = \{a, b, c\}$. B can be put in a one-to-one correspondence with a proper subset of A as follows:

   $\{a, b, c\}$
   ↕ ↕ ↕
   $\{1, 2, 3\}$

39. The statement "If set A is an infinite set and A is not equivalent to the set of counting numbers, then $n(A) = c$" is not always true. For example, A could be the set of all subsets of the set of real numbers. Then, $n(A)$ would be an infinite number greater than c.

40. The statement "If A and B are both countably infinite sets, then $n(A \cup B) = \aleph_0$" is always true. Set A can be put in a one-to-one correspondence with the set of counting numbers. By dropping elements in B that are in A, the remaining elements, whether a finite number or not, can be added to the list of A to produce $A \cup B$. The cardinal number of that set is $\aleph_0$.

41. (a) Use the figure (in the text), where the line segment between 0 and 1 has been bent into a semicircle and positioned above the line, to prove that $\{x \mid x \text{ is a real number between 0 and 1}\}$ is equivalent to $\{x \mid x \text{ is a real number}\}$.

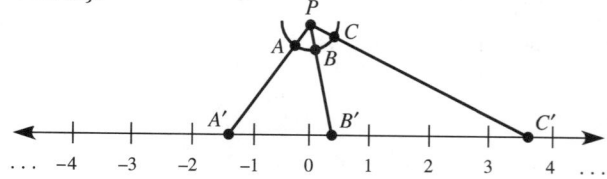

Rays emanating from point P will establish a geometric pairing of the points on the semicircle with the points on the line.

(b) The fact part (a) establishes about the set of real numbers is that the set of real numbers is infinite, having been placed in a one-to-one correspondence with a proper subset of itself.

42. Draw a dashed line through the upper endpoints of the two segments. Label the point where this line intersects the base line as P. Choose any point A on the shorter segment; draw a line through P and A, intersecting the longer segment at a point we label A'. Now choose a point B' on the longer segment. Follow the same procedure to find the corresponding point B on the shorter segment. In this manner, we can set up a one-to-one correspondence between all the points on the two given line segments. This correspondence shows that the sets of points on the two segments have the same cardinal number; that is, the segments have the same number of points.

43. $\{3, \quad 6, \quad 9, \quad 12, \quad \ldots, \quad 3n, \quad \ldots\}$
    $\updownarrow \quad \updownarrow \quad \updownarrow \quad \updownarrow \qquad\qquad \updownarrow$
    $\{6, \quad 9, \quad 12, \quad 15, \quad \ldots, \quad 3n+3, \quad \ldots\}$

44. $\{4, \quad 7, \quad 10, \quad 13, \quad \ldots, \quad 3n+1, \quad \ldots\}$
    $\updownarrow \quad \updownarrow \quad \updownarrow \quad \updownarrow \qquad\qquad \updownarrow$
    $\{7, \quad 10, \quad 13, \quad 16, \quad \ldots, \quad 3n+4, \quad \ldots\}$

45. $\{3/4, \ 3/8, \ 3/12, \ 3/16, \ldots, 3/(4n), \quad \ldots\}$
    $\updownarrow \quad \updownarrow \quad \updownarrow \quad \updownarrow \qquad\qquad \updownarrow$
    $\{3/8, \ 3/12, \ 3/16, \ 3/20, \ldots, 3/(4n+4), \ldots\}$

46. $\{1, \ 4/3, 5/3, \ 2, \quad \ldots, (n+2)/3, \ \ldots\}$
    $\updownarrow \quad \updownarrow \quad \updownarrow \quad \updownarrow \qquad\qquad \updownarrow$
    $\{4/3, 5/3, \ 2, \ 7/3, \ldots, (n+3)/3, \ldots\}$

47. $\{1/9, \quad 1/18, \quad 1/27, \quad \ldots, \quad 1/(9n), \qquad \ldots\}$
    $\updownarrow \quad \updownarrow \quad \updownarrow \qquad\qquad \updownarrow$
    $\{1/18, \ 1/27, \ 1/36, \ \ldots, \ 1/(9n+9), \ \ldots\}$

48. $\{-3, \ -5, \ -9, \ -17, \ \ldots, \ -(2^n+1), \quad \ldots\}$
    $\updownarrow \quad \updownarrow \quad \updownarrow \quad \updownarrow \qquad\qquad \updownarrow$
    $\{-5, \ -9, \ -17, \ -33, \ \ldots, \ -(2^{n+1}+1), \ \ldots\}$

49. Writing exercise
50. Writing exercise
51. Writing exercise
52. Writing exercise

CHAPTER 2 TEST

1. $A \cup C = \{a,b,c,d\} \cup \{a,e\} = \{a,b,c,d,e\}$.

2. $B \cap A = \{b,e,a,d\} \cap \{a,b,c,d\} = \{a,b,d\}$.

3. $B' = \{b,e,a,d\}' = \{c,f,g,h\}$.

4. $A - (B \cap C') = A - (\{b,e,a,d\} \cap \{b,c,d,f,g,h\})$
    $= \{a,b,c,d\} - \{b,d\}$
    $= \{a,c\}$.

5. $b \in A$ is true since b is member of set A.

6. $C \subseteq A$ is false since the element e, which is a member of set C, is not also a member of set A.

7. $B \subset (A \cup C)$ is true since all members of set B are also members of $A \cup C$.

8. $c \notin C$ is true because c is not a member of set C.

9. $n[(A \cup B) - C] = 4$ is false. Because,
    $n[(A \cup B) - C] = n[\{a,b,c,d,e\} - \{a,e\}]$
    $= n(\{b,c,d\})$
    $= 3$.

10. $\emptyset \subset C$ is true. The empty set is considered a subset of any set. C has more elements then $\emptyset$ which makes $\emptyset$ a proper subset of C.

11. $(A \cap B')$ is equivalent to $(B \cap A')$ is true. Because,
    $n(A \cap B') = n(\{c\}) = 1$
    $n(B \cap A') = n(\{e\}) = 1$.

12. $(A \cup B)' = A' \cap B'$ is true by one of De Morgan's laws.

13. $n(A \times C) = n(A) \times n(C) = 4 \times 2 = 8$

14. The number of proper subsets of A is
    $$2^4 - 1 = 16 - 1 = 15.$$

Answers may vary for Exercises 15–18.

15. A word description for $\{-3, -1, 1, 3, 5, 7, 9\}$ is the set of all odd integers between -4 and 10.

16. A word description for $\{$January, February, March, $\ldots$, December$\}$ is the set of months of the year.

17. Set-builder notation for $\{-1, -2, -3, -4, \ldots\}$ would be $\{x \mid x \text{ is a negative integer}\}$.

18. Set-builder notation for $\{24, 32, 40, 48, \ldots, 88\}$ would be $\{x \mid x \text{ is a multiple of 8 between 20 and 90}\}$.

19. $\emptyset$ $\boxed{\subseteq}$ $\{x \mid x \text{ is a counting number between 17 and 18}\}$ since the empty set is a subset of any set.

20. $\{4, 9, 16\}$ $\boxed{\text{neither}}$ $\{4, 5, 6, 7, 8, 9, 10\}$ since the element 16 is not a member of the second set.

21. $X \cup Y'$

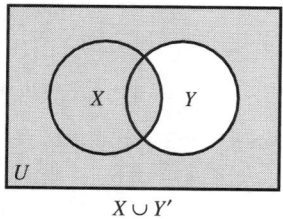

$X \cup Y'$

22. $X' \cap Y'$

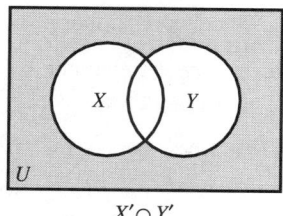

$X' \cap Y'$

23. $(X \cup Y) - Z$

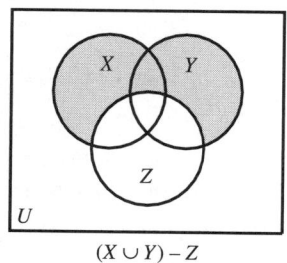

$(X \cup Y) - Z$

24. $[(X \cap Y) \cup (Y \cap Z) \cup (X \cap Z)] - (X \cap Y \cap Z)$

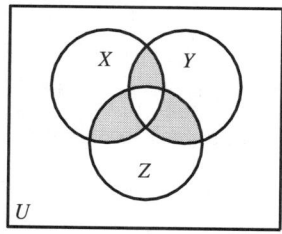

$[(X \cap Y) \cup (Y \cap Z) \cup (X \cap Z)] - (X \cap Y \cap Z)$

25. $A \cap T =$
{Electric razor, Telegraph, Zipper} ∩ {Electric razor, Fiber optics, Geiger counter, Radar} = {Electric Razor}.

26. $(A \cup T)' =$
({Electric razor, Telegraph, Zipper} ∪ {Electric razor, Fiber optics, Geiger counter, Radar})' = {Electric razor, Fiber optics, Geiger counter, Radar, Telegraph, Zipper}' = {Adding machine, Barometer, Pendulum clock, Thermometer}.

27. $A - T' =$ {Electric razor, Telegraph, Zipper} − {Electric razor, Fiber optics, Geiger counter, Radar}' = {Electric razor, Telegraph, Zipper} − {Adding machine, Barometer, Pendulum clock, Telegraph, Thermometer, Zipper} = {Electric razor}.

28. Writing exercise

29. (a) $n(A \cup B) = 12 + 3 + 7 = 22$.

    (b) $n(A \cap B') = n(A - B) = 12$. These are the elements in A but outside of B.

    (c) $n(A \cap B)' = n(\{3\}') = 12 + 7 + 9 = 28$.

30. Let $G =$ set of students who are receiving government grants. Let $S =$ set of students who are receiving private scholarships. Let $A =$ set of students who are receiving aid from the college. Complete a Venn diagram by inserting the appropriate cardinal number for each region in the diagram. Begin with the intersection of all three sets: $n(G \cap S \cap A) = 8$. Since $n(S \cap A) = 28$, $n(S \cap A \cap G') = 28 - 8 = 20$. Since $n(G \cap A) = 18$, $n(G \cap A \cap S') = 18 - 8 = 10$. Since $n(G \cap S) = 23$, $n(G \cap S \cap A') = 23 - 8 = 15$.

Since $n(A) = 43$,
$$n(A \cap (G \cup S)') = 43 - (10 + 8 + 20)$$
$$= 43 - 38 = 5.$$
Since $n(S) = 55$,
$$n(S \cap (G \cup A)') = 55 - (15 + 8 + 20)$$
$$= 55 - 43 = 12.$$
Since $n(G) = 49$,
$$n(G \cap (S \cup A)') = 49 - (10 + 8 + 15)$$
$$= 49 - 33 = 16.$$

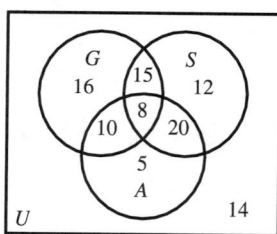

Thus,

(a) $n(G \cap (S \cup A)') = 16$ have a government grant only.

(b) $n(S \cap G') = 32$ have a private scholarship but not a government grant.

(c) $16 + 12 + 5 = 33$ receive financial aid from only one of these sources.

(d) $10 + 15 + 20 = 45$ receive aid from exactly two of these sources.

(e) $n(G \cup S \cup A)' = 14$ receive no financial aid from any of these sources.

(f) $n(S \cap (A \cup G)') + n(A \cup G \cup S)' = 12 + 14 = 26$ received private scholarships or no aid at all

3.1 EXERCISES

1. Because the declarative sentence "December 7, 1941, was a Sunday" has the property of being true or false, it is considered a statement.

2. Because the declarative sentence "The ZIP code for Manistee, MI, is 49660" has the property of being true or false, it is considered a statement.

3. "Listen my children and you shall hear of the midnight ride of Paul Revere" is not a declarative sentence and does not have the property of being true or false. Hence, it is not considered a statement.

4. "Yield to oncoming traffic" is a command, not a declarative sentence and therefore, it is not a statement.

5. "$5 + 8 = 13$ and $4 - 3 = 1$" is a declarative sentence that is true and, therefore, is considered a statement.

6. "$5 + 8 = 12$ or $4 - 3 = 2$" is a declarative sentence that is false and, therefore, is considered a statement.

7. "Some numbers are negative" is a declarative sentence that is true and, therefore, is a statement.

8. "Andrew Johnson was president of the United States in 1867" is a declarative sentence that is false and is, therefore, a statement.

9. "Accidents are the main cause of deaths of children under the age of 8" is a declarative sentence that has the property of being true or false and, therefore, is considered to be a statement.

10. "*Star Wars: Episode I - The Phantom Menace* was the top-grossing movie of 1999" is a declarative sentence that has the property of being either true or false and, therefore, is considered to be a statement.

11. "Where are you going today?" is a question, not a declarative sentence and, therefore, is not considered a statement.

12. "Behave yourself and sit down" is a command, not a declarative sentence and, therefore, is not considered a statement.

13. "Kevin 'Catfish' McCarthy once took a prolonged continuous shower for 340 hours, 40 minutes" is a declarative sentence that has the property of being either true or false and, therefore, is considered to be a statement.

14. "One gallon of milk weighs more than 4 pounds" is a declarative sentence that has the property of being either true or false (probably false) and, therefore, is considered to be a statement.

15. "I read the Chicago Tribune and I read the New York Times" is a compound statement because it consists of two simple statements combined by the connective "and."

16. "My brother got married in London" is not a compound statement because only one assertion is being made.

17. "Tomorrow is Sunday" is a simple statement because only one assertion is being made.

18. "Dara Lanier is younger than 29 years of age, and so is Teri Orr" is a compound statement because it consists of two simple statements combined by the connective "and."

19. "Jay Beckenstein's wife loves Ben and Jerry's ice cream" is not compound because only one assertion is being made.

20. "The sign on the back of the car read 'California or bust!'" is not compound because only one assertion is being made.

21. "If Julie Ward sells her quota, then Bill Leonard will be happy" is a compound statement because it consists of two simple statements combined by the connective "if ... then."

22. "If Mike is a politician, then Jerry is a crook" is a compound statement because it consists of two simple statements combined by the connective "if ... then."

23. The negation of "Her aunt's name is Lucia" is "Her aunt's name is not Lucia."

24. The negation of "The flowers are to be watered" is "The flowers are not to be watered."

25. A negation of "Every dog has its day" is "At least one dog does not have its day."

26. A negation of "No rain fell in southern California today" is "Some rain fell in southern California today."

27. A negation of "Some books are longer than this book" is "No book is longer than this book."

28. A negation of "All students present will get another chance" is "At least one student present will not get another chance."

29. A negation of "No computer repairman can play blackjack" is "At least one computer repairman can play blackjack."

30. A negation of "Some people have all the luck" is "No people have all the luck."

31. A negation of "Everybody loves somebody sometime" is "Someone does not love somebody sometime."

32. A negation of "Everyone loves a winner" is "Someone does not love a winner."

33. A negation for "$y > 12$" (without using a slash sign) would be "$y \leq 12$."

34. A negation for "$x < -6$" (without using a slash sign) would be "$x \geq -6$."

35. A negation for "$q \geq 5$" would be "$q < 5$."

36. A negation for "$r \leq 19$" would be "$r > 19$."
37. Writing exercise
38. Writing exercise

Let p represent the statement "She has green eyes," and let q represent "He is 48 years old." Translate each symbolic compound statement into words.

39. A translation for "$\sim p$" is "She does not have green eyes."
40. A translation for "$\sim q$" is "He is not 48 years old."
41. A translation for "$p \wedge q$" is "She has green eyes and he is 48 years old."
42. A translation for "$p \vee q$" is "She has green eyes or he is 48 years old."
43. A translation for "$\sim p \vee q$" is "She does not have green eyes or he is 48 years old."
44. A translation for "$p \wedge \sim q$" is "She has green eyes and he is not 48 years old."
45. A translation for "$\sim p \vee \sim q$" is "She does not have green eyes or he is not 48 years old."
46. A translation for "$\sim p \wedge \sim q$" is "She does not have green eyes and he is not 48 years old."
47. A translation for "$\sim(\sim p \wedge q)$" is "It is not the case that she does not have green eyes and he is 48 years old."
48. A translation for "$\sim(p \vee \sim q)$" is "It is not the case that she has green eyes or he is not 48 years old."
49. "Chris collects videotapes and Jack does not play the tuba" may be symbolized as $p \wedge \sim q$.
50. "Chris does not collect videotapes or Jack does not play the tuba" may be symbolized as $\sim p \vee \sim q$.
51. "Chris does not collect videotapes or Jack plays the tuba" may be symbolized as $\sim p \vee q$.
52. "Jack plays the tuba and Chris does not collect videotapes" may be symbolized as $q \wedge \sim p$.
53. "Neither Chris collects videotapes nor Jack plays the tuba" may be symbolized as $\sim(p \vee q)$ or equivalently, $\sim p \wedge \sim q$.
54. "Either Jack plays the tuba or Chris collects videotapes, and it is not the case that both Jack plays the tuba and Chris collects videotapes" may be symbolized as $(q \vee p) \wedge [\sim(q \wedge p)]$.
55. Writing exercise
56. Writing exercise

Refer to the sketches labeled A, B, and C in the text, and identify the sketch (or sketches) that is (are) satisfied by the given statement involving a quantifier.

57. The condition that "all pictures have frames" is satisfied by group C.
58. The condition that "No picture has a frame" is satisfied by group B.
59. The condition that "At least one picture does not have a frame" is met by groups A and B.
50. The condition that "Not every picture has a frame" is satisfied by groups A and B. Observe that this statement is equivalent to "At least one picture does not have a frame"
61. The condition that "At least one picture has a frame" is satisfied by groups A and C.
62. The condition that "No picture does not have a frame" is satisfied by group C. Observe that this statement is equivalent to "All pictures have a frame"
63. The condition that "all pictures do not have frames" is satisfied by group B. Observe that this statement is equivalent to "No pictures have a frame."
64. The condition that "Not every picture does not have a frame" is satisfied by groups A and C. Observe that this statement is equivalent to "At least one picture does have a frame."
65. Since all whole numbers are integers, the statement "Every whole number is an integer" is true.
66. Since all natural numbers are integers, the statement "Every natural number is an integer" is true.
67. Since 1/2 is a rational number but not an integer, the statement "There exists a rational number that is not an integer" is true.
68. Since -3 is an integer but not a natural number, the statement "There exists an integer that is not a natural number" is true.
69. Since rational numbers are real numbers, the statement "All rational numbers are real numbers" is true.
70. Since irrational numbers are real numbers, the statement "All irrational numbers are real numbers" is true.
71. Since 1/2 is a rational number but not an integer, the statement "Some rational numbers are not integers" is true.
72. Since whole numbers are rational, the statement "Some whole numbers are not rational numbers" is false.
73. The number 0 is a whole number but not positive. Thus, the statement "Each whole number is a positive number" is false.
74. The number $-1/2$ is a rational number. Thus, the statement "Each rational number is a positive number" is false.
75. Writing exercise
76. Writing exercise

77. We might write the statement "There is no one here who has not done that at one time or another" using the word "every" as "Everyone here has done that at one time or another."

78. A. The statement "For some real number x, $x \not< 0$" is true. Let $x = 10$, for example. This real number satisfies the statement.

   B. The statement "For all real numbers x, $x^3 > 0$" is false. For example let $x = -2$. Observe that $(-2)^3 = -8 \not> 0$.

   C. The statement "For all real numbers x less than 0, x^2 is also less than 0" is false. Let $x = -4$, which is less than 0, but $x^2 = (-4)^2 = 16$, which is not less than 0.

   D. The statement "For some real number x, $x^2 < 0$" is false. If you square any real number (negative, positive, or 0), the result will be ≥ 0.

3.2 EXERCISES

1. If q is false, then $(p \wedge \sim q) \wedge q$ must be false, since both conjuncts (parts of the conjunction) must be true for the compound statement to be true.

2. If q is true, then $q \vee (q \wedge \sim p)$ is true, since only one disjunct (parts of the disjunctive statement) need be true for the compound statement to be true.

3. If $p \wedge q$ is true, and p is true, then q must also be true in order for the conjunctive statement to be true. Observe that both conjuncts must be true for a conjunctive statement to be true.

4. If $p \vee q$ is false, and p is false, then q must also be false. Observe that both disjuncts must be false for a disjunctive statement to be false.

5. If $\sim(p \vee q)$ is true, both components (disjuncts) must be false. Thus, the disjunction itself is false making its negation true.

6. If $\sim(p \wedge q)$ is false, then both p and q must be true. This will assure that the conjunction itself is true making its negation false.

In exercises 7–18, p represents a false statement and q represents a true statement.

7. Since $p = F$,
$$\sim p = \sim F$$
$$= T.$$
That is, replace p by F and determine the truth of $\sim F$.

8. Since $q = T$,
$$\sim q = \sim T$$
$$= F.$$
Thus, $\sim q$ is false.

9. Since p is false and q is true, we may consider the "or" statement as
$$F \vee T$$
$$T,$$
by the logical definition of an "or" statement. That is, $p \vee q$ is true.

10. Since p is false and q is true, we may consider the "and" statement as
$$F \wedge T$$
$$F,$$
by the logical definition of an "and" statement. That is, $p \wedge q$ is false.

11. With the given truth values for p and q, we may consider $p \vee \sim q$ as
$$F \vee \sim T$$
$$F \vee F$$
$$F,$$
by the logical definition of " $\vee$ ".

12. With the given truth values for p and q, we may consider $\sim p \wedge q$ as
$$\sim F \wedge T$$
$$T \wedge T$$
$$T,$$
by the logical definition of " $\wedge$ ". Thus, the compound statement is true.

13. With the given truth values for p and q, we may consider $\sim p \vee \sim q$ as
$$\sim F \vee \sim T$$
$$T \vee F$$
$$T.$$
Thus, the compound statement is true.

14. Replacing p and q with the given truth values, we have
$$F \wedge \sim T$$
$$F \wedge F$$
$$F.$$
Thus, the compound statement $p \wedge \sim q$ is false.

15. Replacing p and q with the given truth values, we have
$$\sim(F \wedge \sim T)$$
$$\sim(F \wedge F)$$
$$\sim F$$
$$T.$$
Thus, the compound statement $\sim(p \wedge \sim q)$ is true.

16. Replacing p and q with the given truth values, we have

$$\sim(\sim F \vee \sim T)$$
$$\sim(T \vee F)$$
$$\sim(T)$$
$$F.$$

Thus, the compound statement $\sim(\sim p \vee \sim q)$ is false.

17. Replacing p and q with the given truth values, we have

$$\sim[\sim F \wedge (\sim T \vee F)]$$
$$\sim[T \wedge (F \vee F)]$$
$$\sim[T \wedge F]$$
$$\sim F$$
$$T.$$

Thus, the compound statement $\sim[\sim p \wedge (\sim q \vee p)]$ is true.

18. Replacing p and q with the given truth values, we have

$$\sim[(\sim F \wedge \sim T) \vee \sim T]$$
$$\sim[(T \wedge F) \vee F]$$
$$\sim[F \vee F]$$
$$\sim F$$
$$T.$$

Thus, the compound statement $\sim[(\sim p \wedge \sim q) \vee \sim q]$ is true.

19. The statement $3 \geq 1$ is a disjunction since it means "$3 > 1$" or "$3 = 1$."

20. The statement "$6 \geq 2$" is true because $6 > 2$.
The statement "$6 \geq 6$" is true because $6 = 6$.

In exercises 21–28, p represents a true statement, and q and r represent false statements.

21. Replacing p, q and r with the given truth values, we have

$$(T \wedge F) \vee \sim F$$
$$F \vee T$$
$$T.$$

Thus, the compound statement $(p \wedge r) \vee \sim q$ is true.

22. Replacing p, q and r with the given truth values, we have

$$(F \vee \sim F) \wedge T$$
$$(F \vee T) \wedge T$$
$$T \wedge T$$
$$T.$$

Thus, the compound statement $(q \vee \sim r) \wedge p$ is true.

23. Replacing p, q and r with the given truth values, we have

$$T \wedge (F \vee F)$$
$$T \wedge F$$
$$F.$$

Thus, the compound statement $p \wedge (q \vee r)$ is false.

24. Replacing p, q and r with the given truth values, we have

$$(\sim T \wedge F) \vee \sim F$$
$$(F \wedge F) \vee T$$
$$F \vee T$$
$$T.$$

Thus, the compound statement $(\sim p \wedge q) \vee \sim r$ is true.

25. Replacing p, q and r with the given truth values, we have

$$\sim(T \wedge F) \wedge (F \vee \sim F)$$
$$\sim F \wedge (F \vee T)$$
$$T \wedge T$$
$$T.$$

Thus, the compound statement $\sim(p \wedge q) \wedge (r \vee \sim q)$ is true.

26. Replacing p, q and r with the given truth values, we have

$$(\sim F \wedge \sim F) \vee (\sim F \wedge F)$$
$$(T \wedge T) \vee (T \wedge F)$$
$$T \vee F$$
$$T.$$

Thus, the compound statement $(\sim r \wedge \sim q) \vee (\sim r \wedge q)$ is true.

27. Replacing p, q and r with the given truth values, we have

$$\sim[(\sim T \wedge F) \vee F]$$
$$\sim[(F \wedge F) \vee F]$$
$$\sim[F \vee F]$$
$$\sim F$$
$$T.$$

Thus, the compound statement $\sim[(\sim p \wedge q) \vee r]$ is true.

28. Replacing p, q and r with the given truth values, we have

$$\sim[F \vee (\sim F \wedge \sim T)]$$
$$\sim[F \vee (T \wedge F)]$$
$$\sim[F \vee F]$$
$$\sim F$$
$$T.$$

Thus, the compound statement $\sim[r \vee (\sim q \wedge \sim p)]$ is true.

Let p represent the statement "$2 > 7$," which is false, let q represent "$8 \not> 6$," which is false and let r represent "$19 \leq 19$," which is true. [E.g. $p = F$, $q = F$ and $r = T$.]

29. Replacing p and r with the given truth values, we have

$$F \wedge T$$
$$F.$$

The compound statement $p \wedge r$ is false.

30. Replacing p and q with the observed truth values, we have
$$F \vee \sim F$$
$$F \vee T$$
$$T.$$
The compound statement $p \vee \sim q$ is true.

31. Replacing q and r with the observed truth values, we have
$$\sim F \vee \sim T$$
$$T \vee F$$
$$T.$$
The compound statement $\sim q \vee \sim r$ is true.

32. Replacing p and r with the observed truth values, we have
$$\sim F \wedge \sim T$$
$$T \wedge F$$
$$F.$$
The compound statement $\sim p \wedge \sim r$ is false.

33. Replacing p, q and r with the observed truth values, we have
$$(F \wedge F) \vee T$$
$$F \vee T$$
$$T.$$
The compound statement $(p \wedge q) \vee r$ is true.

34. Replacing p, q and r with the observed truth values, we have
$$\sim F \wedge (\sim T \vee \sim F)$$
$$T \wedge (F \vee T)$$
$$T \wedge T$$
$$T.$$
The compound statement $\sim p \wedge (\sim r \vee \sim q)$ is true.

35. Replacing p, q and r with the observed truth values, we have
$$(\sim T \wedge F) \vee \sim F$$
$$(F \wedge F) \vee T$$
$$F \vee T$$
$$T.$$
The compound statement $(\sim r \wedge q) \vee \sim p$ is true.

36. Replacing p, q and r with the observed truth values, we have
$$\sim(F \vee \sim F) \vee \sim T$$
$$\sim(F \vee T) \vee F$$
$$\sim T \vee F$$
$$F \vee F$$
$$F.$$
The compound statement $\sim(p \vee \sim q) \vee \sim r$ is false.

37. Since there are two simple statements (p and r), we have $2^2 = 4$ combinations of truth values, or rows in the truth table, to examine.

38. Since there are three simple statements (p, r, and s), we have $2^3 = 8$ combinations of truth values, or rows in the truth table, to examine.

39. Since there are four simple statements (p, q, r, and s), we have $2^4 = 16$ combinations of truth values, or rows in the truth table, to examine.

40. Since there are five simple statements (p, q, r, s, and t), we have $2^5 = 32$ combinations of truth values, or rows in the truth table, to examine.

41. Since there are seven simple statements (p, q, r, s, t, u, and v), we have $2^7 = 128$ combinations of truth values, or rows in the truth table, to examine.

42. Since there are eight simple statements (p, q, r, s, m, n, u, and v), we have $2^8 = 256$ combinations of truth values, or rows in the truth table, to examine.

43. If the truth table for a certain compound statement has 64 rows, then there must be six distinct component statements ($2^6 = 64$).

44. It is not possible for a truth table of a compound statement to have exactly 48 rows, because 48 is not a whole number power of 2.

45. $\sim p \wedge q$

p	q	$\sim p$	$\sim p \wedge q$
T	T	F	F
T	F	F	F
F	T	T	T
F	F	T	F

46. $\sim p \vee \sim q$

p	q	$\sim p$	$\sim q$	$\sim p \vee \sim q$
T	T	F	F	F
T	F	F	T	T
F	T	T	F	T
F	F	T	T	T

47. $\sim(p \wedge q)$

p	q	$p \wedge q$	$\sim(p \wedge q)$
T	T	T	F
T	F	F	T
F	T	F	T
F	F	F	T

48. $p \vee \sim q$

p	q	$\sim q$	$p \vee \sim q$
T	T	F	T
T	F	T	T
F	T	F	F
F	F	T	T

49. $(q \vee \sim p) \vee \sim q$

p	q	$\sim p$	$q \vee \sim p$	$\sim q$	$(q \vee \sim p) \vee \sim q$
T	T	F	T	F	T
T	F	F	F	T	T
F	T	T	T	F	T
F	F	T	T	T	T

50. $(p \wedge \sim q) \wedge p$

p	q	$\sim q$	$p \wedge \sim q$	$(p \wedge \sim q) \wedge p$
T	T	F	F	F
T	F	T	T	T
F	T	F	F	F
F	F	T	F	F

In Exercises 51–59 we are using the alternative method, filling in columns in the order indicated by the numbers. Observe that columns with the same number are combined (by the logical definition of the connective) to get the next numbered column. Note that <u>this is different</u> than the way the numbered columns are used in the textbook. Remember that the last column (highest numbered column) completed yields the truth values for the complete compound statement. Be sure to align truth values under the appropriate logical connective or simple statement.

51. $\sim q \wedge (\sim p \vee q)$

p	q	$\sim q$	$\wedge$	$(\sim p \vee q)$
T	T	F	F	F T T
T	F	T	F	F F F
F	T	F	F	T T T
F	F	T	T	T T F
		2	3	1 2 1

52. $\sim p \vee (\sim q \wedge \sim p)$

p	q	$\sim p$	$\vee$	$(\sim q \wedge \sim p)$
T	T	F	F	F F F
T	F	F	F	T F F
F	T	T	T	F F T
F	F	T	T	T T T
		2	3	1 2 1

53. $(p \vee \sim q) \wedge (p \wedge q)$

p	q	$(p \vee \sim q)$	$\wedge$	$(p \wedge q)$
T	T	T T F	T	T T T
T	F	T T T	F	T F F
F	T	F F F	F	F F T
F	F	F T T	F	F F F
		1 2 1	3	1 2 1

54. $(\sim p \wedge \sim q) \vee (\sim p \vee q)$

p	q	$(\sim p \wedge \sim q)$	$\vee$	$(\sim p \vee q)$
T	T	F F F	T	F T T
T	F	F F T	F	F F F
F	T	T F F	T	T T T
F	F	T T T	T	T T F
		1 2 1	3	1 2 1

55. $(\sim p \wedge q) \wedge r$

p	q	r	$(\sim p \wedge q)$	$\wedge$	r
T	T	T	F F T	F	T
T	T	F	F F T	F	F
T	F	T	F F F	F	T
T	F	F	F F F	F	F
F	T	T	T T T	T	T
F	T	F	T T T	F	F
F	F	T	T F F	F	T
F	F	F	T F F	F	F
			1 2 1	3	2

56. $r \vee (p \wedge \sim q)$

p	q	r	r	$\vee$	$(p \wedge \sim q)$
T	T	T	T	T	T F F
T	T	F	F	F	T F F
T	F	T	T	T	T T T
T	F	F	F	T	T T T
F	T	T	T	T	F F F
F	T	F	F	F	F F F
F	F	T	T	T	F F T
F	F	F	F	F	F F T
			1	3	1 2 1

57. $(\sim p \land \sim q) \lor (\sim r \lor \sim p)$

p	q	r	$(\sim p$	$\land$	$\sim q)$	$\lor$	$(\sim r$	$\lor$	$\sim p)$
T	T	T	F	F	F	F	F	F	F
T	T	F	F	F	F	T	T	T	F
T	F	T	F	F	T	F	F	F	F
T	F	F	F	F	T	T	T	T	F
F	T	T	T	F	F	T	F	T	T
F	T	F	T	F	F	T	T	T	T
F	F	T	T	T	T	T	F	T	T
F	F	F	T	T	T	T	T	T	T
			1	2	1	3	1	2	1

58. $(\sim r \lor \sim p) \land (\sim p \lor \sim q)$

p	q	r	$(\sim r$	$\lor$	$\sim p)$	$\land$	$(\sim p$	$\lor$	$\sim q)$
T	T	T	F	F	F	F	F	F	F
T	T	F	T	T	F	F	F	F	F
T	F	T	F	F	F	F	F	T	T
T	F	F	T	T	F	T	F	T	T
F	T	T	F	T	T	T	T	T	F
F	T	F	T	T	T	T	T	T	F
F	F	T	F	T	T	T	T	T	T
F	F	F	T	T	T	T	T	T	T
			1	2	1	3	1	2	1

59. $\sim(\sim p \land \sim q) \lor (\sim r \lor \sim s)$

p	q	r	s	$\sim(\sim p$	$\land$	$\sim q)$	$\lor$	$(\sim r$	$\lor$	$\sim s)$
T	T	T	T	T	F	F	F	T	F	F F
T	T	T	F	T	F	F	F	T	F	T T
T	T	F	T	T	F	F	F	T	T	T F
T	T	F	F	T	F	F	F	T	T	T T
T	F	T	T	T	F	F	T	T	F	F F
T	F	T	F	T	F	F	T	T	F	T T
T	F	F	T	T	F	F	T	T	T	T F
T	F	F	F	T	F	F	T	T	T	T T
F	T	T	T	T	T	F	F	T	F	F F
F	T	T	F	T	T	F	F	T	F	T T
F	T	F	T	T	T	F	F	T	T	T F
F	T	F	F	T	T	F	F	T	T	T T
F	F	T	T	F	T	T	T	F	F	F F
F	F	T	F	F	T	T	T	F	F	T T
F	F	F	T	F	T	T	T	F	T	T F
F	F	F	F	F	T	T	T	T	T	T T
				3	1	2	1	4	2	3 2

60. $(\sim r \lor s) \land (\sim p \land q)$

p	q	r	s	$(\sim r$	$\lor$	$s)$	$\land$	$(\sim p$	$\land$	$q)$
T	T	T	T	F	T	T	F	F	F	T
T	T	T	F	F	F	F	F	F	F	T
T	T	F	T	T	T	T	F	F	F	T
T	T	F	F	T	T	F	F	F	F	T
T	F	T	T	F	T	T	F	F	F	F
T	F	T	F	F	F	F	F	F	F	F
T	F	F	T	T	T	T	F	F	F	F
T	F	F	F	T	T	F	F	F	F	F
F	T	T	T	F	T	T	T	T	T	T
F	T	T	F	F	F	F	F	T	T	T
F	T	F	T	T	T	T	T	T	T	T
F	T	F	F	T	T	F	T	T	T	T
F	F	T	T	F	T	T	F	T	F	F
F	F	T	F	F	F	F	F	T	F	F
F	F	F	T	T	T	T	F	T	F	F
F	F	F	F	T	T	F	F	T	F	F
				1	2	1	3	1	2	1

61. "You can pay me now or you can pay me later" has the symbolic form $(p \lor q)$. The negation, $\sim(p \lor q)$, is equivalent, by one of De Morgan's laws, to $(\sim p \land \sim q)$. The corresponding word statement is "You can't pay me now and you can't pay me later."

62. "I am not going or she is going" has the symbolic form $\sim p \lor q$. Its negation, $\sim(\sim p \lor q)$, is equivalent, by De Morgan's, to $p \land \sim q$. The word translation for the negation is "I am going and she is not going."

63. "It is summer and there is no snow" has the symbolic form $p \land \sim q$. The negation, $\sim(p \land \sim q)$, is equivalent by De Morgan's, to $\sim p \lor q$. The word translation for the negation is "It is not summer or there is snow."

64. "1/2 is a positive number and -12 is less than zero" is of the form $p \land q$. The negation, $\sim(p \land q)$, is equivalent, by De Morgan's, to $\sim p \lor \sim q$. The word translation for the negation is "1/2 is not a positive number or $-12 \geq$ zero." (Note that the inequality "$\geq$" is equivalent to "not less than.")

65. "I said yes but she said no" is of the form $p \land q$. The negation, $\sim(p \land q)$, equivalent, by De Morgan's, to $\sim p \lor \sim q$. The word translation for the negation is "I did not say yes or she did not say no." (Note that the connective "but" is equivalent to that of "and.")

66. "Pauline Mula tried to sell the book, but she was unable to do so" is of the form $p \land q$. The negation, $\sim(p \land q)$, equivalent, by De Morgan's, to $\sim p \lor \sim q$. The word translation for the negation is "Pauline Mula did not try to sell the book, or she was able to do so."

67. "$5 - 1 = 4$ and $9 + 12 \neq 7$" is of the form $p \wedge \sim q$. The negation, $\sim(p \wedge \sim q)$, equivalent, by De Morgan's, to $\sim p \vee q$. The translation for the negation is "$5 - 1 \neq 4$ or $9 + 12 = 7$."

68. "$3 < 10$ or $7 \neq 2$" is of the form $p \vee \sim q$. The negation, $\sim(p \vee \sim q)$, is equivalent, by De Morgan's, to $\sim p \wedge q$. A translation for the negation is "$3 \geq 10$ and $7 = 2$." (Note that the inequality "$\geq$" is equivalent to "$\not<$").

69. "Cupid or Vixen will lead Santa's sleigh next Christmas" is of the form $p \vee q$. The negation, $\sim(p \vee q)$, is equivalent, by De Morgan's, to $\sim p \wedge \sim q$. A translation for the negation is "Neither Cupid nor Vixen will lead Santa's sleigh next Christmas."

70. "The lawyer and the client appeared in court" is of the form $(p \wedge q)$. The negation, $\sim(p \wedge q)$, is equivalent, by De Morgan's, to $\sim p \vee \sim q$. The word translation for the negation is "The lawyer did not appear in court or the client did not appear in court."

71. "For every real number y, $y < 13$ or $y > 6$" is <u>true</u> since for any real number at least one of the component statements is true.

72. "For every real number t, $t > 9$ or $t < 9$" is <u>false</u> since for the real number $t = 9$, both component statements are false.

73. "For some integer p, $p \geq 4$ and $p \leq 4$" is <u>true</u> since both component statements are true for the integer $p = 4$.

74. "There exists an integer n such that $n > 0$ and $n < 0$" is <u>false</u> since any integer which is true for one of the component statements will be false for the other.

75. $p \underline{\vee} q$

p	q	$p \underline{\vee} q$
T	T	F
T	F	T
F	T	T
F	F	F

Observe that it is only the first line in the truth table that changes for "exclusive disjunction" since the component statements can not both be true at the same time.

76. The phrase "and/or" represents <u>inclusive disjunction</u> since both or either component statement(s) may be true.

77. "$3 + 1 = 4 \underline{\vee} 2 + 5 = 7$" is <u>false</u> since both component statements are true.

78. "$3 + 1 = 4 \underline{\vee} 2 + 5 = 9$" is <u>true</u> since the first component statement is true and the second is false.

79. "$3 + 1 = 7 \underline{\vee} 2 + 5 = 7$" is <u>true</u> since the first component statement is false and the second is true.

80. "$3 + 1 = 7 \underline{\vee} 2 + 5 = 9$" is <u>false</u> since both component statements are false.

3.3 EXERCISES

1. The statement "It must be alive if it is breathing" becomes "If it is breathing, then it must be alive."

2. The statement "You can believe it if you see it on the Internet" becomes "If you see it on the Internet, then you can believe it."

3. The statement "Lorri Morgan visits Ireland every summer" becomes "If it is summer, then Lorri Morgan visits Hawaii."

4. The statement "Tom Shaffer's area code is 216" becomes "If the person is Tom Shaffer, then his area code is 216."

5. The statement "Every picture tells a story" becomes "If it is a picture, then it tells a story."

6. The statement "All marines love boot camp" becomes "If the soldier is a marine, then the soldier loves boot camp."

7. The statement "No guinea pigs are scholars" becomes "If it is a guinea pig, then it is not a scholar."

8. The statement "No koalas live in Texas" becomes "If it is a koala, then it does not live in Texas."

9. The statement "Running Bear loves Little White Dove" becomes "If he is Running Bear, then he loves Little White Dove."

10. The statement "An opium-eater cannot have self-command" becomes "If it is an opium eater, then it has no self-command."

11. The statement "If the antecedent of a conditional statement is false, the conditional statement is true" is <u>true</u>, since a false antecedent will always yield a true conditional statement.

12. The statement "If the consequent of a conditional statement is true, the conditional statement is true" is <u>true</u>, since a true consequent is always associated with a true conditional statement (i.e., it doesn't matter what the truth value of the antecedent is, if the consequent itself is true).

13. The statement "If q is true, then $(p \wedge q) \rightarrow q$ is true" is <u>true</u>, since with a true consequent the conditional statement is always true (even though the antecedent may be false).

14. The statement "If p is true, then $\sim p \rightarrow (q \vee r)$ is true" is <u>true</u> since the antecedent, $\sim p$, is false.

15. The negation of "If pigs fly, I'll believe it" is "If pigs don't fly, I won't believe it." This statement is <u>false</u>. The negation is "Pigs fly and I don't believe it."

16. The statements "If it flies, then it's a bird" and "It does not fly or it's a bird" are logically equivalent. To decide if the above is true, examine the corresponding truth tables for each individual statement.

p	q	$p \to q$	$\sim p \lor q$
T	T	T T T	F T T
T	F	T F F	F F F
F	T	F T T	T T T
F	F	F T F	T T F
		1 2 1	1 2 1

    This statement is <u>true</u> because the truth values (column 2) for each compound statement are the same, showing that the statements equivalent.

17. "Given that $\sim p$ is true and q is false, the conditional $p \to q$ is true" is a <u>true</u> statement since the antecedent, p, must be false.

18. "Given that $\sim p$ is false and q is false, the conditional $p \to q$ is true" is a <u>true</u> statement since the antecedent, p, is false.

19. Writing exercise

20. Writing exercise

21. "F $\to$ (4 $\neq$ 7)" is a <u>true</u> statement, since a false antecedent always yields a conditional statement which is true.

22. "T $\to$ (6 < 3)" is a <u>false</u> statement, since the antecedent is true and the consequent is false.

23. "(6 $\geq$ 6) $\to$ F" is a <u>false</u> statement, since the antecedent is true and the consequent is false.

24. "F $\to$ (3 $\neq$ 3)" is a <u>true</u> statement, since a false antecedent always yields a true conditional statement.

25. "(4 = 11 $-$ 7) $\to$ (8 > 0)" is <u>true</u>, since the antecedent and the consequent are both true.

26. "($4^2 \neq 16$) $\to$ (4 $-$ 4 = 8)" is a <u>true</u> statement, since a false antecedent always yields a true conditional statement.

Let s represent the statement "She has a snake for a pet," let p represent the statement "he trains ponies," and let m represent "they raise monkeys."

27. "$\sim m \to p$" expressed in words, becomes "If they do not raise monkeys, then he trains ponies."

28. "$p \to \sim m$" expressed in words, becomes "If he trains ponies, then they do not raise monkeys."

29. "$s \to (m \land p)$" expressed in words, becomes "If she has a snake for a pet, then they raise monkeys and he trains ponies."

30. "$(s \land p) \to m$" expressed in words, becomes "If she has a snake for a pet and he trains ponies, then they raise monkeys."

31. "$\sim p \to (\sim m \lor s)$" expressed in words, becomes "If he does not train ponies, then they do not raise monkeys or she has a snake for a pet."

32. "$(\sim s \lor \sim m) \to \sim p$" expressed in words, becomes "If she does not have a snake for a pet or they do not raise monkeys, then he does not train ponies."

Let b represent the statement "I ride my bike," let r represent the statement "it rains" and let p represent "the play is cancelled."

33. The statement "If it rains, then I ride my bike" can be symbolized as "$r \to b$."

34. The statement "If I ride my bike, then the play is cancelled." can be symbolized as "$b \to p$."

35. The statement "If I do not ride my bike, then it does not rain" can be symbolized as "$\sim b \to \sim r$."

36. The statement "If the play is cancelled., then it does not rain" can be symbolized as "$p \to \sim r$."

37. The statement "I ride my bike, or if the play is cancelled, then it rains" can be symbolized as "$b \lor (p \to r)$."

38. The statement "The play is cancelled., and if it rains then I do not ride my bike" can be symbolized as "$p \land (r \to \sim b)$."

39. The statement "I'll ride my bike if it doesn't rain" can be symbolized as "$\sim r \to b$."

40. The statement "It rains if the play is cancelled." can be symbolized as "$p \to r$."

Assume that p and r are false, and q is true.

41. Replacing r and q with the given truth values, we have

    $$\sim F \to T$$
    $$T \to T$$
    $$T.$$

    Thus, the compound statement $\sim r \to q$ is true.

42. Replacing p and, r with the given truth values, we have

    $$\sim F \to \sim F$$
    $$T \to T$$
    $$T.$$

    Thus, the compound statement $\sim p \to \sim r$ is true.

43. Replacing p and q with the given truth values, we have

    $$T \to F$$
    $$F.$$

    Thus, the compound statement $q \to p$ is false.

44. Replacing r and p with the given truth values, we have

$$\sim F \rightarrow F$$
$$T \rightarrow F$$
$$F.$$

Thus, the compound statement $\sim r \rightarrow p$ is false.

45. Replacing p and q with the given truth values, we have

$$F \rightarrow T$$
$$T.$$

Thus, the compound statement $p \rightarrow q$ is true.

46. Replacing q and r with the given truth values, we have

$$\sim T \rightarrow F$$
$$F \rightarrow F$$
$$T.$$

Thus, the compound statement $\sim q \rightarrow r$ is true.

47. Replacing p, r and q with the given truth values, we have

$$\sim F \rightarrow (T \wedge F)$$
$$T \rightarrow F$$
$$F.$$

Thus, the compound statement $\sim p \rightarrow (q \wedge r)$ is false.

48. Replacing p, r and q with the given truth values, we have

$$(\sim F \vee F) \rightarrow F$$
$$(T \vee F) \rightarrow F$$
$$T \rightarrow F$$
$$F.$$

Thus, the compound statement $(\sim r \vee p) \rightarrow p$ is false.

49. Replacing p, r and q with the given truth values, we have

$$\sim T \rightarrow (F \wedge F)$$
$$F \rightarrow F$$
$$T.$$

Thus, the compound statement $\sim q \rightarrow (p \wedge r)$ is true.

50. Replacing p, r and q with the given truth values, we have

$$(\sim F \wedge \sim T) \rightarrow (F \wedge \sim F)$$
$$(T \wedge F) \rightarrow (F \wedge T)$$
$$F \rightarrow F$$
$$T.$$

Thus, the compound statement $(\sim p \wedge \sim q) \rightarrow (p \wedge \sim r)$ is true.

51. Replacing p, r and q with the given truth values, we have

$$(F \rightarrow \sim T) \rightarrow (\sim F \wedge \sim F)$$
$$(F \rightarrow F) \rightarrow (T \wedge T)$$
$$T \rightarrow T$$
$$T.$$

Thus, the compound statement $(p \rightarrow \sim q) \rightarrow (\sim p \wedge \sim r)$ is true.

52. Replacing p, r and q with the given truth values, we have

$$(F \rightarrow \sim T) \wedge (F \rightarrow F)$$
$$(F \rightarrow F) \wedge T$$
$$T \wedge T$$
$$T.$$

Thus, the compound statement $(p \rightarrow \sim q) \wedge (p \rightarrow r)$ is true.

53. Writing exercise

54. Answers will vary. One example is $[p \rightarrow (\sim q \wedge r) \vee p]$. Observe that any "If..., then..." statement where the antecedent is either p, q or r will work since, all that is needed to make the statement true, is a false antecedent.

55. $\sim q \rightarrow p$

p	q	$\sim q$	$\rightarrow$	p
T	T	F	T	T
T	F	T	T	T
F	T	F	T	F
F	F	T	F	F
		1	2	1

56. $p \rightarrow \sim q$

p	q	p	$\rightarrow$	$\sim q$
T	T	T	F	F
T	F	T	T	T
F	T	F	T	F
F	F	F	T	T
		1	2	1

57. $(\sim p \rightarrow q) \rightarrow p$

p	q	$(\sim p$	$\rightarrow$	$q)$	$\rightarrow$	p
T	T	F	T	T	T	T
T	F	F	T	F	T	T
F	T	T	T	T	F	F
F	F	T	F	F	T	F
		1	2	1	3	2

58. $(\sim q \rightarrow \sim p) \rightarrow \sim q$

p	q	$(\sim q$	$\rightarrow$	$\sim p)$	$\rightarrow$	$\sim q$
T	T	F	T	F	F	F
T	F	T	F	F	T	T
F	T	F	T	T	F	F
F	F	T	T	T	T	T
		1	2	1	3	2

59. $(p \vee q) \rightarrow (q \vee p)$

p	q	$(p \vee q)$	$\rightarrow$	$(q \vee p)$
T	T	T T T	T	T T T
T	F	T T F	T	F T T
F	T	F T T	T	T T F
F	F	F F F	T	F F F
		1 2 1	3	1 2 1

Since this statement is always true (column 3), it is a tautology.

60. $(p \wedge q) \rightarrow (p \vee q)$

p	q	$(p \wedge q)$	$\rightarrow$	$(p \vee q)$
T	T	T T T	T	T T T
T	F	T F F	T	T T F
F	T	F F T	T	F T T
F	F	F F F	T	F F F
		1 2 1	3	1 2 1

Since this statement is always true (column 3), it is a tautology.

61. $(\sim p \rightarrow \sim q) \rightarrow (p \wedge q)$

p	q	$(\sim p \rightarrow \sim q)$	$\rightarrow$	$(p \wedge q)$
T	T	F T F	T	T T T
T	F	F T T	F	T F F
F	T	T F F	T	F F T
F	F	T T T	F	F F F
		1 2 1	3	1 2 1

62. $r \rightarrow (p \wedge \sim q)$

p	q	r	r	$\rightarrow$	$(p \wedge \sim q)$
T	T	T	T	F	T F F
T	T	F	F	T	T F F
T	F	T	T	T	T T T
T	F	F	F	T	T T T
F	T	T	T	F	F F F
F	T	F	F	T	F F F
F	F	T	T	F	F F T
F	F	F	F	T	F F T
			2	3	1 2 1

63. $[(r \vee p) \wedge \sim q] \rightarrow p$

p	q	r	$[(r \vee p)$	$\wedge$	$\sim q]$	$\rightarrow$	p
T	T	T	T T T	F	F	T	T
T	T	F	F T T	F	F	T	T
T	F	T	T T T	T	T	T	T
T	F	F	F T T	T	T	T	T
F	T	T	T T F	F	F	T	F
F	T	F	F F F	F	F	T	F
F	F	T	T T F	T	T	F	F
F	F	F	F F F	F	T	T	F
			1 2 1	3	2	4	3

64. $(\sim r \rightarrow s) \vee (p \rightarrow \sim q)$

p	q	r	s	$(\sim r \rightarrow s)$	$\vee$	$(p \rightarrow \sim q)$
T	T	T	T	F T T	T	T F F
T	T	T	F	F T F	T	T F F
T	T	F	T	T T T	T	T F F
T	T	F	F	T F F	F	T F F
T	F	T	T	F T T	T	T T T
T	F	T	F	F T F	T	T T T
T	F	F	T	T T T	T	T T T
T	F	F	F	T F F	T	T T T
F	T	T	T	F T T	T	F T F
F	T	T	F	F T F	T	F T F
F	T	F	T	T T T	T	F T F
F	T	F	F	T F F	T	F T F
F	F	T	T	F T T	T	F T T
F	F	T	F	F T F	T	F T T
F	F	F	T	T T T	T	F T T
F	F	F	F	T F F	T	F T T
				1 2 1	3	1 2 1

65. $(\sim p \wedge \sim q) \rightarrow (s \rightarrow r)$

p	q	r	s	$(\sim p \wedge \sim q)$	$\rightarrow$	$(s \rightarrow r)$
T	T	T	T	F F F	T	T T T
T	T	T	F	F F F	T	F T T
T	T	F	T	F F F	T	T F F
T	T	F	F	F F F	T	F T F
T	F	T	T	F F T	T	T T T
T	F	T	F	F F T	T	F T T
T	F	F	T	F F T	T	T F F
T	F	F	F	F F T	T	F T F
F	T	T	T	T F F	T	T T T
F	T	T	F	T F F	T	F T T
F	T	F	T	T F F	T	T F F
F	T	F	F	T F F	T	F T F
F	F	T	T	T T T	T	T T T
F	F	T	F	T T T	T	F T T
F	F	F	T	T T T	F	T F F
F	F	F	F	T T T	T	F T F
				1 2 1	3	1 2 1

66. The statement is not a tautology if only one F appears in the final column of a truth table, since a tautology requires all T's in the final column.

67. The negation of "If that is an authentic Persian rug, I'll be surprised" is "That is an authentic Persian rug and I am not surprised."

68. The negation of "If Ella reaches that note, she will shatter glass" is "Ella reaches that note and she will not shatter glass."

69. The negation of "If the English measures are not converted to metric measures, then the spacecraft will crash on the surface of Mars" is "The English measures are not converted to metric measures and the spacecraft does not crash on the surface of Mars"

70. The negation of "If you say 'I do,' then you'll be happy for the rest of your life" is "Say 'I do' and you are not happy for the rest of your life."

71. The negation of "If you want to be happy for the rest of your life, never make a pretty woman your wife" is "You want to be happy for the rest of your life and you make a pretty woman you wife."

72. The negation of "If loving you is wrong, I don't want to be right" is "Loving you is wrong and I want to be right."

73. An equivalent statement to "If you give your plants tender, loving care, they will flourish" is "You do not give your plants tender, loving care or they flourish."

74. An equivalent statement to "If the check is in the mail, I'll be surprised" is "The check is not in the mail or I'll be surprised."

75. An equivalent statement to "If she doesn't, he will" is "She does or he will."

76. An equivalent statement to "If I say yes, she says no" is "I do not say yes or she says no."

77. An equivalent conditional statement to "All residents of Butte are residents of Montana" is "If you are a resident of Butte, then you are a resident of Montana." An equivalent statement would be "The person is a not a resident of Butte or is a resident of Montana."

78. An equivalent conditional statement to "All women were once girls" is "If you are a women, then you were once a girl." An equivalent statement would be "The person is not a women or was once a girl."

79. The statements $p \rightarrow q$ and $\sim p \vee q$ are equivalent if they have the same truth tables.

p	q	$p \rightarrow q$	$\sim p \vee q$
T	T	T T T	F T T
T	F	T F F	F F F
F	T	F T T	T T T
F	F	F T F	T T F
		1 2 1	1 2 1

Since the truth values in the final columns for each statement are the same, the statements are equivalent.

80.
p	q	$\sim(p \rightarrow q)$	$p \wedge \sim q$
T	T	F T T T	T F F
T	F	T T F F	T T T
F	T	F F T T	F F F
F	F	F F T F	F F T
		3 1 2 1	1 2 1

Since the truth values in the final columns for each statement are the same, the statements are equivalent.

81.
p	q	$p \rightarrow q$	$\sim q \rightarrow \sim p$
T	T	T T T	F T F
T	F	T F F	T F F
F	T	F T T	F T T
F	F	F T F	T T T
		1 2 1	1 2 1

Since the truth values in the final columns for each statement are the same, the statements are equivalent.

82.
p	q	$q \rightarrow p$	$\sim p \rightarrow \sim q$
T	T	T T T	F T F
T	F	F T T	F T T
F	T	T F F	T F F
F	F	F T F	T T T
		1 2 1	1 2 1

Since the truth values in the final columns for each statement are the same, the statements are equivalent.

83.
p	q	$p \rightarrow \sim q$	$\sim p \vee \sim q$
T	T	T F F	F F F
T	F	T T T	F T T
F	T	F T F	T T F
F	F	F T T	T T T
		1 2 1	1 2 1

Since the truth values in the final columns for each statement are the same, the statements are equivalent.

84.

p	q	$p \to q$	$q \to p$
T	T	T T T	T T T
T	F	T F F	F T T
F	T	F T T	T F F
F	F	F T F	F T F
		1 2 1	1 2 1

Since the truth values in the final columns for each statement are not the same, the statements are not equivalent.

85.

p	q	$p \land \sim q$	$\sim q \to \sim p$
T	T	T F F	F T F
T	F	T T T	T F F
F	T	F F F	F T T
F	F	F F T	T T T
		1 2 1	1 2 1

Since the truth values in the final columns for each statement are not the same, the statements are not equivalent. Observe that since they have opposite truth values, each statement is the negation of the other.

86.

p	q	$\sim p \land q$	$\sim p \to q$
T	T	F F T	F T T
T	F	F F F	F T F
F	T	T T T	T T T
F	F	T F F	T F F
		1 2 1	1 2 1

Since the truth values in the final columns for each statement are not the same, the statements are not equivalent.

87. In the diagram, two series circuits are shown, which correspond to $p \land q$ and $p \land \sim q$. These circuits, in turn, form a parallel circuit. Thus, the logical statement is

$$(p \land q) \lor (p \land \sim q).$$

One pair of equivalent statements listed in the text includes

$$(p \land q) \lor (p \land \sim q) \equiv p \land (q \lor \sim q).$$

Since $(q \lor \sim q)$ is always true, $p \land (q \lor \sim q)$ simplifies to

$$p \land \text{T} \equiv p.$$

88. In the diagram, a parallel circuit is shown, which corresponds to $r \lor q$. This circuit, in turn, is in series with p. Thus, the logical statement is

$$p \land (r \lor q).$$

89. In the diagram, a series circuit is shown, which corresponds to $\sim q \land r$. This circuit, in turn, forms a parallel circuit with p. Thus, the logical statement is

$$p \lor (\sim q \land r).$$

90. The diagram shows q in parallel with a series circuit consisting of p and the parallel circuit involving q and $\sim p$. Thus, the logical statement is

$$q \lor [p \land (q \lor \sim p)].$$

One pair of equivalent statements listed in the text includes

$$p \land (q \lor \sim p) \equiv (p \land q) \lor (p \land \sim p).$$

Since $(p \land \sim p)$ is never true, $p \land (q \lor \sim p)$ simplifies to

$$(p \land q) \lor \text{F} \equiv (p \land q). \text{ Thus,}$$

$$q \lor [p \land (q \lor \sim p)] \equiv q \lor (p \land q)$$
$$\equiv (q \lor p) \land (q \land q)$$
$$\equiv (q \lor p) \land q$$
$$\equiv q.$$

91. In the diagram, a parallel circuit corresponds to $p \lor q$. This circuit is parallel to $\sim p$. Thus, the total circuit corresponds to the logical statement

$$\sim p \lor (p \lor q).$$

This statement in turn, is equivalent to

$$(\sim p \lor p) \lor (\sim p \lor q).$$

Since $\sim p \lor p$ is always true, we have

$$\text{T} \lor (\sim p \lor q) \equiv \text{T}.$$

92. The diagram shows two parallel circuits, $\sim p \lor q$ and $\sim p \lor \sim q$ which are parallel to each other. Thus, the total circuit can be represented as

$$(\sim p \lor q) \lor (\sim p \lor \sim q).$$

This circuit can be simplified using the following equivalencies:

$$(\sim p \lor q) \lor (\sim p \lor \sim q)$$
$$\equiv \sim p \lor q \lor \sim p \lor \sim q$$
$$\equiv \sim p \lor q \lor \sim q$$
$$\equiv \sim p \lor (q \lor \sim q)$$
$$\equiv \sim p \lor \text{T}$$
$$\equiv \text{T}.$$

93. The logical statement, $p \land (q \lor \sim p)$, can be represented by the following circuit.

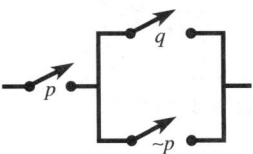

The statement, $p \wedge (q \vee \sim p)$, simplifies to $p \wedge q$ as follows:
$$p \wedge (q \vee \sim p) \equiv (p \wedge q) \vee (p \wedge \sim p)$$
$$\equiv (p \wedge q) \vee F$$
$$\equiv p \wedge q.$$

94. The logical statement, $(\sim p \wedge \sim q) \wedge \sim r$, can be represented by the following circuit.

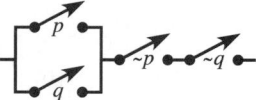

95. The logical statement, $(p \vee q) \wedge (\sim p \wedge \sim q)$, can be represented by the following circuit.

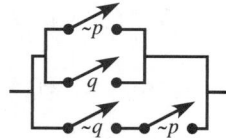

The statement, $(p \vee q) \wedge (\sim p \wedge \sim q)$, simplifies to F as follows:
$$(p \vee q) \wedge (\sim p \wedge \sim q) \equiv p \vee q \wedge \sim p \wedge \sim q$$
$$\equiv p \vee \sim p \wedge q \wedge \sim q$$
$$\equiv (p \vee \sim p) \wedge (q \wedge \sim q)$$
$$\equiv T \wedge F$$
$$\equiv F.$$

96. The logical statement, $(\sim q \wedge \sim p) \vee (\sim p \vee q)$, can be represented by the following circuit.

The statement, $(\sim q \wedge \sim p) \vee (\sim p \vee q)$, simplifies to $\sim p \vee q$ as follows:

$$(\sim q \wedge \sim p) \vee (\sim p \vee q) \equiv [\sim q \vee (\sim p \vee q)] \wedge [\sim p \vee (\sim p \vee q)]$$
$$\equiv [\sim q \vee \sim p \vee q] \wedge [\sim p \vee \sim p \vee q]$$
$$\equiv [(\sim q \vee q) \vee \sim p] \wedge [(\sim p \vee \sim p) \vee q]$$
$$\equiv (T \vee \sim p) \wedge (\sim p \vee q)$$
$$\equiv T \wedge (\sim p \vee q)$$
$$\equiv \sim p \vee q.$$

97. The logical statement, $[(p \vee q) \wedge r] \wedge \sim p$, can be represented by the following circuit.

The statement, $[(p \vee q) \wedge r] \wedge \sim p$, simplifies to $(r \wedge \sim p) \wedge q$ as follows:

$$[(p \vee q) \wedge r] \wedge \sim p \equiv [(p \wedge r) \vee (q \wedge r)] \wedge \sim p$$
$$\equiv [(p \wedge r) \wedge \sim p] \vee [(q \wedge r) \wedge \sim p]$$
$$\equiv [p \wedge r \wedge \sim p] \vee [q \wedge r \wedge \sim p]$$
$$\equiv [(p \wedge \sim p) \wedge r] \vee [(r \wedge \sim p) \wedge q]$$
$$\equiv (F \wedge r) \vee [(r \wedge \sim p) \wedge q]$$
$$\equiv F \vee [(r \wedge \sim p) \wedge q]$$
$$\equiv (r \wedge \sim p) \wedge q.$$

98. The logical statement, $[(\sim p \wedge \sim r) \vee \sim q] \wedge (\sim p \wedge r)$, can be represented by the following circuit.

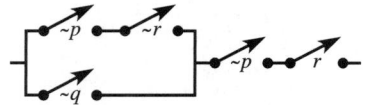

The statement, $[(\sim p \wedge \sim r) \vee \sim q] \wedge (\sim p \wedge r)$, can simplify to $(\sim p \wedge r) \wedge \sim q$ in the following manner.

Both $[(\sim p \wedge \sim r) \vee \sim q]$ and $(\sim p \wedge r)$ must be true. But if $(\sim p \wedge r)$ is true, then $(\sim p \wedge \sim r)$ is false. If $(\sim p \wedge \sim r)$ is false, then $\sim q$ must be true for the original disjunction to be true. Thus,

$$[(\sim p \wedge \sim r) \vee \sim q] \wedge (\sim p \wedge r) \equiv (F \vee \sim q) \wedge (\sim p \wedge r)$$
$$\equiv \sim q \wedge (\sim p \wedge r)$$
$$\equiv (\sim p \wedge r) \wedge \sim q.$$

99. The logical statement, $\sim q \rightarrow (\sim p \rightarrow q)$, can be represented by the following circuit.

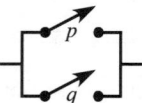

The statement, $\sim q \rightarrow (\sim p \rightarrow q)$, simplifies to $p \vee q$ as follows:

$$\sim q \rightarrow (\sim p \rightarrow q) \equiv \sim q \rightarrow (p \vee q)$$
$$\equiv q \vee (p \vee q)$$
$$\equiv q \vee p \vee q$$
$$\equiv p \vee q \vee q$$
$$\equiv p \vee (q \vee q)$$
$$\equiv p \vee q.$$

100. The logical statement, $\sim p \rightarrow (\sim p \vee \sim q)$, can be represented by the following circuit.

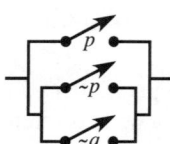

The statement, $\sim p \to (\sim p \vee \sim q)$, simplifies to T as follows:

$$\sim p \to (\sim p \vee \sim q) \equiv p \vee (\sim p \vee \sim q)$$
$$\equiv p \vee \sim p \vee \sim q$$
$$\equiv (p \vee \sim p) \vee \sim q$$
$$\equiv T \vee \sim q$$
$$\equiv T.$$

101. Writing exercise

102. Referring to Figures 5 and 6 of Example 6 in the text:

   Cost per year of the circuit in Figure 5
   = number of switches × $.03 × 24 hrs × 365 days
   = (4) × (.03) × 24 × 365
   = $1051.20.

   Cost per year of the circuit in Figure 6
   = number of switches × $.03 × 24 hrs × 365 days
   = (3) × (.03) × 24 × 365
   = $788.40.

   Thus, the savings is $1051.20 − $788.40 = $262.80.

3.4 EXERCISES

For each given direct statement (symbolically as $p \to q$), write (a) the converse ($q \to p$), (b) the inverse ($\sim p \to \sim q$), and (c) the contrapositive ($\sim q \to \sim p$) in if ... then forms. Wording may vary in the answers to Exercises 1–10.

1. The direct statement: If beauty were a minute, then you would be an hour.

   (a) *Converse*: If you were an hour, then beauty would be a minute.

   (b) *Inverse*: If beauty were not a minute, then you would not be an hour.

   (c) *Contrapositive*: If you were not an hour, then beauty would not be a minute.

2. The direct statement: If you lead, then I will follow.

   (a) *Converse*: If I follow, then you lead.

   (b) *Inverse*: If you do not lead, then I will not follow.

   (c) *Contrapositive*: If I do not follow, then you do not lead.

3. The direct statement: If it ain't broke, don't fix it.

   (a) *Converse*: If you don't fix it, then it ain't broke.

   (b) *Inverse*: If it's broke, then fix it.

   (c) *Contrapositive*: If you fix it, then it's broke.

4. The direct statement: If I had a nickel for each time that happened, I would be rich.

   (a) *Converse*: If I were rich, then I would have a nickel for each time that happened.

   (b) *Inverse*: If I did not have a nickel for each time that happened, then I would not be rich.

   (c) *Contrapositive*: If I were not rich, then I would not have a nickel for each time that happened.

It is helpful to restate the direct statement in an if ... then form for the exercises 5–9.

5. The direct statement: If you walk in front of a moving car, then it is dangerous to your health.

   (a) *Converse*: If it is dangerous to your health, then you walk in front of a moving car.

   (b) *Inverse*: If you do not walk in front of a moving car, then it is not dangerous to your health.

   (c) *Contrapositive*: If it is not dangerous to your health, then you do not walk in front of a moving car.

6. The direct statement: If it's milk, then it contains calcium.

   (a) *Converse*: If it contains calcium, then it's milk.

   (b) *Inverse*: If it's not milk, then it does not contain calcium.

   (c) *Contrapositive*: If it does not contain calcium, then it's not milk.

7. The direct statement: If they are birds of a feather, then they flock together.

   (a) *Converse*: If they flock together, then they are birds of a feather.

   (b) *Inverse*: If they are not birds of a feather, then they do not flock together.

   (c) *Contrapositive*: If they do not flock together, then they are not birds of a feather.

8. The direct statement: If it is a rolling stone, then it gathers no moss.

   (a) *Converse*: If it gathers no moss, then it is a rolling stone.

   (b) *Inverse*: If it is not a rolling stone, then it gathers moss.

   (c) *Contrapositive*: If it gathers moss, then it is not a rolling stone.

9. The direct statement: If you build it, then he will come.

   (a) *Converse*: If he comes, then you built it.

   (b) *Inverse*: If you don't build it, then he won't come.

   (c) *Contrapositive*: If he doesn't come, then you didn't build it.

10. The direct statement: If there's smoke, then there's fire.

    (a) *Converse*: If there's fire, then there's smoke.

    (b) *Inverse*: If there's no smoke, then there's no fire.

(c) *Contrapositive*: If there's no fire, then there's no smoke.

11. The direct statement: $p \to \sim q$.
    (a) *Converse*: $\sim q \to p$.
    (b) *Inverse*: $\sim p \to q$.
    (c) *Contrapositive*: $q \to \sim p$.

12. The direct statement: $\sim p \to q$.
    (a) *Converse*: $q \to \sim p$.
    (b) *Inverse*: $p \to \sim q$.
    (c) *Contrapositive*: $\sim q \to p$.

13. The direct statement: $\sim p \to \sim q$.
    (a) *Converse*: $\sim q \to \sim p$.
    (b) *Inverse*: $p \to q$.
    (c) *Contrapositive*: $q \to p$.

14. The direct statement: $\sim q \to \sim p$.
    (a) *Converse*: $\sim p \to \sim q$.
    (b) *Inverse*: $q \to p$.
    (c) *Contrapositive*: $p \to q$.

15. The direct statement: $p \to (q \lor r)$.
    (a) *Converse*: $(q \lor r) \to p$.
    (b) *Inverse*: $\sim p \to \sim(q \lor r)$ or $\sim p \to (\sim q \land \sim r)$.
    (c) *Contrapositive*: $(\sim q \land \sim r) \to \sim p$.

16. The direct statement: $(r \lor \sim q) \to p$.
    (a) *Converse*: $p \to (r \lor \sim q)$.
    (b) *Inverse*: $\sim(r \lor \sim q) \to \sim p$ or $(\sim r \land q) \to \sim p$.
    (c) *Contrapositive*: $\sim p \to \sim(r \lor \sim q)$ or $\sim p \to (\sim r \land q)$.

17. Writing exercise

18. Writing exercise

Writing the statements, Exercises 19–40, in the form "if p, then q" we arrive at the following results.

19. The statement "If it is muddy, I'll wear my galoshes" becomes "If it is muddy, then I'll wear my galoshes."

20. The statement "If I finish studying, I'll go to the party" becomes "If I finish studying, then I'll go to the party.

21. The statement "'17 is positive' implies that $17 + 1$ is positive" becomes "If 17 is positive, then $17 + 1$ is positive."

22. The statement "'Today is Wednesday' implies that yesterday was Tuesday" becomes "If Today is Wednesday, then yesterday was Tuesday."

23. The statement "All integers are rational numbers" becomes "If a number is an integer, then it is a rational number."

24. The statement "All whole numbers are integers" becomes "If a number is a whole number, then it is an integer."

25. The statement "Doing crossword puzzles is sufficient for driving me crazy" becomes "If I do crossword puzzles, then I am driven crazy."

26. The statement "Being in Fort Lauderdale is sufficient for being in Florida" becomes "If you are in Fort Lauderdale, then you are in Florida."

27. "A day's growth of beard is necessary for Greg Tobin to shave" becomes "If Greg Tobin is to shave, then he must have a day's growth of beard."

28. The statement "Being an environmentalist is necessary for being elected" becomes "If one is elected, then one is an environmentalist."

29. The statement "I can go from Boardwalk to Connecticut Avenue only if I pass GO" becomes "If I go from Boardwalk to Connecticut, then I pass GO."

30. The statement "The principal will hire more teachers only if the school board approves" becomes "If the principal hires more teachers, then the school board approves."

31. The statement "No whole numbers are not integers" becomes "If a number is a whole number, then it is an integer."

32. The statement "No integers are irrational numbers" becomes "If a number is an integer, then it is rational."

33. The statement "The Indians will win the pennant when their pitching improves" becomes "If their pitching improves, then the Indians will win the pennant."

34. The statement "Jesse will be a liberal when pigs fly" becomes "If pigs fly, then Jesse will be a liberal."

35. The statement "A rectangle is a parallelogram with a right angle" becomes "If the figure is a rectangle, then it is a parallelogram with a right angle."

36. The statement "A parallelogram is a four-sided figure with opposite sides parallel" becomes "If the figure is a parallelogram, then it is a four-sided figure with opposite sides parallel."

37. The statement "A triangle with two sides of the same length is isosceles" becomes "If a triangle has two sides of the same length, then it is isosceles."

38. The statement "A square is rectangle with two adjacent sides of equal length" becomes "If the figure is a square, then it is a rectangle with two adjacent sides of equal length."

39. The statement "The square of a two-digit number whose units digit is 5 will end in 25" becomes "If a two-digit number whose units digit is 5 is squared, then it will end in 25."

40. The statement "An integer whose units digit is 0 or 5 is divisible by 5" becomes "If an integer has a units digit of 0 or 5, then it is divisible by 5."

41. Option D is the answer since "r is necessary for s" represents the converse, $s \to r$, of all of the other statements.

42. Writing exercise

43. Writing exercise

44. Writing exercise

45. The statement "$5 = 9 - 4$ if and only if $8 + 2 = 10$" is true, since this is a biconditional composed of two true statements.

46. The statement "$3 + 1 \neq 6$ if and only if $8 \neq 8$" is false since this is a biconditional consisting of a true and a false statement.

47. The statement "$8 + 7 \neq 15$ if and only if $3 \times 5 \neq 9$" is false, since this is a biconditional consisting of a false and a true statement.

48. The statement "$6 \times 2 = 14$ if and only if $9 + 7 \neq 16$" is true, since this is a biconditional consisting of two false statements.

49. The statement "Bill Clinton was president if and only if Jimmy Carter was not president" is false, since this is a biconditional consisting of a true and a false statement.

50. Burger King sells Big Macs if and only if IBM manufactures computers" is false, since this is a biconditional consisting of a false and a true statement.

51. The statements "Elvis is alive" and "Elvis is dead" are contrary, since both cannot be true at the same time.

52. The statements "George W. Bush is a Democrat" and "George W. Bush is a Republican" are contrary, since both cannot be true at the same time.

53. The statements "That animal has four legs" and "That animal is a dog" are consistent, since both statements can be true.

54. The statements "That book is nonfiction" and "That book costs more than $70" are consistent, since both statements can be true.

55. The statements "This number is an integer" and "This number is irrational" are contrary, since both cannot be true at the same time.

56. The statements "This number is positive" and "This number is a natural number" are consistent, since both statements can be true.

57. Answers will vary. One example is: That man is Kent Merrill; That man sells books.

58. Answers will vary. One example is: Jennifer Kerber is 35 years old; Jennifer Kerber is less then 35 years old.

3.5 EXERCISES

1. Draw an Euler diagram where the region representing "boxers" must be inside the region representing "those who wear trunks" so that the first premise is true.

   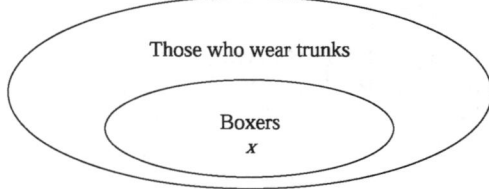

   Let x represent Cris Mader. By premise 2, x must lie in the "boxers" region. Since this forces the conclusion to be true, the argument is valid.

2. Draw an Euler diagram where the region representing "amusement parks" must be inside the region representing "locations that have thrill rides" so that the first premise is true.

   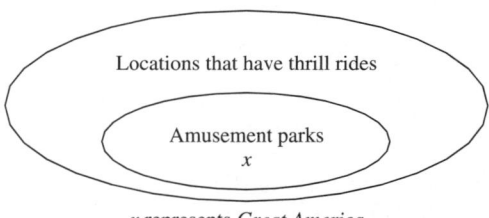

   Let x represent the amusement park "Great America". By the second premise, x must lie in the "amusement parks" region. Since this forces the conclusion to be true, the argument is valid.

3. Draw an Euler diagram where the region representing "Southerners" lies inside the region representing " those who speak with an accent. Let x represent Bill Leonard. By the second premise, Bill Leonard must lie inside the region of those who speak with an accent and hence may lie inside the inner or the outer region. But, for the conclusion to be true, he must lie only inside the inner region. Thus, the argument is invalid.

   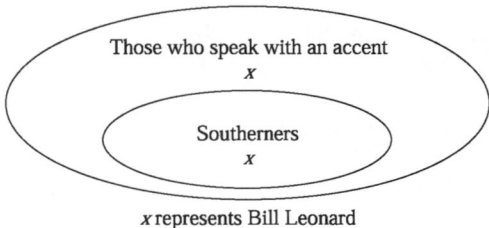

4. Draw an Euler diagram where the region representing "politicians" must be inside the region representing "those who lie, cheat, and steal" so that the first premise is true.

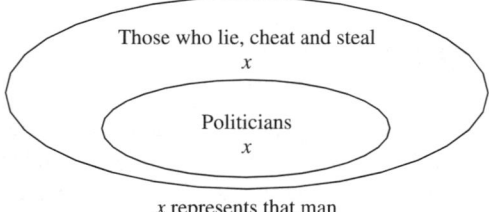

x represents that man

Let x represent "that man." By the second premise, x must lie in the "those who lie, cheat, and steal" region. Thus, he could be inside or outside the inner region. Since this allows for a false conclusion (he doesn't have to be in the "politicians" region for both premises to be true), the argument is <u>invalid</u>.

5. Draw an Euler diagram where the region representing "contractors" must be inside the region representing "those who use cell phones" so that the first premise is true.

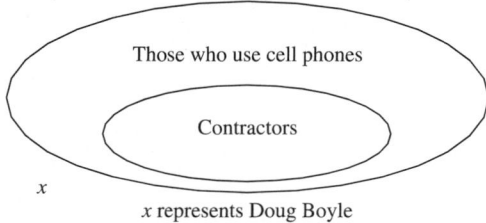

x represents Doug Boyle

Let x represent "Doug Boyle." By the second premise, x must lie outside the region representing "those who use cell phones." Since this forces the conclusion to be true, the argument is <u>valid</u>.

6. Draw an Euler diagram where the region representing "dogs" must be inside the region representing "creatures that love to bury bones" so that the first premise is true.

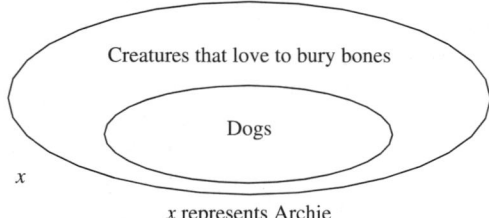

x represents Archie

Let x represent "Archie." By the second premise, x must lie outside the region representing "creatures that love to bury bones." Since this forces the conclusion to be true, the argument is <u>valid</u>.

7. Draw an Euler diagram where the region representing "people who apply for a loan" must be inside the region representing "people who pay for a title search" so that the first premise is true.

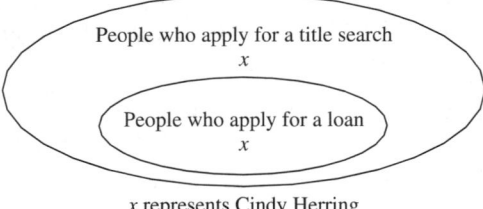

x represents Cindy Herring

Let x represent "Cindy Herring." By the second premise, x must lie in the "people who pay for a title search" region. Thus, she could be inside or outside the inner region. Since this allows for a false conclusion (she doesn't have to be in the "people who apply for a loan" region for both premises to be true), the argument is <u>invalid</u>.

8. Draw an Euler diagram where the region representing "residents of Minnesota" must be inside the region representing "those who know how to live in freezing temperatures" so that the first premise is true.

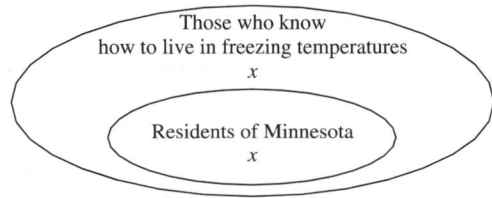

x represents Wendy Rockswold

Let x represent "Wendy Rockswold." By the second premise, x must lie in the "those who know how to live in freezing temperatures" region. Thus, she could be inside or outside the inner region. Since this allows for a false conclusion (she doesn't have to be in the "residents of Minnesota" region for both premises to be true), the argument is <u>invalid</u>.

9. Draw an Euler diagram where the region representing "philosophers" intersects the region representing "those who are absent minded." This keeps the first premise true.

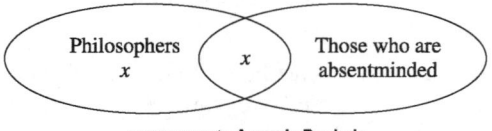

x represents Amanda Perdaris

Let x represent "Amanda Perdaris." By the second premise, x must lie in the region representing "philosophers." Thus, she could be inside or outside the region "representing people who are absent minded." Since this allows for a false conclusion, the argument is <u>invalid</u>.

10. Draw an Euler diagram where the region representing "dinosaurs" intersects the region representing "plant-eaters." This keeps the first premise true.

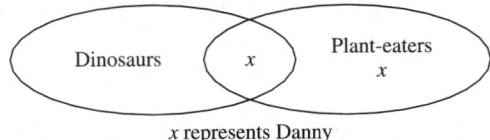

x represents Danny

Let x represent "Danny". By the second premise, x must lie in the region representing "plant-eaters." Thus, he could be inside or outside the region "dinosaurs." Since this allows for a false conclusion, the argument is invalid.

11. Draw an Euler diagram where the region representing "trucks" intersects the region representing "vehicles with sound systems." This keeps the first premise true. There are several ways to represent the 2nd premise as true. One way is as shown below. Examining the diagram, however, it is apparent that the conclusion is false.

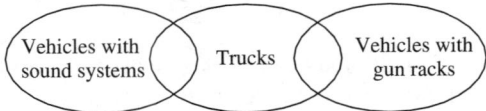

Since the diagram shows true premises but also a false conclusion, the argument is invalid. Note that all ways to draw an Euler diagram representing true premises must also yield a true conclusion for the argument to be valid.

12. Draw an Euler diagram where the region representing "nurses" intersects the region representing "those who wear blue uniforms." This keeps the first premise true.

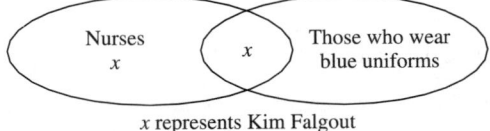

x represents Kim Falgout

Let x represent "Kim Folgout". By the second premise, x must lie in the region representing "nurses." Thus, she could be inside or outside the region "those who wear blue uniforms." Since this allows for a false conclusion, the argument is invalid.

13. Interchanging the second premise and the conclusion of Example 3 (in the text) yields the following argument,

All banana trees have green leaves.
That plant is a banana tree.
That plant has green leaves.

Draw an Euler diagram where the region representing "Banana trees" must be inside the region representing "Things that have green leaves" so that the first premise is true.

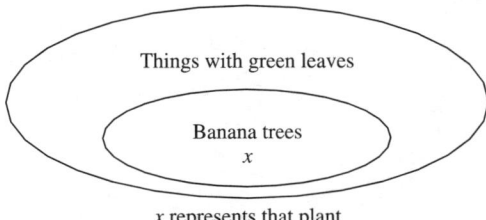

x represents that plant

Let x represent "That plant." By the second premise, x must lie inside the region representing "Banana trees." Since this forces the conclusion to be true, the argument is valid, which makes the answer to the question yes.

14. The (valid) argument of Example 4 (in the text) is,

All expensive things are desirable.
All desirable things make you feel good.
All things that make you feel good make you live longer.
All expensive things make you live longer.

Another possible conclusion, which will keep the argument valid, is "All expensive things make you feel good." The argument remains valid since the premises diagrammed (Figure 13, in the text) force this conclusion to be true also.

15. The following is a valid argument which can be constructed from the given Euler diagram.

People who have major surgery must go to the hospital.
Julianne Peterson is having major surgery.
Julianne Peterson must go to the hospital.

16. The following is a valid argument which can be constructed from the given Euler diagram.

All people with blue eyes have blond hair.
Julie Ward does not have blond hair.
Julie Word does not have blue eyes.

17. The following Euler diagram represents true premises.

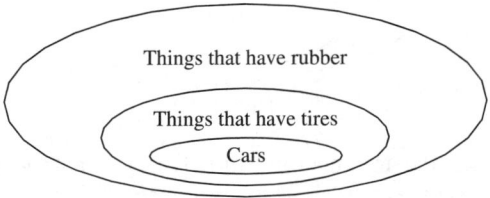

Since the diagram forces the conclusion to be true also, the argument is valid.

18. The following represents one way to diagram the premises so that they are true; however, the arugment is <u>invalid</u> since, according to the diagram, all birds are planes, which is false even though the stated conclusion is true.

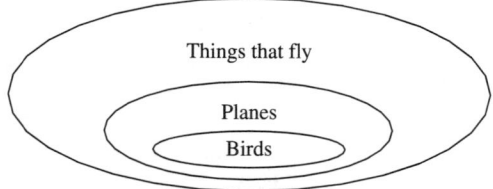

19. The following Euler diagram represents true premises.

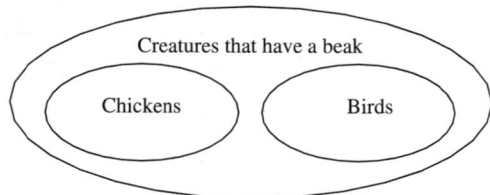

The argument is <u>invalid</u> even though the conclusion is true since the diagram implies that no chickens are birds - a false statement.

20. The following Euler diagram yields true premises. It also forces the conclusion to be true.

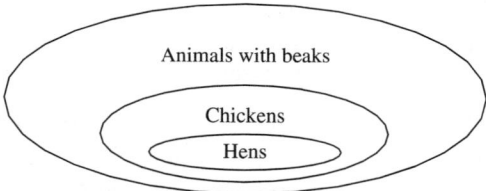

Thus, the argument in <u>valid</u>. Observe that the diagram is the only way to show true premises.

21. The following Euler diagram yields true premises. It also forces the conclusion to be true.

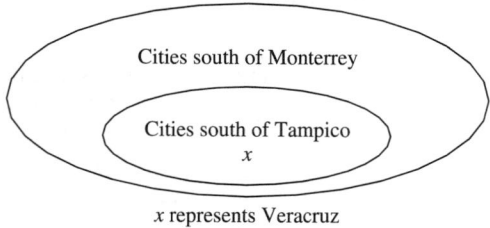

x represents Veracruz

Thus, the argument in <u>valid</u>. Observe that the diagram is the only way to show true premises.

22. The following Euler diagram represents true premises.

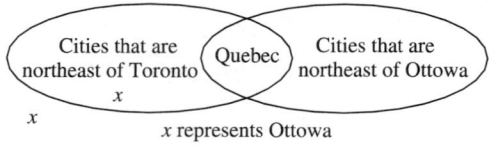

x represents Ottowa

But x can reside inside or outside of the "Cities that are northeast of Toronto" diagram. In the one case (x inside) the conclusion is true. In the other case (x outside) the conclusion is false. Since true premises must always give a true conclusion, the argument is <u>invalid</u>.

23. The following Euler diagram represents true premises.

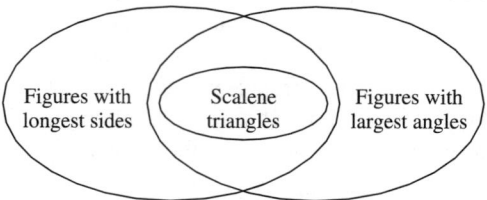

No information, however, is given regarding the relationship between the largest angle and the longest side. The argument is <u>invalid</u> even though the conclusion is true.

24. The following Euler diagram represents the two premises as being true and we are forced into a true conclusion.

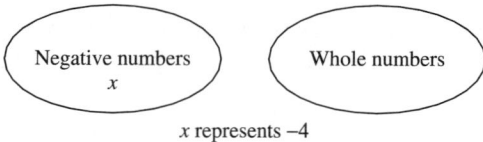

x represents -4

Thus, the argument is <u>valid</u>.

The premises marked A, B, and C are followed by several possible conclusions (Exercises 25–30). Take each conclusion in turn, and check whether the resulting argument is valid or invalid.

A. All people who drive contribute to air pollution.
B. All people who contribute to air pollution make life a little worse.
C. Some people who live in a suburb make life a little worse.
   Diagram the three premises to be true.

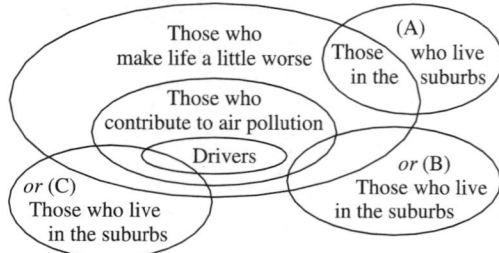

25. We are not forced into the conclusion, "Some people who live in a suburb drive" since diagrams (A) or (B) represent true premises where this conclusion is false. Thus, the argument is <u>invalid</u>.

26. We are not forced into the conclusion, "Some people who live in a suburb contribute to air pollution" since option (A) represents true premises and a false conclusion. Thus, the argument is <u>invalid</u>.

27. We are not forced into the conclusion, "Some people who contribute to air pollution live in a suburb" since option (A) represents true premises and a false conclusion. Thus, the argument is <u>invalid</u>.

28. We are not forced into the conclusion, "Suburban residents never drive" since diagrams (C) represents true premises where this conclusion is false. Thus, the argument is <u>invalid</u>.

29. The conclusion, "All people who drive make life a little worse" yields a <u>valid</u> argument since all three options (A–C) represent true premises and force this conclusion to be true.

30. The conclusion, "Some people who make life a little worse live in a suburb" yields a <u>valid</u> argument since all three options (A–C) represent true premises and force this conclusion to be true.

31–32. No answers

EXTENSION: LOGIC PUZZLES

1. Draw charts as indicated and complete using the initial information given. Use "●" for Yes and "x" for No. For any cell that is assigned "●", mark "x" in the remaining unmarked cells in that row and column.

    (1) *Neither Lauren nor Zach was the child accompanied by Ms. Reed. Tara was more interested in shopping at the country store than in picking pumpkins.* Mark (x)'s in the boxes for Lauren and Zach under Ms. Reed's column. Place (●) into the intersection of Tara and country store since that was her interest. Note: remember to always mark (x)'s in remaining cells of a row or column when marking a cell with (●)

    (2) *Xander and his father, Mr. Morgan, didn't go on the hay ride. Ms. Fedor's child (who isn't Zach or Lauren) was fascinated by the cider-making process.* Place (●) into the cell representing the intersection of Xander and Mr. Morgan since you know that they are father and son. Place (x) into the cell representing the intersection of Xander and hay ride as well as the cell for his father, Mr. Morgan, and hay ride. Place (●) into the cell at the intersection of Ms. Fedor and cider-making. Place (x)'s in the cells representing Lauren and Zach in Ms. Fedor's column. Since neither Lauren nor Zach is Ms. Fedor's child neither can be interested in cider-making. Mark (x)'s in the cells for cider-making in Zach and Lauren's rows.

    (3) *Zach is neither the child who went on the hay ride nor the one who wanted to go apple picking. Mr. Hanson's child didn't go on the hay ride.* Place (x) in the aspect cells for apple picking and hay ride in Zach's row. This leaves feeding animals as Zach's favorite aspect. Place (●) in the feeding animal cell for Zach. Mark (x) in the aspect cell for the hay ride in Mr. Hanson's column.

    Tara and Raven are the only two children that could still belong to Ms. Fedor. Since Ms. Fedor's child liked cider-making check the aspect columns and notice Tara liked the country store. This leaves Raven as Ms. Fedor's child. Place (●) in the cell for Raven in Ms. Fedor's column and (●) in the cider-making cell in Raven's row. The only aspect now available to Xander is apple picking. Place (●) in that cell. The last aspect, the hay ride, is now Lauren's only possible aspect. Place (●) in the appropriate cell.

    At this point *Chart 1* will have been completed.

    Chart 1

		PARENT					ASPECT				
		MS. FEDOR	M. HANSON	MR. MAIER	MR. MORGAN	MS. REED	APPLE PICKING	CIDER MAKING	COUNTRY STR.	FEEDING ANM.	HAY RIDE
CHILD	LAUREN	X			X	X	X	X	X	X	●
	RAVEN	●	X	X	X	X	X	●	X	X	X
	TARA	X			X		X	X	●	X	X
	XANDER	X	X	X	●	X	●	X	X	X	X
	ZACH	X			X	X	X	X	X	●	X
ASPECT	APPLE PICK.	X									
	CIDER MAK.	●	X	X	X	X					
	COUNTRY	X									
	FEEDING	X									
	HAY RIDE	X	X		X						

    Tara is now the only child left in Ms. Reed's column. Place (●) at the intersection of Lauren and Ms. Reed. Notice that Xander's favorite aspect is apple picking and we know that Mr. Morgan is Xander's father. Thus we can place (●) in Mr. Morgan's column in the apple picking cell. This leaves Lauren and Zach with the possibility of Mr. Hanson and Mr. Maier for parents. By comparing Lauren's favorite aspect, the hay ride, to Mr. Hanson and Mr. Maier's aspect columns we find that Mr. Hanson's child did NOT like the hay ride. Thus Mr. Maier must be Lauren's father. Mr. Hanson is left as Zach's father. Now it is a simple matter of matching each child's interest with the aspect cell for his or her parent. Zach likes feeding animals; mark (●) in the feeding animals cell for Mr. Hanson. Lauren likes the hay ride; mark (●) in the hay ride cell for Mr. Maier. The country store is now the only available interest in Ms. Reed's column. Check to see that Ms. Reed's child, Tara, is, in fact, the child that likes the country store.

Place (●) in the appropriate cell and the logic puzzle is complete (*Chart 2*).

Chart 2

		PARENT					ASPECT				
		MS. FEDOR	M. HANSON	MR. MAIER	MR. MORGAN	MS. REED	APPLE PICKING	CIDER MAKING	COUNTRY STR.	FEEDING ANM.	HAY RIDE
CHILD	LAUREN	X	X	●	X	X	X	X	X	X	●
	RAVEN	●	X	X	X	X	X	●	X	X	X
	TARA	X	X	X	X	●	X	X	●	X	X
	XANDER	X	X	X	●	X	●	X	X	X	X
	ZACH	X	●	X	X	X	X	X	X	●	X
ASPECT	APPLE PICK.	X	X	X	●	X					
	CIDER MAK.	●	X	X	X	X					
	COUNTRY	X	X	X	X	●					
	FEEDING	X	●	X	X	X					
	HAY RIDE	X	X	●	X	X					

Thus, Lauren, Mr. Mair, hay ride; Raven, Ms. Fedor, cider making; Tara, Ms. Reed, country store; Xander, Mr. Morgan, apple picking; Zach, Mr. Hanson, feeding animals.

2. Draw charts as indicated and complete using the initial information given. Use "●" for Yes and "x" for No. For any cell that is assigned "●", mark "x" in the remaining unmarked cells in that row and column.

   (1) Arlene had her name imprinted on her new orange bowling ball. Devon bought a bowling bag for his new ball, which is lighter than Silas's. Arlene's ball is orange and, therefore, can't be another color. Devon's ball is lighter than Silas's. Thus Devon's ball cannot weigh the most, 18 lb. Mark (x) appropriately.

   (2) Tina bowls with a 14-pound ball. The pink bowling ball is exactly 6 pounds lighter than the turquoise one. Tina's ball is 14 lbs. and, therefore, can't be any other weight (all other cells x).

   (3) The red bowling ball, which isn't Rosetta's, weighs the most. Rosetta didn't buy the 16-pound ball. The red ball weighs the most, so it must be 18 lbs. Place (●) in the 18 lb. cell in the red column. The red ball (which is 18 lbs.) is not Rosetta's, so (x) the 18 lb. and red columns in Rosetta's row. Also mark (x) in the 16 lb. cell in Rosetta's row.

   The smallest ball is 10 lbs., and the largest ball is 18 lbs. Having already determined that the 18 lb. ball is red, the turquoise ball (which is 6 pounds heavier than the pink ball) must be 16 lb. and the pink ball 10 lb. Put (x)s in all rows and columns that contain a known value (●). At this point you will have completed the following chart.

Chart 1

		COLOR					WEIGHT/POUNDS				
		GRAY	ORANGE	PINK	RED	TURQUOISE	10	12	14	16	18
BOWLER	ARLENE	X	●	X	X	X			X		
	DEVON		X						X		X
	ROSETTA		X						X	X	
	SILAS		X						X		
	TINA		X		X	X	X	X	●	X	X
WEIGHT	10	X	X	●	X	X					
	12		X	X	X						
	14		X	X	X						
	16	X	X	X	X	●					
	18	X	X	X	●	X					

The orange ball cannot be 10, 16, or 18 lbs. By transferring these (x)'s to Arlene's row (Arlene owns the orange ball), we determine that Arlene's ball is 12 lb. Place (●) in the cell at the intersection of 12 lbs. and the orange ball. The only weight cell open in Rosetta's row is 10 lbs. and we know that the 10 lb. ball is pink. Thus Rosetta has the 10 lb. pink ball. This leaves the 16 lb. ball as the only weight available to Devon. Silas's ball must be 18 lb. to be heavier than Devon's. The 16 lb. ball is turquoise; therefore, Devon has the turquoise ball. We know the 18 lb. ball, which belongs to Silas, is red. Thus, Tina must own the gray ball. Place (●) in the appropriate cell. Since Tina's (gray) ball is 14 lbs., mark (●) in the 14 lb. cell in the gray column. This leaves the orange ball at 12 lbs. We can check by noting the orange ball belongs to Arlene, and Arlene's ball is 12 lbs. Your chart should now be completed as in *Chart 2* below.

Chart 2

		COLOR					WEIGHT/POUNDS				
		GRAY	ORANGE	PINK	RED	TURQUOISE	10	12	14	16	18
BOWLER	ARLENE	X	●	X	X	X	X	●	X	X	X
	DEVON	X	X	X	X	●	X	X	X	●	X
	ROSETTA	X	X	●	X	X	●	X	X	X	X
	SILAS	X	X	X	●	X	X	X	X	X	●
	TINA	●	X	X	X	X	X	X	●	X	X
WEIGHT	10	X	X	●	X	X					
	12	X	●	X	X	X					
	14	●	X	X	X	X					
	16	X	X	X	X	●					
	18	X	X	X	●	X					

Thus, Arlene, orange, 12; Devon, turquoise, 16; Rosetta, pink, 10; Silas, red, 18; Tina, gray, 14.

3. Draw charts as indicated and complete using the initial information given. Use "●" for Yes and "x" for No. For any cell that is assigned "●", mark "x" in the remaining unmarked cells in that row and column.

(1) *The tie with the grinning leprechauns wasn't a present from a daughter.* Mark (x) in the cell at the intersection of daughter and grinning leprechaun.

(2) *Mr. Crow's tie features neither the dancing reindeer nor the yellow happy faces.* (x) the dancing reindeer and the yellow happy faces from Mr. Crow's row.

(3) *Mr. Speigler's tie wasn't a present from his uncle.* Place (x) in the uncle cell in Mr. Speigler's row.

(4) *The tie with the yellow happy faces wasn't a gift from a sister.* Mark (x) in the cell at the intersection of yellow happy faces and sister.

(5) *Mr. Evans and Mr. Speigler won the tie with the grinning leprechauns and the tie that was a present from a father-in-law, in some order.* The leprechaun tie could not have come from the father-in-law so mark (x) at that intersection. For either Mr. Evans or Mr. Speigler to receive the tie from his father-in-law no one else could have received a tie from his father-in-law so (x) the father-in-law cells for Mr. Crow and Mr. Hurley. The same logic applies to the leprechaun; neither Mr. Crow nor Mr. Hurley could have received the leprechaun tie. (x) the appropriate cells. Mr. Crow's only tie option is now the tie with cupids so place (●) in the cell at the intersection of cupids and Mr. Crow.

(6) *Mr. Hurley received his flamboyant tie from his sister.* Place (●) for the cell for sister in Mr. Hurley's row. Since Mr. Hurley received the tie from his sister and the sister did NOT give the happy faces tie, (x) the happy faces cell in Mr. Hurley's row. This leaves reindeer as the only choice for Mr. Hurley's row. (●) the cell for reindeer in Mr. Hurley's row, and (●) the cell for reindeer in the sister row because Mr. Hurley received his tie from his sister.

Since Mr. Crow did NOT receive his tie from his father-in-law or sister, (x) those cells in the column under cupids. The father-in-law could now only have given the happy faces tie, so (●) the appropriate cell. This leaves the cupids tie for the daughter and the leprechaun tie for the uncle. Since the cupid tie belongs to Mr. Crow, place (●) in the daughter cell in Mr. Crow's row.

Mr. Speigler's tie could now only have come from his father-in-law, leaving the uncle's tie for Mr. Evans. Now notice that Mr. Evan's tie came from his uncle, and the uncle purchased the leprechaun tie so Mr. Evans received the tie with leprechauns, and Mr. Speigler received the only remaining tie, the tie with happy faces. Your completed chart should look like the following.

Chart

	Cupids	Happy faces	Leprechauns	Reindeer	Daughter	Father-in-law	Sister	Uncle
Mr. Crow	●	X	X	X	●	X	X	X
Mr. Evans	X	X	●	X	X	X	X	●
Mr. Hurley	X	X	X	●	X	X	●	X
Mr. Speigler	X	●	X	X	X	●	X	X
Daughter	●	X	X	X				
Father-in-law	X	●	X	X				
Sister	X	X	X	●				
Uncle	X	X	●	X				

Thus, Mr. Crow, cupids, daughter; Mr. Evans, leprechauns, uncle; Mr. Hurley, reindeer, sister; Mr. Speigler, happy faces, father-in-law.

4. Draw charts as indicated and complete using the initial information given. Use "●" for Yes and "x" for No. For any cell that is assigned "●", mark "x" in the remaining unmarked cells in that row and column.

(1) *The seal (who isn't the creation of either Joanne or Lou) neither rode to the moon in a spaceship nor took a trip around the world on a magic train.* Place (x) in the cells for train and spaceship in the column under seal. Also (x) the seal cell for Joanne and Lou.

(2) *Joanne's imaginary friend (who isn't the grizzly bear) went to the circus.* Place (x) in the cell for grizzly bear in Joanne's row. Mark (●) at the intersection of Joanne and circus. Since the grizzly bear did not go to the circus, (x) the appropriate cell in the grizzly bear column.

(3) *Winnie's imaginary friend is a zebra.* Place (●) for zebra in Winnie's row. Joanne is left with the moose as her character, so place (●) in that cell. This leaves the grizzly bear for Lou and once that is marked, the seal is left for Ralph. Joanne's moose went to the circus, so place (●) in the cell for circus in the moose's column. The seal's only adventure choice is now the rock band, so place (●) appropriately. Since Ralph created the seal, place (●) in the rock band cell in Ralph's row.

(4) *The grizzly bear didn't board the spaceship to the moon.* Mark (x) in the cell at the intersection of grizzly bear and spaceship. The grizzly bear must have taken the train, so (●) the train cell in the grizzly bear column. This leaves the spaceship for the zebra. Lou's character, the grizzly bear, rode the train, so mark (●) in the train cell in Lou's row. This leaves the spaceship for Winnie's character, the zebra. The completed chart would look like the following.

Chart

	Grizzly bear	Moose	Seal	Zebra	Circus	Rock band	Spaceship	Train
Joanne	X	●	X	X	●	X	X	X
Lou	●	X	X	X	X	X	X	●
Ralph	X	X	●	X	X	●	X	X
Winnie	X	X	X	●	X	X	●	X
Circus	X	●	X	X				
Rock band	X	X	●	X				
Spaceship	X	X	X	●				
Train	●	X	X	X				

Thus, Joanne, moose, circus; Lou, grizzly bear, train; Ralph, seal, rock band; Winnie, zebra, spaceship.

3.6 EXERCISES

1. Let p represent "you use binoculars," q represent "you get a glimpse of the comet," and r represent "you will be amazed." The argument is then represented symbolically by:

$$p \to q$$
$$q \to r$$
$$p \to r.$$

This is the <u>valid</u> argument form "reasoning by transitivity."

2. Let p represent "Billy Joel comes to town," q represent "I will go to the concert," and r represent "I'll call in for work." The argument is then represented symbolically by:

$$p \to q$$
$$q \to r$$
$$p \to r.$$

This is the <u>valid</u> argument form "reasoning by transitivity."

3. Let p represent "Frank Steed sells his quota" and q represent "He will get a bonus." The argument is then represented symbolically by:

$$p \to q$$
$$p$$
$$q.$$

This is the <u>valid</u> argument form "modus ponens."

4. Let p represent "Amy McRee works hard enough" and q represent "she will get a promotion." The argument is then represented symbolically by:

$$p \to q$$
$$p$$
$$q$$

This is the <u>valid</u> argument form "modus ponens."

5. Let p represent "She buys another pair of shoes" and q represent "her closet will overflow." The argument is then represented symbolically by:

$$p \to q$$
$$q$$
$$p.$$

Since this is the form "fallacy of the converse," it is invalid and considered a <u>fallacy</u>.

6. Let p represent "He doesn't have to get up at 5:30 a.m." and q represent "he is ecstatic." The argument is then represented symbolically by:

$$p \to q$$
$$q$$
$$p.$$

Since this is the form "fallacy of the converse," it is invalid and considered a <u>fallacy</u>.

7. Let p represent "Patrick Roy plays" and q represent "the opponent gets shut out." The argument is then represented symbolically by:

$$p \to q$$
$$\sim q$$
$$\sim p.$$

This is the <u>valid</u> argument form "modus tollens."

8. Let p represent "Pedro Martinez pitches" and q represent "the Red Sox wins." The argument is then represented symbolically by:

$$p \to q$$
$$\sim q$$
$$\sim p.$$

This is the <u>valid</u> argument form "modus tollens."

9. Let p represent "we evolved a race of Isaac Newtons." and q represent "that would not be progress." The argument is then represented symbolically by:

$$p \to q$$
$$\sim p$$
$$\sim q.$$

Note: that since we let q represent "that <u>would not</u> be progress," then $\sim q$ represents "that <u>is</u> progress." Since this is the form "fallacy of the inverse," it is invalid and considered a <u>fallacy</u>.

10. Let p represent "I have seen farther than others" and q represent "it is because I have stood on the shoulders of giants." The argument is then represented symbolically by:

$$p \to q$$
$$\sim p$$
$$\overline{\sim q.}$$

Since this is the form "fallacy of the inverse," it is invalid and considered a <u>fallacy</u>.

11. Let p represent "Alison Romike jogs" and q represent "Kaare Taylor pumps iron" The argument is then represented symbolically by:

$$p \vee q \quad (\text{or } q \vee p)$$
$$\sim q$$
$$\overline{p.}$$

Since this is the form "disjunctive syllogism," it is a <u>valid</u> argument.

12. Let p represent "She uses e-commerce" and q represent "she pays by credit card." The argument is then represented symbolically by:

$$p \vee q \quad (\text{or } q \vee p)$$
$$\sim q$$
$$\overline{p.}$$

Since this is the form "disjunctive syllogism," it is a <u>valid</u> argument.

To show validity for the arguments in the following exercises, we must show that the conjunction of the premises implies the conclusion. That is, the conditional statement $[P_1 \wedge P_2 \wedge \ldots \wedge P_n] \to C$ must be a tautology. For exercises 13 and 14 we will use the standard (long format) to develop the corresponding truth tables. For the remainder of the exercises we will use the alternate (short format) to create the truth tables.

13. Form the conditional statement

$$[(p \vee q) \wedge p] \to \sim q$$

from the argument. Complete a truth table.

p	q	$p \vee q$	$(p \vee q) \wedge p$	$\sim q$	$[(p \vee q) \wedge p] \to \sim q$
T	T	T	T	F	F
T	F	T	T	T	T
F	T	T	F	F	T
F	F	F	F	T	T

Since the conditional, formed by the conjunction of premises implying the conclusion, is not a tautology, the argument is <u>invalid</u>.

14. Form the conditional statement

$$[(p \wedge \sim q) \wedge p] \to \sim q$$

from the argument. Complete a truth table.

p	q	$\sim q$	$p \wedge \sim q$	$(p \wedge \sim q) \wedge p$	$[(p \wedge \sim q) \wedge p] \to \sim q$
T	T	F	F	F	T
T	F	T	T	T	T
F	T	F	F	F	T
F	F	T	F	F	T

Since the conditional, formed by the conjunction of premises implying the conclusion, is a tautology, the argument is <u>valid</u>.

15. Form the conditional statement

$$[(\sim p \to \sim q) \wedge q] \to p$$

from the argument. Complete a truth table.

p	q	$[(\sim p$	$\to$	$\sim q)$	$\wedge q]$	$\to$	p
T	T	F	T	F	T T	T	T
T	F	F	T	T	F F	T	T
F	T	T	F	F	F T	T	F
F	F	T	T	T	F F	T	F
		1	2	1	3 2	4	3

Since the conditional, formed by the conjunction of premises implying the conclusion, is a tautology, the argument is <u>valid</u>.

16. Form the conditional statement

$$[(p \vee \sim q) \wedge p] \to \sim q$$

from the argument. Complete a truth table.

p	q	$[(p \vee$	$\sim q)$	$\wedge p]$	$\to$	$\sim q$
T	T	T T	F	T T	F	F
T	F	T T	T	T T	T	T
F	T	F F	F	F F	T	F
F	F	F T	T	F F	T	T
		1 2 1		3 2	4	3

Since the conditional, formed by the conjunction of premises implying the conclusion, is not a tautology, the argument is <u>invalid</u>.

17. Form the conditional statement

$$[(p \to q) \wedge (q \to p)] \to (p \wedge q)$$

from the argument. Complete a truth table.

p	q	$[(p \to q)$	$\wedge$	$(q \to p)]$	$\to$	$(p \wedge q)$
T	T	T	T	T	T	T
T	F	F	F	T	T	F
F	T	T	F	F	T	F
F	F	T	T	T	F	F
		1	3	2	4	3

Since the conditional, formed by the conjunction of premises implying the conclusion, is not a tautology, the argument is <u>invalid</u>.

18. Form the conditional statement

$$[(\sim p \to q) \land p] \to \sim q$$

from the argument. Complete a truth table.

p	q	[($\sim p$	$\to$	q)	$\land$	p]	$\to$	$\sim q$
T	T	F	T	T	T	T	F	F
T	F	F	T	F	T	T	T	T
F	T	T	T	T	F	F	T	F
F	F	T	F	F	F	F	T	T
		1	2	1		3	2	4	3

Since the conditional, formed by the conjunction of premises implying the conclusion, is not a tautology, the argument is <u>invalid</u>. Note: If you are completing the truth table along rows (rather than down columns), you could stop after completing the first row, knowing that with a false conditional, the statement will not be a tautology.

19. Form the conditional statement

$$[(p \to \sim q) \land q] \to \sim p$$

from the argument. Complete a truth table.

p	q	[(p	$\to$	$\sim q$)	$\land$	q]	$\to$	$\sim p$
T	T	T	F	F	F	T	T	F
T	F	T	T	T	F	F	T	F
F	T	F	T	F	T	T	T	T
F	F	F	T	T	F	F	T	T
		1	2	1		3	2	4	3

Since the conditional, formed by the conjunction of premises implying the conclusion, is a tautology, the argument is <u>valid</u>.

20. Form the conditional statement

$$[(p \to \sim q) \land \sim p] \to \sim q$$

from the argument. Complete a truth table.

p	q	[(p	$\to$	$\sim q$)	$\land$	$\sim p$]	$\to$	$\sim q$
T	T	T	F	F	F	F	T	F
T	F	T	T	T	F	F	T	T
F	T	F	T	F	T	T	F	F
F	F	F	T	T	T	T	T	T
		1	2	1		3	2	4	3

Since the conditional, formed by the conjunction of premises implying the conclusion, is not a tautology, the argument is <u>invalid</u>.

21. Form the conditional statement

$$[(\sim p \lor q) \land (\sim p \to q) \land p] \to \sim q$$

from the argument. Complete a truth table.

p	q	[($\sim p$	$\lor$	q)	$\land$	($\sim p$	$\to$	q)]	$\land$	p]	$\to$	$\sim q$
T	T	F	T	T	T	F	T	T	T	T	F	F
T	F	F	F	F	F	F	T	F	F	T	T	T
F	T	T	T	T	T	T	T	T	F	F	T	F
F	F	T	T	F	F	T	F	F	F	F	T	T
		1	2	1		3		2		4	3	5	4

Since the conditional, formed by the conjunction of premises implying the conclusion, is not a tautology, the argument is <u>invalid</u>.

22. Form the conditional statement

$$\{[(p \to q) \land (q \to p)] \land p\} \to (p \lor q)$$

from the argument. Complete a truth table.

p	q	{[(p	$\to$	q)	$\land$	(q	$\to$	p)]	$\land$	p}	$\to$	(p	$\lor$	q)
T	T	T	T	T	T	T	T	T	T	T	T	T	T	T
T	F	T	F	F	F	F	T	T	F	T	T	T	T	F
F	T	F	T	T	F	T	F	F	F	F	T	F	T	T
F	F	F	T	F	T	F	T	F	F	F	T	F	F	F
		1	2	1		3	1	2	1		4	3	5	3	4	3

Since the conditional, formed by the conjunction of premises implying the conclusion, is a tautology, the argument is <u>valid</u>.

23. Form the conditional statement

$$\{[(\sim p \land r) \to (p \lor q)] \land (\sim r \to p)\} \to (q \to r)$$

from the argument.

p	q	r	{[($\sim p$	$\land$	r)	$\to$	(p	$\lor$	q)]	$\land$	($\sim r$	$\to$	p)}	$\to$	(q	$\to$	r)
T	T	T	F	F	T	T	T	T	T	T	F	T	T	T	T	T	T
T	T	F	F	F	F	T	T	T	T	T	T	T	T	F	T	F	F
T	F	T	F	F	T	T	T	T	F	T	F	T	T	T	F	T	T
T	F	F	F	F	F	T	T	T	F	T	T	T	T	T	F	T	F
F	T	T	T	T	T	T	F	T	T	T	F	T	F	T	T	T	T
F	T	F	T	F	F	T	F	T	T	F	T	F	F	T	T	F	F
F	F	T	T	T	T	F	F	F	F	F	F	T	F	T	F	T	T
F	F	F	T	F	F	T	F	F	F	F	T	F	F	T	F	T	F
			1	2	1		3		2		4	2	3	2		5		4

The F in the final column 5 shows us that the statement is not a tautology and hence, the argument is <u>invalid</u>.

24. Form the conditional statement

$$\{[(r \wedge p) \to (r \vee q)] \wedge (q \wedge p)\} \to (r \vee p)$$

from the argument.

p	q	r	{[(r ∧ p)	→	(r ∨ q)]	∧	(q ∧ p)}	→	(r ∨ p)
T	T	T	T T T	T	T T T	T	T T T	T	T T T
T	T	F	F F T	T	F T T	T	T T T	T	F T T
T	F	T	T T T	T	T T T	F	F F T	T	T T T
T	F	F	F F T	T	F T T	F	F F T	T	F T T
F	T	T	T F F	T	T T F	F	T F F	T	T T F
F	T	F	F F F	T	F F F	F	T F F	T	F F F
F	F	T	T F F	T	T T F	F	F F F	T	T T F
F	F	F	F F F	T	F F F	F	F F F	T	F F F
			1 2 1	3	1 2 1	4	2 3 2	5	3 4 3

Since the conditional, formed by the conjunction of premises implying the conclusion, is a tautology, the argument is <u>valid</u>.

25. Writing exercise

26. Every time something squeaks, I use WD-40.
Every time I use WD-40, I must go to the hardware store.
Every time something squeaks, I go to the hardware store.

27. Let p represent "Jeff loves to play golf," q represent "Joan likes to sew," and r represent "Brad sings in the choir." The argument is then represented symbolically by:

$$p$$
$$q \to \sim p$$
$$\sim q \to r$$
$$r.$$

Construct the truth table for

$$[p \wedge (q \to \sim p) \wedge (\sim q \to r)] \to r.$$

p	q	r	[p ∧	(q → ~p)	∧	(~q → r)]	→	r
T	T	T	T F	T F F	F	F T T	T	T
T	T	F	T F	T F F	F	F T F	T	F
T	F	T	T T	F T F	T	T T T	T	T
T	F	F	T T	F T F	F	T F F	T	F
F	T	T	F F	T T T	F	F T T	T	T
F	T	F	F F	T T T	F	F T F	T	F
F	F	T	F F	F T T	F	T T T	T	T
F	F	F	F F	F T T	F	T F F	T	F
			2 3	1 2 1	4	2 3 2	5	4

Since the conditional, formed by the conjunction of premises implying the conclusion, is a tautology, the argument is <u>valid</u>.

28. Let p represent "that tree is infested with pine bark beetles," q represent "it will die," and r represent "people plant trees on Arbor Day." The argument is then represented symbolically by:

$$p \to q$$
$$r \wedge \sim q$$
$$r \to \sim p.$$

Construct the truth table for

$$[(p \to q) \wedge (r \wedge \sim q)] \to (r \to \sim p).$$

p	q	r	[(p → q)	∧	(r ∧ ~q)]	→	(r → ~p)
T	T	T	T T T	F	T F F	T	T F F
T	T	F	T T T	F	F F F	T	F T F
T	F	T	T F F	F	T T T	T	T F F
T	F	F	T F F	F	F F T	T	F T F
F	T	T	F T T	F	T F F	T	T T T
F	T	F	F T T	F	F F F	T	F T T
F	F	T	F T F	T	T T T	T	T T T
F	F	F	F T F	F	F F T	T	F T T
			1 2 1	3	1 2 1	4	2 3 2

Since the conditional, formed by the conjunction of premises implying the conclusion, is a tautology, the argument is <u>valid</u>.

29. Let p represent "the Bobble head doll craze continues," q represent "Beanie Babies will remain popular," and r represent "Barbie dolls continue to be favorites." The argument is then represented symbolically by:

$$p \to q$$
$$r \vee q$$
$$\sim r$$
$$\sim p.$$

Construct the truth table for

$$[(p \rightarrow q) \wedge (r \vee q) \wedge (\sim r)] \rightarrow \sim p.$$

Note: we do not have to complete a column under each simple statement $p, q,$ and r, (as we did in exercises above) since it is easy to compare the appropriate index columns to create the truth value for each connective.

p	q	r	$[(p \rightarrow q)$	$\wedge$	$(r \vee q)$	$\wedge$	$(\sim r)]$	$\rightarrow$	$\sim p$
T	T	T	T	T	T	F	F	T	F
T	T	F	T	T	T	T	T	F	F
T	F	T	F	F	T	F	F	T	F
T	F	F	F	F	F	F	T	T	F
F	T	T	T	T	T	F	F	T	T
F	T	F	T	T	T	T	T	T	T
F	F	T	T	T	T	F	F	T	T
F	F	F	T	F	F	F	T	T	T
			1	2	1	3	2	4	3

Since the conditional, formed by the conjunction of premises implying the conclusion, is not a tautology, the argument is <u>invalid</u>. Note: If you are completing the truth table along rows (rather than down columns), you could stop after completing the second row, knowing that with a false conditional, the statement will not be a tautology.

30. Let p represent "Christina Aguilera sings," q represent "Ricky Martin is a teen idol," and r represent "Britney Spears wins an American Music Award." The argument is then represented symbolically by:

$$p \vee \sim q$$
$$\sim q \rightarrow \sim r$$
$$\underline{r}$$
$$\sim p.$$

Construct the truth table for

$$[(p \vee \sim q) \wedge (\sim q \rightarrow \sim r) \wedge r] \rightarrow \sim p.$$

p	q	r	$[(p \vee \sim q)$	$\wedge$	$(\sim q \rightarrow \sim r)$	$\wedge$	$r]$	$\rightarrow$	$\sim p$
T	T	T	T T F	T	F T F	T	T	F	F
T	T	F	T T F	T	F T T	F	F	T	F
T	F	T	T T T	F	T F F	F	T	T	F
T	F	F	T T T	T	T T T	F	F	T	F
F	T	T	F F F	F	F T F	F	T	T	T
F	T	F	F F F	F	F T T	F	F	T	T
F	F	T	F T T	F	T F F	F	T	T	T
F	F	F	F T T	F	T T T	F	F	T	T
			1 2 1	3	1 2 1	4	3	5	4

Since the conditional, formed by the conjunction of premises implying the conclusion, is not a tautology, the argument is <u>invalid</u>. Note: If you are completing the truth table along rows (rather than down columns), you could stop after completing the first row, knowing that with a false conditional, the statement will not be a tautology.

31. Let p represent "I've got you under my skin," q represent "you are deep in the heart of me," and r represent "you are really a part of me." The argument is then represented symbolically by:

$$p \rightarrow q$$
$$q \rightarrow \sim r$$
$$\underline{q \vee r}$$
$$p \rightarrow r.$$

Construct the truth table for

$$[(p \rightarrow q) \wedge (q \rightarrow \sim r) \wedge (q \vee r)] \rightarrow (p \rightarrow r).$$

p	q	r	$[(p \rightarrow q)$	$\wedge$	$(q \rightarrow \sim r)$	$\wedge$	$(q \vee r)]$	$\rightarrow$	$(p \rightarrow r)$
T	T	T	T	F	T F F	F	T	T	T
T	T	F	T	T	T T T	T	T	F	F
T	F	T	F	F	F T F	F	T	T	T
T	F	F	F	F	F T T	F	F	T	F
F	T	T	T	F	T F F	F	T	T	T
F	T	F	T	T	T T T	T	T	T	T
F	F	T	T	F	F T F	F	T	T	T
F	F	F	T	T	T T T	F	F	T	T
			2	3	1 2 1	4	3	5	4

Since the conditional, formed by the conjunction of premises implying the conclusion, is not a tautology, the argument is <u>invalid</u>. Note: If you are completing the truth table along rows (rather than down columns), you could stop after completing the second row, knowing that with a false conditional, the statement will not be a tautology.

32. Let p represent "The Colts will be in the playoffs," q represent "Peyton leads the league in passing," and r represent "Marv loves the Colts." The argument is then represented symbolically by

$$p \leftrightarrow q$$
$$r \vee q$$
$$\underline{\sim r}$$
$$\sim p.$$

Construct the truth table for

$$[(p \leftrightarrow q) \wedge (r \vee q) \wedge (\sim r)] \rightarrow \sim p.$$

p	q	r	$[(p \leftrightarrow q)$	$\wedge$	$(r \vee q)$	$\wedge$	$(\sim r)]$	$\rightarrow$	$\sim p$
T	T	T	T	T	T	F	F	T	F
T	T	F	T	T	T	T	T	F	F
T	F	T	F	F	T	F	F	T	F
T	F	F	F	F	F	F	T	T	F
F	T	T	F	F	T	F	F	T	T
F	T	F	F	F	T	F	T	T	T
F	F	T	T	T	T	F	F	T	T
F	F	F	T	F	F	F	T	T	T
			1	2	1	3	2	4	3

Since the conditional, formed by the conjunction of premises implying the conclusion, is not a tautology, the argument is invalid. Note: If you are completing the truth table along rows (rather than down columns), you could stop after completing the second row, knowing that with a false conditional, the statement will not be a tautology.

33. Let p represent "Otis is a disc jockey," q represent "he lives in Lexington," and r represent "he is a history buff." The argument is then represented symbolically by

$$p \to q$$
$$q \wedge r$$
$$\overline{\sim r \to \sim p.}$$

Construct the truth table for

$$[(p \to q) \wedge (q \wedge r)] \to (\sim r \to \sim p).$$

p	q	r	$[(p$	$\to$	$q)$	$\wedge$	$(q \wedge r)]$	$\to$	$(\sim r$	$\to$	$\sim p)$
T	T	T		T		T		T	F	T	F
T	T	F		T		F		T	T	F	F
T	F	T		F		F		T	F	T	F
T	F	F		F		F		T	T	F	F
F	T	T		T		T		T	F	T	T
F	T	F		T		F		T	T	T	T
F	F	T		T		F		T	F	T	T
F	F	F		T		F		T	T	T	T
				1		2		3	1	2	1

Since the conditional, formed by the conjunction of premises implying the conclusion, is a tautology, the argument is valid.

34. Let p represent "I am your women," q represent "you are my man," and r represent "I stop loving you." The argument is then represented symbolically by:

$$(p \wedge q) \to \sim r$$
$$\underline{r}$$
$$\sim p \vee \sim q.$$

Construct the truth table for

$$\{[(p \wedge q) \to \sim r] \wedge r\} \to (\sim p \vee \sim q).$$

p	q	r	$\{[(p \wedge q)$	$\to$	$\sim r]$	$\wedge$	$r\}$	$\to$	$(\sim p$	$\vee$	$\sim q)$
T	T	T	T	F	F	F	T	T	F	F	F
T	T	F	T	T	T	F	F	T	F	F	F
T	F	T	F	T	F	T	T	T	F	T	T
T	F	F	F	T	T	F	F	T	F	T	T
F	T	T	F	T	F	T	T	T	T	T	F
F	T	F	F	T	T	F	F	T	T	T	F
F	F	T	F	T	F	T	T	T	T	T	T
F	F	F	F	T	T	F	F	T	T	T	T
			1	2	1	3	2	4	2	3	2

Since the conditional, formed by the conjunction of premises implying the conclusion, is a tautology, the argument is valid.

The following exercises involve Quantified arguments and can be analyzed, as such, by Euler diagrams. However, the quantified statements can be represented as conditional statements as well. This allows us to use a truth table – or recognize a valid argument form – to analyze the validity of the argument.

35. Let p represent "you are a man," q represent "you are created equal," and r represent "you are a women." The argument is then represented symbolically by:

$$p \to q$$
$$q \to r$$
$$\overline{p \to r.}$$

This is a "Reasoning by Transitivity" argument form and, hence, is valid.

36. Let p represent "you are a man," q represent "you are mortal," and r represent "you are Socrates." The argument is then represented symbolically by:

$$p \to q$$
$$r \to p$$
$$\overline{r \to q.}$$

By interchanging the first premise with the second premise the argument becomes a "reasoning by transitivity" form and, hence, is valid.

37. We apply reasoning by repeated transitivity to the six premises. A conclusion from this reasoning, which makes the argument valid, is reached by linking the first antecedent to the last consequent. This conclusion is "If I tell you the time, then my life will be miserable."

38. Writing exercise

Answers in Exercises, 39 – 46 may be replaced by their contrapositives.

39. The statement "All my poultry are ducks" becomes "If it is my poultry, then it is a duck."

40. The statement "None of your sons can do logic" becomes "If he is your son, then he can't do logic."

41. The statement "Guinea pigs are hopelessly ignorant of music" becomes "If it is a Guinea pig, then it is hopelessly ignorant of music."

42. The statement "No teetotalers are pawnbrokers" becomes "If the person is a teetotaler, then the person is not a pawnbroker."

43. The statement "No teachable kitten has green eyes" becomes "If it is a teachable kitten, then it does not have green eyes."

44. The statement "Opium-eaters have no self-command" becomes "If it is an opium-eater, then it has no self-command."

45. The statement "I have not filed any of them that I can read" becomes "If I can read it, then I have not filed it."

46. The statement "All of them written on blue paper are filed" becomes "If it is written on blue paper, then it is filed."

47. (a) "No ducks are willing to waltz" becomes "if it is a duck, then it is not willing to waltz."

    (b) "No officers ever decline to waltz" becomes "if one is an officer, then one is willing to waltz."

    (c) "All my poultry are ducks" becomes "if it is my poultry, then it is a duck."

    (d) In symbols, the three premises are
    $$p \rightarrow \sim s$$
    $$r \rightarrow s$$
    $$q \rightarrow p.$$

    Begin with q, which only appears once. Replacing $r \rightarrow s$ with its contrapositive, $\sim s \rightarrow \sim r$, rearrange the three premises.
    $$q \rightarrow p$$
    $$p \rightarrow \sim s$$
    $$\sim s \rightarrow \sim r$$

    By repeated use of reasoning by transitivity, the conclusion which provides a valid argument is
    $$q \rightarrow \sim r.$$

    In words, "If it is my poultry, then it is not an officer," or "none of my poultry are officers."

48. (a) "Everyone who is sane can do logic" becomes "if one is sane, then one is able to do logic."

    (b) "No lunatics are fit to serve on a jury" becomes "if one is a lunatic (or not sane), then one is not fit to serve on a jury."

    (c) "None of your sons can do logic" becomes "if he is your son then, then he can not do logic."

    (d) In symbols, the three premises are
    $$r \rightarrow p$$
    $$\sim r \rightarrow \sim q$$
    $$s \rightarrow \sim p.$$

    Replacing $r \rightarrow p$ with its contrapositive, $\sim p \rightarrow \sim r$, rearrange the three premises.
    $$s \rightarrow \sim p$$
    $$\sim p \rightarrow \sim r$$
    $$\sim r \rightarrow \sim q$$

    By repeated use of reasoning by transitivity, the conclusion which provides a valid argument is
    $$s \rightarrow \sim q.$$

    In words, "If he is your son, then he is not fit to serve on a jury," or "your sons are not fit to serve on a jury."

49. (a) "Promise-breakers are untrustworthy" becomes "if one is a promise-breaker, then one is not trustworthy."

    (b) "Wine-drinkers are very communicative" becomes "if one is a wine-drinker, then one is very communicative."

    (c) "A person who keeps a promise is honest" becomes "if one is not a promise-breaker, then one is honest."

    (d) "No teetotalers are pawnbrokers" becomes "if one is not a wine-drinker, then one is not a pawnbroker."

    (e) "One can always trust a very communicative person" becomes "if one is very communicative, then one is trustworthy."

    (f) In symbols, the statements are
    $$r \rightarrow \sim s$$
    $$u \rightarrow t$$
    $$\sim r \rightarrow p$$
    $$\sim u \rightarrow \sim q$$
    $$t \rightarrow s.$$

    Begin with q, which only appears once. Using the contrapositive of $\sim u \rightarrow \sim q$, $(q \rightarrow u)$, and $r \rightarrow \sim s$, $(s \rightarrow \sim r)$, rearrange the five premises as follows:
    $$q \rightarrow u$$
    $$u \rightarrow t$$
    $$t \rightarrow s$$
    $$s \rightarrow \sim r$$
    $$\sim r \rightarrow p.$$

    By repeated use of reasoning by transitivity, the conclusion which provides a valid argument is
    $$q \rightarrow p.$$

    In words, this conclusion can be stated as "if one is a pawnbroker, then one is honest," or "all pawnbrokers are honest."

50. Let p be "it is a guinea pig," q be "it is hopelessly ignorant of music," r be "it keeps silent while the Moonlight Sonata is being played," and s be "it appreciates Beethoven."

    (a) "Nobody who really appreciates Beethoven fails to keep silent while the Moonlight Sonata is being played" becomes "if one really appreciates Beethoven, then one fails to keep silent while the Moonlight Sonata is being played."

    (b) "Guinea pigs are hopelessly ignorant of music" becomes "If you are a guinea pig, then you are hopelessly ignorant of music."

(c) "No one who is hopelessly ignorant of music ever keeps silent while the Moonlight Sonata is being played," becomes "If one is hopelessly ignorant of music, then one fails to keep silent while the Moonlight Sonata is being played."

(d) In symbols, the statements are

  (a) $s \rightarrow r$
  (b) $p \rightarrow q$
  (c) $q \rightarrow \sim r$.

Using the contrapositive of the premise $s \rightarrow r$, $(\sim r \rightarrow \sim s)$, rearrange the premises as follows:

$$p \rightarrow q$$
$$q \rightarrow \sim r$$
$$\sim r \rightarrow \sim s.$$

By repeated use of reasoning by transitivity, the conclusion which provides a valid argument is

$$p \rightarrow \sim s.$$

In words, this conclusion can be stated as "if you are a guinea pig, then you do not appreciate Beethoven," or equivalently, "Guinea pigs don't appreciate Beethoven."

51. Begin by changing each quantified premise to a conditional statement:

(a) The statement "All the dated letters in this room are written on blue paper" becomes "If it is dated, then it is on blue paper."

(b) The statement "None of them are in black ink, except those that are written in the third person" becomes "If is not in the third person, then it is not in black ink."

(c) The statement "I have not filed any of them that I can read" becomes "If I can read it, then it is not filed."

(d) The statement "None of them that are written on one sheet are undated" becomes "If it is on one sheet, then it is dated."

(e) The statement "All of them that are not crossed are in black ink" becomes "If it is not crossed, then it is in black ink."

(f) The statement "All of them written by Brown begin with 'Dear Doctor'" becomes "If it is written by Brown, then it begins with 'Dear Sir'."

(g) The statement "All of them written on blue paper are filed" becomes "If it is on blue paper, then it is filed."

(h) The statement "None of them written on more than one sheet are crossed" becomes "If it is not on more than one sheet, then it is not crossed."

(i) The statement "None of them that begin with "Dear Sir' are written in the third person" becomes "If it begins with 'Dear Sir,' then it is not written in the third person."

(j) In symbols, the statements are

  (a) $r \rightarrow w$
  (b) $\sim u \rightarrow \sim t$
  (c) $v \rightarrow \sim s$
  (d) $x \rightarrow r$
  (e) $\sim q \rightarrow t$
  (f) $y \rightarrow p$
  (g) $w \rightarrow s$
  (h) $\sim x \rightarrow \sim q$
  (i) $p \rightarrow \sim u$.

Begin with y, which appears only once. Using contrapositives of $v \rightarrow \sim s$ $(s \rightarrow \sim v)$, $\sim q \rightarrow t$ $(\sim t \rightarrow q)$, and $\sim x \rightarrow \sim q$ $(q \rightarrow x)$, rearrange the nine statements:

$$y \rightarrow p$$
$$p \rightarrow \sim u$$
$$\sim u \rightarrow \sim t$$
$$\sim t \rightarrow q$$
$$q \rightarrow x$$
$$x \rightarrow r$$
$$r \rightarrow w$$
$$w \rightarrow s$$
$$s \rightarrow \sim v.$$

By repeated use of reasoning by transitivity, the conclusion that makes the argument valid is

$$y \rightarrow \sim v.$$

In words, the conclusion can be stated as "if it is written by Brown, then I can't read it," or equivalently "I can't read any of Brown's letters."

52. (a) "No one who is going to a party ever fails to brush his hair" becomes "If one is going to a party, then he brushes his hair."

(b) "No one looks fascinating if he is untidy" becomes "If one is untidy, then he does not look fascinating."

(c) "Opium-eaters have no self-command" becomes "If one is an opium-eater; then he has no self control."

(d) "Everyone who has brushed his hair looks fascinating" becomes "If one has brushed his hair, then he looks fascinating."

(e) "No one wears white kid gloves unless he is going to a party" becomes "If he wears white gloves, then he is going to the party."

(f) "A man is always untidy if he has no self-command." becomes" becomes "If a man has no self-command, then a man is not tidy."

(g) In symbols, the statements are

$$
\begin{aligned}
&\text{(a)} && p \to q \\
&\text{(b)} && \sim u \to \sim s \\
&\text{(c)} && t \to \sim r \\
&\text{(d)} && q \to s \\
&\text{(e)} && v \to p \\
&\text{(f)} && \sim r \to \sim u.
\end{aligned}
$$

Begin with t, which only occurs once. Using the contrapositives of $q \to s, (\sim s \to \sim q), p \to q, (\sim q \to \sim p)$, and $v \to p, (\sim p \to \sim v)$, rearrange the five premises as follows:

$$
\begin{aligned}
t &\to \sim r \\
\sim r &\to \sim u \\
\sim u &\to \sim s \\
\sim s &\to \sim q \\
\sim q &\to \sim p \\
\sim p &\to \sim v.
\end{aligned}
$$

By repeated use of reasoning by transitivity, the conclusion which provides a valid argument is

$$ t \to \sim v. $$

In words, this conclusion can be stated as "If he is an opium-eater, then he doesn't wear white gloves" or equivalently, "Opium-eaters do not wear white kid gloves."

Chapter 3 Test

1. The negation of "$6 - 3 = 3$" is "$6 - 3 \neq 3$."

2. The negation of "All men are created equal" is "Some men are not created equal."

3. The negation of "Some members of the class went on the field trip" is "No members of the class went on the field trip." An equivalent answer would be "All members of the class did not go on the field trip."

4. The negation of "If that's the way you feel, then I will accept it" is "That's the way you feel and I won't accept it." Remember that $\sim(p \to q) \equiv (p \wedge \sim q)$.

5. The negation of "She passed GO and collected $200" is "She did not pass GO or did not collect $200." Remember that $\sim(p \wedge q) \equiv (\sim p \vee \sim q)$.

Let p represent "You will love me" and let q represent "I will love you."

6. The symbolic form of "If you won't love me, then I will love you" is "$\sim p \to q$."

7. The symbolic form of "I will love you if you will love me" (or equivalently, "if you will love me, then I will love you") is "$p \to q$."

8. The symbolic form of "I won't love you if and only if you won't love me" is "$\sim q \leftrightarrow \sim p$."

9. Writing the symbolic form "$\sim p \wedge q$" in words, we get "You won't love me and I will love you."

10. Writing the symbolic form "$\sim(p \vee \sim q)$" in words, we get "It is not the case that you will love me or I won't love you" (or equivalently, by DeMorgan's, "you won't love me and I will love you").

Assume that p is true and that q and r are false for Exercises 11–14.

11. Replacing q and r with the given truth values, we have

$$
\begin{aligned}
&\sim F \wedge \sim F \\
&T \wedge T \\
&T.
\end{aligned}
$$

The compound statement $\sim q \wedge \sim r$ is true.

12. Replacing p, q and r with the given truth values, we have

$$
\begin{aligned}
&F \vee (T \wedge \sim F) \\
&F \vee (T \wedge T) \\
&F \vee T \\
&T.
\end{aligned}
$$

The compound statement $r \vee (p \wedge \sim q)$ is true.

13. Replacing r with the given truth value (s not known), we have

$$
\begin{aligned}
&F \to (s \vee F) \\
&F \to \text{not known} \\
&T.
\end{aligned}
$$

The compound statement $r \to (s \vee r)$ is true.

14. Replacing p and q with the given truth values, we have

$$
\begin{aligned}
&T \leftrightarrow (T \to F) \\
&T \leftrightarrow \ (F) \\
&F.
\end{aligned}
$$

The compound statement $p \leftrightarrow (p \to q)$ is false.

15. Writing exercise

16. The necessary condition for

   (a) a conditional statement to be false is that the antecedent must be true and the consequent must be false.

   (b) a conjunction to be true is that both component statements must be true.

   (c) a disjunction to be false is that both component statements must be false.

17.
p	q	p	$\wedge$	$(\sim p$	$\vee$	$q)$
T	T	T	T	F	T	T
T	F	T	F	F	F	F
F	T	F	F	T	T	T
F	F	F	F	T	T	F
		2	3	1	2	1

18.
p	q	$\sim$	$(p \wedge q)$	$\to$	$(\sim p$	$\vee$	$\sim q)$
T	T	F	T	T	F	F	F
T	F	T	F	T	F	T	T
F	T	T	F	T	T	T	F
F	F	T	F	T	T	T	T
		2	1	3	1	2	1

Since the last completed column (3) is all true, the conditional is a tautology.

19. The statement "Some negative integers are whole numbers" is <u>false</u>, since all whole numbers are non-negative.

20. The statement "All irrational numbers are real numbers" is <u>true</u>, because the real numbers are made up of both the rational and irrational numbers.

The wording may vary in the answers in Exercises 21–26.

21. "All integers are rational numbers" can be stated as "If the number is an integer, then it is a rational number."

22. "Being a rhombus is sufficient for a polygon to be a quadrilateral" can be stated as "If a polygon is a rhombus, then it is a quadrilateral."

23. "Being divisible by 3 is necessary for a number to be divisible by 9" can be stated as "If a number is divisible by 9, then it is divisible by 3." Remember that the "necessary" part of the statement becomes the consequent.

24. "She digs dinosaur bones only if she is a paleontologist" can be stated as "If she digs dinosaur bones, then she is a paleontologist." Remember that the "only if" part of the statement becomes the consequent.

25. The direct statement: If a picture paints a thousand words, the graph will help me understand it.

    (a) *Converse*: If the graph will help me understand it, then a picture paints a thousand words.

    (b) *Inverse*: If a picture doesn't paint a thousand words, then the graph won't help me understand it.

    (c) *Contrapositive*: If the graph doesn't help me understand it, then a picture doesn't paint a thousand words.

26. The direct statement: $\sim p \to (q \wedge r)$.

    (a) Converse: $(q \wedge r) \to \sim p$.

    (b) Inverse: $p \to \sim(q \wedge r)$, or $p \to (\sim q \vee \sim r)$.

    (c) Contrapositive: $\sim(q \wedge r) \to p$, or $(\sim q \vee \sim r) \to p$.

27. Complete an Euler diagram as:

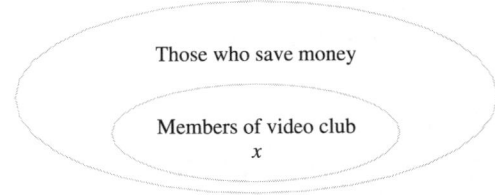

x represents Pat Pearson

Since, when the premises are diagrammed as being true, we are forced into a true conclusion, the argument is <u>valid</u>.

28. (a) Let p represent "he eats liver" and q represent "he will eat anything." The argument is then represented symbolically by:

$$p \to q$$
$$\underline{p\quad\;}$$
$$q.$$

This is the valid argument form "modus ponens," hence the answer is A.

(b) Let p represent "you use your seat belt" and q represent "you will be safer." The argument is then represented symbolically by:

$$p \to q$$
$$\underline{\sim p\;}$$
$$\sim q.$$

The answer is F, a fallacy of the inverse.

(c) Let p represent "I hear *Come Sunday Morning*," q represent "I think of her," and "I get depressed." The argument is then represented symbolically by:

$$p \to q$$
$$\underline{q \to r}$$
$$p \to r.$$

This is the valid argument form "reasoning by transitivity," hence the answer is C.

(d) Let p represent "she sings" and q represent "she dances." The argument is then represented symbolically by:

$$p \vee q$$
$$\underline{\sim p\;}$$
$$q.$$

This is the valid argument form "disjunctive syllogism," hence the answer is D.

29. Let p represent "I write a check," q represent "It will bounce," and "The bank guarantees it." The argument is then represented symbolically by:

$$p \to q$$
$$r \to \sim q$$
$$\underline{r}$$
$$\sim p.$$

Construct the truth table for

$$\{[(p \to q) \wedge (r \to \sim q)] \wedge r\} \to (\sim p).$$

p	q	r	$\{[(p$	$\to$	$q)$	$\wedge$	$(r$	$\to$	$\sim q)]$	$\wedge$	$r\}$	$\to$	$(\sim p)$
T	T	T		T		F		T	F F		T		F
T	T	F		T		T		F	T F		F		F
T	F	T		F		F		F	T T T		F		F
T	F	F		F		F		F	T T		F		F
F	T	T		T		F		T	F F		F		T
F	T	F		T		T		F	T F		F		T
F	F	T		T		T		T	T T		T		T
F	F	F		T		T		F	T T		F		T
				2		3	1	2 1		4	3	5	4

Since the conditional, formed by the conjunction of premises implying the conclusion, is a tautology, the argument is <u>valid</u>.

30. Construct the truth table for

$$[(\sim p \to \sim q) \wedge (q \to p)] \to (p \vee q).$$

p	q	$[(\sim p$	$\to$	$\sim q)$	$\wedge$	$(q \to p)]$	$\to$	$(p \vee q)$
T	T	F	T	F	T	T T T	T	T T T
T	F	F	T	T	T	F T T	T	T T F
F	T	T	F	F	F	T F F	T	F T T
F	F	T	T	T	T	F T F	F	F F F
		1	2	1	3	1 2 1	4	2 3 2

Since the conditional, formed by the conjunction of premises implying the conclusion, is not a tautology, the argument is <u>invalid</u>.

4.1 EXERCISES

For Reference:

EGYPTIAN

Number	Symbol	Description
1	│	Stroke
10	∩	Heel Bone
100	ϙ	Scroll
1000	₤	Lotus Flower
10,000	⌒	Pointing Finger
100,000	⌒⌒	Burbot Fish
1,000,000	⚹	Astonished Person

CHINESE

Number	Symbol
0	零
1	一
2	二
3	三
4	四
5	五
6	六
7	七
8	八
9	九
10	十
100	百
1000	千

1. $(1 \times 10{,}000) + (3 \times 1000) + (0 \times 100) + (3 \times 10) + (6 \times 1) = 13{,}036$

2. $(2 \times 1000) + (4 \times 100) + (1 \times 10) + (2 \times 1) = 2412$

3. $(7 \times 1{,}000{,}000) + (6 \times 100{,}000) + (3 \times 10{,}000) + (0 \times 1000) + (7 \times 100) + (2 \times 10) + (9 \times 1) = 7{,}630{,}729$

4. $(3 \times 1{,}000{,}000) + (0 \times 100{,}000) + (0 \times 10{,}000) + (5 \times 1000) + (2 \times 100) + (3 \times 10) + (1 \times 1) = 3{,}005{,}231$

5. ⌒⌒ ₤₤₤ ϙ ∩∩∩∩ │││││

6. ϙϙϙϙ ∩∩ │││││││

7. ⚹⚹⚹⚹⚹⚹⚹ ⌒⌒⌒⌒⌒⌒ ⌒⌒⌒⌒ ₤₤₤₤₤₤₤

8. ⌒⌒⌒ ₤₤₤₤₤ ∩∩∩∩∩∩∩

9. ⌒⌒⌒⌒ ₤₤₤₤₤₤₤₤ ϙϙϙ

10. ⌒⌒⌒⌒ ₤₤₤₤₤ ϙϙϙϙϙ

11. ⌒⌒⌒⌒⌒ ₤₤₤₤ ϙϙϙϙϙϙ

12. ⌒⌒⌒⌒ ₤₤₤₤₤ ϙϙϙϙϙϙ ∩∩∩∩∩

13. ⌒⌒⌒⌒⌒ ₤₤ ϙϙϙϙϙϙϙ

14. ⌒⌒⌒⌒ ₤₤₤₤ ϙϙϙϙ

15. $(9 \times 100) + (3 \times 10) + (5 \times 1) = 935$

16. $(2 \times 100) + (4 \times 10) + (6 \times 1) = 246$

17. $(3 \times 1000) + (7 \times 1) = 3007$

18. $(4 \times 1000) + (9 \times 100) + (2 \times 1) = 4902$

19. 九百六十

20. 六十三

21. 七十零

22. 七十二四百十六

23. 一十三百六十八 to 一十六百四十四

24. 九百六十 to 一十六二百七十九

25. [Egyptian numeral] to [Egyptian numeral]

26. [Egyptian numeral] to [Egyptian numeral]

27. There is a total of one scroll, eleven heelbones, and six strokes. Group ten heelbones to create a second scroll.
$$(2 \times 100) + (1 \times 10) + (6 \times 1) = 200 + 10 + 6$$
$$= 216$$

28. There is a total of three scrolls, eight heelbones, and twelve strokes. Group ten strokes to create a ninth heelbone, leaving two strokes.
$$(3 \times 100) + (9 \times 10) + (2 \times 1) = 300 + 90 + 2$$
$$= 392$$

29. There is a total of five pointing fingers, three lotus flowers, five scrolls, nine heelbones, and eleven strokes. Group ten strokes to create another heelbone. The ten heelbones then create another scroll.
$$(5 \times 10,000) + (3 \times 1000) + (6 \times 100) + (0 \times 10)$$
$$+ (1 \times 1) = 50000 + 3000 + 600 + 0 + 1$$
$$= 53,601$$

30. There is a total of eight strokes, six heelbones, eleven scrolls, and five lotus flowers. Regroup ten of the scrolls to create another lotus flower.
$$(6 \times 1000) + (1 \times 100) + (6 \times 10) + (8 \times 1)$$
$$= 6000 + 100 + 60 + 8$$
$$= 6168$$

31. After subtracting, there is one scroll, one heelbone, and three strokes.
$$(1 \times 100) + (1 \times 10) + (3 \times 1) = 100 + 10 + 3$$
$$= 113$$

32. Two strokes remain and two heelbones.
$$(2 \times 10) + (2 \times 1) = 20 + 2 = 22$$

33. Regroup the pointing finger to make ten lotus flowers for a total of eleven. Then one lotus flower must be regrouped to make ten scrolls for a total of twelve, and one scroll must be regrouped to make ten heelbones. Regroup one heelbone to make ten strokes. Then ten lotus flowers less three yields seven; eleven scrolls less six yields five; nine heelbones remain; fourteen strokes less six yields eight.

$$(7 \times 1000) + (5 \times 100) + (9 \times 10) + (8 \times 1)$$
$$= 7000 + 500 + 90 + 8$$
$$= 7598$$

34. One scroll must be regrouped to make ten heelbones for a total of thirteen. Then one lotus flower remains; four scrolls less two yields two; thirteen heelbones less seven yields six; six strokes less three yields three.
$$(1 \times 1000) + (2 \times 100) + (6 \times 10) + (3 \times 1)$$
$$= 1000 + 200 + 60 + 3$$
$$= 1263$$

35. Form two columns, headed by 1 and 53. Keep doubling each row until there are numbers in the first column that add up to 26.

	1	53	
→	2	106	←
	4	212	
→	8	424	←
→	16	848	←

$2 + 8 + 16 = 26$. Then add corresponding numbers from the second column:
$$106 + 424 + 848 = 1378.$$

36. Form two columns, headed by 1 and 81. Keep doubling each row until there are numbers in the first column that add up to 33:

	1	81	
→	1	81	←
	2	162	
	4	324	
	8	648	
	16	1296	
→	32	2592	←

$1 + 32 = 33$. Then add corresponding numbers from the second column:
$$81 + 2592 = 2673.$$

37. Form two columns, headed by 1 and 103. Keep doubling each row until there are numbers in the first column that add up to 58:

	1	103	
→	2	206	←
	4	412	
→	8	824	←
→	16	1648	←
→	32	3296	←

$2 + 8 + 16 + 32 = 58$. Then add corresponding numbers from the second column:
$$206 + 824 + 1648 + 3296 = 5974.$$

38. Form two columns, headed by 1 and 115. Keep doubling each row until there are numbers in the first column that add up to 67:

$$\begin{array}{rrl} \to & 1 & 115 \leftarrow \\ \to & 2 & 230 \leftarrow \\ & 4 & 460 \\ & 8 & 920 \\ & 16 & 1840 \\ & 32 & 3680 \\ \to & 64 & 7360 \leftarrow \end{array}$$

$1 + 2 + 64 = 67$. Then add corresponding numbers from the second column:

$$115 + 230 + 7360 = 7705.$$

39. thirty golden basins ∩∩∩
 a thousand silver basins 𝄢
 four hundred ten silver bowls 9999 ∩
 thirty golden bowls ∩∩∩
 3000 shekels 𝄢 𝄢 𝄢
 500 shekels 99999
 50 shekels ∩∩∩∩∩
 400 shekels 9999

30×3000

$$\begin{array}{rrl} & 1 & 3000 \\ \to & 2 & 6000 \leftarrow \\ \to & 4 & 12,000 \leftarrow \\ \to & 8 & 24,000 \leftarrow \\ \to & 16 & 48,000 \leftarrow \end{array}$$

$30 \times 3000 = 6000 + 12,000 + 24,000 + 48,000$
$= 90,000$

500×1000

$$\begin{array}{rrl} & 1 & 1000 \\ & 2 & 2000 \\ \to & 4 & 4000 \leftarrow \\ & 8 & 8000 \\ \to & 16 & 16,000 \leftarrow \\ \to & 32 & 32,000 \leftarrow \\ \to & 64 & 64,000 \leftarrow \\ \to & 128 & 128,000 \leftarrow \\ \to & 256 & 256,000 \leftarrow \end{array}$$

$500 \times 1000 = 4000 + 16,000 + 32,000 + 64,000 + 128,000 + 256,000 = 500,000$

50×410

$$\begin{array}{rrl} & 1 & 410 \\ \to & 2 & 820 \leftarrow \\ & 4 & 1640 \\ & 8 & 3280 \\ \to & 16 & 6560 \leftarrow \\ \to & 32 & 13,120 \leftarrow \end{array}$$

$50 \times 410 = 820 + 6560 + 13,120 = 20,500$

30×400

$$\begin{array}{rrl} & 1 & 400 \\ \to & 2 & 800 \leftarrow \\ \to & 4 & 1600 \leftarrow \\ \to & 8 & 3200 \leftarrow \\ \to & 16 & 6400 \leftarrow \end{array}$$

$30 \times 400 = 800 + 1600 + 3200 + 6400 = 12,000$
Now add 90,000, 500,000, 20,500, and 12,000 using Egyptian symbols.

𝄢𝄢𝄢𝄢𝄢𝄢𝄢 + ⌒⌒⌒⌒⌒ + 𝄢𝄢
99999 + 𝄢𝄢 𝄢𝄢 = ⌒⌒⌒⌒
𝄢𝄢𝄢𝄢𝄢𝄢𝄢𝄢𝄢𝄢 𝄢𝄢 99999

Regroup ten pointing fingers to one burbot fish.

⌒⌒⌒⌒⌒⌒ 𝄢𝄢 𝄢𝄢 99999

The total value of the treasure is

$(6 \times 100,000) + (2 \times 10,000) + (2 \times 1000)$
$+ (5 \times 100) = 622,500$ shekels.

40. $5500 = 𝄢𝄢𝄢𝄢\ 99999$
 $4600 = 𝄢𝄢𝄢𝄢\ 999999$
 $32 = ∩∩∩\ \text{II}$
 $900 = 999999999$
 $16 = ∩\ \text{IIIIII}$

2×5500

$$\begin{array}{rrl} & 1 & 5500 \\ \to & 2 & 11,000 \leftarrow \end{array}$$

$2 \times 5500 = 11,000$

$7 \times 11,000$

$$\begin{array}{rrl} \to & 1 & 11,000 \leftarrow \\ \to & 2 & 22,000 \leftarrow \\ \to & 4 & 44,000 \leftarrow \end{array}$$

$7 \times 11,000 = 11,000 + 22,000 + 44,000 = 77,000.$

3×4600

$$\begin{array}{rrl} \to & 1 & 4600 \leftarrow \\ \to & 2 & 9200 \leftarrow \end{array}$$

$3 \times 4600 = 4600 + 9200 = 13,800$

$32 \times 13,800$

$$\begin{array}{rrl} & 1 & 13,800 \\ & 2 & 27,600 \\ & 4 & 55,200 \\ & 8 & 110,400 \\ & 16 & 220,800 \\ \to & 32 & 441,600 \leftarrow \end{array}$$

$32 \times 13,800 = 441,600$

16×900

1	900
2	1800
4	3600
8	7200
→ 16	14,400 ←

77,000 in Egyptian numerals is

[6 scrolls] [7 lotus flowers].

441,600 in Egyptian numerals is

[4 pointing fingers] [4 scrolls] [1 lotus flower] [6 heels].

14,400 in Egyptian numerals is

[1 scroll] [4 lotus flowers] [4 heels].

Then:

```
        [6 scrolls]   [7 lotus flowers]
[2 fingers] [4 scrolls] [1 lotus]            [6 heels]
[2 fingers]
+           [1 scroll]  [4 lotus]            [4 heels]
―――――――――――――――――――――――――――――――――――――――――――――――――――
[2 fingers] [5 scrolls] [6 lotus flowers]    [5 heels]
[2 fingers] [5 scrolls] [6 lotus flowers]    [5 heels]
```

Regroup ten scrolls to make one lotus flower:

[4 pointing fingers] [5 scrolls] [6 lotus flowers] [5 scrolls] [6 lotus flowers].

Regroup ten lotus flowers to make one pointing finger:

[3 pointing fingers] [6 scrolls] [3 lotus] [5 scrolls].

Regroup ten pointing fingers to make one burbot fish:

[5 burbot fish] [3 scrolls] [3 lotus].

This is equivalent to 533,000 shekels.

41. Writing exercise
42. Writing exercise
43. Writing exercise
44. Writing exercise
45. 99,999. Five distinct symbols allows only five positions.
46. 9,999,999,999.
47. The largest number is $44,444_{\text{five}}$, which is equivalent to 3124_{ten}.
48. The largest number is $4,444,444,444_{\text{five}}$, which is equivalent to $9,765,624_{\text{ten}}$.
49. $10^d - 1$. Examine Exercise 45 to see that $10^5 - 1 = 100,000 - 1 = 99,999$.
50. $5^d - 1$
51. $7^d - 1$.
52. $b^d - 1$.
53. Writing exercise
54. Writing exercise

4.2 EXERCISES

1. $73 = (7 \times 10) + (3 \times 1) = (7 \times 10^1) + (3 \times 10^0)$

2. $925 = (9 \times 100) + (2 \times 10) + (5 \times 1)$
   $= (9 \times 10^2) + (2 \times 10^1) + (5 \times 10^0)$

3. $3774 = (3 \times 1000) + (7 \times 100) + (7 \times 10)$
   $+ (4 \times 1)$
   $= (3 \times 10^3) + (7 \times 10^2) + (7 \times 10^1)$
   $+ (4 \times 10^0)$

4. $12,398 = (1 \times 10,000) + (2 \times 1000) + (3 \times 100)$
   $+ (9 \times 10) + (8 \times 1)$
   $= (1 \times 10^4) + (2 \times 10^3) + (3 \times 10^2)$
   $+ (9 \times 10^1) + (8 \times 10^0)$

5. $4924 = (4 \times 1000) + (9 \times 100) + (2 \times 10)$
   $+ (4 \times 1)$
   $= (4 \times 10^3) + (9 \times 10^2) + (2 \times 10^1)$
   $+ (4 \times 10^0)$.

6. $52,118 = (5 \times 10,000) + (2 \times 1000) + (1 \times 100)$
   $+ (1 \times 10) + (8 \times 1)$
   $= (5 \times 10^4) + (2 \times 10^3) + (1 \times 10^2)$
   $+ (1 \times 10^1) + (8 \times 10^0)$

7. $14,206,040 = (1 \times 10,000,000) + (4 \times 1,000,000)$
   $+ (2 \times 100,000) + (0 \times 10,000)$
   $+ (6 \times 1000) + (0 \times 100)$
   $+ (4 \times 10) + (0 \times 1)$
   $= (1 \times 10^7) + (4 \times 10^6) + (2 \times 10^5)$
   $+ (0 \times 10^4) + (6 \times 10^3)$
   $+ (0 \times 10^2) + (4 \times 10^1)$
   $+ (0 \times 10^0)$

8. $212,011,916 = (2 \times 100,000,000) + (1 \times 10,000,000)$
   $+ (2 \times 1,000,000)$
   $+ (0 \times 100,000) + (1 \times 10,000)$
   $+ (1 \times 1000) + (9 \times 100)$
   $+ (1 \times 10) + (6 \times 1)$
   $= (2 \times 10^8) + (1 \times 10^7) + (2 \times 10^6)$
   $+ (0 \times 10^5) + (1 \times 10^4)$
   $+ (1 \times 10^3) + (9 \times 10^2)$
   $+ (1 \times 10^1) + (6 \times 10^0)$

9. $(4 \times 10) + (2 \times 1) = 42$

10. $(3 \times 100) + (5 \times 10) = 350$

11. $(6 \times 1000) + (2 \times 100) + (9 \times 1) = 6209$

12. $(5 \times 100,000) + (3 \times 1000) + (5 \times 100) + (6 \times 10) + (8 \times 1) = 503,568$

13. $(7 \times 10,000,000) + (4 \times 100,000) + (1 \times 1000) + (9 \times 1) = 70,401,009$

14. $(3 \times 100,000,000) + (8 \times 10,000,000) + (2 \times 100) + (3 \times 1) = 380,000,203.$

15. $\quad 54 = (5 \times 10^1) + (4 \times 10^0)$
    $\underline{+ 35 = (3 \times 10^1) + (5 \times 10^0)}$
    $\quad\quad = (8 \times 10^1) + (9 \times 10^0)$
    $\quad\quad = 80 + 9$
    $\quad\quad = 89$

16. $\quad 782 = (7 \times 10^2) + (8 \times 10^1) + (2 \times 10^0)$
    $\underline{+ 413 = (4 \times 10^2) + (1 \times 10^1) + (3 \times 10^0)}$
    $\quad\quad = (11 \times 10^2) + (9 \times 10^1) + (5 \times 10^0)$
    $\quad\quad = (1 \times 10^3) + (1 \times 10^2) + (9 \times 10^1)$
    $\quad\quad\quad + (5 \times 10^0)$
    $\quad\quad = 1000 + 100 + 90 + 5$
    $\quad\quad = 1195$

17. $\quad 85 = (8 \times 10^1) + (5 \times 10^0)$
    $\underline{- 53 = (5 \times 10^1) + (3 \times 10^0)}$
    $\quad\quad = (3 \times 10^1) + (2 \times 10^0)$
    $\quad\quad = 30 + 2$
    $\quad\quad = 32$

18. $\quad 784 = (7 \times 10^2) + (8 \times 10^1) + (4 \times 10^0)$
    $\underline{- 523 = (5 \times 10^2) + (2 \times 10^1) + (3 \times 10^0)}$
    $\quad\quad = (2 \times 10^2) + (6 \times 10^1) + (1 \times 10^0)$
    $\quad\quad = 200 + 60 + 1$
    $\quad\quad = 261$

19. $\quad 75 = (7 \times 10^1) + (5 \times 10^0)$
    $\underline{+ 34 = (3 \times 10^1) + (4 \times 10^0)}$
    $\quad\quad = (10 \times 10^1) + (9 \times 10^0)$
    $\quad\quad = (1 \times 10^2) + (9 \times 10^0)$
    $\quad\quad = 100 + 9$
    $\quad\quad = 109$

20. $\quad 537 = (5 \times 10^2) + (3 \times 10^1) + (7 \times 10^0)$
    $\underline{+ 278 = (2 \times 10^2) + (7 \times 10^1) + (8 \times 10^0)}$
    $\quad\quad = (7 \times 10^2) + (10 \times 10^1) + (15 \times 10^0)$
    $\quad\quad = (7 \times 10^2) + (10 \times 10^1) + (1 \times 10^1)$
    $\quad\quad\quad + (5 \times 10^0)$
    $\quad\quad = (7 \times 10^2) + (1 \times 10^2) + (1 \times 10^1)$
    $\quad\quad\quad + (5 \times 10^0)$
    $\quad\quad = (8 \times 10^2) + (1 \times 10^1) + (5 \times 10^0)$
    $\quad\quad = 800 + 10 + 5$
    $\quad\quad = 815$

21. $\quad 434 = (4 \times 10^2) + (3 \times 10^1) + (4 \times 10^0)$
    $\underline{+ 299 = (2 \times 10^2) + (9 \times 10^1) + (9 \times 10^0)}$
    $\quad\quad = (6 \times 10^2) + (12 \times 10^1) + (13 \times 10^0)$
    $\quad\quad = (6 \times 10^2) + (12 \times 10^1) + (1 \times 10^1)$
    $\quad\quad\quad + (3 \times 10^0)$
    $\quad\quad = (6 \times 10^2) + (13 \times 10^1) + (3 \times 10^0)$
    $\quad\quad = (6 \times 10^2) + (1 \times 10^2) + (3 \times 10^1)$
    $\quad\quad\quad + (3 \times 10^0)$
    $\quad\quad = (7 \times 10^2) + (3 \times 10^1) + (3 \times 10^0)$
    $\quad\quad = 700 + 30 + 3$
    $\quad\quad = 733$

22. $\quad 6755 = (6 \times 10^3) + (7 \times 10^2) + (5 \times 10^1)$
    $\quad\quad\quad + (5 \times 10^0)$
    $\underline{+ 4827 = (4 \times 10^3) + (8 \times 10^2) + (2 \times 10^1)}$
    $\quad\quad\quad\underline{+ (7 \times 10^0)}$
    $\quad\quad = (10 \times 10^3) + (15 \times 10^2) + (7 \times 10^1)$
    $\quad\quad\quad + (12 \times 10^0)$
    $\quad\quad = (10 \times 10^3) + (15 \times 10^2) + (7 \times 10^1)$
    $\quad\quad\quad + (1 \times 10^1) + (2 \times 10^0)$
    $\quad\quad = (11 \times 10^3) + (5 \times 10^2) + (8 \times 10^1)$
    $\quad\quad\quad + (2 \times 10^0)$
    $\quad\quad = (10 \times 10^3) + (1 \times 10^3) + (5 \times 10^2)$
    $\quad\quad\quad + (8 \times 10^1) + (2 \times 10^0)$
    $\quad\quad = (1 \times 10^4) + (1 \times 10^3) + (5 \times 10^2)$
    $\quad\quad\quad + (8 \times 10^1) + (2 \times 10^0)$
    $\quad\quad = 10,000 + 1000 + 500 + 80 + 2$
    $\quad\quad = 11,582$

23. $\quad 54 = (5 \times 10^1) + (4 \times 10^0)$
    $\underline{- 48 = (4 \times 10^1) + (8 \times 10^0)}$

    Since, in the units position, we cannot subtract 8 from 4, we use the distributive property to modify the top expansion as follows.

    $(4 \times 10^1) + (1 \times 10^1) + (4 \times 10^0)$
    $(4 \times 10^1) + (10 \times 10^0) + (4 \times 10^0)$

    $\quad 54 = (4 \times 10^1) + (14 \times 10^0)$
    $\underline{- 48 = (4 \times 10^1) + (8 \times 10^0)}$
    $\quad\quad = (0 \times 10^1) + (6 \times 10^0)$
    $\quad\quad = 6$

24. $\quad 364 = (3 \times 10^2) + (6 \times 10^1) + (4 \times 10^0)$
    $\underline{- 59 = (0 \times 10^2) + (5 \times 10^1) + (9 \times 10^0)}$

    Since, in the units position, we cannot subtract 9 from 4, we use the distributive property to modify the top expansion as follows.

$$(3 \times 10^2) + (5 \times 10^1) + (1 \times 10^1) + (4 \times 10^0)$$
$$(3 \times 10^2) + (5 \times 10^1) + (10 \times 10^0) + (4 \times 10^0)$$

$$364 = (3 \times 10^2) + (5 \times 10^1) + (14 \times 10^0)$$
$$-\,59 = (0 \times 10^2) + (5 \times 10^1) + (9 \times 10^0)$$
$$= (3 \times 10^2) + (0 \times 10^1) + (5 \times 10^0)$$
$$= 300 + 5$$
$$= 305$$

25. $\quad 645 = (6 \times 10^2) + (4 \times 10^1) + (5 \times 10^0)$
$-\,439 = (4 \times 10^2) + (3 \times 10^1) + (9 \times 10^0)$

Since, in the units position, we cannot subtract 9 from 5, we use the distributive property to modify the top expansion as follows.

$$(6 \times 10^2) + (3 \times 10^1) + (1 \times 10^1) + (5 \times 10^0)$$
$$(6 \times 10^2) + (3 \times 10^1) + (10 \times 10^0) + (5 \times 10^0)$$

$$645 = (6 \times 10^2) + (3 \times 10^1) + (15 \times 10^0)$$
$$-\,439 = (4 \times 10^2) + (3 \times 10^1) + (9 \times 10^0)$$
$$= (2 \times 10^2) + (0 \times 10^1) + (6 \times 10^0)$$
$$= 200 + 6$$
$$= 206$$

26. $\quad 816 = (8 \times 10^2) + (1 \times 10^1) + (6 \times 10^0)$
$-\,335 = (3 \times 10^2) + (3 \times 10^1) + (5 \times 10^0)$

Since, in the tens position, we cannot subtract 3 from 1, we use the distributive property to modify the top expansion as follows.

$$(7 \times 10^2) + (1 \times 10^2) + (1 \times 10^1) + (6 \times 10^0)$$
$$(7 \times 10^2) + (10 \times 10^1) + (1 \times 10^1) + (6 \times 10^0)$$
$$(7 \times 10^2) + (11 \times 10^1) + (6 \times 10^0)$$

$$816 = (7 \times 10^2) + (11 \times 10^1) + (6 \times 10^0)$$
$$-\,335 = (3 \times 10^2) + (3 \times 10^1) + (5 \times 10^0)$$
$$= (4 \times 10^2) + (8 \times 10^1) + (1 \times 10^0)$$
$$= 400 + 80 + 1$$
$$= 481$$

27. Reading the abacus from the right. The number represented by this abacus is
$[(1 \times 5) + (1 \times 1)] + (1 \times 50) + (2 \times 100)$
$= 6 + 50 + 200 = 256.$

28. Reading the abacus from the right
$(3 \times 1) + (2 \times 10) = 3 + 20 = 23.$

29. The number represented by this abacus is
$[(1 \times 5) + (4 \times 1)] + (1 \times 50) + (2 \times 100)$
$+ (3 \times 1000) + [(1 \times 50{,}000) + (1 \times 10{,}000)]$
$= 4 + 5 + 50 + 200 + 3000 + 50{,}000 + 10{,}000$
$= 63{,}259.$

30. The number represented by this abacus is
$[(1 \times 5) + (1 \times 1)] + (3 \times 10) + (1 \times 500)$
$+ (4 \times 1000) = 6 + 30 + 500 + 4000 = 4536.$

31. $38 = (3 \times 10) + [(1 \times 5) + (3 \times 1)]$

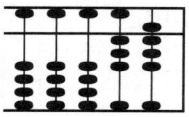

32. $183 = (1 \times 100) + [(1 \times 50) + (3 \times 10)] + (3 \times 1)$

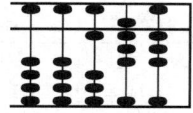

33. $2547 = (2 \times 1000) + (1 \times 500) + (4 \times 10)$
$+ [(1 \times 5) + (2 \times 1)]$

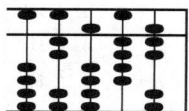

34. $70{,}163 = [(1 \times 50{,}000) + (2 \times 10{,}000)]$
$\phantom{70{,}163}+ (1 \times 100) + [(1 \times 50) + (1 \times 10)] + (3 \times 1)$

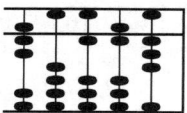

35. 65×29 is written around the top and right side.

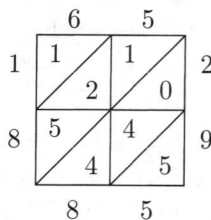

Obtain the numbers inside each box by finding the products of all the pairs of digits on the top and side: $6 \times 2 = 12$, $5 \times 2 = 10$, etc. Then add diagonally starting from the bottom right, placing the sums outside. For example, $0 + 4 + 4 = 8$. Finally, read the answer around the left side and the bottom as 1885.

36. 32 × 741 is written around the top and right side.

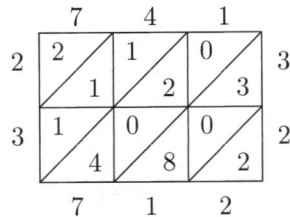

Obtain the numbers inside each box by finding the products of all the pairs of digits on the top and side: 7 × 3 = 21, 4 × 3 = 12, etc. Then add diagonally starting from the bottom right, placing the sums outside. For example, 3 + 0 + 8 = 11. Write the 1 outside the box and carry the one to the next diagonal. Finally, read the answer around the left side and the bottom as 23,712.

37. 525 × 73 is written around the top and right side.

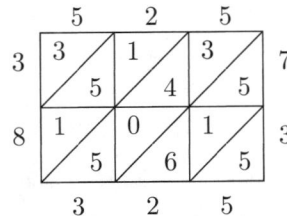

Find each number inside the boxes by finding the product of all the pairs of digits on the top and side. Then add diagonally beginning from the bottom right. For example, 5 + 1 + 6 = 12. Write the 2 outside the box and carry the 1. Now add 1 + 3 + 4 + 5 = 13. Again carry to the next diagonal above. Read the answer around the left side and the bottom as 38,325.

38. 912 × 483 is written around the top and right side.

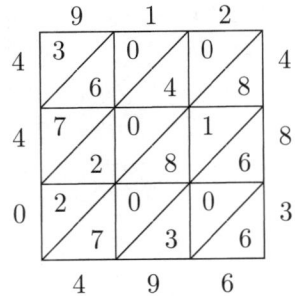

Find each number inside the boxes by finding the product of all the pairs of digits on the top and side. Then add diagonally beginning from the bottom right. For example, 8 + 1 + 8 + 7 = 24. Write the 4 outside the box and carry the 2. Now add 2 + 4 + 2 + 2 = 10. Again carry to the next diagonal above. Read the answer around the left side and the bottom as 440,496.

39. 723 × 4198 is written around the top and right side.

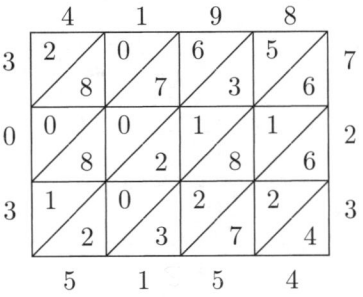

Find each number inside the boxes by finding the product of all the pairs of digits on the top and side. Then add diagonally beginning from the bottom right. For example, 6 + 2 + 7 = 15. Write the 5 outside the box and carry the 1. Now add 1 + 6 + 1 + 8 + 2 + 3 = 21. Again carry the 2 to the next diagonal above. Read the answer around the left side and the bottom as 3,035,154.

40. Writing exercise

41. Select the rods for 6 and 2 and place them side by side. Use the index to locate the row or level for a multiplier of 8.

Index	6	2
1	0/6	0/2
2	1/2	0/4
3	1/8	0/6
4	2/4	0/8
5	3/0	1/0
6	3/6	1/2
7	4/2	1/4
→ 8	4/8	1/6
9	5/4	1/8

The resulting lattice is shown below.

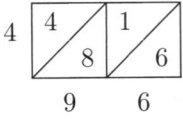

The product of 8 and 62 is 496.

42. Select the rods for 7 and 3 and place them side by side. Use the index to locate the row or level for multipliers of 3 and 2.

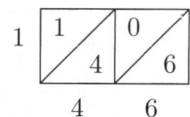

$2 \times 73 = 146$ is shown in the lattice below.

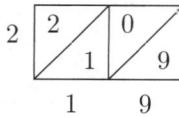

$3 \times 73 = 219$ is shown in the lattice below.

To create the table below, write the multiplicand on the top row. Write the digits of the multiplier in reverse order in the right hand column as shown. Insert the product of 2 and 73 as the first entry. Insert the product of 3 and 73 as the second entry, shifted one column to the left because it is actually 30×73. The final answer is found by addition; $23 \times 73 = 2336$.

```
     73
    146 | 2
    219 | 3
   2336
```

43. Select the rods for 8, 3, 5, and 4 and place them side by side. Use the index to first locate the row or level for multipliers of 2 and 6.

The product $6 \times 8354 = 50,124$.

Find the product of 2×8354 in a similar way, but using the index for a multiplier of 2.

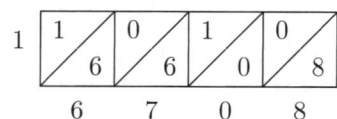

To create the table below, write the multiplicand on the top row and the multiplier in the right hand column as shown. Insert the product of 6 and 8354 as the first entry. Insert the product of 2 and 8354 as the second entry, shifted one column to the left because it is actually 20×8354. The final answer is found by addition; $26 \times 8354 = 217,204$.

```
        8354
      50124 | 6
      16708 | 2
     217204
```

44. 526×4863

Select the rods for 4, 8, 6, and 3 and place them side by side. Use the index to first locate the row or level for multipliers of 5, 2, and 6.

Index	4	8	6	3
1	0/4	0/8	0/6	0/3
→ 2	0/8	1/6	1/2	0/6
3	1/2	2/4	1/8	0/9
4	1/6	3/2	2/4	1/2
→ 5	2/0	4/0	3/0	1/5
→ 6	2/4	4/8	3/6	1/8
7	2/8	5/6	4/2	2/1
8	3/2	6/4	4/8	2/4
9	3/6	7/2	5/4	2/7

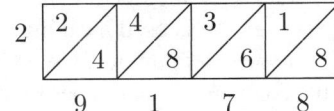

The product $6 \times 4863 = 29{,}178$.

Find the product of 2×4863 in a similar way, but using the index for a multiplier of 2.

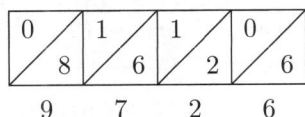

The product $2 \times 4863 = 9726$.

Find the product of 5×4863 in a similar way, but using the index for a multiplier of 5.

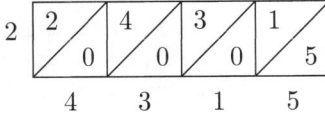

The product $5 \times 4863 = 24{,}315$.

To create the table below, write the multiplicand on the top row and the multiplier in the right hand column as shown. Insert the product of 6 and 4863 as the first entry. Insert the product of 2 and 4863 as the second entry, shifted one column to the left because it is actually 20×4863. Insert the product of 5 and 4863 as the third entry, shifted two columns to the left because it is actually 500×4863. The final answer is found by addition; $526 \times 4863 = 2{,}557{,}938$.

```
      4863
    29178  | 6
     9726  | 2
    24315  | 5
   2557938
```

45. Complete missing place value with 0.

$$283 \\ -\,041$$

Replace digits in subtrahend (041) with the nines complement of each and add.

$$283 \\ +\,958 \\ \overline{1241}$$

Delete the first digit on left and add that 1 to the remaining part of the sum: $241 + 1 = 242$.

46.
$$536 \\ -\,425$$

Replace digits in subtrahend (425) with the nines complement of each and add.

$$536 \\ +\,574 \\ \overline{1110}$$

Delete the first digit on left and add that 1 to the remaining part of the sum: $110 + 1 = 111$.

47. Complete missing place values with 0.

$$50000 \\ -\,00199$$

Replace digits in subtrahend (00199) with the nines complement of each and add.

$$50000 \\ +\,99800 \\ \overline{149800}$$

Delete the first digit on left and add that 1 to the remaining part of the sum: $49{,}800 + 1 = 49{,}801$.

48. Complete missing place value with 0.

$$40002 \\ -\,04846$$

Replace digits in subtrahend (04846) with the nines complement of each and add.

$$40002 \\ +\,95153 \\ \overline{135155}$$

Delete the first digit on left and add that 1 to the remaining part of the sum: $35{,}155 + 1 = 35{,}156$.

49. To multiply 5 and 92 using the Russian peasant method, write each number at the top of a column.

$$\begin{array}{rrl} \rightarrow & 5 & 92 \quad \leftarrow \\ & 2 & 184 \\ \rightarrow & 1 & 368 \quad \leftarrow \end{array}$$

Divide the first column by 2 and double the second column until 1 is obtained in the first column. Ignore the remainders when dividing. Add the numbers in the second column that correspond to the odd numbers in the first: $92 + 368 = 460$.

50. To multiply 41 and 53 using the Russian peasant method, write each number at the top of a column.

$$
\begin{array}{rr}
\rightarrow\ 41 & 53\ \leftarrow \\
20 & 106 \\
10 & 212 \\
\rightarrow\ 5 & 424\ \leftarrow \\
2 & 848 \\
\rightarrow\ 1 & 1696\ \leftarrow
\end{array}
$$

Divide the first column by 2 and double the second column until 1 is obtained in the first column. Ignore the remainders when dividing. Add the numbers in the second column that correspond to the odd numbers in the first: $53 + 424 + 1696 = 2173.51$. To multiply 62 and 529 using the Russian peasant method, write each number at the top of a column.

$$
\begin{array}{rr}
62 & 529 \\
\rightarrow\ 31 & 1058\ \leftarrow \\
\rightarrow\ 15 & 2116\ \leftarrow \\
\rightarrow\ 7 & 4232\ \leftarrow \\
\rightarrow\ 3 & 8464\ \leftarrow \\
\rightarrow\ 1 & 16{,}928\ \leftarrow
\end{array}
$$

Divide the first column by 2 and double the second column until 1 is obtained in the first column. Ignore the remainders when dividing. Add the numbers in the second column that correspond to the odd numbers in the first column.

$1058 + 2116 + 4232 + 8464 + 16{,}928 = 32{,}798$.

51. To multiply 62 and 529 using the Russian peasant method, write each number at the top of a column.

$$
\begin{array}{rr}
62 & 529 \\
\rightarrow\ 31 & 1058\ \leftarrow \\
\rightarrow\ 15 & 2116\ \leftarrow \\
\rightarrow\ 7 & 4232\ \leftarrow \\
\rightarrow\ 3 & 8464\ \leftarrow \\
\rightarrow\ 1 & 16{,}928\ \leftarrow
\end{array}
$$

Divide the first column by 2 and double the second column until 1 is obtained in the first column. Ignore the remainders when dividing. Add the numbers in the second column that correspond to the odd numbers in the first column.

$1058 + 2116 + 4232 + 8464 + 16{,}928 = 32{,}798$.

52. To multiply 63 and 145 using the Russian peasant method, write each number at the top of a column.

$$
\begin{array}{rr}
\rightarrow\ 63 & 145\ \leftarrow \\
\rightarrow\ 31 & 290\ \leftarrow \\
\rightarrow\ 15 & 580\ \leftarrow \\
\rightarrow\ 7 & 1160\ \leftarrow \\
\rightarrow\ 3 & 2320\ \leftarrow \\
\rightarrow\ 1 & 4640\ \leftarrow
\end{array}
$$

Divide the first column by 2 and double the second column until 1 is obtained in the first column. Ignore the remainders when dividing. Add the numbers in the second column that correspond to the odd numbers in the first column.

$145 + 290 + 580 + 1160 + 2320 + 4640 = 9135$.

4.3 EXERCISES

1. 1, 2, 3, 4, 5, and 6 are the first six digits. To represent the number seven, 10 is used meaning $(1 \times 7^1) + (0 \times 7^0)$. The next six numbers would be 11, 12, 13, 14, 15, and 16. To express the number fourteen, 20 is used meaning $(2 \times 7^1) + (0 \times 7^0)$. Continue in this pattern: 21, 22, 23, 24, 25, 26.

2. 1, 2, 3, 4, 5, 6, and 7 are the first seven digits. To represent the number eight, 10 is used meaning $(1 \times 8^1) + (0 \times 8^0)$. The next seven numbers would be 11, 12, 13, 14, 15, 16, 17. To express the number sixteen, 20 is used meaning $(2 \times 8^1) + (0 \times 8^0)$. Continue in this pattern: 21, 22, 23, 24.

3. 1, 2, 3, 4, 5, 6, 7, and 8 are the first eight digits. To represent the number nine, 10 is used which means $(1 \times 9^1) + (0 \times 9^0)$. The next eight numbers are 11, 12, 13, 14, 15, 16, 17, 18. To express the number eighteen, 20 is used which means $(2 \times 9^1) + (0 \times 9^0)$. Continue in this pattern: 21, 22.

4. In base sixteen, we need sixteen "digits." Since the normal ten digits we use for base ten numbers are not sufficient, we add the "digits" A through F to represent the numbers ten through 15. Note that 10 represents the number sixteen. The first twenty numbers in base sixteen are: 1, 2, 3, 4, 5, 6, 7, 8, 9, A, B, C, D, E, F, 10, 11, 12, 13, 14.

5. 13_{five} is the number just before, and 20_{five} is the number just after the given number.

6. 554_{six} is the number just before, and 1000_{six} just after the given number.

7. $\text{B6E}_{\text{sixteen}}$ is the number just before, and $\text{B70}_{\text{sixteen}}$ is the number just after the given number.

8. 10110_{two} is the number just before, and 11000_{two} is the number just after the given number.

9. Three distinct symbols are needed.

10. Seven distinct symbols are needed.

11. Eleven distinct symbols are needed.

12. Sixteen distinct symbols are needed.

13. The smallest four-digit number in base three is 1000, which means $(1 \times 3^3) = 27$. The largest four-digit number in base three is 2222, which means $(2 \times 3^3) + (2 \times 3^2) + (2 \times 3^1) + (2 \times 3^0)$. This is equivalent to $54 + 18 + 6 + 2 = 80$.

14. In base sixteen the smallest four-digit number is 1000_{sixteen}. The decimal equivalent of this number is $1 \times 16^3 = 4096$. The largest four-digit number is $FFFF_{\text{sixteen}}$. The decimal equivalent of this number is:
$(15 \times 16^3) + (15 \times 16^2) + (15 \times 16^1) + (15 \times 16^0)$
$= (15 \times 4096) + (15 \times 256) + (15 \times 16) + (15 \times 1)$
$= 61,440 + 3840 + 240 + 15$
$= 65,535$.

15. $(2 \times 5^1) + (4 \times 5^0) = 10 + 4 = 14$
Using the calculator shortcut: $(2 \times 5) + 4 = 14$.

16. $(6 \times 7^1) + (2 \times 7^0) = 42 + 2 = 44$
Using the calculator shortcut: $(6 \times 7) + 2 = 44$.

17. $2^3 + 2^1 + 2^0 = 8 + 2 + 1 = 11$
Using the calculator shortcut:
$[(1 \times 2 + 0) \times 2 + 1] \times 2 + 1$
$= [5] \times 2 + 1$
$= 11$.

18. $(3 \times 8^1) + (5 \times 8^0) = 24 + 5 = 29$
Using the calculator shortcut: $(3 \times 8) + 5 = 29$.

19. $(3 \times 16^2) + (11 \times 16^1) + (12 \times 16^0)$
$= 3 \times 256 + 11 \times 16 + 12$
$= 956$
Using the calculator shortcut:
$(3 \times 16 + 11) \times 16 + 12 = 956$.

20. $(3 \times 5^4) + (4 \times 5^3) + (4 \times 5^2) + (3 \times 5^1) + (2 \times 5^0)$
$= 3 \times 625 + 4 \times 125 + 4 \times 25 + 3 \times 5 + 2 \times 1$
$= 1875 + 500 + 100 + 15 + 2$
$= 2492$
Using the calculator shortcut:
$\{[(3 \times 5 + 4) \times 5 + 4] \times 5 + 3\} \times 5 + 2 = 2492$.

21. $(2 \times 7^3) + (3 \times 7^2) + (6 \times 7^1) + (6 \times 7^0)$
$= 686 + 147 + 42 + 6$
$= 881$
Using the calculator shortcut:
$[(2 \times 7 + 3) \times 7 + 6] \times 7 + 6 = [125] \times 7 + 6 = 881$.

22. $(1 \times 2^8) + (0 \times 2^7) + (1 \times 2^6) + (1 \times 2^5)$
$+ (0 \times 2^4) + (1 \times 2^3) + (1 \times 2^2) + (1 \times 2^1)$
$+ (0 \times 2^0)$
$= 256 + 64 + 32 + 8 + 4 + 2$
$= 366$

23. $(7 \times 8^4) + (0 \times 8^3) + (2 \times 8^2) + (6 \times 8^1)$
$+ (6 \times 8^0)$
$= 28,672 + 128 + 48 + 6$
$= 28,854$
Using the calculator shortcut:
$\{[(7 \times 8 + 0) \times 8 + 2] \times 8 + 6\} \times 8 + 6 = 28,854$.

24. Remember that in base sixteen, A is the symbol for 10, B is the symbol for 11, C for 12, and D for 13.
$(10 \times 16^3) + (11 \times 16^2) + (12 \times 16^1) + (13 \times 16^0)$
$= 40,960 + 2816 + 192 + 13 = 43,981$
Using the calculator shortcut:
$[(10 \times 16 + 11) \times 16 + 12] \times 16 + 13 = 43,981$.

25. $(2 \times 4^3) + (0 \times 4^2) + (2 \times 4^1) + (3 \times 4^0)$
$= 128 + 8 + 3 = 139$
Using the calculator shortcut:
$[(2 \times 4 + 0) \times 4 + 2] \times 4 + 3 = 139$.

26. $(6 \times 9^3) + (1 \times 9^2) + (8 \times 9^1) + (5 \times 9^0)$
$= 4374 + 81 + 72 + 5$
$= 4532$
Using the calculator method:
$[(6 \times 9 + 1) \times 9 + 8] \times 9 + 5 = 4532$.

27. $(4 \times 6^4) + (1 \times 6^3) + (5 \times 6^2) + (3 \times 6^1)$
$+ (3 \times 6^0)$
$= 5184 + 216 + 180 + 18 + 3$
$= 5601$
Using the calculator method:
$\{[(4 \times 6 + 1) \times 6 + 5] \times 6 + 3\} \times 6 + 3 = 5601$.

28. $(8 \times 9^4) + (8 \times 9^3) + (7 \times 9^2) + (0 \times 9^1)$
$+ (3 \times 9^0)$
$= 52,488 + 5832 + 567 + 3$
$= 58,890$
Using the calculator method:
$\{[(8 \times 9 + 8) \times 9 + 7] \times 9 + 0\} \times 9 + 3 = 58,890$.

29. The base five place values, starting from the right, are 1, 5, 25, 125, and so on. Since 86 is between 25 and 125, we will need some 25's but no 125's. Begin by dividing 86 by 25; then divide the remainder obtained from this division by 5. Finally, divide the remainder obtained from the previous division by 1, giving a remainder of 0.

$$86 \div 25 = 3, \text{ remainder } 11$$
$$11 \div 5 = 2, \text{ remainder } 1$$
$$1 \div 1 = 1, \text{ remainder } 0$$

The digits of the answer are found by reading quotients from the top down.
$86 = 321_{\text{five}}$

Shortcut:

```
5|86       Rem
5|17   ←   1
 5|3   ←   2
  0    ←   3
```

Read the answer from the remainder column, reading from the bottom up.
$$86 = 321_{\text{five}}.$$

30. The base seven place values, starting from the right, are 1, 7, 49, 343, and so on. Since 65 is between 49 and 343, we will need some 49's, but no 343's. Begin by dividing 65 by 49; then divide the remainder obtained from this division by 7. Finally, divide the remainder obtained from the previous division by 1, giving a remainder of 0.

$$65 \div 49 = 1, \text{ remainder } 16$$
$$16 \div 7 = 2, \text{ remainder } 2$$
$$2 \div 1 = 2, \text{ remainder } 0$$

The digits of the answer are found by reading quotients from the top down.
$$65 = 122_{\text{seven}}$$

Repeated division shortcut:

$$
\begin{array}{r|l}
7 & 65 \quad\quad \text{Rem} \\
7 & 9 \leftarrow 2 \\
7 & 1 \leftarrow 2 \\
& 0 \leftarrow 1
\end{array}
$$

Read the answer from the remainder column, reading from the bottom up.
$$65 = 122_{\text{seven}}.$$

31.
$$
\begin{array}{r|l}
2 & 19 \quad\quad \text{Rem} \\
2 & 9 \leftarrow 1 \\
2 & 4 \leftarrow 1 \\
2 & 2 \leftarrow 0 \\
2 & 1 \leftarrow 0 \\
& 0 \leftarrow 1
\end{array}
$$
$19 = 10011_{\text{two}}$

32.
$$
\begin{array}{r|l}
8 & 935 \quad\quad \text{Rem} \\
8 & 116 \leftarrow 7 \\
8 & 14 \leftarrow 4 \\
8 & 1 \leftarrow 6 \\
& 0 \leftarrow 1
\end{array}
$$
$935 = 1647_{\text{eight}}$

33.
$$
\begin{array}{r|l}
16 & 147 \quad\quad \text{Rem} \\
16 & 9 \leftarrow 3 \\
& 0 \leftarrow 9
\end{array}
$$
$147 = 93_{\text{sixteen}}$

34.
$$
\begin{array}{r|l}
16 & 2730 \quad\quad \text{Rem} \\
16 & 170 \leftarrow A \\
16 & 10 \leftarrow A \\
& 0 \leftarrow A
\end{array}
$$
$2730 = AAA_{\text{sixteen}}$

Remember that the symbol "A" means ten in base sixteen.

35.
$$
\begin{array}{r|l}
5 & 36401 \quad\quad \text{Rem} \\
5 & 7280 \leftarrow 1 \\
5 & 1456 \leftarrow 0 \\
5 & 291 \leftarrow 1 \\
5 & 58 \leftarrow 1 \\
5 & 11 \leftarrow 3 \\
5 & 2 \leftarrow 1 \\
& 0 \leftarrow 2
\end{array}
$$
$36401 = 2131101_{\text{five}}$

36.
$$
\begin{array}{r|l}
7 & 70893 \quad\quad \text{Rem} \\
7 & 10127 \leftarrow 4 \\
7 & 1446 \leftarrow 5 \\
7 & 206 \leftarrow 4 \\
7 & 29 \leftarrow 3 \\
7 & 4 \leftarrow 1 \\
& 0 \leftarrow 4
\end{array}
$$
$70893 = 413454_{\text{seven}}$

37.
$$
\begin{array}{r|l}
2 & 586 \quad\quad \text{Rem} \\
2 & 293 \leftarrow 0 \\
2 & 146 \leftarrow 1 \\
2 & 73 \leftarrow 0 \\
2 & 36 \leftarrow 1 \\
2 & 18 \leftarrow 0 \\
2 & 9 \leftarrow 0 \\
2 & 4 \leftarrow 1 \\
2 & 2 \leftarrow 0 \\
2 & 1 \leftarrow 0 \\
& 0 \leftarrow 1
\end{array}
$$
$586 = 1001001010_{\text{two}}$

38.
$$
\begin{array}{r|l}
8 & 12888 \quad\quad \text{Rem} \\
8 & 1611 \leftarrow 0 \\
8 & 201 \leftarrow 3 \\
8 & 25 \leftarrow 1 \\
8 & 3 \leftarrow 1 \\
& 0 \leftarrow 3
\end{array}
$$
$12888 = 31130_{\text{eight}}$

39.
$$\begin{array}{r|l} 3 & 8407 \\ 3 & 2802 \\ 3 & 934 \\ 3 & 311 \\ 3 & 103 \\ 3 & 34 \\ 3 & 11 \\ 3 & 3 \\ 3 & 1 \\ & 0 \end{array} \quad \begin{array}{l} \text{Rem} \\ \leftarrow 1 \\ \leftarrow 0 \\ \leftarrow 1 \\ \leftarrow 2 \\ \leftarrow 1 \\ \leftarrow 1 \\ \leftarrow 2 \\ \leftarrow 0 \\ \leftarrow 1 \end{array}$$

$8407 = 102112101_{\text{three}}$

40.
$$\begin{array}{r|l} 4 & 11028 \\ 4 & 2757 \\ 4 & 689 \\ 4 & 172 \\ 4 & 43 \\ 4 & 10 \\ 4 & 2 \\ & 0 \end{array} \quad \begin{array}{l} \text{Rem} \\ \leftarrow 0 \\ \leftarrow 1 \\ \leftarrow 1 \\ \leftarrow 0 \\ \leftarrow 3 \\ \leftarrow 2 \\ \leftarrow 2 \end{array}$$

$11028 = 2230110_{\text{four}}$

41.
$$\begin{array}{r|l} 6 & 9346 \\ 6 & 1557 \\ 6 & 259 \\ 6 & 43 \\ 6 & 7 \\ 6 & 1 \\ & 0 \end{array} \quad \begin{array}{l} \text{Rem} \\ \leftarrow 4 \\ \leftarrow 3 \\ \leftarrow 1 \\ \leftarrow 1 \\ \leftarrow 1 \\ \leftarrow 1 \end{array}$$

$9346 = 111134_{\text{six}}$

42.
$$\begin{array}{r|l} 9 & 99999 \\ 9 & 11111 \\ 9 & 1234 \\ 9 & 137 \\ 9 & 15 \\ 9 & 1 \\ & 0 \end{array} \quad \begin{array}{l} \text{Rem} \\ \leftarrow 0 \\ \leftarrow 5 \\ \leftarrow 1 \\ \leftarrow 2 \\ \leftarrow 6 \\ \leftarrow 1 \end{array}$$

$99999 = 162150_{\text{nine}}$

43. First convert 43_{five} to base ten.

$$(4 \times 5) + 3 = 23$$

Then convert 23 to base seven.

$$\begin{array}{r|l} 7 & 23 \\ 7 & 3 \\ & 0 \end{array} \quad \begin{array}{l} \text{Rem} \\ \leftarrow 2 \\ \leftarrow 3 \end{array}$$

$43_{\text{five}} = 32_{\text{seven}}$

44. First convert 27_{eight} to base ten.

$$(2 \times 8) + 7 = 23$$

Then convert 23 to base five.

$$\begin{array}{r|l} 5 & 23 \\ 5 & 4 \\ & 0 \end{array} \quad \begin{array}{l} \text{Rem} \\ \leftarrow 3 \\ \leftarrow 4 \end{array}$$

$27_{\text{eight}} = 43_{\text{five}}$

45. First convert 6748_{nine} to base ten.

$$(6 \times 9^3) + (7 \times 9^2) + (4 \times 9) + 8 = 4985$$

Then convert 4985 to base four.

$$\begin{array}{r|l} 4 & 4985 \\ 4 & 1246 \\ 4 & 311 \\ 4 & 77 \\ 4 & 19 \\ 4 & 4 \\ 4 & 1 \\ & 0 \end{array} \quad \begin{array}{l} \text{Rem} \\ \leftarrow 1 \\ \leftarrow 2 \\ \leftarrow 3 \\ \leftarrow 1 \\ \leftarrow 3 \\ \leftarrow 0 \\ \leftarrow 1 \end{array}$$

$6748_{\text{nine}} = 1031321_{\text{four}}$

46. First convert C02 to base ten.

$$(12 \times 16^2) + 2 = 3074$$

Then convert 3074 to base seven.

$$\begin{array}{r|l} 7 & 3074 \\ 7 & 439 \\ 7 & 62 \\ 7 & 8 \\ 7 & 1 \\ & 0 \end{array} \quad \begin{array}{l} \text{Rem} \\ \leftarrow 1 \\ \leftarrow 5 \\ \leftarrow 6 \\ \leftarrow 1 \\ \leftarrow 1 \end{array}$$

$\text{C02}_{\text{sixteen}} = 11651_{\text{seven}}$

47. Replace each octal digit with its 3-digit binary equivalent. Then combine all the binary equivalents into a single binary numeral.

$$\begin{array}{ccc} 3 & 6 & 7 \\ \downarrow & \downarrow & \downarrow \\ 011 & 110 & 111 \end{array}$$

$367_{\text{eight}} = 11110111_{\text{two}}$

48. Replace each octal digit with its 3-digit binary equivalent. Then combine all the binary equivalents into a single binary numeral.

$$\begin{array}{cccc} 2 & 4 & 0 & 6 \\ \downarrow & \downarrow & \downarrow & \downarrow \\ 10 & 100 & 000 & 110 \end{array}$$

$2406_{\text{eight}} = 10100000110_{\text{two}}$

49. Starting at the right, break the digits into groups of three. Then convert the groups to their octal equivalents. (Refer to Table 7.)

$$\begin{array}{ccc} 100 & 110 & 111 \\ \downarrow & \downarrow & \downarrow \\ 4 & 6 & 7 \end{array}$$

$100110111_{\text{two}} = 467_{\text{eight}}$

50. Starting at the right, break the digits into groups of three. Then convert the groups to their octal equivalents. (Refer to Table 7.)

$$\begin{array}{cccc} 11 & 010 & 111 & 101 \\ \downarrow & \downarrow & \downarrow & \downarrow \\ 3 & 2 & 7 & 5 \end{array}$$

$11010111101_{\text{two}} = 3275_{\text{eight}}$

51. Each hexadecimal digit yields a 4-digit binary equivalent. (See Table 8.)

$$\begin{array}{cc} D & C \\ \downarrow & \downarrow \\ 1101 & 1100 \end{array}$$

$DC_{\text{sixteen}} = 11011100_{\text{two}}$

52.
$$\begin{array}{cccc} F & 1 & 1 & 1 \\ \downarrow & \downarrow & \downarrow & \downarrow \\ 1111 & 0001 & 0001 & 0001 \end{array}$$

$F111_{\text{sixteen}} = 1111000100010001_{\text{two}}$

53. Starting at the right, break the digits into groups of four. Then convert the groups to their hexadecimal equivalent. (Refer to Table 8.)

$$\begin{array}{cc} 10 & 1101 \\ \downarrow & \downarrow \\ 2 & D \end{array}$$

$101101_{\text{two}} = 2D_{\text{sixteen}}$

54.
$$\begin{array}{cccc} 101 & 1110 & 1110 & 1000 \\ \downarrow & \downarrow & \downarrow & \downarrow \\ 5 & E & E & 8 \end{array}$$

$101111011101000_{\text{two}} = 5EE8_{\text{sixteen}}$

55. In order to compare these numbers, we need to write them in the same base. Convert each of them to decimal form (base ten).

$42_{\text{seven}} = \quad (4 \times 7^1) + (2 \times 7^0) = 28 + 2$ or 30

$37_{\text{eight}} = \quad (3 \times 8^1) + (7 \times 8^0) = 24 + 7$ or 31

$1D_{\text{sixteen}} = \quad (1 \times 16^1) + (13 \times 16^0) = 16 + 13$ or 29

The largest number is 37_{eight}.

56. In order to compare these numbers, we need to write them in the same base. Convert each of them to decimal form (base ten).

1101110_{two}
$= (1 \times 2^6) + (1 \times 2^5) + (0 \times 2^4) + (1 \times 2^3) +$
$\quad (1 \times 2^2) + (1 \times 2^1) + (0 \times 2^0)$
$= 64 + 32 + 8 + 4 + 2$
$= 110$

407_{five}
$= (4 \times 5^2) + (0 \times 5^1) + (7 \times 5^0)$
$= 100 + 0 + 7$
$= 107$

$6F_{\text{sixteen}}$
$= (6 \times 16^1) + (15 \times 16^0)$
$= 96 + 15$
$= 111$

The largest number is $6F_{\text{sixteen}}$.

57. $(9 \times 12^2) + (10 \times 12) + 11 = 1427$ copies

58. 3 gross, 6 dozen
$= (3 \times 12^2) + (6 \times 12)$
$= 432 + 72$
$= 504$

2 gross, 19 dozen
$= (2 \times 12^2) + (19 \times 12)$
$= 288 + 228$
$= 516$

Then 2 gross, 19 dozen is the larger.

59. Since A is assigned the number 65, C is assigned the number 67. Change 67 from decimal form to binary form.

$$\begin{array}{rll} 2\underline{|67} & & \text{Rem} \\ 2\underline{|33} & \leftarrow & 1 \\ 2\underline{|16} & \leftarrow & 1 \\ 2\underline{|8} & \leftarrow & 0 \\ 2\underline{|4} & \leftarrow & 0 \\ 2\underline{|2} & \leftarrow & 0 \\ 2\underline{|1} & \leftarrow & 0 \\ 0 & \leftarrow & 1 \end{array}$$

$C = 1000011_{\text{two}}$

60. A capital X is equivalent to 88. The binary code for 88 is found by repeated division by 2.

$$\begin{array}{rll} 2\underline{|88} & & \text{Rem} \\ 2\underline{|44} & \leftarrow & 0 \\ 2\underline{|22} & \leftarrow & 0 \\ 2\underline{|11} & \leftarrow & 0 \\ 2\underline{|5} & \leftarrow & 1 \\ 2\underline{|2} & \leftarrow & 1 \\ 2\underline{|1} & \leftarrow & 0 \\ 0 & \leftarrow & 1 \end{array}$$

$X = 1011000_{\text{two}}$

4.3 CONVERSION BETWEEN NUMBER BASES   89

61. Since a is assigned the number 97, k is assigned the number 107. (Since k is the eleventh letter of the alphabet, its corresponding number will be ten more than the number corresponding to a.) Change 107 to binary form.

```
2|107        Rem
  2|53   ←  1
    2|26  ←  1
      2|13  ←  0
        2|6   ←  1
          2|3   ←  0
            2|1   ←  1
              0   ←  1
```

$k = 1101011_{two}$

62. Convert each seven-digit binary number to decimal form; then find the corresponding letters.

Base Two	Base Ten	Letter
1001000	64 + 8 = 72	H
1000101	64 + 4 + 1 = 69	E
1001100	64 + 8 + 4 = 76	L
1010000	64 + 16 = 80	P

The given number represents HELP.

63. Convert each seven-digit binary number to decimal form; then find the corresponding letters.

Base Two	Base Ten	Letter
1000011	64 + 2 + 1 = 67	C
1001000	64 + 8 = 72	H
1010101	64 + 16 + 4 + 1 = 85	U
1000011	64 + 2 + 1 = 67	C
1001011	64 + 8 + 2 + 1 = 75	K

The given number represents CHUCK.

64. To translate the word "New" into an ASCII string of binary digits, find the number corresponding to each letter of the word, translate each of these numbers to binary form, and finally combine these three binary numbers into a single string.

Letter	Base Ten	Base Two
N	78	1001110
e	101	1100101
w	119	1110111

The word "New" is represented by the following ASCII string: $100111011001011110111_{two}$.

65.

Letter	Base Ten	Base Two
O	79	1001111
r	114	1110010
l	108	1101100
e	101	1100101
a	97	1100001
n	110	1101110
s	115	1110011

The word "Orleans" is represented by the following ASCII string:

O   r   l   e
1001111   1110010   1101100   1100101
a   n   s
1100001   1101110 1110011_{two}.

66. Writing exercise

67. The largest base eight number that consists of two digits is 77_{eight}, which is equivalent to 63. In base three this number is 2100_{three}. The smallest base eight number that consists of two nonzero digits is 11_{eight}, which is equivalent to 9. However, in base three this number has only three digits, 100_{three}. In base three the smallest four-digit number is 1000_{three}, which equals 27. In base eight, this number is 33_{eight}. The smallest number, then, is 27 and the largest, 63.

68. (a) The binary ones digit is 1.
    (b) The binary twos digit is 1.
    (c) The binary fours digit is 1.
    (d) The binary eights digit is 1.
    (e) The binary sixteens digit is 1.

69. Writing exercise

70. In order to include all ages up to 63, we will need to add one more column to Table 9. This column will contain the numbers whose binary thirty-twos digit is 1. Thus, 6 columns would be needed.

71. Seven columns would be needed. The trick in Table 9 is that it encodes ages 1–31 in a binary fashion. Ages 1–31 require five bits. In Table 9 the bits are in reverse order with the column weights in the first row the same as binary positional notation. Ages 1–127 require seven bits. Thus the number of columns increases from five to seven and the number of rows increases from 16 to 64.

To better understand Table 9, increase it by one column and 16 rows to encode ages 1–63 using the patterns in Table 9. Alternatively, truncate Table 9 to four columns and eight rows to see that it now only encodes ages 1–15.

72. In base two, every even number has 0 as its ones digit and every odd number has a 1 as its ones digit. Thus, we can distinguish odd and even numbers by looking at their ones digit. The criterion works.

90 CHAPTER 4 NUMERATION AND MATHEMATICAL SYSTEMS

73. No. Consider 12_{three}, which is equivalent to $(1 \times 3) + 2 = 5$. Although the ones digit is a 2, the number is odd.

74. In base four, every even number has 0 or 2 as its ones digit, while every odd number has 1 or 3 as its ones digit. Thus, we can distinguish odd and even numbers by looking at their ones digit. The criterion works.

75. No. Consider 12_{five} which is equivalent to $(1 \times 5) + 2 = 7$, an odd number.

80. Writing exercise

81. Writing exercise

82. The units digit, 4, is the same for base 5 and base 10. Hence the number is not divisible by 5.

83. The units digit, 0, is the same for base 5 and base 10. Hence the number is divisible by 5.

84. The units digit, 0, is the same for base 5 and base 10. Hence the number is divisible by 5.

85. The units digit, 2, is the same for base 5 and base 10. Hence the number is not divisible by 5.

4.4 EXERCISES

1. All properties are satisfied. 1 is the identity element; 1 is its own inverse, as is 2.

2. All properties are satisfied. 1 is the identity element; 1 is its own inverse, as is 4. 2 and 3 are inverses.

3. All properties are satisfied except closure and inverse. 1 is the identity element; 2, 4, and 6 have no inverses. Note that $0 \notin \{1, 2, 3, 4, 5, 6, 7\}$.

4. The commutative, associative, and identity properties are satisfied. 1 is the identity element. 2, 3, and 4 have no inverses.

5. All properties are satisfied except the inverse. 1 is the identity element; 5 has no inverse.

6. All properties are satisfied. 1 is the identity element. 1, 3, 5, and 7 are their own inverses.

7. All properties are satisfied. F is the identity element. A and B are inverses; F is its own inverse.

8. Only the closure and commutative properties are satisfied.

9. All properties are satisfied. t is the identity element. s and r are inverses; t and u are their own inverses.

10. All properties are satisfied. A is the identity element. J and U are inverses; A and T are their own inverses.

11. The letter b represents a rotation of $90°$, and d represents a rotation of $270°$. If both rotations are performed, the

76. In base six, every even number has 0, 2, or 4 as its ones digit, while every odd number has 1, 3, or 5 as its ones digit. The criterion works.

77. No. Consider 14_{seven}, which is equivalent to $(1 \times 7) + 4 = 11$, an odd number.

78. In base eight, every even number has 0, 2, 4, or 6 as its ones digit, while every odd number has 1, 3, 5, or 7 as its ones digit. The criterion works.

79. No. Consider 13_{nine}, which is equivalent to $(1 \times 9) + 3 = 12$, an even number.

square returns to its original position. The answer, then, is $b \square d = a$.

12. The letter b represents a rotation of $90°$. Two $90°$ rotations is equivalent to $180°$. Then $b \square b = c$

13. A rotation of $270°$ (represented by d) followed by a rotation of $90°$ (b), again returns the square to its original position. The answer, then, is $d \square b = a$.

14. This is a $0°$ rotation followed by a $90°$ rotation: $a \square b = b$.

15.

$\square$	a	b	c	d
a	a	b	c	d
b	b	c	$\underline{d}$	a
c	c	$\underline{d}$	a	$\underline{b}$
d	d	a	$\underline{b}$	$\underline{c}$

16. All properties are satisfied. Closure is satisfied because all answers in the body of the table are elements of the set $\{a, b, c, d\}$. The commutative property is satisfied. For example, $b \square c = d$ and $c \square b = d$. The associative property is satisfied. For example, $(a \square c) \square b = c \square b = d$ and $a \square (c \square b) = a \square d = d$. The identity element is a. Each element has an inverse.

17. Call the new set $\{U, \emptyset, \{1\}, \{2\}, \dots\}$ U star.
Closure: Yes. The set U^* is closed because the intersection of any two members in U^* is a member of U^*.
Commutative: Yes. The order of the intersection does not change the outcome.
Associative: Yes. The grouping can be changed without affecting the outcome.
Identity: Yes. For the operation intersection, the identity is U.

18. Call the new set $\{U, \emptyset, \{1\}, \{2\}, \dots\}$ U star.
Closure: Yes. The set U^* is closed because the union of any two members in U^* is a member of U^*.
Commutative: Yes. The order of the union does not change the outcome.
Associative: Yes. The grouping can be changed without affecting the outcome.
Identity: Yes. For the operation union, the identity is $\emptyset$.

19. Here is one possibility.

	a	b	c	d
a	a	b	c	d
b	b	a	d	c
c	c	d	a	b
d	d	c	b	a

20. Try $3 - (5 \times 4)$. If the distributive property of subtraction over multiplication did hold, this expression would equal $(3 - 5) \times (3 - 4)$. The first expression $3 - (5 \times 4) = 3 - 20 = -17$. But $(3 - 5) \times (3 - 4) = -2 \times -1 = 2$. The property does not hold.

21. Try $2 + (6 - 4)$. If the distributive property of addition over subtraction did hold, this expression would equal $(2 + 6) - (2 + 4)$. The first expression $2 + (6 - 4) = 4$. But $(2 + 6) - (2 + 4) = 8 - 6 = 2$. The property does not hold.

22. Try $2 - (6 + 4)$. If the distributive property of subtraction over addition did hold, this expression would equal $(2 - 6) + (2 - 4)$. The first expression $2 - (6 + 4) = 2 - 10 = -8$. But $(2 - 6) + (2 - 4) = -6$. The property does not hold.

23. (a) $a = 2, b = -5, c = 4$
    $$2 + (-5 \times 4) = 2 + (-20) = -18$$
    $$(2 + (-5)) \times (2 + 4) = -3 \times 6 = -18$$
    The equation is true for these values.

    (b) $a = -7, b = 5, c = 3$
    $$-7 + (5 \times 3) = -7 + (15) = 8$$
    $$(-7 + 5) \times (-7 + 3) = -2 \times -4 = 8$$
    The equation is true for these values.

    (c) $a = -8, b = 14, c = -5$
    $$-8 + (14 \times (-5)) = -8 + (-70) = -78$$
    $$(-8 + 14) \times (-8 + (-5)) = 6 \times -13 = -78$$
    The equation is true for these values.

    (d) $a = 1, b = 6, c = -6$
    $$1 + (6 \times (-6)) = 1 + (-36) = -35$$
    $$(1 + 6) \times (1 + (-6)) = 7 \times -5 = -35$$
    The equation is true for these values.

24. $a = 5, b = -7, c = 3$
    $$5 + (-7 \times 3) = 5 + (-21) = -16$$
    $$(5 + (-7)) \times (5 + 3) = -2 \times 8 = -16$$

25. The statement is true when $a + b + c = 1$ or $a = 0$.

26. Writing exercise

27. (a) True only when $a = 0$.
    (b) True only when $a = 0$.

28. (a) Two sets of values that make this equation true are $a = 0, b = 0, c = 0$; and $a = 1, b = 1, c = 1$.
    $$0 - (0 \times 0) = 0$$
    $$(0 - 0) \times (0 - 0) = 0$$
    $$1 - (1 \times 1) = 0$$
    $$(1 - 1) \times (1 - 1) = 0$$

    (b) Writing exercise

29. Use the tables of Example 4 to see that $d \circ e = c$. Then $c \star c = e$. Examine the right side of the equation to see that $c \star d = b$ and $c \star e = d$. Then $b \circ d = e$. Each side simplifies to e.

30. $a \star (a \circ b) = a \star b = a$ and $(a \star a) \circ (a \star b) = a \circ a = a$.
    Since both sides simplify to a, the distributive property has been verified.

31. $d \star (e \circ c) = d \star b = d$ and $(d \star e) \circ (d \star c) = c \circ b = d$. The left side of the equation is equivalent to the right side.

32. $b \star (b \circ b) = b \star c = c$ and $(b \star b) \circ (b \star b) = b \circ b = c$. Since both sides simplify to c, the distributive property has been verified.

33.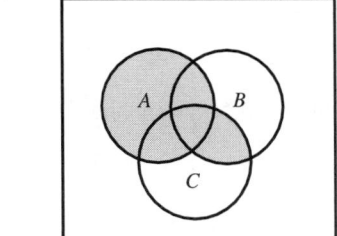

    $A \cup (B \cap C) = (A \cup B) \cap (A \cup C)$

34.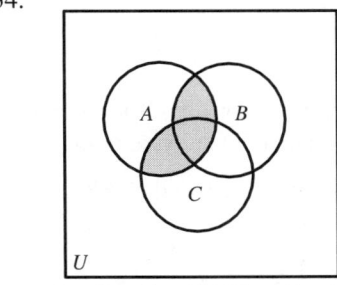

    $A \cap (B \cup C) = (A \cap B) \cup (A \cap C)$

35.

p	q	r	p ∨	(q ∧ r)	(p ∨ q)	∧	(p ∨ r)
T	T	T	T	T	T	T	T
T	T	F	T	F	T	T	T
T	F	T	T	F	T	T	T
T	F	F	T	F	T	T	T
F	T	T	T	T	T	T	T
F	T	F	F	F	T	F	F
F	F	T	F	F	F	F	T
F	F	F	F	F	F	F	F
			2	1	1	2	1

Compare column 2 for each logic statement. Since they are equal row by row, the distributive property is proven to hold.

36.

p	q	r	p ∧	(q ∨ r)	(p ∧ q)	∨	(p ∧ r)
T	T	T	T	T	T	T	T
T	T	F	T	T	T	T	F
T	F	T	T	T	F	T	T
T	F	F	F	F	F	F	F
F	T	T	F	T	F	F	F
F	T	F	F	T	F	F	F
F	F	T	F	T	F	F	F
F	F	F	F	F	F	F	F
			2	1	1	2	1

Compare column 2 for each logic statement. Since they are equal row by row, the distributive property is proven to hold.

4.5 EXERCISES

1. The operation is not specified.

2. The set of elements is not specified.

3. Form the multiplication table for this system and check the necessary four properties.

×	0
0	0

   The set {0} under multiplication forms a group. All properties are satisfied: closure, associative, identity, and inverse.

4.
+	0
0	0

   The set {0} under addition forms a group. All properties are satisfied: closure, associative, identity, and inverse.

5. Form the addition table for this system and check the necessary four properties.

+	0	1
0	0	1
1	1	2

   The set {0, 1} under addition does not form a group because closure is not satisfied, and the inverse property is not satisfied.

6.
−	0
0	0

   The set {0} under subtraction forms a group. All properties are satisfied: closure, associative, identity, and inverse.

7.
÷	−1	1
−1	1	−1
1	−1	1

   This system forms a group. All properties are satisfied: closure, associative, identity, and inverse.

8.
×	0	1
0	0	0
1	0	1

   The set {0, 1} under multiplication does not form a group because the inverse property is not satisfied.

9.
×	−1	0	1
−1	1	0	−1
0	0	0	0
1	−1	0	1

   This set does not form a group because the inverse property is not satisfied.

10.
+	−1	0	1
−1	−2	−1	0
0	−1	0	1
1	0	1	2

    This set does not form a group because closure property is not satisfied.

11. The set of integers under the operation of subtraction is not a group because the associative, identity, and inverse properties are not satisfied. Here are some examples to show why these properties are not satisfied:

    Associative: $(2 - 3) - 1 = -1 - 1 = -2$, but
    $2 - (3 - 1) = 2 - 2 = 0$.

    Identity: $3 - 0 = 3$, but $0 - 3 \neq 3$.

    If there is no identity element, the inverse property cannot be checked.

12. The set of integers under multiplication does not form a group because the inverse property does not hold. For example, the inverse of 3 is 1/3, but 1/3 is not in the set of integers.

13. The closure, associative, and identity properties are satisfied, with 1 as the identity element. However, the inverse property is not satisfied. While -1 and 1 are their own inverses, 0 has no multiplicative inverse, and the multiplicative inverses of the other odd integers are not integers at all. For example, the multiplicative inverse of 3 is 1/3, which is not an element of the set of the system. Therefore, the set of odd integers under multiplication is not a group.

14. The counting numbers are the elements of the set $\{1, 2, 3, 4, \dots\}$. Because zero is not a part of the set, there is no identity element for the operation of addition. Therefore, the inverse property does not hold. This mathematical system is not a group.

15. The closure, associative, and identity properties are all satisfied, with 0 as the identity element. Every rational number p/q has an additive inverse (or "opposite") number $-p/q$ such that
$$\frac{p}{q} + \left(\frac{-p}{q}\right) = \left(\frac{-p}{q}\right) + \frac{p}{q} = 0.$$
For example,
$$\frac{3}{5} + \left(\frac{-3}{5}\right) = \left(\frac{-3}{5}\right) + \frac{3}{5} = 0.$$
Since all of the required properties are satisfied, the set of rational numbers under multiplication is a group.

16. The even integers are the elements of the set $\{\dots, -4, -2, 0, 2, 4, 6, \dots\}$. The sum of two even integers produces an even integer, so closure is satisfied. The associative property holds; an example is $[(-6) + (-2)] + 4 = -4$. At the same time, $(-6) + [(-2) + 4] = -4$. The identity element is 0, and each element has an inverse. For example, -4 and 4 are inverses of each other. Therefore, the set of even integers under addition forms a group.

17. This system is not closed, since the sum of two prime numbers is not always a prime number. For example, $5 + 7 = 12$. The identity property is not satisfied since 0 is not a prime number. Since the identity property is not satisfied, the inverse property cannot be satisfied. Therefore, the set of prime numbers under addition is not a group.

18. The set of nonzero rational numbers contains all the numbers that can be represented on the number line, except for zero. When any two of these numbers are multiplied, the product is always an element of the set. For example,
$$(.5) \times \left(\frac{-1}{4}\right) = (.5) \times (-.25) = -0.125.$$
Closure is satisfied. The associative property also holds. Try adding three different rational numbers; then change the grouping and add again. You should get the same result. The identity element is one. Each number has an inverse, because all nonzero rational numbers have reciprocals. Some examples are: 6 and 1/6, -2.3 and $-1/2.3$, 4/3 and 3/4. All the properties of a group are satisfied.

19. Writing exercise

20. Writing exercise

21. Read the table in the text to see that $RN = S$.

22. Read the table in the text to see that $PR = T$.

23. Read the table in the text to see that $TV = N$.

24. Read the table in the text to see that $VP = S$.

25. Use the table in the text to see that $N(TR) = NP$ or M; also, $(NT)R = VR$ or M.

26. $V(PS) = VV = Q; (VP)S = SS = Q.$

27. $T(VN) = TT = Q; (TV)N = NN = Q.$

28. $S(MR) = SV = M; (SM)R = VR = M.$

29. The identity element is Q. Then the inverse of N is N, because $NN = Q$.

30. The identity element is Q. Since $QQ = Q$, the inverse of Q is Q.

31. The inverse of R is R, because $RR = Q$.

32. The inverse of S is S, because $SS = Q$.

33. The inverse of T is T, because $TT = Q$.

34. The inverse of V is V, because $VV = Q$.

35. The symmetries of a square is not commutative. If the order in which the operation is done is changed, the answer changes. See the table in Example 4 to see that $R \square P = V$, but $P \square R = T$.

36. The subgroup of this system is $\{M, N, P, Q\}$. Refer to the textbook for Example 6. Apply the diagonal line test. Since the part of the table above the diagonal from upper left to lower right is a mirror image of the part below the diagonal, the group is commutative.

37. This group is commutative. For example, $4 + 2 = 6$ and $2 + 4 = 6$.

38. Examine the table for this group. Since $A*B* = D*$, but $B*A* = F*$, this group is not commutative. We may also apply the diagonal line test to see that the group is not commutative.

39. Writing exercise

40. Writing exercise

41. Writing exercise

42. Writing exercise

Chapter 4 Test

1. These are symbols from the ancient Egyptian numeration system. The lotus flower represents 1000; the scroll represents 100; the heelbone represents 10; the stroke represents 1. Therefore, the number is 2536.

2. $(8 \times 1000) + (3 \times 100) + (6 \times 10) + (4 \times 1) = 8364$

3. $(6 \times 10,000) + (0 \times 1000) + (9 \times 100) + (2 \times 10)$
   $\quad + (3 \times 1)$
   $= (6 \times 10^4) + (0 \times 10^3) + (9 \times 10^2) + (2 \times 10^1)$
   $\quad + (3 \times 10^0)$

4. Using the Egyptian method, form two columns, headed by 1 and 54. Keep doubling each row until there are numbers in the first column that add up to 37.

   $$\begin{array}{rrrl} \rightarrow & 1 & 54 & \leftarrow \\ & 2 & 108 & \\ \rightarrow & 4 & 216 & \leftarrow \\ & 8 & 432 & \\ & 16 & 864 & \\ \rightarrow & 32 & 1728 & \leftarrow \end{array}$$

   $1 + 4 + 32 = 37$. Then add corresponding numbers from the second column: $54 + 216 + 1728 = 1998$.

5. 236×94

   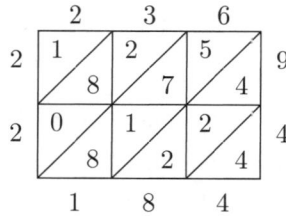

   Read the answer around the left side and bottom: 22,184.

6. $21,325 - 8498$
   Complete missing place value with 0.

   $$\begin{array}{r} 21325 \\ -\ 08498 \end{array}$$

   Replace digits in subtrahend (08498) with the nines complement of each and add.

   $$\begin{array}{r} 21325 \\ +\ 91501 \\ \hline 112826 \end{array}$$

   Delete the first digit on left and add that 1 to the remaining part of the sum: $12,826 + 1 = 12,827$.

7. 424_{five} is equivalent to
   $(4 \times 5^2) + (2 \times 5^1) + (4 \times 5^0)$.
   Then
   $(4 \times 25) + (2 \times 5) + (4 \times 1) = 100 + 10 + 4 = 114$.

8. 100110_{two} is equivalent to
   $(1 \times 2^5) + (0 \times 2^4) + (0 \times 2^3) + (1 \times 2^2) +$
   $(1 \times 2^1) + (0 \times 2^0)$.
   Then $(1 \times 32) + (1 \times 4) + (1 \times 2) = 38$.

9. $A80C_{\text{sixteen}}$ is equivalent to
   $(10 \times 16^3) + (8 \times 16^2) + (0 \times 16^1) \times (12 \times 16^0)$.
   Then $(10 \times 4096) + (8 \times 256) + (12 \times 1)$
   $= 40960 + 2048 + 12$
   $= 43,020$.

10. Using repeated division,

    $$\begin{array}{rll} 2\underline{|58} & & \text{Rem} \\ 2\underline{|29} & \leftarrow & 0 \\ 2\underline{|14} & \leftarrow & 1 \\ 2\underline{|7} & \leftarrow & 0 \\ 2\underline{|3} & \leftarrow & 1 \\ 2\underline{|1} & \leftarrow & 1 \\ 0 & \leftarrow & 1 \end{array}$$

    $58 = 111010_{\text{two}}$.

11. Using repeated division,

    $$\begin{array}{rll} 5\underline{|1846} & & \text{Rem} \\ 5\underline{|369} & \leftarrow & 1 \\ 5\underline{|73} & \leftarrow & 4 \\ 5\underline{|14} & \leftarrow & 3 \\ 5\underline{|2} & \leftarrow & 4 \\ 0 & \leftarrow & 2 \end{array}$$

    $1846 = 24341_{\text{five}}$.

12. Starting at the right, break the digits into groups of three. Then convert the groups to their octal equivalents. (See Table 6.)

    $$\begin{array}{ccc} 10 & 101 & 110 \\ \downarrow & \downarrow & \downarrow \\ 2 & 5 & 6 \end{array}$$

    $10101110_{\text{two}} = 256_{\text{eight}}$

13. Each hexadecimal digit yields a 4-digit binary equivalent. (See Table 6.)

    $$\begin{array}{ccc} B & 5 & 2 \\ \downarrow & \downarrow & \downarrow \\ 1011 & 0101 & 0010 \end{array}$$

    $B52_{\text{sixteen}} = 101101010010_{\text{two}}$

14. The advantage of multiplicative grouping over simple grouping is that there is less repetition of symbols.

15. The advantage, in a positional numeration system, of a smaller base over a larger base, is that there are fewer symbols to learn.

16. The advantage, in a positional numeration system, of a larger base over a smaller base, is that there are fewer digits in the numerals.

17. The identity element is zero for addition of whole numbers.

18. For multiplication of rational numbers, the inverse of 3 is 1/3.

19. $(a + b) + c = (b + a) + c$ illustrates the commutative property of addition. The order of the addends has changed, but not the grouping.

20. Writing exercise

21. (a) There is an identity element.
    (b) o

22. (a) Closure is satisfied.
    (b) Writing exercise

23. (a) The system is not commutative.
    (b) Writing exercise

24. (a) The distributive property is not satisfied.
    (b) Writing exercise

25. (a) The system is not a group.
    (b) Writing exercise

5.1 EXERCISES

1. True. Remember that the natural numbers are also called the counting numbers: 1, 2, 3, 4, ...

2. True. All natural numbers greater than 1 are classified as either prime or composite.

3. False. The even number 2 is a prime number.

4. True. If a number is divisible by 9, it must also be divisible by 3 since 9 is divisible by 3.

5. False. If $n = 5$ it is true that $5|5$ (i.e., 5 divides 5), but it is not true that $10|5$ (i.e., 10 divides 5).

6. False. The smallest prime number is 2. The number 1 is neither prime nor composite.

7. True. Consider an example such as 7. It is a factor of itself because $7 \div 7 = 1$. It is also a multiple of itself because $7 \cdot 1 = 7$.

8. True because 2, 4, and 8 all divide 16 evenly.

9. False. Every composite number has a unique prime factorization.

10. True. The natural number factors are 53 and 1.

11. False. $2^{11} - 1 = 2047$. The factors of 2047 are 23 and 89.

12. True.

13. Remember that all natural number factors are those that divide the given number with remainder 0: 1, 2, 3, 4, 6, 12.

14. 1, 2, 3, 6, 9, 18.

15. 1, 2, 4, 5, 10, 20.

16. 1, 2, 4, 7, 14, 28.

17. 1, 2, 3, 4, 5, 6, 8, 10, 12, 15, 20, 24, 30, 40, 60, and 120.

18. 1, 2, 4, 43, 86, 172.

19. (a) It is not divisible by 2 because it is an odd number.

    (b) It is divisible by 3 because the sum of the digits is 9, a number divisible by 3.

    (c) It is not divisible by 4 because 15, the number formed by the last two digits, is not divisible by 4.

    (d) It is divisible by 5 because the last digit is 5.

    (e) It is not divisible by 6 because although it is divisible by 3, it is not divisible by 2. It must be divisible by both.

    (f) It is not divisible by 8 because the three digits form a number that is not divisible by 8.

    (g) It is divisible by 9 because the sum of the digits is 9.

    (h) It is not divisible by 10 because the last digit is not 0.

    (i) It is not divisible by 12 because, although it is divisible by 3, it is not divisible by 4. It must be divisible by both.

20. (a) It is divisible by 2 because it is an even number.

    (b) It is divisible by 3 because the sum of the digits is 9, a number divisible by 3.

    (c) It is not divisible by 4 because 30, the number formed by the last two digits, is not divisible by 4.

    (d) It is divisible by 5 because the last digit is 0.

    (e) It is divisible by 6 because it is divisible by both 3 and 2.

    (f) It is not divisible by 8 because the three digits form a number that is not divisible by 8.

    (g) It is divisible by 9 because the sum of the digits is 9.

    (h) It is divisible by 10 because the last digit is 0.

    (i) It is not divisible by 12 because, although it is divisible by 3, it is not divisible by 4. It must be divisible by both.

21. (a) It is not divisible by 2 because it is an odd number.

    (b) It is not divisible by 3 because the sum of the digits is 14, which is not divisible by 3.

    (c) It is not divisible by 4 because 25, the number formed by the last two digits, is not divisible by 4.

    (d) It is divisible by 5 because the last digit is 5.

    (e) It is not divisible by 6 because it is not divisible by 3 or 2. It must be divisible by both.

    (f) It is not divisible by 8 because the last three digits form a number that is not divisible by 8.

    (g) It is not divisible by 9 because the sum of the digits is 14, which is not divisible by 9.

    (h) It is not divisible by 10 because the last digit is not 0.

    (i) It is not divisible by 12 because it is not divisible by 3 or 4. It must be divisible by both.

22. (a) It is not divisible by 2 because it is an odd number.

    (b) It is not divisible by 3 because the sum of the digits is 23, which is not divisible by 3.

    (c) It is not divisible by 4 because 15 is not divisible by 4.

    (d) It is divisible by 5 because the last digit is 5.

    (e) It is not divisible by 6 because it is not divisible by 3 or 2. It must be divisible by both.

    (f) It is not divisible by 8 because the last three digits form a number that is not divisible by 8.

    (g) It is not divisible by 9 because the sum of the digits is 23, which is not divisible by 9.

(h) It is not divisible by 10 because the last digit is not 0.

(i) It is not divisible by 12 because it is not divisible by 3 or 4. It must be divisible by both.

23. (a) It is not divisible by 2 because it is odd.

(b) It is divisible by 3 because the sum of the digits is 45, a number divisible by 3.

(c) It is not divisible by 4 because 89, the number formed by the last two digits, is not divisible by 4.

(d) It is not divisible by 5 because it does not end in 5 or 0.

(e) It is not divisible by 6 because, although it is divisible by 3, it is not divisible by 2. It must be divisible by both.

(f) It is not divisible by 8 because the last three digits form a number that is not divisible by 8.

(g) It is divisible by 9 because the sum of the digits is 45, a number divisible by 9.

(h) It is not divisible by 10 because it does not end in 0.

(i) It is not divisible by 12 because, although it is divisible by 3, it is not divisible by 4. It must be divisible by both.

24. (a) It is not divisible by 2 because it is odd.\

(b) It is divisible by 3 because the sum of the digits is 45, a number divisible by 3.

(c) It is not divisible by 4 because 21, the number formed by the last two digits, is not divisible by 4.

(d) It is not divisible by 5 since it does not end in 5 or 0.

(e) It is not divisible by 6 because, although it is divisible by 3, it is not divisible by 2. It must be divisible by both.

(f) It is not divisible by 8 because the last three digits form a number that is not divisible by 8.

(g) It is divisible by 9 because the sum of the digits is 45, a number divisible by 9.

(h) It is not divisible by 10 because it does not end in 0.

(i) It is not divisible by 12 because, although it is divisible by 3, it is not divisible by 4. It must be divisible by both.

25. (a) Writing exercise

(b) The largest prime number whose multiples would have to be considered is 13; the square of 13 is 169, which is less than 200. The next prime number after 13 is 17, but the square of 17 is 289, a number greater than 200.

(c) square root; square root; square root

(d) prime

26. (a) <u>101</u> 102 <u>103</u> 104 105 106 <u>107</u> 108 <u>109</u> 110
111 112 <u>113</u> 114 115 116 117 118 119 120
121 122 <u>123</u> 124 125 126 <u>127</u> 128 129 130
<u>131</u> 132 133 134 135 136 <u>137</u> 138 <u>139</u> 140
141 142 143 144 145 146 147 148 <u>149</u> 150
<u>151</u> 152 153 154 155 156 <u>157</u> 158 159 160
161 162 <u>163</u> 164 165 166 <u>167</u> 168 169 170
171 172 <u>173</u> 174 175 176 177 178 <u>179</u> 180
<u>181</u> 182 183 184 185 186 187 188 189 190
<u>191</u> 192 <u>193</u> 194 195 196 <u>197</u> 198 <u>199</u> 200

There are 21 prime numbers in this table.

(b) $M_7 = 2^7 - 1 = 128 - 1 = 127$. Using divisibility tests, 127 is not divisible by 2 because it is not even. It is not divisible by 3 because the sum of the digits is 10, which is not divisible by 3. It is not divisible by 5 because it does not end in 5. It is not divisible by 7 or 11. We can't stop here because 11^2 is 121 which is less than 127. Checking the next prime, 13, we see that it doesn't divide 127; and 13^2 is 169, which is greater than 127. So we can stop checking larger primes. The number 127 is prime.

27. Two primes that are consecutive natural numbers are 2 and 3; there are no others because one of them would be an even number, which is divisible by 2.

28. Writing exercise

29. The last digit must be zero, because the number must be divisible by 10.

30. The divisibility test for 2 requires that the last single digit be divisible by 2. The divisibility test for 4 requires that the two digits form a number divisible by 4. The divisibility test for 8 requires that the last three digits form a number divisible by 8. By induction, a divisibility test for 16 might be that the last four digits form a number that is divisible by 16. To test the given number: The last four digits $2320 \div 16 = 145$. The number should be divisible by 16.

31.
$$\begin{array}{r|l} 2 & 240 \\ \hline 2 & 120 \\ \hline 2 & 60 \\ \hline 2 & 30 \\ \hline 3 & 15 \\ \hline & 5 \end{array}$$

The prime factorization of 240 is $2^4 \cdot 3 \cdot 5$.

32.
$$\begin{array}{r|l} 2 & 300 \\ \hline 2 & 150 \\ \hline 3 & 75 \\ \hline 5 & 25 \\ \hline & 5 \end{array}$$

The prime factorization is $2^2 \cdot 3 \cdot 5^2$.

33. $\begin{array}{r|r} 2 & 360 \\ 2 & 180 \\ 2 & 90 \\ 3 & 45 \\ 3 & 15 \\ \hline & 5 \end{array}$

The prime factorization is $2^3 \cdot 3^2 \cdot 5$.

34. $\begin{array}{r|r} 5 & 425 \\ 5 & 85 \\ \hline & 17 \end{array}$

The prime factorization is $5^2 \cdot 17$.

35. $\begin{array}{r|r} 3 & 663 \\ 13 & 221 \\ \hline & 17 \end{array}$

The prime factorization is $3 \cdot 13 \cdot 17$.

36. $\begin{array}{r|r} 3 & 885 \\ 5 & 295 \\ \hline & 59 \end{array}$

The prime factorization is $3 \cdot 5 \cdot 59$.

37. To test 142,891 for divisibility by 7:
    (a) and (b)
    $$14,289 - 2 = 14,287$$
    $$1428 - 14 = 1414$$
    $$141 - 8 = 133$$
    $$13 - 6 = 7.$$
    (c) Because 7 is divisible by 7, the given number is also divisible by 7.

38. To test 409,311 for divisibility by 7:
    (a) and (b)
    $$40,931 - 2 = 40,929$$
    $$4092 - 18 = 4074$$
    $$407 - 8 = 399$$
    $$39 - 18 = 21.$$
    (c) Because 21 is divisible by 7, the given number is also divisible by 7.

39. To test 458,485 for divisibility by 7:
    (a) and (b)
    $$45,848 - 10 = 45,838$$
    $$4583 - 16 = 4567$$
    $$456 - 14 = 442$$
    $$44 - 4 = 40.$$
    (c) Because 40 is not divisible by 7, the given number is not divisible by 7.

40. To test 287,824 for divisibility by 7:
    (a) and (b)
    $$28,782 - 8 = 28,774$$
    $$2877 - 8 = 2869$$
    $$286 - 18 = 268$$
    $$26 - 16 = 10.$$
    (c) Because 10 is not divisible by 7, the given number is not divisible by 7.

41. (a) $8 + 9 + 9 + 9 = 35$
    (b) $4 + 3 + 6 = 13$
    (c) $35 - 13 = 22$
    (d) Because 22 is divisible by 11, the given number is also divisible by 11.

42. (a) $8 + 7 + 6 + 9 + 2 = 32$
    (b) $4 + 6 + 7 + 4 = 21$
    (c) $32 - 21 = 11$
    (d) Because 11 is divisible by 11, the given number is also divisible by 11.

43. (a) $4 + 3 + 9 + 2 + 8 = 26$
    (b) $5 + 8 + 6 + 4 = 23$
    (c) $26 - 23 = 3$
    (d) Because 3 is not divisible by 11, the given number is not divisible by 11

44. (a) $5 + 2 + 4 + 9 + 3 = 23$
    (b) $5 + 7 + 9 + 1 = 22$
    (c) $23 - 22 = 1$
    (d) Because 1 is not divisible by 11, the given number is not divisible by 11.

45. Based on the divisibility test for 6, which says that the number must be divisible by both 2 and 3, the divisibility test for 15 is that the number must be divisible by both 3 and 5. That is, the sum of the digits must be divisible by 3 and the last digit must be 5 or 0.

46. Writing exercise

47. 0, 2, 4, 6, or 8 because they are all the even single digit numbers.

48. The sum of the digits must be a number that is divisible by 3. Because $2 + 4 + 5 + 7 + 6 + 5 = 29$, the remaining digit must be either 1, 4, or 7.

49. The last two digits must form a number that is divisible by 4. The possible values for x are 0, 4, and 8.

50. For this number to be divisible by 5, x must be 0 or 5.

51. The number is divisible by 6 if it is divisible by both 2 and 3. Both 985,230 and 985,236 are divisible by 2 and 3. Then $x = 0$ or 6.

52. For this number to be divisible by 8, the three-digit $24x$ must be divisible by 8. The only possible such numbers in which x is a single digit are

$$240 = 8 \cdot 30 \text{ and } 248 = 8 \cdot 31$$

Thus, 0 and 8 are the only possible replacements for x.

53. The sum of the existing digits is 30. The overall sum must be divisible by 9, so x must be 6.

54. To be divisible by 10, a number must end in 0. This number ends in 0 if any digit replaces x.

55. Prime factorization: $36 = 3^2 \cdot 2^2$.

    Add one to each exponent: $\quad 2 + 1 = 3$
    $\qquad\qquad\qquad\qquad\qquad\quad 2 + 1 = 3.$

    Multiply: $3 \cdot 3 = 9$. There are 9 divisors of 36.

56. Prime factorization: $48 = 2^4 \cdot 3^1$.

    Add one to each exponent: $\quad 4 + 1 = 5$
    $\qquad\qquad\qquad\qquad\qquad\quad 1 + 1 = 2.$

    Multiply: $5 \cdot 2 = 10$. There are 10 divisors of 48.

57. Prime factorization: $72 = 2^3 \cdot 3^2$.

    Add one to each exponent: $\quad 3 + 1 = 4$
    $\qquad\qquad\qquad\qquad\qquad\quad 2 + 1 = 3.$

    Multiply: $4 \cdot 3 = 12$. There are 12 divisors of 72.

58. Prime factorization: $144 = 2^4 \cdot 3^2$.

    Add one to each exponent: $\quad 4 + 1 = 5$
    $\qquad\qquad\qquad\qquad\qquad\quad 2 + 1 = 3.$

    Multiply: $5 \cdot 3 = 15$. There are 15 divisors of 144.

59. Prime factorization: $2^8 \cdot 3^2$.

    Add one to each exponent: $\quad 8 + 1 = 9$
    $\qquad\qquad\qquad\qquad\qquad\quad 2 + 1 = 3.$

    Multiply: $9 \cdot 3 = 27$. There are 27 divisors of $2^8 \cdot 3^2$.

60. Prime factorization: $2^4 \cdot 3^4 \cdot 5^2$.

    Add one to each exponent: $\quad 4 + 1 = 5$
    $\qquad\qquad\qquad\qquad\qquad\quad 4 + 1 = 5$
    $\qquad\qquad\qquad\qquad\qquad\quad 2 + 1 = 3.$

    Multiply: $5 \cdot 5 \cdot 3 = 75$. There are 75 divisors of this number.

61. It is divisible by 4 and does not end in two zeros, so it is a leap year.

62. It is divisible by 4 and does not end in two zeros, so it is a leap year.

63. It is not divisible by 4, so it is not a leap year.

64. It is not divisible by 4, so it is not a leap year.

65. Because 2400 is divisible by 4 and 400, it is a leap year.

66. Although 1800 is divisible by 4, it is not divisible by 400; therefore, it is not a leap year.

67. Writing exercise

68. Three consecutive natural numbers: 1, 2, 3.
    Multiply: $1 \cdot 2 \cdot 3 = 6$.
    Divide by 6: $6 \div 6 = 1$.
    Three consecutive natural numbers: 2, 3, 4.
    Multiply: $2 \cdot 3 \cdot 4 = 24$.
    Divide by 6: $24 \div 6 = 4$.
    Three consecutive natural numbers: 3, 4, 5.
    Multiply: $3 \cdot 4 \cdot 5 = 60$.
    Divide by 6: $60 \div 6 = 10$.
    Three consecutive natural numbers: 20, 21, 22.
    Multiply: $20 \cdot 21 \cdot 22 = 9240$.
    Divide by 6: $9240 \div 6 = 1540$.

    Conjecture: The product of any three consecutive natural numbers is divisible by 6.

69. Writing exercise

70. Choose 921,921.
    Divide: $921,921 \div 13 = 70,917$.
    Divide: $921,921 \div 11 = 83,811$.
    Divide: $921,921 \div 7 = 131,703$.
    The number is divisible by 13, 11, and 7. Thus, the number must be divisible by the product of these numbers: $13 \cdot 11 \cdot 7 = 1001$.
    Note that $921,921 \div 1001 = 921$, the number formed by the first three digits. Any 3-digit number multiplied by 1001 produces a 6-digit number consisting of three digits followed by the same three digits. Therefore, the 6-digit number must be divisible by 13, 11, and 7.

71. $41^2 - 41 + 41 = 41^2$, which is not a prime.

72. (a) $42^2 - 42 + 41 = 1764 - 42 + 41 = 1763$

    (b) $43^2 - 43 + 41 = 1849 - 43 + 41 = 1847$

73. For $n > 41$, Euler's formula produces a prime sometimes. In Exercise 72, Euler's formula produces 1763 for $n = 42$. The square root of 1763 is approximately 41.99, so all of the prime numbers less than this number must be tested. If 41 is tested as a divisor of 1763, the quotient is 43. Therefore, the number is not prime. On the other hand, Euler's formula produces 1847 for $n = 43$. The square root of 1847 is approximately 42.98. If all the prime numbers less than this number are tested, none divides 1847 evenly. Therefore, the number is prime.

74. $81^2 - 79 \cdot 81 + 1601 = 1763$
    $82^2 - 79 \cdot 82 + 1601 = 1847$

    See Exercise 73 for the explanation that 1763 is not prime because $41 \cdot 43 = 1763$. The same exercise also shows that 1847 is prime. Therefore, for $n > 80$, Escott's formula produces a prime sometimes.

75. (a) $2^{2^4} + 1 = 2^{16} + 1 = 65,536 + 1 = 65,537$

    (b) The square root of 65,537 is approximately 256. All primes less than 256 must be checked as possible factors of this number; 251 is then the largest potential prime factor.

76. (a) $2^{2^5} + 1 = 2^{32} + 1$
    $= 4,294,967,296 + 1$
    $= 4,294,967,297$

    (b) $4,294,967,297 \div 641 = 6,700,417$

77. Writing exercise

78. Writing exercise

79. Writing exercise

80. $M = 2 \cdot 3 \cdot 5 \cdot 7 \cdot 11 \cdot 13 + 1 = 30,031$
    This number M is composite because it can be factored into $59 \cdot 509$.

81. $M_6 = 2^6 - 1 = 64 - 1 = 63$

82. $2^3 - 1 = 8 - 1 = 7$
    From exercise 81, see that $M_6 = 63$; 7 is a factor of 63.

83. $2^p - 1$

84. $M_{10} = 2^{10} - 1 = 1024 - 1 = 1023$

85. The two prime factors of 10 are 2 and 5.
    $$2^2 - 1 = 4 - 1 = 3$$
    $$2^5 - 1 = 32 - 1 = 31$$
    Two distinct factors of M_{10} are 3 and 31.

86. Writing exercise

5.2 EXERCISES

1. True

2. This statement is false because 2 and 3 do not differ by 2.

3. True

4. True. Euclid showed that if the number $2^n - 1$ is prime then the number $2^{n-1}(2^n - 1)$ is a perfect number.

5. True. By definition, the only factors of a prime number are the number itself and 1. The proper divisors of a number are all the divisors except the number itself. For a prime number, then, the only proper divisor is 1. That means that any prime number must be deficient.

6. The statement is false. Goldbach's conjecture states that every even number greater than 2 can be written as the sum of two primes; however, 51 is not a prime number because it can be factored into $3 \cdot 17$.

7. False. For every new Mersenne prime, there is a perfect number.

8. False. $16^2 - 15^2 = 256 - 225 = 31$

9. True. Since $(2^7 - 1) = 128 - 1 = 127$ is a prime number, then $2^6(2^7 - 1) = 8128$ is a perfect number by Euclid. (See answer to Exercise 4)

10. True because any natural number greater than 1 must either have the sum of its proper divisors either equal to, less than, or greater than the number itself.

11. $1 + 2 + 4 + 8 + 16 + 31 + 62 + 124 + 248 = 496$

12. $1 + 2 + 4 + 8 + 16 + 32 + 64 + 127 + 254 + 508$
    $+ 1016 + 2032 + 4064 = 8128$

13. $2^{13} - 1 = 8191$, which is prime.
    $2^{13-1}(2^{13} - 1) = 4096(8192 - 1) = 33,550,336$
    This number is even and ends in 6.

14. $2^{13,446,916}(2^{13,446,917} - 1)$

15. The divisors of 6 are 1, 2, 3, and 6. The sum of the reciprocals of these numbers is
    $$\frac{1}{1} + \frac{1}{2} + \frac{1}{3} + \frac{1}{6}$$
    $$= \frac{6}{6} + \frac{3}{6} + \frac{2}{6} + \frac{1}{6}$$
    $$= \frac{12}{6}$$
    $$= 2.$$

16. $496 = 1 + 2 + 3 + 4 + 5 + 6 + 7 + 8 + 9 + 10 + 11$
    $+ 12 + 13 + 14 + 15 + 16 + 17 + 18 + 19$
    $+ 20 + 21 + 22 + 23 + 24 + 25 + 26 + 27$
    $+ 28 + 29 + 30 + 31$

17. The proper divisors of 36 are 1, 2, 3, 4, 6, 9, 12, and 18. The sum of these numbers is 55. Because the sum is greater than 36, the number is abundant.

18. The proper divisors of 30 are 1, 2, 3, 5, 6, 10 and 15. The sum of these numbers is 42. Because the sum is greater than 30, the number is abundant.

19. The proper divisors of 75 are 1, 3, 5, 15, and 25. The sum of these numbers is 49. Because the sum is less than 75, the number is deficient.

20. The proper divisors of 95 are 1, 5, and 19. The sum of these numbers is 25. Because the sum is less than 95, the number is deficient.

21. Examine the Sieve of Eratosthenes in Table 5.1 to see that the prime numbers 2, 3, 5, 7, 11, 13, 17, 19, and 23 can be deleted from the search. Examine the remaining numbers: 4 is deficient because $1 + 2 = 3$; 6 is perfect because $1 + 2 + 3 = 6$; 8 is deficient because $1 + 2 + 4 = 7$; 9 is deficient because $1 + 3 = 4$; 10 is deficient because $1 + 2 + 5 = 8$. The number 12 is abundant because the sum of its proper divisors is $1 + 2 + 3 + 4 + 6 = 16$. The numbers 14, 15, and 16 are deficient. Verify this by adding their proper divisors.

The number 18 is the next abundant number; the sum of its proper divisors is 21. Then 20 is the third abundant number; the sum of its proper divisors is 22. The numbers 21 and 22 are both deficient. Verify this. Finally the fourth abundant number is 24; the sum of its proper divisors is 36.

22. Writing exercise
23. The sum of the proper divisors is 975. You can verify this with a calculator. Therefore, the number is abundant.
24. Writing exercise
25. The sum of the proper divisors of 1,184 is 1,210. The sum of the proper divisors of 1,210 is 1,184. Thus, they are amicable.
26. (a) When $n = 2$,
$$x = 3 \cdot 2^{2-1} - 1 \quad = 3 \cdot 2 - 1 \quad = 5$$
$$y = 3 \cdot 2^2 - 1 \quad = 3 \cdot 4 - 1 \quad = 11$$
$$z = 9 \cdot 2^{2 \cdot 2 - 1} - 1 \quad = 9 \cdot 8 - 1 \quad = 71.$$
Then
$$2^n xy = 2^2 \cdot 5 \cdot 11 \quad = 4 \cdot 5 \cdot 11 \quad = 220$$
$$2^n z = 2^2 \cdot 71 \quad = 4 \cdot 71 \quad = 284.$$

(b) When $n = 4$,
$$x = 3 \cdot 2^{4-1} - 1 \quad = 3 \cdot 8 - 1 \quad = 23$$
$$y = 3 \cdot 2^4 - 1 \quad = 3 \cdot 16 - 1 \quad = 47$$
$$z = 9 \cdot 2^{2 \cdot 4 - 1} - 1 \quad = 9 \cdot 128 - 1 \quad = 1151.$$
Then
$$2^n xy = 2^4 \cdot 23 \cdot 47 \quad = 16 \cdot 23 \cdot 47 \quad = 17,296$$
$$2^n z = 2^4 \cdot 1151 \quad = 16 \cdot 1151 \quad = 18,416.$$

27. $3 + 11, 7 + 7$
28. $11 + 11, 3 + 19, 5 + 17$
29. $3 + 23, 7 + 19, 13 + 13$
30. $3 + 29, 13 + 19$
31. If $a = 5$ and $b = 3$, then $5 + 2 \cdot 3 = 11$.
32. $6 = 11 - 5; \; 12 = 17 - 5; \; 18 = 23 - 5$
33. Check the Sieve of Eratosthenes (Table 1) in Section 5.1, to find 71 and 73. These numbers differ by 2.
34. Examine the Sieve of Eratosthenes to find 101 and 103.
35. Examine the Sieve of Eratosthenes to find 137 and 139.
36. (a) $a^{p-1} - 1 = 3^{5-1} - 1$
$$= 3^4 - 1$$
$$= 81 - 1$$
$$= 80$$

Then, $80 \div 5 = 16$.

(b) $a^{p-1} - 1 = 2^{7-1} - 1$
$$= 2^6 - 1$$
$$= 64 - 1$$
$$= 63$$

Then, $63 \div 7 = 9$.

37. (a) $5 = 9 - 4$ or $3^2 - 2^2$.
    (b) $11 = 36 - 25$ or $6^2 - 5^2$.
38. (a) When $k = 1$,
$$4k + 1 = 4 \cdot 1 + 1$$
$$= 5.$$
Then
$$2^2 + 1^2 = 5.$$

(b) When $k = 3$,
$$4k + 1 = 4 \cdot 3 + 1$$
$$= 13.$$
Then
$$2^2 + 3^2 = 13.$$

39. $5^2 + 2 = 3^3; \; 25 + 2 = 27$
40. For $a = 2$ and $b = 2$,
$$a^2 + 4 = b^3$$
$$2^2 + 4 = 2^3$$
$$4 + 4 = 8.$$
For $a = 11$ and $b = 5$,
$$a^2 + 4 = b^3$$
$$11^2 + 4 = 5^3$$
$$121 + 4 = 125.$$

41. Examine the numbers.
42. No; The fifth contains 8 digits and the sixth contains 10 digits.
43. No; for the first six, the sequence is 6, 8, 6, 8, 6, 6.
44. Writing exercise
45. Writing exercise
46. (a) For $p = 7, \; 2 \cdot 3 \cdot 5 \cdot 7 \pm 1 = 209$ or 211.
    (b) The only prime number is 211; $209 = 11 \cdot 19$.
47. B; Sometimes

Reference table for Exercises 48–53

	p	$2p+1$	Sophie Germaine prime?
48.	2	5	yes
49.	3	7	yes
50.	5	11	yes
51.	7	15	no
52.	11	23	yes
53.	13	27	no

48. For $p = 2$, $2p + 1 = 5$, a Sophie Germaine prime.
49. For $p = 3$, $2p + 1 = 7$, a Sophie Germaine prime.
50. For $p = 5$, $2p + 1 = 11$, a Sophie Germaine prime.
51. For $p = 7$, $2p + 1 = 15$, not a Sophie Germaine prime.
52. For $p = 11$, $2p + 1 = 23$, a Sophie Germaine prime.
53. For $p = 13$, $2p + 1 = 27$, not a Sophie Germaine prime.

54. – 56.

	n	$n!$	$n!-1$	$n!+1$	$n!-1$ prime?	$n!+1$ prime?
	2	2	1	3	no	yes
54.	3	6	5	7	yes	yes
55.	4	24	23	25	yes	no
56.	5	120	119	121	no	no

To calculate the underlined numerical values in the table, recall the meaning of n-factorial:
$$n! = n(n-1)(n-2)\ldots(3)(2)(1).$$

n	$n!$	$n!-1$	$n!+1$
3	$3 \cdot 2 \cdot 1 = 6$	5	7
4	$4 \cdot 3 \cdot 2 \cdot 1 = 24$	23	25
5	$5 \cdot 4 \cdot 3 \cdot 2 \cdot 1 = 120$	119	121

The numbers 5 and 7 are both prime. Although 23 is prime, the number 25 is not as it can be written as $5 \cdot 5$. The number 119 can be written as $7 \cdot 17$, and 121 can be written as $11 \cdot 11$.

57. Writing exercise
58. B; Sometimes
59. B; Sometimes
60. B; Sometimes

5.3 EXERCISES

1. True. They would have a common factor of at least 2. Two numbers that are relatively prime have a greatest common factor of 1.

2. True. Two numbers are relatively prime if they are prime relative to each other; that is, their greatest common factor is 1. Since a prime number has only 1 and itself as factors, two different prime numbers must have 1 as their greatest common factor.

3. True. Consider the prime number 5; $5^2 = 25$. The greatest common factor of 5 and 25 is 5.

4. False. Since p is prime, the prime factorizations of p and p^2 are:
$$p = p^1$$
$$p^2 = p^2.$$
Since 2 is the largest exponent of powers indicated in the prime factorizations, p^2 is the least common multiple of p and p^2.

5. False. The variable $p = 2$ proves this statement false.

6. False. Consider an example. Choose 6 and 9 as the natural numbers. The set of multiples of 6 is
$$\{6, 12, 18, 24, 30, 36, \ldots\}$$
while the set of multiples of 9 is
$$\{9, 18, 27, 36, 45, 54, \ldots\}.$$
The set of common multiples of 6 and 9 is
$$\{18, 36, 54, \ldots\},$$
which is precisely the set of multiples of 18, the least common multiple of 6 and 9. Since this set is infinite, the given statement is false.

7. True. All natural numbers have 1 as a common factor.

8. True. The prime factorizations of primes p and q are
$$p = p^1$$
$$q = q^1.$$
Forming the products of the primes that belong to any factorization, we have the least common multiple pq.

9. True. Consider some examples: 25 and 9, 14 and 33, 16 and 39, etc.

10. True. Since the set of factors in the prime factorization of any natural number is finite, the set of common factors of two given natural numbers will also be finite.

11. To find the greatest common factor by the prime factors method, first write the prime factorization of each number:
$$70 = 2 \cdot 5 \cdot 7$$
$$120 = 2^3 \cdot 3 \cdot 5$$
Now find the primes common to all factorizations, with each prime raised to the smallest exponent from either factorization: 2, 5.
Finally, form the product of these numbers:
$$2 \cdot 5 = 10.$$

12. To find the greatest common factor by the prime factors method, first write the prime factorization of each number:
$$180 = 2^2 \cdot 3^2 \cdot 5$$
$$300 = 2^2 \cdot 3 \cdot 5^2.$$

    Now find the primes common to all factorizations, with each prime raised to the smallest exponent from either factorization: 2^2, 3, 5.

    Finally, form the product of these numbers:
    $$2^2 \cdot 3 \cdot 5 = 60.$$

13. $480 = 2^5 \cdot 3 \cdot 5$
    $1800 = 2^3 \cdot 3^2 \cdot 5^2$

    Now find the primes common to all factorizations, with each prime raised to the smallest exponent from either factorization: 2^3, 3, 5.

    Finally, form the product of these numbers:
    $$2^3 \cdot 3 \cdot 5 = 120.$$

14. $168 = 2^3 \cdot 3 \cdot 7$
    $504 = 2^3 \cdot 3^2 \cdot 7$

    Now find the primes common to all factorizations, with each prime raised to the smallest exponent from either factorization: 2^3, 3, 7.

    Finally, form the product of these numbers:
    $$2^3 \cdot 3 \cdot 7 = 168.$$

15. $28 = 2^2 \cdot 7$
    $35 = 5 \cdot 7$
    $56 = 2^3 \cdot 7$

    Now find the primes common to all factorizations, with each prime raised to the smallest exponent from all factorizations: 7. The only factor common to all three numbers is 7.

16. $252 = 2^2 \cdot 3^2 \cdot 7$
    $308 = 2^2 \cdot 7 \cdot 11$
    $504 = 2^3 \cdot 3^2 \cdot 7$

    Now find the primes common to all factorizations, with each prime raised to the smallest exponent from all factorizations: 2^2, 7.
    $$2^2 \cdot 7 = 28$$

17. To find the greatest common factor of 60 and 84 by the method of dividing by prime factors, first write the numbers in a row and divide by their smallest common prime factor, which is 2.

    $$\begin{array}{r|rr} 2 & 60 & 84 \\ \hline & 30 & 42 \end{array}$$

    Now divide 30 and 42 by their next largest prime factor, which is 2.

    $$\begin{array}{r|rr} 2 & 60 & 84 \\ 2 & 30 & 42 \\ \hline & 15 & 21 \end{array}$$

    Divide again by the next largest prime factor, which is 3.

    $$\begin{array}{r|rr} 2 & 60 & 84 \\ 2 & 30 & 42 \\ 3 & 15 & 21 \\ \hline & 5 & 7 \end{array}$$

    Since 5 and 7 have no common prime factors, there are no more divisions. To find the greatest common factor, find the product of the primes on the left, 2, 3 and 3.
    $$2 \cdot 2 \cdot 3 = 12$$

18. To find the greatest common factor of 130 and 455 by the method of dividing by prime factors, first write the numbers in a row and divide by their smallest common prime factor, which is 5.

    $$\begin{array}{r|rr} 5 & 130 & 455 \\ \hline & 26 & 91 \end{array}$$

    Now divide 26 and 91 by their next largest prime factor, which is 13.

    $$\begin{array}{r|rr} 5 & 130 & 455 \\ 13 & 26 & 91 \\ \hline & 2 & 7 \end{array}$$

    Since 2 and 7 have no common prime factors, there are no more divisions. To find the greatest common factor, find the product of the primes on the left, 5 and 13.
    $$5 \cdot 13 = 65$$

19. To find the greatest common factor of 310 and 460 by the method of dividing by prime factors, first write the numbers in a row and divide by their smallest common prime factor, which is 2.

    $$\begin{array}{r|rr} 2 & 310 & 460 \\ \hline & 155 & 230 \end{array}$$

    Now divide 155 and 230 by their next largest prime factor, which is 5.

    $$\begin{array}{r|rr} 2 & 310 & 460 \\ 5 & 155 & 230 \\ \hline & 31 & 46 \end{array}$$

    Since 31 and 46 have no common prime factors, there are no more divisions. To find the greatest common factor, find the product of the primes on the left, 2 and 5.
    $$2 \cdot 5 = 10$$

20. 234 and 470

$$\begin{array}{r|rr} 2 & 234 & 470 \\ \hline & 117 & 235 \end{array}$$

Since 117 and 235 have no common prime factors ($117 = 3^2 \cdot 13$ and $235 = 5 \cdot 47$), the greatest common factor of 234 and 470 is 2.

21. 12, 18, and 30

$$\begin{array}{r|rrr} 2 & 12 & 18 & 30 \\ 3 & 6 & 9 & 15 \\ \hline & 2 & 3 & 5 \end{array}$$

Since 2, 3, and 5 have no common prime factors, there are no more divisions. Find the product of the primes on the left.

$$2 \cdot 3 = 6$$

22. 450, 1500, and 432

$$\begin{array}{r|rrr} 2 & 450 & 1500 & 432 \\ 3 & 225 & 750 & 216 \\ \hline & 75 & 250 & 72 \end{array}$$

Since 75, 250, and 72 have no common prime factors ($75 = 3 \cdot 5^2$; $250 = 2 \cdot 5^3$; $72 = 2^3 \cdot 3^2$), there are no more divisions. Find the product of the primes on the left.

$$2 \cdot 3 = 6$$

23.

	Remainder
$60 \div 36 = 1$	24
$36 \div 24 = 1$	12
$24 \div 12 = 2$	0

Then 12, the last positive remainder, is the greatest common factor.

24.

	Remainder
$70 \div 25 = 2$	20
$25 \div 20 = 1$	5
$20 \div 5 = 4$	0

Then 5, the last positive remainder, is the greatest common factor.

25.

	Remainder
$180 \div 84 = 2$	12
$84 \div 12 = 7$	0

Then 12, the last positive remainder, is the greatest common factor.

26.

	Remainder
$120 \div 72 = 1$	48
$72 \div 48 = 1$	24
$48 \div 24 = 2$	0

Then 24, the last positive remainder, is the greatest common factor.

27.

	Remainder
$560 \div 210 = 2$	140
$210 \div 140 = 1$	70
$140 \div 70 = 2$	0

Then 70, the last positive remainder, is the greatest common factor.

28.

	Remainder
$480 \div 150 = 3$	30
$150 \div 30 = 5$	0

Then 30, the last positive remainder, is the greatest common factor.

29. Writing exercise

30. Writing exercise

31. To find the least common multiple by the prime factors method, first write the prime factorization of each number.

$$24 = 2^3 \cdot 3$$
$$30 = 2 \cdot 3 \cdot 5$$

Choose all primes belonging to any factorization, with each prime raised to the largest exponent that it has in any factorization: 2^3, 3, 5. Finally, form the product of these numbers:

$$2^3 \cdot 3 \cdot 5 = 120.$$

The least common multiple of 24 and 30 is 120.

32. To find the least common multiple by the prime factors method, first write the prime factorization of each number.

$$12 = 2^2 \cdot 3$$
$$32 = 2^5$$

Choose all primes belonging to any factorization, with each prime raised to the largest exponent that it has in any factorization: 2^5, 3. Finally, form the product of these numbers:

$$2^5 \cdot 3 = 96.$$

The least common multiple of 12 and 32 is 96.

33. Prime factorizations:

$$56 = 2^3 \cdot 7$$
$$96 = 2^5 \cdot 3.$$

$$\text{Least common multiple} = 2^5 \cdot 3 \cdot 7$$
$$= 672.$$

5.3 GREATEST COMMON FACTOR AND LEAST COMMON MULTIPLE

34. Prime factorizations:
$$28 = 2^2 \cdot 7$$
$$70 = 2 \cdot 5 \cdot 7.$$
Least common multiple $= 2^2 \cdot 5 \cdot 7$
$$= 140.$$

35. Prime factorizations:
$$30 = 2 \cdot 3 \cdot 5$$
$$40 = 2^3 \cdot 5$$
$$70 = 2 \cdot 5 \cdot 7.$$
Least common multiple $= 2^3 \cdot 3 \cdot 5 \cdot 7$
$$= 840.$$

36. Prime factorizations:
$$24 = 2^3 \cdot 3$$
$$36 = 2^2 \cdot 3^2$$
$$48 = 2^4 \cdot 3.$$
Least common multiple $= 2^4 \cdot 3^2$
$$= 144.$$

37.

2	24	32
2	12	16
2	6	8
	3	4

Therefore, the product $2 \cdot 2 \cdot 2 \cdot 3 \cdot 4 = 96$ is the LCM.

38.

7	35	56
	5	8

Therefore, the product $7 \cdot 5 \cdot 8 = 280$ is the LCM.

39.

5	45	75
3	9	15
	3	5

Therefore, the product $5 \cdot 3 \cdot 3 \cdot 5 = 225$ is the LCM.

40.

2	48	54	60
3	24	27	30
2	8	9	10
	4	9	5

Therefore, the product $2 \cdot 3 \cdot 2 \cdot 4 \cdot 9 \cdot 5 = 2160$ is the LCM.

41.

2	16	120	216
2	8	60	108
2	4	30	54
3	2	15	27
	2	5	9

Therefore, the product $2 \cdot 2 \cdot 2 \cdot 3 \cdot 2 \cdot 5 \cdot 9 = 2160$ is the LCM.

42.

5	210	385	2310
3	42	77	462
2	14	77	154
7	7	77	77
11	1	11	11
	1	1	1

Therefore, the product $5 \cdot 3 \cdot 2 \cdot 7 \cdot 11 = 2310$ is the LCM.

43. Use the result of Exercise 23 and the following formula.
Least common multiple of m and n
$$= \frac{m \cdot n}{\text{Greatest common factor of } m \text{ and } n}.$$
Least common multiple of 36 and 60 $= \dfrac{36 \cdot 60}{12}$
$$= 180.$$

44. Use the result of Exercise 24 and the following formula.
Least common multiple of m and n
$$= \frac{m \cdot n}{\text{Greatest common factor of } m \text{ and } n}.$$
Least common multiple of 25 and 70 $= \dfrac{25 \cdot 70}{5}$
$$= 350.$$

45. Use the result of Exercise 25 and the following formula.
Least common multiple of m and n
$$= \frac{m \cdot n}{\text{Greatest common factor of } m \text{ and } n}.$$
Least common multiple of 84 and 180 $= \dfrac{84 \cdot 180}{12}$
$$= 1260.$$

46. Use the result of Exercise 26 and the following formula.
Least common multiple of m and n
$$= \frac{m \cdot n}{\text{Greatest common factor of } m \text{ and } n}.$$
Least common multiple of 72 and 120 $= \dfrac{72 \cdot 120}{24}$
$$= 360.$$

47. Use the result of Exercise 27 and the following formula.
Least common multiple of m and n
$$= \frac{m \cdot n}{\text{Greatest common factor of } m \text{ and } n}.$$
Least common multiple of 210 and 560 $= \dfrac{210 \cdot 560}{70}$
$$= 1680.$$

48. Use the result of Exercise 28 and the following formula.

    Least common multiple of m and n
    $$= \frac{m \cdot n}{\text{Greatest common factor of } m \text{ and } n}.$$

    Least common multiple of 150 and 480 $= \dfrac{150 \cdot 480}{30}$
    $= 2400.$

49. In creating the GCF choose the smallest exponent on common factors. Note that $a > b > c$. Choose p^b (since $a > b$), q^c (since $a > c$), and r^c (since $b > c$). The GCF is $p^b q^c r^c$.

    (b) In creating the LCM choose the largest exponent on common factors. Since $a > b > c$. Choose p^a (since $a > b$), q^a (since $a > c$), and r^b (since $b > c$). The LCM is $p^a q^a r^b$.

50. (a) Find the primes common to each factorization, with each prime having as exponent the smallest exponent from either factorization.

    Greatest common factor $= 2^{31} \cdot 5^{17} \cdot 7^{13}$.

    (b) Form the product of all the primes that occur in either factorization, with each prime having as exponent the largest exponent from either factorization.

    Least common multiple $= 2^{34} \cdot 5^{22} \cdot 7^{21}$.

51. 150, 210, and 240
    1. $210 \div 150 = 1$ with a remainder of 60.
    2. $150 \div 60 = 2$ with a remainder of 30.
    3. $60 \div 30 = 2$ with remainder of 0.
    4. Then 30, the last positive remainder, is the greatest common factor of 150 and 210.
    5. $240 \div 30 = 8$ with a remainder of 0.
    6. Then 30 is the greatest common factor of the three given numbers.

52. 12, 75, and 120
    1. $75 \div 12 = 6$ with a remainder of 3.
    2. $12 \div 3 = 4$ with a remainder of 0.
    3. Then 3, the last positive remainder, is the greatest common factor of 75 and 12.
    4. $120 \div 3 = 40$ with a remainder of 0.
    5. Then 3 is the greatest common factor of the three given numbers.

53. 90, 105, and 315
    1. $105 \div 90 = 1$ with a remainder of 15.
    2. $90 \div 15 = 6$ with a remainder of 0.
    3. Then 15, the last positive remainder, is the greatest common factor of 90 and 105.
    4. $315 \div 15 = 21$ with a remainder of 0.
    5. Then 15 is the greatest common factor of the three given numbers.

54. 48, 315, and 450
    1. $315 \div 48 = 6$ with a remainder of 27.
    2. $48 \div 27 = 1$ with a remainder of 21.
    3. $27 \div 21 = 1$ with a remainder of 6.
    4. $21 \div 6 = 3$ with a remainder of 3
    5. $6 \div 3 = 2$ with a remainder of 0.
    6. Then 3, the last positive remainder, is the greatest common factor of 315 and 48.
    7. $450 \div 3 = 150$ with a remainder of 0.
    8. Then 3 is the greatest common factor of the three given numbers.

55. 144, 180, and 192
    1. $180 \div 144 = 1$ with a remainder of 36.
    2. $144 \div 36 = 4$ with a remainder of 0.
    3. Then 36, the last positive remainder, is the greatest common factor of 144 and 180.
    4. $192 \div 36 = 5$ with a remainder of 12.
    5. $36 \div 12 = 3$ with a remainder of 0.
    6. Then 12 is the greatest common factor of the three given numbers.

56. 180, 210, and 630
    1. $210 \div 180 = 1$ with a remainder of 30.
    2. $180 \div 30 = 6$ with a remainder of 0.
    3. Then 30, the last positive remainder, is the greatest common factor of 180 and 210.
    4. $630 \div 30 = 21$ with a remainder of 0.
    5. Then 30 is the greatest common factor of the three given numbers.

57. (a) The GCF is found in the intersection: $2 \cdot 3 = 6$.

    (b) The LCM is found in the union: $2^2 \cdot 3^2 = 36$.

58. (a) The GCF is found in the intersection: $3^2 = 9$.

    (b) The LCM is found in the union: $2^2 \cdot 3^3 = 108$.

59. (a) The GCF is found in the intersection: $2 \cdot 3^2 = 18$.

    (b) The LCM is found in the union: $2^3 \cdot 3^3 = 216$.

60. The number p is a factor of q, or q is a multiple of p.

61. The numbers p and q must be relatively prime.

62. The number p is a factor of q, or q is a multiple of p. Consider an example. If $p = 3$ and $q = 21$, the greatest common factor of p and q is 3, or p. We see that the greatest common factor of p and q is p whenever p is a factor of q, or, equivalently, q is a multiple of p.

63. Writing exercise

64. Writing exercise

65. Find the least common multiple of 16 and 36. $16 = 2^4$; $36 = 2^2 \cdot 3^2$. The LCM is $2^4 \cdot 3^2 = 144$. The 144th calculator is the first that they will both inspect.

66. The question may be restated as "What is the smallest number of nights it will take for Paul and Cindy to be off together?" We must find the least common multiple of 6 and 10.
$$6 = 2 \cdot 3$$
$$10 = 2 \cdot 5$$
Least common multiple $= 2 \cdot 3 \cdot 5 = 30$. The next night they will be off together is 30 nights after July 1, or July 31.

67. Find the greatest common factor of 240 and 288.
$240 = 2^4 \cdot 3 \cdot 5$; $288 = 2^5 \cdot 3^2$.
The GCF is $2^4 \cdot 3 = 48$.

68. The question may be restated as "What is the smallest number of seconds it will take for Kathryn and Tami to reach the starting point simultaneously?" We must find the least common multiple of 40 and 45.
$$40 = 2^3 \cdot 5$$
$$45 = 3^2 \cdot 5$$
Least common multiple $= 2^3 \cdot 3^2 \cdot 5 = 360$. They will reach the starting point again in 360 seconds.

69. To answer the first question, find the least common multiple of the two dollar amounts.
$$24 = 2^3 \cdot 3 \text{ and } 50 = 2 \cdot 5^2.$$
The LCM is $2^3 \cdot 3 \cdot 5^2 = 600$. The answer is $600. To answer the second question, divide $600 by $24, the price per book to obtain 25. He would have sold 25 books at this price.

70. We wish to find the largest number that will divide evenly into both 60 and 72. We must find the greatest common factor of 60 and 72.
$$60 = 2^2 \cdot 3 \cdot 5$$
$$72 = 2^3 \cdot 3^2$$
Greatest common factor $= 2^2 \cdot 3 = 12$. The longest Jill can cut each piece is 12 inches.

5.4 EXERCISES

1. To perform the subtraction $8 - 3$ on the 12-hour clock, begin at 8 and move 3 hours counterclockwise, ending at 5. Thus, in this system,
$$8 - 3 = 5.$$

Another way to find this difference is based on the formal definition of subtraction in clock arithmetic:
$$8 - 3 = 8 + (-3).$$
From the table of additive inverses for clock arithmetic given in the textbook, the additive inverse of 3 is 9, so we have
$$8 - 3 = 8 + (-3)$$
$$= 8 + 9$$
$$= 5.$$

2. To perform the subtraction $4 - 9$ on the 12-hour clock, begin at 4 and move 9 hours counterclockwise, ending at 7. Thus, in this system,
$$4 - 9 = 7.$$
Another way to find this difference is based on the formal definition of subtraction in clock arithmetic:
$$4 - 9 = 4 + (-9)$$
From the table of additive inverses for clock arithmetic given in the textbook, the additive inverse of 9 is 3, so we have
$$4 - 9 = 4 + (-9)$$
$$= 4 + 3$$
$$= 7.$$

3. To perform the subtraction $2 - 8$ on the 12-hour clock, begin at 2 and move 8 hours counterclockwise, ending at 6. Thus, in this system,
$$2 - 8 = 6.$$
Another way to find this difference using the formal definition of subtraction in clock arithmetic:
$$2 - 8 = 2 + (-8)$$
$$= 2 + 4$$
$$= 6.$$

4. To perform the subtraction $0 - 3$ on the 12-hour clock, begin at 0 and move 3 hours counterclockwise, ending at 9. Thus, in this system,
$$0 - 3 = 9.$$
We may also use the formal definition of subtraction in clock arithmetic, together with the table of additive inverses:
$$0 - 3 = 0 + (-3)$$
$$= 0 + 9$$
$$= 9.$$

5.

×	0	1	2	3	4	5	6	7	8	9	10	11
0	0	0	0	0	0	0	0	0	0	0	0	0
1	0	1	2	3	4	5	6	7	8	9	10	11
2	0	2	4	6	8	10	0	2	4	6	8	10
3	0	3	6	9	0	3	6	9	0	3	6	9
4	0	4	8	0	4	8	0	4	8	0	4	8
5	0	5	10	3	8	1	6	11	4	9	2	7
6	0	6	0	6	0	6	0	6	0	6	0	6
7	0	7	2	9	4	11	6	1	8	3	10	5
8	0	8	4	0	8	4	0	8	4	0	8	4
9	0	9	6	3	0	9	6	3	0	9	6	3
10	0	10	8	6	4	2	0	10	8	6	4	2
11	0	11	10	9	8	7	6	5	4	3	2	1

6. closure

   Yes: From the multiplication table constructed in Exercise 5, we see that the product of any two of the 12-hour clock numbers is always a member of the set of 12-hour clock numbers, $\{0, 1, 2, 3, \ldots, 11\}$.

7. commutative

   Yes: The commutative property holds because the order of the numbers being multiplied can be changed without changing the product. For example, $2 \times 6 = 6 \times 2$ or 0. Note that is comes from the symmetry with respect to the main diagonal.

8. identity

   Yes: From the second column of the 12-hour clock multiplication table, we see that for every element a of the set $\{0, 1, 2, 3, \ldots, 11\}$,

   $$a \times 1 = 1 \times a = a.$$

9.

+	0	1	2	3	4
0	0	1	2	3	4
1	1	2	3	4	0
2	2	3	4	0	1
3	3	4	0	1	2
4	4	0	1	2	3

10. closure

    Yes: The sum of any two elements of the set $\{0, 1, 2, 3, 4\}$ is an element of this set, so closure holds for 5-hour clock addition.

11. commutative

    Yes: The commutative property is satisfied because the order of the addends can be changed without affecting the sum. For example, $2 + 3 = 3 + 2$, which is 0.

12. identity

    Yes: For every element a belonging to the set $\{0, 1, 2, 3, 4\}$,

    $$a + 0 = 0 + a = a.$$

    Therefore, the identity property holds for the 5-hour clock addition with 0 as the identity element.

13. inverse

    Yes: Inverses exist for the 5-hour clock and addition. The inverse of 0 is 0 because $0 + 0 = 0$. The inverse of 1 is 4, and vice versa, because $1 + 4 = 0$. The sumbers 2 and 3 are inverses of each other because $2 + 3 = 3 + 2 = 0$. Each sum yields the identity element, 0.

14.

×	0	1	2	3	4
0	0	0	0	0	0
1	0	1	2	3	4
2	0	2	4	1	3
3	0	3	1	4	2
4	0	4	3	2	1

15. closure

    Yes: Closure is satisfied because all the products that can be formed are numbers of the system: $\{0, 1, 2, 3, 4\}$.

16. commutative

    Yes: Examine the table for 5-hour clock multiplication that was completed in Exercise 14. Draw a diagonal line from upper left to lower right. We see that the part of the table above the diagonal is a mirror image of the part below the diagonal. For example, $2 \times 3 = 1$ and $3 \times 2 = 1$, also. Therefore, the commutative property is satisfied by the system of 5-hour clock numbers with the operation of multiplication.

17. Yes: The identity element is 1 since $1 \times$ any element is equal to that element.

18. $1 \div 3 = q$ if and only if $3 \times q = 1$. Use the multiplication table from Exercise 14 to find the correct value of q. Locate 3 in the column on the left. Read across this row in the body of the table to locate 1. We see that in this system,

    $$3 \times 2 = 1.$$

    Therefore, in 5-hour clock arithmetic

    $$1 \div 3 = 2.$$

19. $3 \div 1 = q$ if and only if $1 \times q = 3$. Use the multiplication table from Exercise 14 to find the correct value of q. Locate 1 in the column on the left. Read across this row in the body of the table to locate 3. We see that in this system,

    $$1 \times 3 = 3.$$

Therefore, in 5-hour clock arithmetic
$$3 \div 1 = 3.$$

20. $2 \div 3 = q$ if and only if $3 \times q = 2$. From our 5-hour clock multiplication table, we see that
$$3 \times 4 = 2.$$
Therefore, in 5-hour clock arithmetic
$$2 \div 3 = 4.$$

21. $3 \div 2 = q$ if and only if $2 \times q = 3$. From our 5-hour clock multiplication table, we see that
$$2 \times 4 = 3.$$
Therefore, in 5-hour clock arithmetic
$$3 \div 2 = 4.$$

22. Writing exercise
23. Writing exercise
24. $1400 + 500 = 1900$

    (This equation is equivalent to stating that 5 hours after 2 P.M. is 7 P.M.)

25. $1300 + 1800 = 3100$.

    Because 2400 represents two complete revolutions on the clock, the time is then 0700.

26. $0750 + 1630 = 2380$
    $= 2300 + 60 + 20$
    $= 2400 + 20$
    $= 0020.$

27. $1545 + 0815 = 2360$
    $= 2300 + 60$
    $= 2400$
    $= 0000.$

28. In ordinary whole number arithmetic,
    $$1145 + 1135 = 2280.$$
    In 12-hour clock arithmetic,
    $1145 + 1135 = 2280 - 1200$
    $= 1080$
    $= 1000 + 60 + 20$
    $= 1120.$

    In 24-hour clock arithmetic,
    $1145 + 1135 = 2280$
    $= 2260 + 20$
    $= 2300 + 20$
    $= 2320.$

Therefore, all of the statements are true, each one in a different system.

29. $5 \equiv 19 \pmod 3$ is false. The difference $19 - 5 = 14$ is not divisible by 3. $5 \equiv 20 \pmod 3$.

30. $35 \equiv 8 \pmod 9$ is true. The difference $35 - 8 = 27$ is divisible by 9.

31. $5445 \equiv 0 \pmod 3$ is true. The difference $5445 - 0 = 5445$ is divisible by 3.

32. $7021 \equiv 4202 \pmod 6$ is false. The difference $7021 - 4202 = 2819$ is not divisible by 6.

33. $(12 + 7) \pmod 4$
    First add 12 and 7 to get 19. Then divide 19 by 4. The remainder is 3, so $(12 + 7) \equiv 3 \pmod 4$.

34. $(62 + 95) \pmod 9$
    First add 62 and 95 to get 157. Then divide 157 by 9. The remainder is 4, so $(62 + 95) \equiv 4 \pmod 9$.

35. $(35 - 22) \pmod 5$
    First subtract 22 from 35 to get 13. Then divide 13 by 5. The remainder is 3, so $(35 - 22) \equiv 3 \pmod 5$.

36. $(82 - 45) \pmod 3$
    First subtract 45 from 82 to get 37. Then divide 37 by 3. The remainder is 1, so $(82 - 45) \equiv 1 \pmod 3$.

37. $(5 \times 8) \pmod 3$
    $5 \times 8 = 40$. Then divide 40 by 3. The remainder is 1, so $(5 \times 8) \equiv 1 \pmod 3$.

38. $(32 \times 21) \pmod 8$
    $32 \times 21 = 672$. Then divide 672 by 8. The remainder is 0, so $(32 \times 21) \equiv 0 \pmod 8$.

39. $[4 \times (13 + 6)] \pmod{11}$
    $[4 \times (13 + 6)] = 4 \times 19 = 76$. Then divide 76 by 11. The remainder is 10, so $[4 \times (13 + 6)] \pmod{11} \equiv 10 \pmod{11}$.

40. $[(10 + 7) \times (5 + 3)] \pmod{10}$
    $[(10 + 7) \times (5 + 3)] = 17 \times 8 = 136$. Then divide 136 by 10. The remainder is 6, so $[(10 + 7) \times (5 + 3)] \equiv 6 \pmod{10}$.

41. Writing exercise

42. $(3 - 27) \pmod 5$
    $3 - 27 = -24$. Since this number is negative, add as many multiples of 5 as necessary to obtain a positive number.
    $-24 + (5 \times 5) = -24 + 25$
    $= 1$
    $\equiv 1 \pmod 5$

    Therefore, $(3 - 27) \equiv 1 \pmod 5$.

43. $(16 - 60) \pmod 7$
    $16 - 60 = -44$. Since this number is negative, add as many multiples of 7 as necessary to obtain a positive number.
    $$-44 + (7 \times 7) = -44 + 49$$
    $$= 5$$
    $$\equiv 5 \pmod 7$$
    Therefore, $(16 - 60) \equiv 5 \pmod 7$.

44. $[(-8) \times 11] \pmod 3$
    $(-8) \times 11 = -88$. Since this number is negative, add as many multiples of 3 as necessary to obtain a positive number.
    $$-88 + (30 \times 3) = -88 + 90$$
    $$= 2$$
    $$\equiv 2 \pmod 3$$
    Therefore, $[(-8) \times 11] \equiv 2 \pmod 3$.

45. $[2 \times (-23)] \pmod 5$
    $2 \times (-23) = -46$. Since this number is negative, add as many multiples of 5 as necessary to obtain a positive number.
    $$-46 + (10 \times 5) = -46 + 50$$
    $$= 4$$
    $$\equiv 4 \pmod 5$$
    Therefore, $[2 \times (-23)] = 4 \pmod 5$.

46. (a)
+	0	1	2	3
0	0	1	2	3
1	1	2	3	0
2	2	3	0	1
3	3	0	1	2

    (b) All four properties are satisfied.

    (c) Since 0 is the identity element, two elements will be inverses if their sum is 0. From the table we see that 2 is its own inverse and 0 is its own inverse. The numbers 1 and 3 are inverses of each other.

47. (a)
+	0	1	2	3	4	5	6
0	0	1	2	3	4	5	6
1	1	2	3	4	5	6	0
2	2	3	4	5	6	0	1
3	3	4	5	6	0	1	2
4	4	5	6	0	1	2	3
5	5	6	0	1	2	3	4
6	6	0	1	2	3	4	5

    (b) All four properties are satisfied.

    (c) Since 0 is the identity element, two elements will be inverses if their sum is 0. From the table we see that 0 is its own inverse. The numbers 1 and 6 are inverses of each other; 3 and 4 are inverses of each other; 2 and 5 are inverses of each other.

48. (a)
×	0	1
0	0	0
1	0	1

    (b) There is no inverse property since 0 does not have and inverse.

    (c) The number 1 is its own inverse.

49. (a)
×	0	1	2
0	0	0	0
1	0	1	2
2	0	2	1

    (b) There is no inverse property since 0 does not have and inverse.

    (c) The number 1 is its own inverse, as is 2.

50. (a)
×	0	1	2	3
0	0	0	0	0
1	0	1	2	3
2	0	2	0	2
3	0	3	2	1

    (b) There is no inverse property. The number 1 is the identity element; 2 has no inverse.

    (c) The number 1 is its own inverse and 3 is its own inverse; 2 has no inverse.

51. (a)
×	0	1	2	3	4	5	6	7	8
0	0	0	0	0	0	0	0	0	0
1	0	1	2	3	4	5	6	7	8
2	0	2	4	6	8	1	3	5	7
3	0	3	6	0	3	6	0	3	6
4	0	4	8	3	7	2	6	1	5
5	0	5	1	6	2	7	3	8	4
6	0	6	3	0	6	3	0	6	3
7	0	7	5	3	1	8	6	4	2
8	0	8	7	6	5	4	3	2	1

    (b) There is no inverse property.

    (c) The number 1 is its own inverse as is 8; 2 and 5 are inverses of each other as are 4 and 7; 3 has no inverse and 6 has no inverse.

52. A modulo system satisfies the inverse property for multiplication only if the modulo number is a prime number.

53. $\{3, 10, 17, 24, 31, 38, \ldots\}$ is the solution set. To find these numbers, multiply 7 times the whole numbers and add 3 each time.
    $$(7 \times 0) + 3 = 3$$
    $$(7 \times 1) + 3 = 10$$
    $$(7 \times 2) + 3 = 17$$

54. In a modulo 3 system, any integer will be congruent to one of the integers 0, 1, or 2. Try each of these integers as a replacement for x. (Notice that $7 \equiv 1 \pmod 3$.)

Try $x = 0$: Is it true that
$$(2 + 0) \equiv 7 \pmod 3?$$
No, because $2 + 0 = 2$ and $2 \equiv 7 \pmod 3$ is not true.

Try $x = 1$: Is it true that
$$(2 + 1) \equiv 7 \pmod 3?$$
No, because $2 + 1 = 3$ and $3 \equiv 7 \pmod 3$ is not true.

Try $x = 2$: Is it true that
$$(2 + 2) \equiv 7 \pmod 3?$$
Yes, because $2 + 2 = 4$ and $4 \equiv 7 \pmod 3$ is true.

Of the integers 0 through 2, only 2 is a solution. Find other solutions of this mod 3 equation by repeatedly adding 3:
$$2$$
$$2 + 3 = 5$$
$$2 + 3 + 3 = 8$$
$$2 + 3 + 3 + 3 = 11.$$
Therefore, $\{2, 5, 8, 11, 14, 17, \ldots\}$ is the solution set.

55. $6x \equiv 2 \pmod 2$

Try the integers 0 and 1. Both work so this is an identity.

56. $(5x - 3) \equiv 7 \pmod 4$

In a modulo 4 system, any integer will be congruent to 0, 1, 2, or 3. Try each of these integers as a replacement for x. Notice that $7 \equiv 3 \pmod 4$.

Try $x = 0$: Is it true that
$$(5 \cdot 0 - 3) \equiv 7 \pmod 4?$$
No, because $0 - 3 = -3$ and $-3 + 4 \equiv 1 \pmod 4$.

Try $x = 1$: Is it true that
$$(5 \cdot 1 - 3) \equiv 7 \pmod 4?$$
No, because $5 - 3 = 2$ and $2 \equiv 2 \pmod 4$.

Try $x = 2$: Is it true that
$$(5 \cdot 2 - 3) \equiv 7 \pmod 4?$$
Yes, because $10 - 3 = 7$ and $7 \equiv 7 \pmod 4$.

Thus, 2 is a solution. Find other solutions of this mod 4 equation by repeatedly adding 4.

Therefore, $\{2, 6, 10, 14, 18, 22\ldots\}$ is the solution set.

57. The modulus is 100,000. After, the odometer reaches 99,999, it "rolls over" to 000,000.

58. If a car's five-digit whole number odometer shows a reading of 29,306 miles, the car might have traveled
$$29,306 + 100,000k$$
miles, where k is any non negative integer.

59. A value of x must be found that will satisfy the following equations:
$$x \equiv 2 \pmod{10}$$
$$x \equiv 2 \pmod{15}$$
$$x \equiv 2 \pmod{20}$$

The solution set for the first equation is:
$\{12, 22, 32, 42, \ldots, 62, \ldots\}$.
The solution set for the second equation is:
$\{17, 32, 47, 62, \ldots\}$.
The solution set for the third equation is:
$\{22, 42, 62, \ldots\}$. The smallest number that satisfies all three equations is 62.

60. A value of x must be found that will satisfy the following equations:
$$x \equiv 6 \pmod 7$$
$$x \equiv 1 \pmod 8$$
$$x \equiv 3 \pmod{15}$$

The solution set for the first equation is:
$\{6, 13, 20, 27, 34, 41, 48, 55, 62, 69\ldots, \}$.
The solution set for the second equation is:
$\{1, 9, 17, 25, 33, 41, 49, 57, 65, 73\ldots\}$.
The solution set for the third equation is:
$\{3, 18, 33, 48, 63, 78, \ldots\}$. Continue writing the elements of these sets until an element is found which is common to all three sets. The smallest number that satisfies all three equations, which is the only solution less than 200, is 153. Therefore, Mary's collection contains 153 spoons.

61. (a) The next year would contain 365 days.

(b) Divide 365 days by 7 to obtain 52 with a remainder of 1. That means that it will be one day past Thursday, which is Friday.

62. If next year (starting today) does not contain 366 days, today must be earlier in the year than the date when an extra day is added in a leap year. Since "leap day" is inserted between February 28 and March 1, the range of possible dates for today is January 1 through February 28.

63. Because there are 31 days in July and 31 days in August for a total of 62, it is unnecessary to find any integers larger than 62 for the modulo systems. Robin's 21-day schedule indicates modulo 21; Kristyn's 30-day schedule indicates modulo 30.

Chicago
For Robin, $x \equiv 1 \pmod{21}$ has solution set $\{1, 22, 43, \ldots\}$. These integers correspond to July 1, July 22, and August 12.
$x \equiv 2 \pmod{21}$ has solution set $\{2, 23, 44, \ldots\}$. These integers correspond to July 2, July 23, and August 13.
$x \equiv 8 \pmod{21}$ has solution set $\{8, 29, 50, \ldots\}$. These integers correspond to July 8, July 29, and August 19.

For Kristyn, $x \equiv 23 \pmod{30}$ has solution set $\{23, 53, \dots\}$. These integers correspond to July 23 and August 22.

$x \equiv 29 \pmod{30}$ has solution set $\{29, 59, \dots\}$. These integers correspond to July 29 and August 28.

$x \equiv 30 \pmod{30}$ has solution set $\{30, 60, \dots\}$. These integers correspond to July 30 and August 29.

The only days that they will both be in Chicago are July 23 and July 29.

New Orleans
For Robin, $x \equiv 5 \pmod{21}$ has solution set $\{5, 26, 47, \dots\}$. These integers correspond to July 5, July 26, and August 16.

$x \equiv 12 \pmod{21}$ has solution set $\{12, 33, 54, \dots\}$. These integers correspond to July 12, August 2, and August 23.

For Kristyn, $x \equiv 5 \pmod{30}$ has solution set $\{5, 35, \dots\}$. These integers correspond to July 5 and August 4.

$x \equiv 6 \pmod{30}$ has solution set $\{6, 36, \dots\}$. These integers correspond to July 6 and August 5.

$x \equiv 17 \pmod{30}$ has solution set $\{17, 47, \dots\}$. These integers correspond to July 17 and August 16.

The only days that they will both be in New Orleans are July 5 and August 16.

San Francisco
For Robin, $x \equiv 6 \pmod{21}$ has solution set $\{6, 27, 48, \dots\}$. These integers correspond to July 6, July 27, and August 17.

$x \equiv 18 \pmod{21}$ has solution set $\{18, 39, 60, \dots\}$. These integers correspond to July 18, August 9, and August 29.

$x \equiv 19 \pmod{21}$ has solution set $\{19, 40, 61, \dots\}$. These integers correspond to July 19, August 9, and August 30.

For Kristyn, $x \equiv 8 \pmod{30}$ has solution set $\{8, 38, \dots\}$. These integers correspond to July 8 and August 7.

$x \equiv 10 \pmod{30}$ has solution set $\{10, 40, \dots\}$. These integers correspond to July 10 and August 9.

$x \equiv 15 \pmod{30}$ has solution set $\{15, 45, \dots\}$. These integers correspond to July 15 and August 14.

$x \equiv 20 \pmod{30}$ has solution set $\{20, 50, \dots\}$. These integers correspond to July 20 and August 19. Finally, $x \equiv 25 \pmod{30}$ has the solution set $\{25, 55, \dots\}$, which means July 25 and August 24. The only day that they will both be in San Francisco is August 9.

64. i^{16}

The powers of i form a modulo 4 system. Since
$$16 \equiv 0 \pmod 4$$
$$i^{16} = i^0 = 1.$$

65. i^{47}
$$47 \equiv 3 \pmod 4$$
$$i^{47} = i^3 = -i$$

66. i^{98}
$$98 \equiv 2 \pmod 4$$
$$i^{47} = i^2 = -1$$

67. i^{137}
$$137 \equiv 1 \pmod 4$$
$$i^{137} = i^1 = i$$

68. 1812
Substitute 1812 for y in the formula given in the text.
$$\begin{aligned} a &= 1812 + [\![1811/4]\!] - [\![1811/100]\!] + [\![1811/400]\!] \\ &= 1812 + [\![452.75]\!] - [\![18.11]\!] + [\![4.5275]\!] \\ &= 1812 + 452 - 18 + 4 \\ &= 2250 \end{aligned}$$

Since
$$2250 = (7 \times 321) + 3$$
$$2250 \equiv 3 \pmod 7$$

and 3 corresponds to Wednesday, January 1, 1812, was a Wednesday.

69. 1865
Substitute 1865 for y in the formula given in the text.
$$\begin{aligned} a &= 1865 + [\![1864/4]\!] - [\![1864/100]\!] + [\![1864/400]\!] \\ &= 1865 + [\![466]\!] - [\![18.64]\!] + [\![4.66]\!] \\ &= 1865 + 466 - 18 + 4 \\ &= 2317 \end{aligned}$$

Since
$$2317 = (7 \times 331) + 0$$
$$2317 \equiv 0 \pmod 7$$

and 0 corresponds to Sunday, January 1, 1865, was a Sunday.

70. 2002
Substitute 2002 for y in the formula given in the text.
$$\begin{aligned} a &= 2002 + [\![2001/4]\!] - [\![2001/100]\!] + [\![2001/400]\!] \\ &= 2002 + [\![500.25]\!] - [\![20.01]\!] + [\![5.0025]\!] \\ &= 2002 + 500 - 20 + 5 \\ &= 2487 \end{aligned}$$

Since
$$2487 = (7 \times 355) + 2$$
$$2487 \equiv 2 \pmod 7$$

and 2 corresponds to Tuesday, January 1, 2002, was a Tuesday.

71. 2020

Substitute 2020 for y in the formula given in the text.

$a = 2020 + [2019/4] - [2019/100] + [2019/400]$
$= 2020 + [504.75] - [20.19] + [5.0475]$
$= 2020 + 504 - 20 + 5$
$= 2509$

Since
$$2509 = (7 \times 358) + 3$$
$$2509 \equiv 3 \pmod 7$$

and 3 corresponds to Wednesday, January 1, 2020, will be a Wednesday.

72. 2003

Substitute 2003 for y in the formula given in the text for Exercises 68–71.

$a = 2003 + [2002/4] - [2002/100] + [2002/400]$
$= 2003 + [500.5] - [20.02] + [5.005]$
$= 2003 + 500 - 20 + 5$
$= 2488$

Since
$$2488 = (7 \times 355) + 3$$
$$2488 \equiv 3 \pmod 7$$

and 3 corresponds to Wednesday, the first day of the year is Wednesday, and 2003 is not a leap year (not divisible by 4). Therefore, the table in the textbook shows that Friday the thirteenth occurs in June.

73. 2002

As in Exercise 70, substitute 2002 for y in the formula given in the text for Exercises 68–71.

$a = 2002 + [2001/4] - [2001/100] + [2001/400]$
$= 2002 + [500.25] - [20.01] + [5.0025]$
$= 2002 + 500 - 20 + 5$
$= 2487$

Since
$$2487 = (7 \times 355) + 2$$
$$2487 \equiv 2 \pmod 7$$

and 2 corresponds to Tuesday, January 1, 2002, was a Tuesday. Since 2002 is not a leap year (not divisible by 4), the table in the textbook shows that Friday the thirteenth occurs in September and December.

74. 2004

Substitute 2004 for y in the formula given in the text for Exercises 68–71.

$a = 2004 + [2003/4] - [2003/100] + [2003/400]$
$= 2004 + [500.75] - [20.03] + [5.0075]$
$= 2004 + 500 - 20 + 5$
$= 2489$

Since
$$2489 = (7 \times 355) + 4$$
$$2489 \equiv 4 \pmod 7$$

and 4 corresponds to Thursday, the first day of the year is Thursday, and 2004 is a leap year (divisible by 4). Therefore, the table in the textbook shows that Friday the thirteenth occurs in February and August.

75. 2200

Substitute 2200 for y in the formula given in the text for Exercises 68–71.

$a = 2200 + [2199/4] - [2199/100] + [2199/400]$
$= 2200 + [549.75] - [21.99] + [5.4975]$
$= 2200 + 549 - 21 + 5$
$= 2733$

Since
$$2733 = (7 \times 390) + 3$$
$$2733 \equiv 3 \pmod 7$$

and 3 corresponds to Wednesday, the first day of the year is Wednesday, and 2200 is not a leap year (not divisible by 400). Therefore, the table in the textbook shows that Friday the thirteenth occurs in June.

76.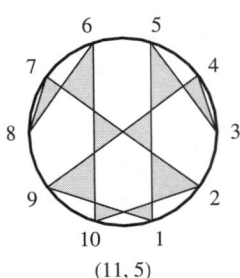

(11, 5)

(a) See figure.

(b) See figure.

(c) $2 \times 5 \equiv \underline{10} \pmod{11}$. Connect 2 and $\underline{10}$.

(d) $3 \times 5 \equiv \underline{4} \pmod{11}$. Connect 3 and $\underline{4}$.

(e) $4 \times 5 \equiv \underline{9} \pmod{11}$. Connect 4 and $\underline{9}$.

(f) $5 \times 5 \equiv \underline{3} \pmod{11}$. Connect 5 and $\underline{3}$.

(g) $6 \times 5 \equiv \underline{8} \pmod{11}$. Connect 6 and $\underline{8}$.

(h) $7 \times 5 \equiv \underline{2} \pmod{11}$. Connect 7 and $\underline{2}$.

(i) $8 \times 5 \equiv \underline{7} \pmod{11}$. Connect 8 and $\underline{7}$.

(j) $9 \times 5 \equiv \underline{1} \pmod{11}$. Connect 9 and $\underline{1}$.

(k) $10 \times 5 \equiv \underline{6} \pmod{11}$. Connect 10 and $\underline{6}$.

(l) See figure.

77. 0–399–13615–4

$(10 \times 0) + (9 \times 3) + (8 \times 9) + (7 \times 9) + (6 \times 1)$
$+ (5 \times 3) + (4 \times 6) + (3 \times 1) + (2 \times 5) = 220$.

Because 220 is a multiple of 11, the check digit is 0; 4 is incorrect.

78. 0–691–02356–5

    $(10 \times 0) + (9 \times 6) + (8 \times 9) + (7 \times 1) + (6 \times 0)$
    $+ (5 \times 2) + (4 \times 3) + (3 \times 5) + (2 \times 6) = 182.$

    Because $182 + 5 = 187$, a multiple of 11, the check digit is 5. The ISBN has the correct check digit.

79. 0–684–81906–__

    $(10 \times 0) + (9 \times 6) + (8 \times 8) + (7 \times 4) + (6 \times 8)$
    $+ (5 \times 1) + (4 \times 9) + (3 \times 0) + (2 \times 6) = 247.$

    Because $247 + 6 = 253$ is a multiple of 11, the check digit is 6.

80. 0–802–71366–__

    $(10 \times 0) + (9 \times 8) + (8 \times 0) + (7 \times 2) + (6 \times 7)$
    $+ (5 \times 1) + (4 \times 3) + (3 \times 6) + (2 \times 6) = 175.$

    Because $175 + 1 = 176$, a multiple of 11, the check digit is 1.

81. 0–679–43247–__

    $(10 \times 0) + (9 \times 6) + (8 \times 7) + (7 \times 9) + (6 \times 4)$
    $+ (5 \times 3) + (4 \times 2) + (3 \times 4) + (2 \times 7) = 246.$

    Because $246 + 7 = 253$, a multiple of 11, the check digit is 7.

82.. 0–374–19066–

    $(10 \times 0) + (9 \times 3) + (8 \times 7) + (7 \times 4) + (6 \times 1)$
    $+ (5 \times 9) + (4 \times 0) + (3 \times 6) + (2 \times 6) = 192.$

    Because $192 + 6 = 198$, a multiple of 11, the check digit is 6.

5.5 EXERCISES

1. Find the sum of the sixteenth and seventeenth terms to obtain the eighteenth:

    $987 + 1597 = 2584.$

2. In the Fibonacci sequence the first number is 1 and the fifth number is 5. $F_n = n$ only when $n = 1$ and $n = 5$.

3. Find the difference between F_{25} and F_{23}:

    $75025 - 28657 = 46,368.$

4. If two successive terms of the Fibonacci sequence are odd, their sum will be even, so the next term is even.

5. $\dfrac{1 + \sqrt{5}}{2}$

6. The approximate value of the golden ratio to the nearest thousandth is found from calculating

    $$\frac{1 + \sqrt{5}}{2} \approx 1.618.$$

Reference table for Exercises 7–12

Term	All	Even	Odd
F_1	1		1
F_2	1	1	
F_3	2		2
F_4	3	3	
F_5	5		5
F_6	8	8	
F_7	13		13
F_8	21	21	
F_9	34		34
F_{10}	55	55	
F_{11}	89		89
F_{12}	144	144	
F_{13}	233		233

7. The pattern on the left side of the nth equation consists of the sum of the first n Fibonacci numbers. The pattern on the right side of the nth equation consists of the $(n + 2)$ number less one. Therefore, the 6th equation is

    $1 + 1 + 2 + 3 + 5 + 8 = 21 - 1,$

    which checks.

8. The pattern on the left side of the equations consists of the beginning consecutive even terms of the Fibonacci sequence. The pattern on the right side of the equations consists of consecutive individual terms of the Fibonacci sequence less one. On both sides add the next consecutive term of the Fibonacci sequence to produce the next equation. The next term on the left side is 144, and the next term on the right side is 233. Therefore, the next equation is

    $1 + 3 + 8 + 21 + 55 + 144 = 233 - 1,$

    which checks.

9. The pattern on the left side of the equations consists of the beginning consecutive odd terms of the Fibonacci sequence. The pattern on the right side of the equations consists of consecutive individual even terms of the Fibonacci sequence. On both sides add the next consecutive odd or even term of the Fibonacci sequence to produce the next equation. The next term on the left side is 89, and the next term on the right side is 144. Therefore, the next equation is

    $1 + 2 + 5 + 13 + 34 + 89 = 144,$

    which checks.

10. The pattern on the left side of the equations consists of the two consecutive terms of the Fibonacci sequence. The pattern on the right side of the equations consists of consecutive individual odd terms of the Fibonacci sequence. It is the sum of the squares of the terms on the left side that equals the term on the right side. Notice also that on the left side, one of the Fibonacci terms in the sum is used again in the next sum. On the left side, then, add the squares of the next consecutive terms of the Fibonacci sequence to produce the next equation. The next terms on the left side are 8 and 13, and the next term on the right side is 233. Therefore, the next equation is

$$8^2 + 13^2 = 233,$$

which checks.

11. The pattern on the left side of the equations consists of the two terms of the Fibonacci sequence in reverse order. The pattern on the right side of the equations consists of consecutive individual even terms of the Fibonacci sequence. It is the difference of the squares of the terms on the left side that equals the term on the right side. Notice also that on the left side, the differences alternate between the squares of consecutive odd terms and consecutive even term. On the left side, then, write the difference of squares of the next consecutive odd terms, in reverse order to produce the next equation. The next terms on the left side are 13^2 and 5^2, and the next term on the right side is 144. Therefore, the next equation is

$$13^2 - 5^2 = 144,$$

which checks.

12. Notice that the cubed numbers on the left side of each equation are descending Fibonacci numbers. Their totals are also Fibonacci numbers. The next set of descending numbers, each one cubed is:

$$13^3 + 8^3 - 5^3 = 2584.$$

13. The left side of each equation is built with alternating terms of the Fibonacci sequence with alternating signs. The right side of each equation consists of Fibonacci numbers squared with alternating signs. The next equation should be $1 - 2 + 5 - 13 + 34 - 89 = -8^2$. The left side of this equation has a sum of -64. The right side of the equations also equals -64.

14. The left side of each equation consists of sequential Fibonacci numbers with alternating signs. The right side of each equation is a bit trickier to describe. The first term alternates signs, beginning with a negative; the second term is always $+1$. Also, ignoring the signs of the first term, they are the Fibonacci numbers in order from 1. Therefore, the next equation is:

$$1 - 1 + 2 - 3 + 5 - 8 + 13 = 8 + 1.$$

Each expression equals 9.

15. (a) $37 = 34 + 3$. Another possibility is $37 = 21 + 8 + 5 + 3$. Can you find more?

(b) $40 = 34 + 3 + 2 + 1$.

(c) $52 = 21 + 13 + 8 + 5 + 3 + 2$.

16. If m divides n, then F_m is a factor of F_n.

(a) $m = 2, n = 6$
Two divides 6, so we wish to show that F_2 is a factor of F_6. $F_2 = 1$ is a factor of $F_6 = 8$.

(b) $m = 3, n = 9$
Three divides 9, so we wish to show that F_3 is a factor of F_9. $F_3 = 2$ is a factor of $F_9 = 34$.

(c) $m = 4, n = 8$
Four divides 8, so we wish to show that F_4 is a factor of F_8. $F_4 = 3$ is a factor of $F_8 = 21$.

17. (a) $m = 10$ and $n = 4$. The greatest common factor of 10 and 4 is 2. Also, $F_{10} = 55$ and $F_4 = 3$; $F_2 = 1$, which is the greatest common factor of 55 and 3.

(b) $m = 12$ and $n = 6$. The greatest common factor of 12 and 6 is 6. Also $F_{12} = 144$ and $F_6 = 8$; The greatest common factor of 144 and 8 is 8.

(c) $m = 14$ and $n = 6$. The greatest common factor of 14 and 6 is 2. Also, $F_{14} = 377$ and $F_6 = 8$; $F_2 = 1$, which is the greatest common factor of 377 and 8.

18. (a) $p = 3$
$$F_{p+1} = F_4 = 3$$
$$F_{p-1} = F_2 = 1$$
$$F_4 = 3 \text{ is divisible by 3.}$$

(b) $p = 7$
$$F_{p+1} = F_8 = 21$$
$$F_{p-1} = F_6 = 8$$
$$F_8 = 21 \text{ is divisible by 7.}$$

(c) $p = 11$
$$F_{p+1} = F_{12} = 144$$
$$F_{p-1} = F_{10} = 55$$
$$F_{10} = 55 \text{ is divisible by 11.}$$

19. (a) Square 13. Multiply the terms of the sequence two positions away from 13 (i.e., 5 and 34). Subtract the smaller result from the larger, and record your answer. $13^2 = 169$; $5 \cdot 34 = 170$; $170 - 169 = 1$.

(b) Square 13. Multiply the terms of the sequence three positions away from 13. Once again subtract the smaller result from the larger, and record your answer. $13^2 = 169$; $3 \cdot 55 = 165$; $169 - 165 = 4$.

(c) Repeat the process, moving four terms away from 13. $13^2 = 169$; $2 \cdot 89 = 178$; $178 - 169 = 9$.

(d) Make a conjecture about what will happen when you repeat the process, moving five terms away. Verify your answer. The results in parts a, b, and c are 1, 4, and 9. These are the squares of the Fibonacci numbers. Because the next Fibonacci number is 5, the result should be $5^2 = 25$. Verification: $13^2 = 169$; $1 \cdot 144 = 144$. $169 - 144 = 25$.

20. Let $x =$ the first number and $y =$ the second number.

    3rd number $= x + y$
    4th number $= y + (x + y) = x + 2y$
    5th number $= (x + y) + (x + 2y) = 2x + 3y$
    6th number $= (x + 2y) + (2x + 3y) = 3x + 5y$
    7th number $= (2x + 3y) + (3x + 5y) = 5x + 8y$
    8th number $= (3x + 5y) + (5x + 8y) = 8x + 13y$
    9th number $= (5x + 8y) + (8x + 13y) = 13x + 21y$
    10th number $= (8x + 13y) + (13x + 21y) = 21x + 34y$

    The sum of the final expressions for each number is $55x + 88y$ or $11(5x + 8y)$. Since $5x + 8y$ is the seventh number in the list, the process is verified.

21. This term is found by adding the ninth and tenth terms: $76 + 123 = 199$.

22. $1, 3, 4, 7, 11, 18, \ldots$

    Square the second number: $3^2 = 9$.
    Multiply the first and third terms: $1 \cdot 4 = 4$.
    Subtract the result: $9 - 4 = 5$.
    Square the third term: $4^2 = 16$.
    Multiply the second and fourth terms: $3 \cdot 7 = 21$.
    Subtract the results: $21 - 16 = 5$.

    In each case, the difference is 5.

23. $1, 3, 4, 7, 11, 18, \ldots$

    Add the first and third terms: $1 + 4 = 5$.
    Add the first, third, and fifth terms: $1 + 4 + 11 = 16$.
    Add the first, third, fifth, and seventh terms:
    $1 + 4 + 11 + 29 = 45$.
    Also, $1 + 4 + 11 + 29 + 76 = 121$.

    Each of the sums is 2 less than a Lucas number.

24. Let L_n be a term of the Lucas sequence.

    $L_2 + L_4 = 3 + 7 = 10$
    $L_2 + L_4 + L_6 = 3 + 7 + 18 = 28$
    $L_2 + L_4 + L_6 + L_8 = 3 + 7 + 18 + 47 = 75$
    $L_2 + L_4 + L_6 + L_8 + L_{10} = 3 + 7 + 18 + 47 + 123$
    $= 198$.

    (Note that $L_{11} = 76 + 123 = 199$.)
    Each sum is 1 less than a Lucas number.

25. (a) $1 \cdot 1 = 1$
    $1 \cdot 3 = 3$
    $2 \cdot 4 = 8$
    $3 \cdot 7 = 21$
    $5 \cdot 11 = 55$

    Notice that reading downward, the first numbers in each equation are the first five members of the Fibonacci sequence, and the second number in each equation is a Lucas number. The products, reading downward, are the 2nd, 4th, 6th, 8th, and 10th terms of the Fibonacci sequence. The next equation should be $8 \cdot 18 = 144$.

    (b) $1 + 2 = 3$
    $1 + 3 = 4$
    $2 + 5 = 7$
    $3 + 8 = 11$
    $5 + 13 = 18$

    Reading downward, the first terms on the left side of the equation are the first five Fibonacci numbers. The second terms on the left side are also ascending Fibonacci numbers, beginning with 2. The results are Lucas numbers. The next equation should be $8 + 21 = 29$.

    (c) $1 + 1 = 2 \cdot 1$
    $1 + 3 = 2 \cdot 2$
    $2 + 4 = 2 \cdot 3$
    $3 + 7 = 2 \cdot 5$
    $5 + 11 = 2 \cdot 8$

    Read downward to see the first five Fibonacci numbers on the extreme left side of the equation added to the first five Lucas numbers. Read downward on the extreme right side to see the second through sixth Fibonacci numbers. The next equation should be $8 + 18 = 2 \cdot 13$.

26. $\dfrac{3}{1} = 3$

    $\dfrac{4}{3} = 1.\overline{3}$

    $\dfrac{7}{4} = 1.75$

    $\dfrac{11}{7} = 1.\overline{571428}$

    $\dfrac{18}{11} = 1.\overline{63}$

    $\dfrac{29}{18} = 1.6\overline{1}$

    $\dfrac{47}{29} = 1.620689655$

    $\dfrac{76}{47} = 1.617021277$

    $\dfrac{123}{76} = 1.618421053$

    These quotients seem to approach the golden ratio.

27. $1, 1, 2, 3$

    Multiply the first and fourth: $1 \cdot 3 = 3$.
    Double the product of the second and third: $2(1 \cdot 2) = 4$.
    Add the squares of the second and third: $1^2 + 2^2 = 5$.
    We have obtained the triple 3, 4, 5. We can substitute

these numbers into the Pythagorean theorem to verify that this is a Pythagorean triple:
$$3^2 + 4^2 = 5^2$$
$$9 + 16 = 25.$$

28. $1, 2, 3, 5$

Multiply the first and fourth: $1 \cdot 5 = 5$.
Double the product of the second and third: $2(2 \cdot 3) = 12$.
Add the squares of the second and third: $2^2 + 3^2 = 13$.
We have obtained the triple 5, 12, 13. We can substitute these numbers into the Pythagorean theorem to verify that this is a Pythagorean triple:
$$5^2 + 12^2 = 13^2$$
$$25 + 144 = 169.$$

29. $2, 3, 5, 8$

Multiply the first and fourth: $2 \cdot 8 = 16$.
Double the product of the second and third: $2(3 \cdot 5) = 30$.
Add the squares of the second and third: $3^2 + 5^2 = 34$.
We have obtained the triple 16, 30, 34. We can substitute these numbers into the Pythagorean theorem to verify that this is a Pythagorean triple:
$$16^2 + 30^2 = 34^2$$
$$256 + 900 = 1156.$$

30. The values of the hypotenuse (c) are:
in Exercise 27,
$$c = 1^2 + 2^2 = 5;$$
in Exercise 28,
$$c = 13;$$
in Exercise 29,
$$c = 3^2 + 5^2 = 9 + 25 = 34.$$

Each of these is a term of the Fibonacci sequence.

31. The sums of the terms on the diagonals are the Fibonacci numbers: $1, 1, 2, 3, 5, 8, 13, \ldots$

32. Writing exercise

33. $\dfrac{1+\sqrt{5}}{2} \approx 1.618033989; \dfrac{1-\sqrt{5}}{2} \approx -0.618033989$

The digits in the decimal positions are the same.

34. $\dfrac{2}{1+\sqrt{5}} \approx .618033989; \dfrac{1+\sqrt{5}}{2} \approx 1.618033989$

The digits in the decimal positions are the same.

35. $n = 14$

$$\dfrac{\left(\dfrac{1+\sqrt{5}}{2}\right)^{14} - \left(\dfrac{1-\sqrt{5}}{2}\right)^{14}}{\sqrt{5}} = 377$$

36. $n = 20$

$$\dfrac{\left(\dfrac{1+\sqrt{5}}{2}\right)^{20} - \left(\dfrac{1-\sqrt{5}}{2}\right)^{20}}{\sqrt{5}} = 6765$$

The 20th Fibonacci number is 6765.

37. $n = 22$

Use a calculator to find
$$\dfrac{\left(\dfrac{1+\sqrt{5}}{2}\right)^{22} - \left(\dfrac{1-\sqrt{5}}{2}\right)^{22}}{\sqrt{5}} = 17,711.$$

The 22nd Fibonacci number is 17,711.

38. $n = 25$

Use a calculator to find
$$\dfrac{\left(\dfrac{1+\sqrt{5}}{2}\right)^{25} - \left(\dfrac{1-\sqrt{5}}{2}\right)^{25}}{\sqrt{5}} = 75,025.$$

The 25th Fibonacci number is 75,025.

Term	All	Even	Odd
F_1	1		1
F_2	1	1	
F_3	2		2
F_4	3	3	
F_5	5		5
F_6	8	8	
F_7	13		13
F_8	21	21	
F_9	34		34
F_{10}	55	55	
F_{11}	89		89
F_{12}	144	144	

Reference table for Exercises 39–42.

39. Observe the pattern in Exercise 9:

$$1 = 1$$
$$1 + 2 = 3$$
$$1 + 2 + 5 = 8$$
$$1 + 2 + 5 + 13 = 21$$
$$1 + 2 + 5 + 13 + 34 = 55$$

Using the Reference Table above, notice that the odd terms of the Fibonacci sequence are added on the left. Also notice that the numbers on the right are the even terms. For n ≥ 1, the subscript "2n" represents an even number and "2n + 1" represents an odd number. Therefore, the formula is:

$$F_1 + F_3 + \ldots + F_{2n-1} = F_{2n}.$$

40. Observe the pattern in Exercise 10:

$$1^2 + 1^2 = 2$$
$$1^2 + 2^2 = 5$$
$$2^2 + 3^2 = 13$$
$$3^2 + 5^2 = 34$$
$$5^2 + 8^2 = 89.$$

Using the Reference Table above, notice that the squares of sequential Fibonacci numbers are added on the left. The sums on the right are the sequential odd Fibonacci numbers with the exception of the first odd number, which is 1. For n ≥ 1, the formula is:

$$F_n^2 + F_{n+1}^2 = F_{2n+1}.$$

41. Observe the pattern in Exercise 11:

$$2^2 - 1^2 = 3$$
$$3^2 - 1^2 = 8$$
$$5^2 - 2^2 = 21$$
$$8^2 = 3^2 = 55.$$

The numbers on the left side of each equation are squares of the Fibonacci numbers, beginning with $F_3^2 - F_{3-2}^2 = F_3^2 - F_1^2$. The next difference of squares is $F_4^2 - F_{4-2}^2 = F_4^2 - F_2^2$, etc. Refer to the Reference Table to see that this pattern continues. The totals on the right are all Fibonacci numbers beginning with F_4, then F_6, F_8, and F_{10}. When $n \geq 3$, these numbers are represented by $2n - 2$. Note that $2 \cdot 3 = 2 = 4$, $2 \cdot 4 - 2 = 6$, etc. The formula then, when $n \geq 3$, is:

$$F_n^2 - F_{n-2}^2 = F_{2n-2}.$$

42. Observe the pattern in Exercise 12:

$$2^3 + 1^3 - 1^3 = 8$$
$$3^3 + 2^3 - 1^3 = 34$$
$$5^3 + 3^3 - 2^3 = 144$$
$$8^3 + 5^3 - 3^3 = 610.$$

If $n = 3$, then the first equation is

$$F_3^3 + F_{3-1}^3 - F_{3-2}^3 = F_3^3 + F_2^3 - F_1^3 = F_6.$$

In other words, the Fibonacci numbers descend from left to right. Notice that the numbers on the right side of the equation are also Fibonacci numbers, but the general pattern is more difficult to discern. The first total, 8, is F_6; the next, 34, is F_9; the third number, 144, if F_{12}; the fourth number, 610, can be shown to be F_{15}. That is, beginning with the ninth number, subsequent totals are every third number. This can be expressed as $3n - 3$ for $n = 3$. The formula for this pattern for n ≥ 3 is:

$$F_n^3 + F_{n-1}^3 - F_{n-2}^3 = F_{3n-3}.$$

EXTENSION: MAGIC SQUARES

1. 180° in a clockwise direction
   Imagine a straight line (180°) from the top left corner to the bottom right corner of the magic square in Figure 10. That moves the 8 from its original position to the bottom right corner, and the other numbers follow.

2	7	6
9	5	1
4	3	8

2. Start with Figure 10. Rotating this magic square 90° in a counterclockwise direction produces the following magic square.

4	9	2
3	5	7
8	1	6

3. 90° in a clockwise direction
   Imagine the top left corner of the box that contains the number 17 as an origin. Rotate the entire square 90° clockwise. Then the 17 will be in the top right box and the 11 will be in the top left. All the other numbers follow.

11	10	4	23	17
18	12	6	5	24
25	19	13	7	1
2	21	20	14	8
9	3	22	16	15

4. Start with Figure 12. Rotating this magic square 180° in a clockwise direction produces the following magic square.

9	2	25	18	11
3	21	19	12	10
22	20	13	6	4
16	14	7	5	23
15	8	1	24	17

5. 90° in a counterclockwise direction

   Use the upper right corner of the box containing 15 as the pivot point. Rotate 90° in a counterclockwise direction. Then the 9 will be in the upper right box, and 15 will be in the upper left.

15	16	22	3	9
8	14	20	21	2
1	7	13	19	25
24	5	6	12	18
17	23	4	10	11

6. A magic square of order 2 would have to be such that for four different natural numbers a, b, c, and d,
   $$a + b = a + c = a + d,$$
   because in a magic square the sum along each row, column, and diagonal is the same. This is impossible. Such a magic square is impossible to construct.

7. Figure 10, multiply by 3

24	9	12
3	15	27
18	21	6

   $$\begin{aligned} \text{MS} &= \frac{n(n^2 + 1)}{2} \cdot 3 \\ &= \frac{3(3^2 + 1)}{2} \cdot 3 \\ &= \frac{3(10)}{2} \cdot 3 \\ &= 45. \end{aligned}$$

8. Start with Figure 10. Adding 7 to each entry produces the following magic square.

15	10	11
8	12	16
13	14	9

   The new magic sum is
   $$\begin{aligned} 15 + 10 + 11 &= 13 + 12 + 11 \\ &= 36. \end{aligned}$$

9. Figure 12, divide by 2

$\frac{17}{2}$	12	$\frac{1}{2}$	4	$\frac{15}{2}$
$\frac{23}{2}$	$\frac{5}{2}$	$\frac{7}{2}$	7	8
2	3	$\frac{13}{2}$	10	11
5	6	$\frac{19}{2}$	$\frac{21}{2}$	$\frac{3}{2}$
$\frac{11}{2}$	9	$\frac{25}{2}$	1	$\frac{9}{2}$

   $$\begin{aligned} \text{MS} &= \frac{n(n^2 + 1)}{2} \div 2 \\ &= \frac{5(5^2 + 1)}{2} \div 2 \\ &= \frac{5(26)}{2} \div 2 \\ &= \frac{65}{2} = 32\frac{1}{2} \end{aligned}$$

10. Start with Figure 12. Subtracting 10 from each entry produces the following magic square.

7	14	−9	−2	5
13	−5	−3	4	6
−6	−4	3	10	12
0	2	9	11	−7
1	8	15	−8	−1

    The new magic sum is
    $$\begin{aligned} &7 + 14 + (-9) + (-2) + 5 \\ &= 1 + 2 + 3 + 4 + 5 \\ &= 15. \end{aligned}$$

11. Using the third row, the magic sum is
    $$281 + 467 + 59 = 807.$$
    Then
    $$807 - (71 + 257) = 479.$$
    The missing entry is 479.

12. Using the first column, the magic sum is
    $$389 + 107 + 311 = 807.$$
    Then
    $$807 - (389 + 227) = 191.$$
    The missing entry is 191.

13. Using the first column, the magic sum is
    $$389 + 71 + 347 = 807.$$
    Then
    $$807 - (191 + 149) = 467.$$
    The missing entry is 467.

14. Using the first column, the magic sum is
    $$401 + 47 + 359 = 807.$$
    Then
    $$807 - (359 + 137) = 311.$$
    The missing entry is 311.

15. Using the first column, the magic sum is
$$401 + 17 + 389 = 807.$$
Then
$$807 - (257 + 281) = 269.$$
The missing entry is 269.

16. The magic sum for each of these five magic squares is 807, so the fort was built in the year 807.

17. Using the second column, the magic sum is
$$68 + 72 + 76 = 216.$$
   (a) $216 - (75 + 68) = 73$
   (b) $216 - (75 + 71) = 70$
   (c) Use the answer from (b) to find
$$216 - (72 + 70) = 74.$$
   (d) $216 - (71 + 76) = 69$

18. Using the diagonal, the magic sum is
$$1 + 14 + 4 + 15 = 34.$$
   (a) Using the first row:
$$1 + 8 + 13 + a = 34$$
$$22 + a = 34$$
$$a = 12.$$
   (b) Using the second row:
$$b + 14 + 7 + 2 = 34$$
$$b + 23 = 34$$
$$b = 11.$$
   (c) Using the third row:
$$16 + 9 + 4 + c = 34$$
$$29 + c = 34$$
$$c = 5.$$
   (d) Using the first column, since $b = 11$:
$$1 + 11 + 16 + d = 34$$
$$28 + d = 34$$
$$d = 6.$$
   (e) Using the second column:
$$8 + 14 + 9 + e = 34$$
$$31 + e = 34$$
$$e = 3.$$

   (f) Using the third column:
$$13 + 7 + 4 + f = 34$$
$$24 + f = 34$$
$$f = 10.$$

19. Using the second column to obtain the magic sum,
$$20 + 14 + 21 + 8 + 2 = 65.$$
   (a) $65 - (3 + 20 + 24 + 11) = 7$
   (b) $65 - (14 + 1 + 18 + 10) = 22$
   (c) $65 - (9 + 21 + 13 + 17) = 5$
   (d) $65 - (16 + 8 + 25 + 12) = 4$
   (e) Use the first column and (b):
$$65 - (3 + 9 + 16 + 22) = 15.$$
   (f) Use the third column and (a):
$$65 - (1 + 13 + 25 + 7) = 19.$$
   (g) Use the fourth column and (c):
$$65 - (24 + 18 + 12 + 5) = 6.$$
   (h) Use the fifth column and (d):
$$65 - (11 + 10 + 17 + 4) = 23.$$

20. Find the magic sum from the first row:
$$3 + 36 + 2 + 35 + 31 + 4 = 111.$$
   (a) Second row:
$$111 - (10 + 12 + 26 + 7 + 27) = 29.$$
   (b) Third row:
$$111 - (21 + 13 + 17 + 14 + 22) = 24.$$
   (c) Second column:
$$111 - (36 + 12 + 13 + 30 + 1) = 19.$$
   (d) Insert the entry for (c) in the fourth row:
$$111 - (16 + 19 + 23 + 18 + 15) = 20.$$
   (e) Fifth row:
$$111 - (28 + 30 + 8 + 25 + 9) = 11.$$
   (f) Sixth row:
$$111 - (1 + 32 + 5 + 6 + 34) = 33.$$

21. Use the "staircase method" to construct a magic square of order 7, containing the entries 1, 2, 3, ...

	31	40	49	2	11	20	
30	39	48	1	10	19	28	30
38	47	7	9	18	27	29	38
46	6	8	17	26	35	37	46
5	14	16	25	34	36	45	5
13	15	24	33	42	44	4	13
21	23	32	41	43	3	12	21
22	31	40	49	2	11	20	

22. Using the first row, we find the magic sum:
$$16 + 3 + 2 + 13 = 34.$$

23. The sum of the entries in the four corners is
$$16 + 13 + 4 + 1 = 34.$$

24. Upper left 2 by 2 square:
$$16 + 3 + 5 + 10 = 34 \geq$$

Upper right 2 by 2 square:
$$2 + 13 + 11 + 8 = 34.$$

Lower left 2 by 2 square:
$$9 + 6 + 4 + 15 = 34.$$

Lower right 2 by 2 square:
$$7 + 12 + 14 + 1 = 34.$$

Each sum is equal to 34.

25. The entries in the diagonals are
$$16 + 10 + 7 + 1 + 13 + 11 + 6 + 4 = 68.$$

The entries not in the diagonals are
$$3 + 2 + 5 + 8 + 9 + 12 + 15 + 14 = 68.$$

26. The sum of the squares of the entries in the diagonals is
$$16^2 + 10^2 + 7^2 + 1^2 + 13^2 + 11^2 + 6^2 + 4^2$$
$$= 256 + 100 + 49 + 1 + 169 + 121 + 36 + 16$$
$$= 748.$$

The sum of the squares of the entries not in the diagonals is
$$3^2 + 2^2 + 5^2 + 8^2 + 9^2 + 12^2 + 15^2 + 14^2$$
$$= 9 + 4 + 25 + 64 + 81 + 144 + 225 + 196$$
$$= 748.$$

Each sum is equal to 748.

27. Sum of cubes of diagonal entries:
$$16^3 + 10^3 + 7^3 + 1^3 + 13^3 + 11^3 + 6^3 + 4^3 =$$
$$4096 + 1000 + 343 + 1 + 2197 + 1331 + 216 + 64 =$$
$$9248.$$

Sum of cubes of entries not in the diagonals:
$$3^3 + 2^3 + 5^3 + 8^3 + 9^3 + 12^3 + 15^3 + 14^3 =$$
$$27 + 8 + 125 + 512 + 729 + 1728 + 3375 + 2744 =$$
$$9248.$$

28. The sum of the squares of the entries in the top two rows is
$$16^2 + 3^2 + 2^2 + 13^2 + 5^2 + 10^2 + 11^2 + 8^2$$
$$= 256 + 9 + 4 + 169 + 25 + 100 + 121 + 64$$
$$= 748.$$

The sum of the squares of the entries in the bottom two rows is
$$9^2 + 6^2 + 7^2 + 12^2 + 4^2 + 15^2 + 14^2 + 1^2$$
$$= 81 + 36 + 49 + 144 + 16 + 225 + 196 + 1$$
$$= 748.$$

Each sum is equal to 748.

29. $16^2 + 3^2 + 2^2 + 13^2 + 9^2 + 6^2 + 7^2 + 12^2 =$
$256 + 9 + 4 + 169 + 81 + 36 + 49 + 144 = 748;$

$5^2 + 10^2 + 11^2 + 8^2 + 4^2 + 15^2 + 14^2 + 1^2 =$
$25 + 100 + 121 + 64 + 16 + 225 + 196 + 1 = 748.$

30. Writing exercise

31.
→	2	3	→
5	→	→	8
9	→	→	12
→	14	15	→

16	2	3	13
5	11	10	8
9	7	6	12
4	14	15	1

The second and third columns are interchanged.

32. $a = 5, b = 1, c = -3$

Replace a, b, and c with these numbers to find the entries in the magic square.

$$a + b = 5 + 1 = 6$$
$$a - b - c = 5 - 1 - (-3) = 7$$
$$a + c = 5 + (-3) = 2$$
$$a - b + c = 5 - 1 + (-3) = 1$$
$$a = 5$$
$$a + b - c = 5 + 1 - (-3) = 9$$
$$a - c = 5 - (-3) = 8$$
$$a + b + c = 5 + 1 + (-3) = 3$$
$$a - b = 5 - 1 = 4$$

The magic square is then

6	7	2
1	5	9
8	3	4

122 CHAPTER 5 NUMBER THEORY

33. $a = 16, b = 2, c = -6$

    Replace a, b, and c with these numbers to find the entries in the magic square.

    $$a + b = 16 + 2 = 18$$
    $$a - b - c = 16 - 2 - (-6) = 20$$
    $$a + c = 16 + (-6) = 10$$
    $$a - b + c = 16 - 2 + (-6) = 8$$
    $$a = 16$$
    $$a + b - c = 16 + 2 - (-6) = 24$$
    $$a - c = 16 - (-6) = 22$$
    $$a + b + c = 16 + 2 + (-6) = 12$$
    $$a - b = 16 - 2 = 14$$

    The magic square is then as follows.

18	20	10
8	16	24
22	12	14

34. $a = 5, b = 4, c = -8$

    Replace a, b, and c with these numbers to find the entries in the magic square.

    $$a + b = 5 + 4 = 9$$
    $$a - b - c = 5 - 4 - (-8) = 9$$
    $$a + c = 5 + (-8) = -3$$
    $$a - b + c = 5 - 4 + (-8) = -7$$
    $$a = 5$$
    $$a + b - c = 5 + 4 - (-8) = 17$$
    $$a - c = 5 - (-8) = 13$$
    $$a + b + c = 5 + 4 + (-8) = 1$$
    $$a - b = 5 - 4 = 1$$

    The magic square is then as follows.

9	9	-3
-7	5	17
13	1	1

35.
39	48	57	10	19	28	37
47	56	16	18	27	36	38
55	15	17	26	35	44	46
14	23	25	34	43	45	54
22	24	33	42	51	53	13
30	32	41	50	52	12	21
31	40	49	58	11	20	29

    $$\text{MS} = \frac{7(2 \cdot 10 + 7^2 - 1)}{2}$$
    $$= \frac{7(20 + 49 - 1)}{2}$$
    $$= \frac{7(68)}{2}$$
    $$= 238$$

36. Use the formula of exercise 35 to find the magic sum. When $n = 4$ and $k = 24$,

    $$\text{MS} = \frac{4(2 \cdot 24 + 4^2 - 1)}{2}$$
    $$= \frac{4(48 + 16 - 1)}{2}$$
    $$= \frac{4(63)}{2}$$
    $$= 126$$

    (a) First row:
    $$126 - (38 + 37 + 27) = 24.$$

    (b) Second row:
    $$126 - (35 + 30 + 32) = 29.$$

    (c) Third row:
    $$126 - (31 + 33 + 28) = 34.$$

    (d) Insert the entry for (a) in the first column:
    $$126 - (24 + 35 + 31) = 36.$$

    (e) Insert the entry for (d) in the fourth row:
    $$126 - (36 + 26 + 25) = 39.$$

37. There are many ways to find the magic sum. One way is by adding the top row:
    $$52 + 61 + 4 + 13 + 20 + 29 + 36 + 45 = 260.$$

38. First half of first row:
    $$52 + 61 + 4 + 13 = 130.$$

    First half of second row:
    $$14 + 3 + 62 + 51 = 130.$$

    First half of third row:
    $$53 + 60 + 5 + 12 = 130.$$

    Continuing in this manner, we find that the sum of the four entries in the first half of each of the remaining rows is also 130.

    Second half of first row:
    $$20 + 29 + 36 + 45 = 130.$$

    Second half of second row:
    $$46 + 35 + 30 + 19 = 130.$$

    Second half of third row:
    $$21 + 28 + 37 + 44 = 130.$$

    Continuing in this manner, we find that the sum of the four entries in the second half of each of the remaining rows is also 130. Thus, the sum of half of each row is equal to 130. This is half the magic sum of 260, which was found in Exercise 37.

39. $52 + 45 + 16 + 17 + 54 + 43 + 10 + 23 = 260$.

40. Start with 50:

    $50 + 8 + 7 + 54 + 43 + 26 + 25 + 47 = 260$.

    Start with 9:

    $9 + 58 + 59 + 12 + 21 + 38 + 39 + 24 = 260$.

    Start with 55:

    $55 + 6 + 5 + 51 + 46 + 28 + 27 + 42 = 260$.

    Start with 11:

    $11 + 60 + 62 + 13 + 20 + 35 + 37 + 22 = 260$.

    In each case we obtain the magic sum, 260.

41. Start by placing 1 in the fourth row, second column. Move up two, right one and place 2 in the second row, third column. If we now move up two, right one again, we will go outside the square. Drop down 5 cells (one complete column) to the bottom of the fourth column to place 3. Move up two, right one, to place 4 in the third row, fifth column. Moving up two, right one, we again go outside the square; move 5 cells to the left to place 5 in the first row, first column. Moving up two, right one, takes us two rows outside the square. Counting downward 5 cells, find that the fourth row, second column is blocked with a 1 already there. The number 6 is then placed just below the entry 5. Continue in this manner until all 24 numbers have been placed. Notice that in trying to enter the number 21, it is blocked by 16 which is already in the cell. Because 20 is already in the bottom cell of the last row, dropping a cell just "below" this one moves it to the top row, third column. The completed magic square is shown here.

		14	22	10	18	
		20	3	16	24	
5	13	21	9	17	5	
6	19	2	15	23	11	
12	25	8	16	4	12	
18	1	14	22	10		
24	7	20	3	11		

42. Start by placing 1 in the third row, third column (the middle cell). Move up 1, right 2, and place 2 in the second row, fifth column. If we now move up 1, right 2 again, we will go outside the square by two columns. Move 5 cells (one complete row) to the left and place 3 in the first row, second column. When we try to enter 4, we will go outside the square again. Drop down 5 cells (one complete column) to the bottom of the fourth column. When we try to enter 5, we will go outside the square again. Move 5 cells to the left to enter 5 in the fourth row, first column. If we move up 1, right 2 from this cell, we are back to the middle cell, which is already occupied by 1. Because we are blocked, we move one cell to the left of the cell containing 5. This puts us outside the magic square, so move 5 cells to the right and enter 6 in row 4, column 5.

    Continue to fill the cells in this manner. Each time you are blocked (this occurs after 5, 10, 15, and 20), move one cell to the left of the previous entry. When entering 17, note that you will need to move 5 cells to the left and then 5 cells down in order to find a position inside the magic square. The completed magic square is shown here.

	10	4	23	17		
9	3	22	16	15	9	3
21	20	14	8	2		20
13	7	1	25	19	13	7
6	5	24	18	12	6	5	24
	17	11	10	4	23	

CHAPTER 5 TEST

1. False; 2 and 3 differ by one and both are prime numbers.

2. True

3. True, because 9 is divisible by 3

4. True

5. True; 1 is a factor of all numbers. The smallest multiple of a given number is found by multiplying that number by 1.

6. 331,153,470

    (a) It is divisible by 2 because it is even.

    (b) It is divisible by 3 because the sum of the digits is 27, a number divisible by 3.

    (c) It is not divisible by 4 because the last two digits taken as the number 70 is not divisible by 4.

    (d) It is divisible by 5 because it ends in 0.

    (e) It is divisible by 6 because it is divisible by both 2 and 3.

    (f) It is not divisible by 8 because the last three digits form the number 470, which is not divisible by 8.

    (g) It is divisible by 9 because the sum of the digits, 27, is divisible by 9.

    (h) It is divisible by 10 because it ends in 0.

    (i) It is not divisible by 12 because it is not divisible by both 3 and 4; it is only divisible by 3.

7. (a) The number 93 is composite because it has factors other than itself and 1. $3 \cdot 31 = 93$.

    (b) The number 1 is neither prime nor composite.

    (c) The number 59 is prime because its only factors are 59 and 1.

8. $\begin{array}{r|r} 2 & 1440 \\ \hline 2 & 720 \\ \hline 2 & 360 \\ \hline 2 & 180 \\ \hline 2 & 90 \\ \hline 3 & 45 \\ \hline 3 & 15 \\ \hline & 5 \end{array}$

   The prime factorization of 1440 is $2^5 \cdot 3^2 \cdot 5$.

9. Writing exercise

10. (a) The only proper factor of 17 is 1, so 17 is deficient.

    (b) The proper factors of 6 are 1, 2, and 3. Because $1 + 2 + 3 = 6$, the number is perfect.

    (c) The proper factors of 24 are 1, 2, 3, 4, 6, 8, and 12. Their sum, $1 + 2 + 3 + 4 + 6 + 8 + 12 = 36$, is greater than 24, so the number is abundant.

11. Statement c is false. Goldbach's Conjecture for the number 8 is verified by the equation $8 = 3 + 5$. The number 1 is not prime.

12. A pair of twin primes between 40 and 50 is 41 and 43.

13. $270 = 2 \cdot 3^3 \cdot 5$
    $450 = 2 \cdot 3^2 \cdot 5^2$

    Then $2 \cdot 3^2 \cdot 5 = 90$, the greatest common factor.

14. $24 = 2^3 \cdot 3$
    $36 = 2^2 \cdot 3^2$
    $60 = 2^2 \cdot 3 \cdot 5$

    Then $2^3 \cdot 3^2 \cdot 5 = 360$, the least common multiple.

15. This exercise is similar to Exercise 66 in section 5.3. If Sherrie is off every 6 days and Della is off every 4 days, this corresponds to modulo 6 and modulo 4. Also, the days of the week, beginning with Sunday, correspond to modulo 7.

    For Sherrie, start with $x = 3 \pmod 6$. This is the set of integers $\{3, 9, 15, 21, \ldots\}$. For Della, $x = 3 \pmod 4$. This is the set of integers $\{3, 7, 11, 15, 19, \ldots\}$.

    The next common number for Sherrie and Della is 15, which corresponds to Monday.

16. $17,711 + 28,657 = 46,368$

17. Notice that the numbers that are being subtracted are four Fibonacci numbers, in order, and each "set" of them deletes the first and adds another as the equations progress. The final values on the right side of the equation are also Fibonacci numbers. Finally, the first number in each equation is also a Fibonacci number, with the first several of them omitted. The next equation should be

    $$89 - (8 + 13 + 21 + 34) = 13.$$

18. Option B Sometimes. See Section 5.1.

19. $(12 + 16)(\text{mod } 5)$

    First add 12 and 16 to get 28. Then divide 28 by 5. The remainder is 3, so $(12 + 16)(\text{mod } 5) \equiv 3$.

20. $(4x - 7) \equiv 2 \pmod 3$

    The value on the left must be congruent to 2, mod 3. Replace x with integer values beginning with zero, to see if it makes the equation true.

    $x = 0$: Is it true that $4 \cdot 0 - 7 \equiv 2 \pmod 3$? No
    $x = 1$: Is it true that $4 \cdot 1 - 7 \equiv 2 \pmod 3$? No
    $x = 2$: Is it true that $4 \cdot 2 - 7 \equiv 2 \pmod 3$? No
    $x = 3$: Is it true that $4 \cdot 3 - 7 \equiv 2 \pmod 3$? Yes

    Then, the set of positive solutions for x is $\{3, 6, 9, 12, \ldots\}$, all the multiples of 3.

21. (a) The sequence is obtained by adding two successive terms into order to obtain the next:
    $1, 5, 6, 11, 17, 28, 45, 73$.

    (b) First let us choose 11.
    $$11^2 = 121$$
    $$6 \cdot 17 = 102$$
    $$121 - 102 = 19$$

    Now let us choose 45.
    $$45^2 = 2025$$
    $$28 \cdot 73 = 2044$$
    $$2044 - 2025 = 19$$

    It appears that this process yields 19 each time.

22. Option A is the exact value of the golden ratio. Options C and D are approximations.

23. Writing exercise

6.1 EXERCISES

1. An integer between 3.5 and 4.5 is 4.
2. A rational number between 3.8 and 3.9 is 3.82. There are many others.
3. A whole number that is not positive and is less than 1 is 0.
4. A whole number that is greater than 4.5 is 6. There are many others.
5. An irrational number that is between $\sqrt{11}$ and $\sqrt{13}$ is $\sqrt{12}$. There are many others.
6. A real number that is neither negative nor positive is 0.
7. It is true that every natural number is positive. The natural numbers consist of $\{1, 2, 3, 4, \ldots\}$.
8. Every whole number is not positive. Zero is a whole number, and it is neither positive nor negative.
9. True. The set of integers is included in the set of rational numbers.
10. True. The set of rational numbers is included in the set of real numbers.
11. (a) 3, 7
    (b) 0, 3, 7
    (c) $-9, 0, 3, 7$
    (d) $-9, -1\frac{1}{4}, -\frac{3}{5}, 0, 3, 5.9, 7$
    (e) $-\sqrt{7}, \sqrt{5}$
    (f) All are real numbers.
12. (a) 3
    (b) 0, 3
    (c) $-5, -1, 0, 3$
    (d) $-5.3, -5, -1, -\frac{1}{9}, 0, 1.2, 1.8, 3$
    (e) $-\sqrt{3}, \sqrt{11}$
    (f) All are real numbers.
13. Writing exercise.
14. The decimal representation of a rational number will either terminate or repeat.
15. 568,488
16. 1450
17. $-30°$
18. 5436
19. $-34,841$
20. 159,376
21. -8
22. 20; 25°; $-3°$
23. (a) Pacific Ocean, Indian Ocean, Caribbean Sea, South China Sea, Gulf of California

    (b) Point Success, Ranier, Matlalcueyetl, Steele, McKinley

    (c) This statement is true because the absolute value of each number is its nonnegative value.

    (d) This statement is false. The absolute value of the depth of the Gulf of California is 2375; the absolute value of the depth of the Caribbean Sea is 8448.
24. (a) Largest: U.S.; smallest: India

    (b) U.S., Germany, and Japan

    (c) Japan: about $300; Italy: about $150

    (d) The difference between Japan and Italy is about $300 - \$150 = \$150$
25. ![number line with points at -6, -4, -2, 0, 2, 4]
26. ![number line with points at -6, -4, -2, 0, 2, 4]
27. ![number line with points at $-3\frac{4}{5}, -1\frac{5}{8}, \frac{1}{4}, 2\frac{1}{2}$]
28. ![number line with points at $-3\frac{2}{5}, -2\frac{1}{3}, 4\frac{5}{9}, 5\frac{1}{4}$]
29. (a) $|-7| = 7$, which is choice A.
    (b) $-(-7) = 7$, which is choice A.
    (c) $-|-7| = -7$, which is choice B.
    (d) $-|-(-7)| = -|7| = -7$, which is choice B.
30. The opposite of -2 is $\underline{2}$, while the absolute value of -2 is $\underline{2}$. The additive inverse of -2 is $\underline{2}$, while the additive inverse of the absolute value of -2 is $\underline{-2}$.
31. -2
    (a) Additive inverse is 2.
    (b) Absolute value is 2.
32. -8
    (a) Additive inverse is 8.
    (b) Absolute value is 8.
33. 6
    (a) Additive inverse is -6.
    (b) Absolute value is 6.

34. 11
    (a) Additive inverse is -11.
    (b) Absolute value is 11.
35. $7 - 4 = 3$
    (a) Additive inverse is -3.
    (b) Absolute value is 3.
36. $8 - 3 = 5$
    (a) Additive inverse is -5.
    (b) Absolute value is 5.
37. $7 - 7 = 0$
    (a) Additive inverse is 0.
    (b) Absolute value is 0.
38. $3 - 3 = 0$
    (a) Additive inverse is 0.
    (b) Absolute value is 0.
39. If $a - b > 0$, then the absolute value of $a - b$ in terms of a and b is $a - b$ because this expression produces a nonnegative number.
40. If $a - b = 0$, then the absolute value of $a - b$ is 0.
41. -12
42. -14
43. -8
44. -16
45. The smaller number is 3 because $|-4| = 4$.
46. $|-2| = 2$, which is the smaller number.
47. $|-3|$ is the smaller number.
48. $|-8|$
49. $-|-6| = -6$, the smaller number.
50. $-|-3| = -3$, the smaller number.
51. $|5 - 3| = |2| = 2$; $|6 - 2| = |4| = 4$. The first is the smaller number.
52. $|7 - 2| = |5| = 5$; $|8 - 1| = |7| = 7$. The first is the smaller number.
53. $6 > -(-2)$ is a true statement because $6 > 2$.
54. $-8 > -(-2)$ is a false statement because $-8 < 2$.
55. $-4 \leq -(-5)$ is true because $-4 \leq 5$.
56. $-6 \leq -(-3)$ is true because $-6 \leq 3$.
57. $|-6| < |-9|$ is true because $6 < 9$.
58. $|-12| < |-20|$ is true because $12 < 20$.
59. $-|8| > |-9|$ is false because $-8 < 9$.
60. $-|12| > |-15|$ is false because $-12 < 15$.
61. $-|-5| \geq -|-9|$ is true because $-5 \geq -9$.
62. $-|-12| \leq -|-15|$ is false because $-12 \geq -15$.
63. $|6 - 5| \geq |6 - 2|$ is false because $1 \leq 4$.
64. $|13 - 8| \leq |7 - 4|$ is false because $5 \geq 3$.
65. The greatest drop is for the commodity video/audio equipment from 1995 to 1996.
66. The commodity apparel from $1995 - 1996$ represents the least change.
67. $1996 - 1997$ since $|-2.2| < |-2.6|$.
68. $1996 - 1997$ since $|-.5| < |.9|$.
69. Computer/data processing services shows the greatest change.
70. Writing exercise

Answers may vary in Exercises $71 - 76$.

71. Three positive real numbers but not integers between -6 and 6 are 5.1, 0.25, and $4\frac{1}{2}$.
72. Three real numbers but not positive between -6 and 6 are -5.1, -4, and $-\frac{2}{3}$.
73. Three real numbers but not whole numbers between -6 and 6 are -5.1, 0.25, and $4\frac{1}{2}$.
74. Three rational numbers but not integers between -6 and 6 are -5.1, 0.25, and $4\frac{1}{2}$.
75. Three real numbers but not rational numbers between -6 and 6 are $-\sqrt{6}$, $\sqrt{2}$, and $\sqrt{5}$.
76. Three rational numbers but not negative numbers between -6 and 6 are 1.1, $3\frac{3}{8}$, and 5.

6.2 EXERCISES

1. The sum of two negative numbers will always be a <u>negative</u> number.
2. The sum of a number and its opposite will always be <u>zero</u>.
3. To simplify the expression $8 + [-2 + (-3 + 5)]$, I should begin by adding $\underline{-3}$ and $\underline{5}$, according to the rule for order of operations.
4. If I am adding a positive number and a negative number, and the negative number has the larger absolute value, the sum will be a <u>negative</u> number.
5. Writing exercise
6. Writing exercise

7. $-12 + (-8) = -20$
8. $-5 + (-2) = -7$
9. $12 + (-16) = -4$
10. $-6 + 17 = 11$
11. $-12 - (-1) = -12 + 1 = -11$
12. $-3 - (-8) = -3 + 8 = 5$
13. $-5 + 11 + 3 = 6 + 3 = 9$
14. $-9 + 16 + 5 = 7 + 5 = 12$
15. $12 - (-3) - (-5) = 12 + 3 + 5 = 20$
16. $15 - (-6) - (-8) = 15 + 6 + 8 = 29$
17. $-9 - (-11) - (4 - 6) = -9 + 11 - (-2)$
    $= -9 + 11 + 2$
    $= 2 + 2$
    $= 4$
18. $-4 - (-13) + (-5 + 10) = -4 + 13 + (5)$
    $= -4 + 18$
    $= 14$
19. $(-12)(-2) = 24$
20. $(-3)(-5) = 15$
21. $9(-12)(-4)(-1)3 = -1296$
22. $-5(-17)(2)(-2)4 = -1360$
23. $\dfrac{-18}{-3} = 6$
24. $\dfrac{-100}{-50} = 2$
25. $\dfrac{36}{-6} = -6$
26. $\dfrac{52}{-13} = -4$
27. $\dfrac{0}{12} = 0$
28. $\dfrac{0}{-7} = 0$
29. $-6 + [5 - (3 + 2)] = -6 + [5 - 5]$
    $= -6 + 0$
    $= -6$
30. $-8[4 + (7 - 8)] = -8[4 + (-1)]$
    $= -8[3]$
    $= -24$
31. $-8(-2) - [(4^2) + (7 - 3)] = 16 - [16 + (4)]$
    $= 16 - [20]$
    $= -4$
32. $-7(-3) - [(2^3) - (3 - 4)] = 21 - [8 - (-1)]$
    $= 21 - [9]$
    $= 12$
33. $-4 - 3(-2) + 5^2 = -4 + 6 + 25$
    $= 2 + 25$
    $= 27$
34. $-6 - 5(-8) + 3^2 = -6 + 40 + 9$
    $= -6 + 49$
    $= 43$
35. $(-8 - 5)(-2 - 1) = (-13)(-3)$
    $= 39$
36. $\dfrac{(-10 + 4) \cdot (-3)}{-7 - 2} = \dfrac{(-6) \cdot (-3)}{-9}$
    $= \dfrac{18}{-9}$
    $= -2$
37. $\dfrac{(-6 + 3) \cdot (-4)}{-5 - 1} = \dfrac{(-3) \cdot (-4)}{-6}$
    $= \dfrac{12}{-6}$
    $= -2$
38. $\dfrac{2(-5 + 3)}{-2^2} - \dfrac{(-3^2 + 2)3}{3 - (-4)} = \dfrac{2(-2)}{-4} - \dfrac{(-9 + 2)3}{3 + 4}$
    $= \dfrac{-4}{-4} - \dfrac{(-7)3}{7}$
    $= 1 - \dfrac{-21}{7}$
    $= 1 - (-3)$
    $= 1 + 3$
    $= 4$
39. $\dfrac{2(-5) + (-3)(-2^2)}{-6 + 5 + 1} = \dfrac{-10 + (-3)(-4)}{-6 + 6}$
    $= \dfrac{-10 + (-3)(-4)}{0}$
    Division by zero is undefined.
40. $\dfrac{3(-4) + (-5)(-2)}{2^3 - 2 + (-6)} = \dfrac{-12 + 10}{8 - 2 - 6}$
    $= \dfrac{-12 + 10}{0}$
    Division by zero is undefined.
41. $-\dfrac{1}{4}[3(-5) + 7(-5) + 1(-2)] = -\dfrac{1}{4}[-15 - 35 - 2]$
    $= -\dfrac{1}{4}[-52]$
    $= 13$

42. $\dfrac{5-3\left(\frac{-5-9}{-7}\right)-6}{-9-11+3\cdot 7} = \dfrac{5-3\left(\frac{-14}{-7}\right)-6}{-20+21}$
$= \dfrac{5-3(2)-6}{1}$
$= \dfrac{5-6-6}{1}$
$= -7$

43. Division by zero is undefined, so A, B, and C are all undefined.

44. Writing exercise

45. Commutative property of addition

46. Commutative property of multiplication

47. Associative property of addition

48. Commutative property of multiplication

49. Inverse property of addition

50. Identity property of addition

51. Identity property of multiplication

52. Inverse property of multiplication

53. Identity property of addition

54. Associative property of multiplication

55. Distributive property

56. Distributive property

57. Closure property of addition

58. Closure property of multiplication

59. (a) $6-8 = -2$ and $8-6 = 2$

    (b) By the results of part (a), we may conclude that subtraction is not a(n) commutative operation.

    (c) When $a = b$, it is a true statement. For example, let $a = b = 5$. Then $5 - 5 = 5 - 5$ or 0.

60. (a) $4 \div 8 = \dfrac{4}{8} = \dfrac{1}{2}$ and $8 \div 4 = 2$

    (b) By the results of part (a), we may conclude that division is not a(n) commutative operation.

    (c) This statement is true when $|a| = b| \neq 0$. For example, let $a = b = 3$. Then $3 \div 3 = 3 \div 3 = 1$.

61. (a) The inverse of cleaning up your room would be messing up your room.

    (b) The inverse of earning money would be spending money.

    (c) The inverse of increasing the volume on your CD player would be decreasing the volume.

62. (a) Putting on your shoes and putting on your socks are not commutative activities. The order of activities affects the outcome.

    (b) Getting dressed and taking a shower are not commutative activities; changing the order affects the outcome.

    (c) Combing your hair and brushing your teeth are commutative activities; the order in which they are done can be changed without affecting the outcome.

63. Jack recognized the identity property for addition.

64. (a) This could be interpreted as a difficult (test question), meaning that the question on the test is difficult. It could also be interpreted as a (difficult test) question, meaning that the question is part of a difficult test.

    (b) This could be interpreted as a woman (fearing husband), meaning that the woman fears her husband. It could also be interpreted as a (woman fearing) husband, meaning that the husband fears women.

    (c) This could be interpreted as a man (biting dog), meaning that the man is biting the dog. It could also be interpreted as a (man biting) dog, meaning that the dog is guilty of biting men.

65. Use the given hint: Let $a = 2$, $b = 3$, $c = 4$. Now test $a + (b \cdot c) = (a+b) \cdot (a+c)$.

$$a + (b \cdot c) = 2 + (3 \cdot 4) = 2 + 12$$
$$= 14$$

but,

$$(a+b) \cdot (a+c) = (2+3) \cdot (2+4) = 5 \cdot 6$$
$$= 30$$

The two expressions are not equivalent. The distributive property for addition with respect to multiplication does not hold.

66. Writing exercise

67. Writing exercise

68. First evaluate the expression using the order of operations:

$$9(11+15) = 9(26) = 234.$$

Then evaluate the expression using the distributive property:

$$9(11+15) = 9 \cdot 11 + 9 \cdot 15$$
$$= 99 + 135$$
$$= 234.$$

69. $-3^4 = -81$ The notation indicates the opposite of 3^4.

70. $-(3^4) = -81$

71. $(-3)^4 = (-3)(-3)(-3)(-3) = 81$

72. $-(-3^4) = -(-81) = 81$

73. $-(-3)^4 = -81$

74. $[-(-3)]^4 = [3]^4 = 81$

75. $-[-(-3)]^4 = -[3]^4 = -81$

76. $-[-(-3^4)] = -[-(-81)] = -[81] = -81$

77. (a) The change in outlay from 1991 to 1992 was
$$298.4 - 273.3 = \$25.1 \text{ billion.}$$
(b) The change in outlay from 1993 to 1994 was
$$279.8 - 291.1 = -\$11.3 \text{ billion.}$$
(c) The change in outlay from 1996 to 1997 was
$$258.3 - 253.3 = \$5.0 \text{ billion.}$$
(d) The change in outlay from 1997 to 1998 was
$$256.1 - 258.3 = -\$2.2 \text{ billion.}$$

78. (a) The difference between the height of Mt. Foraker and the depth of the Philippine Trench is $17,400 - (-32,995) = 50,395$ feet.
(b) The difference between the height of Pikes Peak and the depth of the Java Trench is $14,110 - (-23,376) = 37,486$ feet.
(c) To find how much deeper the Cayman Trench is than the Java Trench:
$-23,376 - (-24,721) = 1345$ feet.
(d) To find how much deeper the Philippine trench is than the Cayman Trench:
$-24,721 - (-32,995) = 8274$ feet.

79. (a) The difference between tax revenue and cost of benefits in the year 2000: $\$538 - 409 = \$129$ billion.
The difference between projected tax revenue and cost of benefits in the year 2010: $\$916 - 710 = \$206$ billion.
The difference between projected tax revenue and cost of benefits in the year 2020: $\$1479 - 1405 = \$74$ billion.
The difference between projected tax revenue and cost of benefits in the year 2030: $\$2041 - 2542 = -\$501$ billion.
(b) The cost of Social Security will exceed revenue in 2030 by $501 billion.

80. (a) 1996: $|19.07 - 21.39| = \$2.32$ million (in the red).
(b) 1997: $|67.41 - 90.29| = \$22.88$ million (in the red).
(c) 1998: $|203.20 - 177.61| = \$25.59$ million (in the black).

81. Shalita's new balance is $54 - 89 = -\$35.00$.

82. $-130 + (-54) = -184$ meters

83. $-4 + 49 = 45°F$

84. $-5 + 117 = 112°F$

85. $-27 - 14 = -41°F$

86. $15 - (-12) = 15 + 12 = 27$ feet

87. $14,494 - (-282) = 14,494 + 282 = 14,776$ feet

88. $535 - (-8) = 535 + 8 = 543$ feet

89. $660 - 2(45) - 205 = 660 - 90 - 205 = 365$ pounds

90. $(-40 + 20) - (-34) = -20 + 34 = 14$ feet

91. $6 - 12 + 43 = -6 + 43 = 37$ yards

92. $34,000 - 2500 + 3000 = 31,500 + 3000 = 34,500$ ft.

93. $-19 + 28 - 5 + 13$ Sometimes it is easier to use the commutative property of addition to change the order of the terms so that the negative numbers can be combined separately from the positive numbers as follows.
$$-19 + 28 - 5 + 13 = -19 - 5 + 28 + 13$$
$$= -24 + 41$$
$$= 17$$

94. Because these years are similar to negative numbers on a number line: $-426 + (-43) = -469$, which is 469 BC.

95.
$$-870 + 35.90 + 150 - 82.50 - 2(10) + 500 - 37.23$$
$$= -870 - 82.50 - 20 - 37.23 + 35.90 + 150 + 500$$
$$= -1009.73 + 685.90$$
$$= -323.83$$
This means that she still owes $323.83.

96. Ignoring for the moment minutes (as these races were won or lost in seconds) we have:
$$(25.86 + 10.07) - 7.04 = 28.89 \text{ seconds.}$$
Thus, the answer is 1 minute, 28.89 seconds.

6.3 EXERCISES

1. $\dfrac{4}{8} = \dfrac{1}{2} = .5 = .5\overline{0}$ which are choices A, C, and D.

The fraction $\frac{4}{8}$ can be simplified to its equivalent fraction $\frac{1}{2}$. It can also be changed to decimal form by dividing 4 by 8. Remember that the overline on zero indicates that this digit repeats indefinitely.

2. $\dfrac{2}{3} = .\overline{6} = \dfrac{20}{30} = .666\ldots$ which are choices B, C, and D.

This fraction can be changed to an equivalent fraction, $\frac{20}{30}$, by multiplying numerator and denominator by 10. It can be represented as a decimal by dividing 2 by 3; the digit 6 repeats.

3. $\dfrac{5}{9} = .\overline{5}$, which is choice C. When 5 is divided by 9, the digit 5 repeats.

4. $\dfrac{1}{4} = .25 = .24\overline{9} = \dfrac{25}{100}$, which are choices A, B, and C.

This fraction can be changed to an equivalent fraction, $\frac{25}{100}$, by multiplying numerator and denominator by 25. It can be represented as a decimal by dividing 1 by 4.

130 CHAPTER 6 THE REAL NUMBERS AND THEIR REPRESENTATIONS

Finally, it can be shown that the repeating decimal $.24\overline{9}$ is equivalent by the method of Example 8 in the text.

5. $\dfrac{16}{48} = \dfrac{16 \cdot 1}{16 \cdot 3} = \dfrac{1}{3}$

6. $\dfrac{21}{28} = \dfrac{7 \cdot 3}{7 \cdot 4} = \dfrac{3}{4}$

7. $-\dfrac{15}{35} = -\dfrac{5 \cdot 3}{5 \cdot 7} = -\dfrac{3}{7}$

8. $-\dfrac{8}{48} = -\dfrac{8 \cdot 1}{8 \cdot 6} = -\dfrac{1}{6}$

9. $\dfrac{3}{8} = \dfrac{3 \cdot 2}{8 \cdot 2} = \dfrac{6}{16}$
    $\dfrac{3}{8} = \dfrac{3 \cdot 3}{8 \cdot 3} = \dfrac{9}{24}$
    $\dfrac{3}{8} = \dfrac{3 \cdot 4}{8 \cdot 4} = \dfrac{12}{32}$

10. $\dfrac{9}{10} = \dfrac{9 \cdot 2}{10 \cdot 2} = \dfrac{18}{20}$
    $\dfrac{9}{10} = \dfrac{9 \cdot 3}{10 \cdot 3} = \dfrac{27}{30}$
    $\dfrac{9}{10} = \dfrac{9 \cdot 4}{10 \cdot 4} = \dfrac{36}{40}$

11. $-\dfrac{5}{7} = -\dfrac{5 \cdot 2}{7 \cdot 2} = -\dfrac{10}{14}$
    $-\dfrac{5}{7} = -\dfrac{5 \cdot 3}{7 \cdot 3} = -\dfrac{15}{21}$
    $-\dfrac{5}{7} = -\dfrac{5 \cdot 4}{7 \cdot 4} = -\dfrac{20}{28}$

12. $-\dfrac{7}{12} = -\dfrac{7 \cdot 2}{12 \cdot 2} = -\dfrac{14}{24}$
    $-\dfrac{7}{12} = -\dfrac{7 \cdot 3}{12 \cdot 3} = -\dfrac{21}{36}$
    $-\dfrac{7}{12} = -\dfrac{7 \cdot 4}{12 \cdot 4} = -\dfrac{28}{48}$

13. (a) $\dfrac{2}{6} = \dfrac{1}{3}$
    (b) $\dfrac{2}{8} = \dfrac{1}{4}$
    (c) $\dfrac{4}{10} = \dfrac{2}{5}$
    (d) $\dfrac{3}{9} = \dfrac{1}{3}$

14. (a) $\dfrac{12}{24} = \dfrac{1}{2}$
    (b) $\dfrac{6}{24} = \dfrac{1}{4}$
    (c) $\dfrac{12}{16} = \dfrac{3}{4}$
    (d) $\dfrac{2}{16} = \dfrac{1}{8}$

15. There are two dots in the intersection of the triangle and the rectangle. This represents 2 out of the 24 dots. As a fraction this is expressed as 2/24 or 1/12.

16. Both records can be expressed as fractions and then converted to decimals as follows.

    Tobin: $\dfrac{8}{20} =$ $8 \div 20 = .400$

    Jordan: $\dfrac{12}{30} =$ $12 \div 30 = .400$

    Their averages are the same.

17. (a) Christine O'Brien had 12 hits out of 36 at-bats. The fraction 12/36 simplifies to 1/3.

    (b) Brenda Bravener had 5 hits out of 11 at-bats. The fraction 5/11 is a little less than 1/2.

    (c) Brenda Bravener had 1 home run out of 11 at-bats. The fraction 1/11 is just less than 1/10.

    (d) Anne Kelly made 9 hits out of 40 times at bat. The fraction 9/40 is just less than 10/40, which equals 1/4.

    (e) Otis Taylor made 8 hits out of 16 times at bat; Carol Britz made 10 hits out of 20 times at bat. The fractions 8/16 and 10/20 both reduce to 1/2.

18. Writing exercise

19. $\dfrac{3}{8} + \dfrac{1}{8} = \dfrac{3+1}{8}$
    $= \dfrac{4}{8}$
    $= \dfrac{4 \cdot 1}{4 \cdot 2}$
    $= \dfrac{1}{2}$

20. $\dfrac{7}{9} + \dfrac{1}{9} = \dfrac{8}{9}$

21. $\dfrac{5}{16} \cdot \dfrac{3}{3} + \dfrac{7}{12} \cdot \dfrac{4}{4} = \dfrac{15}{48} + \dfrac{28}{48} = \dfrac{43}{48}$

22. $\dfrac{1}{15} \cdot \dfrac{6}{6} + \dfrac{7}{18} \cdot \dfrac{5}{5} = \dfrac{6}{90} + \dfrac{35}{90} = \dfrac{41}{90}$

23. $\dfrac{2}{3} \cdot \dfrac{8}{8} - \dfrac{7}{8} \cdot \dfrac{3}{3} = \dfrac{16}{24} - \dfrac{21}{24} = -\dfrac{5}{24}$

24. $\dfrac{13}{20} \cdot \dfrac{3}{3} - \dfrac{5}{12} \cdot \dfrac{5}{5} = \dfrac{39}{60} - \dfrac{25}{60}$
    $= \dfrac{14}{60}$
    $= \dfrac{7 \cdot 2}{30 \cdot 2}$
    $= \dfrac{7}{30}$

25. $\dfrac{5}{8} \cdot \dfrac{7}{7} - \dfrac{3}{14} \cdot \dfrac{4}{4} = \dfrac{35}{56} - \dfrac{12}{56} = \dfrac{23}{56}$

26. $\dfrac{19}{15} \cdot \dfrac{4}{4} - \dfrac{7}{12} \cdot \dfrac{5}{5} = \dfrac{76}{60} - \dfrac{35}{60} = \dfrac{41}{60}$

27. $\dfrac{3}{4} \cdot \dfrac{9}{5} = \dfrac{27}{20}$

6.3 RATIONAL NUMBERS AND DECIMAL REPRESENTATION

28. $\dfrac{3}{8} \cdot \dfrac{2}{7} = \dfrac{6}{56}$
    $= \dfrac{3 \cdot 2}{28 \cdot 2}$
    $= \dfrac{3}{28}$

29. $-\dfrac{2}{3} \cdot -\dfrac{5}{8} = \dfrac{10}{24} = \dfrac{5 \cdot 2}{12 \cdot 2} = \dfrac{5}{12}$

30. $-\dfrac{2}{4} \cdot \dfrac{3}{9} = -\dfrac{6}{36}$
    $= -\dfrac{6 \cdot 1}{6 \cdot 6}$
    $= -\dfrac{1}{6}$

31. $\dfrac{5}{12} \div \dfrac{15}{4} = \dfrac{5}{12} \cdot \dfrac{4}{15}$
    $= \dfrac{20}{180}$
    $= \dfrac{20 \cdot 1}{20 \cdot 9}$
    $= \dfrac{1}{9}$

32. $\dfrac{15}{16} \div \dfrac{30}{8} = \dfrac{15}{16} \cdot \dfrac{8}{30}$
    $= \dfrac{120}{480}$
    $= \dfrac{120 \cdot 1}{120 \cdot 4}$
    $= \dfrac{1}{4}$

33. $-\dfrac{9}{16} \div -\dfrac{3}{8} = -\dfrac{9}{16} \cdot -\dfrac{8}{3}$
    $= \dfrac{72}{48}$
    $= \dfrac{3 \cdot 24}{2 \cdot 24}$
    $= \dfrac{3}{2}$

34. $-\dfrac{3}{8} \div \dfrac{5}{4} = -\dfrac{3}{8} \cdot \dfrac{4}{5}$
    $= -\dfrac{12}{40}$
    $= -\dfrac{3 \cdot 4}{10 \cdot 4}$
    $= -\dfrac{3}{10}$

35. $\left(\dfrac{1}{3} \div \dfrac{1}{2}\right) + \dfrac{5}{6} = \left(\dfrac{1}{3} \cdot \dfrac{2}{1}\right) + \dfrac{5}{6}$
    $= \dfrac{2}{3} + \dfrac{5}{6}$
    $= \dfrac{2}{3} \cdot \dfrac{2}{2} + \dfrac{5}{6}$
    $= \dfrac{4}{6} + \dfrac{5}{6}$
    $= \dfrac{9}{6}$

    The fraction $\dfrac{9}{6}$ can be simplified: $\dfrac{3 \cdot 3}{3 \cdot 2} = \dfrac{3}{2}$.

36. $\dfrac{2}{5} \div \left(-\dfrac{4}{5} \div \dfrac{3}{10}\right) = \dfrac{2}{5} \div \left(-\dfrac{4}{5} \cdot \dfrac{10}{3}\right)$
    $= \dfrac{2}{5} \div \left(-\dfrac{40}{15}\right)$
    $= \dfrac{2}{5} \cdot -\dfrac{15}{40}$
    $= -\dfrac{30}{200}$
    $= -\dfrac{3}{20}$

37. (a) $6 \cdot \dfrac{3}{4} = \dfrac{6}{1} \cdot \dfrac{3}{4} = \dfrac{18}{4}$ or $4\dfrac{1}{2}$ cups

    (b) $\dfrac{1}{2}$ of $\left(\dfrac{3}{4} + 1\right) = \dfrac{1}{2} \cdot \left(\dfrac{3}{4} + \dfrac{4}{4}\right) = \dfrac{1}{2} \cdot \dfrac{7}{4} = \dfrac{7}{8}$ cup

38. (a) Other regions:
    $1 - \left(\dfrac{3}{10} + \dfrac{13}{100} + \dfrac{13}{25}\right) = \dfrac{100}{100} - \left(\dfrac{30}{100} + \dfrac{13}{100} + \dfrac{52}{100}\right)$
    $= \dfrac{100 - 95}{100}$
    $= \dfrac{5}{100} = \dfrac{1}{20}$

    (b) Latin America or Asia:
    $\dfrac{13}{25} + \dfrac{3}{10} = \dfrac{13 \cdot 4}{25 \cdot 4} + \dfrac{3 \cdot 10}{10 \cdot 10}$
    $= \dfrac{52 + 30}{100}$
    $= \dfrac{82}{100} = \dfrac{41}{50}$

    (c) Number from Europe: More than $\dfrac{13}{100} \times 8$ million $= \dfrac{13 \cdot 8}{100}$ million
    $= \dfrac{104}{100}$ million
    $= 1\dfrac{4}{100} = 1\dfrac{1}{25}$ million

39. $4 + \dfrac{1}{3} = \dfrac{4}{1} + \dfrac{1}{3} = \dfrac{12}{3} + \dfrac{1}{3} = \dfrac{13}{3}$

40. $3 + \dfrac{7}{8} = \dfrac{3}{1} + \dfrac{7}{8} = \dfrac{24}{8} + \dfrac{7}{8} = \dfrac{31}{8}$

41. $2\dfrac{9}{10} = 2 + \dfrac{9}{10} = \dfrac{2}{1} + \dfrac{9}{10} = \dfrac{20}{10} + \dfrac{9}{10} = \dfrac{29}{10}$

42. $18 \div 5 = 3\dfrac{3}{5}$

43. $27 \div 4 = 6\dfrac{3}{4}$

44. $19 \div 3 = 6\dfrac{1}{3}$

45. $3\dfrac{1}{4} + 2\dfrac{7}{8} = \dfrac{13}{4} + \dfrac{23}{8}$
$= \dfrac{26}{8} + \dfrac{23}{8}$
$= \dfrac{49}{8}$
$= 6\dfrac{1}{8}$

46. $6\dfrac{1}{5} - 2\dfrac{7}{15} = \dfrac{31}{5} - \dfrac{37}{15}$
$= \dfrac{93}{15} - \dfrac{37}{15}$
$= \dfrac{56}{15}$
$= 3\dfrac{11}{15}$

47. $-4\dfrac{7}{8} \cdot 3\dfrac{2}{3} = -\dfrac{39}{8} \cdot \dfrac{11}{3}$
$= -\dfrac{429}{24}$
$= -17\dfrac{21}{24}$
$= -17\dfrac{7}{8}$

48. $-4\dfrac{1}{6} \div 1\dfrac{2}{3} = -\dfrac{25}{6} \div \dfrac{5}{3}$
$= -\dfrac{25}{6} \cdot \dfrac{3}{5}$
$= -\dfrac{75}{30}$
$= -2\dfrac{1}{2}$

49. $\dfrac{3}{4} - \dfrac{3}{16} = \dfrac{3 \cdot 4}{4 \cdot 4} - \dfrac{3}{16} = \dfrac{12 - 3}{16} = \dfrac{9}{16}$ inch

50. $\dfrac{11}{16} - \dfrac{3}{8} = \dfrac{11}{16} - \dfrac{3 \cdot 2}{8 \cdot 2} = \dfrac{11 - 6}{16} = \dfrac{5}{16}$ inch

51. Using Method 1,

$\dfrac{\tfrac{1}{2} + \tfrac{1}{4}}{\tfrac{1}{2} - \tfrac{1}{4}} = \dfrac{3}{4} \div \dfrac{1}{4}$
$= \dfrac{3}{4} \cdot \dfrac{4}{1}$
$= \dfrac{12}{4}$
$= 3.$

52. Using Method 1,

$\dfrac{\tfrac{2}{3} + \tfrac{1}{6}}{\tfrac{2}{3} - \tfrac{1}{6}} = \dfrac{5}{6} \div \dfrac{1}{2}$
$= \dfrac{5}{6} \cdot \dfrac{2}{1}$
$= \dfrac{10}{6}$
$= \dfrac{5}{3}.$

53. Using Method 1,

$\dfrac{\tfrac{5}{8} - \tfrac{1}{4}}{\tfrac{1}{8} + \tfrac{3}{4}} = \dfrac{3}{8} \div \dfrac{7}{8}$
$= \dfrac{3}{8} \cdot \dfrac{8}{7}$
$= \dfrac{24}{56}$
$= \dfrac{3}{7}.$

54. Using Method 2,

$\dfrac{\tfrac{3}{16} + \tfrac{1}{2}}{\tfrac{5}{16} - \tfrac{1}{8}} \cdot \dfrac{16}{16} = \dfrac{3 + 8}{5 - 2}$
$= \dfrac{11}{3}.$

55. Using Method 2,

$\dfrac{\tfrac{7}{11} + \tfrac{3}{10}}{\tfrac{1}{11} - \tfrac{9}{10}} \cdot \dfrac{110}{110} = \dfrac{70 + 33}{10 - 99}$
$= \dfrac{103}{-89}$
$= -\dfrac{103}{89}.$

56. Using Method 2,

$\dfrac{\tfrac{11}{15} + \tfrac{1}{9}}{\tfrac{13}{15} - \tfrac{2}{3}} \cdot \dfrac{45}{45} = \dfrac{33 + 5}{39 - 30}$
$= \dfrac{38}{9}.$

6.3 RATIONAL NUMBERS AND DECIMAL REPRESENTATION

57. $2 + \dfrac{1}{1 + \frac{1}{3+\frac{1}{2}}} = 2 + \dfrac{1}{1 + \frac{1}{\frac{6}{2}+\frac{1}{2}}}$

$= 2 + \dfrac{1}{1 + \frac{1}{\frac{7}{2}}}$

$= 2 + \dfrac{1}{1 + \frac{2}{7}}$

$= 2 + \dfrac{1}{\frac{7}{7} + \frac{2}{7}}$

$= 2 + \dfrac{1}{\frac{9}{7}}$

$= 2 + \dfrac{7}{9}$

$= \dfrac{18}{9} + \dfrac{7}{9}$

$= \dfrac{25}{9}$

58. $4 + \dfrac{1}{2 + \frac{1}{1+\frac{1}{3}}} = 4 + \dfrac{1}{2 + \frac{1}{\frac{3}{3}+\frac{1}{3}}}$

$= 4 + \dfrac{1}{2 + \frac{1}{\frac{4}{3}}}$

$= 4 + \dfrac{1}{2 + \frac{3}{4}}$

$= 4 + \dfrac{1}{\frac{11}{4}}$

$= 4 + \dfrac{4}{11}$

$= \dfrac{44}{11} + \dfrac{4}{11}$

$= \dfrac{48}{11}$

59. $\dfrac{\frac{1}{2} + \frac{3}{4}}{2} = \dfrac{\frac{2}{4} + \frac{3}{4}}{2}$

$= \dfrac{\frac{5}{4}}{2}$

$= \dfrac{5}{4} \div \dfrac{2}{1}$

$= \dfrac{5}{4} \cdot \dfrac{1}{2}$

$= \dfrac{5}{8}$

60. $\dfrac{\frac{1}{3} + \frac{5}{12}}{2} = \dfrac{\frac{4}{12} + \frac{5}{12}}{2}$

$= \dfrac{\frac{9}{12}}{2}$

$= \dfrac{9}{12} \div \dfrac{2}{1}$

$= \dfrac{9}{12} \cdot \dfrac{1}{2}$

$= \dfrac{9}{24}$

$= \dfrac{3}{8}$

61. $\dfrac{\frac{3}{5} + \frac{2}{3}}{2} = \dfrac{\frac{9}{15} + \frac{10}{15}}{2}$

$= \dfrac{\frac{19}{15}}{2}$

$= \dfrac{19}{15} \div \dfrac{2}{1}$

$= \dfrac{19}{15} \cdot \dfrac{1}{2}$

$= \dfrac{19}{30}$

62. $\dfrac{\frac{7}{12} + \frac{5}{8}}{2} = \dfrac{\frac{14}{24} + \frac{15}{24}}{2}$

$= \dfrac{\frac{29}{24}}{2}$

$= \dfrac{29}{24} \div \dfrac{2}{1}$

$= \dfrac{29}{24} \cdot \dfrac{1}{2}$

$= \dfrac{29}{48}$

63. $\dfrac{-\frac{2}{3} + \left(-\frac{5}{6}\right)}{2} = \dfrac{-\frac{4}{6} + \left(-\frac{5}{6}\right)}{2}$

$= \dfrac{-\frac{9}{6}}{2}$

$= -\dfrac{9}{6} \div \dfrac{2}{1}$

$= -\dfrac{9}{6} \cdot \dfrac{1}{2}$

$= -\dfrac{9}{12}$

$= -\dfrac{3}{4}$

134   CHAPTER 6   THE REAL NUMBERS AND THEIR REPRESENTATIONS

64. $\dfrac{-3+\left(-\frac{5}{2}\right)}{2} = \dfrac{-\frac{6}{2}+\left(-\frac{5}{2}\right)}{2}$

$= \dfrac{-\frac{11}{2}}{2}$

$= -\dfrac{11}{2} \div \dfrac{2}{1}$

$= -\dfrac{11}{2} \cdot \dfrac{1}{2}$

$= -\dfrac{11}{4}$

65. $\dfrac{15.30 + 15.08 + 12.37 + 14.23 + 12.43 + 7.69 + 12.33 + 11.39}{8}$

$= \dfrac{100.82}{8}$

$\approx \$12.60$

66. Find the sum of the housing costs and then divide by the number of cities represented.

$$\dfrac{\$641909}{11} \approx \$58,355.$$

This is the average housing cost.

67. $\dfrac{5+9}{6+13} = \dfrac{14}{19}$

68. $\dfrac{10+13}{11+19} = \dfrac{23}{30}$

69. $\dfrac{4+9}{13+16} = \dfrac{13}{29}$

70. $\dfrac{6+13}{11+14} = \dfrac{19}{25}$

71. $\dfrac{2+3}{1+1} = \dfrac{5}{2}$

72. $\dfrac{3+4}{1+1} = \dfrac{7}{2}$

73. Using the consecutive integers 6 and 7,

$$\dfrac{6+7}{1+1} = \dfrac{13}{2} \text{ or } 6\dfrac{1}{2}.$$

The number will be halfway between the integers.

74. Writing exercise

75. $\dfrac{3}{4} = .75$

76. $\dfrac{7}{8} = .875$

77. $\dfrac{3}{16} = .1875$

78. $\dfrac{9}{32} = .28125$

79. $\dfrac{3}{11} = .\overline{27}$

80. $\dfrac{9}{11} = .\overline{81}$

81. $\dfrac{2}{7} = .\overline{285714}$

82. $\dfrac{11}{15} = .7\overline{3}$

83. $.4 = \dfrac{4}{10} = \dfrac{2 \cdot 2}{2 \cdot 5} = \dfrac{2}{5}$

84. $.9 = \dfrac{9}{10}$

85. $.85 = \dfrac{85}{100} = \dfrac{5 \cdot 17}{5 \cdot 20} = \dfrac{17}{20}$

86. $.105 = \dfrac{105}{1000} = \dfrac{5 \cdot 21}{5 \cdot 200} = \dfrac{21}{200}$

87. $.934 = \dfrac{934}{1000} = \dfrac{2 \cdot 467}{2 \cdot 500} = \dfrac{467}{500}$

88. $.7984 = \dfrac{7984}{10,000} = \dfrac{16 \cdot 499}{16 \cdot 625} = \dfrac{499}{625}$

89. $\dfrac{8}{15}: 15 = 3 \cdot 5$

Because 3 is one of the prime factors of the denominator, the fraction will yield a repeating decimal.

90. $\dfrac{8}{35}: 35 = 5 \cdot 7$

Because 7 is one of the prime factors of the denominator, the fraction will yield a repeating decimal.

91. $\dfrac{13}{125}: 125 = 5^3$

Because 5 is the only prime number that is a factor of the denominator, the fraction will yield a terminating decimal.

92. $\dfrac{3}{24} = \dfrac{1}{8}: 8 = 2^3$

Because 2 is the only prime number that is a factor of the denominator, the fraction will yield a terminating decimal.

93. $\dfrac{22}{55} = \dfrac{2 \cdot 11}{5 \cdot 11} = \dfrac{2}{5}$

Because 5 is the only prime number that is a factor of the denominator, the fraction will yield a terminating decimal.

94. $\dfrac{24}{75} = \dfrac{3 \cdot 8}{3 \cdot 25} = \dfrac{8}{25}: 25 = 5^2$

Because 5 is the only prime number that is a factor of the denominator, the fraction will yield a terminating decimal.

95. (a) The decimal representation for 1/3 is .333....

(b) The decimal representation for 2/3 is .666....

(c) .333... + .666... = .999....

(d) Writing exercise

96. $3 \cdot \dfrac{1}{3} = 3 \cdot .333\ldots$

    $1 = .999\ldots$

6.4 EXERCISES

1. This number is rational because it can be written as the ratio of one integer to another.

2. Rational

3. This number is irrational because it cannot be written as the ratio of one integer to another; only an approximation of the number can be written in this form.

4. Irrational

5. $.37 = \dfrac{37}{100}$, a rational number

6. $.91 = \dfrac{91}{100}$, a rational number

7. $.\overline{41}$, a rational number. Use Example 8 from Section 6.3 to show that it is equivalent to the rational number $\dfrac{41}{99}$.

8. $.\overline{32}$, a rational number. Use Example 8 from Section 6.3 to show that it is equivalent to the rational number $\dfrac{32}{99}$.

9. The number symbolized by π is irrational. Its value is a nonterminating, nonrepeating decimal. The values given in Exercises 13 and 14 are approximations of the value of π.

10. Zero can be written as the ratio of two integers; therefore, it is rational. An example is $\dfrac{0}{4}$.

11. This number is irrational; it is nonterminating and nonrepeating.

12. This number is irrational; it is nonterminating and nonrepeating.

13. This number is rational. It can be written as the ratio of one integer to another

    $3\dfrac{14159}{100000}$ or $\dfrac{314159}{100000}$

14. This number is rational; a ratio of integers.

15. (a)   .272772777277772...

    $+$ .616116111611116...

          .888888888888888...

    (b) Based on the result of part (a), we can conclude that the sum of two <u>irrational</u> numbers may be a(n) <u>rational</u> number.

16. (a)   .010110111011110...

    $+$ .252552555255552...

          .262662666266662...

    (b) Based on the result of part (a), we can conclude that the sum of two <u>irrational</u> numbers may be a(n) <u>irrational</u> number.

17. $\sqrt{39} \approx 6.244997998$

18. $\sqrt{44} \approx 6.633249581$

19. $\sqrt{15.1} \approx 3.885871846$

20. $\sqrt{33.6} \approx 5.796550698$

21. $\sqrt{884} \approx 29.73213749$

22. $\sqrt{643} \approx 25.35744467$

23. First find $9 \div 8 = 1.125$ on your calculator. Then take the square root of the resulting quotient.

    $\sqrt{1.125} \approx 1.060660172$

24. First find $6 \div 5 = 1.2$ on your calculator. Then take the square root of the quotient.

    $\sqrt{1.2} \approx 1.095445115$

25. $\sqrt{50} = \sqrt{25 \cdot 2} = \sqrt{25} \cdot \sqrt{2} = 5\sqrt{2}$

    Using a calculator,

    $\sqrt{50} \approx 7.071067812$

    and

    $5\sqrt{2} \approx 7.071067812.$

26. $\sqrt{32} = \sqrt{16 \cdot 2} = \sqrt{16} \cdot \sqrt{2} = 4\sqrt{2}$

    Using a calculator,

    $\sqrt{32} \approx 5.656854249$

    and

    $4\sqrt{2} \approx 5.656854249.$

27. $\sqrt{75} = \sqrt{25 \cdot 3} = \sqrt{25} \cdot \sqrt{3} = 5\sqrt{3}$

    Using a calculator,

    $\sqrt{75} \approx 8.660254038$

    and

    $5\sqrt{3} \approx 8.660254038.$

28. $\sqrt{150} = \sqrt{25 \cdot 6} = \sqrt{25} \cdot \sqrt{6} = 5\sqrt{6}$

    Using a calculator,
    $$\sqrt{150} \approx 12.24744871$$
    and
    $$5\sqrt{6} \approx 12.24744871.$$

29. $\sqrt{288} = \sqrt{144 \cdot 2} = \sqrt{144} \cdot \sqrt{2} = 12\sqrt{2}$

    Using a calculator,
    $$\sqrt{288} \approx 16.97056275$$
    and
    $$12\sqrt{2} \approx 16.97056275.$$

30. $\sqrt{200} = \sqrt{100 \cdot 2} = \sqrt{100} \cdot \sqrt{2} = 10\sqrt{2}$

    Using a calculator,
    $$\sqrt{200} \approx 14.14213562$$
    and
    $$10\sqrt{2} \approx 14.14213562.$$

31. $\dfrac{5}{\sqrt{6}} = \dfrac{5}{\sqrt{6}} \cdot \dfrac{\sqrt{6}}{\sqrt{6}} = \dfrac{5\sqrt{6}}{6}$

    Using a calculator,
    $$\dfrac{5}{\sqrt{6}} \approx 2.041241452$$
    and
    $$\dfrac{5\sqrt{6}}{6} \approx 2.041241452.$$

32. $\dfrac{3}{\sqrt{2}} = \dfrac{3}{\sqrt{2}} \cdot \dfrac{\sqrt{2}}{\sqrt{2}} = \dfrac{3\sqrt{2}}{2}$

    Using a calculator,
    $$\dfrac{3}{\sqrt{2}} \approx 2.121320344$$
    and
    $$\dfrac{3\sqrt{2}}{2} \approx 2.121320344.$$

33. $\sqrt{\dfrac{7}{4}} = \dfrac{\sqrt{7}}{\sqrt{4}} = \dfrac{\sqrt{7}}{2}$

    Using a calculator,
    $$\sqrt{\dfrac{7}{4}} \approx 1.322875656$$
    and
    $$\dfrac{\sqrt{7}}{2} \approx 1.322875656.$$

34. $\sqrt{\dfrac{8}{9}} = \dfrac{\sqrt{8}}{\sqrt{9}} = \dfrac{\sqrt{8}}{3} = \dfrac{\sqrt{4} \cdot \sqrt{2}}{3} = \dfrac{2\sqrt{2}}{3}$

    Using a calculator,
    $$\sqrt{\dfrac{8}{9}} \approx .9428090416$$
    and
    $$\dfrac{2\sqrt{2}}{3} \approx .9428090416.$$

35. $\sqrt{\dfrac{7}{3}} = \dfrac{\sqrt{7}}{\sqrt{3}} \cdot \dfrac{\sqrt{3}}{\sqrt{3}} = \dfrac{\sqrt{21}}{3}$

    Using a calculator,
    $$\sqrt{\dfrac{7}{3}} \approx 1.527525232$$
    and
    $$\dfrac{\sqrt{21}}{3} \approx 1.527525232.$$

36. $\sqrt{\dfrac{14}{5}} = \dfrac{\sqrt{14}}{\sqrt{5}} \cdot \dfrac{\sqrt{5}}{\sqrt{5}} = \dfrac{\sqrt{70}}{5}$

    Using a calculator,
    $$\sqrt{\dfrac{14}{5}} \approx 1.673320053$$
    and
    $$\dfrac{\sqrt{70}}{5} \approx 1.673320053.$$

37. $\sqrt{17} + 2\sqrt{17} = (1+2)\sqrt{17} = 3\sqrt{17}$

38. $3\sqrt{19} + \sqrt{19} = (3+1)\sqrt{19} = 4\sqrt{19}$

39. $5\sqrt{7} - \sqrt{7} = (5-1)\sqrt{7} = 4\sqrt{7}$

40. $3\sqrt{27} - \sqrt{27} = (3-1)\sqrt{27} = 2\sqrt{27}$

    Notice that $\sqrt{27}$ can be simplified:
    $$\sqrt{27} = \sqrt{9} \cdot \sqrt{3} = 3\sqrt{3}.$$

The final answer is then
$$2 \cdot 3\sqrt{3} = 6\sqrt{3}.$$

41. $3\sqrt{18} + \sqrt{2} = 3\sqrt{9 \cdot 2} + \sqrt{2}$
$= 3\sqrt{9} \cdot \sqrt{2} + \sqrt{2}$
$= 3 \cdot 3\sqrt{2} + \sqrt{2}$
$= 9\sqrt{2} + \sqrt{2}$
$= (9+1)\sqrt{2}$
$= 10\sqrt{2}$

42. $2\sqrt{48} - \sqrt{3} = 2\sqrt{16 \cdot 3} - \sqrt{3}$
$= 2\sqrt{16} \cdot \sqrt{3} - \sqrt{3}$
$= 2 \cdot 4\sqrt{3} - \sqrt{3}$
$= 8\sqrt{3} - \sqrt{3}$
$= (8-1)\sqrt{3}$
$= 7\sqrt{3}$

43. $-\sqrt{12} + \sqrt{75} = -\sqrt{4 \cdot 3} + \sqrt{25 \cdot 3}$
$= -\sqrt{4} \cdot \sqrt{3} + \sqrt{25} \cdot \sqrt{3}$
$= -2\sqrt{3} + 5\sqrt{3}$
$= (-2+5)\sqrt{3}$
$= 3\sqrt{3}$

44. $2\sqrt{27} - \sqrt{300} = 2\sqrt{9 \cdot 3} - \sqrt{100 \cdot 3}$
$= 2\sqrt{9} \cdot \sqrt{3} - \sqrt{100} \cdot \sqrt{3}$
$= 2 \cdot 3\sqrt{3} - 10\sqrt{3}$
$= (6-10)\sqrt{3}$
$= -4\sqrt{3}$

45.

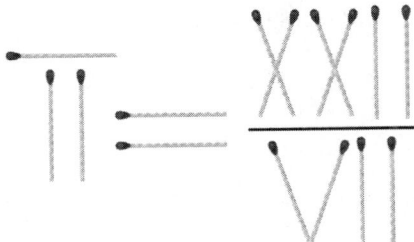

46. $\sqrt{\dfrac{2143}{22}} \approx 9.869604395\ldots$

Then the take the square root of this number.
$$\sqrt{9.869604395\ldots} \approx 3.141592653\ldots$$

The result agrees with the first nine digits of π.

47. $355 \div 113 = 3.1415929$, which agrees with the first seven digits in the decimal for π.

48. The values $22/7$ and 3.14 are rational numbers that are approximations for π.

49. Using 3.14 for π and the given formula:
$$P = 2\pi\sqrt{\dfrac{L}{32}}$$
$$P \approx 2(3.14)\sqrt{\dfrac{5.1}{32}}$$
$$P \approx 6.28\sqrt{.159375}$$
$$P \approx 2.5 \text{ seconds.}$$

50. Substitute the given values into the formula and follow the order of operations.
$$r = \dfrac{-h + \sqrt{h^2 + .64S}}{2}$$
$$r = \dfrac{-12 + \sqrt{12^2 + .64(400)}}{2}$$
$$r = \dfrac{-12 + \sqrt{144 + 256}}{2}$$
$$r = \dfrac{-12 + \sqrt{400}}{2}$$
$$r = \dfrac{-12 + 20}{2}$$
$$r = \dfrac{8}{2}$$
$$r = 4 \text{ inches}$$

51. Add 6 and 28 to get the total height.
$$H = 6 + 28 = 34$$

Substitute the given values into the formula and follow the order of operations.
$$D = \sqrt{2H}$$
$$D = \sqrt{2 \cdot 34}$$
$$D = \sqrt{68}$$
$$D \approx 8.2 \text{ miles}$$

52. Substitute the given values into the formula and follow the order of operations.
$$I = \sqrt{\dfrac{2P}{L}}$$
$$I = \sqrt{\dfrac{2 \cdot 120}{80}}$$
$$I = \sqrt{\dfrac{240}{80}}$$
$$I = \sqrt{3}$$
$$I \approx 1.7 \text{ amps}$$

53. The semiperimeter, s, of the Bermuda triangle is $\tfrac{1}{2}(850 + 925 + 1300)$ or 1537.5 miles.
$$\sqrt{1537.5(1537.5 - 850)(1537.5 - 925)(1537.5 - 1300)}$$
$$= \sqrt{1537.5(687.5)(612.5)(237.5)}$$
$$\approx 392,000 \text{ square miles}$$

138 CHAPTER 6 THE REAL NUMBERS AND THEIR REPRESENTATIONS

54. The semiperimeter, s, of the Vietnam Veterans' Memorial is $\frac{1}{2}(246.75 + 246.75 + 438.14)$ or 465.82 feet.
$$\sqrt{465.82(465.82 - 246.75)(465.82 - 246.75)(465.82 - 438.14)}$$
$$= \sqrt{465.82(219.07)(219.07)(27.68)}$$
$$= \sqrt{261750.8226}$$
$$\approx 24,900 \text{ square feet}$$

55. Substitute the given values into the formula and follow the order of operations with $L = 4$, $W = 3$, and $H = 2$.
$$D = \sqrt{L^2 + W^2 + H^2}$$
$$D = \sqrt{4^2 + 3^2 + 2^2}$$
$$D = \sqrt{16 + 9 + 4}$$
$$D = \sqrt{29}$$
$$D \approx 5.4 \text{ feet}$$

56. Substitute the given values into the formula and follow the order of operations:
$$r = \frac{\sqrt{A} - \sqrt{P}}{\sqrt{P}}$$
$$r = \frac{\sqrt{A} - \sqrt{P}}{\sqrt{P}} \cdot \frac{\sqrt{P}}{\sqrt{P}}$$
$$r = \frac{(\sqrt{A} - \sqrt{P})\sqrt{P}}{P}$$
$$r = \frac{(\sqrt{58,320} - \sqrt{50,000})\sqrt{50,000}}{50,000}$$
$$r = \frac{\sqrt{2916000000} - 50,000}{50,000}$$
$$r = \frac{54,000 - 50,000}{50,000}$$
$$r = \frac{4,000}{50,000}$$
$$r = .08.$$

57. (a) Substitute the given values into the formula and follow the order of operations.
$$s = 30\sqrt{\frac{a}{p}}$$
$$s = 30\sqrt{\frac{862}{156}}$$
$$s \approx 30\sqrt{3.979166667}$$
$$s \approx 70.5 \text{ miles per hour}$$

(b) Substitute the given values into the formula and follow the order of operations.
$$s = 30\sqrt{\frac{a}{p}}$$
$$s = 30\sqrt{\frac{382}{96}}$$
$$s \approx 30\sqrt{3.979166667}$$
$$s \approx 59.8 \text{ miles per hour}$$

(c) Substitute the given values into the formula and follow the order of operations.
$$s = 30\sqrt{\frac{a}{p}}$$
$$s = 30\sqrt{\frac{84}{26}}$$
$$s \approx 30\sqrt{3.230769231}$$
$$s \approx 53.9 \text{ miles per hour}$$

58. (a) Substitute the given values into the formula and follow the order of operations.
$$d = 1.22\sqrt{x}$$
$$d = 1.22\sqrt{15,000}$$
$$d = 1.22 \cdot 122.4744871$$
$$d \approx 149.4 \text{ miles}$$

(b) Substitute the given values into the formula and follow the order of operations.
$$d = 1.22\sqrt{18,000}$$
$$d \approx 163.7 \text{ miles}$$

(c) Substitute the given values into the formula and follow the order of operations.
$$d = 1.22\sqrt{24,000}$$
$$d \approx 189.0 \text{ miles}$$

59. $\sqrt[3]{64} = 4$

60. $\sqrt[3]{125} = 5$

61. $\sqrt[3]{343} = 7$

62. $\sqrt[3]{729} = 9$

63. $\sqrt[3]{216} = 6$

64. $\sqrt[3]{512} = 8$

65. $\sqrt[4]{1} = 1$

66. $\sqrt[4]{16} = 2$

6.5 APPLICATIONS OF DECIMALS AND PERCENTS

67. $\sqrt[4]{256} = 4$

68. $\sqrt[4]{625} = 5$

69. $\sqrt[4]{4096} = 8$

70. $\sqrt[4]{2401} = 7$

71. $\sqrt[3]{43} \approx 3.50339806$

72. $\sqrt[3]{87} \approx 4.431047622$

73. $\sqrt[3]{198} \approx 5.828476683$

74. $\sqrt[3]{2107} \approx 6.775106617$

75. $\sqrt{10265.2} \approx 10.06565066$

76. $\sqrt[3]{863.5} \approx 5.420827475$

77. Substitute the given values into the formula and follow the order of operations.

$$h = 12.3\sqrt[3]{T}$$
$$h = 12.3\sqrt[3]{180}$$
$$h \approx 12.3(5.64622)$$
$$h \approx 69.4 \text{ inches}$$

This is approximately 5.8 feet or about 5 feet, 9 inches.

78. We have:
$$f_1 = f_2\sqrt{\frac{F_1}{F_2}}$$
$$f_1 = 260\sqrt{\frac{300}{60}}$$
$$f_1 = 260\sqrt{5}$$
$$f_1 \approx 581.$$

79. Use a calculator to verify that:

(a) $2^{\frac{1}{2}} = \sqrt{2} \approx 1.414213562.$

(b) $7^{\frac{1}{2}} = \sqrt{7} \approx 2.645751311.$

(c) $13.2^{\frac{1}{2}} = \sqrt{13.2} \approx 3.633180425.$

(d) $25^{\frac{1}{2}} = \sqrt{25} \approx 5.$

80. (a) The expression $a^{\frac{1}{3}}$ as a radical would be $\sqrt[3]{a}$.

(b) $\sqrt[3]{16} \approx 2.5198421$

(c) $16^{\frac{1}{3}} \approx 2.5198421$

81. $(1.1)^{10} \approx 2.59374246$

$(1.01)^{100} \approx 2.704813829$

$(1.001)^{1000} \approx 2.716923932$

$(1.0001)^{10,000} \approx 2.718145927$

$(1.00001)^{100,000} \approx 2.718268273$

The computed values seem to be approaching the value of e.

82. (a) $e^2 \approx 7.389056099$

(b) $e^3 \approx 20.08553692$

(c) $\sqrt{e} \approx 1.648721271$

6.5 EXERCISES

1. True. $3.00(12) = 36$

2. True. $25\% = .25 = \dfrac{1}{4}$

3. False. When 759.367 is rounded to the nearest hundredth, the result is 759.37.

4. False. When 759.367 is rounded to the nearest hundred, the result is 800.

5. True. $50\% = .5 = \dfrac{1}{2}$, and multiplying by one half yields the same result as dividing by 2.

6. True. $\dfrac{12}{12+8} = \dfrac{12}{20} \cdot \dfrac{5}{5} = \dfrac{60}{100} = 60\%$

7. True. $.70(50) = 35$

8. False. $.40(120) = 48$, and 30 is less than 48.

9. False. $.99 \text{¢} = \dfrac{99}{100}$ cent, meaning that it is less than the value of one penny.

10. False. $.10(70) = 7$, indicating the item will be $7 less than the original price of $70. $70 - 7 = \$63$.

11. $8.53 + 2.785 = 11.315$

12. $9.358 + 7.2137 = 16.5717$

13. $8.74 - 12.955 = -4.215$

14. $2.41 - 3.997 = -1.587$

15. $25.7 \times .032 = .8224$

16. $45.1 \times 8.344 = 376.3144$

17. $1019.825 \div 21.47 = 47.5$

18. $-262.563 \div 125.03 = -2.1$

19. $\dfrac{118.5}{1.45 + 2.3} = \dfrac{118.5}{3.75} = 31.6$

20. $2.45(1.2 + 3.4 - 5.6) = 2.45(-1) = -2.45$

21. Philadelphia had the greatest change. The population declined 10.6%.

22. Chicago had the smallest change. The population increased .6%.

23. The indicated price is .33¢, which means $\dfrac{33}{100}$ of a penny. You should be able to purchase three stamps for one cent.

24. The total price should be 33 cents rather than 33 dollars.

25. Since
$$\text{Taxes} = \frac{\text{home value} \times .5485 - 4850}{1000} \times 31.44,$$
substitute given home values and follow order of operations.

   (a) $\text{Taxes} = \dfrac{100000 \times .5485 - 4850}{1000} \times 31.44$
   $= (50) \times 31.44$
   $= \$1572$

   (b) $\text{Taxes} = \dfrac{150000 \times .5485 - 4850}{1000} \times 31.44$
   $= (77.425) \times 31.44$
   $\approx \$2434$

   (c) $\text{Taxes} = \dfrac{200000 \times .5485 - 4850}{1000} \times 31.44$
   $= (104.85) \times 31.44$
   $\approx \$3296$

26. (a) $\$1742.18 + \$9271.94 + \$28.37 = \$11{,}042.49$

   (b) $\$7195.14 + \$511.09 + \$1291.03 = \$8997.26$

   (c) $(\$1856.12 + \$11{,}042.49) - \$8997.26 = \$3901.35$

27. (a) $\text{BAC} = \dfrac{[11.52]}{190} - (.03) \approx .031$

   (b) $\text{BAC} = \dfrac{[10.8]}{135} - (.045) = .035$

28. Substitute given values and follow order of operations.
$$\text{MPH} = \frac{5600 \times 26}{3.12 \times 336}$$
$$\text{MPH} = \frac{145600}{1048.32}$$
$$\text{MPH} \approx 139$$

29. Substitute given values and follow order of operations.
$$\text{Horsepower} = \frac{195 \times 302 \times 4000}{792{,}000}$$
$$\text{Horsepower} = \frac{235{,}560{,}000}{792{,}000}$$
$$\text{Horsepower} \approx 297$$

30. Substitute given values and follow order of operations.
$$\text{Torque} = \frac{5252 \times 400}{4500}$$
$$\text{Torque} = \frac{2{,}100{,}800}{4500}$$
$$\text{Torque} \approx 467$$

31. (a) 78.4
   (b) 78.41

32. (a) 3689.5
   (b) 3689.54

33. (a) .1
   (b) .08

34. (a) .1
   (b) .07

35. (a) 12.7
   (b) 12.69

36. (a) 44.0
   (b) 44.00

37. $.42 = \dfrac{42}{100} = 42\%$

38. $.87 = \dfrac{87}{100} = 87\%$

39. $.365 = \dfrac{365}{1000} \div \dfrac{10}{10} = \dfrac{36.5}{100} = 36.5\%$

40. $.792 = \dfrac{792}{1000} \div \dfrac{10}{10} = \dfrac{79.2}{100} = 79.2\%$

41. $.008 = \dfrac{8}{1000} \div \dfrac{10}{10} = \dfrac{.8}{100} = .8\%$

42. $.0093 = \dfrac{93}{10{,}000} \div \dfrac{100}{100} = \dfrac{.93}{100} = .93\%$

43. $2.1 = 2\dfrac{1}{10} \cdot \dfrac{10}{10} = 2\dfrac{10}{100} = \dfrac{210}{100} = 210\%$

44. $8.9 = 8\dfrac{9}{10} \cdot \dfrac{10}{10} = 8\dfrac{90}{100} = \dfrac{890}{100} = 890\%$

45. $\dfrac{1}{5} = 1 \div 5 = .2$, which is 20%.

46. $\dfrac{2}{5} = 2 \div 5 = .4$, which is 40%.

47. $\dfrac{1}{100} = 1 \div 100 = .01$, which is 1%.

48. $\dfrac{1}{50} = 1 \div 50 = .02$, which is 2%.

49. $\dfrac{3}{8} = 3 \div 8 = .375$, which is 37.5%.

50. $\dfrac{5}{6} = 5 \div 6 = .83\dfrac{1}{3}$, which is $83\dfrac{1}{3}\%$.

51. $\dfrac{3}{2} = 3 \div 2 = 1.5$, which is 150%.

52. $\dfrac{7}{4} = 7 \div 4 = 1.75$, which is 175%.

53. Writing exercise

54. (a) 25% matches letter E, $\dfrac{1}{4}$.

   (b) 10% matches letter D, $\dfrac{1}{10}$.

6.5 APPLICATIONS OF DECIMALS AND PERCENTS

(c) 2% matches letter B, $\frac{1}{50}$.

(d) 20% matches letter F, $\frac{1}{5}$.

(e) 75% matches letter C, $\frac{3}{4}$.

(f) $33\frac{1}{3}$% matches letter A, $\frac{1}{3}$.

55. (a) 5% means $\underline{5}$ in every 100.

(b) 25% means 6 in every $\underline{24}$.

(c) 200% means $\underline{8}$ for every 4.

(d) .5% means $\underline{.5}$ in every 100.

(e) $\underline{600}$ % means 12 for every 2.

56. The total number of elements in all the regions is 40.

(a) 6 out of 40 is 15%.

(b) 16 out of 40 is 40%.

(c) 6 out of 30 is 20%.

(d) 10 out of 40 is 25%.

57. No. If the item is discounted 20%, its new price is $60 - .2 \times $60 = 48. Then, if the new price is marked up 20%, the price becomes $48 + .2 \times $48 = 57.60.

58. (a) $\frac{2}{6} = \frac{1}{3} = .333\ldots = 33\frac{1}{3}\%$

(b) $\frac{2}{8} = \frac{1}{4} = .25 = 25\%$

(c) $\frac{4}{10} = .4 = 40\%$

(d) $\frac{3}{9} = \frac{1}{3} = .333\ldots = 33\frac{1}{3}\%$

59. (a) Minnesota: $\frac{90}{90 + 66} = \frac{90}{156} = .577$

(b) Chicago: $\frac{79}{79 + 78} = \frac{79}{157} = .503$

(c) Cleveland: $\frac{70}{70 + 87} = \frac{70}{157} = .446$

60. (a) Atlanta: $\frac{97}{97 + 58} = \frac{97}{155} = .626$

(b) Florida: $\frac{76}{76 + 81} = \frac{76}{157} = .484$

(c) New York: $\frac{74}{74 + 82} = \frac{74}{156} = .474$

61. Using Method 1: $(.26)(480) = 124.8$.

62. Using Method 1: $(.38)(12) = 4.56$.

63. Using Method 1: $(.105)(28) = 2.94$.

64. Using Method 1: $(.486)(19) = 9.234$.

65. Using Method 2:
$$\frac{x}{100} = \frac{45}{30}$$
$$30x = 4500$$
$$x = \frac{4500}{30}$$
$$x = 150\%.$$

66. Using Method 2:
$$\frac{x}{100} = \frac{20}{48}$$
$$48x = 2000$$
$$x = \frac{2000}{48}$$
$$x = 41.\overline{6}\%.$$

67. Using Method 2:
$$\frac{25}{100} = \frac{150}{x}$$
$$25x = 15000$$
$$x = \frac{15000}{25}$$
$$x = 600.$$

68. Using Method 2:
$$\frac{12}{100} = \frac{3600}{x}$$
$$12x = 360000$$
$$x = \frac{360000}{12}$$
$$x = 30,000.$$

69. Using Method 1:
$$(x)(28) = .392$$
$$x = \frac{.392}{28}$$
$$x = .014$$
$$x = 1.4\%.$$

70. Using Method 1:
$$(x)(292) = 78.84$$
$$x = \frac{78.84}{292}$$
$$x = .27$$
$$x = 27\%.$$

71. Choice A. The difference between $5 and $4 is $1; $1 compared to the original $4 is 1/4 or 25%.

72. Choice D. The difference between $5000 and $2000 is $3000; 3000 compared to the original 5000 is 3/5 or 60%.

73. Choice C. Rounding, the population of Alabama is approximately 4,000,000. About 25% of this number is $1/4 \times 4,000,000 = 1,000,000$.

74. Choice A. 20% of 1,200,000 is 240,000.

75. Price increase was $2.50 - 1.50 = 1$. The percent increase is $\frac{1}{1.50}(100) \approx 67\%$.

76. Price decrease was about $2.50 - 1.55 = .95$ The percent decrease is about $\frac{.95}{2.50}(100) \approx 40\%$.

77. (a) $\frac{20.48}{4131.15} \approx .005$. This is .5%.

    (b) $\frac{71.36}{(11489.36 + 71.36)} = \frac{71.36}{11560.72} \approx .006$, or .6%.

78. The number of members the previous year is
    $$16,500,000 - 265,000 = 16,235,000.$$
    Therefore, the percentage increase is
    $$\frac{265,000}{16,235,000} \approx .0163.$$
    This is 1.63%

79. $0.164(8450) = \$1385.80$

80. $\frac{2500 - 625}{625} = \frac{1875}{625}$
    $= 3$ or 300%

81. $\frac{340}{1500} = .226666...$
    $\approx 22.7\%$

82. The customer was able to save 5.5% of $26,410.
    $$.055(26410) \approx \$1452.55$$
    The amount paid for the car was
    $26410 - 1452.55 = \$24,957.45.$

83. $29.57
    1. Rounded to the nearest dollar, the amount of the bill is $30.
    2. 10% of $30 is $3.
    3. 1/2 of $3 is $1.50. $3 + 1.50 = \$4.50$.

84. $38.32
    1. Rounded to the nearest dollar, the amount of the bill is $38.
    2. 10% of $38 is $3.80.
    3. 1/2 of $3.80 is $1.90. $3.80 + 1.90 = \$5.70$.

85. $5.15
    1. Rounded to the nearest dollar, the amount of the bill is $5.
    2. 10% of $5 is $.50.
    3. 1/2 of $.50 is $.25. $.50 + .25 = \$.75$.

86. $7.89
    1. Rounded to the nearest dollar, the amount of the bill is $8.
    2. 10% of $8 is $.80.
    3. 1/2 of $.80 is $.40. $.80 + .40 = \$1.20$.

87. $59.36 \approx 59$
    10% of $59 = 5.9 \approx 6$
    $2 \times 6 = \$12.00$

88. $40.24 \approx 40$
    10% of $40 = 4$
    $2 \times 4 = \$8.00$

89. $180.43 \approx 180$
    10% of $180 = 18$
    $2 \times 18 = \$36.00$

90. $199.86 \approx 200$
    10% of $200 = 20$
    $2 \times 20 = \$40.00$

91. Writing exercise

92. Writing exercise

EXTENSION: COMPLEX NUMBERS

1. $\sqrt{-144} = i\sqrt{144} = 12i$

2. $\sqrt{-196} = i\sqrt{196} = 14i$

3. $-\sqrt{-225} = -i\sqrt{225} = -15i$

4. $-\sqrt{-400} = -i\sqrt{400} = -20i$

5. $\sqrt{-3} = i\sqrt{3}$

6. $\sqrt{-19} = i\sqrt{19}$

7. $\sqrt{-75} = i\sqrt{25 \cdot 3} = i\sqrt{25} \cdot \sqrt{3} = 5i\sqrt{3}$

8. $\sqrt{-125} = i\sqrt{25 \cdot 5} = i\sqrt{25} \cdot \sqrt{5} = 5i\sqrt{5}$

9. $\sqrt{-5} \cdot \sqrt{-5} = i\sqrt{5} \cdot i\sqrt{5}$
   $= i^2\sqrt{5 \cdot 5}$
   $= i^2\sqrt{25}$
   $= 5i^2$
   $= 5(-1)$
   $= -5$

10. $\sqrt{-3} \cdot \sqrt{-3} = i\sqrt{3} \cdot i\sqrt{3}$
    $= i^2\sqrt{3 \cdot 3}$
    $= i^2\sqrt{9}$
    $= 3i^2$
    $= 3(-1)$
    $= -3$

11. $\sqrt{-9} \cdot \sqrt{-36} = i\sqrt{9} \cdot i\sqrt{36}$
    $= 3i \cdot 6i$
    $= 18i^2$
    $= 18(-1)$
    $= -18$

12. $\sqrt{-4} \cdot \sqrt{-81} = i\sqrt{4} \cdot i\sqrt{81}$
    $= 2i \cdot 9i$
    $= 18i^2$
    $= 18(-1)$
    $= -18$

13. $\sqrt{-16} \cdot \sqrt{-100} = i\sqrt{16} \cdot i\sqrt{100}$
    $= i^2 \cdot 4 \cdot 10$
    $= 40i^2$
    $= 40(-1)$
    $= -40$

14. $\sqrt{-81} \cdot \sqrt{-121} = i\sqrt{81} \cdot i\sqrt{121}$
    $= 9i \cdot 11i$
    $= 99i^2$
    $= 99(-1)$
    $= -99$

15. $\dfrac{\sqrt{-200}}{\sqrt{-100}} = \dfrac{i\sqrt{200}}{i\sqrt{100}}$
    $= \sqrt{\dfrac{200}{100}}$
    $= \sqrt{2}$

    $\dfrac{\sqrt{-200}}{\sqrt{-100}} = \dfrac{i\sqrt{200}}{i\sqrt{100}}$
    $= \sqrt{\dfrac{200}{100}}$
    $= \sqrt{2}$

16. $\dfrac{\sqrt{-50}}{\sqrt{-2}} = \dfrac{i\sqrt{50}}{i\sqrt{2}}$
    $= \sqrt{\dfrac{50}{2}}$
    $= \sqrt{25}$
    $= 5$

17. $\dfrac{\sqrt{-54}}{\sqrt{6}} = \dfrac{i\sqrt{54}}{\sqrt{6}}$
    $= i\sqrt{\dfrac{54}{6}}$
    $= i\sqrt{9}$
    $= 3i$

18. $\dfrac{\sqrt{-90}}{\sqrt{10}} = \dfrac{i\sqrt{90}}{\sqrt{10}}$
    $= i\sqrt{\dfrac{90}{10}}$
    $= i\sqrt{9}$
    $= 3i$

19. $\dfrac{\sqrt{-288}}{\sqrt{-8}} = \dfrac{i\sqrt{288}}{i\sqrt{8}}$
    $= \sqrt{\dfrac{288}{8}}$
    $= \sqrt{36}$
    $= 6$

20. $\dfrac{\sqrt{-48} \cdot \sqrt{-3}}{\sqrt{-2}} = \dfrac{i\sqrt{48} \cdot i\sqrt{3}}{i\sqrt{2}}$
    $= \dfrac{i^2\sqrt{144}}{i\sqrt{2}}$
    $= i\sqrt{\dfrac{144}{2}}$
    $= i\sqrt{72}$
    $= i\sqrt{36 \cdot 2}$
    $= i\sqrt{36} \cdot \sqrt{2}$
    $= 6i\sqrt{2}$

21. Writing exercise

22. Writing exercise

23. $i^8 = i^4 \cdot i^4 = 1 \cdot 1 = 1$

24. $i^{16} = (i^4)^4 = (1)^4 = 1$

25. i^{42}
    $42 \div 4 = 10$, with a remainder of 2. This means that $i^{42} = i^2$, which is -1.

26. i^{86}
    $86 \div 4 = 21$, with a remainder of 2. This means that $i^{86} = i^2$, which is -1.

27. i^{47}
    $47 \div 4 = 11$, with a remainder of 3. This means that $i^{47} = i^3$, which is $-i$.

28. i^{63}
    $63 \div 4 = 15$, with a remainder of 3. This means that $i^{63} = i^3$, which is $-i$.

29. i^{101}
    $101 \div 4 = 25$, with a remainder of 1. This means that $i^{101} = i$.

30. i^{141}
    $141 \div 4 = 35$, with a remainder of 1. This means that $i^{141} = i$.

31. Writing exercise

32. Writing exercise

CHAPTER 6 TEST

1. $\{-4, -\sqrt{5}, -3/2, -.5, 0, \sqrt{3}, 4.1, 12\}$

    (a) The only natural number is 12.

    (b) Whole numbers are 0 and 12.

    (c) Integers are -4, 0, and 12.

    (d) Rational numbers are -4, $-3/2$, $-.5$, 0, 4.1, and 12.

    (e) Irrational numbers are $-\sqrt{5}$ and $\sqrt{3}$.

    (f) All the numbers in the set are real numbers.

2. (a) C

    (b) B

    (c) D

    (d) A

3. (a) False. The absolute value of a number is always nonnegative; the absolute value of zero is zero.

    (b) True. Both sides of the equation are equal to 7.

    (c) True.

    (d) False. Zero, a real number, is neither positive nor negative.

4. $6^2 - 4(9-1) = 36 - 4(8)$
    $= 36 - 32$
    $= 4$

5. $(-3)(-2) - [5 + (8-10)] = 6 - [5 + (-2)]$
    $= 6 - [3]$
    $= 3$

6. $\dfrac{(-8+3) - (5+10)}{7-9} = \dfrac{-5-15}{-2}$
    $= \dfrac{-20}{-2}$
    $= 10$

7. (a) $-6439 - 5039 = -11,478$

    (b) $2284 - 20,060 = -17,776$

    (c) $-5588 - 1377 \approx -6000 - 1000 = -7000$; Option C

    (d) $1377 - (-17929) \approx 1000 - (-18000) = 1000 + 18000$; Option C

8. $(225 + 3852) - (-1299 + 80) = 5296$ feet.
    After   Before

9. $97,069 - 88,140 = 8929.$
    $86,133 - 97,069 = -10,936$
    $71,558 - 86,133 = -14,575$
    $71,128 - 71,558 = -430$
    $71,931 - 71,128 = 803.$

10. (a) E, Associative property

    (b) A, Distributive property

    (c) B, Identity property

    (d) D, Commutative property

    (e) F, Inverse property

    (f) C, Closure property

11. (a) Whitney made 4 out of 6, which is more than 1/2.

    (b) Moura made 13 out of 40; because 13 out of 39 would be 1/3, this is just less than 1/3. Dawkins made 2 out of 7; because 2 out of 6 would be 1/3, this is just less than 1/3.

    (c) Whitney made 4 out of 6, which is the same ratio as 2 out of 3.

    (d) Pritchard made 4 out of 10; Miller made 8 out of 20.
    $$\dfrac{4}{10} = \dfrac{8}{20} = \dfrac{2}{5}$$

12. $\dfrac{3}{16} + \dfrac{1}{2} = \dfrac{3}{16} + \dfrac{1}{2} \cdot \dfrac{8}{8}$
    $= \dfrac{3}{16} + \dfrac{8}{16}$
    $= \dfrac{11}{16}$

13. $\dfrac{9}{20} - \dfrac{3}{32} = \dfrac{9}{20} \cdot \dfrac{8}{8} - \dfrac{3}{32} \cdot \dfrac{5}{5}$
    $= \dfrac{72}{160} - \dfrac{15}{160}$
    $= \dfrac{57}{160}$

14. $\dfrac{3}{8} \cdot \left(-\dfrac{16}{15}\right) = -\dfrac{48}{120}$
    $= -\dfrac{2 \cdot 24}{5 \cdot 24}$
    $= -\dfrac{2}{5}$

15. $\dfrac{7}{9} \div \dfrac{14}{27} = \dfrac{7}{9} \cdot \dfrac{27}{14}$
    $= \dfrac{7 \cdot 3 \cdot 9}{7 \cdot 2 \cdot 9}$
    $= \dfrac{3}{2}$

16. (a) $\dfrac{9}{20} = .45$

    (b) $\dfrac{5}{12} = .41\overline{6}$

17. (a) $.72 = \dfrac{72}{100} = \dfrac{4 \cdot 18}{4 \cdot 25} = \dfrac{18}{25}$

(b) $.\overline{58}$
Let $x = .\overline{58}$
$$100x = 58.585858\ldots$$
$$-x = .585858\ldots$$
$$99x = 58$$
$$x = \dfrac{58}{99}$$

18. (a) $\sqrt{10}$ is irrational because it cannot be written as the ratio of two integers.

(b) $\sqrt{16}$ is rational. Its value is 4, which can be written as 4/1.

(c) $.01 = 1/100$, a rational number.

(d) $.\overline{01}$ can be converted to 1/99, a rational number.

(e) $.0101101110\ldots$ is an irrational number.

19. (a) $\sqrt{150} \approx 12.24744871$

(b) $\sqrt{150} = \sqrt{25} \cdot \sqrt{6} = 5\sqrt{6}$

20. (a) $\dfrac{13}{\sqrt{7}} \approx 4.913538149$

(b) $\dfrac{13}{\sqrt{7}} \cdot \dfrac{\sqrt{7}}{\sqrt{7}} = \dfrac{13\sqrt{7}}{\sqrt{49}}$
$= \dfrac{13\sqrt{7}}{7}$

21. (a) $2\sqrt{32} - 5\sqrt{128} \approx -45.254834$

(b) $2\sqrt{32} - 5\sqrt{128} = 2\sqrt{16 \cdot 2} - 5\sqrt{64 \cdot 2}$
$= 2 \cdot 4\sqrt{2} - 5 \cdot 8\sqrt{2}$
$= 8\sqrt{2} - 40\sqrt{2}$
$= -32\sqrt{2}$

22. Writing exercise

23. (a) $4.6 + 9.21 = 13.81$

(b) $12 - 3.725 - 8.59 = -.315$

(c) $86(.45) = 38.7$

(d) $236.439 \div (-9.73) = -24.3$

24. (a) 9.04

(b) 9.045

25. (a) $.185(90) = 16.65$

(b) $\dfrac{145}{100} = \dfrac{x}{70}$
$100x = 10150$
$x = \dfrac{10150}{100}$
$x = 101.5$

26. (a) 4 out of 15; $4 \div 15 = .26\tfrac{2}{3} = 26\tfrac{2}{3}\%$.

(b) 10 out of 15; $10 \div 15 = .66\tfrac{2}{3} = 66\tfrac{2}{3}\%$.

27. Choice D. Since $300,000 is 100%, $600,000 is 200%, and $900,000 is 300% of the original value.

28. (a) 69% of 500: $.69(500) = 345$.

(b) 42% of 500: $.42(500) = 210$.

(c) What % of 500 is 190: $x \cdot 500 = 190$
$x = 190/500$
$= .38 = 38\%$

(d) What % of 500 is 75: $x \cdot 500 = 75$
$x = 75/500$
$= .15 = 15\%$

29. (a) 8% of 2500 $= .08 \times 2500 = 200$

(b) Since the percentage indicated in the graph is twice as great, we would expect twice as many people to name Charlie Brown, or $2 \times 45 = 90$.

30. The amount of decrease is $11.1 - 9.6 = 1.5$. Thus, the percent of decrease is given by
$$\dfrac{1.5}{11.1} \approx .135 = 13.5\%.$$

7.1 EXERCISES

1. Equations A and C are linear in x (or represent "first-degree equations in x) because the highest power on the variable x is one.

2. Equations B and D are not linear in x. Equation B is not linear because the variable x is squared or of degree 2. Equation D contains an x to the negative one power.

3. $$3(x+4) = 5x$$
$$3(6+4) = 5 \cdot 6$$
$$3(10) = 30$$
$$30 = 30$$

   Six is a solution because the left side of the equation equals the right side when 6 is substituted for x.

4. $$5(x+4) - 3(x+6) = 9(x+1)$$
$$5(-2+4) - 3(-2+6) = 9(-2+1)$$
$$5(2) - 3(4) = 9(-1)$$
$$10 - 12 = -9$$
$$-2 = -9$$

   Negative two is not a solution of this equation because the left side does not equal the right side when it is substituted for x.

5. If two equations are equivalent, they have the same <u>solution set</u>.

6. Suppose your last name is Quincy, which has six letters. Substitute 6 for x:
$$4[x + (2-3x)] = 2(4-4x)$$
$$4[6 + (2-3\cdot6)] = 2(4-4\cdot6)$$
$$4[6 + (2-18)] = 2(4-24)$$
$$4[6 + (-16)] = 2(-20)$$
$$4[-10] = -40$$
$$-40 = -40$$

   This shows that 6 is a solution.

7. $$.06(10-x)(100) = (100).06(10-x)$$
$$= 6(10-x)$$
$$= 6\cdot10 - 6\cdot x$$
$$= 60 - 6x$$

   The left side of the equation is equivalent to choice B.

8. Writing exercise

9. $$7k + 8 = 1$$
$$7k + 8 - 8 = 1 - 8$$
$$7k = -7$$
$$\frac{7k}{7} = \frac{-7}{7}$$
$$k = -1$$

10. $$5m - 4 = 21$$
$$5m - 4 + 4 = 21 + 4$$
$$5m = 25$$
$$\frac{5m}{5} = \frac{25}{5}$$
$$m = 5$$

11. $$8 - 8x = -16$$
$$-8 + 8 - 8x = -8 + (-16)$$
$$-8x = -24$$
$$\frac{-8x}{-8} = \frac{-24}{-8}$$
$$x = 3$$

12. $$9 - 2r = 15$$
$$-9 + 9 - 2r = -9 + 15$$
$$-2r = 6$$
$$\frac{-2r}{-2} = \frac{6}{-2}$$
$$r = -3$$

13. $$7x - 5x + 15 = x + 8$$
$$2x + 15 = x + 8$$
$$2x + 15 - 15 = x + 8 - 15$$
$$2x = x - 7$$
$$2x - x = x - x - 7$$
$$x = -7$$

14. $$2x + 4 - x = 4x - 5$$
$$2x - x + 4 = 4x - 5$$
$$x + 4 = 4x - 5$$
$$x + 4 - 4 = 4x - 5 - 4$$
$$x = 4x - 9$$
$$-4x + x = -4x + 4x - 9$$
$$-3x = -9$$
$$\frac{-3x}{-3} = \frac{-9}{-3}$$
$$x = 3$$

15. $$12w + 15w - 9 + 5 = -3w + 5 - 9$$
$$27w - 4 = -3w - 4$$
$$27w + 3w - 4 = -3w + 3w - 4$$
$$30w - 4 = -4$$
$$30w - 4 + 4 = -4 + 4$$
$$30w = 0$$
$$\frac{30w}{30} = \frac{0}{30}$$
$$w = 0$$

16.
$$-4t + 5t - 8 + 4 = 6t - 4$$
$$t - 4 = 6t - 4$$
$$t - 4 + 4 = 6t - 4 + 4$$
$$t = 6t$$
$$t - 6t = 6t - 6t$$
$$-5t = 0$$
$$\frac{-5t}{-5} = \frac{0}{-5}$$
$$t = 0$$

17.
$$2(x + 3) = -4(x + 1)$$
$$2 \cdot x + 2 \cdot 3 = -4 \cdot x - 4 \cdot 1$$
$$2x + 6 = -4x - 4$$
$$2x + 6 - 6 = -4x - 4 - 6$$
$$2x = -4x - 10$$
$$2x + 4x = -4x + 4x - 10$$
$$6x = -10$$
$$\frac{6x}{6} = \frac{-10}{6}$$
$$x = \frac{-2 \cdot 5}{2 \cdot 3}$$
$$x = -\frac{5}{3}$$

18.
$$4(x - 9) = 8(x + 3)$$
$$4 \cdot x - 4 \cdot 9 = 8 \cdot x + 8 \cdot 3$$
$$4x - 36 = 8x + 24$$
$$4x - 36 + 36 = 8x + 24 + 36$$
$$4x = 8x + 60$$
$$4x - 8x = 8x - 8x + 60$$
$$-4x = 60$$
$$\frac{-4x}{-4} = \frac{60}{-4}$$
$$x = -15$$

19.
$$3(2w + 1) - 2(w - 2) = 5$$
$$3 \cdot 2w + 3 \cdot 1 - 2 \cdot w + 2 \cdot 2 = 5$$
$$6w + 3 - 2w + 4 = 5$$
$$6w - 2w + 3 + 4 = 5$$
$$4w + 7 = 5$$
$$4w + 7 - 7 = 5 - 7$$
$$4w = -2$$
$$\frac{4w}{4} = \frac{-2}{4}$$
$$w = -\frac{2 \cdot 1}{2 \cdot 2}$$
$$w = -\frac{1}{2}$$

20.
$$4(x - 2) + 2(x + 3) = 6$$
$$4 \cdot x - 4 \cdot 2 + 2 \cdot x + 2 \cdot 3 = 6$$
$$4x - 8 + 2x + 6 = 6$$
$$4x + 2x - 8 + 6 = 6$$
$$6x - 2 = 6$$
$$6x - 2 + 2 = 6 + 2$$
$$6x = 8$$
$$\frac{6x}{6} = \frac{8}{6}$$
$$x = \frac{2 \cdot 4}{2 \cdot 3}$$
$$x = \frac{4}{3}$$

21.
$$2x + 3(x - 4) = 2(x - 3)$$
$$2x + 3 \cdot x - 3 \cdot 4 = 2 \cdot x - 2 \cdot 3$$
$$2x + 3x - 12 = 2x - 6$$
$$5x - 12 = 2x - 6$$
$$5x - 12 + 12 = 2x - 6 + 12$$
$$5x = 2x + 6$$
$$5x - 2x = 2x - 2x + 6$$
$$3x = 6$$
$$\frac{3x}{3} = \frac{6}{3}$$
$$x = 2$$

22.
$$6x - 3(5x + 2) = 4(1 - x)$$
$$6x - 3 \cdot 5x - 3 \cdot 2 = 4 \cdot 1 - 4 \cdot x$$
$$6x - 15x - 6 = 4 - 4x$$
$$-9x - 6 = 4 - 4x$$
$$-9x - 6 + 6 = 4 + 6 - 4x$$
$$-9x = 10 - 4x$$
$$-9x + 4x = 10 - 4x + 4x$$
$$-5x = 10$$
$$\frac{-5x}{-5} = \frac{10}{-5}$$
$$x = -2$$

23.
$$6p - 4(3 - 2p) = 5(p - 4) - 10$$
$$6p - 4 \cdot 3 + 4 \cdot 2p = 5 \cdot p - 5 \cdot 4 - 10$$
$$6p - 12 + 8p = 5p - 20 - 10$$
$$6p + 8p - 12 = 5p - 30$$
$$14p - 12 = 5p - 30$$
$$14p - 12 + 12 = 5p - 30 + 12$$
$$14p = 5p - 18$$
$$14p - 5p = 5p - 5p - 18$$
$$9p = -18$$
$$\frac{9p}{9} = \frac{-18}{9}$$
$$x = -2$$

24.
$$-2k - 3(4 - 2k) = 2(k - 3) + 2$$
$$-2k - 3 \cdot 4 + 3 \cdot 2k = 2 \cdot k - 2 \cdot 3 + 2$$
$$-2k - 12 + 6k = 2k - 6 + 2$$
$$-2k + 6k - 12 = 2k - 4$$
$$4k - 12 = 2k - 4$$
$$4k - 12 + 12 = 2k - 4 + 12$$
$$4k = 2k + 8$$
$$4k - 2k = 2k - 2k + 8$$
$$2k = 8$$
$$\frac{2k}{2} = \frac{8}{2}$$
$$k = 4$$

25.
$$-[2z - (5z + 2)] = 2 + (2z + 7)$$
$$-[2z - 5z - 2] = 2 + 2z + 7$$
$$-[-3z - 2] = 2 + 7 + 2z$$
$$-[-3z - 2] = 9 + 2z$$
$$+3z + 2 = 9 + 2z$$
$$+3z + 2 - 2 = 9 - 2 + 2z$$
$$3z = 7 + 2z$$
$$3z - 2z = 7 + 2z - 2z$$
$$z = 7$$

26.
$$-[6x - (4x + 8)] = 9 + (6x + 3)$$
$$-[6x - 4x - 8] = 9 + 6x + 3$$
$$-[2x - 8] = 9 + 3 + 6x$$
$$-2x + 8 = 12 + 6x$$
$$-2x + 8 - 8 = 12 - 8 + 6x$$
$$-2x = 4 + 6x$$
$$-2x - 6x = 4 + 6x - 6x$$
$$-8x = 4$$
$$\frac{-8x}{-8} = \frac{4}{-8}$$
$$x = -\frac{1}{2}$$

27.
$$-3m + 6 - 5(m - 1) = 4m - (2m - 4) - 9m + 5$$
$$-3m + 6 - 5 \cdot m + 5 \cdot 1 = 4m - 2m + 4 - 9m + 5$$
$$-3m + 6 - 5m + 5 = 4m - 2m - 9m + 4 + 5$$
$$-3m - 5m + 6 + 5 = 4m - 2m - 9m + 4 + 5$$
$$-8m + 11 = -7m + 9$$
$$-8m + 11 - 11 = -7m + 9 - 11$$
$$-8m = -7m - 2$$
$$-8m + 7m = -7m + 7m - 2$$
$$-m = -2$$
$$\frac{-m}{-1} = \frac{-2}{-1}$$
$$m = 2$$

28.
$$4(k + 2) - 8k - 5 = -3k + 9 - 2(k + 6)$$
$$4 \cdot k + 4 \cdot 2 - 8k - 5 = -3k + 9 - 2 \cdot k - 2 \cdot 6$$
$$4k + 8 - 8k - 5 = -3k + 9 - 2k - 12$$
$$4k - 8k + 8 - 5 = -3k - 2k + 9 - 12$$
$$-4k + 3 = -5k - 3$$
$$-4k + 3 - 3 = -5k - 3 - 3$$
$$-4k = -5k - 6$$
$$-4k + 5k = -5k + 5k - 6$$
$$k = -6$$

29.
$$-[3x - (2x + 5)] = -4 - [3(2x - 4) - 3x]$$
$$-[3x - 2x - 5] = -4 - [3 \cdot 2x - 3 \cdot 4 - 3x]$$
$$-[x - 5] = -4 - [6x - 12 - 3x]$$
$$-x + 5 = -4 - [6x - 3x - 12]$$
$$-x + 5 = -4 - [3x - 12]$$
$$-x + 5 = -4 - 3x + 12$$
$$-x + 5 = -4 + 12 - 3x$$
$$-x + 5 = 8 - 3x$$
$$-x + 5 - 5 = 8 - 5 - 3x$$
$$-x = 3 - 3x$$
$$-x + 3x = 3 - 3x + 3x$$
$$2x = 3$$
$$\frac{2x}{2} = \frac{3}{2}$$
$$x = \frac{3}{2}$$

30.
$$2[-(x - 1) + 4] = 5 + [-(6x - 7) + 9x]$$
$$2[-x + 1 + 4] = 5 + [-6x + 7 + 9x]$$
$$2[-x + 5] = 5 + [-6x + 9x + 7]$$
$$2[-x + 5] = 5 + [3x + 7]$$
$$-2x + 10 = 5 + 3x + 7$$
$$-2x + 10 = 5 + 7 + 3x$$
$$-2x + 10 = 12 + 3x$$
$$-2x + 10 - 10 = 12 - 10 + 3x$$
$$-2x = 2 + 3x$$
$$-2x - 3x = 2 + 3x - 3x$$
$$-5x = 2$$
$$\frac{-5x}{-5} = \frac{2}{-5}$$
$$x = -\frac{2}{5}$$

31.
$$-(9 - 3a) - (4 + 2a) - 3 = -(2 - 5a) + (-a) + 1$$
$$-9 + 3a - 4 - 2a - 3 = -2 + 5a - a + 1$$
$$3a - 2a - 9 - 4 - 3 = -2 + 1 + 5a - a$$
$$a - 16 = -1 + 4a$$
$$a - 16 + 16 = -1 + 16 + 4a$$
$$a = 15 + 4a$$
$$a - 4a = 15 + 4a - 4a$$
$$-3a = 15$$
$$\frac{-3a}{-3} = \frac{15}{-3}$$
$$a = -5$$

32. $-(-2+4x)-(3-4x)+5 = -(-3+6x)+x+1$
$+2-4x-3+4x+5 = +3-6x+x+1$
$-4x+4x+2-3+5 = +3+1-6x+x$
$0+4 = 4-5x$
$4-4 = 4-4-5x$
$0 = -5x$
$\dfrac{0}{-5} = \dfrac{-5x}{-5}$
$0 = x$

33. $2(-3+m)-(3m-4) = -(-4+m)-4m+6$
$2\cdot-3+2\cdot m-3m+4 = 4-m-4m+6$
$-6+2m-3m+4 = 4-m-4m+6$
$+4-6+2m-3m = +6+4-m-4m$
$-2-m = 10-5m$
$-2+2-m = 10+2-5m$
$-m = 12-5m$
$-m+5m = 12-5m+5m$
$4m = 12$
$\dfrac{4m}{4} = \dfrac{12}{4}$
$m = 3$

34. The smallest power of 10 is 10^2 or 100. This value will move the decimal point two places to the right in the coefficients 0.05 and 0.12.

35. Writing exercise

36. $4\cdot\left(\dfrac{3x}{4}+\dfrac{5x}{2}\right) = (13)\cdot 4$
$\dfrac{4}{1}\cdot\dfrac{3x}{4}+\dfrac{4}{1}\cdot\dfrac{5x}{2} = 52$
$3x+2\cdot 5x = 52$
$3x+10x = 52$
$13x = 52$
$\dfrac{13x}{13} = \dfrac{52}{13}$
$x = 4$

37. $12\cdot\left(\dfrac{8x}{3}-\dfrac{2x}{4}\right) = (-13)\cdot 12$
$\dfrac{12}{1}\cdot\dfrac{8x}{3}-\dfrac{12}{1}\cdot\dfrac{2x}{4} = -156$
$4\cdot 8x-3\cdot 2x = -156$
$32x-6x = -156$
$26x = -156$
$\dfrac{26x}{26} = \dfrac{-156}{26}$
$x = -6$

38. $15\cdot\left(\dfrac{x-8}{5}+\dfrac{8}{5}\right) = -\dfrac{x}{3}\cdot 15$
$\dfrac{15}{1}\cdot\dfrac{x-8}{5}+\dfrac{15}{1}\cdot\dfrac{8}{5} = -\dfrac{x}{3}\cdot\dfrac{15}{1}$
$3\cdot(x-8)+3\cdot 8 = -5\cdot x$
$3x-24+24 = -5x$
$3x = -5x$
$3x+5x = -5x+5x$
$8x = 0$
$\dfrac{8x}{8} = \dfrac{0}{8}$
$x = 0$

39. $21\cdot\left(\dfrac{2r-3}{7}+\dfrac{3}{7}\right) = -\dfrac{r}{3}\cdot 21$
$\dfrac{21}{1}\cdot\dfrac{2r-3}{7}+\dfrac{21}{1}\cdot\dfrac{3}{7} = -\dfrac{r}{3}\cdot\dfrac{21}{1}$
$3\cdot(2r-3)+3\cdot 3 = -7\cdot r$
$6r-9+9 = -7r$
$6r = -7r$
$6r+7r = -7r+7r$
$13r = 0$
$\dfrac{13r}{13} = \dfrac{0}{13}$
$r = 0$

40. $6\cdot\dfrac{4t+1}{3} = \left(\dfrac{t+5}{6}+\dfrac{t-3}{6}\right)\cdot 6$
$\dfrac{6}{1}\cdot\dfrac{4t+1}{3} = \dfrac{6}{1}\cdot\dfrac{t+5}{6}+\dfrac{6}{1}\cdot\dfrac{t-3}{6}$
$2\cdot(4t+1) = (t+5)+(t-3)$
$8t+2 = 2t+2$
$8t+2-2 = 2t+2-2$
$8t = 2t$
$8t-2t = 2t-2t$
$6t = 0$
$\dfrac{6t}{6} = \dfrac{0}{6}$
$t = 0$

41. $10\cdot\dfrac{2+5}{5} = \left(\dfrac{3x+1}{2}+\dfrac{-x+7}{2}\right)\cdot 10$
$\dfrac{10}{1}\cdot\dfrac{2x+5}{5} = \dfrac{10}{1}\cdot\dfrac{3x+1}{2}+\dfrac{10}{1}\cdot\dfrac{-x+7}{2}$
$2\cdot(2x+5) = 5\cdot(3x+1)+5\cdot(-x+7)$
$4x+10 = 15x+5-5x+35$
$4x+10 = 10x+40$
$4x+10-10 = 10x+40-10$
$4x = 10x+30$
$4x-10x = 10x-10x+30$
$-6x = 30$
$\dfrac{-6x}{-6} = \dfrac{30}{-6}$
$x = -5$

42.
$$100 \cdot [.05x + .12(x + 5000)] = 940 \cdot 100$$
$$100 \cdot .05x + 100 \cdot .12 \cdot (x + 5000) = 94,000$$
$$5x + 12 \cdot (x + 5000) = 94,000$$
$$5x + 12x + 60,000 = 94,000$$
$$17x + 60,000 = 94,000$$
$$17x = 34,000$$
$$\frac{17x}{17} = \frac{34,000}{17}$$
$$x = 2000$$

43.
$$100 \cdot [.09k + .13(k + 300)] = 61 \cdot 100$$
$$100 \cdot .09k + 100 \cdot .13 \cdot (k + 300) = 6100$$
$$9k + 13 \cdot (k + 300) = 6100$$
$$9k + 13k + 3900 = 6100$$
$$22k + 3900 = 6100$$
$$22k + 3900 - 3900 = 6100 - 3900$$
$$22k = 2200$$
$$\frac{22k}{22} = \frac{2200}{22}$$
$$k = 100$$

44.
$$100 \cdot [.02(50) + .08r] = .04(50 - r) \cdot 100$$
$$100 \cdot .02(50) + 100 \cdot .08r = 100 \cdot .04(50 + r)$$
$$2(50) + 8r = 4(50 + r)$$
$$100 + 8r = 4 \cdot 50 + 4 \cdot r$$
$$100 + 8r = 200 + 4r$$
$$100 + 8r - 4r = 200 + 4r - 4r$$
$$100 + 4r = 200$$
$$100 - 100 + 4r = 200 - 100$$
$$4r = 100$$
$$\frac{4r}{4} = \frac{100}{4}$$
$$r = 25$$

45.
$$100 \cdot [.20(14,000) + .14t] = .18(14,000 + t) \cdot 100$$
$$100 \cdot .20(14,000) + 100 \cdot .14t = 100 \cdot .18(14,000 + t)$$
$$20(14,000) + 14t = 18(14,000 + t)$$
$$280,000 + 14t = 252,000 + 18t$$
$$280,000 + 14t - 18t = 252,000 + 18t - 18t$$
$$280,000 - 4t = 252,000$$
$$280,000 - 280,000 - 4t = 252,000 - 280,000$$
$$-4t = -28,000$$
$$\frac{-4t}{-4} = \frac{-28,000}{-4}$$
$$t = 7000$$

46.
$$100 \cdot [.05x + .10(200 - x)] = .45x \cdot 100$$
$$100 \cdot .05x + 100 \cdot .10(200 - x) = 45x$$
$$5x + 10(200 - x) = 45x$$
$$5x + 2000 - 10x = 45x$$
$$2000 - 5x = 45x$$
$$2000 - 5x + 5x = 45x + 5x$$
$$2000 = 50x$$
$$\frac{2000}{50} = \frac{50x}{50}$$
$$40 = x$$

47.
$$100 \cdot [.08x + .12(260 - x)] = .48x \cdot 100$$
$$100 \cdot .08x + 100 \cdot .12(260 - x) = 48x$$
$$8x + 12(260 - x) = 48x$$
$$8x + 3120 - 12x = 48x$$
$$3120 - 4x = 48x$$
$$3120 - 4x + 4x = 48x + 4x$$
$$3120 = 52x$$
$$\frac{3120}{52} = \frac{52x}{52}$$
$$60 = x$$

48. The equation $x + 2 = x + 2$ is called a(n) <u>identity</u>, because its solution set is {all real numbers}. The equation $x + 1 = x + 2$ is called a(n) <u>contradiction</u>, because it has no solutions.

49. Option A is the only conditional equation.

B. This is an identity:
$$x = 3x - 2x$$
$$x = x.$$

C. This is a contradiction:
$$2(x + 2) = 2x + 2$$
$$2x + 4 = 2x + 2..$$

D. This is an identity:
$$5x - 3 = 4x + x - 5 + 2$$
$$5x - 3 = 5x - 3.$$

50. Writing exercise

51.
$$-2p + 5p - 9 = 3(p - 4) - 5$$
$$3p - 9 = 3p - 12 - 5$$
$$3p - 9 = 3p - 17$$
$$3p - 9 + 9 = 3p - 17 + 9$$
$$3p = 3p - 8$$
$$3p - 3p = 3p - 3p - 8$$
$$0 = -8$$

This is a contradiction; the solution is the empty set, which can be symbolized by ∅.

52.
$$-6k + 2k - 11 = -2(2k - 3) + 4$$
$$-4k - 11 = -4k + 6 + 4$$
$$-4k - 11 = -4k + 10$$
$$-4k - 11 + 11 = -4k + 10 + 11$$
$$-4k = -4k + 21$$
$$-4k + 4k = -4k + 4k + 21$$
$$0 = 21$$

This is a contradiction; the solution is the empty set, which can be symbolized by ∅.

53. $6x + 2(x - 2) = 9x + 4$
$6x + 2x - 4 = 9x + 4$
$8x - 4 = 9x + 4$
$8x - 4 + 4 = 9x + 4 + 4$
$8x = 9x + 8$
$8x - 9x = 9x - 9x + 8$
$-x = 8$
$\dfrac{-x}{-1} = \dfrac{8}{-1}$
$x = -8.$

This is a conditional equation.

54. $-4(x + 2) = -3(x + 5) - x$
$-4x - 8 = -3x - 15 - x$
$-4x - 8 = -4x - 15$
$-4x - 8 + 8 = -4x - 15 + 8$
$-4x = -4x - 7$
$-4x + 4x = -4x + 4x - 7$
$0 = -7$

This is a contradiction; the solution is the empty set, which can be symbolized by ∅.

55. $-11m + 4(m - 3) + 6m = 4m - 12$
$-11m + 4m - 12 + 6m = 4m - 12$
$-11m + 4m + 6m - 12 = 4m - 12$
$-11m + 4m + 6m - 12 = 4m - 12$
$-m - 12 = 4m - 12$
$-m - 12 + 12 = 4m - 12 + 12$
$-m = 4m$
$-m - 4m = 4m - 4m$
$-5m = 0$
$\dfrac{-5m}{-5} = \dfrac{0}{-5}$
$m = 0$

This is a conditional equation.

56. $3p - 5(p + 4) + 9 = -11 + 15p$
$3p - 5p - 20 + 9 = -11 + 15p$
$-2p - 11 = -11 + 15p$
$-2p - 11 + 11 = -11 + 11 + 15p$
$-2p = 15p$
$-2p - 15p = 0 + 15p - 15p$
$-17p = 0$
$\dfrac{-17p}{-17} = \dfrac{0}{-17}$
$p = 0$

This is a conditional equation.

57. $7[2 - (3 + 4r)] - 2r = -9 + 2(1 - 15r)$
$7[2 - 3 - 4r] - 2r = -9 + 2 - 30r$
$7[-1 - 4r] - 2r = -7 - 30r$
$-7 - 28r - 2r = -7 - 30r$
$-7 - 30r = -7 - 30r$

This equation is an identity. The solution set is {all real numbers}.

58. $4[6 - (1 + 2m)] + 10m = 2(10 - 3m) + 8m$
$4[6 - 1 - 2m] + 10m = 20 - 6m + 8m$
$4[5 - 2m] + 10m = 20 + 2m$
$20 - 8m + 10m = 20 + 2m$
$20 + 2m = 20 + 2m$

This equation is an identity. The solution set is {all real numbers}.

59. The following algebraic manipulations show that letter A, B, and C are all equivalent.

A. $h = 2\left(\dfrac{A}{b}\right)$
$h = \dfrac{2}{1} \cdot \dfrac{A}{b}$
$h = \dfrac{2A}{b}$

B. $h = 2A\left(\dfrac{1}{b}\right)$
$h = \dfrac{2A}{1} \cdot \dfrac{1}{b}$
$h = \dfrac{2A}{b}$

C. $h = \dfrac{A}{\frac{1}{2}b}$
$h = \dfrac{A}{1} \div \dfrac{1}{2}b$
$h = \dfrac{A}{1} \div \dfrac{b}{2}$
$h = \dfrac{A}{1} \cdot \dfrac{2}{b}$
$h = \dfrac{2A}{b}$

D. Here is the simplification of this equation:
$h = \dfrac{\frac{1}{2}A}{b}$
$h = \dfrac{1}{2}A \div b$
$h = \dfrac{A}{2} \div \dfrac{b}{1}$
$h = \dfrac{A}{2} \cdot \dfrac{1}{b}$
$h = \dfrac{A}{2b}.$

This equation is not equivalent to the given equation.

60. These formulas are equivalent; the distributive property justifies their equivalence.

61. $$d = rt$$
$$\frac{d}{r} = \frac{rt}{r}$$
$$\frac{d}{r} = t$$

62. $$I = prt$$
$$\frac{I}{pt} = \frac{prt}{pt}$$
$$\frac{I}{pt} = r$$

63. $$A = bh$$
$$\frac{A}{h} = \frac{bh}{h}$$
$$\frac{A}{h} = b$$

64. $$P = 2L + 2W$$
$$P - 2W = 2L + 2W - 2W$$
$$P - 2W = 2L$$
$$\frac{P - 2W}{2} = \frac{2L}{2}$$
$$\frac{P - 2W}{2} = L$$

The last equation can also be simplified to
$$L = \frac{P - 2W}{2} = \frac{P}{2} - \frac{2 \cdot W}{2 \cdot 1} = \frac{P}{2} - W$$

65. $$P = a + b + c$$
$$P - b = a + b - b + c$$
$$P - b - c = a + c - c$$
$$P - b - c = a$$

66. $$V = LWH$$
$$\frac{V}{LH} = \frac{LWH}{LH}$$
$$\frac{V}{LH} = W$$

67. $$A = \frac{1}{2}bh$$
$$2 \cdot A = 2 \cdot \frac{1}{2}bh$$
$$2A = bh$$
$$\frac{2A}{b} = \frac{bh}{b}$$
$$\frac{2A}{b} = h$$

68. $$C = 2\pi r$$
$$\frac{C}{2\pi} = \frac{2\pi r}{2\pi}$$
$$\frac{C}{2\pi} = r$$

69. $$S = 2\pi rh + 2\pi r^2$$
$$S - 2\pi r^2 = 2\pi rh + 2\pi r^2 - 2\pi r^2$$
$$S - 2\pi r^2 = 2\pi rh$$
$$\frac{S - 2\pi r^2}{2\pi r} = \frac{2\pi rh}{2\pi r}$$
$$\frac{S - 2\pi r^2}{2\pi r} = h$$

The last equation can be simplified further:
$$h = \frac{S - 2\pi r^2}{2\pi r}$$
$$= \frac{S}{2\pi r} - \frac{2\pi r^2}{2\pi r}$$
$$= \frac{S}{2\pi r} - r.$$

70. $$A = \frac{1}{2}(B + b)h$$
$$2 \cdot A = 2 \cdot \frac{1}{2}(B + b)h$$
$$2A = (B + b)h$$
$$2A = Bh + bh$$
$$2A - bh = Bh + bh - bh$$
$$2A - bh = Bh$$
$$\frac{2A - bh}{h} = \frac{Bh}{h}$$
$$\frac{2A - bh}{h} = B$$

The last equation can be simplified further:
$$B = \frac{2A - bh}{h}$$
$$= \frac{2A}{h} - \frac{bh}{h}$$
$$= \frac{2A}{h} - b.$$

71. $$C = \frac{5}{9}(F - 32)$$
$$\frac{9}{5} \cdot C = \frac{9}{5} \cdot \frac{5}{9}(F - 32)$$
$$\frac{9}{5}C = (F - 32)$$
$$\frac{9}{5}C + 32 = F - 32 + 32$$
$$\frac{9}{5}C + 32 = F$$

72.
$$F = \frac{9}{5}C + 32$$
$$F - 32 = \frac{9}{5}C + 32 - 32$$
$$F - 32 = \frac{9}{5}C$$
$$\frac{5}{9}(F - 32) = \frac{5}{9} \cdot \frac{9}{5}C$$
$$\frac{5}{9}(F - 32) = C$$

73.
$$A = 2HW + 2LW + 2LH$$
$$A - 2LW = 2HW + 2LW - 2LW + 2LH$$
$$A - 2LW = 2HW + 2LH$$
$$A - 2LW = H(2W + 2L)$$
$$\frac{A - 2LW}{2W + 2L} = \frac{H(2W + 2L)}{2W + 2L}$$
$$\frac{A - 2LW}{2W + 2L} = H$$

74.
$$V = \frac{1}{3}Bh$$
$$3 \cdot V = \frac{3}{1} \cdot \frac{1}{3}Bh$$
$$3V = Bh$$
$$\frac{3V}{B} = \frac{Bh}{B}$$
$$\frac{3V}{B} = h$$

75. (a) $y = .61x - 48.3$
$y = .61(95) - 48.3$
$= 57.95 - 48.3$
$= 9.65$ million tickets

(b)
$$y = .61x - 48.3$$
$$7.9 = .61x - 48.3$$
$$7.9 + 48.3 = .61x - 48.3 + 48.3$$
$$56.2 = .61x$$
$$\frac{56.2}{.61} = \frac{.61x}{.61}$$
$$92.1 \approx x$$

This corresponds to 1992–1993 season.

76. (a) To estimate the number of youths attending in 1993–1994, substitute 93 for x in the model:
$$y = .1x - 8.5$$
$$y = .1(93) - 8.5$$
$$= 9.3 - 8.5$$
$$= 0.8 \text{ million youths.}$$

(b) To find in which season the ticket sales amounted to .75 million, substitute .75 for y and solve for x:
$$y = .1x - 8.5$$
$$.75 = .1x - 8.5$$
$$.75 + 8.5 = .1x - 8.5 + 8.5$$
$$9.25 = .1x$$
$$\frac{9.25}{.1} = \frac{.1x}{.1}$$
$$x = 92.5.$$

This corresponds to the 1992–1993 season

77. Examine the bar graph to see that the sales increased from approximately 9.4 million during the 95–96 season to approximately 10.5 million during the 96–97 season. This looks to be the greatest jump in sales. Thus the greatest increase appears to be between the 1995–1996 and the 1996–1997 seasons.

78. (a) $y = .61x - 48.3$
$y = .61(94) - 48.3$
$= 57.34 - 48.3$
$= 9.04$ million tickets

(b) Yes, this is a good approximation.

7.2 EXERCISES

1. Expression
2. Expression
3. Equation: $\frac{2}{3} \cdot x = 12$
4. Equation: $7 = x + 3$
5. Expression
6. Equation: $\frac{15}{x} = 3$

7. Let $x=$ the number of five-dollar bills
$x-25=$ the number of ten-dollar bills

Number of bills	Denomination	Dollar Value
x	5	$5x$
$x-25$	10	$10(x-25)$

$$5x+10(x-25)=200$$
$$5x+10x-250=200$$
$$15x-250=200$$
$$15x-250+250=200+250$$
$$15x=450$$
$$\frac{15x}{15}=\frac{450}{15}$$
$$x=30$$

The number of five-dollar bills is 30; the number of ten-dollar bills is $30-25=5$. The answer has not changed.

8. Writing exercise

9. $x-18$

10. $x+12$

11. $(x-9)(x+6)$

12. $\dfrac{x}{6}$ or $x\div 6$

13. $\dfrac{12}{x}$, $(x\neq 0)$

14. $\dfrac{6}{7}\cdot x$

15. Writing exercise

16. Letter D is not a valid translation because there is no symbol for "the number."

17. $2x+\dfrac{x}{6}=x-8$

18. $x-(-4x)=x+9$

19. $12-\dfrac{2}{3}x=10$

20. $6+.75x=x+3$

21. *Step 2*: $x-70=$ the number of shoppers at small chain/independent bookstores.

    Step 3: $x+x-70=442$.

    Step 4:
    $$2x-70=442$$
    $$2x-70+70=442+70$$
    $$2x=512$$
    $$\frac{2x}{2}=\frac{512}{2}$$
    $$x=256.$$

    Step 5: There were 256 large chain bookstore shoppers and $256-70$ or 186 small chain/independent shoppers.

    Step 6: The number of large chain shoppers was 70 more than the number of small chain/independent shoppers, and the total number of these two bookstore types was 442.

22. *Step 1*: Let $x=$ the revenue of rock band U2..

    Step 2: Then $196.5-x=$ the revenue (in millions of dollars) generated by N Sync.

    Step 3: $x-(196.5-x)=22.9$.

    Step 4:
    $$x-(196.5-x)=22.9$$
    $$2x-196.5=22.9$$
    $$2x=219.4$$
    $$x=109.7$$

    Step 5: Thus U2's revenue is \$109.7 million.

    Step 6: Since the revenue of N Sync is \$22.9 million less than that of U2, we arrive at \$86.8 million for N Sync's revenue.

23. Let $x=$ the number of sales for Toyota Camrys in thousands. Then $x-15=$ the number of sales for Honda Accord in thousands.
    $$x+(x-15)=828$$
    $$2x-15=828$$
    $$2x-15+15=828+15$$
    $$2x=843$$
    $$x=421.5$$

    The sale of Camrys then is 421.5 thousand. The sales of the Accords then is $421.5-15=406.5$ thousand.

24. Let $x=$ the number of games lost.
    Let $3x-2=$ the number of games won. Since losses plus wins must equal total number of games, we have
    $$x+(3x-2)=82$$
    $$4x-2=82$$
    $$4x=84$$
    $$x=21.$$

    Thus the Kings had 21 losses and $3x-2=3(21)-2=61$ wins.

25. Let $x=$ the number of base hits for Hornsby.
    Let $x-57=$ the number of base hits for Ruth.
    $$x+(x-57)=5803$$
    $$2x-57=5803$$
    $$2x-57+57=5803+57$$
    $$2x=5860$$
    $$\frac{2x}{2}=\frac{5860}{2}$$
    $$x=2930$$

    The number of hits for Hornsby was 2930; the number of hits for Ruth was $2930-57=2873$ hits.

26. Let x = the number of electoral votes for George W. Bush. Let $x - 5$ = the number of electoral votes for Al Gore.
$$x + (x - 5) = 537$$
$$2x - 5 = 537$$
$$2x = 542$$
$$x = 271$$

The number of votes for Bush 271; the number of votes for Gore was $271 - 5 = 266$ votes.

27. Let x = the revenue from video sales in billions of dollars.
Let $2x + .27$ = the revenue from video rentals in billions of dollars.
$$x + 2x + .27 = 9.81$$
$$3x + .27 - .27 = 9.81 - .27$$
$$3x = 9.54$$
$$\frac{3x}{3} = \frac{9.54}{3}$$
$$x = 3.18$$

The total revenue for video sales was $3.18 billion; the total revenue for rentals was $2(3.18) + .27 = \$6.63$ billion.

28. Let x = the revenue from video sales in billions of dollars.
Let $x + .15$ = the revenue from video rentals in billions of dollars.
$$x + x + .15 = 14.83$$
$$2x + .15 - .15 = 14.83 - .15$$
$$2x = 14.68$$
$$\frac{2x}{2} = \frac{14.68}{2}$$
$$x = 7.34$$

The total revenue for video sales was $7.34 billion; the total revenue for rentals was $7.34 + .15 = \$7.49$ billion. Examine the bar graph to see that this occurred in 1995.

29. Let x = the number of movie screens in New York. Then $x + 11$ = the number of movie screens in California.
$$x + (x + 11) = 107$$
$$2x + 11 = 107$$
$$2x = 96$$
$$\frac{2x}{2} = \frac{96}{2}$$
$$x = 48$$

The number of screens in New York is 48; the number of screens in California is $48 + 11 = 59$.

30. Let x = the number viewers (in millions) on Saturday nights. Then $x + 20$ = the number viewers (in millions) on Thursday nights.
$$x + (x + 20) = 102$$
$$2x + 20 = 102$$
$$2x = 82$$
$$x = 41$$

Thus, there are 41 million viewers on Saturday and $41 + 20 = 61$ million viewers on Thursday.

31. 14% of 250
$$.14(250) = 35 \text{ ml}$$

32. 30% of 150
$$.30(150) = 45 \text{ L}$$

33. 3.5% of 10,000
$$.035(10,000) = \$350$$

34. Use the formula $I = Prt$.
$$I = (25,000)(.03)(2)$$
$$= \$1500$$

35. $283(.05) = \$14.15$

36. $35(.50) = \$17.50$

37. Let x = the number of liters of 20% solution.

Strength	L of solution	L of alcohol
12%	12	.12(12)
20%	x	.20(x)
14%	$12 + x$	.14($12 + x$)

Create an equation by adding the first two algebraic expressions in the last column to total the third:
$$.12(12) + .20(x) = .14(12 + x)$$
$$100[.12(12) + .20(x)] = 100[.14(12 + x)]$$
$$12(12) + 20x = 14(12 + x)$$
$$144 - 144 + 20x = 168 - 144 + 14x$$
$$20x - 14x = 24 + 14x - 14x$$
$$6x = 24$$
$$\frac{6x}{6} = \frac{24}{6}$$
$$x = 4 \text{ liters of 20\% solution.}$$

38. Let x = the number of liters of 10% alcohol solution.

Strength	L of solution	L of alcohol
10%	x	.10(x)
50%	40	.50(40)
40%	$x + 40$	.40($x + 40$)

Create an equation by adding the first two algebraic expressions in the last column to total the third:

$$.10(x) + .50(40) = .40(x + 40)$$
$$100[.10(x) + .50(40)] = 100[.40(x + 40)]$$
$$10(x) + 50(40) = 40(x + 40)$$
$$10x + 2000 = 40x + 1600$$
$$10x + 2000 - 2000 = 40x + 1600 - 2000$$
$$10x = 40x - 400$$
$$10x - 40x = 40x - 40x - 400$$
$$-30x = -400$$
$$\frac{-30x}{-30} = \frac{-400}{-30}$$
$$x = 13\frac{1}{3} \text{ liters of 10\% solution..}$$

39. Let x = the number of liters of pure alcohol.

Strength	L of solution	L of alcohol
70%	50	.70(50)
100%	x	1.00(x)
78%	50 + x	.78(50 + x)

Create an equation by adding the first two algebraic expressions in the last column to total the third:

$$.70(50) + 1.00(x) = .78(50 + x)$$
$$100[.70(50) + 1.00(x)] = 100[.78(50 + x)]$$
$$70(50) + 100(x) = 78(50 + x)$$
$$3500 + 100x = 3900 + 78x$$
$$3500 + 100x - 78x = 3900 + 78x - 78x$$
$$3500 - 3500 + 22x = 3900 - 3500$$
$$22x = 400$$
$$\frac{22x}{22} = \frac{400}{22}$$
$$x = 18\frac{2}{11} \text{ liters of pure alcohol.}$$

40. Let x = the number of gallons of water to be added.

Strength	Gallons of solution	Gallons of insecticide
0%	x	.00(x)
4%	3	.04(3)
3%	$x + 3$	.03($x + 3$)

Create an equation by adding the first two algebraic expressions in the last column to total the third:

$$.00(x) + .04(3) = .03(x + 3)$$
$$100[.00(x) + .04(3)] = 100[.03(x + 3)]$$
$$0 + 4(3) = 3(x + 3)$$
$$12 = 3x + 9$$
$$12 - 9 = 3x + 9 - 9$$
$$3 = 3x$$
$$\frac{3}{3} = \frac{3x}{3}$$
$$1 = x.$$

One gallon of water must be added.

41. Let x = number of liters of 20% solution to be replaced with 100% solution. The amount of 20% solution that is replaced must be subtracted while the amount of 100% solution must be added.

Strength	L of solution	L of chemical
20%	20	.20(20)
20%	x	$-.20(x)$
100%	x	1.00(x)
40%	20	.40(20)

Create an equation by adding the first three algebraic expressions in the last column to total the last. To clear the decimal in this equation, multiply by 10 to move the decimal point only one place to the right:

$$.20(20) - .20(x) + 1.00(x) = .40(20)$$
$$10[.20(20) - .20(x) + 1.00(x)] = 10[.40(20)]$$
$$2(20) - 2x + 10x = 4(20)$$
$$40 - 2x + 10x = 80$$
$$40 + 8x = 80$$
$$40 - 40 + 8x = 80 - 40$$
$$8x = 40$$
$$\frac{8x}{8} = \frac{40}{8}$$
$$x = 5 \text{ liters.}$$

42. Let x = number of liters of 30% solution to be replaced with 70% solution. The amount of 30% solution that is replaced must be subtracted while the amount of 70% solution must be added.

Strength	L of solution	L of chemical
30%	80	.30(80)
30%	x	$-.30(x)$
70%	x	.70(x)
40%	80	.40(80)

Create an equation by adding the first three algebraic expressions in the last column to total the last. To clear the decimal in this equation, multiply by 10 to move the decimal point only one place to the right:

$$.30(80) - .30(x) + .70(x) = .40(80)$$
$$10[.30(80) - .30(x) + .70(x)] = 10[.40(80)]$$
$$3(80) - 3(x) + 7(x) = 4(80)$$
$$240 - 3x + 7x = 320$$
$$240 + 4x = 320$$
$$240 - 240 + 4x = 320 - 240$$
$$4x = 80$$
$$\frac{4x}{4} = \frac{80}{4}$$
$$x = 20 \text{ liters.}$$

43. Let $x =$ the amount invested at 3%.

% as decimal	Amount Invested	Interest in one year
.03	x	$.03x$
.04	$12,000 - x$	$.04(12,000 - x)$
Totals	$12,000$	440

Create an equation by adding the first two algebraic expressions in the last column to total the third:

$$.03x + .04(12,000 - x) = 440$$
$$100[.03x + .04(12,000 - x)] = 100(440)$$
$$3x + 4(12,000 - x) = 44,000$$
$$3x + 48,000 - 4x = 44,000$$
$$48,000 - x = 44,000$$
$$48,000 - 48,000 - x = 44,000 - 48,000$$
$$-x = -4000$$
$$\frac{-x}{-1} = \frac{-4000}{-1}$$
$$x = 4000.$$

The amount invested at 3% is $4000; the amount invested at 4% is: $12,000 - 4000 = \$8000$.

44.

% as decimal	Amount Invested	Interest in one year
.02	x	$.02x$
.03	$60,000 - x$	$.03(60,000 - x)$
Totals	$60,000$	1600

Create an equation by adding the first two algebraic expressions in the last column to total the third:

$$.02x + .03(60,000 - x) = 1600$$
$$100[.02x + .03(60,000 - x)] = 100(1600)$$
$$2x + 3(60,000 - x) = 160,000$$
$$2x + 180,000 - 3x = 160,000$$
$$180,000 - x = 160,000$$
$$180,000 - 180,000 - x = 160,000 - 180,000$$
$$-x = -20,000$$
$$\frac{-x}{-1} = \frac{-20,000}{-1}$$
$$x = 20,000.$$

The amount invested at 2% is $20,000; the amount invested at 3% is: $60,000 - 20,000 = \$40,000$.

45.

% as decimal	Amount Invested	Interest in one year
.045	x	$.045x$
.03	$2x - 1000$	$.03(2x - 1000)$
Total		1020

Create an equation by adding the first two algebraic expressions in the last column to total the third. To clear the decimals, both sides of the equation must be multiplied by 1000 to move the decimal point 3 places to the right:

$$.045x + .03(2x - 1000) = 1020$$
$$1000[.045x + .03(2x - 1000)] = 1000(1020)$$
$$45x + 30(2x - 1000) = 1,020,000$$
$$45x + 60x - 30.000 = 1,020,000$$
$$105x - 30,000 = 1,020,000$$
$$105x - 30,000 + 30,000 = 1,020,000 + 30,000$$
$$105x = 1,050,000$$
$$\frac{105x}{105} = \frac{1,050,000}{105}$$
$$x = 10,000.$$

The amount invested at 4.5% is $10,000; the amount invested at 3% is: $2 \cdot 10,000 - 1000 = \$19,000$.

46.

% as decimal	Amount Invested	Interest in one year
.035	x	$.035x$
.04	$3x + 5000$	$.04(3x + 5000)$
Total		1440

Create an equation by adding the first two algebraic expressions in the last column to total the third. To clear the decimals, both sides of the equation must be multiplied by 1000 to move the decimal point 3 places to the right:

$$.035x + .04(3x + 5000) = 1440$$
$$1000[.035x + .04(3x + 5000)] = 1000(1440)$$
$$35x + 40(3x + 5000) = 1,440,000$$
$$35x + 120x + 200,000 = 1,440,000$$
$$155x + 200,000 = 1,440,000$$
$$155x + 200,000 - 200,000 = 1,440,000 - 200,000$$
$$155x = 1,240,000$$
$$\frac{155x}{155} = \frac{1,240,000}{155}$$
$$x = 8000.$$

The amount invested at 3.5% is $8000; the amount invested at 4% is: $3 \cdot 8000 + 5000 = \$29,000$.

47.

% as decimal	Amount Invested	Interest in one year
.05	$29,000$	$.05(29,000)$
.02	x	$.02x$
.03	$29.000 + x$	$.03(29.000 + x)$

Create an equation by adding the first two algebraic expressions in the last column to total the third:

$$.05(29,000) + .02x = .03(29,000 + x)$$
$$100[.05(29,000) + .02x] = 100[.03(29,000 + x)]$$
$$5(29,000) + 2x = 3(29,000 + x)$$
$$145,000 + 2x = 87,000 + 3x$$
$$145,000 + 2x - 2x = 87,000 + 3x - 2x$$
$$145,000 - 87,000 = 87,000 - 87,000 + x$$
$$58,000 = x.$$

He should invest $58,000 at 2% in order to have an average return of 3%.

48.

% as decimal	Amount Invested	Interest in one year
.06	15,000	.06(15,000)
.04	x	$.04x$
.055	$15,000 + x$	$.055(15,000 + x)$

Create an equation by adding the first two algebraic expressions in the last column to total the third. To clear the decimals, both sides of the equation must be multiplied by 1000 to move the decimal point 3 places to the right:

$$.06(15,000) + .04x = .055(15,000 + x)$$
$$1000[.06(15,000) + .04x] = 1000[.055(15,000 + x)]$$
$$60(15,000) + 40x = 55(15,000 + x)$$
$$900,000 + 40x = 825,000 + 55x$$
$$900,000 + 40x - 40x = 825,000 + 55x - 40x$$
$$900,000 = 825,000 + 15x$$
$$900,000 - 825,000 = 825,000 - 825,000 + 15x$$
$$75,000 = 15x$$
$$\frac{75,000}{15} = \frac{15x}{15}$$
$$5000 = x.$$

Terry should invest $5000 at 4% in order to have an average return of 5.5%.

49.

Number of coins	Denomination	Value
x	.01	$.01x$
x	.10	$.10x$
$44 - 2x$	.25	$.25(44 - 2x)$

Because there are 44 coins altogether, the number of quarters is represented by subtracting the total number of pennies and dimes from 44. Now create an equation by adding the algebraic expressions in the last column; the total value of all the coins is $4.37.

$$.01x + .10x + .25(44 - 2x) = 4.37$$
$$100[.01x + .10x + .25(44 - 2x)] = 100(4.37)$$
$$x + 10x + 25(44 - 2x) = 437$$
$$x + 10x + 1100 - 50x = 437$$
$$x + 10x - 50x + 1100 = 437$$
$$-39x + 1100 - 1100 = 437 - 1100$$
$$-39x = -663$$
$$\frac{-39x}{-39} = \frac{-663}{-39}$$
$$x = 17$$

Sam has 17 pennies, 17 dimes, and $44 - 2 \cdot 17 = 10$ quarters.

50.

Number of coins	Denomination	Value
x	.05	$.05x$
x	.25	$.25x$
$2x$	.50	$.50(2x)$

Now create an equation by adding the algebraic expressions in the last column; the total value of all the coins is $2.60.

$$.05x + .25x + .50(2x) = 2.60$$
$$100[.05x + .25x + .50(2x)] = 100(2.60)$$
$$5x + 25x + 50(2x) = 260$$
$$5x + 25x + 100x = 260$$
$$130x = 260$$
$$\frac{130x}{130} = \frac{260}{130}$$
$$x = 2$$

Roma found 2 nickels, 2 quarters, and $2 \cdot 2 = 4$ half-dollars.

51.

Number of tickets	Value/ticket	Total Value
x	3	$3x$
$410 - x$	7	$7(410 - x)$

Now create an equation by adding the algebraic expressions in the last column; the total value of all the tickets is $1650.

$$3x + 7(410 - x) = 1650$$
$$3x + 2870 - 7x = 1650$$
$$2870 - 2870 - 4x = 1650 - 2870$$
$$-4x = -1220$$
$$\frac{-4x}{-4} = \frac{-1220}{-4}$$
$$x = 305$$

There were 305 students who attended and $410 - 305 = 105$ nonstudents who attended.

52.

Number of tickets	Value/ticket	Total Value
x	40	$40x$
$550 - x$	28	$28(550 - x)$

Now create an equation by adding the algebraic expressions in the last column; the total value of all the tickets is $20,800.

$$40x + 28(550 - x) = 20,800$$
$$40x + 15,400 - 28x = 20,800$$
$$12x + 15,400 = 20,800$$
$$12x + 15,400 - 15,400 = 20,800 - 15,400$$
$$12x = 5400$$
$$\frac{12x}{12} = \frac{5400}{12}$$
$$x = 450$$

There were 450 floor tickets sold and $550 - 450 = 100$ balcony tickets sold.

53.

Number of tickets	Value/ticket	Total Value
x	35	$35x$
$105 - x$	30	$30(105 - x)$

Now create an equation by adding the algebraic expressions in the last column; the total value of all the tickets is $3420.

$$35x + 30(105 - x) = 3420$$
$$35x + 3150 - 30x = 3420$$
$$5x + 3150 = 3420$$
$$5x + 3150 - 3150 = 3420 - 3150$$
$$5x = 270$$
$$\frac{5x}{5} = \frac{270}{5}$$
$$x = 54$$

There were 54 seats sold in Row 1 and $105 - 54 = 51$ seats sold in Row 2.

54.

Number of coins	Denomination	Value
x	.02	$.02x$
$3x$	.03	$.03(3x)$

Now create an equation by adding the algebraic expressions in the last column; the total value of all the coins is $1.21.

$$.02x + .03(3x) = 1.21$$
$$100[.02x + .03(3x)] = 100(1.21)$$
$$2x + 3(3x) = 121$$
$$2x + 9x = 121$$
$$11x = 121$$
$$\frac{11x}{11} = \frac{121}{11}$$
$$x = 11$$

Lee Ann has 11 two-cent pieces and $3 \cdot 11 = 33$ three-cent pieces.

55.

Number of coins	Denomination	Value
x	10	$10x$
$80 - x$	20	$20(80 - x)$

Now create an equation by adding the algebraic expressions in the last column; the total value of all the coins is $1060.

$$10x + 20(80 - x) = 1060$$
$$10x + 1600 - 20x = 1060$$
$$-10x + 1600 = 1060$$
$$-10x + 1600 - 1600 = 1060 - 1600$$
$$-10x = -540$$
$$\frac{-10x}{-10} = \frac{-540}{-10}$$
$$x = 54$$

Dave has 54 $10 coins and $80 - 54 = 26$ $20 coins.

56. A half-cent coin means 1/2 of $.01.

$$\frac{1}{2}(\$.01) = \$.005$$

57. Solve the formula $rt = d$ for t.

$$rt = d$$
$$\frac{rt}{r} = \frac{d}{r}$$
$$t = \frac{d}{r}$$

Now substitute the given values for d and r.

$$t = \frac{500}{131.294} \approx 3.808 \text{ hours}$$

58. Substitute the given values for d and r.

$$t = \frac{500}{161.794} \approx 3.090 \text{ hours}$$

59. Substitute the given values for d and r.

$$t = \frac{255}{148.725} \approx 1.715 \text{ hours}$$

60. Substitute the given values for d and r.

$$t = \frac{435}{149.213} \approx 2.915 \text{ hours}$$

61. Solve the formula $rt = d$ for r.

$$rt = d$$
$$\frac{rt}{t} = \frac{d}{t}$$
$$r = \frac{d}{t}$$

Now substitute the given values for d and t.

$$r = \frac{100}{12.65} \approx 7.91 \text{ meters per second}$$

62. Substitute the given values for d and t.

$$r = \frac{400}{53.02} \approx 7.54 \text{ meters per second.}$$

63. Substitute the given values for d and t.

$$r = \frac{400}{47.50} \approx 8.42 \text{ meters per second}$$

64. Substitute the given values for d and t.

$$r = \frac{400}{43.84} \approx 9.12 \text{ meters per second}$$

65. Use the formula $rt = d$.

$$53 \cdot 10 = 530 \text{ miles}$$

66. Use the formula $rt = d$.

$$164 \cdot 2 = 328 \text{ miles}$$

67. Writing exercise

68. The estimate might be made by mentally dividing 400 by 8, which is 50. Letter A seems to be the best estimate.

69.
	Rate	Time	Distance
First train	85	t	$85t$
Second train	95	t	$95t$

Because the trains are traveling in opposite directions, the sum of their distances will equal the total distance apart of 315 kilometers. Use this information to create an equation.

$$85t + 95t = 315$$
$$180t = 315$$
$$\frac{180t}{180} = \frac{315}{180}$$
$$t = 1.75 \text{ hours}$$

70.
	Rate	Time	Distance
First steamer	22	t	$22t$
Second steamer	22	t	$22t$

Because the steamers are traveling in opposite directions, the sum of their distances will equal the total distance apart of 110 miles. Use this information to create an equation.

$$22t + 22t = 110$$
$$44t = 110$$
$$\frac{44t}{44} = \frac{110}{44}$$
$$t = 2.5 \text{ hours}$$

71.
	Rate	Time	Distance
Nancy	35	t	$35t$
Mark	40	$t - \frac{1}{4}$	$40(t - \frac{1}{4})$

Because Nancy and Mark are traveling in opposite directions, the sum of their distances will equal the total distance apart of 140 miles. Use this information to create an equation.

$$35t + 40\left(t - \frac{1}{4}\right) = 140$$
$$35t + 40t - 10 = 140$$
$$75t - 10 = 140$$
$$75t - 10 + 10 = 140 + 10$$
$$75t = 150$$
$$\frac{75t}{75} = \frac{150}{75}$$
$$t = 2 \text{ hours}$$

The question asks at what time will they be 140 miles apart. The value of t tells us that in two hours they will be 140 miles apart. Because Nancy left the house at 8:00, the time would be 10:00 A.M.

72.
	Rate	Time	Distance
Jeff	5	t	$5t$
Joan	8	$t - \frac{1}{2}$	$8(t - \frac{1}{2})$

When Joan catches up with Jeff, they will have traveled the same distance. Use this information to create an equation:

$$5t = 8\left(t - \frac{1}{2}\right)$$
$$5t = 8t - 4$$
$$5t - 8t = 8t - 8t - 4$$
$$-3t = -4$$
$$\frac{-3t}{-3} = \frac{-4}{-3}$$
$$t = 1\frac{1}{3} \text{ hours}$$

The question asks at what time will Joan catch up with Jeff. The value of t tells us that it will take 1 1/3 hours. Because Jeff left the house at 8:30, the time would be 9:50 A.M., 1 hour and 20 minutes later.

73.
	Rate	Time	Distance
Car	r	$\frac{1}{2}$	$\frac{1}{2}r$
Bus	$r - 12$	$\frac{3}{4}$	$\frac{3}{4}(r - 12)$

The distance to and from work is the same, so set the distances equal to each other.

$$\frac{1}{2}r = \frac{3}{4}(r - 12)$$
$$\frac{4}{1} \cdot \frac{1}{2}r = \frac{4}{1} \cdot \frac{3}{4}(r - 12)$$
$$2r = 3(r - 12)$$
$$2r = 3r - 36$$
$$2r - 3r = 3r - 3r - 36$$
$$-r = -36$$
$$\frac{-r}{-1} = \frac{-36}{-1}$$
$$r = 36 \text{ miles per hour}$$

Tri's rate of speed when driving is 36 mph. The distance to work is $\frac{1}{2} \cdot 36 = 18$ miles.

74.
	Rate	Time	Distance
Bike	r	$\frac{1}{4}$	$\frac{1}{4}r$
Walk	$r - 10$	$\frac{3}{4}$	$\frac{3}{4}(r - 10)$

The distance to and from school is the same, so set the distances equal to each other.

$$\frac{1}{4}r = \frac{3}{4}(r-10)$$
$$\frac{4}{1} \cdot \frac{1}{4}r = \frac{4}{1} \cdot \frac{3}{4}(r-10)$$
$$r = 3(r-10)$$
$$r = 3r - 30$$
$$r - 3r = 3r - 3r - 30$$
$$-2r = -30$$
$$\frac{-2r}{-2} = \frac{-30}{-2}$$
$$r = 15 \text{ miles per hour}$$

Latoya's rate of speed on her bike is 15 mph. The distance to school is $\frac{1}{4} \cdot 15 = \frac{15}{4}$ or 3.75 miles.

75.

	Rate	Time	Distance
First part	10	t	$10t$
Second part	15	t	$15t$

The sum of the distances is 100.
$$10t + 15t = 100$$
$$25t = 100$$
$$\frac{25t}{25} = \frac{100}{25}$$
$$t = 4 \text{ hours.}$$

The time for each part of the trip is 4 hours, so the total time for the trip is 8 hours.

76.

	Rate	Time	Distance
Steve	50	t	$50t$
David	60	$t - \frac{1}{2}$	$60\left(t - \frac{1}{2}\right)$

The sum of the distances is 80.
$$50t + 60\left(t - \frac{1}{2}\right) = 80$$
$$50t + 60t - 30 = 80$$
$$110t - 30 = 80$$
$$110t - 30 + 30 = 80 + 30$$
$$110t = 110$$
$$\frac{110t}{110} = \frac{110}{110}$$
$$t = 1 \text{ hour.}$$

The value of t tells us that Steve will travel for 1 hour before he meets David. The question asks how long after David leaves until they meet. Because David leaves one half hour later than Steve, it will take one half hour for them to meet after David leaves.

7.3 EXERCISES

1. $\dfrac{25}{40} = \dfrac{5 \cdot 5}{5 \cdot 8} = \dfrac{5}{8}$

2. $\dfrac{16}{48} = \dfrac{16 \cdot 1}{16 \cdot 3} = \dfrac{1}{3}$

3. $\dfrac{18}{72} = \dfrac{18 \cdot 1}{18 \cdot 4} = \dfrac{1}{4}$

4. $\dfrac{300}{250} = \dfrac{50 \cdot 6}{50 \cdot 5} = \dfrac{6}{5}$

5. The units of measure must be the same in order to factor the numerator and denominator to simplify the fraction. This simplification can be done by dimensional analysis:
$$\frac{144 \text{ inches}}{6 \text{ feet}} \cdot \frac{1 \text{ foot}}{12 \text{ inches}} = \frac{12 \cdot 12}{12 \cdot 6} = \frac{2}{1}.$$

6. The units of measure must be the same in order to factor the numerator and denominator to simplify the fraction. This simplification can be done by dimensional analysis:
$$\frac{60 \text{ inches}}{2 \text{ yards}} \cdot \frac{1 \text{ yard}}{36 \text{ inches}} = \frac{12 \cdot 5}{2 \cdot 6 \cdot 6} = \frac{5}{6}.$$

7. $\dfrac{5 \text{ days}}{40 \text{ hours}} \cdot \dfrac{24 \text{ hours}}{1 \text{ day}} = \dfrac{5 \cdot 8 \cdot 3}{40} = \dfrac{3}{1}$

8. $\dfrac{75 \text{ minutes}}{2 \text{ hours}} \cdot \dfrac{1 \text{ hour}}{60 \text{ minutes}} = \dfrac{15 \cdot 5}{15 \cdot 8} = \dfrac{5}{8}$

9. (a) $.4 = \dfrac{4}{10} = \dfrac{2}{5}$

   (b) 4 to 10 means $\dfrac{4}{10}$, which is equivalent to $\dfrac{2}{5}$

   (c) 20 to 50 means $\dfrac{20}{50}$, which simplifies to $\dfrac{2}{5}$

   (d) 5 to 2 means $\dfrac{5}{2}$, which is not the same ratio as 2 to 5.

10. Three ratios that are equivalent to 4 to 3 are: 8 to 6, 12 to 9, and 16 to 12.

11. Writing exercise

12. Writing exercise

In exercises 13 through 18, check to see if the cross products are equal to determine if the proportions are true or false.

13. $5 \cdot 56 = 280; 35 \cdot 8 = 280$. The proportion is true.

14. $4 \cdot 21 = 84; 12 \cdot 7 = 84$. The proportion is true.

15. $120 \cdot 10 = 1200; 82 \cdot 7 = 574$. The proportion is false.

16. $27 \cdot 110 = 2970; 160 \cdot 18 = 2880$. The proportion is false.

17. $\dfrac{1}{2} \cdot 10 = 5; 5 \cdot 1 = 5$ The proportion is true.

18. $\dfrac{1}{3} \cdot 18 = 6; 6 \cdot 1 = 6$ The proportion is true.

19. $\dfrac{k}{4} = \dfrac{175}{20}$
$20k = 4 \cdot 175$
$20k = 700$
$\dfrac{20k}{20} = \dfrac{700}{20}$
$k = 35$

20. $\dfrac{49}{56} = \dfrac{z}{8}$
$56z = 49 \cdot 8$
$56z = 392$
$\dfrac{56z}{56} = \dfrac{392}{56}$
$z = 7$

21. $\dfrac{x}{6} = \dfrac{18}{4}$
$4x = 6 \cdot 18$
$4x = 108$
$\dfrac{4x}{4} = \dfrac{108}{4}$
$x = 27$

22. $\dfrac{z}{80} = \dfrac{20}{100}$
$100z = 80 \cdot 20$
$100z = 1600$
$\dfrac{100z}{100} = \dfrac{1600}{100}$
$z = 16$

23. $\dfrac{3y - 2}{5} = \dfrac{6y - 5}{11}$
$5(6y - 5) = 11(3y - 2)$
$30y - 25 = 33y - 22$
$30y - 25 + 25 = 33y - 22 + 25$
$30y - 33y = 33y - 33y + 3$
$-3y = 3$
$\dfrac{-3y}{-3} = \dfrac{3}{-3}$
$y = -1$

24. $\dfrac{2p + 7}{3} = \dfrac{p - 1}{4}$
$3(p - 1) = 4(2p + 7)$
$3p - 3 = 8p + 28$
$3p - 3 + 3 = 8p + 28 + 3$
$3p - 8p = 8p - 8p + 31$
$-5p = 31$
$\dfrac{-5p}{-5} = \dfrac{31}{-5}$
$p = -\dfrac{31}{5}$

25. Let $x =$ number of fluid ounces of oil.
$\dfrac{2.5 \text{ fl oz.}}{1 \text{ gallon}} = \dfrac{x}{2.75 \text{ gallons}}$
$1 \cdot x = (2.5)(2.75)$
$x = 6.875$ ounces

26. Let $x =$ number of gallons of gasoline.
$\dfrac{5.5 \text{ fl oz.}}{1 \text{ gallon}} = \dfrac{22 \text{ fl oz}}{x}$
$5.5x = (1)(22)$
$\dfrac{5.5x}{5.5} = \dfrac{22}{5.5}$
$x = 4$ gallons

27. Let $x =$ amount of U.S. money exchanged.
$\dfrac{1 \text{ pound}}{\$1.6762} = \dfrac{400 \text{ pounds}}{x}$
$1 \cdot x = (1.6762)(400)$
$x = \$670.48$

28. Let $x =$ number of Swiss francs.
$\dfrac{\$3}{4.5204 \text{ francs}} = \dfrac{\$49.20}{x}$
$3 \cdot x = (4.5204)(49.20)$
$3x = 222.40368$
$\dfrac{3x}{3} = \dfrac{222.40368}{3}$
$x \approx 74.13$

29. Let $c =$ cost to fill tank.
$\dfrac{6 \text{ gal}}{\$3.72} = \dfrac{15 \text{ gal}}{c}$
$6 \cdot c = (3.72)(15)$
$6c = 55.8$
$\dfrac{6c}{6} = \dfrac{55.8}{6}$
$x = \$9.30$

30. Let $t =$ sales tax.
$\dfrac{16}{1.32} = \dfrac{120}{t}$
$16 \cdot t = (1.32)(120)$
$16t = 158.4$
$\dfrac{16t}{16} = \dfrac{158.4}{16}$
$t = \$9.90$

31. Let $x =$ number of feet.
$\dfrac{600 \text{ miles}}{2.4 \text{ feet}} = \dfrac{1000 \text{ miles}}{x}$
$600 \cdot x = (2.4)(1000)$
$600x = 2400$
$\dfrac{600x}{600} = \dfrac{2400}{600}$
$x = 4$ feet

32. Let $x =$ number of feet.
$$\frac{3300 \text{ miles}}{11 \text{ inches}} = \frac{7700 \text{ miles}}{x}$$
$$3300 \cdot x = (11)(7700)$$
$$3300x = 84,700$$
$$\frac{3300x}{3300} = \frac{84,700}{3300}$$
$$x = 25\frac{2}{3} \text{ feet}$$

33. Let $x =$ number of fish in lake.
$$\frac{250 \text{ tagged}}{x} = \frac{7 \text{ tagged}}{350 \text{ fish}}$$
$$7 \cdot x = (250)(350)$$
$$7x = 87,500$$
$$\frac{7x}{7} = \frac{87,500}{7}$$
$$x = 12,500 \text{ fish}$$

34. Let $x =$ number of fish in lake.
$$\frac{420 \text{ tagged}}{x} = \frac{9 \text{ tagged}}{500 \text{ fish}}$$
$$9 \cdot x = (420)(500)$$
$$9x = 210,000$$
$$\frac{9x}{9} = \frac{210,000}{9}$$
$$x = 23,333.\overline{3}$$

Rounded to the nearest hundred, there are approximately 23,300 fish in the lake.

35. (a) $$\frac{26}{100} = \frac{x}{350}$$
$$100 \cdot x = 26 \cdot 350$$
$$100x = 9100$$
$$\frac{100x}{100} = \frac{9100}{100}$$
$$x = \$91 \text{ million}$$

(b) $$\frac{32}{100} = \frac{x}{350}$$
$$100 \cdot x = 32 \cdot 350$$
$$100x = 11,200$$
$$\frac{100x}{100} = \frac{11,200}{100}$$
$$x = \$112 \text{ million}$$

The amount provided by sponsorship was $112 million. Next $112 divided equally among 10 sponsors is $11.2 million per sponsor.

(c) $$\frac{34}{100} = \frac{x}{350}$$
$$100 \cdot x = 34 \cdot 350$$
$$100x = 11,900$$
$$\frac{100x}{100} = \frac{11,900}{100}$$
$$x = \$119 \text{ million}$$

36. (a) $$\frac{22}{100} = \frac{x}{350}$$
$$100 \cdot x = 22 \cdot 350$$
$$100x = 7700$$
$$\frac{100x}{100} = \frac{7700}{100}$$
$$x = \$77 \text{ million}$$

(b) $$\frac{67}{100} = \frac{x}{350}$$
$$100 \cdot x = 67 \cdot 350$$
$$100x = 23,450$$
$$\frac{100x}{100} = \frac{23,450}{100}$$
$$x = \$234.5 \text{ million}$$

37. $$\frac{\$3.09}{20 \text{ bags}} = \$0.1545 \text{ per bag}$$

$$\frac{\$4.59}{30 \text{ bags}} = \$0.153 \text{ per bag}$$

The 30-count size is the better buy.

38. $$\frac{\$.99}{1 \text{ oz.}} = \$0.99 \text{ per ounce}$$

$$\frac{\$1.65}{2 \text{ oz.}} = \$0.825 \text{ per ounce}$$

$$\frac{\$4.39}{4 \text{ oz.}} = \$1.0975 \text{ per ounce}$$

The 2-ounce size is the best buy.

39. $$\frac{\$2.99}{15 \text{ oz.}} = \$0.199 \text{ per ounce}$$

$$\frac{\$4.49}{25 \text{ oz.}} = \$0.180 \text{ per ounce}$$

$$\frac{\$5.49}{31 \text{ oz.}} = \$0.177 \text{ per ounce}$$

The 31-ounce size is the best buy.

164 CHAPTER 7 THE BASIC CONCEPTS OF ALGEBRA

40. $\dfrac{\$1.39}{8 \text{ oz.}} \approx \$0.174$ per ounce

$\dfrac{\$2.19}{16 \text{ oz.}} \approx \$0.137$ per ounce

$\dfrac{\$2.99}{32 \text{ oz.}} \approx \$0.093$ per ounce

The 32-ounce size is the best buy.

41. $\dfrac{\$0.89}{14 \text{ oz.}} \approx \$0.064$ per ounce

$\dfrac{\$1.19}{32 \text{ oz.}} \approx \$0.037$ per ounce

$\dfrac{\$2.95}{64 \text{ oz.}} \approx \$0.046$ per ounce

The 32-ounce size is the best buy.

42. $\dfrac{\$0.45}{8 \text{ oz.}} \approx \$0.056$ per ounce

$\dfrac{\$0.49}{16 \text{ oz.}} \approx \$0.031$ per ounce

$\dfrac{\$1.59}{50 \text{ oz.}} \approx \$0.032$ per ounce

The 16-ounce size is the best buy.

43. $\dfrac{12}{x} = \dfrac{9}{3}$

$9 \cdot x = 12 \cdot 3$

$9x = 36$

$\dfrac{9x}{9} = \dfrac{36}{9}$

$x = 4$

44. $\dfrac{2}{x} = \dfrac{3}{12}$

$3 \cdot x = 2 \cdot 12$

$3x = 24$

$\dfrac{3x}{3} = \dfrac{24}{3}$

$x = 8$

45. $\dfrac{3}{x} = \dfrac{6}{2}$

$6 \cdot x = 3 \cdot 2$

$6x = 6$

$\dfrac{6x}{6} = \dfrac{6}{6}$

$x = 1$

46. $\dfrac{3}{x} = \dfrac{3}{2}$

$3 \cdot x = 3 \cdot 2$

$3x = 6$

$\dfrac{3x}{3} = \dfrac{6}{3}$

$x = 2$

47. (a)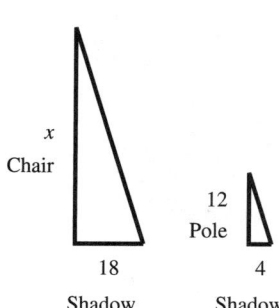

(b) Let $x =$ height of the chair.

$\dfrac{18}{4} = \dfrac{x}{12}$

$4 \cdot x = 18 \cdot 12$

$4x = 216$

$\dfrac{4x}{4} = \dfrac{216}{4}$

$x = 54$ feet

48. (a)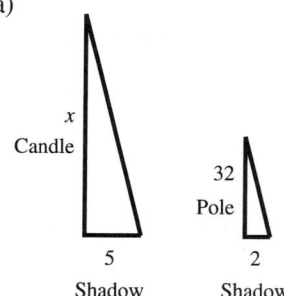

(b) Let $x =$ height of the candle

$\dfrac{5}{2} = \dfrac{x}{32}$

$2 \cdot x = 5 \cdot 32$

$2x = 160$

$\dfrac{2x}{2} = \dfrac{160}{2}$

$x = 80$ feet

49. $\dfrac{225}{130.7} = \dfrac{x}{140.3}$

$130.7x = 225(140.3)$

$130.7x = 31567.5$

$\dfrac{130.7x}{130.7} = \dfrac{31567.5}{130.7}$

$x \approx \$242$

50. $\dfrac{225}{130.7} = \dfrac{x}{148.2}$
$130.7x = 225(148.2)$
$130.7x = 33345$
$\dfrac{130.7x}{130.7} = \dfrac{33345}{130.7}$
$x \approx \$255$

51. $\dfrac{225}{130.7} = \dfrac{x}{156.9}$
$130.7x = 225(156.9)$
$130.7x = 35302.5$
$\dfrac{130.7x}{130.7} = \dfrac{35302.5}{130.7}$
$x \approx \$270$

52. $\dfrac{225}{130.7} = \dfrac{x}{163.0}$
$130.7x = 225(163.0)$
$130.7x = 36675$
$\dfrac{130.7x}{130.7} = \dfrac{36675}{130.7}$
$x \approx \$281$

53. $\dfrac{225}{130.7} = \dfrac{x}{166.6}$
$130.7x = 225(166.6)$
$130.7x = 37485$
$\dfrac{130.7x}{130.7} = \dfrac{37485}{130.7}$
$x \approx \$287$

54. $\dfrac{225}{130.7} = \dfrac{x}{172.2}$
$130.7x = 225(172.2)$
$130.7x = 38745$
$\dfrac{130.7x}{130.7} = \dfrac{38745}{130.7}$
$x \approx \$296$

55. If $x = ky$
$27 = k \cdot 6$
$\dfrac{27}{6} = k,$

then

$x = \dfrac{27}{6} \cdot \dfrac{2}{1} = 9.$

56. If $z = kx$
$30 = k \cdot 8$
$\dfrac{30}{8} = k$
$\dfrac{15}{4} = k,$

then

$z = \dfrac{15}{4} \cdot \dfrac{4}{1} = 15.$

57. If $m = kp^2$
$20 = k \cdot 2^2$
$20 = 4k$
$\dfrac{20}{4} = \dfrac{4k}{4}$
$5 = k,$

then

$= 5 \cdot 5^2 = 5 \cdot 25 = 125.$

58. If $a = kb^2$
$48 = k \cdot 4^2$
$48 = k \cdot 16$
$\dfrac{48}{16} = k$
$3 = k,$

then

$a = 3 \cdot 7^2 = 3 \cdot 49 = 147.$

59. If $p = \dfrac{k}{q^2}$
$4 = \dfrac{k}{\left(\frac{1}{2}\right)^2}$
$4 = \dfrac{k}{\frac{1}{4}}$
$4 = k \div \dfrac{1}{4}$
$4 = k \cdot \dfrac{4}{1}$
$4 = 4k$
$1 = k,$

then

$p = \dfrac{1}{\left(\frac{3}{2}\right)^2}$
$p = 1 \div \dfrac{9}{4}$
$p = 1 \cdot \dfrac{4}{9}$
$p = \dfrac{4}{9}.$

166 CHAPTER 7 THE BASIC CONCEPTS OF ALGEBRA

60. If
$$z = \frac{k}{x^2}$$
$$9 = \frac{k}{\left(\frac{2}{3}\right)^2}$$
$$9 = \frac{k}{\frac{4}{9}}$$
$$9 = k \div \frac{4}{9}$$
$$9 = k \cdot \frac{9}{4}$$
$$9 \cdot \frac{4}{9} = k \cdot \frac{9}{4} \cdot \frac{4}{9}$$
$$4 = k,$$

then
$$z = \frac{4}{\left(\frac{5}{4}\right)^2}$$
$$p = 4 \div \frac{25}{16}$$
$$p = 4 \cdot \frac{16}{25}$$
$$p = \frac{64}{25}.$$

61. Let i = the amount of interest and r = rate of interest.
$$i = kr$$
$$48 = k(.05)$$
$$\frac{48}{.05} = k$$
$$960 = k$$

Then
$$i = 960(.042) = \$40.32.$$

62. Let A = area of triangle and h = height.
$$A = kh$$
$$10 = k \cdot 4$$
$$\frac{10}{4} = k$$
$$\frac{5}{2} = k$$

Then $A = \frac{5}{2} \cdot 6 = 15$ square inches.

63. Let r = speed or rate and t = time.
$$r = \frac{k}{t}$$
$$30 = \frac{k}{.5}$$
$$30(.5) = \frac{k}{.5} \cdot \frac{.5}{1}$$
$$15 = k$$

Then
$$r = \frac{15}{.75} = 20 \text{ miles per hour.}$$

64. Let l = length of rectangle and w = width.
$$l = \frac{k}{w}$$
$$27 = \frac{k}{10}$$
$$27 \cdot 10 = \frac{k}{10} \cdot \frac{10}{1}$$
$$270 = k$$

Then
$$l = \frac{270}{18} = 15 \text{ feet.}$$

65. Let m = the weight of an object on the moon and e = the weight of the object on the earth.
$$m = ke$$
$$59 = k \cdot 352$$
$$\frac{59}{352} = \frac{k \cdot 352}{352}$$
$$\frac{59}{352} = k$$

Then
$$m = \frac{59}{352} \cdot 1800 \approx 302 \text{ pounds.}$$

66. Let d = distance and t = time.
$$d = kt$$
$$2226 = k \cdot 103$$
$$\frac{2226}{103} = \frac{k \cdot 103}{103}$$
$$\frac{2226}{103} = k$$

Then
$$d = \frac{2226}{103} \cdot 120 \approx 2593 \text{ miles.}$$

67. Let p = pressure and d = depth.
$$p = kd$$
$$50 = k \cdot 10$$
$$\frac{50}{10} = \frac{k \cdot 10}{10}$$
$$5 = k$$

Then
$$p = 5 \cdot 20 = 100 \text{ pounds per square inch.}$$

68. Let p = pressure and t = temperature.
$$p = kt$$
$$5 = k \cdot 200$$
$$\frac{5}{200} = \frac{k \cdot 200}{200}$$
$$.025 = k$$

Then
$$p = (.025)300 = 7.5 \text{ pounds per square inch.}$$

69. Let p = pressure and v = volume.

$$p = \frac{k}{v}$$
$$10 = \frac{k}{3}$$
$$10 \cdot 3 = \frac{k}{3} \cdot 3$$
$$30 = k$$

Then
$$p = \frac{30}{1.5} = 20 \text{ pounds per square foot.}$$

70. Let f = force and l = length of spring.

$$f = kl$$
$$12 = k \cdot 3$$
$$\frac{12}{3} = \frac{k \cdot 3}{3}$$
$$4 = k$$

Then
$$f = 4 \cdot 5 = 20 \text{ pounds per square inch.}$$

71. Let d = distance and t = time.

$$d = kt^2$$
$$400 = k \cdot 5^2$$
$$400 = k \cdot 25$$
$$\frac{400}{25} = \frac{k \cdot 25}{25}$$
$$16 = k$$

Then
$$d = 16 \cdot 3^2 = 16 \cdot 9 = 144 \text{ feet.}$$

72. Let I = illumination and d = distance.

$$I = \frac{k}{d^2}$$
$$75 = \frac{k}{4^2}$$
$$75 \cdot 16 = \frac{k}{16} \cdot 16$$
$$1200 = k$$

Then
$$I = \frac{1200}{9^2} = \frac{1200}{81} \approx 14.8 \text{ or } 14\frac{22}{27} \text{ foot-candles.}$$

73. Let B = body mass index and W = a person's weight and H = the person's height.

$$B = \frac{kW}{H^2}$$
$$20 = \frac{k \cdot 118}{64^2}$$
$$64^2 \cdot 20 = \frac{k \cdot 118}{64^2} \cdot 64^2$$
$$81920 = 118k$$
$$k \approx 694.24$$

Then
$$B = \frac{694.24 \times 165}{70^2} = \frac{114549.6}{4900} \approx 23.$$

74. Let f = force, r = radius of curve, w = weight of car, and s = speed of car.

$$f = \frac{kws^2}{r}$$
$$242 = \frac{k \cdot 2000 \times 30^2}{500}$$
$$242 = \frac{k \cdot 1800000}{500}$$
$$242 = 3600k$$
$$k \approx .06722$$

Thus
$$f \approx \frac{.06722ws^2}{r}$$
$$f = \frac{.06722 \times 2000 \times 50^2}{750}$$
$$f \approx 448.1 \text{ pounds}$$

75. Let w = weight of bass, and g = girth, and l = length.

$$w = kgl^2$$
$$22.7 = k \cdot 21 \cdot 36^2$$
$$22.7 = k \cdot 27216$$
$$\frac{22.7}{27216} = \frac{k \cdot 27216}{27216}$$
$$k \approx .000834$$

Thus
$$w = .000834 \cdot 18 \cdot 28^2 \approx 11.8 \text{ pounds.}$$

76. Let w = weight, and g = girth, and l = length of trout.

$$w = klg^2$$
$$10.5 = k \cdot 26 \cdot 18^2$$
$$10.5 = k \cdot 8424$$
$$\frac{10.5}{8424} = \frac{k \cdot 8424}{8424}$$
$$k \approx .001246$$

Thus
$$w = .001246 \cdot 22 \cdot 15^2 \approx 6.2 \text{ pounds.}$$

7.4 EXERCISES

1. $\{x \mid x \leq 3\}$ matches with letter D, $(-\infty, 3]$. The bracket is used in interval notation because 3 is included in the set.

2. $\{x \mid x > 3\}$ matches with letter C, $(3, \infty)$. Both notations mean all values greater than 3.

3. $\{x \mid x < 3\}$ matches with letter B, the number line showing all values less than 3.

4. $\{x \mid x \geq 3\}$ matches with letter A, the number line showing all values that are greater than or equal to 3.

5. $\{x \mid -3 \leq x \leq 3\}$ matches with letter F, $[-3, 3]$. The brackets indicate that the endpoints, -3 and 3, are included.

6. $\{x \mid -3 < x < 3\}$ matches with letter E, $(-3, 3)$. The parentheses indicate that the endpoints are not included in the set.

7. Writing exercise

8. Writing exercise

9. $$4x + 1 \geq 21$$
$$4x + 1 - 1 \geq 21 - 1$$
$$4x \geq 20$$
$$\frac{4x}{4} \geq \frac{20}{4}$$
$$x \geq 5$$

   Interval notation: $[5, \infty)$
   Graph:

10. $$5t + 2 \geq 52$$
$$5t + 2 - 2 \geq 52 - 2$$
$$5t \geq 50$$
$$\frac{5t}{5} \geq \frac{50}{5}$$
$$t \geq 10$$

    Interval notation: $[10, \infty)$
    Graph:

11. $$\frac{3k-1}{4} > 5$$
$$\frac{4}{1} \cdot \frac{3k-1}{4} > 5 \cdot 4$$
$$3k - 1 > 20$$
$$3k - 1 + 1 > 20 + 1$$
$$3k > 21$$
$$\frac{3k}{3} > \frac{21}{3}$$
$$k > 7$$

    Interval notation: $(7, \infty)$
    Graph:

12. $$\frac{5z-6}{8} < 8$$
$$\frac{8}{1} \cdot \frac{5z-6}{8} < 8 \cdot 8$$
$$5z - 6 < 64$$
$$5z - 6 + 6 < 64 + 6$$
$$5z < 70$$
$$\frac{5z}{5} < \frac{70}{5}$$
$$z < 14$$

    Interval notation: $(-\infty, 14)$
    Graph:

13. In this exercise, remember to reverse the inequality sign when multiplying or dividing by a negative number.
$$-4x < 16$$
$$\frac{-4x}{-4} > \frac{16}{-4}$$
$$x > -4$$

    Interval notation: $(-4, \infty)$
    Graph:

14. In this exercise remember to reverse the inequality sign when multiplying or dividing by a negative number.
$$-2m > 10$$
$$\frac{-2m}{-2} < \frac{10}{-2}$$
$$m < -5$$

    Interval notation: $(-\infty, -5)$
    Graph:

15. In this exercise remember to reverse the inequality sign when multiplying or dividing by a negative number.
$$-\frac{3}{4}r \geq 30$$
$$-\frac{4}{3} \cdot -\frac{3}{4}r \leq \frac{30}{1} \cdot -\frac{4}{3}$$
$$r \leq -\frac{30 \cdot 4}{3}$$
$$r \leq -40$$

    Interval notation: $(-\infty, -40]$
    Graph:

16. In this exercise, it is wise to change to all fractional notation or all decimal notation. Here is the exercise in decimal notation:
$$-1.5x \le -4.5$$
$$\frac{-1.5x}{-1.5} \ge \frac{-4.5}{-1.5}$$
$$x \ge 3.$$
Interval notation: $[3, \infty)$
Graph:

17. $-1.3m \ge -5.2$
$$\frac{-1.3m}{-1.3} \le \frac{-5.2}{-1.3}$$
$$y \le 4$$
Interval notation: $(-\infty, 4]$
Graph:

18. $-2.5x \le -1.25$
$$\frac{-2.5x}{-2.5} \ge \frac{-1.25}{-2.5}$$
$$x \ge .5$$
Interval notation: $[.5, \infty)$
Graph:

19. $$\frac{2k-5}{-4} > 5$$
$$\frac{-4}{1} \cdot \frac{2k-5}{4} < 5 \cdot -4$$
$$2k - 5 < -20$$
$$2k - 5 + 5 < -20 + 5$$
$$2k < -15$$
$$\frac{2k}{2} < \frac{-15}{2}$$
$$k < \frac{-15}{2}$$
Interval notation: $(-\infty, \frac{-15}{2})$
Graph:

20. $$\frac{3z-2}{-5} < 6$$
$$\frac{-5}{1} \cdot \frac{3z-2}{-5} > 6 \cdot -5$$
$$3z - 2 > -30$$
$$3z - 2 + 2 > -30 + 2$$
$$3z > -28$$
$$\frac{3z}{3} > \frac{-28}{3}$$
$$z > \frac{-28}{3}$$

Interval notation: $(-\frac{28}{3}, \infty)$
Graph:

21. $x + 4(2x - 1) \ge x$
$$x + 8x - 4 \ge x$$
$$9x - 4 \ge x$$
$$9x - 4 + 4 \ge x + 4$$
$$9x \ge x + 4$$
$$9x - x \ge x - x + 4$$
$$8x \ge 4$$
$$\frac{8x}{8} \ge \frac{4}{8}$$
$$x \ge \frac{4 \cdot 1}{4 \cdot 2}$$
$$x \ge \frac{1}{2}$$
Interval notation: $[\frac{1}{2}, \infty)$
Graph:

22. $m - 2(m - 4) \le 3m$
$$m - 2m + 8 \le 3m$$
$$-m + 8 \le 3m$$
$$-m + 8 - 8 \le 3m - 8$$
$$-m \le 3m - 8$$
$$-m - 3m \le 3m - 3m - 8$$
$$-4m \le -8$$
$$\frac{-4m}{-4} \ge \frac{-8}{-4}$$
$$m \ge 2$$
Interval notation: $[2, \infty)$
Graph:

23. $-(4 + r) + 2 - 3r < -14$
$$-4 - r + 2 - 3r < -14$$
$$-2 - 4r < -14$$
$$-2 + 2 - 4r < -14 + 2$$
$$-4r < -12$$
$$\frac{-4m}{-4} > \frac{-12}{-4}$$
$$m > 3$$
Interval notation: $(3, \infty)$
Graph:

24. $-(9+k)-5+4k \geq 4$
$-9-k-5+4k \geq 4$
$-14+3k \geq 4$
$-14+14+3k \geq 4+14$
$3k \geq 18$
$\dfrac{3k}{3} \geq \dfrac{18}{3}$
$k \geq 6$

Interval notation: $[6, \infty)$
Graph:

25. $-3(z-6) > 2z-2$
$-3z+18 > 2z-2$
$-3z-2z+18 > 2z-2z-2$
$-5z+18 > -2$
$-5z+18-18 > -2-18$
$-5z > -20$
$\dfrac{-5z}{-5} < \dfrac{-20}{-5}$
$z < 4$

Interval notation: $(-\infty, 4)$
Graph:

26. $-2(x+4) \leq 6x+16$
$-2x-8 \leq 6x+16$
$-2x-8+8 \leq 6x+16+8$
$-2x \leq 6x+24$
$-2x-6x \leq 6x-6x+24$
$-8x \leq 24$
$\dfrac{-8x}{-8} \geq \dfrac{24}{-8}$
$x \geq -3$

Interval notation: $[-3, \infty)$
Graph:

27. Clear the fractions by multiplying both sides of the equation by the LCM, 6.

$\dfrac{2}{3}(3k-1) \geq \dfrac{3}{2}(2k-3)$
$\dfrac{6}{1} \cdot \dfrac{2}{3}(3k-1) \geq \dfrac{6}{1} \cdot \dfrac{3}{2}(2k-3)$
$2 \cdot 2(3k-1) \geq 3 \cdot 3(2k-3)$
$4(3k-1) \geq 9(2k-3)$
$12k-4 \geq 18k-27$
$12k-4+4 \geq 18k-27+4$
$12k \geq 18k-23$
$12k-18k \geq 18k-18k-23$
$-6k \geq -23$
$\dfrac{-6k}{-6} \leq \dfrac{-23}{-6}$
$k \leq \dfrac{23}{6}$

Interval notation: $(-\infty, \frac{23}{6}]$
Graph:

28. Clear the fractions by multiplying both sides of the equation by the LCM, 15.

$\dfrac{7}{5}(10m-1) < \dfrac{2}{3}(6m+5)$
$\dfrac{15}{1} \cdot \dfrac{7}{5}(10m-1) < \dfrac{15}{1} \cdot \dfrac{2}{3}(6m+5)$
$3 \cdot 7(10m-1) < 5 \cdot 2(6m+5)$
$21(10m-1) < 10(6m+5)$
$210m-21 < 60m+50$
$210m-21+21 < 60m+50+21$
$210m < 60m+71$
$210m-60m < 60m-60m+71$
$150m < 71$
$\dfrac{150m}{150} < \dfrac{71}{150}$

Interval notation: $(-\infty, \frac{71}{150})$
Graph:

29. Clear the fractions by multiplying both sides of the equation by the LCM, 8.

$$-\frac{1}{4}(p+6) + \frac{3}{2}(2p-5) < 10$$
$$\frac{8}{1} \cdot -\frac{1}{4}(p+6) + \frac{8}{1} \cdot \frac{3}{2}(2p-5) < 10 \cdot 8$$
$$2 \cdot -1(p+6) + 4 \cdot 3(2p-5) < 80$$
$$-2(p+6) + 12(2p-5) < 80$$
$$-2p - 12 + 24p - 60 < 80$$
$$22p - 72 < 80$$
$$22p - 72 + 72 < 80 + 72$$
$$22p < 152$$
$$\frac{22p}{22} < \frac{152}{22}$$
$$p < \frac{76 \cdot 2}{11 \cdot 2}$$
$$p < \frac{76}{11}$$

Interval notation: $(-\infty, \frac{76}{11})$
Graph:

30. Clear the fractions by multiplying both sides of the equation by the LCM, 20.

$$\frac{3}{5}(k-2) - \frac{1}{4}(2k-7) \leq 3$$
$$\frac{20}{1} \cdot \frac{3}{5}(k-2) - \frac{20}{1} \cdot \frac{1}{4}(2k-7) \leq 3 \cdot 20$$
$$4 \cdot 3(k-2) - 5 \cdot 1(2k-7) \leq 60$$
$$12(k-2) - 5(2k-7) \leq 60$$
$$12k - 24 - 10k + 35 \leq 60$$
$$2k + 11 \leq 60$$
$$2k + 11 - 11 \leq 60 - 11$$
$$2k \leq 49$$
$$\frac{2k}{2} \leq \frac{49}{2}$$
$$k \leq \frac{49}{2}$$

Interval notation: $(-\infty, \frac{49}{2}]$
Graph:

31. $$3(2x - 4) - 4x < 2x + 3$$
$$6x - 12 - 4x < 2x + 3$$
$$2x - 12 < 2x + 3$$
$$2x - 2x - 12 < 2x - 2x + 3$$
$$-12 < 3$$

Because it is always true that negative 12 is less than 3, x can be replaced by any real number and the inequality will be true. This is called an identity.

Interval notation: $(-\infty, +\infty)$
Graph:

32. $$7(4 - x) + 5x < 2(16 - x)$$
$$28 - 7x + 5x < 32 - 2x$$
$$28 - 2x < 32 - 2x$$
$$28 - 2x + 2x < 32 - 2x + 2x$$
$$28 < 32$$

Because it is always true that 28 is less than 32, x can be replaced by any real number and the inequality will be true. This is called an identity.

Interval notation: $(-\infty, +\infty)$
Graph:

33. $$8\left(\frac{1}{2}x + 3\right) < 8\left(\frac{1}{2}x - 1\right)$$
$$4x + 24 < 4x - 8$$
$$4x - 4x + 24 < 4x - 4x - 8$$
$$24 < -8$$

Because it is not true that 24 is less than negative 8, there is no real number that can replace x to obtain a true statement. This is called a contradiction and there is no solution. This can be symbolized by the empty set: $\emptyset$.

34. $$10x + 2(x - 4) < 12x - 10$$
$$10x + 2x - 8 < 12x - 10$$
$$12x - 8 < 12x - 10$$
$$12x - 12x - 8 < 12x - 12x - 10$$
$$-8 < -10$$

Because it is not true that negative 8 is less than negative 10, there is no real number that can replace x to obtain a true statement. This is called a contradiction and there is no solution. This can be symbolized by the empty set: $\emptyset$.

35. Writing exercise

36. (a) $(0, \infty)$ matches with letter A, the positive real numbers. The interval notation indicates all the numbers that are greater than zero.

    (b) $[0, \infty)$ matches with letter D, the nonnegative real numbers. The interval notation indicates the set of numbers beginning with zero and continuing in a positive direction. In order to include zero with the positive real numbers, the term "nonnegative" is used.

    (c) $(-\infty, 0]$ matches with letter C, the nonpositive real numbers. The set of numbers includes zero and all negative real numbers.

    (d) $(-\infty, 0)$ matches with letter B, the negative real numbers. Zero is not included in this set of numbers.

37. In a three-part inequality, remember to work toward isolating the variable by applying the same operation to all three parts of the inequality.

$$-4 < x - 5 < 6$$
$$-4 + 5 < x - 5 + 5 < 6 + 5$$
$$1 < x < 11$$

Interval notation: $(1, 11)$
Graph:

38. $\quad -1 < x + 1 < 8$
$\quad -1 - 1 < x + 1 - 1 < 8 - 1$
$\quad -2 < x < 7$

Interval notation: $(-2, 7)$
Graph:

39. $\quad -9 \leq k + 5 \leq 15$
$\quad -9 - 5 \leq k + 5 - 5 \leq 15 - 5$
$\quad -14 \leq k \leq 10$

Interval notation: $[-14, 10]$
Graph:

40. $\quad -4 \leq m + 3 \leq 10$
$\quad -4 - 3 \leq m + 3 - 3 \leq 10 - 3$
$\quad -7 \leq m \leq 7$

Interval notation: $[-7, 7]$
Graph:

41. $\quad -6 \leq 2z + 4 \leq 16$
$\quad -6 - 4 \leq 2z + 4 - 4 \leq 16 - 4$
$\quad -10 \leq 2z \leq 12$
$\quad \dfrac{-10}{2} \leq \dfrac{2z}{2} \leq \dfrac{12}{2}$
$\quad -5 \leq z \leq 6$

Interval notation: $[-5, 6]$
Graph:

42. $\quad -15 < 3p + 6 < -12$
$\quad -15 - 6 < 3p + 6 - 6 < -12 - 6$
$\quad -21 < 3p < -18$
$\quad \dfrac{-21}{3} < \dfrac{3p}{3} < \dfrac{-18}{3}$
$\quad -7 < p < -6$

Interval notation: $(-7, -6)$
Graph:

43. $\quad -19 \leq 3x - 5 \leq 1$
$\quad -19 + 5 \leq 3x - 5 + 5 \leq 1 + 5$
$\quad -14 \leq 3x \leq 6$
$\quad \dfrac{-14}{3} \leq \dfrac{3x}{3} \leq \dfrac{6}{3}$
$\quad \dfrac{-14}{3} \leq x \leq 2$

Interval notation: $\left[\dfrac{-14}{3}, 2\right]$
Graph:

44. $\quad -16 < 3t + 2 < -10$
$\quad -16 - 2 < 3t + 2 - 2 < -10 - 2$
$\quad -18 < 3t < -12$
$\quad \dfrac{-18}{3} < \dfrac{3t}{3} < \dfrac{-12}{3}$
$\quad -6 < t < -4$

Interval notation: $(-6, -4)$
Graph:

45. $\quad -1 \leq \dfrac{2x - 5}{6} \leq 5$
$\quad 6 \cdot -1 \leq \dfrac{6}{1} \cdot \dfrac{2x - 5}{6} \leq 6 \cdot 5$
$\quad -6 \leq 2x - 5 \leq 30$
$\quad -6 + 5 \leq 2x - 5 + 5 \leq 30 + 5$
$\quad -1 \leq 2x \leq 35$
$\quad \dfrac{-1}{2} \leq \dfrac{2x}{2} \leq \dfrac{35}{2}$
$\quad \dfrac{-1}{2} \leq x \leq \dfrac{35}{2}$

Interval notation: $\left[\dfrac{-1}{2}, \dfrac{35}{2}\right]$
Graph:

46. $\quad -3 \leq \dfrac{3m + 1}{4} \leq 3$
$\quad 4 \cdot -3 \leq \dfrac{4}{1} \cdot \dfrac{3m + 1}{4} \leq 4 \cdot 3$
$\quad -12 \leq 3m + 1 \leq 12$
$\quad -12 - 1 \leq 3m + 1 - 1 \leq 12 - 1$
$\quad -13 \leq 3m \leq 11$
$\quad \dfrac{-13}{3} \leq \dfrac{3m}{3} \leq \dfrac{11}{3}$
$\quad \dfrac{-13}{3} \leq m \leq \dfrac{11}{3}$

Interval notation: $\left[\dfrac{-13}{3}, \dfrac{11}{3}\right]$
Graph:

47. Remember to reverse the inequality symbols in this exercise when dividing by -9.

$$4 \leq 5 - 9x < 8$$
$$4 - 5 \leq 5 - 5 - 9x < 8 - 5$$
$$-1 \leq -9x < 3$$
$$\frac{-1}{-9} \geq \frac{-9x}{-9} > \frac{3}{-9}$$
$$\frac{1}{9} \geq x > -\frac{1}{3}$$

To make this more meaningful, restate the inequality as

$$-\frac{1}{3} < x \leq \frac{1}{9}$$

Interval notation: $(-\frac{1}{3}, \frac{1}{9}]$
Graph:

48. $$4 \leq 3 - 2x < 8$$
$$4 - 3 \leq 3 - 3 - 2x < 8 - 3$$
$$1 \leq -2x < 5$$
$$\frac{1}{-2} \geq \frac{-2x}{-2} > \frac{5}{-2}$$
$$-\frac{1}{2} \geq x > -\frac{5}{2}$$

To make this more meaningful, restate the inequality as

$$-\frac{5}{2} < x \leq -\frac{1}{2}$$

Interval notation: $(-\frac{5}{2}, -\frac{1}{2}]$
Graph:

49. The following months show percentages greater than 7.7%: April, 12.9%; May, 22.1%; June, 20.7%, and July, 11.1%.

50. The months in which the percent of tornadoes was 12.9% or greater were: April, 12.9%; May, 22.1%, and June, 20.7%.

51. First find what percent 1500 is of 17,252.
Let r = the percent in decimal form

$$r \cdot 17{,}252 = 1500$$
$$\frac{r \cdot 17{,}252}{17{,}252} = \frac{1500}{17{,}252}$$
$$r = .087$$

This decimal value, .087, is equivalent to 8.7%. Look for the months for which the percent is less than 8.7%. These months are: January, February, March, August, September, October, November, and December.

52. In March, the percentage of tornadoes was 7.0%; in October, the percentage was 3.0%. This is a difference of 4.0%. Now find 4.0% of 17252.

$$.04(17252) \approx 690 \text{ more tornadoes}$$

53. Read the vertical axis of the graph to find the 90°. Drawing a horizontal line (or lining up a horizontal sheet of paper) through this point we see that it intersects the curve between 2:30 P.M. and 6:00 P.M..

54. Read the vertical axis of the graph to find the 70°. Drawing a horizontal line (or lining up a horizontal sheet of paper) through this point, we see that it intersects the curve at about 10:15 P.M and about 9:00 P.M.. This would indicate the range in time values where the temperature is no more that 70°F to be from 8:00 A.M. to 10:15 A.M. and after about 9:00 P.M..

55. Draw a vertical line (or use the side of a sheet of paper) through the point representing the Women's 100 meter. The vertical line intersects the lower boundary at about 84°F and the upper boundary about 91°F.

56. Draw a vertical line (or use the side of a sheet of paper) through the point representing the Men's 10,000 meters. The vertical line intersects the lower boundary at about 65°F and the upper boundary about 67°F.

57. Let d = number of additional 1/5 miles. Dantrell must pay \$1.50 plus \$.25 for each additional 1/5 mile, and the amount cannot exceed \$3.75. Solve the inequality.

$$1.50 + .25d \leq 3.75$$
$$1.50 - 1.50 + .25d \leq 3.75 - 1.50$$
$$.25d \leq 2.25$$
$$\frac{.25d}{.25} \leq \frac{2.25}{.25}$$
$$d \leq 9$$

Then Dantrell can travel $9 \cdot \frac{1}{5} = \frac{9}{5}$ or $1\frac{4}{5}$ miles in addition to the first $\frac{1}{5}$ mile for a total of 2 miles.

58. Let d = number of additional 1/7 miles. Dantrell must pay \$.90 plus \$.10 for each additional 1/7 mile, and the amount cannot exceed \$3.75. Solve the inequality.

$$.90 + .10d \leq 3.75$$
$$.90 - .90 + .10d \leq 3.75 - .90$$
$$.10d \leq 2.85$$
$$\frac{.10d}{.10} \leq \frac{2.85}{.10}$$
$$d \leq 28.5$$

Then Dantrell can travel $28.5 \cdot \frac{1}{7} = \frac{28.5}{7}$ or $4\frac{1}{14}$ miles in addition to the first $\frac{1}{7}$ mile for a total of $4\frac{3}{14}$ miles. This is approximately 4.2 miles. The final answer to the question is the difference between 4.2 miles and 2 miles (from Exercise 57), which is 2.2 miles.

59. Let x = score needed on the third test. Solve the inequality:
$$\frac{90+82+x}{3} \geq 84$$
$$\frac{3}{1} \cdot \frac{90+82+x}{3} \geq 84 \cdot 3$$
$$90+82+x \geq 252$$
$$172+x \geq 252$$
$$172-172+x \geq 252-172$$
$$x \geq 80$$

Margaret must score at least a grade of 80 on the third test.

60. Let x = score needed on the third test. Solve the inequality:
$$\frac{92+96+x}{3} \geq 90$$
$$\frac{3}{1} \cdot \frac{92+96+x}{3} \geq 90 \cdot 3$$
$$92+96+x \geq 270$$
$$188+x \geq 270$$
$$188-188+x \geq 270-188$$
$$x \geq 82$$

Susan must score at least a grade of 82 on the third test.

61. Let m = number of miles, and the cost to rent each vehicle is:
Avis = $35 + .14m$
Downtown Toyota = $34 + .16m$
To determine the number of miles at which the cost to rent from Downtown Toyota will exceed the cost from Avis, solve the inequality:

$$\text{Cost for Downtown Toyota} > \text{Cost for Avis}$$
$$34 + .16m > 35 + .14m$$
$$34 - 34 + .16m > 35 - 34 + .14m$$
$$.16m > 1 + .14m$$
$$.16m - .14m > 1 + .14m - .14m$$
$$.02m > 1$$
$$\frac{.02m}{.02} > \frac{1}{.02}$$
$$m > 50 \text{ miles.}$$

The price to rent from Downtown Toyota would exceed the price to rent from Avis after 50 miles.

62. Let m = number of miles, and the cost to rent each vehicle for a week is:
Avis = $28 \cdot 7 = 196$
Downtown Toyota = $108 + .14m$
To determine the number of miles at which the cost to rent from Avis is less than from Toyota, solve the inequality:

$$196 < 108 + .14m$$
$$196 - 108 < 108 - 108 + .14m$$
$$88 < .14m$$
$$\frac{88}{.14} < \frac{.14m}{.14}$$
$$628.6 < m, \text{ or } m > 628.6 \text{ miles.}$$

They would have to drive more than 628.6 miles before the Avis price would be less than the Toyota price.

63. Solve the following inequalities for weight (w) given each value of height (h).
$$19 < \text{BMI} < 25$$
$$19 < \frac{704 \times w}{h^2} < 25$$

(a) $h = 72$ in.
$$19 < \frac{704 \times w}{72^2} < 25$$
$$72^2 \cdot 19 < 704 \times w < 72^2 \cdot 25$$
$$98496 < 704 \times w < 129600$$
$$\frac{98496}{704} < \frac{704w}{704} < \frac{129600}{704}$$
$$140 < w < 184 \text{ pounds}$$

(b) Answers will vary.

64. Solve the following inequality for THR given the age value, A.
$$.7(220 - A) \leq THR \leq .85(220 - A)$$

(a) Let $A = 35$
$$.7(220-35) \leq THR \leq .85(220-35)$$
$$.7(185) \leq THR \leq .85(185)$$
$$130 \leq THR \leq 157$$

(b) Answers will vary.

For 65 and 66 we want $R > C$ for each company to show a profit.

65. $R > C$
$$24x > 20x + 100$$
$$4x > 100$$
$$\frac{4x}{4} > \frac{100}{4}$$
$$x > 25$$

Thus, they must sell at least 26 tapes to make a profit.

66. $R > C$
$$5.50x > 3x + 2300$$
$$2.5x > 2300$$
$$\frac{2.5x}{2.5} > \frac{2300}{2.5}$$
$$x > 920$$

Thus, they must make at least 921 deliveries to make a profit.

7.5 EXERCISES

1. $\left(\dfrac{5}{3}\right)^2 = \dfrac{25}{9}$, which is choice A.

2. $\left(\dfrac{3}{5}\right)^2 = \dfrac{9}{25}$, which is choice C.

3. $\left(-\dfrac{3}{5}\right)^{-2} = \dfrac{1}{\left(-\frac{3}{5}\right)^2}$
   $= \dfrac{1}{\frac{9}{25}}$
   $= 1 \div \dfrac{9}{25}$
   $= 1 \cdot \dfrac{25}{9}$
   $= \dfrac{25}{9}$, which is choice A

4. $\left(-\dfrac{5}{3}\right)^{-2} = \dfrac{1}{\left(-\frac{5}{3}\right)^2}$
   $= \dfrac{1}{\frac{25}{9}}$
   $= 1 \div \dfrac{25}{9}$
   $= 1 \cdot \dfrac{9}{25}$
   $= \dfrac{9}{25}$, or alternatively
   $\left(-\dfrac{5}{3}\right)^{-2} = \left(-\dfrac{3}{5}\right)^2$
   $= \dfrac{9}{25}$, which is choice C

5. $-\left(-\dfrac{3}{5}\right)^2 = -\left(-\dfrac{3}{5} \cdot -\dfrac{3}{5}\right)$
   $= -\left(\dfrac{9}{25}\right)$
   $= -\dfrac{9}{25}$, which is choice D

6. $-\left(-\dfrac{5}{3}\right)^2 = -\left(-\dfrac{5}{3} \cdot -\dfrac{5}{3}\right)$
   $= -\left(\dfrac{25}{9}\right)$
   $= -\dfrac{25}{9}$, which is choice B

7. $5^4 = 5 \cdot 5 \cdot 5 \cdot 5 = 625$

8. $10^3 = 10 \cdot 10 \cdot 10 = 1000$

9. $(-2)^5 = -2 \cdot -2 \cdot -2 \cdot -2 \cdot -2 = -32$

10. $(-5)^4 = -5 \cdot -5 \cdot -5 \cdot -5 = 625$

11. $-2^3 = -(2 \cdot 2 \cdot 2) = -8$

12. $-3^2 = -(3 \cdot 3) = -9$

13. $-(-3)^4 = -(-3 \cdot -3 \cdot -3 \cdot -3) = -81$

14. $-(-5)^2 = -(-5 \cdot -5) = -25$

15. $7^{-2} = \dfrac{1}{7^2} = \dfrac{1}{49}$

16. $4^{-1} = \dfrac{1}{4}$

17. $-7^{-2} = -\dfrac{1}{7^2} = -\dfrac{1}{49}$

18. $-4^{-1} = -\dfrac{1}{4}$

19. $\dfrac{2}{(-4)^{-3}} = 2 \div (-4)^{-3}$
    $= 2 \div \dfrac{1}{(-4)^3}$
    $= 2 \cdot (-4)^3$
    $= 2 \cdot -64$
    $= -128$

20. $\dfrac{2^{-3}}{3^{-2}} = \dfrac{\frac{1}{2^3}}{\frac{1}{3^2}}$
    $= \dfrac{1}{2^3} \div \dfrac{1}{3^2}$
    $= \dfrac{1}{8} \cdot \dfrac{3^2}{1}$
    $= \dfrac{9}{8}$

21. $\dfrac{5^{-1}}{4^{-2}} = \dfrac{\frac{1}{5}}{\frac{1}{4^2}}$
    $= \dfrac{1}{5} \div \dfrac{1}{4^2}$
    $= \dfrac{1}{5} \cdot \dfrac{4^2}{1}$
    $= \dfrac{16}{5}$

22. $\left(\dfrac{1}{2}\right)^{-3} = \dfrac{1}{\left(\frac{1}{2}\right)^3}$
    $= 1 \div \left(\dfrac{1}{2}\right)^3$
    $= 1 \cdot \dfrac{8}{1}$
    $= 8$

23. $\left(\dfrac{1}{5}\right)^{-3} = \dfrac{1^{-3}}{5^{-3}}$
$= \dfrac{5^3}{1^3}$
$= 125$

24. $\left(\dfrac{2}{3}\right)^{-2} = \dfrac{2^{-2}}{3^{-2}}$
$= \dfrac{3^2}{2^2}$
$= \dfrac{9}{4}$

25. $\left(\dfrac{4}{5}\right)^{-2} = \dfrac{4^{-2}}{5^{-2}}$
$= \dfrac{5^2}{4^2}$
$= \dfrac{25}{16}$

26. $3^{-1} + 2^{-1} = \dfrac{1}{3} + \dfrac{1}{2}$
$= \dfrac{2}{2} \cdot \dfrac{1}{3} + \dfrac{1}{2} \cdot \dfrac{3}{3}$
$= \dfrac{2}{6} + \dfrac{3}{6}$
$= \dfrac{5}{6}$

27. $4^{-1} + 5^{-1} = \dfrac{1}{4} + \dfrac{1}{5}$
$= \dfrac{5}{5} \cdot \dfrac{1}{4} + \dfrac{1}{5} \cdot \dfrac{4}{4}$
$= \dfrac{5}{20} + \dfrac{4}{20}$
$= \dfrac{9}{20}$

28. $8^0 = 1$, by definition.

29. $12^0 = 1$

30. $(-23)^0 = 1$

31. $(-4)^0 = 1$

32. $-2^0 = -1$ The base is 2; it is not negative 2. That is, find the opposite of 2^0.

33. $3^0 - 4^0 = 1 - 1 = 0$

34. $-8^0 - 7^0 = -1 - 1 = -2$

35. In order to raise a fraction to a negative power, we may change the fraction to its reciprocal and change the exponent to additive inverse of the original exponent.

36. Writing exercise

37. (a) Simplify each side of the equation to see if the statement is true:
$$-\dfrac{3}{4} = \left(\dfrac{3}{4}\right)^{-1}$$
$$= \dfrac{3^{-1}}{4^{-1}}$$
$$= \dfrac{4}{3}.$$
This equation is not correct.

(b) $\dfrac{3^{-1}}{4^{-1}} = \left(\dfrac{4}{3}\right)^{-1}$
$\dfrac{4}{3} = \dfrac{4^{-1}}{3^{-1}}$
$= \dfrac{3}{4}$
This equation is not correct.

(c) $\dfrac{3^{-1}}{4} = \dfrac{3}{4^{-1}}$
$\dfrac{1}{3 \cdot 4} = 3 \cdot 4$
$\dfrac{1}{12} = 12$
This equation is not correct.

(d) $\dfrac{3^{-1}}{4^{-1}} = \left(\dfrac{3}{4}\right)^{-1}$
$= \dfrac{3^{-1}}{4^{-1}}$
This equation is correct.

38. (a) One of the exponent laws shows that this equation is correct:
$$(3r)^{-2} = 3^{-2}r^{-2}.$$

(b) This equation is incorrect. See part (a).

(c) This equation is correct by the laws of exponents.

(d) This equation is correct by the laws of exponents.

39. $x^{12} \cdot x^4 = x^{12+4} = x^{16}$

40. $\dfrac{x^{12}}{x^4} = x^{12-4} = x^8$

41. $\dfrac{5^{17}}{5^{16}} = 5^{17-16} = 5$

42. $\dfrac{3^{12}}{3^{13}} = 3^{12-13} = 3^{-1} = \dfrac{1}{3}$

43. $\dfrac{3^{-5}}{3^{-2}} = 3^{-5-(-2)} = 3^{-5+2} = 3^{-3} = \dfrac{1}{3^3} = \dfrac{1}{27}$

7.5 PROPERTIES OF EXPONENTS AND SCIENTIFIC NOTATION

44. $\dfrac{2^{-4}}{2^{-3}} = 2^{-4-(-3)} = 2^{-4+3} = 2^{-1} = \dfrac{1}{2}$

45. $\dfrac{9^{-1}}{9} = 9^{-1-1} = 9^{-2} = \dfrac{1}{9^2} = \dfrac{1}{81}$

46. $\dfrac{12}{12^{-1}} = 12^{1-(-1)} = 12^{1+1} = 12^2 = 144$

47. $t^5 t^{-12} = t^{5+(-12)} = t^{-7} = \dfrac{1}{t^7}$

48. $p^5 p^{-6} = p^{5+(-6)} = p^{-1} = \dfrac{1}{p}$

49. $(3x)^2 = 3^2 \cdot x^2 = 9x^2$

50. $(-2x^{-2})^2 = (-2)^2 \cdot x^{-4} = 4 \cdot \dfrac{1}{x^4} = \dfrac{4}{x^4}$

51. $a^{-3} a^2 a^{-4} = a^{-3+2+(-4)} = a^{-3+2-4} = a^{-5} = \dfrac{1}{a^5}$

52. $k^{-5} k^{-3} k^4 = k^{-5+(-3)+4} = k^{-4} = \dfrac{1}{k^4}$

53. $\dfrac{x^7}{x^{-4}} = x^{7-(-4)} = x^{7+4} = x^{11}$

54. $\dfrac{p^{-3}}{p^5} = p^{-3-5} = p^{-8} = \dfrac{1}{p^8}$

55. $\dfrac{r^3 r^{-4}}{r^{-2} r^{-5}} = \dfrac{r^{3+(-4)}}{r^{-2+(-5)}} = \dfrac{r^{-1}}{r^{-7}} = r^{-1-(-7)} = r^{-1+7} = r^6$

56. $\dfrac{z^{-4} z^{-2}}{z^3 z^{-1}} = \dfrac{z^{-4+(-2)}}{z^{3+(-1)}} = \dfrac{z^{-6}}{z^2} = z^{-6-2} = z^{-8} = \dfrac{1}{z^8}$

57. $7k^2(-2k)(4k^{-5}) = (7 \cdot -2 \cdot 4)k^{2+1+(-5)}$
$= -56k^{-2}$
$= -\dfrac{56}{k^2}$

58. $3a^2(-5a^{-6})(-2a) = (3 \cdot -5 \cdot -2)a^{2+(-6)+1}$
$= 30a^{-3}$
$= \dfrac{30}{a^3}$

59. $(z^3)^{-2} z^2 = z^{3 \cdot -2} z^2 = z^{-6} z^2 = z^{-6+2} = z^{-4} = \dfrac{1}{z^4}$

60. $(p^{-1})^3 p^{-4} = p^{-3} p^{-4} = p^{-3+(-4)} = p^{-7} = \dfrac{1}{p^7}$

61. $-3r^{-1}(r^{-3})^2 = -3r^{-1} r^{-3 \cdot 2}$
$= -3r^{-1} r^{-6}$
$= -3r^{-1+(-6)}$
$= -3r^{-7}$
$= -\dfrac{3}{r^7}$

62. $2(y^{-3})^4 (y^6) = 2y^{-3 \cdot 4} y^6$
$= 2y^{-12} y^6$
$= 2y^{-12+6}$
$= 2y^{-6}$
$= \dfrac{2}{y^6}$

63. $(3a^{-2})^3 (a^3)^{-4} = 3^3 a^{-6} \cdot a^{-12}$
$= 27a^{-6+(-12)}$
$= 27a^{-18}$
$= \dfrac{27}{a^{18}}$

64. $(m^5)^{-2}(3m^{-2})^3 = 3^3 m^{5 \cdot -2} \cdot m^{-2 \cdot 3}$
$= 27m^{-10} \cdot m^{-6}$
$= 27m^{-10+(-6)}$
$= 27m^{-16}$
$= \dfrac{27}{m^{16}}$

65. $(x^{-5} y^2)^{-1} = x^{-5 \cdot -1} y^{2 \cdot -1}$
$= x^5 y^{-2}$
$= \dfrac{x^5}{y^2}$

66. $(a^{-3} b^{-5})^2 = a^{-3 \cdot 2} b^{-5 \cdot 2}$
$= a^{-6} b^{-10}$
$= \dfrac{1}{a^6 b^{10}}$

67. The reciprocal of x is $\dfrac{1}{x}$.

(a) $x^{-1} = \dfrac{1}{x}$, by definition.

(b) $\dfrac{1}{x}$ is the reciprocal.

(c) $\left(\dfrac{1}{x^{-1}}\right)^{-1} = \dfrac{1^{-1}}{x^1} = \dfrac{1}{x}$

(d) $-x$ is the opposite of x. It is not the reciprocal.

68. (a) 6.02×10^{23} is in scientific notation.

(b) 14×10^{-6} is not in scientific notation because there are two digits to the left of the decimal point in 14.

(c) 1.4×10^{-5} is in scientific notation.

(d) 3.8×10^3 is in scientific notation.

69. $230 = 2.3 \times 10^2$

70. $46{,}500 = 4.65 \times 10^4$

71. $.02 = 2 \times 10^{-2}$

72. $.0051 = 5.1 \times 10^{-3}$

73. $6.5 \times 10^3 = 6500$

74. $2.317 \times 10^5 = 231,700$

75. $1.52 \times 10^{-2} = .0152$

76. $1.63 \times 10^{-4} = .000163$

77. $\dfrac{.002 \times 3900}{.000013} = \dfrac{(2 \times 10^{-3}) \times (3.9 \times 10^3)}{1.3 \times 10^{-5}}$
$= \dfrac{(2 \times 3.9) \times (10^{-3} \times 10^3)}{1.3 \times 10^{-5}}$
$= \dfrac{(7.8) \times (10^0)}{1.3 \times 10^{-5}}$
$= \dfrac{7.8}{1.3} \times \dfrac{10^0}{10^{-5}}$
$= 6 \times 10^{0-(-5)}$
$= 6 \times 10^{0+5}$
$= 6 \times 10^5$

78. $\dfrac{.009 \times 600}{.02} = \dfrac{(9 \times 10^{-3}) \times (6 \times 10^2)}{2 \times 10^{-2}}$
$= \dfrac{(9 \times 6) \times (10^{-3} \times 10^2)}{2 \times 10^{-2}}$
$= \dfrac{54 \times 10^{-1}}{2 \times 10^{-2}}$
$= \dfrac{54}{2} \times \dfrac{10^{-1}}{10^{-2}}$
$= 27 \times 10^{-1-(-2)}$
$= 27 \times 10^1$
$= 2.7 \times 10^2$

Note that 27×10^1 is not in scientific notation. As 27 is reduced by a power of ten, the exponent on 10 must increase by a power of ten.

79. $\dfrac{.0004 \times 56,000}{.000112} = \dfrac{(4 \times 10^{-4}) \times (5.6 \times 10^4)}{1.12 \times 10^{-4}}$
$= \dfrac{(4 \times 5.6) \times (10^{-4} \times 10^4)}{1.12 \times 10^{-4}}$
$= \dfrac{22.4 \times 10^0}{1.12 \times 10^{-4}}$
$= \dfrac{22.4}{1.12} \times \dfrac{10^0}{10^{-4}}$
$= 20 \times 10^{0-(-4)}$
$= 20 \times 10^4$
$= 2.0 \times 10^5$

Note that 20×10^4 is not in scientific notation. As 20 is reduced by a power of ten, the exponent on 10 must increase by a power of ten.

80. $\dfrac{.018 \times 20,000}{300 \times .0004} = \dfrac{(1.8 \times 10^{-2}) \times (2 \times 10^4)}{(3 \times 10^2) \times (4 \times 10^{-4})}$
$= \dfrac{(1.8 \times 2) \times (10^{-2} \times 10^4)}{(3 \times 4) \times (10^2 \times 10^{-4})}$
$= \dfrac{3.6 \times 10^2}{12 \times 10^{-2}}$
$= \dfrac{3.6}{12} \times \dfrac{10^2}{10^{-2}}$
$= .3 \times 10^{2-(-2)}$
$= .3 \times 10^4$
$= 3 \times 10^3$

Note that $.3 \times 10^4$ is not in scientific notation. As .3 is increased by a power of ten, the exponent on 10 must decrease by a power of ten.

81. $\dfrac{840,000 \times .03}{.00021 \times 600} = \dfrac{(8.4 \times 10^5) \times (3 \times 10^{-2})}{(2.1 \times 10^{-4}) \times (6 \times 10^2)}$
$= \dfrac{(8.4 \times 3) \times (10^5 \times 10^{-2})}{(2.1 \times 6) \times (10^{-4} \times 10^2)}$
$= \dfrac{25.2 \times 10^3}{12.6 \times 10^{-2}}$
$= 2 \times 10^{3-(-2)}$
$= 2 \times 10^{3+2}$
$= 2 \times 10^5$

82. $\dfrac{28 \times .0045}{140 \times 1500} = \dfrac{(2.8 \times 10^1) \times (4.5 \times 10^{-3})}{(1.4 \times 10^2) \times (1.5 \times 10^3)}$
$= \dfrac{(2.8 \times 4.5) \times (10^1 \times 10^{-3})}{(1.4 \times 1.5) \times (10^2 \times 10^3)}$
$= \dfrac{12.6 \times 10^{-2}}{2.1 \times 10^5}$
$= 6 \times 10^{-2-5}$
$= 6 \times 10^{-7}$

83. $\$13,757,400,000 = \$1.37574 \times 10^{10}$

84. $207,754,000 = 2.07754 \times 10^8$

85. $1,000,000,000 = 1 \times 10^9$;
$1,000,000,000,000 = 1 \times 10^{12}$;
$2,128,000,000,000 = 2.128 \times 10^{12}$;
$144,419 = 1.44419 \times 10^5$

86. $1500 = 1.5 \times 10^3$; $680,000 = 6.8 \times 10^5$;
$1,100,000 = 1.1 \times 10^6$; $2,200,000 = 2.2 \times 10^6$

87. 10 billion $= 10,000,000,000 = 1 \times 10^{10}$

88. 4.7 million $= 4,700,000 = 4.7 \times 10^6$

89. $2 \times 10^9 = 2,000,000,000$.

90. 10^{-13} means $1 \times 10^{-13} = .0000000000001$

7.6 POLYNOMIALS AND FACTORING

91. $\dfrac{3.9 \times 10^8}{15{,}000} = \dfrac{3.9 \times 10^8}{1.5 \times 10^4}$
$= \dfrac{3.9}{1.5} \times 10^{8-4}$
$= 2.6 \times 10^4 \text{ or } 26000$

92. $\dfrac{8.009 \times 10^7}{2000} = \dfrac{8.009 \times 10^7}{2 \times 10^3}$
$= \dfrac{8.009}{2} \times \dfrac{10^7}{10^3}$
$= 4.0045 \times 10^4$

At $1 per ticket, this is $40,045.

93. $\dfrac{1.8 \times 10^7}{19 \times 10^{12}} = \dfrac{1.8}{19} \times 10^{7-12} \approx .09474 \times 10^{-5}$

Written in scientific notation, this is 9.474×10^{-7}.

94. $\dfrac{1}{1.57828 \times 10^{-5}} = \dfrac{1 \times 10^0}{1.57828 \times 10^{-5}}$
$= \dfrac{1}{1.57828} \times 10^{0-(-5)}$
$\approx .63360 \times 10^5$
$= 63{,}360 \text{ inches}$

95. $\dfrac{9 \times 10^{12}}{3 \times 10^{10}} = \dfrac{9}{3} \times 10^{12-10} = 3 \times 10^2$, which is 300 sec.

96. $\dfrac{9.3 \times 10^7}{2.9 \times 10^3} = \dfrac{9.3}{2.9} \times 10^{7-3} \approx 3.2 \times 10^4$

This is 32,000 hours or about 3.7 years.

97. $\dfrac{1.86 \times 10^5 \text{mi}}{1\text{ s}} \cdot \dfrac{60\text{s}}{1\text{min}} \cdot \dfrac{60\text{ min}}{1\text{ hr}} \cdot \dfrac{24\text{ hr}}{1\text{ day}} \cdot \dfrac{365\text{ days}}{1\text{ year}}$.

This simplifies to $58{,}665{,}960 \times 10^5$ miles; in scientific notation this is approximately 5.87×10^{12}.

98. From Exercise 96 the average distance from the earth to the sun is 9.3×10^7 miles. From Exercise 97 the speed of light is 1.86×10^5 miles per second. Use the formula $d = rt$ and solve for t.

$$9.3 \times 10^7 = (1.86 \times 10^5)t$$
$$\dfrac{9.3 \times 10^7}{1.86 \times 10^5} = \dfrac{(1.86 \times 10^5)t}{1.86 \times 10^5}$$
$$\dfrac{9.3}{1.86} \times \dfrac{10^7}{10^5} = t$$
$$5 \times 10^2 = t$$

This time, 500, is in seconds. To convert to minutes,

$\dfrac{500 \text{ seconds}}{1} \cdot \dfrac{1 \text{ minute}}{60 \text{ seconds}} = \dfrac{500}{60} \approx 8.3$ minutes

99. Use the formula $d = rt$. The total distance that the spacecraft must travel is the difference between 6.7×10^7 and 3.6×10^7. The rate of the spacecraft is 1.55×10^3. Insert this information into the formula and solve for t.

$$6.7 \times 10^7 - 3.6 \times 10^7 = (1.55 \times 10^3)t$$
$$3.1 \times 10^7 = (1.55 \times 10^3)t$$
$$\dfrac{3.1 \times 10^7}{1.55 \times 10^3} = \dfrac{(1.55 \times 10^3)t}{1.55 \times 10^3}$$
$$\dfrac{3.1}{1.55} \times \dfrac{10^7}{10^3} = t$$
$$2 \times 10^4 = t$$

The time would be 20,000 hours.

100. $(4.6 \times 10^8) - (4.14 \times 10^8) = 0.46 \times 10^8$. Change this to scientific notation by moving the decimal point one position to the right and decreasing the power of 10 by 1: 4.6×10^7 meters.

101. $123{,}000 = 1.23 \times 10^5$

102. $3000 = 3 \times 10^3$

103. $424{,}000 = 4.24 \times 10^5$

104. $136{,}000 = 1.36 \times 10^5$

105. $440{,}000 = 4.4 \times 10^5$

106. $374{,}000 = 3.74 \times 10^5$

7.6 EXERCISES

1. $(3x^2 - 4x + 5) + (-2x^2 + 3x - 2) =$
$3x^2 - 2x^2 - 4x + 3x + 5 - 2 = x^2 - x + 3$

2. $(4m^3 - 3m^2 + 5) + (-3m^3 - m^2 + 5) =$
$4m^3 - 3m^3 - 3m^2 - m^2 + 5 + 5 = m^3 - 4m^2 + 10$

3. Remember that the negative in front of the parenthesis affects all the terms within the grouping symbol; i. e., all the signs change.

$(12y^2 - 8y + 6) - (3y^2 - 4y + 2) =$
$12y^2 - 3y^2 - 8y + 4y + 6 - 2 = 9y^2 - 4y + 4$

4. $(8p^2 - 5p) - (3p^2 - 2p + 4) =$
$8p^2 - 3p^2 - 5p + 2p - 4 = 5p^2 - 3p - 4$

5. $(6m^4 - 3m^2 + m) - (2m^3 + 5m^2 + 4m) + (m^2 - m) =$
$6m^4 - 3m^2 + m - 2m^3 - 5m^2 - 4m + m^2 - m =$
$6m^4 - 2m^3 - 3m^2 - 5m^2 + m^2 + m - 4m - m =$
$6m^4 - 2m^3 - 7m^2 - 4m$

6. $-(8x^3 + x - 3) + (2x^3 + x^2) - (4x^2 + 3x - 1) =$
$-8x^3 - x + 3 + 2x^3 + x^2 - 4x^2 - 3x + 1 =$
$-8x^3 + 2x^3 + x^2 - 4x^2 - x - 3x + 3 + 1 =$
$-6x^3 - 3x^2 - 4x + 4$

7. $5(2x^2 - 3x + 7) - 2(6x^2 - x + 12) =$
$10x^2 - 15x + 35 - 12x^2 + 2x - 24 =$
$10x^2 - 12x^2 - 15x + 2x + 35 - 24 =$
$-2x^2 - 13x + 11$

8. $8x^2y - 3xy^2 + 2x^2y - 9xy^2 =$
$8x^2y + 2x^2y - 3xy^2 - 9xy^2 =$
$10x^2y - 12xy^2$

9. Use the FOIL method:
$(x+3)(x-8) = x \cdot x + x \cdot (-8) + 3 \cdot x + 3 \cdot (-8)$
$= x^2 - 8x + 3x - 24$
$= x^2 - 5x - 24.$

10. Use the FOIL method:
$(y-3)(y-9) = y \cdot y + y \cdot (-9) + (-3) \cdot y + (-3) \cdot (-9)$
$= y^2 - 9y - 3y + 27$
$= y^2 - 12y + 27.$

11. $(4r-1)(7r+2) = 4r \cdot 7r + 4r \cdot 2 + (-1) \cdot 7r + (-1) \cdot 2$
$= 28r^2 + 8r - 7r - 2$
$= 28r^2 + r - 2$

12. $(5m-6)(3m+4) = 5m \cdot 3m + 5m \cdot 4 + -6 \cdot 3m + -6 \cdot 4$
$= 15m^2 + 20m - 18m - 24$
$= 15m^2 + 2m - 24$

13. Use the distributive property. Also remember to add exponents when multiplying variables that have the same base.
$4x^2(3x^3 + 2x^2 - 5x + 1) =$
$4x^2 \cdot 3x^3 + 4x^2 \cdot 2x^2 + 4x^2 \cdot (-5x) + 4x^2 \cdot 1 =$
$12x^5 + 8x^4 - 20x^3 + 4x^2$

14. $2b^3(b^2 - 4b + 3) =$
$2b^3 \cdot b^2 + 2b^3 \cdot (-4b) + 2b^3 \cdot 3 =$
$2b^5 - 8b^4 + 6b^3$

15. The FOIL method can always be used to multiply two binomials.
$(2m+3)(2m-3) = 2m \cdot 2m + -3 \cdot 2m + 3 \cdot 2m + 3 \cdot -3$
$= 4m^2 - 6m + 6m - 9$
$= 4m^2 - 9$

However, it is helpful to recognize that it is the product of the sum and difference of two terms.

16. $(8s-3t)(8s+3t) = (8s)^2 - (3t)^2 = 64s^2 - 9t^2$

17. It is important to remember that the binomial $4m + 2n$ is the base that is being squared.
$(4m+2n)^2 = (4m+2n)(4m+2n)$
$= 4m \cdot 4m + 4m \cdot 2n + 2n \cdot 4m + 2n \cdot 2n$
$= 16m^2 + 16mn + 4n^2$

It is also helpful to recognize the pattern of a binomial squared.

18. $(a-6b)^2 = (a-6b)(a-6b)$
$= a^2 - 2 \cdot (6ab) + (6b)^2$
$= a^2 - 12ab + 36b^2$

19. It is important to remember that the binomial $5r + 3t^2$ is the base that is being squared:
$(5r+3t^2)^2 = (5r+3t^2)(5r+3t^2)$
$= 5r \cdot 5r + 5r \cdot 3t^2 + 3t^2 \cdot 5r + 3t^2 \cdot 3t^2$
$= 25r^2 + 15rt^2 + 15rt^2 + 9t^4$
$= 25r^2 + 30rt^2 + 9t^4$

It is also helpful to recognize the pattern of a binomial squared.

20. $(2z^4 - 3y)^2 = (2z^4 - 3y)(2z^4 - 3y)$
$= 4z^8 - 2 \cdot (6yz^4) + (3y)^2$
$= 4z^8 - 12yz^4 + 9y^2$

21. Vertical multiplication is often less confusing when multiplying two polynomials of more than two terms. Multiply from the right as in number multiplication; i.e., start with -1 times -4. Line up like terms in columns.

$$\begin{array}{r} -z^2 + 3z - 4 \\ 2z - 1 \\ \hline z^2 - 3z + 4 \\ -2z^3 + 6z^2 - 8z \\ \hline -2z^3 + 7z^2 - 11z + 4 \end{array}$$

22. Using vertical multiplication,

$$\begin{array}{r} 12k^3 - 3k^2 + k + 1 \\ k + 2 \\ \hline 24k^3 - 6k^2 + 2k + 2 \\ 12k^4 - 3k^3 + k^2 + k \\ \hline 12k^4 + 21k^3 - 5k^2 + 3k + 2. \end{array}$$

23. Using vertical multiplication, these polynomials have been multiplied beginning at the right: First multiply $-3k$ from the second polynomial times each term of the first polynomial; then multiply $+2n$ times each term of the first polynomial; finally multiply $+m$ times each term of the first polynomial. Place like terms in columns as you multiply to simplify being added later.

$$\begin{array}{r} m - n + k \\ m + 2n - 3k \\ \hline -3km + 3kn - 3k^2 \\ +2mn + 2kn - 2n^2 \\ -mn + km + m^2 \\ \hline +mn - 2km + 5kn + m^2 - 2n^2 - 3k^2 \end{array}$$

24.
$$\begin{array}{r} r - 3s + t \\ 2r - s + t \\ \hline +rt - 3st + t^2 \\ -rs + 3s^2 - st \\ +2r^2 - 6rs + 2rt \\ \hline 2r^2 - 7rs + 3s^2 + 3rt - 4st + t^2 \end{array}$$

25. Vertical multiplication might be the preferred method as in the previous exercise.

$$\begin{array}{r} a \quad -b \quad +2c \\ a \quad -b \quad +2c \\ \hline +2ac - 2bc + 4c^2 \\ -ab + b^2 \quad\quad - 2bc \\ +a^2 - ab \quad\quad + 2ac \\ \hline a^2 - 2ab + b^2 + 4ac - 4bc + 4c^2 \end{array}$$

26. $(k - y + 3m)^2 = (k - y + 3m)(k - y + 3m)$

    Vertical multiplication might be the preferred method as in the previous exercise.

$$\begin{array}{r} k \quad -y \quad +3m \\ k \quad -y \quad +3m \\ \hline +3km - 3my + 9m^2 \\ -ky + y^2 \quad\quad - 3my \\ +k^2 - ky \quad\quad + 3km \\ \hline k^2 - 2ky + y^2 + 6km - 6my + 9m^2 \end{array}$$

27. The answer is (A). Choices (B) and (C) are trinomials of degree 6; but they are not in descending powers; choice (D) is not a trinomial.

28. One example is $3x^5 - 4x^3 + 7x^2 - 1$

29. Writing exercise

30. Writing exercise

31. $$8m^4 + 6m^3 - 12m^2 =$$
    $$2m^2 \cdot 4m^2 + 2m^2 \cdot 3m - 2m^2 \cdot 6 =$$
    $$2m^2(4m^2 + 3m - 6)$$

32. $$2p^5 - 10p^4 + 16p^3 =$$
    $$2p^3 \cdot p^2 - 2p^3 \cdot 5p + 2p^3 \cdot 8 =$$
    $$2p^3(p^2 - 5p + 8)$$

33. $$4k^2m^3 + 8k^4m^3 - 12k^2m^4 =$$
    $$4k^2m^3 \cdot 1 + 4k^2m^3 \cdot 2k^2 - 4k^2m^3 \cdot 3m =$$
    $$4k^2m^3(1 + 2k^2 - 3m)$$

34. $$28r^4s^2 + 7r^3s - 35r^4s^3 =$$
    $$7r^3s \cdot 4rs + 7r^3s \cdot 1 - 7r^3s \cdot 5rs^2 =$$
    $$7r^3s(4rs + 1 - 5rs^2)$$

35. In this exercise, the greatest common factor is $2(a + b)$.

$$2(a + b) + 4m(a + b) =$$
$$2 \cdot (a + b) + 2m \cdot 2 \cdot (a + b) =$$
$$2(a + b)(1 + 2m)$$

36. In this exercise, the greatest common factor is $(y - 2)$.

$$4(y - 2)^2 + 3(y - 2) =$$
$$(y - 2) \cdot 4(y - 2) + (y - 2) \cdot 3 =$$
$$(y - 2)[4(y - 2) + 3] =$$
$$(y - 2)[4y - 8 + 3] =$$
$$(y - 2)(4y - 5)$$

37. In this exercise, the greatest common factor is $m - 1$.

$$2(m - 1) - 3(m - 1)^2 + 2(m - 1)^3 =$$
$$(m - 1) \cdot 2 - (m - 1) \cdot 3(m - 1)^1 + (m - 1) \cdot 2(m - 1)^2 =$$
$$(m - 1)[2 - 3(m - 1) + 2(m - 1)^2] =$$
$$(m - 1)[2 - 3m + 3 + 2(m - 1)(m - 1)] =$$
$$(m - 1)[2 - 3m + 3 + 2(m^2 - 2m + 1)] =$$
$$(m - 1)[2 - 3m + 3 + 2m^2 - 4m + 2] =$$
$$(m - 1)[2m^2 - 7m + 7]$$

38. In this exercise, the greatest common factor is $a + 3$.

$$5(a + 3)^3 - 2(a + 3) + (a + 3)^2 =$$
$$(a + 3) \cdot 5(a + 3)^2 - (a + 3) \cdot 2 + (a + 3) \cdot (a + 3)^1 =$$
$$(a + 3)[5(a + 3)^2 - 2 + (a + 3)] =$$
$$(a + 3)[5(a^2 + 6a + 9) - 2 + (a + 3)] =$$
$$(a + 3)[5a^2 + 30a + 45 - 2 + a + 3] =$$
$$(a + 3)[5a^2 + 31a + 46]$$

39. $$6st + 9t - 10s - 15 =$$
    $$3t \cdot (2s + 3) - 5 \cdot (2s + 3) =$$
    $$(2s + 3)(3t - 5)$$

    Remember that the binomial factors can be written either as shown above or as $(3t - 5)(2s + 3)$.

40. $$10ab - 6b + 35a - 21 =$$
    $$2b \cdot (5a - 3) + 7 \cdot (5a - 3) =$$
    $$(5a - 3)(2b + 7)$$

41. $$rt^3 + rs^2 - pt^3 - ps^2 =$$
    $$r \cdot (t^3 + s^2) - p \cdot (t^3 + s^2) =$$
    $$(t^3 + s^2)(r - p)$$

42. $$2m^4 + 6 - am^4 - 3a =$$
    $$2 \cdot (m^4 + 3) - a \cdot (m^4 + 3) =$$
    $$(m^4 + 3)(2 - a).$$

43. $$16a^2 + 10ab - 24ab - 15b^2 =$$
    $$2a \cdot (8a + 5b) - 3b \cdot (8a + 5b) =$$
    $$(8a + 5b)(2a - 3b)$$

44. $$15 - 5m^2 - 3r^2 + m^2r^2 =$$
    $$5 \cdot (3 - m^2) - r^2 \cdot (3 - m^2) =$$
    $$(3 - m^2)(5 - r^2)$$

45. $$20z^2 - 8zx - 45zx + 18x^2 =$$
    $$4z \cdot (5z - 2x) - 9x \cdot (5z - 2x) =$$
    $$(5z - 2x)(4z - 9x)$$

46. Each one can be checked by the FOIL method of multiplication. Although the terms of the polynomial are in a different order, it is the same expression for choices (a) through (c).

    (a) $(a-1)(b-1) = ab - a - b + 1$
    (b) $(-a+1)(-b+1) = ab - a - b + 1$
    (c) $(-1+a)(-1+b) = 1 - b - a + ab$
    (d) $(1-a)(b+1) = b + 1 - ab - a$

    Choice (d) does not produce the same polynomial.

47. Recall that the mental process is to think of two numbers whose product is -15 and whose sum is -2. Use these numbers, -5 and $+3$ to rename the middle term of the trinomial. After creating the 4-term polynomial in the second line, factor by grouping.

    $$\begin{aligned} x^2 - 2x - 15 &= x^2 - 5x + 3x - 15 \\ &= x(x-5) + 3(x-5) \\ &= (x-5)(x+3) \end{aligned}$$

48. $\begin{aligned} r^2 + 8r + 12 &= r^2 + 2r + 6r + 12 \\ &= r(r+2) + 6(r+2) \\ &= (r+2)(r+6) \end{aligned}$

49. $\begin{aligned} y^2 + 2y - 35 &= y^2 + 7y - 5y - 35 \\ &= y(y+7) - 5(y+7) \\ &= (y+7)(y-5) \end{aligned}$

50. $\begin{aligned} x^2 - 7x + 6 &= x^2 - x - 6x + 6 \\ &= x(x-1) - 6(x-1) \\ &= (x-1)(x-6) \end{aligned}$

51. First, factor out the greatest common factor 6; then proceed to factor the trinomial.

    $$\begin{aligned} 6a^2 - 48a - 120 &= 6(a^2 - 8a - 20) \\ &= 6[a^2 - 10a + 2a - 20] \\ &= 6[a(a-10) + 2(a-10)] \\ &= 6(a-10)(a+2) \end{aligned}$$

52. $\begin{aligned} 8h^2 - 24h - 320 &= 8(h^2 - 3h - 40) \\ &= 8[h^2 - 8h + 5h - 40] \\ &= 8[h(h-8) + 5(h-8)] \\ &= 8(h-8)(h+5) \end{aligned}$

53. $\begin{aligned} 3m^3 + 12m^2 + 9m &= 3m[m^2 + 4m + 3] \\ &= 3m[m^2 + 3m + m + 3] \\ &= 3m[m(m+3) + 1(m+3)] \\ &= 3m(m+3)(m+1) \end{aligned}$

54. $\begin{aligned} 9y^4 - 54y^3 + 45y^2 &= 9y^2[y^2 - 6y + 5] \\ &= 9y^2[y^2 - 5y - y + 5] \\ &= 9y^2[y(y-5) - 1(y-5)] \\ &= 9y^2(y-5)(y-1) \end{aligned}$

55. When the leading coefficient is not 1, remember that the product to consider is found by multiplying the leading coefficient by the last term. In this exercise multiply 6 times -6, so that a product of -36 is needed along with a sum of $+5$.

    $$\begin{aligned} 6k^2 + 5kp - 6p^2 &= 6k^2 + 9kp - 4kp - 6p^2 \\ &= 3k(2k+3p) - 2p(2k+3p) \\ &= (2k+3p)(3k-2p) \end{aligned}$$

56. $\begin{aligned} 14m^2 + 11mr - 15r^2 &= 14m^2 - 10mr + 21mr - 15r^2 \\ &= 2m(7m-5r) + 3r(7m-5r) \\ &= (7m-5r)(2m+3r) \end{aligned}$

57. $\begin{aligned} 5a^2 - 7ab - 6b^2 &= 5a^2 - 10ab + 3ab - 6b^2 \\ &= 5a(a-2b) + 3b(a-2b) \\ &= (a-2b)(5a+3b) \end{aligned}$

58. $\begin{aligned} 12s^2 + 11st - 5t^2 &= 12s^2 - 4st + 15st - 5t^2 \\ &= 4s(3s-t) + 5t(3s-t) \\ &= (3s-t)(4s+5t) \end{aligned}$

59. $\begin{aligned} 9x^2 - 6x^3 + x^4 &= x^2[9 - 6x + x^2] \\ &= x^2[9 - 3x - 3x + x^2] \\ &= x^2[3(3-x) - x(3-x)] \\ &= x^2(3-x)(3-x) \\ &= x^2(3-x)^2 \end{aligned}$

60. $\begin{aligned} 30a^2 + am - m^2 &= 30a^2 + 6am - 5am - m^2 \\ &= 6a(5a+m) - m(5a+m) \\ &= (5a+m)(6a-m) \end{aligned}$

61. $\begin{aligned} 24a^4 + 10a^3b - 4a^2b^2 &= 2a^2[12a^2 + 5ab - 2b^2] \\ &= 2a^2[12a^2 - 3ab + 8ab - 2b^2] \\ &= 2a^2[3a(4a-b) + 2b(4a-b)] \\ &= 2a^2(4a-b)(3a+2b) \end{aligned}$

62. $\begin{aligned} 18x^5 + 15x^4z - 75x^3z^2 &= 3x^3[6x^2 + 5xz - 25z^2] \\ &= 3x^3[6x^2 - 10xz + 15xz - 25z^2] \\ &= 3x^3[2x(3x-5z) + 5z(3x-5z)] \\ &= 3x^3(3x-5z)(2x+5z) \end{aligned}$

63. Writing exercise

64. Writing exercise

65. $\begin{aligned} 9m^2 - 12m + 4 &= (3m)^2 - 2(3m)(2) + (2)^2 \\ &= (3m-2)^2 \end{aligned}$

66. $\begin{aligned} 16p^2 - 40m + 25 &= (4p)^2 - 2(4p)(5) + (5)^2 \\ &= (4p-5)^2 \end{aligned}$

67. $\begin{aligned} 32a^2 - 48ab + 18b^2 &= 2[16a^2 - 24ab + 9b^2] \\ &= 2[(4a)^2 - 2(4a)(3b) + (3b)^2] \\ &= 2(4a-3b)^2 \end{aligned}$

68. $\begin{aligned} 20p^2 - 100pq + 125q^2 &= 5[4p^2 - 20pq + 25q^2] \\ &= 5[(2p)^2 - 2(2p)(5q) + (5q)^2] \\ &= 5(2p-5q)^2 \end{aligned}$

69. $4x^2y^2 + 28xy + 49 = (2xy)^2 + 2(2xy)(7) + (7)^2$
$= (2xy + 7)^2$

70. $9m^2n^2 - 12mn + 4 = (3mn)^2 - 2(3mn)(2) + (2)^2$
$= (3mn - 2)^2$

71. $x^2 - 36 = (x)^2 - (6)^2 = (x + 6)(x - 6)$

72. $t^2 - 64 = (t)^2 - (8)^2 = (t + 8)(t - 8)$

73. $y^2 - w^2 = (y)^2 - (w)^2 = (y + w)(y - w)$

74. $25 - w^2 = (5)^2 - (w)^2 = (5 + w)(5 - w)$

75. $9a^2 - 16 = (3a)^2 - (4)^2 = (3a + 4)(3a - 4)$

76. $16q^2 - 25 = (4q)^2 - (5)^2 = (4q + 5)(4q - 5)$

77. $25s^4 - 9t^2 = (5s^2)^2 - (3t)^2 = (5s^2 + 3t)(5s^2 - 3t)$

78. First factor out the greatest common factor 9; then factor the difference of squares.
$$36z^2 - 81y^4 = 9(4z^2 - 9y^4)$$
$$= 9\left[(2z)^2 - (3y^2)^2\right]$$
$$= 9(2z + 3y^2)(2z - 3y^2).$$

79. This exercise requires factoring twice.
$$p^4 - 625 = (p^2)^2 - (25)^2$$
$$= (p^2 + 25)(p^2 - 25)$$
$$= (p^2 + 25)\left[(p)^2 - (5)^2\right]$$
$$= (p^2 + 25)(p + 5)(p - 5)$$

80. $m^4 - 81 = (m^2)^2 - (9)^2$
$= (m^2 + 9)(m^2 - 9)$
$= (m^2 + 9)\left[(m)^2 - (3)^2\right]$
$= (m^2 + 9)(m + 3)(m - 3)$

81. $8 - a^3 = (2)^3 - (a)^3$
$= (2 - a)(2^2 + 2 \cdot a + a^2)$
$= (2 - a)(4 + 2a + a^2)$

82. $r^3 + 27 = (r)^3 + (3)^3$
$= (r + 3)(r^2 - r \cdot 3 + 3^2)$
$= (r + 3)(r^2 - 3r + 9)$

83. $125x^3 - 27 = (5x)^3 - (3)^3$
$= (5x - 3)\left[(5x)^2 + 5x \cdot 3 + 3^2\right]$
$= (5x - 3)(25x^2 + 15x + 9)$

84. $8m^3 - 27n^3 = (2m)^3 - (3n)^3$
$= (2m - 3n)\left[(2m)^2 + 2m \cdot 3n + (3n)^2\right]$
$= (2m - 3n)(4m^2 + 6mn + 9n^2)$

85. $27y^9 + 125z^6 = (3y^3)^3 + (5z^2)^3$
$= (3y^3 + 5z^2)\left((3y^3)^2 - 3y^3 \cdot 5z^2 + (5z^2)^2\right)$
$= (3y^3 + 5z^2)(9y^6 - 15y^3z^2 + 25z^4)$

86. First, factor out the greatest common factor, 27; then follow the pattern for factoring the sum of cubes.
$$27z^3 + 729y^3 = 27(z^3 + 27y^3)$$
$$= 27\left[(z)^3 + (3y)^3\right]$$
$$= 27(z + 3y)\left[(z)^2 - z \cdot 3y + (3y)^2\right]$$
$$= 27(z + 3y)(z^2 - 3yz + 9y^2)$$

87. Factor by grouping.
$$x^2 + xy - 5x - 5y =$$
$$x \cdot (x + y) - 5 \cdot (x + y) =$$
$$(x + y)(x - 5)$$

88. Factor this trinomial by multiplying the leading coefficient, 8, by -3 to obtain a product of -24. Then find two numbers whose product is -24 and whose sum is -10.
$$8r^2 - 10rs - 3s^2 =$$
$$8r^2 - 12rs + 2rs - 3s^2 =$$
$$4r \cdot (2r - 3s) + s \cdot (2r - 3s) =$$
$$(2r - 3s)(4r + s)$$

89. Factor out the greatest common factor, $(m - 2n)$.
$$p^4(m - 2n) + q(m - 2n) = (m - 2n)(p^4 + q)$$

90. This is a perfect square trinomial that can be factored by following the pattern.
$$36a^2 + 60a + 25 =$$
$$(6a)^2 + 2(6a)(5) + (5)^2 =$$
$$(6a + 5)^2$$

91. This is a perfect square trinomial that can be factored by following the pattern.
$$4z^2 + 28z + 49 =$$
$$(2z)^2 + 2(2z)(7) + (7)^2 =$$
$$(2z + 7)^2$$

92. Factor this trinomial by multiplying the leading coefficient, 6, by -3 to obtain a product of -18. Then find two numbers whose product is -18 and whose sum is $+7$.
$$6p^4 + 7p^2 - 3 =$$
$$6p^4 - 2p^2 + 9p^2 - 3 =$$
$$2p^2 \cdot (3p^2 - 1) + 3 \cdot (3p^2 - 1) =$$
$$(3p^2 - 1)(2p^2 + 3)$$

93. This is a sum of cubes.

$$1000x^3 + 343y^3 = (10x)^3 + (7y)^3$$
$$= (10x + 7y)\left((10x)^2 - 10x \cdot 7y + (7y)^2\right)$$
$$= (10x + 7y)(100x^2 - 70xy + 49y^2)$$

94. This exercise requires factoring by grouping the first three terms together to create a perfect square trinomial. Then factor the difference of squares.

$$\left(b^2 + 8b + 16\right) - a^2 = (b+4)^2 - a^2$$
$$= [(b+4) + a][(b+4) - a]$$
$$= (b + 4 + a)(b + 4 - a)$$

95. This is the difference of cubes.

$$125m^6 - 216 = \left(5m^2\right)^3 - (6)^3$$
$$= \left(5m^2 - 6\right)\left(\left(5m^2\right)^2 + 5m^2 \cdot 6 + 6^2\right)$$
$$= \left(5m^2 - 6\right)\left(25m^4 + 30m^2 + 36\right)$$

96. This exercise is like number 94.

$$\left(q^2 + 6q + 9\right) - p^2 = (q+3)^2 - p^2$$
$$= [(q+3) + p][(q+3) - p]$$
$$= (q + 3 + p)(q + 3 - p)$$

97. Factor this trinomial by multiplying the leading coefficient, 12, by −35 to obtain a product of −420. Then, find two numbers whose product is −420 and whose sum is +16.

$$12m^2 + 16mn - 35n^2 =$$
$$12m^2 - 14mn + 30mn - 35n^2 =$$
$$2m \cdot (6m - 7n) + 5n \cdot (6m - 7n) =$$
$$(6m - 7n)(2m + 5n)$$

98. This is a sum of cubes.

$$216p^3 + 125q^3 = (6p)^3 + (5q)^3$$
$$= (6p + 5q)\left((6p)^2 - 6p \cdot 5q + (5q)^2\right)$$
$$= (6p + 5q)\left(36p^2 - 30pq + 25q^2\right)$$

99. Replace x with 4 and y with 2 to show that the left side of the equation does not equal the right side.

$$x^2 + y^2 \neq (x + y)(x + y)$$
$$4^2 + 2^2 \neq (4+2)(4+2)$$
$$16 + 4 \neq (6)(6)$$
$$20 \neq 36$$

100. Yes, it can be factored. Only the greatest common factor, 9, however, can be factored out of this expression:

$$9x^2 + 36 = 9\left(x^2 + 4\right).$$

7.7 EXERCISES

1. For the quadratic equation $4x^2 + 5x - 9 = 0$, the values of a, b, and c are respectively <u>4</u>, <u>5</u>, and <u>−9</u>.

2. To solve the equation $3x^2 - 5x = -2$ by the quadratic formula, the first step is to add <u>+2</u> to both sides of the equation.

3. When using the quadratic formula, when $b^2 - 4ac$ is positive, the equation has <u>two</u> real solution(s). Consider the quadratic formula

$$x = \frac{-b \pm \sqrt{b^2 - 4ac}}{2a}$$

If the quantity under the radical is positive, there will be two real solutions:

$$x_1 = \frac{-b + \sqrt{\text{positive number}}}{2a}$$
$$x_2 = \frac{-b - \sqrt{\text{positive number}}}{2a}$$

4. If a, b, and c are integers in $ax^2 + bx + c = 0$ and $b^2 - 4ac = 17$, then the equation has <u>two</u> irrational solutions. See Exercise 3. The solutions will be irrational when the quantity under the radical is not a perfect square.

5. Set each factor equal to zero and solve each equation.

$$(x+3)(x-9) = 0$$
$$\begin{array}{ccc} x + 3 = 0 & \text{or} & x - 9 = 0 \\ x = -3 & & x = 9 \end{array}$$

The solution set is $\{-3, 9\}$.

6. Set each factor equal to zero and solve each equation.

$$(m+6)(m+4) = 0$$
$$\begin{array}{ccc} m + 6 = 0 & \text{or} & m + 4 = 0 \\ m = -6 & & m = -4 \end{array}$$

The solution set is $\{-6, -4\}$.

7. Set each factor equal to zero and solve each equation.

$$(2t-7)(5t+1) = 0$$
$$\begin{array}{ccc} 2t - 7 = 0 & \text{or} & 5t + 1 = 0 \\ 2t = 7 & & 5t = -1 \\ t = \tfrac{7}{2} & & t = -\tfrac{1}{5} \end{array}$$

The solution set is $\left\{\tfrac{7}{2}, -\tfrac{1}{5}\right\}$.

8. Set each factor equal to zero and solve each equation.
$$(7x - 3)(6x + 4) = 0$$

$7x - 3 = 0$ or $6x + 4 = 0$
$7x = 3$ $\qquad$ $6x = -4$
$x = \frac{3}{7}$ $\qquad$ $x = -\frac{4}{6}$

Simplify $-\frac{4}{6}$ to $-\frac{2}{3}$.

The solution set is $\{\frac{3}{7}, -\frac{2}{3}\}$.

9. Factor the trinomial to obtain the two factors; then set each one equal to zero.
$$x^2 - x - 12 = 0$$
$$(x - 4)(x + 3) = 0$$

$x - 4 = 0$ or $x + 3 = 0$
$x = 4$ $\qquad$ $x = -3$

The solution set is $\{4, -3\}$.

10. Factor the trinomial to obtain the two factors; then set each one equal to zero.
$$m^2 + 4m - 5 = 0$$
$$(m - 1)(m + 5) = 0$$

$m - 1 = 0$ or $m + 5 = 0$
$m = 1$ $\qquad$ $m = -5$

The solution set is $\{1, -5\}$.

11. Factor the trinomial to obtain the two factors; then set each one equal to zero.
$$x^2 + 9x + 14 = 0$$
$$(x + 2)(x + 7) = 0$$

$x + 2 = 0$ or $x + 7 = 0$
$x = -2$ $\qquad$ $x = -7$

The solution set is $\{-2, -7\}$.

12. Add -2 to both sides of the equation and then factor the left side.
$$15r^2 + 7r = 2$$
$$15r^2 + 7r - 2 = 0$$
$$(3r + 2)(5r - 1) = 0$$

$3r + 2 = 0$ or $5r - 1 = 0$
$3r = -2$ $\qquad$ $5r = 1$
$r = -\frac{2}{3}$ $\qquad$ $y = \frac{1}{5}$

The solution set is $\{-\frac{2}{3}, \frac{1}{5}\}$.

13. Add -1 to both sides of the equation and then factor the left side.
$$12x^2 + 4x = 1$$
$$12x^2 + 4x - 1 = 0$$
$$(2x + 1)(6x - 1) = 0$$

$2x + 1 = 0$ or $6x - 1 = 0$
$2x = -1$ $\qquad$ $6x = 1$
$x = -\frac{1}{2}$ $\qquad$ $x = \frac{1}{6}$

The solution set is $\{-\frac{1}{2}, \frac{1}{6}\}$.

14. First use the distributive property to multiply the left side of the equation. Then add -4 to both sides to set the equation equal to zero.
$$x(x + 3) = 4$$
$$x^2 + 3x = 4$$
$$x^2 + 3x - 4 = 0$$
$$(x + 4)(x - 1) = 0$$

$x + 4 = 0$ or $x - 1 = 0$
$x = -4$ $\qquad$ $x = 1$

The solution set is $\{-4, 1\}$.

15. FOIL the left side of the equation and combine like terms. Then add 16 to both sides to set the equation equal to zero.
$$(x + 4)(x - 6) = -16$$
$$x^2 - 6x + 4x - 24 = -16$$
$$x^2 - 2x - 24 = -16$$
$$x^2 - 2x - 8 = 0$$
$$(x - 4)(x + 2) = 0$$

$x - 4 = 0$ or $x + 2 = 0$
$x = 4$ $\qquad$ $x = -2$

The solution set is $\{4, -2\}$.

16. FOIL the left side of the equation and combine like terms. Then add $-4w$ to both sides to set the equation equal to zero.
$$(w - 1)(3w + 2) = 4w$$
$$3w^2 + 2w - 3w - 2 = 4w$$
$$3w^2 - w - 2 = 4w$$
$$3w^2 - 5w - 2 = 0$$
$$(3w + 1)(w - 2) = 0$$

$3w + 1 = 0$ or $w - 2 = 0$
$3w = -1$ $\qquad$ $w = 2$
$w = -\frac{1}{3}$

The solution set is $\{-\frac{1}{3}, 2\}$.

17. $x^2 = 64$
$\sqrt{x^2} = \pm\sqrt{64}$
$x = \pm 8$

18. $w^2 = 16$
$\sqrt{w^2} = \pm\sqrt{16}$
$w = \pm 4$

19. $x^2 = 24$
$\sqrt{x^2} = \pm\sqrt{24}$
$x = \pm\sqrt{4 \cdot 6}$
$x = \pm 2\sqrt{6}$

20. $x^2 = 48$
$\sqrt{x^2} = \pm\sqrt{48}$
$x = \pm\sqrt{16 \cdot 3}$
$x = \pm 4\sqrt{3}$

21. $r^2 = -5$. The solution set is $\emptyset$. There is no real number that will produce a negative number when it is squared.

22. $x^2 = -10$. The solution set is $\emptyset$.

23. $(x-4)^2 = 3$
$\sqrt{(x-4)^2} = \pm\sqrt{3}$
$x - 4 = \pm\sqrt{3}$
$x - 4 + 4 = 4 \pm \sqrt{3}$
$x = 4 \pm \sqrt{3}$

24. $(x+3)^2 = 11$
$\sqrt{(x+3)^2} = \pm\sqrt{11}$
$x + 3 = \pm\sqrt{11}$
$x + 3 - 3 = -3 \pm \sqrt{11}$
$x = -3 \pm \sqrt{11}$

25. $(2x-5)^2 = 13$
$\sqrt{(2x-5)^2} = \pm\sqrt{13}$
$2x - 5 = \pm\sqrt{13}$
$2x - 5 + 5 = 5 \pm \sqrt{13}$
$2x = 5 \pm \sqrt{13}$
$x = \dfrac{5 \pm \sqrt{13}}{2}$

26. $(4x+1)^2 = 19$
$\sqrt{(4x+1)^2} = \pm\sqrt{19}$
$4x + 1 = \pm\sqrt{19}$
$4x + 1 - 1 = -1 \pm \sqrt{19}$
$4x = -1 \pm \sqrt{19}$
$x = \dfrac{-1 \pm \sqrt{19}}{4}$

27. For the equation $4x^2 - 8x + 1 = 0$, $a = 4$, $b = -8$, and $c = 1$. Substitute these values into the quadratic formula.

$x = \dfrac{-b \pm \sqrt{b^2 - 4ac}}{2a}$

$x = \dfrac{-(-8) \pm \sqrt{(-8)^2 - 4 \cdot 4 \cdot 1}}{2 \cdot 4}$

$= \dfrac{8 \pm \sqrt{64 - 16}}{8}$

$= \dfrac{8 \pm \sqrt{48}}{8}$

$= \dfrac{8 \pm \sqrt{16 \cdot 3}}{8}$

$= \dfrac{8 \pm 4\sqrt{3}}{8}$

$= \dfrac{4\left(2 \pm 1\sqrt{3}\right)}{8}$

$= \dfrac{2 \pm \sqrt{3}}{2}$

28. For the equation $m^2 + 2m - 5 = 0$, $a = 1$, $b = 2$, and $c = -5$. Substitute these values into the quadratic formula.

$x = \dfrac{-b \pm \sqrt{b^2 - 4ac}}{2a}$

$m = \dfrac{-(2) \pm \sqrt{(2)^2 - 4 \cdot 1 \cdot -5}}{2 \cdot 1}$

$= \dfrac{-2 \pm \sqrt{4 + 20}}{2}$

$= \dfrac{-2 \pm \sqrt{24}}{2}$

$= \dfrac{-2 \pm \sqrt{4 \cdot 6}}{2}$

$= \dfrac{-2 \pm 2\sqrt{6}}{2}$

$= \dfrac{2\left(-1 \pm 1\sqrt{6}\right)}{2}$

$= -1 \pm \sqrt{6}$

29. First write the equation in standard form, which is $2x^2 - 2x - 1 = 0$. Then $a = 2, b = -2, c = -1$. Substitute these values into the quadratic formula.

$$x = \frac{-b \pm \sqrt{b^2 - 4ac}}{2a}$$

$$x = \frac{-(-2) \pm \sqrt{(-2)^2 - 4 \cdot 2 \cdot -1}}{2 \cdot 2}$$

$$= \frac{2 \pm \sqrt{4 + 8}}{4}$$

$$= \frac{2 \pm \sqrt{12}}{4}$$

$$= \frac{2 \pm \sqrt{4 \cdot 3}}{4}$$

$$= \frac{2 \pm 2\sqrt{3}}{4}$$

$$= \frac{2(1 \pm \sqrt{3})}{4}$$

$$= \frac{1 \pm \sqrt{3}}{2}$$

30. First write the equation in standard form, which is $9r^2 + 6r - 1 = 0$. Then, $a = 9, b = 6, c = -1$. Substitute these values into the quadratic formula.

$$x = \frac{-b \pm \sqrt{b^2 - 4ac}}{2a}$$

$$r = \frac{-(6) \pm \sqrt{(6)^2 - 4 \cdot 9 \cdot -1}}{2 \cdot 9}$$

$$= \frac{-6 \pm \sqrt{36 + 36}}{18}$$

$$= \frac{-6 \pm \sqrt{72}}{18}$$

$$= \frac{-6 \pm \sqrt{36 \cdot 2}}{18}$$

$$= \frac{-6 \pm 6\sqrt{2}}{18}$$

$$= \frac{6(-1 \pm 1\sqrt{2})}{18}$$

$$= \frac{-1 \pm \sqrt{2}}{3}$$

31. First write the equation in standard form, which is $q^2 - q - 1 = 0$. Then, $a = 1, b = -1, c = -1$. Substitute these values into the quadratic formula.

$$x = \frac{-b \pm \sqrt{b^2 - 4ac}}{2a}$$

$$q = \frac{-(-1) \pm \sqrt{(-1)^2 - 4 \cdot 1 \cdot -1}}{2 \cdot 1}$$

$$= \frac{1 \pm \sqrt{1 + 4}}{2}$$

$$= \frac{1 \pm \sqrt{5}}{2}$$

32. First write the equation in standard form, which is $2p^2 - 4p - 5 = 0$. Then $a = 2, b = -4, c = -5$. Substitute these values into the quadratic formula.

$$x = \frac{-b \pm \sqrt{b^2 - 4ac}}{2a}$$

$$p = \frac{-(-4) \pm \sqrt{(-4)^2 - 4 \cdot 2 \cdot -5}}{2 \cdot 2}$$

$$= \frac{4 \pm \sqrt{16 + 40}}{4}$$

$$= \frac{4 \pm \sqrt{56}}{4}$$

$$= \frac{4 \pm \sqrt{4 \cdot 14}}{4}$$

$$= \frac{4 \pm 2\sqrt{14}}{4}$$

$$= \frac{2(2 \pm 1\sqrt{14})}{4}$$

$$= \frac{2 \pm \sqrt{14}}{2}$$

33. Write the equation in standard form by expanding the left side of the equation to obtain $4k^2 + 4k = 1$. Then subtract 1 from both sides to obtain $4k^2 + 4k - 1 = 0$. Now $a = 4, b = 4, c = -1$. Substitute these values into the quadratic formula.

$$x = \frac{-b \pm \sqrt{b^2 - 4ac}}{2a}$$

$$k = \frac{-(4) \pm \sqrt{(4)^2 - 4 \cdot 4 \cdot -1}}{2 \cdot 4}$$

$$= \frac{-4 \pm \sqrt{16 + 16}}{8}$$

$$= \frac{-4 \pm \sqrt{32}}{8}$$

$$= \frac{-4 \pm \sqrt{16 \cdot 2}}{8}$$

$$= \frac{-4 \pm 4\sqrt{2}}{8}$$

$$= \frac{4\left(-1 \pm 1\sqrt{2}\right)}{8}$$

$$= \frac{-1 \pm \sqrt{2}}{2}$$

34. Write the equation in standard form by expanding the left side of the equation to obtain $4r^2 - 4r = 19$. Then subtract 19 from both sides to obtain $4r^2 + 4r - 19 = 0$. Now $a = 4, b = 4, c = -19$. Substitute these values into the quadratic formula.

$$x = \frac{-b \pm \sqrt{b^2 - 4ac}}{2a}$$

$$r = \frac{-(4) \pm \sqrt{(4)^2 - 4 \cdot 4 \cdot -19}}{2 \cdot 4}$$

$$= \frac{-4 \pm \sqrt{16 + 304}}{8}$$

$$= \frac{-4 \pm \sqrt{320}}{8}$$

$$= \frac{-4 \pm \sqrt{64 \cdot 5}}{8}$$

$$= \frac{-4 \pm 8\sqrt{5}}{8}$$

$$= \frac{4\left(-1 \pm 2\sqrt{5}\right)}{8}$$

$$= \frac{-1 \pm 2\sqrt{5}}{2}$$

35. FOIL the left side of the equation to obtain $g^2 - g - 6 = 1$. Then subtract 1 from both sides to obtain $g^2 - g - 7 = 0$. Now $a = 1, b = -1, c = -7$. Substitute these values into the quadratic formula.

$$x = \frac{-b \pm \sqrt{b^2 - 4ac}}{2a}$$

$$g = \frac{-(-1) \pm \sqrt{(-1)^2 - 4 \cdot 1 \cdot -7}}{2 \cdot 1}$$

$$= \frac{1 \pm \sqrt{1 + 28}}{2}$$

$$= \frac{1 \pm \sqrt{29}}{2}$$

36. FOIL the left side of the equation to obtain $x^2 - 3x - 10 = 6$. Then subtract 6 from both sides to obtain $x^2 - 3x - 16 = 0$. Now $a = 1, b = -3, c = -16$. Substitute these values into the quadratic formula.

$$x = \frac{-b \pm \sqrt{b^2 - 4ac}}{2a}$$

$$x = \frac{-(-3) \pm \sqrt{(-3)^2 - 4 \cdot 1 \cdot -16}}{2 \cdot 1}$$

$$= \frac{3 \pm \sqrt{9 + 64}}{2}$$

$$= \frac{3 \pm \sqrt{73}}{2}$$

37. Write the equation in standard form by adding 14 to both sides of the equation: $m^2 - 6m + 14 = 0$. Now $a = 1, b = -6, c = 14$. Substitute these values into the quadratic formula.

$$x = \frac{-b \pm \sqrt{b^2 - 4ac}}{2a}$$

$$m = \frac{-(-6) \pm \sqrt{(-6)^2 - 4 \cdot 1 \cdot 14}}{2 \cdot 1}$$

$$= \frac{6 \pm \sqrt{36 - 56}}{2}$$

$$= \frac{6 \pm \sqrt{-20}}{2}.$$

A negative under the radical creates an imaginary number. The final solutions to the equation are not real numbers. This can be symbolized by ∅.

38. Write the equation in standard form: $x^2 - 2x + 2 = 0$. Now $a = 1, b = -2, c = 2$. Substitute these values into the quadratic formula.

$$x = \frac{-b \pm \sqrt{b^2 - 4ac}}{2a}$$

$$x = \frac{-(-2) \pm \sqrt{(-2)^2 - 4 \cdot 1 \cdot 2}}{2 \cdot 1}$$

$$= \frac{2 \pm \sqrt{4 - 8}}{2}$$

$$= \frac{2 \pm \sqrt{-4}}{2}.$$

A negative under the radical creates an imaginary number. The final solutions to the equation are not real numbers. This can be symbolized by ∅.

39. Writing exercise
40. Writing exercise
41. Writing exercise
42. Writing exercise
43. $a = 1, b = 6, c = 9$

$$b^2 - 4ac = 6^2 - 4 \cdot 1 \cdot 9$$
$$= 36 - 36$$
$$= 0$$

(c); The equation has one rational solution because the quantity under the radical is zero. The solution will be

$$x = \frac{-b}{2a}$$

44. $a = 4, b = 20, c = 25$

$$b^2 - 4ac = 20^2 - 4 \cdot 4 \cdot 25$$
$$= 400 - 400$$
$$= 0$$

(c); The equation has one rational solution because the quantity under the radical is zero. The solution will be

$$x = \frac{-b}{2a}$$

45. $a = 6, b = 7, c = -3$

$$b^2 - 4ac = 7^2 - 4 \cdot 6 \cdot -3$$
$$= 49 + 72$$
$$= 121$$

(a); The equation has two different rational solutions. $\sqrt{121} = 11$.

46. $a = 2, b = 1, c = -3$

$$b^2 - 4ac = 1^2 - 4 \cdot 2 \cdot -3$$
$$= 1 + 24$$
$$= 25$$

(a); The equation has two different rational solutions. $\sqrt{25} = 5$.

47. $a = 9, b = -30, c = 15$

$$b^2 - 4ac = (-30)^2 - 4 \cdot 9 \cdot 15$$
$$= 900 - 540$$
$$= 360$$

(b); The equation has two different irrational solutions because 360 is not a perfect square.

48. $a = 2, b = -1, c = 1$

$$b^2 - 4ac = (-1)^2 - 4 \cdot 2 \cdot 1$$
$$= 1 - 8$$
$$= -7$$

(d); The equation has no real solutions.

49. Replace s with 200:

$$200 = -16t^2 + 45t + 400$$
$$200 - 200 = -16t^2 + 45t + 400 - 200$$
$$0 = -16t^2 + 45t + 200.$$

Then $a = -16, b = 45, c = 200$

$$t = \frac{-45 \pm \sqrt{(45)^2 - 4 \cdot -16 \cdot 200}}{2 \cdot -16}$$

$$t = \frac{-45 \pm \sqrt{2025 + 12,800}}{-32}$$

$$t = \frac{-45 \pm \sqrt{14,825}}{-32}$$

$$t \approx \frac{-45 \pm 121.76}{-32}$$

Then $t \approx \frac{-45 + 121.76}{-32}$ or $t \approx \frac{-45 - 121.76}{-32}$. The first equation produces a negative value of t, which is not meaningful. The second equation produces $t \approx 5.2$ seconds.

50. Replace s with 450:

$$450 = -16t^2 + 75t + 407$$
$$450 - 450 = -16t^2 + 75t + 407 - 450$$
$$0 = -16t^2 + 75t - 43.$$

Then $a = -16, b = 75, c = -43$

$$t = \frac{-75 \pm \sqrt{(75)^2 - 4 \cdot -16 \cdot -43}}{2 \cdot -16}$$

$$t = \frac{-75 \pm \sqrt{5625 - 2752}}{-32}$$

$$t = \frac{-75 \pm \sqrt{2873}}{-32}$$

$$t \approx \frac{-75 \pm 53.60}{-32}$$

Then $t \approx \frac{-75 + 53.60}{-32}$ or $t \approx \frac{-75 - 53.60}{-32}$. The first equation produces the result $t \approx 0.7$ sec. The second equation produces $t \approx 4.0$ seconds.

51. At time zero, the object is at ground level. Therefore, by letting $t = 0$, the value of s can be found. This is the height of the building.

52. Consider the starting point to be zero. Let $s = 0$ and factor right side of the resulting equation.

$$0 = 100t^2 - 300t$$
$$0 = 100t(t - 3)$$

Set each factor equal to zero.

$$100t = 0 \quad \text{or} \quad t - 3 = 0$$
$$t = 0 \qquad\qquad t = 3$$

The object will take 3 minutes to return to its starting point.

53. (a) Replace s with 128. Obtain standard form for the equation by adding $16t^2$ and subtracting $144t$ to and from both sides of the equation.

$$s = 144t - 16t^2$$
$$128 = 144t - 16t^2$$
$$16t^2 - 144t + 128 = 0$$

Now solve for t by factoring.

$$16t^2 - 144t + 128 = 0$$
$$16(t^2 - 9t + 8) = 0$$
$$16(t - 1)(t - 8) = 0$$

Set each factor containing t equal to zero.

$$t - 1 = 0 \quad \text{or} \quad t - 8 = 0$$
$$t = 1 \qquad\qquad t = 8$$

The object will be 128 feet above the ground at 1 second and at 8 seconds. As it travels upward it will reach this height at 1 second; as it falls back to earth, it will be at the same height at 8 seconds.

(b) The object will strike the ground when $s = 0$.

$$0 = 144t - 16t^2$$
$$0 = 16t(9 - t)$$

Set each factor equal to zero.

$$16t = 0 \quad \text{or} \quad 9 - t = 0$$
$$t = 0 \qquad\qquad 9 = t$$

The object is on the ground at time zero when it is first projected, and it falls back to the ground 9 seconds later.

54. Substitute 190 in the formula for D. Obtain standard form for the equation by adding $13t^2$ and subtracting $100t$ to and from both sides of the equation.

$$D = 100t - 13t^2$$
$$190 = 100t - 13t^2$$
$$13t^2 - 100t + 190 = 0$$

Now solve for t by using the quadratic formula.

Then $a = 13, b = -100, c = 190$.

$$t = \frac{-(-100) \pm \sqrt{(-100)^2 - 4 \cdot 13 \cdot 190}}{2 \cdot 13}$$

$$t = \frac{100 \pm \sqrt{10,000 - 9880}}{26}$$

$$t = \frac{100 \pm \sqrt{120}}{26}$$

$$t \approx \frac{100 \pm 10.95}{26}$$

Then $t \approx \frac{100 + 10.95}{26} = 4.3$ seconds, or

$$t \approx \frac{100 - 10.95}{26} = 3.4 \text{ seconds.}$$

The second answer of approximately 3.4 seconds is the appropriate answer; the first is extraneous because the time cannot exceed 3.8 seconds.

55. Use the Pythagorean theorem with the legs represented by the algebraic expressions $2m$ and $2m + 3$. The longest side or hypotenuse has the value $5m$.

$$(2m)^2 + (2m + 3)^2 = (5m)^2$$
$$4m^2 + (4m^2 + 12m + 9) = 25m^2$$
$$8m^2 + 12m + 9 = 25m^2$$
$$8m^2 - 25m^2 + 12m + 9 = 0$$
$$-17m^2 + 12m + 9 = 0$$

Now $a = -17$, $b = 12$, $c = 9$.

$$m = \frac{-(12) \pm \sqrt{(12)^2 - 4 \cdot -17 \cdot 9}}{2 \cdot -17}$$

$$= \frac{-12 \pm \sqrt{144 + 612}}{-34}$$

$$= \frac{-12 \pm \sqrt{756}}{-34}$$

$$\approx \frac{-12 \pm 27.495}{-34}$$

Now $m \approx \dfrac{-12 + 27.495}{-34}$ yields a negative number, which is not meaningful; however,

$$m \approx \frac{-12 - 27.495}{-34} \approx 1.16.$$

Then, $2m = 2(1.16)$ or approximately 2.3; $2m + 3 = 2.3 + 3$ or 5.3; and $5m = 5(1.16)$ or 5.8.

56. Let $x =$ the shorter leg,
    $x + 1 =$ the longer leg,
    $x + 4 =$ the hypotenuse.

$$(x)^2 + (x+1)^2 = (x+4)^2$$
$$x^2 + (x^2 + 2x + 1) = x^2 + 8x + 16$$
$$2x^2 + 2x + 1 = x^2 + 8x + 16$$
$$x^2 - 6x - 15 = 0$$

Now $a = 1$, $b = -6$, $c = -15$.

$$x = \frac{-(-6) \pm \sqrt{(-6)^2 - 4 \cdot 1 \cdot -15}}{2 \cdot 1}$$

$$= \frac{6 \pm \sqrt{36 + 60}}{2}$$

$$= \frac{6 \pm \sqrt{96}}{2}$$

$$\approx \frac{6 \pm 9.8}{2}$$

Now $x \approx \dfrac{6 - 9.8}{2}$ yields a negative number, which is not meaningful. However, $x \approx \dfrac{6 + 9.8}{2} \approx 7.9$. Then $x + 1 \approx 8.9$; $x + 4 \approx 11.9$.

57. Let $100 =$ length of the shorter leg,
    $400 =$ length of the longer leg (height of the Mart),
    $c =$ length of the hypotenuse.

Use the Pythagorean theorem: $a^2 + b^2 = c^2$.

$$(100)^2 + (400)^2 = c^2$$
$$10,000 + 160,000 = c^2$$
$$170,000 = c^2$$
$$\sqrt{170,000} = \sqrt{c^2}$$
$$412.3 \approx c$$

The length of the wire is about 412.3 feet.

58. Let $d =$ the distance between the base of the Center and the point on the ground where the wire is attached. Then the length of the wire can be expressed as $2d$. The hypotenuse is $2d$ and one of the legs is d. The other leg is the height of the building, 407 feet. Use the Pythagorean theorem with $a = d$, $b = 407$, and $c = 2d$.

$$d^2 + 407^2 = (2d)^2$$
$$d^2 + 407^2 = 4d^2$$
$$d^2 - d^2 + 165649 = 4d^2 - d^2$$
$$165649 = 3d^2$$
$$\frac{165649}{3} = \frac{3d^2}{3}$$
$$55216.\overline{3} = d^2$$
$$\sqrt{55216.\overline{3}} = \sqrt{d^2}$$
$$235 \approx d$$

The length of the wire is then approximately $2 \cdot 235 = 470$ feet.

59. Examine the figure in the text to see that the two legs of the right triangle can be represented by x and $x + 70$. If the two ships are 170 miles apart, this value is the length of the hypotenuse. Again use the Pythagorean theorem with $a = x$, $b = x + 70$, and $c = 170$.

$$x^2 + (x+70)^2 = (170)^2$$
$$x^2 + (x^2 + 140x + 4900) = 28900$$
$$2x^2 + 140x + 4900 = 28900$$
$$2x^2 + 140x - 24000 = 0$$

Now use the quadratic formula with $a = 2$, $b = 140$, $c = -24,000$.

$$x = \frac{-(140) \pm \sqrt{(140)^2 - 4 \cdot 2 \cdot -24,000}}{2 \cdot 2}$$

$$= \frac{-140 \pm \sqrt{19,600 + 192,000}}{4}$$

$$= \frac{-140 \pm \sqrt{211,600}}{4}$$

$$= \frac{-140 \pm 460}{4}$$

Use $\dfrac{-140 + 460}{4} = 80$ miles. Then the ship traveling due east had traveled 80 miles, and the ship traveling south had traveled $80 + 70$ or 150 miles.

60. Using the Pythagorean theorem, let $a = x$, $b = x + 30$, and $c = 150$.

$$x^2 + (x+30)^2 = (150)^2$$
$$x^2 + (x^2 + 60x + 900) = 22,500$$
$$2x^2 + 60x + 900 = 22,500$$
$$2x^2 + 60x - 21,600 = 0$$

Now use the quadratic formula with $a = 2$; $b = 60$; $c = -21,600$.

$$x = \frac{-(60) \pm \sqrt{(60)^2 - 4 \cdot 2 \cdot -21,600}}{2 \cdot 2}$$
$$= \frac{-60 \pm \sqrt{3600 + 172,800}}{4}$$
$$= \frac{-60 \pm \sqrt{176,400}}{4}$$
$$= \frac{-60 \pm 420}{4}$$

Using only $\dfrac{-60 + 420}{4}$ to obtain a positive answer, $x = 90$. The height of the kite is represented by $x + 30$, so the height is 120 feet.

61. Let a = length of the shorter leg,
$2a + 2$ = length of the longer leg,
$2a + 2 + 1$ = length of the hypotenuse.
Use the Pythagorean theorem: $a^2 + b^2 = c^2$.

$$a^2 + (2a+2)^2 = (2a+2+1)^2$$
$$a^2 + (4a^2 + 8a + 4) = (2a+3)^2$$
$$5a^2 + 8a + 4 = 4a^2 + 12a + 9$$
$$a^2 - 4a - 5 = 0$$

Now use the quadratic formula with $a = 1$, $b = -4$, and $c = -5$.

$$a = \frac{-(-4) \pm \sqrt{(-4)^2 - 4 \cdot 1 \cdot -5}}{2 \cdot 1}$$
$$= \frac{4 \pm \sqrt{16 + 20}}{2}$$
$$= \frac{4 \pm \sqrt{36}}{2}$$
$$= \frac{4 \pm 6}{2}$$

Use $\dfrac{4+6}{2} = 5$ to obtain a positive value for a. Then the shorter leg is 5 cm, the longer leg is $2 \cdot 5 + 2 = 12$ cm, and the hypotenuse is $12 + 1 = 13$ cm.

62. Let a = length of the longer side,
$a + 8$ = length of the hypotenuse,
$(a + 8) - 9$ = length of the shorter side.

The length of the shorter side simplifies to $a - 1$.
Use the Pythagorean theorem: $a^2 + b^2 = c^2$.

$$a^2 + (a-1)^2 = (a+8)^2$$
$$a^2 + (a^2 - 2a + 1) = a^2 + 16a + 64$$
$$2a^2 - 2a + 1 = a^2 + 16a + 64$$
$$a^2 - 18a - 63 = 0$$

Now use the quadratic formula with $a = 1$, $b = -18$, and $c = -63$.

$$a = \frac{-(-18) \pm \sqrt{(-18)^2 - 4 \cdot 1 \cdot -63}}{2 \cdot 1}$$
$$= \frac{18 \pm \sqrt{324 + 252}}{2}$$
$$= \frac{18 \pm \sqrt{576}}{2}$$
$$= \frac{18 \pm 24}{2}$$

Use $\dfrac{18+24}{2} = 21$ to obtain a positive value for a. Then the longer leg is 21 meters; the hypotenuse is $21 + 8 = 29$ meters, and the shorter leg is $21 - 1 = 20$ meters.

63. Let w = width of the rectangle. Then $2w - 1$ = length of rectangle. Use the Pythagorean theorem with the width and length as the legs and 2.5 as the hypotenuse of the right triangle.

$$w^2 + (2w-1)^2 = (2.5)^2$$
$$w^2 + (4w^2 - 4w + 1) = 6.25$$
$$5w^2 - 4w + 1 = 6.25$$
$$5w^2 - 4w - 5.25 = 0$$

Now use the quadratic formula with $a = 5$, $b = -4$, and $c = -5.25$.

$$a = \frac{-(-4) \pm \sqrt{(-4)^2 - 4 \cdot 5 \cdot -5.25}}{2 \cdot 5}$$
$$= \frac{4 \pm \sqrt{16 + 105}}{10}$$
$$= \frac{4 \pm \sqrt{121}}{10}$$
$$= \frac{4 \pm 11}{10}$$

Use $\dfrac{4+11}{10} = 1.5$ to obtain a positive value for w. The second value $\dfrac{4-11}{10}$ is negative, which is not meaningful. The width is 1.5 cm and the length is $2(1.5) - 1 = 2$ cm.

64. See the diagram in the text. Let x = the distance from the top of the ladder to the ground; then $x - 7$ = the distance from the bottom of the ladder to the house. These are the two legs of the right triangle. The hypotenuse is 13. Use the Pythagorean theorem: $a^2 + b^2 = c^2$.

$$x^2 + (x-7)^2 = (13)^2$$
$$x^2 + (x^2 - 14x + 49) = 169$$
$$2x^2 - 14x + 49 = 169$$
$$2x^2 - 14x - 120 = 0$$
$$2(x+5)(x-12) = 0$$

When each factor is set equal to zero, only $x - 12$ produces a positive value for x, which is 12. Then the distance from the bottom of the ladder to the house is $12 - 7 = 5$ feet.

65. The area of the floor is $15 \cdot 20$ or 300 square feet, and the area of the rug is given as 234 square feet. The remaining area of the strip around the rug will be $300 - 234 = 66$ square feet. Draw a sketch of the rug surrounded by the flooring. Let w equal the width of the border; that is, the distance from the rug to the wall.

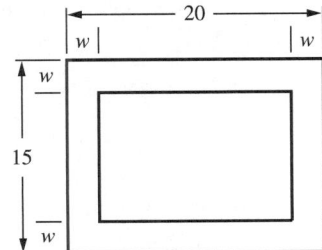

Four rectangles can be formed: The two rectangles on the left and right sides of the figure each have an area of $15 \cdot w$. The rectangles on the top and bottom of the figure each have an area of $(20 - 2w) \cdot w$. Because the total area of these rectangles is 66, an equation can be written:

$$2(15w) + 2[(20 - 2w) \cdot w] = 66$$
$$30w + 2[20w - 2w^2] = 66$$
$$30w + 40w - 4w^2 = 66$$
$$70w - 4w^2 = 66$$
$$-4w^2 + 70w - 66 = 66 - 66$$
$$-4w^2 + 70w - 66 = 0.$$

Before using the quadratic formula, both sides of the equation can be divided by 2 or -2 in order to make the coefficients smaller.

$$\frac{-4w^2}{-2} + \frac{70w}{-2} - \frac{66}{-2} = \frac{0}{-2}$$
$$2w^2 - 35w + 33 = 0$$

Now use the quadratic formula with $a = 2$, $b = -35$, $c = 33$.

$$w = \frac{-(-35) \pm \sqrt{(-35)^2 - 4 \cdot 2 \cdot 33}}{2 \cdot 2}$$
$$= \frac{35 \pm \sqrt{1225 - 264}}{4}$$
$$= \frac{35 \pm \sqrt{961}}{4}$$
$$= \frac{35 \pm 31}{4}$$

The first value of w, $\frac{35+31}{4}$ yields $\frac{66}{4}$ or $16\frac{1}{2}$ feet. The second value of w, $\frac{35-31}{4}$ yields $\frac{4}{4}$ or 1 foot. Only the second value is meaningful in the context of the problem. That is, the width of the border is 1 foot rather than $16\frac{1}{2}$ feet.

66. Let $x =$ the width of the border. Then, the dimensions of the pool plus the border are $40 + 2x$ and $30 + 2x$. The area of the border, 296 square feet, is given by subtracting the area of the pool from the area of the pool plus the border:

$$296 = (40 + 2x)(30 + 2x) - 40 \cdot 30$$
$$296 = 1200 + 40(2x) + (2x)30 + (2x^2) - 1200$$
$$296 = 80x + 60x + 4x^2$$
$$0 = 4x^2 + 140x - 296 \text{ or, dividing by 4,}$$
$$0 = x^2 + 35x - 74. \text{ Solve by factoring:}$$
$$0 = (x - 2)(x + 37).$$

The solutions are $x = 2$, and $x = -37$.

Only the positive solution is meaningful here, so choose $x = 2$ feet as the border width.

67. It is helpful to make a sketch of the proposed garden and grass strip surrounding it. In the figure, the inner rectangle that represents the garden has dimensions $20 - 2x$ meters by $30 - 2x$ meters, where x represents the width of strip of grass surrounding the garden. The outer rectangle, which represents the backyard has dimensions 20 by 30. Also, the area of the garden added to the area of the border will equal the total area of the backyard, or the area of the surrounding rectangle. This can be written as an equation.

$$(30 - 2x)(20 - 2x) + 184 = 30 \cdot 20$$
$$600 - 100x + 4x^2 + 184 = 600$$
$$784 - 100x + 4x^2 = 600$$
$$784 - 600 - 100x + 4x^2 = 600 - 600$$
$$184 - 100x + 4x^2 = 0$$

To make computations with the quadratic formula a little easier, divide both sides of the equation by 4 to give

$$46 - 25x + x^2 = 0$$

Now use the quadratic formula with $a = 1$; $b = -25$; $c = 46$.

$$x = \frac{-(-25) \pm \sqrt{(-25)^2 - 4 \cdot 1 \cdot 46}}{2 \cdot 1}$$
$$= \frac{25 \pm \sqrt{625 - 184}}{2}$$
$$= \frac{25 \pm \sqrt{441}}{2}$$
$$= \frac{25 \pm 21}{2}$$

The first fraction yields 23 and the second fraction yields 2. However, one dimension of the garden, $20 - 2x$, would be a negative number if 23 is used to replace x. When 2 is used to replace x, the dimensions are $20 - 2 \cdot 2 = 16$ meters by $30 - 2 \cdot 2 = 26$ meters, which are reasonable answers.

68. Let $w =$ the width of the sheet of metal. Then the length of the sheet of metal can be represented by $l = 2w - 4$. In creating the box, 2 inches is taken away on each side of the length and on each side of the width. (See the diagram in the text.) The dimensions of the box are:

Length: $L = l - 2 \cdot 2 = 2w - 4 - 2 \cdot 2 = 2w - 8$
Width: $W = w - 2 \cdot 2 = w - 4$
Height: $H = 2$ (the depth or height of the box when the corners have been turned up).

The volume of box is $256 = LWH$. Substitute these algebraic expressions into the formula and solve for w.

$$256 = (2w - 8)(w - 4) \cdot 2$$
$$256 = (2w^2 - 8w - 8w + 32) \cdot 2$$
$$256 = (2w^2 - 16w + 32) \cdot 2$$
$$256 = 4w^2 - 32w + 64$$
$$0 = 4w^2 - 32w + 64 - 256$$
$$0 = 4w^2 - 32w - 192$$

To make computations with the quadratic formula easier, divide both sides of the equation by 4 to give

$$w^2 - 8w - 48 = 0$$

Now use the quadratic formula with $a = 1$, $b = -8$, $c = -48$.

$$w = \frac{-(-8) \pm \sqrt{(-8)^2 - 4 \cdot 1 \cdot -48}}{2 \cdot 1}$$
$$= \frac{8 \pm \sqrt{64 + 192}}{2}$$
$$= \frac{8 \pm \sqrt{256}}{2}$$
$$= \frac{8 \pm 16}{2}$$
$$= 4 \pm 8$$

Since w must be positive, use $w = 4 + 8 = 12$ inches and $l = 2 \cdot 12 - 4 = 20$ inches.

69. Replace the variables in the formula with the given number values and solve for r. The computations will be easier if both sides of the equation are divided by 2000 before expanding $(1 + r)^2$.

$$A = P(1 + r)^2$$
$$2142.25 = 2000(1 + r)^2$$
$$1.071125 = (1 + r)^2$$
$$1.071125 = 1 + 2r + r^2$$
$$0 = r^2 + 2r - .071125$$

Now use the quadratic formula with $a = 1$, $b = 2$, $c = -.071125$.

$$r = \frac{-(2) \pm \sqrt{(2)^2 - 4 \cdot 1 \cdot -.071125}}{2 \cdot 1}$$
$$= \frac{-2 \pm \sqrt{4 + .2845}}{2}$$
$$= \frac{-2 \pm \sqrt{4.2845}}{2}$$
$$\approx \frac{-2 \pm 2.07}{2}$$

The solution $\frac{-2 + 2.07}{2}$ yields approximately .035.

70. Replace V with 432 and solve the equation for x.

$$432 = 3(x - 6)^2$$
$$432 = 3(x - 12x + 36)$$
$$432 = 3x^2 - 36x + 108$$
$$0 = 3x^2 - 36x - 324$$

To make computations with the quadratic formula easier, divide both sides of the equation by 3 to give

$$x^2 - 12x - 108 = 0$$

Now use the quadratic formula with $a = 1$, $b = -12$, $c = -108$.

$$x = \frac{-(-12) \pm \sqrt{(-12)^2 - 4 \cdot 1 \cdot -108}}{2 \cdot 1}$$
$$= \frac{12 \pm \sqrt{144 + 432}}{2}$$
$$= \frac{12 \pm \sqrt{576}}{2}$$
$$= \frac{12 \pm 24}{2}$$

Use $x = \dfrac{12 + 24}{2} = 18$ inches.

71. Set these two expressions equal to each other and solve for p. Begin by multiplying both sides of the equation by p to clear the fractions.

$$\frac{3200}{p} = 3p - 200$$
$$3200 = 3p^2 - 200p$$
$$0 = 3p^2 - 200p - 3200$$

Now use the quadratic formula with $a = 3$, $b = -200$, $c = -3200$.

$$p = \frac{-(-200) \pm \sqrt{(-200)^2 - 4 \cdot 3 \cdot -3200}}{2 \cdot 3}$$
$$= \frac{200 \pm \sqrt{40,000 + 38,400}}{6}$$
$$= \frac{200 \pm \sqrt{78,400}}{6}$$
$$= \frac{200 \pm 280}{6}$$

Use $p = \dfrac{200 + 280}{6} = 80$ cents.

72. As in Exercise 71, set the two algebraic expressions equal to each other and solve for p in dollars.

$$\frac{700}{p} = 5p - 1$$
$$700 = 5p^2 - p$$
$$0 = 5p^2 - p - 700$$

Now use the quadratic formula with $a = 5$, $b = -1$, $c = -700$.

$$p = \frac{-(-1) \pm \sqrt{(-1)^2 - 4 \cdot 5 \cdot -700}}{2 \cdot 5}$$
$$= \frac{1 \pm \sqrt{1 + 14,000}}{10}$$
$$= \frac{1 \pm \sqrt{14,001}}{10}$$
$$\approx \frac{1 \pm 118.33}{10}$$

Use $p = \dfrac{1 + 118.33}{10} \approx \$11.93$.

73. Substitute the given values into the formula for the Froude number and solve for v. Clear the fractions by multiplying both sides of the equation by the lowest common denominator, $(9.8)(1.2)$. Then take the square root of both sides and use only the principal or positive root.

$$F = \frac{v^2}{gl}$$
$$2.57 = \frac{v^2}{(9.8)(1.2)}$$
$$30.2232 = v^2$$
$$5.5 \approx v$$

The value of v is approximately 5.5 meters per second.

74. Substitute the given values into the formula for the Froude number and solve for v. Clear the fractions by multiplying both sides of the equation by the lowest common denominator, $(9.8)(2.8)$. Then take the square root of both sides and use only the principal or positive root.

$$F = \frac{v^2}{gl}$$
$$.16 = \frac{v^2}{(9.8)(2.8)}$$
$$4.3904 = v^2$$
$$2.1 \approx v$$

The value of v is approximately 2.1 meters per second.

75. To find the length of side AC, use proportions to write the following equation:

$$\frac{AC}{DF} = \frac{BC}{EF}$$
$$\frac{3x - 19}{x - 3} = \frac{x - 4}{4}$$
$$4(3x - 19) = (x - 3)(x - 4)$$
$$12x - 76 = x^2 - 7x + 12$$
$$12x - 12x - 76 = x^2 - 7x - 12x + 12$$
$$-76 + 76 = x^2 - 19x + 12 + 76$$
$$0 = x^2 - 19x + 88.$$

Now use the quadratic formula with $a = 1$, $b = -19$, $c = 88$.

$$x = \frac{-(-19) \pm \sqrt{(-19)^2 - 4 \cdot 1 \cdot 88}}{2 \cdot 1}$$
$$= \frac{19 \pm \sqrt{361 - 352}}{2}$$
$$= \frac{19 \pm \sqrt{9}}{2}$$
$$= \frac{19 \pm 3}{2}$$

Using $x = \dfrac{19 + 3}{2} = 11$ or $x = \dfrac{19 - 3}{2} = 8$, there are two possible values for side AC:

$$3 \cdot 11 - 19 = 14 \quad \text{or} \quad 3 \cdot 8 - 19 = 5$$

The possible values for AC are 14 or 5.

76. To find the length of side RQ, use proportions to write the following equation. Then solve by clearing the fractions:

$$\frac{RQ}{UT} = \frac{PQ}{ST}$$
$$\frac{x-5}{3x-11} = \frac{3}{x+3}$$
$$(x-5)(x+3) = 3(3x-11)$$
$$x^2 - 2x - 15 = 9x - 33$$
$$x^2 - 11x + 18 = 0.$$

Now use the quadratic formula with $a = 1$, $b = -11$, $c = 18$.

$$x = \frac{-(-11) \pm \sqrt{(-11)^2 - 4 \cdot 1 \cdot 18}}{2 \cdot 1}$$
$$= \frac{11 \pm \sqrt{121 - 72}}{2}$$
$$= \frac{11 \pm \sqrt{49}}{2}$$
$$= \frac{11 \pm 7}{2}$$

Use $x = \dfrac{11+7}{2} = 9$. The length of side RQ is then $9 - 5 = 4$. If $x = \dfrac{11-7}{2} = 2$ is used, the length of side RQ is a negative number, $2 - 5 = -3$, which is meaningless.

CHAPTER 7 TEST

1. $5x - 3 + 2x = 3(x - 2) + 11$
$7x - 3 = 3x - 6 + 11$
$7x - 3 = 3x + 5$
$7x - 3 + 3 = 3x + 5 + 3$
$7x = 3x + 8$
$7x - 3x = 3x - 3x + 8$
$4x = 8$
$\dfrac{4x}{4} = \dfrac{8}{4}$
$x = 2$

2. $\dfrac{2p-1}{3} + \dfrac{p+1}{4} = \dfrac{43}{12}$
$\dfrac{12}{1} \cdot \dfrac{2p-1}{3} + \dfrac{12}{1} \cdot \dfrac{p+1}{4} = \dfrac{12}{1} \cdot \dfrac{43}{12}$
$4(2p-1) + 3(p+1) = 43$
$8p - 4 + 3p + 3 = 43$
$11p - 1 = 43$
$11p - 1 + 1 = 43 + 1$
$11p = 44$
$\dfrac{11p}{11} = \dfrac{44}{11}$
$p = 4$

3. $3x - (2 - x) + 4x = 7x - 2 - (-x)$
$3x - 2 + x + 4x = 7x - 2 + x$
$8x - 2 = 8x - 2$

This is an identity; the solution set is {all real numbers}.

4. $S = vt - 16t^2$
$S + 16t^2 = vt - 16t^2 + 16t^2$
$S + 16t^2 = vt$
$\dfrac{S + 16t^2}{t} = \dfrac{vt}{t}$
$\dfrac{S + 16t^2}{t} = v$

5. Let $k =$ the area of Kauai; $k + 177 =$ the area of Maui; and $(k + 177) + 3293 =$ the area of Hawaii

$k + k + 177 + (k + 177) + 3293 = 5300$
$3k + 3647 = 5300$
$3k + 3647 - 3647 = 5300 - 3647$
$3k = 1653$
$\dfrac{3k}{3} = \dfrac{1653}{3}$
$k = 551$

Then the area of Kauai is 551 square miles; the area of Maui is $551 + 177 = 728$ square miles; and the area of Hawaii is $728 + 3293 = 4021$ square miles.

6.
Strength	L of solution	L of alcohol
20%	x	$.20(x)$
50%	10	$.50(10)$
40%	$x+10$	$.40(x+10)$

Create an equation by adding the first two algebraic expressions in the last column to total the third:

$$.2(x) + .5(10) = .4(x+10)$$
$$10[.2(x) + .5(10)] = 10[.4(x+10)]$$
$$2x + 50 = 4(x+10)$$
$$2x + 50 = 4x + 40$$
$$2x + 50 - 50 = 4x + 40 - 50$$
$$2x = 4x - 10$$
$$2x - 4x = 4x - 4x - 10$$
$$-2x = -10$$
$$\frac{-2x}{-2} = \frac{-10}{-2}$$
$$x = 5 \text{ liters of 20\% solution.}$$

7.
	Rate	Time	Distance
Passenger train	60	t	$60t$
Freight train	75	t	$75t$

Because the trains are traveling in opposite directions, the sum of their distances will equal the total distance apart of 297 miles. Use this information to create an equation:

$$60t + 75t = 297$$
$$135t = 297$$
$$\frac{135t}{135} = \frac{297}{135}$$
$$t = 2.2 \text{ hours.}$$

8. $\dfrac{\$2.19}{8 \text{ slices}} \approx \$0.274$ per slice

$\dfrac{\$3.30}{12 \text{ bags}} \approx \$0.275$ per slice

The 8-count size is the better buy.

9. Let x = the actual distance between Seattle and Cincinnati.

$$\frac{z}{46} = \frac{1050}{21}$$
$$21z = 46 \cdot 1050$$
$$21z = 48,300$$
$$\frac{21z}{21} = \frac{48,300}{21}$$
$$z = 2300 \text{ miles}$$

10.
$$I = \frac{k}{r}$$
$$80 = \frac{k}{30}$$
$$30 \cdot 80 = k$$
$$2400 = k$$

Then
$$I = \frac{2400}{12}$$
$$I = 200 \text{ amps.}$$

11.
$$-4x + 2(x-3) \geq 4x - (3+5x) - 7$$
$$-4x + 2x - 6 \geq 4x - 3 - 5x - 7$$
$$-2x - 6 \geq -x - 10$$
$$-2x - 6 + 6 \geq -x - 10 + 6$$
$$-2x \geq -x - 4$$
$$-2x + x \geq -x + x - 4$$
$$-x \geq -4$$
$$\frac{-x}{-1} \leq \frac{-4}{-1}$$
$$x \leq 4$$

Interval notation: $(-\infty, 4]$
Graph:

12.
$$-10 < 3k - 4 \leq 14$$
$$-10 + 4 < 3k - 4 + 4 \leq 14 + 4$$
$$-6 < 3k \leq 18$$
$$\frac{-6}{3} < \frac{3k}{3} \leq \frac{18}{3}$$
$$-2 < k \leq 6$$

Interval notation: $(-2, 6]$
Graph:

13. (a)
$$-3x < 9$$
$$\frac{-3x}{-3} > \frac{9}{-3}$$
$$x > -3$$

This is not equivalent.

(b)
$$-3x > -9$$
$$\frac{-3x}{-3} < \frac{-9}{-3}$$
$$x < 3$$

This is not equivalent.

(c)
$$-3x > 9$$
$$\frac{-3x}{-3} < \frac{9}{-3}$$
$$x < -3$$

This is equivalent.

(d)
$$-3x < -9$$
$$\frac{-3x}{-3} > \frac{-9}{-3}$$
$$x > 3$$

This is not equivalent.

14. Let x = the possible scores on the fourth test.
$$\frac{83 + 76 + 79 + x}{4} \geq 80$$
$$\frac{4}{1} \cdot \frac{238 + x}{4} \geq 80 \cdot 4$$
$$238 + x \geq 320$$
$$x \geq 82$$

He must score an 82 or better.

15. $\left(\frac{4}{3}\right)^2 = \frac{4}{3} \cdot \frac{4}{3} = \frac{16}{9}$

16. $-(-2)^6 = -(-2 \cdot -2 \cdot -2 \cdot -2 \cdot -2 \cdot -2)$
$= -(64)$
$= -64$

17. $\left(\frac{3}{4}\right)^{-3} = \frac{3^{-3}}{4^{-3}}$
$= \frac{\frac{1}{3^3}}{\frac{1}{4^3}}$
$= \frac{\frac{1}{27}}{\frac{1}{64}}$
$= \frac{1}{27} \div \frac{1}{64}$
$= \frac{1}{27} \cdot \frac{64}{1}$
$= \frac{64}{27}$

18. $-5^0 + (-5)^0 = -1 + 1 = 0$

19. $9(4p^3)(6p^{-7}) = 9 \cdot 4 \cdot 6 \cdot p^{3+(-7)}$
$= 216p^{-4}$
$= \frac{216}{1} \cdot \frac{1}{p^4}$
$= \frac{216}{p^4}$

20. $\frac{m^{-2}(m^3)^{-3}}{m^{-4}m^7} = \frac{m^{-2} \cdot m^{-9}}{m^{-4+7}}$
$= \frac{m^{-11}}{m^3}$
$= m^{-11-3}$
$= m^{-14}$
$= \frac{1}{m^{14}}$

21. $\frac{(2,500,000)(.00003)}{(.05)(5,000,000)} = \frac{(2.5 \times 10^6)(3 \times 10^{-5})}{(5 \times 10^{-2})(5 \times 10^6)}$
$= \frac{(2.5 \times 3) \times (10^{6-5})}{(5 \times 5) \times (10^{-2+6})}$
$= \frac{7.5 \times 10^1}{25 \times 10^4}$
$= \frac{7.5}{25} \times \frac{10^1}{10^4}$
$= .3 \times 10^{1-4}$
$= .3 \times 10^{-3}$
$= 3 \times 10^{-4}$

Remember that scientific notation has one nonzero digit to the left of the decimal point. As .3 is made larger by a power of ten to become 3, 10^{-3} must become smaller by a power of ten.

22. Solve the formula $D = rt$ for t by dividing both sides of the equation by r.
$$\frac{D}{r} = \frac{rt}{r} = t.$$

Then replace the given distance and rate and simplify.
$$\frac{4.58 \times 10^9}{3.00 \times 10^5} = \frac{4.58}{3.00} \times \frac{10^9}{10^5}$$
$$\approx 1.53 \times 10^{9-5}$$
$$= 1.53 \times 10^4$$
$$= 15{,}300 \text{ seconds}$$

23. *For 1995:*
102.8 billion is 102,800,000,000. In scientific notation this is 1.028×10^{11}.

For 1997:
114.9 billion is 114,900,000,000. In scientific notation this is 1.149×10^{11}.

For 1999:
123.6 billion is 123,600,000,000. In scientific notation this is 1.236×10^{11}.

24.
$$(3k^3 - 5k^2 + 8k - 2) - (3k^3 - 9k^2 + 2k - 12) =$$
$$3k^3 - 5k^2 + 8k - 2 - 3k^3 + 9k^2 - 2k + 12 =$$
$$(3k^3 - 3k^3) + (-5k^2 + 9k^2) + (8k - 2k) + (-2 + 12) =$$
$$4k^2 + 6k + 10$$

25. $(5x + 2)(3x - 4) = 5x \cdot 3x + 5x \cdot -4 + 2 \cdot 3x + 2 \cdot -4$
$= 15x^2 - 20x + 6x - 8$
$= 15x^2 - 14x - 8$

26. $(4x^2 - 3)(4x^2 + 3) = (4x^2)^2 - (3)^2$
$= 16x^4 - 9$

If the pattern is recognized, this multiplication of binomials can be done quickly. Otherwise, use the FOIL method as in Exercise 25.

27. Using vertical multiplication:

$$3x^2 + 8x - 9$$
$$x + 4$$
$$\overline{12x^2 + 32x - 36}$$
$$3x^3 + 8x^2 - 9x$$
$$\overline{3x^3 + 20x^2 + 23x - 36.}$$

28. One of many possibilities is

$$2t^5 + 8t^4 + t^3 - 7t^2 + 2t + 1.$$

29. Find two quantities whose product is $2 \cdot 3q^2 = 6q^2$ and whose sum is $-5q$, the coefficient of the middle term. The two quantities are $-2q$ and $-3q$. Use these two algebraic expressions as coefficients for p in place of the middle term; then factor the four-term polynomial by grouping.

$$2p^2 - 5pq + 3q^2 = 2p^2 - 2pq - 3pq + 3q^2$$
$$= 2p(p-q) - 3q(p-q)$$
$$= (p-q)(2p-3q)$$

Remember that the two factors can also be expressed as

$$(2p-3q)(p-q).$$

30. This is the difference of squares.

$$(10x)^2 - (7y)^2 = (10x+7y)(10x-7y)$$

31. This is the difference of cubes.

$$(3y)^3 - (5x)^3 = (3y-5x)\left[(3y)^2 + 3y \cdot 5x + (5x)^2\right]$$
$$= (3y-5x)(9y^2 + 15xy + 25x^2)$$

32. Factor by grouping.

$$4x + 4y - mx - my = 4(x+y) - m(x+y)$$
$$= (x+y)(4-m)$$

33. In this equation $a = 6$, $b = 7$, $c = -3$. Substitute these values into the quadratic formula.

$$x = \frac{-b \pm \sqrt{b^2 - 4ac}}{2a}$$
$$x = \frac{-(7) \pm \sqrt{(7)^2 - 4 \cdot 6 \cdot -3}}{2 \cdot 6}$$
$$= \frac{-7 \pm \sqrt{49 + 72}}{12}$$
$$= \frac{-7 \pm \sqrt{121}}{12}$$
$$= \frac{-7 \pm 11}{12}$$

The two solutions are

$$\frac{-7+11}{12} = \frac{4}{12} = \frac{1}{3} \text{ or}$$

$$\frac{-7-11}{12} = \frac{-18}{12} = -\frac{3}{2}.$$

The solution set is

$$\left\{-\frac{3}{2}, \frac{1}{3}\right\}.$$

This equation could also be solved by factoring.

34. First write the equation in standard form.

$$x^2 - x - 7 = 0$$

In this equation $a = 1$, $b = -1$, $c = -7$. Substitute these values into the quadratic formula.

$$x = \frac{-b \pm \sqrt{b^2 - 4ac}}{2a}$$
$$x = \frac{-(-1) \pm \sqrt{(-1)^2 - 4 \cdot 1 \cdot -7}}{2 \cdot 1}$$
$$= \frac{1 \pm \sqrt{1 + 28}}{2}$$
$$= \frac{1 \pm \sqrt{29}}{2}$$

The solution set is

$$\left\{\frac{1 \pm \sqrt{29}}{2}\right\}.$$

35. Replace s with 25 and solve for t.

$$25 = 16t^2 + 15t$$
$$25 - 25 = 16t^2 + 15t - 25$$
$$0 = 16t^2 + 15t - 25$$

In this equation $a = 16$, $b = 15$, $c = -25$. Substitute these values into the quadratic formula.

$$x = \frac{-b \pm \sqrt{b^2 - 4ac}}{2a}$$
$$x = \frac{-(15) \pm \sqrt{(15)^2 - 4 \cdot 16 \cdot -25}}{2 \cdot 16}$$
$$= \frac{-15 \pm \sqrt{225 + 1600}}{32}$$
$$= \frac{-15 \pm \sqrt{1825}}{32}$$
$$\approx \frac{-15 \pm 42.7}{32}$$

The first fraction simplifies to about .87 seconds. The second fraction produces a negative number, which is not meaningful.

8.1 EXERCISES

1. (a) The variable, x, represents the year and the variable, y, represents the percent of women in math or computer science professions.

   (b) The percent of women decreased in the decade, 1990–2000.

   (c) The percent of women reached a maximum in 1990.

   (d) Twenty–seven percent of women in the profession occurred in 1980.

2. (a) The variable, x, represents the year and the variable, y, represents the revenue in billions of dollars.

   (b) The revenue was about 1450 billion dollars in 1996.

   (c) The revenue was about 1700 billion dollars in 1998.

3. For any value of x, the point $(x, 0)$ lies on the x-axis.

4. For any value of y, the point $(0, y)$ lies on the y-axis.

5. The circle $x^2 + y^2 = 9$ has the point $(0, 0)$ as its center.

6. The point $(2, 0)$ is the center of the circle $(x - 2)^2 + y^2 = 16$.

7. (a) The point $(1, 6)$ is located in quadrant I since the x-value and the y-value are both positive.

   (b) The point $(-4, -2)$ is located in quadrant III since the x-value and the y-value are both negative.

   (c) The point $(-3, 6)$ is located in quadrant II since the x-value is negative and the y-value is positive.

   (d) The point $(7, -5)$ is located in quadrant IV since the x-value is positive and the y-value is negative.

   (e) The point $(-3, 0)$ is located between quadrants II and III (i.e., on the negative x-axis) since the x-value is negative and the y-value is 0.

8. (a) The point $(-2, -10)$ is located in quadrant III since the x-value and the y-value are both negative.

   (b) The point $(4, 8)$ is located in quadrant I since the x-value and the y-value are both positive.

   (c) The point $(-9, 12)$ is located in quadrant II since the x-value is negative and the y-value is positive.

   (d) The point $(3, -9)$ is located in quadrant IV since the x-value is positive and the y-value is negative.

   (e) The point $(0, -8)$ is located between quadrants III and IV (i.e., on the negative y-axis) since the x-value is 0 and the y-value is negative.

9. (a) If $xy > 0$, then x and y must have same signs. Thus, the point must lie in Quadrant I or III.

   (b) If $xy < 0$, then x and y must have opposite signs. Thus, the point must lie in Quadrant II or IV.

   (c) If $\frac{x}{y} < 0$, then x and y must have opposite signs. Thus, the point must lie in Quadrant II or IV.

   (d) If $\frac{x}{y} > 0$, then x and y must have same signs. Thus, the point must lie in Quadrant I or III.

10. For the coordinates of any point that lie along an axis, one of the coordinates must be zero.

11–20. Note: graph shows even and odd answers.

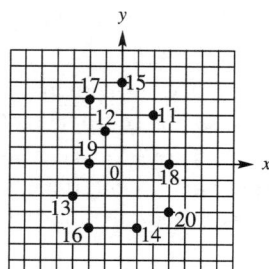

21. (a) $d = \sqrt{(x_2 - x_1)^2 + (y_2 - y_1)^2}$

    $= \sqrt{(-2 - 3)^2 + (1 - 4)^2}$

    $= \sqrt{(-5)^2 + (-3)^2}$

    $= \sqrt{25 + 9}$

    $= \sqrt{34}$

    (b) Using $\left(\dfrac{x_1 + x_2}{2}, \dfrac{y_1 + y_2}{2}\right)$, we have

    $\left(\dfrac{3 + (-2)}{2}, \dfrac{4 + 1}{2}\right) = \left(\dfrac{1}{2}, \dfrac{5}{2}\right)$.

22. (a) $d = \sqrt{(x_2 - x_1)^2 + (y_2 - y_1)^2}$

    $= \sqrt{[3 - (-2)]^2 + (-2 - 1)^2}$

    $= \sqrt{(5)^2 + (-3)^2}$

    $= \sqrt{25 + 9}$

    $= \sqrt{34}$

    (b) Using $\left(\dfrac{x_1 + x_2}{2}, \dfrac{y_1 + y_2}{2}\right)$, we have

    $\left(\dfrac{-2 + 3}{2}, \dfrac{1 + (-2)}{2}\right) = \left(\dfrac{1}{2}, -\dfrac{1}{2}\right)$.

23. (a) $d = \sqrt{(x_2 - x_1)^2 + (y_2 - y_1)^2}$

    $= \sqrt{[3 - (-2)]^2 + (-2 - 4)^2}$

    $= \sqrt{5^2 + (-6)^2}$

    $= \sqrt{25 + 36}$

    $= \sqrt{61}$

(b) Using $\left(\dfrac{x_1+x_2}{2}, \dfrac{y_1+y_2}{2}\right)$, we have

$$\left(\dfrac{(-2)+3}{2}, \dfrac{4+(-2)}{2}\right) = \left(\dfrac{1}{2}, 1\right).$$

24. (a) $d = \sqrt{(x_2-x_1)^2 + (y_2-y_1)^2}$
$= \sqrt{(6-1)^2 + [3-(-5)]^2}$
$= \sqrt{(5)^2 + (8)^2}$
$= \sqrt{25+64}$
$= \sqrt{89}$

(b) Using $\left(\dfrac{x_1+x_2}{2}, \dfrac{y_1+y_2}{2}\right)$,

$$\left(\dfrac{1+6}{2}, \dfrac{-5+3}{2}\right) = \left(\dfrac{7}{2}, -1\right).$$

25. (a) $d = \sqrt{(x_2-x_1)^2 + (y_2-y_1)^2}$
$= \sqrt{[2-(-3)]^2 + (-4-7)^2}$
$= \sqrt{(5)^2 + (-11)^2}$
$= \sqrt{25+121}$
$= \sqrt{146}$

(b) Using $\left(\dfrac{x_1+x_2}{2}, \dfrac{y_1+y_2}{2}\right)$,

$$\left(\dfrac{-3+2}{2}, \dfrac{7+(-4)}{2}\right) = \left(-\dfrac{1}{2}, \dfrac{3}{2}\right).$$

26. (a) $d = \sqrt{(x_2-x_1)^2 + (y_2-y_1)^2}$
$= \sqrt{[0-(-3)]^2 + (5-12)^2}$
$= \sqrt{(3)^2 + (-7)^2}$
$= \sqrt{9+49}$
$= \sqrt{58}$

(b) Using $\left(\dfrac{x_1+x_2}{2}, \dfrac{y_1+y_2}{2}\right)$,

$$\left(\dfrac{0+(-3)}{2}, \dfrac{5+12}{2}\right) = \left(-\dfrac{3}{2}, \dfrac{17}{2}\right).$$

27. $(x-3)^2 + (y-2)^2 = 25$

The equation indicates a graph of a circle with a radius of 5 and centered at $(3, 2)$. Therefore, choose graph B.

28. $(x-3)^2 + (y+2)^2 = 25$
$(x-3)^2 + [y-(-2)]^2 = 25$

The equation indicates a graph of a circle with a radius of 5 and centered at $(3, -2)$. Thus, choose graph C.

29. $(x+3)^2 + (y-2)^2 = 25$
$[x-(-3)]^2 + (y-2)^2 = 25$

The equation indicates a graph of a circle with a radius of 5 and centered at $(-3, 2)$. Thus, choose graph D.

30. $(x+3)^2 + (y+2)^2 = 25$
$[x-(-3)]^2 + [y-(-2)]^2 = 25$

The equation indicates a graph of a circle with a radius of 5 and centered at $(-3, -2)$. Thus, choose graph A.

31. Use the equation of a circle, where $h=0$, $k=0$, and $r=6$.
$$(x-h)^2 + (y-k)^2 = r^2$$
$$(x-0)^2 + (y-0)^2 = 6^2$$
$$x^2 + y^2 = 36$$

32. Use the equation of a circle, where $h=0$, $k=0$, and $r=5$.
$$(x-h)^2 + (y-k)^2 = r^2$$
$$(x-0)^2 + (y-0)^2 = 5^2$$
$$x^2 + y^2 = 25$$

33. Use the equation of a circle, where $h=-1$, $k=3$, and $r=4$.
$$(x-h)^2 + (y-k)^2 = r^2$$
$$[x-(-1)]^2 + (y-3)^2 = 4^2$$
$$(x+1)^2 + (y-3)^2 = 16$$

34. Use the equation of a circle, where $h=2$, $k=-2$, and $r=3$.
$$(x-h)^2 + (y-k)^2 = r^2$$
$$(x-2)^2 + [y-(-2)]^2 = 3^2$$
$$(x-2)^2 + (y+2)^2 = 9$$

35. Use the equation of a circle, where $h=0$, $k=4$, and $r=\sqrt{3}$.
$$(x-h)^2 + (y-k)^2 = r^2$$
$$(x-0)^2 + (y-4)^2 = \left(\sqrt{3}\right)^2$$
$$x^2 + (y-4)^2 = 3$$

36. Use the equation of a circle, where $h=-2$, $k=0$, and $r=\sqrt{5}$.
$$(x-h)^2 + (y-k)^2 = r^2$$
$$[x-(-2)]^2 + (y-0)^2 = \left(\sqrt{5}\right)^2$$
$$(x+2)^2 + y^2 = 5$$

37. The equation, $x^2 + y^2 = r^2$, $r > 0$ or equivalently, $(x-0)^2 + (y-0)^2 = r^2$, $r > 0$, implies that the center is located at $(0,0)$ and the radius is r.

38. (a) There is only one point, $(4, 1)$. The only way for the sum of two squares to equal zero is for both of the squares to equal zero. This is satisfied when $x = 4$ and $y = 1$. Another interpretation is that the circle, centered at $(4, 1)$, has no radius (since $r = \sqrt{0} = 0$) and thus, is only the one point at the center.

    (b) There are no points because the sum of two squares cannot be negative.

39. To find the center and radius, complete the square on x and y.
$$x^2 + y^2 + 4x + 6y + 9 = 0$$
$$\left(x^2 + 4x + \right) + \left(y^2 + 6y + \right) = -9$$
$$\left(x^2 + 4x + 4\right) + \left(y^2 + 6y + 9\right) = -9 + 4 + 9$$
$$(x + 2)^2 + (y + 3)^2 = 4 \text{ or}$$
$$[x - (-2)]^2 + [y - (-3)]^2 = 4$$

    Thus, by inspection, the center is located at $(-2, -3)$ and the radius is given by $r = \sqrt{4} = 2$.

    Remember that the added constants, 4 and 9, come from squaring $\frac{1}{2}$ of the coefficients of each first-degree term $\left(\text{i.e., } \left[\frac{1}{2}(4)\right]^2 = 4 \text{ and } \left[\frac{1}{2}(6)\right]^2 = 9\right)$.

40. To find the center and radius, complete the square on x and y.
$$x^2 + y^2 - 8x - 12y + 3 = 0$$
$$\left(x^2 - 8x + \right) + \left(y^2 - 12y + \right) = -3$$
$$\left(x^2 - 8x + 16\right) + \left(y^2 - 12y + 36\right) = -3 + 16 + 36$$
$$(x - 4)^2 + (y - 6)^2 = 49$$

    Thus, by inspection, the center is located at $(4, 6)$ and the radius is given by $r = \sqrt{49} = 7$.

41. To find the center and radius, complete the square on x and y.
$$x^2 + y^2 + 10x - 14y - 7 = 0$$
$$\left(x^2 + 10x + \right) + \left(y^2 - 14y + \right) = 7$$
$$\left(x^2 + 10x + 25\right) + \left(y^2 - 14y + 49\right) = 7 + 25 + 49$$
$$(x + 5)^2 + (y - 7)^2 = 81 \text{ or}$$
$$[x - (-5)]^2 + (y - 7)^2 = 9^2$$

    Thus, by inspection, the center is located at $(-5, 7)$ and the radius is given by $r = 9$.

42. To find the center and radius, complete the square on x and y.
$$x^2 + y^2 - 2x + 4y - 4 = 0$$
$$\left(x^2 - 2x + \right) + \left(y^2 + 4y + \right) = 4$$
$$\left(x^2 - 2x + 1\right) + \left(y^2 + 4y + 4\right) = 4 + 1 + 4$$
$$(x - 1)^2 + (y + 2)^2 = 9 \text{ or}$$
$$(x - 1)^2 + (y - (-2))^2 = 3^2$$

    Thus, by inspection, the center is located at $(1, -2)$ and the radius is given by $r = 3$.

43. To find the center and radius, complete the square on x and y.
$$3x^2 + 3y^2 - 12x - 24y + 12 = 0$$
$$x^2 + y^2 - 4x - 8y + 4 = 0 \quad \text{Divide by 3}$$
$$\left(x^2 - 4x + \right) + \left(y^2 - 8y + \right) = -4$$
$$\left(x^2 - 4x + 4\right) + \left(y^2 - 8y + 16\right) = -4 + 4 + 16$$
$$(x - 2)^2 + (y - 4)^2 = 16$$

    Thus, by inspection, the center is located at $(2, 4)$ and the radius is given by $r = \sqrt{16} = 4$.

44. To find the center and radius, complete the square on x and y.
$$2x^2 + 2y^2 + 20x + 16y + 10 = 0$$
$$x^2 + y^2 + 10x + 8y + 5 = 0$$
$$\left(x^2 + 10x + \right) + \left(y^2 + 8y + \right) = -5$$
$$\left(x^2 + 10x + 25\right) + \left(y^2 + 8y + 16\right) = -5 + 25 + 16$$
$$(x + 5)^2 + (y + 4)^2 = 36 \text{ or}$$
$$[x - (-5)]^2 + [y - (-4)]^2 = 6^2$$

    Thus, by inspection, the center is located at $(-5, -4)$ and the radius is given by $r = 6$.

45. The equation, $x^2 + y^2 = 36$, is equivalent to
$$(x - 0)^2 + (y - 0)^2 = 6^2.$$

    Thus, the center of the graph is located at $(0, 0)$ and has a radius $r = 6$. The graph is as follows.

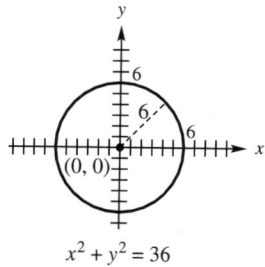

8.1 THE RECTANGULAR COORDINATE SYSTEM AND CIRCLES 203

46. The equation, $x^2 + y^2 = 81$, is equivalent to
$$(x - 0)^2 + (y - 0)^2 = 9^2.$$
Thus, the center of the graph is located at $(0, 0)$ and has a radius $r = 9$. The graph is as follows.

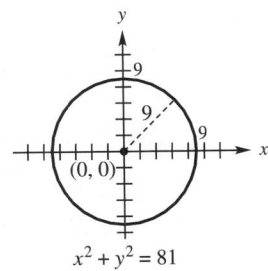

$x^2 + y^2 = 81$

47. The equation, $(x - 2)^2 + y^2 = 36$, is equivalent to
$$(x - 2)^2 + (y - 0)^2 = 6^2.$$
Thus, the center of the graph is located at $(2, 0)$ and has a radius $r = 6$. The graph is as follows.

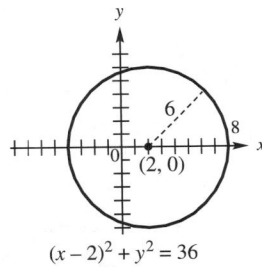

$(x - 2)^2 + y^2 = 36$

48. The equation, $x^2 + (y + 3)^2 = 49$, is equivalent to
$$(x - 0)^2 + [y - (-3)]^2 = 7^2.$$
Thus, the center of the graph is located at $(0, -3)$ and has a radius $r = 7$. The graph is as follows.

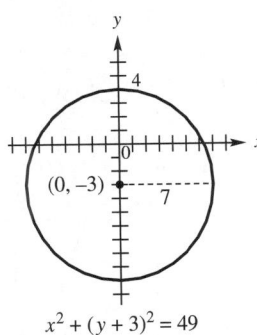

$x^2 + (y + 3)^2 = 49$

49. The equation, $(x + 2)^2 + (y - 5)^2 = 16$, is equivalent to
$$[x - (-2)]^2 + (y - 5)^2 = 4^2.$$
Thus, the center of the graph is located at $(-2, 5)$ and has a radius $r = 4$. The graph is as follows.

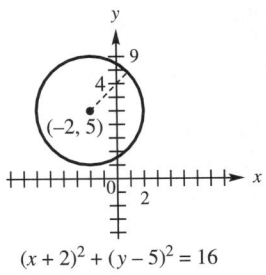

$(x + 2)^2 + (y - 5)^2 = 16$

50. The equation, $(x - 4)^2 + (y - 3)^2 = 25$, is equivalent to
$$(x - 4)^2 + (y - 3)^2 = 5^2.$$
Thus, the center of the graph is located at $(4, 3)$ and has a radius $r = 5$. The graph is as follows.

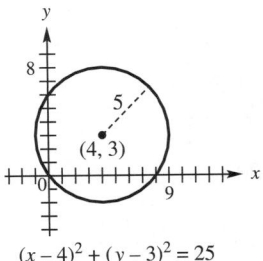

$(x - 4)^2 + (y - 3)^2 = 25$

51. The equation, $(x + 3)^2 + (y + 2)^2 = 36$, is equivalent to
$$[x - (-3)]^2 + [y - (-2)]^2 = 6^2$$
Thus, the center of the graph is located at $(-3, -2)$ and has a radius $r = 6$. The graph is as follows.

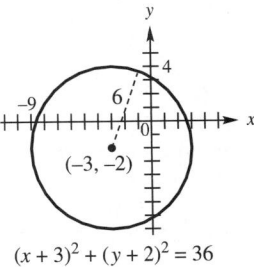

$(x + 3)^2 + (y + 2)^2 = 36$

52. The equation, $(x - 5)^2 + (y + 4)^2 = 49$, is equivalent to
$$(x - 5)^2 + [y - (-4)]^2 = 7^2.$$
Thus, the center of the graph is located at $(5, -4)$ and has a radius $r = 7$. The graph is as follows.

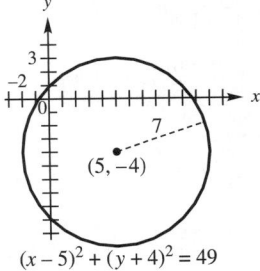

$(x - 5)^2 + (y + 4)^2 = 49$

53. (a) $d = \sqrt{(x_2 - x_1)^2 + (y_2 - y_1)^2}$
$= \sqrt{(2-(-4))^2 + (5-3)^2}$
$= \sqrt{(6)^2 + (2)^2}$
$= \sqrt{36 + 4}$
$= \sqrt{40} = 2\sqrt{10}$

(b) Using $\left(\dfrac{x_1 + x_2}{2}, \dfrac{y_1 + y_2}{2}\right)$, we have
$\left(\dfrac{-4 + 2}{2}, \dfrac{3 + 5}{2}\right) = (-1, 4).$

54. (a) $d = \sqrt{(x_2 - x_1)^2 + (y_2 - y_1)^2}$
$= \sqrt{(6 - (-7))^2 + ((-2) - 4)^2}$
$= \sqrt{(13)^2 + (-6)^2}$
$= \sqrt{169 + 36}$
$= \sqrt{205}$

(b) Using $\left(\dfrac{x_1 + x_2}{2}, \dfrac{y_1 + y_2}{2}\right)$, we have
$\left(\dfrac{-7 + 6}{2}, \dfrac{4 + (-2)}{2}\right) = \left(-\dfrac{1}{2}, 1\right).$

55. Using $\dfrac{y_1 + y_2}{2}$, we have
$\dfrac{10017 + 15380}{2} = 12698.5$ or \$12,698.50.

56. *For 1985:* Using $\dfrac{y_1 + y_2}{2}$, we have
$\dfrac{4.5 + 5.2}{2} = 4.85$ or 4.85 million students.

For 1995: Using $\dfrac{y_1 + y_2}{2}$, we have
$\dfrac{5.2 + 5.8}{2} = 5.5$ or 5.5 million students.

57. *For 1965:* Using $\dfrac{y_1 + y_2}{2}$, we have
$\dfrac{3022 + 3968}{2} = 3495$ dollars.

For 1985: Using $\dfrac{y_1 + y_2}{2}$, we have
$\dfrac{8414 + 13359}{2} \approx 10887$ dollars.

58. Writing exercise

59. Writing exercise

60. Writing exercise

61. Writing exercise

62. Create equations of circles with centers at $(1, 4)$, $(-6, 0)$, and $(5, -2)$ and radii of 4 units, 5 units, and 10 units respectively.

$(x - 1)^2 + (y - 4)^2 = 4^2 = 16$
$(x + 6)^2 + y^2 = 5^2 = 25$
$(x - 5)^2 + (y + 2)^2 = 10^2 = 100$

If we replace x by -3 and y by 4 in each of these equations we see that $(-3, 4)$ is a solution in all three equations. That is, $(-3, 4)$ is the epicenter.

63.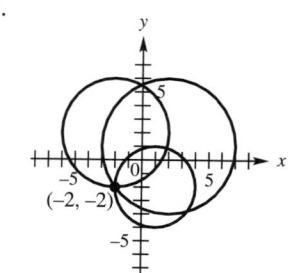

Thus, the epicenter is at $(-2, -2)$.

64. Writing exercise

65. Writing exercise

66. If the endpoints of a line segment have coordinates (x_1, y_1) and (x_2, y_2), then

(a) $d_1 = \sqrt{\left(\dfrac{x_1 + x_2}{2} - x_1\right)^2 + \left(\dfrac{y_1 + y_2}{2} - y_1\right)^2}$
$= \sqrt{\left(\dfrac{x_1 + x_2 - 2x_1}{2}\right)^2 + \left(\dfrac{y_1 + y_2 - 2y_1}{2}\right)^2}$
$= \sqrt{\left(\dfrac{x_2 - x_1}{2}\right)^2 + \left(\dfrac{y_2 - y_1}{2}\right)^2}$

$d_2 = \sqrt{\left(\dfrac{x_1 + x_2}{2} - x_2\right)^2 + \left(\dfrac{y_1 + y_2}{2} - y_2\right)^2}$
$= \sqrt{\left(\dfrac{x_1 + x_2 - 2x_2}{2}\right)^2 + \left(\dfrac{y_1 + y_2 - 2y_2}{2}\right)^2}$
$= \sqrt{\left(\dfrac{x_1 - x_2}{2}\right)^2 + \left(\dfrac{y_1 - y_2}{2}\right)^2}$
$= \sqrt{\left(\dfrac{x_2 - x_1}{2}\right)^2 + \left(\dfrac{y_2 - y_1}{2}\right)^2}.$

Thus, $d_1 = d_2$.

(b) From part (a)

$$d_1 + d_2 = 2d_1$$
$$= 2\sqrt{\left(\frac{x_2 - x_1}{2}\right)^2 + \left(\frac{y_2 - y_1}{2}\right)^2}$$
$$= 2\sqrt{\frac{1}{4}\left[(x_2 - x_1^2) + (y_2 - y_1)^2\right]}$$
$$= \sqrt{(x_2 - x_1)^2 + (y_2 - y_1)^2}$$
$$= \text{distance between } (x_1, y_1) \text{ and } (x_2, y_2).$$

(c) We can conclude that the point given by $\left(\frac{x_1 + x_2}{2}, \frac{y_1 + y_2}{2}\right)$ is the midpoint of the segment whose endpoints are (x_1, y_1) and (x_2, y_2).

67. Let $(x_1, y_1) = (3, -8)$, $(x, y) = (6, 5)$, and (x_2, y_2) be the coordinates of the other endpoint, Then

$$x = \frac{x_1 + x_2}{2}$$
$$6 = \frac{3 + x_2}{2}$$
$$12 = 3 + x_2$$
$$x_2 = 9$$

and

$$y = \frac{y_1 + y_2}{2}$$
$$5 = \frac{-8 + y_2}{2}$$
$$10 = -8 + y_2$$
$$y_2 = 18.$$

Thus, the coordinates of the other endpoint are $(9, 18)$.

68. Only option (B) can be written in the form $x^2 + y^2 = r^2$ where $r^2 > 0$, and hence, is the only equation that will represent a circle.

69. Writing exercise

8.2 EXERCISES

1. The given equation is
$$2x + y = 5.$$

For $(0, \ )$:
$$2 \cdot 0 + y = 5$$
$$y = 5$$

or the ordered pair $(0, 5)$.

For $(\ , 0)$:
$$2x + 0 = 5$$
$$x = \frac{5}{2}$$

or the ordered pair $\left(\frac{5}{2}, 0\right)$.

For $(1, \ )$:
$$2 \cdot 1 + y = 5$$
$$y = 3$$

or the ordered pair $(1, 3)$.

For $(\ , 1)$:
$$2x + 1 = 5$$
$$2x = 4$$
$$x = 2$$

or the ordered pair $(2, 1)$.

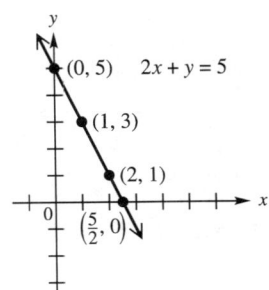

2. The given equation is
$$3x - 4y = 24.$$

For $(0, \ )$:
$$3 \cdot 0 - 4y = 24$$
$$-4y = 24$$
$$y = -6$$

or the ordered pair $(0, -6)$.

For $(\ , 0)$:
$$3x - 4 \cdot 0 = 24$$
$$3x = 24$$
$$x = 8$$

or the ordered pair $(8, 0)$.

For $(6, \ )$:
$$3 \cdot 6 - 4y = 24$$
$$18 - 4y = 24$$
$$-4y = 6$$
$$y = -\frac{3}{2}$$

or the ordered pair $\left(6, -\frac{3}{2}\right)$.

For $(\ , -3)$:
$$3x - 4(-3) = 24$$
$$3x + 12 = 24$$
$$3x = 12$$
$$x = 4$$

or the ordered pair $(4, -3)$.

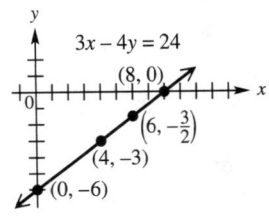

3. The given equation is
$$x - y = 4.$$

For (0,):
$$0 - y = 4$$
$$-y = 4$$
$$y = -4$$

or the ordered pair $(0, -4)$.

For (, 0):
$$x - 0 = 4$$
$$x = 4$$

or the ordered pair $(4, 0)$.

For (2,):
$$2 - y = 4$$
$$-y = 2$$
$$y = -2$$

or the ordered pair $(2, -2)$.

For (, -1):
$$x - (-1) = 4$$
$$x + 1 = 4$$
$$x = 3$$

or the ordered pair $(3, -1)$.

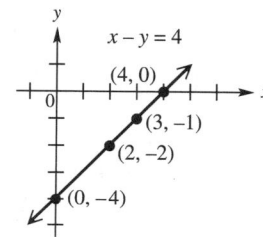

4. The given equation is
$$x + 3y = 12.$$

For (0,):
$$0 + 3y = 12$$
$$y = 4$$

or the ordered pair $(0, 4)$.

For (, 0):
$$x + 3 \cdot 0 = 12$$
$$x = 12$$

or the ordered pair $(12, 0)$.

For (3,):
$$3 + 3y = 12$$
$$3y = 9$$
$$y = 3$$

or the ordered pair $(3, 3)$.

For (, 6):
$$x + 3 \cdot 6 = 12$$
$$x + 18 = 12$$
$$x = -6$$

or the ordered pair $(-6, 6)$.

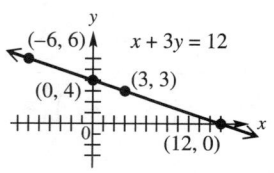

5. The given equation is
$$4x + 5y = 20.$$

For (0,):
$$4 \cdot 0 + 5y = 20$$
$$5y = 20$$
$$y = 4$$

or the ordered pair $(0, 4)$.

For (, 0):
$$4x + 5 \cdot 0 = 20$$
$$4x = 20$$
$$x = 5$$

or the ordered pair $(5, 0)$.

For (3,):
$$4 \cdot 3 + 5y = 20$$
$$12 + 5y = 20$$
$$5y = 8$$
$$y = \frac{8}{5}$$

or the ordered pair $\left(3, \frac{8}{5}\right)$.

For (, 2):
$$4x + 5 \cdot 2 = 20$$
$$4x + 10 = 20$$
$$4x = 10$$
$$x = \frac{5}{2}$$

or the ordered pair $\left(\frac{5}{2}, 2\right)$.

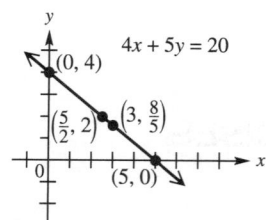

6. The given equation is
$$2x - 5y = 12.$$

For (0,):
$$2 \cdot 0 - 5y = 12$$
$$-5y = 12$$
$$y = -\frac{12}{5}$$

or the ordered pair $\left(0, -\frac{12}{5}\right)$.

For (, 0):
$$2x - 5 \cdot 0 = 12$$
$$2x = 12$$
$$x = 6$$

or the ordered pair $(6, 0)$.

For (, −2): $2x - 5 \cdot (-2) = 12$
$2x + 10 = 12$
$2x = 2$
$x = 1$

or the ordered pair $(1, -2)$.

For (−2,): $2(-2) - 5y = 12$
$-4 - 5y = 12$
$-5y = 16$
$y = -\dfrac{16}{5}$

or the ordered pair $\left(-2, -\dfrac{16}{5}\right)$.

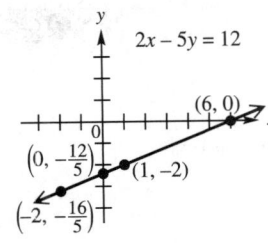

7. The given equation is
$$3x + 2y = 8.$$

From the partially completed table in text we want to complete the evaluation of the ordered pairs (0,), (, 0), (2,), (, −2).

For (0,): $3 \cdot 0 + 2y = 8$
$2y = 8$
$y = 4$

or the ordered pair $(0, 4)$.

For (, 0): $3x + 0 \cdot y = 8$
$3x = 8$
$x = \dfrac{8}{3}$

or the ordered pair $\left(\dfrac{8}{3}, 0\right)$.

For (2,): $3 \cdot 2 + 2y = 8$
$6 + 2y = 8$
$2y = 2$
$y = 1$

or the ordered pair $(2, 1)$.

For (, −2): $3x + 2 \cdot (-2) = 8$
$3x - 4 = 8$
$3x = 12$
$x = 4$

or the ordered pair $(4, -2)$.

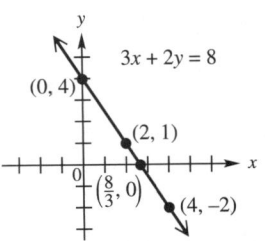

8. The given equation is
$$5x + y = 12.$$

For (0,): $5 \cdot 0 + y = 12$
$y = 12$

or the ordered pair $(0, 12)$.

For (, 0): $5x + 0 = 12$
$5x = 12$
$x = \dfrac{12}{5}$

or the ordered pair $\left(\dfrac{12}{5}, 0\right)$.

For (, −3): $5x + (-3) = 12$
$5x = 15$
$x = 3$

or the ordered pair $(3, -3)$.

For (2,): $5 \cdot 2 + y = 12$
$10 + y = 12$
$y = 2$

or the ordered pair $(2, 2)$.

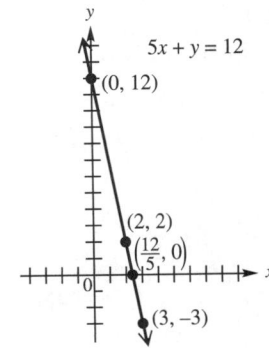

9. Writing exercise

10. Writing exercise

11. Option A is correct since y is constant at $y = 3$.

12. From geometry we know that two points are enough to determine a unique line. Since the graph of every linear equation in two variables is a line, the minimum number of points needed is two.

13. The given equation is
$$3x + 2y = 12.$$
To find the x-intercept, let $y = 0$.
$$3x + 2 \cdot 0 = 12$$
$$3x = 12$$
$$x = 4$$
The x-intercept is $(4, 0)$.

To find the y-intercept, let $x = 0$.
$$3 \cdot 0 + 2y = 12$$
$$2y = 12$$
$$y = 6$$
The y-intercept is $(0, 6)$.

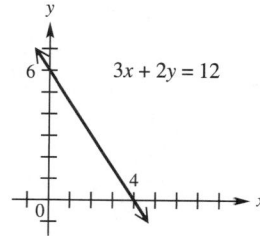

14. The given equation is
$$2x + 5y = 10.$$
To find the x-intercept, let $y = 0$.
$$2x + 5 \cdot 0 = 10$$
$$2x = 10$$
$$x = 5$$
The x-intercept is $(5, 0)$.

To find the y-intercept, let $x = 0$.
$$2 \cdot 0 + 5y = 10$$
$$5y = 10$$
$$y = 2$$
The y-intercept is $(0, 2)$.

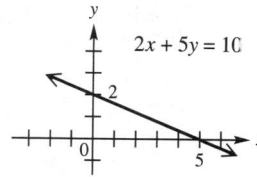

15. The given equation is
$$5x + 6y = 10.$$
To find the x-intercept, let $y = 0$.
$$5x + 6 \cdot 0 = 10$$
$$5x = 10$$
$$x = 2$$

The x-intercept is $(2, 0)$.

To find the y-intercept, let $x = 0$.
$$5 \cdot 0 + 6y = 10$$
$$6y = 10$$
$$y = \frac{5}{3}$$
The y-intercept is $\left(0, \frac{5}{3}\right)$.

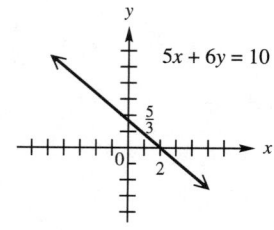

16. The given equation is
$$3y + x = 6.$$
To find the x-intercept, let $y = 0$.
$$3 \cdot 0 + x = 6$$
$$x = 6$$
The x-intercept is $(6, 0)$.

To find the y-intercept, let $x = 0$.
$$3y + 0 = 6$$
$$3y = 6$$
$$y = 2$$
The y-intercept is $(0, 2)$.

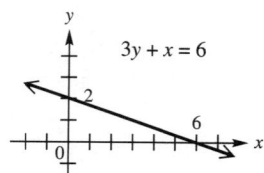

17. The given equation is
$$2x - y = 5.$$
To find the x-intercept, let $y = 0$.
$$2x - 0 = 5$$
$$x = \frac{5}{2}$$
The x-intercept is $\left(\frac{5}{2}, 0\right)$.

To find the y-intercept, let $x = 0$.
$$2 \cdot 0 - y = 5$$
$$-y = 5$$
$$y = -5$$

The y-intercept is $(0, -5)$.

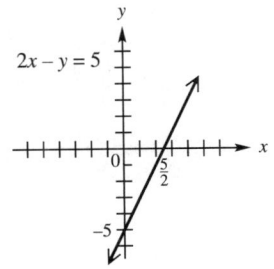

18. The given equation is
$$3x - 2y = 4.$$
To find the x-intercept, let $y = 0$.
$$3x - 2 \cdot 0 = 4$$
$$3x = 4$$
$$x = \frac{4}{3}$$
The x-intercept is $\left(\frac{4}{3}, 0\right)$.

To find the y-intercept, let $x = 0$.
$$3 \cdot 0 - 2y = 4$$
$$-2y = 4$$
$$y = -2$$
The y-intercept is $(0, -2)$.

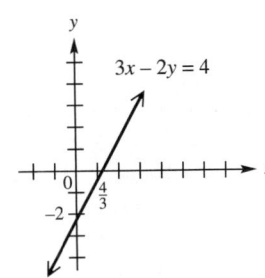

19. The given equation is
$$x - 3y = 2.$$
To find the x-intercept, let $y = 0$.
$$x - 3 \cdot 0 = 2$$
$$x = 2$$
The x-intercept is $(2, 0)$.

To find the y-intercept, let $x = 0$.
$$0 - 3y = 2$$
$$y = -\frac{2}{3}$$

The x-intercept is $\left(0, -\frac{2}{3}\right)$.

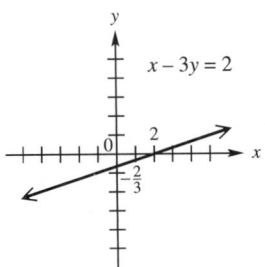

20. The given equation is
$$y - 4x = 3.$$
To find the x-intercept, let $y = 0$.
$$0 - 4 \cdot x = 3$$
$$x = -\frac{3}{4}$$
The x-intercept is $\left(-\frac{3}{4}, 0\right)$.

To find the y-intercept, let $x = 0$.
$$y - 4 \cdot 0 = 3$$
$$y = 3$$
The y-intercept is $(0, 3)$.

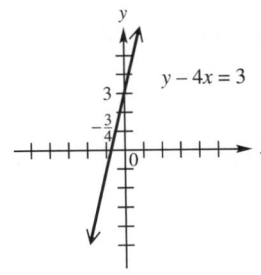

21. The given equation is
$$y + x = 0.$$
To find the y-intercept, let $x = 0$.
$$y + 0 = 0$$
$$y = 0$$

The y-intercept is $(0, 0)$. Observe that this is also the x-intercept. Thus, the graph runs through the origin.

To find a second point let x (or y) take on a value and solve for the other variable, i.e. let $x = 2$.
$$y + 2 = 0$$
$$y = -2$$

Thus, a second point would have coordinates $(2, -2)$.

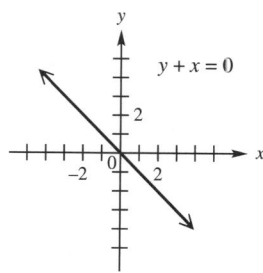

22. The given equation is
$$2x - y = 0.$$
To find the x-intercept, let $y = 0$.
$$2x - 0 = 0$$
$$x = 0$$

The x-intercept is $(0, 0)$. Observe that this is also the y-intercept. Thus, the graph runs through the origin.

To find a second point let x (or y) take on a value and solve for the other variable, i.e. let $y = 2$.
$$2x - 2 = 0$$
$$2x = 2$$
$$x = 1$$

Thus, a second point would have coordinates $(1, 2)$.

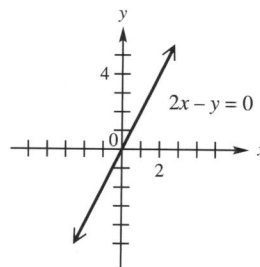

23. The given equation is
$$3x = y.$$
To find the y-intercept, let $x = 0$.
$$3 \cdot 0 = y$$
$$y = 0$$

The y-intercept is $(0, 0)$. Observe that this is also the x-intercept. Thus, the graph runs through the origin.

To find a second point let x (or y) take on a value and solve for the other variable, i.e. let $x = 1$.
$$3 \cdot 1 = y$$
$$y = 3$$

Thus, a second point would have coordinates $(1, 3)$.

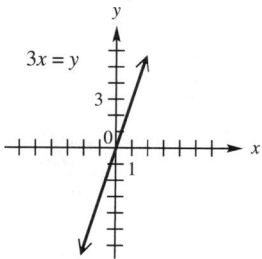

24. The given equation is
$$x = -4y.$$
To find the x-intercept, let $y = 0$.
$$x = -4 \cdot 0$$
$$x = 0$$

The x-intercept (and y-intercept) is $(0, 0)$. Thus, the graph runs through the origin. Solve for one more point to sketch the graph.

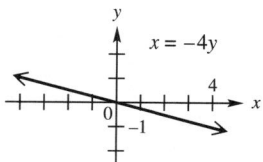

25. The given equation is
$$x = 2.$$
The equation is represented by a vertical line where x is 2 for any value of y. Thus, when $y = 0$, x remains the value 2. The x-intercept is $(2, 0)$. There is no y-intercept.

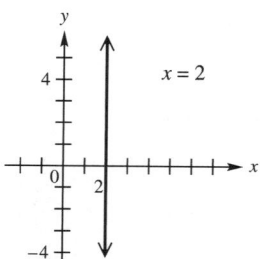

26. The given equation is
$$y = -3.$$
The equation is represented by a horizontal line, where y is -3 for any value of x. Thus, when $x = 0$, y remains the value -3. The y-intercept is $(0, -3)$. There is no x-intercept.

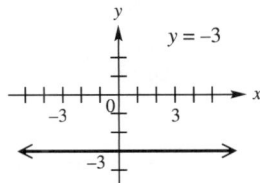

27. The given equation is
$$y = 4.$$
The equation is represented by a horizontal line where y is 4, for any value of x. Thus, when $x = 0$, y remains the value 4. The y-intercept is $(0, 4)$. There is no x-intercept.

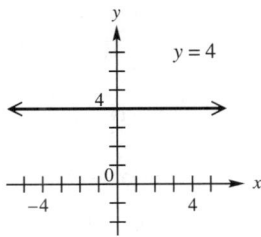

28. The given equation is
$$x = -2.$$
The equation is represented by a vertical line, where x is -2 for any value of y. Thus, x remains the value -2, when $y = 0$. The x-intercept is $(-2, 0)$. There is no y-intercept.

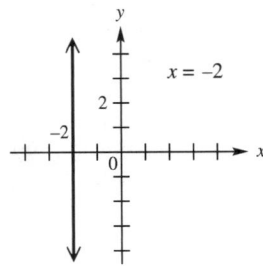

29. The graph of $y + 2 = 0$ is a horizontal line crossing the y-axis at $y = -2$. This fits option C.

30. The graph of $y + 4 = 0$ is a horizontal line crossing the y-axis at $y = -4$. This fits option C.

31. The graph of $x + 3 = 0$ is a vertical line which crosses the x-axis at $x = -3$. This fits option A.

32. The graph of $x + 7 = 0$ is a vertical line which crosses the x-axis at $x = -7$. This fits option A.

33. The graph of $y - 2 = 0$ is a horizontal line crossing the y-axis at $y = 2$. This fits option D.

34. The graph of $y - 4 = 0$ is a horizontal line crossing the y-axis at $y = 4$. This fits option D.

35. The graph of $x - 3 = 0$ is a vertical line which crosses the x-axis at $x = 3$. This fits option B.

36. The graph of $x - 7 = 0$ is a vertical line which crosses the x-axis at $x = 7$. This fits option B.

37. The diagram of the roof indicates a rise (change in y) of 6 feet and a run (change in x) of 20 feet. Thus, the slope (pitch) is given by
$$m = \frac{\text{rise}}{\text{run}} = \frac{6}{20} = \frac{3}{10}.$$

38. The diagram of the hill indicates a rise (change in y) of 32 meters and a run (change in x) of 108 meters. Thus, the slope (grade) is given by
$$m = \frac{\text{rise}}{\text{run}} = \frac{32}{108} = \frac{8}{27}.$$

39. (a) The coordinates of the given points are $(-1, -4)$ and $(3, 2)$. Let $(x_1, y_1) = (-1, -4)$ and $(x_2, y_2) = (3, 2)$. Then,
$$m = \frac{y_2 - y_1}{x_2 - x_1}$$
$$= \frac{2 - (-4)}{3 - (-1)}$$
$$= \frac{3}{2}.$$
Observe that either point may be chosen as (x_1, y_1) or (x_2, y_2).

(b) The coordinates of the given points are $(-3, 5)$ and $(1, -2)$. Let $(x_1, y_1) = (-3, 5)$ and $(x_2, y_2) = (1, -2)$. Then,
$$m = \frac{y_2 - y_1}{x_2 - x_1}$$
$$= \frac{-2 - 5}{1 - (-3)}$$
$$= -\frac{7}{4}.$$

40. (a) The slope is <u>positive</u> since as you move left to right on the graph, the "change in y" and the "change in x" are both positive, keeping the slope ratio
$$m = \frac{\text{change in } y}{\text{change in } x}$$
itself positive.

(b) The slope of any vertical line is <u>undefined</u> since the denominator, "change in x," of the slope ratio,
$$m = \frac{\text{change in } y}{\text{change in } x},$$
is zero (and division by zero is undefined).

(c) The slope is negative since as you move left to right on the graph, the "change in y" is negative while the "change in x" is positive, which makes the slope ratio,

$$m = \frac{\text{change in } y}{\text{change in } x},$$

negative.

(d) The slope of any horizontal line is zero. This is true since the numerator of the slope ratio,

$$m = \frac{\text{change in } y}{\text{change in } x},$$

is zero.

41. Let $(x_1, y_1) = (-2, -3)$ and $(x_2, y_2) = (-1, 5)$.
Then,
$$m = \frac{y_2 - y_1}{x_2 - x_1}$$
$$= \frac{5 - (-3)}{-1 - (-2)}$$
$$= 8.$$

42. Let $(x_1, y_1) = (-4, 3)$ and $(x_2, y_2) = (-3, 4)$.
Then,
$$m = \frac{y_2 - y_1}{x_2 - x_1}$$
$$= \frac{4 - 3}{-3 - (-4)}$$
$$= 1.$$

43. Let $(x_1, y_1) = (8, 1)$ and $(x_2, y_2) = (2, 6)$.
Then,
$$m = \frac{y_2 - y_1}{x_2 - x_1}$$
$$= \frac{6 - 1}{2 - 8}$$
$$= -\frac{5}{6}.$$

44. Let $(x_1, y_1) = (13, -3)$ and $(x_2, y_2) = (5, 6)$.
Then,
$$m = \frac{y_2 - y_1}{x_2 - x_1}$$
$$= \frac{6 - (-3)}{5 - 13}$$
$$= -\frac{9}{8}.$$

45. Let $(x_1, y_1) = (2, 4)$ and $(x_2, y_2) = (-4, 4)$.
Then,
$$m = \frac{y_2 - y_1}{x_2 - x_1}$$
$$= \frac{4 - 4}{(-4) - 2}$$
$$= \frac{0}{-6}$$
$$= 0.$$

46. Let $(x_1, y_1) = (-6, 3)$ and $(x_2, y_2) = (2, 3)$.
Then,
$$m = \frac{y_2 - y_1}{x_2 - x_1}$$
$$= \frac{3 - 3}{2 - (-6)}$$
$$= \frac{0}{8}$$
$$= 0.$$

47. Refer to graph (Figure A, in the text).
(a) Let $(x_1, y_1) = (1990, 11{,}338)$ and $(x_2, y_2) = (2005, 14{,}818)$.
Then,
$$m = \frac{y_2 - y_1}{x_2 - x_1}$$
$$= \frac{14{,}818 - 11{,}338}{2005 - 1990}$$
$$= \frac{3480}{15}$$
$$= 232 \ (\textit{thousand})$$
$$= 232{,}000.$$

(b) The slope of the line in Figure A is positive. This means that during the period represented, enrollment increased.

(c) Since the slope, or *rate of change*, is

$$\frac{232{,}000 \text{ students}}{1 \text{ yr}}$$

the increase in students per year is $232{,}000$.

(d) Refer to graph (Figure B, in the text). Let $(x_1, y_1) = (1990, 22)$ and $(x_2, y_2) = (2000, 5.4)$.
Then,
$$m = \frac{y_2 - y_1}{x_2 - x_1}$$
$$= \frac{5.4 - 22}{2000 - 1990}$$
$$= \frac{-16.6}{10}$$
$$= -1.66.$$

(e) The slope of the line in Figure B is negative. This shows us that during the period represented, the number of students per computer decreased.

(f) Since the slope, or *rate of change*, is

$$\frac{-1.66 \text{ students per computer}}{1 \text{ yr}}$$

the decrease in students per computer *per year* is 1.66 students per computer.

48. Writing exercise

49. Locate the point $(-3, 2)$. The slope is

$$m = \frac{1}{2} = \frac{\text{change in } y}{\text{change in } x}.$$

From $(-3, 2)$ move 1 unit up and 2 units to the right. This brings you to another point, $(-1, 3)$. Draw the line through $(-3, 2)$ and $(-1, 3)$.

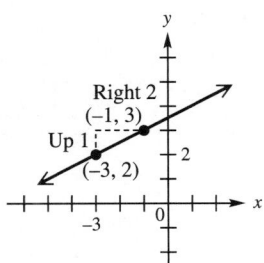

50. Locate the point $(0, 1)$. The slope is

$$m = \frac{2}{3} = \frac{\text{change in } y}{\text{change in } x}.$$

From $(0, 1)$ move 2 units up and 3 units to the right. This brings you to another point, $(3, 3)$. Draw the line through $(0, 1)$ and $(3, 3)$.

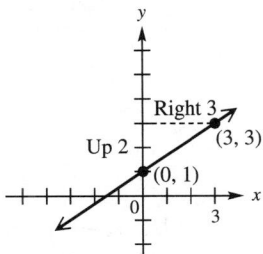

51. Locate the point $(-2, -1)$. The slope is

$$m = \frac{-5}{4} = \frac{\text{change in } y}{\text{change in } x}.$$

From $(-2, -1)$ move 5 units down and 4 units to the right. This brings you to another point, $(2, -6)$. Draw the line through $(-2, -1)$ and $(2, -6)$.

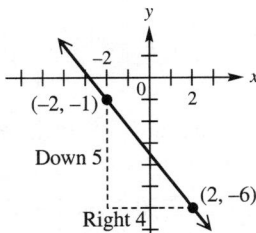

52. Locate the point $(-1, -2)$. The slope is

$$m = \frac{-3}{2} = \frac{\text{change in } y}{\text{change in } x}.$$

From $(-1, -2)$ move 3 units down and 2 units to the right. This brings you to another point, $(1, -5)$. Draw the line through $(-1, -2)$ and $(1, -5)$.

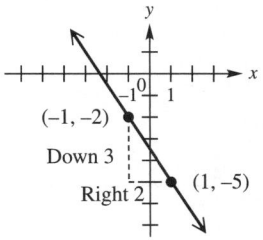

53. Locate the point $(-1, -4)$. The slope is

$$m = -2 = \frac{-2}{1} = \frac{\text{change in } y}{\text{change in } x}.$$

From $(-1, -4)$ move 2 units down and 1 unit to the right. This brings you to another point, $(0, -6)$. Draw the line through $(-1, -4)$ and $(0, -6)$.

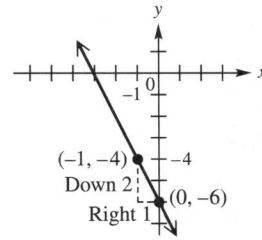

54. Locate the point $(1, 2)$. The slope is

$$m = 3 = \frac{3}{1} = \frac{\text{change in } y}{\text{change in } x}.$$

From $(1, 2)$ move 3 units up and 1 unit to the right. This brings you to another point, $(2, 5)$. Draw the line through $(1, 2)$ and $(2, 5)$.

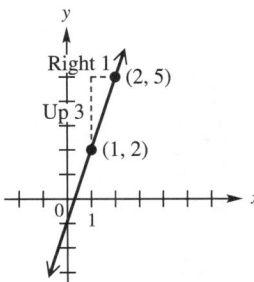

55. First locate the point $(2, -5)$. Use the definition of slope ($m = 0 = \frac{0}{n}$, for any non-zero integer, n) to move up (or down) 0 units (change in y-values) and to the right (or left) n units (change in x-values) to locate another point on the graph. Draw a line through these two points.

Locate the point $(2, -5)$. The slope is

$$m = 0 = \frac{0}{n} = \frac{\text{change in } y}{\text{change in } x}, \text{ for any non-zero integer, } n.$$

From $(2, -5)$ move 0 units up and n units to the left or right to locate another point on the line. Draw the line through these two points.

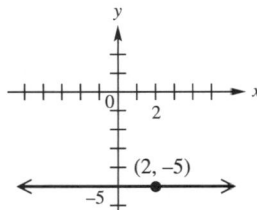

56. Locate $(-3, 1)$. An undefined slope means that the line is vertical. The x-value of every point is -3.

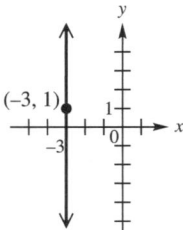

57. L_1 is through $(4, 6)$ and $(-8, 7)$, and L_2 is through $(7, 4)$ and $(-5, 5)$.

For L_1,
$$m = \frac{y_2 - y_1}{x_2 - x_1} = \frac{7 - 6}{-8 - 4} = \frac{-1}{12}.$$

For L_2,
$$m_2 = \frac{y_2 - y_1}{x_2 - x_1} = \frac{5 - 4}{-5 - 7} = \frac{-1}{12}.$$

Since the slopes are equal, the lines are parallel.

58. L_1 is through $(9, 15)$ and $(-7, 12)$, and L_2 is through $(-4, 8)$ and $(-20, 5)$.

For L_1,
$$m = \frac{y_2 - y_1}{x_2 - x_1} = \frac{12 - 15}{-7 - 9} = \frac{-3}{-16} = \frac{3}{16}.$$

For L_2,
$$m_2 = \frac{y_2 - y_1}{x_2 - x_1} = \frac{5 - 8}{-20 - (-4)} = \frac{-3}{-16} = \frac{3}{16}.$$

Since the slopes are equal, the lines are parallel.

59. L_1 is through $(2, 0)$ and $(5, 4)$, and L_2 is through $(6, 1)$ and $(2, 4)$.

For L_1,
$$m = \frac{y_2 - y_1}{x_2 - x_1} = \frac{4 - 0}{5 - 2} = \frac{4}{3}.$$

For L_2,
$$m_2 = \frac{y_2 - y_1}{x_2 - x_1} = \frac{4 - 1}{2 - 6} = \frac{3}{-4}.$$

Since the slopes are negative reciprocals of each other, the lines are perpendicular.

60. L_1 is through $(0, -7)$ and $(2, 3)$, and L_2 is through $(0, -3)$ and $(1, -2)$.

For L_1,
$$m = \frac{y_2 - y_1}{x_2 - x_1} = \frac{3 - (-7)}{2 - 0} = \frac{10}{2} = 5.$$

For L_2,
$$m_2 = \frac{y_2 - y_1}{x_2 - x_1} = \frac{-2 - (-3)}{1 - 0} = \frac{1}{1} = 1.$$

The slopes are not related and hence, the lines are neither parallel nor perpendicular.

61. L_1 is through $(0, 1)$ and $(2, -3)$, and L_2 is through $(10, 8)$ and $(5, 3)$.

For L_1,
$$m = \frac{y_2 - y_1}{x_2 - x_1} = \frac{-3 - 1}{2 - 0} = \frac{-4}{2} = -2.$$

For L_2,
$$m_2 = \frac{y_2 - y_1}{x_2 - x_1} = \frac{3 - 8}{5 - 10} = \frac{-5}{-5} = 1.$$

The slopes are not related and hence, the lines are neither parallel nor perpendicular.

62. L_1 is through $(1, 2)$ and $(-7, -2)$, and L_2 is through $(1, -1)$ and $(5, -9)$.

For L_1,
$$m = \frac{y_2 - y_1}{x_2 - x_1} = \frac{-2 - 2}{-7 - 1} = \frac{-4}{-8} = \frac{1}{2}.$$

For L_2,
$$m_2 = \frac{y_2 - y_1}{x_2 - x_1} = \frac{-9 - (-1)}{5 - 1} = \frac{-8}{4} = \frac{-2}{1}.$$

Since the slopes are negative reciprocals of each other, the lines are perpendicular.

63. Use the two points $(160, 0)$ and $(250, 63)$ to represent the front of the upper deck and the back of the upper deck. The change in x is found by
$$250 \text{ feet} - 160 \text{ feet} = 90 \text{ feet}$$
and the change is y is the height or 63 feet. Thus,
$$m = \frac{\text{change in } y}{\text{change in } x} = \frac{63 \text{ ft}}{90 \text{ ft}} = \frac{7}{10}.$$

64. Since the maximum grade that an elephant can walk is 13%,

$$m = \frac{\text{vertical rise}}{\text{horizontal run}} = 13\%$$
$$= \frac{13}{100}.$$

Thus, $\frac{x}{150} = \frac{13}{100}$,
$$x = 150 \times \frac{13}{100}.$$
$$x = 19.5 \text{ feet}$$

65. (a) The average rate of change is the same value as the slope. Thus,

$$\text{average rate of change} = \frac{y_2 - y_1}{x_2 - x_1}$$
$$= \frac{2080 - 1280}{1999 - 1995}$$
$$= \frac{800}{4} = 200.$$

The average rate of change is $200 million per year.

(b) Writing exercise

66. (a) The average rate of change is the same value as the slope. Thus,

$$\text{average rate of change} = \frac{y_2 - y_1}{x_2 - x_1}$$
$$= \frac{20.8 - 25.9}{1998 - 1995}$$
$$= \frac{-5.1}{3} = -1.7.$$

The average rate of change is -1.7 million recipients per year.

(b) Writing exercise

67. (a) The average rate of change in book sales for the following years:

1995–1996

$$m = \frac{y_2 - y_1}{x_2 - x_1} = \frac{20000 - 19000}{1996 - 1995} = \frac{1000}{1} = 1000,$$

or 1000 million per year.

1995 to 1999

$$m = \frac{y_2 - y_1}{x_2 - x_1} = \frac{23000 - 19000}{1999 - 1995} = \frac{4000}{4} = 1000,$$

or 1000 million per year.

1998 to 2000

$$m = \frac{y_2 - y_1}{x_2 - x_1} = \frac{24000 - 22000}{2000 - 1998} = \frac{2000}{2} = 1000,$$

or 1000 million per year.

Thus, the average rate of change is the same, no matter which two ordered pairs are selected to calculate it. This is true because the data points lie on a straight line.

(b)
1995–1996

$$r = \frac{1000}{19000} \approx 5.3\% \text{ increase per year.}$$

1995 to 1999

$$r = \frac{4000}{19000} \approx 21.05\% \text{ for the four years,}$$

or $21.05\%/4 \approx 5.3\%$ increase per year.

1998 to 2000

$$r = \frac{2000}{22000} \approx 9.09\% \text{ for the two years,}$$

or $9.09\%/2 \approx 4.5\%$ increase per year.

All are about 5%.

68. (a) The average rate of change in subscribers is:

1994–1995

$$m = \frac{y_2 - y_1}{x_2 - x_1} = \frac{33786 - 24134}{1995 - 1994} = \frac{9652}{1} = 9652,$$

or 9652 thousand subscribers per year.

1995 to 1996

$$m = \frac{y_2 - y_1}{x_2 - x_1} = \frac{44043 - 33786}{1996 - 1995} = \frac{10257}{1} = 10257,$$

or 10,257 thousand subscribers per year.

1996 to 1997

$$m = \frac{y_2 - y_1}{x_2 - x_1} = \frac{55312 - 44043}{1997 - 1996} = \frac{11269}{1} = 11269,$$

or 11,269 thousand subscribers per year.

1997 to 1998

$$m = \frac{y_2 - y_1}{x_2 - x_1} = \frac{69209 - 55312}{1998 - 1997} = \frac{13897}{1} = 13897,$$

or 13,897 thousand subscribers per year.

1998 to 1999

$$m = \frac{y_2 - y_1}{x_2 - x_1} = \frac{86047 - 69209}{1999 - 1998} = \frac{16838}{1} = 16838,$$

or 16,838 thousand subscribers per year.

(b) The average rate of change for successive years is not the same. The plot would not approximate a straight line since the line segments have different slopes.

69. Writing exercise

70. Writing exercise

8.3 EXERCISES

1. The slope $= -2$, through the point $(4, 1)$ matches D, $y - 1 = -2(x - 4)$. Use point–slope form of line.

2. The slope $= -2$, y–intercept $(0, 1)$ matches C, $y = -2x + 1$. Use slope–intercept form of line.

3. The line passing through the points $(0, 0)$ and $(4, 1)$ matches B, $y = \frac{1}{4}x$. Find the slope using the two points, then use the slope and either point in the point–slope form of line.

4. The line passing through the points $(0, 0)$ and $(1, 4)$ matches A, $y = 4x$. Find the slope using the two points, then use the slope and either point in the point–slope form of line.

5. Using the two intercept points, the slope is given by
$$m = \frac{y_2 - y_1}{x_2 - x_1}$$
$$= \frac{0 - (-3)}{1 - 0}$$
$$= 3,$$
and $b = -3$ (the ordinate of the y–intercept). Thus, the slope–intercept form of the line is
$$y = mx + b$$
$$= 3x + (-3), \text{ or}$$
$$y = 3x - 3.$$

6. Using the two intercept points, the slope is given by
$$m = \frac{y_2 - y_1}{x_2 - x_1}$$
$$= \frac{0 - (-4)}{2 - 0}$$
$$= \frac{4}{2}$$
$$= 2,$$
and $b = -4$ (the ordinate of the y–intercept). Thus, the slope–intercept form of the line is
$$y = mx + b$$
$$= 2x + (-4), \text{ or}$$
$$y = 2x - 4.$$

7. Using the two intercept points, the slope is given by
$$m = \frac{y_2 - y_1}{x_2 - x_1}$$
$$= \frac{0 - 3}{3 - 0}$$
$$= \frac{-3}{3}$$
$$= -1,$$
and $b = 3$ (the ordinate of the y–intercept). Thus, the slope–intercept form of the line is
$$y = mx + b$$
$$= (-1)x + 3, \text{ or}$$
$$y = -x + 3.$$

8. Using the two intercept points, the slope is given by
$$m = \frac{y_2 - y_1}{x_2 - x_1}$$
$$= \frac{0 - 2}{4 - 0}$$
$$= \frac{-2}{4}$$
$$= -\frac{1}{2}$$
and $b = 2$ (the ordinate of the y–intercept). Thus, the slope–intercept form of the line is
$$y = mx + b$$
$$= \left(-\frac{1}{2}\right)x + 2, \text{ or}$$
$$y = -\frac{1}{2}x + 2.$$

9. The equation $y = 2x + 3$ matches the graph A. Observe that the positive slope, 2, discounts options B, C, D, E, and G. Of the remaining options only A suggests the y–intercept as potentially the value 3.

10. The equation $y = -2x + 3$ matches the graph D. Observe that the negative slope, -2, discounts options A, B, E, F, and H. Of the remaining options only D suggests the y–intercept as potentially the value 3.

11. The equation $y = -2x - 3$ matches the graph C. Observe that the negative slope, -2, discounts options A, B, E, F, and H. Of the remaining options only C suggests the y–intercept as potentially the value -3.

12. The equation $y = 2x - 3$ matches the graph F. Observe that the positive slope, 2, discounts options B, C, D, E, and G. Of the remaining options only F suggests the y–intercept as potentially the value -3.

13. The equation $y = 2x$ matches the graph H. Observe that the y–intercept is the value 0. This means that the graph runs through the origin, $(0, 0)$. Only G and H satisfy this condition and G indicates a negative slope.

14. The equation $y = -2x$ matches the graph G. Observe that the y–intercept is the value 0. This means that the graph runs through the origin, $(0, 0)$. Only G and H satisfy this condition and H indicates a positive slope.

15. The equation $y = 3$ represents a horizontal line with a y–intercept at 3. These conditions match option B.

16. The equation $y = -3$ represents a horizontal line with a y–intercept at -3. These conditions match option E.

8.3 EQUATIONS OF LINES AND LINEAR MODELS

17. Use the point-slope form of the line to write the equation:
$$y - y_1 = m(x - x_1), \text{ or}$$
$$y - 4 = -\frac{3}{4}[x - (-2)].$$

To write the equation in slope–intercept form:
$$y - 4 = -\frac{3}{4}x + \left(-\frac{3}{4}\right) \cdot 2$$
$$y = -\frac{3}{4}x - \frac{3}{2} + 4$$
$$y = -\frac{3}{4}x - \frac{3}{2} + \frac{8}{2}$$
$$y = -\frac{3}{4}x + \frac{5}{2}.$$

18. Use the point-slope form of the line to write the equation:
$$y - y_1 = m(x - x_1), \text{ or}$$
$$y - 6 = -\frac{5}{6}[x - (-1)].$$

To write the equation in slope–intercept form:
$$y - 6 = -\frac{5}{6}x + \left(-\frac{5}{6}\right) \cdot 1$$
$$y = -\frac{5}{6}x - \frac{5}{6} + 6$$
$$y = -\frac{5}{6}x - \frac{5}{6} + \frac{36}{6}$$
$$y = -\frac{5}{6}x + \frac{31}{6}.$$

19. Use the point-slope form of the line to write the equation:
$$y - y_1 = m(x - x_1), \text{ or}$$
$$y - 8 = -2(x - 5).$$

To write the equation in slope–intercept form:
$$y - 8 = -2x + 10$$
$$y = -2x + 10 + 8$$
$$y = -2x + 18.$$

20. Use the point-slope form of the line to write the equation:
$$y - y_1 = m(x - x_1), \text{ or}$$
$$y - 10 = 1 \cdot (x - 12).$$

To write the equation in slope–intercept form:
$$y - 10 = x - 12$$
$$y = x - 12 + 10$$
$$y = x - 2.$$

21. Use the point-slope form of the line to write the equation:
$$y - y_1 = m(x - x_1), \text{ or}$$
$$y - 4 = \frac{1}{2}[x - (-5)].$$

To write the equation in slope–intercept form:
$$y - 4 = \frac{1}{2}x + \left(\frac{1}{2}\right) \cdot 5$$
$$y = \frac{1}{2}x + \frac{5}{2} + 4$$
$$y = \frac{1}{2}x + \frac{5}{2} + \frac{8}{2}$$
$$y = \frac{1}{2}x + \frac{13}{2}.$$

22. Use the point-slope form of the line to write the equation:
$$y - y_1 = m(x - x_1), \text{ or}$$
$$y - (-2) = \frac{1}{4}(x - 7).$$

To write the equation in slope–intercept form:
$$y + 2 = \frac{1}{4}x - \left(\frac{1}{4}\right) \cdot 7$$
$$y = \frac{1}{4}x - \frac{7}{4} - 2$$
$$y = \frac{1}{4}x - \frac{7}{4} - \frac{8}{4}$$
$$y = \frac{1}{4}x - \frac{15}{4}.$$

23. Use the point-slope form of the line to write the equation:
$$y - y_1 = m(x - x_1), \text{ or}$$
$$y - 0 = 4 \cdot (x - 3).$$

This simplifies to slope–intercept form:
$$y = 4x - 12.$$

24. Use the point-slope form of the line to write the equation:
$$y - y_1 = m(x - x_1), \text{ or}$$
$$y - 0 = -5[x - (-2)].$$

This simplifies to slope–intercept form:
$$y = -5x - 10.$$

25. Using the point-slope form of the line to write the equation,
$$y - y_1 = m(x - x_1), \text{ or}$$
$$y - 5 = 0 \cdot (x - 9).$$

This simplifies to slope–intercept form:
$$y - 5 = 0, \text{ or}$$
$$y = 5.$$

Alternatively, with $m = 0$ (horizontal line), the equation takes the form of $y = k$ and by inspection we can recognize $y = 5$ as the equation of the horizontal line.

26. Using the point-slope form of the line to write the equation,
$$y - y_1 = m(x - x_1), \text{ or}$$
$$y - (-2) = 0 \cdot [x - (-4)].$$

This simplifies to slope–intercept form:
$$y + 2 = 0, \text{ or}$$
$$y = -2.$$

Alternatively, with $m = 0$ (horizontal line), the equation takes the form of $y = k$ and by inspection we can recognize $y = -2$ as the equation of the horizontal line.

27. An undefined slope indicates that the line is vertical and therefore, the equation is of the form $x = k$. Thus, by inspection, $x = 9$ is the equation of the vertical line.

28. An undefined slope indicates that the line is vertical and therefore, the equation is of the form $x = k$. Thus, $x = -2$ is the equation of the vertical line.

29. A vertical line is of the form $x = k$. Thus, by inspection, $x = .5$ is the equation of the line.

30. A vertical line is of the form $x = k$. Thus, $x = \frac{5}{8}$ is the equation of the line.

31. A horizontal line is of the form $y = k$. Thus, by inspection, $y = 8$ is the equation of the line.

32. A horizontal line is of the form $y = k$. Thus, $y = 7$ is the equation of the line.

33. Using the two points, the slope is given by
$$m = \frac{y_2 - y_1}{x_2 - x_1}$$
$$= \frac{8 - 4}{5 - 3}$$
$$= \frac{4}{2}$$
$$= 2.$$

Use the point-slope form of the line (and either point) to write the equation:
$$y - y_1 = m(x - x_1), \text{ or}$$
$$y - 4 = 2(x - 3).$$

To write the equation in slope–intercept form:
$$y - 4 = 2x - 6$$
$$y = 2x - 6 + 4$$
$$y = 2x - 2.$$

34. Using the two points, the slope is given by
$$m = \frac{y_2 - y_1}{x_2 - x_1}$$
$$= \frac{14 - (-2)}{(-3) - 5}$$
$$= \frac{16}{-8}$$
$$= -2.$$

Use the point-slope form of the line (and either point) to write the equation:
$$y - y_1 = m(x - x_1), \text{ or}$$
$$y - (-2) = -2(x - 5).$$

To write the equation in slope–intercept form:
$$y + 2 = -2x + 10$$
$$y = -2x + 10 - 2$$
$$y = -2x + 8.$$

35. Using the two points, the slope is given by
$$m = \frac{y_2 - y_1}{x_2 - x_1}$$
$$= \frac{5 - 1}{(-2) - 6}$$
$$= \frac{4}{-8}$$
$$= -\frac{1}{2}.$$

Use the point-slope form of the line (and either point) to write the equation:
$$y - y_1 = m(x - x_1), \text{ or}$$
$$y - 1 = -\frac{1}{2}(x - 6).$$

To write the equation in slope–intercept form:
$$y - 1 = -\frac{1}{2}x - \left(-\frac{1}{2}\right)6$$
$$y - 1 = -\frac{1}{2}x + 3$$
$$y = -\frac{1}{2}x + 3 + 1$$
$$y = -\frac{1}{2}x + 4.$$

36. Using the two points, the slope is given by
$$m = \frac{y_2 - y_1}{x_2 - x_1}$$
$$= \frac{1 - 5}{(-8) - (-2)}$$
$$= \frac{-4}{-6}$$
$$= \frac{2}{3}.$$

Use the point-slope form of the line (and either point) to write the equation:
$$y - y_1 = m(x - x_1), \text{ or}$$
$$y - 5 = \frac{2}{3}[x - (-2)].$$

To write the equation in slope–intercept form:
$$y - 5 = \frac{2}{3}(x + 2)$$
$$y - 5 = \frac{2}{3}x + \left(\frac{2}{3}\right) \cdot 2$$
$$y = \frac{2}{3}x + \frac{4}{3} + 5$$
$$y = \frac{2}{3}x + \frac{4}{3} + \frac{15}{3}$$
$$y = \frac{2}{3}x + \frac{19}{3}.$$

37. Using the two points, the slope is given by
$$m = \frac{y_2 - y_1}{x_2 - x_1}$$
$$= \frac{(2/3) - (2/5)}{(4/3) - (-2/5)}$$
$$= \frac{(10/15) - (6/15)}{(20/15) + (6/15)}$$
$$= \frac{(4/15)}{(26/15)}$$
$$= \frac{4}{15} \cdot \frac{15}{26}$$
$$= \frac{2}{13}.$$

Use the point-slope form of the line (and either point) to write the equation:
$$y - y_1 = m(x - x_1), \text{ or}$$
$$y - \frac{2}{5} = \frac{2}{13}\left[x - \left(-\frac{2}{5}\right)\right].$$

To write the equation in slope-intercept form:
$$y - \frac{2}{5} = \frac{2}{13}\left(x + \frac{2}{5}\right)$$
$$y - \frac{2}{5} = \frac{2}{13}x + \left(\frac{2}{13}\right) \cdot \left(\frac{2}{5}\right)$$
$$y = \frac{2}{13}x + \frac{4}{65} + \frac{2}{5}$$
$$y = \frac{2}{13}x + \frac{4}{65} + \frac{26}{65}$$
$$y = \frac{2}{13}x + \frac{30}{65}$$
$$y = \frac{2}{13}x + \frac{6}{13}.$$

38. Using the two points, the slope is given by
$$m = \frac{y_2 - y_1}{x_2 - x_1}$$
$$= \frac{(2/3) - (8/3)}{(2/5) - (3/4)}$$
$$= \frac{(-6/3)}{(8/20) - (15/20)}$$
$$= \frac{-2}{(-7/20)}$$
$$= (-2) \cdot \left(-\frac{20}{7}\right)$$
$$= \frac{40}{7}.$$

Use the point-slope form of the line (and either point) to write the equation:
$$y - y_1 = m(x - x_1), \text{ or}$$
$$y - \frac{8}{3} = \frac{40}{7}\left[x - \frac{3}{4}\right].$$

To write the equation in slope-intercept form:
$$y - \frac{8}{3} = \frac{40}{7}x - \frac{40}{7} \cdot \frac{3}{4}$$
$$y - \frac{8}{3} = \frac{40}{7}x - \frac{30}{7}$$
$$y = \frac{40}{7}x - \frac{30}{7} + \frac{8}{3}$$
$$y = \frac{40}{7}x - \frac{90}{21} + \frac{56}{21}$$
$$y = \frac{40}{7}x - \frac{34}{21}.$$

39. Using the two points, the slope is given by
$$m = \frac{y_2 - y_1}{x_2 - x_1}$$
$$= \frac{5 - 5}{1 - 2}$$
$$= \frac{0}{-1}$$
$$= 0.$$

Using the point-slope form of the line to write the equation,
$$y - y_1 = m(x - x_1), \text{ or}$$
$$y - 5 = 0 \cdot (x - 2).$$

This simplifies to slope-intercept form:
$$y - 5 = 0, \text{ or}$$
$$y = 5.$$

40. Using the two points, the slope is given by
$$m = \frac{y_2 - y_1}{x_2 - x_1}$$
$$= \frac{2 - 2}{4 - (-2)}$$
$$= \frac{0}{6}$$
$$= 0.$$

Using the point-slope form of the line to write the equation,
$$y - y_1 = m(x - x_1), \text{ or}$$
$$y - 2 = 0 \cdot [x - (-2)].$$

This simplifies to slope-intercept form:
$$y - 2 = 0, \text{ or}$$
$$y = 2.$$

41. These points lie on a vertical line and the slope is undefined (since the denominator in the slope ratio, $x_2 - x_1 = 7 - 7$, is 0). Therefore, one can't use the point-slope form of the line. Rather, the equation of a vertical line is in the form $x = k$. Thus, $x = 7$.

42. These points lie on a vertical line and the slope is undefined (since the denominator in the slope ratio, $x_2 - x_1 = 13 - 13$, is 0). Therefore, one can't use the point-slope form of the line. Rather, the equation of a vertical line is in the form $x = k$. Thus, $x = 13$.

43. Using the two points, the slope is given by
$$m = \frac{y_2 - y_1}{x_2 - x_1}$$
$$= \frac{-3 - (-3)}{-1 - 1}$$
$$= \frac{0}{-2}$$
$$= 0.$$

Instead of using the point-slope form of the line to write the equation, we offer an alternative solution. We know that with $m = 0$, the line is horizontal and is of the form $y = k$. Thus, $y = -3$ is the equation.

44. Using the two points, the slope is given by
$$m = \frac{y_2 - y_1}{x_2 - x_1}$$
$$= \frac{6 - 6}{5 - (-4)}$$
$$= \frac{0}{9}$$
$$= 0.$$

Instead of using the point-slope form of the line to write the equation, we offer an alternative solution. We know that with $m = 0$, the line is horizontal and is of the form $y = k$. Thus, $y = 6$ is the equation.

45. Using the slope-intercept form of the line:
$$y = mx + b, \text{ or}$$
$$y = 5x + 15.$$

46. Using the slope-intercept form of the line:
$$y = mx + b, \text{ or}$$
$$y = -2x + 12.$$

47. Using the slope-intercept form of the line:
$$y = mx + b, \text{ or}$$
$$y = -\frac{2}{3}x + \frac{4}{5}.$$

48. Using the slope-intercept form of the line:
$$y = mx + b, \text{ or}$$
$$y = -\frac{5}{8}x + \left(-\frac{1}{3}\right)$$
$$y = -\frac{5}{8}x - \frac{1}{3}.$$

49. Using the slope-intercept form of the line with $m = \frac{2}{5}$ and $b = 5$:
$$y = mx + b, \text{ or}$$
$$y = \frac{2}{5}x + 5.$$

50. Using the slope-intercept form of the line with $m = -\frac{3}{4}$ and $b = 7$:
$$y = mx + b, \text{ or}$$
$$y = -\frac{3}{4}x + 7.$$

51. Writing exercise

52. The standard form of an equation representing a straight line is $Ax + By = C$. Thus, D is the correct option.

53. (a) To write $x + y = 12$ in slope-intercept form, solve for y:
$$x + y = 12$$
$$y = -x + 12.$$

(b) By inspection, the slope, m, is given by $m = -1$ (the understood coefficient of the x-term).

(c) By inspection, $b = 12$ (the constant), so the y-intercept is $(0, 12)$.

54. (a) To write $x - y = 14$ in slope-intercept form, solve for y:
$$x - y = 14$$
$$-y = -x + 14$$
$$y = x - 14.$$

(b) By inspection, the slope, m, is given by $m = 1$.

(c) By inspection, $b = -14$, so the y-intercept is $(0, -14)$.

55. (a) To write $5x + 2y = 20$ in slope-intercept form, solve for y:
$$5x + 2y = 20$$
$$2y = -5x + 20$$
$$y = -\frac{5}{2}x + 10.$$

(b) By inspection, the slope, m, is given by $m = -\frac{5}{2}$.

(c) By inspection, $b = 10$, so the y-intercept is $(0, 10)$.

56. (a) To write $6x + 5y = 40$ in slope-intercept form, solve for y:
$$6x + 5y = 40$$
$$5y = -6x + 40$$
$$y = -\frac{6}{5}x + 8.$$

(b) By inspection, the slope, m, is given by $m = -\frac{6}{5}$.

(c) By inspection, $b = 8$, so the y-intercept is $(0, 8)$.

57. (a) To write $2x - 3y = 10$ in slope-intercept form, solve for y:
$$2x - 3y = 10$$
$$-3y = -2x + 10$$
$$y = \frac{2}{3}x - \frac{10}{3}.$$

(b) By inspection, the slope, m, is given by $m = \frac{2}{3}$.

(c) By inspection, $b = -\frac{10}{3}$, so the y-intercept is $\left(0, -\frac{10}{3}\right)$.

58. (a) To write $4x - 3y = 10$ in slope-intercept form, solve for y:
$$4x - 3y = 10$$
$$-3y = -4x + 10$$
$$y = \frac{4}{3}x - \frac{10}{3}.$$

(b) By inspection, the slope, m, is given by $m = \frac{4}{3}$.

(c) By inspection, $b = -\frac{10}{3}$, so the y-intercept is $\left(0, -\frac{10}{3}\right)$.

59. Write $3x - y = 8$ in slope-intercept form in order to identify the slope of the given line:
$$3x - y = 8$$
$$-y = -3x + 8$$
$$y = 3x - 8.$$

By inspection, $m = 3$. This is also the slope of the new line, since they are parallel. Using the point-slope form of the line,
$$y - y_1 = m(x - x_1), \text{ or}$$
$$y - 2 = 3(x - 7).$$

This simplifies to slope-intercept form:
$$y - 2 = 3x - 21, \text{ or}$$
$$y = 3x - 19.$$

60. Write $2x + 5y = 10$ in slope-intercept form in order to identify the slope of the given line:
$$2x + 5y = 10$$
$$5y = -2x + 10$$
$$y = -\frac{2}{5}x + 2.$$

By inspection, $m = -\frac{2}{5}$. This is also the slope of the new line, since they are parallel. Using the point-slope form of the line,
$$y - y_1 = m(x - x_1), \text{ or}$$
$$y - 1 = -\frac{2}{5}(x - 4)$$
$$y - 1 = -\frac{2}{5}x + \frac{8}{5}$$
$$y = -\frac{2}{5}x + \frac{8}{5} + 1$$
$$y = -\frac{2}{5}x + \frac{8}{5} + \frac{5}{5}$$
$$y = -\frac{2}{5}x + \frac{13}{5}.$$

61. Write $-x + 2y = 10$ in slope-intercept form in order to identify the slope of the given line:
$$-x + 2y = 10$$
$$2y = x + 10$$
$$y = \frac{1}{2}x + 5.$$

By inspection, $m = \frac{1}{2}$. This is also the slope of the new line, since they are parallel. Using the point, $(-2, -2)$ and point-slope form of the line,
$$y - y_1 = m(x - x_1), \text{ or}$$
$$y - (-2) = \frac{1}{2}[x - (-2)]$$
$$y + 2 = \frac{1}{2}x + \frac{2}{2}$$
$$y = \frac{1}{2}x + 1 - 2$$
$$y = \frac{1}{2}x - 1.$$

62. Write $-x + 3y = 12$ in slope-intercept form in order to identify the slope of the given line:
$$-x + 3y = 12$$
$$3y = x + 12$$
$$y = \frac{1}{3}x + 4.$$

By inspection, $m = \frac{1}{3}$. This is also the slope of the new line, since they are parallel. Using the point $(-1, 3)$, and point-slope form of the line,

$$y - y_1 = m(x - x_1), \text{ or}$$
$$y - 3 = \frac{1}{3}[x - (-1)]$$
$$y - 3 = \frac{1}{3}x + \frac{1}{3}$$
$$y = \frac{1}{3}x + \frac{1}{3} + 3$$
$$y = \frac{1}{3}x + \frac{1}{3} + \frac{9}{3}$$
$$y = \frac{1}{3}x + \frac{10}{3}.$$

63. Write $2x - y = 7$ in slope-intercept form in order to identify the slope of the given line:

$$2x - y = 7$$
$$-y = -2x + 7$$
$$y = 2x - 7.$$

By inspection, $m = 2$. Thus, the slope of any line perpendicular to the given line is $-\frac{1}{2}$.

Using the given point $(8, 5)$ and the point-slope form of the line,

$$y - y_1 = m(x - x_1), \text{ or}$$
$$y - 5 = -\frac{1}{2}(x - 8)$$
$$2y - 10 = -x + 8$$
$$2y = -x + 8 + 10$$
$$y = -\frac{1}{2}x + \frac{18}{2}$$
$$y = -\frac{1}{2}x + 9.$$

64. Write $5x + 2y = 18$ in slope-intercept form in order to identify the slope of the given line:

$$5x + 2y = 18$$
$$2y = -5x + 18$$
$$y = -\frac{5}{2}x + 9.$$

By inspection, $m = -\frac{5}{2}$. Thus, the slope of any line perpendicular to the given line is $\frac{2}{5}$.

Using the given point $(2, -7)$ and the point-slope form of the line,

$$y - y_1 = m(x - x_1), \text{ or}$$
$$y - (-7) = \frac{2}{5}(x - 2)$$
$$5y + 35 = 2(x - 2)$$
$$5y = 2x - 4 - 35$$
$$y = \frac{1}{5}(2x - 39)$$
$$y = \frac{2}{5}x - \frac{39}{5}.$$

65. Since $x = 9$ is a vertical line, any line perpendicular to it will be horizontal and have slope $= 0$. To write the equation of a horizontal line through the point $(-2, 7)$ we can use the form $y = k$ with $k = 7$. Thus, the line is $y = 7$.

66. Since $x = -3$ is a vertical line, any line perpendicular to it will be horizontal and have slope $= 0$. To write the equation of a horizontal line through the point $(8, 4)$ we can use the form $y = k$ with $k = 4$. Thus, the line is $y = 4$

67. (a)

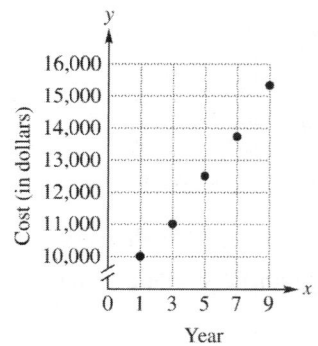

Yes, they do approximate a straight line.

(b) Using the points $(3, 11025)$ and $(9, 15380)$. The slope is given by

$$m = \frac{y_2 - y_1}{x_2 - x_1}$$
$$= \frac{15380 - 11025}{9 - 3}$$
$$= \frac{4355}{6}$$
$$\approx 725.8.$$

Using the point-slope form for the equation, the first point, and the calculated slope we have:

$$y - y_1 = m(x - x_1), \text{ or}$$
$$y - 11025 = 725.8(x - 3)$$
$$y = 11025 + 725.8(x - 3)$$
$$y = 11025 + 725.8x - 2177.4$$
$$y = 725.8x + 8847.6 \text{ or}$$
$$y = 725.8x + 8847.5 \text{ (if other point was used)}.$$

(c) Substituting $x = 13$ (for the year 2003), we have
$$y = 725.8x + 8847.8$$
$$y = 725.8(13) + 8847.8$$
$$y \approx \$18,283.$$

68. (a) Yes, the points do approximate a straight line.

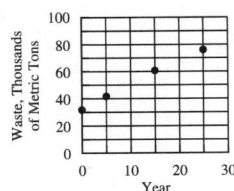

(b) Using the points $(0, 32)$ and $(25, 76)$, the slope is given by
$$m = \frac{y_2 - y_1}{x_2 - x_1}$$
$$= \frac{76 - 32}{25 - 0}$$
$$= \frac{44}{25}$$
$$= 1.76.$$

Use the vertical axis intercept $(0, 32)$ as the point $(0, b)$ and the slope-intercept form for the equation to arrive at
$$y = mx + b, \text{ or}$$
$$y = 1.76x + 32.$$

(c) In the year 2005, $x = 10$. Therefore,
$$y = 1.76(10) + 32$$
$$= 17.6 + 32$$
$$= 49.6.$$

Thus, there will be about 49.6 thousand $= 49.6(1000)$ or 49,600 metric tons produced in the year 2005.

69. Choose the first and last year for the data points, $(4, 28657)$ and $(8, 27952)$, to create the slope. The slope is given by
$$m = \frac{y_2 - y_1}{x_2 - x_1}$$
$$= \frac{27952 - 28657}{8 - 4}$$
$$= \frac{-705}{4}$$
$$= -176.25.$$

Using the slope and any data point, we choose $(4, 28657)$, in conjunction with the point-slope form of the line we arrive at:
$$y - y_1 = m(x - x_1), \text{ or}$$
$$y - 28657 = -176.25(x - 4)$$
$$y = 28657 - 176.25(x - 4)$$
$$y = 28657 - 176.25x + 705$$
$$y = -176.25x + 29362.$$

70. Choose the first and last year for the data points, $(150, 5000)$ and $(1400, 24000)$, to create the slope. The slope is given by
$$m = \frac{y_2 - y_1}{x_2 - x_1}$$
$$= \frac{24000 - 5000}{1400 - 150}$$
$$= \frac{19000}{1250}$$
$$= 15.2.$$

Using the slope and any data point, we choose $(150, 5000)$, in conjunction with the point-slope form of the line we arrive at:
$$y - y_1 = m(x - x_1), \text{ or}$$
$$y - 5000 = 15.2(x - 150)$$
$$y = 5000 + 15.2(x - 150)$$
$$y = 5000 + 15.2x - 2280$$
$$y = 15.2x + 2720.$$

71. (a) Use the data points for years 1993 and 1997, $(3, 21696)$ and $(7, 25050)$, to create the slope. The slope is given by
$$m = \frac{y_2 - y_1}{x_2 - x_1}$$
$$= \frac{25050 - 21696}{7 - 3}$$
$$= \frac{3354}{4}$$
$$= 838.5.$$

Using the slope and any data point, we choose $(7, 25050)$, in conjunction with the point-slope form of the line we arrive at:
$$y - y_1 = m(x - x_1), \text{ or}$$
$$y - 25050 = 838.5(x - 7)$$
$$y = 25050 + 838.5(x - 7)$$
$$y = 25050 + 838.5x - 5869.5$$
$$y = 838.5x + 19180.5.$$

(b) Substituting $x = 5$ (for the year 1995), we have
$$y = 838.5x + 19180.5$$
$$y = 838.5(5) + 19180.5$$
$$y \approx \$23,373.$$

This is close to the actual value.

72. (a) Use the data points for years 1995 and 1997, $(5, 24075)$ and $(7, 26628)$, to create the slope. The slope is given by
$$m = \frac{y_2 - y_1}{x_2 - x_1}$$
$$= \frac{26628 - 24075}{7 - 5}$$
$$= \frac{2553}{2}$$
$$= 1276.5.$$

224 CHAPTER 8 GRAPHS, FUNCTIONS, AND SYSTEMS OF EQUATIONS AND INEQUALITIES

Using the slope and any data point, we choose $(5, 24075)$, in conjunction with the point-slope form of the line we arrive at:

$$y - y_1 = m(x - x_1), \text{ or}$$
$$y - 24075 = 1276.5(x - 5)$$
$$y = 24075 + 1276.5(x - 5)$$
$$y = 24075 + 1276.5x - 6382.5$$
$$y = 1276.5x + 17692.5.$$

(b) Writing exercise

73. (a) Using the points $(0, 5.6)$ and $(2.5, 3.2)$, the slope is given by

$$m = \frac{y_2 - y_1}{x_2 - x_1}$$
$$= \frac{3.2 - 5.6}{2.5 - 0}$$
$$= \frac{-2.4}{2.5}$$
$$= -.96.$$

Use the vertical axis intercept $(0, 5.6)$ as the point $(0, b)$ and the slope-intercept form for the equation to arrive at

$$y = mx + b, \text{ or}$$
$$y = -.96x + 5.6.$$

(b) In the year 2001, $x = 5$, Therefore,

$$y = -.96(5) + 5.6$$
$$= -4.8 + 5.6$$
$$= 0.8.$$

Thus, the landfill capacity, in the year 2001, is predicted to be about .8 million $= .8(1,000,000)$ or $800,000$ tons.

(c) To find the x-intercept, let $y = 0$ in the linear equation. Thus,

$$y = -.96x + 5.6$$
$$0 = -.96x + 5.6$$
$$.96x = 5.6$$
$$x = \frac{5.6}{.96}$$
$$= 5.833333.$$

This answer suggests that late in the 5th year, i.e. 2001, the capacity is predicted to be zero. To find out how late in the year take $.8333333(12) = 9.999996 \approx 10$. In other words, in the tenth month, October, zero capacity is predicted.

74. (a) When $C = 0°$, $F = \underline{32°}$; when $C = 100°$, $F = \underline{212°}$.

(b) The two ordered pairs are $(0, 32)$ and $(100, 212)$.

(c) Using the points $(0, 32)$ and $(100, 212)$, the slope is given by

$$m = \frac{y_2 - y_1}{x_2 - x_1}$$
$$= \frac{212 - 32}{100 - 0}$$
$$= \frac{180}{100}$$
$$= \frac{9}{5}.$$

(d) Use the vertical axis intercept $(0, 32)$ as the point $(0, b)$ and the slope-intercept form for the equation to arrive at

$$F = mC + b, \text{ or}$$
$$F = \frac{9}{5}C + 32.$$

(e) To solve for C in terms of F:

$$F = \frac{9}{5}C + 32$$
$$F - 32 = \frac{9}{5}C$$
$$\frac{5}{9}(F - 32) = C, \text{ or}$$
$$C = \frac{5}{9}(F - 32).$$

(f) When the Celsius temperature is 50°, the Fahrenheit temperature is 122°.

75. (a) Choosing $(0, 1)$ and $(4, 4)$, the slope is given by

$$m = \frac{y_2 - y_1}{x_2 - x_1}$$
$$= \frac{4 - 1}{4 - 0}$$
$$= \frac{3}{4}.$$

Use the y-axis intercept, $(0, 1)$, as the point $(0, b)$ and the slope-intercept form for the equation of the line to arrive at

$$y = mx + b, \text{ or}$$
$$Y_1 = \frac{3}{4}X + 1.$$

(b) Choosing $(0, -2)$ and $(2, -10)$, the slope is given by

$$m = \frac{y_2 - y_1}{x_2 - x_1}$$
$$= \frac{-10 - (-2)}{2 - 0}$$
$$= \frac{-8}{2}$$
$$= -4.$$

Use the y-axis intercept, $(0, -2)$, as the point $(0, b)$ and the slope-intercept form for the equation of the line to arrive at

$$y = mx + b, \text{ or}$$
$$Y_1 = -4X - 2.$$

8.4 EXERCISES

1. Writing exercise

2. Writing exercise

3. The first element in an ordered pair is the independent variable.

4. The relation, $\{(1,1),(1,-1),(2,4),(2,-4),(3,9),(3,-9)\}$, is not a function, since corresponding to the first component, there are more than one second components.
The domain is $\{1,2,3\}$.
The range is $\{-9,-4,-1,1,4,9\}$.

5. The relation, $\{(2,5),(3,7),(4,9),(5,11)\}$ is a function, since corresponding to each first component, there is a unique second component.
The domain is $\{2,3,4,5\}$.
The range is $\{5,7,9,11\}$.

6. The set of ordered pairs, $\{$(US, 83.9), (Japan, 90.91), (Canada, 85.26), (Britain, 83.79), (France, 87.01), (Germany, 83.12), (Italy, 82.26)$\}$, does represent a function since each country is paired with only one outcome. The domain is $\{$US, Japan, Canada, Britain, France, Germany, Italy$\}$.
The range is $\{83.9, 90.91, 85.26, 83.79, 87.01, 83.12, 82.26\}$.

7. The input-output machine would not represent a function. The domain is $(0, \infty)$. The range is $(-\infty, 0) \cup (0, \infty)$.

8. The indicated mapping is a function, since for the domain element (2), there are two associated range elements (15 and 19). The domain is $\{4, 9, 11, 17, 25\}$. The range is $\{-69, 14, 32, 47\}$.

9. The table does represent a function since each race is paired with only one outcome. The domain is $\{$Hispanic, Native American, Asian American, African American, White$\}$. The range (in millions) is $\{21.3, 1.6, 8.2, 24.6, 152.0\}$.

10. The set of ordered pairs $\{(x,y) \mid x = |y|\}$ is not a function since there are two y-values for most x-values, e.g. the ordered pairs $(2,-2)$ and $(2,2)$ satisfy the relation. The domain is $[0, \infty)$. The range is $(-\infty, \infty)$.

11. The graph represents a function, since it passes the "vertical line test" (only one intersection). The domain is $(-\infty, \infty)$. The range is $(-\infty, 4]$.

12. The graph represents a function, since it passes the "vertical line test." The domain is $[-2, 2]$. The range is $[0, 4]$.

13. The graph does not represent a function, since it does not pass the "vertical line test." The domain is $[-4, 4]$. The range is $[-3, 3]$.

14. The graph does not represents a function, since it does not pass the "vertical line test" (more than one intersection). The domain is $[3, \infty)$. The range is $(-\infty, \infty)$.

15. The equation, $y = x^2$, represent a function, since any value for x will yield exactly one value for y (for a number squared, there is only one answer). The domain is the set of reals numbers or $(-\infty, \infty)$.

16. The equation, $y = x^3$, represents a function, since any value for x will yield exactly one value for y. The domain is the set of real numbers or $(-\infty, \infty)$.

17. To determine if the equation, $x = y^2$, is or is not a function, solve for y: $y = \pm \sqrt{x}$. Since any replacement for x will yield two values for y, the equation doesn't represent a function. In order to get "real" number values for y, x can be replaced only with values such that $x \geq 0$. These values represent the domain $[0, \infty)$.

18. To determine if the equation, $x = y^4$, is or is not a function, solve for y: $y = \pm \sqrt[4]{x}$. Since any replacement for x will yield two values for y, the equation doesn't represent a function. In order to get "real" number values for y, x can be replaced only with values such that $x \geq 0$. These values represent the domain $[0, \infty)$.

19. Since ordered pairs such as $(2,0)$, $(2,1)$, and $(2,3)$, etc., all satisfy the inequality, $x + y < 4$, it does not represent a function. Note that the graph of any linear inequality, such as this one, will be a half-plane and, hence will not satisfy the "vertical line test." Any real number can be used for x. Therefore, the domain is $(-\infty, \infty)$.

20. Since ordered pairs such as $(2,0)$, $(2,1)$, and $(2,3)$, etc., all satisfy the inequality, $x - y < 3$, it does not represent a function. Note that the graph of any linear inequality, such as this one, will be a half-plane and, hence will not satisfy the "vertical line test." Any real number can be used for x. Therefore, the domain is $(-\infty, \infty)$.

21. Since any value for x in the domain of the relation, $y = \sqrt{x}$, will yield exactly one value for y, the principal square root of x, the equation represents a function. To keep y "real valued" (i.e., a real number), x must satisfy the inequality $x \geq 0$. Thus, the domain is given by $[0, \infty)$.

22. Since any value for x in the domain of the relation, $y = -\sqrt{x}$, will yield exactly one value for y, the negative square root of x, the equation represents a function. To keep y "real valued" (i.e., a real number), x must satisfy the inequality $x \geq 0$. Thus, the domain is given by $[0, \infty)$.

23. Solve the equation, $xy = 1$, for y: $y = \frac{1}{x}$. Since any value for x, in the domain, will yield exactly one value for y, the equation represents a function. Observe that in order to keep the fraction defined, $x \neq 0$, but all other (real) numbers will work. This implies that the domain is all real numbers except 0 or $(-\infty, 0) \cup (0, \infty)$.

24. Solve the equation, $xy = -3$, for y: $y = \frac{-3}{x}$. Since any value for x, in the domain, will yield exactly one value for y, the equation represents a function. Observe that in order to keep the fraction defined, $x \neq 0$, but all other (real) numbers will work. This implies that the domain is all real numbers except 0 or $(-\infty, 0) \cup (0, \infty)$.

25. The relation, $y = \sqrt{4x + 2}$, is a function, because for any choice of x in the domain, there is exactly one corresponding value of y. The domain is the set of values that satisfies $4x + 2 \geq 0$. Solving the inequality we arrive at $x \geq \frac{-2}{4}$ or $x \geq -\frac{1}{2}$. Thus, the domain is $[-\frac{1}{2}, \infty)$.

26. The relation, $y = \sqrt{9 - 2x}$, is a function, because for any choice of x in the domain, there is exactly one corresponding value of y. The domain is the set of values that satisfy $9 - 2x \geq 0$. Solving the inequality we arrive at $x \leq \frac{9}{2}$. The domain is, therefore, $(-\infty, \frac{9}{2}]$.

27. The relation, $y = \frac{2}{x-9}$, is a function, because for any choice of x in the domain, there is exactly one corresponding value of y. We may use any real value for x except those which make the denominator 0, i.e., $x \neq 9$. Thus, the domain is given by $(-\infty, 9) \cup (9, \infty)$.

28. The relation, $y = \frac{-7}{x-16}$, is a function, because for any choice of x in the domain, there is exactly one corresponding value of y. We may use any real value for x except those which make the denominator 0, i.e., $x \neq 16$. Thus, the domain is given by $(-\infty, 16) \cup (16, \infty)$.

29. (a) The values along the vertical axis, representing the dependent variable, are $[0, 3000]$.

    (b) The water is increasing between 0 and 25 hours, so the water is increasing for a total of $25 - 0 = 25$ hours. The water is decreasing between 50 and 75 hours, so the water is decreasing for a total of $75 - 50 = 25$ hours.

    (c) The graph shows that 2000 gallons of water are left in the pool after 90 hours.

    (d) The value of $f(0) = 0$, represents the amount of water in the pool at 0 hours.

30. (a) This graph is a function, since it passes the "vertical line test."

    (b) The indicated domain is $[0, 24]$.

    (c) The amount of electricity used at 8 A.M. is about 1200 megawatts.

    (d) Most electricity was used at 18 hours or 6 P.M.. Least electricity was used at 4 A.M.

31. One example: The height of a child depends on her age, so height is a function of age.

32. The correct response for the notation "$f(3)$" is B, "the value of the dependent variable when the independent variable is 3."

33. If $f(x) = 3 + 2x$, then $f(1) = 3 + 2(1) = 5$.

34. If $f(x) = 3 + 2x$, then $f(4) = 3 + 2(4) = 11$.

35. If $g(x) = x^2 - 2$, then $g(2) = 2^2 - 2 = 2$.

36. If $g(x) = x^2 - 2$, then $g(0) = 0^2 - 2 = -2$.

37. If $g(x) = x^2 - 2$, then $g(-1) = (-1)^2 - 2 = -1$.

38. If $g(x) = x^2 - 2$, then $g(-3) = (-3)^2 - 2 = 7$.

39. If $f(x) = 3 + 2x$, then $f(-8) = 3 + 2(-8) = -13$.

40. If $f(x) = 3 + 2x$, then $f(-5) = 3 + 2(-5) = -7$.

41. $f(x) = -2x + 5$

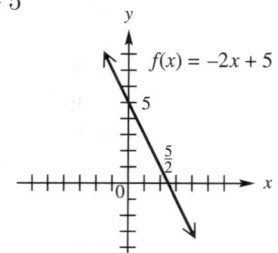

The domain and range is $(-\infty, \infty)$.

42. $f(x) = 4x - 1$

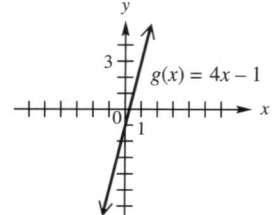

The domain and range is $(-\infty, \infty)$.

43. $h(x) = \frac{1}{2}x + 2$

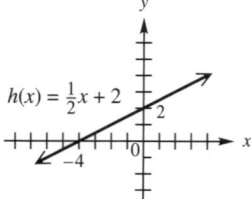

The domain and range is $(-\infty, \infty)$.

44. $F(x) = -\frac{1}{4}x + 1$

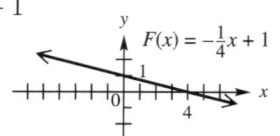

The domain and range is $(-\infty, \infty)$.

45. $G(x) = 2x$

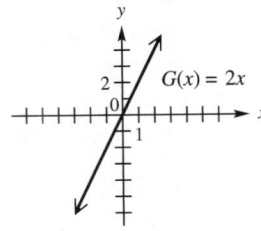

The domain and range is $(-\infty, \infty)$.

46. $H(x) = -3x$

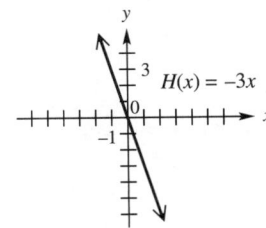

The domain and range is $(-\infty, \infty)$.

47. $f(x) = 5$

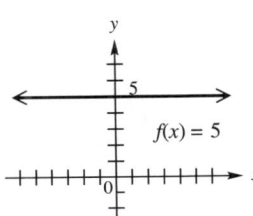

The domain is $(-\infty, \infty)$.
The range is $\{5\}$.

48. $g(x) = -4$

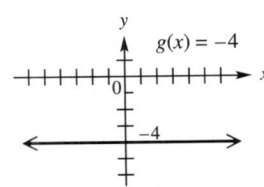

The domain is $(-\infty, \infty)$.
The range is $\{-4\}$.

49. (a) $y + 2x^2 = 3$
$y = 3 - 2x^2$
Thus, $f(x) = 3 - 2x^2$.

(b) $f(3) = 3 - 2(3)^2 = -15$

50. (a) $y - 3x^2 = 2$
$y = 2 + 3x^2$
Thus, $f(x) = 2 + 3x^2$.

(b) $f(3) = 2 + 3(3)^2 + 3 = 29$

51. (a) $4x - 3y = 8$
$-3y = 8 - 4x$
$y = \dfrac{8 - 4x}{-3}$
Thus, $f(x) = \dfrac{8 - 4x}{-3}$.

(b) $f(3) = \dfrac{8 - 4(3)}{-3} = \dfrac{4}{3}$

52. (a) $-2x + 5y = 9$
$5y = 9 + 2x$
$y = \dfrac{9 + 2x}{5}$
Thus, $f(x) = \dfrac{9 + 2x}{5}$.

(b) $f(3) = \dfrac{9 + 2(3)}{5} = 3$

53. The equation $2x + y = 4$ has a straight <u>line</u> as its graph. One point that lies on the line is $(3, \underline{-2})$. If we solve the equation for y and use function notation, we have a linear function $f(x) = \underline{-2x + 4}$. For this function, $f(3) = -2(3) + 4 = \underline{-2}$, meaning that the point $(3, -2)$ lies on the graph of the function.

54. Only A, $y = \dfrac{x - 5}{4}$, can be written in the form of a linear function, $f(x) = \frac{1}{4}x - \frac{5}{4}$.

55. (a)

x	$f(x)$
0	$0
1	$1.50
2	$3.00
3	$4.50

(b) The linear function that gives a rule for the amount charged is $f(x) = \underline{1.50x}$.

(c) The graph of this function for x, where x is an element of $\{0, 1, 2, 3\}$ is:

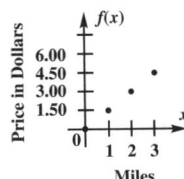

56. (a) Using $f(x) = 2.75x$, then
$$f(3) = 2.75(3) = 8.25 \text{ (dollars)}.$$

(b) Three (3) is the value of the independent variable, which represents a package weight of 3 pounds; $f(3)$ is the value of the dependent variable representing the cost to mail a 3-pound package.

(c) Using $f(x) = 2.75x$, then
$$f(5) = 2.75(5) = \$13.75.$$

57. (a) Using $h(r) = 69.09 + 2.24r$, then
$$h(56) = 69.09 + 2.24(56) = 194.53 \text{ cm}.$$

(b) Using $h(t) = 81.69 + 2.39t$, then
$$h(40) = 81.69 + 2.39(40) = 177.29 \text{ cm}.$$

(c) Using $h(r) = 61.41 + 2.32r$, then
$$h(50) = 61.41 + 2.32(50) = 177.41 \text{ cm}.$$

(d) Using $h(t) = 72.57 + 2.53t$, then
$$h(36) = 72.57 + 2.53(36) = 163.65 \text{ cm}.$$

58. Using $f(x) = (.91)(3.14)x^2$, then

(a) $f(0.8) = (.91)(3.14)(0.8)^2 = 1.83 \text{ m}^3$.

(b) $f(1.0) = (.91)(3.14)(1.0)^2 = 2.86 \text{ m}^3$.

(c) $f(1.2) = (.91)(3.14)(1.2)^2 = 4.11 \text{ m}^3$.

(d) $f(1.5) = (.91)(3.14)(1.5)^2 = 6.43 \text{ m}^3$.

59. Using $f(x) = -183x + 40034$, then

(a) $f(1) = -183(1) + 40034 = 39,851$.

(b) $f(3) = -183(3) + 40034 = 39,485$.

(c) $f(5) = -183(5) + 40034 = 39,119$.

(d) In 1992, there were 39,668 post offices in the US

60. Using $f(x) = -6324x + 305294$, then

(a) $f(3) = -6324(3) + 305294 = \$286,322$ million.

(b) $f(5) = -6324(5) + 305294 = \$273,674$ million.

(c) $f(6) = -6324(6) + 305294 = \$267,350$ million.

(d) In 1994, the defense budget was $\$279,998$ million.

61. Let x represent the number of envelopes stuffed. Then,

(a) $C(x) = .02x + 200$.

(b) $R(x) = .04x$.

(c) Set $C(x) = R(x)$ and solve for x.
$$.02x + 200 = .04x$$
$$200 = .04x - .02x$$
$$200 = .02x$$
$$x = 10,000$$

(d)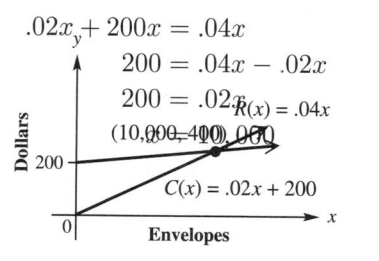

For $x < 10,000$, a loss
For $x > 10,000$, a profit

62. Let x represent the number of copies he makes. Then,

(a) $C(x) = .01x + 3500$.

(b) $R(x) = .05x$.

(c) Set $C(x) = R(x)$ and solve for x.
$$.01x + 3500 = .05x$$
$$3500 = .05x - .01x$$
$$3500 = .04x$$
$$x = 87,500$$

(d)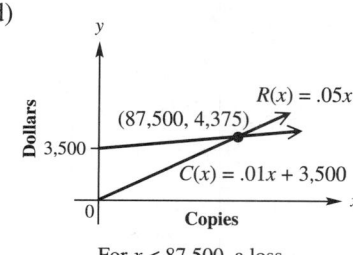

For $x < 87,500$, a loss
For $x > 87,500$, a profit

63. Let x represent the number of deliveries he makes. Then,

(a) $C(x) = 3x + 2300$.

(b) $R(x) = 5.50x$.

(c) Set $C(x) = R(x)$ and solve for x.
$$3.00x + 2300 = 5.50x$$
$$2300 = 5.50x - 3.00x$$
$$2300 = 2.50x$$
$$x = 920$$

(d)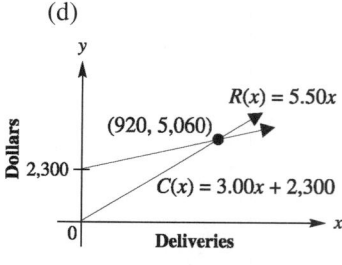

For $x < 920$, a loss
For $x > 920$, a profit

64. Let x represent the number of cakes sold. Then,

(a) $C(x) = 2.50x + 40$.

(b) $R(x) = 6.50x$.

(c) Set $C(x) = R(x)$ and solve for x.
$$2.5x + 40 = 6.50x$$
$$40 = 6.50x - 2.5x$$
$$40 = 4x$$
$$x = 10.$$

(d)

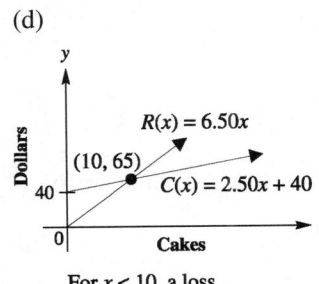

For $x < 10$, a loss
For $x > 10$, a profit

8.5 EXERCISES

1. The equation $g(x) = x^2 - 5$ matches F with a vertex $(0, -5)$ and opening up since the leading coefficient (1) is positive.

2. The equation $h(x) = -x^2 + 4$ matches B with a vertex $(0, 4)$ and opening down since the leading coefficient (-1) is less than 0.

3. The equation $F(x) = (x - 1)^2$ matches C with a vertex $(1, 0)$ and opening up since the leading coefficient (1) is positive.

4. The equation $G(x) = (x + 1)^2$ matches A with a vertex $(-1, 0)$ and opening up since the leading coefficient (1) is positive.

5. The equation $H(x) = (x - 1)^2 + 1$ matches E with a vertex $(1, 1)$ and opening up since the leading coefficient (1) is positive.

6. The equation $K(x) = (x + 1)^2 + 1$ matches D with a vertex $(-1, 1)$ and opening up since the leading coefficient (1) is positive.

7. Writing exercise

8. Writing exercise

9. Write the function, $f(x) = -3x^2$, in the form $f(x) = a(x - h)^2 + k$:
$$f(x) = -3(x - 0)^2 + 0.$$
Thus, the vertex, (h, k), is given by $(0, 0)$.

10. Write the function, $f(x) = -.5x^2$, in the form $f(x) = a(x - h)^2 + k$:
$$f(x) = -.5(x - 0)^2 + 0.$$
Thus, the vertex, (h, k), is given by $(0, 0)$.

11. Write the function, $f(x) = x^2 + 4$, in the form $f(x) = a(x - h)^2 + k$:
$$f(x) = (x - 0)^2 + 4.$$
Thus, the vertex, (h, k), is given by $(0, 4)$.

12. Write the function, $f(x) = x^2 - 4$, in the form $f(x) = a(x - h)^2 + k$:
$$f(x) = (x - 0)^2 + (-4).$$
Thus, the vertex, (h, k), is given by $(0, -4)$.

13. Write the function, $f(x) = (x - 1)^2$, in the form $f(x) = a(x - h)^2 + k$:
$$f(x) = (x - 1)^2 + 0.$$
Thus, the vertex, (h, k), is given by $(1, 0)$.

14. Write the function, $f(x) = (x + 3)^2$, in the form $f(x) = a(x - h)^2 + k$:
$$f(x) = [x - (-3)]^2 + 0.$$
Thus, the vertex, (h, k), is given by $(-3, 0)$.

15. Write the function, $f(x) = (x + 3)^2 - 4$, in the form $f(x) = a(x - h)^2 + k$:
$$f(x) = [x - (-3)]^2 + (-4).$$
Thus, the vertex, (h, k), is given by $(-3, -4)$.

16. Write the function, $f(x) = (x - 5)^2 - 8$, in the form $f(x) = a(x - h)^2 + k$:
$$f(x) = (x - 5)^2 + (-8).$$
Thus, the vertex, (h, k), is given by $(5, -8)$.

17. Writing exercise

18. The graph $f(x) = -2x^2$ opens downward since the leading coefficient (-2) is negative. It is narrower since $|-2| > 1$.

19. The graph $f(x) = -3x^2 + 1$ opens downward since the leading coefficient (-3) is negative. It is narrower since $|-3| > 1$.

20. The graph $f(x) = .5x^2$ opens upward since the leading coefficient $(.5)$ is positive. It is wider since $|.5| < 1$.

21. The graph $f(x) = \frac{2}{3}x^2 - 4$ opens upward since the leading coefficient $\left(\frac{2}{3}\right)$ is positive. It is wider since $\left|\frac{2}{3}\right| < 1$.

22. If $|a| > 1$, the graph of $f(x) = a(x - h)^2 + k$ is narrower than the graph of $y = x^2$. If $0 < |a| < 1$, the graph is wider than the graph of $y = x^2$. If a is negative, the graph opens downward while the graph of $y = x^2$ opens upward.

23. (a) With $h > 0$, $k > 0$, both coordinates of the vertex are positive, $(+, +)$, which puts the vertex in quadrant I.

    (b) With $h > 0$, $k < 0$, we have $(+, -)$ and the vertex is in quadrant IV.

(c) With $h < 0$, $k > 0$, we have $(-, +)$ and the vertex is in quadrant II.

(d) With $h < 0$, $k < 0$, we have $(-, -)$ and the vertex is in quadrant III.

24. (a) With the vertex on the y-axis, $x = 0$. Hence, $h = 0$.

(b) With the vertex on the x-axis, $y = 0$. Hence, $k = 0$.

25. Write the function, $f(x) = 3x^2$, in the form $f(x) = a(x - h)^2 + k$:

$$f(x) = 3(x - 0)^2 + 0.$$

Thus, the vertex, (h, k), is given by $(0, 0)$. Since $|a| = 3 > 1$, the graph has narrower branches than $f(x) = x^2$ and opens upward. To find two other points:

$$f(\pm 1) = 3(\pm 1)^2 = 3.$$

Thus, $(-1, 3)$ and $(1, 3)$ also lie on the graph.

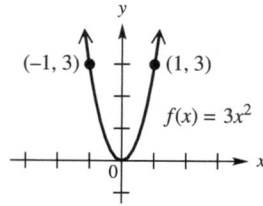

26. Write the function, $f(x) = -2x^2$, in the form $f(x) = a(x - h)^2 + k$:

$$f(x) = -2(x - 0)^2 + 0.$$

Thus, the vertex, (h, k), is given by $(0, 0)$. Since $|a| = 2 > 1$, the graph has narrower branches than $f(x) = x^2$. It opens downward, since $a < 0$. To find two other points:

$$f(\pm 1) = -2(\pm 1)^2 = -2.$$

Thus, $(-1, -2)$ and $(1, -2)$ also lie on the graph.

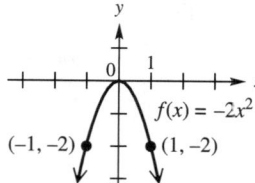

27. Write the function, $f(x) = -\frac{1}{4}x^2$, in the form $f(x) = a(x - h)^2 + k$:

$$f(x) = -\frac{1}{4}(x - 0)^2 + 0.$$

Thus, the vertex, (h, k), is given by $(0, 0)$. Since $|a| = \frac{1}{4} < 1$, the graph has wider branches than $f(x) = x^2$.

It opens downward, since $a = -\frac{1}{4} < 0$. To find two other points:

$$f(\pm 4) = -\frac{1}{4}(\pm 4)^2 = -4.$$

Thus, $(-4, -4)$ and $(4, -4)$ also lie on the graph.

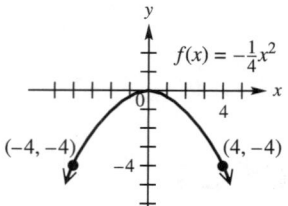

28. Write the function, $f(x) = \frac{1}{3}x^2$, in the form $f(x) = a(x - h)^2 + k$:

$$f(x) = \frac{1}{3}(x - 0)^2 + 0.$$

Thus, the vertex, (h, k), is given by $(0, 0)$. Since $|a| = \frac{1}{3} < 1$, the graph has wider branches than $f(x) = x^2$. It opens upward, since $a = \frac{1}{3} > 0$. To find two other points:

$$f(\pm 3) = \frac{1}{3}(\pm 3)^2 = 3.$$

Thus, $(-3, 3)$ and $(3, 3)$ also lie on the graph.

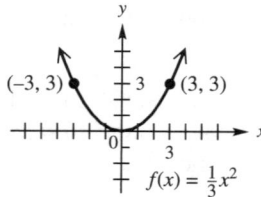

29. Write the function, $f(x) = x^2 - 1$, in the form $f(x) = a(x - h)^2 + k$:

$$f(x) = 1 \cdot (x - 0)^2 + (-1).$$

Thus, the vertex, (h, k), is given by $(0, -1)$. Since $|a| = 1$, the graph has the same branches as $f(x) = x^2$. It opens upward, since $a = 1 > 0$. To find two other points:

$$f(\pm 2) = 1 \cdot (\pm 2)^2 - 1 = 3.$$

Thus, $(-2, 3)$ and $(2, 3)$ also lie on the graph.

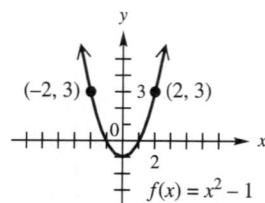

30. Write the function, $f(x) = x^2 + 3$, in the form $f(x) = a(x-h)^2 + k$:

$$f(x) = 1 \cdot (x-0)^2 + 3.$$

Thus, the vertex, (h, k), is given by $(0, 3)$. Since $|a| = 1$, the graph has the same branches as $f(x) = x^2$. It opens upward, since $a = 1 > 0$. To find two other points:

$$f(\pm 1) = 1 \cdot (\pm 1)^2 + 3 = 4.$$

Thus, $(-1, 4)$ and $(1, 4)$ also lie on the graph.

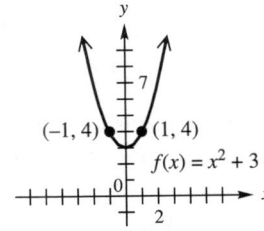

31. Write the function, $f(x) = -x^2 + 2$, in the form $f(x) = a(x-h)^2 + k$:

$$f(x) = -1 \cdot (x-0)^2 + 2.$$

Thus, the vertex, (h, k), is given by $(0, 2)$. Since $|a| = 1$, the graph has the same branches as $f(x) = x^2$. It opens downward since, $a = -1 < 0$. To find two other points:

$$f(\pm 2) = -1 \cdot (\pm 2)^2 + 2 = -2.$$

Thus, $(-2, -2)$ and $(2, -2)$ also lie on the graph.

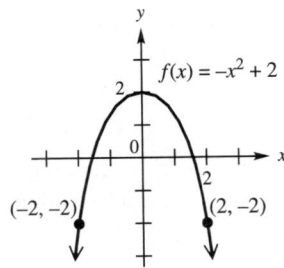

32. Write the function, $f(x) = -x^2 - 4$, in the form $f(x) = a(x-h)^2 + k$:

$$f(x) = -1 \cdot (x-0)^2 + (-4).$$

Thus, the vertex, (h, k), is given by $(0, -4)$. Since $|a| = 1$, the graph has the same branches as $f(x) = x^2$. It opens downward, since $a = -1 < 0$. To find two other points:

$$f(\pm 1) = -1 \cdot (\pm 1)^2 - 4 = -5.$$

Thus, $(-1, -5)$ and $(1, -5)$ also lie on the graph.

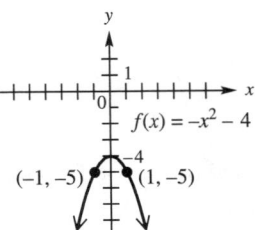

33. Write the function, $f(x) = 2x^2 - 2$, in the form $f(x) = a(x-h)^2 + k$:

$$f(x) = 2(x-0)^2 + (-2).$$

Thus, the vertex, (h, k), is given by $(0, -2)$. Since $|a| = 2 > 1$, the graph has narrower branches than $f(x) = x^2$. It opens upward, since $a > 0$. To find two other points:

$$f(\pm 1) = 2(\pm 1)^2 - 2 = 0.$$

Thus, $(-1, 0)$ and $(1, 0)$ also lie on the graph.

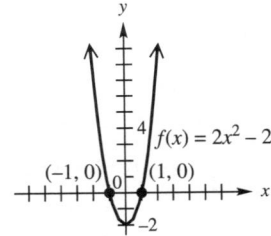

34. Write the function, $f(x) = -3x^2 + 1$, in the form $f(x) = a(x-h)^2 + k$:

$$f(x) = -3(x-0)^2 + (1).$$

Thus, the vertex, (h, k), is given by $(0, 1)$. Since $|a| = 3 > 1$, the graph has the branches which are narrower than $f(x) = x^2$. It opens downward, since $a < 0$. To find two other points:

$$f(\pm 1) = -3(\pm 1)^2 + 1 = -2.$$

Thus, $(-1, -2)$ and $(1, -2)$ also lie on the graph.

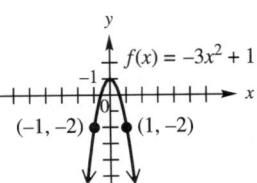

35. Write the function, $f(x) = (x-4)^2$, in the form $f(x) = a(x-h)^2 + k$:

$$f(x) = 1 \cdot (x-4)^2 + 0.$$

Thus, the vertex, (h, k), is given by $(4, 0)$. It opens upward, since $a > 0$. Find two other points, e.g. let $x = 3$ and $x = 5$:

$$f(3) = (3 - 4)^2 = 1.$$
$$f(5) = (5 - 4)^2 = 1.$$

Thus, $(3, 1)$ and $(5, 1)$ also lie on the graph.

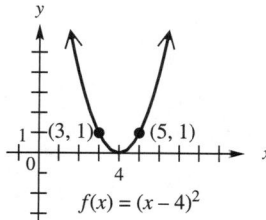

36. Write the function, $f(x) = (x - 3)^2$, in the form $f(x) = a(x - h)^2 + k$:

$$f(x) = 1 \cdot (x - 3)^2 + 0.$$

Thus, the vertex, (h, k), is given by $(3, 0)$. It opens upward, since $a > 0$. Find two other points, e.g. let $x = 2$ and $x = 4$:

$$f(2) = (2 - 3)^2 = 1,$$
$$f(4) = (4 - 3)^2 = 1.$$

Thus, $(2, 1)$ and $(4, 1)$ also lie on the graph.

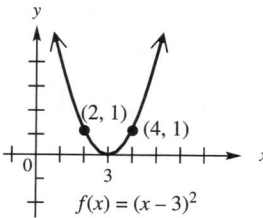

37. Write the function, $f(x) = 3(x + 1)^2$, in the form $f(x) = a(x - h)^2 + k$:

$$f(x) = 3[x - (-1)]^2 + 0.$$

Thus, the vertex, (h, k), is given by $(-1, 0)$. It opens upward, since $a > 0$. Find two other points, e.g. let $x = -2$ and $x = 0$:

$$f(-2) = 3(-2 + 1)^2 = 3,$$
$$f(0) = 3(0 + 1)^2 = 3.$$

Thus, $(-2, 3)$ and $(0, 3)$ also lie on the graph.

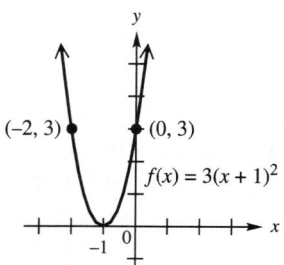

38. Write the function, $f(x) = -2(x + 1)^2$, in the form $f(x) = a(x - h)^2 + k$:

$$f(x) = -2[x - (-1)]^2 + 0.$$

Thus, the vertex, (h, k), is given by $(-1, 0)$. It opens downward, since $a = -2 < 0$. Find two other points, e.g. let $x = -2$ and $x = 0$:

$$f(-2) = -2(-2 + 1)^2 = -2,$$
$$f(0) = -2(0 + 1)^2 = -2.$$

Thus, $(-2, -2)$ and $(0, -2)$ also lie on the graph.

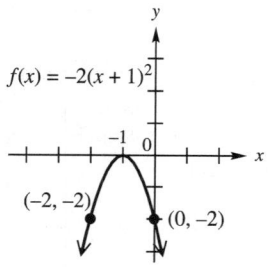

39. Write the function, $f(x) = (x + 1)^2 - 2$, in the form $f(x) = a(x - h)^2 + k$:

$$f(x) = 1 \cdot [x - (-1)]^2 - 2.$$

Thus, the vertex, (h, k), is given by $(-1, -2)$. It opens upward, since $a > 0$. Find two other points, e.g. let $x = -2$ and $x = 0$:

$$f(-2) = (-2 + 1)^2 - 2 = -1,$$
$$f(0) = (0 + 1)^2 - 2 = -1.$$

Thus, $(-2, -1)$ and $(0, -1)$ also lie on the graph.

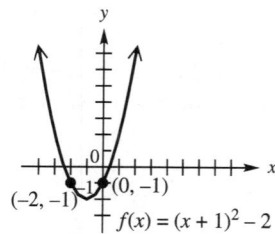

40. Write the function, $f(x) = (x-2)^2 + 3$, in the form $f(x) = a(x-h)^2 + k$:

$$f(x) = 1 \cdot (x-2)^2 + 3.$$

Thus, the vertex, (h, k), is given by $(2, 3)$. It opens upward, since $a > 0$. Find two other points, e.g. let $x = 1$ and $x = 3$:

$$f(1) = (1-2)^2 + 3 = 4,$$
$$f(3) = (3-2)^2 + 3 = 4.$$

Thus, $(1, 4)$ and $(3, 4)$ also lie on the graph.

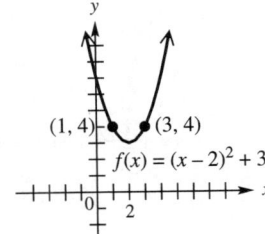

In exercises 41–46, we are finding the vertex by writing each function in the standard form of a quadratic function in order to identify directly the vertex, (h, k). Completing the square is the technique used. However, an alternate technique, using the formula $\left(-\dfrac{b}{2a}, f\left(-\dfrac{b}{2a}\right)\right)$ will yield the coordinates of each vertex.

41. Write the function, $f(x) = x^2 + 8x + 14$, in the form $f(x) = a(x-h)^2 + k$ by completing the square on x.

$$f(x) = x^2 + 8x + 14$$
$$= \left[x^2 + 8x + \left(\frac{8}{2}\right)^2\right] + 14 - \left(\frac{8}{2}\right)^2$$
$$= (x^2 + 8x + 16) + 14 - 16$$
$$= (x+4)^2 - 2$$
$$f(x) = [x - (-4)]^2 - 2$$

Thus, the vertex, (h, k), is given by $(-4, -2)$. It opens upward, since $a > 0$.

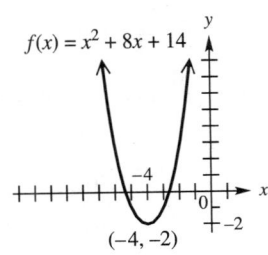

42. Write the function, $f(x) = x^2 + 10x + 23$, in the form $f(x) = a(x-h)^2 + k$ by completing the square on x.

$$f(x) = x^2 + 10x + 23$$
$$= \left[x^2 + 10x + \left(\frac{10}{2}\right)^2\right] + 23 - \left(\frac{10}{2}\right)^2$$
$$= [x^2 + 10x + 25] + 23 - 25$$
$$= (x+5)^2 - 2$$
$$f(x) = [x - (-5)]^2 - 2$$

Thus, the vertex, (h, k), is given by $(-5, -2)$. It opens upward, since $a > 0$.

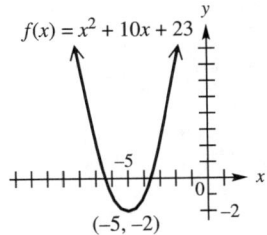

43. Write the function, $f(x) = x^2 + 2x - 4$, in the form $f(x) = a(x-h)^2 + k$ by completing the square on x.

$$f(x) = x^2 + 2x - 4$$
$$= \left[x^2 + 2x + \left(\frac{2}{2}\right)^2\right] - 4 - \left(\frac{2}{2}\right)^2$$
$$= [x^2 + 2x + 1] - 4 - 1$$
$$= (x+1)^2 - 5$$
$$f(x) = [x - (-1)]^2 - 5$$

Thus, the vertex, (h, k), is given by $(-1, -5)$. It opens upward, since $a > 0$.

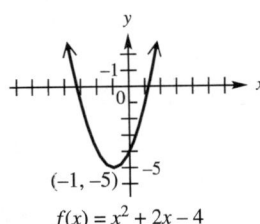

44. Write the function, $f(x) = 3x^2 - 9x + 8$, in the form $f(x) = a(x-h)^2 + k$ by completing the square on x.

$$f(x) = 3x^2 - 9x + 8$$
$$= (3x^2 - 9x) + 8$$
$$= 3(x^2 - 3x) + 8$$
$$= 3\left[x^2 - 3x + \left(\frac{3}{2}\right)^2\right] + 8 - 3\left(\frac{3}{2}\right)^2$$
$$= 3\left(x - \frac{3}{2}\right)^2 + 8 - \frac{27}{4}$$
$$= 3\left(x - \frac{3}{2}\right)^2 + \frac{32}{4} - \frac{27}{4}$$
$$f(x) = 3\left(x - \frac{3}{2}\right)^2 + \frac{5}{4}$$

Thus, the vertex, (h, k), is given by $\left(\frac{3}{2}, \frac{5}{4}\right)$. It opens upward, since $a > 0$.

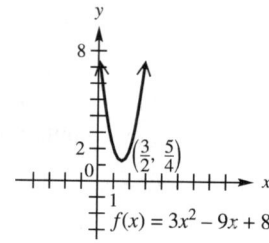

45. Write the function, $f(x) = -2x^2 + 4x + 5$, in the form $f(x) = a(x-h)^2 + k$ by completing the square on x.

$$f(x) = -2x^2 + 4x + 5$$
$$= -2(x^2 - 2x) + 5$$
$$= -2\left[x^2 - 2x + \left(\frac{2}{2}\right)^2\right] + 5 - (-2)\left(\frac{2}{2}\right)^2$$
$$= -2(x-1)^2 + 5 + 2$$
$$f(x) = -2(x-1)^2 + 7$$

Thus, the vertex, (h, k), is given by $(1, 7)$. It opens downward, since $a < 0$.

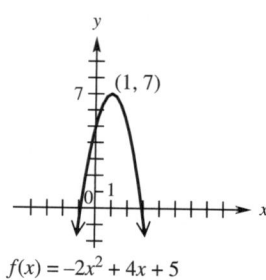

46. Write the function, $f(x) = -5x^2 - 10x + 2$, in the form $f(x) = a(x-h)^2 + k$ by completing the square on x.

$$f(x) = -5x^2 - 10x + 2$$
$$= -5(x^2 + 2x) + 2$$
$$= -5(x^2 + 2x + 1) + 2 - (-5)(1)$$
$$= -5(x+1)^2 + 7$$
$$f(x) = -5[x - (-1)]^2 + 7$$

Thus, the vertex, (h, k), is given by $(-1, 7)$. It opens downward, since $a < 0$.

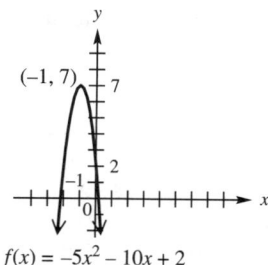

47. If we let x represent the width dimension, then $50 - x$ represents the length dimension (length plus width dimensions will equal half of the fencing needed for perimeter). Create a function representing the area.

$$A(x) = \text{length} \cdot \text{width}$$
$$= (50 - x) \cdot x$$
$$= -x^2 + 50x$$

Use the formula $\dfrac{-b}{2a}$ to create the x-value of the turning point. Thus,

$$x = \frac{-50}{2(-1)} = 25 \text{ meters.}$$

48. If we let x represent the width dimension, then $280 - 2x$ represents the length dimension (length plus two width dimensions will equal the fencing needed for perimeter). Create a function representing the area.

$$A(x) = \text{length} \cdot \text{width}$$
$$= (280 - 2x) \cdot x$$
$$= -2x^2 + 280x$$

Use the formula $\dfrac{-b}{2a}$ to create the x-value of the turning point. Thus,

$$x = \frac{-280}{2(-2)} = 70 \text{ feet.}$$

The length dimension is given by

$$280 - 2(70) = 140 \text{ feet.}$$

The area is given by
$$A(x) = -2(70)^2 + 280(70)$$
$$= 9800 \text{ square feet.}$$

49. Begin by writing the equation in standard form.
$$h = -16t^2 + 32t$$

Observe that the vertex $(t, h(t))$ of the parabola represents the time, t, when the object reaches its maximum and the maximum height, $h(t)$. Use the formula $\frac{-b}{2a}$ to find at what time the object reaches the maximum height. Thus,
$$t = \frac{-b}{2a} = \frac{-32}{2(-16)} = 1 \text{ sec.}$$

Then,
$$h(1) = -16(1)^2 + 32(1) = 16 \text{ feet}$$

is the maximum height. To find when the object hits the ground let $h = 0$. Then, solving for t, we have
$$32t - 16t^2 = 0$$
$$16t(2 - t) = 0$$
$$t = 0 \text{ and } t = 2.$$

Therefore, $t = 0$ seconds or $t = 2$ seconds. It is at ground level at $t = 0$ seconds (when it is thrown) and $t = 2$ seconds (when it hits the ground). Notice that it takes 1 second to reach maximum height and 1 more second to hit the ground.

50. The answer to both questions is given by the vertex (highest point) of the parabola suggested by the equation, $s(t) = -16t^2 + 400t$. Use the formula $\frac{-b}{2a}$ to find at what time the object reaches the maximum height. Thus,
$$t = \frac{-b}{2a} = \frac{-400}{2(-16)} = 12.5 \text{ seconds.}$$

Then,
$$s(12.5) = -16(12.5)^2 + 400(12.5)$$
$$= 2500 \text{ feet}$$

is the maximum height.

51. The answer to both questions is given by the vertex (highest point) of the parabola suggested by the equation $s(t) = -4.9t^2 + 40t$. To find the vertex, use the formula, $\left(-\frac{b}{2a}, s\left(-\frac{b}{2a}\right)\right)$, or complete the square to reach standard form. Using the formula $\frac{-b}{2a}$ to find at what time the object reaches the maximum height, we have

$$\frac{-b}{2a} = \frac{-40}{2(-4.9)} \approx 4.1 \text{ seconds.}$$

Then,
$$s(4.1) = -4.9(4.1)^2 + 40(4.1)$$
$$\approx 81.6 \text{ meters}$$

is the maximum height.

52. The answer to both questions is given by the vertex (highest point) of the parabola suggested by the equation $f(x) = 1.727x - .0013x^2$. Begin by writing the function in standard form:
$$f(x) = -.0013x^2 + 1.727x.$$

To find the vertex, use the formula, $\left(-\frac{b}{2a}, f\left(-\frac{b}{2a}\right)\right)$, or complete the square to reach standard form. Using the formula $\frac{-b}{2a}$ to find at what time the object reaches the maximum height, we have

$$t = \frac{-b}{2a} = \frac{-1.727}{2(-.0013)} \approx 644.23 \text{ seconds. Then,}$$

$$f(644.2) = -.0013(644.23)^2 + 1.727(644.23)$$
$$\approx 573.04 \text{ feet}$$

is the maximum height.

53. For the function $f(x) = .228x^2 - 2.57x + 8.97$:

    (a) The coefficient, $.228$ of x^2 is positive so the parabola opens upward causing the y-value of the vertex value to be a minimum.

    (b) To find the vertex, use the formula,
    $$\left(-\frac{b}{2a}, f\left(-\frac{b}{2a}\right)\right).$$

    $$-\frac{b}{2a} = -\frac{-2.57}{2(.228)}$$
    $$\approx 5.6. \text{ Thus, the minimum accrued in 1995.}$$

    $$f\left(-\frac{b}{2a}\right) = f(5.6)$$
    $$= .228(5.6)^2 - 2.57(5.6) + 8.97$$
    $$\approx 1.7\%.$$

54. For the function $f(x) = -.566x^2 + 5.08x + 29.2$:

    (a) The coefficient, $-.566$, of x^2 is negative, so the parabola opens downward causing the y-value of the vertex to be a maximum.

(b) To find the vertex, use the formula,
$$\left(-\frac{b}{2a}, f\left(-\frac{b}{2a}\right)\right).$$
$$-\frac{b}{2a} = -\frac{5.08}{2(-.566)}$$
$$\approx 4.5. \text{ Thus, the maximum accrued in 1994.}$$
$$f\left(-\frac{b}{2a}\right) = f(4.5)$$
$$= -.566(4.5)^2 + 5.08(4.5) + 29.2$$
$$\approx 41 \text{ million.}$$

55. For the function $f(x) = -20.57x^2 + 758.9x - 3140$:

    (a) The coefficient is negative since the parabola opens downward.

    (b) To find the vertex, use the formula,
    $$\left(-\frac{b}{2a}, f\left(-\frac{b}{2a}\right)\right).$$
    $$-\frac{b}{2a} = -\frac{758.9}{2(-20.57)}$$
    $$\approx 18.45.$$
    $$f\left(-\frac{b}{2a}\right) = f(18.45)$$
    $$= -20.57(18.45)^2 + 758.9(18.45) - 3140$$
    $$\approx 3860.$$

    Thus, the vertex is $(18.45, 3860)$.

    (c) In 2018 Social Security assets will reach their maximum value of $3860 billion.

56. (a) No, the graph is not a function since, for example, two ordered pairs correspond to the same x-value of about 15.9%. That is, it doesn't pass the "vertical line test."

    (b) The investment mix shown at the vertex is 80% U.S. and 20% foreign. This point shows a relative risk of about 15.8% (horizontal axis) and a return on investment of about 14.2% (vertical axis).

    (c) The point corresponding to 100% foreign stocks represents the riskiest investment mixture. The return on investment is about 16%.

57. (a) $R(x) = $ (no. of seats)(cost per seat)
    $$= (100 - x)(200 + 4x)$$
    $$= 20,000 + 200x - 4x^2.$$

    (b)
    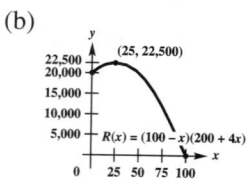

(c) The maximum revenue will occur at the vertex, since the parabola opens downward ($a < 0$). The formula, $-\frac{b}{2a}$, yields the number of unsold seats that will produce the maximum revenue while $R\left(-\frac{b}{2a}\right)$ yields the maximum revenue. Thus,
$$-\frac{b}{2a} = -\frac{200}{2(-4)}$$
$$= 25 \text{ seats.}$$

(d) The maximum revenue is
$$R(25) = 20,000 + 200(25) - 4(25)^2$$
$$= \$22,500.$$

58. (a) $R(x) = $ (no. of seats)(cost per seat)
    $$= (42 - x)(48 + 2x)$$
    $$= 2016 + 36x - 2x^2$$
    $$= -2x^2 + 36x + 2016.$$

    (b)
    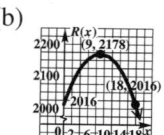

    (c) The maximum revenue will occur at the vertex, since the parabola opens downward ($a < 0$). The formula, $-\frac{b}{2a}$, yields the number of unsold seats that will produce the maximum revenue while $R\left(-\frac{b}{2a}\right)$ yields the maximum revenue. Thus,
    $$-\frac{b}{2a} = -\frac{36}{2(-2)}$$
    $$= 9 \text{ seats.}$$

    The maximum revenue is
    $$R(9) = 2016 + 36(9) - 2(9)^2$$
    $$= \$2178.$$

8.6 EXERCISES

1. For an exponential function $f(x) = a^x$, if $a > 1$, the graph <u>rises</u> from left to right. If $0 < a < 1$, the graph <u>falls</u> from left to right.

2. The y-intercept of the graph of $y = a^x$ is $\underline{(0, 1)}$.

3. The graph of the exponential function $y = a^x$ <u>does not</u> have an x-intercept, since $a^x \neq 0$ for any value of x.

4. The point $(2, \underline{243})$ is on the graph of $y = 3^{4x-3}$, since $3^{4(2)-3} = 3^5 = 243$.

8.6 EXPONENTIAL AND LOGARITHMIC FUNCTIONS, APPLICATIONS, AND MODELS

5. For a logarithmic function $g(x) = \log_a x$, if $a > 1$, the graph <u>rises</u> from left to right. If $0 < a < 1$, the graph <u>falls</u> from left to right.

6. The x-intercept of the graph of $y = \log_a x$ is $\underline{(1, 0)}$, since $a^0 = 1$ (equivalent to $0 = \log_a 1$).

7. The graph of the exponential function $g(x) = \log_a x$ <u>does not</u> have a y-intercept, since $\log_a 0$ does not exist.

8. The point $(98, 2)$ lies on the graph of $g(x) = \log_{10}(x + 2)$, since

$$\begin{aligned} g(98) &= \log_{10}(98 + 2) \\ &= \log_{10}(100) \\ &= \log_{10}(10)^2 \\ &= 2 \cdot \log_{10}(10) \\ &= 2 \cdot 1 \\ &= 2. \end{aligned}$$

9. $9^{\frac{3}{7}} \approx 2.56425419972$

10. $14^{\frac{2}{7}} \approx 2.12551979078$

11. $(.83)^{-1.2} \approx 1.25056505582$

12. $(.97)^{3.4} \approx .901620746784$

13. $\left(\sqrt{6}\right)^{\sqrt{5}} \approx 7.41309466897$

14. $\left(\sqrt{7}\right)^{\sqrt{3}} \approx 5.39357064307$

15. $\left(\dfrac{1}{3}\right)^{9.8} \approx 2.10965628481 \times 10^{-5}$
    $= .0000210965628481$

16. $\left(\dfrac{2}{5}\right)^{8.1} \approx 5.97978996117 \times 10^{-4}$
    $= .000597978996117$

17. Generate several ordered pairs that satisfy the function $f(x) = 3^x$ (e.g., $\left(-1, \frac{1}{3}\right)$, $(0, 1)$, $(1, 3)$, etc.) and plot these values. Sketch a smooth curve through these points. Remember that the graph will rise from left to right since $b > 1$ and that the x-axis acts as an asymptote.

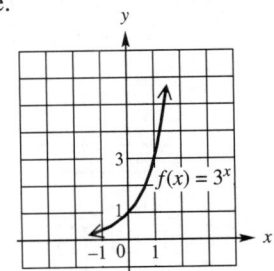

18. Generate several ordered pairs that satisfy the function $f(x) = 5^x$ (e.g., $\left(-1, \frac{1}{5}\right)$, $(0, 1)$, $(1, 5)$, etc.) and plot these values. Sketch a smooth curve through these points. Remember that the graph will rise from left to right since $b > 1$ and that the x-axis acts as an asymptote.

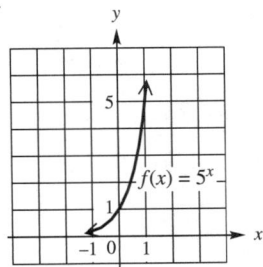

19. Generate several ordered pairs that satisfy the function $f(x) = \left(\frac{1}{4}\right)^x$ (e.g., $(-1, 4)$, $(0, 1)$, $\left(1, \frac{1}{4}\right)$, etc.) and plot these values. Sketch a smooth curve through these points. Remember that the graph will fall from left to right since $b < 1$ and that the x-axis acts as an asymptote.

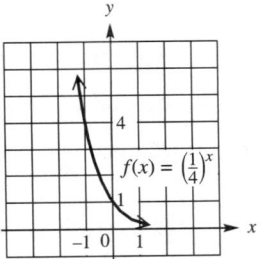

20. Generate several ordered pairs that satisfy the function $f(x) = \left(\frac{1}{3}\right)^x$ (e.g., $(-1, 3)$, $(0, 1)$, $\left(1, \frac{1}{3}\right)$, etc.) and plot these values. Sketch a smooth curve through these points. Remember that the graph will fall from left to right since $b < 1$ and that the x-axis acts as an asymptote.

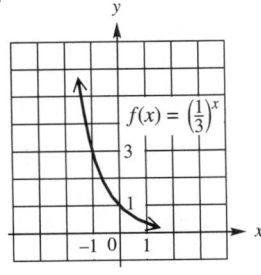

21. $e^3 \approx 20.0855369232$

22. $e^4 \approx 54.5981500331$

23. e^{-4} or $1/e^4 \approx .018315638889$

24. e^{-3} or $1/e^3 \approx .049787068368$

25. $4^2 = 16$ is equivalent to $2 = \log_4 16$.

26. $5^3 = 125$ is equivalent to $3 = \log_5 125$.

27. $\left(\frac{2}{3}\right)^{-3} = \frac{27}{8}$ is equivalent to $-3 = \log_{2/3}\left(\frac{27}{8}\right)$.

28. $\left(\frac{1}{10}\right)^{-4} = 10000$ is equivalent to $-4 = \log_{1/10}(10000)$.

29. $5 = \log_2 32$ is equivalent to $2^5 = 32$.

30. $3 = \log_4 64$ is equivalent to $4^3 = 64$.

31. $1 = \log_3 3$ is equivalent to $3^1 = 3$.

32. $0 = \log_{12} 1$ is equivalent to $12^0 = 1$.

33. $\ln 4 \approx 1.38629436112$

34. $\ln 6 \approx 1.79175946923$

35. $\ln .35 \approx -1.0498221245$

36. $\ln 2.45 \approx .896088024557$

37. By inspecting the graph, the year 2000 corresponds

    (a) to an approximate $.5°C$ increase on the exponential curve, and

    (b) to an approximate $.35°C$ increase on the linear graph.

38. By inspecting the graph, the year 2010 corresponds

    (a) to an approximate $1.0°C$ increase on the exponential curve, and

    (b) to an approximate $.4°C$ increase on the linear graph.

39. By inspecting the graph, the year 2020 corresponds

    (a) to an approximate $1.6°C$ increase on the exponential curve, and

    (b) to an approximate $.5°C$ increase on the linear graph.

40. By inspecting the graph, the year 2040 corresponds

    (a) to an approximate $3.0°C$ increase on the exponential curve, and

    (b) to an approximate $.7°C$ increase on the linear graph.

41. Since $g(x) = \log_3 x$ is the inverse of $f(x) = 3^x$ (Exercise 17), we can reflect the graph of $f(x)$ across the line $y = x$ to get the graph of g(x). This may be accomplished by interchanging the roll of the x and y values that were generated to graph $f(x)$. Thus, some generated ordered pairs for g(x) would include $\left(\frac{1}{3}, -1\right)$, $(1, 0)$, $(3, 1)$, etc.

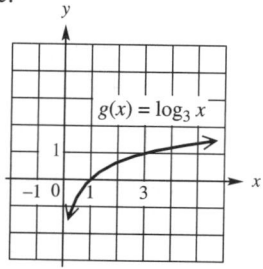

42. Since $g(x) = \log_5 x$ is the inverse of $f(x) = 5^x$ (Exercise 18), we can reflect the graph of $f(x)$ across the line $y = x$ to get the graph of g(x). This may be accomplished by interchanging the roll of the x and y values that were generated to graph $f(x)$. Thus, some generated ordered pairs for g(x) would include $\left(\frac{1}{5}, -1\right)$, $(1, 0)$, $(5, 1)$, etc. Observe that this will now give an asymptote along the y-axis.

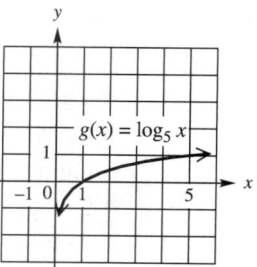

43. Since $g(x) = \log_{1/4} x$ is the inverse of $f(x) = \left(\frac{1}{4}\right)^x$ (Exercise 19), we can reflect the graph of $f(x)$ across the line $y = x$ to get the graph of $g(x)$. This may be accomplished by interchanging the roll of the x and y values that were generated to graph $f(x)$. Thus, some generated ordered pairs for $g(x)$ would include $\left(\frac{1}{4}, 1\right)$, $(1, 0)$, $(4, -1)$, etc. Observe that this will now give an asymptote along the y-axis.

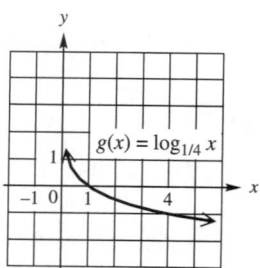

44. Since $g(x) = \log_{1/3} x$ is the inverse of $f(x) = \left(\frac{1}{3}\right)^x$ (Exercise 20), we can reflect the graph of $f(x)$ across the line $y = x$ to get the graph of $g(x)$. This may be accomplished by interchanging the roll of the x and y values that were generated to graph $f(x)$. Thus, some generated ordered pairs for $g(x)$ would include $\left(\frac{1}{3}, 1\right)$, $(0, 1)) (-1, 3))$ etc. Observe that this will now give an asymptote along the y-axis.

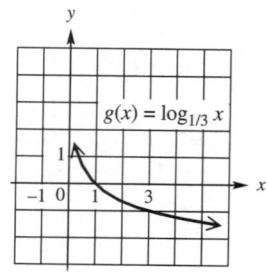

8.6 EXPONENTIAL AND LOGARITHMIC FUNCTIONS, APPLICATIONS, AND MODELS

45. Using the compound interest formula $A = P\left(1 + \frac{r}{n}\right)^{nt}$ where $P = \$4{,}292$, $r = 6\% = .06$, $n = 1$ and $t = 10$ years, we have

$$A = 4292\left(1 + \frac{.06}{1}\right)^{1 \cdot 10}$$
$$= 4292(1.06)^{10}$$
$$\approx 4292 \cdot (1.79085)$$
$$\approx \$7686.32.$$

46. Using the compound interest formula $A = P\left(1 + \frac{r}{n}\right)^{nt}$ where $P = \$8906.54$, $r = 5\% = .05$, $n = 2$ and $t = 9$ years, we have

$$A = 8906.54\left(1 + \frac{.05}{2}\right)^{2 \cdot 9}$$
$$= 8906.54(1.025)^{18}$$
$$\approx 8906.54 \cdot (1.55966)$$
$$\approx \$13{,}891.16.$$

47. Use the compound interest formula $A = P\left(1 + \frac{r}{n}\right)^{nt}$ where $P = \$56{,}780$, $r = 5.3\% = .053$, $n = 4$ and $t = \frac{23}{4}$ years. Note that the total number of compoundings (nt) is 23.

$$A = 56780\left(1 + \frac{.053}{4}\right)^{23}$$
$$= 56780(1.01325)^{23}$$
$$\approx 56780 \cdot (1.353574255)$$
$$\approx \$76{,}855.95$$

48. Using the compound interest formula $A = P\left(1 + \frac{r}{n}\right)^{nt}$ where $P = \$45{,}788$, $r = 6\% = .06$, $n = 365$ and $t = 11$ years,

$$A = 45788\left(1 + \frac{.06}{365}\right)^{365 \cdot 11}$$
$$= 45788(1.000164384)^{4015}$$
$$\approx 45788 \cdot (1.93487393)$$
$$\approx \$88{,}585.47.$$

49. Using the continuous compounding formula $A = Pe^{rt}$ where $P = \$25{,}000$ and $r = 5\% = .05$,

(a) For $n = 1$:
$$A = 25000e^{.05(1)}$$
$$\approx 25000 \cdot (1.05127)$$
$$\approx \$26{,}281.78.$$

(b) For $n = 5$:
$$A = 25000e^{.05(5)}$$
$$= 25000e^{.25}$$
$$\approx 25000 \cdot (1.284025)$$
$$\approx \$32{,}100.64.$$

(c) For $n = 10$:
$$A = 25000e^{.05(10)}$$
$$= 25000e^{.5}$$
$$\approx 25000 \cdot (1.64872)$$
$$\approx \$41{,}218.03.$$

50. For the first plan we use the compounding formula $A = P\left(1 + \frac{r}{n}\right)^{nt}$ with $P = \$60{,}000$, $n = 4$, $r = 7\% = .07$, and $t = 5$ years.

$$A = 60000\left(1 + \frac{.07}{4}\right)^{4 \cdot 5}$$
$$= 60000(1.0175)^{20}$$
$$\approx 60000 \cdot (1.414778196)$$
$$\approx \$84{,}886.69$$

For the second plan we use the compounding formula $A = Pe^{rt}$ (since we are compounding continuously) with $P = \$60{,}000$, $r = 6.75\% = .0675$, and $t = 5$ years.

$$A = 60000e^{.0675(5)}$$
$$= 60000e^{.3375}$$
$$\approx 60000 \cdot (1.401339608)$$
$$\approx \$84{,}086.38$$

The difference is $\$84{,}886.69 - \$84{,}086.38 = \$800.31$. Thus, the 7% plan earns $\$800.31$ more in the five years.

51. Using the model
$$f(x) = 132359(1.0124)^{-x},$$
the approximate number of emissions in

(a) 1970, where $x = 0$, is given by
$$f(0) = 132359(1.0124)^{-0}$$
$$= 132359(1)$$
$$= 132{,}359 \text{ thousand tons.}$$

(b) 1995, where $x = 25$, is given by
$$f(25) = 132359(1.0124)^{-25}$$
$$\approx 132359(.7348)$$
$$\approx 97{,}264 \text{ thousand tons.}$$

(c) 1998, where $x = 28$, is given by
$$f(28) = 132359(1.0124)^{-28}$$
$$\approx 132359(.70817)$$
$$\approx 93{,}733 \text{ thousand tons.}$$

The actual amount of emissions in 1998 was 89,454 thousand tons, or somewhat less than that predicted by the model.

52. Using the model
$$f(x) = 157.28(1.0204)^x,$$
the approximate gross waste in

(a) 1980, where $x = 0$, is given by
$$f(0) = 157.28(1.0204)^0$$
$$= 157.28(1)$$
$$= 157.28 \text{ million tons.}$$

(b) 1985, where $x = 5$, is given by
$$f(5) = 157.28(1.0204)^5$$
$$\approx 157.28(1.06247)$$
$$\approx 173.99 \text{ million tons.}$$

(c) 1999, where $x = 19$, is given by
$$f(33) = 157.28(1.0204)^{19}$$
$$\approx 157.28(1.4677)$$
$$\approx 230.84 \text{ million tons.}$$

The actual amount of emissions in 1993 was 229.9 million tons. This is a slightly less amount that the model provides.

53. Using the model
$$P(x) = 176000(2)^{.008x},$$
the approximate total population in

(a) 2000, where $x = 0$, is given by
$$P(0) = 176000(2)^{.008(0)}$$
$$\approx 176000(1)$$
$$\approx 176,000 \text{ people.}$$

(b) 2025, where $x = 25$, is given by
$$P(25) = 176000(2)^{.008(25)}$$
$$\approx 176000(1.1487)$$
$$\approx 202,171 \text{ people.}$$

(c) The population increased by
$$202171 - 176000 = 26171.$$

Thus, the percentage increase is given by
$$\frac{26171}{176000} \approx .1487 \approx 15\%.$$

54. Using the model
$$f(x) = 146250(2)^{.0176x},$$
the approximate total population in

(a) 2000, where $x = 0$, is given by
$$f(0) = 146250(2)^{.0176(0)}$$
$$= 146250(1)$$
$$= 146,250 \text{ people.}$$

(b) 2025, where $x = 25$, is given by
$$f(25) = 146250(2)^{.0176(25)}$$
$$\approx 146250(1.3566)$$
$$\approx 198,403 \text{ people.}$$

(c) The population increased by
$$198403 - 146250 = 52153.$$

Thus, the percentage increase is given by
$$\frac{52153}{146250} \approx .3566 \approx 36\%.$$

55. Use the model,
$$B(x) = 8768e^{.072x},$$
to find the approximate consumer expenditure on all types of books in the US in 1998.

For the year 1998, $x = 18$, and
$$B(18) = 8768e^{.072(18)}$$
$$\approx 8768(3.65468796)$$
$$\approx 32,044 \text{ million dollars.}$$

56. Use the model,
$$D(x) = 815{,}427e^{.0137x},$$
to find the approximate total number of bachelor degrees earned in the US For the year 1994, $x = 24$, thus
$$D(24) = 815427e^{.0137(24)}$$
$$\approx 815427(1.389299968)$$
$$\approx 1,132,873 \text{ degrees.}$$

57. Use the model,
$$A(t) = 100(3.2)^{-.5t},$$
to approximate the amount of radioactive material present at any time t.

(a) The initial measurement amount, $t = 0$, is given by
$$A(0) = 100(3.2)^{-.5(0)}$$
$$= 100(1)$$
$$= 100 \text{ grams.}$$

(b) The amount 2 months later, $t = 2$, is given by
$$A(0) = 100(3.2)^{-.5(2)}$$
$$= 100(.3125)$$
$$= 31.25 \text{ grams.}$$

(c) The amount 10 months later, $t = 10$, is given by
$$A(0) = 100(3.2)^{-.5(10)}$$
$$\approx 100(.0029802322)$$
$$\approx .30 \text{ grams.}$$

8.6 EXPONENTIAL AND LOGARITHMIC FUNCTIONS, APPLICATIONS, AND MODELS

58. Use the model,
$$V(t) = 5000(2)^{-.15t},$$
to approximate the value at any time t.

(a) The original value, $t = 0$, is given by
$$V(0) = 5000(2)^{-.15(0)}$$
$$= 5000(1)$$
$$= \$5000.$$

(b) Five years later, $t = 5$, the value is
$$V(5) = 5000(2)^{-.15(5)}$$
$$\approx 5000(.5946035575)$$
$$\approx \$2973.$$

(c) Ten years later, $t = 10$, the value is
$$V(10) = 5000(2)^{-.15(10)}$$
$$\approx 5000(.3535533906)$$
$$\approx \$1768.$$

59. Using the model,
$$f(x) = 11.34 + 317.01 \log_2 x,$$
to the approximate number of hazardous waste sites in the US:

(a) In 1984, where $x = 4$, the number is given by
$$f(4) = 11.34 + 317.01 \log_2 4.$$

Note that you will want to

(1) Use the change of base formula to write the $\log_2$ either ln (base e) or log (base 10) in order to use your calculator: Choosing base 10:

$$\log_2 4 = \frac{\log_{10} 4}{\log_{10} 2}$$
$$\approx \frac{.6020599913}{.3010299957}$$
$$= 2.$$

Or (2) simplify $\log_2 4$ by properties of logarithms as follows:
$$\log_2 4 = \log_2 (2^2)$$
$$= 2\log_2 (2)$$
$$= 2(1)$$
$$= 2.$$

Thus,
$$f(4) = 11.34 + 317.01(2)$$
$$= 645 \text{ sites}.$$

(b) In 1988, where $x = 8$, the number is given by
$$f(8) = 11.34 + 317.01 \log_2 8.$$

As above, we will want to choose a method to evaluate or simplify $\log_2 8$. Simplifying by properties of logarithms results in the following:
$$\log_2 8 = \log_2 (2^3)$$
$$= 3\log_2 (2)$$
$$= 3(1)$$
$$= 3.$$

Thus,
$$f(8) = 11.34 + 317.01(3)$$
$$= 962 \text{ sites}.$$

60. Using the model,
$$f(x) = 51.47 + 6.044 \log_2 x,$$
for the approximate world wide consumption of natural gas:

(a) In 1980, where $x = 1$, the number is given by
$$f(1) = 51.47 + 6.044 \log_2 1$$
$$= 51.47 + 6.044(0) \quad \textit{Remember, } \log_b(1) = 0$$
$$= 51.47 \text{ trillion cubic feet}.$$

(b) In 1987, where $x = 8$, the number is given by
$$f(8) = 51.47 + 6.044 \log_2 8.$$
$$= 51.47 + 6.044 \log_2 (2)^3$$
$$= 51.47 + 6.044(3) \quad \textit{As in Exercise 59}$$
$$= 69.602 \text{ trillion cubic feet}.$$

61. Using the exponential form of the Richter scale model,
$$x = 10^R x_0:$$
For the Lander's earthquake, the intensity is given by
$$x = 10^{7.3} x_0.$$
For the Northbridge earthquake, the intensity is given by
$$x = 10^{6.7} x_0.$$
Thus, the Lander's earthquake is about
$$\frac{10^{7.3} x_0}{10^{6.7} x_0} = \frac{10^{7.3}}{10^{6.7}}$$
$$= 10^{7.3 - 6.7}$$
$$= 10^{.6}$$
$$\approx 4 \text{ times as powerful}.$$

62. Using the exponential form of the Richter scale model,
$$x = 10^R x_0:$$
For the Lander's earthquake, the intensity is given by
$$x = 10^{7.3} x_0.$$

For the smallest rated earthquake ($R = 4.8$), the intensity is given by

$$x = 10^{4.8}x_0.$$

Thus, the Lander's earthquake is about

$$\frac{10^{7.3}x_0}{10^{4.8}x_0} = \frac{10^{7.3}}{10^{4.8}}$$
$$= 10^{7.3-4.8}$$
$$= 10^{2.5}$$
$$\approx 300 \text{ times as powerful.}$$

8.7 EXERCISES

1. (a) ABC dominated evening news from 1989 until 1997 since their graph was above the others.

   (b) ABC's dominance ended in 1997. NBC equaled ABC's share that same year in 1997 at 17%.

   (c) ABC and CBS had an equal share in 1989 of 20% and in 1998 of 16%.

   (d) NBC and ABC had a most recent equal share in 2000 of 16% or (2000, 16).

   (e) The viewership of the three networks has generally declined during these years.

2. (a) In 1991 they both reached the level of about 350 million.

   (b) (1987, 100 million)

   (c) Cassette production stabilized and remained constant from 1988 until 1990.

   (d) CD production generally increased during these years. The slope would be positive.

   (e) The slope would be negative since the graph of cassette production is decreasing.

3. $x + y = 6$
   $x - y = 4$

   To decide if $(5, 1)$ is a solution of the system, substitute 5 for x and 1 for y in each equation to see if the results are true statements.

   $$5 + 1 = 6$$
   $$6 = 6 \quad True$$
   $$5 - 1 = 4$$
   $$4 = 4 \quad True$$

   Therefore, $(5, 1)$ is a solution to the above system.

4. $x - y = 17$
   $x + y = -1$

   To decide if $(8, -9)$ is a solution of the system, substitute 8 for x and -9 for y in each equation to see if the results are true statements.

   $$8 - (-9) = 17$$
   $$17 = 17 \quad True$$
   $$8 + -9 = -1$$
   $$-1 = -1 \quad True$$

   Therefore, $(8, -9)$ is a solution to the above system.

5. $2x - y = 8$
   $3x + 2y = 20$

   To decide if $(5, 2)$ is a solution of the system, substitute 5 for x and 2 for y in each equation to see if the results are true statements.

   $$2(5) - 2 = 8$$
   $$8 = 8 \quad True$$
   $$3(5) + 2(2) = 20$$
   $$19 = 20 \quad False$$

   Therefore, $(5, 2)$ is not a solution to the system.

6. $3x - 5y = -12$
   $x - y = 1$

   To decide if $(-1, 2)$ is a solution of the system, substitute -1 for x and 2 for y in each equation to see if the results are true statements.

   $$3(-1) - 5(2) = -12$$
   $$-13 = -12 \quad False$$
   $$(-1) - 2 = 1$$
   $$-3 = 1 \quad False$$

   Therefore, $(-1, 2)$ is not a solution to the system.

7. $x + y = 4$
   $2x - y = 2$

   Graph the line $x + y = 4$ through its intercepts $(0, 4)$ and $(4, 0)$ and the line $2x - y = 2$ through its intercepts $(0, -2)$ and $(1, 0)$.

   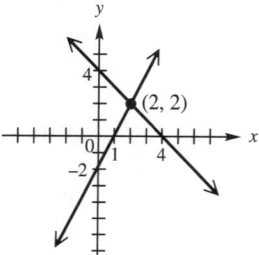

   The lines appear to intersect at $(2, 2)$. Check this ordered pair in the system.

   $$2 + 2 = 4$$
   $$4 = 4 \quad True$$
   $$2(2) - 2 = 2$$
   $$2 = 2 \quad True$$

   Thus, $\{(2, 2)\}$ is the solution.

8. $x + y = -5$
$-2x + y = 1$

Graph the line $x + y = -5$ through its intercepts $(0, -5)$ and $(-5, 0)$ and the line $-2x + y = 1$ through its intercepts $(0, 1)$ and $\left(\frac{-1}{2}, 0\right)$.

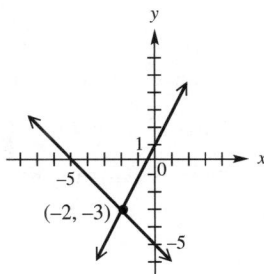

The lines appear to intersect at $(-2, -3)$. Check this ordered pair in the system.
$$-2 + (-3) = -5$$
$$-5 = -5 \quad \textit{True}$$
$$-2(-2) + -3 = 1$$
$$1 = 1 \quad \textit{True}$$

The solution set is $\{(-2, -3)\}$.

9. $2x - 5y = 11$
$3x + y = 8$

Multiply the second equation by 5 then add to the first equation.
$$2x - 5y = 11$$
$$15x + 5y = 40$$
$$\overline{17x = 51}$$
$$x = 3$$

Substitute this value for x into the first equation and solve for y.
$$2(3) - 5y = 11$$
$$-5y = 11 - 6$$
$$-5y = 5$$
$$y = -1$$

Since the ordered pair $(3, -1)$ satisfies both equations, it checks. The solution set is $\{(3, -1)\}$.

10. $-2x + 3y = 1$
$-4x + y = -3$

Multiply the second equation by -3 then add to the first equation.
$$-2x + 3y = 1$$
$$12x - 3y = 9$$
$$\overline{10x = 10}$$
$$x = 1$$

Substitute this value for x into the first equation and solve for y.
$$-2(1) + 3y = 1$$
$$3y = 1 + 2$$
$$3y = 3$$
$$y = 1$$

Since the ordered pair $(1, 1)$ satisfies both equations, it checks. The solution set is $\{(1, 1)\}$.

11. $3x + 4y = -6$
$5x + 3y = 1$

Multiply the first equation by 3 and the second by -4 then add.
$$9x + 12y = -18$$
$$-20x - 12y = -4$$
$$\overline{-11x = -22}$$
$$x = 2$$

Substitute this value for x into the first equation and solve for y.
$$3(2) + 4y = -6$$
$$4y = -6 - 6$$
$$4y = -12$$
$$y = -3$$

Since the ordered pair $(2, -3)$ satisfies both equations, it checks. The solution set is $\{(2, -3)\}$.

12. $4x + 3y = 1$
$3x + 2y = 2$

Multiply the first equation by 3 and the second by -4 then add.
$$12x + 9y = 3$$
$$-12x - 8y = -8$$
$$\overline{y = -5}$$

Substitute this value for y into the first equation and solve for x.
$$4x + 3(-5) = 1$$
$$4x = 1 + 15$$
$$4x = 16$$
$$x = 4$$

Since the ordered pair $(4, -5)$ satisfies both equations, it checks. The solution set is $\{(4, -5)\}$.

13. $3x + 3y = 0$
$4x + 2y = 3$

Simplify the first equation by dividing both sides by 3.
$$x + y = 0$$
$$4x + 2y = 3$$

Multiply the first equation by -2 then add. Solve the resulting equation for x.

$$-2x - 2y = 0$$
$$4x + 2y = 3$$
$$\overline{2x = 3}$$
$$x = \frac{3}{2}$$

Substitute this value for x into the equation $x + y = 0$ and solve for y.

$$\frac{3}{2} + y = 0$$
$$y = -\frac{3}{2}$$

Since the ordered pair $\left(\frac{3}{2}, -\frac{3}{2}\right)$ satisfies both equations, it checks. The solution set is $\left\{\left(\frac{3}{2}, -\frac{3}{2}\right)\right\}$.

14. $8x + 4y = 0$
    $4x - 2y = 2$

To simplify the first and second equation, divide the first by 4 and the second by 2.

$$2x + y = 0$$
$$2x - y = 1$$

Add the two equations to eliminate the y-term. Solve the resulting equation for x.

$$4x = 1$$
$$x = \frac{1}{4}$$

Substitute this value for x into the equation $2x + y = 0$ and solve for y.

$$2\left(\frac{1}{4}\right) + y = 0$$
$$\frac{1}{2} + y = 0$$
$$y = -\frac{1}{2}$$

Since the ordered pair $\left(\frac{1}{4}, -\frac{1}{2}\right)$ satisfies both equations, it checks. The solution set is $\left\{\left(\frac{1}{4}, -\frac{1}{2}\right)\right\}$.

15. $7x + 2y = 6$
    $-14x - 4y = -12$

To simplify the second equation, divide by 2. Then add both equations.

$$7x + 2y = 6$$
$$-7x - 2y = -6$$
$$\overline{0 = 0 \ \text{True}}$$

The equations are dependent and hence have an infinite number of solutions. Solve the first equation for x in term of y.

$$7x + 2y = 6$$
$$7x = 6 - 2y$$
$$x = \frac{6 - 2y}{7}$$

We will leave the answer as the ordered pair $\left(\frac{6-2y}{7}, y\right)$. The solution set is $\left\{\left(\frac{6-2y}{7}, y\right)\right\}$. Note that this allows one to create a specific solution as an ordered pair, for any real number replacement of y.

16. $x - 4y = 2$
    $4x - 16y = 8$

To simplify the second equation, divide by 4. Then multiply second equation by -1 and add both equations.

$$x - 4y = 2$$
$$-x + 4y = -2$$
$$\overline{0 = 0 \ \text{True}}$$

The equations are dependent and hence have an infinite number of solutions. Solve the first equation for x in term of y.

$$x - 4y = 2$$
$$x = 2 + 4y$$

We will leave the answer as the ordered pair $(2 + 4y, y)$. The solution set is $\{(2 + 4y, y)\}$. Note that this allows one to create a specific solution as an ordered pair, for any real number replacement of y.

17. $\frac{x}{2} + \frac{y}{3} = -\frac{1}{3}$
    $\frac{x}{2} + 2y = -7$

Multiply the first equation by 6 and the second by -6 in order to eliminate fractions and to get opposite coefficients for the x-terms. Add the two resulting equations and solve for y.

$$3x + 2y = -2$$
$$-3x - 12y = 42$$
$$\overline{-10y = 40}$$
$$y = -4$$

Substitute this value for y in the equation $3x + 2y = -2$ and solve for x.

$$3x + 2(-4) = -2$$
$$3x - 8 = -2$$
$$3x = 6$$
$$x = 2$$

Since the ordered pair $(2, -4)$ satisfies both equations, it checks. The solution set is $\{(2, -4)\}$.

18. $\dfrac{x}{5} + y = \dfrac{6}{5}$

$\dfrac{x}{10} + \dfrac{y}{3} = \dfrac{5}{6}$

Multiply the first equation by 10 and the second by -30 in order to eliminate fractions and to get opposite coefficients for the y-terms. Add the two resulting equations and solve for x.
$$2x + 10y = 12$$
$$-3x - 10y = -25$$
$$\overline{-x = -13}$$
$$x = 13$$

Substitute this value for x in the equation $2x + 10y = 12$ or the equivalent simpler equation $x + 5y = 6$ (resulting from division by 2) and solve for y.
$$x + 5y = 6$$
$$13 + 5y = 6$$
$$5y = -7$$
$$y = -\dfrac{7}{5}$$

Since the ordered pair $\left(13, -\dfrac{7}{5}\right)$ satisfies both equations, it checks. The solution set is $\left\{\left(13, -\dfrac{7}{5}\right)\right\}$.

19. $5x - 5y = 3$

$x - y = 12$

Multiply the second equation by -5 then add both equations.
$$5x - 5y = 3$$
$$-5x + 5y = 60$$
$$\overline{0 = 63 \quad \text{False}}$$

The equations are inconsistent and, thus, have no solutions. They are parallel lines with no points of intersection. The solution set is the empty set $\emptyset$.

20. $2x - 3y = 7$

$-4x + 6y = 14$

Multiply the first equation by 2 then add both equations.
$$4x - 6y = 14$$
$$-4x + 6y = 14$$
$$\overline{0 = 28 \quad \text{False}}$$

The equations are inconsistent and, thus, have no solutions. They are parallel lines with no points of intersection. The solution set is the empty set $\emptyset$.

21. $4x + y = 6$

$y = 2x$

Substitute the value for y from the second equation into the first equation and solve for x.
$$4x + (2x) = 6$$
$$6x = 6$$
$$x = 1$$

Substitute this value of x into the equation $y = 2x$ and solve for y.
$$y = 2(1)$$
$$y = 2$$

Since the ordered pair $(1, 2)$ satisfies both equations, it checks. The solution set is $\{(1, 2)\}$.

22. $2x - y = 6$

$y = 5x$

Substitute the value for y from the second equation into the first equation and solve for x.
$$2x - (5x) = 6$$
$$-3x = 6$$
$$x = -2$$

Substitute this value of x into the equation $y = 5x$ and solve for y.
$$y = 5(-2)$$
$$y = -10$$

Since the ordered pair $(-2, -10)$ satisfies both equations, it checks. The solution set is $\{(-2, -10)\}$.

23. $3x - 4y = -22$

$-3x + y = 0$

Solve the second equation for y.
$$-3x + y = 0$$
$$y = 3x$$

Substitute this value for y into the first equation and solve for x.
$$3x - 4(3x) = -22$$
$$3x - 12x = -22$$
$$-9x = -22$$
$$x = \dfrac{22}{9}$$

Replace $x = \dfrac{22}{9}$ in the equation $y = 3x$ and evaluate for y.
$$y = 3\left(\dfrac{22}{9}\right)$$
$$y = \dfrac{22}{3}$$

Since the ordered pair $\left(\dfrac{22}{9}, \dfrac{22}{3}\right)$ satisfies both equations, it checks. The solution set is $\left\{\left(\dfrac{22}{9}, \dfrac{22}{3}\right)\right\}$.

24. $-3x + y = -5$
$x + 2y = 0$

Solve the second equation for x.

$$x + 2y = 0$$
$$x = -2y$$

Substitute this value for x into the first equation and solve for y.

$$-3(-2y) + y = -5$$
$$6y + y = -5$$
$$7y = -5$$
$$y = -\frac{5}{7}$$

Replace $y = -\frac{5}{7}$ in the equation $x = -2y$ and evaluate x.

$$x = -2\left(-\frac{5}{7}\right)$$
$$x = \frac{10}{7}$$

Since the ordered pair $\left(\frac{10}{7}, -\frac{5}{7}\right)$ satisfies both equations, it checks. The solution set is $\left\{\left(\frac{10}{7}, -\frac{5}{7}\right)\right\}$.

25. $-x - 4y = -14$
$2x = y + 1$

Solve the second equation for y.

$$2x = y + 1$$
$$2x - 1 = y$$

Substitute this value for y into the first equation and solve for x.

$$-x - 4(2x - 1) = -14$$
$$-x - 8x + 4 = -14$$
$$-9x = -18$$
$$x = 2$$

Substitute $x = 2$ into the equation $y = 2x - 1$ and evaluate y.

$$y = 2(2) - 1$$
$$y = 3$$

Since the ordered pair $(2, 3)$ satisfies both equations, it checks. The solution set is $\{(2, 3)\}$.

26. $-3x - 5y = -17$
$4x = y - 8$

Solve the second equation for y.

$$4x = y - 8$$
$$4x + 8 = y$$

Substitute this value of y into the first equation and solve for x.

$$-3x - 5(4x + 8) = -17$$
$$-3x - 20x - 40 = -17$$
$$-23x = 23$$
$$x = -1$$

Substitute $x = -1$ into the equation $y = 4x + 8$ and evaluate for y.

$$y = 4(-1) + 8$$
$$y = 4$$

Since the ordered pair $(-1, 4)$ satisfies both equations, it checks. The solution set is $\{(-1, 4)\}$.

27. $5x - 4y = 9$
$3 - 2y = -x$

Solve the second equation for x by multiplication of both sides by -1.

$$3 - 2y = -x$$
$$-3 + 2y = x$$

Substitute this value of x into the first equation and solve for y.

$$5(-3 + 2y) - 4y = 9$$
$$-15 + 10y - 4y = 9$$
$$6y = 24$$
$$y = 4$$

Substitute this value of y into the equation $x = -3 + 2y$ and evaluate for x.

$$x = -3 + 2(4)$$
$$x = 5$$

Since the ordered pair $(5, 4)$ satisfies both equations, it checks. The solution set is $\{(5, 4)\}$.

28. $6x - y = -9$
$4 + 7x = -y$

Solve the second equation for y through multiplication of both sides by -1.

$$4 + 7x = -y$$
$$-4 - 7x = y$$

Substitute this value for y into the first equation and solve for x.

$$6x - (-4 - 7x) = -9$$
$$6x + 4 + 7x = -9$$
$$13x = -13$$
$$x = -1$$

Substitute $x = -1$ into the equation $y = -4 - 7x$ and evaluate for y.

$$y = -4 - 7(-1)$$
$$y = 3$$

Since the ordered pair $(-1, 3)$ satisfies both equations, it checks. The solution set is $\{(-1, 3)\}$.

29. $x = 3y + 5$
    $x = \dfrac{3}{2}y$

Replace x in the first equation by $\frac{3}{2}y$, the value of x in the second equation. Multiply both sides by 2 in order to eliminate the fraction and solve for y.

$$\dfrac{3}{2}y = 3y + 5$$
$$3y = 6y + 10$$
$$-3y = 10$$
$$y = -\dfrac{10}{3}$$

Substitute this value for y into the second equation (or the first equation) and evaluate for x.

$$x = \dfrac{3}{2}y$$
$$x = \dfrac{3}{2}\left(-\dfrac{10}{3}\right)$$
$$x = -5$$

Since the ordered pair $\left(-5, -\frac{10}{3}\right)$ satisfies both equations, it checks. The solution set is $\left\{\left(-5, -\frac{10}{3}\right)\right\}$.

30. $x = 6y - 2$
    $x = \dfrac{3}{4}y$

Replace x in the first equation by $\frac{3}{4}y$, the value of x in the second equation. Multiply both sides by 4 in order to eliminate the fraction and solve for y.

$$\dfrac{3}{4}y = 6y - 2$$
$$3y = 24y - 8$$
$$-21y = -8$$
$$y = \dfrac{8}{21}$$

Substitute this value for y into the second equation (or the first equation) and evaluate for x.

$$x = \dfrac{3}{4}y$$
$$x = \dfrac{3}{4}\left(\dfrac{8}{21}\right)$$
$$x = \dfrac{2}{7}$$

Since the ordered pair $\left(\frac{2}{7}, \frac{8}{21}\right)$ satisfies both equations, it checks. The solution set is $\left\{\left(\frac{2}{7}, \frac{8}{21}\right)\right\}$.

31. $\dfrac{1}{2}x + \dfrac{1}{3}y = 3$
    $y = 3x$

Substitute the value for y from the second equation into the first equation and solve for x.

$$\dfrac{1}{2}x + \dfrac{1}{3}(3x) = 3$$
$$\dfrac{1}{2}x + x = 3$$
$$\dfrac{3}{2}x = 3$$
$$x = \left(\dfrac{2}{3}\right)3$$
$$x = 2$$

Substitute this value of x into the equation $y = 3x$ and solve for y.

$$y = 3(2)$$
$$y = 6$$

Since the ordered pair $(2, 6)$ satisfies both equations, it checks. The solution set is $\{(2, 6)\}$.

32. $\dfrac{1}{4}x - \dfrac{1}{5}y = 9$
    $y = 5x$

Substitute the value for y from the second equation into the first equation and solve for x.

$$\dfrac{1}{4}x - \dfrac{1}{5}(5x) = 9$$
$$\dfrac{1}{4}x - x = 9$$
$$-\dfrac{3}{4}x = 9$$
$$x = \left(-\dfrac{4}{3}\right)9$$
$$x = -12$$

Substitute this value of x into the equation $y = 5x$ and solve for y.

$$y = 5(-12)$$
$$y = -60$$

Since the ordered pair $(-12, -60)$ satisfies both equations, it checks. The solution set is $\{(-12, -60)\}$.

33. Writing exercise

34. Start with the solution set $\{(3, 1, 2)\}$. Write an expression involving x, y, and z. Then substitute the components of the ordered triple to complete the equation.

$$x + y + z = 3 + 1 + 2 = 6$$

A first equation might be
$$x + y + z = 6.$$
A second equation might be
$$\begin{aligned}2x + 3y - z &= 2(3) + 3(1) - 2 \\ &= 6 + 3 - 2 \\ &= 7.\end{aligned}$$
A third equation might be
$$\begin{aligned}3x - y - z &= 3(3) - 1 - 2 \\ &= 9 - 1 - 2 \\ &= 6.\end{aligned}$$
Solve the system:
$$\begin{aligned}x + y + z &= 6 \\ 2x + 3y - z &= 7 \\ 3x - y - z &= 6.\end{aligned}$$

Eliminate z from the first and second equations by adding.
$$\begin{aligned}x + y + z &= 6 \\ 2x + 3y - z &= 7 \\ \hline 3x + 4y &= 13\end{aligned}$$

Eliminate z from the first and third equations by adding.
$$\begin{aligned}x + y + z &= 6 \\ 3x - y - z &= 6 \\ \hline 4x &= 12 \\ x &= 3\end{aligned}$$

Substitute this value for x into the equation $3x + 4y = 13$ and solve for y.
$$\begin{aligned}3(3) + 4y &= 13 \\ 4y &= 13 - 9 \\ y &= \frac{4}{4} \\ y &= 1\end{aligned}$$

Substitute these values for x and y into the first equation and solve for z.
$$\begin{aligned}3 + (1) + z &= 6 \\ 4 + z &= 6 \\ z &= 2\end{aligned}$$

The solution set is $\{(3, 1, 2)\}$, our starting solution set.

35. $3x + 2y + z = 8$
$2x - 3y + 2z = -16$
$x + 4y - z = 20$

Eliminate z from the first and third equations by adding.
$$\begin{aligned}3x + 2y + z &= 8 \\ x + 4y - z &= 20 \\ \hline 4x + 6y &= 28\end{aligned}$$

Eliminate z from the second and third equations by adding 2 times the third to the second.
$$\begin{aligned}2x - 3y + 2z &= -16 \\ 2x + 8y - 2z &= 40 \\ \hline 4x + 5y &= 24\end{aligned}$$

We are left with two equations in x and y. Multiply the second by -1 and add to the first to eliminate the x-term.
$$\begin{aligned}4x + 6y &= 28 \\ -4x - 5y &= -24 \\ \hline y &= 4\end{aligned}$$

Substitute this value of y into the equation $4x + 5y = 24$ (Either equation in x and y may be used.) Solve for x.
$$\begin{aligned}4x + 5(4) &= 24 \\ 4x &= 4 \\ x &= 1\end{aligned}$$

Replace x and y by these values in the equation $x + 4y - z = 20$ (Any one of the original 3 equations may be used.) Solve for z.
$$\begin{aligned}(1) + 4(4) - z &= 20 \\ 17 - z &= 20 \\ -z &= 3 \\ z &= -3\end{aligned}$$

Since the ordered triple $(1, 4, -3)$ satisfies all three equations, it checks. The solution set is $\{(1, 4, -3)\}$.

36. $-3x + y - z = -10$
$-4x + 2y + 3z = -1$
$2x + 3y - 2z = -5$

Eliminate z from the first and second equations by multiplying the first by 3 and adding.
$$\begin{aligned}-9x + 3y - 3z &= -30 \\ -4x + 2y + 3z &= -1 \\ \hline -13x + 5y &= -31\end{aligned}$$

Eliminate z from the first and third equations by multiplying the first by -2 then adding.
$$\begin{aligned}6x - 2y + 2z &= 20 \\ 2x + 3y - 2z &= -5 \\ \hline 8x + y &= 15\end{aligned}$$

We are left with two equations in x and y. Multiply the second by -5 and add to the first to eliminate the y-term. Solve for x.
$$\begin{aligned}-13x + 5y &= -31 \\ -40x - 5y &= -75 \\ \hline -53x &= -106 \\ x &= 2\end{aligned}$$

Substitute this value of x into the equation $8x + y = 15$ (Either equation in x and y may be used.) Solve for y.

$$8(2) + y = 15$$
$$y = -1$$

Replace x and y by these values in the equation $-3x + y - z = -10$ (Any one of the original 3 equations may be used.) Solve for z.

$$-3(2) + (-1) - z = -10$$
$$-7 - z = -10$$
$$-z = -3$$
$$z = 3$$

The solution set is $\{(2, -1, 3)\}$.

37. $2x + 5y + 2z = 0$
$4x - 7y - 3z = 1$
$3x - 8y - 2z = -6$

Eliminate z from the first and third equations by adding.

$$2x + 5y + 2z = 0$$
$$3x - 8y - 2z = -6$$
$$\overline{5x - 3y = -6}$$

Eliminate z from first and second equations by adding 3 times the first to 2 times the second.

$$6x + 15y + 6z = 0$$
$$8x - 14y - 6z = 2$$
$$\overline{14x + y = 2}$$

We are left with two equations in x and y. Multiply the second equation by 3 and add to the first to eliminate the y-term.

$$5x - 3y = -6$$
$$42x + 3y = 6$$
$$\overline{47x = 0}$$
$$x = 0$$

Substitute this value of x into the equation $14x + y = 2$ (Either equation in x and y may be used.) Solve for y.

$$14(0) + y = 2$$
$$y = 2$$

Replace x and y by these values in the equation $2x + 5y + 2z = 0$ (Any one of the original 3 equations may be used.) Solve for z.

$$2(0) + 5(2) + 2z = 0$$
$$10 + 2z = 0$$
$$2z = -10$$
$$z = -5$$

The solution set is $\{(0, 2, -5)\}$.

38. $5x - 2y + 3z = -9$
$4x + 3y + 5z = 4$
$2x + 4y - 2z = 14$

Eliminate y from the first and second equations by multiplying the first by 3 and the second by 2, then adding.

$$15x - 6y + 9z = -27$$
$$8x + 6y + 10z = 8$$
$$\overline{23x + 19z = -19}$$

Eliminate y from first and third equations by adding 2 times the first to the third.

$$10x - 4y + 6z = -18$$
$$2x + 4y - 2z = 14$$
$$\overline{12x + 4z = -4}$$

We are left with two equations in x and y. Multiply the first by 4 and the second by -19, then add.

$$92x + 76z = -76$$
$$-228x - 76z = 76$$
$$\overline{-136x = 0}$$
$$x = 0$$

Substitute this value of x into the equation $12x + 4z = -4$. Solve for z.

$$12(0) + 4z = -4$$
$$4z = -4$$
$$z = -1$$

Replace x and z by these values in the equation $5x - 2y + 3z = -9$. Solve for y.

$$5(0) - 2y + 3(-1) = -9$$
$$-2y - 3 = -9$$
$$-2y = -6$$
$$y = 3$$

The solution set is $\{(0, 3, -1)\}$.

39. $x + y - z = -2$
$2x - y + z = -5$
$-x + 2y - 3z = -4$

Add the first and second equations and solve for x.

$$x + y - z = -2$$
$$2x - y + z = -5$$
$$\overline{3x = -7}$$
$$x = -\frac{7}{3}$$

Add -3 times the first equation to the third.

$$-3x - 3y + 3z = 6$$
$$-x + 2y - 3z = -4$$
$$\overline{-4x - y = 2}$$

Substitute $x = -\frac{7}{3}$ into the equation $-4x - y = 2$.

$$-4\left(-\frac{7}{3}\right) - y = 2$$

$$\frac{28}{3} - y = 2$$

$$-y = \frac{6}{3} - \frac{28}{3}$$

$$-y = -\frac{22}{3}$$

$$y = \frac{22}{3}$$

Replace x and y by these values in the equation $x + y - z = -2$. Solve for z.

$$\left(-\frac{7}{3}\right) + \left(\frac{22}{3}\right) - z = -2$$

$$-z = -2 - \frac{15}{3}$$

$$-z = -\frac{6}{3} - \frac{15}{3}$$

$$-z = -\frac{21}{3}$$

$$z = 7$$

The solution set is $\left\{\left(-\frac{7}{3}, \frac{22}{3}, 7\right)\right\}$.

40. $x + 2y + 3z = 1$
$-x - y + 3z = 2$
$-6x + y + z = -2$

Add the first and second equations to eliminate x.

$$x + 2y + 3z = 1$$
$$\underline{-x - y + 3z = 2}$$
$$y + 6z = 3$$

Add 6 times the first to the third equation to eliminate x.

$$6x + 12y + 18z = 6$$
$$\underline{-6x + y + z = -2}$$
$$13y + 19z = 4$$

Multiply the equation $y + 6z = 3$ by -13 and add to the above equation.

$$-13y - 78z = -39$$
$$\underline{13y + 19z = 4}$$
$$-59z = -35$$

$$z = \frac{35}{59}$$

Substitute $z = \frac{35}{59}$ in the equation $y + 6z = 3$ and evaluate for y.

$$y + 6\left(\frac{35}{59}\right) = 3$$

$$y + \frac{210}{59} = 3$$

$$y = \frac{177}{59} - \frac{210}{59}$$

$$y = -\frac{33}{59}$$

Substitute $y = -\frac{33}{59}$ and $z = \frac{35}{59}$ into the equation $x + 2y + 3z = 1$. Solve for x.

$$x + 2\left(-\frac{33}{59}\right) + 3\left(\frac{35}{59}\right) = 1$$

$$x - \frac{66}{59} + \frac{105}{59} = 1$$

$$x + \frac{39}{59} = 1$$

$$x = \frac{59}{59} - \frac{39}{59}$$

$$x = \frac{20}{59}$$

The solution set is $\left\{\left(\frac{20}{59}, -\frac{33}{59}, \frac{35}{59}\right)\right\}$.

41. $2x - 3y + 2z = -1$
$x + 2y + z = 17$
$2y - z = 7$

Add -2 times the second equation to the first equation.

$$2x - 3y + 2z = -1$$
$$\underline{-2x - 4y - 2z = -34}$$
$$-7y = -35$$
$$y = 5$$

Substitute $y = 5$ into the equation $2y - z = 7$.

$$2(5) - z = 7$$
$$10 - z = 7$$
$$-z = -3$$
$$z = 3$$

Substitute these values of y and z into the equation $x + 2y + z = 17$ and solve for x.

$$x + 2(5) + 3 = 17$$
$$x + 13 = 17$$
$$x = 4$$

The solution set is $\{(4, 5, 3)\}$.

42. $2x - y + 3z = 6$
$x + 2y - z = 8$
$2y + z = 1$

Add 3 times the second equation to the first equation.

$$2x - y + 3z = 6$$
$$\underline{3x + 6y - 3z = 24}$$
$$5x + 5y = 30$$

Add the second equation to the third equation.

$$x + 2y - z = 8$$
$$\underline{ 2y + z = 1}$$
$$x + 4y = 9$$

Add -5 times this equation to $5x + 5y = 30$

$$5x + 5y = 30$$
$$\underline{-5x - 20y = -45}$$
$$-15y = -15$$
$$y = 1$$

Substitute $y = 1$ into $x + 4y = 9$ and solve for x.

$$x + 4(1) = 9$$
$$x + 4 = 9$$
$$x = 5$$

Substitute these values for x and y into $2x - y + 3z = 6$ and solve for z.

$$2(5) - 1 + 3z = 6$$
$$9 + 3z = 6$$
$$3z = -3$$
$$z = -1$$

The solution set is $\{(5, 1, -1)\}$.

43. $4x + 2y - 3z = 6$
$x - 4y + z = -4$
$-x + 2z = 2$

Add the second equation and third equation.

$$x - 4y + z = -4$$
$$\underline{-x + 2z = 2}$$
$$-4y + 3z = -2$$

Add -4 times the second equation to the first equation.

$$4x + 2y - 3z = 6$$
$$\underline{-4x + 16y - 4z = 16}$$
$$18y - 7z = 22$$

We are left with two equations in y and z. Multiply the first equation by 7 and the second equation by 3.

$$-28y + 21z = -14$$
$$\underline{54y - 21z = 66}$$
$$26y = 52$$
$$y = 2$$

Replace this value of y in the equation $-4y + 3z = -2$.

$$-4(2) + 3z = -2$$
$$3z = -2 + 8$$
$$3z = 6$$
$$z = 2$$

Substitute this value for z into $-x + 2z = 2$ and solve for x.

$$-x + 2(2) = 2$$
$$-x + 4 = 2$$
$$-x = -2$$
$$x = 2$$

The solution set is $\{(2, 2, 2)\}$.

44. $2x + 3y - 4z = 4$
$x - 6y + z = -16$
$-x + 3z = 8$

Since the third equation has no y-term, eliminate y from the first or second equation. To do this, add 2 times the first equation to the second equation.

$$4x + 6y - 8z = 8$$
$$\underline{x - 6y + z = -16}$$
$$5x - 7z = -8$$

Eliminate the x-term from the two remaining equations in x and z. Do this by multiplying the first by 5 and adding to the second. Solve for z.

$$-5x + 15z = 40$$
$$\underline{5x - 7z = -8}$$
$$8z = 32$$
$$z = 4$$

Substitute this value for z into the equation $-x + 3z = 8$ and solve for x.

$$-x + 3(4) = 8$$
$$-x = 8 - 12$$
$$-x = -4$$
$$x = 4$$

Substitute the values $x = 4$ and $z = 4$ into the first equation $2x + 3y - 4z = 4$ and solve for x.

$$2(4) + 3y - 4(4) = 4$$
$$3y - 8 = 4$$
$$3y = 12$$
$$y = 4$$

The solution set is $\{(4, 4, 4)\}$.

45. $2x + y = 6$
$3y - 2z = -4$
$3x - 5z = -7$

Multiply the first equation by -3 and add to the second equation in order to eliminate the y-term.

$$-6x - 3y = -18$$
$$\underline{3y - 2z = -4}$$
$$-6x - 2z = -22 \text{ or}$$
$$-3x - z = -11$$

Eliminate the x-term from this equation and the third original equation. Solve for z.

$$3x - 5z = -7$$
$$\underline{-3x - z = -11}$$
$$-6z = -18$$
$$z = 3$$

Substitute this value of z into the original equation $3x - 5z = -7$ to find x.

$$3x - 5(3) = -7$$
$$3x - 15 = -7$$
$$3x = 8$$
$$x = \frac{8}{3}$$

Substitute the value $z = 3$ into the equation $3y - 2z = -4$ and solve for y.

$$3y - 2z = -4$$
$$3y - 2(3) = -4$$
$$3y - 6 = -4$$
$$3y = 2$$
$$y = \frac{2}{3}$$

The solution set is $\left\{\left(\frac{8}{3}, \frac{2}{3}, 3\right)\right\}$.

46. $4x - 8y = -7$
    $4y + z = 7$
    $-8x + z = -4$

Multiply the second equation by -1 and add to the third equation in order to eliminate the z-term.

$$-4y - z = -7$$
$$\underline{-8x + z = -4}$$
$$-8x - 4y = -11$$

Multiply the original first equation by 2 and add to this equation.

$$8x - 16y = -14$$
$$\underline{-8x - 4y = -11}$$
$$-20y = -25$$
$$y = \frac{5}{4}$$

Substitute this value of y into the original first equation $4x - 8y = -7$ to find x.

$$4x - 8\left(\frac{5}{4}\right) = -7$$
$$4x - 10 = -7$$
$$4x = 3$$
$$x = \frac{3}{4}$$

Substitute the value $y = \frac{5}{4}$ into the equation $4y + z = 7$ and solve for z.

$$4\left(\frac{5}{4}\right) + z = 7$$
$$5 + z = 7$$
$$z = 2$$

The solution set is $\left\{\left(\frac{3}{4}, \frac{5}{4}, 2\right)\right\}$.

47. Let $x =$ the number of baseball games won and
    $y =$ the number of baseball games lost.

    Write an equation to represent the total number of games played.

    $$x + y = 162$$

    Write a second equation to represent the relationship between games won and games lost.

    $$x = y + 28$$

    Substituting this value of x into the first equation, we arrive at a new equation in y only.

    $$(y + 28) + y = 162$$

    Solve for y.

    $$(y + 28) + y = 162$$
    $$2y + 28 = 162$$
    $$2y = 134$$
    $$y = 67$$

    Substitute this value for y into the equation $x = y + 28$ and solve for x.

    $$x = (67) + 28$$
    $$x = 95$$

    There were 95 wins and 67 losses for the season.

48. Let $x =$ the number of baseball games won and
    $y =$ the number of baseball games lost.

    Write an equation to represent the total number of games played.

    $$x + y = 162$$

    Write a second equation to represent the relationship between games won and games lost.

    $$y = x + 32$$

    Substituting this value of y into the first equation, we arrive at a new equation in y only.

    $$x + (x + 32) = 162$$

8.7 SYSTEMS OF EQUATIONS AND APPLICATIONS

Solve for x.
$$x + (x + 32) = 162$$
$$2x + 32 = 162$$
$$2x = 130$$
$$x = 65$$

Substitute this value for x into the equation $y = x + 32$ and solve for y.
$$y = (65) + 32$$
$$y = 97$$

There were 65 wins and 97 losses for the season.

49. Let l = the length of the tennis court and
    w = the width of the tennis court.

    Use the relationship of the perimeter to the side lengths, $P = 2l + 2w$, to write the following equation.
    $$228 = 2l + 2w \quad \text{or}$$
    $$l + w = 114$$

    Write a second equation to represent the relationship between the length and the width dimensions.
    $$w = l - 42 \quad \text{or}$$
    $$l - w = 42$$

    Solve the resulting system of equations by adding to eliminate the w-term.
    $$l + w = 114$$
    $$l - w = 42$$
    $$\overline{2l = 156}$$
    $$l = 78$$

    Substituting this value of l into the equation $w = l - 42$, we find the width dimension.
    $$w = 78 - 42$$
    $$= 36$$

    The dimensions are: length = 78 feet and width = 36 feet.

50. Let l = the length of the basketball court and w = the width of the basketball court.

    Use the relationship of the perimeter to the side lengths, $P = 2l + 2w$, to write the following equation.
    $$2l + 2w = 288 \quad \text{or}$$
    $$l + w = 144$$

    Write a second equation to represent the relationship between the length and the width dimensions.
    $$w = l - 44 \quad \text{or}$$
    $$l - w = 44$$

Solve the resulting system of equations by adding to eliminate the w-term.
$$l - w = 44$$
$$l + w = 144$$
$$\overline{2l = 188}$$
$$l = 94$$

Substituting this value of l into the equation $w = l - 44$, we find the width dimension.
$$w = 94 - 44$$
$$= 50$$

The dimensions are: length = 94 feet and width = 50 feet.

51. Let x = the revenue for ExxonMobile and
    y = the revenue for General Motors.

    Write an equation to represent the total revenue of the two companies.
    $$x + y = 399$$

    Write a second equation to represent the relationship between the revenues of the two companies.
    $$x = y + 29$$

    Solve the resulting system of equations by substituting this value of x in the first equation.
    $$x + y = 399$$
    $$(y + 29) + y = 399$$
    $$2y + 29 = 399$$
    $$2y = 370$$
    $$y = 185$$

    Substituting this value of y into the equation $x = y + 29$, we arrive at a new equation in x only.
    $$x = 185 + 29$$
    $$x = 214$$

    ExxonMobil's revenue was $214 billion, and General Motors revenue was $185 billion.

52. Let x = the exports and imports with Canada and
    y = the exports and imports with Mexico.

    Write an equation to represent the total exports and imports of the two countries
    $$x + y = 211$$

    Write a second equation to represent the relationship between the two countries exports and imports.
    $$x = y + 57$$

Solve the resulting system of equations by substituting this value of x in the first equation.

$$x + y = 211$$
$$(y + 57) + y = 211$$
$$2y + 57 = 211$$
$$2y = 154$$
$$y = 77$$

Substituting this value of y into the equation $x = y + 57$, we arrive at a new equation in x only.

$$x = 77 + 57$$
$$x = 134$$

US exports and imports with Canada was $134 billion and with Mexico was $77 billion.

53. Let $l =$ the length of the rectangle and
    $w =$ the width of the rectangle.

Write an equation to represent the relationship between the length and the width dimensions.

$$l = w + 7$$

Use the relationship of the perimeter to the side lengths, $P = 2l + 2w$, to write the following equation.

$$2(l - 3) + 2(w + 2) = 32 \text{ or}$$
$$(l - 3) + (w + 2) = 16 \text{ or}$$
$$l + w = 17$$

Substituting this value of l from the first equation into the second, we arrive at a new equation in w only.

$$(w + 7) + w = 17$$

Solve for w.

$$(w + 7) + w = 17$$
$$2w + 7 = 17$$
$$2w = 10$$
$$w = 5$$

Substitute this value for w into the equation $l = w + 7$ and solve for l.

$$l = 5 + 7$$
$$l = 12$$

The dimensions are: length $= 12$ feet and width $= 5$ feet.

54. Let $x =$ the side length of the equilateral triangle and
    $y =$ the side length of the square.

Write an equation to represent the relationship between the side lengths.

$$y = x + 4 \text{ or}$$
$$-x + y = 4$$

The perimeter of the square may be expressed as $4y$.
The perimeter of the triangle may be expressed as $3x$.

Write a second equation to represent the relationship between the perimeters.

$$4y = 3x + 24 \text{ or}$$
$$-3x + 4y = 24$$

Solve the resulting system of equations by multiplying the first equation $-x + y = 4$ by -3 and adding to the second equation $-3x + 4y = 24$.

$$3x - 3y = -12$$
$$-3x + 4y = 24$$
$$\overline{y = 12}$$

Substituting this value of y into the equation $-x + y = 4$ gives the value for x.

$$-x + 12 = 4 \text{ or}$$
$$-x = -8$$
$$x = 8$$

The side dimension for the square is 12 centimeters, and the side dimension for the triangle is 8 centimeters.

55. Let $x =$ the cost of the colored paper and
    $y =$ the cost of the marker pens.

Write two equations to represent the two sets of purchase costs.

$$8x + 3y = 6.50$$
$$2x + 2y = 3.00 \text{ or}$$
$$x + y = 1.50$$

Multiply the two equations by 10 to eliminate the decimal point.

$$80x + 30y = 65$$
$$10x + 10y = 15$$

Multiply the second equation by -3 and add to the first equation in order to eliminate the y-term.

$$80x + 30y = 65$$
$$-30x - 30y = -45$$
$$\overline{50x = 20}$$
$$x = .40$$

Substitute this value of x into the equation $x + y = 1.50$ and solve for y.

$$.40 + y = 1.50$$
$$y = 1.10$$

The cost of the marker pen is $1.10 and that of the colored paper is $.40.

56. Let $x =$ the cost of a CGA monitor and
    $y =$ the cost of a VGA monitor.

Write two equations to represent the two sets of purchase prices.

$$4x + 6y = 4600$$
$$6x + 4y = 4400$$

Simplify both equations by dividing out the greatest common factor, 2.

$$2x + 3y = 2300$$
$$3x + 2y = 2200$$

Multiply the first equation by 2 and the second equation by -3 and add to eliminate the y-term.

$$4x + 6y = 4600$$
$$-9x - 6y = -6600$$
$$\overline{-5x = -2000}$$
$$x = 400$$

Substitute this value of x into the equation $2x + 3y = 2300$ and solve for y.

$$2(400) + 3y = 2300$$
$$800 + 3y = 2300$$
$$3y = 1500$$
$$y = 500$$

The CGA monitors cost $400 for each monitor, and the VGA monitors cost $500 for each monitor.

57. Let $x =$ the FCI for the National Hockey League and $y =$ the FCI for the National Basketball Association.

Write an equation to represent the total FCI prices.

$$x + y = 423.12$$

Write an equation to represent the relationship of the FCI prices for the two groups.

$$x = y + 16.36$$

Substituting this value of x from the second equation into the first equation, we arrive at a new equation in y only.

$$(y + 16.36) + y = 423.12$$

Solve for y.

$$(y + 16.36) + y = 423.12$$
$$2y + 16.36 = 423.12$$
$$2y = 406.76$$
$$y = 203.38$$

Substitute this value for y into the equation $x = y + 16.36$ and solve for x.

$$x = (203.38) + 16.36$$
$$x = 219.74$$

The FCI for the National Hockey League was $219.74. The FCI for the National Basketball Association was $203.38.

58. Let $x =$ the FCI for Major League Baseball and $y =$ the FCI for the National Football League.

Write an equation to represent the total FCI prices.

$$x + y = 311.03$$

Write an equation to represent the relationship of the FCI prices for the two groups.

$$y = x + 105.87$$

Substituting this value of y from the second equation into the first equation, we arrive at a new equation in x only.

$$x + (x + 105.87) = 311.03$$

Solve for x.

$$x + (x + 105.87) = 311.03$$
$$2x + 105.87 = 311.03$$
$$2x = 205.16$$
$$x = 102.58$$

Substitute this value for x into the equation $y = x + 105.87$ and solve for y.

$$y = (102.58) + 105.87$$
$$y = 208.45$$

The FCI for Major League Baseball was $102.58, and the FCI for the National Football League was $208.45.

59. Let $x =$ the cost of a single hamburger and $y =$ the cost of a double hamburger.

Write two equations to represent the two sets of purchase costs.

$$15x + 10y = 63.25$$
$$30x + 5y = 78.65$$

Multiply the second equation by -2 and add to the first equation in order to eliminate the y-term.

$$15x + 10y = 63.25$$
$$-60x - 10y = -157.30$$
$$\overline{-45x = -94.05}$$
$$x = 2.09$$

Substitute this value of x into the equation $30x + 5y = 78.65$ and solve for y.

$$30(2.09) + 5y = 78.65$$
$$62.70 + y = 78.65$$
$$5y = 15.95$$
$$y = 3.19$$

The cost of a single is $2.09 and that of a double is $3.19.

60. Let x = the average cost of a day in Tokyo and
    y = the average cost of a day in New York.

    Write two equations to represent the two sets of total costs
    $$2x + 3y = 2015$$
    $$4x + 2y = 2490 \text{ or}$$
    $$2x + y = 1245$$

    Multiply the second equation by -3 and add to the first equation in order to eliminate the y-term.
    $$2x + 3y = 2015$$
    $$\underline{-6x - 3y = -3735}$$
    $$-4x = -1720$$
    $$x = 430$$

    Substitute this value of x into the equation $2x + y = 1245$ and solve for y.
    $$2(430) + y = 1245$$
    $$860 + y = 1245$$
    $$y = 385$$

    The average day in Tokyo costs $430, and the average day in New York costs $385.

61. Let x = the number of units of yarn and
    y = the number of units of thread.

    Write two equations representing the number of hours per day each machine runs.
    $$\text{Machine A:} \quad 1x + 1y = 8$$
    $$\text{Machine B:} \quad 2x + 1y = 14$$

    Multiply the first equation by -1 and add to the second equation in order to eliminate the y-term.
    $$-x - y = -8$$
    $$\underline{2x + y = 14}$$
    $$x = 6$$

    Substitute this value of x into the equation $x + y = 8$ and solve for y.
    $$(6) + y = 8$$
    $$y = 2$$

    Thus, 6 units of yarn and 2 units of thread will keep both machines running to capacity.

62. Let x = the cost of the dark clay and
    y = the cost of the light clay.

    Write two equations to represent the two sets of purchase costs.
    $$2x + 3y = 22$$
    $$1x + 2y = 13$$

    Multiply the second equation by -2 and add to the first equation in order to eliminate the x-term.
    $$2x + 3y = 22$$
    $$\underline{-2x - 4y = -26}$$
    $$-y = -4$$
    $$y = 4$$

    Substitute this value of y into the equation $x + 2y = 13$ and solve for x.
    $$x + 2(4) = 13$$
    $$x + 8 = 13$$
    $$x = 5$$

    The dark clay cost $5 per kilogram, and the light clay cost $4 per kilogram.

63. (a) $(.10)(60) = 6$ oz

    (b) $(.25)(60) = 15$ oz

    (c) $(.40)(60) = 24$ oz

    (d) $(.50)(60) = 30$ oz

64. (a) $(\$5000)(.02) = \$100$

    (b) $(\$5000)(.03) = \$150$

    (c) $(\$5000)(.04) = \$200$

    (d) $(\$5000)(.035) = \$175$

65. The cost of x turkeys is $.99x$.

66. $8y$ is collected from the sale of y tickets.

67. Let x = the number of gallons of 25% alcohol needed and y = the number of gallons of 35% alcohol needed.

    Write an equation to represent the total gallons of solution.
    $$x + y = 20$$

    Write another equation to represent the total amount of alcohol.
    $$.25x + .35y = .32(20), \text{ or equivalently}$$
    $$25x + 35y = 32(20)$$
    $$5x + 7y = 128$$

    Solve the resulting system of equations.
    $$x + y = 20$$
    $$5x + 7y = 128$$

    Multiply the first equation by -5 and add to the second equation in order to eliminate the x-term.
    $$-5x - 5y = -100$$
    $$\underline{5x + 7y = 128}$$
    $$2y = 28$$
    $$y = 14$$

Substitute this value of y into the equation $x + y = 20$ and solve for x.
$$x + (14) = 20$$
$$x = 6$$

It requires 6 gallons of 25% alcohol and 14 gallons of 35% alcohol to obtain 20 gallons of 32% alcohol.

68. Let $x = $ the number of liters of 15% acid and
$y = $ the number of liters of 33% acid needed.

Write an equation to represent the total liters of solution.
$$x + y = 40$$

Write another equation to represent the total amount of acid.
$$.15x + .33y = .21(40), \text{ or equivalently}$$
$$15x + 33y = 21(40)$$
$$5x + 11y = 7(40)$$
$$5x + 11y = 280$$

Solve the resulting system of equations.
$$x + y = 40$$
$$5x + 11y = 280$$

Multiply the first equation by -5 and add to the second equation in order to eliminate the x-term.
$$-5x - 5y = -200$$
$$\underline{5x + 11y = 280}$$
$$6y = 80$$
$$y = \frac{40}{3} = 13\frac{1}{3}$$

Substitute this value of y into the equation $x + y = 40$ and solve for x.
$$x + \left(\frac{40}{3}\right) = 40$$
$$x = 40 - \frac{40}{3}$$
$$x = \frac{120}{3} - \frac{40}{3}$$
$$x = \frac{80}{3} = 26\frac{2}{3}$$

It requires $26\frac{2}{3}$ liters of 15% acid and $13\frac{1}{3}$ liters of 33% acid to make 40 liters of 21% acid.

69. Let $x = $ the number of liters of 100% (pure) acid and
$y = $ the number of liters of 10% acid needed.

Write an equation to represent the total liters of solution.
$$x + y = 27$$

Write another equation to represent the total amount of acid.
$$1.00x + .10y = .20(27)$$
$$10.0x + 1.0y = 2.0(27)$$
$$10x + 1y = 54$$

Solve the resulting system of equations.
$$x + y = 27$$
$$10x + y = 54$$

Multiply the first equation by -1 and add to the second equation in order to eliminate the y-term.
$$-x - y = -27$$
$$\underline{10x + y = 54}$$
$$9x = 27$$
$$x = 3$$

Substitute this value of x into the equation $x + y = 27$ and solve for y.
$$(3) + y = 27$$
$$y = 24$$

Three liters of pure acid and 24 liters of 10% acid should be used.

70. Let $x = $ the number of liters of 100% antifreeze and
$y = $ the number of liters of 4% antifreeze needed.

Write an equation to represent the total liters of solution
$$x + y = 18$$

Write (and simplify) another equation to represent the total amount of antifreeze.
$$1.00x + .04y = .20(18)$$
$$100x + 4y = 20(18)$$
$$100x + 4y = 360$$

Solve the resulting system of equations.
$$x + y = 18$$
$$100x + 4y = 360$$

Multiply the first equation by -4 and add to the second equation in order to eliminate the y-term.
$$-4x - 4y = -72$$
$$\underline{100x + 4y = 360}$$
$$96x = 288$$
$$x = 3$$

There are 3 liters of pure antifreeze needed.

71. Let $x = $ the number of pounds of \$3.60 pecan clusters and $y = $ the number of pounds of \$7.20 chocolate truffles.

Write an equation to represent the total number of pounds of the mixture.
$$x + y = 80$$

Write (and simplify) another equation to represent the total value of the candy.
$$3.60x + 7.20y = 4.95(80)$$
$$360x + 720y = 495(80)$$
$$x + 2y = 110$$

Solve the resulting system of equations.

$$x + y = 80$$
$$x + 2y = 110$$

Multiply the first equation by -1 and add to the second equation in order to eliminate the x-term.

$$-x - y = -80$$
$$\underline{x + 2y = 110}$$
$$y = 30$$

Substitute this value of y into the equation $x + y = 80$ and solve for x.

$$x + (30) = 80$$
$$x = 50$$

Thus, 50 pounds of \$3.60 pecan clusters and 30 pounds of \$7.20 chocolate truffles are to be used.

72. Let $x =$ the number of liters of 50% juice and
    $y =$ the number of liters of 30% juice.

Write an equation to represent the total amount of juice mixture.

$$x + y = 200$$

Write (and simplify) another equation to represent the total amount of pure juice.

$$.50x + .30y = .45(200)$$
$$50x + 30y = 45(200)$$
$$5x + 3y = 900$$

Solve the resulting system of equations.

$$x + y = 200$$
$$5x + 3y = 900$$

Multiply the first equation by -3 and add to the second equation in order to eliminate the y-term.

$$-3x - 3y = -600$$
$$\underline{5x + 3y = 900}$$
$$2x = 300$$
$$x = 150$$

Substitute this value of x into the equation $x + y = 200$ and solve for y.

$$(150) + y = 200$$
$$y = 50$$

Thus, 150 liters of 50% juice and 50 liters of 30% juice are to be used.

73. Let $x =$ the number of general admission (\$2.50) tickets sold and $y =$ the number of student admission (\$2.00) tickets sold.

Write an equation to represent the total number of people who saw the performance.

$$x + y = 184$$

Write (and simplify) another equation to represent the total amount of money collected.

$$2.50x + 2.00y = 406$$
$$25x + 20y = 4060$$
$$5x + 4y = 812$$

Solve the resulting system of equations.

$$x + y = 184$$
$$5x + 4y = 812$$

Multiply the first equation by -4 and add to the second equation in order to eliminate the y-term.

$$-4x - 4y = -736$$
$$\underline{5x + 4y = 812}$$
$$x = 76$$

Substitute this value of x into the equation $x + y = 184$ and solve for y.

$$(76) + y = 184$$
$$y = 108$$

Thus, 76 with general admission tickets and 108 tickets with student identification were sold.

74. Let $x =$ the amount of \$1.20 per pound candy and
    $y =$ the amount of \$2.40 per pound candy.

Write an equation to represent the total pounds of candy mix.

$$x + y = 80$$

Write (and simplify) another equation to represent the total value of the candy.

$$1.20x + 2.40y = 1.65(80)$$
$$12x + 24y = 1320$$
$$x + 2y = 110$$

Solve the resulting system of equations.

$$x + y = 80$$
$$x + 2y = 110$$

Multiply the first equation by -1 and add to the second equation in order to eliminate the x-term.

$$-x - y = -80$$
$$\underline{x + 2y = 110}$$
$$y = 30$$

Substitute this value of y into the equation $x + y = 80$ and solve for x.

$$x + (30) = 80$$
$$x = 50$$

He wants to mix 50 pounds of the \$1.20 per pound candy with 30 pounds of the \$2.40 per pound candy.

75. Let $x =$ the amount of money invested at 2% and
    $y =$ the amount of money invested at 4%.

    Write an equation to represent the total number of dollars invested.
    $$x + y = 3000$$

    Write another equation to represent the total return (interest earned) on the investments.
    $$.02x + .04y = 100$$
    $$2x + 4y = 10000$$
    $$x + 2y = 5000$$

    Solve the resulting system of equations.
    $$x + y = 3000$$
    $$x + 2y = 5000$$

    Multiply the first equation by -1 and add to the second equation in order to eliminate the x-term.
    $$-x - y = -3000$$
    $$\underline{x + 2y = 5000}$$
    $$y = 2000$$

    Substitute this value of y into the equation $x + y = 3000$ and solve for x.
    $$x + (2000) = 3000$$
    $$x = 1000$$

    There was $1000 deposited at 2% and $2000 deposited at 4%.

76. Let $x =$ the amount of money invested at 4% and
    $y =$ the amount of money invested at 3%.

    Write an equation to represent the total number of dollars invested
    $$x + y = 15000$$

    Write another equation to represent the total return (interest earned) on the investments.
    $$.04x + .03y = 550$$
    $$4x + 3y = 55000$$

    Solve the resulting system of equations.
    $$x + y = 15000$$
    $$4x + 3y = 55000$$

    Multiply the first equation by -3 and add to the second equation in order to eliminate the y-term.
    $$-3x - 3y = -45000$$
    $$\underline{4x + 3y = 55000}$$
    $$x = 10000$$

    Substitute this value of x into the equation $x + y = 15000$ and solve for y.
    $$(10000) + y = 15000$$
    $$y = 5000$$

    He should invest $10,000 at 4% and $5000 at 3%.

77. Let $x =$ the speed of the freight train and
    $y =$ the speed of the express train.

    Write an equation to show the relationship of the speed of the freight train compared to that of the express train.
    $$x = y - 30$$

    The distance traveled, $d = rt$, by the freight train during the 3 hours is $3x$ and that of the express train is $3y$. Since the total distance traveled by both trains is 390 kilometers, construct an equation showing total distance.
    $$3x + 3y = 390$$
    $$x + y = 130$$

    Solve the resulting system of equations.
    $$x = y - 30$$
    $$x + y = 130$$

    Substitute the value for x into the second equation $x + y = 130$ and solve for y
    $$(y - 30) + y = 130$$
    $$2y - 30 = 130$$
    $$2y = 160$$
    $$y = 80$$

    Substitute this value of y into the equation $x = y - 30$ and solve for x.
    $$x = (80) - 30$$
    $$x = 50$$

    Thus, the speed of the freight train was 50 kilometers per hour and that of the express train was 80 kilometers per hour.

78. Let $x =$ the speed of the train and
    $y =$ the speed of the plane.

    Write an equation to showing the relationship of the speed of the plane compared to that of the train.
    $$y = 3x - 20$$

    Because the time values are the same for the train and the plane, solve $d = rt$ for t.

    The time for the train can be represented as
    $$t = \frac{d}{r} = \frac{150}{x}.$$

    The time for the plane can be represented as
    $$t = \frac{d}{r} = \frac{400}{y}.$$

    Thus,
    $$\frac{150}{x} = \frac{400}{y}$$
    $$150y = 400x.$$

Substituting the value $y = 3x - 20$ into the resulting equation gives

$$150(3x - 20) = 400x$$
$$450x - 3000 = 400x$$
$$50x = 3000$$
$$x = 60.$$

Substituting this value of x into $y = 3x - 20$ gives

$$y = 3(60) - 20$$
$$y = 160.$$

Thus, the speed of the train is 60 kilometers per hour, and that of the plane is 160 kilometers per hour.

79. Let $x =$ the top speed of the snow speeder and
$y =$ the speed of the wind.

Remembering the relationship $d = rt$, the distance traveled by the snow speeder into the wind is given by $3600 = (x - y)(2)$ and the same distance returning with the wind is $(x + y)(1.5)$. The resulting two equations are

$$3600 = (x - y)(2)$$
$$3600 = (x + y)(1.5).$$

Write the equations in standard form.

$$2x - 2y = 3600$$
$$1.5x + 1.5y = 3600$$

Observe that the second equation may be simplified as follows before solving the system of equations.

$$1.5x + 1.5y = 3600$$
$$15x + 15y = 36000$$
$$x + y = 2400$$

The resulting system is

$$2x - 2y = 3600$$
$$x + y = 2400.$$

Solve by multiplying the second equation by 2 and adding to the first.

$$2x - 2y = 3600$$
$$2x + 2y = 4800$$
$$\overline{4x = 8400}$$
$$x = 2100$$

Substitute this value of x into the equation $x + y = 2400$ and solve for y.

$$(2100) + y = 2400$$
$$y = 300$$

Thus, the top speed is 2100 miles per hour, and the wind speed is 300 miles per hour.

80. Let $x =$ the speed of the boat in still water and
$y =$ the speed of the current.

Using the relationship $d = rt$, the distance traveled by boat upstream into the current is given by $36 = (x - y)(2)$ and the same distance returning with the current is $(x + y)(1.5)$. The resulting two equations are

$$36 = (x - y)(2)$$
$$36 = (x + y)(1.5)$$

Write the equations in standard form.

$$2x - 2y = 36$$
$$1.5x + 1.5y = 36$$

Observe that the second equation may be simplified as follows before solving the system of equations.

$$1.5x + 1.5y = 36$$
$$15x + 15y = 360$$
$$x + y = 24$$

The resulting system is

$$2x - 2y = 36$$
$$x + y = 24.$$

Solve by multiplying the second equation by 2 and adding to the first.

$$2x - 2y = 36$$
$$2x + 2y = 48$$
$$\overline{4x = 84}$$
$$x = 21$$

Substitute this value of x into the equation $x + y = 24$ and solve for y.

$$(21) + y = 24$$
$$y = 3$$

Thus, the speed of the boat is 21 miles per hour, and the speed of the current is 3 miles per hour.

81. Let $x =$ the number Independents,
$y =$ the number of Democrats, and
$z =$ the number of Republicans.

Write an equation to represent the total number in the sample.

$$x + y + z = 100$$

Write a second equation to represent the relationship between the number of Independents and Republicans.

$$x = z + 10$$
$$x - z = 10 \quad \text{in standard form}$$

Write a third equation to represent the relationship between the Republicans and Democrats.

$$y = z + 6$$
$$y - z = 6 \quad \text{in standard form}$$

The resulting system equations is as follows.

$$x + y + z = 100$$
$$x - z = 10$$
$$y - z = 6$$

Add the first equation to the second equation to eliminate the z-term.

$$x + y + z = 100$$
$$\underline{x - z = 10}$$
$$2x + y = 110$$

Add the second equation to (-1) times the third equation to eliminate the z-term.

$$x - z = 10$$
$$\underline{-y + z = -6}$$
$$x - y = 4$$

Add the resulting equations in x and y to eliminate the y-term.

$$2x + y = 110$$
$$\underline{x - y = 4}$$
$$3x = 114 \text{ or}$$
$$x = 38$$

Substitute this value of x into the equation $x - z = 10$ to find z.

$$(38) - z = 10$$
$$-z = -28$$
$$z = 28$$

Substitute the values for x and z into the equation $x + y + z = 100$ to find the value for y.

$$(38) + y + (28) = 100$$
$$y + 66 = 100$$
$$y = 34$$

There are 38 Independents, 34 Democrats, and 28 Republicans in the sample.

82. Let $x =$ the number gold medals,
$y =$ the number of silver medals, and
$z =$ the number of bronze medals won.
Write an equation to represent the total number of medals won by the United States.

$$x + y + z = 97$$

Write a second equation to represent the relationship between the number of gold and silver medals.

$$x = y + 14$$
$$x - y = 14 \quad \text{in standard form}$$

Write a third equation to represent the relationship between the number of bronze and silver medals.

$$2y - 17 = z$$
$$2y - z = 17 \quad \text{in standard form}$$

The resulting system equations is as follows.

$$x + y + z = 97$$
$$x - y = 14$$
$$2y - z = 17$$

Solve for z in the third equation to get

$$z = 2y - 17$$

Solve for y in the second equation to get

$$y = x - 14$$

Substitute this value of y in the equation $z = 2y - 17$

$$z = 2(x - 14) - 17$$
$$z = 2x - 28 - 17$$
$$z = 2x - 45$$

This leaves the two variables y and z, in terms of x. Replace y and z in the first equation to gain a single equation in x.

$$x + (x - 14) + (2x - 45) = 97$$
$$4x - 59 = 97$$
$$4x = 156$$
$$x = 39$$

Substitute this value of x into the equation $y = x - 14$ to find y.

$$y = (39) - 14$$
$$y = 25$$

Substitute the values for x and y into the equation $x + y + z = 97$ to find the value for z.

$$(39) + (25) + z = 97$$
$$64 + z = 97$$
$$z = 33$$

The United States won 39 gold medals, 25 silver medals, and 33 bronze medals.

83. Let $x =$ the length of the shortest side,
$y =$ the length of the middle side, and
$z =$ the length of the longest side.

Write an equation to represent the perimeter.

$$x + y + z = 70$$

Write a second equation to represent the relationship between the longest side and the others.

$$z = (x + y) - 4, \text{ or}$$
$$-x - y + z = -4$$

Write a third equation to represent the relationship between the longest and shortest sides.
$$z = 2x + 9, \text{ or}$$
$$-2x + z = 9$$

The resulting system equations is as follows.
$$x + y + z = 70$$
$$-x - y + z = -4$$
$$-2x + z = 9$$

Add the first and second equation to eliminate the x, and y-terms.
$$x + y + z = 70$$
$$\underline{-x - y + z = -4}$$
$$2z = 66$$
$$z = 33$$

Substitute this value of z into the equation $-2x + z = 9$ to find x.
$$-2x + (33) = 9$$
$$-2x = -24$$
$$x = 12$$

Substitute the values for x and z into the equation $x + y + z = 70$ to find the value for y.
$$(12) + y + (33) = 70$$
$$y + 45 = 70$$
$$y = 25$$

The shortest side is 12 centimeters, the middle side is 25 centimeters, and the longest side is 33 centimeters.

84. Let $x =$ the length of the shortest side,
$y =$ the length of the middle side, and
$z =$ the length of the longest side.

Write an equation to represent the perimeter.
$$x + y + z = 56$$

Write a second equation to represent the relationship between the longest side and the others.
$$z = (x + y) - 4$$
$$-x - y + z = -4 \quad \text{in standard form}$$

Write a third equation to represent the relationship between the longest and shortest sides.
$$z = 3x - 4$$
$$-3x + z = -4 \quad \text{in standard form}$$

The resulting system equations is as follows.
$$x + y + z = 56$$
$$-x - y + z = -4$$
$$-3x + z = -4$$

Add the first and second equation to eliminate the x and y-terms.
$$x + y + z = 56$$
$$\underline{-x - y + z = -4}$$
$$2z = 52$$
$$z = 26$$

Substitute this value of z into the equation $-3x + z = -4$ to find x.
$$-3x + (26) = -4$$
$$-3x = -30$$
$$x = 10$$

Substitute the values for x and z into the equation $x + y + z = 56$ to find the value for y.
$$(10) + y + (26) = 56$$
$$y + 36 = 56$$
$$y = 20$$

The shortest side is 10 inches, the middle side is 20 inches, and the longest side is 26 inches.

85. Let $x =$ the number of cases sent to wholesaler A,
$y =$ the number of cases sent to wholesaler B, and
$z =$ the number of cases sent to wholesaler C.

Write an equation to represent the total cases sent.
$$x + y + z = 320$$

Write a second equation to represent the relationship between the number of wholesaler A and wholesale B's cases of trinkets.
$$x = 3y, \text{ or}$$
$$x - 3y = 0$$

Write a third equation to represent the relationship between the number of cases sent to wholesaler C and those to wholesalers A and B.
$$z = (x + y) - 160, \text{ or}$$
$$-x - y + z = -160$$

The resulting system of equations is as follows:
$$x + y + z = 320$$
$$x - 3y = 0$$
$$-x - y + z = -160.$$

Add the first and third equation to eliminate the x- and y-terms.
$$x + y + z = 320$$
$$\underline{-x - y + z = -160}$$
$$2z = 160$$
$$z = 80$$

Substitute this value of z into the equation $x+y+z=320$ and use with the other equation $x-3y=0$ in x and y.

$$x+y+(80)=320$$
$$x+y=240$$
$$x-3y=0$$

Multiply the equation $x-3y=0$ by -1 and add to the equation $x+y=240$.

$$x+y=240$$
$$\underline{-x+3y=0}$$
$$4y=240$$
$$y=60$$

Substitute the values for y and z into the equation $x+y+z=320$ and solve for x.

$$x+(60)+(80)=320$$
$$x+140=320$$
$$x=180$$

She must send 180 cases to wholesaler A, 60 cases to wholesaler B, and 80 cases to wholesaler C.

86. Let $x=$ the number of units of type A clamps,
 $y=$ the number of units of type B clamps, and
 $z=$ the number of units of type C clamps.

Write an equation to represent the total number of units produced.

$$x+y+z=490$$

Write a second equation to represent the relationship between the number of units of type C clamps and units of type A and type B clamps.

$$z=(x+y)+10, \text{ or}$$
$$-x-y+z=10$$

Write a third equation to to represent the relationship between the number type C and type B clamps.

$$y=2x, \text{ or}$$
$$-2x+y=0$$

The resulting system of equations is as follows:

$$x+y+z=490$$
$$-x-y+z=10$$
$$-2x+y=0.$$

Add the first and second equations to eliminate the x- and y-terms.

$$x+y+z=490$$
$$\underline{-x-y+z=10}$$
$$2z=500$$
$$z=250$$

Substitute this value of z into the equation $x+y+z=490$ and use with the other equation $-2x+y=0$, in x and y.

$$x+y+(250)=490$$
$$x+y=240$$
$$-2x+y=0$$

Multiply the equation $-2x+y=0$ by -1 and add to the equation $x+y=240$.

$$2x-y=0$$
$$x+y=240$$
$$3x=240$$
$$x=80$$

Substitute the values for x and z into the equation $x+y+z=490$ and solve for y.

$$(80)+y+(250)=490$$
$$y+330=490$$
$$y=160$$

The shop must make 80 type A, 160 type B, and 250 type C clamps per day.

87. Let $x=$ the number of $10 tickets,
 $y=$ the number of $18 tickets, and
 $z=$ the number of VIP tickets.

Write an equation to represent the total sale of all tickets.

$$10x+18y+30z=9500$$

Write a second equation to represent the relationship between the number of $18 tickets and VIP tickets sold.

$$y=5z$$

Write a third equation to represent the relationship between the number of $10, $18 tickets, and VIP tickets sold.

$$x=y+2z$$

Replacing y with $5z$ in the last equation leaves x in terms of z.

$$x=(5z)+2z$$
$$x=7z$$

Replacing x and z in the first equation leaves a single equation in z to solve.

$$10x+18y+30z=9500$$
$$10(7z)+18(5z)+30z=9500$$
$$70z+90z+30z=9500$$
$$190z=9500$$
$$z=50$$

Replace z with 50 in the two equations for x and y. Solve the resulting equations in x and y.

$$x = 7z$$
$$x = 7(50)$$
$$x = 350 \text{ and}$$
$$y = 5z$$
$$y = 5(50)$$
$$y = 250$$

Thus 350 - $10 tickets, 250 - $18, and 50 - VIP ($30) tickets have been sold.

88. Let $x =$ the price of the "up close" tickets, $y =$ the price of the "middle" tickets, and $z =$ price of the "far out" tickets.

Write an equation to represent the relative costs.

$$x = y + 10$$
$$y = z + 10$$
$$2x = 3z + 20$$

Write these equations in standard form.

$$x - y = 10$$
$$y - z = 10$$
$$2x - 3z = 20$$

Add the first two equations to eliminate the y-term.

$$\begin{aligned} x - y &= 10 \\ y - z &= 10 \\ \hline x - z &= 20 \end{aligned}$$

Solve the two equations in x and z for z by multiplying the above equation $x - z = 20$ by -2 and adding to the other equation $2x - 3z = 20$.

$$\begin{aligned} 2x - 3z &= 20 \\ -2x + 2z &= -40 \\ \hline -z &= -20 \\ z &= 20 \end{aligned}$$

Use this value of z in the equation $x - z = 20$ to find the value for x.

$$x - (20) = 20$$
$$x = 40$$

In the same manner replace the value for z in the equation $y - z = 10$ to find the corresponding value for y.

$$y - (20) = 10$$
$$y = 30$$

The price of the "up close" tickets is $40, the price of the "middle" tickets is $30, and the price of the "far out" tickets is $20.

EXTENSION: USING MATRIX ROW OPERATIONS TO SOLVE SYSTEMS

1. $x + y = 5$
   $x - y = -1$

   Write the augmented matrix.

   $$\begin{bmatrix} 1 & 1 & | & 5 \\ 1 & -1 & | & -1 \end{bmatrix}$$

   Multiply row 1 by -1 and add to row 2.

   $$\begin{bmatrix} 1 & 1 & | & 5 \\ 0 & -2 & | & -6 \end{bmatrix}$$

   Multiply row 2 by $-1/2$.

   $$\begin{bmatrix} 1 & 1 & | & 5 \\ 0 & 1 & | & 3 \end{bmatrix}$$

   Multiply row 2 by -1 and add to row 1.

   $$\begin{bmatrix} 1 & 0 & | & 2 \\ 0 & 1 & | & 3 \end{bmatrix}$$

   The resulting matrix represents the following system of equations.

   That is,
   $$1x + 0y = 2$$
   $$0x + 1y = 3$$
   $$x = 2$$
   $$y = 3.$$

   Thus, $\{(2, 3)\}$ is the solution set.

2. $x + 2y = 5$
   $2x + y = -2$

   Form the augmented matrix.

   $$\begin{bmatrix} 1 & 2 & | & 5 \\ 2 & 1 & | & -2 \end{bmatrix}$$

   Multiply row 1 by -2 and add to row 2.

   $$\begin{bmatrix} 1 & 2 & | & 5 \\ 0 & -3 & | & -12 \end{bmatrix}$$

   Multiply row 2 by $-1/3$.

   $$\begin{bmatrix} 1 & 2 & | & 5 \\ 0 & 1 & | & 4 \end{bmatrix}$$

   Multiply row 2 by -2 and add to row 1.

   $$\begin{bmatrix} 1 & 0 & | & -3 \\ 0 & 1 & | & 4 \end{bmatrix}$$

   That is,
   $$x = -3$$
   $$y = 4.$$

   Thus, $\{(-3, 4)\}$ is the solution set.

3. $x + y = -3$
   $2x - 5y = -6$

   Form the augmented matrix.
   $$\begin{bmatrix} 1 & 1 & | & -3 \\ 2 & -5 & | & -6 \end{bmatrix}$$

   Multiply row 1 by -2 and add to row 2.
   $$\begin{bmatrix} 1 & 1 & | & -3 \\ 0 & -7 & | & 0 \end{bmatrix}$$

   Multiply row 2 by $-1/7$.
   $$\begin{bmatrix} 1 & 1 & | & -3 \\ 0 & 1 & | & 0 \end{bmatrix}$$

   Multiply row 2 by -1 and add to row 1.
   $$\begin{bmatrix} 1 & 0 & | & -3 \\ 0 & 1 & | & 0 \end{bmatrix}$$

   That is,
   $$x = -3$$
   $$y = 0.$$

   Thus, $\{(-3, 0)\}$ is the solution set.

4. $3x - 2y = 4$
   $3x + y = -2$

   Form the augmented matrix.
   $$\begin{bmatrix} 3 & -2 & | & 4 \\ 3 & 1 & | & -2 \end{bmatrix}$$

   Multiply row 1 by -1 and add to row 2.
   $$\begin{bmatrix} 3 & -2 & | & 4 \\ 0 & 3 & | & -6 \end{bmatrix}$$

   Multiply row 2 by $1/3$.
   $$\begin{bmatrix} 3 & -2 & | & 4 \\ 0 & 1 & | & -2 \end{bmatrix}$$

   Multiply row 2 by 2 and add to row 1.
   $$\begin{bmatrix} 3 & 0 & | & 0 \\ 0 & 1 & | & -2 \end{bmatrix}$$

   Multiply row 1 by $1/3$.
   $$\begin{bmatrix} 1 & 0 & | & 0 \\ 0 & 1 & | & -2 \end{bmatrix}$$

   That is,
   $$x = 0$$
   $$y = -2.$$

   Thus, $\{(0, -2)\}$ is the solution set.

5. $2x - 3y = 10$
   $2x + 2y = 5$

   Form the augmented matrix.
   $$\begin{bmatrix} 2 & -3 & | & 10 \\ 2 & 2 & | & 5 \end{bmatrix}$$

   Multiply row 1 by $1/2$.
   $$\begin{bmatrix} 1 & -3/2 & | & 5 \\ 2 & 2 & | & 5 \end{bmatrix}$$

   Multiply row 1 by -2 and add to row 2.
   $$\begin{bmatrix} 1 & -3/2 & | & 5 \\ 0 & 5 & | & -5 \end{bmatrix}$$

   Multiply row 2 by $1/5$.
   $$\begin{bmatrix} 1 & -3/2 & | & 5 \\ 0 & 1 & | & -1 \end{bmatrix}$$

   Multiply row 2 by $3/2$ and add to row 1.
   $$\begin{bmatrix} 1 & 0 & | & 7/2 \\ 0 & 1 & | & -1 \end{bmatrix}$$

   That is,
   $$x = 7/2$$
   $$y = -1.$$

   Thus, $\{(7/2, -1)\}$ is the solution set.

6. $4x + y = 5$
   $2x + y = 3$

   Form the augmented matrix.
   $$\begin{bmatrix} 4 & 1 & | & 5 \\ 2 & 1 & | & 3 \end{bmatrix}$$

   Multiply row 1 by $1/4$.
   $$\begin{bmatrix} 1 & 1/4 & | & 5/4 \\ 2 & 1 & | & 3 \end{bmatrix}$$

   Multiply row 1 by -2 and add to row 2.
   $$\begin{bmatrix} 1 & 1/4 & | & 5/4 \\ 0 & 1/2 & | & 1/2 \end{bmatrix}$$

   Multiply row 2 by 2.
   $$\begin{bmatrix} 1 & 1/4 & | & 5/4 \\ 0 & 1 & | & 1 \end{bmatrix}$$

   Multiply row 2 by $-1/4$ and add to row 1.
   $$\begin{bmatrix} 1 & 0 & | & 1 \\ 0 & 1 & | & 1 \end{bmatrix}$$

   That is,
   $$x = 1$$
   $$y = 1.$$

   Thus, $\{(1, 1)\}$ is the solution set.

7. $3x - 7y = 31$
   $2x - 4y = 18$

   Form the augmented matrix.
   $$\begin{bmatrix} 3 & -7 & | & 31 \\ 2 & -4 & | & 18 \end{bmatrix}$$

   Multiply row 1 by 1/3.
   $$\begin{bmatrix} 1 & -7/3 & | & 31/3 \\ 2 & -4 & | & 18 \end{bmatrix}$$

   Multiply row 1 by -2 and add to row 2.
   $$\begin{bmatrix} 1 & -7/3 & | & 31/3 \\ 0 & 2/3 & | & -8/3 \end{bmatrix}$$

   Multiply row 2 by 3/2.
   $$\begin{bmatrix} 1 & -7/3 & | & 31/3 \\ 0 & 1 & | & -4 \end{bmatrix}$$

   Multiply row 2 by 7/3 and add to row 1.
   $$\begin{bmatrix} 1 & 0 & | & 1 \\ 0 & 1 & | & -4 \end{bmatrix}$$

   That is,
   $x = 1$
   $y = -4.$

   Thus, $\{(1, -4)\}$ is the solution set.

8. $5x - y = 14$
   $x + 8y = 11$

   Form the augmented matrix.
   $$\begin{bmatrix} 5 & -1 & | & 14 \\ 1 & 8 & | & 11 \end{bmatrix}$$

   Interchange row 1 and row 2.
   $$\begin{bmatrix} 1 & 8 & | & 11 \\ 5 & -1 & | & 14 \end{bmatrix}$$

   Multiply row 1 by -5 and add to row 2.
   $$\begin{bmatrix} 1 & 8 & | & 11 \\ 0 & -41 & | & -41 \end{bmatrix}$$

   Multiply row 2 by $-1/41$.
   $$\begin{bmatrix} 1 & 8 & | & 11 \\ 0 & 1 & | & 1 \end{bmatrix}$$

   Multiply row 2 by -8 and add to row 1.
   $$\begin{bmatrix} 1 & 0 & | & 3 \\ 0 & 1 & | & 1 \end{bmatrix}$$

   That is,
   $x = 3$
   $y = 1.$

   Thus, $\{(3, 1)\}$ is the solution set.

9. $x + y - z = 6$
   $2x - y + z = -9$
   $x - 2y + 3z = 1$

   Form the augmented matrix.
   $$\begin{bmatrix} 1 & 1 & -1 & | & 6 \\ 2 & -1 & 1 & | & -9 \\ 1 & -2 & 3 & | & 1 \end{bmatrix}$$

   Multiply row 1 by -1 and add to row 3.
   Multiply row 1 by -2 and add to row 2.
   $$\begin{bmatrix} 1 & 1 & -1 & | & 6 \\ 0 & -3 & 3 & | & -21 \\ 0 & -3 & 4 & | & -5 \end{bmatrix}$$

   Multiply row 2 by -1 and add to row 3.
   $$\begin{bmatrix} 1 & 1 & -1 & | & 6 \\ 0 & -3 & 3 & | & -21 \\ 0 & 0 & 1 & | & 16 \end{bmatrix}$$

   Multiply row 2 by $-1/3$.
   $$\begin{bmatrix} 1 & 1 & -1 & | & 6 \\ 0 & 1 & -1 & | & 7 \\ 0 & 0 & 1 & | & 16 \end{bmatrix}$$

   Multiply row 2 by -1 and add to row 1.
   $$\begin{bmatrix} 1 & 0 & 0 & | & -1 \\ 0 & 1 & -1 & | & 7 \\ 0 & 0 & 1 & | & 16 \end{bmatrix}$$

   Add row 3 to row 2.
   $$\begin{bmatrix} 1 & 0 & 0 & | & -1 \\ 0 & 1 & 0 & | & 23 \\ 0 & 0 & 1 & | & 16 \end{bmatrix}$$

   That is,
   $x = -1$
   $y = 23$
   $z = 16.$

   Thus, $\{(-1, 23, 16)\}$ is the solution set.

10. $x + 3y - 6z = 7$
    $2x - y + 2z = 0$
    $x + y + 2z = -1$

    Form the augmented matrix.
    $$\begin{bmatrix} 1 & 3 & -6 & | & 7 \\ 2 & -1 & 2 & | & 0 \\ 1 & 1 & 2 & | & -1 \end{bmatrix}$$

    Multiply row 1 by -1 and add to row 3.
    Multiply row 1 by -2 and add to row 2.
    $$\begin{bmatrix} 1 & 3 & -6 & | & 7 \\ 0 & -7 & 14 & | & -14 \\ 0 & -2 & 8 & | & -8 \end{bmatrix}$$

Multiply row 2 by $-1/7$ and row 3 by $1/2$.
$$\begin{bmatrix} 1 & 3 & -6 & | & 7 \\ 0 & 1 & -2 & | & 2 \\ 0 & -1 & 4 & | & -4 \end{bmatrix}$$

Add row 2 to row 3.
$$\begin{bmatrix} 1 & 3 & -6 & | & 7 \\ 0 & 1 & -2 & | & 2 \\ 0 & 0 & 2 & | & -2 \end{bmatrix}$$

Multiply row 3 by $1/2$.
$$\begin{bmatrix} 1 & 3 & -6 & | & 7 \\ 0 & 1 & -2 & | & 2 \\ 0 & 0 & 1 & | & -1 \end{bmatrix}$$

Multiply row 3 by 2 and add to row 2.
$$\begin{bmatrix} 1 & 3 & -6 & | & 7 \\ 0 & 1 & 0 & | & 0 \\ 0 & 0 & 1 & | & -1 \end{bmatrix}$$

Multiply row 2 by -3 and add to row 1.
$$\begin{bmatrix} 1 & 0 & -6 & | & 7 \\ 0 & 1 & 0 & | & 0 \\ 0 & 0 & 1 & | & -1 \end{bmatrix}$$

Multiply row 3 by 6 and add to row 1.
$$\begin{bmatrix} 1 & 0 & 0 & | & 1 \\ 0 & 1 & 0 & | & 0 \\ 0 & 0 & 1 & | & -1 \end{bmatrix}$$

That is,
$$x = 1$$
$$y = 0$$
$$z = -1.$$

Thus, $\{(1, 0, -1)\}$ is the solution set.

11. $2x - y + 3z = 0$
    $x + 2y - z = 5$
    $2y + z = 1$

Form the augmented matrix.
$$\begin{bmatrix} 2 & -1 & 3 & | & 0 \\ 1 & 2 & -1 & | & 5 \\ 0 & 2 & 1 & | & 1 \end{bmatrix}$$

Interchange row 1 and row 2.
$$\begin{bmatrix} 1 & 2 & -1 & | & 5 \\ 2 & -1 & 3 & | & 0 \\ 0 & 2 & 1 & | & 1 \end{bmatrix}$$

Multiply row 1 by -2 and add to row 2.
$$\begin{bmatrix} 1 & 2 & -1 & | & 5 \\ 0 & -5 & 5 & | & -10 \\ 0 & 2 & 1 & | & 1 \end{bmatrix}$$

Multiply row 2 by $-1/5$.
$$\begin{bmatrix} 1 & 2 & -1 & | & 5 \\ 0 & 1 & -1 & | & 2 \\ 0 & 2 & 1 & | & 1 \end{bmatrix}$$

Multiply row 2 by -2 and add to row 3.
$$\begin{bmatrix} 1 & 2 & -1 & | & 5 \\ 0 & 1 & -1 & | & 2 \\ 0 & 0 & 3 & | & -3 \end{bmatrix}$$

Multiply row 3 by $1/3$.
$$\begin{bmatrix} 1 & 2 & -1 & | & 5 \\ 0 & 1 & -1 & | & 2 \\ 0 & 0 & 1 & | & -1 \end{bmatrix}$$

Add row 3 to row 1 and to row 2.
$$\begin{bmatrix} 1 & 2 & 0 & | & 4 \\ 0 & 1 & 0 & | & 1 \\ 0 & 0 & 1 & | & -1 \end{bmatrix}$$

Add -2 times row 2 to row 1.
$$\begin{bmatrix} 1 & 0 & 0 & | & 2 \\ 0 & 1 & 0 & | & 1 \\ 0 & 0 & 1 & | & -1 \end{bmatrix}$$

That is,
$$x = 2$$
$$y = 1$$
$$z = -1.$$

Thus, $\{(2, 1, -1)\}$ is the solution set.

12. $4x + 2y - 3z = 6$
    $x - 4y + z = -4$
    $-x + 2z = 2$

Form the augmented matrix.
$$\begin{bmatrix} 4 & 2 & -3 & | & 6 \\ 1 & -4 & 1 & | & -4 \\ -1 & 0 & 2 & | & 2 \end{bmatrix}$$

Interchange row 1 and row 2.
$$\begin{bmatrix} 1 & -4 & 1 & | & -4 \\ 4 & 2 & -3 & | & 6 \\ -1 & 0 & 2 & | & 2 \end{bmatrix}$$

Multiply row 1 by -4 and add to row 2.
$$\begin{bmatrix} 1 & -4 & 1 & | & -4 \\ 0 & 18 & -7 & | & 22 \\ -1 & 0 & 2 & | & 2 \end{bmatrix}$$

Add row 1 to row 3.
$$\begin{bmatrix} 1 & -4 & 1 & | & -4 \\ 0 & 18 & -7 & | & 22 \\ 0 & -4 & 3 & | & -2 \end{bmatrix}$$

Multiply row 3 by -1 and add to row 1.
$$\begin{bmatrix} 1 & 0 & -2 & | & -2 \\ 0 & 18 & -7 & | & 22 \\ 0 & -4 & 3 & | & -2 \end{bmatrix}$$
Multiply row 3 by 4 and add to row 2.
$$\begin{bmatrix} 1 & 0 & -2 & | & -2 \\ 0 & 2 & 5 & | & 14 \\ 0 & -4 & 3 & | & -2 \end{bmatrix}$$
Multiply row 2 by 2 and add to row 3.
$$\begin{bmatrix} 1 & 0 & -2 & | & -2 \\ 0 & 2 & 5 & | & 14 \\ 0 & 0 & 13 & | & 26 \end{bmatrix}$$
Multiply row 3 by 1/13.
$$\begin{bmatrix} 1 & 0 & -2 & | & -2 \\ 0 & 2 & 5 & | & 14 \\ 0 & 0 & 1 & | & 2 \end{bmatrix}$$
Multiply row 3 by -5 and add to row 2.
$$\begin{bmatrix} 1 & 0 & -2 & | & -2 \\ 0 & 2 & 0 & | & 4 \\ 0 & 0 & 1 & | & 2 \end{bmatrix}$$
Multiply row 2 by 1/2.
$$\begin{bmatrix} 1 & 0 & -2 & | & -2 \\ 0 & 1 & 0 & | & 2 \\ 0 & 0 & 1 & | & 2 \end{bmatrix}$$
Multiply row 3 by 2 and add to row 1.
$$\begin{bmatrix} 1 & 0 & 0 & | & 2 \\ 0 & 1 & 0 & | & 2 \\ 0 & 0 & 1 & | & 2 \end{bmatrix}$$
That is,
$$x = 2$$
$$y = 2$$
$$z = 2.$$
Thus, $\{(2, 2, 2)\}$ is the solution set.

13. $-x + y = -1$
$y - z = 6$
$x + z = -1$

Form the augmented matrix.
$$\begin{bmatrix} -1 & 1 & 0 & | & -1 \\ 0 & 1 & -1 & | & 6 \\ 1 & 0 & 1 & | & -1 \end{bmatrix}$$
Multiply row 1 by -1.
$$\begin{bmatrix} 1 & -1 & 0 & | & 1 \\ 0 & 1 & -1 & | & 6 \\ 1 & 0 & 1 & | & -1 \end{bmatrix}$$

Multiply row 1 by -1 and add to row 3.
$$\begin{bmatrix} 1 & -1 & 0 & | & 1 \\ 0 & 1 & -1 & | & 6 \\ 0 & 1 & 1 & | & -2 \end{bmatrix}$$
Add row 2 to row 1.
$$\begin{bmatrix} 1 & 0 & -1 & | & 7 \\ 0 & 1 & -1 & | & 6 \\ 0 & 1 & 1 & | & -2 \end{bmatrix}$$
Multiply row 2 by -1 and add to row 3.
$$\begin{bmatrix} 1 & 0 & -1 & | & 7 \\ 0 & 1 & -1 & | & 6 \\ 0 & 0 & 2 & | & -8 \end{bmatrix}$$
Multiply row 3 by 1/2.
$$\begin{bmatrix} 1 & 0 & -1 & | & 7 \\ 0 & 1 & -1 & | & 6 \\ 0 & 0 & 1 & | & -4 \end{bmatrix}$$
Add row 3 to row 1.
$$\begin{bmatrix} 1 & 0 & 0 & | & 3 \\ 0 & 1 & -1 & | & 6 \\ 0 & 0 & 1 & | & -4 \end{bmatrix}$$
Add row 3 to row 2.
$$\begin{bmatrix} 1 & 0 & 0 & | & 3 \\ 0 & 1 & 0 & | & 2 \\ 0 & 0 & 1 & | & -4 \end{bmatrix}$$
That is,
$$x = 3$$
$$y = 2$$
$$z = -4.$$
Thus, $\{(3, 2, -4)\}$ is the solution set.

14. $x + y = 1$
$2x - z = 0$
$y + 2z = -2$

Form the augmented matrix.
$$\begin{bmatrix} 1 & 1 & 0 & | & 1 \\ 2 & 0 & -1 & | & 0 \\ 0 & 1 & 2 & | & -2 \end{bmatrix}$$
Multiply row 1 by -2 and add to row 2.
$$\begin{bmatrix} 1 & 1 & 0 & | & 1 \\ 0 & -2 & -1 & | & -2 \\ 0 & 1 & 2 & | & -2 \end{bmatrix}$$
Interchange row 2 and row 3.
$$\begin{bmatrix} 1 & 1 & 0 & | & 1 \\ 0 & 1 & 2 & | & -2 \\ 0 & -2 & -1 & | & -2 \end{bmatrix}$$

Multiply row 2 by 2 and add to row 3.
$$\begin{bmatrix} 1 & 1 & 0 & | & 1 \\ 0 & 1 & 2 & | & -2 \\ 0 & 0 & 3 & | & -6 \end{bmatrix}$$

Multiply row 3 by 1/3.
$$\begin{bmatrix} 1 & 1 & 0 & | & 1 \\ 0 & 1 & 2 & | & -2 \\ 0 & 0 & 1 & | & -2 \end{bmatrix}$$

Multiply row 3 by -2 and add to row 2.
$$\begin{bmatrix} 1 & 1 & 0 & | & 1 \\ 0 & 1 & 0 & | & 2 \\ 0 & 0 & 1 & | & -2 \end{bmatrix}$$

Multiply row 2 by -1 and add to row 1.
$$\begin{bmatrix} 1 & 0 & 0 & | & -1 \\ 0 & 1 & 0 & | & 2 \\ 0 & 0 & 1 & | & -2 \end{bmatrix}$$

That is,
$$x = -1$$
$$y = 2$$
$$z = -2.$$

Thus, $\{(-1, 2, -2)\}$ is the solution set.

15. $2x - y + 4z = -1$
$-3x + 5y - z = 5$
$2x + 3y + 2z = 3$

Form the augmented matrix.
$$\begin{bmatrix} 2 & -1 & 4 & | & -1 \\ -3 & 5 & -1 & | & 5 \\ 2 & 3 & 2 & | & 3 \end{bmatrix}$$

Add row 2 to row 1.
$$\begin{bmatrix} -1 & 4 & 3 & | & 4 \\ -3 & 5 & -1 & | & 5 \\ 2 & 3 & 2 & | & 3 \end{bmatrix}$$

Multiply row 1 by 2 and add to row 3.
$$\begin{bmatrix} -1 & 4 & 3 & | & 4 \\ -3 & 5 & -1 & | & 5 \\ 0 & 11 & 8 & | & 11 \end{bmatrix}$$

Multiply row 1 by -1.
$$\begin{bmatrix} 1 & -4 & -3 & | & -4 \\ -3 & 5 & -1 & | & 5 \\ 0 & 11 & 8 & | & 11 \end{bmatrix}$$

Multiply row 1 by 3 and add to row 2.
$$\begin{bmatrix} 1 & -4 & -3 & | & -4 \\ 0 & -7 & -10 & | & -7 \\ 0 & 11 & 8 & | & 11 \end{bmatrix}$$

Multiply row 2 by $-1/7$.
$$\begin{bmatrix} 1 & -4 & -3 & | & -4 \\ 0 & 1 & 10/7 & | & 1 \\ 0 & 11 & 8 & | & 11 \end{bmatrix}$$

Multiply row 2 by 4 and add to row 1.
$$\begin{bmatrix} 1 & 0 & 19/7 & | & 0 \\ 0 & 1 & 10/7 & | & 1 \\ 0 & 11 & 8 & | & 11 \end{bmatrix}$$

Multiply row 2 by -11 and add to row 3.
$$\begin{bmatrix} 1 & 0 & 19/7 & | & 0 \\ 0 & 1 & 10/7 & | & 1 \\ 0 & 0 & -54/7 & | & 0 \end{bmatrix}$$

Multiply row 3 by $-7/54$.
$$\begin{bmatrix} 1 & 0 & 19/7 & | & 0 \\ 0 & 1 & 10/7 & | & 1 \\ 0 & 0 & 1 & | & 0 \end{bmatrix}$$

Multiply row 3 by $-19/7$ and add to row 1.
Multiply row 3 by $-10/7$ and add to row 2.
$$\begin{bmatrix} 1 & 0 & 0 & | & 0 \\ 0 & 1 & 0 & | & 1 \\ 0 & 0 & 1 & | & 0 \end{bmatrix}$$

That is,
$$x = 0$$
$$y = 1$$
$$z = 0.$$

Thus, $\{(0, 1, 0)\}$ is the solution set.

16. $5x - 3y + 2z = -5$
$2x + 2y - z = 4$
$4x - y + z = -1$

Form the augmented matrix.
$$\begin{bmatrix} 5 & -3 & 2 & | & -5 \\ 2 & 2 & -1 & | & 4 \\ 4 & -1 & 1 & | & -1 \end{bmatrix}$$

Multiply row 2 by -2 and add to row 3.
$$\begin{bmatrix} 5 & -3 & 2 & | & -5 \\ 2 & 2 & -1 & | & 4 \\ 0 & -5 & 3 & | & -9 \end{bmatrix}$$

Multiply row 2 by 1/2.
$$\begin{bmatrix} 5 & -3 & 2 & | & -5 \\ 1 & 1 & -1/2 & | & 2 \\ 0 & -5 & 3 & | & -9 \end{bmatrix}$$

Interchange row 1 and row 2.
$$\begin{bmatrix} 1 & 1 & -1/2 & | & 2 \\ 5 & -3 & 2 & | & -5 \\ 0 & -5 & 3 & | & -9 \end{bmatrix}$$

Multiply row 1 by -5 and add to row 2.
$$\begin{bmatrix} 1 & 1 & -1/2 & | & 2 \\ 0 & -8 & 9/2 & | & -15 \\ 0 & -5 & 3 & | & -9 \end{bmatrix}$$

Multiply row 2 by $1/8$.
$$\begin{bmatrix} 1 & 1 & -1/2 & | & 2 \\ 0 & -1 & 9/16 & | & -15/8 \\ 0 & -5 & 3 & | & -9 \end{bmatrix}$$

Add row 2 to row 1.
$$\begin{bmatrix} 1 & 0 & 1/16 & | & 1/8 \\ 0 & -1 & 9/16 & | & -15/8 \\ 0 & -5 & 3 & | & -9 \end{bmatrix}$$

Multiply row 2 by -5 and add to row 3.
$$\begin{bmatrix} 1 & 0 & 1/16 & | & 1/8 \\ 0 & -1 & 9/16 & | & -15/8 \\ 0 & 0 & 3/16 & | & 3/8 \end{bmatrix}$$

Multiply row 2 by -1 and row 3 by $16/3$.
$$\begin{bmatrix} 1 & 0 & 1/16 & | & 1/8 \\ 0 & 1 & -9/16 & | & 15/8 \\ 0 & 0 & 1 & | & 2 \end{bmatrix}$$

Multiply row 3 by $-1/16$ and add to row 1.
Multiply row 3 by $9/16$ and add to row 2.
$$\begin{bmatrix} 1 & 0 & 0 & | & 0 \\ 0 & 1 & 0 & | & 3 \\ 0 & 0 & 1 & | & 2 \end{bmatrix}$$

That is,
$$x = 0$$
$$y = 3$$
$$z = 2.$$

Thus, $\{(0, 3, 2)\}$ is the solution set.

8.8 EXERCISES

1. The answer is C since this represents the region where $x \geq 5$ and (at the same time) $y \leq -3$.

2. The answer is A since this represents the region where $x \leq 5$ and (at the same time) $y \geq -3$.

3. The answer is B since this represents the region where $x > 5$ and (at the same time) $y < -3$.

4. The answer is D since this represents the region where $x < 5$ and (at the same time) $y > -3$.

5. $x + y \leq 2$
   Graph the boundary line
   $$x + y = 2.$$
   Since the inequality is of the form "$\leq$," i.e., includes the equality, the graph is a solid line. Next, try a test point not on the line, such as $(0, 0)$, in the inequality.
   $$x + y \leq 2$$
   $$0 + 0 \leq 2 \quad \text{True}$$
   Thus, shade the region containing the point $(0, 0)$ and all the other points in the region below and including the line itself.

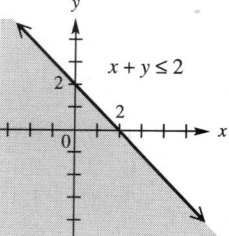

6. $x - y \geq -3$
   Graph the boundary line
   $$x - y = -3.$$
   Since the inequality is of the form "$\geq$," i.e., includes the equality, the graph is a solid line. Next, try a test point not on the line, such as $(0, 0)$, in the inequality.
   $$x - y \geq -3$$
   $$0 - 0 \geq -3 \quad \text{True}$$
   Thus, shade the region containing the point $(0, 0)$, and, all the other points in the region below and including the line.

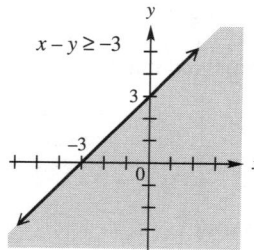

7. $4x - y \leq 5$
   Graph the boundary line
   $$4x - y = 5.$$
   Try a test point, such as $(0, 0)$, in the inequality.
   $$4x - y \leq 5$$
   $$4(0) - 0 \leq 5 \quad \text{True}$$
   Thus, the half-plane including $(0, 0)$ is to be shaded. Observe that the line itself is a part of the solution set.

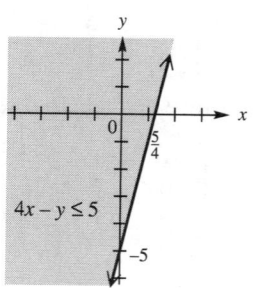

8. $3x + y \geq 6$
   Graph the boundary line
   $$3x + y = 6.$$
   Try a test point, such as $(0, 0)$, in the inequality.
   $$3x + y \geq 6$$
   $$3(0) + (0) \geq 6$$
   $$0 \geq 6 \quad \text{False}$$

   Since the point does not work, the half-plane not including $(0, 0)$ is to be shaded. That is, shade the other half-plane. Observe that the line itself is a part of the solution set.

   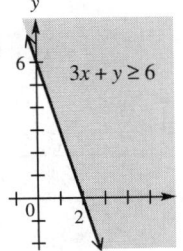

9. $x + 3y \geq -2$
   Graph the boundary line
   $$x + 3y = -2.$$
   Try a test point, such as $(0, 0)$, in the inequality.
   $$x + 3y \geq -2$$
   $$0 + 3(0) \geq -2 \quad \text{True}$$

   Thus, the half-plane including $(0, 0)$ is to be shaded. The line itself is also a part of the solution set.

   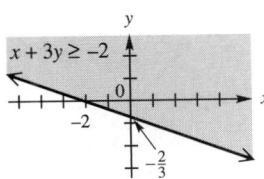

10. $4x + 6y \leq -3$
    Graph the boundary line
    $$4x + 6y = -3.$$
    Try a test point, such as $(0, 0)$, in the inequality.
    $$4x + 6y \leq -3$$
    $$4(0) + 6(0) \leq -3$$
    $$0 \leq -3 \quad \text{False}$$

    Since the point does not work, the other half-plane is to be shaded. The line itself is a part of the solution set.

    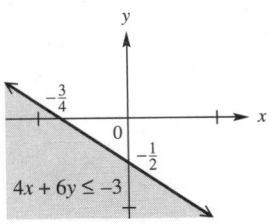

11. $x + 2y \leq -5$
    Graph the boundary line
    $$x + 2y = -5.$$
    Try a test point, such as $(0, 0)$, in the inequality.
    $$x + 2y \leq -5$$
    $$0 + 2(0) \leq -5$$
    $$0 \leq -5 \quad \text{False}$$

    Since the point does not work, the other half-plane represents the solution and is to be shaded. The line itself is a part of the solution set.

12. $2x - 4y \leq 3$
    Graph the boundary line
    $$2x - 4y = 3.$$
    Try a test point, such as $(0, 0)$, in the inequality.
    $$2x - 4y \leq 3$$
    $$2(0) - 4(0) \leq 3$$
    $$0 \leq 3 \quad \text{True}$$

    Thus, the half-plane including $(0, 0)$ is to be shaded. The line itself is a part of the solution set.

    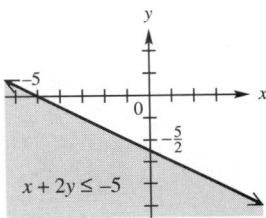

13. $4x - 3y < 12$
    Graph the boundary line
    $$4x - 3y = 12.$$
    Try a test point, such as $(0, 0)$, in the inequality.
    $$4x - 3y < 12$$
    $$4(0) - 3(0) < 12$$
    $$0 < 12 \quad \text{True}$$

Thus, the half-plane including $(0, 0)$ is to be shaded. The line itself is not a part of the solution set and, therefore, is indicated with a dashed line.

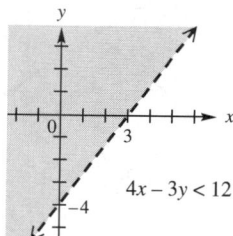

14. $5x + 3y > 15$

    Graph the boundary line

    $$5x + 3y = 15.$$

    Try a test point, such as $(0, 0)$, in the inequality.

    $$5x + 3y > 15$$
    $$5(0) + 3(0) > 15$$
    $$0 > 15 \quad False$$

    Thus, the half-plane which does not include $(0, 0)$ is to be shaded. The line itself is not at part of the solution.

    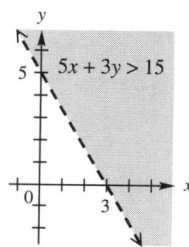

15. $y > -x$

    Use the alternate equivalent form, $x + y > 0$. Graph the boundary line

    $$x + y = 0.$$

    Try a test point, such as $(1, 1)$, which does not lie on the line.

    $$1 + 1 > 0$$
    $$2 > 0 \quad True$$

    Thus, the half-plane to be shaded is that above the line and includes the point $(1, 1)$. The line must be dashed since "=" is not included in the strict inequality " $>$."

    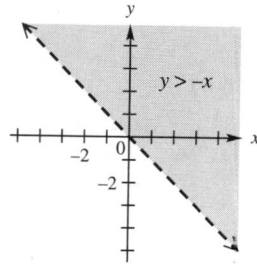

16. $y < x$

    Use the alternate equivalent form, $-x + y < 0$. Graph the boundary line

    $$-x + y = 0.$$

    Try a test point, such as $(2, 3)$, which lies above the line and hence does not lie on the line.

    $$-2 + 3 < 0$$
    $$1 < 0 \quad False$$

    Thus, the half-plane to be shaded is that below the line and does not include the point $(2, 3)$. The line must be dashed since "=" is not included in the strict inequality " $<$."

    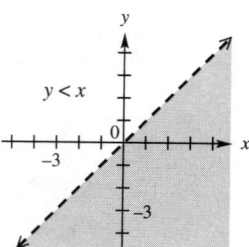

17. $x + y \leq 1$
    $x \geq 0$

    Using the boundary equations

    $$x + y = 1 \quad \text{and}$$
    $$x = 0, \ (y\text{-axis})..$$

    sketch the graph of each individual inequality, shading the appropriate half-planes. The intersection of these regions (area in common) represents the solution set as below.

    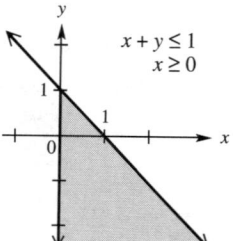

18. $3x - 4y \leq 6$
    $y \geq 1$

    Using the boundary equations

    $$3x - 4y = 6 \quad \text{and}$$
    $$y = 1,$$

    sketch the graph of each individual inequality, shading the appropriate half-planes. The intersection of these regions (area in common) represents the solution set as below.

19. $2x - y \geq 1$
    $3x + 2y \geq 6$

    Using the boundary equations
    $$2x - y = 1 \quad \text{and}$$
    $$3x + 2y = 6,$$
    sketch the graph of each individual inequality, shading the appropriate half-planes. The intersection of these regions represents the solution set as shown below.

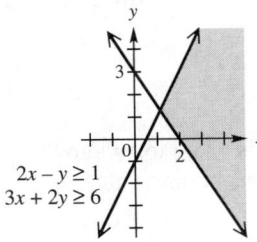

20. $x + 3y \geq 6$
    $3x - 4y \leq 12$

    Using the boundary equations
    $$x + 3y = 6 \quad \text{and}$$
    $$3x - 4y = 12,$$
    sketch the graph of each individual inequality, shading the appropriate half-planes. The intersection of these regions represents the solution set as shown below.

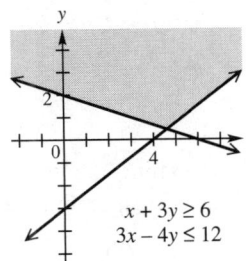

21. $-x - y < 5$
    $x - y \leq 3$

    Use the boundary equations
    $$-x - y = 5 \quad \text{and}$$
    $$x - y = 3$$
    to sketch the graph of each individual inequality, shading the appropriate half-planes. Note that the line itself is not included with the first inequality. The intersection of these regions represents the solution set as shown below.

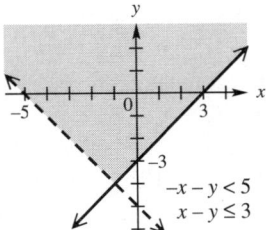

22. $6x - 4y < 8$
    $x + 2y \geq 4$

    Use the boundary equations
    $$6x - 4y = 8 \quad \text{and}$$
    $$x + 2y = 4$$
    to sketch the graph of each individual inequality, shading the appropriate half-planes. Note that the line itself is not included with the first inequality. The intersection of these regions represents the solution set as shown below.

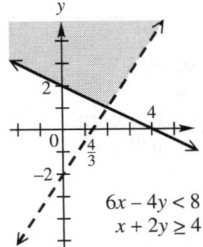

23. Evaluate $3x + 5y$ at each vertex.

Point	Value = $3x + 5y$	
(1, 1)	$3(1) + 5(1) = 8$	←minimum
(6, 3)	$3(6) + 5(3) = 33$	
(5, 10)	$3(5) + 5(10) = 65$	←maximum
(2, 7)	$3(2) + 5(7) = 41$	

    Thus, there is a maximum of 65 at $(5, 10)$ and a minimum of 8 at $(1, 1)$.

24. Evaluate $40x + 75y$ at each vertex.

Point	Value = $40x + 75y$	
(0, 0)	$40(0) + 75(0) = 0$	←minimum
(0, 12)	$40(0) + 75(12) = 900$	←maximum
(4, 8)	$40(4) + 75(8) = 760$	
(7, 3)	$40(7) + 75(3) = 505$	
(8, 0)	$40(8) + 75(0) = 320$	

    Thus, there is a maximum of 900 at $(0, 12)$ and a minimum of 0 at $(0, 0)$.

25. Find $x \geq 0$ and $y \geq 0$ such that
    $$2x + 3y \leq 6$$
    $$4x + y \leq 6$$
    and $5x + 2y$ is maximized.

To find the vertex points, solve the system of boundary equations.
$$2x + 3y = 6$$
$$4x + y = 6$$

Multiply the second equation by -3 and add to the first equation.
$$2x + 3y = 6$$
$$-12x - 3y = -18$$
$$\overline{-10x = -12}$$
$$x = \frac{-12}{-10} = \frac{6}{5}$$

Substitute this value for x in the second equation $4x + y = 6$.
$$4\left(\frac{6}{5}\right) + y = 6$$
$$y = 6 - \frac{24}{5}$$
$$y = \frac{30}{5} - \frac{24}{5}$$
$$y = \frac{6}{5}$$

Thus, a corner point is $\left(\frac{6}{5}, \frac{6}{5}\right)$.

Sketch the graph of the feasible region representing the intersection of all of the constraints (i.e., system of inequalities).

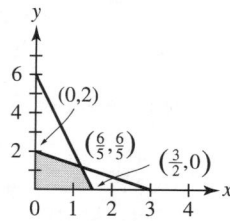

Evaluate $5x + 2y$ at each vertex.

Point	Value $= 5x + 2y$	
$(0, 2)$	$5(0) + 2(2) = 4$	
$\left(\frac{6}{5}, \frac{6}{5}\right)$	$5\left(\frac{6}{5}\right) + 2\left(\frac{6}{5}\right) = \frac{42}{5}$	←maximum
$\left(\frac{3}{2}, 0\right)$	$5\left(\frac{3}{2}\right) + 2(0) = \frac{15}{2}$	
$(0, 0)$	$5(0) + 2(0) = 0$	

Thus, the maximum value occurs at the vertex point $\left(\frac{6}{5}, \frac{6}{5}\right)$ and has the value $\frac{42}{5}$.

26. Find $x \geq 0$ and $y \geq 0$ such that
$$x + y \leq 10$$
$$5x + 2y \geq 20$$
$$2y \geq x$$

and $x + 3y$ is minimized.

Begin by sketching a graph of the solution to the system (feasible region).

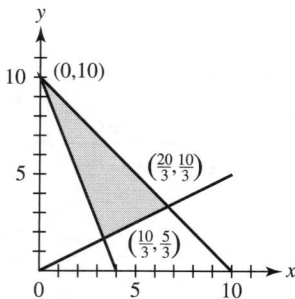

The vertices of the feasible region are the intersection points of each pair of boundary lines. The intersection of the lines $x + y = 10$ and $5x + 2y = 20$ is their common y-intercept $(0, 10)$. The other two vertices can be found by solving the following systems of equations.
$$x + y = 10$$
$$2y = x$$

Substitute $2y$ for x in the first equation and solve for y.
$$(2y) + y = 10$$
$$3y = 10$$
$$y = \frac{10}{3}$$

Substitute this value for y into the second equation to find the corresponding value for x.
$$2\left(\frac{10}{3}\right) = x$$
$$x = \frac{20}{3}$$

This gives the vertex $\left(\frac{20}{3}, \frac{10}{3}\right)$.

To find the remaining vertex solve the following system.
$$5x + 2y = 20$$
$$2y = x$$

Substitute $2y$ for x in the first equation and solve for y.
$$5(2y) + 2y = 20$$
$$12y = 20$$
$$y = \frac{20}{12} = \frac{5}{3}$$

Substitute this value for y into the second equation $2y = x$ to find the corresponding value for x.
$$2\left(\frac{5}{3}\right) = x$$
$$x = \frac{10}{3}$$

This gives the vertex $\left(\frac{10}{3}, \frac{5}{3}\right)$.

Evaluate $x + 3y$ at each vertex.

Point	Value $= x + 3y$
$(0, 10)$	$(0) + 3(10) = 30$
$\left(\frac{20}{3}, \frac{10}{3}\right)$	$\left(\frac{20}{3}\right) + 3\left(\frac{10}{3}\right) = \frac{50}{3}$
$\left(\frac{10}{3}, \frac{5}{3}\right)$	$\left(\frac{10}{3}\right) + 3\left(\frac{5}{3}\right) = \frac{25}{3}$ ←minimum

Thus, the minimum value occurs at the corner point $\left(\frac{10}{3}, \frac{5}{3}\right)$ and has the value $\frac{25}{3}$.

27. Find $x \geq 2$ and $y \geq 5$ such that
$$3x - y \geq 12$$
$$x + y \leq 15$$
and $2x + y$ is minimized.

Sketch a graph of the solution to the system (feasible region).

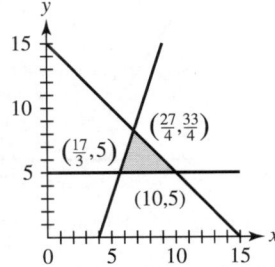

The vertices of the feasible region are the intersection points of each pair of boundary lines. The top vertex point is the intersection of the lines $3x - y = 12$ and $x + y = 15$.

Solve this system.
$$3x - y = 12$$
$$x + y = 15$$

Add the equations to eliminate y.
$$3x - y = 12$$
$$x + y = 15$$
$$\overline{4x = 27}$$
$$x = \frac{27}{4}$$

Substitute this value of x into the second equation $x + y = 15$ and solve for y.
$$\left(\frac{27}{4}\right) + y = 15$$
$$y = 15 - \frac{27}{4}$$
$$y = \frac{60}{4} - \frac{27}{4} = \frac{33}{4}$$

The resulting vertex point is $\left(\frac{27}{4}, \frac{33}{4}\right)$.

The bottom left vertex can be found by solving the system.
$$3x - y = 12$$
$$y = 5$$

Substitute this value of y into the first equation $3x - y = 12$ and solve for x.
$$3x - (5) = 12$$
$$3x = 17$$
$$x = \frac{17}{3}$$

This gives the vertex $\left(\frac{17}{3}, 5\right)$.

To find the remaining vertex solve the following system.
$$x + y = 15$$
$$y = 5$$

Substitute this value of y into the first equation $x + y = 15$ and solve for x.
$$x + (5) = 15$$
$$x = 10$$

This gives the vertex $(10, 5)$.

Evaluate $2x + y$ at each vertex.

Point	Value $= 2x + y$
$\left(\frac{27}{4}, \frac{33}{4}\right)$	$2\left(\frac{27}{4}\right) + \left(\frac{33}{4}\right) = \frac{87}{4} = 21.75$
$\left(\frac{17}{3}, 5\right)$	$2\left(\frac{17}{3}\right) + (5) = \frac{49}{3} \approx 16.33$ ←minimum
$(10, 5)$	$2(10) + (5) = 25$

Thus, the vertex point $\left(\frac{17}{3}, 5\right)$ gives the minimum value, $\frac{49}{3}$.

28. Find $x \geq 10$ and $y \geq 20$ such that
$$2x + 3y \leq 100$$
$$5x + 4y \leq 200$$
and $x + 3y$ is maximized.

Sketch a graph of the feasible region.

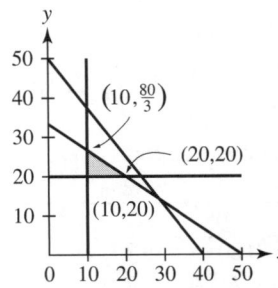

Evaluate $x + 3y$ at each vertex.

Point	Value $= x + 3y$
$(10, 20)$	$(10) + 3(20) = 70$
$(20, 20)$	$(20) + 3(20) = 80$
$\left(10, \frac{80}{3}\right)$	$(10) + 3\left(\frac{80}{3}\right) = 90$ ←maximum

Thus, the maximum value occurs at the corner point $\left(10, \frac{80}{3}\right)$ and has the value 90.

Note: In the following exercises, $x \geq 0$ and $y \geq 0$ are understood constraints.

29. Let $x =$ the number of refrigerators shipped to warehouse A and

    $y =$ the number of refrigerators shipped to Warehouse B.

    Since at least 100 refrigerators must be shipped,
    $$x + y \geq 100.$$

    Since warehouse A holds a maximum of 100 refrigerators and has 25 already, it has room for at most 75 more, so
    $$0 \leq x \leq 75.$$

    Similarly, warehouse B has room for at most 80 more refrigerators, so
    $$0 \leq y \leq 80.$$

    The cost function is given by $12x + 10y$.

    The linear programming problem may be stated as follows:

    Find $x \geq 0$ and $y \geq 0$ such that
    $$x + y \geq 100$$
    $$x \leq 75$$
    $$y \leq 80$$

    and $12x + 10y$ is minimized.

    Graph the feasible region.

    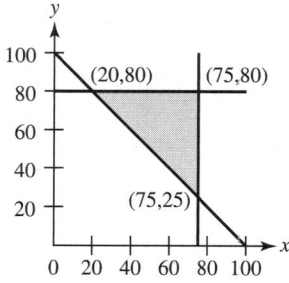

    Evaluate the objective function at each vertex.

Point	Cost = $12x + 10y$	
$(20, 80)$	$12(20) + 10(80) = 1040$	←minimum
$(75, 80)$	$12(75) + 10(80) = 1700$	
$(75, 25)$	$12(75) + 10(25) = 1150$	

    The minimum value of the objective function is 1040 occurring at $(20, 80)$. Therefore, 20 refrigerators should be shipped to warehouse A and 80 to warehouse B, for a minimum cost of $1040.

30. Let $x =$ number of servings of product A and
    $y =$ number of servings of product B.

    Then, the following constraints apply.

    Find $x \geq 0$ and $y \geq 0$ such that
    $$3x + 2y \geq 15 \quad \text{(supplement I)}$$
    $$2x + 4y \geq 15 \quad \text{(supplement II)}$$

    where the Cost $= .25x + .40y = \frac{1}{4}x + \frac{2}{5}y$ is the objective function to be minimized.

    Solve the corresponding equations to find the vertex points, and graph the feasible region.

    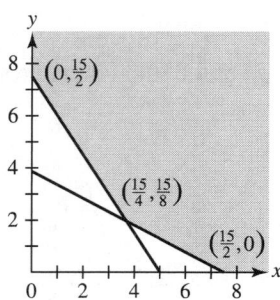

    Evaluate the objective function at each vertex.

Point	Value = $12x + 10y$	
$\left(0, \frac{15}{2}\right)$	$\frac{1}{4}(0) + \frac{2}{5}\left(\frac{15}{2}\right) = \$3.00$	
$\left(\frac{15}{4}, \frac{15}{8}\right)$	$\frac{1}{4}\left(\frac{15}{4}\right) + \frac{2}{5}\left(\frac{15}{8}\right) = \frac{27}{16} \approx \$1.69$	←minimum
$\left(\frac{15}{2}, 0\right)$	$\frac{1}{4}\left(\frac{15}{2}\right) + \frac{2}{5}(0) = \frac{15}{8} \approx \$1.88$	

    Thus, the vertex point $\left(\frac{15}{4}, \frac{15}{8}\right)$ gives the minimum value, $1.69. The answer, then is to use $3\frac{3}{4}$ servings of A and $1\frac{7}{8}$ servings of B for a minimum cost of $1.69.

31. Let $x =$ the number of red pills and
    $y =$ the number of blue pills.

    Use a table to summarize the given information.

	Vitamin A	Vitamin B_1	Vitamin C
Red pills	8	1	2
Blue pills	2	1	7
Daily requirement	16	5	20

    The linear programming problem may be stated as follows:

    Find $x \geq 0$ and $y \geq 0$ such that
    $$8x + 2y \geq 16 \quad \text{(Vitamin A)}$$
    $$x + y \geq 5 \quad \text{(Vitamin } B_1\text{)}$$
    $$2x + 7y \geq 20 \quad \text{(Vitamin C)}$$

    and the Cost $= .1x + .2y$ is minimized.

Solve the corresponding equations to find the vertex points, and graph the feasible region.

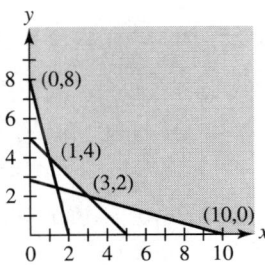

Evaluate the objective function at each vertex.

Point	Cost = $.1x + .2y$	
$(0, 8)$	$.1(0) + .2(8) = 1.6$	
$(1, 4)$	$.1(1) + .2(4) = .9$	
$(3, 2)$	$.1(3) + .2(2) = .7$	←minimum
$(10, 0)$	$.1(10) + .2(0) = 1$	

The minimum value of the objective function is .7 (i.e., 70¢) occurring at $(3, 2)$. Therefore, she should take 3 red pills and 2 blue pills, for a minimum cost of 70¢ per day.

32. Let x = number of type A bolts and
    y = number of type B bolts.

The linear programming problem may be stated as follows:

Find $x \geq 0$ and $y \geq 0$ such that

$.10x + .10y \leq 240$ (machine group I)
$.10x + .40y \leq 720$ (machine group II)
$.10x + .50y \leq 160$ (machine group III).

These constraints may be written in the simpler equivalent form:

$$x + y \leq 2400$$
$$x + 4y \leq 7200$$
$$x + 5y \leq 1600,$$

where Revenue = $.10x + .12y$ and is to be maximized.

Solve the corresponding system of boundary equations for their points of intersection to find the vertices, and sketch the graph representing the intersection of the constraints.

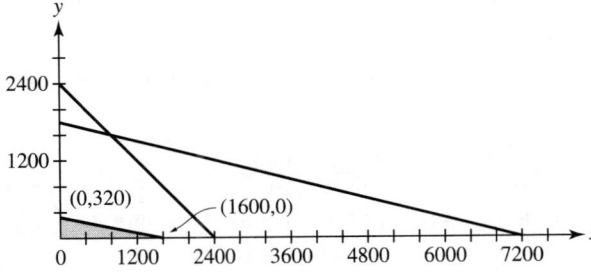

The only vertex points of interest come from the x-intercept, $(1600, 0)$, and the y-intercept, $(0, 320)$ of the boundary equation $x + 5y = 1600$.

Evaluate the objective function at each of these vertex points.

Point	Revenue = $.10x + .12y$	
$(1600, 0)$	$.10(1600) + .12(0) = \$160$	←maximum
$(0, 320)$	$.10(0) + .12(320) = \$38.40$	
$(0, 0)$	$.10(0) + .12(0) = \$0$	

Thus, manufacturing 1600 Type A and 0 Type B bolts yields a maximum revenue of $160.

33. Let x = number of barrels of gasoline and
    y = number of barrels of oil.

The linear programming problem may be stated as follows:

Find $x \geq 0$ and $y \geq 0$ with the following constraints:

$$x \geq 2y$$
$$y \geq 3 \text{ million}$$
$$x \leq 6.4 \text{ million},$$

where Revenue = $1.9x + 1.5y$ and is to be maximized.

Solve the corresponding system of boundary equations for their points of intersection to find the vertices, and sketch the graph representing the intersection of the constraints.

The vertices of the feasible region are

$$(6.4, 3), (6, 3) \text{ and } (6.4, 3.2).$$

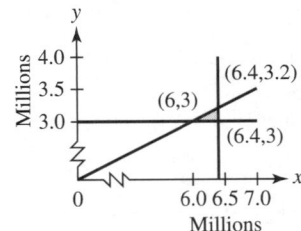

Evaluate the objective function at each vertex.

Point	Revenue = $1.9x + 1.5y$	(in millions)
$(6.4, 3)$	$1.9(6.4) + 1.5(3) = 16.66$	
$(6, 3)$	$1.9(6) + 1.5(3) = 15.9$	
$(6.4, 3.2)$	$1.9(6.4) + 1.5(3.2) = 16.96$	←maximum

Thus, producing 6.4 million barrels of gasoline and 3.2 million barrels of fuel oil should yield the maximum revenue of $16,960,000.

34. Let x = number of batches of cakes and
    y = number of batches of cookies.

Then, the following constraints apply:

For $x \geq 0$ and $y \geq 0$,

$$2x + 1\frac{1}{2}y \leq 15 \quad \text{(Oven)}$$
$$3x + \frac{2}{3}y \leq 13. \quad \text{(Decorating room)}$$

These constraints may be written in the simpler equivalent form:

$$4x + 3y \leq 30$$
$$9x + 2y \leq 39,$$

where Profit $= 30x + 20y$ and is to be maximized.

Solve the corresponding system of boundary equations for their points of intersection to find the vertices, and sketch the graph representing the intersection of the constraints.

The vertices of the feasible region are

$$(0,10), (3,6) \text{ and } \left(\frac{13}{3}, 0\right).$$

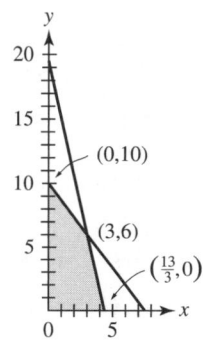

Evaluate the objective function at each vertex.

Point	Profit $= 30x + 20y$
$(0, 10)$	$30(0) + 20(10) = 200$
$(3, 6)$	$30(3) + 20(6) = 210$ ←maximum
$\left(\frac{13}{3}, 0\right)$	$30\left(\frac{13}{3}\right) + 20(0) = 130$

Thus, 3 batches of cakes and 6 batches of cookies yield a maximum profit of \$210.

35. Let x = number of medical kits and
y = number of containers of water.

Use a table to summarize the given information.

	Volume	Weight
Medical Kits	1	10
Containers of water	1	20
Maximum allowed	6000	80000

Weight constraint:

$$10x + 20y \leq 80000.$$

Volume constraint:

$$(1)x + (1)y \leq 6000.$$

Objective function:

$$6x + 10y$$

The problem may be stated as follows.

For $x \geq 0$ and $y \geq 0$,

$$10x + 20y \leq 80000 \quad \text{(Pounds)}$$
$$x + y \leq 6000. \quad \text{(Cubic feet)}$$

These constraints may be written in the simpler equivalent form:

$$x + 2y \leq 8000$$
$$x + y \leq 6000.$$

The number of people aided $= 6x + 10y$ and is to be maximized.

Solve the corresponding equations for their points of intersection to find the vertices, and sketch the graph representing the intersection of the constraints.

The vertex points are

$$(0, 4000), (4000, 2000) \text{ and } (6000, 0).$$

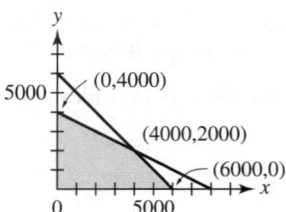

Evaluate the objective function at each vertex.

Point	people aided $= 6x + 10y$
$(0, 4000)$	$6(0) + 10(4000) = 40000$
$(4000, 2000)$	$6(4000) + 10(2000) = 44000$ ←maximum
$(6000, 0)$	$6(6000) + 10(0) = 36000$

Thus, ship 4000 medical kits and 2000 containers of water in order to maximize the number of people aided.

36. Let x = number of medical kits and
y = number of containers of water.

Then, the constraints and feasible region of the plane are exactly the same as Exercise 35 above. The objective function is changed, however.

The number of people aided $= 4x + 10y$ and is to be maximized.

Evaluate the objective function at each vertex. (See Exercise 35.)

Point	people aided $= 4x + 10y$
$(0, 4000)$	$4(0) + 10(4000) = 40000$ ←maximum
$(4000, 2000)$	$4(4000) + 10(2000) = 36000$
$(6000, 0)$	$4(6000) + 10(0) = 24000$

Thus, ship 0 medical kits and 4,000 containers of water in order to maximize the number of people.

Observe that the only difference made by the change of medical kits aiding 4 people instead of 6, in terms of the problem solution, is a change to the objective function.

CHAPTER 8 TEST

1. $d = \sqrt{(x_2 - x_1)^2 + (y_2 - y_1)^2}$
   $= \sqrt{(2 - (-3))^2 + (1-5)^2}$
   $= \sqrt{(5)^2 + (-4)^2}$
   $= \sqrt{25 + 16}$
   $= \sqrt{41}$

2. Use the equation of a circle, where $h = -1$, $k = 2$, and $r = 3$.
   $$(x - h)^2 + (y - k)^2 = r^2$$
   $$(x - (-1))^2 + (y - 2)^2 = 3^2$$
   $$(x + 1)^2 + (y - 2)^2 = 9$$

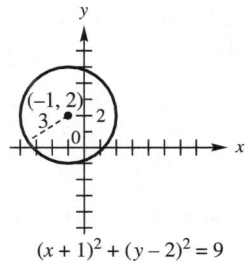

3. $3x - 2y = 8$

   To find the x-intercept, let $y = 0$.
   $$3x - 2(0) = 8$$
   $$3x = 8$$
   $$x = \frac{8}{3}$$

   The x-intercept is $\left(\frac{8}{3}, 0\right)$.

   To find the y-intercept, let $x = 0$.
   $$3(0) - 2y = 8$$
   $$-2y = 8$$
   $$y = -4$$

   The y-intercept is $(0, -4)$.

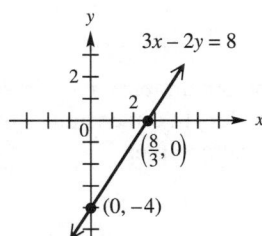

4. Let $(x_1, y_1) = (6, 4)$ and $(x_2, y_2) = (-1, 2)$. Then,
   $$m = \frac{y_2 - y_1}{x_2 - x_1}$$
   $$= \frac{2 - 4}{-1 - 6}$$
   $$= \frac{-2}{-7}$$
   $$= \frac{2}{7}.$$

5. The slope-intercept form of the equation of the line
   (a) passing through the point $(-1, 3)$, with slope $-2/5$.

   Use the point-slope form of the line to write the equation:
   $$y - y_1 = m(x - x_1)$$
   $$y - 3 = -\frac{2}{5}[x - (-1)]$$
   $$y = -\frac{2}{5}x - \frac{2}{5} + 3$$
   $$y = -\frac{2}{5}x + \frac{13}{5}.$$

   (b) passing through $(-7, 2)$ and perpendicular to $y = 2x$.

   Use the slope of the given line, 2, to find the slope of any perpendicular line, $-\frac{1}{2}$. Use the point-slope form of the line to write the equation:
   $$y - y_1 = m(x - x_1), \text{ or}$$
   $$y - 2 = -\frac{1}{2}[x - (-7)]$$
   $$y = -\frac{1}{2}x - \frac{7}{2} + 2$$
   $$y = -\frac{1}{2}x - \frac{3}{2}.$$

   (c) the line shown in the textbook displays.

   The line displays show two points on the line, $(-2, 3)$ and $(6, -1)$. Use these points to create the slope of the line:
   $$m = \frac{y_2 - y_1}{x_2 - x_1}$$
   $$= \frac{-1 - 3}{6 - (-2)}$$
   $$= \frac{-4}{8}$$
   $$= -\frac{1}{2}.$$

Use this slope and either known point to write the equation:

$$y - y_1 = m(x - x_1)$$
$$y - 3 = -\frac{1}{2}[x - (-2)]$$
$$y = -\frac{1}{2}x - 1 + 3$$
$$y = -\frac{1}{2}x + 2.$$

6. Option B shows the graph of a line with a positive slope and a negative y-intercept.

7. (a) Choose the data points represented by the two years, $(3, 21696)$ and $(7, 25050)$ to create slope.

$$m = \frac{y_2 - y_1}{x_2 - x_1}$$
$$= \frac{27910 - 22393}{9 - 5}$$
$$= \frac{5517}{4}$$
$$= 1379.25$$

Use the slope, the point-slope form of the line, and either data point to write the equation of the line.

$$y - y_1 = m(x - x_1), \text{ or}$$
$$y - 22393 = 1379.25(x - 5)$$
$$y - 22393 = 1379.25x - 1379.25(5)$$
$$y = 1379.25x - 6896.25 + 22393$$
$$y = 1379.25x + 15496.75$$

(b) To approximate median income for 1997, let $x = 7$ and solve for y.

$$y = 1379.25x + 15496.75$$
$$= 1379.25(7) + 15496.75$$
$$= 9654.75 + 15496.75$$
$$\approx 25152$$

The predicted value, $25,152 is close to the actual value, $25,050.

8. Let $x =$ the number of days the book is overdue and $y =$ the total fine. Then

$$y = .05x + .50$$

will model the situation. To find three ordered pairs, evaluate the equation at $x = 1, 5,$ and 10.

$$y = .05(1) + .50 = .55 \quad \text{or} \quad (1, .55)$$
$$y = .05(5) + .50 = .75 \quad \text{or} \quad (5, .75)$$
$$y = .05(10) + .50 = 1.00 \quad \text{or} \quad (10, 1.00)$$

9. Estimate the values for two points on the line such as $(0, 1)$ and $(3, 3)$. Use these values to create the slope.

$$m = \frac{y_2 - y_1}{x_2 - x_1}$$
$$= \frac{3 - 1}{3 - 0}$$
$$= \frac{2}{3}$$

Because one of our points is the y-intercept, $(0, 1)$, we may choose to use the slope-intercept form of the line to write the equation.

$$y = mx + b$$
$$y = \frac{2}{3}x + 1$$

10. Given the function $f(x) = x^2 - 3x + 12$, then

(a) the domain is $(-\infty, \infty)$, and

(b) $f(-2) = (-2)^2 - 3(-2) + 12 = 22.$

11. Given the function $f(x) = \dfrac{2}{x - 3}$, then the domain is $(-\infty, 3) \cup (3, \infty)$.

12. To find the break-even point with $C(x) = 50x + 5000$ and $R(x) = 60x$, set $C(x) = R(x)$ and solve for x.

$$50x + 5000 = 60x$$
$$5000 = 10x$$
$$x = 500$$

Evaluate $R(x)$ at $x = 500$ to find the corresponding revenue.

$$R(500) = 60(500) = 30000$$

Thus, 500 calculators, which produces $30,000 in revenue, is the break even point.

13. $f(x) = -(x + 3)^2 + 4$

By inspection of the given standard form of the equation, the axis is at $x = -3$ and the vertex is at $(-3, 4)$. The domain is $(-\infty, \infty)$ and the range is $(-\infty, 4]$.

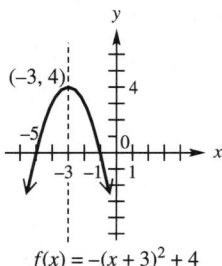

14. Let $x =$ the width dimension of the lot and $(320 - 2x) =$ length dimension of the lot. If $y =$ area function, then

$$y = l \cdot w$$
$$= (320 - 2x) \cdot x$$
$$= -2x^2 + 320x$$

Use the formula for the x-value of the vertex. This value of x will give the largest value for y, the area.

$$x = \frac{-b}{2a}$$
$$= \frac{-320}{2(-2)}$$
$$= 80$$

Evaluate $(320 - 2x)$ at this value of x to determine the length dimension.

$$\text{length} = 320 - 2(80)$$
$$= 320 - 160$$
$$= 160$$

Thus, the dimensions are 80 feet by 160 feet.

15. (a) $5.1^{4.7} \approx 2116.31264888$

    (b) $e^{-1.85} \approx .157237166314$

    (c) $\ln 23.56 \approx 3.15955035878$

16. A is a false statement since the domain of the log function is $(0, \infty)$.

17. $P = \$12{,}000$, $n = 4$, $r = 4\% = .04$, and $t = 3$ years.
    (a) Use the compounding formula
    $A = P\left(1 + \frac{r}{n}\right)^{nt}$:

    $$A = 12000\left(1 + \frac{.04}{4}\right)^{4 \cdot 3}$$
    $$= 12000(1.01)^{12}$$
    $$\approx \$13{,}521.90.$$

    (b) Use the compounding formula $A = Pe^{rt}$ for continuous compounding:

    $$A = 12000e^{(.04)3}$$
    $$= 12000e^{.12}$$
    $$\approx \$13{,}529.96.$$

18. $A(t) = 2.00e^{-.053t}$

    (a) The amount present in 4 years is given by

    $$A(4) = 2.00e^{-.053(4)}$$
    $$= 2.00e^{-.212}$$
    $$\approx 1.62 \text{ grams}.$$

    (b) The amount present in 10 years is given by

    $$A(10) = 2.00e^{-.053(10)}$$
    $$= 2.00e^{-.53}$$
    $$\approx 1.18 \text{ grams}.$$

    (c) The amount present in 20 years is given by

    $$A(20) = 2.00e^{-.053(20)}$$
    $$= 2.00e^{-1.06}$$
    $$\approx .69 \text{ grams}.$$

    (d) The initial amount present given by

    $$A(0) = 2.00e^{-.053(0)}$$
    $$= 2.00e^0$$
    $$\approx 2.00 \text{ grams}.$$

19. $2x + 3y = 2$
    $3x - 4y = 20$

    Multiply the first equation by 4 and the second by 3 then add.

    $$8x + 12y = 8$$
    $$9x - 12y = 60$$
    $$\overline{17x = 68}$$
    $$x = 4$$

    Substitute this value for x into the first equation and solve for y.

    $$2(4) + 3y = 2$$
    $$3y = 2 - 8$$
    $$3y = -6$$
    $$y = -2$$

    Since the ordered pair $(4, -2)$ satisfies both equations, it checks. The solution set is $\{(4, -2)\}$.

20. $2x + y + z = 3$
    $x + 2y - z = 3$
    $3x - y + z = 5$

    Eliminate z from the first and second equations by adding.

    $$2x + y + z = 3$$
    $$x + 2y - z = 3$$
    $$\overline{3x + 3y = 6}$$
    $$x + y = 2$$

    Eliminate z from first and third equations by multiplying the first equation by -1 and adding to the third.

    $$-2x - y - z = -3$$
    $$3x - y + z = 5$$
    $$\overline{x - 2y = 2}$$

    We are left with two equations in x and y. Multiply the first by 2 and add to the second in order to eliminate the y-term.

    $$2x + 2y = 4$$
    $$x - 2y = 2$$
    $$\overline{3x = 6}$$
    $$x = 2$$

    Substitute this value of x into the equation $x + y = 2$ (Either equation in x and y may be used.) Solve for y.

    $$(2) + y = 2$$
    $$y = 0$$

Replace x and y by these values in the first equation $2x + y + z = 3$ (Any one of the original 3 equations may be used.) Solve for z.

$$2(2) + (0) + z = 3$$
$$z = 3 - 4$$
$$z = -1$$

The solution set is $\{(2, 0, -1)\}$.

21. $2x + 3y - 6z = 11$
$x - y + 2z = -2$
$4x + y - 2z = 7$

Eliminate y and z from the second and third equations by adding.
$$\begin{array}{r} x - y + 2z = -2 \\ 4x + y - 2z = 7 \\ \hline 5x = 5 \\ x = 1 \end{array}$$

It is not possible to find a unique value for y and z corresponding to the value $x = 1$. This means that the system is dependent and has an infinite number of answers.

Therefore, replace $x = 1$ into the third equation (this may be done in any of the original three equations) and solve for y in terms of z.

$$4(1) + y - 2z = 7$$
$$y - 2z = 7 - 4$$
$$y = 2z + 3$$

The solution set may now be expressed as $\{(1, 2z + 3, z)\}$.

22. Let x = gross receipts for *Pretty Woman* and y = the gross receipts for *Runaway Bride*.

Write an equation representing the gross receipts for both movies
$$x + y = 330.7$$

Write a second equation that relates gross receipts for both movies.
$$y = x - 26.1$$

Substitute this value for y in the first equation
$$x + (x - 26.1) = 330.7$$
$$2x - 26.1 = 330.7$$
$$2x = 356.8$$
$$x = 178.4$$

Substitute this value of x into the equation $y = x - 26.1$ and solve for y.
$$y = (178.4) - 26.1$$
$$y = 152.3$$

The gross receipts for *Pretty Woman* was $178.4 million and for *Runaway Bride*, $152.3 million.

23. Let x = the sale price with a 10% commission, y = the sale price with a 6% commission and z = the sale price with a 5% commission.

The total property sold is then given by
$$x + y + z = 280,000.$$

The total commission is given by
$$.10x + .06y + .05z = 17,000.$$

In addition, there is the following relationship between the sales:
$$z = x + y.$$

Simplifying the second equation and writing all in standard form results in the system
$$x + y + z = 280,000$$
$$10x + 6y + 5z = 1,700,000$$
$$-x - y + z = 0.$$

Adding the first and last equation results in the elimination of x and y terms.

$$\begin{array}{r} x + y + z = 280,000 \\ -x - y + z = 0 \\ \hline 2z = 280,000 \\ z = 140,000 \end{array}$$

Multiply the first equation $x + y + z = 280000$ by -6 and add to the second equation to eliminate the y terms.

$$\begin{array}{r} -6x - 6y - 6z = -1,680,000 \\ 10x + 6y + 5z = 1,700,000 \\ \hline 4x - z = 20,000 \end{array}$$

Replace the value for z in this last equation and solve for x.
$$4x - 140,000 = 20,000$$
$$4x = 160,000$$
$$x = 40,000$$

Replace x and z in the equation $z = x + y$ by their values and solve for y.
$$140,000 = 40,000 + y$$
$$140,000 - 40,000 = y$$
$$y = 100,000$$

Thus, Keshon Grant sold a property for $40,000 with a 10% commission, another property for $100,000 with a 6% commission, and a third for $140,000 with a 5% commission.

24. $x + y \leq 6$
$2x - y \geq 3$

Use the boundary equations
$$x + y = 6 \quad \text{and}$$
$$2x - y = 3$$

to sketch the graph of each individual inequality, shading the appropriate half-planes. The intersection of these regions represents the solution set as shown below.

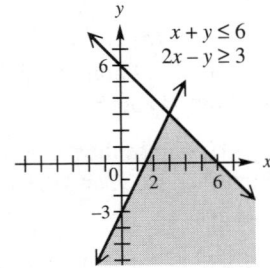

25. Let $x =$ the number of VIP rings produced, and $y =$ the number of SST rings produced.

Since the company can produce no more than 24 rings in a day, the constraint is represented by:

$$x + y \leq 24.$$

Since the company can afford no more than 60 hours of labor in a day, the constraint is represented by

$$3x + 2y \leq 60.$$

The profit function is given by $30x + 40y$.

The linear programming problem may be stated as follows:

Find $x \geq 0$ and $y \geq 0$ such that

$$x + y \leq 24$$
$$3x + 2y \leq 60$$

and Profit $= 30x + 40y$ is maximized.

Solve the corresponding equations for their points of intersection to find the vertices, and sketch the graph representing the intersection of the constraints.

To find the vertex points, solve the system of boundary equations.

$$x + y = 24$$
$$3x + 2y = 60$$

Multiply the first equation by -2 and add to the second equation.

$$-2x - 2y = -48$$
$$3x + 2y = 60$$
$$\overline{\,\, x = 12}$$

Substitute this value for x in the first equation $x + y = 24$.

$$12 + y = 24$$
$$y = 12$$

Thus, a corner point (or vertex) is $(12, 12)$.

The vertices of the feasible region are

$$(0, 24), (12, 12) \text{ and } (20, 0).$$

Sketch the graph of the feasible region representing the intersection of all of the constraints.

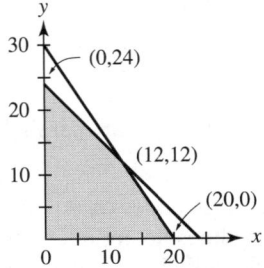

Evaluate the objective function at each vertex.

Point	Profit $= 30x + 40y$	
$(0, 24)$	$30(0) + 40(24) = 960$	←maximum
$(12, 12)$	$30(12) + 40(12) = 840$	
$(20, 0)$	$30(20) + 40(0) = 600$	

Thus, to maximize its profit, the Alessic company should design and sell 0 VIP rings and 24 SST rings for a profit of $960.

9.1 EXERCISES

1. The sum of the measures of two complementary angles is <u>90</u> degrees.

2. The sum of the measures of two supplementary angles is <u>180</u> degrees.

3. The measures of two vertical angles are <u>equal</u>.

4. The measures of <u>two</u> right angles add up to the measure of a straight angle.

5. It is true that a line segment has two endpoints.

6. It is true that a ray has one endpoint.

7. It is false that if A and B are distinct points on a line, then ray AB and ray BA represent the same set of points. The initial point of ray AB is point A; the initial point of ray BA is point B.

8. It is true that if two lines intersect, they lie in the same plane.

9. It is true that if two lines are parallel, they lie in the same plane.

10. It is false that if two lines do not intersect, they must be parallel. Skew lines do not intersect and are not parallel; skew lines do not lie in the same plane.

11. It is true that segment AB and segment BA represent the same set of points. A and B are the endpoints of the line segment and can be named in either order.

12. It is false that there is no angle that is its own complement. A 45° angle is the complement of itself.

13. It is false that there is no angle that is its own supplement. A 90° or right angle is supplementary to a 90° angle.

14. It is false that the origin of the use of the degree as a unit of measure of an angle goes back to the Egyptians. It dates back to the Babylonians.

15. (a) $\overleftrightarrow{AB}$ (b) A———B

16. (a) $\overrightarrow{BC}$ (b) B—C—D→

17. (a) $\overleftarrow{CB}$ (b) ←A—B—C

18. (a) $\overline{AD}$ (b) A—B—C—D

19. (a) $\overset{\circ\!\!\rightarrow}{BC}$ (b) B∘—C—D→

20. (a) $\overset{\circ\!\!\rightarrow}{AD}$ (b) A∘—B—C—D

21. (a) $\overleftrightarrow{BA}$ (b) ←—A—B—→

22. (a) $\overrightarrow{DA}$ (b) ←A—B—C—D

23. (a) $\overline{CA}$ (b) A—B—C

24. (a) $\overleftrightarrow{DA}$ (b) A—B—C—D

25. Letter F. Line segment PQ is the same as line segment QP.

26. Letter A. Half-line QR names the same set of points as half-line QS. The initial point is Q, and the set of points passes through both R and S.

27. Letter D. Ray QR names the same set of points as ray QS. The initial point is Q, and the set of points passes through both R and S.

28. Letter G. Line PQ is the same as line RS.

29. Letter B. Ray RP is the same as ray RQ. The initial point for both is R.

30. Letter C. Half-line SQ is the same as half-line SR.

31. Letter E. Line segment PS is the same as line segment SP.

32. Letter H. Half-line PS does not match.

33. $\overline{MN} \cup \overline{NO}$ names the same set of points as $\overline{MO}$. The union symbol joins the two line segments.

34. $\overline{MN} \cap \overline{NO}$ is the same as point N.

35. $\overline{MO} \cap \overline{OM}$ indicates the intersection or overlap of the same line segment. Therefore a simpler way is either $\overline{MO}$ or $\overline{OM}$.

36. $\overline{MO} \cup \overline{OM}$ is the same set of points as $\overleftrightarrow{MO}$.

37. $\overset{\circ\!\!\rightarrow}{OP} \cap O$ have no points in common because point O is not part of the half-line $\overset{\circ\!\!\rightarrow}{OP}$. Therefore, the intersection is the empty set, symbolized by ∅.

38. $\overset{\circ\!\!\rightarrow}{OP} \cup O$ indicates the union of the half-line with the point. The final result is a ray: $\overrightarrow{OP}$.

39. $\overrightarrow{NP} \cap \overrightarrow{OP}$ indicates the same set of points as $\overrightarrow{OP}$.

40. $\overrightarrow{NP} \cup \overrightarrow{OP}$ indicates the same of set of points as $\overrightarrow{NP}$.

41. $90 - 28 = 62°$

42. $90 - 32 = 58°$

43. $90 - 89 = 1°$

44. $90 - 45 = 45°$

45. $(90 - x)°$

46. $90 - (90 - x) = 90 - 90 + x = x°$

47. $180 - 132 = 48°$

48. $180 - 105 = 75°$

49. $180 - 26 = 154°$

50. $180 - 90 = 90°$

51. $(180 - y)°$

52. $180 - (180 - y) = 180 - 180 + y = y°$

53. $\angle ABE$ and $\angle CBD$; $\angle ABD$ and $\angle CBE$

54. $\angle TQR$ and $\angle PQS$; $\angle TQS$ and $\angle PQR$

55. (a) $52°$. They are vertical angles.

    (b) $180 - 52 = 128°$. They are supplementary angles.

56. (a) $126°$. They are vertical angles.

    (b) $180 - 126 = 54°$. They are supplementary angles.

57. The designated angles are supplementary; their sum is $180°$.
$$(10x + 7) + (7x + 3) = 180$$
$$17x + 10 = 180$$
$$17x = 170$$
$$x = 10$$

    Then, one angle measure is $10 \cdot 10 + 7 = 107°$, and the other angle measure is $7 \cdot 10 + 3 = 73°$.

58. The designated angles are supplementary; their sum is $180°$.
$$(x + 1) + (4x - 56) = 180$$
$$5x - 55 = 180$$
$$5x = 235$$
$$x = 47$$

    Then, one angle measure is $47 + 1 = 48°$, and the other angle measure is $4 \cdot 47 - 56 = 132°$.

59. The angles are vertical, so they have the same measurement. Set the algebraic expressions each to each other.
$$3x + 45 = 7x + 5$$
$$-4x = -40$$
$$x = 10$$

    Then, one angle measure is $3 \cdot 10 + 45 = 75°$, and the other angle measure is $7 \cdot 10 - 5 = 75°$.

60. The angles are vertical, so they have the same measurement. Set the algebraic expressions each to each other.
$$5x - 129 = 2x - 21$$
$$3x = 108$$
$$x = 36$$

    Then, one angle measure is $5 \cdot 36 - 129 = 51°$, and the other angle measure is $2 \cdot 36 - 21 = 51°$.

61. The angles are vertical, so they have the same measurement. Set the algebraic expressions each to each other.
$$11x - 37 = 7x + 27$$
$$4x = 64$$
$$x = 16$$

    Then, one angle measure is $11 \cdot 16 - 37 = 139°$, and the other angle measure is $7 \cdot 16 + 27 = 139°$.

62. The angles are vertical, so they have the same measurement. Set the algebraic expressions each to each other.
$$10x + 15 = 12x - 3$$
$$-2x = -18$$
$$x = 9$$

    Then, one angle measure is $10 \cdot 9 + 15 = 105°$, and the other angle measure is $12 \cdot 9 - 3 = 105°$.

63. The designated angles are supplementary; their sum is $180°$.
$$(3x + 5) + (5x + 15) = 180$$
$$8x + 20 = 180$$
$$8x = 160$$
$$x = 20$$

    Then, one angle measure is $3 \cdot 20 + 5 = 65°$, and the other angle measure is $5 \cdot 20 + 15 = 115°$.

64. The designated angles are complementary; their sum is $90°$.
$$(5x - 1) + 2x = 90$$
$$7x - 1 = 90$$
$$7x = 91$$
$$x = 13$$

    Then, one angle measure is $5 \cdot 13 - 1 = 64°$, and the other angle measure is $2 \cdot 13 = 26°$.

65. The designated angles are complementary; their sum is $90°$.
$$(5k + 5) + (3x + 5) = 90$$
$$8k + 10 = 90$$
$$8k = 80$$
$$k = 10$$

    Then, one angle measure is $5 \cdot 10 + 5 = 55°$, and the other angle measure is $3 \cdot 10 + 5 = 35°$.

66. Alternate interior angles have equal measures.
$$2x - 5 = x + 22$$
$$x = 27$$

    Then, the measure of each angle is $2 \cdot 27 - 5 = 49°$, and $27 + 29 = 49°$.

67. Alternate exterior angles have equal measures.
$$2x + 61 = 6x - 51$$
$$-4x = -112$$
$$x = 28$$

    Then, the measure of each angle is $2 \cdot 28 + 61 = 117°$, and $6 \cdot 28 - 51 = 117°$.

286 CHAPTER 9 GEOMETRY

68. The angles are supplementary; their sum is 180°.
$$(x+1) + (4x - 56) = 180$$
$$5x - 55 = 180$$
$$5x = 235$$
$$x = 47$$

Then, one angle measure is $47 + 1 = 48°$, and the other angle measure is $4 \cdot 47 - 56 = 132°$.

69. Alternate exterior angles have equal measures.
$$10x + 11 = 15x - 54$$
$$-5x = -65$$
$$x = 13$$

Then the measure of each angle is $10 \cdot 13 + 11 = 141°$, and $15 \cdot 13 - 54 = 141°$.

70. Let x = measure of the angle,
$180 - x$ = measure of its supplement,
$90 - x$ = measure of its complement.
$$180 - x = 25 + 2(90 - x)$$
$$180 - x = 25 + 180 - 2x$$
$$180 - x = 205 - 2x$$
$$180 - x + 2x = 205 - 2x + 2x$$
$$180 + x = 205$$
$$180 - 180 + x = 205 - 180$$
$$x = 25°$$

71. Let x = measure of the angle,
$180 - x$ = measure of its supplement,
$90 - x$ = measure of its complement.
$$90 - x = \frac{1}{5}(180 - x) - 10$$
$$5 \cdot (90 - x) = \frac{5}{1} \cdot \frac{1}{5}(180 - x) - 5 \cdot 10$$
$$450 - 5x = (180 - x) - 50$$
$$450 - 5x = 130 - x$$
$$450 - 5x + x = 130 - x + x$$
$$450 - 4x = 130$$
$$450 - 450 - 4x = 130 - 450$$
$$-4x = -320$$
$$x = 80°$$

72. Let x = the measure of the angle,
$180 - x$ = measure of its supplement,
$90 - x$ = measure its complement.
$$(180 - x) + (90 - x) = 210$$
$$270 - 2x = 210$$
$$270 - 270 - 2x = 210 - 270$$
$$-2x = -60$$
$$\frac{-2x}{-2} = \frac{-60}{-2}$$
$$x = 30°$$

73. Let x = measure of the angle,
$180 - x$ = measure of its supplement,
$90 - x$ = measure of its complement.
$$\frac{1}{2}(180 - x) = 2(90 - x) - 12$$
$$\frac{2}{1} \cdot \frac{1}{2}(180 - x) = 2 \cdot 2(90 - x) - 2 \cdot 12$$
$$180 - x = 4(90 - x) - 24$$
$$180 - x = 360 - 4x - 24$$
$$180 - x = 336 - 4x$$
$$180 - x + 4x = 336 - 4x + 4x$$
$$180 + 3x = 336$$
$$180 - 180 + 3x = 336 - 180$$
$$3x = 156$$
$$\frac{3x}{3} = \frac{156}{3}$$
$$x = 52°$$

74. (a) Measure of $\angle 2$ = measure of $\angle 3$, since they are vertical angles.

    (b) Measure of $\angle 3$ = measure of $\angle 6$, since they are alternate interior angles.

    (c) Measure of $\angle 6$ = measure of $\angle 7$, since they are vertical angles.

    (d) By the results of parts (a), (b), and (c), the measure of $\angle 2$ must equal the measure of $\angle 7$, showing that alternate <u>exterior</u> angles have equal measures.

75. Some of the unknown angles must be solved before other unknown angles. Here is one order that they can be solved.
    $\angle 1 = 55°$; vertical angle to 55°.
    $\angle 8 = 180 - 120 = 60°$; supplementary angle to 120°.
    $\angle 6 = 180 - 60 = 120°$; supplementary angle to $\angle 8$.
    $\angle 7 = 60°$; vertical angle to $\angle 8$.
    $\angle 5 = 60°$; alternate interior angle to $\angle 7$.
    $\angle 3 = 60°$; vertical angle to $\angle 5$.
    $\angle 2 = 180 - (55 + 60) = 180 - 55 - 60 = 65°$.
    Angles 2, 3, and 55 all add to 180°.
    $\angle 4 = 180 - (55 + 60) = 180 - 55 - 60 = 65°$;
    straight angle composed of $\angle 4$, $\angle 5$, and 55°.
    $\angle 10 = 55°$; alternate interior angle to the 55° angle.
    $\angle 9 = 55°$; vertical angle to $\angle 10$.

76. (a) $m(\angle 1) + m(\angle 2) = \underline{180}°$. They are supplementary.

    (b) $m(\angle 2) + m(\angle 3) = \underline{180}°$. They are supplementary.

    (c) Subtract the equation in part (b) from the equation in part (a) to get
    $[m(\angle 1) + m(\angle 2)] - [m(\angle 2) + m(\angle 3)] = \underline{180}° - \underline{180}°$.

    (d) $m(\angle 1) + m(\angle 2) - m(\angle 2) - m(\angle 3) = \underline{0}°$

    (e) $m(\angle 1) - m(\angle 3) = \underline{0}$

    (f) $m(\angle 1) = m(\angle \underline{3})$

77. Refer to the figure for Exercise 74 (in the text). The interior angles are ∠3, ∠4, ∠5, and ∠6. The interior angles on the same side of the transversal are ∠4 and ∠6, and ∠3 and ∠5. (a) ∠3 = ∠2; vertical angles.

(b) Measure of ∠3 = measure of ∠6; alternate interior angles.

(c) Measure of ∠2 = measure of ∠6; by parts (a) and (b) above.

(d) ∠2 and ∠4 are supplementary, by definition.

(e) ∠6 and ∠4 are supplementary, by parts (c) and (d) above.

The result can be shown similarly for ∠3 and ∠5. Therefore, interior angles on the same side of a transversal are supplementary.

78. Given $x + y = 40$ means that $x = 40 - y$. Replace x with $40 - y$ in each expression. Because the angles are supplementary, their sum is 180.

$$(x + 2y) + 11x = 180$$
$$[(40 - y) + 2y] + 11(40 - y) = 180$$
$$[40 + y] + 440 - 11y = 180$$
$$480 - 10y = 180$$
$$-10y = -300$$
$$y = 30$$

If $y = 30$, then $x = 40 - 30 = 10$.

9.2 EXERCISES

1. A segment joining two points on a circle is called a(n) <u>chord</u>.

2. A segment joining the center of a circle and a point on the circle is called a(n) <u>radius</u>.

3. A regular triangle is called a(n) <u>equilateral</u> triangle.

4. A chord that contains the center of a circle is called a(n) <u>diameter</u>.

5. False. A rhombus does not have equal angle measures.

6. True. An isosceles triangle has two sides with the same measure. A scalene triangle has all three sides of unequal measure.

7. False. The sum of the angle measures of a triangle always equals 180°. If a triangle had two obtuse angles, then the sum of the measures of the angles would exceed 180° which is impossible. Therefore, a triangle has at most one obtuse angle.

8. True. A square has opposite sides parallel and of equal measure, as does a rectangle. A square also has all four sides of equal measure, as does a rhombus.

9. True. A square is a rhombus with four 90° angles.

10. False. A square must have four 90° angles; a rhombus does not have this requirement.

11. Writing exercise

12. A STOP sign has the shape of an octagon.

13. Both. It is closed, and there are no intersecting curves.

14. Simple

15. Closed

16. Both

17. Closed

18. Not simple; closed

19. Neither

20. Neither

21. Convex

22. Not convex

23. Convex

24. Convex

25. Not convex

26. Not convex

27. Right; scalene

28. Obtuse; scalene

29. Acute; equilateral

30. Acute; isosceles

31. Right; scalene

32. Obtuse; isosceles

33. Right; isosceles

34. Right; scalene

35. Obtuse; scalene

36. Acute; equilateral

37. Acute; isosceles

38. Right; scalene

39. Writing exercise

40. Writing exercise

41. Writing exercise

42. See the isosceles triangle in Exercise 30.

$$\sqrt{9} + \sqrt{9} \neq \sqrt{6}$$

That is, $3 + 3 \neq \sqrt{6}$.

43. The sum of the measures of the three angles of any triangle is always 180°.

$$x + (x+20) + (210-3x) = 180$$
$$-x + 230 = 180$$
$$-x = -50$$
$$x = 50$$

The measure of angle A is 50°; angle B is $50 + 20 = 70°$; angle C is $210 - 3 \cdot 50 = 60°$.

44. $(x+15) + (x+5) + (10x-20) = 180$
$$12x = 180$$
$$x = 15$$

The measure of angle A is $10 \cdot 15 - 20 = 130°$; angle B is $15 + 15 = 30°$; angle C is $15 + 5 = 20°$.

45. $(x-30) + (2x-120) + \left(\frac{1}{2}x + 15\right) = 180$
$$3\frac{1}{2}x - 135 = 180$$
$$\frac{7}{2}x = 315$$
$$\frac{2}{7} \cdot \frac{7}{2}x = 315 \cdot \frac{2}{7}$$
$$x = 90$$

The measure of angle A is $90 - 30 = 60°$; angle B is $2 \cdot 90 - 120 = 60°$; angle C is $\frac{1}{2} \cdot 90 + 15 = 60°$.

46. This exercise can be done algebraically or by analysis. If each of the angle measures is the same, divide 180° by 3 to obtain 60°.

47. Let x = the angle measure of A (or B), $x + 24$ = the angle measure of C.
$$x + x + (x+24) = 180$$
$$3x + 24 = 180$$
$$3x = 156$$
$$x = 52$$

The measure of angle A (or B) is 52°; the measure of angle C is $52 + 24 = 76°$.

48. Let x = the angle measure of B (or C), $x + 30$ = the angle measure of A.
$$x + x + (x+30) = 180$$
$$3x + 30 = 180$$
$$3x = 150$$
$$x = 50$$

The measure of angle B (or C) is 50°; the measure of angle A is $50 + 30 = 80°$.

49. The measure of the exterior angle of a triangle is equal to the sum of the measures of the two opposite interior angles.

$$10x + (15x - 10) = 20x + 25$$
$$25x - 10 = 20x + 25$$
$$25x - 20x - 10 = 20x - 20x + 25$$
$$5x - 10 = 25$$
$$5x - 10 + 10 = 25 + 10$$
$$5x = 35$$
$$x = 7$$

Then $(20 \cdot 7) + 25 = 165°$.

50. $13x + (10x + 15) = 26x$
$$23x + 15 = 26x$$
$$23x - 23x + 15 = 26x - 23x$$
$$15 = 3x$$
$$5 = x$$

Then $26 \cdot 5 = 130°$.

51. $(2 - 7x) + (100 - 10x) = 90 - 20x$
$$-17x + 102 = 90 - 20x$$
$$-17x + 102 - 102 = 90 - 102 - 20x$$
$$-17x = -12 - 20x$$
$$3x = -12$$
$$x = -4$$

Then $90 - 20 \cdot (-4) = 90 + 80 = 170°$.

52. $2x + 4x = 100 - 4x$
$$6x = 100 - 4x$$
$$10x = 100$$
$$x = 10$$

Then $100 - 4 \cdot 10 = 100 - 40 = 60°$.

53. (a) The center is at point O.

(b) There are four line segments that are radii: $\overline{OA}$, $\overline{OC}$, $\overline{OB}$, and $\overline{OD}$.

(c) There are two diameters: $\overline{AC}$ and $\overline{BD}$.

(d) There are four chords: $\overline{AC}$, $\overline{BD}$, $\overline{BC}$, and $\overline{AB}$.

(e) There are two secants: $\overleftrightarrow{AB}$ and $\overleftrightarrow{BC}$.

(f) There is one tangent: $\overleftrightarrow{AE}$

54. According to figure 19, angles 3 and 6 are supplementary; the sum of the measures of their angles is equal to 180°. Also the sum of the interior measures of the angles of a triangle is always equal to 180°. Therefore, the sum of the measures of angles 1 and 2 must be equivalent to the measure of angle 6.

55. Writing exercise

EXTENSION: GEOMETRIC CONSTRUCTIONS

1. With the radius of the compasses greater than one-half the length PQ, place the point of the compasses at P and swing arcs above and below line r. Then with the same radius and the point of the compasses at Q, swing two more arcs above and below line r. Locate the two points of intersections of the arcs above and below, and call them A and B. With a straightedge, join A and B. AB is the perpendicular bisector of PQ.

2. With radius of the compasses greater than one-half the length PQ, place the point of the compasses at P and swing arcs to the left and right of line r. Then with the same radius and the point of the compasses at Q, swing two more arcs to the left and right of line r. Locate the two points of intersections of the arcs to the left and right, and call them A and B. With a straightedge, join A and B. AB is the perpendicular bisector of PQ.

3. With the radius of the compasses greater than the distance from P to r, place the point of the compasses at P and swing an arc intersecting line r in two points. Call these points A and B. Swing arcs of equal radius to the left of line r, with the point of the compasses at A and at B, intersecting at point Q. With a straightedge, join P and Q. PQ is the perpendicular from P to r.

4. With the radius of the compasses greater than the distance from P to r, place the point of the compasses at P and swing an arc intersecting line r in two points. Call these points A and B. Swing arcs of equal radius above line r, with the point of the compasses at A and at B, intersecting at point Q. With a straightedge, join P and Q. PQ is the perpendicular from P to line r.

5. With any radius, place the point of the compasses at P and swing arcs to the left and right, intersecting line r in two points. Call these points A and B. With an arc of sufficient length, place the point of the compasses first at A and then at B, and swing arcs either both above or both below line r, intersecting at point Q. With a straightedge, join P and Q. PQ is perpendicular to line r at P.

6. With any radius, place the point of the compasses at P and swing arcs above and below, intersecting line r in two points. Call these points A and B. With an arc of sufficient length, place the point of the compasses first at A and then at B, and swing arcs either both to the left or both to the right of line r, intersecting at point Q. With a straightedge, join P and Q. PQ is perpendicular to line r at P.

7. With any radius, place the point of the compasses at vertex A and swing an arc intersecting both sides of the angle. Call these points of intersection P and Q. With an arc of sufficient length, place the point of the compasses first at P and then at Q, and swing arcs in the interior of the angle, intersecting each other at point B. With a straightedge, join A and B. AB is the bisector of the angle.

8. With any radius, place the point of the compasses at vertex P and swing an arc intersecting both sides of the angle. Call these points of intersection A and B. With an arc of sufficient length, place the point of the compasses first at A and then at B, and swing arcs in the interior of the angle, intersecting each other at point Q. With a straightedge, join P and Q. PQ is the bisector of the angle.

9. Use Construction 3 to construct a perpendicular to a line at a point. Then use Construction 4 to bisect one of the right angles formed. This yields a 45° angle.

10. Writing exercise

9.3 EXERCISES

1. The perimeter of an equilateral triangle with side length equal to 12 inches is the same as the perimeter of a rectangle with length 10 inches and width 8 inches. The perimeter of the rectangle is $2 \cdot 10 + 2 \cdot 8 = 36$. The perimeter of the triangle is also 36. If all three sides must have the same length, then one side has length $36 \div 3 = 12$ inches.

2. A square with an area 16 square cm has perimeter 16 cm. The area of a square is found by the formula $A = s^2$. If the area is 16, the length of one side is 4 cm. Perimeter is found by $P = 4s$; then, $P = 4 \cdot 4 = 16$ cm.

3. If the area of a certain triangle is 24 square inches, and the base measures 8 inches, then the height must measure 6 inches. The formula for the area of a triangle is $A = \frac{1}{2}bh$. Substitute the given values into the formula and solve for h.

$$24 = \frac{1}{2} \cdot 8h$$
$$24 = 4h$$
$$6 = h$$

4. If the radius of a circle is doubled, then its area is multiplied by a factor of 4. The formula for the area of a circle if $A = \pi r^2$. Replace r with $2r$ and compute the new area: $A = \pi(2r)^2 = \pi \cdot 4r^2 = 4\pi r^2$.

5. The area of an equilateral triangle with side length 6 inches is $9\sqrt{3}$ square inches. Use Heron's area formula with $a = b = c = 6$. Then $s = \frac{1}{2}(6 + 6 + 6) = 9$; and

$$A = \sqrt{s(s-a)(s-b)(s-c)}$$
$$= \sqrt{9(9-6)(9-6)(9-6)}$$
$$= \sqrt{9(3)(3)(3)}$$
$$= \sqrt{243}$$
$$= \sqrt{81 \cdot 3}$$
$$= 9\sqrt{3}.$$

6. If the length and the width of a rectangle are doubled, then the area is multiplied by a factor of 4. In the formula for the area of a rectangle, replace l and w by $2l$ and $2w$:

   Formula for area. $A = lw$
   Replace l and w with $2l$ and $2w$. $A = 2l \cdot 2w$
   Area is multiplied by 4. $A = 4lw$

7. $A = lw$
   $A = 4 \cdot 3$
   $A = 12 \text{ cm}^2$

8. $A = s^2$
   $A = 3^2$
   $A = 9 \text{ cm}^2$

9. $A = lw$
   $A = 2\frac{1}{2} \cdot 2$
   $A = 5 \text{ cm}^2$

10. $A = lw$
    $A = 1 \cdot 3$
    $A = 3 \text{ cm}^2$

11. $A = bh$
    $A = 4 \cdot 2$
    $A = 8 \text{ in.}^2$

12. $A = bh$
    $A = 4 \cdot 2\frac{1}{2}$
    $A = \frac{4}{1} \cdot \frac{5}{2}$
    $A = 10 \text{ in.}^2$

13. $A = bh$
    $A = 3 \cdot 1.5$
    $A = 4.5 \text{ cm}^2$

14. $A = \frac{1}{2}bh$
    $A = \frac{1}{2} \cdot 52 \cdot 36$
    $A = 936 \text{ mm}^2$

15. $A = \frac{1}{2}bh$
    $A = \frac{1}{2} \cdot 22 \cdot 38$
    $A = \frac{1}{2} \cdot \frac{22}{1} \cdot \frac{38}{1}$
    $A = 418 \text{ mm}^2$

16. $A = \frac{1}{2}bh$
    $A = \frac{1}{2} \cdot 5 \cdot 3$
    $A = \frac{1}{2} \cdot \frac{5}{1} \cdot \frac{3}{1}$
    $A = \frac{15}{2}$
    $A = 7.5 \text{ m}^2$

17. $A = \frac{1}{2}h(b + B)$
    $A = \frac{1}{2} \cdot 2(3 + 5)$
    $A = 1(8)$
    $A = 8 \text{ cm}^2$

18. $A = \frac{1}{2}h(b + B)$
    $A = \frac{1}{2} \cdot 3(4 + 5)$
    $A = \frac{3}{2} \cdot (9)$
    $A = \frac{27}{2}$
    $A = 13.5 \text{ cm}^2$

19. $A = \pi r^2$
    $A = (3.14)(1)^2$
    $A = 3.14 \text{ cm}^2$

20. $A = \pi r^2$
    $A = (3.14)(15)^2$
    $A = (3.14)(225)$
    $A = 706.5 \text{ cm}^2$

21. The diameter is 36, so the radius is 18 m.
    $$A = \pi r^2$$
    $$A = (3.14)(18)^2$$
    $$A = (3.14)(324)$$
    $$A = 1017.36 \text{ m}^2$$

22. The diameter is 12, so the radius is 6 m.
    $$A = \pi r^2$$
    $$A = (3.14)(6)^2$$
    $$A = (3.14)(36)$$
    $$A = 113.04 \text{ m}^2$$

23. Let $s =$ length of a side of the window. Use the formula $P = 4s$. Replace P with $7s - 12$ and solve for s.

$$P = 4s$$
$$7s - 12 = 4s$$
$$7s - 7s - 12 = 4s - 7s$$
$$-12 = -3s$$
$$\frac{-12}{-3} = \frac{-3s}{-3}$$
$$4 = s$$

The length of a side of the window is 4 m.

24. Use the formula $P = 2l + 2w$ and $l = 20 + w$. Replace P with 176, and replace l with $20 + w$. Solve for w.

$$P = 2l + 2w$$
$$176 = 2(20 + w) + 2w$$
$$176 = 40 + 2w + 2w$$
$$176 = 40 + 4w$$
$$136 = 4w$$
$$\frac{136}{4} = \frac{4w}{4}$$
$$34 = w$$

The width is 34 inches; the length is $20 + 34 = 54$ inches.

25. The formula for perimeter of a triangle is $P = a + b + c$. Translating the problem, let a be the shortest side. Then $b = 100 + a$ and $c = 200 + a$. Replace a, b, and c in the formula with these expressions; replace P with 1200 and solve for a.

$$P = a + b + c$$
$$1200 = a + (100 + a) + (200 + a)$$
$$1200 = 3a + 300$$
$$900 = 3a$$
$$300 = a$$

Side a is 300 ft.; $b = 100 + 300 = 400$ ft.; side $c = 200 + 300 = 500$ ft.

26. Let $a =$ the length of the shorter side of the triangle. Then the other two sides each have a length of $a + 18$. Substitute these values into the formula for the perimeter of a triangle; $P = 54$ inches.

$$P = a + b + c$$
$$54 = a + (a + 18) + (a + 18)$$
$$54 = 3a + 36$$
$$18 = 3a$$
$$6 = a$$

Side a is 6 in.; the other two equal sides each have a length of $6 + 18 = 24$ inches.

27. One formula for circumference is $C = 2\pi r$. Translating the second sentence of the problem, $C = 6r + 12.88$. Equate these two expressions for C and solve the equation for r.

$$6r + 12.88 = 2\pi r$$
$$6r + 12.88 = 2(3.14)r$$
$$6r + 12.88 = 6.28r$$
$$6r - 6r + 12.88 = 6.28r - 6r$$
$$12.88 = 0.28r$$
$$\frac{12.88}{0.28} = \frac{0.28r}{0.28}$$
$$46 = r$$

The radius is 46 ft.

28. Use the formula $C = 2\pi r$. Translating the problem indicates $C = 3r + 8.2$. Set these expressions equal to each other and solve for r.

$$3r + 8.2 = 2\pi r$$
$$3r + 8.2 = 2(3.14)r$$
$$3r + 8.2 = 6.28r$$
$$3r - 3r + 8.2 = 6.28r - 3r$$
$$8.2 = 3.28r$$
$$\frac{8.2}{3.28} = \frac{3.28r}{3.28}$$
$$2.5 = r$$

The radius is 2.5 cm.

29. The formula for the area of a trapezoid is $A = \frac{1}{2}h(b + B)$. Substitute the numerical values given in the problem and compute to find area.

$$A = \frac{1}{2}h(b + B)$$
$$A = \frac{1}{2}(165.97)(115.80 + 171.00)$$
$$A = \frac{1}{2}(165.97)(286.8)$$
$$A = \frac{47600.196}{2}$$
$$A = 23,800.098$$

Rounded to the nearest hundredth, the area is 23,800.10 sq ft.

30. The formula for the area of a trapezoid is $A = \frac{1}{2}h(b + B)$. Substitute the numerical values given in the problem and compute to find area.

$$A = \frac{1}{2}h(b + B)$$
$$A = \frac{1}{2}(165.97)(26.84 + 82.05)$$
$$A = \frac{1}{2}(165.97)(108.89)$$
$$A = \frac{18072.473}{2}$$
$$A = 9036.2365$$

Rounded to the nearest hundredth, the area is 9036.24 sq ft.

31. You would need to use perimeter. Fencing is sold by linear measure.

32. You would need to use area.

Table for Exercises 33–40

	r	d	C	A
33.	6 in.	12 in.	12π in.	36π in.^{2}
34.	9 in.	18 in.	18π in.	81π in.^{2}
35.	5 ft	10 ft	10π ft	25π ft^2
36.	20 ft	40 ft	40π ft	400π ft^2
37.	6 cm	12 cm	12π cm	36π cm^2
38.	9 cm	18 cm	18π cm	81π cm^2
39.	10 in.	20 in.	20π in.	100π in.^{2}
40.	16 in.	32 in.	32π in.	256π in.^{2}

33. $d = 2r = 2 \cdot 6 = 12$ in.
    $C = 2\pi r = 2\pi \cdot 6 = 12\pi$ in.
    $A = \pi r^2 = \pi \cdot 6^2 = 36\pi$ in.^{2}

34. $d = 2r = 2 \cdot 9 = 18$ in.
    $C = 2\pi r = 2\pi \cdot 9 = 18\pi$ in.
    $A = \pi r^2 = \pi \cdot 9^2 = 81\pi$ in.^{2}

35. $r = \dfrac{1}{2} \cdot 10 = 5$ ft
    $C = \pi d = \pi \cdot 10 = 10\pi$ ft
    $A = \pi r^2 = \pi \cdot 5^2 = 25\pi$ ft^2

36. $r = \dfrac{1}{2} \cdot 40 = 20$ ft
    $C = \pi d = \pi \cdot 40 = 40\pi$ ft
    $A = \pi r^2 = \pi \cdot 20^2 = 400\pi$ ft^2

37. $d = \dfrac{C}{\pi} = \dfrac{12\pi}{\pi} = 12$ cm
    $r = \dfrac{1}{2} \cdot 12 = 6$ cm
    $A = \pi \cdot 6^2 = 36\pi$ cm^2

38. $d = \dfrac{C}{\pi} = \dfrac{18\pi}{\pi} = 18$ cm
    $r = \dfrac{1}{2} \cdot 18 = 9$ cm
    $A = \pi \cdot 9^2 = 81\pi$ cm^2

39. $r^2 = \dfrac{A}{\pi} = \dfrac{100\pi}{\pi} = 100$. Then $r = \sqrt{100} = 10$ in.
    $d = 2r = 2 \cdot 10 = 20$ in.
    $C = 2\pi r = 2 \cdot \pi \cdot 10 = 20\pi$ in.

40. $r^2 = \dfrac{A}{\pi} = \dfrac{256\pi}{\pi} = 256$. Then $r = \sqrt{256} = 16$ in.
    $d = 2r = 2 \cdot 16 = 32$ in.
    $C = 2\pi r = 2 \cdot \pi \cdot 16 = 32\pi$ in.

41. Use the formula $P = 4s$, replacing s with x and P with 58.
    $$P = 4x$$
    $$58 = 4x$$
    $$\dfrac{58}{4} = \dfrac{4x}{4}$$
    $$14.5 = x$$

42. Use the formula $P = a + b + c$, replacing a, b, and c with the expressions in x and replacing P with 42.
    $$42 = x + (x+2) + (x+7)$$
    $$42 = 3x + 9$$
    $$33 = 3x$$
    $$\dfrac{33}{3} = \dfrac{3x}{3}$$
    $$11 = x$$

43. Use the formula $P = 2l + 2w$, replacing l and w with the expressions in x and replacing P with 38.
    $$38 = 2(2x - 3) + 2(x+1)$$
    $$38 = 4x - 6 + 2x + 2$$
    $$38 = 6x - 4$$
    $$42 = 6x$$
    $$\dfrac{42}{6} = \dfrac{6x}{6}$$
    $$7 = x$$

44. Use the formula $P = 2l + 2w$, replacing l and w with the expressions in x and P with 278.
    $$278 = 2(5x+1) + 2(x)$$
    $$278 = 10x + 2 + 2x$$
    $$278 = 12x + 2$$
    $$276 = 12x$$
    $$\dfrac{276}{12} = \dfrac{12x}{12}$$
    $$23 = x$$

45. Use the formula $A = s^2$, replacing s with x and A with 26.01.
    $$26.01 = x^2$$
    $$\sqrt{26.01} = \sqrt{x^2}$$
    $$5.1 = x$$

46. Use the formula $A = lw$, replacing l and w with the expressions in x and replacing A with 28.
    $$28 = x(x+3)$$
    $$28 = x^2 + 3x$$
    $$0 = x^2 + 3x - 28$$

    Now factor the trinomial and set each factor equal to zero.
    $$0 = (x+7)(x-4)$$
    $$x + 7 = 0 \quad \text{or} \quad x - 4 = 0$$
    $$x = -7 \qquad\qquad x = 4$$

    The solution -7 is not meaningful because length or width of a rectangle must be positive numbers. The answer then is 4.

9.3 PERIMETER, AREA, AND CIRCUMFERENCE

47. Use the formula $A = \frac{1}{2}bh$, replacing b and h with the expressions in x and replacing A with 15.

$$15 = \frac{1}{2}x(x+1)$$
$$2 \cdot 15 = \frac{2}{1} \cdot \frac{1}{2}x(x+1)$$
$$30 = x(x+1)$$
$$30 = x^2 + x$$
$$0 = x^2 + x - 30$$

Now factor the trinomial and set each factor equal to zero.

$$0 = (x+6)(x-5)$$

$x + 6 = 0$ or $x - 5 = 0$
$x = -6$ $x = 5$

The solution -6 is not meaningful because the base of a triangle must be positive a number. The answer then is 5.

48. Use the formula $A = \frac{1}{2}h(b + B)$, replacing h, b, and B with the expressions in x and replacing A with 30.

$$30 = \frac{1}{2}3[x + (x+4)]$$
$$2 \cdot 30 = \frac{2}{1} \cdot \frac{1}{2}3[x + (x+4)]$$
$$60 = 3[x + (x+4)]$$
$$\frac{60}{3} = \frac{3[x + (x+4)]}{3}$$
$$20 = [x + (x+4)]$$
$$20 = 2x + 4$$
$$16 = 2x$$
$$8 = x$$

49. Use the formula $C = 2\pi r$, replacing C with 37.68 and r with the expression in x.

$$37.68 = 2(3.14)(x+1)$$
$$37.68 = 6.28(x+1)$$
$$\frac{37.68}{6.28} = \frac{6.28(x+1)}{6.28}$$
$$6 = x + 1$$
$$5 = x$$

50. Use the formula $C = \pi d$, replacing d with $3x - 5$ and C with 54.95.

$$54.95 = (3.14)(3x - 5)$$
$$\frac{54.95}{3.14} = \frac{3.14(3x-5)}{3.14}$$
$$17.5 = 3x - 5$$
$$22.5 = 3x$$
$$7.5 = x$$

51. Use the formula $A = \pi r^2$, replacing A with 18.0864 and r with x.

$$18.0864 = 3.14x^2$$
$$\frac{18.0864}{3.14} = \frac{3.14x^2}{3.14}$$
$$5.76 = x^2$$
$$\sqrt{5.76} = \sqrt{x^2}$$
$$2.4 = x$$

52. Use the formula $A = \pi r^2$, replacing A with 28.26. The diameter is $4x$, so this expression must be divided in half for r.

$$28.26 = 3.14(2x)^2$$
$$\frac{28.26}{3.14} = \frac{3.14(2x)^2}{3.14}$$
$$9 = (2x)^2$$
$$9 = 4x^2$$
$$\frac{9}{4} = \frac{4x^2}{4}$$
$$\frac{9}{4} = x^2$$
$$\sqrt{\frac{9}{4}} = \sqrt{x^2}$$
$$\frac{3}{2} = x$$

53. (a) $A = 4 \cdot 5 = 20$ cm^2

(b) $A = 8 \cdot 10 = 80$ cm^2

(c) $A = 12 \cdot 15 = 180$ cm^2

(d) $A = 16 \cdot 20 = 320$ cm^2

(e) The rectangle in part (b) had sides twice as long as the sides of the rectangle in part (a). Divide the larger area by the smaller $(80 \div 20 = 4)$. By doubling the sides, the area increased <u>4</u> times.

(f) To get the rectangle in part (c) each side of the rectangle of part (a) was multiplied by <u>3</u>. This made the larger area <u>9</u> times the smaller area $(180 \div 20 = 9)$.

(g) To get the rectangle of part (d) each side of the rectangle of part (a) was multiplied by <u>4</u>. This made the area increase to <u>16</u> times what it was originally $(320 \div 20 = 16)$.

(h) In general, if the length of each side of a rectangle is multiplied by n, the area is multiplied by <u>n^2</u>.

54. Because each measurement is multiplied by 2, the area will increase by $2^2 = 4$. Then $4 \cdot 60 = \$240$.

55. Because each measurement is multiplied by 2, the area will increase by $2^2 = 4$. Then $4 \cdot 200 = \$800$.

56. Because each measurement is multiplied by 3, the area will increase by $3^2 = 9$. Then $9 \cdot 80 = \$720$.

57. If the radius of a circle is multiplied by n, then the area of the circle is multiplied by $\underline{n^2}$.

58. If the height of a triangle is multiplied by n and the base length remains the same, then the area of the triangle is multiplied by $\underline{n}$. Only one dimension is multiplied by n, so the area is multiplied by n only once.

59. Find the area of the parallelogram and the area of the triangle. Then add the two area values.

    Parallelogram $\quad A = 6 \cdot 10 = 60$
    Triangle $\quad A = \dfrac{1}{2}(10)(4) = 20$
    Total area $\quad 60 + 20 = 80$

60. Find the area of the triangle, rectangle, and parallelogram. Then add the three area values.

    Triangle $\quad A = \dfrac{1}{2}(10)(9) = 45$
    Rectangle $\quad A = 4 \cdot 10 = 40$
    Parallelogram $\quad A = 10 \cdot 3 = 30$
    Total area $\quad 45 + 40 + 30 = 115$

61. There are 2 semicircles or equivalently 1 full circle with radius of 3. Find the area of this circle and of the rectangle.

    Rectangle $\quad A = 8 \cdot 6 = 48$
    Circle $\quad A = (3.14) \cdot 3^2 = 28.26$
    Total area $\quad 48 + 28.26 = 76.26$

62. The four semicircles create two full circles, each with a radius of 4. Find the sum of the areas of the two circles and the square.

    Square $\quad A = 8 \cdot 8 = 64$
    Circle $\quad A = (3.14) \cdot 4^2 = 50.24$
    Total area $\quad 64 + 2(50.24) = 164.48$

63. Find the area of the trapezoid that surrounds the triangle. Subtract the area of the triangle.

    Trapezoid $\quad A = \dfrac{1}{2}(12)(18 + 11) = 174$
    Triangle $\quad A = \dfrac{1}{2}(12)(7) = 42$
    Shaded area $\quad 174 - 42 = 132$ ft^2

64. Find the area of the trapezoid that surrounds the triangle. Subtract the area of the triangle. The length of the lower base of the trapezoid is $19 + 38 = 57$.

    Trapezoid $\quad A = \dfrac{1}{2}(24)(28 + 57) = 1020$
    Triangle $\quad A = \dfrac{1}{2}(19)(16) = 152$
    Shaded area $\quad 1020 - 152 = 868$ cm^2

65. Find the area of the rectangle that surrounds the triangles. Subtract the areas of the triangles. The length of the rectangle is $48 + 48 = 96$.

    Rectangle $\quad A = 74 \cdot 96 = 7104$
    One triangle $\quad A = \dfrac{1}{2}(48)(36) = 864$
    Shaded area $\quad 7104 - 2(864) = 5376$ cm^2.

66. Find the area of the rectangle that surrounds the semicircle. Subtract the area of the semicircle. The diameter is 21; therefore, the radius is $21 \div 2 = 10.5$.

    Rectangle $\quad A = 21 \cdot 23 = 483$
    Semicircle $\quad A = \dfrac{1}{2}(3.14)(10.5)^2 = 173.0925$
    Shaded area $\quad 483 - 173.0925 = 309.9075$

    Rounded to the nearest hundredth, the shaded area is 309.91 ft.^{2}.

67. Find the area of the square that surrounds the circle. Subtract the area of the circle. The diameter is 26; therefore, the radius is $26 \div 2 = 13$.

    Square $\quad A = 26^2 = 676$
    Circle $\quad A = (3.14)(13)^2 = 530.66$
    Shaded area $\quad 676 - 530.66 = 145.34$ m^2

68. Find the area of the large circle that surrounds the two smaller circles. Subtract the area of the smaller circles. The radius of each of the smaller circles is 4; the radius of the large circle is 8.

    Large circle $\quad A = (3.14)(8)^2 = 200.96$
    Small circle $\quad A = (3.14)(4)^2 = 50.24$
    Shaded area $\quad 200.96 - 2(50.24) = 100.48$ cm^2

69. The best buy is the pizza with the lowest cost per square inch or unit price if you have enough money and you can eat all of it!

    10" pizza $\quad A = (3.14)(5^2) = 78.5$ in.^{2}
    Unit price $\quad \dfrac{5.99}{78.5} \approx \$.076$
    12" pizza $\quad A = (3.14)(6^2) = 113.04$ in.^{2}
    Unit price $\quad \dfrac{7.99}{113.04} \approx \$.071$
    14" pizza $\quad A = (3.14)(7^2)$ or 153.86 in.^{2}
    Unit price $\quad \dfrac{8.99}{153.86} \approx \$.058$

    The best buy is the 14" pizza.

70. The best buy is the pizza with the lowest cost per square inch or unit price.

    10" pizza $A = (3.14)(5^2) = 78.5$ in^2
    Unit price $\dfrac{7.99}{78.5} \approx \$.102$
    12" pizza $A = (3.14)(6^2) = 113.04$ in^2
    Unit price $\dfrac{9.99}{113.04} \approx \$.088$
    14" pizza $A = (3.14)(7^2)$ or 153.86 in^2
    Unit price $\dfrac{10.99}{153.86} \approx \$.071$

    The best buy is the 14" pizza.

71. The best buy is the pizza with the lowest cost per square inch or unit price.

    10" pizza $A = (3.14)(5^2) = 78.5$ in^2
    Unit price $\dfrac{9.99}{78.5} \approx \$.127$
    12" pizza $A = (3.14)\sqrt{6^2} = 113.04$ in^2
    Unit price $\dfrac{11.99}{113.04} \approx \$.106$
    14" pizza $A = (3.14)(7^2)$ or 153.86 in^2
    Unit price $\dfrac{12.99}{153.86} \approx \$.084$

    The best buy is the 14" pizza.

72. The best buy is the pizza with the lowest cost per square inch or unit price.

    10" pizza $A = (3.14)(5^2) = 78.5$ in^2
    Unit price $\dfrac{11.99}{78.5} \approx \$.153$
    12" pizza $A = (3.14)(6^2) = 113.04$ in^2
    Unit price $\dfrac{13.99}{113.04} \approx \$.124$
    14" pizza $A = (3.14)(7^2)$ or 153.86 in^2
    Unit price $\dfrac{14.99}{153.86} \approx \$.097$

    The best buy is the 14" pizza

73. The key is to construct OB and to realize that the diagonals of a rectangle are equal in length. So by inspection, $OB = AC = 13$ in. OB is a radius. Therefore, the diameter $= 2 \cdot 13 = 26$ in.

74. In the figure, $EF = ED$ and $FB = BC$.
    The perimeter of $\triangle AEB$ is

    $AE + EB + AB = AE + (EF + FB) + AB$.

    Because $EF = ED$,

    $AE + EF = AE + ED = AD = 20$ in.

    Also because $FB = BC$,

    $FB + AB = BC + AB = 34$ in.

    By substitution, the perimeter of $\triangle AEB$ is

    $20 + 34 = 54$ in.

75. The key is to construct TV and UW to create more triangles. By inspection, all the small triangles are equal. $PQRS$ has 8 triangles. $TUVW$ has 4 triangles. Therefore $TUVW$ has half the area of $PQRS$, which is 625 ft^2. Otherwise, find the area by first solving for the length of one side using the Pythagorean theorem.

76. In rectangle $ABCD$, $l = 2w$. Since the perimeter of $ABCD$ is 96 inches,

    $P = 2l + 2w$
    $96 = 2(2w) + 2w$
    $96 = 6w$
    $16 = w$ and $l = 2w = 32$.

    A midpoint divides a segment into two parts of equal length. Since P, Q, R, and S are midpoints of the sides, AP, PB, DR, and RC are each equivalent to $\frac{1}{2}l$, which is 16. Also, AS, SD, BQ, and QC are each equivalent to $\frac{1}{2}w$, which is 8.

    The area of rectangle $ABCD = lw = 32 \cdot 16 = 512$ inches^2. Notice that the triangles are right triangles, so the legs can be used as their bases and heights.

    Area of $\triangle APS$ $\dfrac{1}{2}(16)(8) = 64$
    Area of $\triangle PBS$ $\dfrac{1}{2}(16)(8) = 64$
    Area of $\triangle SDR$ $\dfrac{1}{2}(16)(8) = 64$
    Area of $\triangle QCR$ $\dfrac{1}{2}(16)(8) = 64$

    The area of quadrilateral $PQRS$ is equal to the area of $ABCD$ minus the sum of the areas of the four triangles.

    $512 - 4(64) = 256$ in.^{2}

77. The key is to construct a perpendicular line from E to side DC. Let the point of intersection of side DC be labeled point F. By inspection there are two sets of equal triangles: $\triangle DAE$ and $\triangle EFD$; $\triangle CBE$ and $\triangle EFC$. Then the area of the shaded region is half the area of the square. Therefore,

    $\text{Area} = \dfrac{36^2}{2} = \dfrac{1296}{2} = 648$ in.^{2}

78. Yes, the perimeter can be found. Draw lines to complete the large rectangle. These lines are equal in measure to the short vertical and horizontal lines that create the cutout portion of the large rectangle.

    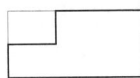

Therefore, the perimeter is
$$P = 2 \cdot 13 + 2 \cdot 7 = 40 \text{ in.}$$

79. The key is to construct a square using two radii from O and bounding the shaded region. The area of the small square is r^2, and the area of the quarter circle is $\pi r^2/4$. Therefore, the area of the shaded region is
$$r^2 - \frac{\pi r^2}{4} = r^2\left(1 - \frac{\pi}{4}\right) = \frac{(4-\pi)r^2}{4}.$$

80. First find the length of AB. In the right triangle ABE, using $\overline{AB}$ as the base and $\overline{EA}$ as the corresponding height, we have
$$A = \frac{1}{2}bh$$
$$30 = \frac{1}{2}(\overline{AB})(6)$$
$$\frac{2(30)}{6} = \overline{AB}$$
$$10 = \overline{AB}$$

Now find the area of the trapezoid with $h = 6$, $b = 10$, and $B = 14$.
$$A = \frac{1}{2}h(b+B)$$
$$A = \frac{1}{2}6(10+14)$$
$$A = 3(24)$$
$$A = 72 \text{ in.}^2$$

9.4 EXERCISES

1. | STATEMENTS | REASONS |
    |---|---|
    | 1. $AC = BD$ | 1. Given |
    | 2. $AD = BC$ | 2. Given |
    | 3. $AB = AB$ | 3. Reflexive property |
    | 4. $\triangle ABD \cong \triangle BAC$ | 4. SSS Congruence Property |

2. | STATEMENTS | REASONS |
    |---|---|
    | 1. $AC = BC$ | 1. Given |
    | 2. $\angle ACD = \angle BCD$ | 2. Given |
    | 3. $CD = CD$ | 3. Reflexive property |
    | 4. $\triangle ADC \cong \triangle BDC$ | 4. SAS Congruence Property |

3. | STATEMENTS | REASONS |
    |---|---|
    | 1. $\overleftrightarrow{DB}$ is perpendicular to $\overleftrightarrow{AC}$ | 1. Given |
    | 2. $AB = BC$ | 2. Given |
    | 3. $\angle ABD = \angle CBD$ | 3. Both are right angles by definition of perpendicularity. |
    | 4. $DB = DB$ | 4. Reflexive property |
    | 5. $\triangle ABD \cong \triangle CBD$ | 5. SAS congruence property |

4. | STATEMENTS | REASONS |
    |---|---|
    | 1. $BC = BA$ | 1. Given |
    | 2. $\angle 1 = \angle 2$ | 2. Given |
    | 3. $\angle DBC = \angle DBA$ | 3. Supplements of equal angles are equal to each other. |
    | 4. $DB = DB$ | 4. Reflexive property |
    | 5. $\triangle DBC \cong \triangle DBA$ | 5. SAS congruence property |

5. | STATEMENTS | REASONS |
    |---|---|
    | 1. $\angle BAC = \angle DAC$ | 1. Given |
    | 2. $\angle BCA = \angle DCA$ | 2. Given |
    | 3. $AC = AC$ | 3. Reflexive property |
    | 4. $\triangle ABC \cong \triangle ADC$ | 4. ASA congruence property |

6. | STATEMENTS | REASONS |
    |---|---|
    | 1. $BO = OE$ | 1. Given |
    | 2. $\overleftrightarrow{OB}$ is perpendicular to $\overleftrightarrow{AC}$. | 2. Given |
    | 3. $\overleftrightarrow{OE}$ is perpendicular to $\overleftrightarrow{DF}$. | 3. Given |
    | 4. $\angle ABO = \angle FEO$ | 4. Both are right angles by definition of perpendicularity. |
    | 5. $\angle AOB = \angle FOE$ | 5. Vertical angles are equal. |
    | 6. $\triangle AOB \cong \triangle FOE$ | 6. ASA congruence property |

7. If $\angle B$ measures 46°, then $\angle A$ measures <u>67°</u> and $\angle C$ measures <u>67°</u>. In an isosceles triangle, the angles opposite the equal sides are also equal in measure. Thus, $\angle A = \angle C$. The sum of the angles of the triangle is 180. Then $180 - 46 = 134$ and $134 \div 2 = 67°$.

8. Because $\angle C = 52°$, $\angle A = 52°$ because the angles opposite the equal sides are also equal in measure. Then, because the sum of the three angles is 180°,
$$180 - (52 + 52) = 76°.$$

9. The length of side $AB = 12$ because it has the same measure as BC. Then $30 - 2 \cdot 12 = 6$ in.

10. Subtract 10 from 40 to find the total length of the two remaining sides. Then because they are equal in length, divide by 2. $40 - 10 = 30$ and $30 \div 2 = 15$ in.

11. Writing exercise

12. Writing exercise

13. Corresponding angles are equal in measure.
$$\angle A \text{ and } \angle P$$
$$\angle B \text{ and } \angle Q$$
$$\angle C \text{ and } \angle R$$

Corresponding sides are proportional in length.
$$\overleftrightarrow{AB} \text{ and } \overleftrightarrow{PQ}$$
$$\overleftrightarrow{AC} \text{ and } \overleftrightarrow{PR}$$
$$\overleftrightarrow{CB} \text{ and } \overleftrightarrow{RQ}$$

14. Corresponding angles are equal in measure.

    $\angle C$ and $\angle R$
    $\angle B$ and $\angle Q$
    $\angle A$ and $\angle P$

    Corresponding sides are proportional in length.

    $\overline{AB}$ and $\overline{PQ}$
    $\overline{AC}$ and $\overline{PR}$
    $\overline{CB}$ and $\overline{RQ}$

15. Sometimes it is helpful to sketch the triangles, drawing them side by side. It is easier to determine the corresponding sides.

    Corresponding angles are equal in measure.

    $\angle HGK$ and $\angle EGF$ because they are vertical
    $\angle H$ and $\angle F$
    $\angle K$ and $\angle E$

    Corresponding sides are proportional in length.

    $\overline{HK}$ and $\overline{EF}$
    $\overline{GH}$ and $\overline{GF}$
    $\overline{GK}$ and $\overline{GE}$

16. Sometimes it is helpful to re-sketch the triangles, drawing them side by side. It is easier to determine the corresponding sides.

    Corresponding angles are equal in measure.

    $\angle ABE$ and $\angle CBD$ because they are vertical
    $\angle A$ and $\angle C$
    $\angle D$ and $\angle E$

    Corresponding sides proportional in length.

    $\overline{AE}$ and $\overline{CD}$
    $\overline{AB}$ and $\overline{BC}$
    $\overline{BE}$ and $\overline{BD}$

17. $\angle P = \angle C = 78°$
    $\angle M = \angle B = 46°$
    $\angle N = \angle A = 180 - (78 + 46) = 180 - 124 = 56°$
    because the sum of the angle measures must equal 180°.

18. $\angle P = \angle C = 90°$
    $\angle Q = \angle A = 42°$
    $\angle R = \angle B = 180 - (90 + 42) = 180 - 132 = 48°$
    because the sum of the angle measures must equal 180°.

19. $\angle T = \angle X = 74°$
    $\angle V = \angle Y = 28°$
    $\angle W = \angle Z = 180 - (74 + 28) = 180 - 102 = 78°$
    because the sum of the angle measures must equal 180°.

20. $\angle B = \angle K = 106°$
    $\angle C = \angle N = 30°$ (This is given.)
    $\angle A = \angle M = 180 - (106 + 30) = 180 - 136 = 44°$
    because the sum of the angle measures must equal 180°.

21. $\angle T = \angle P = 20°$
    $\angle V = \angle Q = 64°$
    $\angle U = \angle R = 180 - (20 + 64) = 180 - 84 = 96°$
    because the sum of the angle measures must equal 180°.

22. $\angle N = \angle Y = 90°$ (This is given.)
    $\angle P = \angle Z = 38°$ (This is given.)
    $\angle M = \angle X = 180 - (90 + 38) = 180 - 128 = 52°$,
    because the sum of the angle measures must equal 180°.

23. Corresponding sides must be proportional.

    $$\frac{a}{8} = \frac{25}{10}$$
    $$10 \cdot a = 8 \cdot 25$$
    $$\frac{10a}{10} = \frac{200}{10}$$
    $$a = 20$$

    and

    $$\frac{b}{6} = \frac{25}{10}$$
    $$10 \cdot b = 6 \cdot 25$$
    $$\frac{10b}{10} = \frac{150}{10}$$
    $$b = 15$$

24. Corresponding sides must be proportional.

    $$\frac{a}{10} = \frac{75}{25}$$
    $$25 \cdot a = 10 \cdot 75$$
    $$\frac{25a}{25} = \frac{750}{25}$$
    $$a = 30$$

    and

    $$\frac{b}{20} = \frac{75}{25}$$
    $$25 \cdot b = 20 \cdot 75$$
    $$\frac{25b}{25} = \frac{1500}{25}$$
    $$b = 60$$

25. Corresponding sides must be proportional.

    $$\frac{a}{12} = \frac{6}{12}$$
    $a = 6$ because the denominators are equal

And, $\dfrac{b}{15} = \dfrac{6}{12}$

$12 \cdot b = 15 \cdot 6$

$\dfrac{12b}{12} = \dfrac{90}{12}$

$b = \dfrac{6 \cdot 15}{6 \cdot 2}$

$b = \dfrac{15}{2}$.

26. Corresponding sides must be proportional.

    $\dfrac{a}{6} = \dfrac{3}{9}$

    $9 \cdot a = 3 \cdot 6$

    $\dfrac{9a}{9} = \dfrac{18}{9}$

    $a = 2$

27. Corresponding sides must be proportional.

    $\dfrac{x}{4} = \dfrac{9}{6}$

    $6 \cdot x = 4 \cdot 9$

    $\dfrac{6x}{6} = \dfrac{36}{6}$

    $x = 6$

28. Corresponding sides must be proportional.

    $\dfrac{m}{12} = \dfrac{21}{14}$

    $14 \cdot m = 12 \cdot 21$

    $\dfrac{14m}{14} = \dfrac{252}{14}$

    $m = 18$

29. Corresponding sides must be proportional.

    $\dfrac{x}{50} = \dfrac{220}{100}$

    $100 \cdot x = 50 \cdot 220$

    $\dfrac{100x}{100} = \dfrac{11000}{100}$

    $x = 110$

30. Corresponding sides must be proportional.

    $\dfrac{y}{60} = \dfrac{40}{160 + 40}$

    $\dfrac{y}{60} = \dfrac{40}{200}$

    $200 \cdot y = 60 \cdot 40$

    $\dfrac{200y}{200} = \dfrac{2400}{200}$

    $y = 12$

31. Corresponding sides must be proportional. In the third step, reduce the fraction on the right to lowest terms to make computations easier.

    $\dfrac{c}{100} = \dfrac{10 + 90}{90}$

    $\dfrac{c}{100} = \dfrac{100}{90}$

    $\dfrac{c}{100} = \dfrac{10}{9}$

    $9 \cdot c = 100 \cdot 10$

    $\dfrac{9c}{9} = \dfrac{1000}{9}$

    $c = 111\dfrac{1}{9}$

32. Corresponding sides must be proportional.

    $\dfrac{m}{80} = \dfrac{75 + 5}{75}$

    $\dfrac{m}{80} = \dfrac{80}{75}$

    $75 \cdot m = 80 \cdot 80$

    $\dfrac{75m}{75} = \dfrac{6400}{75}$

    $m = 85\dfrac{1}{3}$

33. Let $h =$ height of the tree.

    $\dfrac{h}{45} = \dfrac{2}{3}$

    $3 \cdot h = 45 \cdot 2$

    $\dfrac{3h}{3} = \dfrac{90}{3}$

    $h = 30$ m

34. Let $h =$ height of the tower.

    $\dfrac{h}{180} = \dfrac{9}{15}$

    $15 \cdot h = 180 \cdot 9$

    $\dfrac{15h}{15} = \dfrac{1620}{15}$

    $h = 108$ feet

35. Let $x =$ length of the mid-length side.

    $\dfrac{5}{x} = \dfrac{4}{400}$

    $4 \cdot x = 5 \cdot 400$

    $\dfrac{4x}{4} = \dfrac{2000}{4}$

    $x = 500$ m

    Now let $y =$ the longest side.

    $\dfrac{7}{y} = \dfrac{4}{400}$

    $4 \cdot y = 7 \cdot 400$

    $\dfrac{4y}{4} = \dfrac{2800}{4}$

    $y = 700$ m

    The lengths of the other two sides are 500 meters and 700 meters.

9.4 THE GEOMETRY OF TRIANGLES: CONGRUENCE, SIMILARITY, AND THE PYTHAGOREAN THEOREM

36. Let h = height of the lighthouse keeper.
$$\frac{h}{3.5} = \frac{14}{28}$$
$$28 \cdot h = 3.5 \cdot 14$$
$$\frac{28h}{28} = \frac{49}{28}$$
$$h = 1.75 \text{ m}$$

37. Let h = height of the building. In step 2 the fraction on the right is reduced to make further computations easier.
$$\frac{h}{15} = \frac{300}{40}$$
$$\frac{h}{15} = \frac{15}{2}$$
$$2 \cdot h = 15 \cdot 15$$
$$\frac{2h}{2} = \frac{225}{2}$$
$$h = 112.5 \text{ ft}$$

38. A sketch may help to see the following proportion.
Let x = the distance from Phoenix to Tucson
$$\frac{x}{8} = \frac{230}{12}$$
$$12 \cdot x = 8 \cdot 230$$
$$\frac{12x}{12} = \frac{1840}{12}$$
$$h \approx 153.3 \text{ km}$$

Now let y = the distance from Yuma to Tucson.
$$\frac{y}{17} = \frac{230}{12}$$
$$12 \cdot y = 17 \cdot 230$$
$$\frac{12y}{12} = \frac{3910}{12}$$
$$y = 325.8 \text{ km}$$

39. Let x = length of carved body.
$$\frac{x}{\text{actual body height}} = \frac{\text{height of carved head}}{\text{actual head height}}$$
$$\frac{x}{6\frac{1}{3}} = \frac{60}{\frac{3}{4}}$$
$$\frac{3}{4} \cdot x = 6\frac{1}{3} \cdot 60$$
$$\frac{3}{4}x = \frac{19}{3} \cdot \frac{60}{1}$$
$$\frac{3}{4}x = 380$$
$$\frac{4}{3} \cdot \frac{3}{4}x = 380 \cdot \frac{4}{3}$$
$$x = 506\frac{2}{3} \text{ feet}$$

40. Let h = height of Robert Wadlow.
$$\frac{h}{35.7} = \frac{6}{24}$$
$$24 \cdot h = 35.7 \cdot 6$$
$$\frac{24h}{24} = \frac{214.2}{24}$$
$$h = 8.925 \text{ feet}$$

Also, $0.925(12) = 11.1$. This means Robert Wadlow was about 8 feet 11 inches tall.

41. Use the Pythagorean theorem $a^2 + b^2 = c^2$ with $a = 8$ and $b = 15$.
$$8^2 + 15^2 = c^2$$
$$64 + 225 = c^2$$
$$289 = c^2$$
$$\sqrt{289} = \sqrt{c^2}$$
$$17 = c$$

42. Use the Pythagorean theorem $a^2 + b^2 = c^2$ with $a = 7$ and $c = 25$.
$$7^2 + b^2 = 25^2$$
$$49 + b^2 = 625$$
$$b^2 = 576$$
$$\sqrt{b^2} = \sqrt{576}$$
$$b = 24$$

43. Use the Pythagorean theorem $a^2 + b^2 = c^2$ with $b = 84$ and $c = 85$.
$$a^2 + 84^2 = 85^2$$
$$a^2 + 7056 = 7225$$
$$a^2 = 169$$
$$\sqrt{a^2} = \sqrt{169}$$
$$a = 13$$

44. Use the Pythagorean theorem $a^2 + b^2 = c^2$ with $a = 24$ and $c = 25$.
$$24^2 + b^2 = 25^2$$
$$576 + b^2 = 625$$
$$b^2 = 49$$
$$\sqrt{b^2} = \sqrt{49}$$
$$b = 7 \text{ cm}$$

45. Use the Pythagorean theorem $a^2 + b^2 = c^2$ with $a = 14$ and $b = 48$.
$$14^2 + 48^2 = c^2$$
$$196 + 2304 = c^2$$
$$2500 = c^2$$
$$\sqrt{2500} = \sqrt{c^2}$$
$$50 \text{ m} = c$$

46. Use the Pythagorean theorem $a^2 + b^2 = c^2$ with $a = 28$ and $c = 100$.
$$28^2 + b^2 = 100^2$$
$$784 + b^2 = 10000$$
$$b^2 = 9216$$
$$\sqrt{b^2} = \sqrt{9216}$$
$$b = 96 \text{ km}$$

47. Use the Pythagorean theorem $a^2 + b^2 = c^2$ with $b = 21$ and $c = 29$.
$$a^2 + 21^2 = 29^2$$
$$a^2 + 441 = 841$$
$$a^2 = 400$$
$$\sqrt{a^2} = \sqrt{400}$$
$$a = 20 \text{ in.}$$

48. Use the Pythagorean theorem $a^2 + b^2 = c^2$ with $b = 120$ and $c = 169$.
$$a^2 + 120^2 = 169^2$$
$$a^2 + 14400 = 28561$$
$$a^2 = 14161$$
$$\sqrt{a^2} = \sqrt{14161}$$
$$a = 119 \text{ ft}$$

49. The sum of the squares of the two shorter sides of a right triangle is equal to the square of the longest side.

50. Use a right triangle with $a = 6$, $b = 8$, and $c = 10$. Although it is true that $6^2 + 8^2 = 10^2$, it is not true that $6 + 8 = 10$.

51. Given $r = 2$ and $s = 1$,
$$a = r^2 - s^2 = 2^2 - 1^2 = 4 - 1 = 3$$
$$b = 2rs = 2 \cdot 2 \cdot 1 = 4$$
$$c = r^2 + s^2 = 2^2 + 1^2 = 4 + 1 = 5$$

The Pythagorean triple is $(3, 4, 5)$.

52. Given $r = 3$ and $s = 2$,
$$a = r^2 - s^2 = 3^2 - 2^2 = 9 - 4 = 5$$
$$b = 2rs = 2 \cdot 3 \cdot 2 = 12$$
$$c = r^2 + s^2 = 3^2 + 2^2 = 9 + 4 = 13$$

The Pythagorean triple is $(5, 12, 13)$.

53. Given $r = 4$ and $s = 3$,
$$a = r^2 - s^2 = 4^2 - 3^2 = 16 - 9 = 7$$
$$b = 2rs = 2 \cdot 4 \cdot 3 = 24$$
$$c = r^2 + s^2 = 4^2 + 3^2 = 16 + 9 = 25$$

The Pythagorean triple is $(7, 24, 25)$.

54. Given $r = 3$ and $s = 1$,
$$a = r^2 - s^2 = 3^2 - 1^2 = 9 - 1 = 8$$
$$b = 2rs = 2 \cdot 3 \cdot 1 = 6$$
$$c = r^2 + s^2 = 3^2 + 1^2 = 9 + 1 = 10$$

The Pythagorean triple is $(8, 6, 10)$.

55. Given $r = 4$ and $s = 2$,
$$a = r^2 - s^2 = 4^2 - 2^2 = 16 - 4 = 12$$
$$b = 2rs = 2 \cdot 4 \cdot 2 = 16$$
$$c = r^2 + s^2 = 4^2 + 2^2 = 16 + 4 = 20$$

The Pythagorean triple is $(12, 16, 20)$.

56. Given $r = 4$ and $s = 1$,
$$a = r^2 - s^2 = 4^2 - 1^2 = 16 - 1 = 15$$
$$b = 2rs = 2 \cdot 4 \cdot 1 = 8$$
$$c = r^2 + s^2 = 4^2 + 1^2 = 16 + 1 = 17$$

The Pythagorean triple is $(15, 8, 17)$.

57. Substitute the expressions in r and s for a and b.
$$a^2 + b^2 = (r^2 - s^2)^2 + (2rs)^2$$
$$= r^4 - 2r^2s^2 + s^4 + 4r^2s^2$$
$$= r^4 + 2r^2s^2 + s^2$$
$$= (r^2 + s^2)^2$$
$$= c^2$$

58. Let $x = 3$ and $y = 5$. The original Pythagorean triple is $(x, x+1, y) = (3, 4, 5)$. Another triple can be found by
$$3x + 2y + 1 = 3 \cdot 3 + 2 \cdot 5 + 1 = 20$$
$$3x + 2y + 2 = 3 \cdot 3 + 2 \cdot 5 + 2 = 21$$
$$4x + 3y + 2 = 4 \cdot 3 + 3 \cdot 5 + 2 = 29$$

The triple is $(20, 21, 29)$.

Use this triple to generate another.
$$3x + 2y + 1 = 3 \cdot 20 + 2 \cdot 29 + 1 = 119$$
$$3x + 2y + 2 = 3 \cdot 20 + 2 \cdot 29 + 2 = 120$$
$$4x + 3y + 2 = 4 \cdot 20 + 3 \cdot 29 + 2 = 169$$

The triple is $(119, 120, 169)$.

Use this triple to generate another.
$$3x + 2y + 1 = 3 \cdot 119 + 2 \cdot 169 + 1 = 696$$
$$3x + 2y + 2 = 3 \cdot 119 + 2 \cdot 169 + 2 = 697$$
$$4x + 3y + 2 = 4 \cdot 119 + 3 \cdot 169 + 2 = 985$$

The triple is $(696, 697, 985)$.

59. When $m = 3$,
$$\frac{m^2 - 1}{2} = \frac{3^2 - 1}{2} = \frac{8}{2} = 4$$
$$\frac{m^2 + 1}{2} = \frac{3^2 + 1}{2} = \frac{10}{2} = 5.$$

The Pythagorean triple is $(3, 4, 5)$.

9.4 THE GEOMETRY OF TRIANGLES: CONGRUENCE, SIMILARITY, AND THE PYTHAGOREAN THEOREM

60. When $m = 5$,
$$\frac{m^2 - 1}{2} = \frac{5^2 - 1}{2} = \frac{24}{2} = 12$$
$$\frac{m^2 + 1}{2} = \frac{5^2 + 1}{2} = \frac{26}{2} = 13.$$
The Pythagorean triple is $(5, 12, 13)$.

61. When $m = 7$,
$$\frac{m^2 - 1}{2} = \frac{7^2 - 1}{2} = \frac{48}{2} = 24$$
$$\frac{m^2 + 1}{2} = \frac{7^2 + 1}{2} = \frac{50}{2} = 25.$$
The Pythagorean triple is $(7, 24, 25)$.

62. When $m = 9$,
$$\frac{m^2 - 1}{2} = \frac{9^2 - 1}{2} = \frac{80}{2} = 40$$
$$\frac{m^2 + 1}{2} = \frac{9^2 + 1}{2} = \frac{82}{2} = 41.$$
The Pythagorean triple is $(9, 40, 41)$.

63. Replace a^2 with m^2, replace b^2 with $\left(\frac{m^2-1}{2}\right)^2$, and show that their sum simplifies to c^2.
$$a^2 + b^2 = m^2 + \left(\frac{m^2 - 1}{2}\right)^2$$
$$= m^2 + \frac{m^4 - 2m^2 + 1}{4}$$
$$= \frac{4m^2}{4} + \frac{m^4 - 2m^2 + 1}{4}$$
$$= \frac{m^4 + 2m^2 + 1}{4}$$
$$= \frac{(m^2 + 1)^2}{4}$$
$$= \left(\frac{m^2 + 1}{2}\right)^2$$
$$= c^2$$

64. Consecutive even numbers take the form x, $x + 2$, and $x + 4$ where x is an even number. For the Pythagorean triple $(x, x + 2, x + 4)$,
$$x^2 + (x + 2)^2 = (x + 4)^2.$$
Solve for x to find the possible consecutive even numbers that are Pythagorean triples.
$$x^2 + (x^2 + 4x + 4) = x^2 + 8x + 16$$
$$2x^2 + 4x + 4 = x^2 + 8x + 16$$
$$x^2 - 4x - 12 = 0$$
$$(x - 6)(x + 2) = 0$$
$$x - 6 = 0 \quad \text{or} \quad x + 2 = 0$$
$$x = 6 \qquad\qquad x = -2$$

Pythagorean triples must be positive numbers, so $(6, 8, 10)$ is the only Pythagorean triple consisting of consecutive even numbers.

65. For $n = 2$,
$$2n = 2 \cdot 2 = 4$$
$$n^2 - 1 = 2^2 - 1 = 3$$
$$n^2 + 1 = 2^2 + 1 = 5.$$
The Pythagorean triple is $(4, 3, 5)$.

66. For $n = 3$,
$$2n = 2 \cdot 3 = 6$$
$$n^2 - 1 = 3^2 - 1 = 8$$
$$n^2 + 1 = 3^2 + 1 = 10.$$
The Pythagorean triple is $(6, 8, 10)$.

67. For $n = 4$,
$$2n = 2 \cdot 4 = 8$$
$$n^2 - 1 = 4^2 - 1 = 15$$
$$n^2 + 1 = 4^2 + 1 = 17.$$
The Pythagorean triple is $(8, 15, 17)$.

68. For $n = 5$,
$$2n = 2 \cdot 5 = 10$$
$$n^2 - 1 = 5^2 - 1 = 24$$
$$n^2 + 1 = 5^2 + 1 = 26.$$
The Pythagorean triple is $(10, 24, 26)$.

69. Replace a with $2n$ and b with $n^2 - 1$ in the Pythagorean theorem, and show that the expression simplifies to c^2.
$$a^2 + b^2 = (2n)^2 + (n^2 - 1)^2$$
$$= 4n^2 + n^4 - 2n^2 + 1$$
$$= n^4 + 2n^2 + 1$$
$$= (n^2 + 1)^2$$
$$= c^2$$

70. No, since if the integer a is the length of one of the equal sides, then $a\sqrt{2}$ must be the length of the hypotenuse, and $a\sqrt{2}$ is not an integer.
Let $a =$ the integer length of one of the equal sides. Then by the Pythagorean theorem,
$$a^2 + a^2 = c^2$$
$$2a^2 = c^2$$
$$\sqrt{2a^2} = \sqrt{c^2}$$
$$a\sqrt{2} = c.$$

71. Let $b =$ length of longer leg,
    $b + 1 =$ length of hypotenuse c,
    $7 =$ length of shorter leg, a.
    Substitute these expressions into the Pythagorean theorem, $a^2 + b^2 = c^2$, and solve for b.
    $$7^2 + b^2 = (b+1)^2$$
    $$49 + b^2 = b^2 + 2b + 1$$
    $$49 + b^2 - b^2 = b^2 - b^2 + 2b + 1$$
    $$49 = 2b + 1$$
    $$48 = 2b$$
    $$24 = b$$

    The longer leg is 24 m.

72. Let $a =$ the length of the shorter leg,
    $2a + 1 =$ the length of the hypotenuse, c,
    $3a - 9 =$ the length of the leg, b.
    Substitute these expressions into the Pythagorean theorem, $a^2 + b^2 = c^2$, and solve for a.
    $$a^2 + (3a-9)^2 = (2a+1)^2$$
    $$a^2 + 9a^2 - 54a + 81 = 4a^2 + 4a + 1$$
    $$10a^2 - 54a + 81 = 4a^2 + 4a + 1$$
    $$6a^2 - 58a + 80 = 0$$

    Now solve for a by factoring.
    $$6a^2 - 58a + 80 = 0$$
    $$2(3a^2 - 29a + 40) = 0$$
    $$2(3a - 5)(a - 8) = 0$$
    $$3a - 5 = 0 \quad \text{or} \quad a - 8 = 0$$
    $$a = \frac{5}{3} \qquad\qquad a = 8$$

    If $a = \frac{5}{3}$, then $3 \cdot \frac{5}{3} - 9 = -4$, which is not reasonable. Therefore, the lengths of the sides are 8 cm, $2 \cdot 8 + 1 = 17$ cm, and $3 \cdot 8 - 9 = 15$ cm.

73. Let $h =$ the height of the tree, one of the legs of the triangle,
    $2h + 2 =$ the hypotenuse, c,
    $a = 30$, another leg of the triangle.
    Substitute these expressions into the Pythagorean theorem, $a^2 + b^2 = c^2$, and solve for h.
    $$30^2 + h^2 = (2h+2)^2$$
    $$900 + h^2 = 4h^2 + 8h + 4$$
    $$900 + h^2 - h^2 = 4h^2 - h^2 + 8h + 4$$
    $$900 = 3h^2 + 8h + 4$$
    $$900 - 900 = 3h^2 + 8h + 4 - 900$$
    $$0 = 3h^2 + 8h - 896$$
    $$0 = (3h + 56)(h - 16)$$
    $$3h + 56 = 0 \quad \text{or} \quad h - 16 = 0$$
    $$h = -\frac{56}{3} \qquad\qquad h = 16$$

    A negative height is not meaningful. The height is 16 feet.

74. Let $w =$ the width of the rectangle, one of the legs of the triangle,
    $2w - 2 =$ the length of the rectangle, another leg of the triangle,
    The hypotenuse is 5.
    Substitute these expressions into the Pythagorean theorem, $a^2 + b^2 = c^2$, and solve for w.
    $$w^2 + (2w-2)^2 = 5^2$$
    $$w^2 + 4w^2 - 8w + 4 = 25$$
    $$5w^2 - 8w + 4 = 25$$
    $$5w^2 - 8w + 4 - 25 = 25 - 25$$
    $$5w^2 - 8w - 21 = 0$$
    $$(5w + 7)(w - 3) = 0$$
    $$5w + 7 = 0 \quad \text{or} \quad w - 3 = 0$$
    $$w = -\frac{7}{5} \qquad\qquad w = 3$$

    A negative width is not meaningful. The width is 3 inches, and the length is $2 \cdot 3 - 2 = 4$ inches.

75. Let $h =$ the height of the break,
    $a = 3$ ft, one leg of the triangle,
    $c = 10 - h$, the hypotenuse of the triangle.
    Substitute these expressions into the Pythagorean theorem, $a^2 + b^2 = c^2$, and solve for h.
    $$3^2 + h^2 = (10-h)^2$$
    $$9 + h^2 = 100 - 20h + h^2$$
    $$9 + h^2 - h^2 = 100 - 20h + h^2 - h^2$$
    $$9 = 100 - 20h$$
    $$9 - 100 = 100 - 100 - 20h$$
    $$-91 = -20h$$
    $$4.55 = h$$

    The height of the break is 4.55 ft.

76. The image creates a right triangle with the length of the reed as the hypotenuse. The radius of the circle is one leg and the depth of the pond is the other leg. A sketch might help.
    Let $d =$ the depth of the pond.
    another leg $= 5$, the radius.
    $d + 1 =$ the hypotenuse because the reed projects 1 foot.
    Substitute these expressions into the Pythagorean theorem, $a^2 + b^2 = c^2$, and solve for d.
    $$d^2 + 5^2 = (d+1)^2$$
    $$d^2 + 25 = d^2 + 2d + 1$$
    $$d^2 - d^2 + 25 = d^2 - d^2 + 2d + 1$$
    $$25 = 2d + 1$$
    $$24 = 2d$$
    $$12 = d$$

    The depth of the pond is 12 feet.

77. Let $c=$ the length of the diagonal.
$$12^2 + 15^2 = c^2$$
$$144 + 225 = c^2$$
$$369 = c^2$$
$$\sqrt{369} = \sqrt{c^2}$$
$$19.21 \approx c$$

Then $.21(12) = 2.52$, which is 3 inches to the nearest inch. The diagonal should be 19 feet, 3 inches.

78. Let $c=$ the length of the diagonal.
$$14^2 + 20^2 = c^2$$
$$196 + 400 = c^2$$
$$596 = c^2$$
$$\sqrt{596} = \sqrt{c^2}$$
$$24.41 \approx c$$

Then $.41(12) = 4.92$, which is 5 inches to the nearest inch. The diagonal should be 24 feet, 5 inches.

79. Let $c=$ the length of the diagonal.
$$16^2 + 24^2 = c^2$$
$$256 + 576 = c^2$$
$$832 = c^2$$
$$\sqrt{832} = \sqrt{c^2}$$
$$28.84 \approx c$$

Then $.84(12) = 10.08$, which is 10 inches to the nearest inch. The diagonal should be 28 feet, 10 inches.

80. Let $c=$ the length of the diagonal.
$$20^2 + 32^2 = c^2$$
$$400 + 1024 = c^2$$
$$1424 = c^2$$
$$\sqrt{1424} = \sqrt{c^2}$$
$$37.74 \approx c$$

Then $.74(12) = 8.88$, which is 9 inches to the nearest inch. The diagonal should be 37 feet, 9 inches.

81. (a) The formula for the area of a trapezoid is $A = \frac{1}{2}h(b+B)$. If ZY is the base B, it can be expressed as b. Base b is then WX, which can also be expressed as a. The height of the trapezoid is WZ, which can also be expressed as $b+a$. Substitute these expressions for B, b, and h.
$$A = \frac{1}{2}(b+a)(a+b) = \frac{1}{2}(a+b)(a+b).$$

(b) The area of $\triangle PWX$ is $\frac{1}{2}ab$. The area of $\triangle PZY$ is $\frac{1}{2}ab$. The area of $\triangle PXY$ is $\frac{1}{2}c^2$.

(c)
$$\frac{1}{2}(a+b)(a+b) = \frac{1}{2}ab + \frac{1}{2}ab + \frac{1}{2}c^2$$
$$\frac{1}{2}(a^2+2ab+b^2) = \frac{1}{2}ab + \frac{1}{2}ab + \frac{1}{2}c^2$$
$$\frac{1}{2}a^2 + ab + \frac{1}{2}b^2 = ab + \frac{1}{2}c^2$$
$$\frac{1}{2}a^2 + \frac{1}{2}b^2 = \frac{1}{2}c^2$$
$$a^2 + b^2 = c^2$$

82. (a) By proportion, we have $c/b = \underline{b/j}$.

(b) By proportion, we also have $c/a = a/\underline{k}$.

(c) From part (a), $b^2 = \underline{cj}$. Cross multiply to obtain this equation.

(d) From part (b), $a^2 = \underline{ck}$. Again, cross multiply to obtain this equation.

(e) From the results of parts (c) and (d) and factoring, $a^2 + b^2 = c\underline{(j+k)}$. But since $\underline{j+k = c}$, it follows that $\underline{a^2 + b^2 = c^2}$. Obtain the first underlined statement from
$$a^2 + b^2 = cj + ck = c(j+k).$$

83. Draw a line connecting B and D to form two right triangles. The area of $\triangle DAB$ is $\frac{1}{2} \cdot 6 \cdot 8 = 24$. The area of $\triangle BCD$ cannot be found until the length of CD is known. First find the length of the hypotenuse BD for $\triangle DAB$: $6^2 + 8^2 = 36 + 64 = 100$. The hypotenuse is 10. To find the length of side CD, use the Pythagorean Theorem again:
$$2^2 + (CD)^2 = 10^2$$
$$4 + (CD)^2 = 100$$
$$(CD)^2 = 96$$
$$\sqrt{(CD)^2} = \sqrt{96}$$
$$CD = \sqrt{16 \cdot 6}$$
$$CD = 4\sqrt{6}.$$

Then the area of $\triangle BCD$ is $\frac{1}{2} \cdot 2 \cdot 4\sqrt{6} = 4\sqrt{6}$. Add the areas of the two triangles to find the area of the quadrilateral: $24 + 4\sqrt{6}$.

84. The perimeter of $\triangle ABC$ is 128 in. The altitude BD is 48 in. Since the triangle is isosceles, $AB = BC$ and $AD = DC$. Add these equations to obtain
$$AB + AD = BC + DC.$$

Since the perimeter of $\triangle ABC$ is equal to
$$AB + AD + BC + DC = 2(AB + AD),$$

one-half the perimeter of $\triangle ABC$ equals
$$AB + AD = \frac{1}{2}(128)$$
$$AB + AD = 64$$
$$AB = 64 - AD.$$

Use $BD = 48$ and the Pythagorean theorem to find AD:
$$(AD)^2 + 48^2 = (64 - AD)^2$$
$$(AD)^2 + 2304 = 4096 - 128(AD) + (AD)^2$$
$$128(AD) = 1792$$
$$AD = 14.$$

Then $AC = AD + DC = 2 \cdot 14 = 28$ in. The area of $\triangle ABC$ is
$$\frac{1}{2}(BD)(AC) = \frac{1}{2}(48)(28)$$
$$= 672 \text{ in.}^2.$$

85. Draw a line from the upper vertex that is perpendicular to the base of 24. Two right triangles are created, each with a base of 12. Use the Pythagorean theorem to find the height, x.
$$x^2 + 12^2 = 13^2$$
$$x^2 + 144 = 169$$
$$x^2 = 25$$
$$x = 5$$

Now separate the two triangles and rearrange them so that the height of each, x, lie next to each other to form the base of a triangle. Then $2x = 10$.

86. Right triangle ABC had $AD = DB + 8$, $AC = 12$ and $AB = 20$. By the Pythagorean theorem
$$(AC)^2 + (BC)^2 = (AB)^2$$
$$12^2 + (BC)^2 = 20^2$$
$$144 + (BC)^2 = 400$$
$$(BC)^2 = 256$$
$$BC = 16.$$

Therefore, $CD = CB - DB = 16 - DB$. We may also apply the Pythagorean theorem to right triangle ACD.
$$(AC)^2 + (CD)^2 = (AD)^2$$

Substitute 12 for AC, $16 - DB$ for CD, and $DB + 8$ for AD in the preceding equation.
$$12^2 + (16 - DB)^2 = (DB + 8)^2$$
$$144 + 256 - 32(DB) + (DB)^2 = DB^2 + 16DB + 64$$
$$336 = 48(DB)$$
$$7 = DB$$

$CD + DB = CB$, so $CD = CB - DB$, or
$$DC = 16 - 7 = 9.$$

87. The area of the pentagon can be found by adding the area of the square and the area of the triangle. If all sides of the pentagon are equal and the perimeter is 80, the length of each side is $80 \div 5 = 16$. Then the area of the square is $16^2 = 256$. Find the height of the triangle in order to use the area formula. One way this can be done is to draw a line from Q to base SR that forms a right angle with the base. This will divide the base in half so that both of the smaller triangles are right triangles. The Pythagorean theorem can be used to find the height.
$$8^2 + h^2 = 16^2$$
$$64 + h^2 = 256$$
$$h^2 = 192$$
$$h = \sqrt{192}$$
$$h = \sqrt{64 \cdot 3}$$
$$h = 8\sqrt{3}$$

Use the formula for the area of a triangle.
$$A = \frac{1}{2}bh$$
$$A = \frac{1}{2} 16 \cdot 8\sqrt{3}$$
$$A = 64\sqrt{3}$$

The area of the pentagon is $256 + 64\sqrt{3}$.

88. Since $\angle A = 50°$,
$$50 + \angle ABC + \angle ACB = 180$$
$$\angle ABC + \angle ACB = 130°.$$

Since OB bisects $\angle ABC$ and OC bisects $\angle ACB$, in $\triangle OBC$,
$$\angle BOC + \angle OBC + \angle OCB = 180°.$$

Substitute $\frac{1}{2}(\angle ABC)$ for $\angle OBC$ and $\frac{1}{2}(\angle ACB)$ for $\angle OCB$.
$$\angle BOC + \frac{1}{2}(\angle ABC) + \frac{1}{2}(\angle ACB) = 180$$
$$\angle BOC + \frac{1}{2}(\angle ABC + \angle ACB) = 180°$$

Substitute 130° for $(\angle ABC + \angle ACB)$.
$$\angle BOC + \frac{1}{2}(130) = 180$$
$$\angle BOC + 65 = 180$$
$$\angle BOC = 115°$$

89. In section 9.2 it is shown that any angle inscribed in a semicircle must be a right angle; $\angle ACB$ is then a right angle. Use the Pythagorean theorem to find the length of the hypotenuse AB, which is also the diameter of the circle. Finally, divide the diameter by two to obtain the length of the radius.
$$8^2 + 6^2 = (AB)^2$$
$$64 + 36 = (AB)^2$$
$$100 = (AB)^2$$
$$10 = AB$$

Then $10 \div 2 = 5$ in.

90. $\angle ACB$ is inscribed in a semicircle, so it is a right angle. $AB = 13$ cm and $BC = AC + 7$. Use the Pythagorean theorem.

$$(AC)^2 + (BC)^2 = (AB)^2$$
$$(AC)^2 + (AC + 7)^2 = (13)^2$$
$$(AC)^2 + (AC)^2 + 14(AC) + 49 = 169$$
$$2(AC)^2 + 14(AC) - 120 = 0$$
$$(AC)^2 + 7(AC) - 60 = 0$$
$$(AC + 12)(AC - 5) = 0$$

$AC + 12 = 0$ or $AC - 5 = 0$
$AC = -12$ $AC = 5$

The negative value is not meaningful. Therefore, the length of AC is 5 cm. By the Pythagorean theorem, the length of BC must be 12 cm.

91. Writing exercise
92. Writing exercise
93. Writing exercise
94. Writing exercise

9.5 EXERCISES

1. True. If the volume is 64 cubic inches, one side of the cube is 4 inches because $4 \cdot 4 \cdot 4 = 64$. Then the area of one face is $4 \cdot 4 = 16$. A cube has six faces so that $6 \cdot 16 = 96$ square inches is the total surface area.

2. True. A tetrahedron has 4 faces and 4 vertices.

3. True. A dodecahedron has 12 faces.

4. False. Each face of an octahedron is triangle.

5. False. The new cube will have eight times the volume of the original cube.

6. True. The formula for the volume of a sphere is $V = \frac{4}{3}\pi r^3$, and the formula for the surface area is $S = 4\pi r^2$. Multiply $r/3$ times the expression for surface area:

$$\frac{r}{3} \cdot 4\pi r^2 = \frac{4}{3}\pi r^3$$

7. (a) $V = lwh$
$= 2 \cdot 1\frac{1}{2} \cdot 1\frac{1}{4}$
$= 2 \cdot \frac{3}{2} \cdot \frac{5}{4}$
$= \frac{15}{4}$
$= 3\frac{3}{4}$ m^3

(b) $S = 2lh + 2hw + 2lw$
$= 2 \cdot 2 \cdot \frac{5}{4} + 2 \cdot \frac{5}{4} \cdot \frac{3}{2} + 2 \cdot 2 \cdot \frac{3}{2}$
$= 5 + \frac{15}{4} + 6$
$= 5 + 3\frac{3}{4} + 6$
$= 14\frac{3}{4}$ m^2

8. (a) $V = lwh$
$= 4 \cdot 6 \cdot 4$
$= 96$ cm^3

(b) $S = 2lh + 2hw + 2lw$
$= 2 \cdot 4 \cdot 4 + 2 \cdot 4 \cdot 6 + 2 \cdot 4 \cdot 6$
$= 32 + 48 + 48$
$= 128$ cm^2

9. It may be helpful to use parentheses to indicate multiplication when working with decimal values. Otherwise, it is possible to confuse the multiplication dot and the decimal points.

(a) $V = lwh$
$= (6)(5)(3.2)$
$= 96$ in^3

(b) $S = 2lh + 2hw + 2lw$
$= 2(6)(3.2) + 2(3.2)(5) + 2(6)(5)$
$= 38.4 + 32 + 60$
$= 130.4$ in^2

10. (a) $V = \frac{4}{3}\pi r^3$
$= \frac{4}{3}(3.14)(7)^3$
$= \frac{4}{3}(3.14)(343)$
≈ 1436.03 m^3

(b) $S = 4\pi r^2$
$= 4(3.14)(7)^2$
$= 4(3.14)(49)$
$= 615.44$ m^2

11. (a) $V = \frac{4}{3}\pi r^3$
$= \frac{4}{3}(3.14)(40)^3$
$= \frac{4}{3}(3.14)(64000)$
$\approx 267,946.67$ ft^3

(b) $S = 4\pi r^2$
$= 4(3.14)(40)^2$
$= 4(3.14)(1600)$
$= 20,096$ ft^2

12. The diameter is 14.8 cm, so the radius is 7.4 cm.

(a) $V = \dfrac{4}{3}\pi r^3$

$= \dfrac{4}{3}(3.14)(7.4)^3$

$= \dfrac{4}{3}(3.14)(405.224)$

$\approx 1696.54 \text{ cm}^3$

(b) $S = 4\pi r^2$

$= 4(3.14)(7.4)^2$

$= 4(3.14)(54.76)$

$\approx 687.79 \text{ cm}^2$

13. (a) $V = \pi r^2 h$

$= (3.14)(5)^2(7)$

$= (3.14)(25)(7)$

$= 549.5 \text{ cm}^3$

(b) $S = 2\pi r^2 + 2\pi r h$

$= 2(3.14)(5)^2 + 2(3.14)(5)(7)$

$= 2(3.14)(25) + 2(3.14)(5)(7)$

$= 157 + 219.8$

$= 376.8 \text{ cm}^2$

14. (a) $V = \pi r^2 h$

$= (3.14)(12)^2(4)$

$= (3.14)(144)(4)$

$= 1808.64 \text{ m}^3$

(b) $S = 2\pi r^2 + 2\pi r h$

$= 2(3.14)(12)^2 + 2(3.14)(12)(4)$

$= 2(3.14)(144) + 2(3.14)(12)(4)$

$= 904.32 + 301.44$

$= 1205.76 \text{ m}^2$

15. (a) $V = \dfrac{1}{3}\pi r^2 h$

$= \dfrac{1}{3}(3.14)(3)^2(7)$

$= \dfrac{1}{3}(3.14)(9)(7)$

$= 65.94 \text{ m}^3$

(b) $S = \pi r\sqrt{r^2 + h^2} + \pi r^2$

$= (3.14)(3)\sqrt{3^2 + 7^2} + (3.14)(3)^2$

$= (3.14)(3)\sqrt{9 + 49} + (3.14)(9)$

$= (3.14)(3)\sqrt{58} + 28.6$

$= 9.42\sqrt{58} + 28.6$

$\approx 100.00 \text{ m}^2$

16. (a) $V = \dfrac{1}{3}\pi r^2 h + \pi r^2$

$= \dfrac{1}{3}(3.14)(4)^2(6)$

$= \dfrac{1}{3}(3.14)(16)(6)$

$= 100.48 \text{ cm}^3$

(b) $S = \pi r\sqrt{r^2 + h^2} + \pi r^2$

$= (3.14)(4)\sqrt{4^2 + 6^2} + (3.14)(4)^2$

$= (3.14)(4)\sqrt{16 + 36} + (3.14)(16)$

$= (3.14)(4)\sqrt{52} + 50.24$

$= 12.56\sqrt{52} + 50.24$

$\approx 140.81 \text{ cm}^2$

17. Remember that B represents the area of the base.

$V = \dfrac{1}{3}Bh$

$= \dfrac{1}{3}(8 \cdot 9) \cdot 7$

$= \dfrac{504}{3}$

$= 168 \text{ in}^3$

18. Remember that B represents the area of the base.

$V = \dfrac{1}{3}Bh$

$= \dfrac{1}{3}(12 \cdot 4) \cdot 10$

$= \dfrac{480}{3}$

$= 160 \text{ ft}^3$

19. $V = \pi r^2 h$

$= (3.14)(6.3)^2(15.8)$

$= (3.14)(36.69)(15.8)$

$= 1969.10 \text{ cm}^3$

20. $V = \pi r^2 h$

$= (3.14)(3.2)^2(9.5)$

$= (3.14)(10.24)(9.5)$

$\approx 305.46 \text{ cm}^3$

21. First find the radius by taking half of the diameter: $r = \tfrac{1}{2}(7.2) = 3.6$. Then use the formula for volume of a right circular cylinder.

$V = \pi r^2 h$

$= (3.14)(3.6)^2(10.5)$

$= (3.14)(12.96)(10.5)$

$\approx 427.29 \text{ cm}^3$

22. First find the radius by taking half of the diameter: $r = \frac{1}{2}(2) = 1$. Then use the formula for volume of a right circular cylinder.

$$V = \pi r^2 h$$
$$= (3.14)(1)^2(40)$$
$$= (3.14)(1)(40)$$
$$= 125.6 \text{ in}^3$$

23. First find the radius by taking half of the diameter: $r = \frac{1}{2}(9) = 4.5$. Then use the formula for volume of a right circular cylinder.

$$V = \pi r^2 h$$
$$= (3.14)(4.5)^2(8)$$
$$= (3.14)(20.25)(8)$$
$$= 508.68 \text{ cm}^3$$

24. First find the radius by taking half of the diameter: $r = \frac{1}{2}(3) = 1.5$. Then use the formula for volume of a right circular cylinder.

$$V = \pi r^2 h$$
$$= (3.14)(1.5)^2(4.3)$$
$$= (3.14)(2.25)(4.3)$$
$$\approx 30.38 \text{ cm}^3$$

25. Remember that B represents the area of the base.

$$V = \frac{1}{3}Bh$$
$$= \frac{1}{3}(230)^2 \cdot 137$$
$$= \frac{1}{3}(52900) \cdot 137$$
$$= \frac{7,247,300}{3}$$
$$\approx 2,415,766.67 \text{ m}^3$$

26. $V = \pi r^2 h$
$$= (3.14)(46)^2(220)$$
$$= (3.14)(2116)(220)$$
$$= 1,461,732.8 \text{ m}^3$$

27. Change 1/2 to the decimal value .5 for ease of computation.

$$V = \frac{1}{3}\pi r^2 h$$
$$= \frac{1}{3}(3.14)(.5)^2(2)$$
$$= \frac{1}{3}(3.14)(.25)(2)$$
$$= \frac{1.57}{3}$$
$$= .52 \text{ m}^3$$

28. $V = \frac{1}{3}\pi r^2 h$
$$= \frac{1}{3}(3.14)(4)^2(12)$$
$$= \frac{1}{3}(3.14)(16)(12)$$
$$= \frac{602.88}{3}$$
$$= 200.96 \text{ in}^3$$

Table for Exercises 29–36

	r	d	V	S
29.	6 in	12 in	288π in^3	144π in^2
30.	9 in	18 in	972π in^3	324π in^2
31.	5 ft	10 ft	$\frac{500}{3}\pi$ ft^3	100π ft^2
32.	20 ft	40 ft	$\frac{32,000}{3}\pi$ ft^3	1600π ft^2
33.	2 cm	4 cm	$\frac{32}{3}\pi$ cm^3	16π cm^2
34.	4 cm	8 cm	$\frac{256}{3}\pi$ cm^3	64π cm^2
35.	1 m	2 m	$\frac{4}{3}\pi$ m^3	4π m^2
36.	6 m	12 m	288π m^3	144π m^2

29. $d = 2r = 2 \cdot 6 = 12$
$$V = \frac{4}{3}\pi r^3 = \frac{4}{3}\pi(6)^3 = \frac{4}{3}\pi(216) = 288\pi$$
$$S = 4\pi r^2 = 4\pi(6)^2 = 4\pi(36) = 144\pi$$

30. $d = 2r = 2 \cdot 9 = 18$
$$V = \frac{4}{3}\pi r^3 = \frac{4}{3}\pi(9)^3 = \frac{4}{3}\pi(729) = 972\pi$$
$$S = 4\pi r^2 = 4\pi(9)^2 = 4\pi(81) = 324\pi$$

31. $r = \frac{1}{2}d = \frac{1}{2} \cdot 10 = 5$
$$V = \frac{4}{3}\pi r^3 = \frac{4}{3}\pi(5)^3 = \frac{4}{3}\pi(125) = \frac{500}{3}\pi$$
$$S = 4\pi r^2 = 4\pi(5)^2 = 4\pi(25) = 100\pi$$

32. $r = \frac{1}{2}d = \frac{1}{2} \cdot 40 = 20$
$$V = \frac{4}{3}\pi r^3 = \frac{4}{3}\pi(20)^3 = \frac{4}{3}\pi(8000) = \frac{32000}{3}\pi$$
$$S = 4\pi r^2 = 4\pi(20)^2 = 4\pi(400) = 1600\pi$$

33. Use the formula for the volume of a sphere to solve for r, by replacing V with the given value, $\frac{32}{3}\pi$.

$$V = \frac{4}{3}\pi r^3$$
$$\frac{32}{3}\pi = \frac{4}{3}\pi r^3$$
$$\frac{32}{3}\pi \div \frac{4}{3}\pi = \frac{4}{3}\pi r^3 \div \frac{4}{3}\pi$$
$$\frac{32\pi}{3} \cdot \frac{3}{4\pi} = r^3$$
$$8 = r^3$$
$$2 = r$$
$$d = 2r = 2 \cdot 2 = 4$$
$$S = 4\pi r^2 = 4\pi(2)^2 = 4\pi(4) = 16\pi$$

34. Use the formula for the volume of a sphere to solve for r, by replacing V with the given value, $\frac{256}{3}\pi$.

$$V = \frac{4}{3}\pi r^3$$
$$\frac{256}{3}\pi = \frac{4}{3}\pi r^3$$
$$\frac{256}{3}\pi \div \frac{4}{3}\pi = \frac{4}{3}\pi r^3 \div \frac{4}{3}\pi$$
$$\frac{256\pi}{3} \cdot \frac{3}{4\pi} = r^3$$
$$64 = r^3$$
$$4 = r$$
$$d = 2r = 2 \cdot 4 = 8$$
$$S = 4\pi r^2 = 4\pi(4)^2 = 4\pi(16) = 64\pi$$

35. Use the formula for the surface area of a sphere to solve for r, by replacing S with the given value, 4π.

$$S = 4\pi r^2$$
$$4\pi = 4\pi r^2$$
$$\frac{4\pi}{4\pi} = r^2$$
$$1 = r^2$$
$$1 = r$$
$$d = 2r = 2 \cdot 1 = 2$$
$$V = \frac{4}{3}\pi r^3 = \frac{4}{3}\pi(1)^3 = \frac{4}{3}\pi(1) = \frac{4}{3}\pi$$

36. Use the formula for the surface area of a sphere to solve for r, by replacing S with the given value, 144π.

$$S = 4\pi r^2$$
$$144\pi = 4\pi r^2$$
$$\frac{144\pi}{4\pi} = r^2$$
$$36 = r^2$$
$$6 = r$$

$$d = 2r = 2 \cdot 6 = 12$$
$$V = \frac{4}{3}\pi r^3 = \frac{4}{3}\pi(6)^3 = \frac{4}{3}\pi(216) = 288\pi$$

37. Volume is a measure of capacity.

38. Surface area

39. The volume of the original cube is x^3. Let the length of the side of the new cube be represented by y. Then $y^3 = 2x^3$. Solve for y by taking the cube root of both sides of the equation.

$$y^3 = 2x^3$$
$$\sqrt[3]{y^3} = \sqrt[3]{2x^3}$$
$$y = \sqrt[3]{2x^3}$$
$$y = x\sqrt[3]{2}$$

40. (a) $V = \frac{4}{3}\pi r^3 = \frac{4}{3}\pi(1)^3 = \frac{4}{3}\pi$ m^3

   (b) $V = \frac{4}{3}\pi(2)^3 = \frac{4}{3}\pi(8) = \frac{32}{3}\pi$ m^3

   (c) $\frac{32}{3}\pi \div \frac{4}{3}\pi = \frac{32\pi}{3} \cdot \frac{3}{4\pi} = 8$ times

   (d) $V = \frac{4}{3}\pi(3)^3 = \frac{4}{3}\pi(27) = 36\pi$ m^3

   (e) $\frac{108}{3}\pi \div \frac{4}{3}\pi = \frac{108\pi}{3} \cdot \frac{3}{4\pi} = 27$ times

   (f) In general, if the radius of a sphere is multiplied by n, the volume is multiplied by $\underline{n^3}$.

41. If the new diameter is 3 times the old diameter, then the new volume will be 3^3 or 27 times greater. Therefore, the cost will also be 27 times greater, or $27 \cdot 300 = \$8100$.

42. If the new diameter is 4 times the old diameter, then the new volume will be 4^3 or 64 times greater. Therefore, the cost will also be 64 times greater, or $64 \cdot 300 = \$19,200$.

43. If the new diameter is 5 times the old diameter, then the new volume will be 5^3 or 125 times greater. Therefore, the cost will also be 125 times greater, or $125 \cdot 300 = \$37,500$.

44. If the radius of the sphere is 1 cm, the surface area is

$$S = 4\pi(1)^2 = 4\pi \text{ cm}^2.$$

If the radius is multiplied by n, $r = n \cdot 1 = n$ cm, so the new surface area is

$$S = 4\pi(n)^2 = 4n^2\pi \text{ cm}.$$

Then $4n^2\pi \div 4\pi = n^2$.

If the radius of a sphere is multiplied by n, then the surface area of the sphere is multiplied by $\underline{n^2}$.

9.5 SPACE FIGURES, VOLUME, AND SURFACE AREA

45. $V = lwh$
$60 = 6 \cdot 4 \cdot x$
$60 = 24x$
$2.5 = x$

46. $V = \dfrac{1}{3}Bh$
$450 = \dfrac{1}{3}x(x+1) \cdot 15$
$450 = 5x(x+1)$
$450 = 5x^2 + 5x$
$0 = 5x^2 + 5x - 450$
$0 = x^2 + x - 90$
$0 = (x+10)(x-9)$
$x + 10 = 0 \quad \text{or} \quad x - 9 = 0$
$x = -10 \qquad\qquad x = 9$

The first value of x is not meaningful. The value of x is 9.

47. In this exercise x = the diameter of the sphere. Therefore $r = \dfrac{x}{2}$.
$V = \dfrac{4}{3}\pi r^3$
$36\pi = \dfrac{4}{3}\pi \left(\dfrac{x}{2}\right)^3$
$36\pi = \dfrac{4}{3}\pi \cdot \dfrac{x^3}{8}$
$36 = \dfrac{4}{3} \cdot \dfrac{x^3}{8}$
$36 = \dfrac{x^3}{6}$
$216 = x^3$
$6 = x$

48. $V = \dfrac{1}{3}\pi r^2 h$
$245\pi = \dfrac{1}{3}\pi (x)^2 \cdot 15$
$245\pi = 5\pi(x)^2$
$49 = (x)^2$
$7 = x$

49. Look at the figure and try some values for the edges of each side that will create the given areas. One side has edges 6 in. and 5 in.; the adjacent side has edges 5 in. and 7 in.; and the third side has edges 7 in. and 6 in. Write these values on the edges of the rectangular box to verify that it can be done to create the given areas. The three dimensions of the box, then, are 6, 7, and 5. The volume of the box is $6 \cdot 7 \cdot 5 = 210$ in.^{3}.

50. A hemisphere is a half sphere, so if its radius is r, the volume is
$$V_{\text{hemisphere}} = \dfrac{1}{2}\left(\dfrac{4}{3}\pi r^3\right) = \dfrac{2}{3}\pi r^3.$$

The inscribed right circular cone has radius r and height r, so its volume is
$$V_{\text{cone}} = \dfrac{1}{3}\pi r^2(r) = \dfrac{1}{3}\pi r^3.$$

The ratio of the volume of the cone to the volume of the hemisphere is
$$\dfrac{V_{\text{cone}}}{V_{\text{hemisphere}}} = \dfrac{\frac{1}{3}\pi r^3}{\frac{2}{3}\pi r^3} = \dfrac{1}{2} \text{ or, 1 to 2.}$$

51. The formula for the volume of a sphere is $V = \frac{4}{3}\pi r^3$, so the radius of the sphere must be found. To find the radius of the circle that is formed by the intersection, set 576π equal to πr^2, the formula for the area of a circle, and solve for r.
$576\pi = \pi r^2$
$\dfrac{576\pi}{\pi} = \dfrac{\pi r^2}{\pi}$
$576 = r^2$
$24 = r$

Now use the Pythagorean theorem to find the length of the hypotenuse, which is also the radius of the sphere.
$7^2 + 24^2 = c^2$
$49 + 576 = c^2$
$625 = c^2$
$25 = c$

Now compute the volume of the sphere.
$V = \dfrac{4}{3}\pi r^3$
$= \dfrac{4}{3}\pi(25)^3$
$= \dfrac{4}{3}\pi(15625)$
$= \dfrac{62500}{3}\pi$ in.^{3}

52. Suppose $h = 2$ ft and $d = 2$ ft. Then $r = 1$ ft.
$V_1 = \pi r^2 h$
$= \pi(1)^2(2)$
$= 2\pi$ ft^3

If the height is halved and the diameter is tripled,

$$h = \frac{2}{2} = 1 \text{ ft}$$
$$d = 3(2) = 6 \text{ ft}$$
$$r = \frac{6}{2} = 3 \text{ ft}$$

Then

$$V_2 = \pi(3)^2(1)$$
$$= 9\pi \text{ ft}^3.$$

Compare the volume of the second cylinder with that of the first.

$$\frac{V_2}{V_1} = \frac{9\pi}{2\pi} = \frac{9}{2} = 4.5$$

The volume is multiplied by 4.5.

53. Rotate the inscribed square 45° so that one of its diagonals is horizontal and the other vertical. Notice that the length of the diagonal is the same length as the side of the circumscribed square. This length is $2r$. That means that the area of the circumscribed square is $A = 4r^2$. Returning to the inscribed square, the length $2r$ is the length of the hypotenuse of a right triangle. Use the Pythagorean theorem to find the length of a side of this square.

Let $x = $ the length of each leg.

$$x^2 + x^2 = (2r)^2$$
$$2x^2 = 4r^2$$
$$x^2 = 2r^2$$

Because the area of this square is equal to x^2, the ratio of the two areas can be determined

$$\frac{\text{area of the circumscribed square}}{\text{area of the inscribed square}} = \frac{4r^2}{2r^2} = \frac{2}{1}$$

The ratio is 2 to 1.

54. Since $\angle ABC$ is a right angle and is inscribed in the circle, arc ABC is a semicircle. Hence AC is a diameter with a length of 8 inches. In right triangle ABC,

$$(AB)^2 + (BC)^2 = 8^2.$$

Since $AB = BC$,

$$2(AB)^2 = 64$$
$$(AB)^2 = 32$$
$$AB = \sqrt{32}$$
$$AB = \sqrt{16 \cdot 2}$$
$$AB = 4\sqrt{2}$$

The sides of the square are equal in length so,

$$P = 4s$$
$$= 4(4\sqrt{2})$$
$$= 16\sqrt{2} \text{ in.}$$

55. Draw a line connecting one diameter RT; draw a line connecting another diameter QS. Recall from section 9.2 that any angle inscribed in a semicircle is a right angle, which means that $\angle RPT$ and $\angle QPS$ are both right angles. From the Pythagorean theorem, $PR^2 + PT^2$ equals the square of the diameter, 12^2. Also, $PQ^2 + PS^2$ equals the square of the diameter. Finally,

$$PR^2 + PT^2 + PQ^2 + PS^2 = 12^2 + 12^2 = 288.$$

56.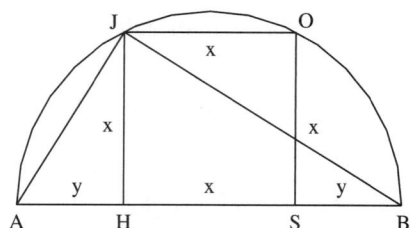

Label the endpoints of the diameter through side HS, A and B. Draw chords at AJ and JB. Angle AJB is a right angle because it is inscribed in a semicircle. Angle JHB is a right angle because it is an angle of square $JOSH$. Angle AHJ is a right angle because it is the supplement of a right angle. Angle JBA in $\triangle AJB$ is the same angle as $\angle JBH$ in $\triangle JHB$. These right angles all have equal measure. Angle JBA in $\triangle AJB$ is the same angle as $\angle JBH$ in $\triangle JHB$. Angle JAB in $\triangle AJB$ is the same angle as $\angle JAH$ in $\triangle AHJ$. Thus, $\triangle AJB$ is similar to $\triangle JHB$ because of the angle-angle similarity property. Likewise $\triangle AJB$ is similar to $\triangle AHJ$ and their corresponding sides are proportional.

$$\frac{x}{y} = \frac{x+y}{x}$$

In Chapter 5 we found that this is the golden ratio or

$$\frac{x}{y} = \frac{1+\sqrt{5}}{2}.$$

9.6 EXERCISES

Exercises 1–8 represent reflection transformations. There is a one-to-one correspondence between each point in the original figure and each corresponding point in the image figure. The original figure and image figure are congruent hence preserving collinearity and distance.

9.6 TRANSFORMATIONAL GEOMETRY 311

1.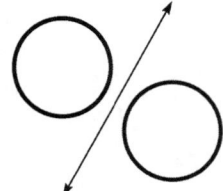

2.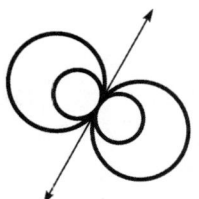

3.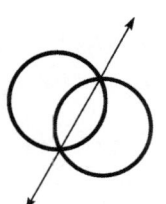

4.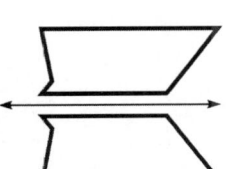

5.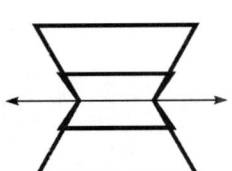

6.

7. The figure is its own reflection image.
8. The figure is its own reflection image.

Exercises 9–12 represent figures that are their own reflections across the lines of symmetry.

9.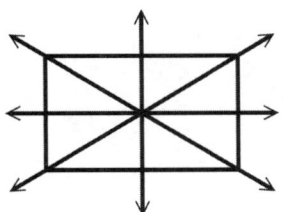

10.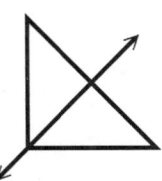

11.

12.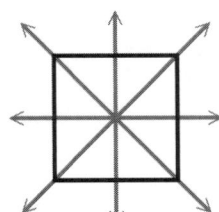

Exercises 13–20 represent the composition or product transformations of r_m followed by r_n, or $r_n \cdot r_m$.

13.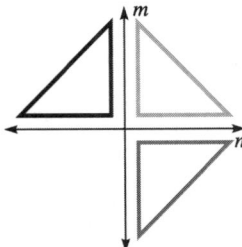

14.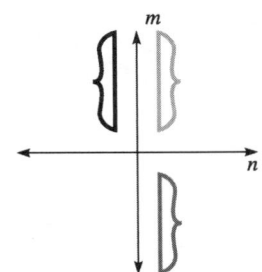

312 CHAPTER 9 GEOMETRY

15.
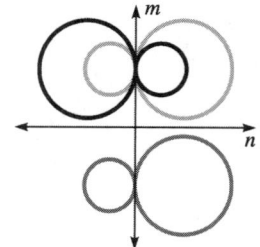

16.
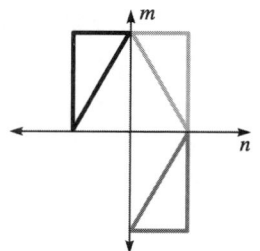

17.
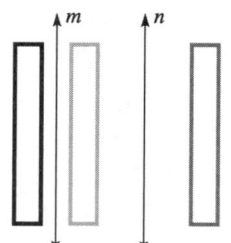

18.
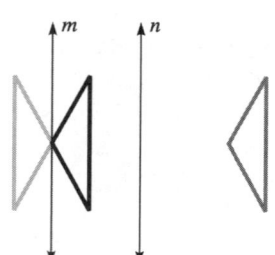

19.
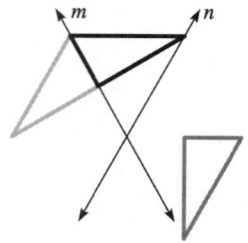

20.
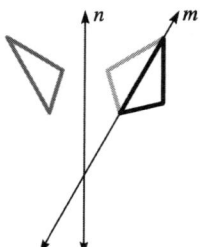

21. r_m
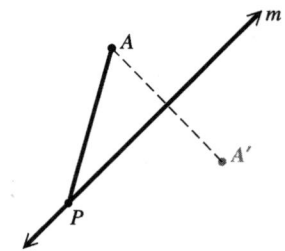

22. R_p
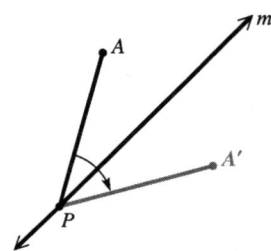

23. T
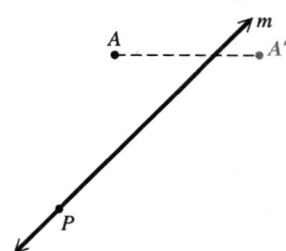

24. $r_m \cdot r_m$
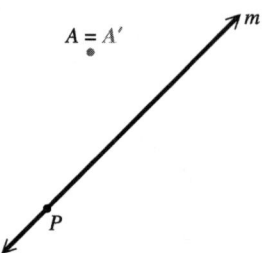

25. $T \cdot T$

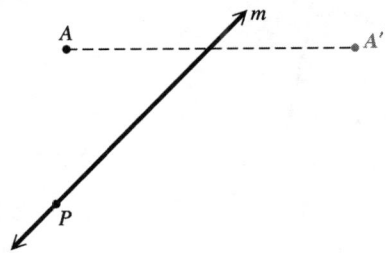

26. $R_p \cdot R_p$

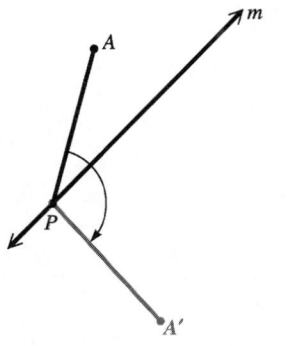

27. $T \cdot R_p$

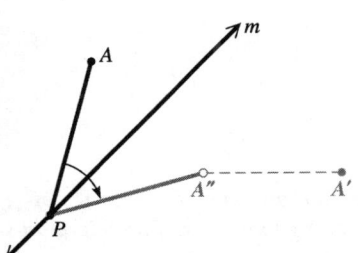

28. $T \cdot r_m$

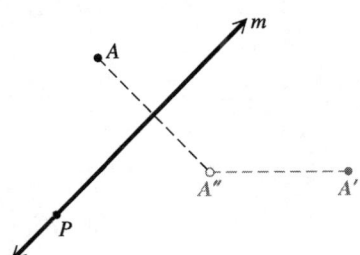

29. $r_m \cdot T$

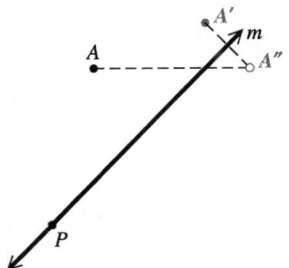

30. $R_p \cdot r_m$

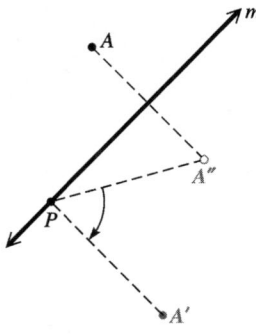

31. $r_m \cdot R_p$

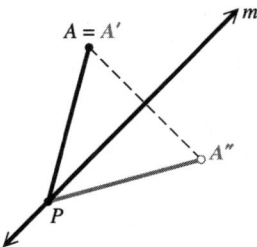

32. $R_p \cdot T$

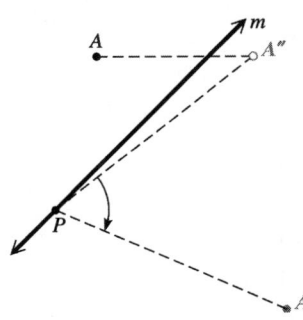

33. Yes, $T \cdot r_m$ is a glide reflection since T is a translation having non zero magnitude.

34. No, $T \cdot r_m \neq r_m \cdot T$. See Exercises 29 and 30 above.

35.

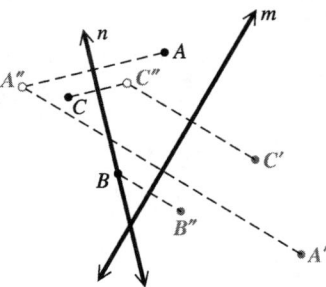

36. (a) Yes, because a glide transformation preserves the shape of the original figure.

    (b) Yes, because a glide transformation preserves the size of the original figure.

37.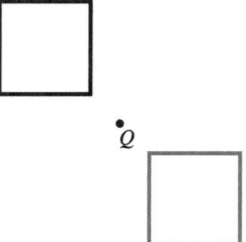

38.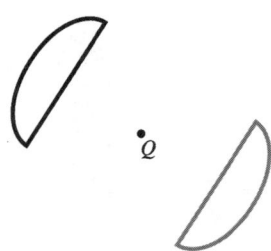

39.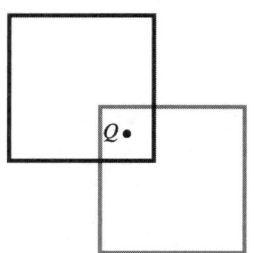

40.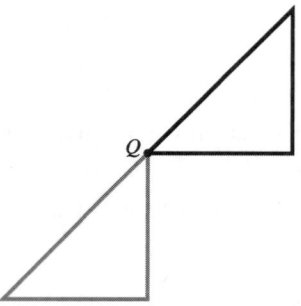

41.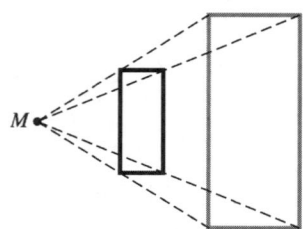

42.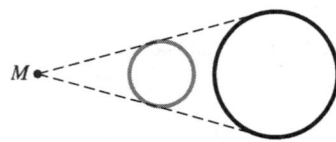

43.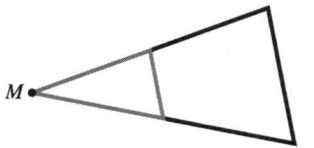

44.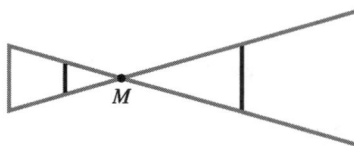

45.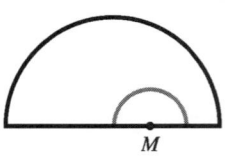

46.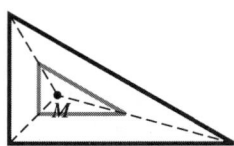

9.7 EXERCISES

The chart in the text characterizes certain properties of Euclidean and non-Euclidean geometries. Study it and use it to respond to Exercises 1–10.

1. Euclidean
2. Riemannian
3. Lobachevskian
4. less than
5. greater than
6. Lobachevskian
7. Riemannian
8. Riemannian
9. Euclidean
10.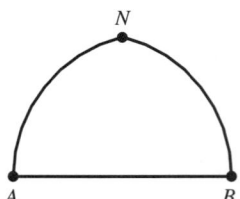

11. Writing exercise

12.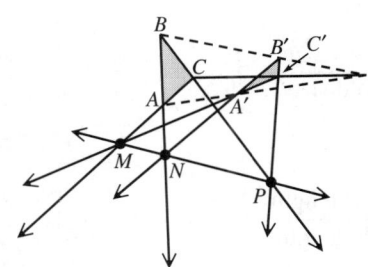

13. C. Both are of genus 2, meaning they have two holes.

14. A carrot has no holes. It is topologically equivalent to the ruler in A and to the nail in E.

15. A and E. All are of genus 0, having no holes.

16. A nut like the one shown has one hole. It is topologically equivalent to the ring in B and to the CD in D.

17. A and E. All three are of genus 0, having no holes.

18. A coin has no holes. It is topologically equivalent to the ruler in A and to the nail in E.

19. None of them

20. A needle has one hole. It is topologically equivalent to the ring in B and to the CD in D.

21. No, both have no holes. They are of genus 0.

22. A mixing bowl has no holes and a colander has many holes, so they do not have the same genus. A topologist would know the difference between them.

23. Yes, the slice of American cheese is of genus 0, and the slice of Swiss cheese is of genus 1 or more.

24. A compact disc has one hole and a phonograph record has one hole. A topologist would not know the difference.

25. A compact disc has one hole, so it is of genus 1.

26. A phonograph record is of genus 1.

27. A sheet of loose-leaf paper made for a three-ring binder has three holes, so it is of genus 3.

28. A sheet of loose-leaf paper made for a two-ring binder has two holes, so it is of genus 2.

29. A wedding band has one hole, so it is of genus 1.

30. A postage stamp has no holes, so it is of genus 0.

31. A, C, D, and F are even vertices because each has two paths leading to or from the vertex; B and E are odd because each has three paths leading to or from the vertex.

32. Each of the vertices A, D, E, and H has two paths to it, so these vertices are all even. Each of the vertices B, C, F, and G has three paths to it, so these vertices are all odd.

33. A, B, C, and F are odd because each has three paths leading to or from the vertex; D, E, and G are even. D and E each have two paths leading to or from the vertex; G has four.

34. Each of the vertices at A and C has four paths to it, and vertex D has two paths to it, so these vertices are even. Each of the vertices at B and E has three paths to it, so these vertices are odd.

35. A, B, C, and D are odd vertices because each has three paths leading to or from the vertex; E is even because it has four.

36. Each of the vertices A through F has three paths leading to it, so all of the vertices are odd.

37. There are two odd vertices at the extremities and the rest are even. Therefore, the network is traversable.

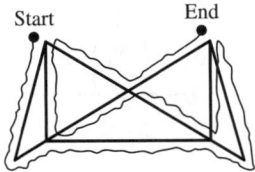

38. Not traversable

39. Not traversable. It has more than two odd vertices.

40. Traversable

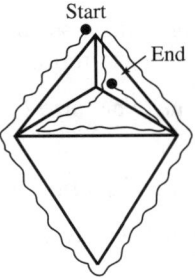

41. The network has exactly 2 odd vertices. It is traversable.

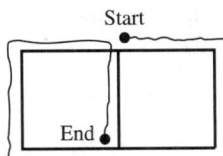

42. Traversable

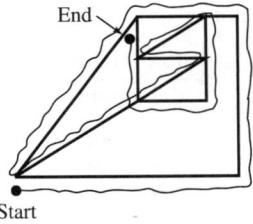

43. Yes. There are exactly two rooms (vertices) with an odd number of doors (paths), the top left room and the bottom left room. The rest of the rooms and the exterior of the house have an even number of doors. Therefore, the house (network) is traversable.

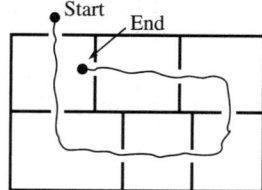

44. Traversable

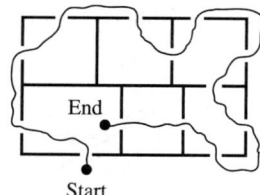

45. No. There are more than two rooms with an odd number of doors (paths).

46. No, not traversable

9.8 EXERCISES

1. The least number of these squares that can be put together edge to edge to form a larger square is 4.

2. If the original square has size 1, the new square has size 4.

3. The length of each edge of the new square is 2.

4. The scale factor between two similar figures is
$$\frac{\text{new length}}{\text{old length}}.$$
The scale factor between the large square and the small square is
$$\frac{2}{1} = 2.$$

5. $\dfrac{\text{new size}}{\text{old size}} = \dfrac{4}{1} = 4$

6. The scale factor is
$$\frac{\text{new length}}{\text{old length}} = \frac{3}{1} = 3.$$
The ratio of
$$\frac{\text{new size}}{\text{old size}} = \frac{9}{1} = 9.$$

7. The scale factor is
$$\frac{\text{new length}}{\text{old length}} = \frac{4}{1} = 4.$$
The ratio of
$$\frac{\text{new size}}{\text{old size}} = \frac{16}{1} = 16.$$

8.
Scale factor	2	3	4	5	6	10
Ratio of new size to old size	4	9	16	25	36	100

9. Each ratio in the bottom row is the square of the scale factor in the top row.

10. The least number of equilateral triangles that can be put together edge to edge to form a similar larger triangle is 4.

11.
Scale factor	2	3	4	5	6	10
Ratio of new size to old size	4	9	16	25	36	100

12. Each ratio in the bottom row is again the square of the scale factor in the top row.

13. Some examples are: $3^d = 9$ and $d = 2$; $4^d = 16$ and $d = 2$; $5^d = 25$ and $d = 2$.

14. The least number of cubes that can be put together face to face to form a larger cube is 8.

15. The scale factor between these two cubes is
$$\frac{\text{new length}}{\text{old length}} = \frac{2}{1} = 2.$$
The ratio of
$$\frac{\text{new size}}{\text{old size}} = \frac{8}{1} = 8.$$

16.
Scale factor	2	3	4	5	6	10
Ratio of new size to old size	8	27	64	125	216	1000

17. Each ratio in the bottom row is the cube of the scale factor in the top row.

18. Since $2^3 = 8$, the value of d in $2^d = 8$ must be 3.

19. The scale factor between stage 1 and stage 2 is $3/1 = 3$.

20. Old size $= 1$, new size $= 4$.

21. $3^d = 4$

    Use trial and error:
    $$3^{1.5} = 5.196\ldots$$
    $$3^{1.25} = 3.948\ldots$$
    $$3^{1.26} = 3.992\ldots$$
    $$3^{1.27} = 4.036\ldots$$
    $$3^{1.261} = 3.996\ldots$$
    $$3^{1.262} = 4.001\ldots$$
    $$3^{1.263} = 4.005\ldots$$
    $$d = 1.262 \text{ to three decimal places,}$$

    or solve using logarithms.
    $$3^d = 4$$
    $$\ln 3^d = \ln 4$$
    $$d \ln 3 = \ln 4$$
    $$d = \frac{\ln 4}{\ln 3}$$
    $$d = 1.262 \text{ to three decimal places.}$$

22. In stage 1, length = 1.
    In stage 2, length = 2.
    Scale factor:
    $$\frac{\text{new length}}{\text{old length}} = \frac{2}{1} = 2.$$

23. Old size = 1, new size = $\underline{3}$

24. Between 1 and 2

25. $2^d = 3$

    Use trial and error:
    $$2^{1.5} = 2.828\ldots$$
    $$2^{1.6} = 3.031\ldots$$
    $$2^{1.55} = 2.928\ldots$$
    $$2^{1.58} = 2.990\ldots$$
    $$2^{1.59} = 3.010\ldots$$
    $$2^{1.584} = 2.998\ldots$$
    $$2^{1.585} = 3.000\ldots$$
    $$d = 1.585 \text{ to three decimal places}$$

    Or solve using logarithms.
    $$2^d = 3$$
    $$\ln 2^d = \ln 3$$
    $$d \ln 2 = \ln 3$$
    $$d = \frac{\ln 3}{\ln 2}$$
    $$d = 1.585 \text{ to three decimal places}$$

26. $y = kx(1-x); k = 3.25$ and $x = .7$

    Begin with $x = .7$ and iterate with a calculator to produce the following sequence. The numbers here are rounded to three decimal places, but keep all digits in the calculator for each next step. Stop when values are repeated at least twice.

    .683, .704, .677, .711, .668, .721, .654, .735, .632, .755, .601, .779, .559, .801, .517, .812, .497, .812, .495, .812, .495, …

    The two attractors are .812 and .495.

27. Given $k = 3.4$, $x = .8$ and formula $y = kx(1-x)$.

    Note that rounded values are used below.
    $$y = 3.4(.8)(1 - .8)$$
    $$= 3.4(.8)(.2)$$
    $$= .544$$

    $$y = 3.4(.544)(1 - .544)$$
    $$= 3.4(.544)(.456)$$
    $$\approx .843$$

    $$y = 3.4(.843)(1 - .843)$$
    $$= 3.4(.843)(.157)$$
    $$\approx .450$$

    $$y = 3.4(.450)(1 - .450)$$
    $$= 3.4(.450)(.550)$$
    $$\approx .842$$

    $$y = 3.4(.842)(1 - .842)$$
    $$= 3.4(.842)(.158)$$
    $$\approx .452$$

    $$y = 3.4(.452)(1 - .452)$$
    $$= 3.4(.452)(.548)$$
    $$\approx .842$$

    $$y = 3.4(.842)(1 - .842)$$
    $$= 3.4(.842)(.158)$$
    $$\approx .452$$

    The attractors are evidently .842 and .452.

28. $y = kx(1-x); k = 3.55$ and $x = .7$

    Begin with $x = .7$ and iterate with a calculator to produce the following sequence. The numbers here are rounded to three decimal places, but keep all digits in the calculator for each next step. Stop when values are repeated at least twice.

    .746, .674, .781, .608, .846, .462, .882, .368, .826, .511, .887, .356, .813, .539, .882, .369, .827, .509, .887, .355, .813, .540, .882, .370, .827, .508, .887, .355, .813, .540, .882, .370, .827, .507, .887, .355, .813, .540, .882, .370, .827, .506, .887, .355, .813, .540, .882, .370, .828, .506, .887, .355, .813, .540, .882, .370, .828, .506, .887, .355, .813, …

The eight attractors are .540, .882, .370, .828, .506, .887, .355 and .813.

CHAPTER 9 TEST

1. (a) The measure of its complement is $90 - 38 = 52°$.

    (b) The measure of its supplement is $180 - 38 = 142°$.

    (c) It is an acute angle because it is less than 90°.

2. The designated angles are supplementary; their sum is 180°.
    $$(2x + 16) + (5x + 80) = 180$$
    $$7x + 96 = 180$$
    $$7x = 84$$
    $$x = 12$$

    Then one angle measure is $2 \cdot 12 + 16 = 40°$ and the other angle measure is $5 \cdot 12 + 80 = 140°$. A check is that their sum is indeed 180°.

3. The angles are vertical so they have the same measurement. Set the algebraic expressions equal to each other.
    $$7x - 25 = 4x + 5$$
    $$3x = 30$$
    $$x = 10$$

    Then one angle measure is $7 \cdot 10 - 25 = 45°$ and the other angle measure is $4 \cdot 10 + 5 = 45°$.

4. The designated angles are complementary; their sum is 90°.
    $$(4x + 6) + 10x = 90$$
    $$14x + 6 = 90$$
    $$14x = 84$$
    $$x = 6$$

    Then one angle measure is $4 \cdot 6 + 6 = 30°$ and the other angle measure is $10 \cdot 6 = 60°$.

5. The designated angles are supplementary.
    $$(7x + 11) + (3x - 1) = 180$$
    $$10x + 10 = 180$$
    $$10x = 170$$
    $$x = 17$$

    Then one angle measure is $7 \cdot 17 + 11 = 130°$ and the other angle measure is $3 \cdot 17 - 1 = 50°$. A check is that their sum is indeed 180°.

6. These are alternate interior angles, which are equal to each other.
    $$13y - 26 = 10y + 7$$
    $$3y = 33$$
    $$y = 11$$

    Then one angle measure is $13 \cdot 11 - 26 = 117°$ and the other angle measure is $10 \cdot 11 + 7 = 117°$.

7. Writing exercise

8. Letter C is false because a triangle cannot have both a right angle and an obtuse angle. A right angle measures 90° and an obtuse angle measures greater than 90°; however, the sum of all three angles of any triangle is exactly 180°.

9. The curve is simple and closed.

10. The curve is neither simple nor closed.

11. The sum of the three angle measures is 180°.
    $$(3x + 9) + (6x + 3) + 21(x - 2) = 180$$
    $$(3x + 9) + (6x + 3) + 21x - 42 = 180$$
    $$30x - 30 = 180$$
    $$30x = 210$$
    $$x = 7$$

    Then one angle measure is $3 \cdot 7 + 9 = 30°$, a second angle measure is $6 \cdot 7 + 3 = 45°$, and the third angle measure is $21(7 - 2) = 21(5) = 105°$. A check is that their sum is indeed 180°.

12. $A = lw$
    $= 6 \cdot 12$
    $= 72 \text{ cm}^2$

13. $A = bh$
    $= 12 \cdot 5$
    $= 60 \text{ in}^2$

14. $A = \dfrac{1}{2}bh$
    $= \dfrac{1}{2} \cdot 17 \cdot 8$
    $= 68 \text{ m}^2$

15. $A = \dfrac{1}{2}h(b + B)$
    $= \dfrac{1}{2} \cdot 9(16 + 24)$
    $= \dfrac{9}{2}(40)$
    $= 180 \text{ m}^2$

16. Replace A in the formula for the area of a circle and solve for r.
    $$A = \pi r^2$$
    $$144\pi = \pi r^2$$
    $$\frac{144\pi}{\pi} = \frac{\pi r^2}{\pi}$$
    $$144 = r^2$$
    $$12 = r$$

Now replace r in the formula for the circumference.

$$C = 2\pi r$$
$$= 2\pi \cdot 12$$
$$= 24\pi \text{ in.}$$

17. Use the formula for the circumference $C = \pi d$.

$$C = (3.14) \cdot 630$$
$$\approx 1978 \text{ ft.}$$

18. Subtract the area of the triangle from the area of the semicircle. First, the area of the semicircle is:

$$A = \frac{1}{2}\pi r^2$$
$$= \frac{1}{2}(3.14)(10)^2$$
$$= \frac{1}{2}(3.14)(100)$$
$$= 157$$

Now find the area of the triangle.

$$A = \frac{1}{2}bh$$
$$= \frac{1}{2} \cdot 20 \cdot 10$$
$$= 100$$

Finally $157 - 100 = 57 \text{ cm}^2$.

19.
STATEMENTS	REASONS
1. $\angle CAB = \angle DBA$	1. Given
2. $DB = CA$	2. Given
3. $AB = AB$	3. Reflexive property
4. $\triangle ABD \cong \triangle BAC$	4. SAS Congruence Property

20. Let $h = $ height of the pole.

$$\frac{h}{30} = \frac{30}{45}$$
$$45 \cdot h = 30 \cdot 30$$
$$\frac{45h}{45} = \frac{900}{45}$$
$$h = 20 \text{ feet}$$

21. Use the Pythagorean theorem to find c.

$$a^2 + b^2 = c^2$$
$$20^2 + 21^2 = c^2$$
$$400 + 441 = c^2$$
$$841 = c^2$$
$$\sqrt{841} = \sqrt{c^2}$$
$$29 = c$$

The length of the diagonal is 29 m.

22.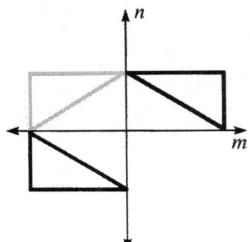

23.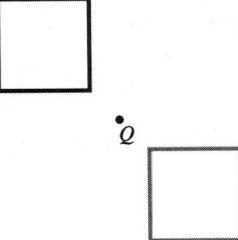

24. (a) $V = \frac{4}{3}\pi r^3$
$$= \frac{4}{3}(3.14)(6)^3$$
$$= \frac{4}{3}(3.14)(216)$$
$$\approx 904.32 \text{ in.}^3$$

(b) $S = 4\pi r^2$
$$= 4(3.14)(6)^2$$
$$= 4(3.14)(36)$$
$$= 452.16 \text{ in.}^2$$

25. (a) $V = lwh$
$$= 12 \cdot 9 \cdot 8$$
$$= 864 \text{ ft}^3$$

(b) $S = 2lh + 2hw + 2lw$
$$= 2 \cdot 12 \cdot 8 + 2 \cdot 8 \cdot 9 + 2 \cdot 12 \cdot 9$$
$$= 192 + 144 + 216$$
$$= 552 \text{ ft}^2$$

26. (a) $V = \pi r^2 h$
$$= (3.14)(6)^2(14)$$
$$= (3.14)(36)(14)$$
$$= 1582.56 \text{ m}^3$$

(b) $S = 2\pi r^2 + 2\pi rh$
$$= 2(3.14)(6)^2 + 2(3.14)(6)(14)$$
$$= 2(3.14)(36) + 2(3.14)(6)(14)$$
$$= 226.08 + 572.52$$
$$= 753.60 \text{ m}^2$$

27. Writing exercise

28. (a) A page of a book and the cover of the same book are topologically equivalent because they both have no holes; they are of genus 1.

(b) A pair of glasses with the lenses removed and the Mona Lisa are not topologically equivalent. The glasses have two holes, but the Mona Lisa has none.

29. (a) Yes.

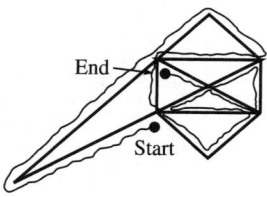

(b) No, because the network has more than two odd vertices.

30. Given $x = .6$ and formula $y = 2.1x(1-x)$.
$y = 2.1(.6)(1-.6)$
$= 2.1(.6)(.4)$
$= .504$

$y = 2.1(.504)(1-.504)$
$= 2.1(.504)(.496)$
$\approx .5249664$

Continuing the itterations using all digits in the calculator for each next step we arrive at the following sequence.

.5236910256, .5238213441, .5238213441, .5238083415, .523809642, .523809512, .523809525, .523809237, .5238095238, .5238095238

The only attractor is .5238095238.

10.1 EXERCISES

1. (a) The complement of 30° is
   $$90° - 30° = 60°,$$
   (b) The supplement of 30° is
   $$180° - 30° = 150°.$$

2. (a) The complement of 60° is
   $$90° - 60° = 30°,$$
   (b) The supplement of 60° is
   $$180° - 60° = 120°.$$

3. (a) The complement of 45° is
   $$90° - 45° = 45°,$$
   (b) The supplement of 45° is
   $$180° - 45° = 135°.$$

4. (a) The complement of 55° is
   $$90° - 55° = 35°,$$
   (b) The supplement of 55° is
   $$180° - 55° = 125°.$$

5. (a) The complement of 89° is
   $$90° - 89° = 1°,$$
   (b) The supplement of 89° is
   $$180° - 89° = 91°.$$

6. (a) The complement of 2° is
   $$90° - 2° = 88°,$$
   (b) The supplement of 2° is
   $$180° - 2° = 178°.$$

7. Given an angle measures x degrees and two angles are complementary if their sum is 90°, then the complement of an angle of $x°$ is $(90 - x)°$.

8. Given an angle measures x degrees and two angles are supplementary if their sum is 180°, then the supplement of an angle of $x°$ is $(180 - x)°$.

9. $62°18' + 21°41' = 83°59'$

10. $75°15' + 83°32' = 158°47'$

11. $71°58' + 47°29' = 118°87'$
    $$= 118°(60 + 27)'$$
    $$= 119°27'$$

12. $90° - 73°48' = 89°60' - 73°48' = 16°12'$

13. $90° - 51°28' = 89°60' - 51°28' = 38°32'$

14. $180° - 124°51' = 179°60' - 124°51' = 55°9'$

15. $90° - 72°58'11'' = 89°59'60'' - 72°58'11''$
    $$= 17°1'49''$$

16. $90° - 36°18'47'' = 89°59'60'' - 36°18'47''$
    $$= 53°41'13''$$

Remember that
$$1' = \frac{1°}{60} \text{ and } 1'' = \frac{1°}{3600}$$
for Exercises 17–22.

17. $20°54' = 20° + \dfrac{54°}{60}$
    $$= 20° + .900°$$
    $$= 20.900°$$

18. $38°42' = 38° + \dfrac{42°}{60}$
    $$= 38.700°$$

19. $91°35'54'' = 91° + \dfrac{35°}{60} + \dfrac{54°}{3600}$
    $$\approx 91.598°$$

20. $34°51'35'' = 34° + \dfrac{51°}{60} + \dfrac{35°}{3600}$
    $$\approx 34.860°$$

21. $274°18'59'' = 274° + \dfrac{18°}{60} + \dfrac{59°}{3600}$
    $$\approx 274.316°$$

22. $165°51'9'' = 165° + \dfrac{51°}{60} + \dfrac{9°}{3600}$
    $$\approx 165.853°$$

23. $31.4296° = 31° + .4296°$
    $$= 31° + (.4296)(60')$$
    $$= 31° + 25.776'$$
    $$= 31° + 25' + (.776)(60'')$$
    $$\approx 31° + 25' + 47''$$
    $$= 31°25'47''$$

24. $59.0854° = 59° + .0854°$
    $$= 59° + (.0854)(60')$$
    $$= 59° + 5.124'$$
    $$= 59° + 5' + (.124)(60'')$$
    $$\approx 59° + 5' + 7''$$
    $$= 59°5'7''$$

25. $89.9004° = 89° + .9004°$
    $= 89° + (.9004)(60')$
    $= 89° + 54.024'$
    $= 89° + 54' + (.024)(60'')$
    $\approx 89° + 54' + 1''$
    $= 89°54'1''$

26. $102.3771° = 102° + .3771°$
    $= 102° + (.3771)(60')$
    $= 102° + 22.626$
    $= 102° + 22' + (.626)(60'')$
    $\approx 102° + 22' + 38''$
    $= 102°22'38''$

27. $178.5994° = 178° + .5994°$
    $= 178° + (.5994)(60')$
    $= 178° + 35.964'$
    $= 178° + 35' + (.964)(60'')$
    $\approx 178° + 35' + 58''$
    $= 178°35'58''$

28. $122.6853° = 122° + .6853°$
    $= 122° + (.6853)(60')$
    $= 122° + 41.118'$
    $= 122° + 41' + (.118)(60'')$
    $\approx 122° + 41' + 7''$
    $= 122°41'7''$

29. $-40°$ is coterminal with $-40° + 360° = 320°$.

30. $-98°$ is coterminal with $-98° + 360° = 262°$.

31. $-125°$ is coterminal with $-125° + 360° = 235°$.

32. $-203°$ is coterminal with $-203° + 360° = 157°$.

33. $539°$ is coterminal with $539° - 360° = 179°$.

34. $699°$ is coterminal with $699° - 360° = 339°$.

35. $850°$ is coterminal with $850° - 2(360°) = 130°$.

36. $1000°$ is coterminal with $1000° - 2(360°) = 280°$.

37. $30°$
    A coterminal angle can be obtained by adding an integer multiple of $360°$ or
    $$30° + n \cdot 360°.$$

38. $45°$
    A coterminal angle can be obtained by adding an integer multiple of $360°$ or
    $$45° + n \cdot 360°.$$

39. $60°$
    A coterminal angle can be obtained by adding an integer multiple of $360°$:
    $$60° + n \cdot 360°.$$

40. $90°$
    A coterminal angle can be obtained by adding an integer multiple of $360°$:
    $$90° + n \cdot 360°.$$

41. A positive angle coterminal with $75°$ is
    $$75° + 360° = 435°.$$
    A negative angle coterminal with $75°$ is
    $$75° - 360° = -285°.$$
    These angles are in quadrant I.

    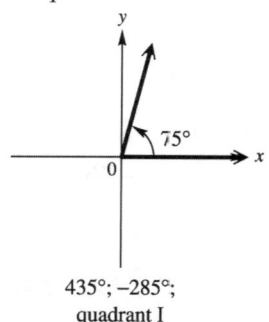

    $435°; -285°;$
    quadrant I

42. A positive angle coterminal with $89°$ is
    $$89° + 360° = 449°.$$
    A negative angle coterminal with $89°$ is
    $$89° - 360° = -271°.$$
    These angles are in quadrant I.

    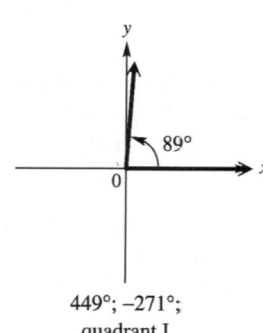

    $449°; -271°;$
    quadrant I

43. A positive angle coterminal with 174° is
$$174° + 360° = 534°.$$
A negative angle coterminal with 174° is
$$174° - 360° = -186°.$$
These angles are in quadrant II.

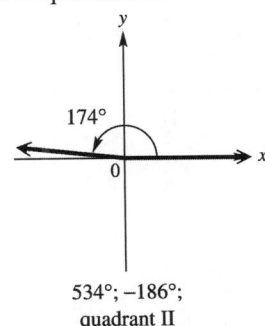

534°; −186°;
quadrant II

44. A positive angle coterminal with 234° is
$$234° + 360° = 594°.$$
A negative angle coterminal with 234° is
$$234° - 360° = -126°.$$
These angles are in quadrant III.

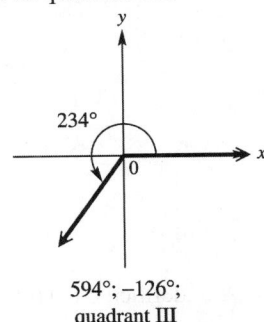

594°; −126°;
quadrant III

45. A positive angle coterminal with 300° is
$$300° + 360° = 660°.$$
A negative angle coterminal with 300° is
$$300° - 360° = -60°.$$
These angles are in quadrant IV.

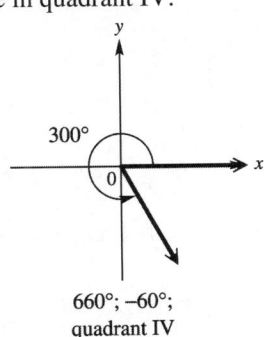

660°; −60°;
quadrant IV

46. A positive angle coterminal with 512° is
$$512° - 360° = 152°.$$
A negative angle coterminal with 512° is
$$512° - 2(360)° = -208°.$$
These angles are in quadrant II.

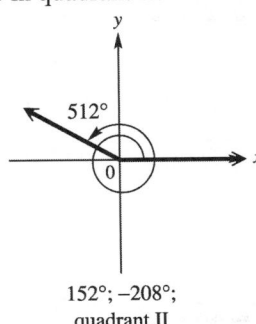

152°; −208°;
quadrant II

47. A positive angle coterminal with −61° is
$$-61° + 360° = 299°.$$
A negative angle coterminal with −61° is
$$-61° - 360° = -421°.$$
These angles are in quadrant IV.

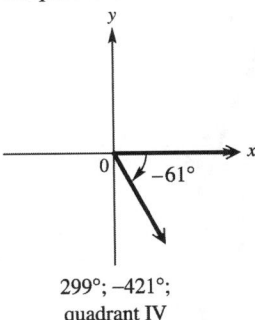

299°; −421°;
quadrant IV

48. A positive angle coterminal with −159° is
$$-159° + 360° = 201°.$$
A negative angle coterminal with −159° is
$$-159° - 360° = -519°.$$
These angles are in quadrant III

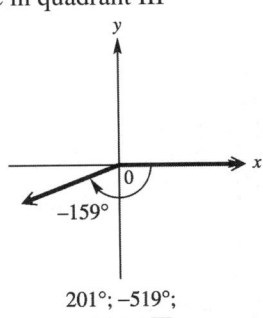

201°; −519°;
quadrant III

10.2 EXERCISES

1.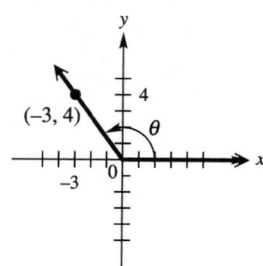

2.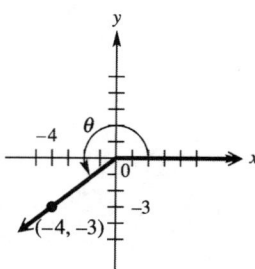

3.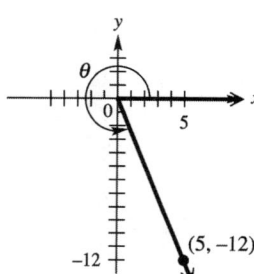

4.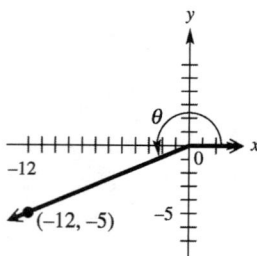

5. $(-3, 4)$

   $x = -3,\ y = 4$

   $r = \sqrt{(-3)^2 + 4^2}$
   $= \sqrt{9 + 16}$
   $= \sqrt{25}$
   $= 5$

   $\sin\theta = \dfrac{y}{r} = \dfrac{4}{5}$ $\qquad \cos\theta = \dfrac{x}{r} = \dfrac{-3}{5} = -\dfrac{3}{5}$

   $\tan\theta = \dfrac{y}{x} = \dfrac{4}{-3} = -\dfrac{4}{3}$ $\qquad \cot\theta = \dfrac{x}{y} = \dfrac{-3}{4} = -\dfrac{3}{4}$

   $\sec\theta = \dfrac{r}{x} = \dfrac{5}{-3} = -\dfrac{5}{3}$ $\qquad \csc\theta = \dfrac{r}{y} = \dfrac{5}{4}$

6. $(-4, -3)$

   $x = -4,\ y = -3$

   $r = \sqrt{(-4)^2 + (-3)^2}$
   $= \sqrt{16 + 9}$
   $= \sqrt{25}$
   $= 5$

   $\sin\theta = \dfrac{y}{r} = \dfrac{-3}{5} = -\dfrac{3}{5}$ $\qquad \cos\theta = \dfrac{x}{r} = \dfrac{-4}{5} = -\dfrac{4}{5}$

   $\tan\theta = \dfrac{y}{x} = \dfrac{-3}{-4} = \dfrac{3}{4}$ $\qquad \cot\theta = \dfrac{x}{y} = \dfrac{-4}{-3} = \dfrac{4}{3}$

   $\sec\theta = \dfrac{r}{x} = \dfrac{5}{-4} = -\dfrac{5}{4}$ $\qquad \csc\theta = \dfrac{r}{y} = \dfrac{5}{-3} = -\dfrac{5}{3}$

7. $(0, 2)$

   $x = 0,\ y = 2$

   $r = \sqrt{(0)^2 + 2^2}$
   $= \sqrt{0 + 4}$
   $= \sqrt{4}$
   $= 2$

   $\sin\theta = \dfrac{y}{r} = \dfrac{2}{2} = 1$ $\qquad \cos\theta = \dfrac{x}{r} = \dfrac{0}{2} = 0$

   $\tan\theta = \dfrac{y}{x} = \dfrac{2}{0}$; undefined $\qquad \cot\theta = \dfrac{x}{y} = \dfrac{0}{2} = 0$

   $\sec\theta = \dfrac{r}{x} = \dfrac{2}{0}$; undifined $\qquad \csc\theta = \dfrac{r}{y} = \dfrac{2}{2} = 1$

8. $(-4, 0)$

   $x = -4,\ y = 0$

   $r = \sqrt{(-4)^2 + 0^2}$
   $= \sqrt{16 + 0}$
   $= \sqrt{16}$
   $= 4$

   $\sin\theta = \dfrac{y}{r} = \dfrac{0}{4} = 0$ $\qquad \cos\theta = \dfrac{x}{r} = \dfrac{-4}{4} = -1$

   $\tan\theta = \dfrac{y}{x} = \dfrac{0}{-4} = 0$ $\qquad \cot\theta = \dfrac{x}{y} = \dfrac{-4}{0}$; undefined

   $\sec\theta = \dfrac{r}{x} = \dfrac{4}{-4} = -1$ $\qquad \csc\theta = \dfrac{r}{y} = \dfrac{4}{0}$; undefined

9. $(1, \sqrt{3})$

$x = 1, y = \sqrt{3}$

$r = \sqrt{(1)^2 + (\sqrt{3})^2}$
$= \sqrt{1+3}$
$= \sqrt{4}$
$= 2$

$\sin\theta = \dfrac{y}{r} = \dfrac{\sqrt{3}}{2}$ $\cos\theta = \dfrac{x}{r} = \dfrac{1}{2}$

$\tan\theta = \dfrac{y}{x} = \dfrac{\sqrt{3}}{1} = \sqrt{3}$ $\cot\theta = \dfrac{x}{y} = \dfrac{1}{\sqrt{3}} = \dfrac{\sqrt{3}}{3}$

$\sec\theta = \dfrac{r}{x} = \dfrac{2}{1} = 2$ $\csc\theta = \dfrac{r}{y} = \dfrac{2}{\sqrt{3}} = \dfrac{2\sqrt{3}}{3}$

Note that for $\cot\theta$ and $\csc\theta$ answers the denominator is "rationalized." For example,

$\cot\theta = = \dfrac{1}{\sqrt{3}} \cdot \dfrac{\sqrt{3}}{\sqrt{3}} = \dfrac{\sqrt{3}}{3}$.

10. $(-2\sqrt{3}, -2)$

$x = -2\sqrt{3}, y = -2$

$r = \sqrt{(-2\sqrt{3})^2 + (-2)^2}$
$= \sqrt{12 + 4}$
$= \sqrt{16}$
$= 4$

$\sin\theta = \dfrac{y}{r} = \dfrac{-2}{4} = -\dfrac{1}{2}$

$\cos\theta = \dfrac{x}{r} = \dfrac{-2\sqrt{3}}{4} = -\dfrac{\sqrt{3}}{2}$

$\tan\theta = \dfrac{y}{x} = \dfrac{-2}{-2\sqrt{3}} = \dfrac{\sqrt{3}}{3}$

$\cot\theta = \dfrac{x}{y} = \dfrac{-2\sqrt{3}}{-2} = \sqrt{3}$

$\sec\theta = \dfrac{r}{x} = \dfrac{4}{-2\sqrt{3}} = -\dfrac{2\sqrt{3}}{3}$

$\csc\theta = \dfrac{r}{y} = \dfrac{4}{-2} = -2$

11. $(3, 5)$

$x = 3, y = 5$

$r = \sqrt{3^2 + 5^2}$
$= \sqrt{9 + 25}$
$= \sqrt{34}$

$\sin\theta = \dfrac{y}{r} = \dfrac{5}{\sqrt{34}} = \dfrac{5\sqrt{34}}{34}$

$\cos\theta = \dfrac{x}{r} = \dfrac{3}{\sqrt{34}} = \dfrac{3\sqrt{34}}{34}$

$\tan\theta = \dfrac{y}{x} = \dfrac{5}{3}$

$\cot\theta = \dfrac{x}{y} = \dfrac{3}{5}$

$\sec\theta = \dfrac{r}{x} = \dfrac{\sqrt{34}}{3}$

$\csc\theta = \dfrac{r}{y} = \dfrac{\sqrt{34}}{5}$

12. $(-2, 7)$

$x = -2, y = 7$

$r = \sqrt{(-2)^2 + 7^2}$
$= \sqrt{4 + 49}$
$= \sqrt{53}$

$\sin\theta = \dfrac{y}{r} = \dfrac{7}{\sqrt{53}} = \dfrac{7\sqrt{53}}{53}$

$\cos\theta = \dfrac{x}{r} = \dfrac{-2}{\sqrt{53}} = -\dfrac{2\sqrt{53}}{53}$

$\tan\theta = \dfrac{y}{x} = \dfrac{7}{-2} = -\dfrac{7}{2}$

$\cot\theta = \dfrac{x}{y} = \dfrac{-2}{7} = -\dfrac{2}{7}$

$\sec\theta = \dfrac{r}{x} = \dfrac{\sqrt{53}}{-2} = -\dfrac{\sqrt{53}}{2}$

$\csc\theta = \dfrac{r}{y} = \dfrac{\sqrt{53}}{7}$

13. $(-8, 0)$

$x = -8, y = 0$

$r = \sqrt{(-8)^2 + 0^2}$
$= \sqrt{64 + 0}$
$= \sqrt{64}$
$= 8$

$\sin\theta = \dfrac{y}{r} = \dfrac{0}{8} = 0$ $\cos\theta = \dfrac{x}{r} = \dfrac{-8}{8} = -1$

$\tan\theta = \dfrac{y}{x} = \dfrac{0}{-8} = 0$ $\cot\theta = \dfrac{x}{y} = \dfrac{-8}{0}$; undefined

$\sec\theta = \dfrac{r}{x} = \dfrac{8}{-8} = -1$ $\csc\theta = \dfrac{r}{y} = \dfrac{8}{0}$; undefined

14. $(0, 9)$

    $x = 0, y = 9$

    $r = \sqrt{0^2 + (9)^2}$
    $= \sqrt{0 + 81}$
    $= \sqrt{81}$
    $= 9$

    $\sin\theta = \dfrac{y}{r} = \dfrac{9}{9} = 1$ $\cos\theta = \dfrac{x}{r} = \dfrac{0}{9} = 0$

    $\tan\theta = \dfrac{y}{x} = \dfrac{9}{0}$; undefined $\cot\theta = \dfrac{x}{y} = \dfrac{0}{9} = 0$

    $\sec\theta = \dfrac{r}{x} = \dfrac{9}{0}$; undefined $\csc\theta = \dfrac{r}{y} = \dfrac{9}{9} = 1$

15. Writing exercise

16. Since $\cot\theta$ is undefined and

    $\cot\theta = \dfrac{x}{y}$, where (x, y) is a point on the terminal side of θ,

    y must equal 0 and x can be any nonzero number.

    $\tan\theta = \dfrac{y}{x} = \dfrac{0}{\text{nonzero number}} = 0$,

    Therefore, $\tan\theta = 0$.

17. The value r is the distance from a point (x, y) on the terminal side of the angle to the origin.

18. Since the terminal side of angle θ is in quadrant III and (x, y) is a pont on that terminal side, both x and y must be negative. Since $r = \sqrt{x^2 + y^2}$, r is always considered positive.

    $\sin\theta = \dfrac{y}{r}$

    Since y is negative and r is positive, $\sin\theta$ is negative.

    $\cos\theta = \dfrac{x}{r}$

    Since x is negative and r is positive, $\cos\theta$ is negative.

    $\tan\theta = \dfrac{y}{x}$

    Since x and y are negative, $\tan\theta$ is positive.

    $\cot\theta = \dfrac{x}{y}$

    Since x and y are negative, $\cot\theta$ is positive

    $\sec\theta = \dfrac{r}{x}$.

    Since x is negative and r is positive, $\sec\theta$ is negative.

    $\csc\theta = \dfrac{y}{r}$

    Since y is negative and r is positive, $\csc\theta$ is negative.

    That is, $\tan\theta$ and $\cot\theta$ are positive and all other function values are negative.

In Exersizes 19–26, $r = \sqrt{x^2 + y^2}$, which is positive.

19. In quadrant II, y is positive, so

    $\dfrac{y}{r}$ is positive.

20. In quadrant II, x is negative, so

    $\dfrac{x}{r}$ is negative.

21. In quadrant III, y is negative, so

    $\dfrac{y}{r}$ is negative.

22. In quadrant III, x is negative, so

    $\dfrac{x}{r}$ is negative.

23. In quadrant IV, x is positive, so

    $\dfrac{x}{r}$ is positive.

24. In quadrant IV, y is negative, so

    $\dfrac{y}{r}$ is negative.

25. In quadrant IV, x is positive and y is negative, so

    $\dfrac{y}{x}$ is negative.

26. In quadrant IV, x is positive and y is negative, so

    $\dfrac{x}{y}$ is negative.

For Exercises 27–32 choose any point on the terminal side of a $90°$ angle such as $(0, 1)$. Then

$$r = \sqrt{x^2 + y^2} = \sqrt{0^2 + 1^2} = 1.$$

27. Since $x = 0, r = 1$, and

    $\cos\theta = \dfrac{x}{r}$, $\cos 90° = \dfrac{0}{1} = 0$.

28. Since $y = 1, r = 1$, and

    $\sin\theta = \dfrac{y}{r}$, $\sin 90° = \dfrac{1}{1} = 1$.

29. Since $x = 0, y = 1$, and

    $\tan\theta = \dfrac{y}{x}$, $\cos 90° = \dfrac{1}{0}$, which is undefined.

30. Since $x = 0, y = 1$, and
$$\cot \theta = \frac{x}{y}, \tan 90° = \frac{0}{1} = 0.$$

31. Since $x = 0, r = 1$, and
$$\sec \theta = \frac{r}{x}, \sec 90° = \frac{1}{0}, \text{ which is undefined.}$$

32. Since $y = 1, r = 1$, and
$$\csc \theta = \frac{r}{y}, \csc 90° = \frac{1}{1} = 1.$$

33. Selecting a point such as $(-1, 0)$ on the terminal side of the angle 180°, we have
$$r = \sqrt{x^2 + y^2} = \sqrt{(-1)^2 + 0^2} = 1.$$
Since $y = 0, r = 1$, and
$$\sin \theta = \frac{y}{r}, \sin 180° = \frac{0}{1} = 0.$$

34. Selecting a point such as $(0, -1)$ on the terminal side of the angle 270°, we have
$$r = \sqrt{x^2 + y^2} = \sqrt{0^2 + (-1)^2} = 1.$$
Since $y = -1, r = 1$, and
$$\sin \theta = \frac{y}{r}, \sin 270° = \frac{-1}{1} = -1.$$

35. Selecting a point such as $(-1, 0)$ on the terminal side of the angle 180°, we have
$$r = \sqrt{x^2 + y^2} = \sqrt{(-1)^2 + 0^2} = 1.$$
Since $x = -1, y = 0$, and
$$\tan \theta = \frac{y}{x}, \tan 180° = \frac{0}{-1} = 0.$$

36. Selecting a point such as $(0, -1)$ on the terminal side of the angle 270°, we have
$$r = \sqrt{x^2 + y^2} = \sqrt{0^2 + (-1)^2} = 1.$$
Since $x = 0, y = -1$, and
$$\cot \theta = \frac{x}{y}, \cot 270° = \frac{0}{-1} = 0.$$

37. Selecting a point such as $(0, 1)$ on the terminal side of the angle $-270°$, we have
$$r = \sqrt{x^2 + y^2} = \sqrt{0^2 + (1)^2} = 1.$$
Since $y = 1, r = 1$, and
$$\sin \theta = \frac{y}{r}, \sin(-270°) = \frac{1}{1} = 1.$$

38. Selecting a point such as $(0, 1)$ on the terminal side of the angle $-270°$, we have
$$r = \sqrt{x^2 + y^2} = \sqrt{0^2 + (1)^2} = 1.$$
Since $x = 0, r = 1$, and
$$\cos \theta = \frac{x}{r}, \cos(-270°) = \frac{0}{1} = 0.$$

39. Selecting a point such as $(1, 0)$ on the terminal side of the angle 0°, we have
$$r = \sqrt{x^2 + y^2} = \sqrt{1^2 + 0^2} = 1.$$
Since $x = 1, y = 0$, and
$$\tan \theta = \frac{y}{x}, \tan 0° = \frac{0}{1} = 0.$$

40. Selecting a point such as $(-1, 0)$ on the terminal side of the angle $-180°$ we have
$$r = \sqrt{x^2 + y^2} = \sqrt{(-1)^2 + 0^2} = 1.$$
Since $x = -1, r = 1$, and
$$\sec \theta = \frac{r}{x}, \sec(-180°) = \frac{1}{-1} = -1.$$

10.3 EXERCISES

1. Since
$$\tan \theta = \frac{1}{\cot \theta} \text{ and } \cot \theta = -3,$$
$$\tan \theta = \frac{1}{-3} = -\frac{1}{3}.$$

2. Since
$$\cot \theta = \frac{1}{\tan \theta} \text{ and } \tan \theta = 5,$$
$$\cot \theta = \frac{1}{5}.$$

3. Since
$$\sin\theta = \frac{1}{\csc\theta} \text{ and } \csc\theta = 3,$$
$$\sin\theta = \frac{1}{3}.$$

4. Since
$$\cos\alpha = \frac{1}{\sec\theta} \text{ and } \sec\alpha = -\frac{5}{2},$$
$$\cos\alpha = \frac{1}{-5/2} = -\frac{2}{5}.$$

5. Since
$$\cot\beta = \frac{1}{\tan\beta} \text{ and } \tan\beta = -\frac{1}{5},$$
$$\cot\beta = \frac{1}{-1/5} = -5.$$

6. Since
$$\sin\alpha = \frac{1}{\csc\theta} \text{ and } \csc\alpha = \sqrt{15},$$
$$\sin\alpha = \frac{1}{\sqrt{15}} = \frac{\sqrt{15}}{15}.$$

7. Since
$$\csc\alpha = \frac{1}{\sin\theta} \text{ and } \sin\alpha = \frac{\sqrt{2}}{4},$$
$$\csc\alpha = \frac{1}{\sqrt{2}/4} = \frac{4}{\sqrt{2}} = 2\sqrt{2}.$$

8. Since
$$\sec\beta = \frac{1}{\cos\beta} \text{ and } \cos\beta = -\frac{1}{\sqrt{7}},$$
$$\sec\beta = \frac{1}{-1/\sqrt{7}} = -\sqrt{7}.$$

9. Since
$$\tan\theta = \frac{1}{\cot\theta} \text{ and } \cot\theta = -\frac{\sqrt{5}}{3},$$
$$\tan\theta = \frac{1}{-\sqrt{5}/3} = -\frac{3}{\sqrt{5}} = -\frac{3\sqrt{5}}{5}.$$

10. Since
$$\cot\theta = \frac{1}{\tan\theta} \text{ and } \tan\theta = \frac{\sqrt{11}}{5},$$
$$\cot\theta = \frac{1}{\sqrt{11}/5} = \frac{5}{\sqrt{11}} = \frac{5\sqrt{11}}{11}.$$

11. Since
$$\sin\theta = \frac{1}{\csc\theta} \text{ and } \csc\theta = 1.5 = 1\frac{1}{2} = \frac{3}{2},$$
$$\sin\theta = \frac{1}{3/2} = \frac{2}{3}.$$

12. Since
$$\cos\alpha = \frac{1}{\sec\theta} \text{ and } \sec\alpha = 7.5 = \frac{15}{2},$$
$$\cos\alpha = \frac{1}{15/2} = \frac{2}{15}.$$

13. Since the sine function is positive and the cosine function is negative only for angles in quadrant II, α must be an angle in quadrant II.

14. Since the cosine function is positive and the tangent function is positive only for angles in quadrant I, β must be an angle in quadrant I.

15. Since the tangent function is positive and the sine function is positive only for angles in quadrant I, γ must be an angle in quadrant I.

16. Since the sine function is negative and the cosine function is positive only for angles in quadrant IV, β must be an angle in quadrant IV.

17. Since the tangent function and the cosine function are negative only for angles in quadrant II, ω must be an angle in quadrant II.

18. Since the cosecant function and the cosine function is negative only for angles in quadrant III, θ must be an angle in quadrant III.

19. Since the cosine function is negative for angles in quadrants II and III, β may be an angle in quadrant II or III.

20. Since the tangent function is positive for angles in quadrants I and III, θ may be an angle in quadrant I or III.

For Exercises 21–30 each angle is assumed to be in standard position (initial ray along the positive x–axis).

21. Since 74° is an acute angle, its terminal ray lies in quadrant I. Since all trig functions are positive for angles in quadrant I, the trigonometric functions of 74° are all positive.

22. Since 129° is an angle with terminal ray in quadrant II where the sine and cosecant functions are positive, $\sin 129° > 0$ and $\csc 129° > 0$. All remaining trigonometric functions of 129° are negative.

23. Since 183° is an angle in quadrant III, for any point (x, y) on the terminal ray, $x < 0$ and $y < 0$. Note that r is always positive. The signs of the trigonometric functions for 183° are

$$\sin 183°: \frac{y}{r} = \frac{-}{+} = -$$
$$\csc 183°: \frac{r}{y} = \frac{+}{-} = -$$
$$\cos 183°: \frac{x}{r} = \frac{-}{+} = -$$
$$\sec 183°: \frac{r}{x} = \frac{+}{-} = -$$
$$\tan 183°: \frac{y}{x} = \frac{-}{-} = +$$
$$\cot 183°: \frac{x}{y} = \frac{-}{-} = +.$$

24. Since 298° is an angle with terminal ray in quadrant IV where the cosine and secant functions are positive, $\cos 298° > 0$ and $\sec 298° > 0$. All remaining trigonometric functions of 298° are negative.

25. Since 302° is an angle in quadrant IV, for any point (x, y) on the terminal ray, $x > 0$ and $y < 0$. Note that r is always positive. The signs of the trigonometric functions for 302° are

$$\sin 302°: \frac{y}{r} = \frac{-}{+} = -$$
$$\csc 302°: \frac{r}{y} = \frac{+}{-} = -$$
$$\cos 302°: \frac{x}{r} = \frac{+}{+} = +$$
$$\sec 302°: \frac{r}{x} = \frac{+}{+} = +$$
$$\tan 302°: \frac{y}{x} = \frac{-}{+} = -$$
$$\cot 302°: \frac{x}{y} = \frac{+}{-} = -.$$

26. Since 406° is an angle with terminal ray in quadrant I, all trigonometric functions of 406° are positive.

27. Since 412° is an angle whose terminal ray lies in quadrant I, all trigonometric functions of 412° are positive.

28. Since −82° is an angle with terminal ray in quadrant IV where the cosine and secant functions are positive, $\cos (-82)° > 0$ and $\sec (-82)° > 0$. All remaining trigonometric functions of −82° are negative.

29. Since −14° is an angle with terminal ray in quadrant IV, for any point (x, y) on the terminal ray, $x > 0$ and $y < 0$. Note that r is always positive. The signs of the trigonometric functions for −14° are

$$\sin (-14°): \frac{y}{r} = \frac{-}{+} = -$$
$$\csc (-14°): \frac{r}{y} = \frac{+}{-} = -$$
$$\cos (-14°): \frac{x}{r} = \frac{+}{+} = +$$
$$\sec (-14°): \frac{r}{x} = \frac{+}{+} = +$$
$$\tan (-14°): \frac{y}{x} = \frac{-}{+} = -$$
$$\cot (-14°): \frac{x}{y} = \frac{+}{-} = -.$$

30. Since −121° is an angle with terminal ray in quadrant III where the tangent and cotangent functions are positive, $\tan (-121°) > 0$ and $\cot (-121°) > 0$. All remaining trigonometric functions of −121° are negative.

31. Since $\tan^2 \alpha + 1 = \sec^2 \alpha$ and $\sec \alpha = 3$,

$$\tan^2 \alpha + 1 = 3^2 = 9$$
$$\tan^2 \alpha = 9 - 1 = 8$$
$$\tan \alpha = \pm \sqrt{8} = \pm 2\sqrt{2}$$

Since α is in quadrant IV, $\tan \alpha$ is negative and

$$\tan \alpha = -2\sqrt{2}.$$

32. Since $\sin^2 \theta + \cos^2 \theta = 1$ and $\sin \theta = 2/3$,

$$\left(\frac{2}{3}\right)^2 + \cos^2 \theta = 1$$
$$\frac{4}{9} + \cos^2 \theta = 1$$
$$\cos^2 \theta = 1 - \frac{4}{9} = \frac{5}{9}$$
$$\cos \theta = \pm \frac{\sqrt{5}}{3}.$$

Since θ is in quadrant II, $\cos \theta$ is negative and

$$\cos \theta = -\frac{\sqrt{5}}{3}.$$

33. Since $\sin^2 \alpha + \cos^2 \alpha = 1$ and $\cos \alpha = -1/4$,

$$\sin^2 \alpha + \left(-\frac{1}{4}\right)^2 = 1$$
$$\sin^2 \alpha + \frac{1}{16} = 1$$
$$\sin^2 \alpha = 1 - \frac{1}{16} = \frac{15}{16}$$
$$\sin \alpha = \pm \frac{\sqrt{15}}{4}.$$

Since α is in quadrant II, $\sin \alpha$ is positive and

$$\sin \alpha = \frac{\sqrt{15}}{4}.$$

34. Since $1 + \cot^2\beta = \csc^2\beta$ and $\cot\beta = -1/2$,

$$1 + \left(-\frac{1}{2}\right)^2 = \csc^2\beta$$
$$1 + \frac{1}{4} = \csc^2\beta$$
$$\csc^2\beta = \frac{5}{4}$$
$$\csc\beta = \pm\frac{\sqrt{5}}{2}.$$

Since β is in quadrant IV, $\csc\beta$ is negative and

$$\csc\beta = -\frac{\sqrt{5}}{2}.$$

35. Since $\sin^2\theta + \cos^2\theta = 1$ and $\cos\theta = 1/3$,

$$\sin^2\theta + \left(\frac{1}{3}\right)^2 = 1$$
$$\sin^2\theta + \frac{1}{9} = 1$$
$$\sin^2\theta = 1 - \frac{1}{9} = \frac{8}{9}$$
$$\sin\theta = \pm\frac{\sqrt{8}}{3}.$$

Since θ is in quadrant IV, $\sin\theta$ is negative and

$$\sin\theta = -\frac{\sqrt{8}}{3}.$$

Since $\tan\theta = \sin\theta/\cos\theta$,

$$\tan\theta = \frac{-\sqrt{8}/3}{1/3}$$
$$\tan\theta = -\frac{\sqrt{8}}{3}\cdot\frac{3}{1}$$
$$= -\sqrt{8}.$$
$$= -2\sqrt{2}.$$

36. Since $\tan^2\theta + 1 = \sec^2\theta$ and $\tan\theta = \sqrt{7}/3$,

$$\left(\frac{\sqrt{7}}{3}\right)^2 + 1 = \sec^2\theta$$
$$\frac{7}{9} + 1 = \sec^2\theta$$
$$\sec^2\theta = \frac{16}{9}$$
$$\sec\theta = \pm\sqrt{\frac{16}{9}} = \pm\frac{4}{3}.$$

Since θ is in quadrant III, $\sec\theta$ is negative and

$$\sec\theta = -\frac{4}{3}.$$

37. Since $\sin\beta = 1/\csc\beta$ and $\csc\beta = -4$

$$\sin\beta = \frac{1}{-4} = -\frac{1}{4}.$$

Since $\sin^2\beta + \cos^2\beta = 1$ and $\sin\beta = -1/4$,

$$\left(-\frac{1}{4}\right)^2 + \cos^2\beta = 1$$
$$\frac{1}{16} + \cos^2\beta = 1$$
$$\cos^2\beta = 1 - \frac{1}{16} = \frac{15}{16}$$
$$\cos\beta = \pm\frac{\sqrt{15}}{4}.$$

Since β is in quadrant III, $\cos\beta$ is negative,

$$\cos\beta = -\frac{\sqrt{15}}{4}.$$

38. Since $\cos\theta = 1/\sec\theta$ and $\sec\theta = 2$,

$$\cos\theta = \frac{1}{2}.$$

Since $\sin^2\theta + \cos^2\theta = 1$ and $\cos\theta = 1/2$,

$$\sin^2\theta + \left(\frac{1}{2}\right)^2 = 1$$
$$\sin^2\theta = 1 - \frac{1}{4}$$
$$\sin^2\theta = \frac{3}{4}$$
$$\sin\theta = \pm\frac{\sqrt{3}}{2}.$$

Since θ is in quadrant IV, $\sin\theta$ is negative and

$$\sin\theta = -\frac{\sqrt{3}}{2}.$$

39. $\tan\alpha = -15/8$, with α in quadrant II

Since $\tan\alpha = y/x$, let $x = -8$ and $y = 15$.

$$r = \sqrt{x^2 + y^2}$$
$$= \sqrt{(-8)^2 + 15^2}$$
$$= \sqrt{64 + 225}$$
$$= \sqrt{289} = 17$$

10.3 TRIGONOMETRIC IDENTITIES

$\sin \alpha = \dfrac{y}{r} = \dfrac{15}{17}$

$\cos \alpha = \dfrac{x}{r} = \dfrac{-8}{17} = -\dfrac{8}{17}$

$\tan \alpha = \dfrac{y}{x} = \dfrac{15}{-8} = -\dfrac{15}{8}$

$\cot \alpha = \dfrac{x}{y} = \dfrac{-8}{15} = -\dfrac{8}{15}$

$\sec \alpha = \dfrac{r}{x} = \dfrac{17}{-8} = -\dfrac{17}{8}$

$\csc \alpha = \dfrac{r}{y} = \dfrac{17}{15}$

40. $\cos \alpha = -3/5$, with α in quadrant III

Since $\cos \alpha = x/r$, let $x = -3$ and $r = 5$.

$x^2 + y^2 = r^2$
$y = \pm \sqrt{r^2 - x^2}$
$= \pm \sqrt{5^2 - (-3)^2}$
$= \pm \sqrt{25 - 9}$
$= \pm \sqrt{16} = -4$ since y is negative in quadrant III..

$\sin \alpha = \dfrac{y}{r} = \dfrac{-4}{5} = -\dfrac{4}{5}$

$\cos \alpha = \dfrac{x}{r} = \dfrac{-3}{5} = -\dfrac{3}{5}$

$\tan \alpha = \dfrac{y}{x} = \dfrac{-4}{-3} = \dfrac{4}{3}$

$\cot \alpha = \dfrac{x}{y} = \dfrac{-3}{-4} = \dfrac{3}{4}$

$\sec \alpha = \dfrac{r}{x} = \dfrac{5}{-3} = -\dfrac{5}{3}$

$\csc \alpha = \dfrac{r}{y} = \dfrac{5}{-4} = -\dfrac{5}{4}$

41. $\cot \gamma = 3/4$, with γ in quadrant III

Since $\cot \gamma = x/y$, let $x = -3$ and $y = -4$.

$r = \sqrt{x^2 + y^2}$
$= \sqrt{(-3)^2 + (-4)^2}$
$= \sqrt{9 + 16}$
$= \sqrt{25} = 5$

$\sin \gamma = \dfrac{y}{r} = \dfrac{-4}{5} = -\dfrac{4}{5}$

$\cos \gamma = \dfrac{x}{r} = \dfrac{-3}{5} = -\dfrac{3}{5}$

$\tan \gamma = \dfrac{y}{x} = \dfrac{-4}{-3} = \dfrac{4}{3}$

$\cot \gamma = \dfrac{x}{y} = \dfrac{-3}{-4} = \dfrac{3}{4}$

$\sec \gamma = \dfrac{r}{x} = \dfrac{5}{-3} = -\dfrac{5}{3}$

$\csc \gamma = \dfrac{r}{y} = \dfrac{5}{-4} = -\dfrac{5}{4}$

42. $\sin \beta = 7/25$, with β in quadrant II

Since $\sin \beta = y/r$, let $y = 7$ and $r = 25$.

$x^2 + y^2 = r^2$
$x = \pm \sqrt{r^2 - y^2}$
$= \pm \sqrt{25^2 - 7^2}$
$= \pm \sqrt{625 - 49}$
$= \pm \sqrt{576} = -24$ since x is negative in quadrant II..

$\sin \beta = \dfrac{y}{r} = \dfrac{7}{25}$

$\cos \beta = \dfrac{x}{r} = \dfrac{-24}{25} = -\dfrac{24}{25}$

$\tan \beta = \dfrac{y}{x} = \dfrac{7}{-24} = -\dfrac{7}{24}$

$\cot \beta = \dfrac{x}{y} = \dfrac{-24}{7} = -\dfrac{24}{7}$

$\sec \beta = \dfrac{r}{x} = \dfrac{25}{-24} = -\dfrac{25}{24}$

$\csc \beta = \dfrac{r}{y} = \dfrac{25}{7}$

43. $\tan \beta = \sqrt{3}$, with β in quadrant III

Since $\tan \beta = y/x$, let $y = -\sqrt{3}$ and $x = -1$.

$x^2 + y^2 = r^2$
$(-1)^2 + \left(-\sqrt{3}\right)^2 = r^2$
$\sqrt{1 + 3} = r$
$2 = r$

$\sin \beta = \dfrac{y}{r} = \dfrac{-\sqrt{3}}{2} = -\dfrac{\sqrt{3}}{2}$

$\cos \beta = \dfrac{x}{r} = \dfrac{-1}{2} = -\dfrac{1}{2}$

$\tan \beta = \dfrac{y}{x} = \dfrac{-\sqrt{3}}{-1} = \sqrt{3}$

$\cot \beta = \dfrac{x}{y} = \dfrac{-1}{-\sqrt{3}} = \dfrac{\sqrt{3}}{3}$

$\sec \beta = \dfrac{r}{x} = \dfrac{2}{-1} = -2$

$\csc \beta = \dfrac{r}{y} = \dfrac{2}{-\sqrt{3}} = -\dfrac{2\sqrt{3}}{3}$

44. $\csc \theta = 2$, with θ in quadrant II

Since $\csc \theta = r/y$, let $y = 1$ and $r = 2$.

$x^2 + y^2 = r^2$
$x = \pm \sqrt{r^2 - y^2}$
$= \pm \sqrt{2^2 - 1^2}$
$= \pm \sqrt{4 - 1}$
$= \pm \sqrt{3} = -\sqrt{3}$ since x is negative in quadrant II.

$$\sin\theta = \frac{y}{r} = \frac{1}{2}$$

$$\cos\theta = \frac{x}{r} = \frac{-\sqrt{3}}{2} = -\frac{\sqrt{3}}{2}$$

$$\tan\theta = \frac{y}{x} = \frac{1}{-\sqrt{3}} = -\frac{\sqrt{3}}{3}$$

$$\cot\theta = \frac{x}{y} = \frac{-\sqrt{3}}{1} = -\sqrt{3}$$

$$\sec\theta = \frac{r}{x} = \frac{2}{-\sqrt{3}} = -\frac{2\sqrt{3}}{3}$$

$$\csc\theta = \frac{r}{y} = \frac{2}{1} = 2$$

45. $\sin\beta = \sqrt{5}/7$, with $\tan\beta > 0$

   For sine and tangent to both be positive, β must lie in quadrant I.

   Since $\sin\beta = y/r$, let $y = \sqrt{5}$ and $r = 7$.

   $x^2 + y^2 = r^2$

   $x = \pm\sqrt{r^2 - y^2}$

   $= \pm\sqrt{7^2 - \left(\sqrt{5}\right)^2}$

   $= \pm\sqrt{49 - 5}$

   $= \sqrt{44} = 2\sqrt{11}$ since β is in quadrant I.

   $\sin\beta = \frac{y}{r} = \frac{\sqrt{5}}{7}$

   $\cos\beta = \frac{x}{r} = \frac{2\sqrt{11}}{7}$

   $\tan\beta = \frac{y}{x} = \frac{\sqrt{5}}{2\sqrt{11}} \cdot \frac{\sqrt{11}}{\sqrt{11}} = \frac{\sqrt{55}}{22}$

   $\cot\beta = \frac{x}{y} = \frac{2\sqrt{11}}{\sqrt{5}} \cdot \frac{\sqrt{5}}{\sqrt{5}} = \frac{2\sqrt{55}}{5}$

   $\sec\beta = \frac{r}{x} = \frac{7}{2\sqrt{11}} \cdot \frac{\sqrt{11}}{\sqrt{11}} = \frac{7\sqrt{11}}{22}$

   $\csc\beta = \frac{r}{y} = \frac{7}{\sqrt{5}} \cdot \frac{\sqrt{5}}{\sqrt{5}} = \frac{7\sqrt{5}}{5}$

46. $\cot\alpha = \sqrt{3}/8$, with $\sin\alpha > 0$

   Since cotangent and sine are both positive, α must lie in quadrant I.

   Since $\cot\alpha = x/y$, let $x = \sqrt{3}$ and $y = 8$.

   $r = \sqrt{x^2 + y^2}$

   $= \sqrt{\left(\sqrt{3}\right)^2 + 8^2}$

   $= \sqrt{3 + 64}$

   $= \sqrt{67}$

$$\sin\alpha = \frac{y}{r} = \frac{8}{\sqrt{67}} = \frac{8\sqrt{67}}{67}$$

$$\cos\alpha = \frac{x}{r} = \frac{\sqrt{3}}{\sqrt{67}} = \frac{\sqrt{201}}{67}$$

$$\tan\alpha = \frac{y}{x} = \frac{8}{\sqrt{3}} = \frac{8\sqrt{3}}{3}$$

$$\cot\alpha = \frac{x}{y} = \frac{\sqrt{3}}{8} = \frac{\sqrt{3}}{8}$$

$$\sec\alpha = \frac{r}{x} = \frac{\sqrt{67}}{\sqrt{3}} = \frac{\sqrt{201}}{3}$$

$$\csc\alpha = \frac{r}{y} = \frac{\sqrt{67}}{8}$$

47. Writing exercise

48. Writing exercise

10.4 EXERCISES

1. $\sin A = \dfrac{\text{side opposite}}{\text{hypotenuse}} = \dfrac{3}{5}$

   $\cos A = \dfrac{\text{side adjacent}}{\text{hypotenuse}} = \dfrac{4}{5}$

   $\tan A = \dfrac{\text{side opposite}}{\text{side adjacent}} = \dfrac{3}{4}$

   $\cot A = \dfrac{\text{side adjacent}}{\text{side opposite}} = \dfrac{4}{3}$

   $\sec A = \dfrac{\text{hypotenuse}}{\text{side adjacent}} = \dfrac{5}{4}$

   $\csc A = \dfrac{\text{hypotenuse}}{\text{side opposite}} = \dfrac{5}{3}$

2. $\sin A = \dfrac{\text{side opposite}}{\text{hypotenuse}} = \dfrac{8}{17}$

   $\cos A = \dfrac{\text{side adjacent}}{\text{hypotenuse}} = \dfrac{15}{17}$

   $\tan A = \dfrac{\text{side opposite}}{\text{side adjacent}} = \dfrac{8}{15}$

   $\cot A = \dfrac{\text{side adjacent}}{\text{side opposite}} = \dfrac{15}{8}$

   $\sec A = \dfrac{\text{hypotenuse}}{\text{side adjacent}} = \dfrac{17}{15}$

   $\csc A = \dfrac{\text{hypotenuse}}{\text{side opposite}} = \dfrac{17}{8}$

3. $\sin A = \dfrac{\text{side opposite}}{\text{hypotenuse}} = \dfrac{21}{29}$

$\cos A = \dfrac{\text{side adjacent}}{\text{hypotenuse}} = \dfrac{20}{29}$

$\tan A = \dfrac{\text{side opposite}}{\text{side adjacent}} = \dfrac{21}{20}$

$\cot A = \dfrac{\text{side adjacent}}{\text{side opposite}} = \dfrac{20}{21}$

$\sec A = \dfrac{\text{hypotenuse}}{\text{side adjacent}} = \dfrac{29}{20}$

$\csc A = \dfrac{\text{hypotenuse}}{\text{side opposite}} = \dfrac{29}{21}$

4. $\sin A = \dfrac{\text{side opposite}}{\text{hypotenuse}} = \dfrac{45}{53}$

$\cos A = \dfrac{\text{side adjacent}}{\text{hypotenuse}} = \dfrac{28}{53}$

$\tan A = \dfrac{\text{side opposite}}{\text{side adjacent}} = \dfrac{45}{28}$

$\cot A = \dfrac{\text{side adjacent}}{\text{side opposite}} = \dfrac{28}{45}$

$\sec A = \dfrac{\text{hypotenuse}}{\text{side adjacent}} = \dfrac{53}{28}$

$\csc A = \dfrac{\text{hypotenuse}}{\text{side opposite}} = \dfrac{53}{45}$

5. $\sin A = \dfrac{\text{side opposite}}{\text{hypotenuse}} = \dfrac{n}{p}$

$\cos A = \dfrac{\text{side adjacent}}{\text{hypotenuse}} = \dfrac{m}{p}$

$\tan A = \dfrac{\text{side opposite}}{\text{side adjacent}} = \dfrac{n}{m}$

$\cot A = \dfrac{\text{side adjacent}}{\text{side opposite}} = \dfrac{m}{n}$

$\sec A = \dfrac{\text{hypotenuse}}{\text{side adjacent}} = \dfrac{p}{m}$

$\csc A = \dfrac{\text{hypotenuse}}{\text{side opposite}} = \dfrac{p}{n}$

6. $\sin A = \dfrac{\text{side opposite}}{\text{hypotenuse}} = \dfrac{k}{z}$

$\cos A = \dfrac{\text{side adjacent}}{\text{hypotenuse}} = \dfrac{y}{z}$

$\tan A = \dfrac{\text{side opposite}}{\text{side adjacent}} = \dfrac{k}{y}$

$\cot A = \dfrac{\text{side adjacent}}{\text{side opposite}} = \dfrac{y}{k}$

$\sec A = \dfrac{\text{hypotenuse}}{\text{side adjacent}} = \dfrac{z}{y}$

$\csc A = \dfrac{\text{hypotenuse}}{\text{side opposite}} = \dfrac{z}{k}$

7. $a = 5, b = 12$

$c^2 = a^2 + b^2$

$c^2 = 5^2 + 12^2$

$c^2 = 25 + 144$

$c^2 = 169$

$c = 13$

$\sin B = \dfrac{\text{side opposite}}{\text{hypotenuse}} = \dfrac{12}{13}$

$\cos B = \dfrac{\text{side adjacent}}{\text{hypotenuse}} = \dfrac{5}{13}$

$\tan B = \dfrac{\text{side opposite}}{\text{side adjacent}} = \dfrac{12}{5}$

$\cot B = \dfrac{\text{side adjacent}}{\text{side opposite}} = \dfrac{5}{12}$

$\sec B = \dfrac{\text{hypotenuse}}{\text{side adjacent}} = \dfrac{13}{5}$

$\csc B = \dfrac{\text{hypotenuse}}{\text{side opposite}} = \dfrac{13}{12}$

8. $a = 3, b = 5$

$c^2 = a^2 + b^2$

$c^2 = 3^2 + 5^2$

$c^2 = 9 + 25$

$c^2 = 34$

$c = \sqrt{34}$

$\sin B = \dfrac{\text{side opposite}}{\text{hypotenuse}} = \dfrac{5}{\sqrt{34}} = \dfrac{5\sqrt{34}}{34}$

$\cos B = \dfrac{\text{side adjacent}}{\text{hypotenuse}} = \dfrac{3}{\sqrt{34}} = \dfrac{3\sqrt{34}}{34}$

$\tan B = \dfrac{\text{side opposite}}{\text{side adjacent}} = \dfrac{5}{3}$

$\cot B = \dfrac{\text{side adjacent}}{\text{side opposite}} = \dfrac{3}{5}$

$\sec B = \dfrac{\text{hypotenuse}}{\text{side adjacent}} = \dfrac{\sqrt{34}}{3}$

$\csc B = \dfrac{\text{hypotenuse}}{\text{side opposite}} = \dfrac{\sqrt{34}}{5}$

9. $a = 6, c = 7$

$c^2 = a^2 + b^2$

$7^2 = 6^2 + b^2$

$b^2 = 49 - 36$

$b^2 = 13$

$b = \sqrt{13}$

$$\sin B = \frac{\text{side opposite}}{\text{hypotenuse}} = \frac{\sqrt{13}}{7}$$
$$\cos B = \frac{\text{side adjacent}}{\text{hypotenuse}} = \frac{6}{7}$$
$$\tan B = \frac{\text{side opposite}}{\text{side adjacent}} = \frac{\sqrt{13}}{6}$$
$$\cot B = \frac{\text{side adjacent}}{\text{side opposite}} = \frac{6}{\sqrt{13}} = \frac{6\sqrt{13}}{13}$$
$$\sec B = \frac{\text{hypotenuse}}{\text{side adjacent}} = \frac{7}{6}$$
$$\csc B = \frac{\text{hypotenuse}}{\text{side opposite}} = \frac{7}{\sqrt{13}} = \frac{7\sqrt{13}}{13}$$

10. $b = 7, c = 12$
$$c^2 = a^2 + b^2$$
$$12^2 = a^2 + 7^2$$
$$a^2 = 144 - 49$$
$$a^2 = 95$$
$$a = \sqrt{95}$$

$$\sin B = \frac{\text{side opposite}}{\text{hypotenuse}} = \frac{7}{12}$$
$$\cos B = \frac{\text{side adjacent}}{\text{hypotenuse}} = \frac{\sqrt{95}}{12}$$
$$\tan B = \frac{\text{side opposite}}{\text{side adjacent}} = \frac{7}{\sqrt{95}} = \frac{7\sqrt{95}}{95}$$
$$\cot B = \frac{\text{side adjacent}}{\text{side opposite}} = \frac{\sqrt{95}}{7}$$
$$\sec B = \frac{\text{hypotenuse}}{\text{side adjacent}} = \frac{12}{\sqrt{95}} = \frac{12\sqrt{95}}{95}$$
$$\csc B = \frac{\text{hypotenuse}}{\text{side opposite}} = \frac{12}{7}$$

11. $\tan 50° = \cot(90° - 50°) = \cot 40°$
12. $\cot 73° = \tan(90° - 73°) = \tan 17°$
13. $\csc 47° = \sec(90° - 47°) = \sec 43°$
14. $\sec 39° = \csc(90° - 39°) = \csc 51°$
15. $\tan 25.4° = \cot(90° - 25.4°) = \cot 64.6°$
16. $\sin 38.7° = \cos(90° - 38.7°) = \cos 51.3°$
17. $\cos 13°30' = \sin(90° - 13°30')$
$ = \sin(89°60' - 13°30')$
$ = \sin 76°30'$
18. $\tan 26°10' = \cot(90° - 26°10')$
$ = \cot(89°60' - 26°10')$
$ = \cot 63°50'$

19. Sketch a 30°-60°-90° right triangle.

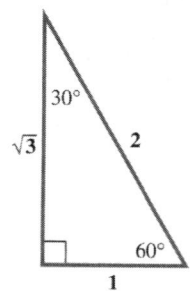

$$\tan 30° = \frac{\text{side opposite}}{\text{side adjacent}}$$
$$= \frac{1}{\sqrt{3}} \cdot \frac{\sqrt{3}}{\sqrt{3}}$$
$$= \frac{\sqrt{3}}{3}$$

20. Sketch a 30°-60°-90° right triangle.

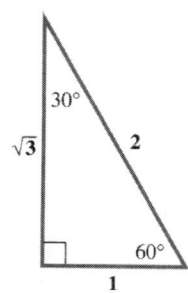

$$\cot 30° = \frac{\text{side adjacent}}{\text{side opposite}}$$
$$= \frac{\sqrt{3}}{1}$$
$$= \sqrt{3}$$

21. Sketch a 30°-60°-90° right triangle.

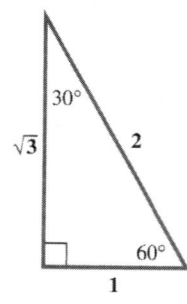

$$\sin 30° = \frac{\text{side opposite}}{\text{hypotenuse}}$$
$$= \frac{1}{2}$$

22. Sketch a 30°-60°-90° right triangle.

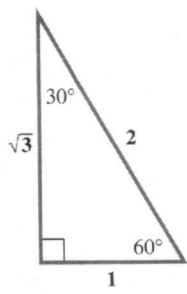

$$\cos 30° = \frac{\text{side adjacent}}{\text{hypotenuse}}$$
$$= \frac{\sqrt{3}}{2}$$

23. Sketch a 45°-45°-90° right triangle.

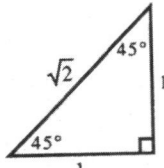

$$\csc 45° = \frac{\text{hypotenuse}}{\text{side opposite}}$$
$$= \frac{\sqrt{2}}{1}$$
$$= \sqrt{2}$$

24. Sketch a 45°-45°-90° right triangle.

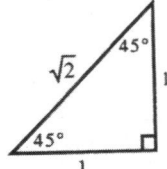

$$\sec 45° = \frac{\text{hypotenuse}}{\text{side adjacent}}$$
$$= \frac{\sqrt{2}}{1}$$
$$= \sqrt{2}$$

25. Sketch a 45°-45°-90° right triangle.

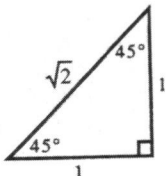

$$\cos 45° = \frac{\text{side adjacent}}{\text{hypotenuse}}$$
$$= \frac{1}{\sqrt{2}} \cdot \frac{\sqrt{2}}{\sqrt{2}}$$
$$= \frac{\sqrt{2}}{2}$$

26. Sketch a 45°-45°-90° right triangle.

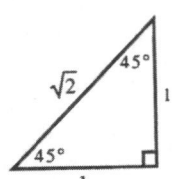

$$\sin 45° = \frac{\text{side opposite}}{\text{hypotenuse}}$$
$$= \frac{1}{\sqrt{2}} \cdot \frac{\sqrt{2}}{\sqrt{2}}$$
$$= \frac{\sqrt{2}}{2}$$

27. Sketch a 30°-60°-90° right triangle.

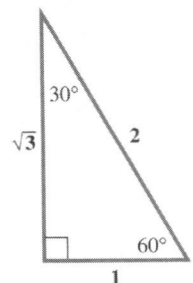

$$\sin 60° = \frac{\text{side opposite}}{\text{hypotenuse}}$$
$$= \frac{\sqrt{3}}{2}$$

28. Sketch a 30°-60°-90° right triangle.

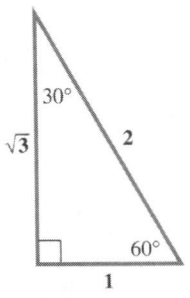

$$\cos 60° = \frac{\text{side adjacent}}{\text{hypotenuse}}$$
$$= \frac{1}{2}$$

29. Sketch a 30°-60°-90° right triangle.

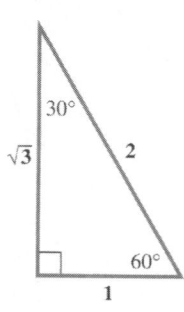

$\tan 60° = \dfrac{\text{side opposite}}{\text{side adjacent}}$

$= \dfrac{\sqrt{3}}{1}$

30. Sketch a 30°-60°-90° right triangle.

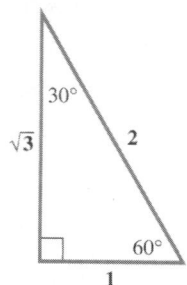

$\cot 60° = \dfrac{\text{side adjacent}}{\text{side opposite}}$

$= \dfrac{1}{\sqrt{3}} \cdot \dfrac{\sqrt{3}}{\sqrt{3}}$

$= \dfrac{\sqrt{3}}{3}$

31. $180° - 98° = 82°$

32. $212° - 180° = 32°$

33. $-135° + 360° = 225°$
    $225° - 180° = 45°$

34. $-60° + 360° = 300°$
    $360° - 300° = 60°$

35. $750° - 2(360°) = 30°$

36. $480° - 360° = 120°$
    $180° - 120° = 60°$

37. 120°

The reference angle is 60°. Since 120° is in quadrant II, the cosine, tangent, cotangent, and secant are negative.

$\sin 120° = \sin 60° = \dfrac{\sqrt{3}}{2}$

$\cos 120° = -\cos 60° = -\dfrac{1}{2}$

$\tan 120° = -\tan 60° = -\sqrt{3}$

$\cot 120° = -\cot 60° = \dfrac{-\sqrt{3}}{3}$

$\sec 120° = -\sec 60° = -2$

$\csc 120° = \csc 60° = \dfrac{2\sqrt{3}}{3}$

38. 135°

The reference angle is 45°. Since 135° is in quadrant II, the cosine, tangent, cotangent, and secant are negative.

$\sin 135° = \sin 45° = \dfrac{\sqrt{2}}{2}$

$\cos 135° = -\cos 45° = -\dfrac{\sqrt{2}}{2}$

$\tan 135° = -\tan 45° = -1$

$\cot 135° = -\cot 45° = -1$

$\sec 135° = -\sec 45° = -\sqrt{2}$

$\csc 135° = \csc 45° = \sqrt{2}$

39. 150°

The reference angle is 30°. Since 150° is in quadrant II, the cosine, tangent, cotangent, and secant are negative.

$\sin 150° = \sin 30° = \dfrac{1}{2}$

$\cos 150° = -\cos 30° = -\dfrac{\sqrt{3}}{2}$

$\tan 150° = -\tan 30° = -\dfrac{\sqrt{3}}{3}$

$\cot 150° = -\cot 30° = -\sqrt{3}$

$\sec 150° = -\sec 30° = -\dfrac{2\sqrt{3}}{3}$

$\csc 150° = \csc 30° = 2$

40. 225°

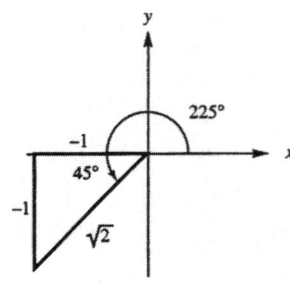

$x = -1, \; y = -1, \; r = \sqrt{2}$

$\sin 225° = \dfrac{y}{r} = \dfrac{-1}{\sqrt{2}} = -\dfrac{\sqrt{2}}{2}$

$\cos 225° = \dfrac{x}{r} = \dfrac{-1}{\sqrt{2}} = -\dfrac{\sqrt{2}}{2}$

$\tan 225° = \dfrac{y}{x} = \dfrac{-1}{-1} = 1$

$\cot 225° = \dfrac{x}{y} = \dfrac{-1}{-1} = 1$

$\sec 225° = \dfrac{r}{x} = \dfrac{\sqrt{2}}{-1} = -\sqrt{2}$

$\csc 225° = \dfrac{r}{y} = \dfrac{\sqrt{2}}{-1} = -\sqrt{2}$

41. 240°

The reference angle is 60°. Since 240° is in quadrant III, the sine, cosine, secant, and cosecant are negative.

$\sin 240° = -\sin 60° = -\dfrac{\sqrt{3}}{2}$

$\cos 240° = -\cos 60° = -\dfrac{1}{2}$

$\tan 240° = \tan 60° = \sqrt{3}$

$\cot 240° = \cot 60° = \dfrac{\sqrt{3}}{3}$

$\sec 240° = -\sec 60° = -2$

$\csc 240° = -\csc 60° = -\dfrac{2\sqrt{3}}{3}$

42. 300°

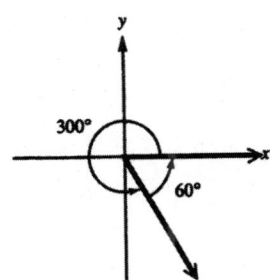

The reference angle is 60°. Since 300° is in quadrant IV, the sine, tangent, cotangent, and cosecant are negative.

$\sin 300° = -\sin 60° = -\dfrac{\sqrt{3}}{2}$

$\cos 300° = \cos 60° = \dfrac{1}{2}$

$\tan 300° = -\tan 60° = -\sqrt{3}$

$\cot 300° = -\cot 60° = \dfrac{-\sqrt{3}}{3} = -\dfrac{\sqrt{3}}{3}$

$\sec 300° = \sec 60° = 2$

$\csc 300° = -\csc 60° = -\dfrac{2\sqrt{3}}{3}$

43. 315°

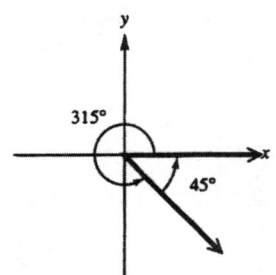

The reference angle is 45°. Since 315° is in quadrant IV, the sine, tangent, cotangent, and cosecant are negative.

$\sin 315° = -\sin 45° = -\dfrac{\sqrt{2}}{2}$

$\cos 315° = \cos 45° = \dfrac{\sqrt{2}}{2}$

$\tan 315° = -\tan 45° = -1$

$\cot 315° = -\cot 45° = -1$

$\sec 315° = \sec 45° = \sqrt{2}$

$\csc 315° = -\csc 45° = -\sqrt{2}$

44. 405°

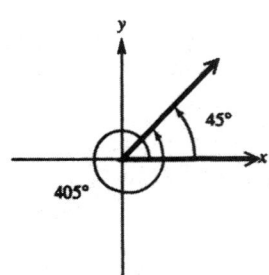

The reference angle is 45°. Since 405° is in quadrant I, all functions are posititve.

$\sin 405° = \sin 45° = \dfrac{\sqrt{2}}{2}$

$\cos 405° = \cos 45° = \dfrac{\sqrt{2}}{2}$

$\tan 405° = \tan 45° = 1$

$\cot 405° = \cot 45° = 1$

$\sec 405° = \sec 45° = \sqrt{2}$

$\csc 405° = \csc 45° = \sqrt{2}$

45. 420°

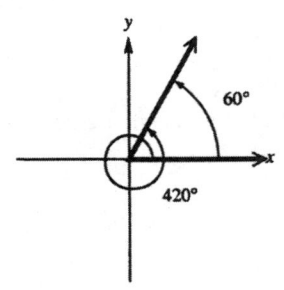

The reference angle is 60°. Since 420° is in quadrant I, all functions are posititve.

$$\sin 420° = \sin 60° = \frac{\sqrt{3}}{2}$$
$$\cos 420° = \cos 60° = \frac{1}{2}$$
$$\tan 420° = \tan 60° = \sqrt{3}$$
$$\cot 420° = \cot 60° = \frac{\sqrt{3}}{3}$$
$$\sec 420° = \sec 60° = 2$$
$$\csc 420° = \csc 60° = \frac{2\sqrt{3}}{3}$$

46. 480°

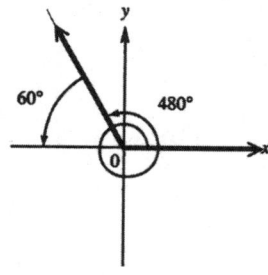

The reference angle is 60°. Since 480° is in quadrant II, the cosine, tangent, cotangent, and secant are negative.

$$\sin 480° = \sin 60° = \frac{\sqrt{3}}{2}$$
$$\cos 480° = -\cos 60° = -\frac{1}{2}$$
$$\tan 480° = -\tan 60° = -\sqrt{3}$$
$$\cot 480° = -\cot 60° = -\frac{\sqrt{3}}{3}$$
$$\sec 480° = -\sec 60° = -2$$
$$\csc 480° = \csc 60° = \frac{2\sqrt{3}}{3}$$

47. 495°

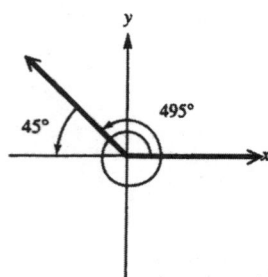

The reference angle is 45°. Since 495° is in quadrant II, the cosine, tangent, cotangent, and secant are negative.

$$\sin 495° = \sin 45° = \frac{\sqrt{2}}{2}$$
$$\cos 495° = -\cos 45° = -\frac{\sqrt{2}}{2}$$
$$\tan 495° = -\tan 45° = -1$$
$$\cot 495° = -\cot 45° = -1$$
$$\sec 495° = -\sec 45° = -\sqrt{2}$$
$$\csc 495° = \csc 45° = \sqrt{2}$$

48. 570°

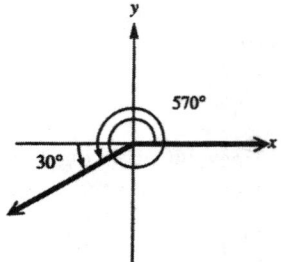

The reference angle is 30°. Since 570° is in quadrant III, the sine, cosine, secant, and cosecant are negative.

$$\sin 570° = \sin 30° = -\frac{1}{2}$$
$$\cos 570° = -\cos 30° = -\frac{\sqrt{3}}{2}$$
$$\tan 570° = \tan 30° = \frac{\sqrt{3}}{3}$$
$$\cot 570° = \cot 30° = \sqrt{3}$$
$$\sec 570° = -\sec 30° = -\frac{2\sqrt{3}}{3}$$
$$\csc 570° = \csc 30° = -2$$

49. 750° is coterminal with
$750° - 2(360°) = 750° - 720° = 30°$.

$$\sin 750° = \sin 30° = \frac{1}{2}$$
$$\cos 750° = \cos 30° = \frac{\sqrt{3}}{2}$$
$$\tan 750° = \tan 30° = \frac{\sqrt{3}}{3}$$
$$\cot 750° = \cot 30° = \sqrt{3}$$
$$\sec 750° = \sec 30° = \frac{2\sqrt{3}}{3}$$
$$\csc 750° = \csc 30° = 2$$

50. 1305° is coterminal with
$1305° - 3(360°) = 1305° - 1080° = 225°$.

The reference angle is 45°. Since 1305° is in quadrant III, the sine, cosine, secant and cosecant are negative.

$\sin 1305° = -\sin 45° = -\dfrac{\sqrt{2}}{2}$

$\cos 1305° = -\cos 45° = -\dfrac{\sqrt{2}}{2}$

$\tan 1305° = \tan 45° = 1$

$\cot 1305° = \cot 45° = 1$

$\sec 1305° = -\sec 45° = -\sqrt{2}$

$\csc 1305° = -\csc 45° = -\sqrt{2}$

51. 1500° is coterminal with
$1500° - 4(360°) = 1500° - 1440° = 60°$.

$\sin 1500° = \sin 60° = \dfrac{\sqrt{3}}{2}$

$\cos 1500° = \cos 60° = \dfrac{1}{2}$

$\tan 1500° = \tan 60° = \sqrt{3}$

$\cot 1500° = \cot 60° = \dfrac{\sqrt{3}}{3}$

$\sec 1500° = \sec 60° = 2$

$\csc 1500° = \csc 60° = \dfrac{2\sqrt{3}}{3}$

52. 2670° is coterminal with
$2670° - 7(360°) = 150°$.

The reference angle is 30°. Since 2670° is in quadrant II, the cosine, tangent, cotangent, and secant are negative.

$\sin 2670° = \sin 30° = \dfrac{1}{2}$

$\cos 2670° = -\cos 30° = -\dfrac{\sqrt{3}}{2}$

$\tan 2670° = -\tan 30° = -\dfrac{\sqrt{3}}{3}$

$\cot 2670° = -\cot 30° = -\sqrt{3}$

$\sec 2670° = -\sec 30° = -\dfrac{2\sqrt{3}}{3}$

$\csc 2670° = \csc 30° = 2$

53. $-390°$ is coterminal with
$-390° + (2)(360°) = 330°$.

The reference angle is 30°. Since $-390°$ is in quadrant IV, the sine, tangent, cotangent, and cosecant are negative.

$\sin(-390)° = -\sin 30° = -\dfrac{1}{2}$

$\cos(-390)° = \cos 30° = \dfrac{\sqrt{3}}{2}$

$\tan(-390)° = -\tan 30° = -\dfrac{\sqrt{3}}{3}$

$\cot(-390)° = -\cot 30° = -\sqrt{3}$

$\sec(-390)° = \sec 30° = \dfrac{2\sqrt{3}}{3}$

$\csc(-390)° = -\csc 30° = -2$

54. $-510°$ is coterminal with
$-510° + (2)(360°) = -510° + 720° = 210°$.

The reference angle is 30°. Since $-510°$ is in quadrant III, the sine, cosine, secant, and cosecant are negative.

$\sin(-510°) = -\sin 30° = -\dfrac{1}{2}$

$\cos(-510°) = -\cos 30° = -\dfrac{\sqrt{3}}{2}$

$\tan(-510°) = \tan 30° = \dfrac{\sqrt{3}}{3}$

$\cot(-510°) = \cot 30° = \sqrt{3}$

$\sec(-510°) = -\sec 30° = -\dfrac{2\sqrt{3}}{3}$

$\csc(-510°) = -\csc 30° = -2$

55. $-1020°$ is coterminal with
$-1020° + (3)(360°) = -1020° + 1080° = 60°$.

$\sin(-1020°) = \sin 60° = \dfrac{\sqrt{3}}{2}$

$\cos(-1020°) = \cos 60° = \dfrac{1}{2}$

$\tan(-1020°) = \tan 60° = \sqrt{3}$

$\cot(-1020°) = \cot 60° = \dfrac{\sqrt{3}}{3}$

$\sec(-1020°) = \sec 60° = 2$

$\csc(-1020°) = \csc 60° = \dfrac{2\sqrt{3}}{3}$

56. $-1290°$ is coterminal with
$-1290° + (4)(360°) = -1290° + 1440° = 150°$.

The reference angle is 30°. Since $-1290°$ is in quadrant II, the cosine, tangent, cotangent, and secant are negative.

$\sin(-1290°) = \sin 30° = \dfrac{1}{2}$

$\cos(-1290°) = -\cos 30° = -\dfrac{\sqrt{3}}{2}$

$\tan(-1290°) = -\tan 30° = -\dfrac{\sqrt{3}}{3}$

$\cot(-1290°) = -\cot 30° = -\sqrt{3}$

$\sec(-1290°) = -\sec 30° = -\dfrac{2\sqrt{3}}{3}$

$\csc(-1290°) = \csc 30° = 2$

57. 30°

$\tan 30° = \dfrac{\sqrt{3}}{3}$

$\cot 30° = \sqrt{3}$

58. 45°

$$\sin 45° = \frac{\sqrt{2}}{2}$$
$$\cos 45° = \frac{\sqrt{2}}{2}$$
$$\sec 45° = \sqrt{2}$$
$$\csc 45° = \sqrt{2}$$

59. 60°

$$\sin 60° = \frac{\sqrt{3}}{2}$$
$$\cot 60° = \frac{\sqrt{3}}{3}$$
$$\csc 60° = \frac{2\sqrt{3}}{3}$$

60. 120°

$$\cos(120°) = -\cos 60° = -\frac{1}{2}$$
$$\cot(120°) = -\cot 60° = -\frac{\sqrt{3}}{3}$$
$$\sec(120°) = -\sec 60° = -2$$

61. 135°

$$\tan 135° = -\tan 45° = -1$$
$$\cot 135° = -\cot 45° = -1$$

62. 150°

$$\sin(150°) = \sin 30° = \frac{1}{2}$$
$$\cot(150°) = -\cot 30° = -\sqrt{3}$$
$$\sec(150°) = -\sec 30° = -\frac{2\sqrt{3}}{3}$$

63. 210°

$$\cos(210°) = -\cos 30° = -\frac{\sqrt{3}}{2}$$
$$\sec(210°) = -\sec 30° = -\frac{2\sqrt{3}}{3}$$

64. 240°

$$\tan(240°) = \tan 60° = \sqrt{3}$$
$$\cot(240°) = \cot 60° = \frac{\sqrt{3}}{3}$$

65. Since the sine is negative, θ must lie in quadrant III or IV. Since the absolute value of $\sin\theta$ is $\frac{1}{2}$, the reference angle must be 30°. The quadrant III angle θ is $180° + 30° = 210°$, and the quadrant IV angle θ is $360° - 30° = 330°$.

66. Since the cosine is negative, θ must lie in quadrant II or III. Since the absolute value of $\cos\theta$ is $\frac{1}{2}$, the reference angle must be 60°. The quadrant II angle θ is $180° - 60° = 120°$, and the quadrant III angle θ is $180° + 60° = 240°$.

67. Since the tangent is positive, θ must lie in quadrant I or III. Since $\tan\theta = 1$, the reference angle must be 45°. The quadrant I angle θ is 45°, and the quadrant III angle θ is $180° + 45° = 225°$.

68. Since the cotangent is positive, θ must lie in quadrant I or III. Since $\cot\theta = \sqrt{3}$, the reference angle must be 30°. The quadrant I angle θ is 30°, and the quadrant III angle θ is $180° + 30° = 210°$.

69. Since the sine is positive, θ must lie in quadrant I or II. Since the value of $\sin\theta$ is $\frac{\sqrt{3}}{2}$, the reference angle must be 60°. The quadrant I angle θ is 60°, and the quadrant II angle θ is $180° - 60° = 120°$.

70. Since the cosine is positive, θ must lie in quadrant I or IV. Since the value of $\cos\theta$ is $\frac{\sqrt{3}}{2}$, the reference angle must be 30°. The quadrant I angle θ is 30°, and the quadrant IV angle θ is $360° - 30° = 330°$.

71. Since the secant is negative, θ must lie in quadrant II or III. Since the absolute value of $\sec\theta$ is 2, the reference angle must be 60°. The quadrant II angle θ is $180° - 60° = 120°$, and the quadrant III angle θ is $180° + 60° = 240°$.

72. Since the cosecant is negative, θ must lie in quadrant III or IV. Since the absolute value of $\csc\theta$ is 2, the reference angle must be 30°. The quadrant III angle θ is $180° + 30° = 210°$, and the quadrant IV angle θ is $360° - 30° = 330°$.

73. Since the sine is negative, θ must lie in quadrant III or IV. Since the absolute value of $\sin\theta$ is $\frac{\sqrt{2}}{2}$, the reference angle must be 45°. The quadrant III angle θ is $180° + 45° = 225°$, and the quadrant IV angle θ is $360° - 45° = 315°$.

74. Since the cosine is negative, θ must lie in quadrant II or III. Since the absolute value of $\cos\theta$ is $\frac{\sqrt{2}}{2}$, the reference angle must be 45°. The quadrant II angle θ is $180° - 45° = 135°$, and the quadrant III angle θ is $180° + 45° = 225°$.

75. Since the tangent is negative, θ must lie in quadrant II or IV. Since the $\tan\theta = -\sqrt{3}$, the reference angle must be 60°. The quadrant II angle θ is $180° - 60° = 120°$, and the quadrant IV° angle θ is $360° - 60° = 300°$.

76. Since the cotangent is negative, θ must lie in quadrant II or IV. Since $\cot\theta = -1$, the reference angle must be 45°. The quadrant II angle θ is $180° - 45° = 135°$, and the quadrant IV° angle θ is $360° - 45° = 315°$.

10.5 EXERCISES

For the following exercises, be sure you calculator is in degree mode. If your calculator accepts angles in degrees, minutes, and seconds, it is not necessary to change angles to decimal degrees. Keystroke sequences may vary on the type and/or model of calculator being used.

1. $\tan 29°\, 30'$

    $\tan 29°\, 30' = \tan\left(29 + \dfrac{30}{60}\right)°$
    $= \tan 29.5°$
    $\approx .5657728$

2. $\sin 38°\, 42'$

    $\sin 38°\, 42' = \sin\left(38 + \dfrac{42}{60}\right)°$
    $= \sin 38.7°$
    $\approx .6252427$

3. $\cot 41°\, 24'$

    $\cot 41°\, 24' = 1/\tan 41°\, 24'$
    $= 1/\tan\left(41 + \dfrac{24}{60}\right)°$
    $= 1/\tan 41.4°$
    $\approx 1/.8816186$
    ≈ 1.1342773

4. $\cos 27°\, 10'$

    $\cos 27°\, 10' = \cos\left(27 + \dfrac{10}{60}\right)°$
    $\approx \cos 27.2°$
    $\approx .8896822$

5. $\sec 13°\, 15'$

    $\sec 13°\, 15' = 1/\cos\left(13 + \dfrac{15}{60}\right)°$
    $= 1/\cos 13.25°$
    $\approx 1/.9733793$
    ≈ 1.0273488

6. $\csc 44°\, 30'$

    $\csc 44°\, 30' = 1/\sin\left(44 + \dfrac{30}{60}\right)°$
    $= 1/\sin 44.5°$
    $\approx 1/.7009093$
    ≈ 1.4267182

7. $\sin 39°\, 40'$

    $\sin 39°\, 40' = \sin\left(39 + \dfrac{40}{60}\right)°$
    $\approx \sin(39.6666667)°$
    $\approx .6383201$

8. $\tan 17°\, 12'$

    $\tan 17°\, 12' = \tan\left(17 + \dfrac{12}{60}\right)°$
    $= \tan 17.2°$
    $= .3095517$

9. $\csc 145°\, 45'$

    $\csc 145°\, 45' = 1/\sin\left(145 + \dfrac{45}{60}\right)°$
    $= 1/\sin 145.75°$
    $\approx 1/.5628049$
    ≈ 1.7768146

10. $\cot 183°\, 48'$

    $\cot 183°\, 48' = 1/\tan 183°\, 48'$
    $= 1/\tan\left(183 + \dfrac{48}{60}\right)°$
    $= 1/\tan 183.8°$
    $\approx 1/.0664199$
    ≈ 15.055723

11. $\cos 421°\, 30'$

    $\cos 421°\, 30' = \cos\left(421 + \dfrac{30}{60}\right)°$
    $= \cos 421.5°$
    $\approx .4771588$

12. $\sec 312°\, 12'$

    $\sec 312°\, 12' = 1/\cos\left(312 + \dfrac{12}{60}\right)°$
    $= 1/\cos 312.2°$
    $\approx 1/.6717206$
    ≈ 1.4887142

13. $\tan(-80°\, 6')$

    $\tan(-80°\, 6') = \tan\left(-\left(80 + \dfrac{6}{60}\right)\right)°$
    $= \tan(-80.1)°$
    ≈ -5.7297416

14. $\sin(-317°\, 36')$

    $\sin -317°\, 36' = \sin\left(-\left(317 + \dfrac{36}{60}\right)\right)°$
    $= \sin(-317.6)°$
    $\approx .6743024$

15. $\cot(-512°\,20')$

$\cot(-512°\,20') = 1/\tan(-512°\,20')$
$= 1/\tan\left(-512 + \dfrac{20}{60}\right)°$
$= 1/\tan(-512.33333°)$
$\approx 1/.5242698$
≈ 1.9074147

16. $\cos(-15')$

$\cos(-15') = \cos\left(-\dfrac{15}{60}\right)°$
$= \cos(-.25)°$
$\approx .9999905$

Depending upon your calculator use "arc," or "INV" followed by trig function. For some calculators you may use "$\sin^{-1}$" for arc sin, etc.

17. $\sin\theta = .84802194$

$\theta = \text{arc sin}(.84802194)$ or
$= \sin^{-1}(.84802194)$ or
$= \text{INV sin}(.84802194)$
$\approx 57.997172°$

18. $\tan\theta = 1.4739716$

$\theta = \tan^{-1}(1.4739716)$
$\approx 55.845496°$

19. $\sec\theta = 1.1606249$

Since $\cos\theta = 1/\sec\theta$,
$\cos\theta = 1/1.1606249$
$\approx .8616048$
$\theta = \text{arc cos}(.8616048)$ or
$= \cos^{-1}(.8616048)$ or
$= \text{INV cos}(.8616048)$
$\approx 30.502748°.$

20. $\cot\theta = 1.2575516$

Since $\tan\theta = 1/\cot\theta$,
$\tan\theta = 1/1.2575516$
$\approx .795195998$
$\theta = \tan^{-1}(.79519600)$
$\approx 38.491580°.$

21. $\sin\theta = .72144101$

$\theta = \text{arc sin}(.72144101)$ or
$= \sin^{-1}(.72144101)$ or
$= \text{INV sin}(.72144101)$
$\approx 46.173581°$

22. $\sec\theta = 2.7496222$

Since $\cos\theta = 1/\sec\theta$,
$\cos\theta = 1/2.7496222$
$\approx .3636863$
$\theta = \cos^{-1}(.3636863)$ or
$\approx 68.673241°.$

23. $\tan\theta = 6.4358841$

$\theta = \tan^{-1}(6.4358841)$
$\approx 81.168073°$

24. $\sin\theta = .27843196$

$\theta = \sin^{-1}(.27843196)$
$\approx 16.166641°$

25. $A = 36°20'$, $c = 964\,\text{m}$

$A + B = 90°$
$B = 90° - A$
$= 90° - 36°20'$
$= 89°60' - 36°20'$
$= 53°40'.$

$\sin A = \dfrac{a}{c}$
$a = c\sin A$
$a = 964\sin 36°20'$
Use a calculator and round answer to three significant digits.
$a \approx 571\,\text{m}$

$\cos A = \dfrac{b}{c}$
$b = c\cos A$
$b = 964\cos 36°20'$
Use a calculator and round answer to three significant digits.
$b \approx 777\,\text{m}$

10.5 APPLICATIONS FO RIGHT TRIANGLES 343

26. $A = 31°40'$, $a = 35.9$ km
$$B = 90° - A$$
$$= 90° - 31°40'$$
$$= 58°20'.$$

$$\tan A = \frac{a}{b}$$
$$b = \frac{a}{\tan A}$$
$$= \frac{35.9}{\tan 31°40'}$$

Use a calculator and round answer to three significant digits.
$$b \approx 58.2 \text{ km}$$

$$\sin A = \frac{a}{c}$$
$$c = \frac{a}{\sin A}$$
$$= \frac{35.9}{\sin 31°40'}$$

Use a calculator and round answer to three significant digits.
$$c \approx 68.4 \text{ km}$$

27. $N = 51.2°$, $m = 124$ m
$$M + N = 90°$$
$$M = 90° - N$$
$$= 90° - 51.2°$$
$$= 38.8°$$

$$\tan N = \frac{n}{m}$$
$$n = m \tan N$$
$$= 124 \cdot \tan 51.2°$$

Use a calculator and round answer to three significant digits.
$$n \approx 154 \text{ m}$$

$$\cos N = \frac{m}{p}$$
$$p = \frac{m}{\cos N}$$
$$= \frac{124}{\cos 51.2°}$$

Use a calculator and round answer to three significant digits.
$$p \approx 198 \text{ m}$$

28. $X = 47.8°$, $b = 89.6$ cm
$$Y = 90° - 47.8°$$
$$= 42.2°$$

$$\frac{x}{89.6} = \sin 47.8°$$
$$x = 89.6 \sin 47.8°$$
$$\approx 66.4 \text{ cm}$$

$$\frac{y}{89.6} = \cos 47.8°$$
$$y = 89.6 \cos 47.8°$$
$$\approx 60.2 \text{ cm}$$

29. $B = 42.0892°$, $z = 56.851$ cm
$$A + B = 90°$$
$$A = 90° - B$$
$$= 90° - 42.0892°$$
$$\approx 47.9108°$$

$$\sin B = \frac{b}{c}$$
$$c = \frac{b}{\sin B}$$
$$= \frac{56.851}{\sin 42.0892°}$$
$$\approx 84.816 \text{ cm}$$

$$\tan B = \frac{b}{a}$$
$$a = \frac{b}{\tan B}$$
$$= \frac{56.851}{\tan 42.0892°}$$
$$\approx 62.942 \text{ cm}$$

30. $B = 68.5142°$, $c = 3579.42$ m
$$A = 90° - 68.5142°$$
$$= 21.4858°$$

$$\frac{a}{3579.42} = \cos 68.5142°$$
$$a = 3579.42 \cos 68.5142°$$
$$\approx 1311.04 \text{ m}$$

$$b = 3579.42 \sin 68.5142°$$
$$\approx 3330.68 \text{ m}$$

31. $A = 28.00°$, $c = 17.4$ ft

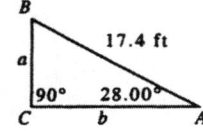

$A + B = 90°$
$B = 90° - A$
$ = 90° - 28.00°$
$ = 62.00°$

$\sin A = \dfrac{a}{c}$
$a = c \sin A$
$ = 17.4 \sin 28.00°$
$ \approx 8.17$ ft

$\cos A = \dfrac{b}{c}$
$b = c \cos A$
$ = 17.4 \cos 28.00°$
$ \approx 15.4$ ft

32. $B = 46.00°$, $c = 29.7$ m

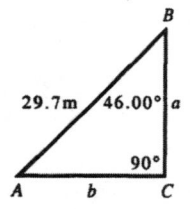

$A = 90° - 46.00°$
$ = 44.00°$

$\cos 46.00° = \dfrac{a}{29.7}$
$a = 29.7 \cos 46.00°$
$ \approx 20.6$ m

$\sin 46.00° = \dfrac{b}{29.7}$
$b = 29.7 \sin 46.00°$
$ = 17.4 \cos 28.00°$
$ \approx 21.4$ m

33. $B = 73.00°$, $b = 128$ in.

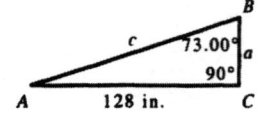

$A = 90° - 73.00°$
$ = 17.00°$

$\tan 73.00° = \dfrac{128}{a}$
$a = \dfrac{128}{\tan 73.00°}$
$ \approx 39.1$ in.

$\sin 73.00° = \dfrac{128}{c}$
$c = \dfrac{128}{\sin 73.00°}$
$ \approx 134$ in.

34. $A = 61°00'$, $b = 39.2$ cm

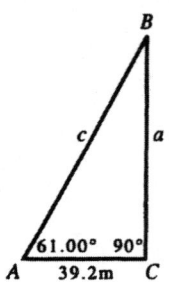

$B = 90° - 61°00'$
$ = 29°00'$

$\tan 61°00' = \dfrac{a}{39.2}$
$a = 39.2 \tan 61°00'$
$ \approx 70.7$ cm

$\sec 61°00' = \dfrac{c}{39.2}$
$c = 39.2 \sec 61°00'$
$ \approx 80.9$ cm

35. $a = 76.4$ yd, $b = 39.3$ yd

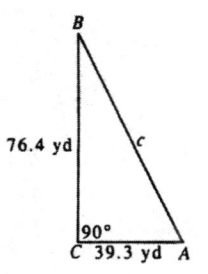

$c^2 = a^2 + b^2$
$c = \sqrt{a^2 + b^2}$
$= \sqrt{(76.4)^2 + (39.3)^2}$
$= 85.9$ yd

$\tan A = \dfrac{76.4}{39.3}$
$A \approx 62°50'$

$\tan B = \dfrac{39.3}{76.4}$
$B \approx 27°10'$

36. $a = 958$ m, $b = 489$ m

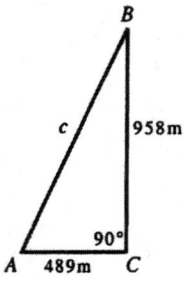

$\tan A = \dfrac{958}{489}$
$A \approx 63.0° = 63°00'$

$\tan B = \dfrac{489}{958}$
$B \approx 27.0° = 27°00'$

Check: $A + B = 90°$
$27° + 63° = 90°$

To find c from the given information, use the Pythagorean theorem.

$c^2 = a^2 + b^2$
$c = \sqrt{a^2 + b^2}$
$= \sqrt{(958)^2 + (489)^2}$
$= \sqrt{1,156,885}$
≈ 1080 m

Another way to find c is to use the given information together with the result for one of the acute angles.

$\csc 63.0° = \dfrac{c}{958}$
$c = 958 \csc 63.0°$
$= 1080$ m

37. $a = 18.9$ cm, $c = 46.3$ cm

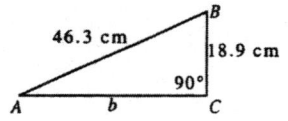

$c^2 = a^2 + b^2$
$b = \sqrt{c^2 - a^2}$
$= \sqrt{(46.3)^2 - (18.9)^2}$
≈ 42.3 cm

$\sin A = \dfrac{18.9}{46.3}$
$A \approx 24°10'$

$\cos B = \dfrac{18.9}{46.3}$
$B \approx 65°50'$

38. $b = 219$ m, $c = 647$ m

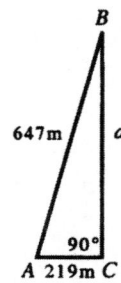

$$a = \sqrt{c^2 - b^2}$$
$$= \sqrt{647^2 - 219^2}$$
$$= \sqrt{418,609 - 47,961}$$
$$= \sqrt{370,648}$$
$$\approx 609 \text{ m}$$

$$\cos A = \frac{219}{647}$$
$$A \approx 70.2° = 70°10'$$

$$\sin B = \frac{219}{647}$$
$$B \approx 19.8° = 19°50'$$

Check that $A + B = 90°$

39. $A = 53°24'$, $c = 387.1$ ft

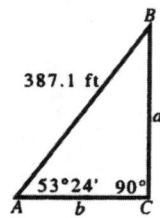

$$A + B = 90°$$
$$B = 90° - A$$
$$= 89°60' - 53°24'$$
$$\approx 36°36'$$

$$\sin 53°24' = \frac{a}{387.1}$$
$$a = 387.1 \sin 53°24'$$
$$\approx 310.8 \text{ ft}$$

$$\cos 53°24' = \frac{b}{387.1}$$
$$b = 387.1 \cos 53°24'$$
$$= 17.4 \cos 28.00°$$
$$\approx 230.8 \text{ ft}$$

40. $A = 13°47'$, $c = 1285$ m

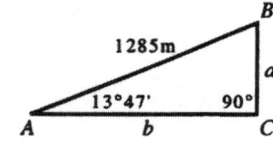

$$B = 90° - A$$
$$= 89°60' - 13°47'$$
$$\approx 76°13'$$

$$\sin 13°47' = \frac{a}{1285}$$
$$a = 1285 \sin 13°47'$$
$$\approx 306.2 \text{ m}$$

$$\cos 13°47' = \frac{b}{1285}$$
$$b = 1285 \cos 13°47'$$
$$\approx 1248 \text{ m}$$

41. $B = 39°9'$, $c = .6231$ m

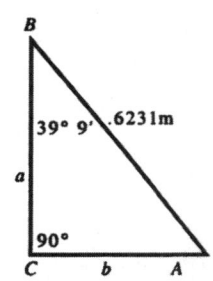

$$A = 90° - 39°9'$$
$$= 50°51'$$

$$\cos 39°9' = \frac{a}{.6231}$$
$$a = (.6231)\cos 39°9'$$
$$\approx .4832 \text{ m}$$

$$\sin 39°9' = \frac{b}{.6231}$$
$$b = (.6231)\sin 39°9'$$
$$\approx .3934 \text{ m}$$

42. $B = 82°51'$, $c = 4.825$ cm

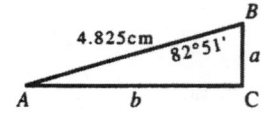

$$A = 90° - 82°51'$$
$$= 7°9'$$

$$\cos 82°51' = \frac{a}{4.825}$$
$$a = (4.825)\cos 82°51'$$
$$\approx .6006 \text{ cm}$$

$$\sin 82°51' = \frac{b}{4.825}$$
$$b = (4.825)\sin 82°51'$$
$$\approx 4.787 \text{ cm}$$

10.5 APPLICATIONS FO RIGHT TRIANGLES   347

43. Let $h = $ the distance the ladder goes up the wall.

$$\sin 43°50' = \frac{h}{13.5}$$
$$h = 13.5 \sin 43°50'$$
$$\approx 9.3496000$$

The ladder goes up the wall 9.35 m.

44. Let $\theta = $ the angle the guy wire makes with the ground.

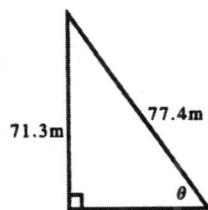

$$\sin \theta = \frac{71.3}{77.4}$$
$$\theta \approx 67.100475°$$
$$\approx 67° + (.100475)(60')$$
$$\approx 67°10'$$

The guy wire makes an angle of $67°10'$ with the ground.

45. Let $x = $ the length of the guy wire.

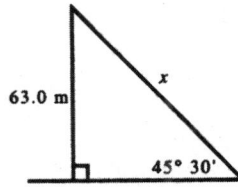

$$\sin 45°30' = \frac{63.0}{x}$$
$$x \sin 45°30' = 63.0$$
$$x = \frac{63.0}{\sin 45°30'}$$
$$\approx 88.328020$$

The length of the guy wire is 88.3 m.

46. Since angle $T = 32°10'$ and angle $S = 57°50'$, $S + T = 90°$, and triangle RST is a right triangle.

$$\tan 32°10' = \frac{RS}{53.1}$$
$$RS = 53.1 \tan 32°10'$$
$$\approx 33.395727$$

The distance across the lake is 33.4 m.

47. The two right triangles are congruent. So corresponding sides are congruent. Since the sides that compose the base of the isosceles triangle are congruent, each side is $\frac{1}{2}(42.36)$, or 21.18 in. Let $x = $ the length of each of the two equal sides of the isosceles triangle.

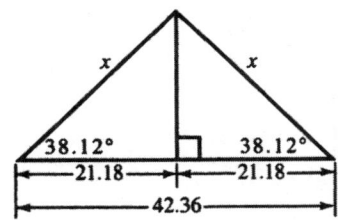

$$\cos 38.12° = \frac{21.18}{x}$$
$$x\cos 38.12° = 21.18$$
$$x = \frac{21.18}{\cos 38.12°}$$
$$\approx 26.921918$$

The length of each of the two equal sides of the triangle is 26.92 in.

48. The altitude of an isosceles triangle bisects the base as well as the angle opposite the base. Let $h = $ the altitude. The base is 184.2 cm so the altitude forms two congruent right triangles each with a leg of $\frac{1}{2}(184.2)$, or 92.1 cm and with an opposite angle of $\frac{1}{2}(68°44') = 34°22'$.

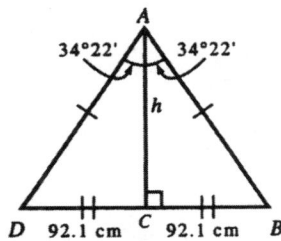

In triangle ABC,

$$\tan 34°22' = \frac{92.1}{h}$$
$$h\tan 34°22' = 92.1$$
$$h = \frac{92.1}{\tan 34°22'}$$
$$\approx 134.67667.$$

The altitude of the triangle is 134.7 cm.

49. $$\tan 30.0° = \frac{y}{1000}$$
$$y = 1000 \tan 30.0°$$
$$\approx 577$$

However, the observer's eye-height is 6 feet from the ground, so the cloud ceiling is $577 + 6 = 583$ feet.

50. let $x = $ the length of the shadow.

$$\tan 23.4° = \frac{5.75}{x}$$
$$x \tan 23.4° = 5.75$$
$$x = \frac{5.75}{\tan 23.4°}$$
$$\approx 13.287466$$

The length of the shadow is 13.3 ft.

51. Let $h =$ height of the tower.

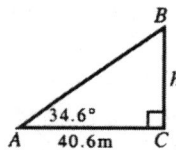

In triangle ABC,

$$\tan 34.6° = \frac{h}{40.6}$$
$$h = 40.6 \tan 34.6°$$
$$h \approx 28.0.$$

The height of the tower is 28.0 m.

52. Let $\theta =$ the angle of elevation of the sun.

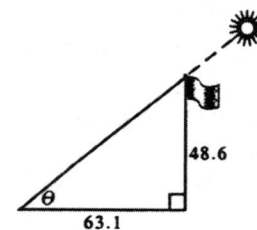

$$\tan \theta = \frac{48.6}{63.1}$$
$$\theta \approx 37.603681°$$
$$\approx 37° + (.603681)(60')$$
$$\approx 37°40'$$

The angle of elevatgion of the sun is $37°40'$.

53. Let $d =$ the distance from the top B of the building to the point on the ground A.

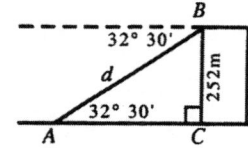

In triangle ABC,

$$\sin 32°30' = \frac{252}{d}$$
$$d = \frac{252}{\sin 32°30'}$$
$$\approx 469.$$

The distance from the top of the building to the point on the ground is 469 m.

54. Let $x =$ the horizontal distance that the plane must fly to be directly over the tree.

$$\tan 13°50' = \frac{10500}{x}$$
$$x \tan 13°50' = 10500$$
$$x = \frac{10500}{\tan 13°50'}$$
$$\approx 42600$$

The horizontal distance that the plane must fly to be directly over the tree is 42,600 ft.

55. Let $x =$ the height of the taller building,
    $h =$ the difference in height between the shorter and taller buildings, and
    $d =$ the distance between the buldings along the ground.

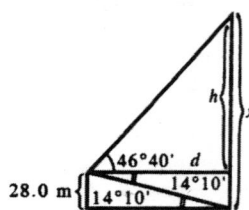

$$\frac{d}{28.0} = \cot 14°10'$$
$$d = (28.0)\cot 14°10'$$
$$= 110.9262493$$
$$\approx 111 \text{ m}$$

$$\frac{h}{d} = \tan 46°40'$$
$$h = d \tan 46°40'$$
$$h = (110.9262493) \tan 46°40'$$
$$\approx 118 \text{ m}$$

$$x = h + 28.0$$
$$= 118 + 28.0$$
$$= 146 \text{ m}$$

The height of the taller building is 146 m.

56. Let $a =$ the angle of depression that the lens makes with the horizontal, and
    $\theta =$ the angle that the lens makes with the vertical.

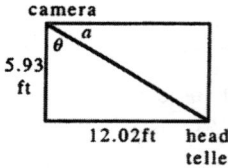

$$\tan \theta = \frac{12.02}{5.93}$$
$$\theta \approx 63°40'$$

$$a = 90° - \theta$$
$$= 90° - 63°40'$$
$$\approx 26°20'$$

The lens should make an angle of $26°20'$ with the horizontal.

57. Let $x =$ the distance from the assigned target.

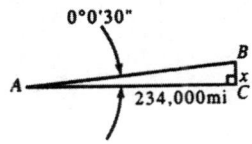

In triangle ABC,
$$\tan 0°0'30'' = \frac{x}{234000}$$
$$x = 234000 \tan 0°0'30''$$
$$x \approx 34.0$$

The distnce from the assigned target is 34.0 mi.

58. (a) The height h of the peak above 14,545 is
$$\sin 5.82° = \frac{h}{(27.0134)5280}$$
$$h = (27.0134)5280 \sin 5.82°$$
$$\approx 14463 \text{ ft}$$

The total height is about
$14,545 \text{ ft} + 14,463 \text{ ft} = 29,008 \text{ ft}.$

(b) The curvature of the earth would make the peak appear <u>shorter</u> than it actually is. Initially, the surveyors did not think Mt. Everest was the tallest peak in the Himalayas. It did not look like the tallest peak because it was farther away than the other large peaks.

59. $a = \frac{1}{2}(24)$, so $a = 12$.
$b = a\sqrt{3}$, so $b = 12\sqrt{3}$.
$d = b$, so $d = 12\sqrt{3}$.
$c = d\sqrt{2}$, so $c = \left(12\sqrt{3}\right)\left(\sqrt{2}\right) = 12\sqrt{6}$.

60. $y = \frac{1}{2}(9)$, so $y = \frac{9}{2}$.
$x = y\sqrt{3}$, so $x = \frac{9\sqrt{3}}{2}$.
$y = z\sqrt{3}$, so
$z = \frac{y}{\sqrt{3}} = \frac{\frac{9}{2}}{\sqrt{3}} = \frac{9\sqrt{3}}{6} = \frac{3\sqrt{3}}{2}.$
$w = 2z$, so $w = 2\left(\frac{3\sqrt{3}}{2}\right) = 3\sqrt{3}.$

10.6 EXERCISES

1. $A = 37°$, $B = 48°$, $c = 18$ m
$C = 180° - A - B$
$= 180° - 37° - 48°$
$= 95°$

$\frac{b}{\sin B} = \frac{c}{\sin C}$
$b = \frac{c \sin B}{\sin C}$
$= \frac{18 \sin 48°}{\sin 95°}$
≈ 13 m

$\frac{a}{\sin A} = \frac{c}{\sin C}$
$a = \frac{c \sin A}{\sin C}$
$= \frac{18 \sin 37°}{\sin 95°}$
≈ 11 m

2. $B = 52°$, $C = 29°$, $a = 43$ cm
$A = 180° - B - C$
$= 180° - 52° - 29°$
$= 99°$

$\frac{a}{\sin A} = \frac{b}{\sin B}$
$b = \frac{a \sin B}{\sin A}$
$= \frac{43 \sin 52°}{\sin 99°}$
≈ 34 cm

$\frac{a}{\sin A} = \frac{c}{\sin C}$
$c = \frac{a \sin C}{\sin A}$
$= \frac{43 \sin 29°}{\sin 99°}$
≈ 21 cm

3. $A = 27.2°$, $C = 115.5°$, $c = 76.0$ ft
$B = 180° - A - C$
$= 180° - 27.2° - 115.5°$
$= 37.3°$

$\frac{a}{\sin A} = \frac{c}{\sin C}$
$a = \frac{c \sin A}{\sin C}$
$= \frac{76.0 \sin 27.2°}{\sin 115.5°}$
≈ 38.5 ft

$\frac{b}{\sin B} = \frac{c}{\sin C}$
$b = \frac{c \sin B}{\sin C}$
$= \frac{76.0 \sin 37.3°}{\sin 115.5°}$
≈ 51.0 ft

4. $C = 124.1°, B = 18.7°, b = 94.6$ m
$$A = 180° - B - C$$
$$= 180° - 18.7° - 124.1°$$
$$= 37.2°$$
$$\frac{a}{\sin A} = \frac{b}{\sin B}$$
$$a = \frac{b \sin A}{\sin B}$$
$$= \frac{94.6 \sin 37.2°}{\sin 18.7°}$$
$$\approx 178 \text{ m}$$
$$\frac{c}{\sin C} = \frac{b}{\sin B}$$
$$c = \frac{b \sin C}{\sin B}$$
$$= \frac{94.6 \sin 124.1°}{\sin 18.7°}$$
$$\approx 244 \text{ m}$$

5. $A = 68.41°, B = 54.23°, a = 12.75$ ft
$$C = 180° - A - B$$
$$= 180° - 68.41° - 54.23°$$
$$\approx 57.36°$$
$$\frac{a}{\sin A} = \frac{b}{\sin B}$$
$$b = \frac{a \sin B}{\sin A}$$
$$= \frac{12.75 \sin 54.23°}{\sin 68.41°}$$
$$\approx 11.13 \text{ ft}$$
$$\frac{a}{\sin A} = \frac{c}{\sin C}$$
$$c = \frac{a \sin C}{\sin A}$$
$$= \frac{12.75 \sin 57.36°}{\sin 68.41°}$$
$$\approx 11.55 \text{ ft}$$

6. $C = 74.08°, B = 69.38°, c = 45.38$ m
$$A = 180° - B - C$$
$$= 180° - 69.38° - 74.08°$$
$$\approx 36.54°$$

$$\frac{a}{\sin A} = \frac{c}{\sin C}$$
$$a = \frac{c \sin A}{\sin C}$$
$$= \frac{45.38 \sin 36.54°}{\sin 74.08°}$$
$$\approx 28.10 \text{ m}$$
$$\frac{b}{\sin B} = \frac{c}{\sin C}$$
$$b = \frac{c \sin B}{\sin C}$$
$$= \frac{45.38 \sin 69.38°}{\sin 74.08°}$$
$$\approx 44.17 \text{ m}$$

7. $A = 87.2°, b = 75.9$ yd, $C = 74.3°$
$$B = 180° - A - C$$
$$= 180° - 87.2° - 74.3°$$
$$\approx 18.5°$$
$$\frac{a}{\sin A} = \frac{b}{\sin B}$$
$$a = \frac{b \sin A}{\sin B}$$
$$= \frac{75.9 \sin 87.2°}{\sin 18.5°}$$
$$\approx 239 \text{ yd}$$
$$\frac{b}{\sin B} = \frac{c}{\sin C}$$
$$c = \frac{b \sin C}{\sin B}$$
$$= \frac{75.9 \sin 74.3°}{\sin 18.5°}$$
$$\approx 230 \text{ yd}$$

8. $B = 38°40', a = 19.7$ cm, $C = 91°40'$
$$A = 180° - B - C$$
$$= 180° - 38°40' - 91°40'$$
$$= 49°40'$$
$$\frac{b}{\sin B} = \frac{a}{\sin A}$$
$$b = \frac{a \sin B}{\sin A}$$
$$= \frac{19.7 \sin 38°40'}{\sin 49°40'}$$
$$\approx 16.1 \text{ cm}$$
$$\frac{c}{\sin C} = \frac{a}{\sin A}$$
$$c = \frac{a \sin C}{\sin A}$$
$$= \frac{19.7 \sin 91°40'}{\sin 49°40'}$$
$$\approx 25.8 \text{ cm}$$

9. $B = 20°50'$, $AC = 132$ ft, $C = 103°10'$
$$A = 180° - B - C$$
$$= 180° - 20°50' - 103°10'$$
$$= 56°00'$$
$$\frac{AC}{\sin B} = \frac{AB}{\sin C}$$
$$AB = \frac{AC \sin C}{\sin B}$$
$$= \frac{132 \sin 103°10'}{\sin 20°50'}$$
$$\approx 361 \text{ ft}$$
$$\frac{BC}{\sin A} = \frac{AC}{\sin B}$$
$$BC = \frac{AC \sin A}{\sin B}$$
$$= \frac{132 \sin 56°00'}{\sin 20°50'}$$
$$\approx 308 \text{ ft}$$

10. $A = 35.3°$, $B = 52.8°$, $AC = 675$ ft
$$C = 180° - A - B$$
$$= 180° - 35.3° - 52.8°$$
$$= 91.9°$$
$$\frac{BC}{\sin A} = \frac{AC}{\sin B}$$
$$BC = \frac{AC \sin A}{\sin B}$$
$$= \frac{675 \sin 35.3°}{\sin 52.8°}$$
$$\approx 490 \text{ ft}$$
$$\frac{AB}{\sin C} = \frac{AC}{\sin B}$$
$$AB = \frac{AC \sin C}{\sin B}$$
$$= \frac{675 \sin 91.9°}{\sin 52.8°}$$
$$\approx 847 \text{ ft}$$

11. $A = 39.70°$, $C = 30.35°$, $b = 39.74$ m
$$B = 180° - A - C$$
$$= 180° - 39.70° - 30.35° = 109.95°$$
$$\approx 110.0°$$

$$\frac{a}{\sin A} = \frac{b}{\sin B}$$
$$a = \frac{b \sin A}{\sin B}$$
$$= \frac{39.74 \sin 39.70°}{\sin 109.95°}$$
$$\approx 27.01 \text{ m}$$
$$\frac{b}{\sin B} = \frac{c}{\sin C}$$
$$c = \frac{b \sin C}{\sin B}$$
$$= \frac{39.74 \sin 30.35°}{\sin 109.95°}$$
$$\approx 21.36 \text{ m}$$

12. $C = 71.83°$, $B = 42.57°$, $a = 2.614$ cm
$$A = 180° - B - C$$
$$= 180° - 42.57° - 71.83°$$
$$= 65.60°$$
$$\frac{b}{\sin B} = \frac{a}{\sin A}$$
$$b = \frac{a \sin B}{\sin A}$$
$$= \frac{2.614 \sin 42.57°}{\sin 65.60°}$$
$$\approx 1.942 \text{ cm}$$
$$\frac{c}{\sin C} = \frac{a}{\sin A}$$
$$c = \frac{a \sin C}{\sin A}$$
$$= \frac{2.614 \sin 71.83°}{\sin 65.60°}$$
$$\approx 2.727 \text{ cm}$$

13. $B = 42.88°$, $C = 102.40°$, $b = 3974$ ft
$$A = 180° - B - C$$
$$= 180° - 42.88° - 102.40°$$
$$\approx 34.72°$$
$$\frac{a}{\sin A} = \frac{b}{\sin B}$$
$$a = \frac{b \sin A}{\sin B}$$
$$= \frac{3974 \sin 34.72°}{\sin 42.88°}$$
$$\approx 3326 \text{ ft}$$
$$\frac{b}{\sin B} = \frac{c}{\sin C}$$
$$c = \frac{b \sin C}{\sin B}$$
$$= \frac{3974 \sin 102.40°}{\sin 42.88°}$$
$$\approx 5704 \text{ ft}$$

14. $A = 18.75°, B = 51.53°, c = 2798$ yd
$C = 180° - A - B$
$= 180° - 18.75° - 51.53°$
$= 109.72°$

$\dfrac{a}{\sin A} = \dfrac{c}{\sin C}$
$a = \dfrac{c \sin A}{\sin C}$
$= \dfrac{2798 \sin 18.75°}{\sin 109.72°}$
≈ 955.4 yd

$\dfrac{b}{\sin B} = \dfrac{c}{\sin C}$
$b = \dfrac{c \sin B}{\sin C}$
$= \dfrac{2798 \sin 51.53°}{\sin 109.72°}$
≈ 2327 yd

15. $A = 39°54', a = 268.7$m, $B = 42°32'$
$C = 180° - A - B$
$= 180° - 39°54' - 42°32'$
$= 97°34'$

$\dfrac{a}{\sin A} = \dfrac{b}{\sin B}$
$b = \dfrac{a \sin B}{\sin A}$
$= \dfrac{268.7 \sin 42°32'}{\sin 39°54'}$
≈ 283.2 m

$\dfrac{a}{\sin A} = \dfrac{c}{\sin C}$
$c = \dfrac{a \sin C}{\sin A}$
$= \dfrac{268.7 \sin 97°34'}{\sin 39°54'}$
≈ 415.2 m

16. $C = 79°18', c = 39.81$ mm, $A = 32°57'$
$B = 180° - A - C$
$= 180° - 32°57' - 79°18'$
$= 67°45'$

$\dfrac{a}{\sin A} = \dfrac{c}{\sin C}$
$a = \dfrac{c \sin A}{\sin C}$
$= \dfrac{39.81 \sin 32°57'}{\sin 79°18'}$
≈ 22.04 mm

$\dfrac{b}{\sin B} = \dfrac{c}{\sin C}$
$b = \dfrac{c \sin B}{\sin C}$
$= \dfrac{39.81 \sin 67°45'}{\sin 79°18'}$
≈ 37.50 mm

17.

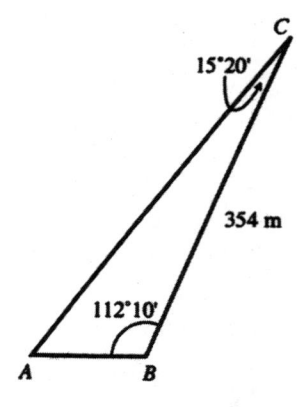

$A = 180° - B - C$
$= 180° - 112°10' - 15°20'$
$= 52°30'$

$\dfrac{BC}{\sin A} = \dfrac{AB}{\sin C}$
$AB = \dfrac{BC \sin C}{\sin A}$
$= \dfrac{354 \sin 15°20'}{\sin 52°30'}$
≈ 118 m

18.

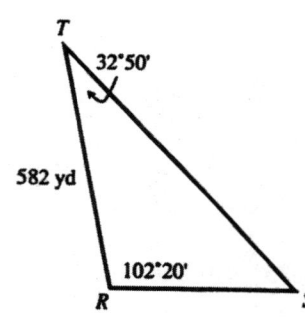

$S = 180° - 32°50' - 102°20'$
$= 44°50'$

$\dfrac{RS}{\sin 32°50'} = \dfrac{582}{\sin 44°50'}$
$RS \approx 448$ yd

19. $S = 180° - 32°50' - 102°20'$
$= 44°50'$
$$\frac{x}{\sin 54.8°} = \frac{12.0}{\sin 70.4°}$$
$x \approx 10.4$ in.

20. Label α in the triangle as shown.

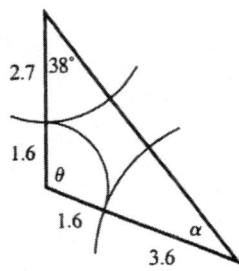

$$\frac{1.6 + 3.6}{\sin 38°} = \frac{1.6 + 2.7}{\sin \alpha}$$
$$\sin a = \frac{(1.6 + 2.7)\sin 38°}{1.6 + 3.6}$$
$\alpha = 31°$
$\theta = 180° - 38° - \alpha$
$= 180° - 38° - 31°$
$\theta = 111°$

21. Label the centers of the atoms A, B, amd C.

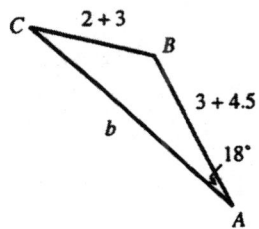

$a = 2.0 + 3.0 = 5.0$
$c = 3.0 + 4.5 = 7.5$
$$\frac{\sin C}{c} = \frac{\sin A}{a}$$
$$\sin C = \frac{(7.5)\sin 18°}{5}$$
$C \approx 28°$

$B = 180° - 18° - 28°$
$= 134°$
$$\frac{b}{\sin B} = \frac{a}{\sin A}$$
$$b = \frac{(5.0)\sin 134°}{\sin 18°}$$
≈ 12

The distance between the centers of atoms A and C is 12.

22. Let $A = $ the location of the ballon;
$B = $ the location of the farther town;
$C = $ the location of the closer town.

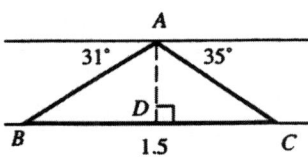

Angle $ABC = 31°$ and angle $ACB = 35°$ because the angles of depression are alternate interior angles with the angles of the triangle.

Angle $BAC = 180° - 31° - 35° = 114°$
$$\frac{\sin BAC}{1.5} = \frac{\sin ACB}{AB}$$
$$AB = \frac{(1.5)\sin 35°}{\sin 114°}$$
$= .9417863636$
$\approx .94$

$$\sin ABC = \frac{AD}{AB}$$
$AD = (.9417863636)\sin 31°$
$\approx .49$

The balloon is .49 mi above the ground.

23. By the law of cosines,
$a^2 = 3^2 + 8^2 - 2(3)(8)\cos 60°$
$= 9 + 64 - 48\left(\frac{1}{2}\right)$
$= 73 - 24$
$a^2 = 49$
$a = 7$.

24. By the law of cosines,
$a^2 = 1^2 + \left(4\sqrt{2}\right)^2 - 2(1)\left(4\sqrt{2}\right)\cos 45°$
$= 1 + 32 - 8\sqrt{2}\left(\frac{\sqrt{2}}{2}\right)$
$= 33 - 8$
$a^2 = 25$
$a = 5$.

25. $\cos \theta = \dfrac{b^2 + c^2 - a^2}{2bc}$
$= \dfrac{1^2 + \left(\sqrt{3}\right)^2 - 1^2}{2(1)\left(\sqrt{3}\right)}$
$= \dfrac{1 + 3 - 1}{2\sqrt{3}}$
$= \dfrac{3}{2\sqrt{3}}$
$= \dfrac{\sqrt{3}}{2}$
$\theta = 30°$

26. $\cos\theta = \dfrac{b^2+c^2-a^2}{2bc}$
$= \dfrac{3^2+5^2-7^2}{2(3)(5)}$
$= \dfrac{9+25-49}{30}$
$= \dfrac{-15}{30}$
$= -\dfrac{1}{2}$
$\theta = 120°$

27. $C = 28.3°$, $b = 5.71$ in., $a = 4.21$ in.
$c^2 = a^2 + b^2 - 2ab\cos C$
$= (4.21)^2 + (5.71)^2 - 2(4.21)(5.71)\cos 28.3°$
$c \approx \sqrt{7.9964337}$
≈ 2.83 in.

Keep all digits of $\sqrt{c^2}$ in the calculator for use in the next calculation. If 2.83 is used, the answer will vary sightly due to round-off error. Find angle A next, since it is the smaller angle and must be acute.

$\sin A = \dfrac{a\sin C}{c} = \dfrac{(4.21)\sin 28.3°}{\sqrt{c^2}}$
$\approx .70581857$
$A = 44.9°$
$B = 180° - 44.9° - 28.3°$
$= 106.8°$

28. $A = 41.4°$, $b = 2.78$ yd, $c = 3.92$ yd
$a^2 = b^2 + c^2 - 2bc\cos A$
$= (2.78)^2 + (3.92)^2 - 2(2.78)(3.92)\cos 41.4°$
$a \approx \sqrt{6.7459792}$
≈ 2.60 yd

Keep all digits of $\sqrt{a^2}$ in the calculator for use in the next calculation. If 2.60 is used, the answer will vary sightly due to round-off error. Find angle B next, since it is the smaller angle and must be acute.

$\dfrac{\sin B}{2.78} = \dfrac{\sin 41.4°}{\sqrt{a^2}}$
$B = 45.1°$
$C = 180° - 41.4° - 45.1°$
$= 93.5°$

29. $C = 45.6°$, $b = 8.94$ m, $a = 7.23$ m
$c^2 = a^2 + b^2 - 2ab\cos C$
$= (7.23)^2 + (8.94)^2 - 2(7.23)(8.94)\cos 45.6°$
$c \approx \sqrt{41.74934078}$
≈ 6.46 m

Find angle A next, since it is the smaller angle and must be acute.

$\sin A = \dfrac{a\sin C}{c} = \dfrac{(7.23)\sin 45.6°}{\sqrt{c^2}}$
$\approx .79946437$
$A = 53.1°$
$B = 180° - 53.1° - 45.6°$
$= 81.3°$

30. $A = 67.3°$, $b = 37.9$ km, $c = 40.8$ km
$a^2 = b^2 + c^2 - 2bc\cos A$
$= (37.9)^2 + (40.8)^2 - 2(37.9)(40.8)\cos 67.3°$
$a \approx \sqrt{1907.581537}$
≈ 43.7 km

Find angle B next, since it is the smaller angle and must be acute.

$\dfrac{\sin B}{37.9} = \dfrac{\sin 67.3°}{\sqrt{a^2}}$
$B = 53.2°$
$C = 180° - 67.3° - 53.2°$
$= 59.5°$

31. $A = 80°40'$, $b = 143$ cm, $c = 89.6$ cm
$a^2 = b^2 + c^2 - 2bc\cos A$
$= 143^2 + (89.6)^2 - 2(143)(89.6)\cos 80°40'$
$a \approx \sqrt{24321.25341}$
≈ 156 cm

Find angle C next, since it is the smaller angle and must be acute.

$\sin C = \dfrac{c\sin A}{a} = \dfrac{(89.6)\sin 80°40'}{\sqrt{a^2}}$
$\approx .56692713$
$C = 34°30'$
$B = 180° - 80°40' - 34°30'$
$= 64°50'$

32. $C = 72°40'$, $a = 327$ ft, b = 251 ft
$c^2 = a^2 + b^2 - 2ab\cos C$
$= 327^2 + 251^2 - 2(327)(251)\cos 72°40'$
$c \approx 348$ ft

Find angle B next, since it is the smaller angle and must be acute.

$\dfrac{\sin B}{251} = \dfrac{\sin 72°40'}{\sqrt{c^2}}$
$B = 43°30'$
$A = 180° - B - C$
$= 63°50'$

10.6 THE LAWS OF SINES AND COSINES; AREA FORMULAS

33. $B = 74.80°$, $a = 8.919$ in., $c = 6.427$ in.

$b^2 = a^2 + c^2 - 2ac\cos B$
$= (8.919)^2 + (6.427)^2 - 2(8.919)(6.427)\cos 74.80°$
$b \approx 9.529$ in.

Find angle C next, since it is the smaller angle and must be acute.

$\sin C = \dfrac{c\sin B}{b} = \dfrac{(6.427)\sin 74.80°}{\sqrt{b^2}}$
$\approx .65089219$
$C \approx 40.61°$
$A = 180° - 74.80° - 40.61°$
$= 64.59°$

34. $C = 59.70°$, $a = 3.725$ mi, $b = 4.698$ mi

$c^2 = a^2 + b^2 - 2ab\cos C$
$= (3.725)^2 + (4.698)^2 - 2(3.725)(4.698)\cos 59.70°$
$c \approx 4.276$ mi

Find angle A next, since it is the smaller angle and must be acute.

$\dfrac{\sin A}{3.725} = \dfrac{\sin 59.70°}{\sqrt{c^2}}$
$A = 48.77°$
$B = 180° - 48.77° - 59.70°$
$= 71.53°$

35. $A = 112.8°$, $b = 6.28$ m, $c = 12.2$ m

$a^2 = b^2 + c^2 - 2bc\cos A$
$= (6.28)^2 + (12.2)^2 - 2(6.28)(12.2)\cos 112.8°$
$a \approx 15.7$ m

Angle A is obtuse, so both B and C are acute. Find either angle next.

$\sin B = \dfrac{b\sin A}{a} = \dfrac{(6.28)\sin 112.8°}{\sqrt{a^2}}$
$\approx .36787456$
$B = 21.6°$
$C = 180° - 112.8° - 21.6°$
$= 45.6°$

36. $B = 168.2°$, $a = 15.1$ cm, $c = 19.2$ cm

$b^2 = a^2 + c^2 - 2ac\cos B$
$= (15.1)^2 + (19.2)^2 - 2(15.1)(19.2)\cos 168.2°$
$b \approx 34.1$ cm

Find angle A next, since it is the smaller angle and must be acute.

$\dfrac{\sin A}{15.1} = \dfrac{\sin 168.2°}{\sqrt{b^2}}$
$A = 5.2°$
$C = 180° - 5.2° - 168.2°$
$= 6.6°$

37. $a = 3.0$ ft, $b = 5.0$ ft, $c = 6.0$ ft

Angle C is the largest, so find it first.

$c^2 = a^2 + b^2 - 2ab\cos C$
$\cos C = \dfrac{a^2 + b^2 - c^2}{2ab}$
$= \dfrac{3.0^2 + 5.0^2 - 6.0^2}{2(3.0)(5.0)}$
$\approx -.06666667$
$C = 94°$
$\sin B = \dfrac{b\sin C}{c}$
$\approx .83147942$
$B = 56°$
$A = 180° - 56° - 94°$
$= 30°$

38. $a = 4.0$ ft, $b = 5.0$ ft, $c = 8.0$ ft

Find the largest angle first, in case it is obtuse. Since c is the longest side, angle C will be the largest angle.

$c^2 = a^2 + b^2 - 2ab\cos C$
$\cos C = \dfrac{a^2 + b^2 - c^2}{2ab}$
$= \dfrac{16 + 25 - 64}{40}$
$= -.575$
$C = 125°$
$\dfrac{\sin B}{5.0} = \dfrac{\sin 125°}{8.0}$
$B = 31°$
$A = 180° - 31° - 125°$
$= 24°$

39. $a = 9.3$ cm, $b = 5.7$ cm, $c = 8.2$ cm

Angle A is largest, so find it first.

$a^2 = b^2 + c^2 - 2bc\cos A$
$\cos A = \dfrac{5.7^2 + 8.2^2 - 9.3^2}{2(5.7)(8.2)}$
$\approx .14163457$
$A = 82°$
$\sin B = \dfrac{b\sin A}{a}$
$\approx .60672455$
$B = 37°$
$C = 180° - 82° - 37°$
$= 61°$

40. $a = 28\,\text{ft}$, $b = 47\,\text{ft}$, $c = 58\,\text{ft}$

Find the largest angle first, in case it is obtuse. Since c is the longest side, angle C will be the largest angle.

$$c^2 = a^2 + b^2 - 2ab\cos C$$
$$\cos C = \frac{a^2 + b^2 - c^2}{2ab}$$
$$= \frac{28^2 + 47^2 - 58^2}{2(28)(47)}$$
$$= -.14095745$$
$$C = 98°$$
$$\frac{\sin A}{28} = \frac{\sin 98°}{58}$$
$$A = 29°$$
$$B = 180° - 29° - 98°$$
$$= 53°$$

41. $a = 42.9\,\text{m}$, $b = 37.6\,\text{m}$, $c = 62.7\,\text{m}$

Angle C is the largest, so find it first.

$$c^2 = a^2 + b^2 - 2ab\cos C$$
$$\cos C = \frac{a^2 + b^2 - c^2}{2ab}$$
$$= \frac{42.9^2 + 37.6^2 - 62.7^2}{2(42.9)(37.6)}$$
$$\approx -.20988940$$
$$C = 102.1°$$
$$\sin B = \frac{b\sin C}{c}$$
$$\approx .58632321$$
$$B = 35.9°$$
$$A = 180° - 35.9° - 102.1°$$
$$= 42.0°$$

42. $a = 187\,\text{yd}$, $b = 214\,\text{yd}$, $c = 325\,\text{yd}$

Angle C is the largest, so find it first.

$$c^2 = a^2 + b^2 - 2ab\cos C$$
$$\cos C = \frac{a^2 + b^2 - c^2}{2ab}$$
$$= \frac{187^2 + 214^2 - 325^2}{2(187)(214)}$$
$$\approx -.31061022$$
$$C = 108.1°$$
$$\frac{\sin B}{214} = \frac{\sin 108.1°}{325}$$
$$B = 38.7°$$
$$A = 180° - 38.7° - 108.1°$$
$$= 33.2°$$

43. Find AB, or c, in the following triangle.

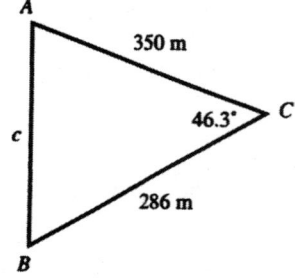

$$c^2 = a^2 + b^2 - 2ab\cos C$$
$$c = \sqrt{286^2 + 350^2 - 2(286)(350)\cos 46.3°}$$
$$c \approx 257$$
$$AB = 257\,\text{m}$$

44. Find the diagonals, BD and AC, of the following parallelogram.

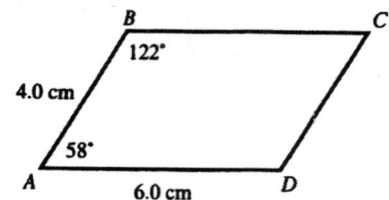

$$BD^2 = AB^2 + AD^2 - 2(AB)(AD)\cos A$$
$$\approx 16 + 36 - 48(.52991926)$$
$$= 26.5638755$$
$$BD \approx 5.2\,\text{cm}$$

$$AC^2 = AB^2 + BC^2 - 2(AB)(BC)\cos B$$
$$\approx 16 + 36 - 48(-.52991926)$$
$$= 77.4361245$$
$$AC \approx 8.8\,\text{cm}$$

The lengths of the diagonals are 5.2 cm and 8.8 cm.

45. Solve for x.

$$x^2 = 25^2 + 25^2 - 2(25)(25)\cos 52°$$
$$\approx 480$$
$$x \approx 22\,\text{ft}$$

46. $AB^2 = 10^2 + 10^2 - 2(10)(10)\cos 128°$
$$\approx 323$$
$$AB \approx 18\,\text{ft}$$

47. Let A = the angle between the beam and the 45-ft cable.

$$\cos A = \frac{45^2 + 90^2 - 60^2}{2(45)(90)}$$
$$A = 36°$$

Let B = the angle between the beam and the 60-ft cable.

$$\cos B = \frac{90^2 + 60^2 - 45^2}{2(90)(60)}$$
$$B = 26°$$

10.6 THE LAWS OF SINES AND COSINES; AREA FORMULAS

48. Using the law of cosines,
$$c^2 = (14.0)^2 + (13.0)^2 - 2(14.0)(13.0)\cos 70°$$
$$\approx 240.5047$$
$$c \approx 15.5 \text{ ft}$$

The length of the property line is approximately $18.0 + 15.5 + 14.0 = 47.5$ feet.

49. Using $A = \frac{1}{2}bh$,
$$A = \frac{1}{2}(1)\left(\sqrt{3}\right)$$
$$= \frac{\sqrt{3}}{2}.$$

Using $A = \frac{1}{2}ab \sin C$,
$$A = \frac{1}{2}\left(\sqrt{3}\right)(1)\sin 90°$$
$$= \frac{1}{2}\left(\sqrt{3}\right)(1)1$$
$$= \frac{\sqrt{3}}{2}.$$

50. Using $A = \frac{1}{2}bh$,
$$A = \frac{1}{2}(2)\left(\sqrt{3}\right)$$
$$= \sqrt{3}.$$

Using $A = \frac{1}{2}ab \sin C$,
$$A = \frac{1}{2}(2)(2)\sin 60°$$
$$= \frac{1}{2}(2)(2)\left(\frac{\sqrt{3}}{2}\right)$$
$$= \sqrt{3}.$$

51. Using $A = \frac{1}{2}bh$,
$$A = \frac{1}{2}(1)\left(\sqrt{2}\right)$$
$$= \frac{\sqrt{2}}{2}.$$

Using $A = \frac{1}{2}ab \sin C$,
$$A = \frac{1}{2}(2)(1)\sin 45°$$
$$= \frac{1}{2}(2)(1)\left(\frac{\sqrt{2}}{2}\right)$$
$$= \frac{\sqrt{2}}{2}.$$

52. Using $A = \frac{1}{2}bh$,
$$A = \frac{1}{2}(2)(1)$$
$$= 1.$$

Using $A = \frac{1}{2}ab \sin C$,
$$A = \frac{1}{2}(2)\left(\sqrt{2}\right)\sin 45°$$
$$= \frac{1}{2}(2)\left(\sqrt{2}\right)\left(\frac{\sqrt{2}}{2}\right)$$
$$= 1.$$

53. $A = 42.5°$, $b = 13.6$ m, $c = 10.1$ m

Angle A is included between sides b and c.
$$\text{Area} = \frac{1}{2}bc \sin A,$$
$$= \frac{1}{2}(13.6)(10.1)\sin 42.5°$$
$$\approx 46.4 \text{ m}^2$$

54. $C = 72.2°$, $b = 43.8$ ft, $a = 35.1$ ft

Angle C is included between sides a and b.
$$\text{Area} = \frac{1}{2}ab \sin C,$$
$$= \frac{1}{2}(35.1)(43.8)\sin 72.2°$$
$$\approx 732 \text{ ft}^2$$

55. $B = 124.5°$, $a = 30.4$ cm, $c = 28.4$ cm

Angle B is included between sides a and c.
$$\text{Area} = \frac{1}{2}ac \sin B,$$
$$= \frac{1}{2}(30.4)(28.4)\sin 124.5°$$
$$\approx 356 \text{ cm}^2$$

56. $C = 142.7°$, $a = 21.9$ km, $b = 24.6$ km

Angle C is included between sides a and b.
$$\text{Area} = \frac{1}{2}ab \sin C,$$
$$= \frac{1}{2}(21.9)(24.6)\sin 142.7°$$
$$\approx 163 \text{ km}^2$$

57. $A = 56.80°$, $b = 32.67$ in., $c = 52.89$ in.

Angle A is included between sides b and c.
$$\text{Area} = \frac{1}{2}bc \sin A,$$
$$= \frac{1}{2}(32.67)(52.89)\sin 56.80°$$
$$\approx 722.9 \text{ in.}^2$$

58. $A = 34.97°$, $b = 35.29$ m, $c = 28.67$ m

    Angle A is included between sides b and c.

    $\text{Area} = \frac{1}{2}bc \sin A$,

    $= \frac{1}{2}(35.29)(28.67)\sin 34.97°$

    $\approx 289.9 \text{ m}^2$

59. Using $\mathcal{A} = \frac{1}{2}bh$,

    $\mathcal{A} = \frac{1}{2}(16)(3\sqrt{3})$

    $= 24\sqrt{3} \approx 41.57$.

    To use Heron's Formula, first find the semiperimeter.

    $s = \frac{1}{2}(a+b+c)$

    $= \frac{1}{2}(6+14+16)$

    $= \frac{1}{2}(36) = 18$

    Now find the area of the triangle.

    $\text{Area} = \sqrt{s(s-a)(s-b)(s-c)}$

    $= \sqrt{18(18-6)(18-14)(18-16)}$

    $= \sqrt{18(12)(4)(2)}$

    $= \sqrt{1728} = \sqrt{576 \cdot 3} = 24\sqrt{3} \approx 41.57$

    Both formulas give the same area.

60. Using $\mathcal{A} = \frac{1}{2}bh$,

    $\mathcal{A} = \frac{1}{2}(10)(3\sqrt{3})$

    $= 15\sqrt{3} \approx 25.98$.

    To use Heron's Formula, first find the semiperimeter.

    $s = \frac{1}{2}(a+b+c)$

    $= \frac{1}{2}(10+6+14)$

    $= \frac{1}{2}(30) = 15$

    Now find the area of the triangle.

    $\text{Area} = \sqrt{s(s-a)(s-b)(s-c)}$

    $= \sqrt{15(15-10)(15-6)(15-14)}$

    $= \sqrt{15(5)(9)(1)}$

    $= \sqrt{675} = \sqrt{225 \cdot 3} = 15\sqrt{3} \approx 25.98$.

    Both formulas give the same area.

61. $a = 12$ m, $b = 16$ m, $c = 25$ m

    $s = \frac{1}{2}(a+b+c)$

    $= \frac{1}{2}(12+16+25)$

    $= \frac{1}{2}(53) = 26.5$

    $\text{Area} = \sqrt{s(s-a)(s-b)(s-c)}$

    $= \sqrt{(26.5)(14.5)(10.5)(1.5)}$

    $\approx 78 \text{ m}^2$

62. $a = 22$ in., $b = 45$ in., $c = 31$ in.

    $s = \frac{1}{2}(a+b+c)$

    $= \frac{1}{2}(22+45+31)$

    $= \frac{1}{2}(98) = 49$

    $\text{Area} = \sqrt{s(s-a)(s-b)(s-c)}$

    $= \sqrt{(49)(27)(4)(18)}$

    $\approx 310 \text{ in.}^2$

63. $a = 154$ cm, $b = 179$ cm, $c = 183$ cm

    $s = \frac{1}{2}(a+b+c)$

    $= \frac{1}{2}(154+179+183)$

    $= \frac{1}{2}(516) = 258$

    $\text{Area} = \sqrt{s(s-a)(s-b)(s-c)}$

    $= \sqrt{(258)(104)(79)(75)}$

    $\approx 12,600 \text{ cm}^2$

64. $a = 25.4$ yd, $b = 38.2$ yd, $c = 19.8$ yd

    $s = \frac{1}{2}(a+b+c)$

    $= \frac{1}{2}(25.4+38.2+19.8)$

    $= 41.7$

    $\text{Area} = \sqrt{s(s-a)(s-b)(s-c)}$

    $= \sqrt{(41.7)(16.3)(3.5)(21.9)}$

    $\approx 228 \text{ yd}^2$

65. $a = 76.3$ ft, $b = 109$ ft, $c = 98.8$ ft

    $s = \frac{1}{2}(a+b+c)$

    $= \frac{1}{2}(76.3+109+98.8)$

    $= 142.05$

    $\text{Area} = \sqrt{s(s-a)(s-b)(s-c)}$

    $= \sqrt{(142.05)(65.75)(33.05)(43.25)}$

    $\approx 3650 \text{ ft}^2$

66. $a = 15.89$ in., $b = 21.74$ in., $c = 10.92$ in.

$s = \dfrac{1}{2}(a+b+c)$

$= \dfrac{1}{2}(15.89 + 21.74 + 10.92)$

$= 24.275$

Area $= \sqrt{s(s-a)(s-b)(s-c)}$

≈ 83.01 in.^{2}

67. Area $= \dfrac{1}{2}ab\sin C$,

$= \dfrac{1}{2}(16.1)(15.2)\sin 125°$

≈ 100 m^2

68. Area $= \dfrac{1}{2}ab\sin C$,

$= \dfrac{1}{2}(52.1)(21.3)\sin 42.2°$

≈ 373 m^2

69. Find the area of the region.

$s = \dfrac{1}{2}(a+b+c)$

$= \dfrac{1}{2}(75+68+85)$

$= 114$

Area $= \sqrt{s(s-a)(s-b)(s-c)}$

$= \sqrt{(114)(39)(46)(29)}$

≈ 2435.35706 m^2

Let $n =$ number of cans needed.

$n = \dfrac{(\text{area in m}^2)}{(\text{m}^2 \text{ per can})}$

$= \dfrac{2435.35761}{75}$

≈ 32.47

She will need to open 33 cans.

70. Find the area of the Bermuda Triangle.

$s = \dfrac{1}{2}(a+b+c)$

$= \dfrac{1}{2}(850+925+1300)$

$= 1537.5$

Area $= \sqrt{s(s-a)(s-b)(s-c)}$

$= \sqrt{(1537.5)(687.5)(612.5)(237.5)}$

$\approx 392,000$

The are of the Bermuda Triangle is 392,000 mi^2.

10.7 EXERCISES

1. $y = \sin x$

   The graph is sinusoidal curve with an amplitude of 1 and a period of 2π. Since $\sin 0 = 0$, the point $(0,0)$ is on the graph. This matches with graph G.

2. $y = \cos x$

   The graph is sinusoidal curve with an amplitude of 1 and a period of 2π. Since $\cos 0 = 1$, the point $(0,1)$ is on the graph. This matches with graph A.

3. $y = -\sin x$

   The graph is sinusoidal curve with an amplitude of 1 and a period of 2π. Because $a = -1$, the graph is a reflection of $y = \sin x$ across the x-axis. This matches with graph E.

4. $y = -\cos x$

   The graph is sinusoidal curve with an amplitude of 1 and a period of 2π. Because $a = -1$, the graph is a reflection of $y = \cos x$ across the x-axis. This matches with graph D.

5. $y = \sin 2x$

   The graph is sinusoidal curve with an amplitude of 1 and a period of π. Since $\sin 2(0) = 0$, the point $(0,0)$ is on the graph. This matches with graph B.

6. $y = \cos 2x$

   The graph is sinusoidal curve with an amplitude of 1 and a period of π. Since $\cos 2(0) = 1$, the point $(0,1)$ is on the graph. This matches with graph H.

7. $y = 2\sin x$

   The graph is sinusoidal curve with an amplitude of 2 and a period of 2π. Since $2\sin 0 = 0$, the point $(0,0)$ is on the graph. This matches with graph F.

8. $y = 2\cos x$

   The graph is sinusoidal curve with an amplitude of 2 and a period of 2π. Since $2\cos 0 = 2$, the point $(0,2)$ is on the graph. This matches with graph C.

9. $y = 2\cos x$

   Amplitude: $|2| = 2$, period $= 2\pi$.

x	0	$\dfrac{\pi}{2}$	π	$\dfrac{3\pi}{2}$	2π
$2\cos x$	2	0	-2	0	2

   This table shows five values for graphing one period of the function. Repeat this cycle for the interval $[-2\pi, 0]$.

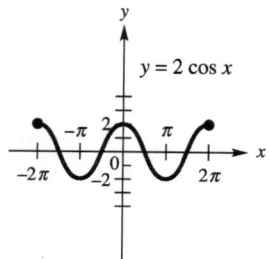

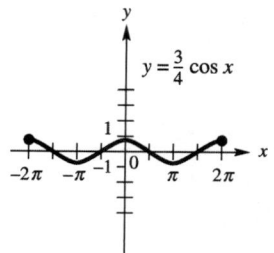

10. $y = 3\sin x$

    Amplitude: $|3| = 3$, period $= 2\pi$.

x	0	$\frac{\pi}{2}$	π	$\frac{3\pi}{2}$	2π
$3\sin x$	0	3	0	-3	0

    This table shows five values for graphing one period of the function. Repeat this cycle for the interval $[-2\pi, 0]$.

13. $y = -\cos x$

    Amplitude: $|-1| = 1$, period $= 2\pi$.

x	0	$\frac{\pi}{2}$	π	$\frac{3\pi}{2}$	2π
$-\cos x$	-1	0	1	0	-1

    This table shows five values for graphing one period of the function. Repeat this cycle for the interval $[-2\pi, 0]$.

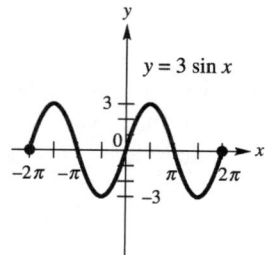

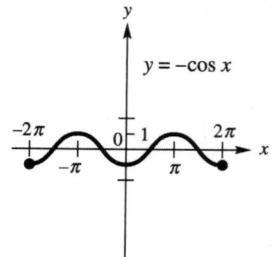

11. $y = \frac{2}{3}\sin x$

    Amplitude: $|\frac{2}{3}| = \frac{2}{3}$, period $= 2\pi$.

x	0	$\frac{\pi}{2}$	π	$\frac{3\pi}{2}$	2π
$\frac{2}{3}\sin x$	0	.7	0	$-.7$	0

    This table shows five values for graphing one period of the function. Repeat this cycle for the interval $[-2\pi, 0]$.

14. $y = -\sin x$

    Amplitude: $|-1| = 1$, period $= 2\pi$.

x	0	$\frac{\pi}{2}$	π	$\frac{3\pi}{2}$	2π
$-\sin x$	0	-1	0	1	0

    This table shows five values for graphing one period of the function. Repeat this cycle for the interval $[-2\pi, 0]$.

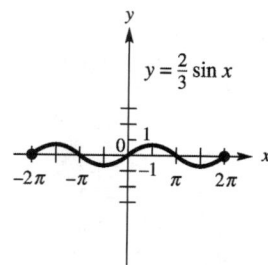

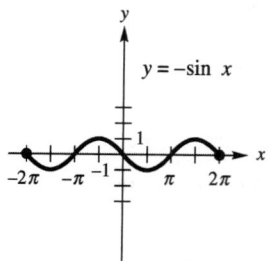

12. $y = \frac{3}{4}\cos x$

    Amplitude: $|\frac{3}{4}| = \frac{3}{4}$, period $= 2\pi$.

x	0	$\frac{\pi}{2}$	π	$\frac{3\pi}{2}$	2π
$\frac{3}{4}\cos x$	$\frac{3}{4}$	0	$-\frac{3}{4}$	0	$\frac{3}{4}$

    This table shows five values for graphing one period of the function. Repeat this cycle for the interval $[-2\pi, 0]$.

15. $y = -2\sin x$

    Amplitude: $|-2| = 2$, period $= 2\pi$.

x	0	$\frac{\pi}{2}$	π	$\frac{3\pi}{2}$	2π
$-2\sin x$	0	-2	0	2	0

    This table shows five values for graphing one period of the function. Repeat this cycle for the interval $[-2\pi, 0]$.

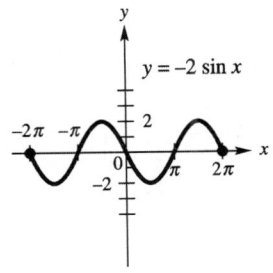

16. $y = -3\cos x$

    Amplitude: $|-3| = 3$, period $= 2\pi$.

x	0	$\frac{\pi}{2}$	π	$\frac{3\pi}{2}$	2π
$-3\cos x$	-3	0	3	0	-3

    This table shows five values for graphing one period of the function. Repeat this cycle for the interval $[-2\pi, 0]$.

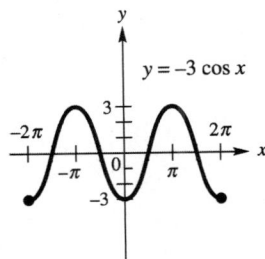

17. $y = \sin \frac{1}{2}x$

    Period: $\frac{2\pi}{\frac{1}{2}} = 4\pi$

    Amplitude: $|1| = 1$

    Divide the interval $[0, 4\pi]$ into four equal parts to get the x-values that will yield minimum and maximum points and x-intercepts. Then make a table.

x	0	π	2π	3π	4π
$\frac{1}{2}x$	0	$\frac{\pi}{2}$	π	$\frac{3\pi}{2}$	2π
$\sin \frac{1}{2}x$	0	1	0	-1	0

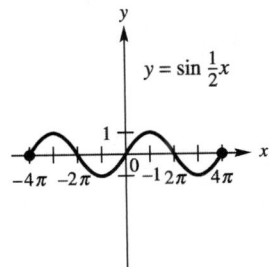

18. $y = \sin \frac{2}{3}x$

    Period: $\frac{2\pi}{\frac{2}{3}} = 3\pi$

    Amplitude: $|1| = 1$

Divide the interval $[0, 3\pi]$ into four equal parts to get the x-values that will yield minimum and maximum points and x-intercepts. Then make a table.

x	0	$\frac{3\pi}{4}$	$\frac{3\pi}{2}$	$\frac{9\pi}{4}$	3π
$\frac{2}{3}x$	0	$\frac{\pi}{2}$	π	$\frac{3\pi}{2}$	2π
$\sin \frac{2}{3}x$	0	1	0	-1	0

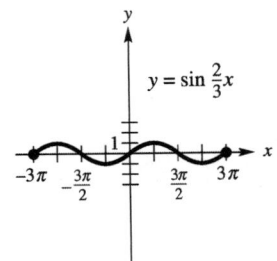

19. $y = \cos \frac{1}{3}x$

    Period: $\frac{2\pi}{\frac{1}{3}} = 6\pi$

    Amplitude: $|1| = 1$

x	0	$\frac{3\pi}{2}$	3π	$\frac{9\pi}{2}$	6π
$\frac{1}{3}x$	0	$\frac{\pi}{2}$	π	$\frac{3\pi}{2}$	2π
$\cos \frac{1}{3}x$	1	0	-1	0	1

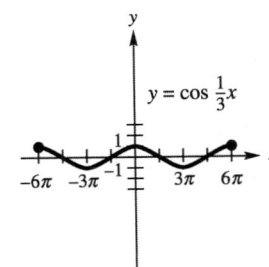

20. $y = \cos \frac{3}{4}x$

    Period: $\frac{2\pi}{\frac{3}{4}} = \frac{8\pi}{3}$

    Amplitude: $|1| = 1$

x	0	$\frac{2\pi}{3}$	$\frac{4\pi}{3}$	2π	$\frac{8\pi}{3}$
$\frac{3}{4}x$	0	$\frac{\pi}{2}$	π	$\frac{3\pi}{2}$	2π
$\cos \frac{3}{4}x$	1	0	-1	0	1

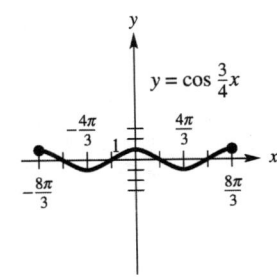

21. $y = \sin 3x$

 Period: $\frac{2\pi}{3}$

 Amplitude: $|1| = 1$

x	0	$\frac{\pi}{6}$	$\frac{\pi}{3}$	$\frac{\pi}{2}$	$\frac{2\pi}{3}$
$3x$	0	$\frac{\pi}{2}$	π	$\frac{3\pi}{2}$	2π
$\sin 3x$	0	1	0	-1	0

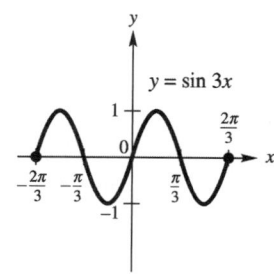

22. $y = \cos 2x$

 Period: $\frac{2\pi}{2} = \pi$

 Amplitude: $|1| = 1$

x	0	$\frac{\pi}{4}$	$\frac{\pi}{2}$	$\frac{3\pi}{4}$	π
$2x$	0	$\frac{\pi}{2}$	π	$\frac{3\pi}{2}$	2π
$\cos 2x$	1	0	-1	0	1

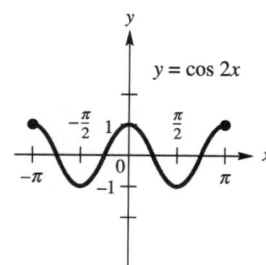

23. $y = 2\sin \frac{1}{4}x$

 Period: $\frac{2\pi}{\frac{1}{4}} = 8\pi$

 Amplitude: $|2| = 2$

x	0	2π	4π	6π	8π
$\frac{1}{4}x$	0	$\frac{\pi}{2}$	π	$\frac{3\pi}{2}$	2π
$2\sin \frac{1}{4}x$	0	2	0	-2	0

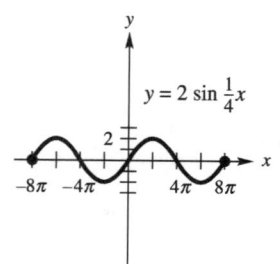

24. $y = 3\sin 2x$

 Period: $\frac{2\pi}{2} = \pi$

 Amplitude: $|3| = 3$

x	0	$\frac{\pi}{4}$	$\frac{\pi}{2}$	$\frac{3\pi}{4}$	π
$3\sin 2x$	0	3	0	-3	0

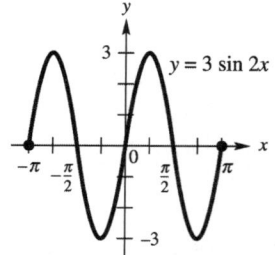

25. $y = -2\cos 3x$

 Period: $\frac{2\pi}{3}$

 Amplitude: $|-2| = 2$

x	0	$\frac{\pi}{6}$	$\frac{\pi}{3}$	$\frac{\pi}{2}$	$\frac{2\pi}{3}$
$3x$	0	$\frac{\pi}{2}$	π	$\frac{3\pi}{2}$	2π
$-2\cos 3x$	-2	0	2	0	-2

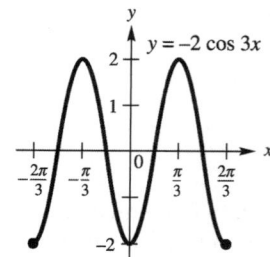

26. $y = -5\cos 2x$

 Period: $\frac{2\pi}{2} = \pi$

 Amplitude: $|-5| = 5$

x	0	$\frac{\pi}{4}$	$\frac{\pi}{2}$	$\frac{3\pi}{4}$	π
$-5\cos 2x$	-5	0	5	0	-5

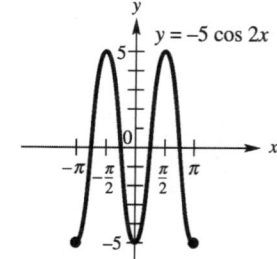

27. $y = \cos \pi x$

Period: $\frac{2\pi}{\pi} = 2$

Amplitude: $|1| = 1$

x	0	$\frac{1}{2}$	1	$\frac{3}{2}$	2
$\cos \pi x$	1	0	-1	0	1

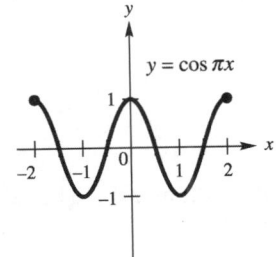

28. $y = -\sin \pi x$

Period: $\frac{2\pi}{\pi} = 2$

Amplitude: $|-1| = 1$

x	0	$\frac{1}{2}$	1	$\frac{3}{2}$	2
πx	0	$\frac{\pi}{2}$	π	$\frac{3\pi}{2}$	2π
$-\sin \pi x$	0	-1	0	1	0

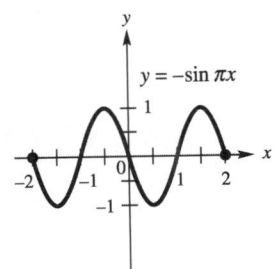

29. (a) The highest temperature is 80°; the lowest is 50°.

(b) The amplitude is $\frac{1}{2}(80° - 50°) = 15°$.

(c) The period is about 35,000 years.

(d) The trend of the temperature now is downward.

30. (a) The latest time that the animals begin their evening activity is 8:00 P.M., the earliest time is 4:00 P.M. $4:00 \le y \le 8:00$

Since there is a difference of 4 hours in these times, the amplitude is $\frac{1}{2}(4) = 2$ hr.

(b) The length of this period is 1 year.

31. (a) The amplitude is $\frac{1}{2}(120 - 80) = \frac{1}{2}(40) = 20$.

(b) Since the period is .8 sec, there are $\frac{1}{.8} = 1.25$ beats per sec and the pulse rate is $60(1.25) = 75$ beats per min.

32. (a) $-1 \le y \le 1$

Amplitude: 1

Period: 8 squares $= 8\left(\frac{\pi}{6}\right) = \frac{4\pi}{3}$.

(b) $-1 \le y \le 1$

Amplitude: 1

Period: 4 squares $= 4\left(\frac{\pi}{6}\right) = \frac{2\pi}{3}$.

CHAPTER 10 TEST

1. $74°17'54'' = 74° + \frac{17°}{60} + \frac{54°}{3600}$
$= 74.2983°$

2. $360° + (-157°) = 203°$

3. $(2, -5)$, $x = 2$, $y = -5$

$r = \sqrt{x^2 + y^2}$
$= \sqrt{2^2 + (-5)^2}$
$= \sqrt{29}$

$\sin \theta = \frac{y}{r} = \frac{-5}{\sqrt{29}} \cdot \frac{\sqrt{29}}{\sqrt{29}} = \frac{-5\sqrt{29}}{29}$

$\cos \theta = \frac{x}{r} = \frac{2}{\sqrt{29}} \cdot \frac{\sqrt{29}}{\sqrt{29}} = \frac{2\sqrt{29}}{29}$

$\tan \theta = \frac{y}{x} = \frac{-5}{2} = -\frac{5}{2}$

4. If $\cos \theta < 0$, then θ is in quadrant II or III. If $\cot \theta > 0$, then θ is in quadrant I or III. Therefore, θ terminates in quadrant III.

5. If $\cos \theta = \frac{4}{5} = \frac{x}{r}$, let $x = 4$ and $r = 5$

$x^2 + y^2 = r^2$
$4^2 + y^2 = 5^2$
$y^2 = \pm\sqrt{9}$
$y = \pm 3$

Since θ is in quadrant IV, y is negative, so $y = -3$.

$\sin \theta = \frac{y}{r} = \frac{-3}{5} = -\frac{3}{5}$

$\tan \theta = \frac{y}{x} = \frac{-3}{4} = -\frac{3}{4}$

$\cot \theta = \frac{x}{y} = \frac{4}{-3} = -\frac{4}{3}$

$\sec \theta = \frac{r}{x} = \frac{5}{4}$

$\csc \theta = \frac{r}{y} = \frac{5}{-3} = -\frac{5}{3}$

6. $\sin A = \dfrac{\text{side opposite}}{\text{hypotenuse}} = \dfrac{12}{13}$

   $\cos A = \dfrac{\text{side adjacent}}{\text{hypotenuse}} = \dfrac{5}{13}$

   $\tan A = \dfrac{\text{side opposite}}{\text{side adjacent}} = \dfrac{12}{5}$

   $\cot A = \dfrac{\text{side adjacent}}{\text{side opposite}} = \dfrac{5}{12}$

   $\sec \theta = \dfrac{\text{hypotenuse}}{\text{side adjacent}} = \dfrac{13}{5}$

   $\csc \theta = \dfrac{\text{hypotenuse}}{\text{side opposite}} = \dfrac{13}{12}$

7. Use a 30°- 60°- 90° right triangle (or the Unit Circle) to find the exact values.

   (a) $\cos 60° = \dfrac{1}{2}$

   (b) $\tan 45° = 1$

   (c) $\tan(-270°) = \tan(-270° + 360°)$
   $= \tan 90°$
   $= \dfrac{1}{0}$; undefind

   (d) $\sec 210° = \sec(210° - 180°)$
   $= -\sec 30°$
   $= -\dfrac{2}{\sqrt{3}} \cdot \dfrac{\sqrt{3}}{\sqrt{3}}$
   $= -\dfrac{2\sqrt{3}}{3}$

   (e) $\csc(-180°) = \dfrac{1}{\sin(-180°)}$
   $= \dfrac{1}{0}$, by Unit Circle; undefind.

   (f) $\sec 135° = -\sec(180° - 135°)$
   $= -\sec 45°$
   $= -\dfrac{\sqrt{2}}{1}$
   $= -\sqrt{2}$

8. Check to see that your calculator is in degree mode.

   (a) $\sin 78°21'$

   $\sin 78°21' = \sin\left(78 + \dfrac{21}{60}\right)°$
   $= \sin 78.35°$
   $\approx .97939940$

   (b) $\tan 11.7689°$

   $\tan 11.7689° \approx .20834446$

   (c) $\sec 58.9041°$

   $\sec 58.9041° = \dfrac{1}{\cos 58.9041°}$
   $\approx \dfrac{1}{.516472054}$
   ≈ 1.9362132

9. $\sin \theta = .27843196$

   $\theta = \sin^{-1}(.27843196)$ or
   $= \text{INV sin}(.27843196)$
   $\approx 16.16664145°$

10. $\cos \theta = \dfrac{-\sqrt{2}}{2}$

   Cosine is negative in quadrants II and III. The reference angle θ' is $45°$. In quadrant II θ is $180° - 45° = 135°$. In quadrant III θ is $180° + 45° = 225°$.

11. $A = 58°30'$, $c = 748$, $C = 90°$

   $\cos A = \dfrac{b}{c}$
   $b = c \cos A$
   $= (748)\cos 58°30'$
   ≈ 391

   $\sin A = \dfrac{a}{c}$
   $a = c \sin A$
   $= (748)\sin 58°30'$
   ≈ 638

   $A + B = 90°$
   $B = 90° - A$
   $= 90° - 58°30'$
   $= 31°30'$

12.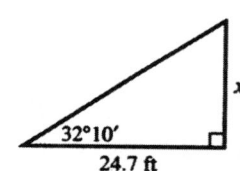

   Let x be the height of the flag pole.

   $\tan 32°10' = \dfrac{x}{24.7}$
   $x = (24.7)\tan 32°10'$
   $\approx 15.5 \text{ ft}$

13. Use the law of sines.

   $\dfrac{\sin 25.2°}{6.92} = \dfrac{\sin B}{4.82}$

   $\sin B = \dfrac{(\sin 25.2°)(4.82)}{6.92}$

   $B = \sin^{-1}\left(\dfrac{(\sin 25.2°)(4.82)}{6.92}\right)$
   $\approx 17.3°$

Use the fact that the sum of the interior angles of a triangle add to 180°.

$$C = 180° - (A + B)$$
$$= 180° - (25.2° + 17.3°)$$
$$= 137.5°$$

14. Use the law of cosines.

$$c^2 = a^2 + b^2 - 2ab\cos C$$
$$= 75^2 + 130^2 - 2(75)(130)\cos 118°$$
$$\approx 31679$$
$$c = \sqrt{31679}$$
$$\approx 180 \text{ km}$$

15. Use the law of cosines.

$$b^2 = a^2 + c^2 - 2ac\cos B$$
$$2ac\cos B = a^2 + c^2 - b^2$$
$$\cos B = \frac{a^2 + c^2 - b^2}{2ac}$$
$$B = \cos^{-1}\left(\frac{a^2 + c^2 - b^2}{2ac}\right)$$
$$= \cos^{-1}\left(\frac{(17.3)^2 + (29.8)^2 - (22.6)^2}{2(17.3)(29.8)}\right)$$
$$\approx 49.0°$$

16. $A = \frac{1}{2}ab\sin C,$
$$= \frac{1}{2}(75)(130)\sin 118°$$
$$\approx 4300 \text{ km}^2$$

(Note: Since c was found in Exercise 14, Heron's formula can also be used.)

17. The semi-perimeter s is:

$$s = \frac{1}{2}(a + b + c)$$
$$= \frac{1}{2}(22 + 26 + 40)$$
$$= 44.$$

Using Heron's formula:

$$\text{Area} = \sqrt{s(s-a)(s-b)(s-c)}$$
$$= \sqrt{44(44-22)(44-26)(44-40)}$$
$$= \sqrt{69696}$$
$$\approx 264 \text{ square units.}$$

18. Find angle C.

$$C = 180° - (47°20' + 24°50')$$
$$= 180° - (72°10')$$
$$= 107°50'$$

Use this result and the law of sines to find AC.

$$\frac{8.4}{\sin 107°50'} = \frac{AC}{\sin 47°20'}$$
$$AC = \frac{(8.4)(\sin 47°20')}{\sin 107°50'}$$
$$= \frac{(8.4)(\sin 47.333°)}{\sin 107.833°}$$
$$\approx 6.5 \text{ mi}$$

Drop a perpendicular line from C to segment AB.

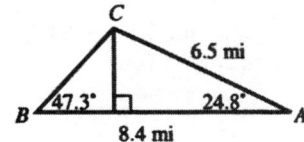

From the resulting triangle(s),

$$\sin 24.833° = \frac{h}{6.5}$$
$$h = (6.5)(\sin 24.833°)$$
$$\approx 2.7 \text{ mi}$$

The balloon is about 2.7 miles above the ground.

19. Let $c = $ the length of the tunnel.

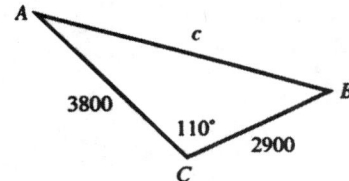

Use the law of cosines to find c.

$$c^2 = a^2 + b^2 - 2ab\cos C$$
$$= 2900^2 + 3800^2 - 2(2900)(3800)\cos 110°$$
$$\approx 30,388,124$$
$$c = \sqrt{30,388,124}$$
$$\approx 5500 \text{ m}$$

The tunnel is about 5500 meters long.

20. Let $A = $ home plate, $B = $ first base, $C = $ second base, $D = $ third base, $P = $ pitcher's rubber. Draw AC through P, draw PB and PD.

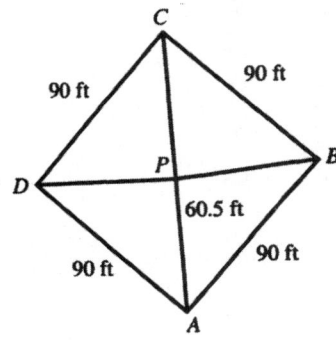

In triangle ABC, angle $B = 90°$, angle $A = $ angle $C = 45°$.

$AC = \sqrt{90^2 + 90^2} = 90\sqrt{2}$

$PC = 90\sqrt{2} - 60.5 = 66.8\,\text{ft}$

In triangle APB, angle $A = 45°$.

$PB^2 = AP^2 + AB^2 - 2(AP)(AB)\cos 45°$
$= (60.5)^2 + (90)^2 - 2(60.5)(90)\cos 45°$
$PB \approx 63.7\,\text{ft}$

Since triangles APB and APD are congruent, $PB = PD = 63.7\,\text{ft}$

The distance to both first and third base is 63.7 feet, and the distance to second base is 66.8 feet.

21. $y = 3\sin x$

Period: 2π

Amplitude: $|3| = 3$

x	0	$\frac{\pi}{2}$	π	$\frac{3\pi}{2}$	2π
$3\sin x$	0	3	0	-3	0

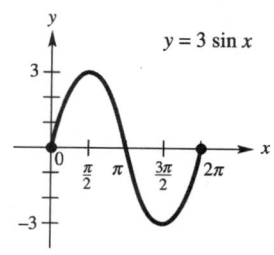

22. $y = -\cos 2x$

Period: $\frac{2\pi}{2} = \pi$

Amplitude: $|-1| = 1$

x	0	$\frac{\pi}{4}$	$\frac{\pi}{2}$	$\frac{3\pi}{4}$	π
$2x$	0	$\frac{\pi}{2}$	π	$\frac{3\pi}{2}$	2π
$-\cos 2x$	-1	0	1	0	-1

This table shows five values for graphing one period of the function. Repeat this cycle for the interval $[-2\pi, 0]$.

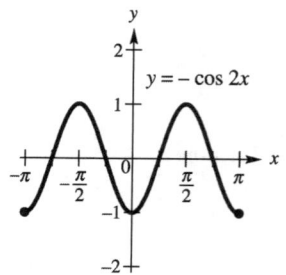

11.1 EXERCISES

In Exercises 1–8 consider the set $N = \{A, B, C, D, E\}$ for {Andy (a man), Bill (a man), Cathy (a woman), David (a man), and Evelyn (a woman)}. List and count the different ways of electing each of the following slates of officers.

1. A president and a treasurer.

   Agreeing that the first letter represents the president and that the second represents the treasurer, we can generate systematically the following symbolic list and count the resulting possibilities: AB, AC, AD, AE; BA, BC, BD, BE; CA, CB, CD, CE; DA, DB, DC, DE; EA, EB, EC, ED. By counting, there are 20 ways to elect a president and treasurer.

2. A president and a treasurer if the president must be female.

   This is a two-part task. Choose the president first. Since Cathy and Evelyn are the only women in the club, there are only two possible results: C and E. For each choice of president, any of the four remaining members may be chosen as secretary. Thus, the president and secretary can be elected in the following ways, with the president listed first: CA, CB, CD, CE; EA, EB, EC, ED. Therefore, the officers can be elected in 8 different ways.

3. A president and a treasurer if the two officers must not be the same sex.

   Since the men include A, B, and D, and the women are C and E, we must also eliminate those doubles from the same list that represent all men (AB, BA, AD, DA, BD, and DB) and those representing all women (CE and EC). We have lost a total of 8 from the original list of 20 (Exercise 1), leaving 12 ways to elect a president and treasurer. The results include: AC, AE, BC, BE, CA, CB, CD, DC, DE, EA, EB, ED. Note that a new list with the above restrictions could be generated rather than subtracting from the original list.

4. A president, a secretary, and a treasurer, if the president and treasurer must be women.

   This is a three-part task. Since Cathy and Evelyn are the only women, we must have either Cathy as president and Evelyn as treasurer, or Evelyn as president and Cathy as treasurer. In either case, any of the three men may be elected secretary. Thus, we obtain the following list of possibilities, using the order president, secretary, treasurer:

   CAE, CBE, CDE, EAC, EBC, EDC.

   The officers may be elected in 6 different ways.

5. A president, a secretary, and a treasurer, if the president must be a man and the other two must be women.

   Generating a new symbolic list where the first member must be a man and the second and third, women, we get

   ACE, AEC, BCE, BEC, DCE, and DEC.

   The officers may be elected in 6 different ways.

6. A president, a secretary, and a treasurer, if the secretary must be a woman and the other two must be men.

   This is a three-part task. The president and treasurer must be selected from A, B, and D while the secretary must be selected from C and E. A tree diagram is helpful. No repeats are allowed!

   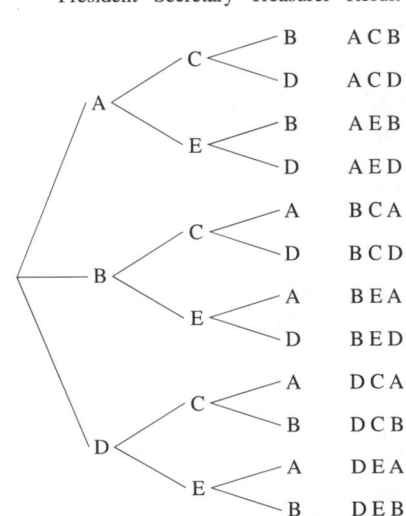

   The officers can be elected in 12 different ways.

List and count the ways club N could appoint a committee of three members under the following conditions.

7. There are no restrictions.

   One method would be to list all triples. Remembering, however, that ABC is the same committee as BAC or CAB, cross out all triples with the same three letters. We are left with: ABC, ABD, ABE, ACD, ACE, ADE, BCD, BCE, BDE, CDE. Therefore, there are 10 ways to select the 3-member committees with no restrictions.

8. The committee must include more men than women.

   Since the committee has three members and there must be more men than women, it must have either 3 men or 2 men and 1 woman. Since there are only 3 men to choose from, there is only 1 committee with 3 men. The number of committees with 2 men and 1 woman is 6. The committees are ABD (all men), ABC, ABE, ACD, ADE, BCD, BDE. The total number of committees is 7.

For Exercises 9–25, refer to Table 2 (the product table for rolling two dice) in the text.

9. Only 1 member of the product table (1, 1) represents an outcome where the sum of the dice is two.

10. Only 2 members of the product table (1, 2), (2, 1) represent an outcome where the sum of the dice is three.

11. Only 3 members of the product table $(3, 1), (2, 2), (1, 3)$ represent outcomes where the sum of the dice is four.

12. There are 4 members of the product table, $(1, 4), (2, 3), (3, 2),$ and $(4, 1)$, which represent outcomes where the sum of the dice is five.

13. There are 5 members of the product table, $(5, 1), (4, 2), (3, 3), (2, 4),$ and $(1, 5)$, which represent outcomes where the sum is six.

14. There are 6 members of the product table, $(1, 6), (2, 5), (3, 4), (4, 3), (5, 2),$ and $(6, 1)$, which represent outcomes where the sum is seven.

15. There are 5 members of the product table, $(6, 2), (5, 3), (4, 4), (3, 5),$ and $(2, 6)$, which represent outcomes where the sum is 8.

16. The sum is nine for the four outcomes $(3, 6), (4, 5), (5, 4),$ and $(6, 3)$.

17. There are only 3 members of the product table, $(6, 4), (5, 5),$ and $(4, 6)$, which represent outcomes where the sum of the dice is ten.

18. The sum is eleven for 2 pairs, $(5, 6)$ and $(6, 5)$.

19. Only 1 member, $(6, 6)$, of the product table yields an outcome where the sum is twelve.

20. The sum is odd for 18 pairs,
    $(1, 2), (1, 4), (1, 6), (2, 1), (2, 3), (2, 5),$
    $(3, 2), (3, 4), (3, 6), (4, 1), (4, 3), (4, 5),$
    $(5, 2), (5, 4), (5, 6), (6, 1), (6, 3), (6, 5).$

    These were obtained by selecting appropriate pairs from each row of Table 2. Some students may find it more natural to work with each odd sum separately, first listing the pairs with a sum of 3, then the pairs with a sum of 5, etc. The same pairs would occur with both procedures.

21. Half of all 36 outcomes suggested by the product table should represent a sum which is even; the other half, odd. Thus, there are 18 outcomes which will be even. They are:
    $(1, 1), (3, 1), (2, 2), (1, 3), (5, 1), (4, 2),$
    $(3, 3), (2, 4), (1, 5), (6, 2), (5, 3), (4, 4),$
    $(3, 5), (2, 6), (6, 4), (5, 5), (4, 6), (6, 6).$

22. To find the sums from 6 through 8 inclusive, we must count all pairs for which the sum is 6, 7, or 8.

    Sum is 6: $(1, 5), (2, 4), (3, 3), (4, 2), (5, 1).$

    Sum is 7: $(1, 6), (2, 5), (3, 4), (4, 3), (5, 2), (6, 1).$

    Sum is 8: $(2, 6), (3, 5), (4, 4), (5, 3), (6, 2).$

    Since there are five pairs with a sum of 6, six pairs with a sum of 7, and five pairs with a sum of 8, there are:
    $$5 + 6 + 5 = 16$$
    pairs with a sum of 6 through 8 inclusive.

23. To find the sums between 6 and 10, we must count pairs in which the sum is 7, 8, or 9.

    Sum is 7: $(1, 6), (2, 5), (3, 4), (4, 3), (5, 2), (6, 1).$

    Sum is 8: $(2, 6), (3, 5), (4, 4), (5, 3), (6, 2).$

    Sum is 9: $(6, 3), (5, 4), (4, 5), (3, 6).$

    Since there are six pairs with a sum of 7, five pairs with a sum of 8, and 4 pairs with a sum of 9, there are:
    $$6 + 5 + 4 = 15$$
    pairs with a sum of 6 through 8 inclusive.

24. Count all pairs in which the sum is 2, 3, or 4.

    Sum is 2: $(1, 1).$

    Sum is 3: $(1, 2), (2, 1).$

    Sum is 4: $(1, 3), (2, 2), (3, 1).$

    There are 6 pairs in which the sum is less than 5.

25. Construct a product table showing all possible two-digit numbers using digits from the set $\{1, 2, 3, 4, 5, 6\}$.

		\multicolumn{6}{c}{2nd digit}
		1	2	3	4	5	6
	1	11	12	13	14	15	16
	2	21	22	23	24	25	26
1st	3	31	32	33	34	35	36
digit	4	41	42	43	44	45	46
	5	51	52	53	54	55	56
	6	61	62	63	64	65	66

26. A number is odd if its units digit is odd. The odd numbers in the table for Exercise 25 are those ending in 1, 3, and 5, so we list the numbers in the column headed by 1, 3, and 5. They are:

    11, 13, 15,
    21, 23, 25,
    31, 33, 35,
    41, 43, 45,
    51, 53, 55,
    61, 63, 65.

27. The following numbers in the table are numbers with repeating digits:

    11, 22, 33, 44, 55, 66.

28. The following numbers in the table are multiples of 6:

    12, 24, 36, 42, 54, 66.

29. A counting number larger than 1 is prime if it is divisible by itself and 1 only. The following numbers in the table are *prime numbers*:

    11, 13, 23, 31, 41, 43, 53, 61.

30. In Chapter 1, we saw that all *triangular numbers* can be written in the form

$$T_n = \frac{n(n+1)}{2}.$$

The smallest triangular number in the table is

$$T_5 = \frac{5(6)}{2} = 15.$$

The following are triangular numbers which are greater than 15 and less than or equal to 66:

$$T_6 = \frac{6(7)}{2} = 21,$$
$$T_7 = \frac{7(8)}{2} = 28 \text{ (but not a member of table)},$$
$$T_8 = \frac{8(9)}{2} = 36,$$
$$T_9 = \frac{9(10)}{2} = 45,$$
$$T_{10} = \frac{10(11)}{2} = 55,$$
$$T_{11} = \frac{11(12)}{2} = 66.$$

31. The following are (perfect) *square* numbers:

$$16, 25, 36, 64.$$

32. Recall from Chapter 5 that *Fibonacci numbers* are the terms of the Fibonacci sequence:

$$1, 1, 2, 3, 5, 8, 13, 21, 34, 55, 89, \ldots.$$

After the first two terms, both 1, each term is obtained by adding the two previous terms. The Fibonacci numbers in the product table from Exercise 25 are:

$$13, 21, 34, \text{ and } 55.$$

33. The numbers 2^4, 2^3, and 2^6 are powers of 2 found in the product table from Exercise 25. That is,

$$16, 32, \text{ and } 64.$$

34. Draw a tree diagram for three fair coins. Then list the ways of getting the following results.

First coin	Second coin	Third coin	Result
h	h	h	h h h
h	h	t	h h t
h	t	h	h t h
h	t	t	h t t
t	h	h	t h h
t	h	t	t h t
t	t	h	t t h
t	t	t	t t t

(a) At least two heads

List the results which contain 3 heads or 2 heads and 1 tail:

$$hhh, hht, hth, thh.$$

(b) More than two heads

All three coins must be heads, so there is only one way to get more than two heads:

$$hhh.$$

(c) No more than two heads

This includes all results which contain 0, 1, or 2 heads:

$$hht, hth, htt, thh, tht, tth, ttt.$$

Note that this includes all results from the tree diagram except for hhh.

(d) Fewer than two heads

List the results which contain either one head or no heads:

$$htt, tht, tth, ttt.$$

35. Extend the tree diagram of Exercise 34 for four fair coins. Then list the ways of getting the following results.

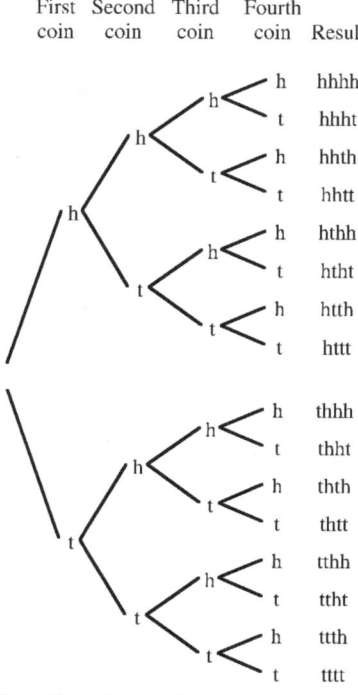

(a) More than three tails

There is only one such outcome:

tttt.

(b) Fewer than three tails

List those outcomes with 0, 1 or 2 tails:

hhhh, hhht, hhth, hhtt,
hthh, htht, htth, thhh
thht, thth, tthh.

(c) At least three tails

List those outcomes with 3 or 4 tails:

httt, thtt, ttht, ttth, tttt.

(d) No more than three tails

List those outcomes with 0, 1, 2, or 3 tails:

hhhh, hhht, hhth, hhtt, hthh,
htht, htth, httt, thhh, thht,
thth, thtt, tthh, ttht, ttth.

36. Label the figure as shown below.

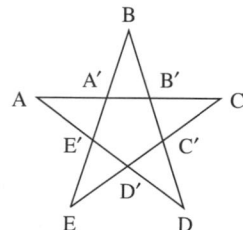

There are 5 small triangles, one at each point of the star:

△AA′E, △BA′B′, △CB′C′, △DC′D′, and △ED′E′.

There are also 5 larger triangles:

△ACD′, △BDE′, △CEA′, △DAB′, and △EBC′.

The figure contains 10 triangles.

37. Begin with the largest triangles which have the long diagonals as their bases. There is 1 triangle on each side of the (2) diagonals. This gives 4 large triangles. Count the intermediate size triangles, each with a base along the outside edge of the large square. There are 4 of these. Furthermore, each of these intermediate sized triangles contain two right triangles within. There are a total of 8 of these. Thus, the total number of triangles is

$$4 + 4 + 8 = 16.$$

38. Begin by counting the two large triangles (on either side of the vertical diagonal). Within each of these there are 4 smaller triangles. Thus, the figure contains a total of 10 triangles.

39. Begin with the larger right triangle at the center square. There are 4. Pairing two of these triangles with each other forms 4 isosceles triangles within the center box. Within each of the four right triangles in the square are two smaller right triangles for a total of 8. Associated with each exterior side of the octagon are 8 triangles, each containing two other right triangles (one of which has already been counted) for a total of 16. There are 4 more isosceles triangles which have their two equal side lengths as exterior edges of the octagon.

Thus, the number of triangles contained in the figure is

$$4 + 4 + 8 + 16 + 4 = 36.$$

40. Label the figure as shown below, so that we can refer to the small squares by number.

1	2	3	4
5	6	7	8
9	10	11	12
13	14	15	16

Find the number of squares of each size and add the results.

There are sixteen 1×1 squares, which are labeled 1 through 16.

Name the 2×2 squares by listing the small squares they contain:

1, 2, 5, 6	2, 3, 6, 7	3, 4, 7, 8
5, 6, 9, 10	6, 7, 10, 11	7, 8, 11, 12
9, 10, 13, 14	10, 11, 14, 15	11, 12, 15, 16

There are nine 2×2 squares.

Name the 3 × 3 squares in a similar manner:

1, 2, 3, 5, 6, 7, 9, 10, 11
2, 3, 4, 6, 7, 8, 10, 11, 12
5, 6, 7, 9, 10, 11, 13, 14, 15
6, 7, 8, 10, 11, 12, 14, 15, 16

There are four 3 × 3 squares.

There is one 4 × 4 square, containing all 16 small squares. Add the results.

Size	Number of squares
1 × 1	16
2 × 2	9
3 × 3	4
4 × 4	1
	30

Thus, the figure contains 30 squares.

41. Label the figure as shown below, so that we can refer to the small squares by number.

	1	2	
3	4	5	6
7	8	9	10
	11	12	

Find the number of squares of each size and add the results.

There are twelve 1 × 1 squares, which are labeled 1 through 12.

Name the 2 × 2 squares by listing the small squares they contain:

1, 2, 4, 5	5, 6, 9, 10
8, 9, 11, 12	3, 4, 7, 8
4, 5, 8, 9	

There are five 2 × 2 squares. There are no squares larger than 2 × 2.

Thus, there are a total of

$$12 + 5 = 17$$

squares contained in the figure.

42. Examine carefully the figure in the text.

There are twelve 1 × 1 squares.

There are five 2 × 2 squares.
(Be sure not to count any of them twice.)

This is a small square in the center whose sides are the diagonals of the four 1 × 1 squares in the center of the figure. Since each side of this square is the hypotenuse of an isosceles right triangle with legs of length 1, the side(s) of this square is(are) $\sqrt{2}$ (Since, $s^2 = 1^2 + 1^2$). The larger square in the middle has sides composed of two diagonals, each of which are $\sqrt{2}$ units long. Thus, the length of each side of this square is $2\sqrt{2}$ units.

Add the results:

Size	Number of squares
1 × 1	12
2 × 2	5
$\sqrt{2} \times \sqrt{2}$	1
$2\sqrt{2} \times 2\sqrt{2}$	1
	19

Thus, the figure contains 19 squares.

43. Examine carefully the figure in the text.

There are sixteen 1 × 1 squares with horizontal bases.

There are three 2 × 2 squares in the each of the first and second rows, the second and third rows, as well as the third and fourth rows with horizontal bases. Thus, there are a total of nine 2 × 2 squares with horizontal bases.

There are two 3 × 3 squares with horizontal bases found in the first, second, and third rows as well as two 3 × 3 squares with horizontal base found in the second, third and fourth rows. Thus, there are a total of four 3 × 3 squares with horizontal bases.

There is one 4 × 4 square (the large square itself).

Visualize the squares along the diagonals (at a slant).

There are twenty-four 1 × 1 squares with bases along diagonals.

There are thirteen 2 × 2 squares with bases along diagonals.

There are four 3 × 3 squares with bases along diagonals.

There is only one 4 × 4 square with bases along diagonals. Add the results:

Size	Number of squares
1 × 1 (horizontal)	16
1 × 1 (slant)	24
2 × 2 (horizontal)	9
2 × 2 (slant)	13
3 × 3 (horizontal)	4
3 × 3 (slant)	4
4 × 4 (horizontal)	1
4 × 4 (slant)	1
	72

There are 72 squares in the figure.

44. The dimensions of the stack are $5 \times 3 \times 2$, so there are 30 small cubes altogether. Of these, the cubes listed in the following table are visible:

Location	Number of cubes
Top layer	15
Bottom layer, along side	5
Bottom layer, along front (exclude cube in corner which has already been counted)	2
	22

    Thus, the number of cubes in the stack that are not visible is: $30 - 22 = 8$.

45. There are $3 \times 3 = 9$ cubes in each of the bottom two layers. This gives a total of 18 in the bottom two layers.

    There are $3 \times 2 = 6$ cubes in each of the middle to layers. This gives a total of 12 in the middle two layers

    There are $3 \times 1 = 3$ cubes in the top two layers. This gives a total of 6 in the top two layers. Altogether, there are

    $$18 + 12 + 6 = 36$$

    $(1 \times 1 \times 1)$ cubes.

    The visible cubes are:

Location	Number of cubes
Top two layers	6
Middle two layers	8
Bottom two layers (exclude cubes in corners which have already been counted)	10
	24

    Thus, the number of cubes in the stack that are not visible is: $36 - 24 = 12$. One could ignore the top two levels since each cube is visible.

46. There are $4 \times 4 = 16$ small cubes on the bottom layer; $3 \times 3 = 9$ on the second layer; $2 \times 2 = 4$ on the third layer, and 1 on the top layer. This gives a total of

    $$16 + 9 + 4 + 1 = 30$$

    $(1 \times 1 \times 1)$ cubes. Of these, the following are visible.

Location	Number of cubes
Bottom layer	7
Second layer	5
Third layer	3
Top layer	1
	16

    Thus, the number of cubes in the stack that are not visible is: $30 - 16 = 14$.

47. There are 4 cubes along each edge of the bottom layer for a total of 10 cubes. There are 3 cubes along each edge of the second layer for a total of 6 cubes. There are 2 cubes along each edge of the third layer for a total of 3 cubes. Remember not to count the back corner cube twice. The top layer cube is visible, so ignore it. Thus, there are a total of

    $$10 + 6 + 3 = 19$$

    $(1 \times 1 \times 1)$ cubes in the bottom three layers. Of these, the following are visible.

Location	Number of cubes
Bottom layer	4
Second layer	3
Third layer	2
	9

    Thus, the number of cubes in the stack that are not visible is: $19 - 9 = 10$.

48. Label the figure as shown below.

    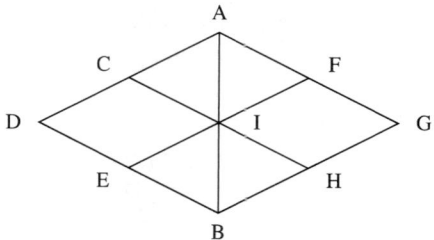

    List all the paths in a systematic way.

    Path down middle of figure: 1;

    AIB.

    Paths down left half of figure: 4;

    ACIB, AIEB, ACDEB, ACIEB.

    Paths down right half of figure: 4;

    AFIB, AIHB, AFGHB, AFIHB.

    Paths crossing between left and right halves of figure: 2;

    ACIHB, AFIEB.

    Thus, there are: $1 + 4 + 4 + 2 = 11$ downward paths from A to B.

49. Label the figure as shown below.

    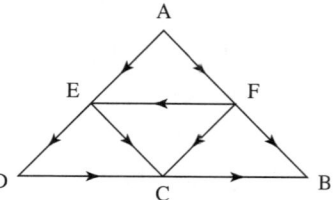

    List all the paths in a systematic way.

    AFB, AFCB, AECB, AEDCB, AFECB, and AFEDCB represent all paths with the indicated restrictions. Thus, there are 6 paths.

50. To determine the number of ways in which 30 can be written as the sum of two primes, use trial and error in a systematic manner. Test each prime, starting with 2, as a possibility for the smaller prime.

    (Since $30 - 2 = 28$, and 28 is not a prime, 2 will not work.) We obtain the following list:
    $$30 = 7 + 23$$
    $$30 = 11 + 19$$
    $$30 = 13 + 17.$$

    Thus, 30 can be written as the sum of two primes in 3 different ways.

51. To determine the number of ways in which 40 can be written as the sum of two primes, use trial and error in a systematic manner. Test each prime, starting with 2, as a possibility for the smaller prime.

    (Since $30 - 2 = 28$, and 28 is not a prime, 2 will not work.) We obtain the following list:
    $$40 = 3 + 37$$
    $$40 = 11 + 29$$
    $$40 = 17 + 23.$$

    Thus, 40 can be written as the sum of two primes in 3 different ways.

52. To determine the number of ways in which 95 can be written as the sum of two primes, follow the same procedure that was used in Exercise 50. We need to test each prime from 2 through 47, since 47 is the largest prime that is less than half of 95. Testing all these possibilities will show that there are no ways in which 95 can be written as the sum of two primes.

    Another approach to this problem is less time-consuming. If 95 can be written as the sum of two primes, we would have
    $$95 = p + q$$
    where p and q are primes.

    If $p = 2$, $q = 95 - 2 = 93 = 3 \cdot 31$, which is not a prime.

    If $p > 2$, p must be odd, since 2 is the only even prime.

    Thus, $q = 95 - p$ must be even, since the difference between two odd numbers is even. But if q is even and is not 2, q cannot be prime.

    We conclude that it is not possible to write 95 as the sum of two primes.

53. A group of twelve strangers sat in a circle, and each one got acquainted only with the person to the left and the person to the right. Then all twelve people stood up and each one shook hands (once) with each of the others who was still a stranger. How many handshakes occurred?

    One strategy is to place and number each person as on a 12-hour clock. Beginning with person #1: he will shake hands with 9 people. This is also true for #2. But person #3 will only shake hands with 8 people, having already shaken hands with #1. Similarly, #4 will shake hands with 7 people, #5 with 6 people, #6 with 5 people, #7 with 4 people, #8 with 3 people, #9 with 2 people, #10 with 1 person.

    Adding, we get 54 handshakes.

54. If there are no ties, each time a game is played, the loser is eliminated. In order to determine the champion, 49 people (all but the champion) must be eliminated. Thus, it will take 49 games to determine the champion.

55. How many of the numbers from 10 through 100 have the sum of their digits equal to a perfect square?

    One strategy would be to list the numbers from 10 to 100. Then check each by adding digits: e.g., 10, yes (since $1 + 0 = 1$, a perfect square number). Similarly, 13, yes (since $1 + 3 = 4$) and so on ... until 100, yes (since $1 + 0 + 0 = 1$). In total there are 18 such numbers: $\{10, 13, 18, 22, 27, 31, 36, 40, 45, 54, 63, 72, 81, 90, 79, 88, 97, 100\}$.

56. Make a systematic list or table of all three-digit numbers that have the sum of their digits equal to 22. Notice that since the largest possible sum of two digits is $9 + 9 = 18$, the smallest possible third digit in any of these number is 4.

499					
589	598				
679	688	697			
769	778	787	796		
859	868	877	886	895	
949	958	967	976	985	994

    This table shows that there are
    $$1 + 2 + 3 + 4 + 5 + 6 = 21$$
    three-digit numbers that have the sum of their digits equal to 22.

57. Make a systematic list, or table, of numbers between 100 and 400 which contain the digit 2. As you are listing the numbers look for patterns.

		102	112					
120–129								
			132	142	152	162	172	182	192
200–299								
			302	312					
320–329								
			332	342	352	362	372	382	392

This table shows that there are

$2 + 10 + 7 + 100 + 2 + 10 + 7 = 138$ numbers

between 100 and 400 which contain the digit 2.

58. Abbreviate each of the choices as follows:

Soups	CC	= clam chowder
	MN	= minestrone
Salads	FS	= fresh spinach
	SH	= shrimp
Breads	SR	= sourdough roll
	BN	= bran muffin
Entrees	LS	= lasagna
	LB	= lobster
	RT	= roast turkey

    The seafood choices are CC, SH, and LB. No more than one of these can be included in the same meal. Also, if the soup choice is MN, the entree choice must be LS, and FS and BN cannot be included in the same meal.

    Construct a tree diagram which takes into account all of these restrictions.

Soup or Salad	Bread	Entree	Meal
CC	SR	LS	CC SR LS
		RT	CC SR RT
	BN	LS	CC BN LS
		RT	CC BN RT
MN	SR	LS	MN SR LS
	BN	LS	MN BN LS
FS	SR	LS	FS SR LS
		LB	FS SR LB
		RT	FS SR RT
SH	SR	LS	SH SR LS
		RT	SH SR RT
	BN	LS	SH BN LS
		RT	SH BN RT

    Counting the meals listed in the final column of the tree diagram, we see that Jim has 13 choices of a complete dinner.

59. Draw a tree diagram showing all possible switch settings.

First Switch	Second Switch	Third Switch	Fourth Switch	Switch Settings
0	0	0	0	0 0 0 0
			1	0 0 0 1
		1	0	0 0 1 0
			1	0 0 1 1
	1	0	0	0 1 0 0
			1	0 1 0 1
		1	0	0 1 1 0
			1	0 1 1 1
1	0	0	0	1 0 0 0
			1	1 0 0 1
		1	0	1 0 1 0
			1	1 0 1 1
	1	0	0	1 1 0 0
			1	1 1 0 1
		1	0	1 1 1 0
			1	1 1 1 1

    Thus, Michelle can choose 16 different switch settings.

60. Refer to Exercise 59 above. If no two adjacent switches can be off *and* no two adjacent switches can be on, the 0's and 1's must alternate. This happens only in the following switch settings:

    0101 and 1010

    Thus, Michelle can choose only 2 different switch settings.

61. There are five switches rather than four, and no two adjacent switches can be on. If no two adjacent switches can be on, the tree diagram that is constructed will not have two "1"s in succession.

11.1 COUNTING BY SYSTEMATIC LISTING

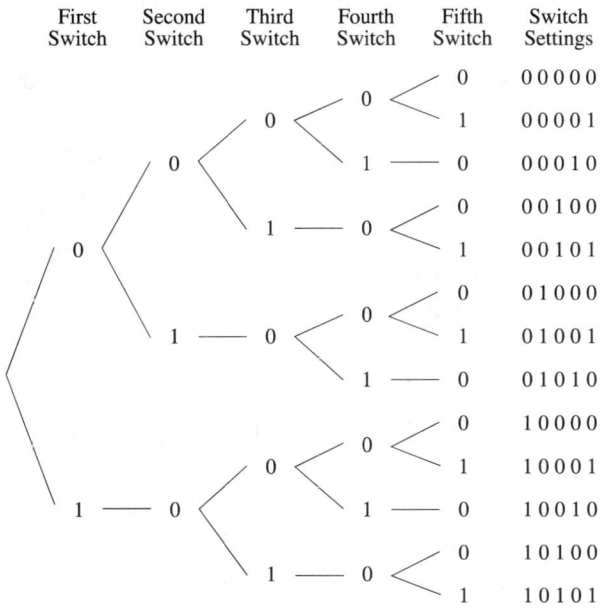

Thus, Michelle can choose 13 different switch settings.

62. Make a systematic list, taking into account the following restrictions:

    1. The first digit cannot be zero.
    2. There may be no repeated digits.
    3. The last digit must be 1 or 3, so that the number will be odd.

    We obtain the following list. (A tree diagram may also be used.)

    $$103, \ 123,$$
    $$201, \ 231, \ 203, \ 213,$$
    $$301, \ 321$$

    Thus, there are 8 possible numbers.

63. A line segment joins the points $(8, 12)$ and $(53, 234)$ in the Cartesian plane. Including its endpoints, how many lattice points does this line segment contain? (A lattice point is a point with integer coordinates.)

    Any point (x, y) on the line segment, when used with either endpoint, must yield the same slope as that of the segment using both endpoints. Therefore, find the slope of the segment.

    $$m = \frac{y_2 - y_1}{x_2 - x_1}$$
    $$= \frac{234 - 12}{53 - 8}$$
    $$= \frac{222}{45}$$
    $$= \frac{74}{15}$$

Set up the slope using the unknown point (x, y) and the known endpoint $(8, 12)$.

$$m = \frac{y_2 - y_1}{x_2 - x_1}$$
$$= \frac{y - 12}{x - 8}$$

Since the slope is the same for all points on the line segment, set these equal to each other and solve for y (in terms of x).

$$\frac{y - 12}{x - 8} = \frac{74}{15}$$
$$y - 12 = \frac{74}{15}(x - 8)$$
$$y = \frac{74}{15}(x - 8) + 12$$

All points on the line segment must be solutions for this equation. For the solutions to be integers (with x between 8 and 53), the denominator 15 will have to divide the value $(x - 8)$ evenly so that y remains an integer. That is, the number "$x - 8$" must be a multiple of 15 for values of x. Systematically trying integers for x from and including 8 to 53 will yield the following results. All other values for x between 8 and 53 would not.

x	$x - 8$	Divisible by 15?
8	$8 - 8 = 0$	Yes
23	$23 - 8 = 15$	Yes
38	$38 - 8 = 30$	Yes
53	$53 - 8 = 45$	Yes

Thus, including the endpoints, there are 4 lattice points.

Note that using a graphing calculator with a table feature would provide a quicker solution. Set up the function

$$y = \frac{74}{15}(x - 8) + 12$$

in the calculator. Adjust the "table set" feature to begin at $x = 8$ with $\triangle x$ set to increase by 1 and create the table (set of ordered pairs) for x and y. Scanning the table for those y-values which are whole numbers will yield the same answers as above.

64. (a) In each row, the 5 dots form 4 segments of length 1. Since there are 5 rows, we can draw $5 \times 4 = 20$ horizontal segments of length 1. Likewise, we can draw 20 vertical segments of length 1. Thus, there are 40 segments of length 1.

    (b) In each row, we can draw 3 segments of length 2. (We may start at the first, second, or third dot in the row and draw a segment extending two dots to the right.) Since there are 5 rows, this gives $5 \times 3 = 15$ horizontal segments of length 2. Likewise, we can draw 15 vertical segments of length 2. Thus, there are 30 segments of length 2.

(c) In each row, we can draw 2 segments of length 3, starting at the first and second dots. Since there are 5 rows, this gives $5 \times 2 = 10$ horizontal segments of length 3. Likewise, there are 10 vertical segments of length 3. Thus, there are 20 segments of length 3.

(d) Connect all of the dots in each row with a horizontal segment of length 4. This gives $5 \times 1 = 5$ horizontal segments of length 1. Likewise, connect the dots in each column to form 5 vertical segments of length 4. Thus, there are 10 segments of length 4.

(e) By the Pythagorean theorem, we know that any right triangle with legs of lengths 3 and 4 will have a hypotenuse of length 5, since

$$3^2 + 4^2 = 5^2.$$

You might recall from Chapter 9, that we call (3,4, 5) a *Pythagorean triple*. Beginning in the upper left corner we can create two right triangles with legs of lengths 3 and 4. Using 4 segments across the top (connecting all of the dots) and 3 segments down the left side will give one desired right triangle whose hypotenuse would be 5 units long. The second may be formed by using 3 segments across the top and 4 segments down the left side, again creating the hypotenuse 5 units long. Since corresponding right triangles can be formed at each of the four corners, there will be $4 \times 2 = 8$ such triangles that can be formed. Therefore, there are 8 segments of length 5.

65. Uniform length matchsticks are used to build a rectangular grid as shown in the figure (in the text). If the grid is 15 matchsticks high and 28 matchsticks wide, how many matchsticks are used?

Each row will contain 28 matchsticks. If the grid is 15 matchsticks high there will be 16 rows of matchsticks (including the top and bottom rows). Therefore, there are $16 \times 28 = 448$ matchsticks in all of the rows. Each column contains 15 matchsticks. Thus, there are $15 \times 29 = 435$ matchsticks in the rows. If the grid is 28 matchsticks wide, then there will be 29 columns of matchsticks counting the first and last columns. Thus, there are $29 \times 15 = 435$ matchsticks in the columns. Altogether, the number of matchsticks are

$$448 + 435 = 883.$$

66. (a) Let us call Square 1 (in the center) the smallest possible square that uses blue and red tiles and has two diagonals that have a common blue tile at the center. Square 1 has 4 blue tiles in the corners and 1 blue tile at the center, or $4 + 1 = 5$ blue tiles. Also, Square 1 has $2 + 1 = 3$ tiles on a side and $3^2 = 9$ tiles altogether. Then, there are $9 - 5 = 4$ red tiles in Square 1.

Now moving outward from the center, consider the next square, Square 2. It has $4 \cdot 2 + 1 = 9$ blue tiles. (Count them!) Also, Square 2 has $2 \cdot 2 + 1 = 5$ tiles on a side

and $5^2 = 25$ tiles altogether. Then Square 2 has $25 - 9 = 16$ red tiles.

Make a table and notice the pattern that emerges in each column.

Name of square	Number of blue tiles	Total number of tiles
Square 1	$4 \cdot 1 + 1$	$(2 \cdot 1 + 1)^2$
Square 2	$4 \cdot 2 + 1$	$(2 \cdot 2 + 1)^2$
Square 3	$4 \cdot 3 + 1$	$(2 \cdot 3 + 1)^2$
$\vdots$	$\vdots$	$\vdots$
Square k	$4 \cdot k + 1$	$(2 \cdot k + 1)^2$

If 81 blue tiles are used, find the number of the square, k.
$$4 \cdot k + 1 = 81$$
$$4k = 80$$
$$k = 20$$

The total number of tiles in Square 20 is

$$(2 \cdot 20 + 1)^2 = (41)^2 = 1681.$$

The number of red tiles is the difference between the total number of tiles and the number of blue tiles,

$$1681 - 81 = 1600.$$

(b) The numbers in place of 81 that would make the problem solvable are the numbers of blue tiles, that is, numbers of the form $4k + 1$, where k is any positive integer.

(c) A formula expressing the number of red tiles required in general is found by subtracting the expression for the number of blue tiles from the expression for the total number of tiles:

$$(2k + 1)^2 - (4k + 1)$$
$$= 4k^2 + 4k + 1 - 4k - 1$$
$$= 4k^2.$$

67. There are 8 people. Let J represent Jeff and 0–7 the other 7 people who meet 0–6 others respectively. One of these is Jeff's son. When we find how many people Jeff had to meet, we will know who his son is since the son met the same number of people as Jeff. One method to solve the problem is to use sets. Each set contains the people that the designated person met.

$0 : \{\}$; he met nobody.
$6 : \{1, 2, 3, 4, 5, J\}$; 6 met all but 0.
$1 : \{6\}$; 5 met 1 (from 6's set) so 6 is the only person in the set.
$5 : \{2, 3, 4, 6, J\}$; 5 met everyone but 0 and 1.
$2 : \{5, 6\}$; 5 and 6 met 2 (from their sets) so 5 and 6 are the only people 2 met.
$4 : \{3, 5, 6, J\}$: 4 met everyone but 0, 1, and 2.
$3 : \{4, 5, 6\}$; 4, 5, and 6 met 3 (from their sets) so 4, 5,

and 6 are the only people that 3 met.
J: {4, 5, 6}; same reasoning as 3, above.

Therefore, Jeff met three people and shook three hands, and "3" is his son.

Wording may vary in answers for Exercises 68–71.

68. (a) Determine the number of two-digit numbers that can be formed using digits from the set $\{1, 2, 3\}$ if repetition of digits is allowed.

    (b) Determine the number of two-digits numbers that can be formed using digits from the set $\{1, 2, 3\}$ if the selection is done with replacement.

69. (a) Determine the number of ordered pairs of digits that can be selected from the set $\{1, 2, 3, 4, 5, 6\}$ if repetition of digits is allowed.

    (b) Determine the number of ordered pairs of digits that can be selected from the set $\{1, 2, 3, 4, 5, 6\}$ if the selection is done with replacement.

70. (a) Find the number of ways to select an ordered pair of letters from the set $\{A, B, C, D, E\}$ if repetition of letters is not allowed.

    (b) Find the number of ways to select an ordered pair of letters from the set $\{A, B, C, D, E\}$ if the selection is done without replacement.

71. (a) Find the number of ways to select three letters from the set $\{A, B, C, D, E\}$ if repetition of letters is not allowed.

    (b) Find the number of ways to select three letters from the set $\{A, B, C, D, E\}$ if the selection is done without replacement.

11.2 EXERCISES

1. Writing exercise
2. Writing exercise
3. (a) No, $(n + m)! \neq n! + m!$
   (b) Writing exercise
4. (a) No, $(n \cdot m)! \neq n! \cdot m!$
   (b) Writing exercise

Evaluate each expression without using a calculator.

5. $4! = 4 \cdot 3 \cdot 2 \cdot 1 = 24$

6. $7! = 7 \cdot 6 \cdot 5 \cdot 4 \cdot 3 \cdot 2 \cdot 1 = 5040$

7. $\dfrac{8!}{5!} = \dfrac{8 \cdot 7 \cdot 6 \cdot \cancel{5} \cdot \cancel{4} \cdot \cancel{3} \cdot \cancel{2} \cdot \cancel{1}}{\cancel{5} \cdot \cancel{4} \cdot \cancel{3} \cdot \cancel{2} \cdot \cancel{1}} = 336$

8. $\dfrac{16!}{14!} = \dfrac{16 \cdot 15 \cdot \cancel{14!}}{\cancel{14!}}$
   $= 16 \cdot 15$
   $= 240$

9. $\dfrac{5!}{(5-2)!} = \dfrac{5!}{3!} = \dfrac{5 \cdot 4 \cdot \cancel{3} \cdot \cancel{2} \cdot \cancel{1}}{\cancel{3} \cdot \cancel{2} \cdot \cancel{1}} = 20$

10. $\dfrac{6!}{(6-4)!} = \dfrac{6!}{2!}$
    $= \dfrac{6 \cdot 5 \cdot 4 \cdot 3 \cdot \cancel{2!}}{\cancel{2!}}$
    $= 6 \cdot 5 \cdot 4 \cdot 3$
    $= 360$

11. $\dfrac{9!}{6!(6-3)!} = \dfrac{9!}{6!3!}$
    $= \dfrac{9 \cdot 8 \cdot 7 \cdot \cancel{6} \cdot \cancel{5} \cdot \cancel{4} \cdot \cancel{3} \cdot \cancel{2} \cdot \cancel{1}}{\cancel{6} \cdot \cancel{5} \cdot \cancel{4} \cdot \cancel{3} \cdot \cancel{2} \cdot \cancel{1} \cdot 3 \cdot 2 \cdot 1}$
    $= \dfrac{\cancel{9}^3 \cdot \cancel{8}^4 \cdot 7}{\cancel{3} \cdot \cancel{2}} = \dfrac{3 \cdot 4 \cdot 7}{1}$
    $= 84$

12. $\dfrac{10!}{4!(10-4)!} = \dfrac{10!}{4!6!}$
    $= \dfrac{10 \cdot 9 \cdot 8 \cdot 7 \cdot \cancel{6!}}{4 \cdot 3 \cdot 2 \cdot 1 \cdot \cancel{6!}}$
    $= 5 \cdot 3 \cdot 2 \cdot 7$
    $= 210$

13. Evaluate
    $\dfrac{n!}{(n-r)!}$, where $n = 7$ and $r = 4$.
    $\dfrac{7!}{(7-4)!} = \dfrac{7!}{3!}$
    $= \dfrac{7 \cdot 6 \cdot 5 \cdot 4 \cdot \cancel{3} \cdot \cancel{2} \cdot \cancel{1}}{\cancel{3} \cdot \cancel{2} \cdot \cancel{1}}$
    $= 840$

14. Evaluate
    $\dfrac{n!}{r!(n-r)!}$, where $n = 12$ and $r = 4$.
    $\dfrac{12!}{4!(12-4)!} = \dfrac{\cancel{12} \cdot 11 \cdot 10 \cdot 9 \cdot \cancel{8!}}{\cancel{4} \cdot \cancel{3} \cdot 2 \cdot 1 \cdot \cancel{8!}}$
    $= \dfrac{11 \cdot 10 \cdot 9}{2}$
    $= 11 \cdot 5 \cdot 9$
    $= 495$

Evaluate each expression using a calculator. (Some answers may not be exact.) For Exercises 15–24, use the factorial key on a calculator, which is labeled $\boxed{x!}$ or $\boxed{n!}$ or, if using a graphing, calculator find "!" in the "Math" menu.

15. $11! = 39,916,800$

16. $17! = 3.556874281 \times 10^{14}$

17. $\dfrac{12!}{7!} = 95,040$

18. $\dfrac{15!}{9!} = 3,603,600$

19. $\dfrac{13!}{(13-3)!} = \dfrac{13!}{10!} = 1716$

20. $\dfrac{16!}{(16-6)!} = \dfrac{16!}{10!} = 5,765,760$

21. $\dfrac{20!}{10! \cdot 10!} = 184,756$

22. $\dfrac{18!}{6! \cdot 12!} = 18,564$

23. Evaluate
$$\dfrac{n!}{(n-r)!}, \text{ where } n = 23 \text{ and } r = 10.$$
$$\dfrac{23!}{(23-10)!} = \dfrac{23!}{13!}$$
$$= 4.151586701 \times 10^{12}$$

24. Evaluate
$$\dfrac{n!}{r!(n-r)!}, \text{ where } n = 28 \text{ and } r = 15.$$
$$\dfrac{28!}{15! \cdot (28-15)!} = \dfrac{28!}{15!13!}$$
$$= 37,442,160$$

25. Since there are two possible outcomes for each switch (on/off), we have $2 \cdot 2 \cdot 2 = 2^3 = 8$.

26.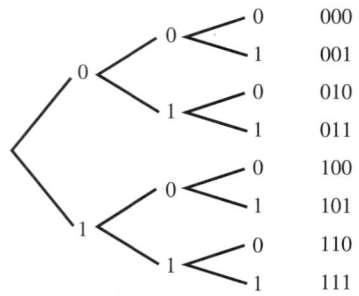

27. Writing exercise

28.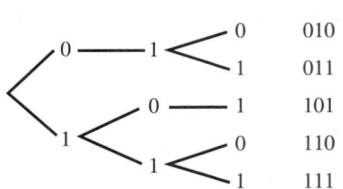

29. Using the fundamental counting principle, this may be considered as a three–part task. There would be
$$\underline{6} \cdot \underline{6} \cdot \underline{6} = 6^3 = 216.$$

30. Selecting a five digit number is a five–part task. There are nine ways to choose the first digit (1 through 9), and then ten ways to choose each of the remaining four digits (0 through 9). By the fundamental counting principle, the total number of possibilities is
$$\underline{9} \cdot \underline{10} \cdot \underline{10} \cdot \underline{10} \cdot \underline{10} = 9 \cdot 10^4 = 90,000.$$

Recall the club $N = \{$Andy, Bill, Cathy, David, Evelyn$\}$.

31. This is a 5-part task. Use the fundamental counting principle to count the number of ways of lining up all five members for a photograph. There would be
$$\underline{5} \cdot \underline{4} \cdot \underline{3} \cdot \underline{2} \cdot \underline{1} = 120 \text{ possibilities}.$$
Similarly, one could use $n!$, where $n = 5$, to arrive at the total number or arrangements of a set of n objects.

32. This is a 5-part task, which involves selecting one member to work on each of the five days. Since there are 5 members to choose from and repeats are allowed, the number of possibilities is
$$\underline{5} \cdot \underline{5} \cdot \underline{5} \cdot \underline{5} \cdot \underline{5} \cdot = 5^5 = 3125.$$

33. This is a 2-part task. Since there are three males to choose from and two females to choose from, the number of possibilities is
$$\underline{3} \cdot \underline{2} = 6.$$

34. This is a 2-part task. Since Bill will not be present, there are only 4 members who can be selected to open the meeting. Since the same person cannot be appointed to close the meeting, there will only be 3 choices for that position. Using the fundamental counting principle, the number of ways to appoint the two members is
$$\underline{4} \cdot \underline{3} = 12.$$

In the following exercises, counting numbers are to be formed using only digits from the set {3, 4, 5}.

35. Choosing two-digit numbers may be considered a 2-part task. Since we can use any of the three given digits for each choice, there are
$$\underline{3}\cdot\underline{3} = 9$$
different numbers that can be obtained.

36. This is a 3-part task where the restriction that the number be odd allows one to choose only an odd number for the unit's place. Any of the three given numbers may be used for the hundred's place and the ten's place. Thus, there are
$$\underline{3}\cdot\underline{3}\cdot\underline{2} = 18$$
ways to choose these numbers.

37. Using the textbook hint, this may be considered a 3-part task. (1) Since there are only 3 positions that the two adjacent 4's can take (1st and 2nd, 2nd and 3rd, and 3rd and 4th positions), (2) two remaining positions that the 3 can take, and (3) the one last remaining digit must filled by the 5, there are
$$\underline{3}\cdot\underline{2}\cdot\underline{1} = 6$$
different numbers that may be created.

38. The choice of five-digit numbers beginning and ending with 3 and with unlimited repetitions allowed may be considered a 5 part task, choosing each digit in succession. There is only one choice for the first and last digit.
$$\underline{1}\cdot\underline{3}\cdot\underline{3}\cdot\underline{3}\cdot\underline{1} = 27.$$

39. Choosing from each of the three food categories is a 3 part task. There are four choices from the soup and salad category, two from the bread category, and three from the entree category. Applying the fundamental counting principle gives
$$\underline{4}\cdot\underline{2}\cdot\underline{3} = 24$$
different dinners that may be selected.

40. There are two choices of soup and three choices of entree, so the number of available dinners is
$$\underline{2}\cdot\underline{3} = 6.$$

41. Since there are 2 choices (T or F) for each question, we have
$$2^6 = 64$$
possible ways.

42. This is a 20-part task. Since there are five answer choices for each of the 20 questions, the number of possibilities is
$$5^{20} = 9.536743164 \times 10^{13}.$$
Note: Your calculator may not display all of these digits.

For each situation in Exercises 43–48, use the table in the text to determine the number of different sets of classes Tiffany can take.

43. All classes shown are available. Choose the number of possible courses from each category and apply the fundamental theorem:
$$\underline{3}\cdot\underline{3}\cdot\underline{4}\cdot\underline{5} = 180.$$

44. She is not eligible for Modern Poetry or for C Programming. This reduces the number of choices in two of the categories, resulting in
$$\underline{2}\cdot\underline{3}\cdot\underline{3}\cdot\underline{5} = 90$$
different sets of classes that she may take.

45. All sections of Minorities in America and Women in American Culture are filled already. The filled classes reduce the options in the Sociology category by 2. Thus, there are
$$\underline{3}\cdot\underline{3}\cdot\underline{4}\cdot\underline{3} = 108$$
possible class schedules.

46. She does not have the prerequisites for Medieval Literature, Finite Mathematics, or C Programming. Due to her lack of prerequisites, Tiffany is restricted to 2 choices in English, 2 choices in Mathematics, and 3 choices in Computer Information Science. She may take any of the 5 Sociology courses. The number of different sets of courses she can take is
$$\underline{2}\cdot\underline{2}\cdot\underline{3}\cdot\underline{5} = 60.$$

47. Funding has been withdrawn for three of the computer courses and for two of the Sociology courses. The reductions to the computer and Sociology categories leave only 1 computer course and 3 sociology courses to choose from. Thus, there are
$$\underline{3}\cdot\underline{3}\cdot\underline{1}\cdot\underline{3} = 27$$
possible class schedules.

48. She must complete English Composition and Aging in America next semester to fulfill her degree requirements. This reduces the number of choices for English and Sociology to one for the next semester. Thus, the number of different sets of courses she can take is
$$\underline{1}\cdot\underline{3}\cdot\underline{4}\cdot\underline{1} = 12.$$

49. This is a 3-part task. Applying the fundamental counting principle, there are
$$\underline{2}\cdot\underline{5}\cdot\underline{6} = 60$$
different outfits that Sean may wear.

50. This is a 4-part task. The number of different complete setups Lionel could choose is
$$\underline{10}\cdot\underline{4}\cdot\underline{6}\cdot\underline{3} = 720.$$

51. The number of different zip codes that can be formed using all of those same five digits, 86726, would be the number of arrangements of the 5 digits. This is given by

$$\underline{5} \cdot \underline{4} \cdot \underline{3} \cdot \underline{2} \cdot \underline{1} = 5! = 120.$$

52. (a) With the men first, the following choices can be made for each of the seven memory locations in his digital phone memory:

$$\underline{3} \cdot \underline{2} \cdot \underline{1} \cdot \underline{4} \cdot \underline{3} \cdot \underline{2} \cdot \underline{1} = 3! \cdot 4! = 144.$$

(b) This may be considered a 3-part task. With the men listed together, make the first task the choice of position in the 7 memory locations for the men. There are 5 locations for the men (1st three memory locations, 2nd three memory locations and so on, until the last three memory locations). The second task will be to select the number of choices for the relative position for each male person, followed the 3rd task, selecting the number of choices for the women in the remaining memory locations. Applying the fundamental counting principle there are

$$\underline{5} \cdot \underline{3} \cdot \underline{2} \cdot \underline{1} \cdot \underline{4} \cdot \underline{3} \cdot \underline{2} \cdot \underline{1} = 5 \cdot 3! \cdot 4! = 720$$

different ways to list them.

(c) Similar to (b), this may be considered a 3-part task. The first task is to locate the positions of the men. It is helpful to underscore seven places side by side and place M in the allowable positions (we are making use of the systematic methods suggested in the last section of the text). Begin with a man in the first location. Then count the number of other places to put men. There are 6 ways to leave a man in position 1 and choose other positions so that the condition (no men are listed together) is satisfied. A second set of groupings is possible beginning with a man in position number two. There are 3 ways to leave a man in position 2 and choose other positions so that the condition is satisfied. Finally, beginning with a man in position 3, there is only 1 way (they must alternate with the women). Thus, there are

$$6 + 3 + 1 = 10$$

ways to position the men to satisfy the condition, no men are listed together. The second task is to count the number of ways to choose the relative position of the three men. To find the number of ways to arrange the three men positions use 3! The 3rd task is to choose the number of ways to arrange the women once the men's locations are fixed. There are 4! arrangements for the women. Thus, by the fundamental counting principle, there are

$$10 \cdot 3! \cdot 4! = 1440$$

ways for John to keep his friends numbers so that no two men are together.

Aaron (A), Bobbette (B), Chuck (C), Deirdre (D), Ed (E), and Fran (F) have reserved six seats in a row at the theater, starting at an aisle seat.

53. Using the textbook hint, divide the task into the series of six parts shown below, performed in order.

(a) If A is seated first, how many seats are available for him?

Six seats are available.

(b) Now, how many are available for B?

Five are available since A is already seated.

(c) Now, how many are available for C?

Four are available since A and B are already seated.

(d) Now, how many for D?

Three, since A, B, and C are already seated.

(e) Now, how many for E?

Two, since A, B, C, and D are already seated.

(f) Now, how many for F?

One, since all other seats have been taken.

Thus, by the fundamental counting principle, there are

$$6 \cdot 5 \cdot 4 \cdot 3 \cdot 2 \cdot 1 = 720$$

arrangements.

54. Using the textbook hint, think of the problem as a six-part task, with the six parts described in (a)–(f) below.

(a) From the sketch in the textbook, we see that there are 5 pairs of adjacent seats that Aaron and Bobbette can occupy.

(b) Given the two seats for A and B, they can be seated in 2 orders (A to the left of B, or B to the left of A).

(c) C may occupy any of the 4 seats which are not taken by A or B.

(d) Once A, B, and C have been seated, there are 3 remaining seats available for D.

(e) Once A, B, C, and D have been seated, there are 2 remaining seats available for E.

(f) Once all of the others have been seated, there is only 1 seat left for F.

Applying the fundamental counting principle, we conclude that the number of ways in which the six people can arrange themselves so that Aaron and Bobbette will be next to each other is

$$5 \cdot 2 \cdot 4 \cdot 3 \cdot 2 \cdot 1 = 240.$$

55. In how many ways can they arrange themselves if the men and women are to alternate seats and a man must sit on the aisle? Using the textbook hint, first answer the following series of questions:

    (a) How many choices are there for the person to occupy the first seat, next to the aisle? (It must be a man.)

    Three, since there are only 3 men.

    (b) Now, how many choices of people may occupy the second seat from the aisle? (It must be a woman.)

    Three, since there are only 3 women.

    (c) Now, how many for the third seat (one of the remaining men)?

    Two, since one of the men has already been seated.

    (d) Now, how many for the fourth seat (a woman)?

    Two, since one of the women has already been seated.

    (e) Now, how many for the fifth seat (a man)?

    One, since only one man remains.

    (f) Now, how many for the sixth seat (a woman)?

    One, since only one woman remains.

    Thus, there are
    $$3 \cdot 3 \cdot 2 \cdot 2 \cdot 1 \cdot 1 = 36$$
    arrangements.

56. Using the text book hint, think of the problem as a six-part task, with the six parts described in (a)–(f) below.

    (a) Any of the 6 people may sit on the aisle, so there are 6 choices.

    (b) There are 3 men and 3 women. For whoever sits on the aisle, any of the 3 people of the opposite sex may occupy the second seat.

    (c) For the third seat, we may choose either of the 2 remaining people of the first sex chosen.

    (d) For the fourth seat, we may choose either of the 2 remaining people of the second sex chosen.

    (e) For the fifth seat, there is only 1 choice, the remaining person of the first sex chosen.

    (f) For the sixth seat, there is only one person left, so there is just 1 choice.

    Multiplying the answers from (a)–(f), we conclude that the number of ways the people can arrange themselves if the men and women are to alternate with either a man or woman on the aisle is
    $$6 \cdot 3 \cdot 2 \cdot 2 \cdot 1 \cdot 1 = 72.$$

57. Writing exercise

58. Writing exercise

59. Repeat Example 4. This time, allow repeated digits. Does the order in which digits are considered matter in this case?

    There are 9 choices for the first digit since it can't be 0. There are 10 choices for the second digit. Finally, there are only 5 choices for the third (unit's) digit since it must be odd. Thus, by the fundamental counting principle there are
    $$9 \cdot 10 \cdot 5 = 450.$$
    The order doesn't matter.

11.3 EXERCISES

Evaluate each of the following expressions.

1. Begin at 6, use four factors.
   $$P(6,4) = 6 \cdot 5 \cdot 4 \cdot 3 = 360$$
   Alternatively, use the factorial formula for permutations
   $$P(n,r) = \frac{n!}{(n-r)!}$$
   where $n = 6$ and $r = 4$.
   $$P(6,4) = \frac{6!}{(6-4)!}$$
   $$= \frac{6!}{(6-4)!}$$
   $$= \frac{6!}{2!}$$
   $$= 360$$

2. Begin at 12, use two factors.
   $$P(12,2) = 12 \cdot 11 = 132$$

3. $C(15,3) = \dfrac{P(15,3)}{3!}$
   $$= \frac{P(15,3)}{3!}$$
   $$= \frac{15 \cdot 14 \cdot 13}{3 \cdot 2 \cdot 1}$$
   $$= \frac{5 \cdot 7 \cdot 13}{1}$$
   $$= 455$$

   Alternatively, use the factorial formula for combinations
   $$C(n,r) = \frac{n!}{r!(n-r)!}$$
   where $n = 15$ and $r = 3$.

$$C(15,3) = \frac{15!}{3!(15-3)!}$$
$$= \frac{15 \cdot 14 \cdot 13 \cdot 12!}{(3 \cdot 2 \cdot 1) \cdot 12!}$$
$$= \frac{15 \cdot 14 \cdot 13}{3 \cdot 2}$$
$$= 5 \cdot 7 \cdot 13$$
$$= 455$$

4. $C(13,6) = \dfrac{P(13,6)}{6!}$
$$= \frac{13 \cdot 12 \cdot 11 \cdot 10 \cdot 9 \cdot 8}{6 \cdot 5 \cdot 4 \cdot 3 \cdot 2 \cdot 1}$$
$$= 1716$$

Alternatively, you may choose to use the factorial formula for combinations.

5. To find the number of permutations of 21 things taken 4 at a time, evaluate $P(21,4)$. Begin at 21, use four factors.
$$P(21,4) = 21 \cdot 20 \cdot 19 \cdot 18 = 143640$$

Thus, the answer is $143{,}640$.

6. To find the number of permutations of 14 things taken 5 at a time, evaluate $P(14,5)$. Begin at 14, use 5 factors.
$$P(14,5) = 14 \cdot 13 \cdot 12 \cdot 11 \cdot 10 = 240240$$

Thus, the answer is $240{,}240$.

Determine the number of combinations (subsets) of each of the following.

7. To find the number of combinations of 10 things taken 5 at a time, evaluate $C(10,5)$.
$$C(10,5) = \frac{P(10,5)}{5!}$$
$$= \frac{10 \cdot 9 \cdot 8 \cdot 7 \cdot 6}{5 \cdot 4 \cdot 3 \cdot 2 \cdot 1}$$
$$= 252$$

Alternatively, use the factorial formula for combinations
$$C(n,r) = \frac{n!}{r!(n-r)!}$$
where $n = 10$ and $r = 5$.
$$C(10,5) = \frac{10!}{5!(10-5)!}$$
$$= \frac{10 \cdot 9 \cdot 8 \cdot 7 \cdot 6 \cdot 5!}{(5 \cdot 4 \cdot 3 \cdot 2 \cdot 1) \cdot 5!}$$
$$= \frac{10 \cdot 9 \cdot 8 \cdot 7 \cdot 6}{5 \cdot 4 \cdot 3 \cdot 2 \cdot 1}$$
$$= 252$$

8. To find the number of combinations of 13 things taken 4 at a time, evaluate $C(13,4)$.
$$C(13,4) = \frac{P(13,4)}{4!}$$
$$= \frac{13 \cdot 12 \cdot 11 \cdot 10}{4 \cdot 3 \cdot 2 \cdot 1}$$
$$= 715$$

Use a calculator to evaluate each expression.

9. Use the nPr or $P(n,r)$ button on a scientific calculator in the following order: 25 $\boxed{nPr}$ 8. Or, with a graphing calculator, find nPr. It is usually found in the probability menu. Insert, in the same order, 25 $\boxed{nPr}$ 8.
$$P(25,8) = 4.3609104 \times 10^{10}$$

10. Use the nCr or $C(n,r)$ button on a scientific calculator in the following order: 36 $\boxed{nCr}$ 14. Or, with a graphing calculator, find nCr. It is usually found in the probability menu. Insert, in the same order, 36 $\boxed{nCr}$ 14.
$$C(36,14) = 3{,}796{,}297{,}200$$

11. Writing exercise

12. Writing exercise

13. Writing exercise

14. Writing exercise

15. (a) Permutation, since the order of the digits is important.

    (b) Permutation, since the order of the digits is important.

    (c) Combination, since the order is unimportant.

    (d) Combination, since the order is unimportant.

    (e) Permutation, since the order of the digits is important.

    (f) Combination, since the order is unimportant.

    (g) Permutation, since the order of the digits is important.

16. Since a first place winner can not also be a second or third place winner in the same race, repetitions are not allowed. Also order is important. Therefore, we are able to use permutations. The number of ways in which 1st, 2nd, and 3rd place winners can occur in a race with six runners competing is given by
$$P(6,3) = 6 \cdot 5 \cdot 4 = 120.$$

17. Since 5 different models will be built, items cannot be repeated. Also, order is important. Therefore, we use permutations. The number of ways in which Jeff can place the homes on the lots is given by
$$P(8,5) = \frac{8!}{3!} = 6720.$$

18. With no repeated digits, there would be
$$P(10,4) = 10 \cdot 9 \cdot 8 \cdot 7 = 5040$$
possible PIN numbers.

19. Since no repetitions are allowed (one person will not be both) and the order of selection is important, the number of ways to choose a president and vice president is
$$P(12,2) = 12 \cdot 11 = 132.$$

20. Since no repetitions are allowed and the order of selection is important, the number of ways to choose the three prize winners is
$$P(10,3) = 10 \cdot 9 \cdot 8 = 720.$$

21. Since no repetitions are allowed and the order of selection is important, the number of ways for the teacher to give the five different prizes to her students is
$$P(25,5) = 25 \cdot 24 \cdot 23 \cdot 22 \cdot 21 = 6,375,600.$$

22. Since no repetitions are allowed and the order of selection is important, the number of ways for the security team to visit the twelve offices is
$$P(12,12) = 12! = 479,001,600.$$

23. (a) To get a sum of 10 we must use the digits $\{1,2,3,4\}$ since, $1 + 2 + 3 + 4 = 10$. They are not repeated and order is important. Thus, there are
$$P(4,4) = 4! = 24$$
such numbers.

(b) To get a sum of 11 we must use the digits $\{1,2,3,5\}$ since, $1 + 2 + 3 + 5 = 11$. They are not repeated and order is important. Thus, there are
$$P(4,4) = 4! = 24$$
such numbers.

24. Since the order of testing each sample is not important (samples are subsets), use combinations to arrive at
$$C(24,5) = 42,504$$
ways.

25. Samples are subsets, so use combinations. There are $24 - 6 = 18$ non-defective players, so we are to select 5 players from a set of 18. The number of samples which contain no defective players is
$$C(18,5) = \frac{18!}{5!13!} = 8568.$$

26. Consider the five-member committees as subsets of the 100 US Senators.
$$C(100,5) = \frac{100!}{5! \cdot 95!} = 75,287,520.$$
Alternatively, we could use
$$C(100,5) = \frac{P(100,5)}{5!} = 75,287,520.$$

27. (a) Any hand represents a combination (or subset) of cards. Here we are choosing 5 cards from the 13 diamonds available. Thus, there are
$$C(13,5) = \frac{13!}{5!8!} = 1287$$
five-card hands.

(b) Since there are 26 black cards in the deck (13 spades and 13 clubs), the number of hands containing all black cards is
$$C(26,5) = \frac{26!}{5!21!} = 65,780.$$

(c) There are only 4 aces making it impossible to draw such a hand. There are 0 ways to do so.

28. Since order is not important and repetition is not allowed, use combinations. The number of different ways to select the lottery numbers is
$$C(39,7) = 15,380,937.$$

29. (a) He has 6 lots to choose from. From the six, he can choose any three to build his standard homes on. Since the standard homes are all the same, the order is not important, and we have
$$C(6,3) = 20$$
possible combinations or choices. Once these have been chosen, the remaining 3 lots will contain the deluxe models.

(b) Jeff may choose the positions for the two deluxe models. Once this has been done, the four standard models must go in the four remaining positions, which can only be done in one way. The number of different positions is therefore
$$C(6,2) = \frac{6!}{2!4!} = 15.$$

Notice that the result will be the same if the four standard homes are positioned first. In this case, we obtain

$$C(6,4) = \frac{6!}{4!2!} = 15.$$

30. Use the hint furnished in the text. Calculate the number of ways of selecting the downward paths. For any selection of the downward steps the horizontal paths are fixed. Thus, there are

$$C(10,3) = 120$$

different paths that may be followed to get from A to B.

31. The baby's last name, Zahrndt, means the last initial of her name is "Z," the last letter of the alphabet. Thus, there are 25 letters, from which the other 2 will be chosen. Order does not matter because they will automatically be put in alphabetical order. The number of monograms is

$$C(25,2) = 300.$$

32. (a) The number of different lotto combinations is

$$C(47,5) = \frac{47!}{5!47!} = 1,533,939.$$

The number of MEGA ticket possibilities is

$$C(27,1) = 27.$$

By the fundamental counting principle the total number of 6-number sets is given by

$$C(47,5) \cdot C(27,1) = 1,533,939 \cdot 27.$$
$$= 41,416,353.$$

(b) Since two of Eleen's lottery numbers are already determined, she only needs to select 3 numbers from the remaining 45 numbers. The number of ways in which Eleen can complete her list is

$$C(45,3) \cdot C(27,1) = 383,130.$$

33. (a) The worst-case scenario is that the first four cards selected will be of different suits, then the 5th card chosen will be the same as one of the earlier choices. Any other scenario will require fewer cards to be drawn. Thus, there is a minimum of 5 cards to be drawn to obtain two cards of the same suit.

(b) By the 5th drawing we are guaranteed 2 cards are of the same suit. The 6th, 7th, and 8th drawings may give results that represent just 2 cards of the same suit for all 4 suits. But the 9th card must then must be of one of the 4 suits, adding a 3rd card of that same suit. There must be a minimum of 9 cards drawn to guarantee three cards of the same suit.

34. Use the fundamental counting principle, where the first task is to choose which suit. There are four possible suits to choose from. Once the choice of suit is made, the 2nd task is to choose how many ways one can select 5 cards from the 13 in any particular suit. Thus, the number of ways to choose five card hands of the same suit is

$$4 \cdot C(13,5) = 5148.$$

35. For each of the first and third groups, there are $P(26,3)$ possible arrangements of letters. For the second group, there are $P(10,3)$ possible arrangements for the digits. Thus, by the fundamental counting principle, there are

$$P(26,3) \cdot P(10,3) \cdot P(26,3) = 175,219,200,000$$

identification numbers.

36. Consider this to be a two-part problem: The number of choices for the first call letter, 2, followed by the number of arrangements of the following two letters can represent, $P(25,2)$. Observe that permutations are used since no letter may be repeated and that we are looking for arrangements (order is important). Choose the two letters from the remaining 25 that have not been selected. Apply the fundamental counting principle to arrive at

$$2 \cdot P(25,2) = 1200$$

sets of call letters.

37. Consider choosing the call letters for a station as a two-part task. The first part consists of choosing the first letter. Since the first letter must be K or W, this may be done in 2 ways. For the remaining three letters, we use permutations because order is important and repetition is not allowed. These three letters may be chosen from any of the 25 letters which were not used as the first letter, so the number of possibilities is $P(25,3)$. Use the fundamental counting principle to combine the results from the two parts of the task. The number of possible call letters is

$$2 \cdot P(25,3) = 27,600.$$

38. There are $C(8,2)$ different games to be played if each team plays another only once. Since the teams are scheduled to play each other three times we have

$$3 \cdot C(8,2) = 3 \cdot 28 = 84$$

games altogether.

39. This is a two-part task. First choose the pitcher. This can be done in 7 ways. Now choose the players and batting order for the rest of the team. Since order is important and repetition is not allowed, use permutations. The number of choices is $P(12,8)$. Use the fundamental counting principle to combine the

results from the two parts of the task. The number of different batting orders is

$$7 \cdot P(12,8) = 139,708,800.$$

40. (a) Use the fundamental counting principle where the first task is to choose the girls and the 2nd is to choose the boys. Since the order is important and there are no repetitions, use permutations to calculate each. The number of different programs is

$$P(8,8) \cdot P(7,7) = 8! \cdot 7! = 203,212,800.$$

(b) Consider the problem as a three-part task: Choosing the girl to perform first, the boy to perform last, followed by choosing the remaining 13 performers. There are 8 possibilities for the girl, 7 possibilities for the boy, and $P(13,13) = 13!$ possibilities for the other 13 performers. Thus, by the fundamental counting principle the number of programs is given by

$$8 \cdot 7 \cdot 13! = 348,713,164,800.$$

(c) Since the first and the last are fixed, we have only 1 choice for the first and last performer. Using the fundamental counting principle, the number of programs is given by

$$1 \cdot 13! \cdot 1 = 6,227,020,800.$$

(d) The first must be a girl–8 possibilities, followed by a boy–7 possibilities, followed by a girl–7 possibilities, etc. By use of the fundamental counting principle, there are

$$(8 \cdot 7) \cdot (7 \cdot 6) \cdot (6 \cdot 5) \cdot (5 \cdot 4) \cdot (4 \cdot 3) \cdot (3 \cdot 2) \cdot (2 \cdot 1)$$

possibilities. This is equivalent, by reordering the factors, to

$$8! \cdot 7! = 203,212,800.$$

(e) Choosing the first, eighth, and fifteenth positions first, we have $8 \cdot 7 \cdot 6$ ways. Follow this by choosing the remaining positions, 12!. Altogether, the number of different program arrangements is given by

$$(8 \cdot 7 \cdot 6) \cdot 12! = 160,944,537,600.$$

41. (a) The number of ways she can choose a book with replacement is found by use to the fundamental counting principle

$$7 \cdot 7 \cdot 7 \cdot 7 \cdot 7 \cdot 7 \cdot 7 = 7^7 = 823,543.$$

(b) The number of ways she can choose a book without replacement can be found by permutations.

$$P(7,7) = 7! = 5040$$

Alternatively, one can apply the fundamental counting principle as well to get

$$7 \cdot 6 \cdot 5 \cdot 4 \cdot 3 \cdot 2 \cdot 1 = 7! = 5040.$$

42. (a) Since there are 13 clubs and 39 non-clubs in the deck, the four clubs can be chosen in $C(13,4)$ ways, and the one non-club can be chosen in 39 ways. Use the fundamental counting principle to combine the two tasks. The number of possible hands is

$$C(13,4) \cdot 39 = 715 \cdot 39$$
$$= 27,885.$$

(b) The two face cards can be chosen from the twelve face cards $C(12,2)$ ways. The three non-face cards can be chosen from the forty non-face cards in $C(40,3)$ ways. Using the fundamental counting principle to combine both tasks, the number of hands is

$$C(12,2) \cdot C(40,3) = 652,080.$$

(c) The two red cards can be chosen in $C(26,2)$ ways, the two clubs can be chosen in $C(13,2)$ ways, and the spade can be chosen in 13 ways. Applying the fundamental counting principle, the number of possible hands is

$$C(26,2) \cdot C(13,2) \cdot 13 = \ = 329,550.$$

43. The number of ways each of the five groupings can be chosen is given by $C(n,r)$ since the order is not important in each grouping. Apply the fundamental counting principle to find the total number of ways to create the 5 groupings altogether.

$$C(25,3) \cdot C(22,4) \cdot C(18,5) \cdot C(13,6) \cdot C(7,7)$$
$$= 2.473653743 \times 10^{14}$$

Note that the last factor $C(7,7) = 1$.

44. Since each pair of points determines a different line, we are looking for the number of 2-element subsets of a set of 7 elements. Subsets are combinations, so the number of lines determined by seven points in a plane, no three of which are collinear, is

$$C(7,2) = \frac{7!}{2!5!} = 21.$$

45. Since each group of three noncollinear points determines a triangle, we are looking for the number of 3-element subsets of a set of 20 elements. Since subsets are combinations, the number of triangles determined by 20 points in a plane, no three of which are collinear, is

$$C(20,3) = \frac{20!}{3!17!} = 1140.$$

46. Use the fundamental counting principle since repetitions are allowed. We have 40 choices for the first, second, and third number. The total number of combinations is

$$40 \cdot 40 \cdot 40 = 40^3 = 64,000.$$

47. (a) Since any pair of the 7 people may drive and the order of selection is important, the number of choices is

$$P(7,2) = \frac{7!}{5!} = 42.$$

(b) Consider choosing the drivers as a two-part task. There are only 3 choices for the driver of the sports car. Once the driver of the first sports car has been chosen, any of the remaining 6 people can be chosen to drive the second car. By the fundamental counting principle, the number of choices for drivers is

$$3 \cdot 6 = 18.$$

(c) Choose the drivers first. Of the two, the first could be the driver of the sports car; and the second, the driver of the other vehicle. There are $P(7,2)$ choices. The second task is to pick the passenger for the sports car. There are five to choose from. By the fundamental counting principle, the total number of choices is

$$P(7,2) \cdot 5 = 210.$$

48. Use the fundamental counting principle. The number of ways to select a winner from the first and from the second race is

$$6 \cdot 8 = 48.$$

Thus, buying 48 different tickets will assure a winner.

49. The number of ways for three of the eight horses running to be 1st, 2nd, and 3rd place winners is given by

$$P(8,3) = 336.$$

Thus, buying 336 different tickets will assure a winner.

50. Follow the textbook hint and answer the 6 questions to solve the exercise.

(a) There are $C(9,2)$ ways to select two-person committees.

(b) There are $C(7,3)$ ways to select three-person committee.

Note that there is only one way to choose the 4 members of the four-person committee. Since there are only 4 left after the other committees are selected, we must choose these 4.

(c) There are 2 ways to select the chair of the two-person committee.

(d) There are 3 ways to select the chair of the three-person committee.

(e) There are 4 ways to select the chair of the four-person committee.

Thus, the total number of ways to make these selections is

$$C(9,2) \cdot C(7,3) \cdot 2 \cdot 3 \cdot 4 = 30,240.$$

51. (a) Altogether on the three-block stretch, how many different choices does Jeff have for positioning the eighteen homes? (Hint: Consider the three blocks separately and use the fundamental counting principle.)

On the first block, he can choose any 2 of the 6 home sites for his deluxe models with $C(6,2)$ choices. On his second block, he can choose any 3 of the sites for his deluxe models or $C(6,3)$ choices. On the third block, he can choose any 4 of the six sites for his deluxe homes or $C(6,4)$ choices. The remaining sites will all be standard homes. Thus, in total, the number of choices he will have is

$$C(6,2) \cdot C(6,3) \cdot C(6,4) = 15 \cdot 20 \cdot 15 = 4500.$$

(b) From the last exercise, the number of choices for the three blocks considered individually are $C(6,2)$, $C(6,3)$, and $C(6,4)$. The number of ways to arrange the three blocks is 3!. Using the fundamental counting principle, we see that the number of choices is now

$$3! \cdot C(6,2) \cdot C(6,3) \cdot C(6,4)$$
$$= 6 \cdot 15 \cdot 20 \cdot 15$$
$$= 27,000.$$

52. (a) How many numbers can be formed using all six digits 4, 5, 6, 7, 8, and 9?

Since the order or arrangement of these digits is important (each being a different number), we can consider the answer to the question to be

$$P(6,6) = 6! = 720$$

different numbers.

(b) The first number is 456,789; the second number is 456,798; the third number is 456,879 and so forth.

The number of arrangements (permutations) of the last five digits is given by $5! = 120$. Thus, the 121st number is 546,789 (where we have moved to the sixth digit from right and interchanged the fifth and sixth digit–4 and 5). There are another 120 permutations (with this change on the fifth and sixth digits) bringing us to the 241st number: 645,789. In a similar manner, numbers beginning with the new first digit–6, we have another 120 permutations using the new set of numbers. For the 361st number, we change the first digit to seven, giving us 745,689. The 362nd number is 745,698. The 363rd number is 745,869 and the 364th number is 745,896

53. (a) Without restrictions, the number of arrangements is

$$8! = 40,320.$$

(b) There are 2 ways that the bride and groom can stand last in line. There are 6! ways the remaining people can stand in line. Thus, by the fundamental counting principle, the total number of arrangements is

$$2 \cdot 6! = 1440.$$

(c) There is only 1 way that the groom can stand in line with the bride next to him. There are 6! ways the remaining people can stand in line. Thus, by the fundamental counting principle, the total number of arrangements is
$$1 \cdot 6! = 720.$$

54. Five percent of the 60 students is $(.05)60$ or 3 students. Five percent of the 40 students is $(.05)40$ or 2 students. The number of ways that he can assign A-grades to the first class is $C(60, 3)$. The number of ways that he can assign A-grades to the second class is $C(40, 2)$. Apply the fundamental counting principle to find the total number of ways the professor may assign A-grades to his students.
$$C(60, 3) \cdot C(40, 2) = 26,691,600$$

55. (a) There are only two sets of distinct digits that add to 12. They are $\{1, 2, 3, 6\}$ and $\{1, 2, 4, 5\}$. Try to find others. There are 4! distinct permutations (which lead to a different counting number) for each set of digits. Thus, using the fundamental counting principle, the total number of counting numbers whose sum of digits is 12 is
$$2 \cdot 4! = 48.$$

(b) There are only three sets of distinct digits that add to 13. They are $\{1, 2, 3, 7\}$, $\{1, 2, 4, 6\}$ and $\{1, 3, 4, 5\}$. Try to find others. There are 4! distinct permutations (which lead to a different counting number) for each set of digits. Therefore, using the fundamental counting principle, the total number of counting numbers whose sum of digits is 13 is
$$3 \cdot 4! = 72.$$

56. This question is the same as asking how many 5-element subsets can be formed from a set with 30 elements. The number of different samples is thus,
$$C(30, 5) = 142,506.$$

57. $C(12, 9) = \dfrac{12!}{9!(12-9)!}$
$= \dfrac{12!}{9!3!}$
$= 220$

$C(12, 3) = \dfrac{12!}{3!(12-3)!}$
$= \dfrac{12!}{3!9!}$
$= 220$

Thus, $C(12, 9) = C(12, 3)$.

58. By the factorial formula for combinations,
$$C(n, r) = \dfrac{n!}{r!(n-r)!}$$
and
$$C(n, n-r) = \dfrac{n!}{(n-r)![n-(n-r)]!}$$
$= \dfrac{n!}{(n-r)!(n-n+r)!}$
$= \dfrac{n!}{(n-r)!r!}$
$= \dfrac{n!}{r!(n-r)!}.$

Thus, $C(n, r) = C(n, n-r)$.

59. (a) Since $P(n, r) = \dfrac{n!}{(n-r)!}$,
$$P(n, 0) = \dfrac{n!}{(n-0)!}$$
$= \dfrac{n!}{n!}$
$= 1.$

(b) Writing exercise

60. (a) Since $C(n, r) = \dfrac{n!}{r!(n-r)!}$,
$$C(n, 0) = \dfrac{n!}{0!(n-0)!}$$
$= \dfrac{n!}{1 \cdot n!}$
$= 1.$

(b) Writing exercise

11.4 EXERCISES

Read the following combination values directly from Pascal's triangle. For exercises 1–8, refer to Table 5 in the text.

1. To find from the value of $C(4, 3)$ from Pascal's triangle, read entry number 3 in row 4 (remember that the top row is row "0" and that in row 4 the "1" is entry "0").
$$C(4, 3) = 4$$

2. To find the value of $C(5, 2)$ from Pascal's triangle, read entry number 2 in row 5.
$$C(5, 2) = 10$$

3. To find the value of $C(6, 4)$ from Pascal's triangle, read entry number 4 in row 6.
$$C(6, 4) = 15$$

4. To find the value of $C(7,5)$ from Pascal's triangle, read entry number 5 in row 7.
$$C(7,5) = 21$$

5. To find the value of $C(8,2)$ from Pascal's triangle, read entry number 2 in row 8.
$$C(8,2) = 28$$

6. To find the value of $C(9,4)$ from Pascal's triangle, read entry number 4 in row 9.
$$C(9,4) = 126$$

7. To find the value of $C(9,7)$ from Pascal's triangle, read entry number 7 in row 9.
$$C(9,7) = 36$$

8. To find the value of $C(10,6)$ from Pascal's triangle, read entry number 6 in row 10.
$$C(10,6) = 210$$

9. Selecting the committee is a two-part task. There are $C(7,1)$ ways of choosing the one Democrat and $C(3,3)$ way of choosing the remaining 3 Republicans. The combination values can be read from Pascal's triangle. By the fundamental counting principle, the total number of ways is
$$C(7,1) \cdot C(3,3) = 7 \cdot 1 = 7.$$

10. A committee with exactly two Democrats will consist of two Democrats and two Republicans. Selecting the committee is a two-part task. The two Democrats can be selected from seven Democrats in $C(7,2)$ ways, while the two Republicans can be selected from three Republicans in $C(3,2)$ ways. The combination values can be read from Pascal's triangle. Using the fundamental counting principle, the number of ways to select the committee is
$$C(7,2) \cdot C(3,2) = 21 \cdot 3 = 63.$$

11. A committee with exactly three Democrats will consist of three Democrats and one Republican. Selecting the committee is a two-part task. There are $C(7,3)$ ways of choosing three Democrats and $C(3,1)$ ways to choose the one remaining Republican. Hence, there are
$$C(7,3) \cdot C(3,1) = 35 \cdot 3 = 105 \text{ ways in total.}$$

12. The number of ways to select four Democrats and no Republicans is
$$C(7,4) \cdot C(3,0) = 35 \cdot 1$$
$$= 35.$$

13. There are $C(8,3) = 56$ ways to choose three different positions for heads. Using Pascal's triangle, find row 8 entry 3. Remember to count first row and first entry as 0. The remaining positions will automatically be tails.

14. The number of ways of obtaining exactly four heads (and thus exactly four tails) can be found from entry 4 in row 8 of Pascal's triangle, that is
$$C(8,4) = 70.$$

15. There are $C(8,5) = 56$ ways to choose exactly five different positions for heads. Using Pascal's triangle, this would be found in row 8, entry 5.

16. The number of ways of obtaining exactly six heads is
$$C(8,6) = 28.$$

Using Pascal's triangle, this would be found in row 8, entry 6.

17. The number of selections for four rooms is given by
$$C(9,4) = 126.$$

Using Pascal's triangle, this would be found in row 9, entry 4.

18. Since only one of the classrooms contains the economics class, there are eight classrooms that do not. The number of ways to select four classrooms that do not contain the class is
$$C(8,4) = 70.$$

19. The number of selections that succeed in locating the class is given by total number of selections (Exercise 17) minus the number of ways which will fail to locate the classroom (Exercise 18), or
$$C(9,4) - C(8,4) = 126 - 70 = 56 \text{ ways.}$$

20. From Exercise 17, there is a total of $C(9,4) = 126$ different selections. From Exercise 18, $C(8,4) = 70$ selections will fail to locate the class. From Exercise 19, the number of selections which succeed in locating the class is $126 - 70 = 56$. Therefore, the fraction of the possible selections that will lead to "success" is
$$\frac{56}{126} \approx .444.$$

21. The number of 0-element subsets for a set of five elements is entry 0 (the first entry) in row 5 of Pascal's triangle. This number is 1.

22. The number of 1-element subsets for a set of five elements is entry 1 (the second entry) in row 5 of Pascal's triangle. This number is 5.

23. The number of 2-element subsets for a set of five elements is entry 2 (the third entry) in row 5 of Pascal's triangle. This number is 10.

24. The number of 3-element subsets for a set of five elements is entry 3 (the fourth entry) in row 5. This number is 10.

25. The number of 4-element subsets for a set of five elements is entry 4 (the fifth entry) in row 5. This number is 5.

26. The number of 5-element subsets of a set of five elements is entry 5 (the last entry) in row 3. This number is 1.

27. The total number of subsets is given by
$$C(5,0) + C(5,1) + C(5,2) + C(5,3) + C(5,4) + C(5,5)$$
$$= 1 + 5 + 10 + 10 + 5 + 1 = 32.$$

    This is the sum of elements in the fifth row of Pascal's triangle.

28. Writing exercise

29. (a) All are multiples of the row number.

    (b) The same pattern holds.

    (c) Row 11:

    1 11 55 165 330 462 462 330 165 55 11 1

    All are multiples of 11. Thus, the same pattern holds.

30. Name the next five numbers of the diagonal sequence indicated in the figure shown in the text. What special name applies to the numbers of this sequence?

    $\ldots, 15, 21, 28, 36, 45, \ldots$ These are the triangular numbers.

31. Following the indicated sums

    $$1, 1, 2, 3, 5,$$

    the sequence continues

    $$8, 13, 21, 34, \ldots .$$

    A number in this sequence comes from the sum of the two preceding terms. This is the Fibonacci sequence.

32.
```
                    1
                  1   1
                1   0   1
              1   1   1   1
            1   0   0   0   1
          1   1   0   0   1   1
```

    In rows 2 and 4, every entry, except for the beginning and ending 1's, is 0. This is because the corresponding entries in the original triangle were all even.

33. Row 8 would be the next row to begin and end with 1, with all other entries 0 (each internal entry in row 8 of Pascal's triangle is even).

34. From Exercises 32 and 33, we see a pattern. In Pascal's triangle, every entry, except for the beginning and ending 1's, will be even whenever the row number is a power of 2. Since $256 = 2^8$ is a power of 2, this pattern will apply to row number 256.

    In row number 256, there are $256 + 1 = 257$ entries. The first and last will be 1's, while all the others will be even. Thus, the number of even numbers will be

    $$257 - 2 = 255.$$

35. The sum of the squares of the entries across the top row equals the entry at the bottom vertex. Choose, for example, the second triangle from the bottom.

    $$1^2 + 3^2 + 3^2 + 1^2$$
    $$= 1 + 9 + 9 + 1$$
    $$= 20 \text{ (the vertex value)}$$

36. The rows of Tartaglia's rectangle correspond to the diagonals of Pascal's triangle.

37. The sum $= N$; Any entry in the array equals the sum of the two entries immediately above it and immediately to its left.

38. The sum $= N$; Any entry in the array equals the sum of the column of entries from its immediate left upward to the top of the array.

39. The sum $= N$; any entry in the array equals the sum of the row of entries from the cell immediately above it to the left boundary of the array.

40. The sum $= N - 1$; Any entry in the array equals 1 more than the sum of all entries whose cells comprise the largest rectangle entirely to the left and above that entry.

41. Reading the coefficients from row 6 of Pascal's triangle and applying the binomial theorem, we obtain

    $$(x+y)^6 = x^6 + 6x^5y + 15x^4y^2 + 20x^3y^3$$
    $$+ 15x^2y^4 + 6xy^5 + y^6.$$

42. Reading the coefficients from row 8 of Pascal's triangle and applying the binomial theorem, we obtain

    $$(x+y)^8 = x^8 + 8x^7y + 28x^6y^2 + 56x^5y^3 + 70x^4y^4$$
    $$+ 56x^3y^5 + 28x^2y^6 + 8xy^7 + y^8.$$

43. Reading the coefficients from row 3 of Pascal's triangle and applying the binomial theorem, we obtain

    $$(z+2)^3 = z^3 + 3z^2(2) + 3z(2^2) + 2^3$$
    $$= z^3 + 6z^2 + 12z + 8.$$

44. Reading the coefficients from row 5 of Pascal's triangle and applying the binomial theorem, we obtain

$$(w+3)^5 = w^5 + 5w^4(3) + 10w^3(3^2) + 10w^2(3^3)$$
$$+ 5w(3^4) + 3^5$$
$$= w^5 + 15w^4 + 90w^3 + 270w^2$$
$$+ 405w + 243.$$

45. Reading the coefficients from row 4 of Pascal's triangle and applying the binomial theorem, we obtain

$$(2a+5b)^4 = (2a)^4 + 4(2a)^3(5b) + 6(2a)^2(5b)^2$$
$$+ 4(2a)(5b)^3 + (5b)^4$$
$$= 16a^4 + 160a^3b + 600a^2b^2 + 1000ab^3$$
$$+ 625b^4.$$

46. Reading the coefficients from row 4 of Pascal's triangle and applying the binomial theorem, we obtain

$$(3d+5f)^4 = (3d)^4 + 4(3d)^3(5f) + 6(3d)^2(5f)^2$$
$$+ 4(3d)(5f)^3 + (5f)^4$$
$$= 81d^4 + 540d^3f + 1350d^2f^2 + 1500df^3$$
$$+ 625f^4.$$

47. Reading the coefficients from row 6 of Pascal's triangle and applying the binomial theorem, we obtain

$$(b-h)^6 = [b+(-h)]^6$$
$$= b^6 + 6b^5(-h) + 15b^4(-h)^2 + 20b^3(-h)^3$$
$$+ 15b^2(-h)^4 + 6b(-h)^5 + (-h)^6$$
$$= b^6 - 6b^5h + 15b^4h^2 - 20b^3h^3 + 15b^2h^4$$
$$- 6bh^5 + h^6.$$

48. Reading the coefficients from row 5 of Pascal's triangle and applying the binomial theorem, we obtain
$$(2n-4m)^5 = [2n+(-4m)]^5$$
$$= (2n)^5 + 5(2n)^4(-4m) + 10(2n)^3(-4m)^2$$
$$+ 10(2n)^2(-4m)^3 + 5(2n)(-4m)^4$$
$$+ (-4m)^5$$
$$= 32n^5 - 320n^4m + 1280n^3m^2$$
$$- 2560n^2m^3 + 2560nm^4 - 1024m^5.$$

49. For the expansion $(x+y)^n$, there will be $n+1$ terms.

50. The exponent on y is $r-1$. Since the sum of the exponents on x and y must be n, the exponent on x is $n-(r-1) = n-r+1$. Thus, the variable part of the rth term is

$$x^{n-r+1}y^{r-1}.$$

Each term of the expansion is the product of a coefficient and its corresponding variable part. Thus, the rth term (the general term) is

$$\frac{n!}{(n-r+1)!(r-1)!}x^{n-r+1}y^{r-1}$$

51. Here $n=14$ and $r=5$. Then $r-1=4$ and $n-r+1=10$. Substituting these values into the result of Exercise 50, we find that the 5th term of $(x+y)^{14}$ is

$$\frac{14!}{10!4!}x^{10}y^4 = 1001x^{10}y^4.$$

52. Here $n=18$ and $r=16$. Then $r-1=15$ and $n-r+1=3$. Substituting these values into the result of Exercise 50, we find that the 16th term of $(a+b)^{18}$ is

$$\frac{18!}{3!15!}a^3b^{15} = 816a^3b^{15}.$$

53. Prove $C(n,r) = C(n-1,r-1) + C(n-1,r)$.

$$C(n-1,r-1) + C(n-1,r)$$
$$= \frac{(n-1)!}{(r-1)![(n-1)-(r-1)]!} + \frac{(n-1)!}{r![(n-1)-r]!}$$
$$= \frac{(n-1)!}{(r-1)!(n-r)!} + \frac{(n-1)!}{r!(n-r-1)!}$$
$$= \frac{n}{n} \cdot \frac{(n-1)!}{(r-1)!(n-r)!} \cdot \frac{r}{r} + \frac{n}{n} \cdot \frac{(n-1)!}{r!(n-r-1)!} \cdot \frac{(n-r)}{(n-r)}$$
$$= \frac{n! \cdot r}{n \cdot r! \cdot (n-r)!} + \frac{n! \cdot (n-r)}{n \cdot r! \cdot (n-r)!}$$
$$= \frac{n! \cdot r + n! \cdot (n-r)}{n \cdot r! \cdot (n-r)!}$$
$$= \frac{n! \cdot [r+(n-r)]}{n \cdot r! \cdot (n-r)!}$$
$$= \frac{n! \cdot \cancel{n}}{\cancel{n} \cdot r! \cdot (n-r)!}$$
$$= \frac{n!}{r! \cdot (n-r)!}$$
$$= C(n,r)$$

11.5 EXERCISES

1. Writing exercise

2. Writing exercise

3. The total number of subsets is 2^4 for a set with 4 elements. The only subset which is not a proper subset is the given set itself. Thus, by the complements principle, the number of proper subsets is

$$2^4 - 1 = 16 - 1 = 15.$$

4. The total number of subsets is 2^5. The only subset which is not a proper subset is the given set. Thus, by the complements principle, the number of proper subsets is

$$2^5 - 1 = 32 - 1 = 31.$$

5. By the fundamental counting principle, there are 2^7 different outcomes if seven coins are tossed. There is only one way to get no heads (all tails); thus, by the complements principle, there are
$$2^7 - 1 = 128 - 1 = 127$$
outcomes with at least one head.

6. By the fundamental counting principle, there are 2^7 different outcomes if seven coins are tossed. Of these, there is 1 way to get zero heads, and there are 7 ways to get one head. Since "at least two" is the complement of "zero or one," the number of ways to get at least two heads is
$$2^7 - (1 + 7) = 128 - 8 = 120.$$

7. "At least two" tails is the complement of "one or none." There are $2^7 = 128$ different outcomes if seven coins are tossed, 1 way of getting no tails (all heads), and 7 ways of getting one tail (tail on 1st coin, or tail on 2nd coin, etc.). Thus, by the complements principle, the number of ways to get at least two tails is
$$2^7 - (1 + 7) = 128 - 8 = 120.$$

8. As in Exercises 3–4, 2^7 or 128 different results are possible. Of these, there is one way to get no heads (seven tails) and one to get no tails (seven heads).

   All of the other results have at least one of each. Thus, the number of ways to obtain at least one of each is
$$2^7 - 2 = 128 - 2 = 126.$$

Refer to Table 2 in the first section of the textbook chapter.

9. Counting outcomes with a 2 on the red die (row 2), there are a total of 6.

10. The smallest possible sum is 2, the complement of "a sum of at least 3." Since there are 36 possible outcomes and only one, the roll represented by $(1, 1)$, with a sum of 2, the number of ways to roll a sum of at least 3 is
$$36 - 1 = 35.$$

11. Counting the number of outcomes in row 4 (4 on red die) + those in column 4 (4 on green die) and subtracting the outcome counted twice, the number of outcomes with "a 4 on at least one of the dice" is
$$6 + 6 - 1 = 11.$$

12. The complement of "a different number on each die" is the "the same number on both dice," which is usually called "doubles." Since there are 6 ways to roll "doubles" (which are found on the main diagonal of Table 2), the number of ways to obtain a different number on each dice is
$$36 - 6 = 30.$$

13. There are nine two digit multiples of 10 (10, 20, 30, ..., 90). Altogether there are $9 \cdot 10 = 90$ two-digit numbers by the fundamental counting principle. Thus, by the complements principle, the number of "two-digit numbers which are not multiples of ten" is
$$90 - 9 = 81.$$

14. There are $99 - 70 = 29$ two-digit numbers greater than 70. There are 9 two-digit multiples of 10. There are 2 two-digit numbers greater than 70 that are also multiples of 10 (be sure not to count these twice). Thus, the number of two-digit numbers satisfying the given conditions is
$$29 + 9 - 2 = 36.$$

15. (a) The number of different sets of three albums she could choose is
$$C(8, 3) = 56.$$

   (b) The number which would not include *Unchained* is
$$C(7, 3) = 35.$$

   (c) The number that would contain *Unchained* is
$$56 - 35 = 21.$$

16. (a) The number of ways that the four can be chosen is
$$C(10, 4) = 210.$$

   (b) The number that would include neither of the productions mentioned is
$$C(8, 4) = 70.$$

   (c) The number that would include at least one of the two productions mentioned is
$$210 - 70 = 140.$$

17. The total number of ways of choosing any three days of the week is $C(7, 3)$. The number of ways of choosing three days of the week that do not begin with S is $C(5, 3)$. Thus, the number of ways of choosing any three days of the week such that "at least one of them begin with S" is
$$C(7, 3) - C(5, 3) = 25.$$

18. The number of ways that Chalon can complete 2 assignments tonight is $C(9, 2)$. The number of ways she can complete her assignment tonight without doing an essay assignment is $C(7, 2)$. Thus, the number of ways

she could complete her assignment tonight, including at least 1 essay assignment, is

$$C(9,2) - C(7,2) = 15.$$

19. If the order of selection is important, the number of choices of restaurants is $P(8,3)$. The number of choices of restaurants that would not include seafood is $P(6,3)$. Thus, the number of choices such that at least one of the three will serve seafood is

$$P(8,3) - P(6,3) = 216.$$

20. If the order of selection is not important, the number of choices of restaurants is $C(8,3)$. The number of choices of restaurants that would not include seafood is $C(6,3)$. Thus, the number of choices such that at least one of the three will serve seafood is

$$C(8,3) - C(6,3) = 36.$$

21. The total number of ways to arrange 3 people among ten seats is $P(10,3)$. The number of ways to arrange three people among the seven (non-aisle) seats is $P(7,3)$. Therefore, by the complements principle, the number of arrangements with at least one aisle seat is

$$P(10,3) - P(7,3) = 510.$$

22. "At least one zero" is the complement of "contains no zeros."
    (a) If repeated digits are allowed, the total number of four-digit numbers is $9 \cdot 10^3$ by the fundamental counting principle. The number of four-digit numbers with no zeros is 9^4. Therefore, by the complements principle, "the number of four-digit numbers containing at least one zero" is

$$9 \cdot 10^3 - 9^4 = 2439.$$

    (b) If repeated digits are not allowed, the total number of four-digit numbers is $9 \cdot 9 \cdot 8 \cdot 7$ by the fundamental counting principle. The number of four-digit numbers with no zeros is $9 \cdot 8 \cdot 7 \cdot 6$. Therefore, by the complements principle, "the number of four-digit numbers containing at least one zero" is

$$9 \cdot 9 \cdot 8 \cdot 7 - 9 \cdot 8 \cdot 7 \cdot 6 = 1512.$$

23. "At least one of these faculty members" is the complement of "none of these faculty members." There is a total of $C(25,4)$ ways of choosing 4-person committees and $C(23,4)$ ways to choose 4-person committees, excluding the two professors. Thus, applying the complements principle, the number of ways of choosing four faculty committees "with at least one of these two professors" is

$$C(25,4) - C(23,4) = 3795.$$

24. "At least one officer" is the complement of "no officers." The number of ways of choosing 4-member search teams is $C(12,4)$. The number of ways to choose 4-member search teams with no officers included is $C(8,4)$. The total number of ways to choose the search team with at least one officer included is

$$C(12,4) - C(8,4) = 425.$$

25. Let C = the set of clubs and J = the set of jacks. Then, $C \cup J$ is the set of cards which are face cards or jacks, and $C \cap J$ is the set of cards which are both clubs and jacks, that is, the jack of clubs. Using the general additive counting principle, we obtain

$$n(C \cup J) = n(C) + n(J) - n(C \cap J)$$
$$= 13 + 4 - 1$$
$$= 16.$$

26. Let F = the set of face cards and B = the set of black cards. Then $F \cup B$ is the set of cards which are face cards or black cards, and $F \cap B$ is the set of cards which are both face cards and black, that is, the black face cards. Using the general additive counting principle, we obtain

$$n(F \cup B) = n(F) + n(B) - N(F \cap B)$$
$$= 12 + 26 - 6$$
$$= 32.$$

27. Let M = the set of students who enjoy music and L = the set of students who enjoy literature. Then $M \cup L$ is the set students who enjoy music or literature, and $M \cap L$ is the set of students who enjoy both music and literature. Using the general additive counting principle, we obtain

$$n(M \cup L) = n(M) + n(L) - n(M \cap L)$$
$$= 30 + 15 - 10$$
$$= 35.$$

28. In Exercise 27, using the general additive property, it was found that the number of students who enjoy at least one of the two subjects is

$$30 + 15 - 10 = 35.$$

The complement, the number of students who like "neither of the subjects," is

$$50 - 35 = 15.$$

29. There are $C(13,5)$ 5-card hands of clubs. Thus, by the complements principle, the number of hands containing "at least one card that is not a club" is

$$2,598,960 - C(13,5) = 2,597,673.$$

30. There are $C(13,5)$ 5-card hands of the same suit. Since there are 4 different suits in a deck, there are $4 \cdot C(13,5)$ 5-card hands of the same suit. Thus, by the complements principle, the number of hands "containing more than one suite" is

$$2{,}598{,}960 - 4 \cdot C(13,5) = 2{,}593{,}812.$$

31. The number of 5-card hands drawn from the 40 non-face cards in the deck is given by $C(40,5)$. Thus, by the complements principle the number of 5-card hands with "at least one face card" is

$$2{,}598{,}960 - C(40,5) = 1{,}940{,}952.$$

32. Using the special addition principle, the number of 5-card hands with no diamonds or all diamonds is $C(39,5) + C(13,5)$. Thus, by the complements principle, the number of 5-card hands with "at least one diamond, but not all diamonds" is

$$2{,}598{,}960 - [C(39,5) + C(13,5)] = 2{,}021{,}916.$$

For Exercises 33–36, the given original set has 12 elements.

33. "At most two elements" is the same as 0, 1, or 2 elements. Thus, the number of subsets is

$$C(12,0) + C(12,1) + C(12,2) = 1 + 12 + 66 = 79.$$

34. "At least ten" means "ten or more." The largest possible subset of a set of twelve elements will itself have 12 elements. Add to this the number of 10-element, and 11-element subsets to obtain

$$\begin{aligned} C(12,10) + C(12,11) + C(12,12) \\ = 66 + 12 + 1 \\ = 79. \end{aligned}$$

35. "More than two elements" is the complement of "at most two elements." Find the number of subsets with "at most two elements" by adding the number of 0-member subsets, 1-member subsets and 2-member subsets.

$$\begin{aligned} C(12,0) + C(12,1) + C(12,2) \\ = 1 + 12 + 66 \\ = 79. \end{aligned}$$

There are 2^{12} subsets altogether. Thus, by the complements principle, the number of subsets of more than two elements is

$$\begin{aligned} 2^{12} - 79 &= 4096 - 79 \\ &= 4017. \end{aligned}$$

36. A set of twelve elements has a total of 2^{12} subsets. The complement of "three through nine elements" is "zero, one, two, ten, eleven, or twelve elements." "Zero, one, or two" is the same as "at most two elements." Exercise 33 shows that there are 79 of these subsets. "Ten, eleven, or twelve" is the same as "at least ten." Exercise 34 shows that there are also 79 of these subsets. Therefore, the number of subsets with three through nine elements is

$$\begin{aligned} 2^{12} - (79 + 79) &= 4096 - 158 \\ &= 3938. \end{aligned}$$

37. The complement of "at least one letter or digit repeated" is "no letters or digits repeated." There are $P(26,3) \cdot P(10,3)$ license plates with no digits repeated, and using the fundamental counting principle we have $26 \cdot 26 \cdot 26 \cdot 10 \cdot 10 \cdot 10 = 26^3 \cdot 10^3$ license plates where any digit can be repeated. By the complements principle, the number of different license numbers with at least one letter or digit repeated is

$$26^3 \cdot 10^3 - P(26,3) \cdot P(10,3) = 6{,}344{,}000.$$

38. Using the fundamental counting principle, we see that the number of ways to draw the king of hearts and then another heart is $1 \cdot 12$, and the number of ways to draw a non-heart king and then a heart is $3 \cdot 13$. Since these components are disjoint, we may apply the special additive principle. The number of different ways in which it is possible to obtain a king on the first draw and a heart on the second is

$$1 \cdot 12 + 3 \cdot 13 = 12 + 39 = 51.$$

39. To choose sites in only one state works as follows: The number of ways to choose three monuments in New Mexico is $C(4,3)$, the number of ways to chose three monuments in Arizona is $C(3,3)$, and the number of ways to pick three monuments in California is $C(5,3)$. Since these components are disjoint, we may use the special additive principle. The number of ways to choose the monuments is

$$\begin{aligned} C(4,3) + C(3,3) + C(5,3) &= 4 + 1 + 10 \\ &= 15. \end{aligned}$$

40. "At least one site not in California" is the complement of "all sites only in California." There are a total of $C(12,3)$ ways of seeing three sites from all states and a total of $C(5,3)$ ways of seeing sites just in California.

Thus, there are

$$C(12,3) - C(5,3) = 210$$

ways of visiting sites with one or more not in California.

41. "Sites in fewer than all three states" is the complement of choosing "sites in all three states." Since there are 12 monuments altogether, the total number of ways to select three monuments (with no restrictions) is $C(12,3)$. Choosing sites in all three states requires choosing one site in each state, which can be done in $4 \cdot 3 \cdot 5$ ways.

Using the complements principles, the number of ways to choose sites in fewer than all three states is
$$C(12,3) - 4 \cdot 3 \cdot 5 = 220 - 60$$
$$= 160.$$

42. "Sites in exactly two of the three states" is the complement of "sites in only one state or sites in all three states." Sites in only one state (Exercise 39) are found by $C(4,3) + C(3,3) + C(5,3) = 15$ and sites in all three states means one site in each state, or, by the fundamental counting principle, $4 \cdot 3 \cdot 5 = 60$. Since there are in total $C(12,3)$ ways of seeing three sites with no limitations, we have
$$C(12,3) - (15 + 60) = 145$$
ways with all three sites being in exactly two states.

43. Writing exercise
44. Writing exercise
45. Writing exercise
46. Writing exercise
47. Writing exercise
48. Writing exercise
49. Writing exercise
50. "Not divisible by 2, 3 or 5" is the complement of "divisible by 2, 3 or 5." There are $300 \div 2 = 150$ even (divisible by 2) numbers between and including 1 and 300. There are $300 \div 3 = 100$ numbers divisible by 3 (every third number) and $300 \div 5 = 60$ divisible by 5 (every fifth number). There are $300 \div 6 = 50$ numbers that are both even and divisible by 3 (must be divisible by $2 \times 3 = 6$; e.g., 6, 12, 18, ... 300). There are 30 numbers that are both even and divisible by 5 (that is, they must be divisible by $2 \times 5 = 10$; e.g., 10, 20, ... 300). There are $300 \div 15 = 20$ numbers that are divisible by 3 and 5 (must be divisible by $3 \times 5 = 15$). Finally, there are $300 \div 30 = 10$ numbers which are divisible by 2, 3 and 5 (divisible by $2 \times 3 \times 5 = 30$). Thus, by the complements principle and by the extension of the general additive principle (Exercise 46), the number of counting numbers 1 through 300 that are not divisible by 2, 3, is
$$300 - (150 + 100 + 60 - 50 - 30 - 20 + 10) = 80.$$

CHAPTER 11 TEST

1. To find three-digit numbers from the set $\{0, 1, 2, 3, 4, 5, 6\}$, use the fundamental counting principle:
$$\underline{6} \cdot \underline{7} \cdot \underline{7} = 294.$$

2. To find even three-digit numbers from the set $\{0, 1, 2, 3, 4, 5, 6\}$, use the fundamental counting principle:
$$\underline{6} \cdot \underline{7} \cdot \underline{4} = 168.$$

3. To find three-digit numbers without repeated digits from the set $\{0, 1, 2, 3, 4, 5, 6\}$, use the fundamental counting principle:
$$\underline{6} \cdot \underline{6} \cdot \underline{5} = 180.$$

4. To find three-digit multiples of five without repeated digits from the set $\{0, 1, 2, 3, 4, 5, 6\}$, use the fundamental counting principle and the special additive principle: Multiples of 5 end in "0" or "5." There are
$$\underline{6} \cdot \underline{5} \cdot \underline{1} = 30$$
multiples that end in 0 and
$$\underline{5} \cdot \underline{5} \cdot \underline{1} = 25$$
that end in 5. Thus, the number of three-digit multiples of five without repeated digits is
$$30 + 25 = 55.$$

5. Make a systematic listing of triangles. Beginning with the smaller inside triangle, there are 4 right triangles off the horizontal bisector of the larger triangle. These triangles may be combined to create 4 larger isosceles triangles. There are 2 isosceles triangles – inside the upper left and lower left corners of the largest triangle and 1 larger right triangles above and 1 below the horizontal bisector of the larger triangle. Of course, count the largest isosceles triangle itself. The total number of triangles is
$$4 + 4 + 2 + 1 + 1 + 1 = 13.$$

6.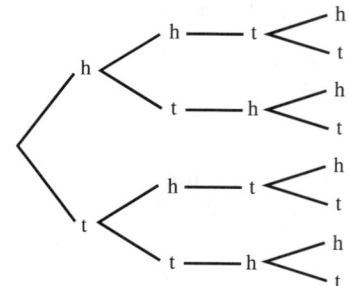

7. There is only one set of 4 digits that add to 30 $(9 + 8 + 7 + 6 = 30)$. The number of arrangements of the digits 9876 is given by
$$P(4, 4) = 4! = 24.$$

8. To find the number of 3-digit numbers with no repeated odd digits from $\{0, 1, 2\}$, use the fundamental counting principle. The only digit which can not be repeated is the odd digit "1", so look (a) at the cases where 1 is positioned in each of the three possible places and (b) those numbers that do not contain the odd digit 1.

   (a) There are
   $$\underline{1} \cdot \underline{2} \cdot \underline{2} = 4$$
   3-digit numbers that begin with 1 and
   $$\underline{1} \cdot \underline{1} \cdot \underline{2} = 2$$
   3-digit numbers whose middle digit is 1, and
   $$\underline{1} \cdot \underline{2} \cdot \underline{1} = 2$$
   3-digit numbers whose last digit is 1. Thus, using the special additive principle, the number of 3-digit numbers without repeated odd digits which contain the odd digit 1 is
   $$4 + 2 + 2 = 8.$$

   (b) The condition "no repeated odds digit" is automatically satisfied by any number that contains "no odd digits." Using only the digits $\{0, 2\}$ and the fundamental counting principle, there are
   $$\underline{1} \cdot \underline{2} \cdot \underline{2} = 4$$
   such numbers.

   There are, therefore,
   $$8 + 4 = 12$$
   numbers that satisfy the condition "no repeated odds digit." The numbers are
   $$100, 102, 120, 122, 200, 201,$$
   $$202, 210, 212, 220, 221, 222.$$

9. $5! = 5 \cdot 4 \cdot 3 \cdot 2 \cdot 1 = 120$

10. $\dfrac{8!}{5!} = \dfrac{8 \cdot 7 \cdot 6 \cdot 5!}{5!}$
    $= \dfrac{8 \cdot 7 \cdot 6}{1}$
    $= 336$

11. $P(12, 4) = 12 \cdot 11 \cdot 10 \cdot 9$
    $= 11,880$

12. $C(7, 3) = \dfrac{P(7, 3)}{3!}$
    $= \dfrac{7 \cdot \cancel{6} \cdot 5}{\cancel{3} \cdot \cancel{2} \cdot 1}$
    $= 35$

13. Since the arrangement of the letters is important and no repetitions are allowed, use $P(26, 5)$.
    $$P(26, 5) = 26 \cdot 25 \cdot 24 \cdot 23 \cdot 22 = 7,893,600$$

14. Since repetitions are allowed, use the fundamental counting principle.
    $$32^5 = 33,554,432$$

15. Since order is not important, use $C(12, 4)$.
    $$C(12, 4) = \dfrac{P(12, 4)}{4!}$$
    $$= \dfrac{12 \cdot 11 \cdot 10 \cdot 9}{4 \cdot 3 \cdot 2 \cdot 1}$$
    $$= 495$$

16. Use the fundamental counting principle along with $C(12, 2)$ and $C(10, 2)$.
    $$C(12, 2) \cdot C(10, 2) = 66 \cdot 45$$
    $$= 2970$$

17. Use the fundamental counting principle along with $C(12, 5)$ and $C(7, 5)$.
    $$C(12, 5) \cdot C(7, 5) = 792 \cdot 21$$
    $$= 16,632$$

18. The complement of "a set of three or more of the players" is "zero, one, or two of the players." The total number of subsets of the 12 players is
    $$2^{12} = 4096.$$
    The total number of 0-member subsets and 1-member subsets and 2-member subsets is
    $$C(12, 0) + C(12, 1) + C(12, 2)$$
    $$= 1 + 12 + 66$$
    $$= 79.$$
    By the complements principle, the number of ways that the coach has in selecting a set of three or more of the players is
    $$4096 - 79 = 4017.$$

19. With no restrictions, use the fundamental counting principle to determine the number of positions that a row of four switches may be set.
    $$\underline{2} \cdot \underline{2} \cdot \underline{2} \cdot \underline{2} = 2^4 = 16$$

20. Use the fundamental counting principle to determine the number of positions that a row of four switches may be set if the first and fourth positions must be on.
    $$\underline{1} \cdot \underline{2} \cdot \underline{2} \cdot \underline{1} = 2^2 = 4$$

21. Use the fundamental counting principle to determine the number of positions that a row of four switches may be set if the first and fourth position must be set the same.
    $$\underline{2} \cdot \underline{2} \cdot \underline{2} \cdot \underline{1} = 2^3 = 8$$

Complete a tree diagram like Exercise 59 in 11.1 to answer questions 22–23.

```
First    Second   Third    Fourth   Switch
Switch   Switch   Switch   Switch   Settings
                    0      0        0 0 0 0
              0
                    0      1        0 0 0 1
         0
                    1      0        0 0 1 0
                           1        0 0 1 1
   0
                    0      0        0 1 0 0
              0
         1                 1        0 1 0 1
                    1      0        0 1 1 0
                           1        0 1 1 1

                    0      0        1 0 0 0
              0
         0                 1        1 0 0 1
                    1      0        1 0 1 0
                           1        1 0 1 1
   1
                    0      0        1 1 0 0
              0
         1                 1        1 1 0 1
                    1      0        1 1 1 0
                           1        1 1 1 1
```

Alternatively, you may also choose to create separate tree diagrams instead, according to the restrictions imposed by each exercise.

22. The following represent switch settings where "no two adjacent switches can be off. Remember that "0" typically represents an "off" switch.

    0101   0110   0111   1010
    1011   1101   1110   1111

    There are 8 switch settings that satisfy the restriction.

23. The following represent switch settings where "no two adjacent switches can be set the same."

    0101   1010

    There are 2 switch settings that satisfy the restriction.

24. The complement of "at least two switches must be on" is "zero or 1 switch is on." Without restrictions there are $2^4 = 16$ different switch settings. There is only 1 way for zero switches to be on and 4 ways for only 1 switch to be on. Thus, by the complements principle, the number of ways for at least two switches to be on is

    $$16 - (1 + 4) = 11.$$

    Alternatively, you may count those settings in the above tree diagram which reflect the given restriction.

Use the set {A, B, C, D, E, F, G} to answer Exercises 25–29.

25. Since the letter "D" must be a member, all that is necessary is to choose three members from the six remaining members. The number of 4-member subsets is then

    $$C(6, 3) = 20.$$

26. Since the letters "A" and "E" must be members, all that is necessary is to choose two members from the five remaining members. The number of 4-member subsets is then

    $$C(5, 2) = 10.$$

27. $C(5, 3)$ represents the number of 4-member subsets that contain exactly one of the letters but not the other. There will be twice this number if we do the same for the 2nd letter. Thus, the number of ways to select 4-member subsets that contain exactly one of the letters but not the other is

    $$2 \cdot C(5, 3) = 20.$$

28. There are two vowels in the set (A and E). The remaining five letters are consonants. For the subsets chosen to have 4 members and for there to be an equal number of consonants and vowels, there must be two of each. The two vowels must be chosen (only one way to do this exists) and the number of ways to choose the two consonants is $C(5, 2)$. Thus, the total number of subsets satisfying the restriction is

    $$C(5, 2) = 10.$$

29. "More consonants then vowels" can happen two ways. One, with 4 consonants and no vowels, can be found by $C(5, 4)$. The second, with 3 consonants and 1 vowel, can be found by choosing one of the two vowels first, $C(2, 1)$ ways, followed by choosing the 3 consonants $C(5, 3)$ ways. Since one "or" the other will satisfy the restrictions (and they cannot both happen at the same time), apply the special addition principle to find the total number of subset choices for "more consonants then vowels."

    $$C(5, 4) + C(2, 1) \cdot C(5, 3)$$
    $$= 5 + 2 \cdot 10$$
    $$= 25$$

30. Reading the coefficients from row 5 of Pascal's triangle and applying the binomial theorem, we obtain

    $$(x + 2)^5 = x^5 + 5x^4(2) + 10x^3(2^2) + 10x^2(2^3)$$
    $$+ 5x(2^4) + 2^5$$
    $$= x^5 + 10x^4 + 40x^3 + 80x^2$$
    $$+ 80x + 32.$$

31. Because $C(n,r)$ and $C(n, r+1)$ are the two entries just above $C(n+1, r+1)$, evaluate $C(n+1, r+1)$ by adding their values.
$$\begin{aligned} C(n+1, r+1) &= C(n,r) + C(n, r+1) \\ &= 495 + 220 \\ &= 715 \end{aligned}$$

32. The sequence of numbers obtained is
$$1, 2, 3, 4, 5, \ldots$$
that is, the counting numbers.

33. Writing exercise

12.1 EXERCISES

The sample space is $\{1, 2, 3\}$.

1. The number of regions in the sample space is $n(S) = 3$. Each region has the same area and thus, has the same likelihood of occurring. Therefore,

   (a) $P(\text{red}) = \dfrac{n(\text{red regions})}{n(S)}$

   $= \dfrac{1}{3}.$

   (b) $P(\text{yellow}) = \dfrac{n(\text{yellow regions})}{n(S)}$

   $= \dfrac{1}{3}.$

   (c) $P(\text{blue}) = \dfrac{n(\text{blue regions})}{n(S)}$

   $= \dfrac{1}{3}.$

2. The number of regions in the sample space is $n(S) = 4$. Each region (piece of the pie) has the same area and thus, has the same likelihood of occurring. However, since two of the regions are yellow, the likelihood of landing on yellow is twice that of landing on red or blue regions.

   (a) Let $E = \{\text{red}\}$. Then

   $P(\text{red}) = \dfrac{n(\text{red regions})}{n(S)}$

   $= \dfrac{1}{4}.$

   (b) Let $E = \{\text{yellow}\}$. Then

   $P(\text{yellow}) = \dfrac{n(\text{yellow regions})}{n(S)}$

   $= \dfrac{2}{4} = \dfrac{1}{2}.$

   (c) Let $E = \{\text{blue}\}$. Then

   $P(\text{blue}) = \dfrac{n(\text{blue regions})}{n(S)}$

   $= \dfrac{1}{4}.$

3. The number of regions in the sample space is $n(S) = 6$. Each region (piece of the pie) has the same area and thus, has the same likelihood of occurring. The probability of landing on any one of the six regions is 1/6, but we must account for the fact that some colors shade more than one region. Therefore,

   (a) $P(\text{red}) = \dfrac{n(\text{red regions})}{n(S)}$

   $= \dfrac{3}{6} = \dfrac{1}{2}.$

   (b) $P(\text{yellow}) = \dfrac{n(\text{yellow regions})}{n(S)}$

   $= \dfrac{2}{6} = \dfrac{1}{3}.$

   (c) $P(E) = \dfrac{n(\text{blue regions})}{n(S)}$

   $= \dfrac{1}{6}.$

4. The number of regions in the sample space is $n(S) = 6$. Each region (piece of the pie) has the same area and thus, has the same likelihood of occurring. The probability of landing on any one of the six regions is 1/6, but we must account for the fact that some colors shade more than one region. Therefore,

   (a) $P(\text{red}) = \dfrac{n(\text{red regions})}{n(S)}$

   $= \dfrac{3}{6} = \dfrac{1}{2}.$

   (b) $P(\text{yellow}) = \dfrac{n(\text{yellow regions})}{n(S)}$

   $= \dfrac{3}{6} = \dfrac{1}{2}.$

   (c) $P(E) = \dfrac{n(\text{blue regions})}{n(S)}$

   $= \dfrac{0}{6} = 0.$

5. (a) The sample space is $\{1, 2, 3\}$.

   (b) The number of favorable outcomes is 2.

   (c) The number of unfavorable outcomes 1.

   (d) The total number of possible outcomes is 3.

   (e) The probability of an odd number is given by

   $P(\text{odd number}) = P(E)$

   $= \dfrac{\text{number of favorable outcomes}}{\text{total number of outcomes}}$

   $= \dfrac{2}{3}.$

   (f) The odds in favor of an odd number is given by

   $\text{Odds in favor} = \dfrac{\text{number of favorable outcomes}}{\text{number of unfavorable outcomes}}$

   $= \dfrac{2}{1}, \text{ or 2 to 1.}$

6. (a) The total number of possible outcomes is 13.

   (b) The number of favorable outcomes is 8.

   (c) The number of unfavorable outcomes 5.

(d) The probability of selecting a girl is given by
$$P(\text{girl}) = P(E)$$
$$= \frac{\text{number of favorable outcomes}}{\text{total number of outcomes}}$$
$$= \frac{8}{13}.$$

(e) The odds in favor of selecting a girl is given by
$$\text{Odds in favor} = \frac{\text{number of favorable outcomes}}{\text{number of unfavorable outcomes}}$$
$$= \frac{8}{5}, \text{ or 8 to 5}.$$

7. (a) The sample space is
$$\{11, 12, 13, 21, 22, 23, 31, 32, 33\}.$$

(b) The probability of an odd number is given by
$$P(\text{odd number}) = P(E)$$
$$= \frac{\text{number of favorable outcomes}}{\text{total number of outcomes}}$$
$$= \frac{6}{9} = \frac{2}{3}.$$

(c) The probability of a number with repeated digit is given by
$$P(\text{number with repeated digits}) = \frac{3}{9} = \frac{1}{3}.$$

(d) the probability of a number greater than 30 is given by
$$P(\text{number greater than 30}) = \frac{3}{9} = \frac{1}{3}.$$

(e) The primes are $\{11, 13, 23, 31\}$. Thus, the probability for a prime number is given by
$$P(\text{prime number}) = \frac{4}{9}.$$

8. (a) the sample space
$$\{hh, ht, th, tt\}$$

(b) The probability of heads on the dime is given by
$$P(\text{heads on the dime}) = \frac{2}{4} = \frac{1}{2}.$$

(c) The probability of heads on the quarter is given by
$$P(\text{heads on the quarter}) = \frac{2}{4} = \frac{1}{2}.$$

(d) The probability of getting both heads is given by
$$P(\text{two heads}) = \frac{1}{4}.$$

(e) The probability of getting the same thing on both coins is given by
$$P(\text{same thing on both coins}) = \frac{2}{4} = \frac{1}{2}.$$

9. (a) The odds in favor of selecting a red ball is given by
$$\text{Odds in favor} = \frac{\text{number of favorable outcomes}}{\text{number of unfavorable outcomes}}$$
$$= \frac{4}{7}, \text{ or 4 to 7}.$$

(b) The odds in favor of selecting a yellow ball is given by
$$\text{Odds in favor} = \frac{\text{number of favorable outcomes}}{\text{number of unfavorable outcomes}}$$
$$= \frac{5}{6}, \text{ or 5 to 6}.$$

(c) The odds in favor of selecting a blue ball is given by
$$\text{Odds in favor} = \frac{\text{number of favorable outcomes}}{\text{number of unfavorable outcomes}}$$
$$= \frac{2}{9}, \text{ or 2 to 9}.$$

10. $N = \{A, B, C, D, E\}$.

(a) The odds against choosing David is found by
$$\text{Odds against} = \frac{\text{number of unfavorable outcomes}}{\text{number of favorable outcomes}}$$
$$= \frac{4}{1}, \text{ or 4 to 1}.$$

(b) The odds against choosing a woman is given by
$$\text{Odds against} = \frac{\text{number of unfavorable outcomes}}{\text{number of favorable outcomes}}$$
$$= \frac{3}{2}, \text{ or 3 to 2}.$$

(c) The odds against choosing a person whose name begins with a vowel is given by
$$\text{Odds against} = \frac{\text{number of unfavorable outcomes}}{\text{number of favorable outcomes}}$$
$$= \frac{3}{2}, \text{ or 3 to 2}.$$

11. (a) $P(\text{Buddy Holly}) = \frac{1}{50}$

(b) $P(\text{The Drifters}) = \frac{2}{50} = \frac{1}{25}$

(c) $P(\text{Bobby Darin}) = \frac{3}{50}$

(d) $P(\text{The Coasters}) = \frac{4}{50} = \frac{2}{25}$

(e) $P(\text{Fats Domino}) = \frac{5}{50} = \frac{1}{10}$

The sample space for three fair coin tosses is $\{hhh, hht, hth, htt, thh, tht, tth, ttt\}$.

12. The sample space is

    $$\{hhh, hht, hth, htt, thh, tht, tth, ttt\}.$$

    (a) If the event of interest is "no heads," then there is only one favorable outcome, $\{ttt\}$.

    $$P(\text{no heads}) = \frac{1}{8}.$$

    (b) Let E represent the event "exactly one head." Then

    $$E = \{htt, tht, tth\}, \text{ and } P(E) = \frac{3}{8}.$$

    (c) Let F represent the event "exactly two heads." Then

    $$F = \{hht, hth, thh\}, \text{ and } P(F) = \frac{3}{8}.$$

    (d) Let G represent the event "exactly three heads." Then

    $$G = \{hhh\}, \text{ and } P(G) = \frac{1}{8}.$$

13. Product table for "sum"

		2nd die				
+	1	2	3	4	5	6
1	2	3	4	5	6	7
2	3	4	5	6	7	8
1st 3	4	5	6	7	8	9
die 4	5	6	7	8	9	10
5	6	7	8	9	10	11
6	7	8	9	10	11	12

    (a) Of the 36 possible outcomes, one gives a sum of 2, so

    $$P(\text{sum is 2}) = \frac{1}{36}.$$

    (b) The sum of 3 appears 2 times in the body of the table. Thus,

    $$P(\text{sum of 3}) = \frac{2}{36} = \frac{1}{18}.$$

    (c) The sum of 4 appears 3 times in the body of the table. Thus,

    $$P(\text{sum of 4}) = \frac{3}{36} = \frac{1}{12}.$$

    (d) The sum of 5 appears 4 times in the table. Thus,

    $$P(\text{sum of 5}) = \frac{4}{36} = \frac{1}{9}.$$

    (e) The sum of 6 appears 5 times in the table. Thus,

    $$P(\text{sum of 6}) = \frac{5}{36}.$$

    (f) The sum of 7 appears 6 times in the table. Thus,

    $$P(\text{sum of 7}) = \frac{6}{36} = \frac{1}{6}.$$

    (g) The sum of 8 appears 5 times in the table. Thus,

    $$P(\text{sum of 8}) = \frac{5}{36}.$$

    (h) The sum of 9 appears 4 times in the table. Thus,

    $$P(\text{sum of 9}) = \frac{4}{36} = \frac{1}{9}.$$

    (i) The sum of 10 appears 3 times in the table. Thus,

    $$P(\text{sum of 10}) = \frac{3}{36} = \frac{1}{12}.$$

    (j) The sum of 11 appears 2 times in the table. Thus,

    $$P(\text{sum of 11}) = \frac{2}{36} = \frac{1}{18}.$$

    (k) The sum of 12 appears 1 time in the table. Thus,

    $$P(\text{sum of 12}) = \frac{1}{36}.$$

In Exercises 14–15, answers are computed to three decimal places.

14. The empirical probability that any particular seed of this type will germinate is given by

    $$P(\text{germination}) = \frac{170}{200} = .850.$$

15. The total number of children born was

    $$38{,}550 + 36{,}770 = 75{,}320.$$

    (a) The empirical probability that one of these births chosen at random is a boy is given by

    $$P(\text{boy}) = \frac{38{,}550}{75{,}320} \approx .512.$$

    (b) The empirical probability that one of these births chosen at random is a girl is given by

    $$P(\text{girl}) = \frac{36{,}770}{75{,}320} \approx .488.$$

16. In Example 3, what would be Michelle's probability of having exactly two daughters if she were to have four children altogether?

    Extend the tree diagram, Figure 2 (in the text) to four children. The resulting sample space is given by:

    $S = \{gggg, gggb, ggbg, ggbb, gbgg, gbgb, gbbg, gbbb,$
    $\quad\quad bggg, bggb, bgbg, bgbb, bbgg, bbgb, bbbg, bbbb\}.$

    Let E be the favorable event of having exactly two girls. $n(E) = 6$ and $n(S) = 16$. Thus, the probability of having two girls is given by

$$P(E) = \frac{n(E)}{n(S)}$$
$$= \frac{6}{16} = \frac{3}{8}..$$

17. Writing exercise

18. Writing exercise

19. Since there is no dominance, only RR will result in red flowers. Thus,
$$P(\text{red}) = \frac{1}{4}.$$

20. A pink snapdragon is produced if there is 1 red gene and 1 white gene. Two outcomes of the four, namely Rr and rR, produce pink. Thus,
$$P(\text{pink}) = \frac{2}{4}$$
$$= \frac{1}{2}.$$

21. Since only rr will result in white flowers
$$P(\text{white}) = \frac{1}{4}.$$

22. Writing exercise

23. (a) Since round peas are dominant over wrinkled peas, the combinations RR, Rr, and rR will all result in round peas. Thus,
$$P(\text{round}) = \frac{3}{4}.$$

    (b) Since wrinkled peas are recessive, only rr will result in wrinkled peas. Thus,
$$P(\text{wrinkled}) = \frac{1}{4}.$$

24. Cystic fibrosis occurs in 1 of every 2000 Caucasian births, so the empirical probability that cystic fibrosis will occur in a randomly selected Caucasian birth is
$$P = \frac{1}{2000} = .0005.$$

25. Cystic fibrosis occurs in 1 of every 250,000 non-Caucasian births, so the empirical probability that cystic fibrosis will occur in a randomly selected non-Caucasian birth is
$$P = \frac{1}{250,000} = .000004.$$

26. One would expect .0005 of the 150,000, or
$$.0005 \times 150,000 = 75$$
births to yield occurrences of cystic fibrosis.

Construct a chart similar to Table 2 in the textbook and determine the probability of each of the following events.

	+	Second parent C	c
First Parent	C	CC	Cc
	c	cC	cc

27. C represents the normal (disease-free gene) and c represents the cystic fibrosis gene. Since c is a recessive gene, only the combination cc results in a child with the disease. Thus, the probability that their first child will have the disease is given by
$$P = \frac{1}{4}.$$

28. Since Cc and cC result in children who are carriers, the probability that their first child will be a carrier is given by
$$P = \frac{2}{4} = \frac{1}{2}.$$

29. Only the combination CC results in a child who neither has nor carries the disease, so the required probability is given by
$$P = \frac{1}{4}.$$

Create a table that gives the possibilities when one parent is a carrier and the other is a non-carrier to answer Exercises 30–32.

	+	Second parent C	c
First Parent	C	CC	Cc
	C	CC	Cc

30. Since no outcomes will yield cc, having a child with cystic fibrosis, the required probability is
$$P = 0.$$

31. The combination Cc results in a child who is a healthy cystic fibrosis carrier. This combination occurs twice in the table (while the other combination that gives a carrier, cC, does not occur). Thus, the required probability is given by
$$P = \frac{2}{4} = \frac{1}{2}.$$

32. Since there are only 2–CC outcomes, the probability that the child will neither have nor carry the disease is given by
$$P = \frac{2}{4} = \frac{1}{2}.$$

33. Sickle-cell anemia occurs in about 1 of every 500 black baby births, so the empirical probability that a randomly selected black baby will have sickle-cell anemia is
$$P = \frac{1}{500} = .002.$$

34. The probability that a randomly selected non-black baby will have sickle-cell anemia is given by
$$P = \frac{1}{160{,}000} \approx .000006.$$

35. From Exercise 33, the probability that a particular black baby will have sickle-cell anemia is .002. Therefore, among 80,000 black baby births, about
$$.002(80{,}000) = 160$$
occurrences of sickle-cell anemia would be expected.

For Exercises 36–38 let S represent the normal gene and s represent the sickle-cell gene. The possibilities for a child with parents who both have sickle-cell trait are given in the following table that gives the possibilities when one parent is a carrier and the other is a non-carrier.

	+	Second parent S	Second parent s
First Parent	S	SS	Ss
	s	sS	ss

36. Since "ss" represents the first child will have sickle-cell anemia the probability that this will occur is given by
$$P = \frac{1}{4}.$$

37. Since the combinations Ss and sS result in sickle-cell trait, the probability that the child will have sickle-cell trait is given by
$$P = \frac{1}{4} = \frac{1}{2}.$$

38. The child will be healthy.
$$P(\text{will be healthy}) = \frac{1}{4}$$

39. (a) The empirical probability formula is
$$P(E) = \frac{\text{number of times event occurred}}{\text{number of times experiment performed}}.$$
Since the number of times that the event described in the exercise has occurred is 0, the probability fraction has a numerator of 0. The denominator is some natural number n. Thus,
$$P(E) = \frac{0}{n}.$$
(b) There is no basis for establishing a theoretical probability for this event.

(c) A woman may break the 10 second barrier at any time in the future, so it is possible that this event will occur.

40. Writing exercise

41. Writing exercise

One approach to Exercises 42–44 is to consider the following:
Odds in favor = $\frac{a}{b}$; Odds against = $\frac{b}{a}$; Probability of same event = $\frac{a}{a+b}$

42. If $P(E) = .37 = 37/100$, find

(a) The odds in favor of E are given by
$$\frac{37}{100 - 37} = \frac{37}{63}, \text{ or 37 to 63.}$$

(b) The odds against E is the reciprocal of the odds in favor
$$\frac{63}{37}, \text{ or 63 to 37.}$$

43. The odds in favor of event E are 12 to 19, where $a = 12$ and $b = 19$. Since
$$P(E) = \frac{a}{a+b},$$
we have
$$P(E) = \frac{12}{12 + 19} = \frac{12}{31}.$$

44. If the odds against event E are 10 to 3, then the odds in favor of event E are 3 to 10, where $a = 3$ and $b = 10$. Since
$$P(E) = \frac{a}{a+b},$$
we have
$$P(E) = \frac{3}{3 + 10} = \frac{3}{13}.$$

45. From Table 1 in the text, there are 2,598,960 5-card poker hands. Of these 36 are straight flushes. Thus,
$$P(\text{straight flush}) = \frac{36}{2{,}598{,}960} \approx .00001385.$$

Refer to Table 1 in the text. Answers are given to eight decimal places.

46. The probability of being dealt two pairs is
$$\frac{123{,}552}{2{,}598{,}960} = .04753902.$$

47. Since there are 13 different possible 4-of-a-kind hands, the probability of being dealt four queens is 1/13 of the total number of 4-of-a-kind hands. Thus, the probability is
$$\left(\frac{1}{13}\right) \cdot \left(\frac{624}{2{,}598{,}960}\right) \approx .00001847.$$

48. Since there are 4 different suits, a hearts flush is 1/4 of all possible non-royal and non-straight flushes. Thus, the probability is
$$\left(\frac{1}{4}\right) \cdot \left(\frac{5108}{2,598,960}\right) \approx .00049135.$$

49. If a dart hits the square target shown in the text at random, what is the probability that it will hit in a colored region? (Hint: Compare the area of the colored regions to the total area of the target.)

    The area of the colored regions is given by
    $$(6cm)^2 - (4cm)^2 + (2cm)^2 = 24\ cm^2.$$
    The total area is given by
    $$(8cm)^2 = 64\ cm^2.$$
    Thus, the probability of hitting a colored region is given by
    $$P = \frac{24}{64} = \frac{3}{8} = .375.$$

50. (a) There are 26 black cards and 26 red cards in the deck. You could draw a maximum of 26 red cards and 2 black cards, a total of 28, and still get fewer than three black cards.

    (b) There are 13 spades and 39 non-spades in the deck. You could draw a maximum of 39 non-spades and 5 spades, a total of 44 cards, and still get fewer than six spades.

    (c) There are 12 face cards (the jack, queen, and king of each suit) and 40 non-face cards in the deck. You could draw a maximum of 40 non-face cards and 3 face cards, a total of 43, and still get fewer than four face cards.

    (d) There are 4 kings and 48 non-kings in the deck. You could draw a maximum of 48 non-kings and 1 king, a total of 49 cards, and still get fewer than two kings.

51. Use the fundamental counting principle to determine the number of favorable seating arrangements where each man will sit immediately to the left of his wife.

    The first seat can be occupied by one of the three men, the second by his wife, the third by one of the two remaining men, etc. or, $3 \cdot 1 \cdot 2 \cdot 1 \cdot 1 \cdot 1 = 6$; since there are $6! = 720$ possible arrangements of the six people in the six seats, we have
    $$P = \frac{6}{720} = \frac{1}{120} \approx .0083.$$

52. Use the fundamental counting principle to find the number of ways in which the 6 people can be seated so that each man will sit immediately to the left of a woman.

    Starting at the left, any of the three men can occupy the first seat, and any of the three women can occupy the second seat. Then, any of the remaining two men can occupy the third seat, and any of the two remaining women can occupy the fourth seat. Finally, the one remaining man must take the fifth seat, and the one remaining woman must take the last (sixth) seat. The number of favorable outcomes is
    $$3 \cdot 3 \cdot 2 \cdot 2 \cdot 1 \cdot 1 = 36.$$
    Thus, the probability that each man will sit immediately to the left of a woman is
    $$P = \frac{36}{720} = \frac{1}{20} = .05.$$

53. Use the fundamental counting principle to determine the number of ways the women can sit in three adjacent seats.

    The first task is to decide in which seat the first woman is to sit. There are 4 choices (seats 1, 2, 3 or 4). Once this is decided, a second task would be to decide how many arrangements the three women can make sitting together (3!). The last task is to decide how many arrangements the three men could make sitting in the remaining three seats (3!). Thus, there are
    $$4 \cdot 3! \cdot 3! = 144$$
    ways to accommodate the three women sitting together.

    The probability of this occurring is given by
    $$P = \frac{144}{720} = \frac{1}{5} = .2,$$
    where $6! = 720$ is the total number of seating arrangements possible.

54. If the women are to be in three adjacent seats and the men are also to be in three adjacent seats, the women must sit in the first three seats (on the left) or the last three seats (on the right). There are 2 choices for the group of seats to be occupied by the women and the men. For each of these choices, the women may be arranged among themselves in 3! ways, and the men may be arranged among themselves in 3! ways. By the fundamental counting principle, the number of favorable outcomes is
    $$2 \cdot 3! \cdot 3! = 2 \cdot 6 \cdot 6 = 72.$$
    Thus, the probability that the women will be in the three adjacent seats, as will the men, is
    $$P = \frac{72}{720} = \frac{1}{10} = .1.$$

55. Two distinct numbers are chosen randomly from the set

$$\{-2, -4/3, -1/2, 0, 1/2, 3/4, 3\}.$$

To evaluate the probability that they will be the slopes of two perpendicular lines, find the size of the sample space and the size of the event of interest. The size of the sample space, $n(S)$, is given by $C(7,2) = 21$, since we are choosing 2 items from a set of 7. The size of the event of interest, $n(E)$, is 2 (remember that perpendicular lines must have slopes that are the negative reciprocals of each other). These are either -2 and $1/2$ (either order) or $-4/3$ and $3/4$. Thus,

$$P = \frac{n(E)}{n(S)} = \frac{2}{21} \approx .095.$$

56. Since repetitions are not allowed and order is important in the finish of a horse race, use permutations. The number of possible ways in which the first, second, and third place winners can occur is $P(5,3)$. Of these ways, only one is the actual result, that is, the correct pick for winning the "trifecta." Let E represent the event of interest. Then the probability of winning is given by

$$P(E) = \frac{\text{number of favorable outcomes}}{\text{total number of outcomes}}$$
$$= \frac{1}{P(5,3)}$$
$$= \frac{1}{60} \approx .017.$$

57. Since repetitions are not allowed and order is not important in selecting courses, use combinations. The number of ways of choosing any three courses from the list of twelve is $C(12,3)$. Let F be the event of interest "all three courses selected are science courses." Then

$$P(F) = \frac{\text{number of favorable outcomes}}{\text{total number of outcomes}}$$
$$= \frac{C(5,3)}{C(12,3)} = \frac{10}{220}$$
$$= \frac{1}{22} \approx .045.$$

58. The total number of ways to make the three selections is given by

$$C(10,3) = 120.$$

Only 1 of these ways represents a success, all three Beethoven. Thus,

$$P = \frac{1}{120} \approx 0.008.$$

59. The total number of ways to make the three selections, in order, is given by

$$P(36,3) = 42840.$$

Only 1 of these ways represents a success. Thus,

$$P = \frac{1}{42840} \approx 0.000023.$$

60. The total number of ways to make the three selections, in any order, is given by

$$C(36,3) = 7140.$$

Only 1 of these ways represents a success. Thus,

$$P = \frac{1}{7140} \approx 0.000140.$$

61. The first eight primes are $2, 3, 5, 7, 11, 13, 17, 19$. The sample space consists of all combinations of the set of 8 elements taken 2 at a time, the size of which is given by $C(8,2) = 28$. Thus,

$$E = \{19+5, 17+7, 13+11\} \text{ and}$$

$$P = \frac{n(E)}{n(S)} = \frac{3}{28} \approx .107.$$

62. Let S be the sample space, which is made up of all the five-digit numbers which can be formed using the digits $1, 2, 3, 4, 5$. Since all five digits must be used, repeats are not possible, so $n(S) = 5!$.

(a) Let E be the event "the number is even." In order for the number to be even, its last (units) digit must be 2 or 4. Because of this restriction, select the units digit first. There are 2 choices. The first four digits will be the remaining four digits (all but the one used as the units digit) arranged in any order. There are 4! possible arrangements. By the fundamental counting principle,

$$n(E) = 2 \cdot 4!.$$

Thus,

$$P(E) = \frac{n(E)}{n(S)} = \frac{2 \cdot 4!}{5!}$$
$$= \frac{2 \cdot 24}{120}$$
$$= \frac{2}{5}.$$

(b) Let G be the event "first and last digits are both even." If the first and last digits are both even, there are two choices for the first digit (2 or 4) and one choice for the last (units) digit (the even digit not already used). The three odd digits can be place in three middle positions in 3! ways.

By the fundamental counting principle,

$$n(G) = 2 \cdot 1 \cdot 3!$$
$$= 2 \cdot 3!.$$

Thus,

$$P(G) = \frac{n(G)}{n(S)}$$
$$= \frac{2 \cdot 3!}{5!}$$
$$= \frac{2 \cdot 6}{120}$$
$$= \frac{1}{10}.$$

63. Two integers are randomly selected from the set $\{1, 2, 3, 4, 5, 6, 7, 8, 9\}$ and are added together. Find the probability that their sum is 11 if they are selected as follows:

(a) With replacement, the event of interest is $E = \{2+9, 9+2, 3+8, 8+3, 4+7, 7+4, 5+6, 6+5.\}$ Thus, $n(E) = 8$. Since there are $9 \cdot 9 = 9^2 = 81$ ways of selecting the two digits to add together, we have $n(S) = 81$ and

$$P(\text{sum is eleven}) = \frac{8}{81} \approx .099.$$

(b) Without replacement, the event of interest is

$$E = \{2+9, 3+8, 4+7, 5+6\}$$

and $n(E) = 4$. However (without replacement),

$$n(S) = C(9, 2) = 36.$$

Thus,

$$P(\text{sum is eleven}) = \frac{4}{36} = \frac{1}{9} \approx .111.$$

64. Let S be the sample space, which is the set of all two-digit numbers. Since the tens digit can be any of the nine digits 1 through 9 and the units digit can be any of the ten digits 0 through 9,

$$n(S) = 9 \cdot 10 = 90.$$

Let E be the event "a palindromic two-digit number is chosen." A two-digit number will be palindromic only if both digits are the same. This repeated digit can be any of the nine digits 1 through 9, that is,

$$E = \{11, 22, 33, \ldots 99\}.$$

Thus,

$$n(E) = 9,$$

and

$$P(E) = \frac{n(E)}{n(S)} = \frac{9}{90} = \frac{1}{10}.$$

65. Let S be the sample space, which is the set of all three-digit numbers. The total number of three-digit numbers is given by the fundamental counting principle,

$$n(S) = 9 \cdot 10 \cdot 10 = 900.$$

Let E be the event "a palindromic three-digit number is chosen." The number of three-digit palindromic numbers is given by

$$n(E) = 9 \cdot 10 \cdot 1,$$

again using the fundamental counting principle. Remember that there are only 9 ways $\{1, 2, \ldots 9\}$ to choose the first digit and 10 ways $\{0, 1, \ldots 9\}$ to choose the second. The third is fixed since it must be the same as the first. Thus,

$$P(E) = \frac{n(E)}{n(S)} = \frac{9 \cdot 10}{9 \cdot 10^2} = \frac{9}{9 \cdot 10} = \frac{1}{10}.$$

12.2 EXERCISES

1. Yes, since event A and event B cannot happen at the same time.

2. For all of the given events to be mutually exclusive, no two of them can happen simultaneously. Since Patricia's alma mater could be a private college and in the East, etc., the events are not all mutually exclusive.

3. Writing exercise

Use the sample space $S = \{1, 2, 3, 4, 5, 6\}$ for Exercises 4–9.

4. Let E be the event "not less than 2." The complementary event E' is "less than 2." Since 1 is the only the outcome less than 2,

$$P(E') = \frac{1}{6}.$$

By the complements rule of probability,

$$P(E) = 1 - \frac{1}{6} = \frac{5}{6}.$$

Thus,

$$P(\text{not less than 2}) = \frac{5}{6}.$$

5. Let E be the event "not prime." Then $E = \{1, 4, 6\}$, the non-prime numbers in S. Thus,

$$P(E) = \frac{3}{6} = \frac{1}{2}.$$

6. Let E be the event "odd or less than 5." Then E' is the event $\{6\}$, getting a 6. Therefore, by the complements rule of probability,
$$P(E) = 1 - P\left(E'\right) = 1 - \frac{1}{6} = \frac{5}{6}.$$

7. Let $E = \{2, 4, 6\}$ and $F = \{2, 3, 5\}$.
$$\begin{aligned}P(E \text{ or } F) &= P(E) + P(F) - P(E \text{ and } F) \\ &= \frac{3}{6} + \frac{3}{6} - \frac{1}{6} \\ &= \frac{5}{6},\end{aligned}$$
by the general addition rule of probability.

8. All of the numbers are "either odd or even" in the set S. Thus,
$$P(\text{odd or even}) = \frac{6}{6} = 1.$$
This is a certain event.

9. Let A be the event "less than 3" and B be the event "greater than 4." Thus, $A = \{1, 2\}$ and $B = \{5, 6\}$.

   Since A and B are mutually exclusive events, use the special addition rule of probability:
$$\begin{aligned}P(A \text{ or } B) &= P(A) + P(B) \\ &= \frac{2}{6} + \frac{2}{6} \\ &= \frac{4}{6} = \frac{2}{3}.\end{aligned}$$

10. (a) Since the complement of "not an ace" is "is an ace," use the complements rule. There are 4 aces, so
$$P(\text{ace}) = \frac{4}{52} = \frac{1}{13}.$$
Then,
$$\begin{aligned}P(\text{not an ace}) &= 1 - P(\text{ace}) \\ &= 1 - \frac{1}{13} \\ &= \frac{12}{13}.\end{aligned}$$

    (b) There are 48 favorable outcomes (not an ace) and 4 unfavorable outcomes, so the odds in favor of drawing a card that is not an ace are 48 to 4, or 12 to 1.

11. (a) Since the two events, drawing a king (K) and drawing a queen (Q) are mutually exclusive, use the special addition rule:

$$\begin{aligned}P(K \text{ or } Q) &= P(K) + P(Q) \\ &= \frac{4}{52} + \frac{4}{52} = \frac{8}{52} = \frac{2}{13}.\end{aligned}$$

    (b) Use the formula for finding odds in favor of an event E:

    Odds in favor of $E = \dfrac{P(E)}{P(E')}$. Thus,

$$\begin{aligned}\frac{P(E)}{P(E')} &= \frac{P(K \text{ or } Q)}{P(\text{not }(K \text{ or } Q))} \\ &= \frac{P(K \text{ or } Q)}{1 - P(K \text{ or } Q)} \\ &= \frac{2/13}{1 - (2/13)} \\ &= \frac{2/13}{11/13} \\ &= \frac{2}{13} \cdot \frac{13}{11} \\ &= \frac{2}{11}, \text{ or } 2 \text{ to } 11.\end{aligned}$$

12. (a) There are 13 clubs and 13 hearts. Since the events "club" and "heart" are mutually exclusive, use the special additional rule for probability:
$$\begin{aligned}P(\text{club or heart}) &= P(club) + P(heart) \\ &= \frac{13}{52} + \frac{13}{52} \\ &= \frac{26}{52} = \frac{1}{2}.\end{aligned}$$

    (b) There are 26 favorable outcomes (13 clubs and 13 hearts) and 26 unfavorable outcomes (13 spades and 13 diamonds). Thus, the odds in favor of drawing a club or a heart are 26 to 26, or 1 to 1.

13. (a) Let S be the event of "drawing a spade" and F be the event of "drawing a face card." Then, $n(S) = 13$ and $n(F) = 12$. (There are 3 face cards in each of the 4 suits.) There are 3 face cards that are also spades so that $n(S \text{ and } F) = 3$. Thus, by the general additive rule for probability,
$$\begin{aligned}P(S \text{ or } F) &= P(S) + P(F) - P(S \text{ and } F) \\ &= \frac{13}{52} + \frac{12}{52} - \frac{3}{52} \\ &= \frac{11}{26}.\end{aligned}$$

(b) Use the formula for finding odds in favor of an event E:

$$\text{Odds in favor of } E = \frac{P(E)}{P(E')}. \text{ Thus,}$$

$$\frac{P(E)}{P(E')} = \frac{P(S \text{ or } F)}{P(\text{not } (S \text{ or } F))}$$
$$= \frac{P(S \text{ or } F)}{1 - P(S \text{ or } F)}$$
$$= \frac{11/26}{1 - (11/26)}$$
$$= \frac{11/26}{15/26}$$
$$= \frac{11}{26} \cdot \frac{26}{15}$$
$$= \frac{11}{15}, \text{ or 11 to 15.}$$

14. (a) Let H be the event "a heart is drawn" and N be the event "a seven is drawn." We want to find $P(H' \text{ or } N)$, or $P(H' \cup N)$. First, use the complements rules of probability to find the probability of drawing a card that is not a heart, or $P(H')$. Since there are 13 hearts in the deck,

$$P(H) = \frac{13}{52}$$
$$P(H') = 1 - P(H)$$
$$= 1 - \frac{13}{52}$$
$$= \frac{39}{52}.$$

There are four 7's in the deck, so

$$P(N) = \frac{4}{52}.$$

Since H' and N are not mutually exclusive events, use the general addition rule of probability. There are three cards which are not hearts but are 7's (club, spade, or diamond), so $n(H' \cap N) = 3$, and

$$P(H' \cap N) = \frac{3}{52}.$$

$$P(H' \cup N) = P(H') + P(N) - P(H' \cap N)$$
$$= \frac{39}{52} + \frac{4}{52} - \frac{3}{52}$$
$$= \frac{40}{52} = \frac{10}{13}.$$

(b) The number of non-hearts totals 39 and the only 7 that is not in that set is the 7 of hearts, so the number of favorable outcomes is $39 + 1 = 40$. The number of unfavorable outcomes is $52 - 40 = 12$. The odds are 40 to 12, or 10 to 3.

15. (a) Let H be the event "a heart is drawn" and S be the event "a seven is drawn." We want to find $P(H \text{ or } S)'$, or $P(H \cup S)'$. Since there are 13 hearts in the deck and 3 other sevens which are not hearts, $n(H \cup N) = 16$ and

$$P(H \cup S) = \frac{16}{52} = \frac{4}{13}.$$

Thus, $P(H \cup S)' = 1 - P(H \cup S)$
$$= 1 - \frac{4}{13}$$
$$= \frac{9}{13}.$$

(b) The number of cards which are hearts or are sevens totals 16. Thus the number of cards which are not hearts nor sevens is $52 - 16 = 36$. The odds in favor of not hearts nor sevens are 36 to 16, or 9 to 4.

Construct a table showing the sum for each of the 36 equally likely outcomes.

		\multicolumn{6}{c}{2nd die}
	+	1	2	3	4	5	6
	1	2	3	4	5	6	7
	2	3	4	5	6	7	8
1st	3	4	5	6	7	8	9
die	4	5	6	7	8	9	10
	5	6	7	8	9	10	11
	6	7	8	9	10	11	12

16. Since getting a sum of 11 and getting a sum of 12 are mutually exclusive events, use the special addition rule of probability:

$$P(11 \text{ or } 12) = P(11) + P(12).$$

The sum of 11 occurs twice and the sum of 12 occurs once. Thus, the probability fractions are

$$P(11 \text{ or } 12) = P(11) + P(12)$$
$$= \frac{2}{36} + \frac{1}{36}$$
$$= \frac{3}{36} = \frac{1}{12}.$$

17. Let E be the event of getting a sum which is an even number. Counting the number of occurrences in the sum table for these even outcomes represents the numerator of the probability fraction.

Then,

$$P(E) = \frac{18}{36}.$$

Let M be the event of getting sums which are multiples of three, $\{3, 6, 9, 12\}$. Counting the number of occurrences in the sum table for these outcomes represents the numerator of the probability fraction.

Then,
$$P(M) = \frac{12}{36}$$
and
$$P(E \text{ and } M) = \frac{6}{36}.$$

Thus, by the general addition rule,
$$\begin{aligned}P(E \cup M) &= P(E) + P(M) - P(E \text{ and } M)\\ &= \frac{18}{36} + \frac{12}{36} - \frac{6}{36}\\ &= \frac{24}{36} = \frac{2}{3}.\end{aligned}$$

18. The set of numbers (possible sums) that are "odd or greater than 9" is $\{3, 5, 7, 9, 10, 11, 12\}$. Use the "sum" table to count the number of times each of these sums occur. These values will be the numerators of the probability fractions. Use the special addition rule to find the probability that the sum is one of these values:

$$\begin{aligned}&P(\text{odd or greater than 9})\\ &= P(3) + P(5) + P(7) + P(9) + P(10) + P(11) + P(12)\\ &= \frac{2}{36} + \frac{4}{36} + \frac{6}{36} + \frac{4}{36} + \frac{3}{36} + \frac{2}{36} + \frac{1}{36}\\ &= \frac{22}{36} = \frac{11}{18}.\end{aligned}$$

19. Since these are mutually exclusive events, use the special additive rule. Since there is only one sum less than 3 (the sum of 2),
$$P(\text{sum less than 3}) = \frac{1}{36},$$
and since there are six sums greater than 9,
$$P(\text{sum greater than 9}) = \frac{6}{36}.$$
Thus,
$$P(\text{sum less than 3 or greater than 9}) = \frac{1}{36} + \frac{6}{36} = \frac{7}{36}.$$

20. (a) There are 5 prime numbers in the given set, $\{2, 3, 5, 7, 11\}$. Thus,
$$P(\text{prime number}) = \frac{5}{12}.$$

(b) Count the number of occurrences for the sums which are prime numbers in table of sums. The number of occurrences represent the numerators for the corresponding probability fractions. Thus, the probability of rolling a prime sum is

$$\begin{aligned}&P(2) + P(3) + P(5) + P(7) + P(11)\\ &= \frac{1}{36} + \frac{2}{36} + \frac{4}{36} + \frac{6}{36} + \frac{2}{36}\\ &= \frac{15}{36} = \frac{5}{12}.\end{aligned}$$

21. $\begin{aligned}P(S) &= P(A \cup B \cup C \cup D)\\ &= P(A) + P(B) + P(C) + P(D)\\ &= 1\end{aligned}$

Refer to Table 1 (12.1 in the textbook) and give answers to six decimal places.

22. These events are mutually exclusive, so use the special addition rule:

$$\begin{aligned}&P(\text{a flush or three of a kind})\\ &= P(\text{flush}) + P(\text{three of a kind})\\ &= \frac{5108}{2,598,960} + \frac{54,912}{2,598,960}\\ &= \frac{60,020}{2,598,960} \approx .023094.\end{aligned}$$

23. Let F be the event of drawing a full house and S be the event of drawing a straight. Using Table 1 in the text to determine
$$n(F) = 3744$$
and
$$n(S) = 10,200,$$
we have
$$P(F) = \frac{3744}{2,598,960} \text{ and } P(S) = \frac{10,200}{2,598,960}.$$
Since the events are mutually exclusive,
$$\begin{aligned}P(F \text{ or } S) &= P(F) + P(S)\\ &= \frac{3,744}{2,598,960} + \frac{10,200}{2,598,960}\\ &= \frac{13,944}{2,598,960} \approx .005365.\end{aligned}$$

24. The events are mutually exclusive, so use the special addition rule. Since 2 of the four suits are black, the number of ways to get a black flush is
$$\left(\frac{1}{2}\right)(5108) = 2554.$$

$$\begin{aligned}&P(\text{a black flush or two pairs})\\ &= P(\text{black flush}) + P(\text{two pairs})\\ &= \frac{2554}{2,598,960} + \frac{123,552}{2,598,960}\\ &= \frac{126,106}{2,598,960} \approx .048522\end{aligned}$$

25. The events are mutually exclusive, so use the special addition rule:

    P(nothing any better than two pairs)
    $= P$(no pair or one pair or two pairs)
    $= P$(no pair) $+ P$(one pair) $+ P$(two pairs)
    $= \dfrac{1,302,540}{2,598,960} + \dfrac{1,098,240}{2,598,960} + \dfrac{123,552}{2,598,960}$
    $= \dfrac{2,524,332}{2,598,960} \approx .971285.$

26. $P(95 \text{ or higher}) = P(95\text{–}99) + P(100 \text{ or above})$
    $= .03 + .01$
    $= .04$

27. "Par or above" is represented by all categories from 70 up. Since these are mutually exclusive, use the special addition rule:

    P(Par or above)
    $= 0.30 + 0.23 + 0.09 + 0.06 + 0.04 + 0.03 + 0.01$
    $= 0.76.$

28. $P(\text{in the } 80's) = P(80\text{–}84) + P(85\text{–}89)$
    $= .09 + .06$
    $= .15$

29. "Less than 90" is represented by all categories under the 90–94 category. Since these are mutually exclusive, use the special addition rule:

    P(Less than 90)
    $= 0.04 + 0.06 + 0.14 + 0.30 + 0.23 + 0.09 + 0.06$
    $= 0.92.$

30. $P(\text{not in the } 70's, 80's, \text{ or } 90's)$
    $= 1 - P(\text{in the } 70's, 80's, \text{ or } 90's)$
    $= 1 - (.30 + .23 + .09 + .06 + .04 + .03)$
    $= 1 - .75 = .25$

31. Odds of Sue's shooting below par $= \dfrac{P(\text{below par})}{P(\text{par or above})}$

    P(par or above) $= .76$ (Exercise 27), and
    P(below par) $= 1 - P$(par or above)
    $= 1 - .76 = .24,$

    by the complements rule of probability.

    Thus,

    odds in favor $= \dfrac{.24}{.76} = \dfrac{6}{19}$, or 6 to 19.

32. There are 11 balls, of which 4 are red, 5 are yellow, and 2 are blue. Since each ball has an equal probability of being chosen, the probability distribution is as follows.

x	$P(x)$
red	4/11
yellow	5/11
blue	2/11

33. Let x denote the sum of two distinct numbers selected randomly from the set $\{1, 2, 3, 4, 5\}$. Construct the probability distribution for the random variable x.

    Create a "sum table" to list the elements in the sample space. Note: can't use $1 + 1 = 2$, etc. Why?

+	1	2	3	4	5
1	–	3	4	5	6
2	3	–	5	6	7
3	4	5	–	7	8
4	5	6	7	–	9
5	6	7	8	9	–

Thus, the probability distribution is as follows.

x	$P(x)$
3	$2/20 = .1$
4	$2/20 = .1$
5	$4/20 = .2$
6	$4/20 = .2$
7	$4/20 = .2$
8	$2/20 = .1$
9	$2/20 = .1$

34. Writing exercise

For Exercises 35–38, let A be an event within the sample space S, and let $n(A) = a$ and $n(S) = s$.

35. $n(A') + n(A) = n(S)$
    $n(A') + a = s$
    Thus, $n(A') = s - a.$

36. By the theoretical probability formula,

    $P(A) = \dfrac{n(A)}{n(S)}$
    $= \dfrac{a}{s},$

    and

    $P(A') = 1 - P(A)$
    $= 1 - \dfrac{a}{s}$
    $= \dfrac{s}{s} - \dfrac{a}{s}$
    $= \dfrac{s - a}{s}.$

37. $P(A) + P(A') = \dfrac{a}{s} + \dfrac{s - a}{s}$
    $= \dfrac{a + s - a}{s}$
    $= \dfrac{s}{s} = 1$

38. In Exercise 37, you have proved the "complements rule of probability."

We want to form three-digit numbers using the set of digits $\{0, 1, 2, 3, 4, 5\}$. For example, 501 and 224 are such numbers but 035 is not.

39. The number of three-digit numbers is, by the fundamental counting principle,

    $$5 \cdot 6 \cdot 6 = 180.$$

    Remember that we can't choose "0" for the first digit.

40. There are 5 choices for the first digit (any digit but 0), 6 choices for the second digit (no restrictions), and 2 choices for the third digit (must be 0 or 5 for the number to be a multiple of 5). By the fundamental principle of counting, the number of possibilities is

    $$5 \cdot 6 \cdot 2 = 60.$$

41. If one three-digit number is chosen at random from all those that can be made from the above set of digits, find the probability that the one chosen is not a multiple of 5.

    The number of three-digit numbers is, by the fundamental counting principle,

    $$5 \cdot 6 \cdot 6 = 180 \text{ (Exercise 39)}.$$

    There are

    $$5 \cdot 6 \cdot 2 = 60$$

    three-digit numbers that are multiples of 5 (Exercise 40). Thus,

    $$P(\text{multiple of 5}) = \frac{60}{180} = \frac{1}{3}.$$

    By the complements rule,

    $$\begin{aligned} P(\text{not a multiple of 5}) &= 1 - P(\text{multiple of 5}) \\ &= 1 - \frac{60}{180} \\ &= 1 - \frac{1}{3} \\ &= \frac{2}{3}. \end{aligned}$$

42. The product of two numbers is even if either of the numbers is even (or both are even). There are 4 spins in which both are even: $(2, 8)$, $(2, 10)$, $(4, 8)$, and $(4, 10)$ and a total of $4 \cdot 3 = 12$ spins to complete the entire sample space (that is, all possible products). So

    $$P(\text{both are even}) = \frac{4}{12} = \frac{1}{3}.$$

    $$\begin{aligned} P(\text{product is even}) &= P(\text{1st or 2nd spinner is even}) \\ &= P(\text{1st even}) + P(\text{2nd even}) - P(\text{both are even}) \\ &= \frac{1}{2} + \frac{2}{3} - \frac{1}{3} \\ &= \frac{3}{6} + \frac{4}{6} - \frac{2}{6} \\ &= \frac{5}{6}. \end{aligned}$$

43. Since box C contains the same number of green marbles as blue and the number of green and blue marbles was the same to begin with, then box A and box B must contain exactly the same number of marbles after all the marbles are drawn. Because this is certain, the probability is 1.

12.3 EXERCISES

1. The events are independent since the outcome on the first toss has no effect on the outcome of the second toss.

2. Informally, we can see that these events are independent because the occurrence of one roll of a die has no effect on the probability of another roll of a die. Formally, we verify this result as follows:

   Let A be the event "even on the first" and B be the event "odd on the second." For the unconditional probability of B, we get

   $$P(B) = \frac{3}{6} = \frac{1}{2}.$$

   For the conditional probability of B, given A, we also have

   $$P(B \mid A) = \frac{3}{6} = \frac{1}{2}.$$

   Since

   $$P(B \mid A) = P(B),$$

   the events are independent.

3. The two planets are selected, with replacement, from the list in Table 5. The events "first is closer than Earth" and "second is farther than Uranus" are independent since the first selection has no affect on the second. This is because the first selection was replaced before the second was made.

4. The two planets are selected, without replacement, from the list in Table 5. The events "first is closer than Jupiter" and "second is farther than Neptune" are not independent. This is because the first selection was not replaced, which may affect the outcome of the second choice.

5. The answers are all guessed on a twenty-question multiple choice test. Let A be the event "first answer correct" and let B be the event "last answer correct." The events A and B are independent since the first answer choice does not affect the last answer choice.

6. Let A be the event "first member selected is a Republican" and B be the event "second member is a Republican." Since the same member can not be selected twice, this process is equivalent to choosing two elements of a set without replacement. Thus, the events are not independent. (If all the Senators were Republican, the events would be independent.)

7. The probability that the student selected is "female" is
$$\frac{42}{100} = \frac{21}{50}.$$

8. Since a total of 34 of the 100 students are "motivated primarily by creativity," the probability is
$$\frac{34}{100} = \frac{17}{50}.$$

9. Since a total of 68 of the 100 students are "not motivated primarily by money," the probability is
$$\frac{68}{100} = \frac{17}{25}.$$

10. There are 18 male students who are motivated primarily by money, so the probability is
$$\frac{18}{100} = \frac{9}{50}.$$

11. Given that the student selected is motivated primarily by "sense of giving to society," the sample space is reduced to 34 students. Of these 19 are male, so the probability is
$$\frac{19}{34}.$$

12. The given condition that the student is female reduces the sample space to 42 students. Of these, 14 are motivated primarily by money and 13 are motivated primarily by creativity. Thus, the probability is
$$\frac{27}{42} = \frac{9}{14}.$$

In Exercises 13–16 the first puppy chosen is replaced before the second is chosen. Note that "with replacement" means that the events may be considered independent and we can apply the special multiplication rule of probability.

13. The probability that both select a "poodle" is
$$P(P_1 \text{ and } P_2) = P(P_1) \cdot P(P_2)$$
$$= \left(\frac{4}{7}\right) \cdot \left(\frac{4}{7}\right)$$
$$= \frac{16}{49}.$$

14. The probability that "Rebecka selects a retriever and Aaron selects a terrier" is given by
$$P(R_1 \text{ and } T_2) = P(R_1) \cdot P(T_2)$$
$$= \left(\frac{1}{7}\right) \cdot \left(\frac{2}{7}\right)$$
$$= \frac{2}{49}.$$

15. The probability that "Rebecka selects a terrier and Aaron selects a retriever" is given by
$$P(T_1 \text{ and } R_2) = P(T_1) \cdot P(R_2)$$
$$= \left(\frac{2}{7}\right) \cdot \left(\frac{1}{7}\right)$$
$$= \frac{2}{49}.$$

16. The probability that "both select a retriever" is given by
$$P(R_1 \text{ and } R_2) = P(R_1) \cdot P(R_2)$$
$$= \left(\frac{1}{7}\right) \cdot \left(\frac{1}{7}\right)$$
$$= \frac{1}{49}.$$

In Exercises 17–22, the first puppy chosen is not replaced before the second is chosen. Thus, the events are not independent. Therefore, apply the general multiplication rule of probability.

17. The probability that "both select a poodle" is given by
$$P(P_1 \text{ and } P_2) = P(P_1) \cdot P(P_2 \mid P_1)$$
$$= \left(\frac{4}{7}\right) \cdot \left(\frac{3}{6}\right)$$
$$= \frac{2}{7}.$$

Remember that the second probability is conditional to the first event as having occurred and hence, the sample space is reduced.

18. The probability that "Aaron selects a terrier, given Rebecka selects a poodle" is found by
$$P(T_2 \mid P_1) = \frac{2}{6} = \frac{1}{3}.$$

Note that Rebecka's choice decreased the sample space by one dog.

19. The probability that "Aaron select a retriever," given "Rebecka selects a poodle" is found by
$$P(R_2 \mid P_1) = \frac{1}{6}.$$
Note that Rebecka's choice decreased the sample space by one dog.

20. The probability that "Rebecka selects a retriever is found by
$$P(R_1) = \frac{1}{7}.$$

21. The probability that "Aaron selects a retriever," given "Rebecka selects a retriever" is found by
$$P(R_2 \mid R_1) = \frac{0}{6} = 0.$$
Note that after Rebecka's selection, there are no remaining retrievers for Aaron to select.

22. The probability that "both select a retriever" is found by
$$\begin{aligned}(R_1 \text{ and } R_2) &= P(R_1) \cdot P(R_2 \mid R_1) \\ &= \left(\frac{1}{7}\right) \cdot \left(\frac{0}{6}\right) \\ &= \frac{0}{42} = 0.\end{aligned}$$

23. Since the cards are dealt without replacement, when the second card is drawn, there will be 51 cards left, of which 12 are spades. Thus,
$$P(S_2 \mid S_1) = \frac{12}{51} = \frac{4}{17}.$$
Note that both the event of interest (numerator) and the sample space (denominator) were reduced by the selection of the first card, a spade.

24. Since the cards are dealt without replacement, when the second card is drawn, there will be 51 cards left, of which 13 are clubs. Thus,
$$P(\text{club second} \mid \text{diamond first}) = \frac{13}{51}.$$

25. Since the cards are dealt without replacement, the events "first is face card" and "second is face card" are not independent, so be sure to use the general multiplication rule of probability. Let F_1 be the event "first is a face card" and F_2 be the event "second is a face card." Then,
$$\begin{aligned}P(\text{two face cards}) &= P(F_1 \text{ and } F_2) \\ &= P(F_1) \cdot P(F_2 \mid F_1) \\ &= \frac{12}{52} \cdot \frac{11}{51} \\ &= \frac{3}{13} \cdot \frac{11}{51} \\ &= \frac{11}{221}.\end{aligned}$$

26. The deck contains 12 face cards and $52 - 12 = 40$ non-face cards. Let N_1 be the event "first not face card" and N_2 be the event "second not face card."
$$\begin{aligned}P(\text{no face cards}) &= P(N_1 \text{ and } N_2) = P(N_1) \cdot P(N_2 \mid N_1) \\ &= \frac{40}{52} \cdot \frac{39}{51} \\ &= \frac{10}{13} \cdot \frac{13}{17} \\ &= \frac{10}{17}.\end{aligned}$$

27. The probability that the "first card dealt is a jack and the second is a face card" is found by
$$\begin{aligned}P(J_1 \text{ and } F_2) &= P(J_1) \cdot (F_2 \mid J_1) \\ &= \frac{4}{52} \cdot \frac{11}{51} \\ &= \frac{11}{663}.\end{aligned}$$
Remember that there are only 11 face cards left once the jack is drawn.

28. (a) $P(A \text{ and } B) = P(A) \cdot P(B \mid A)$

    This is the general multiplication rule of probability.

    (b) Therefore,
    $$P(B \mid A) = \frac{P(A \text{ and } B)}{P(A)}.$$
    Divide both sides of the equation in (a) by $P(A)$ and reverse the equality.

    (c) Therefore,
    $$P(B \mid A) = \frac{n(A \text{ and } B)/n(S)}{n(A)/n(S)}.$$
    Apply the theoretical probability formula to $P(A \text{ and } B)$ and $P(A)$.

    (d) Therefore,
    $$P(B \mid A) = \frac{n(A \text{ and } B)}{n(A)}.$$
    Multiply the numerator and denominator of the expression in (c) by $n(S)$, as follows:
    $$\begin{aligned}P(B \mid A) &= \frac{[n(A \text{ and } B)/n(S)]}{[n(A)/n(S)]} \cdot \frac{n(S)}{n(S)} \\ &= \frac{n(A \text{ and } B)}{n(A)}.\end{aligned}$$

Use the results of Exercise 28 to find each of the following probabilities when a single card is drawn from a standard 52-card deck.

29. $P(\text{queen} \mid \text{face card}) = \dfrac{n(F \text{ and } Q)}{n(F)}$
    $= \dfrac{4}{12} = \dfrac{1}{3}$

30. $P(\text{face card} \mid \text{queen}) = \dfrac{n(Q \text{ and } F)}{n(Q)}$
    $= \dfrac{4}{4} = 1$

31. $P(\text{red} \mid \text{diamond}) = \dfrac{n(D \text{ and } R)}{n(D)}$
    $= \dfrac{13}{13} = 1$

32. $P(\text{diamond} \mid \text{red}) = \dfrac{n(R \text{ and } D)}{n(R)}$
    $= \dfrac{13}{26} = \dfrac{1}{2}$

33. From the integers 1 through 10, the set of primes are $\{2, 3, 5, 7\}$. The set of odds are $\{1, 3, 5, 7, 9\}$. Since there only three primes in the set of 4 odd numbers, the second probability fraction become 2/4 and
    $$P(\text{prime}) \cdot P(\text{odd} \mid \text{prime}) = \dfrac{4}{10} \cdot \dfrac{3}{4}$$
    $$= \dfrac{3}{10}.$$

    This is the same value as computed in the text for $P(\text{odd}) \cdot P(\text{prime} \mid \text{odd})$, that is, the probability of selecting an integer from the set which is "odd and prime."

34. Writing exercise

35. Since the birth of a boy (or girl) is independent of previous births, the probability of having three boys successively is
    $$\left(\dfrac{1}{2}\right) \cdot \left(\dfrac{1}{2}\right) \cdot \left(\dfrac{1}{2}\right) = \dfrac{1}{8}.$$

36. Since the birth of a boy or girl is independent of previous births, the probability of having a boy, then a girl, then a boy is
    $$\left(\dfrac{1}{2}\right) \cdot \left(\dfrac{1}{2}\right) \cdot \left(\dfrac{1}{2}\right) = \dfrac{1}{8}.$$

37. Let S represent a sale purchase for more than \$100. Then,
    $$P(\text{both sales more than \$100}) = P(S_1 \text{ and } S_2)$$
    $$= P(S_1) \cdot P(S_2)$$
    $$= (.80) \cdot (.80)$$
    $$= .640.$$

38. Let S represent a sale purchase for more than \$100. Then,
    $$P(\text{first 3 sales more than \$100})$$
    $$= P(S_1 \text{ and } S_2 \text{ and } S_3)$$
    $$= P(S_1) \cdot P(S_2) \cdot P(S_3)$$
    $$= (.80) \cdot (.80) \cdot (.80)$$
    $$= .512.$$

39. Let S represent a sale purchase for more than \$100. Then, S' represents a sale purchase that is not more than \$100. Since $P(S) = .80$,
    $$P(S') = 1 - .80$$
    $$= .20,$$

    by the complements principle.
    $$P(\text{none of the first 3 sales more than \$100})$$
    $$= P(\text{not } S_1 \text{ and not } S_2 \text{ and not } S_3)$$
    $$= P(S_1') \cdot P(S_2') \cdot P(S_3')$$
    $$= (.20) \cdot (.20) \cdot (.20)$$
    $$= .008.$$

40. Let S represent a sale purchase for more than \$100. Then, S' represents a sale purchase that is not more than \$100. Since $P(S) = .80$,
    $$P(S') = 1 - .80$$
    $$= .20,$$

    by the complements principle. There are three possible orderings of the purchases
    $$S_1 S_1' S_1',$$
    $$S_1' S_1 S_1', \text{ or}$$
    $$S_1' S_1' S_1.$$

    In each case, the probability is given by the product
    $$(.8)(.2)(.2) = .032,$$

    in some order. The final probability is thus,
    $$3(.032) = .096.$$

    Note that the fundamental counting principle is used here, where the first task is to decide how many different orderings there are for the sales. The second task is to decide the probability for any one of the orderings.

41. Since the probability the cloud will move in the critical direction is .05, the probability that it will not move in the critical direction is
    $$1 - .05 = .95,$$

    by the complements formula.

42. Since we are assuming that the probabilities for a particular launch in no way depends on the probabilities for other launches, we may treat the five launches as independent events and apply the special multiplication

rule. The probability that there will not be cloud movement in the critical direction during a particular launch is $1 - .05 = .95$. Therefore, the probability that there will be no cloud movement in the critical direction during *any* of the 5 launches will be

$$(.95)^5 \approx .77.$$

43. The probability that the cloud would not move in the critical direction for each launch is $1 - .05 = .95$. The probability that the cloud would not move in the critical direction for any 5 launches is $(.95)^5$ by the special multiplication rule. The probability that any 5 launches will result in at least one cloud movement in the critical direction is the complement of the probability that a cloud would not move in the critical direction for any 5 launches, or

$$1 - (.95)^5 \approx .23.$$

44. Use the complements rule of probability. Let E represent the given event. Then its complement, E', is the event "no cloud movement in the critical direction will occur in any of 10 launches." By the same reasoning that was used in Exercise 43,

$$P(E') = (.95)^{10}$$
$$\text{and} \quad P(E) = 1 - P(E')$$
$$= 1 - (.95)^{10}$$
$$\approx .40.$$

Four men and three women are waiting to be interviewed for jobs. If they are all selected in random order, find the probability of each of the following events.

45. $P(\text{all women first})$
$= P(W_1) \cdot P(W_2 \mid W_1) \cdot P(W_3 \mid W_1 \text{and } W_2)$
$= \dfrac{3}{7} \cdot \dfrac{2}{6} \cdot \dfrac{1}{5}$
$= \dfrac{1}{35}$

46. The total number of orders in which the seven people can be interviewed is 7!, so $n(S) = 7!$. Consider choosing the order for the interviews as a two-part task, since all of the men are to be interviewed first. The number of orders in which the men may be arranged is 4!, and the number of orders in which the women may be arranged is 3!. Using the fundamental principle of counting, the number of orders in which the people can be selected with all the men first is $4! \cdot 3!$, and the required probability is given by

$$P(\text{all the men will be interviewed first})$$
$$= \dfrac{4! \cdot 3!}{7!}$$
$$= \dfrac{4 \cdot 3 \cdot 2 \cdot 1 \cdot 3 \cdot 2 \cdot 1}{7 \cdot 6 \cdot 5 \cdot 4 \cdot 3 \cdot 2 \cdot 1} = \dfrac{1}{35}.$$

47. The probability that the first person interviewed will be a woman is

$$P(W_1) = \dfrac{3}{7}.$$

48. This event is made up of two cases: a woman is interviewed first and second (W_1 and W_2) and a man is interviewed first and a women second (M_1 and W_2). Since these events are mutually exclusive,

$$P(W_2) = P(W_1 \text{ and } W_2) + P(M_1 \text{ and } W_2)$$
$$= P(W_1) \cdot P(W_2 \mid W_1)$$
$$\quad + P(M_1) \cdot P(W_2 \mid M_1)$$
$$= \dfrac{3}{7} \cdot \dfrac{2}{6} + \dfrac{4}{7} \cdot \dfrac{3}{6}$$
$$= \dfrac{18}{42} = \dfrac{3}{7}.$$

49. Consider choosing the last to be interviewed first. Then the probability that the "last person interviewed will be a woman" is

$$P(W_7) = \dfrac{3}{7}.$$

50. Example 7 in the text gives the probability

$$\dfrac{4}{11} \cdot \dfrac{5}{10} \cdot \dfrac{2}{9} = \dfrac{4}{99}$$

for the case where the first ball drawn is red, the second ball is yellow and the third is blue. For the case where the first ball is red, the second blue, and the third yellow the probability is

$$\dfrac{4}{11} \cdot \dfrac{2}{10} \cdot \dfrac{5}{9} = \dfrac{4}{99}.$$

A similar set of probability fractions will occur for each arrangement of the order in which different colored balls are drawn. The number of different arrangements is 3! In each case the resulting probability is 4/99. Using the fundamental counting principle the total number of ways is

$$3! \cdot \dfrac{4}{99} = (3 \cdot 2 \cdot 1) \cdot \dfrac{4}{99}$$
$$= \dfrac{8}{33}.$$

51. (a) "At least three" is the complement of "one or two" and we can't get two girls from one birth. Thus, find only the probability of having 2 girls with two births,

$$P(gg) = P(g) \cdot P(g)$$
$$= \dfrac{1}{2} \cdot \dfrac{1}{2}$$
$$= \dfrac{1}{4}.$$

Therefore, by the complements principle,

$$P(\text{at least three births}) = 1 - \frac{1}{4} = \frac{3}{4}.$$

(b) "At least four births" is the complement of "two or three births." From (a) above

$$P(\text{two births to get gg}) = \frac{1}{4}.$$

To calculate $P(\text{three births to get gg})$ examine Figure 2 in the text. The three successes are ggb, gbg, and bgg. Don't count, however, the outcome ggb as a success since this has already been counted when computing the probability associated with two births.

$$P(\text{three births to get gg}) = \frac{2}{8} = \frac{1}{4}.$$

$P(\text{two or three births to get gg})$
$= P(\text{two births to get gg}) + P(\text{three births to get gg})$
$= \frac{1}{4} + \frac{1}{4}$
$= \frac{2}{4} = \frac{1}{2}$

Finally, the probability of "at least four births" may be calculated by the complements rule:

$$P(\text{at least four births to get gg}) = 1 - \frac{1}{2} = \frac{1}{2}.$$

(c) "At least five births" is the complement of "two or three or four births."

$P(\text{two or three or four births to obtain gg})$
$= P(\text{two or three births to get gg}) + P(\text{four births to get gg})$

$P(\text{two or three births to get gg}) = 1/2$ was calculated in (b) above. To calculate $P(\text{four births to get gg})$ extend the tree diagram (Figure 2 in the text). Count all of the outcomes with gg (two girls) as successes except bggb and ggbb since they have already been used in calculating the earlier probability. There are then 3 outcomes which may be considered as successes. Thus, $P(\text{four births to get gg})) = 3/16$.

$P(\text{two or three or four births to obtain gg})$
$= P(\text{two or three births to get gg}) + P(\text{four births to get gg})$
$= \frac{1}{2} + \frac{3}{16} = \frac{8}{16} + \frac{3}{16}$
$= \frac{11}{16}.$

Finally, the probability of "at least five births to obtain gg" may be calculated by the complements rule:

$$P(\text{at least five births to obtain gg}) = 1 - \frac{11}{16} = \frac{5}{16}.$$

52. (a) Since there are 12 face cards in the deck, the probability that the first card drawn will be a face card is

$$\frac{12}{52} = \frac{3}{13}.$$

Since we are interested in probabilities for drawing at least one face card, we can stop when the first face card is drawn. Let F_1 represent the event "a face card is obtained on the first draw," N_2 represent the event "a non-face card is obtained on the second draw," and so on. Then,

$$P(F_1) = \frac{3}{13} \approx .23,$$
$$P(F_2) = P(N_1 \cap F_2)$$
$$= P(N_1) \cdot P(F_2 \mid N_1)$$
$$= \frac{40}{52} \cdot \frac{12}{51} \approx .18,$$
$$P(F_3) = P(N_1 \cap N_2 \cap F_3)$$
$$= P(N_1) \cdot P(N_2 \mid N_1)$$
$$\cdot P(F_3 \mid N_1 \text{ and } N_2)$$
$$= \frac{42}{52} \cdot \frac{39}{51} \cdot \frac{12}{50} \approx .14.$$

The probability of drawing a face card on the first draw is about .23. The probability of drawing the first face card on the first or second draw is about

$$.23 + .18 = .42.$$

The probability of drawing the first face card on the third draw is about

$$.24 + .18 + .14 = .55 > \frac{1}{2}.$$

Thus, 3 cards must be drawn before the probability of drawing a face card is greater than $1/2$.

(b) Since there are 4 kings in the deck, the probability that the first card drawn will be a king is

$$\frac{4}{52} = \frac{1}{13}.$$

Follow the same procedure used in part (a). Let K_1 represent the event "a king is obtained on the first draw and N_2 represent the event "a non-king is obtained on the second draw," and so on. Then

$$P(K_1) = \frac{1}{13} \approx .077,$$
$$P(K_2) = P(N_1 \cap K_2)$$
$$= \frac{48}{52} \cdot \frac{4}{51} \approx .072, \text{ and}$$
$$P(K_3) = P(N_1 \cap N_2 \cap K_3)$$
$$= \frac{48}{52} \cdot \frac{47}{51} \cdot \frac{4}{50} \approx .068.$$

The probability of obtaining a king after 3 cards is drawn is about

$$.077 + .072 + .068 = .217.$$

Continue in this manner until the *cumulative* probability exceeds $1/2$ or $.5$. It will be found that 9 cards must be drawn before the probability of obtaining a king is greater than $1/2$.

A coin, biased so that $P(h) = .5200$ and $P(t) = .4800$, it tossed twice. Give answers to four decimal places.

53. Since the two tosses are independent events, use the special multiplication rule:
$$P(hh) = P(h) \cdot P(h)$$
$$= (.5200) \cdot (.5200)$$
$$= .2704.$$

54. Since the two tosses are independent events, use the special multiplication rule:
$$P(ht) = P(h) \cdot P(t)$$
$$= (.5200) \cdot (.4800)$$
$$= .2496.$$

55. Since the two tosses are independent events, use the special multiplication rule:
$$P(th) = P(t) \cdot P(h)$$
$$= (.4800) \cdot (.5200)$$
$$= .2496.$$

56. Since the two tosses are independent events, use the special multiplication rule:
$$P(tt) = P(t) \cdot P(t)$$
$$= (.4800)(.4800)$$
$$= .2304.$$

57. Writing exercise

58. Review "For Further Thought" in this section. Since there are 6 independent switches, the number of different ways in which the switches can be set is,
$$2^6 = 64.$$

59. Using the fundamental counting principle, there are
$$2 \cdot 2 \cdot 2 \cdot 2 \cdot 2 \cdot 2 = 2^6 = 64$$
different switch settings. The probability of randomly getting 1 of the 64 possible settings is
$$\frac{1}{64} \approx .0156.$$

60. Since there are 64 different possible settings, (see Exercises 58 and 59) there are 63 ways in which the second person can choose a different setting from the first, 62 ways in which the third person can choose a different setting from either of the first two, and so on. Use the same procedure as is shown in the text for the birthday problem, which uses both the complements rule and the general multiplication rule of probability.

For the five neighbors, the probability is given by

$$P(\text{at least one duplication of settings})$$
$$= 1 - P(\text{no duplications})$$
$$= 1 - \frac{63}{64} \cdot \frac{62}{64} \cdot \frac{61}{64} \cdot \frac{60}{64}$$
$$= 1 - \frac{14,295,960}{16,777,216}$$
$$= 1 - .8521 = .1479.$$

61. $P(\text{at least one duplication of switch settings})$
$$= 1 - P(\text{no duplication})$$
$$= 1 - \frac{63}{64}$$
$$\approx .016 \text{ for two neighbors;}$$

$$1 - \frac{63}{64} \cdot \frac{62}{64} = 1 - \frac{63 \cdot 62}{(64^2)}$$
$$\approx .046 \text{ for three neighbors;}$$

$$\ldots$$

$$1 - \frac{63 \cdot 62 \cdot 61 \cdot 60 \cdot 59 \cdot 58 \cdot 57 \cdot 56}{(64)^8}$$
$$\approx .445 \text{ for nine neighbors;}$$
and
$$1 - \frac{63 \cdot 62 \cdot 61 \cdot 60 \cdot 59 \cdot 58 \cdot 57 \cdot 56 \cdot 55}{(64)^9}$$
$$\approx .523 > \frac{1}{2} \text{ for ten neighbors.}$$

Thus, the minimum number of neighbors who must use this brand of opener before the probability of at least one duplication of settings is greater than $1/2$ is ten.

62. The three cards have the six sides shown below.

1st card	red	red
2nd card	green	red
3rd card	green	green

Selecting one card at random and placing it on the table is equivalent to selecting at random one of these six sides. Since we know that a red side is up, the reduced sample space made up of the three red sides. Of these three, two belong to the first card and one belongs to the second card. If the red side which is up belongs to the first card, the other side is also red. Thus, the probability that the card is also red on the other side is 2/3.

We can also use conditional probability. Let R_1 be the event "red side up" and R_2 be the event "other side red." We need to find the conditional probability

$$P(R_2 \mid R_1).$$

Select one of the three cards at random. Since one of the three cards has both sides red, the probability that both sides are red, which is $P(R_1 \mid R_2)$, is 1/3.

Since three of the six sides are red, the probability that the side which is up is red, which is

$$P(R_1), \text{ is } 1/2.$$

Using the formula for conditional probability,

$$P(R_2 \mid R_1) = \frac{P(R_1 \cap R_2)}{P(R_1)}$$
$$= \frac{1/3}{1/2}$$
$$= \frac{1}{3} \cdot \frac{2}{1} = \frac{2}{3}.$$

63. Since the events are not independent, use the general multiplication rule. Let R_1 represent rain on the first day, R_2 represent rain on the second day.

$P(\text{rain on two consecutive days in November})$
$= P(R_1 \cap R_2)$
$= P(R_1) \cdot P(R_2 \mid R_1)$
$= (.500) \cdot (.800)$
$= .400$

64. Since the events are not independent, use the general multiplication rule. Let R_1 represent rain on the first day, R_2 represent rain on the second day, and R_3 represent rain on the third day.

$P(\text{rain on three consecutive days in November})$
$= P(R_1 \cap R_2 \cap R_3)$
$= P(R_1) \cdot P(R_2 \mid R_1) \cdot P(R_3 \mid R_2)$
$= (.500)(.800)(.800)$
$= .320$

65. Since the events are not independent, use the general multiplication rule. Let R_1 represent rain on November 1, R_2 represent rain on November 2, and R_3 represent rain on November 3.

$P(\text{rain on November 1st and 2nd, but not on the 3rd})$
$P(R_1 \cap R_2 \cap \text{not } R_3)$
$= P(R_1) \cdot P(R_2 \mid R_1) \cdot P(\text{not } R_3 \mid R_2)$
$= (.500) \cdot (.800) \cdot (1 - .800)$
$= .080$

Note that the probability of not raining after a rainy day is the complement of the probability that it does rain.

66. Since the events are not independent, use the general multiplication rule. Let C_{31} represent a clear day on October 31, R_1 represent rain on November 1, R_2 represent rain November 2, and R_3 represent rain on November 3, and R_4 represent rain November 4.

$P(R_1 \cap R_2 \cap R_3 \cap R_4)$
$= P(R_1 \mid C_{31}) \cdot P(R_2 \mid R_1) \cdot P(R_3 \mid R_2) \cdot P(R_4 \mid R_3)$
$= (.300)(.800)(.800)(.800)$
$\approx .154$

67. To find the probability of "no engine failures", begin by letting F represent a failed engine. The probability that a given engine will fail, $P(F)$, is .10. This means that, by the complements principle, the probability that an engine will not fail is given by

$$P(\text{not } F) = 1 - .10 = .90.$$

Since engines "not failing" are independent events, use the special product rule.

$P(\text{no engine failures})$
$= P(\text{not } F_1 \cap \text{not } F_2 \cap \text{not } F_3 \cap \text{not } F_4)$
$= P(\text{not } F_1) \cdot P(\text{not } F_2) \cdot P(\text{not } F_3) \cdot P(\text{not } F_4)$
$= (.90)^4 = .6561$

68. The probability of "exactly one engine failure" can be found by applying the fundamental counting principle. The probability that a given engine will fail is .10, and the probability of each of the other three engines failing if one engine fails is .20. This means that if one engine fails, the probability that a different engine does *not* fail is $1 - .20 = .80$. There are $C(4, 1)$ possibilities for which an engine will fail. Thus,

$P(\text{exactly one engine failure})$
$= C(4, 1) \cdot (.10)(.80)^3$
$= 4(.10)(.5120) = .2048.$

69. The probability of "exactly two engine failures" can be found by applying the fundamental counting principle, where the first task is to decide which two engines fail $[C(4, 2)]$; the second task, find the probability that one of these engines fail (.10); followed by the second engine failing (.20); followed by finding the probability the third engine does not fail $(1 - .30 = .70)$; followed by the probability that last engine does not fail $(1 - .30 = .70)$. Thus,

$$P = C(4, 2) \cdot (.10) \cdot (.20) \cdot (.70)^2$$
$$= .0588.$$

70. Since the aircraft can fly with as few as two operating engines, a failed flight occurs when a third engine fails, leaving only one operating engine. Using the complements rule and the results of Exercises 67–69, the probability is given by

$$P(\text{flight failure})$$
$$= 1 - P(0 \text{ or } 1 \text{ or } 2 \text{ engines fail})$$
$$= 1 - (.6561 + .2048 + .0588)$$
$$= 1 - .9197$$
$$= .0803.$$

Refer to text discussion of the rules for "one-and-one" basketball. Susan Dratch, a basketball player, has a 70% foul shot record. (She makes 70% of her foul shots.) Find the probability that, on a given one-and-one foul shooting opportunity, Susan will score the following number of points.

71. A one-and-one foul shooting opportunity means that Susan gets a second shot only if she makes her first shot. The probability of scoring "no points" means that she missed her first shot. Thus,

$$P(\text{scoring no points})$$
$$= 1 - P(\text{scoring at least one point})$$
$$= 1 - .70$$
$$= .30.$$

72. A one-and-one foul shooting opportunity means that Susan gets a second shot only if she makes her first shot. She can score one point only by making the first shot and missing the second. Thus,

$$P(\text{one point})$$
$$= P(\text{making the 1st shot and missing 2nd shot})$$
$$= P(\text{making the 1st shot}) \cdot P(\text{missing 2nd shot})$$
$$= (.70)(.30)$$
$$= .21.$$

73. The probability of "scoring two points" is given by

$$P(\text{scoring two points})$$
$$= P(\text{scoring the 1st shot and scoring the 2nd shot})$$
$$= P(\text{scoring on 1st shot}) \cdot P(\text{scoring on 2nd shot})$$
$$= (.70) \cdot (.70)$$
$$= .49.$$

74. (a) Writing exercise

    (b) "At least three rolls" is the complement of "fewer than three rolls."

$$P(\text{at least three rolls}) = 1 - P(\text{fewer than three rolls})$$
$$= 1 - P(\text{one or two rolls})$$
$$= 1 - [P(\text{one roll}) + P(\text{two rolls})]$$
$$= 1 - \{P(7 \text{ on first roll}) + P[(\text{non-7 on first roll})$$
$$\text{ and } (7 \text{ on second roll})]\}$$
$$= 1 - \left[\left(\frac{6}{36}\right) + \left(\frac{30}{36}\right)\left(\frac{6}{36}\right)\right]$$
$$= 1 - \left[\left(\frac{6}{36}\right) + \left(\frac{5}{6}\right)\left(\frac{1}{6}\right)\right]$$
$$= 1 - \left[\left(\frac{6}{36}\right) + \left(\frac{5}{36}\right)\right]$$
$$= 1 - \left(\frac{11}{36}\right)$$
$$= \frac{25}{36} \approx .69$$

75. Writing exercise

12.4 EXERCISES

1. Let heads be "success."
   Then $n = 3, p = q = \frac{1}{2}$, and $x = 0$.

   By the binomial probability formula,

   $$P(0) = C(3, 0) \left(\frac{1}{2}\right)^0 \left(\frac{1}{2}\right)^{3-0}$$
   $$= \frac{3!}{0! \, (3-0)!} \cdot 1 \cdot \left(\frac{1}{2}\right)^3$$
   $$= \frac{3!}{1 \cdot 3!} \cdot \frac{1}{2^3}$$
   $$= 1 \cdot \frac{1}{8} = \frac{1}{8}.$$

   Note $C(3, 0) = 1$, and we could easily reason this result without using the combination formula since there is only one way to choose 0 things from a set of 3 things– take none out.

2. Let heads be "success."
   Then $n = 3, p = q = \frac{1}{2}$, and $x = 1$.

   By the binomial probability formula,

   $$P(1) = C(3, 1) \left(\frac{1}{2}\right)^1 \left(\frac{1}{2}\right)^2$$
   $$= 3 \cdot \frac{1}{2} \cdot \frac{1}{4}$$
   $$= \frac{3}{8}.$$

12.4 BINOMIAL PROBABILITY

3. Let heads be "success."
   Then $n = 3, p = q = \frac{1}{2}$, and $x = 2$.
   By the binomial probability formula,
   $$P(2 \text{ heads}) = C(3,2) \cdot (\tfrac{1}{2})^2 \cdot (\tfrac{1}{2})^{3-2}$$
   $$= \frac{3!}{2!(3-2)!} \cdot \frac{1}{4} \cdot \frac{1}{2}$$
   $$= \frac{3 \cdot 2 \cdot 1}{2 \cdot 1 \cdot 4 \cdot 2} = \frac{3}{8}.$$

4. Here $n = 3, p = q = \frac{1}{2}$, and $x = 3$.
   $$P(3) = C(3,3)\left(\frac{1}{2}\right)^3\left(\frac{1}{2}\right)^0$$
   $$= 1 \cdot \frac{1}{8} \cdot 1$$
   $$= \frac{1}{8}$$

5. Use the special addition rule for calculating the probability of "1 or 2 heads."
   $$P(1 \text{ or } 2 \text{ heads}) = P(1) + P(2)$$
   $$= C(3,1) \cdot (\tfrac{1}{2})^1 \cdot (\tfrac{1}{2})^{3-1}$$
   $$\quad + C(3,2) \cdot (\tfrac{1}{2})^2 \cdot (\tfrac{1}{2})^{3-2}$$
   $$= 3 \cdot \frac{1}{2} \cdot \frac{1}{4} + 3 \cdot \frac{1}{4} \cdot \frac{1}{2}$$
   $$= \frac{6}{8} = \frac{3}{4}$$

6. The complement of "at least 1 head" is "no heads."
   $$P(\text{at least 1 head}) = 1 - P(0)$$
   $$= 1 - C(3,0)\left(\frac{1}{2}\right)^0\left(\frac{1}{2}\right)^3$$
   $$= 1 - 1 \cdot 1 \cdot \frac{1}{8}$$
   $$= \frac{7}{8}$$

7. "No more than 1" is the same as "0 or 1."
   $$P(0 \text{ or } 1) = P(0) + P(1)$$
   $$= C(3,0)\left(\frac{1}{2}\right)^0\left(\frac{1}{2}\right)^3$$
   $$\quad + C(3,1)\left(\frac{1}{2}\right)^1\left(\frac{1}{2}\right)^2$$
   $$= 1 \cdot 1 \cdot \frac{1}{8} + 3 \cdot \frac{1}{2} \cdot \frac{1}{4}$$
   $$= \frac{1}{8} + \frac{3}{8}$$
   $$= \frac{1}{2}$$

8. The complement of "fewer than 3 heads" is "3 heads." The probability of 3 heads was found in Exercise 4.
   $$P(\text{fewer than 3 heads}) = 1 - P(3)$$
   $$= 1 - \frac{1}{8}$$
   $$= \frac{7}{8}$$

9. Assuming boy and girl babies are equally likely, find the probability that a family with three children will have exactly two boys.
   $$P(2 \text{ boys}) = C(3,2)\left(\frac{1}{2}\right)^2\left(\frac{1}{2}\right)^1$$
   $$= 3 \cdot \frac{1}{4} \cdot \frac{1}{2}$$
   $$= \frac{3}{8}$$

10. Writing exercise

11. If n fair coins are tossed, the probability of exactly x heads is the fraction whose numerator is entry number x of row number n in Pascal's triangle, and whose denominator is the sum of the entries in row number n. That is, $x; n; n$.

For Exercises 12–19, refer to Pascal's triangle. Since seven coins are tossed, we will use row number 7 of the triangle. (Recall that the first row is row number 0 and that the first entry in each row is entry number 0.) The sum of the numbers in row 7 is $2^7 = 128$, which will be the denominator in each of the probability fractions.

```
                    1
                  1   1
                1   2   1
              1   3   3   1
            1   4   6   4   1
          1   5  10  10   5   1
        1   6  15  20  15   6   1
      1   7  21  35  35  21   7   1
```

12. For the probability "0 heads", the numerator will be entry number 0 in row number 7, which is 1 and the denominator is the sum of the elements in row 7, or 128. Thus, $P(0 \text{ heads}) = \dfrac{1}{128}$.

13. For the probability "1 head," the numerator of the probability fraction is entry 1 of row 7 of Pascal's triangle, or 7, and the denominator is the sum of the elements in row 7, or
    $$1 + 7 + 21 + 35 + 35 + 21 + 7 + 1 = 128.$$
    Thus, $P(1 \text{ head}) = \dfrac{7}{128}$.

14. For the probability "2 heads," the numerator is entry number 2 in row 7, which is 21. Thus,
$$P(2 \text{ heads}) = \frac{21}{128}.$$

15. For the probability "3 heads," the numerator of the probability fraction is entry number 3 of row 7 of Pascal's triangle, or 35, and the denominator is the sum of the elements in row 7, or 128. Thus,
$$P(3 \text{ heads}) = \frac{35}{128}.$$

16. For the probability "4 heads," the numerator is entry number 4 in row 7, which is 35. Thus,
$$P(4 \text{ heads}) = \frac{35}{128}.$$

17. For the probability "5 heads," the numerator of the probability fraction is entry number 5 of row 7 of Pascal's triangle, or 21, and the denominator is the sum of the elements in row 7, or 128. Thus,
$$P(5 \text{ heads}) = \frac{21}{128}.$$

18. For the probability "6 heads," the numerator is entry number 6 in row 7, which is 7. Thus,
$$P(6 \text{ heads}) = \frac{7}{128}.$$

19. For the probability "7 heads," the numerator of the probability fraction is entry 7 of row 7 of Pascal's triangle, or 1, and the denominator is the sum of the elements in row 7, or 128. Thus,
$$P(7 \text{ heads}) = \frac{1}{128}.$$

For Exercises 20–23, a fair die is rolled three times and a 4 is considered "success," while all other outcomes are "failures."

20. For "0 successes" there are no 4's.
Here $n = 3, p = \frac{1}{6}, q = \frac{5}{6}$, and $x = 0$.
$$P(0) = C(3,0)\left(\frac{1}{6}\right)^0 \left(\frac{5}{6}\right)^3$$
$$= 1 \cdot 1 \cdot \frac{125}{216}$$
$$= \frac{125}{216}$$

21. Here $n = 3, p = \frac{1}{6}, q = \frac{5}{6}$, and $x = 1$.
$$P(1) = C(3,1)\left(\frac{1}{6}\right)^1 \left(\frac{5}{6}\right)^2$$
$$= 3 \cdot \frac{1}{6} \cdot \frac{25}{36}$$
$$= \frac{25}{72}$$

22. Here $n = 3, p = \frac{1}{6}, q = \frac{5}{6}$, and $x = 2$.
$$P(2) = C(3,2)\left(\frac{1}{6}\right)^2 \left(\frac{5}{6}\right)^1$$
$$= 3 \cdot \frac{1}{36} \cdot \frac{5}{6}$$
$$= \frac{5}{72}$$

23. Here $n = 3, p = \frac{1}{6}, q = \frac{5}{6}$, and $x = 3$.
$$P(3) = C(3,3)\left(\frac{1}{6}\right)^3 \left(\frac{5}{6}\right)^0$$
$$= 1 \cdot \frac{1}{216} \cdot 1$$
$$= \frac{1}{216}$$

24. Writing exercise

Answers are rounded to three decimal places.

25. Here $n = 5, p = \frac{1}{3}$, and $x = 4$.
Since $p = \frac{1}{3}$,
$$q = 1 - p$$
$$= 1 - \frac{1}{3}$$
$$= \frac{2}{3}.$$

Substitute these values into the binomial probability formula:
$$P(4) = C(5,4)\left(\frac{1}{3}\right)^4 \left(\frac{2}{3}\right)^1$$
$$= 5 \cdot \frac{2}{3^5}$$
$$\approx .041.$$

26. Here $n = 10, p = .7$, and $x = 5$.
Since $p = .7$,
$$q = 1 - p$$
$$= 1 - .7$$
$$= .3.$$

Substitute these values into the binomial probability formula:
$$P(5) = C(10,5)(.7)^5(.3)^5$$
$$= \frac{10!}{5!5!}(.7)^5(.3)^5$$
$$= 252(.16807)(.00243)$$
$$\approx .103.$$

27. Here $n = 20, p = \frac{1}{8}$, and $x = 2$.
Since $p = \frac{1}{8}$,
$$q = 1 - p$$
$$= 1 - \frac{1}{8}$$
$$= \frac{7}{8}.$$

Substitute these values into the binomial probability formula:

$$P(2) = C(20, 2)\left(\frac{1}{8}\right)^2\left(\frac{7}{8}\right)^{18}$$

$$= \frac{20!}{2!\,18!} \cdot \frac{7^{18}}{8^2 8^{18}}$$

$$= 190 \cdot \frac{7^{18}}{8^{20}}$$

$$\approx .268.$$

28. Here $n = 30, p = .6,$ and $x = 22$.
Since $p = .6$,

$$q = 1 - p$$
$$= 1 - .6$$
$$= .4.$$

Thus,

$$P(22) = C(30, 22)(.6)^{22}(.4)^8$$
$$= 5,852,925(.6)^{22}(.4)^8$$
$$\approx .050.$$

29. Writing exercise

30. When Scott goes up to bat for the tenth time in the series, he will have already batted 9 times in the series, giving him a career total of 4009 times at bat. His greatest possible average would occur if he got hits all nine of his previous times at bat in the series,

$$\frac{1209}{4009} \approx .302.$$

The least his average could be would occur if, in the previous 9 times at bat in the series, he go no hits. His average would then be

$$\frac{1200}{4009} \approx .299.$$

31. Writing exercise

For Exercises 32–35, let a correct answer be a "success." Then $n = 10, p = 2/6 = 1/3,$ and $q = 1 - p = 2/3$.

32. The probability of getting "exactly 4 correct answers" is given by

$$P(4) = C(10, 4)\left(\frac{1}{3}\right)^4\left(\frac{2}{3}\right)^6$$
$$= 210\left(\frac{1}{81}\right)\left(\frac{64}{729}\right)$$
$$\approx .228.$$

33. The probability of getting "exactly 7 correct answers" is given by

$$P(7 \text{ correct answers}) = C(10, 7)\left(\frac{1}{3}\right)^7\left(\frac{2}{3}\right)^3$$
$$= 120 \cdot \frac{2^3}{3^{10}}$$
$$\approx .016.$$

34. The probability of getting "fewer than 3 correct answers" is given by

$$P(x < 3) = P(x = 0 \text{ or } 1 \text{ or } 2)$$
$$= P(0) + P(1) + P(2)$$
$$= C(10, 0)\left(\frac{1}{3}\right)^0\left(\frac{2}{3}\right)^{10}$$
$$+ C(10, 1)\left(\frac{1}{3}\right)^1\left(\frac{2}{3}\right)^9$$
$$+ C(10, 2)\left(\frac{1}{3}\right)^2\left(\frac{2}{3}\right)^8$$
$$\approx 1(.01734) + 10(.00867)$$
$$+ 45(.00434)$$
$$\approx .299.$$

35. "At least seven" means seven, eight, nine, or ten correct answers.

$$P(7 \text{ or } 8 \text{ or } 9 \text{ or } 10)$$
$$= P(7) + P(8) + P(9) + P(10)$$
$$= C(10, 7)\left(\frac{1}{3}\right)^7\left(\frac{2}{3}\right)^3 + C(10, 8)\left(\frac{1}{3}\right)^8\left(\frac{2}{3}\right)^2$$
$$+ C(10, 9)\left(\frac{1}{3}\right)^9\left(\frac{2}{3}\right)^1 + C(10, 10)\left(\frac{1}{3}\right)^{10}\left(\frac{2}{3}\right)^0$$
$$\approx .01626 + .00305 + .00034 + .00002$$
$$\approx .01967 \approx .020$$

36. For "none to have undesirable side effects," $x = 0$ and $n = 8, p = .30, q = 1 - p = .70$. Thus,

$$P(0) = C(8, 0)(.3)^0(.7)^8$$
$$= 1 \cdot 1 \cdot (.05764801)$$
$$\approx .058.$$

37. For "exactly 1 to have undesirable side effects," $x = 1$ and $n = 8, p = .30, q = 1 - p = .70$. Thus,

$$P(1) = C(8, 1)(.30)^1(1 - .30)^{8-1}$$
$$= 8(.30)(.70)^7$$
$$\approx .198.$$

38. For "exactly 2 to have undesirable side effects" $x = 2$ and $n = 8, p = .30, q = 1 - p = .70$. Thus,

$$P(2) = C(8, 2)(.30)^2(.70)^6$$
$$= 28(.09)(.117649)$$
$$\approx .296.$$

39. "More than two" is the complement of 0, 1, or 2. Thus,

$$P(\text{more than two}) = 1 - P(0, 1 \text{ or } 2)$$
$$= 1 - [P(0) + P(1) + P(2)]$$
$$= 1 - [C(8,0)(.30)^0(.70)^8$$
$$\cdot C(8,1)(.30)^1(.70)^7$$
$$+ C(8,2)(.30)^2(.70)^6]$$
$$\approx 1 - [1 \cdot 1 \cdot (.057648)$$
$$+ 8(.30)(.08235)$$
$$+ (28)(.09)(.11765)]$$
$$\approx .448.$$

40. For the probability that "exactly 4 enroll in college," $x = 4, n = 9, p = .50,$ and $q = 1 - p = .50.$ Thus,

$$P(4) = C(9,4)(.50)^4(.50)^5$$
$$= 126(.50)^9$$
$$= 126(.001953125)$$
$$\approx .246.$$

41. For the probability that "from 4 through 6" will attend college, $n = 9, p = .50,$ and $q = 1 - p = .50.$ Thus,

$$P(4 \text{ or } 5 \text{ or } 6) = P(4) + P(5) + P(6)$$
$$= C(9,4)(.50)^4(.50)^5$$
$$+ C(9,5)(.50)^5(.50)^4$$
$$+ C(9,6)(.50)^6(.50)^3$$
$$\approx .246 + .246 + .164$$
$$\approx .656.$$

42. For the probability that "none enroll in college," $x = 0, n = 9, p = .50,$ and $q = 1 - p = .50.$ Thus,

$$P(0) = C(9,0)(.5)^0(.5)^9$$
$$= 1 \cdot 1 \cdot (.001953125)$$
$$\approx .002.$$

43. For the probability that "all 9 enroll in college," $x = 9, n = 9, p = .50,$ and $q = 1 - p = .50.$ Thus,

$$P(\text{all } 9) = P(9)$$
$$= C(9,9)(.50)^9(.50)^0$$
$$= 1 \cdot (.001953125) \cdot 1$$
$$\approx .002.$$

44. For the probability that exactly 3 of the 5 students will have their own computers, $x = 3\ n = 5, p = .80,$ and $q = 1 - .80 = .20.$ Thus,

$$P(3) = C(5,3)(.8)^3(.2)^2$$
$$= 10(.512)(.04)$$
$$\approx .205.$$

45. "At least half of the 6 trees" is the complement of "0, 1, or 2 trees." Here $p = .65, q = 1 - .65 = .35$ and $n = 6.$ Using the complements rule, the probability is found by

$$P(\text{at least half}) = 1 - P(0 \text{ or } 1 \text{ or } 2)$$
$$= 1 - [P(0) + P(1) + P(2)]$$
$$= 1 - [C(6,0)(.65)^0(.35)^6$$
$$+ C(6,1)(.65)^1(.35)^5$$
$$+ C(6,2)(.65)^2(.35)^4]$$
$$\approx 1 - [.0018 + .0205 + .0951]$$
$$\approx 1 - .1174 \approx .883.$$

46. Here $n = 15, p = .64, q = 1 - p = .36,$ and $x = 10.$ Thus,

$$P(10) = C(15,10)(.64)^{10}(.36)^5$$
$$= 3003(.011529215)(.0060466176)$$
$$\approx .209.$$

See discussion in text concerning the probability of a first success on the xth trial which follows $x - 1$ trials which are failures. The formula is as follows:

$$P(F_1 \text{ and } F_2 \text{ and } \ldots \text{ and } F_{x-1} \text{ and } S_x) = q^{x-1} \cdot p.$$

47. Writing exercise

48. Let a girl be a "success."

Then $p = .5, q = 1 - .5 = .5,$ and $x = 4.$

The probability that a family's fourth child will be their first daughter is

$$P(F_1 \text{ and } F_2 \text{ and } F_3 \text{ and } S_4)$$
$$= (.5)^3(.5)$$
$$= (.5)^4$$
$$\approx .063.$$

49. Let $p = 0.38, q = 1 - 0.38 = 0.62,$ and $x = 4.$ Thus,

$$P(F_1 \text{ and } F_2 \text{ and } F_3 \text{ and } S_4)$$
$$= q^{4-1} \cdot p = q^3 \cdot p$$
$$= (.62)^3(.38)$$
$$= (.238328)(.38)$$
$$\approx .091.$$

50. Let a union member be a "success."

Then $p = .30, q = 1 - .20 = .70,$ and $x = 3.$

The probability that the first union member will occur on the third selection is

$$P(F_1 \text{ and } F_2 \text{ and } S_3)$$
$$= (.70)^2(.30)$$
$$= (.49)(.3)$$
$$\approx .147.$$

51. Let an aborted launching be a "success."

    Then $p = .04, q = 1 - .04 = .96,$ and $x = 20$. Thus,

    $$\begin{aligned}P(\text{20th launch is first launch aborted}) &= q^{19} \cdot p \\ &= (.96)^{19} \cdot (.04) \\ &\approx (.460419) \cdot (.04) \\ &\approx .018.\end{aligned}$$

Heads means walk one block north and tails means walk one block south. In each case, ask how many successes, say heads, would be required and use the binomial formula. See discussion in text about a random walk.

52. To end up 10 blocks north of his corner, Harvey must go 10 blocks north and 0 blocks south, so he must toss 10 heads and no tails. Use the binomial probability formula with

    $$n = 10, p = \frac{1}{2}, q = \frac{1}{2}, \text{ and } x = 10.$$

    Then,

    $$\begin{aligned}P(\text{10 heads, 0 tails}) &= P(10) \\ &= C(10, 10)\left(\frac{1}{2}\right)^{10}\left(\frac{1}{2}\right)^{0} \\ &= 1 \cdot \frac{1}{1024} \cdot 1 \\ &= \frac{1}{1024} \approx .001.\end{aligned}$$

53. To end up 6 blocks north of his corner, Harvey must go 8 blocks north and 2 blocks south, so he must toss 8 heads and 2 tails. Use the binomial probability formula with

    $$n = 10, p = \frac{1}{2}, q = \frac{1}{2}, \text{ and } x = 8.$$

    Then,

    $$\begin{aligned}P(\text{8 heads, 2 tails}) &= (8) \\ &= C(10, 8)\left(\frac{1}{2}\right)^{8}\left(\frac{1}{2}\right)^{2} \\ &= 45 \cdot \left(\frac{1}{2}\right)^{10} \\ &= \frac{45}{1024} \approx .044.\end{aligned}$$

54. To end up 6 blocks south of his corner, Harvey must go 2 blocks north and 8 blocks south, so he must toss 2 heads and 8 tails. Use the binomial probability formula with

    $$n = 10, p = \frac{1}{2}, q = \frac{1}{2}, \text{ and } x = 2.$$

    $$\begin{aligned}P(\text{2 heads, 8 tails}) &= P(2) \\ &= C(10, 2)\left(\frac{1}{2}\right)^{2}\left(\frac{1}{2}\right)^{8} \\ &= 45\left(\frac{1}{2}\right)^{10} \\ &= 45\left(\frac{1}{1024}\right) \\ &= \frac{45}{1024} \approx .044.\end{aligned}$$

55. It will be impossible for Harvey to end up 5 blocks south of his corner since the number of heads + the number of tails must = 10 and at the same time the number of tails − the number of heads must = 5.

    To show this, solve the system of equations:

    $$\begin{aligned}h + t &= 10 \\ t - h &= 5\end{aligned}$$ or $$\begin{aligned}2t &= 15 \text{ (adding equations)} \\ t &= 7.5 \text{ and } h = 2.5.\end{aligned}$$

    Thus, $P(\text{5 blocks south}) = 0$.

56. To end up 2 blocks north of his corner, Harvey must go 6 blocks north and 4 blocks south (since $6 + 4 = 10$ and $6 - 4 = 2$, so he must toss 6 heads and 4 tails. Use the binomial probability formula with

    $$n = 10, p = \frac{1}{2}, q = \frac{1}{2}, \text{ and } x = 6.$$

    Then,

    $$\begin{aligned}P(\text{6 heads, 4 tails}) &= P(6) \\ &= C(10, 6)\left(\frac{1}{2}\right)^{6}\left(\frac{1}{2}\right)^{4} \\ &= 210\left(\frac{1}{1024}\right) \\ &= \frac{210}{1024} = \frac{105}{512} \approx .205.\end{aligned}$$

57. $P(\text{at least 2 blocks north})$
    $= P(2 \text{ or } 3 \text{ or } \ldots \text{ or } 10 \text{ blocks north})$
    $= P(2 \text{ blocks north}) + P(3 \text{ blocks north}) \ldots$
    $\quad + P(10 \text{ blocks north})$
    $= P(6 \text{ heads and 4 tails})$
    $\quad + 0[\text{since 3 blocks north is impossible} -$
    $\quad \quad \text{similar to Exercise 55}]$
    $\quad + P(7 \text{ heads and 3 tails})$
    $\quad + 0 \, [5 \text{ blocks north is impossible}]$
    $\quad + P(8 \text{ heads and 2 tails})$
    $\quad + 0[7 \text{ blocks north is impossible}]$
    $\quad + P(9 \text{ heads and 1 tail})$
    $\quad + 0 \, [9 \text{ blocks north is impossible}]$
    $\quad + P(10 \text{ heads})$
    $= P(6 \text{ heads}) + 0 + P(7 \text{ heads}) + 0$
    $\quad + P(8 \text{ heads}) + P(9 \text{ heads}) + 0 + P(10 \text{ heads})$
    $= C(10, 6)\left(\frac{1}{2}\right)^6 \left(\frac{1}{2}\right)^4$
    $\quad + C(10, 7)\left(\frac{1}{2}\right)^7 \left(\frac{1}{2}\right)^3$
    $\quad + C(10, 8)\left(\frac{1}{2}\right)^8 \left(\frac{1}{2}\right)^2$
    $\quad + C(10, 9)\left(\frac{1}{2}\right)^9 \left(\frac{1}{2}\right)^1$
    $\quad + C(10, 10)\left(\frac{1}{2}\right)^{10} \left(\frac{1}{2}\right)^0$
    $= 210 \cdot \left(\frac{1}{2}\right)^{10} + 120 \cdot \left(\frac{1}{2}\right)^{10} + 45 \cdot \left(\frac{1}{2}\right)^{10}$
    $\quad + 10 \cdot \left(\frac{1}{2}\right)^{10} + 1 \cdot \left(\frac{1}{2}\right)^{10}$
    $= (210 + 120 + 45 + 10 + 1) \cdot \left(\frac{1}{2}\right)^{10}$
    $= \dfrac{210 + 120 + 45 + 10 + 1}{1024}$
    $= \dfrac{193}{512} \approx .377$

58. To find the probability that Harvey will end up at least 2 blocks from his corner, use the complements rule. The complement of "at least 2 blocks from his corner" is "less than 2 blocks from his corner" or "1 block from his corner." However, it is impossible for Harvey to end up 1 block from his corner (either north or south). In order for two integers to have a sum of 10, they must be both even or both odd. In either case, their difference will be even. Thus, Harvey can never end up an odd number of blocks from his corner.

    In order to end up *on* his corner, Harvey must go 5 blocks north and 5 blocks south. Let $p = \frac{1}{2}, q = \frac{1}{2}$, and $x = 5$:

    $P(5 \text{ heads}, 5 \text{ tails}) = P(5)$
    $\quad = C(10, 5)\left(\frac{1}{2}\right)^5 \left(\frac{1}{2}\right)^5$
    $\quad = 252\left(\frac{1}{2}\right)^{10}$
    $\quad = 252\left(\frac{1}{1024}\right)$
    $\quad = \dfrac{252}{1024} = \dfrac{63}{256}.$

    Then, the probability that Harvey ends up at least 2 blocks from his corner is

    $1 - P(5) = 1 - \dfrac{63}{256}$
    $\quad = \dfrac{193}{256} \approx .754.$

59. In order to end up *on* his corner, Harvey must go 5 blocks north and 5 blocks south. Let $p = \frac{1}{2}, q = \frac{1}{2}$, and $x = 5$:

    $P(5 \text{ heads}, 5 \text{ tails}) = P(5)$
    $\quad = C(10, 5)\left(\frac{1}{2}\right)^5 \left(\frac{1}{2}\right)^5$
    $\quad = 252\left(\frac{1}{2}\right)^{10}$
    $\quad = 252\left(\frac{1}{1024}\right)$
    $\quad = \dfrac{252}{1024} = \dfrac{63}{256} \approx .246.$

12.5 EXERCISES

1. Writing exercise

2. Writing exercise

3. Five fair coins are tossed. A tree diagram may be helpful to create the following sample space. Use the following sample space to create the individual probabilities.

hhhhh	hhhht	hhhth	hhhtt
hhthh	hhtht	hhtth	hhttt
hthhh	hthht	hthth	hthtt
htthh	httht	htth	htttt
thhhh	thhht	thhth	thhtt
ththh	ththt	thtth	thttt
tthhh	tthht	tthth	tthtt
ttthh	tttht	ttth	ttttt

number of heads, x	Probability $P(x)$	Product $x \cdot P(x)$
0	1/32	0
1	5/32	5/32
2	10/32	20/32
3	10/32	30/32
4	5/32	20/32
5	1/32	5/32

Thus, the expected value is given by:

Expected number of heads
$$= 0 + \frac{5}{32} + \frac{20}{32} + \frac{30}{32} + \frac{20}{32} + \frac{5}{32}$$
$$= \frac{80}{32} = \frac{5}{2}.$$

4. Two cards are drawn, with replacement, from a standard 52-card deck. Find the expected number of diamonds.

Number of diamonds	Probability $P(x)$	Product $x \cdot P(x)$
0	$(3/4) \cdot (3/4) = 9/16$	$0 \cdot (9/16) = 0$
1	$(1/4) \cdot (3/4) + (3/4) \cdot (1/4) = 3/8$	$1 \cdot (3/8) = 3/8$
2	$(1/4) \cdot (1/4) = 1/16$	$2 \cdot (1/16) = 1/8$

Expected value: $4/8 = 1/2$

5. List the given information in a table. Then calculate $P(x)$, the product $x \cdot P(x)$, and their total.

Number Rolled	Payoff	Probability	Product
6	$3	1/6	$(3/6)
5	$2	1/6	$(2/6)
4	$1	1/6	$(1/6)
1–3	$0	3/6	$0

Expected value: $(6/6) = $1

6. Is this game fair, or unfair against the player, or unfair in favor of the player? From Exercise 5, the expected winnings of the game are $1. Thus, a fair price to pay to play this game is $1.

7. List the given information in a table. Then complete the table as follows.

Number Rolled	Payoff	Probability	Product
1	−$1	1/6	−$(1/6)
2	$2	1/6	$(2/6)
3	−$3	1/6	−$(3/6)
4	$4	1/6	$(4/6)
5	−$5	1/6	−$(5/6)
6	$6	1/6	$(6/6)

Expected value: $(3/6) = 50¢

The expected net winnings for this game are 50¢.

8. This game is unfair in favor of the player since expected value is greater than $0 for the player.

9. List the given information in a table, and complete the probability and product columns. Remember that the expected value is the sum of the product column.

Number of heads	Payoff	Probability $P(x)$	Product $x \cdot P(x)$
3	10¢	1/8	(10/8)¢
2	5¢	3/8	(15/8)¢
1	3¢	3/8	(9/8)¢
0	0¢	1/8	0¢

Expected value: $(34/8)¢ = (17/4)¢$

Since it costs 5¢ to play, the expected net winnings are

$$\frac{17}{4}¢ - 5¢ = \frac{17}{4}¢ - \frac{20}{4}¢ = -\frac{3}{4}¢.$$

Because the expected net winnings are not zero, 5¢ is not a fair price to pay to play this game.

10. The pay off is $2 if the ball lands on red. The probability of landing on red is 18/37. The probability of not landing on red is $1 - (18/37) = 19/37$ and the payoff is $0. The cost per play is $1.

Thus, the expect net winnings are

$$(\$2 - \$1) \cdot \frac{18}{37} + (\$0 - \$1) \cdot \frac{19}{37} = \$\left(\frac{18}{37} - \frac{19}{37}\right)$$
$$= -\$\frac{1}{37} \approx -2.7¢.$$

11. The expected number of absences on a given day is

$$x_1 \cdot P(x_1) + x_2 \cdot P(x_2) + x_3 \cdot P(x_3) + x_4 \cdot P(x_4) + x_5 \cdot P(x_5)$$
$$= 0(.12) + 1(.32) + 2(.35) + 3(.14) + 4(.07)$$
$$= 1.72.$$

12. An insurance company will insure a $100,000 home for its total value for an annual premium of $300. If the probability of total loss for such a home in a given year is 0.002, and you assume that either total loss or no loss will occur, what is the company's expected annual gain (or profit) on such a policy?

$P(\text{no loss}) = 1 - 0.002 = 0.998$

Thus, the expected value is

$$= (\$330 - \$100,000)(.002) + \$330(.998)$$
$$= \$130.$$

Since it costs the company $20 per year to service the policy, their expected annual gain (profit) on such a policy is

$$\$130 - \$20 = \$110.$$

A college foundation raises funds by selling raffle tickets for a new car worth $36,000.

13. (a) Since 600 tickets are sold, a person who buys one ticket will have a probability of $1/600 \approx .00167$ of winning the car and a $1 - .0017 = .9983$ probability of not winning anything. For this person, the expected value is

$$\$36,000(.00167) + \$0(.9983) \approx \$60,$$

and the expected *net* winnings (since the ticket costs $120) are

$$\$60 - \$120 = -\$60.$$

(b) By selling 600 tickets at $120 each, the foundation takes in

$$500(\$120) = \$72,000.$$

Since they had to spend $36,000 for the car, the total profit for the foundation is

$$\text{revenue} - \text{cost} = \text{profit}$$
$$\$72,000 - \$36,000 = \$36,000.$$

(c) Without having to pay for the car, the foundation's total profit will be all of the revenue from the ticket sales, which is $72,000.

14. (a) Since 720 tickets are sold, a person who buys one ticket will have a probability of $1/720 \approx .001389$ of winning and a probability of $1 - .001389 = .9986$ of not winning anything. For this person, the expected value is

$$\$36,000(.001389) + \$0(.9986) \approx \$50,$$

and the expected *net* winnings (since the ticket costs $120) are

$$\$50 - \$120 = -\$70.$$

(b) By selling 720 tickets at $120 each, the foundation takes in

$$720(\$120) = \$86,400.$$

Since they had to spend $36,000 for the car, the total profit for the foundation is

$$\text{revenue} - \text{cost} = \text{profit}$$
$$\$86,400 - \$36,000 = \$50,400.$$

(c) Without having to pay for the car, the foundation's total profit will be all of the revenue from the ticket sales, which is $86,400.

Five thousand raffle tickets are sold. One first prize of $1,000, two second prizes of $500 each, and five third prizes of $100 each are to be awarded, with all winners selected randomly.

15. The associated probabilities are

$$P(\text{1st prize}) = \frac{1}{5000},$$
$$P(\text{2nd prize}) = \frac{2}{5000}, \text{ and}$$
$$P(\text{3rd prize}) = \frac{5}{5000}.$$

The expected winnings, ignoring the cost of the raffle ticket, are given by

$$\$1000\left(\frac{1}{5000}\right) + \$500\left(\frac{2}{5000}\right) + \$100\left(\frac{5}{5000}\right)$$
$$= \$.20 + \$.20 + \$.10$$
$$= \$.50, \text{ or } 50¢.$$

16. The associated probabilities are

$$P(\text{1st prize}) = 2 \cdot \frac{1}{5000} = \frac{2}{5000},$$
$$P(\text{2nd prize}) = 2 \cdot \frac{2}{5000} = \frac{4}{5000}, \text{ and}$$
$$P(\text{3rd prize}) = 2 \cdot \frac{5}{5000} = \frac{10}{5000}.$$

The expected winnings, ignoring the cost of the raffle ticket, are given by

$$\$1000 \cdot \frac{2}{5000} + \$500 \cdot \frac{4}{5000} + \$1000 \cdot \frac{10}{5000}$$
$$= \$\left(\frac{2}{5} + \frac{4}{10} + \frac{10}{50}\right) = \$1.$$

17. Since 5000 tickets were sold for $1 each, the sponsor's revenue was $5000 (\$1) = \$5000$. The sponsor's cost was the sum of all the prizes:

$$1(\$1000) + 2(\$500) + 5(\$100)$$
$$= \$2500.$$

Therefore, the sponsor's profit is

$$\text{revenue} - \text{cost} = \text{profit}$$
$$\$5000 - \$2500 = \$2500.$$

18. Remember that the expected value is the sum of the product column.

Number of children snackers	Probability $P(x)$	Product $x \cdot P(x)$
65	65/163	$1 \cdot (65/163) = 65/163$
40	40/163	$2 \cdot (40/163) = 80/163$
26	26/163	$3 \cdot (26/163) = 78/163$
14	14/163	$4 \cdot (14/163) = 56/163$
18	18/163	$0 \cdot (18/163) = 0$

Expected value: $279/163 \approx 1.7$

19. List the given information in a table, and complete the probability and product columns. Remember that the expected value is the sum of the product column.

Number of families	Probability $P(x)$	Product $x \cdot P(x)$
1020	1020/10000	$1 \cdot (1020/10000) = 1020/10000$
3370	3370/10000	$2 \cdot (3370/10000) = 6740/10000$
3510	3510/10000	$3 \cdot (3510/10000) = 10530/10000$
1340	1340/10000	$4 \cdot (1340/10000) = 5360/10000$
510	510/10000	$5 \cdot (510/10000) = 2550/10000$
80	80/10000	$6 \cdot (80/10000) = 480/10000$
170	170/10000	$0 \cdot (170/10000) = 0$

Expected value: $26680/10000 \approx 2.7$

20. The following table gives the sum for each possible pair of cards. Since the cards are drawn without replacement, the two cards cannot have the same number.

		\multicolumn{5}{c}{Second Card}
	+	1	2	3	4	5
	1	–	3	4	5	6
	2	3	–	5	6	7
First	3	4	5	–	7	8
card	4	5	6	7	–	9
	5	6	7	8	9	–

The table shows the ways in which each of the possible sums, 3 through 9, can be obtained. Use this information to compute the frequency of each sum. Note that the frequency will become the numerator of the probability function.

Sum of cards	Frequency f	Probability $P(x)$	Product $x \cdot P(x)$
3	2	2/20	$3 \cdot (2/20) = 6/20$
4	2	2/20	$4 \cdot (2/20) = 8/20$
5	4	4/20	$5 \cdot (4/20) = 20/20$
6	4	4/20	$6 \cdot (4/20) = 24/20$
7	4	4/20	$7 \cdot (4/20) = 28/20$
8	2	2/20	$8 \cdot (2/20) = 16/20$
9	2	2/20	$9 \cdot (2/20) = 18/20$

Expected value: $120/20 \approx 6$

21. The expected value is

$$1200(.2) + 500(.5) + (-800)(.3)$$
$$+ 240 + 250 - 240$$
$$= 250.$$

Since this expected value is positive, the expected change in the number of electronics jobs is an increase of 250.

22. The expected net winnings for this game is given by

$$(\$3.20 - \$1) \cdot \frac{20}{80} - (\$1) \cdot \frac{60}{80}$$
$$= -\$.20$$
$$= -20¢.$$

23. Writing exercise

24. Apply the expected value formula.

Project A: $\$40000(.10) + \$180000(.60) + \$250000(.30) = \$187{,}000$.

Project B: $\$0(.20) + \$210000(.35) + \$290000(.45) = \$204{,}000$.

Project C: $\$60000(.65) + \$340000(.35) = \$158{,}000$.

Using expected value analysis, Kimberli should choose option B.

25. The optimist viewpoint would ignore the probabilities and hope for the best possible outcome, which is Project C since it may return up to $340,000.

26. The pessimist viewpoint will assume the worst case possible, which is also Project C since it could potentially make $60,000, representing the biggest loss from the potential profit.

27. If the contestant takes a chance on the other two prizes, the expected winnings will be

$$\$5000(.20) + \$8000(.15)$$
$$= \$1000 + \$1200$$
$$= \$2200.$$

28. Purely in terms of monetary value, the best choice is to accept the computer, since $2300 > $2200 (expected value from Exercise 27).

29. Compute the remaining values in Column 5 (Expected Value).

Row 5: $(50{,}000)(.5) = 25{,}000$
Row 6: $(100{,}000)(.6) = 60{,}000$
Row 7: $(20{,}000)(.8) = 16{,}000$

30. Compute the remaining values in Column 6 (Existing Volume plus Expected Value of Additional Volume.

Row 3: $20{,}000 + 2{,}000 = 22{,}000$
Row 4: $50{,}000 + 1000 = 51{,}000$
Row 5: $5000 + 25{,}000 = 30{,}000$
Row 6: $0 + 60{,}000 = 60{,}000$
Row 7: $30{,}000 + 16{,}000 = 46{,}000$

31. Classify each amount in column 6.

Row 1: Class C ($17,500 < $45,000)
Row 2: Class C ($40,000 < $45,000)
Row 3: Class C ($22,000 < $45,000)
Row 4: Class B ($45,000 ≤ $51,000 < $55,000)
Row 5: Class C ($30,000 < $45,000)
Row 6: Class A ($55,000 ≤ $60,000)
Row 7: Class B ($45,000 ≤ $46,000 < $55,000)

32. Considering all seven of this salesman's accounts, compute the total additional volume he can "expect" to get. The amount he can "expect" to get is found by adding together the values in Column 5:

$$\$2500 + \$2000 + \$1000 \\ + \$25,000 + \$60,000 + \$16,000 \\ = \$106,500.$$

33. $P(\text{matching 3 numbers})$
$$= \frac{C(20,3)C(60,3)}{C(80,6)} = .1298$$
$P(\text{matching 4 numbers})$
$$= \frac{C(20,4) \cdot C(60,2)}{C(80,6)} = .0285$$
P(matching 5 numbers)
$$= \frac{C(20,5) \cdot C(60,1)}{C(80,6)} = .0031$$
P(matching 6 numbers)
$$= \frac{C(20,6) \cdot C(60,0)}{C(80,6)} = .000129$$
Expected value
$$= (\$.35)(.1298) + (\$2.00)(.0285) \\ + (\$60.00)(.0031) \\ + (\$1250.00)(.000129) \\ = \$.44968 \\ \approx 45¢$$

Since the player pays 60¢ for his ticket, his expected net winnings are about

$$45¢ - 60¢ = -15¢.$$

12.6 EXERCISES

1. Writing exercise

2. There are $50 - 3 = 47$ sets of four successive colors, since a set of four cannot start with any of the last three colors. Count the number of sequences (or "strings") which read "red-red-red-red." Note that the sequence of 8 consecutive reds contains $8 - 3 = 5$ strings of four consecutive reds, since the last three reds cannot begin strings. In this manner, we find that there are

$$5 + 7 + 1 + 2 = 15$$

strings of four consecutive reds. Therefore, the probability that four successive offspring will all have red flowers is

$$\frac{15}{47} \approx .319.$$

3. No, since the probability of an individual girl's birth is (nearly) the same as that for a boy.

4. Answers will vary.

5. Let each of the 50 numbers correspond to one family. For example, the first number, 51592, with middle digits—1(boy), 5(girl), 9(girl)—represents a family with 2 girls and 1 boy. The last number whose middle digits are 800 represents the 50th family which has 1 girl and 2 boys—a success, and so on. Examining each number, we count (tally) 18 successes. Therefore,

$$P(\text{2 boys and 1 girl}) = \frac{18}{50} = .36.$$

Observe that this is quite close to the .375, predicted by the theoretical value.

Refer to discussion in text regarding foul shooting in basketball. After completing the indicated tally, find the empirical probability that, on a given opportunity, Karin will score as follows.

To construct the tally for Exercises 6–8, begin as follows: Since the first number in the table of random digits is 5, which represents a hit, Susan will get a second shot. The second digit is 7, representing a miss on the second shot. Record the results of the first two shots (the first one-and-one opportunity) as "one point." The second and third opportunities correspond to the pairs 3, 4 and 0, 5. Record each of these results as "two points." For the fourth opportunity, the digit 9 indicates that the first shot was missed, so Susan does not get a second shot. In this case only one digit is used. Record this result as "zero points." Continue in this manner until 50 one-and-one opportunities are obtained. The results of the tally is as follows. Note that this, in effect, is a frequency distribution as discussed in Chapter 12.

Number of Points	Tally - frequency
0	15
1	11
2	24
Total	50

6. From the tally, we see that 0 point shots occur 15 times. Thus,

$$P(\text{no points}) = \frac{15}{50} = .30.$$

7. From the tally, we see that 1 point shots occur 11 times. Thus,

$$P(\text{1 point}) = \frac{11}{50} = .22.$$

8. From the tally, we see that 2 point shots occur 24 times. Thus,

$$P(\text{2 points}) = \frac{24}{50} = .48.$$

9. Answers will vary.

10. Answers will vary.

11. Writing exercise

12. If we eliminate all digits other than 1 through 6, we obtain the following list:

    1, 2, 2, 5, 5, 5, 5, 4, 1, 4, 6,
    1, 2, 2, 6, 5, 6, 1, 2, 6, 5, 5, 6.

    Create a table to keep track of results.

Start	Go one block north
1	Turn left: go one block west.
2	Continue one more block west.
2	Continue one more block west.
5	Turn right: go one block north.
5	Turn right: go one block east.
5	Turn right: go one block south.
5	Turn right: go one block west.
4	Turn right: go one block north.
1	Turn left: go one block west.
4	Turn right: go one block north.
6	Turn right: go one block east.
1	Turn left: go one block north.
2	Continue one more block north.
2	Continue one more block north.
6	Turn right: go one block east.
5	Turn right: go one block south.
6	Turn right: go one block west.
1	Turn left: go one block south.
2	Continue one more block south.
6	Turn right: go one block west.
5	Turn right: go one block north.
5	Turn right: go one block east.
6	Turn right: go one block south.

    Tally the blocks walked in each direction.

Direction	Frequency
West	7
North	8
East	4
South	5
	24

    Since north and south are opposite direction, the net result of going 8 block north and 5 blocks south is 3 blocks north of your starting point.

    Thus, the walk ends 3 blocks north and 3 blocks west of the starting point. Since segments 3 block long going north and west form a right triangle, the distance d from start to end is the hypotenuse. By the Pythagorean theorem,

    $$d^2 = 3^2 + 3^2$$
    $$d^2 = 9 + 9 = 18$$
    $$d = \sqrt{18} = \sqrt{9} \cdot \sqrt{2}$$
    $$= 3\sqrt{2}.$$

    Thus, the walk ends $3\sqrt{2}$ blocks northwest of the starting point. That is, 3 blocks west and 3 blocks north of the starting point.

CHAPTER 12 TEST

1. Writing exercise

2. Writing exercise

3. There are 39 non-hearts and 13 hearts, so the odds against getting a heart are 39 to 13, or 3 to 1 (when reduced).

4. There are 50 non-red queens and 2 red queens, so the odds against getting a red queen are 50 to 2, or 25 to 1.

5. There are 12 face cards altogether. Of these, there are 6 black face cards and 2 more non-black kings for a total of 8 cards. There are $52 - 8 = 44$ other cards in the deck. Thus, the odds against getting a black face card or king are 44 to 8, or 11 to 2.

6.
		Second parent	
		C	c
First	C	CC	Cc
Parent	c	cC	cc

7. There are two outcomes (cC) indicating that the next child will be a carrier. Thus,

   $$P(\text{carrier}) = \frac{2}{4} = \frac{1}{2}.$$

8. There are 3 favorable outcomes $(CC, Cc, \text{and } cC)$. Thus, the odds against a child getting the disease (cc) are 3 to 1.

9. Use the fundamental counting principle where the first task is to calculate the probability of the initial employee choosing any day of the week (7/7), the second task is the probability for the second employee to choose any other days of the week (6/7). In a similar manner, the third employee's probability must involve a choice from one of the five remaining days with a resulting probability of (5/7). Thus,

   $$P = \frac{7}{7} \cdot \frac{6}{7} \cdot \frac{5}{7} = \frac{30}{49}.$$

10. Use the fundamental counting principle where the first task is to calculate the probability of the initial employee choosing any day of the week $(7/7)$. For the second and third task the employees must choose the same day. The probability in each case is $(1/7)$. Thus,

$$P = \frac{7}{7} \cdot \frac{1}{7} \cdot \frac{1}{7} = \frac{1}{49}.$$

11. The complement of "exactly two choose the same day" is "all three choose different days or all three choose the same day." These complement probabilities were calculated in Exercises 9 and 10 above. Thus, the probability of "exactly two choosing the same day" is given by

$$P = 1 - \left(\frac{30}{49} + \frac{1}{49}\right) = \frac{18}{49}.$$

Observe that the calculation involves both the complements rule and the special addition rule.

Two numbers are randomly selected without replacement from the set $\{1, 2, 3, 4, 5\}$.

12. To find the probability that "both numbers are even", use combinations to select the number of successes – ways of selecting the two even numbers, $C(2, 2)$ and to calculate the total number of ways of selecting two of the numbers from the 5, $C(5, 2)$. Thus,

$$P(\text{selecting two even numbers}) = \frac{C(2,2)}{C(5,2)} = \frac{1}{10}.$$

As in many of the exercises an alternate solution may be considered here: Let E_1 represent the event of selecting an even number as the first selection and E_2, selecting an even number as the second selection. Using the general multiplication rule the probability is given by

$$P(E_1 \text{ and } E_2) = P(E_1) \cdot P(E_2|E_1)$$
$$= \frac{2}{5} \cdot \frac{1}{4}$$
$$= \frac{1}{10}.$$

13. To find the probability that "both numbers are prime" use combinations. Since there are three prime numbers $\{2, 3, 5\}$, use $C(3, 2)$ to calculate the number of successes. To calculate the total number of ways of selecting two of the numbers from the 5, use $C(5, 2)$. Thus,

$$P(\text{selecting two even numbers}) = \frac{C(3,2)}{C(5,2)} = \frac{3}{10}.$$

14. Create a "product (sum) table" to list the elements in the sample space and the successes (event of interest). Note that "without replacement", one can only use a selected number once. Hence, there are no diagonal values to the table.

	+	1	2	3	4	5
	1	–	3	4	5	6
1st	2	3	–	5	6	7
number	3	4	5	–	7	8
	4	5	6	7	–	9
	5	6	7	8	9	–

(2nd number across top)

There are 12 odd sums in the table and a total of 20 sums in the sample space. Thus,

$$P(\text{sum is odd}) = \frac{12}{20} = \frac{3}{5}.$$

15. Similar to Exercise 14, create a "product table" to list the elements in the sample space and the successes (event of interest). Note that "without replacement", one can only use a selected number once. Hence, there are no diagonal values to the table.

	×	1	2	3	4	5
	1	–	2	3	4	5
1st	2	2	–	6	8	10
number	3	3	6	–	12	15
	4	4	8	12	–	20
	5	5	10	15	20	–

(2nd number across top)

There are 6 odd products in the table and a total of 20 products in the sample space. Thus,

$$P(\text{product is odd}) = \frac{6}{20} = \frac{3}{10}.$$

A three-member committee is selected randomly from a group consisting of three men and two women.

16. Let x represent the number of men on the committee. Then,

x	$P(x)$
0	0
1	3/10
2	6/10
3	1/10

Where,

$$P(1) = \frac{C(3,1) \cdot C(2,2)}{C(5,3)}$$

$$P(2) = \frac{C(3,2) \cdot C(2,1)}{C(5,3)}$$

$$P(3) = \frac{C(3,3) \cdot C(2,0)}{C(5,3)}.$$

Why is $P(0) = 0$?

17. The probability that the "committee members are not all men" is the complement of the "committee are all men," $P(3)$. Hence, use the complements rule,

$$P(\text{committee members are not all men})$$
$$= 1 - P(3)$$
$$= 1 - \frac{1}{10}$$
$$= \frac{9}{10}.$$

18. Complete the table begun in Exercise 16 as an aid to calculating the expected number of men (sum of product column).

x	$P(x)$	$x \cdot P(x)$
0	0	0
1	3/10	3/10
2	6/10	12/10
3	1/10	3/10

Expected number: $18/10 = 9/5$

Create a "product (sum) table" such as below for the "sum" of rolling two dice.

		\multicolumn{6}{c}{2nd die}
	+	1	2	3	4	5	6
	1	2	3	4	5	6	7
	2	3	4	5	6	7	8
1st	3	4	5	6	7	8	9
die	4	5	6	7	8	9	10
	5	6	7	8	9	10	11
	6	7	8	9	10	11	12

Use for Exercises 19–22.

19. There are 30 non-doubles values and 6 doubles values. The odds against doubles are therefore, 30 to 6, or 5 to 1.

20. A "sum greater than 2" is the complement of a "sum equal to or smaller than 2." There is only one sum which satisfies this condition ("snake eyes"). Thus, by the complements rule

$$P(\text{sum greater than 2}) = 1 - \frac{1}{36} = \frac{35}{36}.$$

21. To find the odds against a "sum of 7 or 11" count the sums that satisfy the condition "sum of 7 or 11." There are 8 such sums. It follows that there are $36 - 8 = 28$ sums that are not 7 or 11. Thus, the odds against a "sum of 7 or 11" are 28 to 8, or 7 to 2.

22. Since there are 4 sums that are even and less than 5,

$$P(\text{sum that is even and less than 5}) = \frac{4}{36} = \frac{1}{9}.$$

For Exercises 23–26, the chance of making par on any one hole is .78.

23. By the special multiplication rule,

$$P(\text{making par on all three holes}) = .78^3 \approx .475.$$

24. Use the fundamental counting principle, where the first task is to find the number of ways to choose the two holes he scores par on followed by the tasks of assigning a probability for each hole. Note that since .78 is the probability of scoring par, $1 - .78 = .22$ is the probability of not scoring par on a hole.

$$P(\text{makes par on exactly 2 holes}) = C(3,2)(.78)^2(.22)$$
$$= (3)(.6084)(.22)$$
$$\approx .402$$

25. "At least one of the three holes" is the complement of "none of the three holes." Since the probability of not making par on any of the three holes is $.22^3$,

$$P(\text{at least one of the three holes}) = 1 - (.22)^3$$
$$= 1 - .010648$$
$$\approx .989.$$

26. Use the special multiplication rule since these probabilities are independent. The probability that he makes par on the first and third holes but not on the second is found by

$$P = (.78)(.22)(.78)$$
$$\approx .134.$$

Two cards are drawn, without replacement, from a standard 52-card deck for Exercises 27–30.

27. Let R_1 and R_2 represent the two red cards. Since the cards are not replaced, the events are not independent. Use the general multiplication rule:

$$P(R_1 \text{ and } R_2) = P(R_1) \cdot P(R_2|R_1)$$
$$= \frac{26}{52} \cdot \frac{25}{51}$$
$$= \frac{1}{2} \cdot \frac{25}{51}$$
$$= \frac{25}{102}.$$

28. Let C_1 and C_2 represent two cards of the same color. Since the cards are not replaced, the events are not independent. Use the general multiplication rule. Note that since it doesn't matter what color the first card is, its probability is 1. Thus,

$$P(C_1 \text{ and } C_2) = P(C_1) \cdot P(C_2|C_1)$$
$$= \frac{52}{52} \cdot \frac{25}{51}$$
$$= \frac{25}{51}.$$

29. The first card drawn limits the sample space to 51 cards where all 4 queens are still in the deck. The probability then is given by

$$P(\text{queen given the first card is an ace}) = P(Q_2|A_1)$$
$$= \frac{4}{51}.$$

30. In the event "the first card is a face card and the second is black", the first (face) card may be red or black. Since this affects the probability associated with the second card, look at both cases and use the special addition rule to add (since we are "or–ing") the results.

Case 1 (first card is a red face card):

$$P(F_1 \text{ and } B_2) = P(F_1) \cdot P(B_2|F_1) \backslash$$
$$= \frac{6}{52} \cdot \frac{26}{51}$$
$$= \frac{3}{51} = \frac{1}{17}.$$

Case 2 (first card is a black face card):

$$P(F_1 \text{ and } B_2) = P(F_1) \cdot P(B_2|F_1)$$
$$= \frac{6}{52} \cdot \frac{25}{51}$$
$$= \frac{3}{26} \cdot \frac{25}{51}$$
$$= \frac{1}{26} \cdot \frac{25}{17} = \frac{25}{26 \cdot 17}.$$

The probability is

$$P(\text{Case 1 or Case 2}) = P(\text{Case 1}) + P(\text{Case 2})$$
$$= \frac{1}{17} + \frac{25}{26 \cdot 17}$$
$$= \frac{26 \cdot 1}{26 \cdot 17} + \frac{25}{26 \cdot 17}$$
$$= \frac{26 + 25}{26 \cdot 17} = \frac{51}{26 \cdot 17}$$
$$= \frac{3 \cdot 17}{26 \cdot 17} = \frac{3}{26}.$$

Exercises 31–33 refer to coin sequence in Example 2 of Section 12.6.

31. The number of 3 successive births is 38. It is helpful to set up a tally to count these.

32. Of the 38 only 1 represents a triple consisting of all girls.

33. The empirical probability that 3 successive births will be all girls is found, using the results of Exercises 31 and 32 above, as

$$\frac{1}{38} \approx .026.$$

13.1 EXERCISES

1. (a) Remember that f represents the frequency of each data value, and f/n is a comparison of each frequency to the overall number of data values.

x	f	f/n
0	10	$10/30 \approx 33\%$
1	7	$7/30 \approx 23\%$
2	6	$6/30 \approx 20\%$
3	4	$4/30 \approx 13\%$
4	2	$2/30 \approx 7\%$
5	1	$1/30 \approx 3\%$

    (b)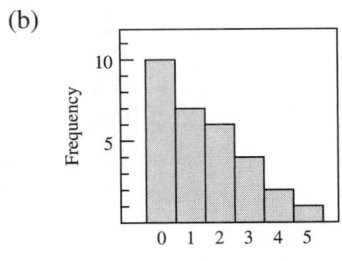

    (c)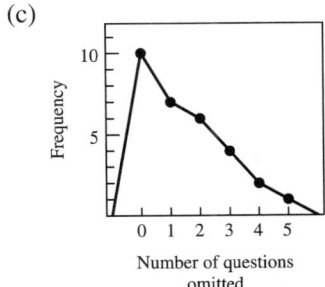

2. (a)
x	f	f/n
1	1	$1/28 \approx 4\%$
2	2	$2/28 \approx 7\%$
3	1	$1/28 \approx 4\%$
4	4	$4/28 \approx 14\%$
5	3	$3/28 \approx 11\%$
6	5	$5/28 \approx 18\%$
7	4	$4/28 \approx 14\%$
8	3	$3/28 \approx 11\%$
9	3	$3/28 \approx 11\%$
10	2	$2/28 \approx 7\%$

    (b)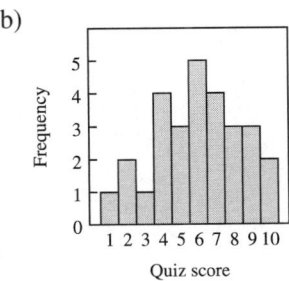

    (c)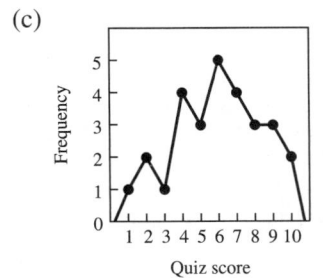

3. (a) In this Exercise, the tally column has been omitted; however, it is a useful tool to use when creating a frequency distribution by hand.

Class Limits	Frequency f	Relative frequency f/n
45–49	3	$3/54 \approx 5.6\%$
50–54	14	$14/54 \approx 25.9\%$
55–59	16	$16/54 \approx 29.6\%$
60–64	17	$17/54 \approx 31.5\%$
65–69	4	$4/54 \approx 7.4\%$

    (b)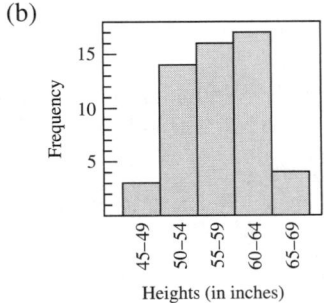

    (c)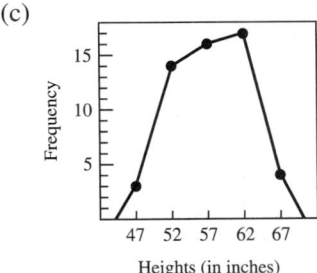

4. (a)
Class Limits	Frequency f	Relative frequency f/n
40–59	3	$3/50 = 6\%$
60–79	11	$11/50 = 22\%$
80–99	10	$10/50 = 20\%$
100–119	12	$12/50 = 24\%$
120–139	8	$8/50 = 16\%$
140–159	3	$3/50 = 6\%$
160–179	3	$3/50 = 6\%$

(b)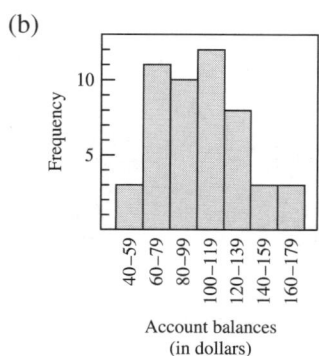
Account balances (in dollars)

(c)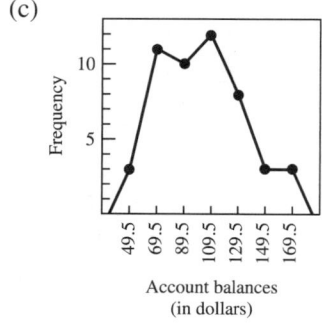
Account balances (in dollars)

5. (a) In this Exercise, the tally column has been omitted; however, it is a useful tool to use when creating a frequency distribution by hand.

Class Limits	Frequency f	Relative frequency f/n
70–74	2	$2/30 \approx 6.7\%$
75–79	1	$1/30 \approx 3.3\%$
80–84	3	$3/30 = 10\%$
85–89	2	$2/30 \approx 6.7\%$
90–94	5	$5/30 \approx 16.7\%$
95–99	5	$5/30 \approx 16.7\%$
100–104	6	$6/30 = 20\%$
105–109	4	$4/30 \approx 13.3\%$
110–114	2	$2/30 \approx 6.7\%$

(b)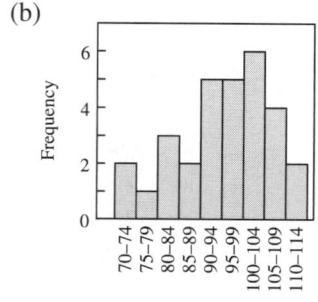
Temperature

(c)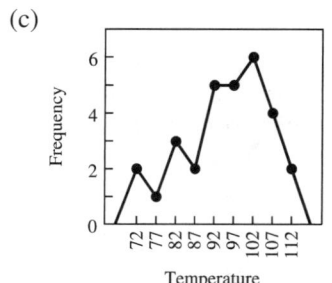
Temperature

6. (a)

Class Limits	Frequency f	Relative frequency f/n
91–95	1	$1/50 = 2\%$
96–100	3	$3/50 = 6\%$
101–105	5	$5/50 = 10\%$
106–110	7	$7/50 = 14\%$
111–115	12	$12/50 = 24\%$
116–120	9	$9/50 = 18\%$
121–125	7	$7/50 = 14\%$
126–130	4	$4/50 = 8\%$
131–135	2	$2/50 = 4\%$

(b)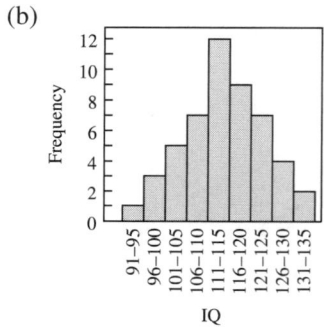
IQ

(c)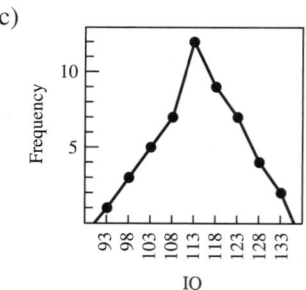
IQ

7.
```
0 | 7 9 8
1 | 1 1 2 8 9 4 3 1 0 5 0 5 5
2 | 0 9 6 6 2 5 2 3 4 4
3 | 1
```

8.
```
0 | 4 8
1 | 2 3 2 5 7 1 3 8 4 9 7 7 0
2 | 1 2 4 6 5 1 0
3 | 3 2 8
4 | 2 2
5 | 3 4
6 | 2
```

9.
```
0 | 8 5 4 9 6 9 4 8
1 | 2 0 1 8 8 2 4 0 8 8 6 3
2 | 6 6 2 5 1 3
3 | 0 4 6
4 | 4
```

10.
```
0  | 7 3
1  | 9 2 8 6
2  | 8 2 4 9 9 5
3  | 6 7 3 9 0 6 3 2
4  | 5 3 3 0 2 1 9 6
5  | 4 8 1 5
6  | 7 0 2
7  | 3 9 3
8  | 8 6
9  | 4
10 | 2
11 | 2
12 | 3
```

11. Read the vertical axis and total the values: $10.1 + 10.6 + 10.7 = 31.4$, or about 31%.

12. $7.8 + 8 + 8.4 = 24.2$, or about 24%.

13. CBS dropped the most, from a high of about 10.7% at the beginning of the time interval to a low of about 7.8%, which is about a 2.9% drop.

14. All the networks increased their ratings during the 1991–92 to the 1992–93 time period.

15. Examine the graph for the 1994–95 time period in which NBC and ABC seem to be farthest apart; the difference is $10 - 8.4 = 1.6\%$, and ABC attained the highest ratings.

16. Examine the Gross Domestic Product graph for 2001. The shorter bar is about one third of the way between 9000 and 10000, which is in billions of dollars. The amount then is about $9200 billion or $9,200,000,000,000.

17. Examine the graph for Consumer Price Index. The tallest bar is about 3.4% level in the years 2000.

18. From the Unemployment Rate graph, the greatest change was the increase was from 2000 to 2001 of about .7%.

19. Writing exercise

20. Writing exercise

21. Take 52% of 360° to find the central angle or
$$(.52)(360) \approx 187°.$$

22. Dividing the 2% contributed by the "other" category by five gives:
$$\frac{2\%}{5} = .4\%.$$

23. To calculate the number of degrees in each sector of the circle, multiply each percentage in decimal form times 360°. Here are a few examples:
$$.33(360) \approx 119°$$
$$.25(360) = 90°$$
$$.12(360) \approx 43°$$

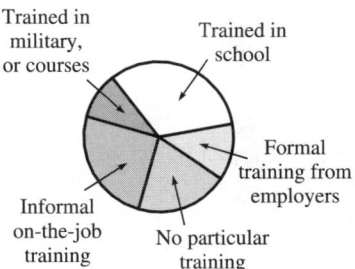

24.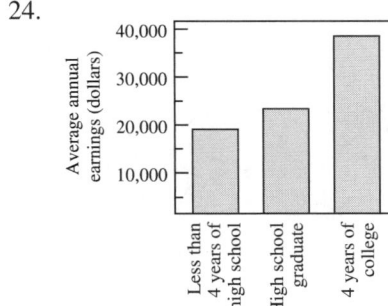

25. Examine the graph to see that he would be about 79.

26. Look at the highest peak of the curve that represents 12%. The dollar amount is about $400,000. Read the corresponding age along the horizontal axis—about 76 years old.

27. (a) Examine the 6% curve to see that Dave's money will run out at age 76. If he reaches age 70, his money would last for about
$$76 - 70 = 6 \text{ years.}$$

   (b) Writing exercise

28. His net worth would be about $400,000 if he averages 12%; it would be about $100,000 if he averages 9%, about four times as much.

29. Writing exercise

30. Writing exercise

31. Writing exercise

32.
(100–104)	10	0 0 4
(105–109)	10	6 9 5 9 7 6 8 5
(110–114)	11	2 0 2 4 4 2 4 0 0 3 1 1 3 0 2 4 4
(115–119)	11	8 8 7 6 8 5 8 7 8 9 7 8
(120–124)	12	0 1 5 2 1 0 0 0
(125–129)	12	8 5
(130–134)	13	4

33. Writing exercise

34.
Letter	Probability
E	.13
T	.09
A, O	.08
N	.07
I, R	.065
S, H	.06
D	.04
L	.035
C, M, U	.03
F, P, Y	.02
W, G, B	.015
V	.01
K, X, J	.005
Q, Z	.002

35.
Letter	Probability
A	$\frac{.08}{.385} \approx .208$
E	$\frac{.13}{.385} \approx .338$
I	$\frac{.065}{.385} \approx .169$
O	$\frac{.08}{.385} \approx .208$
U	$\frac{.03}{.385} \approx .078$

36.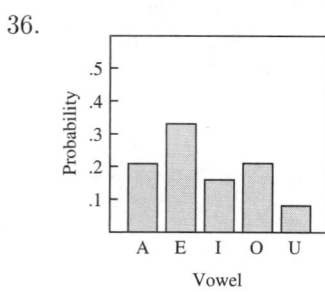

37.
Letter	Probability
A	$\frac{31}{118} \approx .263$
E	$\frac{34}{118} \approx .288$
I	$\frac{20}{118} \approx .169$
O	$\frac{23}{118} \approx .195$
U	$\frac{10}{118} \approx .085$

38. Writing exercise

39.
Class Limits	Probability
10–19	$\frac{6}{40} = .150$
20–29	$\frac{11}{40} = .275$
30–39	$\frac{9}{40} = .225$
40–49	$\frac{7}{40} = .175$
50–59	$\frac{4}{40} = .100$
60–69	$\frac{2}{40} = .050$
70–79	$\frac{1}{40} = .025$

40. (a) Read the table in Exercise 39 to see that the probability is .225 that a given student studied 30–39 hours.

(b) The student would fall into either the 40–49 hour range or the 50–59 hour range. Add the probabilities: $.175 + .100 = .275$

(c) Fewer than 30 hours means either 20–29 hours or 10–19 hours. Add the probabilities: $.275 + .150 = .425$

(d) At least 50 hours means 50 or more; the categories included are 50–59, 60–69, and 70–79. Add the probabilities: $.100 + .050 + .025 = .175$

41. The probability is $\frac{7}{40}$.

42.

Class Limits	Probability
Sailing	$\frac{9}{40} = .225$
Hang gliding	$\frac{5}{40} = .125$
Bungee jumping	$\frac{7}{40} = .175$
Sky diving	$\frac{3}{40} = .075$
Canoeing	$\frac{12}{40} = .300$
Rafting	$\frac{4}{40} = .100$

43. (a) Empirical

   (b) Writing exercise

44. Writing exercise

13.2 EXERCISES

1. (a) $\bar{x} = \dfrac{3+7+12+16+23}{5} = \dfrac{61}{5} = 12.2$

   (b) The data is given in order from smallest to largest; the middle number is 12.

   (c) No mode.

2. (a) $\bar{x} = \dfrac{21+25+32+48+53+62}{6} = \dfrac{241}{6} \approx 40.2$

   (b) The data is given in order from smallest to largest; the median is $\dfrac{32+48}{2} = 40$.

   (c) No mode.

3. (a) $\bar{x} = \dfrac{128+230+196+224+196+233}{6}$
   $= \dfrac{1207}{6}$
   ≈ 201.2

   (b) Arrange the values from smallest to largest or vice versa: 128, 196, 196, 224, 230, 233. Find the mean of the two middle numbers: $\dfrac{196+224}{2} = 210$.

   (c) The value that occurs most frequently is 196.

4. (a) $\bar{x} = \dfrac{26+31+46+31+26+29+31}{7} = \dfrac{220}{7} \approx 31.4$

   (b) Arrange the values from smallest to largest or vice versa: 26, 26, 29, 31, 31, 31, 46. The median is 31.

   (c) The mode is 31.

5. (a) $\bar{x} = \dfrac{3.1+4.5+6.2+7.1+4.5+3.8+6.2+6.3}{8}$
   $= \dfrac{41.7}{8}$
   ≈ 5.2

   (b) Arrange the values from smallest to largest or vice versa: 3.1, 3.8, 4.5, 4.5, 6.2, 6.2, 6.3, 7.1. Find the mean of the two middle numbers: $\dfrac{4.5+6.2}{2} = 5.35$.

   (c) The set of values is bimodal: 4.5 and 6.2.

6. (a) The sum of the data is 64,070. $\dfrac{64070}{4} = 16,017.5$.

   (b) Arrange the values from smallest to largest or vice versa: 14,320, 15,470, 16,950, 17,330. Find the mean of the two middle numbers: $\dfrac{15470+16950}{2} = 16210$.

   (c) No mode.

7. (a) $\bar{x} = \dfrac{.78+.93+.66+.94+.87+.62+.74+.81}{8}$
   $= \dfrac{6.35}{8}$
   $\approx .8$

   (b) Arrange the values from smallest to largest or vice versa: .62, .66, .74, .78, .81, .87, .93, .94. Find the mean of the two middle numbers: $\dfrac{.78+.81}{2} = .795$.

   (c) There is no mode.

8. (a) The sum of the data is 2.09. $\dfrac{2.09}{9} \approx .2$.

   (b) When arranged in order from smallest to largest, the median is .22.

   (c) The mode is .28, which occurs twice.

9. (a) The sum of the data is 330.4. $\bar{x} = \dfrac{330.4}{7} = 47.2$.

   (b) Arrange the values from smallest to largest or vice versa: 1.2, 12.3, 34.5, 45.6, 67.8, 78.9, 90.1. The median is 45.6.

   (c) No mode.

10. (a) The sum of the data is 5.7. $\bar{x} = \dfrac{5.7}{11} \approx .5$.

    (b) Arrange the values from smallest to largest or vice versa: .1, .2, .3, .3, .4, .5, .6, .7, .8, .9, .9. The median is .5.

    (c) There are two modes: .3 and .9.

11. (a) The sum of the data is 1032. $\bar{x} = \dfrac{1032}{8} = 129$.

(b) Arrange the values from smallest to largest or vice versa: 125, 125, 127, 128, 128, 131, 132, 136. The two middle numbers are both 128. The median is

$$\frac{128 + 128}{2} = 128.$$

(c) There are two modes: 125 and 128.

12. (a) The sum of the data is 29.97. $\bar{x} = \frac{29.97}{6} \approx 5.0$.

    (b) Arrange the values from smallest to largest or vice versa. The two middle numbers are 4.35 and 5.64. The median is $\frac{4.35 + 5.64}{2} = 4.995$.

    (c) There is no mode.

13. (a) The sum of the visitors for sites 1 through 5 is 234,540 thousand. Then

    $$\bar{x} = \frac{234540000}{5} \approx 42,779,000.$$

    (b) The median or middle value is 46,908,000.

14. (a) The sum of the visitors for sites 1 through 10 is 362,443 thousand. Then

    $$\bar{x} = \frac{3624430000}{10} \approx 36,244,000.$$

    (b) The median or middle value is

    $$\frac{36802000 + 31854000}{2} \approx 34,328,000.$$

15. (a) The sum of the visitors for sites 11 through 20 is 147,699 thousand ≈ 147,700 thousand. Then

    $$\bar{x} = \frac{147700000}{10} = 14,770,000.$$

    (b) The median or middle value is

    $$\frac{14636000 + 14517000}{2} = 14576500 \approx 14,577,000.$$

16. (a) From Exersizes 14 and 15 the sum of visitors for sites 1 through 20 is

    $362,440,000 + 147,700,000 = 510,140,000$. Then

    $$\bar{x} = \frac{510140000}{20} \approx 25,507,000.$$

    (b) The median or middle value is

    $$\frac{19553000 + 19096000}{2} = 19,324,500 \approx 19,325,000.$$

17. The sum of the yearly losses is 19875. Then

    $$\bar{x} = \frac{19875}{5} = 3975.$$ Rounded to the nearest 100 square miles, this is 4000 square miles.

18. The sum of the emissions is 351,000,000. Then

    $$\bar{x} = \frac{351000000}{5} = 70,200,000.$$ Rounded to the nearest 1 million, this is 70,000,000 metric tons.

19. The sum of the data is 36.92. Then $\bar{x} = \frac{36.92}{7} = 5.27$.

20. The median is the middle number after arranging the values in rank order: 2.39.

21. The sum for the new list is 16.94. Then

    $$\bar{x} = \frac{16.94}{7} = 2.42.$$

22. The median for the new list is 2.28.

23. The mean was affected more.

24. The median is affected less by the presence of an extreme value in the data.

25. $\bar{x} = \frac{79 + 81 + 44 + 89 + 79 + 90}{6} = \frac{462}{6} = 77.$

    Arrange in order from smallest to largest; 44, 79, 79, 81, 89, 90. The median is $\frac{79 + 81}{2} = 80.$

    (c) The mode is 79.

26. The median is probably the best indicator.

27. Let $x =$ the score he must make. Replace the score of 44 with x.

    $$\frac{x + 79 + 79 + 81 + 89 + 90}{6} = 85$$

    $$\frac{6}{1} \cdot \frac{x + 79 + 79 + 81 + 89 + 90}{6} = 85 \cdot 6$$

    $$x + 418 = 510$$

    $$x = 92.$$

28. (a)

Value	Frequency	Value · Frequency
2	5	10
4	1	4
6	8	48
8	4	32
Totals	18	94

    Then $\bar{x} = \frac{94}{18} \approx 5.22$.

(b) From part (a), there are 18 items. The formula for the position of the median is

$$\frac{\Sigma f + 1}{2} = \frac{18 + 1}{2} = 9.5.$$

This means that the median is halfway between the ninth and tenth item. A chart showing cumulative frequency shows the value.

Value	Frequency	Cumulative Frequency
2	5	5
4	1	6
6	8	14
8	4	18

The value 6 is the median.

(c) Examine the frequency column to see that the mode is also 6.

29. (a) Calculate a value · frequency column in order to evaluate the mean.

Value	Frequency	Value · Frequency
603	13	7839
597	8	4776
589	9	5301
598	12	7176
601	6	3606
592	4	2368
Totals	52	31066

Then $\bar{x} = \dfrac{31066}{52} \approx 597.42.$

(b) From part (a), there are 52 items. The formula for the position of the median is

$$\frac{\Sigma f + 1}{2} = \frac{52 + 1}{2} = 26.5.$$

This means that the median is halfway between the 26th and 27th item. A chart showing cumulative frequency shows the value.

Value	Frequency	Cumulative Frequency
589	9	9
592	4	13
597	8	21
598	12	33
601	6	39
603	13	52

The value 598 is the median.

(c) Examine the frequency column to see that the mode is 603.

30.

Value	Frequency	Value · Frequency
19,500	6	117,000
23,000	8	184,000
28,300	4	113,200
34,500	2	69,000
36,900	7	258,300
145,500	1	145,500
Totals	28	887,000

Then $\bar{x} = \dfrac{887000}{28} \approx \$31,700$, to the nearest hundred.

31.

Units	Grade	Units · Grade Value
4	C	4 · 2 = 8
7	B	7 · 3 = 21
3	A	3 · 4 = 12
3	F	3 · 0 = 0
17		41

Then $\bar{x} = \dfrac{41}{17} \approx 2.41$, to the nearest hundredth.

32.

Units	Grade	Units · Grade Value
2	A	2 · 4 = 8
6	B	6 · 3 = 18
5	C	5 · 2 = 10
13		36

Then $\bar{x} = \dfrac{36}{13} \approx 2.77$, to the nearest hundredth.

33. The sum of the area values is 8,506,300. Then

$$\bar{x} = \frac{8506300}{11} = 773,300 \text{ square miles.}$$

34. Writing exercise

35. The sum of the populations values is 278,327,413. Then

$$\bar{x} = \frac{278327413}{11} \approx 25,302,000 \text{ to the nearest } 1000.$$

36. Writing exercise

37. Remember that negative values are deducted from the sum. The sum of the values is 7,164,889. Then

$$\bar{x} = \frac{7164889}{10} \approx 716,489.$$

38. The sum of the seven positive values is 7,651,831. Then

$$\bar{x} = \frac{7651831}{7} = 1,093,119.$$

39. Find the sum of the three negative values: $-486,942$. Then
$$\bar{x} = \frac{-486942}{3} = -162,314.$$

40. $\bar{x} = \frac{[(1093119 \cdot 7) + (-162314 \cdot 3)]}{7+3} \approx 716,489$

    This is the same value that was calculated in Exercise 37.

41. The sum of the dollar values is $2,062,000. Then
$$\bar{x} = \frac{\$2062000}{3} = \$687,333.33.$$

42. The sum of the "Gold" column is 61. $\bar{x} = \frac{61}{10} = 6.1$

43. There are 10 scores. Then the position of the median is given by
$$\frac{10+1}{2} = 5.5.$$

    This means that the score of interest is between the 5th and 6th score. After putting the scores in ascending order, we see that the 5th score is 5 and the 6th score is 5. The mean of these two scores (median score) is
$$\frac{5+5}{2} = 5.$$

44. Examine the "Bronze" column to see that the most often occuring scores, 4 and 6, occur two times; the modes are 4 and 6.

45. (a) The sum of the "Totals" column is 184. Then
$$\bar{x} = \frac{184}{10} = 18.4.$$

    (b) The median is the mean of the two middle numbers:
$$\frac{16+16}{2} = 16.$$

    (c) The modes are 16 and 11, each occuring twice.

46. (a) $\bar{x} = \frac{97 + 102 + 104 + 105 + \ldots + 138}{23} = \frac{2675}{23} \approx \$116.30$

    (b) The median is the 12th piece of data which is $115.

    (c) There is no mode; several numbers occur two times.

47. (a) $\bar{x} = \frac{47 + 51 + 53 + 56 + \ldots + 96}{34} = \frac{2544}{34} \approx 74.8$

    (b) The scores are listed from smallest to the largest. Because there is an even number of scores, the median is the mean of the two middle numbers, the 17th and 18th scores: $\frac{77+78}{2} = 77.5$.

    (c) The most frequently occurring score is 78.

48. Analyze the given information to see that three scores are certain: 92, 92, and 87. If these are arranged in order, it can be seen that the missing test scores must be lower than 87 and they must be different from each other because the data is not bimodal. If one of the scores is 86, this would allow the last score to be as low as possible. Set up the following inequality to determine the lowest possible value of the missing tests:
$$\frac{92 + 92 + 87 + 86 + x}{5} \geq 88$$
$$\frac{5}{1} \cdot \frac{92 + 92 + 87 + 86 + x}{5} \geq 88 \cdot 5$$
$$92 + 92 + 87 + 86 + x \geq 440$$
$$357 + x \geq 440$$
$$x \geq 83.$$

49. Writing exercise

50. (a) $\bar{x} = \frac{5 \cdot 1 + 5 \cdot 2 + 3 \cdot 3 + 6 \cdot 4 + 3 \cdot 6}{22} = \frac{66}{22} = 3$

    Position of the median is $\frac{22+1}{2} = 11.5$. The position is between the 11th and 12th piece of data.

Value	Frequency	Cumulative Frequency
1	5	5
2	5	10
3	3	13
4	6	19
6	3	22

    Examine the table to see that the median is located in the row for the value of 3.

    The mode is 4 with a frequency of 6.

    (b) The company should use the mode because they would want to satisfy most of their customers.

    (c) Writing exercise

51. (a)

Value	Frequency	Value · Frequency
0	1	0
1	1	1
3	1	3
14	2	28
15	1	15
16	2	32
17	2	34
18	3	54
19	1	19
20	1	20
Totals	15	206

    $$\bar{x} = \frac{206}{15} \approx 13.7.$$

Position of the median is $\frac{15+1}{2} = 8$. The position is the 8th piece of data,

Value	Frequency	Cumulative Frequency
0	1	1
1	1	2
3	1	3
14	2	5
15	1	6
16	2	8
17	2	10
18	3	13
19	1	14
20	1	15

Examine the table to see that the median is located in the row for the value of 16.

The mode is 18 with a frequency of 3.

(b) The median, 16, is most representative of the data.

(c) Writing exercise

52. 1, 2, 3, 4, 5, 6, 7. The median is 4.

   (a) There are 7 numbers listed.

   (b) $\bar{x} = \frac{1+2+3+4+5+6+7}{7} = \frac{28}{7} = 4$

53. 2, 3, 5, 7. The median is $\frac{3+5}{2} = 4$.

   (a) There are 4 numbers listed.

   (b) $\bar{x} = \frac{2+3+5+7}{4} = \frac{17}{4} = 4.25$

54. 1, 1, 2, 3, 5, 8, 13, 21. The median is $\frac{3+5}{2} = 4$.

   (a) There are 8 numbers listed.

   (b) $\bar{x} = \frac{1+1+2+3+5+8+13+21}{8}$
   $= \frac{54}{8}$
   $= 6.75.$

55. 1, 3, 6, 10, 15, 21. The median is $\frac{6+10}{2} = 8$.

   (a) There are 6 numbers listed.

   (b) $\bar{x} = \frac{1+3+6+10+15+21}{6} = \frac{56}{6} \approx 9.33$

56. Consecutive whole numbers can be represented as follows: x is the smallest; $x+1$ is the next; $x+2$, the next, etc. Set up an algebraic equation showing the sum of these whole numbers on the left and 147 on the right.

$$x + (x+1) + (x+2) + \ldots (x+6) = 147$$
$$7x + 21 = 147$$
$$7x = 126$$
$$x = 18$$

Because there are 7 numbers, the median or middle number would be the fourth, which is represented by $x+3$ or $18+3 = 21$. The mean of the numbers is

$$\frac{147}{7} = 21.$$

Therefore, $21 - 21 = 0$.

57. Arrange the numbers from smallest to largest. At this point it is uncertain where x will lie. However, if a single number must be the mean, median, and mode, then one of the given numbers must be that number. Because the median is the middle number and five values are given, the median must be 70 or 80. Try each of these as a mean to see which one works.

$$\frac{60+70+80+110+x}{5} = 70$$
$$\frac{5}{1} \cdot \frac{320+x}{5} = 70 \cdot 5$$
$$320 + x = 350$$
$$x = 30$$

This value does not work.

$$\frac{60+70+80+110+x}{5} = 80$$
$$\frac{5}{1} \cdot \frac{320+x}{5} = 80 \cdot 5$$
$$320 + x = 400$$
$$x = 80$$

This value works. The set of numbers is $\{60, 70, 80, 80, 110\}$. The value 80 is the mean, median, and mode.

58. The locations for the number n are before or after the 2, following the 5, following the 8, or following the 9. If the location is before or after the 2, the median is 5. If the mean, then, must also equal 5, find one value of n by the following equation.

$$\frac{n+2+5+8+9}{5} = 5$$
$$\frac{5}{1} \cdot \frac{n+24}{5} = 5 \cdot 5$$
$$n + 24 = 25$$
$$n = 1$$

Now if the location of n is following the 5, then n will be the median and the mean. To find this value of n solve the following equation.

$$\frac{2+5+n+8+9}{5} = n$$

$$\frac{5}{1} \cdot \frac{n+24}{5} = n \cdot 5$$

$$n + 24 = 5n$$

$$24 = 4n$$

$$6 = n$$

If the location of n is following the 8 or the 9, the median is 8. If the mean, then, must also equal 8, find the value of n by the following equation.

$$\frac{2+5+8+n+9}{5} = 8$$

$$\frac{5}{1} \cdot \frac{n+24}{5} = 8 \cdot 5$$

$$n + 24 = 40$$

$$n = 16$$

The three choices for n are 1, 6, and 16.

59. Writing exercise

60. No, because the table gives only how many pieces of data occur within a given interval.

61. No

62. Writing exercise

13.3 EXERCISES

1. The sample standard deviation will be larger because the denominator is $n-1$ instead of n.

2. Writing exercise

3. (a) The range is $19 - 2 = 17$.

   (b) To find the standard deviation:
   1. First find the mean. $\bar{x} = \frac{75}{8} = 9.375$

   2 and 3. Find each deviation from the mean $(x - \bar{x})$ and square each deviation, $(x - \bar{x})^2$. These steps are shown in the table.

Data	Deviations	Squared Deviations
2	$2 - 9.375 = -7.375$	$(-7.375)^2 = 54.3906$
5	$5 - 9.375 = -4.375$	$(-4.375)^2 = 19.1406$
6	$6 - 9.375 = -3.375$	$(-3.375)^2 = 11.3906$
8	$8 - 9.375 = -1.375$	$(-1.375)^2 = 1.8906$
9	$9 - 9.375 = -0.375$	$(-0.375)^2 = 0.1406$
11	$11 - 9.375 = 1.625$	$(1.625)^2 = 2.6406$
15	$15 - 9.375 = 5.625$	$(5.625)^2 = 31.6406$
19	$19 - 9.375 = 9.625$	$(9.625)^2 = 92.6406$
Total		213.8748

4. Sum the squared deviations. The sum is 213.8748.

5. Divide by $n - 1$.

$$\frac{213.8748}{8-1} \approx 30.5535$$

6. Take the square root. $\sqrt{30.5484} \approx 5.53$

4. (a) The range is $21 - 3 = 18$.

   (b) To find the standard deviation:
   1. First find the mean. $\bar{x} = \frac{97}{9} \approx 10.778$.

   2 and 3. Find each deviation from the mean $(x - \bar{x})$ and square each deviation, $(x - \bar{x})^2$. These steps are shown in the table.

Data	Deviations	Squared Deviations
8	$8 - 10.778 = -2.778$	$(-2.778)^2 \approx 7.717$
5	$5 - 10.778 = -5.778$	$(-5.778)^2 \approx 33.385$
12	$12 - 10.778 = 1.222$	$(1.222)^2 \approx 1.493$
8	$8 - 10.778 = -2.778$	$(-2.778)^2 \approx 7.717$
9	$9 - 10.778 = -1.778$	$(-1.778)^2 \approx 3.161$
15	$15 - 10.778 = 4.222$	$(4.222)^2 \approx 17.825$
21	$21 - 10.778 = 10.222$	$(10.222)^2 \approx 104.489$
16	$16 - 10.778 = 5.222$	$(5.222)^2 \approx 27.269$
3	$3 - 10.778 = -7.778$	$(-7.778)^2 \approx 60.497$
Total		263.553

4. Sum the squared deviations. The sum is 263.553.

5. Divide by $n - 1$.

$$\frac{263.553}{9-1} \approx 32.944$$

6. Take the square root. $\sqrt{32.944} \approx 5.74$

13.3 MEASURES OF DISPERSION

5. (a) The range is $41 - 22 = 19$.

   (b) To find the standard deviation:
   1. First find the mean. $\bar{x} = \dfrac{210}{7} = 30$.
   2 and 3. Find each deviation from the mean $(x - \bar{x})$ and square each deviation, $(x - \bar{x})^2$. These steps are shown in the table.

Data	Deviations	Squared Deviations
25	$25 - 30 = -5$	$(-5)^2 = 25$
34	$34 - 30 = 4$	$(4)^2 = 16$
22	$22 - 30 = -8$	$(-8)^2 = 64$
41	$41 - 30 = 11$	$(11)^2 = 121$
30	$30 - 30 = 0$	$(0)^2 = 0$
27	$27 - 30 = -3$	$(-3)^2 = 9$
31	$31 - 30 = 1$	$(1)^2 = 1$
Total		236

   4. Sum the squared deviations. The sum is 236.
   5. Divide by $n - 1$.
   $$\dfrac{236}{7-1} \approx 39.3333$$
   6. Take the square root. $\sqrt{39.3333} \approx 6.27$

6. (a) The range is $84 - 55 = 29$.

   (b) To find the standard deviation:
   1. First find the mean. $\bar{x} = \dfrac{634}{9} \approx 70.444$

   2 and 3. Find each deviation from the mean $(x - \bar{x})$ and square each deviation, $(x - \bar{x})^2$. These steps are shown in the table.

Data	Deviations	Squared Deviations
67	$67 - 70.444 = -3.444$	$(-3.444)^2 = 11.861$
83	$83 - 70.444 = 12.556$	$(12.556)^2 = 157.653$
55	$55 - 70.444 = -15.444$	$(-15.444)^2 = 238.517$
68	$68 - 70.444 = -2.444$	$(-2.444)^2 = 5.973$
77	$77 - 70.444 = 6.556$	$(6.556)^2 = 42.981$
63	$63 - 70.444 = -7.444$	$(-7.444)^2 = 55.413$
84	$84 - 70.444 = 13.556$	$(13.556)^2 = 183.765$
72	$72 - 70.444 = 1.556$	$(1.556)^2 = 2.421$
65	$65 - 70.444 = -5.444$	$(-5.444)^2 = 29.637$
Total		728.221

   4. Sum the squared deviations. The sum is 728.221.
   5. Divide by $n - 1$.
   $$\dfrac{728.221}{9-1} \approx 91.0276$$
   6. Take the square root. $\sqrt{91.0276} \approx 9.54$

Some of the details are omitted in the following exercises. See Exercises 3–6 for details of computing standard deviation. A spreadsheet is a very useful tool in obtaining the intermediate calculations.

7. (a) The range is $331 - 308 = 23$.

   (b) To find the standard deviation:
   1. First find the mean. $\bar{x} = \dfrac{3204}{10} = 320.4$

   2 and 3. Find each deviation from the mean $(x - \bar{x})$ and square each deviation, $(x - \bar{x})^2$.

   4. Sum the squared deviations. The sum is 540.4.
   5. Divide by $n - 1$.
   $$\dfrac{540.4}{10-1} \approx 60.044$$
   6. Take the square root. $\sqrt{60.044} \approx 7.75$

8. (a) The range is $8.5 - 5.7 = 2.8$.

   (b) To find the standard deviation:
   1. First find the mean. $\bar{x} = \dfrac{103.3}{14} = 7.379$

   2 and 3. Find each deviation from the mean $(x - \bar{x})$ and square each deviation, $(x - \bar{x})^2$.
   4. Sum the squared deviations. The sum is 7.584.
   5. Divide by $n - 1$. $\dfrac{7.584}{14-1} \approx .583$
   6. Take the square root. $\sqrt{.583} \approx .76$

9. (a) The range is $85.62 - 84.48 = 1.14$.

   (b) To find the standard deviation:
   1. First find the mean. $\bar{x} = \dfrac{763.62}{9} = 84.84\overline{6}$

   2 and 3. Find each deviation from the mean $(x - \bar{x})$ and square each deviation, $(x - \bar{x})^2$.

   4. Sum the squared deviations. The sum is 1.091401.
   5. Divide by $n - 1$.
   $$\dfrac{1.091401}{9-1} \approx .136425$$
   6. Take the square root. $\sqrt{.136425} \approx .37$

10. (a) The range is $214.2 - 206.3 = 7.9$.

    (b) To find the standard deviation:
    1. First find the mean. $\bar{x} = \dfrac{2109.7}{10} = 210.97$

    2 and 3. Find each deviation from the mean $(x - \bar{x})$ and square each deviation, $(x - \bar{x})^2$.

4. Sum the squared deviations. The sum is 44.681.

5. Divide by $n-1$. $\dfrac{44.681}{10-1} \approx 4.965$

6. Take the square root. $\sqrt{4.965} \approx 2.23$

11. (a) The range is $9 - 1 = 8$.

    (b) To find the standard deviation:
    1. First find the mean.

Value	Frequency	Value · Frequency
9	3	27
7	4	28
5	7	35
3	5	15
1	2	2
Totals	21	107

    Then $\bar{x} = \dfrac{107}{21} \approx 5.095$.

    2 and 3. Find each deviation from the mean $(x - \bar{x})$ and square each deviation, $(x - \bar{x})^2$. These steps are shown in the table.

Value	Deviations	Squared Deviations	Freq · $(x - \bar{x})^2$
9	3.905	15.249	3(15.249)
7	1.905	3.629	4(3.629)
5	−.095	.009	7(.009)
3	−2.095	4.389	5(4.389)
1	−4.095	16.769	2(16.769)
Total			115.809

    4. The fourth column shows the frequency of each value multiplied by the squared deviation. After multiplying each of these, find the sum: 115.809.

    5. Divide by $n - 1$. Remember that the total number of values is 21.

    $$\dfrac{115.809}{21 - 1} \approx 5.790$$

    6. Take the square root. $\sqrt{5.790} \approx 2.41$

12. (a) The range is $26 - 14 = 12$.

    (b) To find the standard deviation:
    1. First find the mean.

Value	Frequency	Value · Frequency
14	8	112
16	12	192
18	15	270
20	14	280
22	10	220
24	6	144
26	3	78
Totals	68	1296

    Then $\bar{x} = \dfrac{1296}{68} \approx 19.059$.

    2 and 3. Find each deviation from the mean $(x - \bar{x})$ and square each deviation, $(x - \bar{x})^2$. These steps are shown in the table.

Value	Deviations	Squared Deviations	Freq · $(x - \bar{x})^2$
14	−5.059	25.593	8(25.593)
16	−3.059	9.357	12(9.357)
18	−1.059	1.121	15(1.121)
20	.941	.885	14(.885)
22	2.941	8.649	10(8.649)
24	4.941	24.413	6(24.413)
26	6.941	48.177	3(48.177)
Total			723.732

    4. The fourth column shows the frequency of each value multiplied by the squared deviation. After multiplying each of these, find the sum: 723.732

    5. Divide by $n - 1$. Remember that the total number of values is 68.

    $$\dfrac{723.732}{68 - 1} \approx 10.802$$

    6. Take the square root. $\sqrt{10.802} \approx 3.29$

13. According to Chebyshev's theorem, the fraction of scores that lie within 2 standard deviations of the mean is at least

    $$1 - \dfrac{1}{2^2} = 1 - \dfrac{1}{4} = \dfrac{3}{4}.$$

14. According to Chebyshev's theorem, the fraction of scores that lie within 4 standard deviations of the mean is at least

    $$1 - \dfrac{1}{4^2} = 1 - \dfrac{1}{16} = \dfrac{15}{16}.$$

15. According to Chebyshev's theorem, the fraction of scores that lie within 7/2 standard deviations of the mean is at least
$$1 - \frac{1}{\left(\frac{7}{2}\right)^2} = 1 - \frac{1}{\frac{49}{4}} = 1 - \frac{4}{49} = \frac{45}{49}.$$

16. According to Chebyshev's theorem, the fraction of scores that lie within 11/4 standard deviations of the mean is at least
$$1 - \frac{1}{\left(\frac{11}{4}\right)^2} = 1 - \frac{1}{\frac{121}{16}} = 1 - \frac{16}{121} = \frac{105}{121}.$$

17. According to Chebyshev's theorem, the fraction of scores that lie within 3 standard deviations of the mean is at least
$$1 - \frac{1}{(3)^2} = 1 - \frac{1}{9} = \frac{8}{9}.$$
Divide 8 by 9 and change the decimal $.\overline{8}$ to 88.9%.

18. According to Chebyshev's theorem, the fraction of scores that lie within 5 standard deviations of the mean is at least
$$1 - \frac{1}{(5)^2} = 1 - \frac{1}{25} = \frac{24}{25}.$$
Divide 24 by 25 and change the decimal .96 to 96%.

19. According to Chebyshev's theorem, the fraction of scores that lie within 5/3 standard deviations of the mean is at least
$$1 - \frac{1}{\left(\frac{5}{3}\right)^2} = 1 - \frac{1}{\frac{25}{9}} = 1 - \frac{9}{25} = \frac{16}{25}.$$
Divide 16 by 25 and change the decimal .64 to 64%.

20. According to Chebyshev's theorem, the fraction of scores that lie within 5/2 standard deviations of the mean is at least
$$1 - \frac{1}{\left(\frac{5}{2}\right)^2} = 1 - \frac{1}{\frac{25}{4}} = 1 - \frac{4}{25} = \frac{21}{25}.$$
Divide 21 by 25 and change the decimal .84 to 84%.

21. Since 54 is 2 standard deviations below the mean $(70 - 2 \cdot 8 = 54)$ and 86 is 2 standard deviations above the mean $(70 + 2 \cdot 8 = 86)$, find the minimum fraction of values that lie within 2 standard deviations of the mean. See Exercise 13 for the answer 3/4.

22. Since 46 is 3 standard deviations below the mean $(70 - 3 \cdot 8 = 46)$ and 94 is 3 standard deviations above the mean $(70 + 3 \cdot 8 = 94)$, find the minimum fraction of values that lie within 3 standard deviations of the mean. See Exercise 17 for the answer 8/9.

23. Since 38 is 4 standard deviations below the mean $(70 - 4 \cdot 8 = 38)$ and 102 is 4 standard deviations above the mean $(70 + 4 \cdot 8 = 102)$, find the minimum fraction of values that lie within 4 standard deviations of the mean. See Exercise 14 for the answer 15/16.

24. Since 30 is 5 standard deviations below the mean $(70 - 5 \cdot 8 = 30)$ and 110 is 5 standard deviations above the mean $(70 + 5 \cdot 8)$, find the minimum fraction of values that lie within 5 standard deviations of the mean. See Exercise 18 for the answer 24/25.

25. This is equivalent to finding the largest fraction of values that lie outside 2 standard deviations from the mean. There are at least $1 - \frac{1}{2^2} = 1 - \frac{1}{4} = \frac{3}{4}$ of the values within 2 standard deviations of the mean. Thus, the largest fraction of values that lie outside this range would be: $1 - \frac{3}{4} = \frac{1}{4}.$

26. To find how many standard deviations below the mean 50 is, use:
$$\frac{50 - 70}{8} = -2.5.$$
Also the value 90 is 2.5 standard deviations above the mean. Then we must find the largest fraction of values that lie outside 2.5 or 2 1/2 standard deviations from the mean. There are at least
$$1 - \frac{1}{\left(\frac{5}{2}\right)^2} = 1 - \frac{1}{\left(\frac{25}{4}\right)} = 1 - \frac{4}{25} = \frac{21}{25}$$
of the values within 2 1/2 standard deviations of the mean. Thus, the largest fraction of values that lie outside this range of values would be: $1 - \frac{21}{25} = \frac{4}{25}.$

27. To find how many standard deviations below the mean 42 is, use:
$$\frac{42 - 70}{8} = -3.5.$$
Also the value 98 is 3.5 standard deviations above the mean. Then we must find the largest fraction of values that lie outside 3.5 or 3 1/2 standard deviations from the mean. There are at least
$$1 - \frac{1}{\left(\frac{7}{2}\right)^2} = 1 - \frac{1}{\left(\frac{49}{4}\right)} = 1 - \frac{4}{49} = \frac{45}{49}$$
of the values within 3 1/2 standard deviations of the mean. Thus, the largest fraction of values that lie outside this range of values would be: $1 - \frac{45}{49} = \frac{4}{49}.$

28. To find how many standard deviations below the mean 52 is, use:

$$\frac{52 - 70}{8} = -2.25.$$

Also the value 88 is 2.25 standard deviations above the mean. Then we must find the largest fraction of values that lie outside 2.25 or 2 1/4 standard deviations from the mean. There are at least

$$1 - \frac{1}{\left(\frac{9}{4}\right)^2} = 1 - \frac{1}{\left(\frac{81}{16}\right)} = 1 - \frac{16}{81} = \frac{65}{81}$$

of the values within 2 1/4 standard deviations of the mean. Thus, the largest fraction of values that lie outside this range of values would be: $1 - \frac{65}{81} = \frac{16}{81}$.

29. The sum of the values is $2430. Then
$$x = \frac{2430}{12} = \$202.50$$

30. To find the standard deviation:

   1. Use the mean from Exercise 29. $\bar{x} = \$202.50$

   2 and 3. Find each deviation from the mean $(x - \bar{x})$ and square each deviation, $(x - \bar{x})^2$. These steps are shown in the table.

Data	Deviations	Squared Deviations
80	$80 - 202.50 = -122.5$	$(-122.5)^2 = 15006.25$
105	$105 - 202.50 = -97.5$	$(-97.5)^2 = 9506.25$
120	$120 - 202.50 = -82.5$	$(-82.5)^2 = 6806.25$
⋮	⋮	⋮
325	$325 - 202.50 = 122.5$	$(122.5)^2 = 15006.25$
Total		71,075

   4. Sum the squared deviations. The sum is 71,075.

   5. Divide by $n - 1$.

   $$\frac{71075}{12 - 1} \approx 6461.364$$

   6. Take the square root. $\sqrt{6461.364} \approx \$80.38$

31. The standard deviation is $80.38. Then $202.50 - 80.38 = \$122.12$, and $202.50 + 80.38 = \$282.88$. There are six bonus amounts that fall within these two boundaries: $175, $185, $190, $205, $210, and $215.

32. Calculate the values for 2 standard deviations above and below the mean.

    $$\$202.50 - 2(80.38) = \$41.74$$
    $$\$202.50 + 2(80.38) = \$363.26$$

    All of the values are within these two amounts.

33. $1 - \frac{1}{2^2} = 1 - \frac{1}{4} = \frac{3}{4}$. Then 3/4 of the data is

    $$\frac{3}{4} \cdot 12 = 9.$$

    There should be at least 9 amounts.

34. Writing exercise

35. To find the mean average life of Brand B, find the sum of the amounts given in the text: 963,510. Then

    $$\bar{x} = \frac{963510}{20} \approx 48,176.$$

    This mean is greater than the mean for Brand A, which is 43,560; Brand B has the longer average life.

36. To answer this question, the standard deviation values must be compared. The standard deviation for Brand A is given as 2116 miles. To calculate the standard deviation for Brand B modules, follow the procedure given in the text or use a calculator or spreadsheet. The value of the standard deviation is 2235 miles. Brand A is more consistent because the value of the standard deviation is smaller.

37. Writing exercise

38. $\bar{x} = \frac{13 + 14 + 16 + 18 + 20 + 22 + 25}{7} \approx 18.29$

    To find the standard deviation:

    1. The mean is 18.29.

    2 and 3. Find each deviation from the mean $(x - \bar{x})$ and square each deviation, $(x - \bar{x})^2$. These steps are shown in the table.

Data	Deviations	Squared Deviations
13	$13 - 18.29 = -5.29$	$(-5.29)^2 = 27.9841$
14	$14 - 18.29 = -4.29$	$(-4.29)^2 = 18.4041$
16	$16 - 18.29 = -2.29$	$(-2.29)^2 = 5.2441$
⋮	⋮	⋮
25	$25 - 18.29 = 6.71$	$(6.71)^2 = 45.0241$
Total		113.429

    4. Sum the squared deviations. The sum is 113.429.

    5. Divide by $n - 1$.

    $$\frac{113.429}{7 - 1} \approx 18.905$$

    6. Take the square root. $\sqrt{18.905} \approx 4.35$

39. $\bar{x} = \dfrac{18 + 19 + 21 + 23 + 25 + 27 + 30}{7} \approx 23.29$

To find the standard deviation:

1. The mean is 23.29.

2 and 3. Find each deviation from the mean $(x - \bar{x})$ and square each deviation, $(x - \bar{x})^2$. These steps are shown in the table.

Data	Deviations	Squared Deviations
18	$18 - 23.29 = -5.29$	$(-5.29)^2 = 27.9841$
19	$19 - 23.29 = -4.29$	$(-4.29)^2 = 18.4041$
21	$21 - 23.29 = -2.29$	$(-2.29)^2 = 5.2441$
$\vdots$	$\vdots$	$\vdots$
30	$30 - 23.29 = 6.71$	$(6.71)^2 = 45.0241$
Total		113.4281

4. Sum the squared deviations. The sum is 113.4281.

5. Divide by $n - 1$.

$$\dfrac{113.4281}{7 - 1} \approx 18.905$$

6. Take the square root. $\sqrt{18.905} \approx 4.35$

40. $\bar{x} = \dfrac{3 + 4 + 6 + 8 + 10 + 12 + 15}{7} \approx 8.29$

To find the standard deviation:

1. The mean is 8.29.

2 and 3. Find each deviation from the mean $(x - \bar{x})$ and square each deviation, $(x - \bar{x})^2$. These steps are shown in the table.

Data	Deviations	Squared Deviations
3	$3 - 8.29 = -5.29$	$(-5.29)^2 = 27.9841$
4	$4 - 8.29 = -4.29$	$(-4.29)^2 = 18.4041$
6	$6 - 8.29 = -2.29$	$(-2.29)^2 = 5.2441$
$\vdots$	$\vdots$	$\vdots$
15	$15 - 8.29 = 6.71$	$(6.71)^2 = 45.0241$
Total		113.429

4. Sum the squared deviations. The sum is 113.429.

5. Divide by $n - 1$.

$$\dfrac{113.429}{7 - 1} \approx 18.905$$

6. Take the square root. $\sqrt{18.905} \approx 4.35$

41. Writing exercise

42. $\bar{x} = \dfrac{39 + 42 + 48 + 54 + 60 + 66 + 75}{7} \approx 54.86$

To find the standard deviation:

1. The mean is 54.86.

2 and 3. Find each deviation from the mean $(x - \bar{x})$ and square each deviation, $(x - \bar{x})^2$. These steps are shown in the table.

Data	Deviations	Squared Deviations
39	$39 - 54.86 = -15.86$	$(-15.86)^2 = 251.5396$
42	$42 - 54.86 = -12.86$	$(-12.86)^2 = 165.3796$
48	$48 - 54.86 = -6.86$	$(-6.86)^2 = 47.0596$
$\vdots$	$\vdots$	$\vdots$
75	$75 - 54.86 = 20.14$	$(20.14)^2 = 405.6196$
Total		1020.857

4. Sum the squared deviations. The sum is 1020.857.

5. Divide by $n - 1$.

$$\dfrac{1020.857}{7 - 1} \approx 170.143$$

6. Take the square root. $\sqrt{170.143} \approx 13.04$

43. Writing exercise

44. Writing exercise

45. (a) Test cities: $\bar{x} = \dfrac{18 + 15 + 7 + 10}{4} = \dfrac{50}{4} = 12.5$.

(b) Control cities:

$$\bar{x} = \dfrac{+1 + (-8) + (-5) + 0}{4} = \dfrac{-12}{4} = -3.0.$$

(c) To find the standard deviation for the test cities:

Steps 1, 2, and 3. Find each deviation from the mean for the test cities, $(x - \bar{x})$ and square each deviation, $(x - \bar{x})^2$. These steps are shown in the table.

Data	Deviations	Squared Deviations
18	$18 - 12.5 = 5.5$	$(5.5)^2 = 30.25$
15	$15 - 12.5 = 2.5$	$(2.5)^2 = 6.25$
7	$7 - 12.5 = -5.5$	$(-5.5)^2 = 30.25$
10	$10 - 12.5 = -2.5$	$(-2.5)^2 = 6.25$
Total		73

4. Sum the squared deviations. The sum is 73.

5. Divide by $n - 1$.

$$\dfrac{73}{4 - 1} \approx 24.333$$

6. Take the square root. $\sqrt{24.333} \approx 4.9$

(d) To find the standard deviation for the control cities:

Steps 1, 2, and 3. Find each deviation from the mean for the control cities, $(x - \bar{x})$ and square each deviation, $(x - \bar{x})^2$. These steps are shown in the table.

Data	Deviations	Squared Deviations
+1	$1 - (-3.0) = 4.0$	$(4.0)^2 = 16$
-8	$-8 - (-3.0) = -5.0$	$(-5.0)^2 = 25$
-5	$-5 - (-3.0) = -2.0$	$(-2.0)^2 = 4$
0	$0 - (-3.0) = 3.0$	$(3.0)^2 = 9$
Total		54

4. Sum the squared deviations. The sum is 54.

5. Divide by $n - 1$.
$$\frac{54}{4-1} \approx 18$$

6. Take the square root. $\sqrt{18} \approx 4.2$

(e) $12.5 - (-3.0) = 12.5 + 3.0 = 15.5$

(f) $15.5 - 7.95 = 7.55$; $15.5 + 7.95 = 23.45$

46. (a) The skewness coefficient will be positive when the mean is greater than the median, when the distribution is skewed to the right.

    (b) The skewness coefficient will be negative when the mean is less than the median, when the distribution is skewed to the left.

47. Writing exercise

48.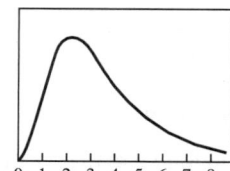

49. No, because the table gives only how many pieces of data occur within a given interval.

50. No, because the individual data items cannot be identified.

51. Writing exercise

52. Writing exercise

53. (a) The smallest item would be the mode because the highest peak of the curve is to the left.

    (b) The largest item would be the mode because the highest peak of the curve is to the right.

    (c) Writing exercise

13.4 EXERCISES

For each of Exercises 1–4, make use of z-scores.

1. The z-score for Chris:
$$z = \frac{5 - 4.6}{2.1} = 0.19.$$

   The z-score for Lynn:
$$z = \frac{6 - 4.6}{2.3} = 0.48.$$

   Lynn's score is greater than Chris's score, so her score is better.

2. All the units must be the same. Let's convert all units to feet. Then Ryan's jump is $6\frac{3}{12}$ or 6.25 ft; Michael's jump is $18\frac{4}{12} \approx 18.3$ ft. The mean is 6 ft, and the standard deviation is $\frac{3.5}{12} \approx .29$ ft for the high jump. The mean is $16\frac{6}{12} = 16.5$ ft and the standard deviation is $1\frac{10}{12} \approx 1.83$ ft for the long jump.

   The z-score for Ryan:
$$z = \frac{6.25 - 6.0}{.29} = 0.86.$$

   The z-score for Michael:
$$z = \frac{18.3 - 16.5}{1.83} \approx 1.00.$$

   Michael's score is greater than Ryan's score, so his score is better.

3. The z-score for Jutta's Brand A tires:
$$z = \frac{37{,}000 - 45{,}000}{4500} \approx -1.78.$$

   The z-score for Arvind's Brand B tires:
$$z = \frac{35{,}000 - 38{,}000}{2080} \approx -1.44.$$

   Arvind's score is greater than Jutta's score, so that score is better.

4. The z-score for Imelda's trout:
$$z = \frac{18 - 12}{2.75} \approx 2.18.$$

   The z-score for Timothy's bass:
$$z = \frac{6 - 4}{.8} \approx 2.50.$$

   Timothy's bass is the better trophy because his z-score is the largest.

5. Find 15% of 40 items: $.15(40) = 6$. Select the 7th item in the data set, which is 58.

6. Find 75% of 40 items: $.75(40) = 30$. Select the 31st item in the data set, which is 70.

7. The third decile is the same as 30%. Find 30% of 40 items: $.30(40) = 12$. Select the 13th item in the data set, which is 62.

8. The eighth decile is the same as 80%. Find 80% of 40 items: $.80(40) = 32$. Select the 33rd item in the data set, which is 71.

9. Find the mean by dividing the sum of scores in the Games Played column by 10, the number of players.

$$\bar{x} = \frac{763}{10} = 76.3$$

Find s by using the six step process described in Section 13.3 or by entering the data into a calculator or spreadsheet that the contains the function to calculate s. The standard deviation is approximately 4.35. Use the formula $z = \frac{x - \bar{x}}{s}$, where x is Mutombo's games played.

$$z = \frac{75 - 76.3}{4.35} \approx -.30$$

10. The sum of the Offensive Rebounds column is 2577.

$$\bar{x} = \frac{2577}{10} = 257.7$$

Find s by using the six step process described in Section 13.3 or by entering the data into a calculator or spreadsheet that the contains the function to calculate s. The standard deviation is approximately 42.2. Use the formula $z = \frac{x - \bar{x}}{s}$, where x is Shaquille O'Neal's offensive rebounds.

$$z = \frac{291 - 257.7}{42.2} \approx 0.79$$

11. The sum of the data in the Average Rebounds Per Game column is 117.1 and there are 10 players. Then

$$\bar{x} = \frac{117.1}{10} = 11.71.$$

Find s by using the six step process described in Section 13.3 or by entering the data into a calculator or spreadsheet that the contains the function to calculate s. The standard deviation is approximately 1.22. Use the formula $z = \frac{x - \bar{x}}{s}$, where x is Chris Webber's average rebounds per game.

$$z = \frac{11.1 - 11.71}{1.22} = -.50$$

12. The sum of the data in the Total Rebounds column is 8928 and there are 10 players. Then

$$\bar{x} = \frac{8928}{10} = 892.8.$$

Find s by using the six step process described in Section 13.3 or by entering the data into a calculator or spreadsheet that the contains the function to calculate s. The standard deviation is approximately 107.8. Use the formula $z = \frac{x - \bar{x}}{s}$, where x is Kevin Garnett's total rebounds.

$$z = \frac{921 - 892.8}{107.8} \approx .26$$

13. Find 35% of 10 items: $.35(10) = 3.5$. Select the 4th item in the data set for defensive rebounds after ranking scores: 461, 513, 598, 605, 628, 649, 702, 708, 738, 749. The 4th value is 605 defensive rebounds made by Antonio McDyess.

14. Find 73% of 10 items: $.73(10) = 7.3$. Select the 8th item after ranking scores in the total rebounds data set: 746, 777, 787, 845, 848, 921, 940, 997, 1017, and 1052. The 8th value is 997 total rebounds made by Tim Duncan.

15. The sixth decile is the same as the 60th percentile. Find 60% of 10 items: $.60(10) = 6$. Select the 7th item (remember we want 60% to be less than this number) after ranking scores in the games played data set: 70, 70, 74, 74, 75, 78, 79, 80, 81, and 82. The 7th value is 79 games played made by Shawn Marion.

16. The third decile is the same as the 30th percentile. Find 30% of 10 items: $.30(10) = 3$. Select the 4th item in the data set for average rebounds. They are already in order from largest to smallest. The 4th value from the bottom is 11.1 average rebounds made by Chris Webber.

17. See Exercise 10 for the values of $\bar{x}$, s and z-score. The z-score for Shaquille O'Neal is

$$z = \frac{291 - 257.7}{42.2} \approx 0.79.$$

Find the mean for Defensive Rebounds by summing the data and dividing by 10. The value of s is aproximately 94.9.

$$\bar{x} = \frac{6351}{10} = 635.1.$$

The z-score for Tim Duncan is

$$z = \frac{738 - 635.1}{94.9} \approx 1.08.$$

Tim Duncan was relatively higher because his z-score of 1.08 is greater than Shaquille O'Neal's z-score of 0.79.

18. The following values are needed in order to draw the box plot for average rebounds per game: Maximum value of 13.5 and minimum value of 10.1. Also the three quartiles must be identified.

Q_2 is the median, which is $\dfrac{11.4 + 12.1}{2} = 11.8$.

Q_1 is the median of the items below Q_2, which is 10.7.
Q_3 is the median of the items above Q_2, which is 12.7.

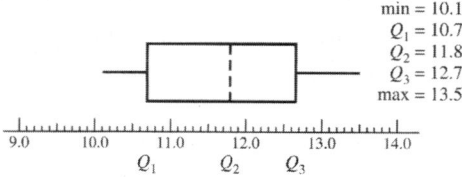

19. (a) The median is 11.8.

    (b) The range is $13.5 - 10.1 = 3.4$.

    (c) The location of the middle half of the items extend from 10.7 to 12.7.

20. Writing exercise
21. Writing exercise
22. Writing exercise
23. Writing exercise
24. Writing exercise
25. Writing exercise
26. Writing exercise
27. The "skewness coefficient" is a measure of the overall distribution.
28. If Jennifer scored at the 92nd percentile, she outscored 92% of those who took the test. Therefore, $100\% - 92\% = 8\%$ scored higher than Jennifer. Since 67,500 individuals took the test, find 8% of this number.

    $.08(67500) = 5400$ scored higher than Jennifer.

29. (a) No; this would only be true if Q_1 and Q_3 are symmetric about Q_2.

    (b) Writing exercise

30. Writing exercise
31. Writing exercise
32. Writing exercise
33. Writing exercise
34. Using the data for Omer, we have

$$z = \dfrac{x - \bar{x}}{s}$$
$$.69 = \dfrac{60 - \bar{x}}{s}$$
$$.69s = 60 - \bar{x}. \quad (1)$$

Using the data for Alessandro, we have

$$z = \dfrac{x - \bar{x}}{s}$$
$$1.67 = \dfrac{72 - \bar{x}}{s}$$
$$1.67s = 72 - \bar{x}. \quad (2)$$

Equations (1) and (2) form a system of equations.

$$.69s = 60 - \bar{x}$$
$$1.67s = 72 - \bar{x}$$

To solve this system by the elimination method, multiply equation (1) by -1 and add the result to equation (2).

$$-.69s = -60 + \bar{x}$$
$$1.67s = 72 - \bar{x}$$
$$\overline{.98s = 12}$$
$$s = 12.24$$

Substitute 12.24 in either of the equations to find $\bar{x}$.

$$.69(12.24) = 60 - \bar{x}$$
$$8.45 = 60 - \bar{x}$$
$$-51.55 = -\bar{x}$$
$$51.55 = \bar{x}$$

The mean is 51.55, and the standard deviation is 12.24.

35. Warner's season rating consists of the following:

Numerator terms	Substitute values	Simplify
$+\left(250 \times \dfrac{C}{A}\right)$	$=\left(250 \times \dfrac{375}{546}\right)$	≈ 171.70
$+\left(1000 \times \dfrac{T}{A}\right)$	$=\left(1000 \times \dfrac{36}{546}\right)$	≈ 65.93
$+\left(12.5 \times \dfrac{Y}{A}\right)$	$=\left(12.5 \times \dfrac{4830}{546}\right)$	≈ 110.58
$+\,6.25$	$+\,6.25$	$+\,6.25$
$-\left(1250 \times \dfrac{I}{A}\right)$	$-\left(1250 \times \dfrac{22}{546}\right)$	≈ -50.37
	TOTAL	≈ 304.09

$$\text{Rating} = \dfrac{\text{TOTAL}}{3} = \dfrac{304.09}{3} \approx 101.4$$

36. Gannon's season rating consists of the following:

Numerator terms	Substitute values	Simplify
$+\left(250 \times \dfrac{C}{A}\right)$	$=\left(250 \times \dfrac{361}{549}\right)$	≈ 164.40
$+\left(1000 \times \dfrac{T}{A}\right)$	$=\left(1000 \times \dfrac{27}{549}\right)$	≈ 49.18
$+\left(12.5 \times \dfrac{Y}{A}\right)$	$=\left(12.5 \times \dfrac{3828}{549}\right)$	≈ 87.16
$+\,6.25$	$+\,6.25$	$=\,+6.25$
$\left(1250 \times \dfrac{I}{A}\right)$	$=\left(1250 \times \dfrac{9}{549}\right)$	$\approx\,-20.49$
	TOTAL	≈ 286.5

$$\text{Rating} = \frac{\text{TOTAL}}{3} = \frac{286.5}{3} \approx 95.5$$

37. Q_2 is the median (mean average of two middle scores):
$$\frac{84.1 + 83.3}{2} = 83.7.$$

Q_1 is the median of the lower half (3rd of 5 scores from bottom), which is 77.7.

Q_3 is the median of the upper half (3rd of 5 scores from top), which is 94.1.

38. Q_2 is the median (mean average of two middle scores):
$$\frac{81.7 + 84.1}{2} = 82.9.$$

Q_1 is the median of the lower half, which is 78.5.

Q_3 is the median of the upper half, which is 86.5.

39. The eighth decile is the 80th percentile. First find 80% of 20 items: $.80(20) = 16$. Then locate the 17th item after arranging all items in order from smallest to largest: 75.3, 76.4, 77.8, 77.1, 77.7, 78.5, 79.6, 80.3, 81.7, 83.3, 84.1, 84.1, 84.1, 84.3, 86.5, 90.2, 94.1, 94.8, 95.5, 101.4. The eighth decile is 94.1.

40. Find 19% of 20 items: $.90(20) = 18$. Then locate the 19th item after arranging all items in order from smallest to largest: 75.3, 76.4, 77.8, 77.1, 77.7, 78.5, 79.6, 80.3, 81.7, 83.3, 84.1, 84.1, 84.1, 84.3, 86.5, 90.2, 94.1, 94.8, 95.5, 101.4. The ninetieth percentile is 95.5.

41.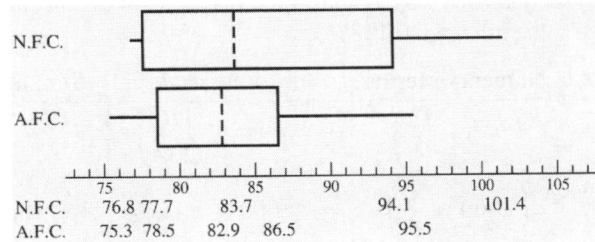

42. The median is greater for the N.F.C. $83.7 - 82.9 \approx$ one unit.

43. The midquartile is found by dividing the sum of Q_1 and Q_3 by 2. For the N.F.C.:
$$\frac{Q_1 + Q_3}{2} = \frac{77.7 + 94.1}{2} = 85.9.$$

For the A.F.C.:
$$\frac{Q_1 + Q_3}{2} = \frac{78.5 + 86.5}{2} = 82.5.$$

The midquartile for the N.F.C. is greater by about 3 points.

44. The range for the N.F.C. is
$$101.4 - 76.8 = 24.6,$$
while the range for the A.F.C. is
$$95.5 - 75.3 = 20.2.$$
The range for the N.F.C. is greater by about 4 units.

45. The interquartile range for the N.F.C. is
$$Q_3 - Q_1 = 94.1 - 77.7 = 16.4.$$
The interquartile range for the A.F.C. is
$$Q_3 - Q_1 = 86.5 - 78.5 = 8.$$
The interquartile range for the N.F.C. is greater by about 8 units.

46. Luckman's rating consists of the following values. In the second row, 28/202 must be replaced with the maximum value allowed, 0.11875.

Numerator terms	Substitute values	Simplify
$\left(250 \times \dfrac{C}{A}\right)$	$= \left(250 \times \dfrac{110}{202}\right)$	≈ 136.14
$\left(1000 \times \dfrac{T}{A}\right)$	$= (1000 \times .11875)$	≈ 118.75
$\left(12.5 \times \dfrac{Y}{A}\right)$	$= \left(12.5 \times \dfrac{2194}{202}\right)$	≈ 135.77
SUM		390.66
		$+\, 6.25$
$\left(1250 \times \dfrac{I}{A}\right)$	$= \left(1250 \times \dfrac{12}{202}\right)$	$-\, 74.26$
	TOTAL	$= 322.65$

$$\text{Rating} = \dfrac{\text{TOTAL}}{3} = \dfrac{322.65}{3} \approx 107.6 \quad \text{Note}$$

that answer is 107.5 using non-rounded numbers.

47. Repeat the process from Exercise 46, using 28/202 for T/A instead of replacing it with the lower value of 0.11875.

Numerator terms	Substitute values	Simplify
$\left(250 \times \dfrac{C}{A}\right)$	$= \left(250 \times \dfrac{110}{202}\right)$	≈ 136.14
$\left(1000 \times \dfrac{T}{A}\right)$	$= \left(1000 \times \dfrac{28}{202}\right)$	≈ 138.61
$\left(12.5 \times \dfrac{Y}{A}\right)$	$= \left(12.5 \times \dfrac{2194}{202}\right)$	≈ 135.77
SUM		410.52
		$+\, 6.25$
$\left(1250 \times \dfrac{I}{A}\right)$	$= \left(1250 \times \dfrac{12}{202}\right)$	$-\, 74.26$
	TOTAL	$= 342.51$

$$\text{Rating} = \dfrac{\text{TOTAL}}{3} = \dfrac{342.51}{3} \approx 114.2$$

48. Luckman had a high percentage of touchdown passes, about $\dfrac{28}{202} \approx 13.9\%$.

49. Writing exercise

13.5 EXERCISES

1. Discrete because the variable can take on only fixed number values such as 1, 2, 3, etc., up to and including 50.

2. Discrete

3. Continuous because the variable is not limited to fixed values. It is measurable rather than countable.

4. Continuous

5. Discrete because the variable is limited to fixed values.

6. Continuous

7. This represents all values to the right of the mean, which is 50% of the total number of values or 50% of 100 students: $.50(100) = 50$ students.

8. The value 85 is one standard deviation below the mean of 86. By the empirical rule there would be $\dfrac{.68(100)}{2} = 34$ of the students reporting between 85 grams and 86 grams. Half of the students report readings more than 86. Therefore, the number of students reporting more than 85 would be $34 + 50 = 84$.

9. These values are 1 standard deviation above and 1 standard deviation below the mean, respectively: (86 ± 1). By the empirical rule, 68% of all scores lie within 1 standard deviation of the mean. Then $.68(100) = 68$.

10. The value 84 is two standard deviations below the mean and 87 is one standard deviation above the mean. By the empirical rule, 95% of all scores lie within 2 standard deviations of the mean, so half of 95% will lie between 84 and 86: $.50(95) = 47.5\%$. Also by the empirical rule, 68% of all scores lie within 1 standard deviation of the mean. Then half of 68% will lie between 86 and 87: $.50(68) = 34\%$. The total percentage is $47.5 + 34 = 81.5\%$. Finally $.815(100) = 81.5$. Approximately 81 or 82 students report these values.

11. Less than 100 represents all values below the mean. This is 50% of the area under the curve or 50% of all the data.

12. The score 115 is one standard deviation above the mean. Find half of 68% and add that value to the 50% that lie below the mean: $34 + 50 = 84$. Subtract 84% from 100% to find the percentage of scores that are greater than 115: $100 - 84 = 16\%$.

13. The score 70 lies 2 standard deviations below the mean, and 130 lies 2 standard deviations above the mean. According to the empirical rule, 95% of the data lies within 2 standard deviations of the mean.

14. The score 145 is three standard deviations above the mean. Find half of 99.7% and add that value to the 50% that lie below the mean: $49.85 + 50 = 99.85\%$. Subtract 99.85% from 100% to find the percentage of scores that are greater than 145: $100 - 99.85 = .15\%$.

15. To find the percent of area between the mean and 1.5 standard deviations, use Table 10 to locate a z-score of 1.5. Read the value of A in the next column. This is the area from the mean to the corresponding value of z. The area is .433, which is 43.3%.

16. To find the percent of area between the mean and .92 standard deviations, use Table 10 to locate a z-score of .92. Read the value of A in the next column. This is the area from the mean to the corresponding value of z. The area is .321, which is 32.1%.

17. Because of the symmetry of the normal curve, the area between the mean and a z value of -1.08 is the same as that from the mean to a z value of $+1.08$. Use Table 10 to locate a z-score of 1.08. Read the value of A in the next column. This is the area from the mean to the corresponding value of z. The area is .360, which is 36.0%.

18. Because of the symmetry of the normal curve, the area between the mean and a z value of -2.25 is the same as that from the mean to a z value of $+2.25$. Use Table 10 to locate a z-score of 2.25. Read the value of A in the next column. This is the area from the mean to the corresponding value of z. The area is .488, which is 48.8%.

19. It is helpful to sketch the area under the normal curve.

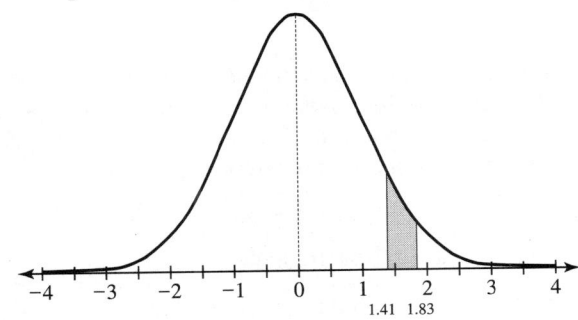

To find the area between $z = 1.41$ and $z = 1.83$, use the z-table to find the area between the mean and $z = 1.83$. The area is .466. Find the area between the mean and $z = 1.41$, which is .421. Subtracting, $.466 - .421 = .045$, which is 4.5%.

20. Use the symmetry of the normal curve to find the area of the positive z-scores. To find the area between $z = 1.74$ and $z = 1.14$, use the z-table to find the area between the mean and $z = 1.74$. The area is .459. Find the area between the mean and $z = 1.14$, which is .373. Subtracting, $.459 - .373 = .086$, which is 8.6%.

21. It is helpful to sketch the area under the normal curve.

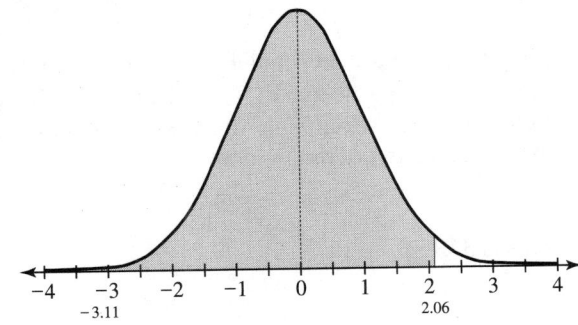

Add the areas under the curve from the mean to $z = -3.11$ to that from the mean to $z = 2.06$. The first area is the same as that from the mean to $z = +3.11$, which is .499. The area from the mean to $z = 2.06$ is .480. Find the total area: $.499 + .480 = .979$ or 97.9%.

22. Add the areas under the curve from the mean to $z = -1.98$ to that from the mean to $z = 1.02$. The first area is the same as that from the mean to $z = +1.98$, which is .476. The area from the mean to $z = 1.02$ is .346. Find the total area: $.476 + .346 = .822$ or 82.2%.

23. If 10% of the total area is to the right of the z-score, there is $50\% - 10\% = 40\%$ of the area between the mean and the value of z. From the z-table, find .40 in the A column. This area under the curve of 40% or .40 yields a z-score of 1.28.

24. If 4% of the total area is to the left of z, the z-score is below the mean, so the answer will be negative. (A sketch helps to see this.) Subtract 4% from 50% to find the amount of area between the mean and the value of z: $50\% - 4\% = 46\%$. Now find .46 in the A column to locate the appropriate value for z: 1.75. The z-score is -1.75.

25. If 9% of the total area is to the left of z, the z-score is below the mean, so the answer will be negative. (A sketch helps to see this.) Subtract 9% from 50% to find the amount of area between the mean and the value of z: $50\% - 9\% = 41\%$. Now find .41 in the A column to locate the appropriate value for z: 1.34. The z-score is -1.34.

26. If 23% of the total area is to the right of the z-score, there is $50 - 23 = 27\%$ of the area between the mean and the value of z. From the z-table, find .27 in the A column. This area under the curve of 27% or .27 yields a z-score of .74

27. Since the mean is 600 hr, we can expect half of the bulbs or 5000 bulbs to last at least 600 hrs.

28. First find the z-score for 675:
$$z = \frac{675 - 600}{50} = 1.5.$$

Then find the amount of area between the mean and $z = 1.5$ from Table 10. The amount is .433 or 43.3%. Finally find 43.3% of 10,000, the number of lightbulbs.
$$.433(10000) = 4330$$

29. Find the z-score for 675:

$$z = \frac{675 - 600}{50} = 1.5.$$

Find the z-score for 740:

$$z = \frac{740 - 600}{50} = 2.8.$$

It is helpful to sketch the area under the normal curve.

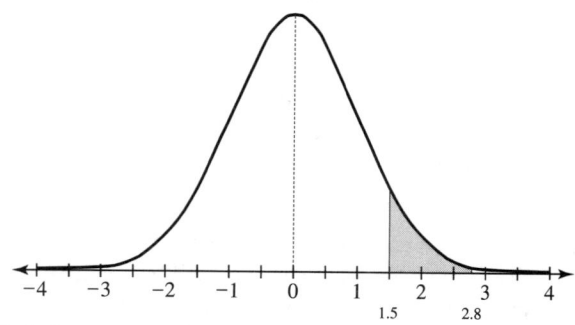

Find the amount of area under the normal curve between 1.5 and 2.8 by finding the corresponding values for A and then subtracting the smaller from the larger:

$$.497 - .433 = .064.$$

Finally find 64% of 10,000:

$$.064(10000) = 640.$$

30. Find the z-score for 490:

$$z = \frac{490 - 600}{50} = -2.2.$$

Find the z-score for 720:

$$z = \frac{720 - 600}{50} = 2.4.$$

Find the amount of area under the normal curve between -2.2 and 2.4 by finding the corresponding values for A and then adding the values:

$$.486 + .492 = .978.$$

Finally find 97.8% of 10,000:

$$.978(10000) = 9780.$$

31. Find the z-score for 740:

$$z = \frac{740 - 600}{50} = 2.8.$$

It is helpful to sketch the area under the normal curve.

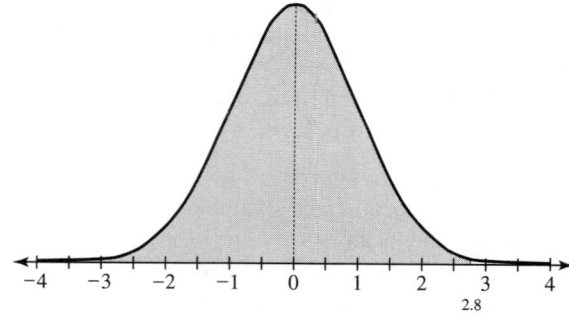

Find the amount of area under the normal curve between the mean and 2.8 by finding the corresponding value for A: .497. Add this area value to .5:

$$.497 + .5 = .997.$$

Finally find 99.7% of 10,000:

$$.997(10000) = 9970.$$

32. Find the z-score for 510:

$$z = \frac{510 - 600}{50} = -1.8.$$

Find the amount of area under the normal curve between the mean and the z-score of -1.8; the area is .464. Subtract this area from .5 to find the amount of area less than or to the left of this z-score:

$$.5 - .464 = .036.$$

Finally find 3.6% of 10,000:

$$.036(10000) = 360.$$

33. Because the mean is 1850 and the standard deviation is 150, the value 1700 corresponds to $z = -1$. Use the empirical rule to evaluate the corresponding area under the curve. The area between $z = -1$ and the mean is half of 68% or 34%. The area to the right of the mean is 50%. The total area to the right of $z = -1$ is $34 + 50 = 84\%$. If the z-table is used, the answer is 84.1%, because the area value in the table for $z = 1$ is .341.

34. Find the z-score for 1800:

$$z = \frac{1800 - 1850}{150} = -0.\overline{3}.$$

Find the amount of area under the normal curve between the mean and the z-score of $-0.\overline{3}$; the area is .129. Subtract this area from .5 to find the amount of area less than or to the left of this z-score:

$$.5 - .129 = .371.$$

This is 37.1%.

35. Find the z-score for 1750:
$$z = \frac{1750 - 1850}{150} \approx -0.67.$$

Find the z-score for 1900:
$$z = \frac{1900 - 1850}{150} = 0.33.$$

It is helpful to sketch the area under the normal curve.

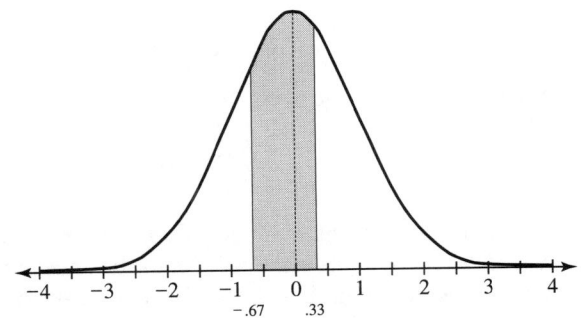

Find the amount of area under the normal curve between -0.67 and 0.33 by finding the corresponding values for Area and then adding the values:
$$.249 + .129 = .378.$$

This is 37.8%.

36. Find the z-score for 1600:
$$z = \frac{1600 - 1850}{150} \approx -1.67.$$

Find the z-score for 2000:
$$z = \frac{2000 - 1850}{150} = 1.00.$$

Find the amount of area under the normal curve between -1.67 and 1.00 by finding the corresponding values for Area and then adding the values:
$$.453 + .341 = .794.$$

This is 79.4%.

37. The z-score corresponding to 24 oz when $s = .5$ is found by
$$z = \frac{24 - 24.5}{.5} = -1.$$

The fraction of boxes that are underweight is equivalent to the area under the curve to the left of -1.

It is helpful to sketch the area under the normal curve.

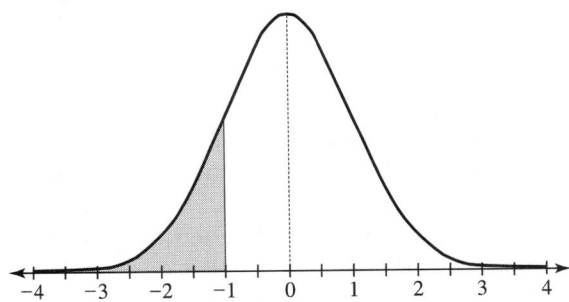

From the z-table, the area under the curve between the mean and $+1$ is .341. Subtract .341 from .5 to obtain the area under the curve to the right of $+1$: $.5 - .341 = .159$. Because of symmetry, this is also the amount of area under the curve to the left of -1. The answer is .159 or 15.9%.

38. The z-score corresponding to 24 oz when $s = .4$ is found by
$$z = \frac{24 - 24.5}{.4} = -1.25.$$

The fraction of boxes that are underweight is equivalent to the area under the curve to the left of -1.25. From the z-table, the area under the curve between the mean and $+1.25$ is .394. Subtract .394 from .5 to obtain the area under the curve to the right of $+1.25$: $.5 - .394 = .106$. Because of symmetry, this is also the amount of area under the curve to the left of -1.25. The answer is .106 or 10.6%.

39. The z-score corresponding to 24 oz when $s = .3$ is found by
$$z = \frac{24 - 24.5}{.3} \approx -1.67.$$

The fraction of boxes that are underweight is equivalent to the area under the curve to the left of -1.67.

It is helpful to sketch the area under the normal curve.

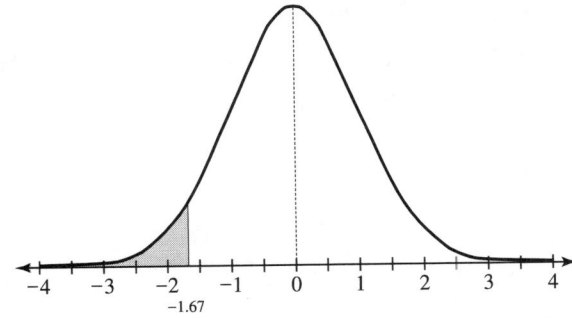

From the z-table, the area under the curve between the mean and $+1.67$ is .453. Subtract .453 from .5 to obtain the area under the curve to the right of $+1.67$: $.5 - .453 = .047$. Because of symmetry, this is also the amount of area under the curve to the left of -1.67. The answer is .047 or 4.7%.

40. The z-score corresponding to 24 oz when $s = .2$ is found by
$$z = \frac{24 - 24.5}{.2} \approx -2.50.$$

   The fraction of boxes that are underweight is equivalent to the area under the curve to the left of -2.50. From the z-table, the area under the curve between the mean and $+2.50$ is .494. Subtract .494 from .5 to obtain the area under the curve to the right of $+2.50$: $.5 - .494 = .006$. Because of symmetry this is also the amount of area under the curve to the left of -2.50. The answer is .006 or 0.6%.

41. The mean plus 2.5 times the standard deviation, $\bar{x} + 2.5s$, corresponds to $z = 2.5$ no matter what the values of $\bar{x}$ and s are:
$$z = \frac{\text{value} - \text{mean}}{\text{standard deviation}}$$
$$= \frac{(\bar{x} + 2.5s) - \bar{x}}{s}$$
$$= \frac{2.5s}{s}$$
$$= 2.5.$$

   The fraction of the population between the mean and $z = 2.5$ is .494, by the z-table. The fraction below the mean is .5. The sum of these is $.5 + .494 = .994$ or 99.4%.

42. The RDA is the value corresponding to $z = 2.5$. Use the z-score formula to find x:
$$z = \frac{x - \bar{x}}{s}$$
$$2.5 = \frac{x - 1800}{140}$$
$$(140) \cdot 2.5 = \frac{x - 1800}{140} \cdot \frac{140}{1}$$
$$350 = x - 1800$$
$$2150 = x.$$

   The RDA is 2150 units.

43. The RDA is the value corresponding to $z = 2.5$. Use the z-score formula to find x:
$$z = \frac{x - \bar{x}}{s}$$
$$2.5 = \frac{x - 159}{12}$$
$$(12) \cdot 2.5 = \frac{x - 159}{12} \cdot \frac{12}{1}$$
$$30 = x - 159$$
$$189 = x.$$

   The RDA is 189 units.

44. Find the z-score for 32:
$$z = \frac{32 - 32.2}{1.2} \approx -0.17.$$

   Find the area between the mean and a z-score of -0.17 in the table: 0.067. Subtract this value from 0.5 to find the amount of area to the left of the z value: $.5 - .067 = .433$.

45. Find the z-score for 7:
$$z = \frac{7 - 7.47}{3.6} \approx -0.13.$$

   Find the z-score for 8:
$$z = \frac{8 - 7.47}{3.6} \approx 0.15.$$

   The area under the curve to the left of $-.13$ and the area under the curve to the right of $.15$ must be added to answer the question. Use the z-table to obtain .052 for the area between the mean and $+.13$. The area to the right of .13 is then $.5 - .052 = .448$. Because of symmetry, this is the amount of area to the left of -1.3. Use the z-table to obtain .060 for the area between the mean and .15. The area to the right of .15 is then $.5 - .060 = .440$. The sum of these two areas is $.448 + .440 = 0.888$.

46. Find the z-score for 18.0:
$$z = \frac{18.0 - 12.3}{4.1} \approx 1.39.$$

   Find the area under the normal curve between the mean and 1.39, which is .418. Subtract .418 from .5 to obtain the area under the curve to the right of $z = 1.39$: $.5 - .418 = .082$.

47. Find the z-score for 2.2:
$$z = \frac{2.2 - 1.5}{.4} = 1.75.$$

   Find the area under the normal curve between the mean and 1.75, which is .460. Subtract .460 from .5 to obtain the area under the curve to the right of $z = 1.75$: $.5 - .460 = .040$ or 4.0%. Find 4.0% of five dozen: $(.04)(5 \cdot 12) = 2.4$. The answer is between 2 and 3 eggs.

48. Find the area as a percent for $\bar{x} + (3/2)s$. This is the area under the curve to the right of $z = 3/2$ or 1.5. The area between the mean and a z-score of 1.5 is .433. Subtract this value from .5: $.5 - .433 = .067$ or 6.7%.

49. Find the area as a percent between $\bar{x} + (1/2)s$ and $\bar{x} + (3/2)s$. This is the area between $z = 1/2$ or .5 and $z = 3/2$ or 1.5. The area under the curve from the mean to $z = .5$ is .191; the area under the curve from the mean to $z = 1.5$ is .433. Subtract: $.433 - .191 = .242$ or 24.2%.

50. Find the area as a percent between $\bar{x} - (1/2)s$ and $\bar{x} + (1/2)s$. This is the area between $z = -1/2$ or $-.5$ and $z = 1/2$ or $+.5$. The area under the curve from the mean to $z = .5$ is $.191$; the area under the curve from the mean to $z = -.5$ is the same. Add these two areas or multiply by 2: $.191 + .191 = .382$ or 38.2%.

51. Writing exercise

52. To find the cutoff grade, subtract 8% from 50% to obtain the percentage or amount of area between the mean and the z-score: $.50 - .08 = .42$. Find z from this known area value. The value of z is between 1.40 and 1.41. Use 1.405. Then use the z-score formula to find x:

$$z = \frac{x - \bar{x}}{s}$$
$$1.405 = \frac{x - 75}{5}$$
$$(5) \cdot (1.405) = \frac{x - 75}{5} \cdot \frac{5}{1}$$
$$7.025 = x - 75$$
$$82.025 = x.$$

Rounded to the nearest whole number, the cutoff score should be 82.

53. To find the bottom cutoff grade, $.15 + .08 = .23$ must be the amount of area under the normal curve to the right of the grade. That means that the amount of area between the mean and this cutoff grade is $.5 - .23 = .27$. Locate this area value in the A column; it corresponds to a z-score of $.74$. Then use the z-score formula to find x:

$$z = \frac{x - \bar{x}}{s}$$
$$.74 = \frac{x - 75}{5}$$
$$(5) \cdot (.74) = \frac{x - 75}{5} \cdot \frac{5}{1}$$
$$3.7 = x - 75$$
$$78.7 = x.$$

Rounded to the nearest whole number, the cutoff score should be 79.

54. If all other students get a grade of C, then find the sum of all the percentages for the other grades and subtract from 100%: $100 - (8 + 8 + 15 + 15) = 100 - 46 = 54\%$. The middle 54% will receive a grade of C. Divide 54% in half because half of the area will be on either side of the mean: $.54 \div 2 = .27$. The cutoff score on the low end will be negative. Locate this area value in the A column; it corresponds to a z-score of $.74$. Replace z with $-.74$ and solve for x:

$$z = \frac{x - \bar{x}}{s}$$
$$-.74 = \frac{x - 75}{5}$$
$$(5) \cdot (-.74) = \frac{x - 75}{5} \cdot \frac{5}{1}$$
$$-3.7 = x - 75$$
$$71.3 = x.$$

Rounded to the nearest whole number, the score should be 71.

55. To find the bottom cutoff grade, $.08$ must be the amount of area under the normal curve to the left of the grade. That means that the amount of area between the mean and this cutoff grade is $.5 - .08 = .42$. Locate this area value in the A column; it corresponds to a z-score of 1.40 or 1.41. Use 1.405. However, the negative z-score must be used because the grade is below the mean.

$$z = \frac{x - \bar{x}}{s}$$
$$-1.405 = \frac{x - 75}{5}$$
$$(5) \cdot (-1.405) = \frac{x - 75}{5} \cdot \frac{5}{1}$$
$$-7.025 = x - 75$$
$$68 \approx x$$

The cutoff grade should be 68.

56. Replace $\bar{x}$, s and z in the z-score formula and solve for x:

$$z = \frac{x - \bar{x}}{s}$$
$$.72 = \frac{x - 76.8}{9.42}$$
$$(9.42) \cdot (.72) = \frac{x - 76.8}{9.42} \cdot \frac{9.42}{1}$$
$$6.7824 = x - 76.8$$
$$83.6 \approx x.$$

57. Replace $\bar{x}$, s and z in the z-score formula and solve for x:

$$z = \frac{x - \bar{x}}{s}$$
$$1.44 = \frac{x - 76.8}{9.42}$$
$$(9.42) \cdot (1.44) = \frac{x - 76.8}{9.42} \cdot \frac{9.42}{1}$$
$$13.5648 = x - 76.8$$
$$90.4 \approx x.$$

58. Replace $\bar{x}$, s and z in the z-score formula and solve for x:

$$z = \frac{x - \bar{x}}{s}$$
$$-2.39 = \frac{x - 76.8}{9.42}$$
$$(9.42) \cdot (-2.39) = \frac{x - 76.8}{9.42} \cdot \frac{9.42}{1}$$
$$-22.5138 = x - 76.8$$
$$54.3 \approx x.$$

59. Replace $\bar{x}$, s and z in the z-score formula and solve for x:

$$z = \frac{x - \bar{x}}{s}$$
$$-3.87 = \frac{x - 76.8}{9.42}$$
$$(9.42) \cdot (-3.87) = \frac{x - 76.8}{9.42} \cdot \frac{9.42}{1}$$
$$-36.4554 = x - 76.8$$
$$40.3 \approx x.$$

60. (a) By Chebyshev's theorem, at least $1 - 1/k^2$ will lie within k standard deviations of the mean.

$$1 - \frac{1}{(1.25)^2} = 1 - \frac{1}{1.5625} = 1 - .64 = .36$$

This is 36%.

(b) Using Table 10 find the area between the mean and $z = 1.25$. The area is .394. Then $2(.394) = .788$ or 78.8%.

61. Writing exercise

EXTENSION: HOW TO LIE WITH STATISTICS

1. There is no explanation of what the solid line represents.

2. We can't tell because there is no scale.

3. The dashed line rises and then falls.

4. We can't tell because there is no scale.

5. Percent of decrease $= \dfrac{\text{old value} - \text{new value}}{\text{old value}}$

$$\frac{2.40 - 1.72}{2.40} \approx .28 \text{ or } 28\%.$$

6. The actual decrease is about 28%, but the artist for the magazine reduced *each* dimension of the original figure by 28%. This causes the area to decrease by about 50%, thus giving a false impression.

7. How long have Toyotas been sold in the United States? How do other makes compare?

8. The tobacco is fresher than what?

9. The dentists preferred Trident Sugarless Gum to what? Which and how many dentists were surveyed? What percentage responded?

10. A Volvo is a smaller car than a Continental and should have a smaller turning radius.

11. How quiet is a glider?

12. There is no scale. We can't tell if the increase is substantial or not.

13. The maps convey their impressions in terms of *area* distributions, whereas personal income distributions may be quite different. The map on the left probably implies too high a level of government spending, while that on the right implies too low a level.

14. It turns out that there were *three* women students, and just *one* had married a faculty member.

15. By the time the figures were used, circumstances may have changed greatly. (The Navy was much larger.) Also, New York City was most likely not typical of the nation as a whole.

16. When Huff's book was written, those three states had the best system for reporting these diseases.

17. (b) Change the fractions to percents and compare:

$$\frac{6}{50} = 12\% \quad \frac{15}{50} = 30\% \quad \frac{29}{50} = 58\%.$$

These values are very close to the overall population values.

18. (c) Change the fractions to percents and compare:

$$\frac{28}{80} = 35\% \quad \frac{23}{80} \approx 29\% \quad \frac{16}{80} = 20\% \quad \frac{13}{80} \approx 16\%.$$

These values are very close to the overall population values.

19. (c) Change the fractions to percents and compare:

$$\frac{50}{120} \approx 42\% \quad \frac{31}{120} \approx 26\% \quad \frac{39}{120} \approx 33\%.$$

These values are very close to the overall population values.

20. (a) Change the fractions to percents and compare:

$$\frac{41}{75} \approx 55\% \quad \frac{21}{75} = 28\% \quad \frac{13}{75} \approx 17\%.$$

These values are very close to the overall population values.

21. (a) The first equation implies that $m = c - 2$. Replace m and a in the following equation and solve for c:

$$10 = m + a + c$$
$$10 = (c - 2) + 2c + c$$
$$10 = 4c - 2$$
$$12 = 4c$$
$$c = 3.$$

Then $m = 3 - 2 = 1$ and $a = 2 \cdot 3 = 6$.

(b) The total number of staff is $7 + 25 + 18 = 50$. The fractions of the entire office staff are: managers, 7/50; agents, 25/50; and clerical, 18/50. Now use the number in the sample as the denominator to solve for each variable using the technique of solving proportions from Section 7.3.

$$\frac{m}{10} = \frac{7}{50}$$
$$50 \cdot m = 7 \cdot 10$$
$$m = \frac{70}{50} \text{ or } 1.4 \text{ managers}$$

$$\frac{a}{10} = \frac{25}{50}$$
$$50 \cdot a = 25 \cdot 10$$
$$a = \frac{250}{50} \text{ or } 5 \text{ agents}$$

$$\frac{c}{10} = \frac{18}{50}$$
$$50 \cdot c = 18 \cdot 10$$
$$c = \frac{180}{50} \text{ or } 3.6 \text{ clerical}$$

22. (a) Substitute $2d$ for f in the second equation and solve for a in terms of d.

$$a = f + d + 1$$
$$a = 2d + d + 1$$
$$a = 3d + 1$$

Because there are 2 deans, substitute 2 for d to find a numerical value for a.

$$a = 3d + 1$$
$$a = 3 \cdot 2 + 1$$
$$a = 7$$

So $d = 2$, $a = 7$, $s = 7$, and $f = 2 \cdot 2 = 4$.

(b) The total number of staff is $12 + 24 + 39 + 45 = 120$. The fractions of the entire office staff are: deans, 12/120; associate professors, 39/120; assistant professors, 45/120; and full professors, 24/120. Now use the number in the sample as the denominator to solve for each variable using the technique of solving proportions from Section 7.3.

$$\frac{d}{20} = \frac{12}{120}$$
$$120 \cdot d = 20 \cdot 12$$
$$d = \frac{240}{120} \text{ or } 2 \text{ deans}$$

$$\frac{a}{20} = \frac{39}{120}$$
$$120 \cdot a = 20 \cdot 39$$
$$a = \frac{780}{120} \text{ or } 6.5 \text{ associate professors}$$

$$\frac{f}{20} = \frac{24}{120}$$
$$120 \cdot f = 20 \cdot 24$$
$$f = \frac{480}{120} \text{ or } 4 \text{ full professors}$$

$$\frac{s}{20} = \frac{45}{120}$$
$$120 \cdot s = 20 \cdot 45$$
$$s = \frac{900}{120} \text{ or } 7.5 \text{ assistant professors}$$

13.6 EXERCISES

1. The equation of the least squares line is $y' = ax + b$ where a and b are found as follows:

$$a = \frac{n(\Sigma xy) - (\Sigma x)(\Sigma y)}{n(\Sigma x^2) - (\Sigma x)^2}$$
$$= \frac{10(75) - (30)(24)}{10(100) - (30)^2}$$
$$= \frac{750 - 720}{1000 - 900}$$
$$= \frac{30}{100}$$
$$= .3, \text{ and}$$

$$b = \frac{\Sigma y - a(\Sigma x)}{n}$$
$$= \frac{24 - .3(30)}{10}$$
$$= \frac{24 - 9}{10}$$
$$= 1.5.$$

Then the equation for the least squares line is

$$y' = .3x + 1.5.$$

2. $$r = \frac{n(\Sigma xy) - (\Sigma x)(\Sigma y)}{\sqrt{n(\Sigma x^2) - (\Sigma x)^2} \cdot \sqrt{n(\Sigma y^2) - (\Sigma y)^2}}$$
$$= \frac{10(75) - (30)(24)}{\sqrt{10(100) - (30)^2} \cdot \sqrt{10(80) - (24)^2}}$$
$$= \frac{750 - 720}{\sqrt{1000 - 900} \cdot \sqrt{800 - 576}}$$
$$= \frac{30}{\sqrt{100} \cdot \sqrt{224}}$$
$$\approx \frac{30}{10 \cdot (14.97)}$$
$$\approx .20$$

3. The regression equation is $y' = .3x + 1.5$. Find y' when $x = 3$ tons.
$$y' = .3(3) + 1.5$$
$$= .9 + 1.5$$
$$= 2.4 \text{ decimeters}$$

4. The equation of the least squares line is $y' = ax + b$ where a and b are found as follows:
$$a = \frac{n(\Sigma xy) - (\Sigma x)(\Sigma y)}{n(\Sigma x^2) - (\Sigma x)^2}$$
$$= \frac{5(28050) - (376)(120)}{5(62522) - (376)^2}$$
$$= \frac{140250 - 45120}{312610 - 141376}$$
$$= \frac{95130}{171234}$$
$$\approx .556, \text{ and}$$
$$b = \frac{\Sigma y - a(\Sigma x)}{n}$$
$$= \frac{120 - .556(376)}{5}$$
$$= \frac{120 - 209.056}{5}$$
$$= -17.8.$$

Then the equation for the least squares line is
$$y' = .556x - 17.8.$$

5. The regression equation is $y' = .556x - 17.8$. Find y' when $x = 120°$.
$$y' = .556(120) - 17.8$$
$$= 66.72 - 17.8$$
$$= 48.92°$$

6. $$r = \frac{n(\Sigma xy) - (\Sigma x)(\Sigma y)}{\sqrt{n(\Sigma x^2) - (\Sigma x)^2} \cdot \sqrt{n(\Sigma y^2) - (\Sigma y)^2}}$$
$$= \frac{5(28050) - (376)(120)}{\sqrt{5(62522) - (376)^2} \cdot \sqrt{5(13450) - (120)^2}}$$
$$= \frac{140250 - 45120}{\sqrt{312610 - 141376} \cdot \sqrt{67250 - 14400}}$$
$$= \frac{95130}{\sqrt{171234} \cdot \sqrt{52850}}$$
$$= \frac{95130}{95130} = 1$$

7. The table shows how to calculate all the sums that are needed in the formula for the least squares line.

x	y	x^2	y^2	xy
62	120	3844	14400	7440
62	140	3844	19600	8680
63	130	3969	16900	8190
65	150	4225	22500	9750
66	142	4356	20164	9372
67	130	4489	16900	8710
68	135	4624	18225	9180
68	175	4624	30625	11900
70	149	4900	22201	10430
72	168	5184	28224	12096
$\Sigma x = 663$	$\Sigma y = 1439$	$\Sigma x^2 = 44059$	$\Sigma y^2 = 209739$	$\Sigma xy = 95748$

The equation of the least squares line is $y' = ax + b$ where a and b are found as follows:
$$a = \frac{n(\Sigma xy) - (\Sigma x)(\Sigma y)}{n(\Sigma x^2) - (\Sigma x)^2}$$
$$= \frac{10(95748) - (663)(1439)}{10(44059) - (663)^2}$$
$$= \frac{957480 - 954057}{440590 - 439569}$$
$$= \frac{3423}{1021}$$
$$\approx 3.35, \text{ and}$$
$$b = \frac{\Sigma y - a(\Sigma x)}{n}$$
$$= \frac{1439 - 3.35(663)}{10}$$
$$= \frac{1439 - 2221.05}{10}$$
$$= -78.2.$$

The value for b calculated here differs slightly from the one in the text because of rounding error. It is highly recommended that a scientific calculator or spreadsheet be used, because then rounding is only done at the very end of all the calculations. Here the value of a was rounded before using it in the calculation for b. Then the equation for the least squares line is
$$y' = 3.35x - 78.2.$$

The value for b in the text is -78.4.

13.6 REGRESSION AND CORRELATION

8. The regression equation is $y' = 3.35x - 78.4$. Find y' when $x = 60$ in.

$$y' = 3.35(60) - 78.4$$
$$= 201 - 78.4$$
$$\approx 123 \text{ pounds}$$

9. The regression equation is $y' = 3.35x - 78.4$. Find y' when $x = 70$ in.

$$y' = 3.35(70) - 78.4$$
$$= 234.5 - 78.4$$
$$\approx 156 \text{ pounds}$$

10. Substitute the sums from Exercise 7 into the formula for the coefficient of correlation.

$$r = \frac{n(\Sigma xy) - (\Sigma x)(\Sigma y)}{\sqrt{n(\Sigma x^2) - (\Sigma x)^2} \cdot \sqrt{n(\Sigma y^2) - (\Sigma y)^2}}$$
$$= \frac{10(95748) - (663)(1439)}{\sqrt{10(44059) - (663)^2} \cdot \sqrt{10(209739) - (1439)^2}}$$
$$= \frac{957480 - 954057}{\sqrt{440590 - 439569} \cdot \sqrt{2097390 - 2070721}}$$
$$= \frac{3423}{\sqrt{1021} \cdot \sqrt{26669}}$$
$$\approx \frac{3423}{31.95 \cdot 163.31}$$
$$\approx .66$$

11.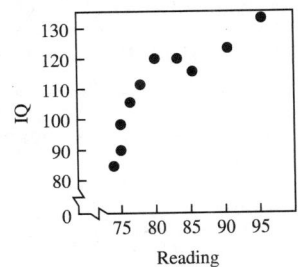

12. Use a scientific calculator or a spreadsheet to calculate the various sums needed in the formula. Actually, a calculator or spreadsheet can calculate the appropriate values for a and b needed for the regression line equation.

$\Sigma x = 811$; $\Sigma y = 1104$; $\Sigma x^2 = 66225$; $\Sigma y^2 = 124060$; $\Sigma xy = 90437$.

The equation of the least squares line is $y' = ax + b$ where a and b are found as follows:

$$a = \frac{n(\Sigma xy) - (\Sigma x)(\Sigma y)}{n(\Sigma x^2) - (\Sigma x)^2}$$
$$= \frac{10(90437) - (811)(1104)}{10(66225) - (811)^2}$$
$$= \frac{9026}{4529}$$
$$\approx 1.99, \text{ and}$$

$$b = \frac{\Sigma y - a(\Sigma x)}{n}$$
$$= \frac{1104 - 1.99(811)}{10}$$
$$= -51.$$

Then the equation for the regression line, rounding a and b to the nearest whole number, is

$$y' = 2x - 51.$$

13. The regression equation is $y' = 2x - 51$. Find y' when $x = 65$.

$$y' = 2(65) - 51$$
$$= 130 - 51$$
$$= 79$$

14. Use a scientific calculator or a spreadsheet to calculate the various sums needed in the formula. Actually, a calculator or spreadsheet can calculate the appropriate values for a and b needed for the regression line equation.

$\Sigma x = 15$; $\Sigma y = 418$; $\Sigma x^2 = 55$; $\Sigma y^2 = 30266$; $\Sigma xy = 1186$.

The equation of the least squares line is $y' = ax + b$ where

$$a = \frac{n(\Sigma xy) - (\Sigma x)(\Sigma y)}{n(\Sigma x^2) - (\Sigma x)^2}$$
$$= \frac{6(1186) - (15)(418)}{6(55) - (15)^2}$$
$$= \frac{7116 - 6270}{330 - 225}$$
$$= \frac{846}{105}$$
$$\approx 8.06, \text{ and}$$

$$b = \frac{\Sigma y - a(\Sigma x)}{n}$$
$$= \frac{418 - 8.06(15)}{6}$$
$$= \frac{418 - 120.9}{6}$$
$$\approx 49.52.$$

Then the equation for the least squares line

$$y' = 8.06x + 49.52.$$

15. Use a scientific calculator or a spreadsheet to calculate the various sums needed in the formula, or see Exercise 7 for the detailed process. Actually, a calculator or spreadsheet can calculate the value of r.

$\Sigma x = 15$; $\Sigma y = 418$; $\Sigma x^2 = 55$; $\Sigma y^2 = 30266$; $\Sigma xy = 1186$; and $n = 6$.

Substitute the sums into the formula for the coefficient of correlation.

$$r = \frac{n(\Sigma xy) - (\Sigma x)(\Sigma y)}{\sqrt{n(\Sigma x^2) - (\Sigma x)^2} \cdot \sqrt{n(\Sigma y^2) - (\Sigma y)^2}}$$

$$= \frac{6(1186) - (15)(418)}{\sqrt{6(55) - (15)^2} \cdot \sqrt{6(30266) - (418)^2}}$$

$$= \frac{7116 - 6270}{\sqrt{330 - 225} \cdot \sqrt{181596 - 174724}}$$

$$= \frac{846}{\sqrt{105} \cdot \sqrt{6872}}$$

$$\approx \frac{846}{10.25 \cdot 82.90}$$

$$\approx .996$$

16. The regression equation is $y' = 8.06x + 49.52$. Find y' when $x = 7$.

$$y' = 8.06(7) + 49.52$$
$$= 56.42 + 49.52$$
$$= 105.94$$

Because $y =$ sales in thousands of dollars, the sales amount would be about \$106,000.

17.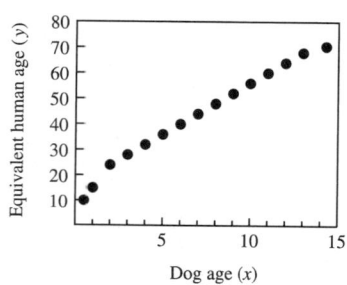

18. Use a scientific calculator or a spreadsheet to calculate the various sums needed in the formula. Actually, a calculator or spreadsheet can calculate the appropriate values for a and b needed for the regression line equation.

$\Sigma x = 105.5$; $\Sigma y = 647.5$; $\Sigma x^2 = 1015.25$; $\Sigma y^2 = 32975.25$; $\Sigma xy = 5719$ and with $n = 15$.

The equation of the least squares line is $y' = ax + b$ where

$$a = \frac{n(\Sigma xy) - (\Sigma x)(\Sigma y)}{n(\Sigma x^2) - (\Sigma x)^2}$$

$$= \frac{15(5719) - (105.5)(647.5)}{15(1015.25) - (105.5)^2}$$

$$= \frac{85785 - 68311.25}{15228.75 - 11130.25}$$

$$= \frac{17473.75}{4098.5}$$

$$\approx 4.26, \text{ and}$$

$$b = \frac{\Sigma y - a(\Sigma x)}{n}$$
$$= \frac{647.5 - 4.26(105.5)}{15}$$
$$= \frac{647.5 - 449.43}{15}$$
$$\approx 13.20.$$

Again, it is recommended that a scientific calculator or spreadsheet be used to calculate these values in order to avoid rounding error. (The value of a has been rounded here.) The value for b that is calculated in the text is 13.18. The equation for the least squares line, then, is

$$y' = 4.26x + 13.18.$$

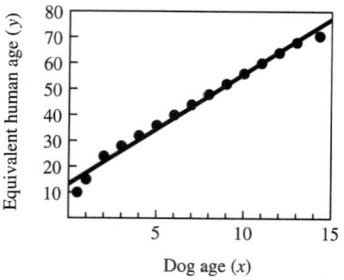

19. Writing exercise

20. Use a scientific calculator or a spreadsheet to calculate the various sums needed in the formula, or see Exercise 7 for the detailed process. Actually, a calculator or spreadsheet can calculate the value of r.

$\Sigma x = 105.5$; $\Sigma y = 647.5$; $\Sigma x^2 = 1015.25$; $\Sigma y^2 = 32975.25$; $\Sigma xy = 5719$; and $n = 15$.

Substitute the sums into the formula for the coefficient of correlation.

$$r = \frac{n(\Sigma xy) - (\Sigma x)(\Sigma y)}{\sqrt{n(\Sigma x^2) - (\Sigma x)^2} \cdot \sqrt{n(\Sigma y^2) - (\Sigma y)^2}}$$

$$= \frac{15(5719) - (105.5)(647.5)}{\sqrt{15(1015.25) - (105.5)^2} \cdot \sqrt{15(32975.25) - (647.5)^2}}$$

$$= \frac{85785 - 68311.25}{\sqrt{15228.75 - 11130.25} \cdot \sqrt{494628.75 - 419256.25}}$$

$$= \frac{17473.75}{\sqrt{4098.5} \cdot \sqrt{75372.5}}$$

$$\approx \frac{17473.75}{64.020 \cdot 274.541}$$

$$\approx .99$$

21. Use a scientific calculator or a spreadsheet to calculate the various sums needed in the formula. Actually, a calculator or spreadsheet can calculate the appropriate values for a and b needed for the regression line equation. See Exercise 7 for the details of calculating the following sums.

$\Sigma x = 56$; $\Sigma y = 77.7$; $\Sigma x^2 = 560$; $\Sigma y^2 = 1110.43$; $\Sigma xy = 786.4$ and with $n = 8$.

The equation of the least squares line is $y' = ax + b$ where

$$a = \frac{n(\Sigma xy) - (\Sigma x)(\Sigma y)}{n(\Sigma x^2) - (\Sigma x)^2}$$
$$= \frac{8(786.4) - (56)(77.7)}{8(560) - (56)^2}$$
$$= \frac{6291.2 - 4351.2}{4480 - 3136}$$
$$= \frac{1940}{1344}$$
$$\approx 1.44, \text{ and}$$

$$b = \frac{\Sigma y - a(\Sigma x)}{n}$$
$$= \frac{77.7 - 1.44(56)}{8}$$
$$= \frac{77.7 - 80.64}{8}$$
$$\approx -.37.$$

Again, it is recommended that a scientific calculator or spreadsheet be used to calculate these values in order to avoid rounding error. (The value of a has been rounded here.) The value for b that is calculated in the text is $-.39$. Then the equation for the least squares line

$$y' = 1.44x - .39.$$

22. Use the values from Exercise 21.

$\Sigma x = 56; \Sigma y = 77.7; \Sigma x^2 = 560; \Sigma y^2 = 1110.43; \Sigma xy = 786.4;$ and $n = 8$.

Substitute the sums into the formula for the coefficient of correlation.

$$r = \frac{n(\Sigma xy) - (\Sigma x)(\Sigma y)}{\sqrt{n(\Sigma x^2) - (\Sigma x)^2} \cdot \sqrt{n(\Sigma y^2) - (\Sigma y)^2}}$$
$$= \frac{8(786.4) - (56)(77.7)}{\sqrt{8(560) - (56)^2} \cdot \sqrt{8(1110.43) - (77.7)^2}}$$
$$= \vdots$$
$$\approx .99$$

23. A correlation of .99 is very strong.

24. The value of x is found by subtracting 1850 from 2010, and dividing by 10 to find the number of decades.

$$x = \frac{2010 - 1850}{10} = 16$$

The regression equation is $y' = 1.44x - .39$. Find y' when $x = 16$.

$$y' = 1.44(16) - .39$$
$$= 23.04 - .39$$
$$\approx 22.7\%.$$

25. Use a scientific calculator or a spreadsheet to calculate the various sums needed in the formulas. Actually, a calculator or spreadsheet can calculate the appropriate values for a and b needed for the regression line equation. See Exercise 7 for the details of calculating the following sums.

$\Sigma x = 152, 273; \Sigma y = 1357; \Sigma x^2 = 2, 868, 146, 155; \Sigma y^2 = 194, 069; \Sigma xy = 21, 747, 588$ and with $n = 10$.

The equation of the least squares line is $y' = ax + b$ where

$$a = \frac{n(\Sigma xy) - (\Sigma x)(\Sigma y)}{n(\Sigma x^2) - (\Sigma x)^2}$$
$$= \frac{10(21747588) - (152273)(1357)}{10(2868146155) - (152273)^2}$$
$$= \vdots$$
$$\approx .00197, \text{ and}$$

$$b = \frac{\Sigma y - a(\Sigma x)}{n}$$
$$= \frac{1357 - .00197(152273)}{10}$$
$$\approx 105.7.$$

Again, it is recommended that a scientific calculator or spread sheet be used to calculate these values in order to avoid rounding error. Then the equation for the regression line

$$y' = .00197x + 105.7.$$

26. Use the values from Exercise 25.

$\Sigma x = 152, 273; \Sigma y = 1357; \Sigma x^2 = 2, 868, 146, 155; \Sigma y^2 = 194, 069; \Sigma xy = 21, 747, 588;$ and $n = 10$.

Substitute the sums into the formula for the coefficient of correlation.

$$r = \frac{n(\Sigma xy) - (\Sigma x)(\Sigma y)}{\sqrt{n(\Sigma x^2) - (\Sigma x)^2} \cdot \sqrt{n(\Sigma y^2) - (\Sigma y)^2}}$$
$$= \frac{10(21747588) - (152273)(1357)}{\sqrt{10(2868146155) - (152273)^2} \cdot \sqrt{10(194, 069) - (1357)^2}}$$
$$\approx \vdots$$
$$\approx .46$$

27. The linear correlation is moderate.

28. The values of x are in thousands. Converting 15 million to thousands yields $x = 15, 000$ thousand. The regression equation from Exercise 25 is $y' = .00197x + 105.7$. Find y' when $x = 15, 000$.

$$y' = .00197(15000) + 105.7$$
$$= 29.55 + 105.7$$
$$= 135.25$$

This is $135.25 thousand, which is approximately $135,000.

CHAPTER 13 TEST

1. Examine the first graph to see that in 1880 about 57.1% of total U.S. workers were in farm occupations. In 1900 about 37.5% were in farm occupations. Somewhere between 1880 and 1900, about 1890, the percentage would have been 50%.

2. No, the percentage was 4 times as great, not the number of workers.

3. Examine the line graph. Find 1970 along the horizontal axis and the value of about 3 million along the vertical axis.

4. Use the bottom two graphs to answer this question. Find the number of farms (in millions) in the graph on the left for each year: about 6.3 million farms in 1940 and about 2.1 million farms in 2000. Multiply each of these values by the size of the average farm; the sizes of the average farm are found in the bar graph on the right: 174 acres for 1940 and 434 acres for 2000.

$$6.3(174) = 1096.2 \text{ and } 2.1(434) = 911.4$$

Find the percentage:

$$\frac{1096.2 - 911.4}{1096.2} \times 100\% \approx 17\%.$$

5. A tally column is not shown here, but it is useful when creating a frequency distribution by hand.

Class Limits	Frequency f	Relative frequency f/n
6–10	3	$3/22 \approx .14$
11–15	6	$6/22 \approx .27$
16–20	7	$7/22 \approx .32$
21–25	4	$4/22 \approx .18$
26–30	2	$2/22 \approx .09$

6. (a)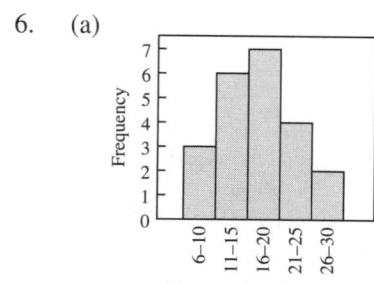

(b)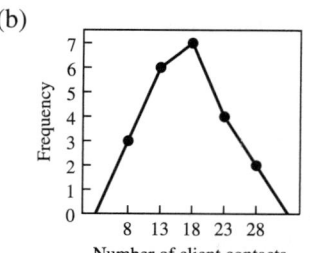

7. If the first class had limits of 7–9, this would be a width of 3. The smallest value is 8, and the largest value is 30. All the data must be included within the classes that are constructed. The classes would be:

7–9
10–12
13–15
16–18
19–21
22–24
25–27
28–30.

This is a total of 8 classes.

8.
Value	Frequency	Value · Frequency
8	3	24
10	8	80
12	10	120
14	8	112
16	5	80
18	1	18
Totals	35	434

Then $\bar{x} = \dfrac{434}{35} = 12.4$.

9. There are 35 items. The formula for the position of the median is

$$\frac{\Sigma f + 1}{2} = \frac{35 + 1}{2} = 18.$$

This means that the median is 18th item when the values are arranged from smallest to largest. A chart showing cumulative frequency shows that the value is in the row for a cumulative frequency of 21.

Value	Frequency	Cumulative Frequency
8	3	3
10	8	11
12	10	21
14	8	29
16	5	34
18	1	35

The value 12 is the median.

10. Examine the frequency column to see that the mode is 12, with a frequency of 10.

11. The range of the data is $18 - 8 = 10$.

12. The leaves are arranged in rank order (from smallest to largest).

    ```
    3 | 3 8
    4 | 3 5 8 9
    5 | 0 2 5
    6 | 1 1 4 5 6 7 7 8
    7 | 0 1 2 3 7 7 8 9 9
    8 | 0 4 4
    9 | 1
    ```

13. There are 33 items. The formula for the position of the median is
    $$\frac{\Sigma f + 1}{2} = \frac{33 + 1}{2} = 17.$$
    This means that the median is the 17th item when the values are arranged from smallest to largest. Count the values in the stem-and-leaf display in the text to see that the 17th value is 35.

14. The mode is 33; it occurs 3 times.

15. The range of the data is $60 - 23 = 37$.

16. The third decile is the same as the 30th percentile. First,
    $$.30(33) = 9.9.$$
    Then the 10th value is the location of the third decile. This value is 31.

17. The eighty-fifth percentile is located by first taking 85% of the 33 data items:
    $$.85(33) = 28.05.$$
    Then the 29th value is located at the 85th percentile. This value is 49.

18. The values shown in the boxplot are: Minimum value, 23; Maximum value, 60; Q_1, 29.5; Q_2, 35, and Q_3, 43.

    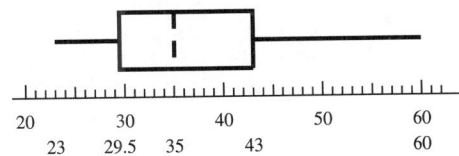

19. The scores 70 and 90 are each two standard deviations away from the mean, because the mean is 80 and the value of one standard deviation is 5. According to the empirical rule, approximately 95% of the data lie within two standard deviations of the mean.

20. The score of 95 is three standard deviations above the mean; the score of 65 is three standard deviations below the mean. According to the empirical rule, approximately 99.7% of the data lie within this interval. Therefore, $100 - 99.7 = .3\%$ lie outside this interval.

21. The score of 75 is one standard deviation below the mean. Using the empirical rule, if 68% of the scores lie within one standard deviation of the mean, then 34% of the scores lie between 75 and 80. Subtract 34% from 50% to find the percentage of scores that are less than 75: $50 - 34 = 16\%$.

22. The score of 85 is one standard deviation above the mean; the score of 90 is two standard deviations above the mean. From Exercise 21 we know that about 34% of the scores lie between 80 and 85. Also from the empirical rule, approximately 95% of the scores lie within two standard deviations. Half of 95% is 47.5%. Subtract 34% from 47.5% to find the percentage of scores between 85 and 90: $47.5 - 34 = 13.5\%$.

23. First find the z-score for 6.5:
    $$z = \frac{6.5 - 5.5}{2.1} \approx .48.$$
    It is always helpful to sketch the area under the normal curve that is being sought. Then find the amount of area between the mean and $z = .48$ from Table 10 in Section 13.5. The amount is .184. This area must be added to .5, the amount of area under the normal curve below the mean:
    $$.184 + .5 = .684$$

24. First find the z-scores for 6.2 and 9.4:
    $$z = \frac{6.2 - 5.5}{2.1} \approx .33$$
    $$z = \frac{9.4 - 5.5}{2.1} \approx 1.86.$$
    It is always helpful to sketch the area under the normal curve that is being sought.

    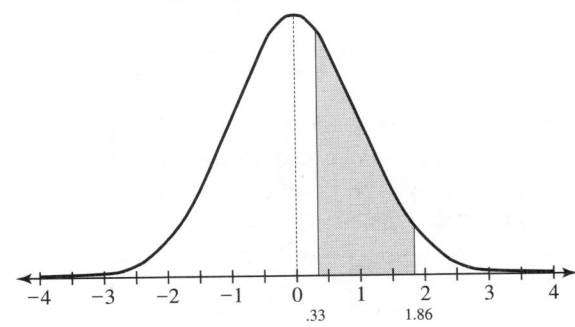

We are trying to find the amount of area under the curve between a z-score of .33 and a z-score of 1.86. Find the amount of area between the mean and $z = 1.86$ from Table 10 in Section 13.5. The amount is .469. Find the amount of area between the mean and $z = .33$; it is .129. The smaller value must be subtracted from the larger.

$$.469 - .129 = .340$$

25. Find the means of the three groups.

$$\text{Eastern teams} \quad \frac{76.4 + 80.0}{2} = 78.2$$

$$\text{Central teams} \quad \frac{78.0 + 78.3}{2} = 78.15$$

$$\text{Western teams} \quad \frac{91.5 + 78.3}{2} = 84.9$$

The West teams had the highest winning average.

26. Find the means of the standard deviations.

$$\text{Eastern teams} \quad \frac{13.9 + 8.1}{2} = 11$$

$$\text{Central teams} \quad \frac{11.8 + 14.5}{2} = 13.15$$

$$\text{Western teams} \quad \frac{21.0 + 14.5}{2} = 17.75$$

The East teams were the most "consistent" because their standard deviation (variation) is the smallest.

27. The average number of games won for all 30 teams is

$$\frac{5(76.4) + 5(78.0) + 4(91.5) + 5(80.0) + 6(78.3) + 5(78.3)}{5 + 5 + 4 + 5 + 6 + 5}$$

$$= \frac{2399.3}{30}$$

$$\approx 80.0.$$

28.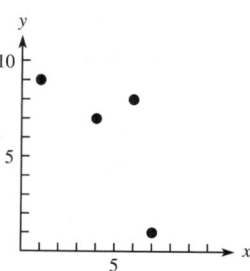

29. It is more efficient to use a calculator or a spreadsheet to find the various sums that are needed in the formulas for a and b. However, if these calculations are done by hand, make a table as follows:

x	y	x^2	y^2	xy
1	9	1	81	9
4	7	16	49	28
6	8	36	64	48
7	1	49	1	7
$\Sigma x = 18$	$\Sigma y = 25$	$\Sigma x^2 = 102$	$\Sigma y^2 = 195$	$\Sigma xy = 92$

The equation of the least squares line is $y' = ax + b$, where a and b are found as follows:

$$a = \frac{n(\Sigma xy) - (\Sigma x)(\Sigma y)}{n(\Sigma x^2) - (\Sigma x)^2}$$

$$= \frac{4(92) - (18)(25)}{4(102) - (18)^2}$$

$$= \frac{368 - 450}{408 - 324}$$

$$= \frac{-82}{84}$$

$$\approx -.97619, \text{ and}$$

$$b = \frac{\Sigma y - a(\Sigma x)}{n}$$

$$= \frac{25 - (-.97619)(18)}{4}$$

$$\approx \frac{25 + 17.5714}{4}$$

$$= 10.64.$$

Then the equation for the least squares line, with a and b rounded to the nearest hundredth, is

$$y' = -.98x + 10.64.$$

30. The regression equation is $y' = -.98x + 10.64$. Find y' when $x = 3$.

$$y' = -.98(3) + 10.64$$
$$= -2.94 + 10.64$$
$$= 7.70.$$

31. $r = \dfrac{n(\Sigma xy) - (\Sigma x)(\Sigma y)}{\sqrt{n(\Sigma x^2) - (\Sigma x)^2} \cdot \sqrt{n(\Sigma y^2) - (\Sigma y)^2}}$

$$= \frac{4(92) - (18)(25)}{\sqrt{4(102) - (18)^2} \cdot \sqrt{4(195) - (25)^2}}$$

$$= \frac{368 - 450}{\sqrt{408 - 324} \cdot \sqrt{780 - 625}}$$

$$= \frac{-82}{\sqrt{84} \cdot \sqrt{155}}$$

$$= \frac{-82}{\sqrt{13020}}$$

$$\approx -.72$$

32. Writing exercise

33. Writing exercise

14.1 EXERCISES

Note: Except where it is stated that a table has been used, exercises in this chapter have been completed with a calculator or spreadsheet. To ensure as much accuracy as possible, rounded values have been avoided in intermediate steps as much as possible. When rounded values have been necessary, several decimal places have been carried throughout the exercise; only the final answer has been rounded to 1 or 2 decimal places. In most cases, answers involving money are rounded to the nearest cent.

1. Use the formula $I = Prt$ with $P = \$1400$; $r = .08$; and $t = 1$ yr.
$$I = Prt$$
$$= (\$1400)(.08)(1)$$
$$= \$112$$

2. Use the formula $I = Prt$ with $P = \$8000$; $r = .05$; and $t = 1$ yr.
$$I = Prt$$
$$= (\$8000)(.05)(1)$$
$$= \$400$$

3. Use the formula $I = Prt$ with $P = \$650$; $r = .06$; and $t = 9/12$ yr. Remember that t must be expressed in years.
$$I = Prt$$
$$= (\$650)(.06)\left(\frac{9}{12}\right)$$
$$= \$29.25$$

4. Use the formula $I = Prt$ with $P = \$6000$; $r = .07$; and $t = 4/12$ yr.
$$I = Prt$$
$$= (\$6000)(.07)\left(\frac{4}{12}\right)$$
$$= \$140$$

5. Use the formula $I = Prt$ with $P = \$2675$; $r = .082$; and $t = 2\,1/2$ or 2.5 yr. Remember that t must be expressed in years.
$$I = Prt$$
$$= (\$2675)(.082)(2.5)$$
$$\approx \$548.38$$

6. Use the formula $I = Prt$ with $P = \$1460$; $r = .0782$; and $t = 22/12$ yr. Remember that t must be expressed in years.

$$I = Prt$$
$$= (\$1460)(.0782)\left(\frac{22}{12}\right)$$
$$\approx \$209.32$$

7. (a) Use the formula $A = P(1 + rt)$, with $P = \$700$; $r = .03$; and $t = 6$.
$$A = P(1 + rt)$$
$$= \$700(1 + .03 \cdot 6)$$
$$= \$700(1 + .18)$$
$$= \$700(1.18)$$
$$= \$826$$

   (b) Use the formula $A = P\left(1 + \frac{r}{m}\right)^n$, with $m =$ number of periods per year and $n =$ total number of periods.
$$A = P\left(1 + \frac{r}{m}\right)^n$$
$$= \$700\left(1 + \frac{.03}{1}\right)^6$$
$$= \$700(1.03)^6$$
$$= \$700(1.194052297)$$
$$\approx \$835.84$$

8. (a) Use the formula $A = P(1 + rt)$, with $P = \$2000$; $r = .04$; and $t = 5$.
$$A = P(1 + rt)$$
$$= \$2000(1 + .04 \cdot 5)$$
$$= \$2000(1 + .2)$$
$$= \$2000(1.2)$$
$$= \$2400$$

   (b) Use the formula $A = P\left(1 + \frac{r}{m}\right)^n$, with $m =$ number of periods per year and $n =$ total number of periods.
$$A = P\left(1 + \frac{r}{m}\right)^n$$
$$= \$2000\left(1 + \frac{.04}{1}\right)^5$$
$$= \$2000(1.04)^5$$
$$= \$2000(1.216652902)$$
$$\approx \$2433.31$$

9. (a) Use the formula $A = P(1 + rt)$, with $P = \$2500$; $r = .02$; and $t = 3$.
$$A = P(1 + rt)$$
$$= \$2500(1 + .02 \cdot 3)$$
$$= \$2500(1 + .06)$$
$$= \$2500(1.06)$$
$$= \$2650$$

(b) Use the formula $A = P\left(1 + \frac{r}{m}\right)^n$, with $m = $ number of periods per year and $n = $ total number of periods.

$$A = P\left(1 + \frac{r}{m}\right)^n$$
$$= \$2500\left(1 + \frac{.02}{1}\right)^3$$
$$= \$2500(1.02)^3$$
$$= \$2500(1.061208)$$
$$\approx \$2653.02$$

10. (a) Use the formula $A = P(1 + rt)$, with $P = \$5000$; $r = .08$; and $t = 5$ yr.

$$A = P(1 + rt)$$
$$= \$5000(1 + .08 \cdot 5)$$
$$= \$5000(1 + .40)$$
$$= \$5000(1.40)$$
$$\approx \$7000$$

(b) Use the formula $A = P\left(1 + \frac{r}{m}\right)^n$, with $m = $ number of periods per year and $n = $ total number of periods.

$$A = P\left(1 + \frac{r}{m}\right)^n$$
$$= \$5000\left(1 + \frac{.08}{1}\right)^5$$
$$= \$5000(1.08)^5$$
$$= \$5000(1.469328)$$
$$\approx \$7346.64$$

11. Use the formula $I = Prt$ with $P = \$7500$; $r = .08$; and $t = 4/12$ yr. Remember that t must be expressed in years.

$$I = Prt$$
$$= \$(7500)(.08)\left(\frac{4}{12}\right)$$
$$= \$200$$

12. Use the formula $I = Prt$ with $P = \$815$; $r = .09$; and $t = 5/12$ yr. Remember that t must be expressed in years.

$$I = Prt$$
$$= \$(815)(.09)\left(\frac{5}{12}\right)$$
$$\approx \$30.56$$

13. First use the formula $I = Prt$ with $P = \$14,800$; $r = .09$; and $t = 10/12$. Remember that t must be expressed in years.

$$I = Prt$$
$$= \$(14800)(.09)\left(\frac{10}{12}\right)$$
$$= \$1110$$

Now the interest amount must be added to the loan amount to find the total amount Chris must repay.

$$\$1110 + \$14800 = \$15,910$$

14. First use the formula $I = Prt$ with $P = \$260$; $r = .06$; and $t = 8/12$ yr. Remember that t must be expressed in years.

$$I = Prt$$
$$= \$(260)(.06)\left(\frac{8}{12}\right)$$
$$= \$10.40$$

Now the interest amount must be added to the loan amount to find the total amount he will receive.

$$\$10.40 + \$260 = \$270.40$$

15. To find the compound interest that was earned, subtract the principal from the final amount.

$$\$1143.26 - \$975 = \$168.26$$

16. To find the compound interest that was earned, subtract the principal from the final amount.

$$\$1737.73 - \$1150 = \$587.73$$

17. To find the final amount add the principal to the compound interest.

$$\$480 + \$337.17 = \$817.17.$$

18. To find the final amount, use the formula for future value $A = P\left(1 + \frac{r}{m}\right)^n$, with $P = \$2370$, $r = .10$, $m = 4$ and $n = 4 \cdot 5 = 20$.

$$A = P\left(1 + \frac{r}{m}\right)^n$$
$$= \$2370\left(1 + \frac{.10}{4}\right)^{20}$$
$$= \$2370(1.025)^{20}$$
$$= \$2370(1.63861)$$
$$\approx \$3883.52$$

Then to find the compound interest, subtract the principal from this final amount.

$$\$3883.52 - \$2370 = \$1513.52$$

19. To find the final amount, use the formula for future value $A = P\left(1 + \frac{r}{m}\right)^n$, with $P = \$7500$, $r = .035$, $m = 1$ and $n = 1 \cdot 25 = 25$.

$$A = P\left(1+\frac{r}{m}\right)^n$$
$$= \$7500\left(1+\frac{.035}{1}\right)^{25}$$
$$= \$7500(1.035)^{25}$$
$$= \$7500(2.363244984)$$
$$\approx \$17,724.34$$

Then to find the compound interest, subtract the principal from this final amount.
$$\$17724.34 - \$7500 = \$10224.34$$

20. To find the final amount, use the formula for future value $A = P\left(1+\frac{r}{m}\right)^n$, with $P = \$3450$, $r = .024$, $m = 2$ and $n = 2 \cdot 10 = 20$.
$$A = P\left(1+\frac{r}{m}\right)^n$$
$$= \$3450\left(1+\frac{.024}{2}\right)^{20}$$
$$= \$3450(1.012)^{20}$$
$$= \$3450(1.269434362)$$
$$\approx \$4379.55$$

Then to find the compound interest, subtract the principal from this final amount.
$$\$4379.55 - \$3450 = \$929.55$$

21. To find the final amount, use the formula for future value $A = P\left(1+\frac{r}{m}\right)^n$.

(a) $A = P\left(1+\frac{r}{m}\right)^n$
$$= \$1000\left(1+\frac{.10}{1}\right)^{3}$$
$$= \$1000(1.10)^3$$
$$= \$1000(1.331)$$
$$\approx \$1331$$

(b) $A = P\left(1+\frac{r}{m}\right)^n$
$$= \$1000\left(1+\frac{.10}{2}\right)^{2\cdot 3}$$
$$= \$1000(1.05)^6$$
$$= \$1000(1.340095641)$$
$$= \$1340.10$$

(c) $A = P\left(1+\frac{r}{m}\right)^n$
$$= \$1000\left(1+\frac{.10}{4}\right)^{4\cdot 3}$$
$$= \$1000(1.025)^{12}$$
$$= \$1000(1.34488824)$$
$$\approx \$1344.89$$

22. (a) $A = P\left(1+\frac{r}{m}\right)^n$
$$= \$3000\left(1+\frac{.05}{1}\right)^{7}$$
$$= \$3000(1.05)^7$$
$$= \$3000(1.407100423)$$
$$\approx \$4221.30$$

(b) $A = P\left(1+\frac{r}{m}\right)^n$
$$= \$3000\left(1+\frac{.05}{2}\right)^{2\cdot 7}$$
$$= \$3000(1.025)^{14}$$
$$= \$3000(1.4129738)$$
$$\approx \$4238.92$$

(c) $A = P\left(1+\frac{r}{m}\right)^n$
$$= \$3000\left(1+\frac{.05}{4}\right)^{4\cdot 7}$$
$$= \$3000(1.0125)^{28}$$
$$= \$3000(1.41599)$$
$$\approx \$4247.98$$

23. To find the final amount, use the formula for future value $A = P\left(1+\frac{r}{m}\right)^n$.

(a) $A = P\left(1+\frac{r}{m}\right)^n$
$$= \$12000\left(1+\frac{.08}{1}\right)^{5}$$
$$= \$12000(1.08)^5$$
$$= \$12000(1.469328077)$$
$$\approx \$17,631.94$$

(b) $A = P\left(1+\frac{r}{m}\right)^n$
$$= \$12000\left(1+\frac{.08}{2}\right)^{2\cdot 5}$$
$$= \$12000(1.04)^{10}$$
$$= \$12000(1.480244)$$
$$\approx \$17,762.93$$

(c) $A = P\left(1 + \dfrac{r}{m}\right)^n$

$= \$12000\left(1 + \dfrac{.08}{4}\right)^{4 \cdot 5}$

$= \$12000(1.02)^{20}$

$= \$12000(1.485947)$

$\approx \$17,831.37$

24. (a) $A = P\left(1 + \dfrac{r}{m}\right)^n$

$= \$15000\left(1 + \dfrac{.07}{1}\right)^9$

$= \$15000(1.07)^9$

$= \$15000(1.838459)$

$\approx \$27,576.89$

(b) $A = P\left(1 + \dfrac{r}{m}\right)^n$

$= \$15000\left(1 + \dfrac{.07}{2}\right)^{2 \cdot 9}$

$= \$15000(1.035)^{18}$

$= \$15000(1.857489)$

$\approx \$27,862.34$

(c) $A = P\left(1 + \dfrac{r}{m}\right)^n$

$= \$15000\left(1 + \dfrac{.07}{4}\right)^{4 \cdot 9}$

$= \$15000(1.0175)^{36}$

$= \$15000(1.867407)$

$\approx \$28,011.11$

25. (a) First find the future value by using the formula $A = P\left(1 + \dfrac{r}{m}\right)^n$ with $P = \$1040$, $r = .076$, $m = 2$, and $n = 2 \cdot 4 = 8$.

$A = P\left(1 + \dfrac{r}{m}\right)^n$

$= \$1040\left(1 + \dfrac{.076}{2}\right)^8$

$= \$1040(1.038)^8$

$= \$1040(1.347655)$

$\approx \$1401.56$

Then subtract the principal from this amount.

$\$1401.56 - \$1040 = \$361.56$

(b) Find the future value, changing to $m = 4$, and $n = 4 \cdot 4 = 16$.

$A = P\left(1 + \dfrac{r}{m}\right)^n$

$= \$1040\left(1 + \dfrac{.076}{4}\right)^{16}$

$= \$1040(1.019)^{16}$

$= \$1040(1.351409)$

$\approx \$1405.47$

Then subtract the principal from this amount

$\$1405.47 - \$1040 = \$365.47.$

(c) Find the future, changing to $m = 12$, and $n = 12 \cdot 4 = 48$.

$A = P\left(1 + \dfrac{r}{m}\right)^n$

$= \$1040\left(1 + \dfrac{.076}{12}\right)^{48}$

$= \$1040(1.00 \overline{63})^{48}$

$= \$1040(1.35397)$

$\approx \$1408.13$

Then subtract the principal from this amount.

$\$1408.13 - \$1040 = \$368.13$

(d) Find the future, changing to $m = 365$, and $n = 365 \cdot 4 = 1460$.

$A = P\left(1 + \dfrac{r}{m}\right)^n$

$= \$1040\left(1 + \dfrac{.076}{365}\right)^{1460}$

$= \$1040(1.000208219)^{1460}$

$= \$1040(1.355226)$

$\approx \$1409.44$

Then subtract the principal from this amount.

$\$1409.44 - \$1040 = \$369.44$

(e) Use the formula for continuous growth $A = Pe^{rt}$ with $P = \$1040$; $r = .076$; and $t = 4$

$A = Pe^{rt}$

$= \$1040 \cdot e^{.076 \cdot 4}$

$= \$1040 \cdot e^{.304}$

$= \$1040(1.3551269056)$

$\approx \$1409.48$

Then subtract the principal from this amount.

$\$1409.48 - \$1040 = \$369.48$

26. (a) First find the future value by using the formula $A = P\left(1 + \frac{r}{m}\right)^n$ with $P = \$1550$, $r = .028$, $m = 2$, and $n = 2 \cdot \frac{33}{12} = 5.5$.

$$A = P\left(1 + \frac{r}{m}\right)^n$$
$$= \$1550\left(1 + \frac{.028}{2}\right)^{5.5}$$
$$= \$1550(1.014)^{5.5}$$
$$= \$1550(1.079465465)$$
$$\approx \$1673.17$$

Then subtract the principal from this amount.

$$\$1673.17 - \$1550 = \$123.17$$

(b) Find the future value, changing to $m = 4$, and $n = 4 \cdot \frac{33}{12} = 11$.

$$A = P\left(1 + \frac{r}{m}\right)^n$$
$$= \$1550\left(1 + \frac{.028}{4}\right)^{11}$$
$$= \$1550(1.007)^{11}$$
$$= \$1550(1.079752395)$$
$$\approx \$1673.62$$

Then subtract the principal from this amount.

$$\$1673.62 - \$1550 = \$123.62$$

(c) Find the future value, changing to $m = 12$, and $n = 12 \cdot \frac{33}{12} = 33$.

$$A = P\left(1 + \frac{r}{m}\right)^n$$
$$= \$1550\left(1 + \frac{.028}{12}\right)^{33}$$
$$= \$1550\left(1.002\overline{3}\right)^{33}$$
$$= \$1550(1.079945208)$$
$$\approx \$1673.92$$

Then subtract the principal from this amount.

$$\$1673.92 - \$1550 = \$123.92$$

(d) Find the future value, changing to $m = 365$, and $n = 1003$.

$$A = P\left(1 + \frac{r}{m}\right)^n$$
$$= \$1550\left(1 + \frac{.028}{365}\right)^{1003}$$
$$= \$1550(1.000076712)^{1003}$$
$$= \$1550(1.079976752)$$
$$\approx \$1673.96$$

Then subtract the principal from this amount.

$$\$1673.96 - \$1550 = \$123.96$$

(e) Use the formula for continuous growth $A = Pe^{rt}$ with $P = \$1550$; $r = .098$; and $t = \frac{1003}{365} = 2.74795$ yr.

$$A = Pe^{rt}$$
$$= \$1550 \cdot e^{(.028)2.74795}$$
$$= \$1550(1.079980084)$$
$$\approx \$1673.97$$

Then subtract the principal from this amount.

$$\$1673.97 - \$1550 = \$123.97$$

27. Writing exercise

28. Use the present value formula for simple interest with $A = \$815.63$, $r = .07$, and $t = 15/12 = 1.25$ yr.

$$P = \frac{A}{1 + rt}$$
$$= \frac{\$815.63}{1 + (.07)(1.25)}$$
$$\approx \$750.00$$

29. Use the present value formula for simple interest with $A = \$1500$, $r = .08$, and $t = 4/12 = .\overline{3}$.

$$P = \frac{A}{1 + rt}$$
$$= \frac{\$1500}{1 + (.08)(.\overline{3})}$$
$$\approx \$1461.04$$

30. Writing exercise

31. Use the present value formula for compound interest with $A = \$1000$, $r = .06$, $m = 1$, and $n = 1 \cdot 5$.

$$P = \frac{A}{\left(1 + \frac{r}{m}\right)^n}$$
$$= \frac{\$1000}{\left(1 + \frac{.06}{1}\right)^5}$$
$$= \frac{\$1000}{(1.06)^5}$$
$$= \frac{\$1000}{1.3382256}$$
$$\approx \$747.26$$

32. Use the present value formula for compound interest with $A = \$14,000$, $r = .04$, $m = 4$, and $n = 4 \cdot 3 = 12$.

$$P = \frac{A}{\left(1 + \frac{r}{m}\right)^n}$$
$$= \frac{\$14000}{\left(1 + \frac{.04}{4}\right)^{12}}$$
$$= \frac{\$14000}{(1.01)^{12}}$$
$$= \frac{\$14000}{1.126825}$$
$$\approx \$12,424.29$$

33. Use the present value formula for compound interest with $A = \$9860$, $r = .08$, $m = 2$, and $n = 2 \cdot 10 = 20$.

$$P = \frac{A}{\left(1 + \frac{r}{m}\right)^n}$$
$$= \frac{\$9860}{\left(1 + \frac{.08}{2}\right)^{20}}$$
$$= \frac{\$9860}{(1.04)^{20}}$$
$$= \frac{\$9860}{2.1911231}$$
$$\approx \$4499.98$$

34. Use the present value formula for compound interest with $A = \$15080$, $r = .05$, $m = 12$, and $n = 12 \cdot 4 = 48$.

$$P = \frac{A}{\left(1 + \frac{r}{m}\right)^n}$$
$$= \frac{\$15080}{\left(1 + \frac{.05}{12}\right)^{48}}$$
$$= \frac{\$15080}{(1.0041\overline{6})^{48}}$$
$$= \frac{\$15080}{1.2208953}$$
$$\approx \$12,351.59$$

35. Use the present value formula for compound interest with $A = \$500,000$, $r = .08$, $m = 4$, and $n = 4 \cdot 30 = 120$.

$$P = \frac{A}{\left(1 + \frac{r}{m}\right)^n}$$
$$= \frac{\$500000}{\left(1 + \frac{.08}{4}\right)^{120}}$$
$$= \frac{\$500000}{(1.02)^{120}}$$
$$= \frac{\$500000}{10.76516303}$$
$$\approx \$46,446.11$$

36. Use the present value formula for compound interest with $A = \$500,000$, $r = .06$, $m = 4$, and $n = 4 \cdot 30 = 120$.

$$P = \frac{A}{\left(1 + \frac{r}{m}\right)^n}$$
$$= \frac{\$500000}{\left(1 + \frac{.06}{4}\right)^{120}}$$
$$= \frac{\$500000}{(1.015)^{120}}$$
$$= \frac{\$500000}{5.969322872}$$
$$\approx \$83,761.59$$

37. Use the present value formula for compound interest with $A = \$500,000$, $r = .08$, $m = 365$, and $n = 365 \cdot 30 = 10950$.

$$P = \frac{A}{\left(1 + \frac{r}{m}\right)^n}$$
$$= \frac{\$500000}{\left(1 + \frac{.08}{365}\right)^{10950}}$$
$$\approx \$45,370.91$$

38. Use the present value formula for compound interest with $A = \$500,000$, $r = .06$, $m = 365$, and $n = 365 \cdot 30 = 10950$.

$$P = \frac{A}{\left(1 + \frac{r}{m}\right)^n}$$
$$= \frac{\$500000}{\left(1 + \frac{.06}{365}\right)^{10950}}$$
$$\approx \$82,661.67$$

39. Use the formula $Y = \left(1 + \frac{r}{m}\right)^m - 1$, with $r = .05$ and $m = 1$.

$$Y = \left(1 + \frac{.05}{1}\right)^1 - 1$$
$$= (1.05) - 1$$
$$= .05, \text{ which is } 5\%$$

40. Use the formula $Y = \left(1 + \frac{r}{m}\right)^m - 1$, with $r = .05$ and $m = 2$.

$$Y = \left(1 + \frac{.05}{2}\right)^2 - 1$$
$$= (1.025)^2 - 1$$
$$= 1.050625 - 1$$
$$= .05063, \text{ which is } 5.063\%$$

41. Use the formula $Y = \left(1 + \frac{r}{m}\right)^m - 1$, with $r = .05$ and $m = 4$.

$$Y = \left(1 + \frac{.05}{4}\right)^4 - 1$$
$$= (1.0125)^4 - 1$$
$$= 1.0509453 - 1$$
$$= .05095, \text{ which is } 5.095\%$$

42. Use the formula $Y = \left(1 + \frac{r}{m}\right)^m - 1$, with $r = .05$ and $m = 12$.

$$Y = \left(1 + \frac{.05}{12}\right)^{12} - 1$$
$$= \left(1.0041\overline{6}\right)^{12} - 1$$
$$= 1.0511619 - 1$$
$$= .0511619, \text{ which is } 5.116\%$$

43. Use the formula $Y = \left(1 + \frac{r}{m}\right)^m - 1$, with $r = .05$ and $m = 365$.

$$Y = \left(1 + \frac{.05}{365}\right)^{365} - 1$$
$$= (1.000136986)^{365} - 1$$
$$= 1.05127 - 1$$
$$= .05127, \text{ which is } 5.127\%$$

44. Use the formula $Y = \left(1 + \frac{r}{m}\right)^m - 1$, with $r = .05$ and $m = 1000$.

$$Y = \left(1 + \frac{.05}{1000}\right)^{1000} - 1$$
$$= (1.00005)^{1000} - 1$$
$$= 1.05127 - 1$$
$$= .05127, \text{ which is } 5.127\%$$

45. Use the formula $Y = \left(1 + \frac{r}{m}\right)^m - 1$, with $r = .05$ and $m = 10,000$.

$$Y = \left(1 + \frac{.05}{10000}\right)^{10000} - 1$$
$$= (1.000005)^{10000} - 1$$
$$= 1.05127 - 1$$
$$= .05127, \text{ which is } 5.127\%$$

46. Writing exercise

47. Because the deposit is more than $2500, use the yield of 1.96% as the rate in the formula $A = P + I$.

$$A = P + I$$
$$= \$20000 + (.0196)(\$20000)$$
$$= \$20,392$$

48. Use trial and error to determine how often compounding occurs. Using the nominal rate of 1.85%, semiannual compounding would give an effective annual yield of

$$Y = \left(1 + \frac{.0185}{2}\right)^2 - 1 = .0185855625, \text{ or about } 1.859\%.$$

This is less than the posted yield of 1.86%. In order to increase the effective yield, interest must be compounded more frequently, so try quarterly compounding. This gives an effective yield of

$$Y = \left(1 + \frac{.0185}{4}\right)^4 - 1 = .0186287399, \text{ or about } 1.86\%.$$

This closely matches the posted yield for the 12-month CD with a minimum balance of $1000, so we conclude that compounding occurs quarterly. (The same result can be obtained by using the rates for the other 3 posted CD's.)

49. Use the formula $A = P\left(1 + \frac{r}{m}\right)^n$ with A having the value $2P$ because we want the amount to double the principal.

$$2P = P\left(1 + \frac{.06}{4}\right)^{4n}$$
$$2 = (1 + .015)^{4n}$$
$$\ln 2 = 4n \cdot \ln(1.015)$$
$$\frac{\ln 2}{\ln(1.015)} = 4n$$
$$\frac{46.55552563}{4} = n$$
$$11.64 \approx n$$

This is 11 years and .64 of a year: $.64(365) = 274$. The answer then is about 11 years and 233 days.

50. At the time when the interest earned is equal to the original principal, the amount in the account will be twice the original principal. Use the future value formula for compound interest with $A = 2P$, $r = .06$, and $m = 365$.

$$2P = P\left(1 + \frac{.06}{365}\right)^n$$
$$2 = \left(1 + \frac{.06}{365}\right)^n$$
$$\ln 2 = n \cdot \ln\left(1 + \frac{.06}{365}\right)$$
$$\frac{\ln 2}{\ln\left(1 + \frac{.06}{365}\right)} = n$$
$$4216.9919 = n$$

Since n represents the number of periods, which in this case is the number of days, divide this number by 365 to obtain the required time period.

$$\frac{4216.9919}{365} = 11.5534 \text{ years, and}$$

$$.5534(365) = 202.$$

This is 11 years, 202 days.

51. Use the formula $Y = \left(1 + \frac{r}{m}\right)^m - 1$ where $r =$ nominal interest rate; $m =$ number of times interest is compounded per year; $Y =$ effective annual yield. Substitute values and solve for r.

$$Y = \left(1 + \frac{r}{m}\right)^m - 1$$

$$.0295 = \left(1 + \frac{r}{12}\right)^{12} - 1$$

$$1.0295 = \left(1 + \frac{r}{12}\right)^{12}$$

$$\ln(1.0295) = 12 \cdot \ln\left(1 + \frac{r}{12}\right)$$

$$\frac{\ln(1.0295)}{12} = \ln\left(1 + \frac{r}{12}\right)$$

$$.0024227706 \approx \ln\left(1 + \frac{r}{12}\right)$$

$$e^{.0024227706} \approx 1 + \frac{r}{12}$$

$$1.00245708 \approx 1 + \frac{r}{12}$$

$$.00245708 \approx \frac{r}{12}$$

$$.0291 \approx r$$

The nominal rate is 2.91%.

52. (a) Use the formula $Y = \left(1 + \frac{r}{m}\right)^m - 1$ where $r = .038$ and $m = 365$ to find Y for the Bank A.

$$Y = \left(1 + \frac{.038}{365}\right)^{365} - 1$$

$$= (1.00010411)^{365} - 1$$

$$\approx 1.038729178 - 1$$

$$\approx .038729178$$

This is a yield of about 3.87%

Now replace Y with this value in decimal form, m with 4, and solve for r for Bank B.

$$.038729178 = \left(1 + \frac{r}{4}\right)^4 - 1$$

$$1.038729178 = \left(1 + \frac{r}{4}\right)^4$$

$$(1.038729178)^{\frac{1}{4}} = \left[\left(1 + \frac{r}{4}\right)^4\right]^{\frac{1}{4}}$$

$$1.009544769 \approx 1 + \frac{r}{4}$$

$$.009544769 \approx \frac{r}{4}$$

$$.0381790761 \approx r$$

The Bank B has a nominal rate of about 3.818%.

(b) Bank A

Use the formula $A = P\left(1 + \frac{r}{m}\right)^n$ with $P = \$2000$, $r = .038$, $m = 365$, and $n = \frac{10}{12}(365)$.

$$A = \$2000\left(1 + \frac{.038}{365}\right)^{\frac{10}{12}(365)}$$

$$\approx \$2064.34$$

Bank B

Use the formula $A = P\left(1 + \frac{r}{m}\right)^n$ with $P = \$2000$, $r = .03818$, $m = 4$, and $n = 3$. Because the time period is 10 months, she earns interest for 3 quarters (or nine months).

$$A = \$2000\left(1 + \frac{.03818}{4}\right)^3$$

$$\approx \$2057.82$$

She should choose the Bank A. She earns \$6.52 more.

(c) Bank A

Use the formula $A = P\left(1 + \frac{r}{m}\right)^n$ with $P = \$6000$, $r = .038$, $m = 365$, and $n = 365$.

$$A = P\left(1 + \frac{r}{m}\right)^n$$

$$A = \$6000\left(1 + \frac{.038}{365}\right)^{365}$$

$$\approx \$6232.38$$

Bank B

$$A = P\left(1 + \frac{r}{m}\right)^n$$

$$A = \$6000\left(1 + \frac{.03818}{4}\right)^4$$

$$\approx \$6232.38$$

There is no difference in the amount for the banks. The interest for both is \$232.38.

53.
$$Y = \left(1 + \frac{r}{m}\right)^m - 1$$

$$Y + 1 = \left(1 + \frac{r}{m}\right)^m$$

$$(Y+1)^{\frac{1}{m}} = \left[\left(1 + \frac{r}{m}\right)^m\right]^{\frac{1}{m}}$$

$$(Y+1)^{\frac{1}{m}} = \left(1 + \frac{r}{m}\right)$$

$$(Y+1)^{\frac{1}{m}} - 1 = \frac{r}{m}$$

$$m\left[(Y+1)^{\frac{1}{m}} - 1\right] = r$$

54. Years to double $= \dfrac{70}{\text{Annual inflation rate}}$

Years to double $= \dfrac{70}{1}$

Years to double $= 70$ years

55. Years to double $= \dfrac{70}{\text{Annual inflation rate}}$

Years to double $= \dfrac{70}{2}$

Years to double $= 35$ years

56. Years to double $= \dfrac{70}{\text{Annual inflation rate}}$

    Years to double $= \dfrac{70}{8}$

    Years to double $= 8.75$ or about 9 years

57. Years to double $= \dfrac{70}{\text{Annual inflation rate}}$

    Years to double $= \dfrac{70}{9}$

    Years to double $= 7.8$ or about 8 years

58. Let $r =$ the inflation rate as a percent.

    $\text{Years to double} = \dfrac{70}{\text{Annual inflation rate}}$

    $5 = \dfrac{70}{r}$

    $5r = 70$

    $r = 14.0\%$

59. Let $r =$ the inflation rate as a percent.

    $\text{Years to double} = \dfrac{70}{\text{Annual inflation rate}}$

    $7 = \dfrac{70}{r}$

    $7r = 70$

    $r = 10.0\%$

60. Let $r =$ the inflation rate as a percent.

    $\text{Years to double} = \dfrac{70}{\text{Annual inflation rate}}$

    $16 = \dfrac{70}{r}$

    $16r = 70$

    $r \approx 4.4\%$

61. Let $r =$ the inflation rate as a percent.

    $\text{Years to double} = \dfrac{70}{\text{Annual inflation rate}}$

    $22 = \dfrac{70}{r}$

    $22r = 70$

    $r \approx 3.2\%$

62. Use the formula for continuous compounding, replacing A with $3P$, representing three times the present value. Solve for t. In the last step, multiplying numerator and denominator by 100 gives t in terms of r, the annual inflation rate as a percent.

    $A = Pe^{rt}$
    $3P = Pe^{rt}$
    $3 = e^{rt}$
    $\ln 3 = \ln e^{rt}$
    $\ln 3 = rt$
    $\dfrac{\ln 3}{r} = t$
    $\dfrac{100 \cdot \ln 3}{100 \cdot r} = t$
    $\dfrac{100 \cdot (1.098612)}{100 \cdot r} \approx t$
    $\dfrac{110}{100r} = t$
    $\dfrac{110}{\text{annual inflation rate}} = t$

63. Writing exercise

Table for Exercises 64–67

Item	2003 Price	2010 2%	2020 2%	2010 10%	2020 10%
64. House	$165,000	$190,000	$232,000	$332,000	$903,000
65. Meal	$4.89	$5.62	$6.87	$9.85	$26.77
66. Gas	$1.65	$1.90	$2.32	$3.32	$9.03
67. Car	$13,500	$15,500	$19,000	$27,000	$73,900

64. Use the formula for continuous compounding $A = Pe^{rt}$, changing r and t as appropriate.

    $A = \$165000 e^{.02(7)}$
    $A \approx \$190,000$

    $A = \$165000 e^{.02(17)}$
    $A \approx \$232,000$

    $A = \$165000 e^{.10(7)}$
    $A \approx \$332,000$

    $A = \$165000 e^{.10(17)}$
    $A \approx \$903,000$

65. Use the formula for continuous compounding $A = Pe^{rt}$, changing r and t as appropriate.

    $A = \$4.89 e^{.02(7)}$
    $A \approx \$5.62$

    $A = \$4.89 e^{.02(17)}$
    $A \approx \$6.87$

    $A = \$4.89 e^{.10(7)}$
    $A \approx \$9.85$

    $A = \$4.89 e^{.10(17)}$
    $A \approx \$26.77$

66. Use the formula for continuous compounding $A = Pe^{rt}$, changing r and t as appropriate.

$$A = \$1.65e^{.02(7)}$$
$$A \approx \$1.90$$

$$A = \$1.65e^{.02(17)}$$
$$A \approx \$2.32$$

$$A = \$1.65e^{.10(7)}$$
$$A \approx \$3.32$$

$$A = \$1.65e^{.10(17)}$$
$$A \approx \$9.03$$

67. Use the formula for continuous compounding $A = Pe^{rt}$, changing r and t as appropriate.

$$A = \$13500e^{.02(7)}$$
$$A \approx \$15,500$$

$$A = \$13500e^{.02(17)}$$
$$A \approx \$19,000$$

$$A = \$13500e^{.10(7)}$$
$$A \approx \$27,200$$

$$A = \$13500e^{.10(17)}$$
$$A \approx \$73,900$$

Table for Exercises 68–71

	Item	Price	Year 1999	Price in 2002
68.	Sofa	$650	Dec. 99	$702
69.	Roses	$78	Dec. 96	$89
70.	Skis	$149	Dec. 95	$176
71.	Motorcycle	$6388	Dec. 94	$7759

68. $\$650 \times 1.034 \times 1.028 \times 1.016 \approx \$702$

69. Find the percent change in Table 2 for each of the six consecutive years from 1996 to 2002. These are (beginning with 1997) 2.3%, 1.6%, 2.2%, 3.4%, 2.8%, and 1.6%; as decimals they are .023, .016, and .022, .034, .028, and .016. Using the compounding interest concept, the problem can be done efficiently by multiplying the beginning amount, $78, by 1 plus the percent increase for each year:

$$\$78 \times 1.023 \times 1.016 \times 1.022 \times 1.034$$
$$\times 1.028 \times 1.016 \approx \$89.$$

70. $\$149 \times 1.030 \times 1.023 \times 1.016 \times 1.022$
$\times 1.034 \times 1.028 \times 1.016 \approx \$176$

71. Find the percent change in Table 2 for each of the eight consecutive years from 1994 to 2002. These are 2.8%, 3.0%, 2.3%, 1.6%, 2.2%, 3.4%, 2.8%, and 1.6%; as decimals they are .028, .030, .023, .016, .022, .034, .028, and .016. Using the compounding interest concept, the problem can be done efficiently by multiplying the beginning amount, $6388, by 1 plus the percent increase for each year:

$$\$6388 \times 1.028 \times 1.030 \times 1.023 \times 1.016 \times 1.022$$
$$\times 1.034 \times 1.028 \times 1.016 \approx \$7759.$$

72. Use the formula $P = \dfrac{A}{\left(1 + \frac{r}{m}\right)^n}$ with $A = \$750,000$; $r = .03$; $m = 4$; $n = 4 \cdot 3 = 12$.

$$P = \frac{A}{\left(1 + \frac{r}{m}\right)^n}$$
$$P = \frac{\$750000}{\left(1 + \frac{.03}{4}\right)^{12}}$$
$$= \$685,678.62$$

73. Use the formula $P = \dfrac{A}{\left(1 + \frac{r}{m}\right)^n}$ with $A = 300 \cdot \$300,000 = \$90,000,000$; $r = .07$; $m = 4$; $n = 4 \cdot \frac{18}{12} = 6$.

$$P = \frac{A}{\left(1 + \frac{r}{m}\right)^n}$$
$$P = \frac{\$90000000}{\left(1 + \frac{.07}{4}\right)^6}$$
$$= \$81,102,828.75$$

74. Writing exercise

EXTENSION: ANNUITIES

In Exercises 1–6, for part (a) use the future value formula, replacing R, r, m, and n with the appropriate values.

1. (a) $$V = \frac{R\left[\left(1 + \frac{r}{m}\right)^n - 1\right]}{\frac{r}{m}}$$
$$= \frac{\$1000\left[\left(1 + \frac{.085}{1}\right)^{10} - 1\right]}{\frac{.085}{1}}$$
$$= \$14,835.10$$

(b) To find the total of all deposits, multiply the amount of the regular deposit by the number of years in the Accumulation Period.

$$10(\$1000) = \$10,000$$

(c) To find the total interest earned, subtract the total deposits from the total accumulation.

$$\$14835.10 - \$10000 = \$4835.10$$

2. (a) $V = \dfrac{R\left[\left(1+\frac{r}{m}\right)^n - 1\right]}{\frac{r}{m}}$

$= \dfrac{\$2000\left[\left(1+\frac{.045}{1}\right)^{20} - 1\right]}{\frac{.045}{1}}$

$= \$62{,}742.85$

(b) To find the total of all deposits, multiply the amount of the regular deposit by the number of years in the Accumulation Period.

$$20(\$2000) = \$40{,}000$$

(c) To find the total interest earned, subtract the total deposits from the total accumulation

$$\$62742.85 - \$40000 = \$22{,}742.85$$

3. (a) $V = \dfrac{R\left[\left(1+\frac{r}{m}\right)^n - 1\right]}{\frac{r}{m}}$

$= \dfrac{\$50\left[\left(1+\frac{.06}{12}\right)^{5\cdot 12} - 1\right]}{\frac{.06}{12}}$

$= \$3488.50$

(b) To find the total of all deposits, multiply the amount of the regular deposit by the total number of deposits, $5 \cdot 12 = 60$.

$$60(\$50) = \$3000$$

(c) To find the total interest earned, subtract the total deposits from the total accumulation

$$\$3488.50 - \$3000 = \$488.50$$

4. (a) $V = \dfrac{R\left[\left(1+\frac{r}{m}\right)^n - 1\right]}{\frac{r}{m}}$

$= \dfrac{\$75\left[\left(1+\frac{.072}{12}\right)^{10\cdot 12} - 1\right]}{\frac{.072}{12}}$

$= \$13{,}125.23$

(b) To find the total of all deposits, multiply the amount of the regular deposit by the total number of deposits, $10 \cdot 12 = 120$.

$$120(\$75) = \$9000$$

(c) To find the total interest earned, subtract the total deposits from the total accumulation

$$\$13125.23 - \$9000 = \$4125.23$$

5. (a) $V = \dfrac{R\left[\left(1+\frac{r}{m}\right)^n - 1\right]}{\frac{r}{m}}$

$= \dfrac{\$20\left[\left(1+\frac{.052}{52}\right)^{3\cdot 52} - 1\right]}{\frac{.052}{52}}$

$= \$3374.70$

(b) To find the total of all deposits, multiply the amount of the regular deposit by the total number of deposits, $3 \cdot 52 = 156$.

$$156(\$20) = \$3120$$

(c) To find the total interest earned, subtract the total deposits from the total accumulation.

$$3374.70 - 3120 = \$254.70$$

6. (a) $V = \dfrac{R\left[\left(1+\frac{r}{m}\right)^n - 1\right]}{\frac{r}{m}}$

$= \dfrac{\$30\left[\left(1+\frac{.078}{52}\right)^{5\cdot 52} - 1\right]}{\frac{.078}{52}}$

$= \$9530.99$

(b) To find the total of all deposits, multiply the amount of the regular deposit by the total number of deposits, $5 \cdot 52 = 260$.

$$260(\$30) = \$7800$$

(c) To find the total interest earned, subtract the total deposits from the total accumulation

$$\$9530.99 - \$7800 = \$1730.99$$

7. Writing exercise

8. Writing exercise

For Exercises 9–12, use the formula

$$V = \dfrac{R\left[\left(1+\frac{r}{m}\right)^{n+1} - 1\right]}{\frac{r}{m}}$$

9. $V = \dfrac{R\left[\left(1+\frac{r}{m}\right)^{n+1} - 1\right]}{\frac{r}{m}}$

$= \dfrac{\$50\left[\left(1+\frac{.06}{12}\right)^{5\cdot 12+1} - 1\right]}{\frac{.06}{12}}$

$= \$3555.94$

10. $V = \dfrac{R\left[\left(1+\frac{r}{m}\right)^{n+1} - 1\right]}{\frac{r}{m}}$

$= \dfrac{\$75\left[\left(1+\frac{.072}{12}\right)^{10\cdot 12+1} - 1\right]}{\frac{.072}{12}}$

$= \$13{,}278.98$

11. $V = \dfrac{R\left[\left(1+\frac{r}{m}\right)^{n+1}-1\right]}{\frac{r}{m}}$

$= \dfrac{\$20\left[\left(1+\frac{.052}{52}\right)^{3\cdot 52+1}-1\right]}{\frac{.052}{52}}$

$= \$3398.08$

12. $V = \dfrac{R\left[\left(1+\frac{r}{m}\right)^{n+1}-1\right]}{\frac{r}{m}}$

$= \dfrac{\$30\left[\left(1+\frac{.078}{52}\right)^{5\cdot 52+1}-1\right]}{\frac{.078}{52}}$

$= \$9575.28$

13. Writing exercise
14. Writing exercise
15. Writing exercise
16. Writing exercise
17. Writing exercise
18. Writing exercise

14.2 EXERCISES

1. Subtract the down payment amount from $2150.

$$\$2150 - \$500 = \$1650$$

2. To find the interest due on this amount, use the simple interest formula with $P = \$1650$, $r = .12$, and $t = 2$.

$$\begin{aligned}I &= Prt \\ &= (\$1650)(.12)(2) \\ &= \$396\end{aligned}$$

3. Referring to Exercise 2, the payment period is 2 years. The total interest is $(\$1650)(.12)(2) = \$396$. Find the sum of the total amount financed and the total interest owed.

$$\$1650 + \$396 = \$2046$$

4. From Exercise 3, the total amount owed on the loan is $2046. The number of monthly payments is $2 \cdot 12 = 24$, so the amount of each monthly payment is

$$\dfrac{\$2046}{24} = \$85.25.$$

5. Subtract the value of the old car from the cost of the new car.

$$\$16,500 - \$3000 = \$13,500$$

6. Use the simple interest formula with $P = \$13,500$, $r = .09$, and $t = 3$.

$$\begin{aligned}I &= Prt \\ &= (\$13,500)(.09)(3) \\ &= \$3645\end{aligned}$$

7. The total interest from Exercise 6 is $(13,500)(.09)(3) = \$3645$. Then the total amount owed is the sum of the principal and the interest.

$$\$13,500 + \$3645 = \$17,145$$

8. From Exercise 7, the total amount owed on the loan is $17,145. The number of monthly payments is $3 \cdot 12 = 36$, so the amount of each monthly payment is

$$\dfrac{\$17145}{36} = \$476.25.$$

9. The interest charge is:

$$\begin{aligned}I &= Prt \\ &= (\$4500)(.09)(3) \\ &= \$1215.\end{aligned}$$

The total amount owed is:

$$\begin{aligned}P + I &= \$4500 + \$1215 \\ &= \$5715.\end{aligned}$$

There are $3 \cdot 12 = 36$ monthly payments, so the amount of each monthly payment is:

$$\dfrac{\$5715}{36} = \$158.75.$$

10. The interest charge is:

$$\begin{aligned}I &= Prt \\ &= (\$2700)(.08)(2) \\ &= \$432.\end{aligned}$$

The total amount owed is:

$$\begin{aligned}P + I &= \$2700 + \$432 \\ &= \$3132.\end{aligned}$$

There are $2 \cdot 12 = 24$ monthly payments, so the amount of each monthly payment is:

$$\dfrac{\$3132}{24} = \$130.50.$$

11. The interest charge is:

$$\begin{aligned}I &= Prt \\ &= (\$750)(.074)\left(\dfrac{18}{12}\right) \\ &= \$83.25.\end{aligned}$$

The total amount owed is:
$$P + I = \$750 + \$83.25$$
$$= \$833.25.$$

There are 18 monthly payments, so the amount of each monthly payment is:
$$\frac{\$833.25}{18} = \$46.29.$$

12. The interest charge is:
$$I = Prt$$
$$= (\$2450)(.092)\left(\frac{30}{12}\right)$$
$$= \$563.50.$$

The total amount owed is:
$$P + I = \$2450 + \$563.50$$
$$= \$3013.50.$$

There are 30 monthly payments, so the amount of each monthly payment is:
$$\frac{\$3013.50}{30} = \$100.45.$$

13. The interest charge is:
$$I = Prt$$
$$= (\$1580)(.06)\left(\frac{10}{12}\right)$$
$$= \$79.$$

The total amount owed is:
$$P + I = \$1580 + \$79$$
$$= \$1659.$$

There are 10 monthly payments, so the amount of each monthly payment is
$$\frac{\$1659}{10} = \$165.90.$$

14. The interest charge is:
$$I = Prt$$
$$= (\$2100)(.12)\left(\frac{14}{12}\right)$$
$$= \$294.$$

The total amount owed is:
$$P + I = \$2100 + \$294$$
$$= \$2394.$$

There are 14 monthly payments, so the amount of each monthly payment is:
$$\frac{\$2394}{14} = \$171.$$

15. The interest charge is:
$$I = Prt$$
$$= (\$535)(.111)\left(\frac{16}{12}\right)$$
$$= \$79.18.$$

The total amount owed is:
$$P + I = \$535 + \$79.18$$
$$= \$614.18.$$

There are 16 monthly payments, so the amount of each monthly payment is:
$$\frac{\$614.18}{16} = \$38.39.$$

16. The interest charge is:
$$I = Prt$$
$$= (\$798)(.103)\left(\frac{29}{12}\right)$$
$$= \$198.64.$$

The total amount owed is:
$$P + I = \$798 + \$198.64$$
$$= \$996.64.$$

There are 29 monthly payments, so the amount of each monthly payment is:
$$\frac{\$996.64}{29} = \$34.37.$$

17. First subtract the down payment from the purchase price of the furniture to find the value of P.
$$\$8500 - \$3000 = \$5500.$$

Now use the formula for simple interest with $P = \$5500$, $r = .10$, and $t = 2.5$.
$$I = Prt$$
$$= (\$5500)(.10)(2.5)$$
$$= \$1375$$

The total amount owed is:
$$P + I = \$5500 + \$1375$$
$$= \$6875.$$

There are $2.5(12) = 30$ monthly payments, so the amount of each monthly payment is:

$$\frac{\$6875}{30} = \$229.17.$$

18. The interest charge is:

$$\begin{aligned} I &= Prt \\ &= (\$9780)(.093)(3) \\ &= \$2728.62. \end{aligned}$$

The total amount owed is:

$$\begin{aligned} P + I &= \$9780 + \$2728.62 \\ &= \$12,508.62. \end{aligned}$$

There are $3 \cdot 12 = 36$ monthly payments, so the amount of each monthly payment is:

$$\frac{\$12508.62}{36} = \$347.46.$$

19. First subtract the down payment from the purchase price of the furniture to find the value of P.

$$\$14240 - \$2900 = \$11340.$$

Now use the formula for simple interest with $P = \$11340$, $r = .10$, and $t = \frac{48}{12} = 4$.

$$\begin{aligned} I &= Prt \\ &= (\$11340)(.10)(4) \\ &= \$4536 \end{aligned}$$

The total amount owed is:

$$\begin{aligned} P + I &= \$11340 + \$4536 \\ &= \$15876. \end{aligned}$$

There are 48 monthly payments, so the amount of each monthly payment is:

$$\frac{\$15876}{48} = \$330.75.$$

20. The interest charge is:

$$\begin{aligned} I &= Prt \\ &= (\$1680)(.09)\left(\frac{10}{12}\right) \\ &= \$126. \end{aligned}$$

The total amount owed is:

$$\begin{aligned} P + I &= \$1680 + \$126 \\ &= \$1806. \end{aligned}$$

There are 10 monthly payments, so the amount of each monthly payment is:

$$\frac{\$1806}{10} = \$180.60.$$

21. First find the total amount to be paid.

$$(48)(314.65) = \$15,103.20$$

Then the rate is .098 and the time is 4 years to correspond to the 48 months.
Let $x =$ amount borrowed.

Simple interest $+$ Amount borrowed $= \$15103.20$
$$\begin{aligned} (.098)(4)x + x &= \$15103.20 \\ .392x + x &= \$15103.20 \\ 1.392x &= \$15103.20 \\ x &= \$10,850 \end{aligned}$$

22. The number of monthly payments is $(3.5)(12) = 42$, so the total amount paid is:

$$(42)(\$207.31) = \$8707.02.$$

This is the total amount owed which is given by $P + Prt$, so we have:

$$P + Prt = \$8707.02.$$

Substitute $P = \$6400$ and $t = 3.5$, and solve for r.

$$\begin{aligned} \$6400 + (\$6400)(3.5)r &= \$8707.02 \\ \$6400 + \$22400r &= \$8707.02 \\ \$22400r &= \$2307.02 \\ \frac{\$22400r}{\$22400} &= \frac{\$2307.02}{\$22400} \\ r &= .103 \end{aligned}$$

This is 10.3%.

23. The total number of payments will be $12 \cdot t$, where $t =$ the time in years. The amount to be paid per year is equivalent to multiplying the monthly payment by 12. This yields:

$$(\$172.44) \cdot 12 = \$2069.28.$$

Substitute $P = \$8000$, $r = .092$, into the formula $P + Prt = A$, and solve for t.

$$\begin{aligned} \$8000 + (\$8000)(.092)t &= \$2069.28t \\ \$8000 + \$736t &= \$2069.28t \\ \$8000 + \$736t - \$736t &= \$2069.28t - \$736t \\ \$8000 &= \$1333.28t \\ \frac{\$8000}{\$1333.28} &= \frac{\$1333.28t}{\$1333.28} \\ 6 \text{ years} &= t \end{aligned}$$

24. Let $t =$ the number of years needed to pay off the loan.

The total amount owed is:

$$\begin{aligned} P + Prt &= \$10000 + (\$10000)(.1015)t \\ &= \$10000 + \$1015t. \end{aligned}$$

The number of monthly payments is $12t$, so the amount of each monthly payment is:

$$\frac{\$10000 + \$1015t}{12t} = \$417.92.$$

Solve this equation for t.

$$\frac{12t}{1} \cdot \frac{\$10000 + \$1015t}{12t} = \$417.92 \cdot 12t$$

$$\$10000 + 1015t = \$5015.04t$$

$$\$10000 + \$1015t - \$1015t = \$5015.04t - \$1015t$$

$$\$10000 = \$4000.04t$$

$$\frac{\$10000}{\$4000.04} = \frac{\$4000.04t}{\$4000.04}$$

$$2.5 = t$$

The time is about 2.5 years, which is $2 \cdot 12 + 6 = 30$ months.

25. $(\$325.50)(.01) = \$3.26$
26. $(\$450.25)(.012) = \$5.40$
27. $(\$242.88)(.0112) = \$2.72$
28. $(\$655.33)(.0121) = \$7.93$

29.

Month	Unpaid Balance at Beginning of Month	Finance Charge	Purchases During Month	Returns	Payments	Unpaid Balance at End of Month
February	$319.10	$3.51	$86.14	0	$50	$358.75
March	358.75	3.95	109.83	$15.75	60	396.78
April	396.78	4.36	39.74	0	72	368.88
May	368.88	4.06	56.29	18.09	50	361.14

Here are explanations of how each underlined value was found:

1. Finance charge = Unpaid balance at beginning of month · Interest rate as decimal.
2. Unpaid balance at end of month = Unpaid balance at beginning of month + Finance charge + Purchases − Returns − Payments.
3. Unpaid balance at beginning of month = Unpaid balance at end of previous month.

February

Unpaid balance at beginning of month: $319.10.
Finance charge: $(\$319.10)(.011) = \$3.51$.
Unpaid balance at end of month: $319.10 + $3.51 + $86.14 − $50 = $358.75.

March

Unpaid balance at beginning of month: $358.75.
Finance charge: $(\$358.75)(.011) = \$3.95$.
Unpaid balance at end of month: $358.75 + $3.95 + $109.83 − $15.75 − $60 = $396.78.

April

Unpaid balance at beginning of month: $396.78.
Finance charge: $(\$396.78)(.011) = \$4.36$.
Unpaid balance at end of month: $396.78 + $4.36 + $39.74 − $72 = $368.88.

May

Unpaid balance at beginning of month: $368.80.
Finance charge: $(\$368.80)(.011) = \$4.06$.
Unpaid balance at end of month: $368.88 + $4.06 + $56.29 − $18.09 − $50 = $361.14.

30.

Month	Unpaid Balance at Beginning of Month	Finance Charge	Purchases During Month	Returns	Payments	Unpaid Balance at End of Month
October	$828.63	$9.11	$128.72	$23.15	$125	$818.31
November	818.31	9.00	291.64	0	170	948.95
December	948.95	10.44	147.11	17.15	150	939.35
January	939.35	10.33	27.84	139.82	200	637.70

Here are explanations of how each underlined value was found:

1. Finance charge = Unpaid balance at beginning of month · Interest rate as decimal.
2. Unpaid balance at end of month = Unpaid balance at beginning of month + Finance charge + Purchases − Returns − Payments.
3. Unpaid balance at beginning of month = Unpaid balance at end of previous month.

October

Unpaid balance at beginning of month: $828.63.
Finance charge: ($828.63)(.011) = $9.11.
Unpaid balance at end of month: $828.63 + $9.11 + $128.72 − $23.15 − $125 = $818.31.

November

Unpaid balance at beginning of month: $818.31.
Finance charge: ($818.31)(.011) = $9.00.
Unpaid balance at end of month: $818.31 + $9.00 + $291.64 − $170 = $948.95.

December

Unpaid balance at beginning of month: $948.95.
Finance charge: ($948.95)(.011) = $10.44.
Unpaid balance at end of month: $948.95 + $10.44 + $147.11 − $17.15 − $150 = $939.35.

January

Unpaid balance at beginning of month: $939.35.
Finance charge: ($939.35)(.011) = $10.33.
Unpaid balance at end of month: $939.35 + $10.33 + $27.84 − $139.82 − $200 = $637.70.

31.

Month	Unpaid Balance at Beginning of Month	Finance Charge	Purchases During Month	Returns	Payment	Unpaid Balance at End of Month
August	$684.17	$7.53	$155.01	$38.11	$100	$708.60
September	708.60	7.79	208.75	0	75	850.14
October	850.14	9.35	56.30	0	90	825.79
November	825.79	9.08	190.00	83.57	150	791.30

Here are explanations of how each underlined value was found:

1. Finance charge = Unpaid balance at beginning of month · Interest rate as decimal.
2. Unpaid balance at end of month = Unpaid balance at beginning of month + Finance charge + Purchases − Returns − Payments.
3. Unpaid balance at beginning of month = Unpaid balance at end of previous month.

August

Unpaid balance at beginning of month: $684.17.
Finance charge: $(\$684.17)(.011) = \$7.53$.
Unpaid balance at end of month: $\$684.17 + \$7.53 + \$155.01 - \$38.11 - \$100 = \$708.60$.

September

Unpaid balance at beginning of month: $708.60.
Finance charge: $(\$708.60)(.011) = \$7.79$.
Unpaid balance at end of month: $\$708.60 + \$7.79 + \$208.75 - \$75 = \$850.14$.

October

Unpaid balance at beginning of month: $850.14.
Finance charge: $(\$850.14)(.011) = \$9.35$.
Unpaid balance at end of month: $\$850.14 + \$9.35 + \$56.30 - \$90 = \$825.79$.

November

Unpaid balance at beginning of month: $825.79.
Finance charge: $(\$825.79)(.011) = \$9.08$.
Unpaid balance at end of month: $\$825.79 + \$9.08 + \$190.00 - \$83.57 - \$150 = \$791.30$.

32.

Month	Unpaid Balance at Beginning of Month	Finance Charge	Purchases During Month	Returns	Payment	Unpaid Balance at End of Month
March	$1230.30	$13.53	$308.13	$74.88	$250	$1227.08
April	$1227.08	13.50	488.35	0	350	1378.93
May	1378.93	15.17	134.99	18.12	175	1335.97
June	1335.97	14.70	157.72	0	190	1318.39

Here are explanations of how each underlined value was found:

1. Finance charge = Unpaid balance at beginning of month · Interest rate as decimal.
2. Unpaid balance at end of month = Unpaid balance at beginning of month + Finance charge + Purchases − Returns − Payments.
3. Unpaid balance at beginning of month = Unpaid balance at end of previous month.

March

Unpaid balance at beginning of month: $1230.30.
Finance charge: $(\$1230.30)(.011) = \$13.53$.
Unpaid balance at end of month: $\$1230.30 + \$13.53 + \$308.13 - \$74.88 - \$250 = \$1227.08$.

April

Unpaid balance at beginning of month: $1227.08.
Finance charge: $(\$1227.08)(.011) = \$13.50$.
Unpaid balance at end of month: $\$1227.08 + \$13.50 + \$488.35 - \$350 = \$1378.93$.

May

Unpaid balance at beginning of month: $1378.93.
Finance charge: $(\$1378.93)(.011) = \$15.17$.
Unpaid balance at end of month: $\$1378.93 + \$15.17 + \$134.99 - \$18.12 - \$175 = \$1335.97$.

484 CHAPTER 14 CONSUMER MATHEMATICS

June

Unpaid balance at beginning of month: $1335.97.
Finance charge: ($1335.97)(.011) = $14.70.
Unpaid balance at end of month: $1335.97 + $14.70 + $157.72 − $190 = $1318.39.

33. ($249.94)(.01) = $2.50

34. ($450.21)(.0106) = $4.77

35. ($1073.40)(.01125) = $12.08

36. ($1320.42)(.01375) = $18.16

37. First, make a table showing all the transactions, and compute the balance due on each date.

Date	Balance due
May 9	$728.36
May 17	$728.36 − $200 = $528.36
May 30	$528.36 + $46.11 = $574.47
June 3	$574.47 + $64.50 = $638.97

    Next, tabulate the balance due figures, along with the number of days until the balance changed. Use this data to calculate the sum of the daily balances.

Date	Balance due	Number of days until balance changed	(Balance due) × (Number of days)
May 9	$728.36	8	$728.36 × 8 = $5826.88
May 17	$528.36	13	$528.36 × 13 = $6868.68
May 30	$574.47	4	$574.47 × 4 = $2297.88
June 3	$638.97	6	$638.97 × 6 = $3833.82
Totals		31	$18,827.26

    (a) Average daily balance = $\dfrac{\text{Sum of daily balances}}{\text{Days in billing period}} = \dfrac{\$18827.26}{31} = \$607.33$

    (b) Finance charge = (.012)($607.33) = $7.29

    (c) Add the finance charge to the June 3 balance of $638.97 to get account balance for next billing:

    $638.97 + $7.29 = $646.26.

38. First, make a table showing all the transactions, and compute the balance due on each date.

Date	Balance due
January 27	$514.79
February 9	$514.79 + $11.08 = $525.87
February 13	$525.87 − $26.54 = $499.33
February 20	$499.33 − $59.00 = $440.33
February 25	$440.33 + $71.19 = $511.52

    Next, tabulate the balance due figures, along with the number of days until the balance changed. Use this data to calculate the sum of the daily balances.

Date	Balance due	Number of days until balance changed	(Balance due) × (Number of days)
January 27	$514.79	13	$514.79 × 13 = $6692.27
February 9	$525.87	4	$525.87 × 4 = $2103.48
February 13	$499.33	7	$499.33 × 7 = $3495.31
February 20	$440.33	5	$440.33 × 5 = $2201.65
February 25	$511.52	2	$511.52 × 2 = $1023.04
Totals		31	$15,515.75

(a) Average daily balance = $\dfrac{\text{Sum of daily balances}}{\text{Days in billing period}} = \dfrac{\$15515.75}{31} = \$500.51$

(b) Finance charge $= (.012)(\$500.51) = \$6.01$

(c) Add the finance charge to the last balance of $511.52 to get account balance for next billing:

$$\$511.52 + \$6.01 = \$517.53.$$

39. First, make a table showing all the transactions, and compute the balance due on each date.

Date	Balance due
June 11	$462.42
June 15	$462.42 − $106.45 = $355.97
June 20	$355.97 + $115.73 = $471.70
June 24	$471.70 + $74.19 = $545.89
July 3	$545.89 − $115.00 = $430.89
July 6	$430.89 + $68.49 = 499.38

Next, tabulate the balance due figures, along with the number of days until the balance changed. Use this data to calculate the sum of the daily balances.

Date	Balance due	Number of days until balance changed	$\left(\begin{array}{c}\text{Balance}\\\text{due}\end{array}\right) \times \left(\begin{array}{c}\text{Number}\\\text{of days}\end{array}\right)$
June 11	$462.42	4	$462.42 × 4 = $1849.68
June 15	$355.97	5	$355.97 × 5 = $1779.85
June 20	$471.70	4	$471.70 × 4 = $1886.80
June 24	$545.89	9	$545.89 × 9 = $4913.01
July 3	$430.89	3	$430.89 × 3 = $1292.67
July 6	$499.38	5	$499.38 × 5 = $2496.9
Totals		30	$14,218.91

(a) Average daily balance = $\dfrac{\text{Sum of daily balances}}{\text{Days in billing period}} = \dfrac{\$14218.91}{30} = \$473.96$

(b) Finance charge $= (.012)(\$607.33) = \$5.69$

(c) Add the finance charge to the June 3 balance of $499.38 to get account balance for next billing:

$$\$499.38 + \$5.69 = \$505.07.$$

40. First, make a table showing all the transactions, and compute the balance due on each date.

Date	Balance due
August 17	$983.25
August 21	$983.25 + $14.92 = $998.17
August 23	$998.17 − $25.41 = $972.76
August 27	$972.76 + $31.82 = $1004.58
August 31	$1004.58 − $108.00 = $896.58
September 9	$896.58 − $71.14 = $825.44
September 11	$825.44 + $110 = $935.44
September 14	$935.44 + $100 = $1035.44

Next, tabulate the balance due figures, along with the number of days until the balance changed. Use this data to calculate the sum of the daily balances.

Date	Balance due	Number of days until balance changed	$\begin{pmatrix}\text{Balance}\\\text{due}\end{pmatrix} \times \begin{pmatrix}\text{Number}\\\text{of days}\end{pmatrix}$
August 17	$983.25	4	$983.25 \times 4 = \$3933.00$
August 21	$998.17	2	$998.17 \times 2 = \$1996.34$
August 23	$972.76	4	$972.76 \times 4 = \$3891.04$
August 27	$1004.58	4	$1004.58 \times 4 = \$4018.32$
August 31	$896.58	9	$896.58 \times 9 = \$8069.22$
September 9	$825.44	2	$825.44 \times 2 = \$1650.88$
September 11	$935.44	3	$935.44 \times 3 = \$2806.32$
September 14	$1035.44	3	$1035.44 \times 3 = \$3106.32$
Totals		31	$29,471.44

(a) Average daily balance $= \dfrac{\text{Sum of daily balances}}{\text{Days in billing period}} = \dfrac{\$29471.44}{31} = \$950.69$

(b) Finance charge $= (.012)(\$950.69) = \$11.41$

(c) Add the finance charge to the June 3 balance of $1035.44 to get account balance for next billing:

$$\$1035.44 + \$11.41 = \$1046.85.$$

41. (a) Using the unpaid balance method to find the finance charge, multiply the previous unpaid balance by the interest rate.

$$(.01)(\$720) = \$7.20$$

(b) Using the average daily balance method, she has a balance of $720 for the first 28 days of the billing period. For the last 3 days of the billing period the balance drops to $120 because of her $600 payment. Then,

Average daily balance $= \dfrac{\$720 \cdot 28 + \$120 \cdot 3}{31} = \$661.94,$

and her finance charge is

$$(.01)(\$661.94) = \$6.62.$$

42. (a) Using the unpaid balance method to find the finance charge, multiply the previous unpaid balance by the interest rate.

$$(.011)(\$1070) = \$11.77$$

(b) Using the average daily balance method, she has a balance of $1070 for the first 25 days of the billing period. For the last 5 days of the billing period the balance drops to $170 because of her $900 payment. Then,

Average daily balance $= \dfrac{\$1070 \cdot 25 + \$170 \cdot 5}{30} = \$920,$

and her finance charge is

$$(.011)(\$920.00) = \$10.12.$$

43. (a) $(\$2900)(.013167) = \$38.18$

(b) At the end of the second month, the interest is first added to the balance before being multiplied by the interest rate.

$$(\$2900 + \$38.18)(.013167) = \$38.69$$

(c) At the end of the third month, both interest amounts are added to the balance before being multiplied by the interest rate.

$$(\$2900 + \$38.18 + \$38.69)(.013167) = \$39.20$$

44. Add the interest amounts from Exercise 43:

$$\$38.18 + \$38.69 + \$39.20 = \$116.07.$$

45. From Exercises 43 and 44, the total interest is $116.07. Use the formula $I = Prt$, with $I = \$116.07$, $P = \$2900$, and $t = 1/4$ or .25.

$$\$116.07 = (\$2900)(r)(.25)$$
$$116.07 = 725r$$
$$.160 = r, \text{ which is } 16.0\%$$

46. $.011(\$1846) = \$20.31$

47. There is a 2% charge for each cash advance of $100. Because she had six cash advances of $100 each, the total charge is

$$6 \times .02(100) = \$12.$$

Add this amount to the late payment fee of $15 and the over-the-credit-limit fee of $5:

$$\$12 + \$15 + \$5 = \$32.$$

48. Writing exercise

49. Writing exercise
50. Writing exercise
51. Writing exercise
52. Writing exercise
53. Writing exercise
54. Writing exercise
55. (a) If they choose Bank A, their estimated total yearly cost will be
$$(\$900)(.0118)(12) = \$127.44.$$
    (b) If they choose Bank B, their estimated total yearly cost will be
$$\$30 + (\$900)(.0101)(12) = \$139.08.$$
56. The card from Bank A is the better choice.

14.3 EXERCISES

1. Find the finance charge per $100 of the amount financed.
$$\frac{75}{1000} \times 100 = \$7.50$$
   In Table 3, find the "12 payments" row and read across to find the number closest to 7.50, which is 7.46. Read up to find the APR, which is 13.5%.

2. Find the finance charge per $100 of the amount financed.
$$\frac{202}{1700} \times 100 = \$11.88$$
   In Table 3, find the "24 payments" row and read across to find the number closest to 11.88, which is 11.86. Read up to find the APR, which is 11.0%.

3. Find the finance charge per $100 of the amount financed.
$$\frac{750}{6600} \times 100 = \$11.36$$
   In Table 3, find the "30 payments" row and read across to find the number closest to 11.36, which is 11.35. Read up to find the APR, which is 8.5%.

4. Find the finance charge per $100 of the amount financed.
$$\frac{1150}{5900} \times 100 = \$19.49$$
   In Table 3, find the "48 payments" row and read across to find the number closest to 19.49, which is 19.45. Read up to find the APR, which is 9.0%.

5. First find the amount financed by subtracting the down payment from the purchase price.
$$\$3000 - \$500 = \$2500$$
   Now find the total payments by adding the amount financed to the finance charge.
$$\$2500 + \$250 = \$2750$$
   The monthly payment is
$$\frac{\$2750}{24} = \$114.58.$$

6. First find the amount financed by subtracting the down payment from the purchase price.
$$\$4280 - \$450 = \$3830$$
   Now find the total payments by adding the amount financed to the finance charge.
$$\$3830 + \$700 = \$4530$$
   The monthly payment is
$$\frac{\$4530}{36} = \$125.83.$$

7. First find the amount financed by subtracting the down payment from the purchase price.
$$\$3950 - \$300 = \$3650$$
   Now find the total payments by adding the amount financed to the finance charge.
$$\$3650 + \$800 = \$4450$$
   The monthly payment is
$$\frac{\$4450}{48} = \$92.71.$$

8. First find the amount financed by subtracting the down payment from the purchase price.
$$\$8400 - \$2500 = \$5900$$
   Now find the total payments by adding the amount financed to the finance charge.
$$\$5900 + \$1300 = \$7200$$
   The monthly payment is
$$\frac{\$7200}{60} = \$120.00.$$

9. First find the amount financed by subtracting the down payment from the purchase price.
$$\$4190 - \$390 = \$3800$$
   Now find the finance charge. The interest rate of 6% will be charged on the amount financed for 1 year (12 payments).
$$\begin{aligned} I &= Prt \\ &= (3800)(.06)(1) \\ &= \$228 \end{aligned}$$

Next find the finance charge per $100 of the amount financed.

$$\frac{228}{3800} \times 100 = \$6.00$$

In Table 3, the number closest to 6.00 in the "12 payments" row is 6.06. Read up to find the APR, which is 11.0%.

10. First find the amount financed by subtracting the down payment from the purchase price.

$$\$3250 - \$750 = \$2500$$

Now find the finance charge. The interest rate of 7% will be charged on the amount financed for 3 years (36 payments).

$$I = Prt$$
$$= (2500)(.07)(3)$$
$$= \$525$$

Next find the finance charge per $100 of the amount financed.

$$\frac{525}{2500} \times 100 = \$21.00$$

In Table 3, the number closest to 21.00 in the "36 payments" row is 21.30. Read up to find the APR, which is 13.0%.

11. First find the amount financed by subtracting the down payment from the purchase price.

$$\$7480 - \$2200 = \$5280$$

Now find the finance charge. The interest rate of 5% will be charged on the amount financed for 1.5 years (18 payments).

$$I = Prt$$
$$= (5280)(.05)(1.5)$$
$$= \$396$$

Next find the finance charge per $100 of the amount financed.

$$\frac{396}{5280} \times 100 = \$7.50$$

In Table 3, the number closest to 7.50 in the "18 payments" row is 7.69. Read up to find the APR, which is 9.5%.

12. First find the amount financed by subtracting the down payment from the purchase price.

$$\$12800 - \$4500 = \$8300$$

Now find the finance charge. The interest rate of 6% will be charged on the amount financed for 4 years (48 payments).

$$I = Prt$$
$$= (8300)(.06)(4)$$
$$= \$1992$$

Next find the finance charge per $100 of the amount financed.

$$\frac{1992}{8300} \times 100 = \$24.00$$

In Table 3, the number closest to 24.00 in the "48 payments" row is 24.06. Read up to find the APR, which is 11.0%

13. (a) Find the intersection of the row for 18 payments and the column for 11.0% to find the value of h: $8.93.

(b) Use the formula $U = kR\left(\dfrac{h}{100+h}\right)$ to find unearned interest, with k = remaining number of payments, R = regular monthly payment, and h = finance charge per $100.

$$U = (18)(346.70)\left(\frac{8.93}{100+8.93}\right)$$
$$= (18)(346.70)\left(\frac{8.93}{108.93}\right)$$
$$= \$511.60$$

(c) The payoff amount is equal to the current payment plus the sum of the scheduled remaining payments minus the unearned interest.

$$\$346.70 + (18)(\$346.70) - 511.60 = \$6075.70$$

14. (a) Find the intersection of the row for 12 payments and the column for 8.5% to find the value of h: $4.66.

(b) Use the formula $U = kR\left(\dfrac{h}{100+h}\right)$ to find unearned interest, with k = remaining number of payments, R = regular monthly payment, and h = finance charge per $100.

$$U = (12)(783.50)\left(\frac{4.66}{100+4.66}\right)$$
$$= (12)(783.50)\left(\frac{4.66}{104.66}\right)$$
$$= \$418.63$$

(c) The payoff amount is equal to the current payment plus the sum of the scheduled remaining payments minus the unearned interest.

$$\$783.50 + (12)(\$783.50) - 418.63 = \$9766.87$$

15. (a) Find the intersection of the row for 6 payments and the column for 9.5% to find the value of h: $2.79.

(b) Use the formula $U = kR\left(\dfrac{h}{100+h}\right)$ to find unearned interest, with k = remaining number of payments, R = regular monthly payment, and h = finance charge per $100.

$$U = (6)(\$595.80)\left(\frac{\$2.79}{\$100 + \$2.79}\right)$$
$$= (6)(\$595.80)\left(\frac{\$2.79}{\$102.79}\right)$$
$$= \$97.03$$

(c) The payoff amount is equal to the current payment plus the sum of the scheduled remaining payments minus the unearned interest.

$$\$595.80 + (6)(\$595.80) - \$97.03 = \$4073.57$$

16. (a) Find the intersection of the row for 24 payments and the column for 10.0% to find the value of h: $10.75.

(b) Use the formula $U = kR\left(\dfrac{h}{100+h}\right)$ to find unearned interest, with k = remaining number of payments, R = regular monthly payment, and h = finance charge per $100.

$$U = (24)(\$314.50)\left(\frac{\$10.75}{\$100 + \$10.75}\right)$$
$$= (24)(\$314.50)\left(\frac{\$10.75}{\$110.75}\right)$$
$$= \$732.65$$

(c) The payoff amount is equal to the current payment plus the sum of the scheduled remaining payments minus the unearned interest.

$$\$314.50 + (24)(\$314.50) - \$732.65 = \$7129.85$$

17. (a) The amount of the total payments is

$$(24)(\$91.50) = \$2196.$$

The finance charge is the difference between this amount and the purchase price.

$$\$2196 - \$1990 = \$206$$

(b) Find the finance charge per $100 of the amount financed.

$$\frac{206}{1990} \times 100 = \$10.35$$

In Table 3, the number closest to 10.35 in the "24 payments" row is 10.19. Read up to find the APR, which is 9.5%.

18. (a) First find the amount financed by subtracting the down payment from the purchase price.

$$\$5090 - \$1240 = \$3850$$

The amount of the total payments is

$$(30)(\$152.70) = \$4581$$

because 2 1/2 years is equivalent to 30 months.

The finance charge is the difference between this amount and amount financed.

$$\$4581 - \$3850 = \$731$$

(b) Find the finance charge per $100 of the amount financed.

$$\frac{731}{3850} \times 100 = \$18.99$$

In Table 3, the number closest to 18.99 in the "30 payments" row is 19.10. Read up to find the APR, which is 14.0%.

19. (a) To find the finance charge use the simple interest formula. The interest rate of 6% will be charged on the amount financed for 1.5 years.

$$I = Prt$$
$$= (\$2000)(.06)(1.5)$$
$$= \$180$$

(b) Find the finance charge per $100 of the amount financed.

$$\frac{180}{2000} \times 100 = \$9.00$$

In Table 3, the number closest to 9.00 in the "18 payments" row is 8.93. Read up to find the APR, which is 11.0%.

20. (a) The amount of the total payments is

$$(36)(\$487.54) = \$17,551.44,$$

because 3 years is equivalent to 36 months.
The finance charge is the difference between this amount and amount financed.

$$\$17551.44 - \$15000 = \$2551.44$$

(b) Find the finance charge per $100 of the amount financed.

$$\frac{2551.44}{15000} \times 100 = \$17.01$$

In Table 3, the number closest to 17.01 in the "36 payments" row, which is exactly 17.01. Read up to find the APR, which is 10.5%.

21. (a) Using the actuarial method, first find the APR. If the loan were not paid off early, the total payments would be

$$(18)(\$201.85) = \$3633.30,$$

and the finance charge would be

$$\$3633.30 - \$3310 = \$323.30.$$

The finance charge per $100 of the amount financed would be

$$\frac{\$323.30}{\$3310} \times 100 = \$9.77.$$

In Table 3, the number 9.77 appears in the "18 payments" row and the 12.0% column, so the APR is 12%. Because the total number of payments remaining is 6, move up to the "6 payments" row to find $h = \$3.53$, the value of h needed in the actuarial method.

Use the formula $U = kR\left(\dfrac{h}{100+h}\right)$ to find unearned interest, with $k =$ remaining number of payments, $R =$ regular monthly payment, and $h =$ finance charge per $100.

$$U = (6)(\$201.85)\left(\frac{\$3.53}{\$100 + \$3.53}\right)$$
$$= (6)(\$201.85)\left(\frac{\$3.53}{\$103.53}\right)$$
$$= \$41.29$$

(b) From part (a), the original finance charge is $323.30. Use the rule of 78 to find unearned interest, with $k = 6$, $n = 18$, and $F = \$323.30$.

$$U = \frac{k(k+1)}{n(n+1)} \times F$$
$$= \frac{6(6+1)}{18(18+1)} \times \$323.30$$
$$= \$39.70$$

22. (a) Using the actuarial method, first find the APR. If the loan were not paid off early, the total payments would be

$$(48)(\$277.00) = \$13296.00,$$

and the finance charge would be

$$\$13296.00 - \$10230 = \$3066.00.$$

The finance charge per $100 of the amount financed would be

$$\frac{\$3066.00}{\$10230} \times 100 = \$29.97.$$

In Table 3, the number 29.97 appears in the "48 payments" row and the 13.5% column, so the APR is 13.5%. Because the total number of payments remaining is 12, move up to the "12 payments" row to find $h = \$7.46$, the value of h needed in the actuarial method.

Use the formula $U = kR\left(\dfrac{h}{100+h}\right)$ to find unearned interest, with $k =$ remaining number of payments, $R =$ regular monthly payment, and $h =$ finance charge per $100.

$$U = (12)(\$277.00)\left(\frac{\$7.46}{\$100 + \$7.46}\right)$$
$$= (12)(\$277.00)\left(\frac{\$7.46}{\$107.46}\right)$$
$$= \$230.76$$

(b) From part (a), the original finance charge is $3066.00. Use the rule of 78 to find unearned interest, with $k = 12$, $n = 48$, and $F = \$3066.00$.

$$U = \frac{k(k+1)}{n(n+1)} \times F$$
$$= \frac{12(12+1)}{48(48+1)} \times \$3066.00$$
$$= \$203.36$$

23. (a) Using the actuarial method, first find the APR. If the loan were not paid off early, the total payments would be

$$(60)(\$641.58) = \$38,494.80,$$

and the finance charge would be

$$\$38494.80 - \$29850 = \$8644.80.$$

The finance charge per $100 of the amount financed would be

$$\frac{\$8644.80}{\$29850} \times 100 = \$28.96.$$

In Table 3, the number 28.96 appears in the "60 payments" row and the 10.5% column, so the APR is 10.5%. Because the total number of payments remaining is 12, move up to the "12 payments" row to find $h = \$5.78$, the value of h needed in the actuarial method.

Use the formula $U = kR\left(\dfrac{h}{100+h}\right)$ to find unearned interest, with $k =$ remaining number of payments, $R =$ regular monthly payment, and $h =$ finance charge per $100.

$$U = (12)(\$641.58)\left(\frac{\$5.78}{\$100 + \$5.78}\right)$$
$$= (12)(\$641.58)\left(\frac{\$5.78}{\$105.78}\right)$$
$$= \$420.68$$

(b) From part (a), the original finance charge is $8644.80. Use the rule of 78 to find unearned interest, with $k = 12$, $n = 60$, and $F = \$8644.80$.

$$U = \frac{k(k+1)}{n(n+1)} \times F$$
$$= \frac{12(12+1)}{60(60+1)} \times \$8644.80$$
$$= \$368.47$$

24. (a) Using the actuarial method, first find the APR. If the loan were not paid off early, the total payments would be

$$(36)(\$539.82) = \$19{,}433.52,$$

and the finance charge would be

$$\$19433.52 - \$16730 = \$2703.52.$$

The finance charge per $100 of the amount financed would be

$$\frac{\$2703.52}{\$16730} \times 100 = \$16.16.$$

In Table 3, the number 16.16 appears in the "36 payments" row and the 10.0% column, so the APR is 10.0%. Because the total number of payments remaining is 18, move up to the "18 payments" row to find $h = \$8.10$, the value of h needed in the actuarial method.

Use the formula $U = kR\left(\dfrac{h}{100+h}\right)$ to find unearned interest, with $k =$ remaining number of payments, $R =$ regular monthly payment, and $h =$ finance charge per \$100.

$$U = (18)(\$539.82)\left(\frac{\$8.10}{\$100 + \$8.10}\right)$$

$$= (18)(\$539.82)\left(\frac{\$8.10}{\$108.10}\right)$$

$$= \$728.08$$

(b) From part (a), the original finance charge is \$2703.52. Use the rule of 78 to find unearned interest, with $k = 18$, $n = 36$, and $F = \$2703.52$.

$$U = \frac{k(k+1)}{n(n+1)} \times F$$

$$= \frac{18(18+1)}{36(36+1)} \times \$2703.52$$

$$= \$694.15$$

25. (a) Use the finance charge formula with $n = 4$ and APR $= .086$.

$$h = \frac{n \times \frac{APR}{12} \times \$100}{1 - \left(1 + \frac{APR}{12}\right)^{-n}} - \$100$$

$$= \frac{4 \times \frac{.086}{12} \times \$100}{1 - \left(1 + \frac{.086}{12}\right)^{-4}} - \$100$$

$$= \$1.80$$

(b) Use the actuarial formula for unearned interest with $k = 4$, $R = \$212$, and $h = \$1.80$.

$$U = kR\left(\frac{h}{\$100 + h}\right)$$

$$= 4(\$212)\left(\frac{\$1.80}{\$100 + \$1.80}\right)$$

$$= \$14.99$$

(c) The payoff amount is equal to the current payment added to the sum of the remaining payments minus the unearned interest.

$$\$212 + 4(\$212) - \$14.99 = \$1045.01$$

26. (a) Use the finance charge formula with $n = 8$ and APR $= .0933$.

$$h = \frac{n \times \frac{APR}{12} \times \$100}{1 - \left(1 + \frac{APR}{12}\right)^{-n}} - \$100$$

$$= \frac{8 \times \frac{.0933}{12} \times \$100}{1 - \left(1 + \frac{.0933}{12}\right)^{-8}} - \$100$$

$$= \$3.53$$

(b) Use the actuarial formula for unearned interest with $k = 8$, $R = \$575$, and $h = \$3.53$.

$$U = kR\left(\frac{h}{\$100 + h}\right)$$

$$= 8(\$575)\left(\frac{\$3.53}{\$100 + \$3.53}\right)$$

$$= \$156.84$$

(c) The payoff amount is equal to the current payment added to the sum of the remaining payments minus the unearned interest.

$$\$575 + 8(\$575) - \$156.84 = \$5018.16$$

27. *Finance Company*

Find the sum of the amount borrowed and the interest that will be charged.

$$\$5000 + (\$5000)(.065)(3) = \$5975$$

Then, the finance charge is

$$\$5975 - \$5000 = \$975.$$

The finance charge per \$100 of the amount financed would be

$$\frac{\$975}{\$5000} \times 100 = \$19.50.$$

In Table 3, the number closest to 19.50 in the "36 payments" row is 19.57. Read up to find the APR, which is 12.0%.

Credit Union

Find the total amount she would pay.

$$36(\$164.50) = \$5922.00$$

Then, the finance charge is

$$\$5922 - \$5000 = \$922.$$

The finance charge per $100 of the amount financed would be

$$\frac{\$922}{\$5000} \times 100 = \$18.44.$$

In Table 3, the number closest to 18.44 in the "36 payments" row is 18.71. Read up to find the APR, which is 11.5%.

The credit union offers the better choice.

28. First calculate the original finance charge. If the loan were not paid off early, the total payments would be

$$36(\$164.50) = \$5922.00,$$

and the finance charge would be

$$\$5922 - \$5000 = \$922.$$

Use the rule of 78 to find the unearned interest if the loan is paid off early. Use $k = 36 - 30 = 6$, $n = 36$, and $F = \$922$.

$$U = \frac{k(k+1)}{n(n+1)} \times F$$
$$= \frac{6(6+1)}{36(36+1)} \times \$922$$
$$= \$29.07$$

29. Use the finance charge formula with $n = 6$ and APR $= .115$. (See Exercise 27.)

$$h = \frac{n \times \frac{APR}{12} \times \$100}{1 - \left(1 + \frac{APR}{12}\right)^{-n}} - \$100$$
$$= \frac{6 \times \frac{.115}{12} \times \$100}{1 - \left(1 + \frac{.115}{12}\right)^{-6}} - \$100$$
$$= \$3.38$$

Then, use the actuarial formula for unearned interest with $k = 6$, $R = \$164.50$, and $h = \$3.38$.

$$U = kR\left(\frac{h}{\$100 + h}\right)$$
$$= 6(\$164.50)\left(\frac{\$3.38}{\$100 + \$3.38}\right)$$
$$= \$32.27$$

30. In Exercise 29, we found that the unearned interest by the actuarial method is $32.27. The payoff amount is equal to the current payment added to the sum of the scheduled remaining payments minus the unearned interest.

$$\$164.50 + 6(\$164.50) - \$32.27 = \$1119.23$$

31. Writing exercise

32. $\text{APR} = \dfrac{2n}{n+1} \times r$
$$= \frac{2 \cdot 24}{24 + 1} \times .07$$
$$= .1345$$

To the nearest half percent, this is 13.5%.

33. Writing exercise

34. Use the rule of 78 to set up an inequality with $n = 24$.

$$\frac{k(k+1)}{n(n+1)} \times F \geq .10F$$
$$\frac{k(k+1)}{n(n+1)} \geq .10$$
$$k(k+1) \geq .10 \times n(n+1)$$
$$k(k+1) \geq .10 \times 24(24+1)$$
$$k^2 + k \geq 60$$

This could be solved as a quadratic inequality; however, trial and error could also be applied. Because k must be an integer, try $k = 7$. Test the inequality:

$$7^2 + 7 \geq 60. \text{ Not true.}$$

Try $k = 8$. Test the inequality:

$$8^2 + 8 \geq 60. \text{ True.}$$

The value of k must be at least 8.

35. Use the rule of 78 to set up an inequality with $n = 36$.

$$\frac{k(k+1)}{n(n+1)} \times F \geq .10F$$
$$\frac{k(k+1)}{n(n+1)} \geq .10$$
$$k(k+1) \geq .10 \times n(n+1)$$
$$k(k+1) \geq .10 \times 36(36+1)$$
$$k^2 + k \geq 133.2$$

This could be solved as a quadratic inequality; however, trial and error could also be applied. Because k must be an integer, try $k = 11$. Test the inequality:

$$11^2 + 11 \geq 133.2. \text{ Not true.}$$

Try $k = 12$. Test the inequality:

$$12^2 + 12 \geq 133.2. \text{ True.}$$

The value of k must be at least 12.

36. Use the rule of 78 to set up an inequality with $n = 48$.

$$\frac{k(k+1)}{n(n+1)} \times F \geq .10F$$
$$\frac{k(k+1)}{n(n+1)} \geq .10$$
$$k(k+1) \geq .10 \times n(n+1)$$
$$k(k+1) \geq .10 \times 48(48+1)$$
$$k^2 + k \geq 235.2$$

This could be solved as a quadratic inequality; however, trial and error could also be applied. Because k must be an integer, try $k = 14$. Test the inequality:

$$14^2 + 14 \geq 235.2. \text{ Not true.}$$

Try $k = 15$. Test the inequality:

$$15^2 + 15 \geq 235.2. \text{ True.}$$

The value of k must be at least 15.

37. Writing exercise
38. Writing exercise
39. Writing exercise
40. Writing exercise
41. Writing exercise
42. Writing exercise

14.4 EXERCISES

1. In Table 4, find the 10.0% row and read across to the column for 20 years to find entry $9.65022. Since this is the monthly payment amount needed to amortize a loan of $1000 and this loan is for $70,000, the required monthly payment is

$$70 \times \$9.65022 = \$675.52.$$

2. In Table 4, find the 11.0% row and read across to the column for 15 years to find entry $11.36597. Since this is the monthly payment amount needed to amortize a loan of $1000 and this loan is for $50,000, the required monthly payment is

$$50 \times \$11.36597 = \$568.30.$$

3. Because the interest rate of 8.7 is not in Table 4, use the formula for regular monthly payment with $P = \$57,300$, $r = .087$, and $t = 25$.

$$R = \frac{P\left(\frac{r}{12}\right)\left(1 + \frac{r}{12}\right)^{12t}}{\left(1 + \frac{r}{12}\right)^{12t} - 1}$$
$$= \frac{\$57300\left(\frac{.087}{12}\right)\left(1 + \frac{.087}{12}\right)^{12(25)}}{\left(1 + \frac{.087}{12}\right)^{12(25)} - 1}$$
$$= \$469.14$$

4. Because the interest rate of 7.9 is not in Table 4, use the formula for regular monthly payment with $P = \$85,000$, $r = .079$, and $t = 30$.

$$R = \frac{P\left(\frac{r}{12}\right)\left(1 + \frac{r}{12}\right)^{12t}}{\left(1 + \frac{r}{12}\right)^{12t} - 1}$$
$$= \frac{\$85000\left(\frac{.079}{12}\right)\left(1 + \frac{.079}{12}\right)^{12(30)}}{\left(1 + \frac{.079}{12}\right)^{12(30)} - 1}$$
$$= \$617.78$$

5. In Table 4, find the 12.5% row and read across to the column for 25 years to find entry $10.90354. Since this is the monthly payment amount needed to amortize a loan of $1000, and this loan is for $227,750, the required monthly payment is

$$\frac{\$227750}{\$1000} \times \$10.90354 = \$2483.28.$$

6. Because the interest rate of 15.5% is not in Table 4, use the formula for regular monthly payment with $P = \$95,450$, $r = .155$, and $t = 5$.

$$R = \frac{P\left(\frac{r}{12}\right)\left(1 + \frac{r}{12}\right)^{12t}}{\left(1 + \frac{r}{12}\right)^{12t} - 1}$$
$$= \frac{\$95450\left(\frac{.155}{12}\right)\left(1 + \frac{.155}{12}\right)^{12(5)}}{\left(1 + \frac{.155}{12}\right)^{12(5)} - 1}$$
$$= \$2295.88$$

7. Because the interest rate of 7.6 and the term of the loan, 22 years, are not in Table 4, use the formula for regular monthly payment with $P = \$132,500$, $r = .076$, and $t = 22$.

$$R = \frac{P\left(\frac{r}{12}\right)\left(1 + \frac{r}{12}\right)^{12t}}{\left(1 + \frac{r}{12}\right)^{12t} - 1}$$
$$= \frac{\$132500\left(\frac{.076}{12}\right)\left(1 + \frac{.076}{12}\right)^{12(22)}}{\left(1 + \frac{.076}{12}\right)^{12(22)} - 1}$$
$$= \$1034.56$$

8. In Table 4, find the 5.5% row and read across to the column for 10 years to find entry $10.85263. Since this is the monthly payment amount needed to amortize a loan of $1000 and this loan is for $205,000, the required monthly payment is

$$\frac{\$205000}{\$1000} \times \$10.85263 = \$2224.79.$$

9. (a) To find the total payment, use Table 4. Find the row for 10.0% interest; read over to the 30 year column to find 8.77572. Since this is the monthly payment amount needed to amortize a loan of $1000 and this loan is for $58,500, the required monthly payment is

$$\frac{\$58500}{\$1000} \times \$8.77572 = \$513.38.$$

(b) This total payment includes both principal and interest. For the first month, interest is charged on the full amount of the mortgage, so use the formula $I = Prt$, with $P = 58500$, $r = .10$, and $t = 1/12$.

$$(\$58500)(.10)\left(\frac{1}{12}\right) = \$487.50$$

(c) The remainder of the total payment is applied to the principal, so the principal payment is

$$\$513.38 - \$487.50 = \$25.88.$$

(d) The balance of the principal is

$$\$58500 - \$25.88 = \$58{,}474.12.$$

10. (a) To find the total payment, use Table 4. Find the row for 8.5% interest; read over to the 20 year column to find $8.67823. Since this is the monthly payment amount needed to amortize a loan of $1000, and this loan is for $87000, the required monthly payment is

$$\frac{\$87000}{\$1000} \times \$8.67823 = \$755.01.$$

(b) This total payment includes both principal and interest. For the first month, interest is charged on the full amount of the mortgage, so use the formula $I = Prt$, with $P = \$87{,}000$, $r = .085$, and $t = 1/12$.

$$(\$87000)(.085)\left(\frac{1}{12}\right) = \$616.25$$

(c) The remainder of the total payment is applied to the principal, so the principal payment is

$$\$755.01 - \$616.25 = \$138.76.$$

(d) The balance of the principal is

$$\$87000 - \$138.76 = \$86{,}861.24.$$

11. (a) To find the total payment, use Table 4. Find the row for 6.5% interest; read over to the 15 year column to find $8.71107. Since this is the monthly payment amount needed to amortize a loan of $1000, and this loan is for $143,200, the required monthly payment is

$$\frac{\$143{,}200}{\$1000} \times \$8.71107 = \$1247.43.$$

(b) This total payment includes both principal and interest. For the first month, interest is charged on the full amount of the mortgage, so use the formula $I = Prt$, with $P = \$143{,}200$, $r = .065$, and $t = 1/12$.

$$(\$143200)(.065)\left(\frac{1}{12}\right) = \$775.67.$$

(c) The remainder of the total payment is applied to the principal, so the principal payment is

$$\$1247.43 - \$775.67 = \$471.76.$$

(d) The balance of the principal is

$$\$143200 - \$471.76 = \$142{,}728.24.$$

(e) Every monthly payment is the same, so the second monthly payment is $1247.43.

(f) The interest payment for the second month is

$$(\$142728.24)(.065)\left(\frac{1}{12}\right) = \$773.11.$$

(g) The principal payment for the second month is

$$\$1247.43 - \$773.11 = \$474.32.$$

(h) The balance of principal after the second month is

$$\$142{,}728.24 - \$474.32 = \$142{,}253.92$$

12. (a) To find the total payment, use Table 4. Find the row for 9% interest; read over to the 25 year column to find $8.39196. Since this is the monthly payment amount needed to amortize a loan of $1000 and this loan is for $124,750, the required monthly payment is

$$\frac{\$124750}{\$1000} \times \$8.39196 = \$1046.90.$$

(b) This total payment includes both principal and interest. For the first month, interest is charged on the full amount of the mortgage, so use the formula $I = Prt$, with $P = \$124750$, $r = .09$, and $t = 1/12$.

$$(\$124750)(.09)\left(\frac{1}{12}\right) = \$935.63$$

(c) The remainder of the total payment is applied to the principal, so the principal payment is

$$\$1046.90 - \$935.63 = \$111.27.$$

(d) The balance of the principal is

$$\$124750 - \$111.27 = \$124{,}638.73.$$

(e) Every monthly payment is the same, so the second monthly payment is $1046.90.

(f) The interest payment for the second month is

$$(\$124,638.73)(.09)\left(\frac{1}{12}\right) = \$934.79.$$

(g) The principal payment for the second month is

$$\$1046.90 - \$934.79 = \$112.11.$$

(h) The balance of principal after the second month is

$$\$124,638.73 - \$112.11 = \$124,526.62.$$

13. (a) Because the interest rate of 8.2 is not in Table 4, use the formula for regular monthly payment with $P = \$113,650$, $r = .082$, and $t = 10$.

$$R = \frac{P\left(\frac{r}{12}\right)\left(1+\frac{r}{12}\right)^{12t}}{\left(1+\frac{r}{12}\right)^{12t} - 1}$$

$$= \frac{\$113650\left(\frac{.082}{12}\right)\left(1+\frac{.082}{12}\right)^{12(10)}}{\left(1+\frac{.082}{12}\right)^{12(10)} - 1}$$

$$= \$1390.93$$

(b) This total payment includes both principal and interest. For the first month, interest is charged on the full amount of the mortgage, so use the formula $I = Prt$, with $P = \$113,650$, $r = .082$, and $t = 1/12$.

$$(\$113650)(.082)\left(\frac{1}{12}\right) = \$776.61$$

(c) The remainder of the total payment is applied to the principal, so the principal payment is

$$\$1390.93 - \$776.61 = \$614.32.$$

(d) The balance of the principal is

$$\$113650 - \$614.32 = \$113,035.68.$$

(e) Every monthly payment is the same, so the second monthly payment is $1390.93.

(f) The interest payment for the second month is

$$(\$113035.68)(.082)\left(\frac{1}{12}\right) = \$772.41.$$

(g) The principal payment for the second month is

$$\$1390.93 - \$772.41 = \$618.52.$$

(h) The balance of principal after the second month is

$$\$113,035.68 - \$618.52 = \$112,417.16.$$

14. (a) Because the term of the loan, 16 years, is not in Table 4, use the formula for regular monthly payment with $P = \$150,000$, $r = .0625$, and $t = 16$.

$$R = \frac{P\left(\frac{r}{12}\right)\left(1+\frac{r}{12}\right)^{12t}}{\left(1+\frac{r}{12}\right)^{12t} - 1}$$

$$= \frac{\$150000\left(\frac{.0625}{12}\right)\left(1+\frac{.0625}{12}\right)^{12(16)}}{\left(1+\frac{.0625}{12}\right)^{12(16)} - 1}$$

$$= \$1237.79$$

(b) This total payment includes both principal and interest. For the first month, interest is charged on the full amount of the mortgage, so use the formula $I = Prt$, with $P = \$150,000$, $r = .0625$, and $t = 1/12$.

$$(\$150000)(.0625)\left(\frac{1}{12}\right) = \$781.25$$

(c) The remainder of the total payment is applied to the principal, so the principal payment is

$$\$1237.79 - \$781.25 = \$456.54.$$

(d) The balance of the principal is

$$\$150000 - \$456.54 = \$149,543.46.$$

(e) Every monthly payment is the same, so the second monthly payment is $1237.79.

(f) The interest payment for the second month is

$$(\$149543.46)(.0625)\left(\frac{1}{12}\right) = \$778.87.$$

(g) The principal payment for the second month is

$$\$1237.79 - \$778.87 = \$458.92.$$

(h) The balance of principal after the second month is

$$\$149543.46 - \$458.92 = \$149,084.54.$$

15. Use Table 4 to find the monthly amortization payment (principal and interest).

$$\frac{\$62300}{\$1000} \times \$7.75299 = \$483.01$$

The monthly tax and insurance payment is

$$\frac{\$610 + \$220}{12} = \$69.17.$$

The total monthly payment, including taxes and insurance, is

$$\$483.01 + \$69.17 = \$552.18.$$

16. Use Table 4 to find the monthly amortization payment (principal and interest).

$$\frac{\$51800}{\$1000} \times \$9.08701 = \$470.71$$

The monthly tax and insurance payment is

$$\frac{\$570 + \$145}{12} = \$59.58.$$

The total monthly payment, including taxes and insurance, is

$$\$470.71 + \$59.58 = \$530.29.$$

17. Use Table 4 to find the monthly amortization payment (principal and interest).

$$\frac{\$89560}{\$1000} \times \$11.35480 = \$1016.94$$

The monthly tax and insurance payment is

$$\frac{\$915 + \$409}{12} = \$110.33.$$

The total monthly payment, including taxes and insurance, is

$$\$1016.94 + \$110.33 = \$1127.27.$$

18. Use Table 4 to find the monthly amortization payment (principal and interest).

$$\frac{\$72890}{\$1000} \times \$8.17083 = \$595.57$$

The monthly tax and insurance payment is

$$\frac{\$1850 + \$545}{12} = \$199.58.$$

The total monthly payment, including taxes and insurance, is

$$\$595.57 + \$199.58 = \$795.15.$$

19. Because the interest rate is not in Table 4, use the formula for regular monthly payment with $P = \$115,400$, $r = .088$, and $t = 20$.

$$R = \frac{P\left(\frac{r}{12}\right)\left(1 + \frac{r}{12}\right)^{12t}}{\left(1 + \frac{r}{12}\right)^{12t} - 1}$$

$$= \frac{\$115400\left(\frac{.088}{12}\right)\left(1 + \frac{.088}{12}\right)^{12(20)}}{\left(1 + \frac{.088}{12}\right)^{12(20)} - 1}$$

$$= \$1023.49$$

The monthly tax and insurance payment is

$$\frac{\$1295.16 + \$444.22}{12} = \$144.95.$$

The total monthly payment, including taxes and insurance, is

$$\$1023.49 + \$144.95 = \$1168.44.$$

20. Because the interest rate is not in Table 4, use the formula for regular monthly payment with $P = \$128,100$, $r = .113$, and $t = 30$.

$$R = \frac{P\left(\frac{r}{12}\right)\left(1 + \frac{r}{12}\right)^{12t}}{\left(1 + \frac{r}{12}\right)^{12t} - 1}$$

$$= \frac{\$128100\left(\frac{.113}{12}\right)\left(1 + \frac{.113}{12}\right)^{12(30)}}{\left(1 + \frac{.113}{12}\right)^{12(30)} - 1}$$

$$= \$1249.05$$

The monthly tax and insurance payment is

$$\frac{\$1476.53 + \$565.77}{12} = \$170.19.$$

The total monthly payment, including taxes and insurance, is

$$\$1249.05 + \$170.19 = \$1419.24.$$

21. $12 \times 30 = 360$

22. There are 360 monthly payments of $1076.48 each for principal and interest, so the total amount that will be paid for principal and interest is

$$360 \times \$1076.48 = \$387,532.80.$$

23. The total interest is

$$\$387,532.80 - \$140,000 = \$247,532.80.$$

24. The total interest is greater than the amount financed.

$$\$247,532.80 - \$140,000 = \$107,532.80$$

25. (a) In Table 5 read the heading of the column on the left to see that the monthly payment is $304.01.

    (b) Read the heading of the column on the right to see that the monthly payment is $734.73.

26. (a) Compare the principal payment to the monthly payment in percent form.

$$\frac{\$79.01}{\$304.01} \times 100 = 26\%$$

    (b) Compare the principal payment to the monthly payment in percent form.

$$\frac{\$9.73}{\$734.73} \times 100 = 1.3\%$$

27. Payment number 12 is the last payment for the year.

    (a) Balance of principal is $59,032.06.

    (b) Balance of principal is $59,875.11.

28. Payment number 240 is the last payment for the 20th year.

    (a) Balance of principal is $29,333.83.

    (b) Balance of principal is $46,417.87.

29. Compare the Interest Payment column with the Principal Payment column.

    (a) The first payment in which the principal payment is higher is number 176.

    (b) The first payment in which the principal payment is higher is number 304

30. Find the last entry in the Interest Payment column, Payment Number 360.

    (a) $1.14

    (b) $8.77

31. Use Table 4 and an annual rate of 7.5% interest. Read across to the 10-year column to find 11.87018; this is the monthly payment for a $1000 mortgage. Multiply by 60 to obtain the monthly payment for a $60,000 mortgage:

    $$60 \times \$11.87018 = \$712.21.$$

    There would be $10 \times 12 = 120$ monthly payments:

    $$120 \times \$712.21 = \$85,465.20.$$

    Then, the interest is the difference between this amount and the loan amount.

    $$\$85,465.20 - \$60000 = \$25,465.20$$

32. Use Table 4 and an annual rate of 7.5% interest. Read across to the 20-year column to find 8.05593; this is the monthly payment for a $1000 mortgage. Multiply by 60 to obtain the monthly payment for a $60,000 mortgage:

    $$60 \times \$8.05593 = \$483.36.$$

    There would be $20 \times 12 = 240$ monthly payments:

    $$240 \times \$483.36 = \$116,006.40.$$

    Then, the interest is the difference between this amount and the loan amount.

    $$\$116006.40 - \$60000 = \$56,006.40$$

33. Use Table 4 and an annual rate of 7.5% interest. Read across to the 30-year column to find 6.99215; this is the monthly payment for a $1000 mortgage. Multiply by 60 to obtain the monthly payment for a $60,000 mortgage:

    $$60 \times \$6.99215 = \$419.53.$$

    There would be $30 \times 12 = 360$ monthly payments:

    $$360 \times \$419.53 = \$151,030.80.$$

    Then, the interest is the difference between this amount and the loan amount is

    $$\$151030.80 - \$60000 = \$91,030.80.$$

34. Because the length of the loan is not in Table 4, use the formula for regular monthly payment with $P = \$60,000$, $r = .075$, and $t = 40$.

    $$R = \frac{P\left(\frac{r}{12}\right)\left(1+\frac{r}{12}\right)^{12t}}{\left(1+\frac{r}{12}\right)^{12t} - 1}$$

    $$= \frac{\$60000\left(\frac{.075}{12}\right)\left(1+\frac{.075}{12}\right)^{12(40)}}{\left(1+\frac{.075}{12}\right)^{12(40)} - 1}$$

    $$= \$394.84$$

    There would be $40 \times 12 = 480$ monthly payments:

    $$480 \times \$394.84 = \$189,523.20.$$

    Then, the interest is the difference between this amount and the loan amount is

    $$\$189523.20 - \$60000 = \$129,523.20.$$

35. (a) Add the initial index rate and the margin to obtain the ARM interest rate.

    $$6.5 + 2.5 = 9.0\%$$

    In Table 4 find 9.0% in the annual rate column and read across to the column for a 20-year mortgage to find 8.99726. Multiply this figure by 75 to obtain the initial monthly payment.

    $$75 \times \$8.99726 = \$674.79$$

    (b) The interest rate for the second adjustment period is given by the ARM interest rate.

    $$8.0 + 2.5 = 10.5\%$$

    Use the formula for Regular monthly payment with $P = \$73,595.52$ (the Adjusted Balance), $r = .105$, and $t = 19$.

    $$R = \frac{P\left(\frac{r}{12}\right)\left(1+\frac{r}{12}\right)^{12t}}{\left(1+\frac{r}{12}\right)^{12t} - 1}$$

    $$= \frac{\$73595.52\left(\frac{.105}{12}\right)\left(1+\frac{.105}{12}\right)^{12(19)}}{\left(1+\frac{.105}{12}\right)^{12(19)} - 1}$$

    $$= \$746.36$$

    (c) The change in monthly payment is the difference in the two amounts from parts (a) and (b).

    $$\$746.36 - \$674.79 = \$71.57$$

36. (a) Add the initial index rate and the margin to obtain the ARM interest rate.

    $$7.2 + 2.75 = 9.95\%$$

Because this interest is not in Table 4, use the formula for regular monthly payment with $P = \$44,500$, $r = .0995$, and $t = 30$.

$$R = \frac{P\left(\frac{r}{12}\right)\left(1+\frac{r}{12}\right)^{12t}}{\left(1+\frac{r}{12}\right)^{12t}-1}$$

$$= \frac{\$44,500\left(\frac{.0995}{12}\right)\left(1+\frac{.0995}{12}\right)^{12(30)}}{\left(1+\frac{.0995}{12}\right)^{12(30)}-1}$$

$$= \$388.88$$

(b) The interest rate for the second adjustment period is given by the ARM interest rate.

$$6.6 + 2.75 = 9.35\%$$

Use the formula for Regular monthly payment with $P = \$43,669.14$ (the Adjusted Balance), $r = .0935$, and $t = 30 - 3 = 27$.

$$R = \frac{P\left(\frac{r}{12}\right)\left(1+\frac{r}{12}\right)^{12t}}{\left(1+\frac{r}{12}\right)^{12t}-1}$$

$$= \frac{\$43,669.14\left(\frac{.0935}{12}\right)\left(1+\frac{.0935}{12}\right)^{12(27)}}{\left(1+\frac{.0935}{12}\right)^{12(27)}-1}$$

$$= \$370.20$$

(c) The change in monthly payment is the difference in the two amounts from parts (a) and (b).

$$\$388.88 - \$370.20 = \$18.68$$

37. (a) The ARM interest rate is 2% plus 7.5% or 9.5%. Then,

$$\frac{.095 \times \$50000}{12} = \$395.83.$$

(b) To find the first monthly payment first add 2% and 7.5% to obtain 9.5% as the ARM interest rate. In Table 4 find 9.5% in the annual rate column and read across to the 20-year mortgage column to find 9.32131. Multiply this figure by 50 to obtain the monthly payment for the $50,000 mortgage.

$$50 \times \$9.32131 = \$466.07$$

38. (a) The interest rate for the second year is the sum of the interest rate and margin.

$$10.0\% + 2\% = 12.0\%$$

However, this rate would be a 2.5% increase, and the periodic rate cap limits the increase to 2%, so the interest rate for the second year is only 11.5%. The balance of principal at the end of the first year is $49,119.48. Interest for the first payment of the second year is based on this balance. Use the formula for simple interest with $P = \$49,119.48$, $r = .115$, and $t = 1/12$.

$$I = Prt$$

$$= (\$49119.48)(.115)\left(\frac{1}{12}\right)$$

$$= \$470.73$$

(b) The remaining term of the mortgage is 19 years. Since Table 4 does not include this term, use the regular monthly payment formula with $P = \$49,119.48$, $r = .115$, and $t = 19$.

$$R = \frac{P\left(\frac{r}{12}\right)\left(1+\frac{r}{12}\right)^{12t}}{\left(1+\frac{r}{12}\right)^{12t}-1}$$

$$= \frac{\$49119.48\left(\frac{.115}{12}\right)\left(1+\frac{.115}{12}\right)^{12(19)}}{\left(1+\frac{.115}{12}\right)^{12(19)}-1}$$

$$= \$531.09$$

39. From Exercise 38, the monthly payment for the first month of the second year is $531.09. The monthly adjustment at the end of the second year is the difference between this amount and the monthly payment amount at the end of the first year.

$$\$531.09 - \$466.07 = \$65.02$$

40. Writing exercise

41. The down payment is 20% of the purchase price of the house.

$$.20 \times \$175,000 = \$35,000$$

Then, the mortgage amount is the difference between the purchase price of the house and this figure.

$$\$175,5000 - \$35,0C0 = \$140,000$$

42. From Exercise 41, the mortgage amount is $140,000. The loan fee is 2 points, which means 2% of the mortgage amount. Therefore, the loan fee is

$$.02 \times \$140,000 = \$2800.$$

43. From Exercise 42, the Loan fee is $.02 \times \$140,000 = \$2800$. Add this figure to the other closing costs listed in the text to obtain the total closing costs of $4275.

44. From Exercise 43, the total closing costs are $4275. The down payment is 20% of the purchase price or $35,000. Therefore, the total amount of cash required of the buyer at closing is

$$\$4275 + \$35,000 = \$39,275.$$

45. Writing exercise

46. Writing exercise

47. Writing exercise

48. Writing exercise

49. Writing exercise
50. Writing exercise
51. Writing exercise
52. Writing exercise
53. Writing exercise

14.5 EXERCISES

1. $29.75
2. 438,600 shares
3. 7¢ per share lower
4. $15.60
5. $47.48
6. $17.40
7. $1.72 per share
8. $2.23 per share
9. 7600 shares
10. 74¢ per share higher
11. 10.6% higher
12. 40.1% lower
13. 35
14. 18
15. $200 \times \$25.70 = \$5140.00$
16. $400 \times \$17.62 = \$17,372.00$
17. $300 \times \$3.87 = \$1161.00$
18. $700 \times \$15.45 = \$10,815.00$
19. The basic cost of stock (Principal Amount) is given by
   (Price per share) × (Number of shares) = $36.69 × 60
   = $2201.40.

   Since this principal amount falls in the first tier of the commission structure (see table in text) the broker's commission is $35 plus 1.7% of this amount:

   Broker's commission = $35 + .017 × $2201.40
   = $72.42.

   The total cost of the shares is
   $2201.40 + $72.42 = $2273.82.

20. Principal Amount = $81.38 × 70 = $5696.60
    Commission = $65 + .0066 × $5696.60 = $102.60
    Total Cost = $5799.20

21. The basic cost of stock (Principal Amount) is given by
    (Price per share) × (Number of shares)
    = $50.12 × 355 = $17792.60.

    Since commission in this exercise is automated and the number of shares is less than 1000, the broker's commission is $29.95 (see table in text).

    The total cost of the shares is
    $17792.60 + $29.95 = $17,822.55.

22. Principal Amount = $38 × 585 = $22230
    Commission = $29.95
    Total Cost = $22,259.95

23. The basic cost of stock (Principal Amount) is given by
    $9.86 × 2500 = $24650.

    Since this principal amount falls in the fourth tier of the commission structure (see table in text) the broker's commission is $100 plus .22% of this amount:

    Broker's commission = $100 + .0022 × $24650
    = $154.23.

    The total cost of the shares is
    $24650 + $154.23 = $24,804.23.

24. Principal Amount = $11.38 × 1500 = $17070
    Commission = $.03 × 1500 = $45
    Total Cost = $17,115

25. The basic cost of stock (Principal Amount) is given by
    (Price per share) × (Number of shares)
    = $23.22 × 2400 = $55728.

    Since commission in this exercise is automated and the number of shares is more than 1000, the broker's commission is $.03 × 2400 = $72 (see table in text).

    The total cost of the shares is
    $55728 + $72 = $55,800.

26. Principal Amount = $5.65 × 10000 = $56500
    Commission = $155 + .0011 × 56500 = $217.15
    Total Cost = $56,717.15

27. The basic cost (Principal Amount) of stock is given by
    $31.36 × 400 = $12544.

    Since this principal amount falls in the third tier of the commission structure (see table in text), the broker's commission is $76 plus .34% of the principal:

    Broker's commission = $76 + .0034 × $12544
    = $118.65.

The SEC fee is

$$\frac{\$12544}{\$1000} \times 3.01¢ = 37.7574¢ = \$.38.$$

The seller receives

$$\$12544 - \$118.65 - \$.38 = \$12,424.97.$$

28. The basic cost (Principal Amount) of stock is given by

    $\$1.07 \times 600 = \$642$

    Since this principal amount falls in the first tier of the commission structure (see table in text), the broker's commission is $35 plus 1.7% of the principal:

    Broker's commission $= \$35 + .017 \times \$642$
    $= \$45.91.$

    The SEC fee is

    $$\frac{\$642}{\$1000} \times 3.01¢ = 1.93242¢ = \$.02.$$

    The seller receives

    $$\$642 - \$45.91 - \$.02 = \$596.07.$$

29. The basic cost of stock (Principal Amount) is given by

    (Price per share) × (Number of shares)
    $= \$11.48 \times 500 = \$5740.$

    Since commission in this exercise is automated and the number of shares is less than 1000, the broker's commission is $29.95 (see table in text).

    The SEC fee is

    $$\frac{\$5740}{\$1000} \times 3.01¢ = 17.2774¢ = \$.17.$$

    The seller receives

    $$\$5740 - \$29.95 - \$.17 = \$5709.88.$$

30. The basic cost of stock (Principal Amount) is given by

    (Price per share) × (Number of shares)
    $= \$56.72 \times 700 = \$39704.$

    Since commission in this exercise is automated and the number of shares is less than 1000, the broker's commission is $29.95 (see table in text).

    The SEC fee is

    $$\frac{\$39704}{\$1000} \times 3.01¢ = 119.50904¢ = \$1.20.$$

    The seller receives

    $$\$39704 - \$29.95 - \$1.20 = \$39,672.85.$$

31. The basic cost of stock (Principal Amount) is given by

    (Price per share) × (Number of shares)
    $= \$12.95 \times 1350 = \$17482.50.$

    Since commission in this exercise is automated and the number of shares is more than 1000, the commission is $\$.03 \times 1350 = \$40.50$ (see table in text).

    The SEC fee is

    $$\frac{\$17482.50}{\$1000} \times 3.01¢ = 52.622325¢ = \$.53.$$

    The seller receives

    $$\$17482.50 - \$40.50 - \$.53 = \$17,441.47.$$

32. The basic cost of stock (Principal Amount) is given by

    (Price per share) × (Number of shares)
    $= \$22.71 \times 2740 = \$62225.40.$

    Since commission in this exercise is automated and the number of shares is more than 1000, the commission is $\$.03 \times 2740 = \$82.20$ (see table in text).

    The SEC fee is

    $$\frac{\$62225.40}{\$1000} \times 3.01¢ = 187.298454¢ = \$1.87.$$

    The seller receives

    $$\$62225.40 - \$82.20 - \$1.87 = \$62141.33.$$

33. The basic cost (Principal Amount) of stock is given by

    $\$38 \times 1480 = \$56240.$

    Since this principal amount falls in the fifth tier of the commission structure (see table in text), the broker's commission is $155 plus .11% of the principal:

    Broker's commission $= \$155 + .0011 \times \$56240$
    $= \$216.86.$

    The SEC fee is

    $$\frac{\$56240}{\$1000} \times 3.01¢ = 169.2824¢ = \$1.69.$$

    The seller receives

    $$\$56240 - \$216.86 - \$1.69 = \$56,021.45.$$

34. The basic cost (Principal Amount) of stock is given by

    $\$28.50 \times 1270 = \$36195.$

    Since this principal amount falls in the fourth tier of the commission structure (see table in text), the broker's commission is $100 plus .22% of the principal:

Broker's commission = $100 + .0022 \times \$36195$
$= \$179.63.$

The SEC fee is

$$\frac{\$36195}{\$1000} \times 3.01¢ = 108.94695¢ = \$1.09.$$

The seller receives

$$\$36195 - \$179.63 - \$1.09 = \$36,014.28.$$

35. Purchase cost

   Principal Amount $= \$1.12 \times 100 = \$112$
   Commission $= \$35 + .017 \times 112 = \$36.90$
   Total Cost $= \$112 + \$36.90 = 148.90$

   Sales profit

   Principal Amount $= \$7.80 \times 20 = \$156$
   Commission $= \$35 + .017 \times 156 = \$37.65$

   The SEC fee $= \dfrac{\$156}{\$1000} \times 3.01¢ = .46956¢ = \$.01$

   He receivers $\$156 - \$37.65 - \$.01 = \$118.34$

   This results in a $\$148.90 - \$118.34 = \$30.56$ payout.

36. Purchase cost

   Principal Amount $= \$23.40 \times 800 = \$18720$
   Commission (Automated) $= \$29.95$
   Total Cost $= \$18720 + \$29.95 = \$18749.95$

   Sales profit

   Principal Amount $= \$28.94 \times 1200 = \$34728$
   Commission(Automated) $= \$.03 \times 1200 = \$36$

   The SEC fee $= \dfrac{\$34728}{\$1000} \times 3.01¢ = 104.53128¢ = \$1.05$

   He receivers $\$34728 - \$36 - \$1.05 = \$34690.95$

   This results in $\$34690.95 - \$18749.95 = \$15,941.00$ net taken in.

37. (a) The total purchase price is

   $(\$20 \text{ per share}) \times (40 \text{ shares}) = \$800.$

   (b) The total dividend amount is

   $(\$2 \text{ per share}) \times (40 \text{ shares}) = \$80.$

   (c) The capital gain is found by multiplying the change in price per share times the number of shares.

   $(\$44 - \$20) \times (40 \text{ shares}) = \$960$

   (d) The total return is the sum of the dividends and the capital gain.

   $\$80 + \$960 = \$1040$

   (e) The percentage return is the quotient of the total return and the total cost as a percent.

   $$\frac{\$1040}{\$800} \times 100 = 130\%$$

38. (a) The total purchase price is

   $(\$25 \text{ per share}) \times (20 \text{ shares}) = \$500.$

   (b) The total dividend amount is

   $(\$1 \text{ per share}) \times (20 \text{ shares}) = \$20.$

   (c) The capital gain is found by multiplying the change in price per share times the number of shares.

   $(\$22 - \$25) \times (20 \text{ shares}) = -\$60$

   (d) The total return is the sum of the dividends and the capital gain.

   $\$20 + (-\$60) = -\$40$

   (e) The percentage return is the quotient of the total return and the total cost as a percent.

   $$\frac{-\$40}{\$500} \times 100 = -8\%$$

39. (a) The total purchase price is

   $(\$12.50 \text{ per share}) \times (100 \text{ shares}) = \$1250.$

   (b) The total dividend amount is

   $(\$1.08 \text{ per share}) \times (100 \text{ shares}) = \$108.$

   (c) The capital gain is found by multiplying the change in price per share times the number of shares.

   $(\$10.15 - \$12.50) \times (100 \text{ shares}) = -\$235$

   (d) The total return is the sum of the dividends and the capital gain.

   $\$108 + (-\$235) = -\$127$

   (e) The percentage return is the quotient of the total return and the total cost as a percent.

   $$\frac{-\$127}{\$1250} \times 100 = -10.16\%$$

40. (a) The total purchase price is

   $(\$8.80 \text{ per share}) \times (200 \text{ shares}) = \$1760.$

   (b) The total dividend amount is

   $(\$1.12 \text{ per share}) \times (200 \text{ shares}) = \$224.$

   (c) The capital gain is found by multiplying the change in price per share times the number of shares.

   $(\$11.30 - \$8.80) \times (200 \text{ shares}) = \$500$

(d) The total return is the sum of the dividends and the capital gain.

$$\$224 + \$500 = \$724$$

(e) The percentage return is the quotient of the total return and the total cost as a percent.

$$\frac{\$724}{\$1760} \times 100 = 41.14\%$$

41. Using the formula for simple interest with $P = \$1000$, $r = .055$, and $t = 5$ the total return is:

$$\begin{aligned} I &= Prt \\ &= \$1000 \times .055 \times 5 \\ &= \$275. \end{aligned}$$

42. Using the formula for simple interest with $P = \$5000$, $r = .064$, and $t = 10$ the total return is:

$$\begin{aligned} I &= Prt \\ &= \$5000 \times .064 \times 10 \\ &= \$3200. \end{aligned}$$

43. Using the formula for simple interest with $P = \$10,000$, $r = .0711$, and $t = 3/12$ the total return is:

$$\begin{aligned} I &= Prt \\ &= \$10000 \times .0711 \times \frac{3}{12} \\ &= \$177.75. \end{aligned}$$

44. Using the formula for simple interest with $P = \$50,000$, $r = .0488$, and $t = 6/12$ the total return is:

$$\begin{aligned} I &= Prt \\ &= \$50000 \times .0488 \times \frac{6}{12} \\ &= \$1220. \end{aligned}$$

45. (a) Use the formula for net asset value with $A = \$875$ million, $L = \$36$ million, and $N = 80$ million.

$$\begin{aligned} NAV &= \frac{A - L}{N} \\ &= \frac{875 - 36}{80} \\ &= \$10.49 \end{aligned}$$

(b) Find the number of shares purchased by dividing the amount invested by the net asset value. Round to the nearest share.

$$\frac{\$3500}{\$10.49} = 334 \text{ shares}$$

46. (a) Use the formula for net asset value with $A = \$643$ million, $L = \$102$ million, and $N = 50$ million.

$$\begin{aligned} NAV &= \frac{A - L}{N} \\ &= \frac{643 - 102}{50} \\ &= \$10.82 \end{aligned}$$

(b) Find the number of shares purchased by dividing the amount invested by the net asset value. Round to the nearest share.

$$\frac{\$1800}{\$10.82} = 166 \text{ shares}$$

47. (a) Use the formula for net asset value with $A = \$2.31$ billion ($\$2,310$ million), $L = \$135$ million, and $N = 263$ million.

$$\begin{aligned} NAV &= \frac{A - L}{N} \\ &= \frac{2310 - 135}{263} \\ &= \$8.27 \end{aligned}$$

(b) Find the number of shares purchased by dividing the amount invested by the net asset value. Round to the nearest share.

$$\frac{\$25470}{\$8.27} = 3080 \text{ shares}$$

48. (a) Use the formula for net asset value with $A = \$1.48$ billion ($\$1480$ million), $L = \$84$ million, and $N = 112$ million.

$$\begin{aligned} NAV &= \frac{A - L}{N} \\ &= \frac{1480 - 84}{112} \\ &= \$12.46 \end{aligned}$$

(b) Find the number of shares purchased by dividing the amount invested by the net asset value. Round to the nearest share.

$$\frac{\$83250}{\$12.46} = 6681 \text{ shares}$$

49. (a) Find the monthly return by multiplying the monthly percentage return, as a decimal, times the amount invested.

$$.013 \times \$645 = \$8.39$$

(b) Find the annual return by multiplying the monthly return by 12.

$$12 \times \$8.39 = \$100.68$$

(c) The annual percentage return is the ratio of the annual return to the amount invested, expressed as a percent.

$$\frac{\$100.68}{\$645} \times 100 = 15.6\%$$

50. (a) Find the monthly return by multiplying the monthly percentage return, as a decimal, times the amount invested.

$$.009 \times \$895 = \$8.06$$

(b) Find the annual return by multiplying the monthly return by 12.

$$12 \times \$8.06 = \$96.72$$

(c) The annual percentage return is the ratio of the annual return to the amount invested, expressed as a percent.

$$\frac{\$96.72}{\$895} \times 100 = 10.8\%$$

51. (a) Find the monthly return by multiplying the monthly percentage return, as a decimal, times the amount invested.

$$.023 \times \$2498 = \$57.45$$

(b) Find the annual return by multiplying the monthly return by 12.

$$12 \times \$57.45 = \$689.40$$

(c) The annual percentage return is the ratio of the annual return to the amount invested, expressed as a percent

$$\frac{\$689.40}{\$2498} \times 100 = 27.6\%$$

52. (a) Find the monthly return by multiplying the monthly percentage return, as a decimal, times the amount invested.

$$.018 \times \$4983 = \$89.69$$

(b) Find the annual return by multiplying the monthly return by 12.

$$12 \times \$89.69 = \$1076.28$$

(c) The annual percentage return is the ratio of the annual return to the amount invested, expressed as a percent.

$$\frac{\$1076.28}{\$4983} \times 100 = 21.6\%$$

53. (a) The beginning value of the investment is the product of the beginning net asset value and the number of shares purchased.

$$\$9.63 \times 125 = \$1203.75$$

(b) To find the first monthly return, multiply the monthly percentage return, in decimal form, times the beginning value from part (a).

$$.015 \times \$1203.75 = \$18.06$$

(c) Using Example 5 from the text as a guide, find the effective annual rate of return as follows. First find the return relative.

$$1 + 1.5\% = 1 + .015 = 1.015$$

Raise this value to the 12th power to get the annual return relative and subtract 1 to get the percentage rate.

$$(1.015)^{12} - 1 \approx .1956, \text{ which is } 19.56\%$$

54. (a) The beginning value of the investment is the product of the beginning net asset value and the number of shares purchased.

$$\$12.40 \times 185 = \$2294$$

(b) To find the first monthly return, multiply the monthly percentage return, in decimal form, times the beginning value from part (a).

$$.023 \times \$2294 = \$52.76$$

(c) Using Example 5 from the text as a guide, find the effective annual rate of return as follows. First find the return relative.

$$1 + 2.3\% = 1 + .023 = 1.023$$

Raise this value to the 12th power to get the annual return relative and subtract 1 to get the percentage rate.

$$(1.023)^{12} - 1 \approx .3137, \text{ which is } 31.37\%$$

55. (a) The beginning value of the investment is the product of the beginning net asset value and the number of shares purchased.

$$\$11.94 \times 350 = \$4179$$

(b) To find the first monthly return, multiply the monthly percentage return, in decimal form, times the beginning value from part (a).

$$.0183 \times \$4179 = \$76.48$$

(c) Using Example 5 from the text as a guide, find the effective annual rate of return as follows. First find the return relative.

$$1 + 1.83\% = 1 + .0183 = 1.0183$$

Raise this value to the 12th power to get the annual return relative and subtract 1 to get the percentage rate.

$$(1.0183)^{12} - 1 \approx .2431, \text{ which is } 24.31\%$$

56. (a) The beginning value of the investment is the product of the beginning net asset value and the number of shares purchased.

    $$\$18.54 \times 548 = \$10,159.92$$

    (b) To find the first monthly return, multiply the monthly percentage return, in decimal form, times the beginning value from part (a).

    $$.0222 \times \$10159.92 = \$225.55$$

    (c) Using Example 5 from the text as a guide, find the effective annual rate of return as follows. First find the return relative.

    $$1 + 2.22\% = 1 + .0222 = 1.0222$$

    Raise this value to the 12th power to get the annual return relative and subtract 1 to get the percentage rate.

    $$(1.0222)^{12} - 1 \approx .3015, \text{ which is } 30.15\%$$

57. (a) Principal Amount

    $$10 \times \$1.00 = \$10.00$$

    (b) Broker-assisted Commission

    $$\$35 + .017 \times \$10.00 = \$35.17$$

    (c) Automated Commission = \$29.95

58. (a) Principal Amount

    $$10 \times \$100.00 = \$1000.00$$

    (b) Broker-assisted Commission

    $$\$35 + .017 \times \$1000.00 = \$52$$

    (c) Automated Commission = \$29.95

59. (a) Principal Amount

    $$400 \times \$1.00 = \$400.00$$

    (b) Broker-assisted Commission

    $$\$35 + .017 \times \$400.00 = \$41.80$$

    (c) Automated Commission = \$29.95

60. (a) Principal Amount

    $$400 \times \$100.00 = \$40,000.00$$

    (b) Broker-assisted Commission

    $$\$100 + .0022 \times \$40000.00 = \$188.00$$

    (c) Automated Commission = \$29.95

61. (a) Principal Amount

    $$4000 \times \$1.00 = \$4000.00$$

    (b) Broker-assisted Commission

    $$\$65 + .0066 \times \$4000.00 = \$91.40$$

    (c) Automated Commission

    $$\$.03 \times 4000.00 = \$120.00$$

62. (a) Principal Amount

    $$4000 \times \$100.00 = \$400,000.00$$

    (b) Broker-assisted Commission

    $$\$155 + .0011 \times \$400000.00 = \$595.00$$

    (c) Automated Commission

    $$\$.03 \times 4000.00 = \$120.00$$

63. A broker-assisted purchase is cheaper than an automated purchase only when a relatively *large* number of shares are purchased for a relatively *low* price per share.

64. Writing exercise

65. Writing exercise

66. The annual report was available free to *Wall Street Journal* readers.

67. One example is Pinnacle Systems (PCLE).

68. Writing exercise

69. *Aggressive Growth*

    $$7\% \text{ of } \$20,000 = .07 \times \$20000 = \$1400$$

    Growth

    $$43\% \text{ of } \$20,000 = .43 \times \$20000 = \$8600$$

    Growth & Income

    $$31\% \text{ of } \$20,000 = .31 \times \$20000 = \$6200$$

    Income

    $$14\% \text{ of } \$20,000 = .14 \times \$20000 = \$2800$$

    Cash

    $$5\% \text{ of } \$20,000 = .05 \times \$20000 = \$1000$$

70. *Aggressive Growth*

    $$6\% \text{ of } \$250,000 = .06 \times \$250000 = \$15,000$$

    Growth

    $$29\% \text{ of } \$250,000 = .29 \times \$250000 = \$72,500$$

Growth & Income

$$36\% \text{ of } \$250{,}000 = .36 \times \$250000 = \$90{,}000$$

Income

$$24\% \text{ of } \$250{,}000 = .24 \times \$250000 = \$60{,}000$$

Cash

$$5\% \text{ of } \$250{,}000 = .05 \times \$250000 = \$12{,}500$$

71. *Aggressive Growth*

$$2\% \text{ of } \$400{,}000 = .02 \times \$400000 = \$8000$$

Growth

$$28\% \text{ of } \$400{,}000 = .28 \times \$400000 = \$112{,}000$$

Growth & Income

$$36\% \text{ of } \$400{,}000 = .36 \times \$400000 = \$144{,}000$$

Income

$$29\% \text{ of } \$400{,}000 = .29 \times \$400000 = \$116{,}000$$

Cash

$$5\% \text{ of } \$400{,}000 = .05 \times \$400000 = \$20{,}000$$

72. *Aggressive Growth*

$$2\% \text{ of } \$845{,}000 = .02 \times \$845000 = \$16{,}900$$

Growth

$$15\% \text{ of } \$845{,}000 = .15 \times \$845000 = \$126{,}750$$

Growth & Income

$$44\% \text{ of } \$845{,}000 = .44 \times \$845000 = \$371{,}800$$

Income

$$34\% \text{ of } \$845{,}000 = .34 \times \$845000 = \$287{,}300$$

Cash

$$5\% \text{ of } \$845{,}000 = .05 \times \$845000 = \$42{,}250$$

73. First find the difference of 100% and 25%.

$$100\% - 25\% = 75\%$$

Then find 75% of 5%.

$$.75 \times .05 = .0375$$

The tax-exempt rate of return is 3.75%.

74. First find the difference of 100% and 30%.

$$100\% - 30\% = 70\%$$

Then find 70% of 7%.

$$.70 \times .07 = .049$$

The tax-exempt rate of return is 4.9%.

75. First find the difference of 100% and 35%.

$$100\% - 35\% = 65\%$$

Then find 65% of 8%.

$$.65 \times .08 = .052$$

The tax-exempt rate of return is 5.2%.

76. First find the difference of 100% and 40%.

$$100\% - 40\% = 60\%$$

Then find 60% of 10%.

$$.60 \times .10 = .06$$

The tax-exempt rate of return is 6.0%.

77. The real rate of return (RRR) is

$$14.2\% - 8.9\% = 5.3\%.$$

78. The real rate of return (RRR) is

$$6.6\% - 1.3\% = 5.3\%.$$

79. Writing exercise
80. Writing exercise
81. Writing exercise
82. Writing exercise
83. Writing exercise
84. Writing exercise
85. Writing exercise
86. Writing exercise

CHAPTER 14 TEST

1. Use the formula $A = P(1 + rt)$, with $P = \$100$; $r = .06$; and $t = 5$.

$$\begin{aligned} A &= P(1+rt) \\ &= 100(1 + .06 \cdot 5) \\ &= 100(1 + .30) \\ &= 100(1.30) \\ &= \$130 \end{aligned}$$

2. To find the final amount, use the formula for future value $A = P\left(1 + \frac{r}{m}\right)^n$, with $P = \$50$, $r = .08$, $m = 4$ and $n = 4 \cdot 2 = 8$ (compounded 4 times per year for 2 years).

$$A = P\left(1 + \frac{r}{m}\right)^n$$
$$= 50\left(1 + \frac{.08}{4}\right)^8$$
$$= 50(1.02)^8$$
$$= 50(1.171659)$$
$$= \$58.58$$

3. Use the formula $Y = \left(1 + \frac{r}{m}\right)^m - 1$, with $r = .03$ and $m = 12$.

$$Y = \left(1 + \frac{.03}{12}\right)^{12} - 1$$
$$= (1.0025)^{12} - 1$$
$$\approx .0304159569$$

This is 3.04%, to the nearest hundredth of a percent.

4. Years to double $= \dfrac{70}{\text{Annual inflation rate}}$

Years to double $= \dfrac{70}{5}$

Years to double $= 14$ years

5. Use the present value formula for compound interest with $A = \$100{,}000$, $r = .04$, $m = 2$, and $n = 2 \cdot 10 = 20$.

$$P = \frac{A}{\left(1 + \frac{r}{m}\right)^n}$$
$$= \frac{100000}{\left(1 + \frac{.04}{2}\right)^{20}}$$
$$= \frac{100000}{(1.02)^{20}}$$
$$\approx \$67297.13$$

6. The interest due is the 1.6% of $680.

$$.016 \times \$680 = \$10.88$$

7. Use the simple interest formula with $P = \$3000$, $r = .08$, and $t = 2$. Because he made a down payment of $1000, he will pay interest on only $3000. In this formula, t must be expressed in years, so 24 months is equivalent to 2 years.

$$I = Prt$$
$$= (3000)(.08)(2)$$
$$= \$480$$

8. The total amount owed is

$$P + I = \$3000 + \$480$$
$$= \$3480.$$

There are 24 monthly payments, so the amount of each monthly payment is

$$\frac{\$3480}{24} = \$145.$$

9. Find the finance charge per $100 of the amount financed.

$$\frac{\$480}{\$3000} \times 100 = \$16$$

In Table 3, find the "24 payments" row and read across to find the number closest to 16. Since 15.23 (associated with an APR of 14.0%) is the last entry in the table, choose the next step up, 14.5%, for an approximate APR.

10. Use the formula $U = kR\left(\dfrac{h}{100 + h}\right)$ to find unearned interest, with $k =$ remaining number of payments, $R =$ regular monthly payment, and $h =$ finance charge per $100. In Table 3, the number 15.80, which is closest to 16, appears in the "24 payments" row. Because the total number of payments remaining is 6, move up to the "6 payments" row to find $h = \$4.27$, the value of h needed in the actuarial method.

$$U = (6)(\$145)\left(\frac{\$4.27}{100 + \$4.27}\right)$$
$$= (6)(\$145)\left(\frac{\$4.27}{\$104.27}\right)$$
$$= \$35.63$$

11. First find the finance charge by multiplying $5 by 6.

$$6 \times \$5 = \$30$$

Find the finance charge per $100 of the amount financed.

$$\frac{\$5}{\$150} \times \$100 \approx \$3.33$$

In Table 3, the number closest to $3.33 in the "6 payments" row is $3.38. Read up to find the APR, which is 11.5%.

12. Writing exercise

13. In Table 4, find the 7.0% row and read across to the column for 20 years to find entry $7.75299. Since this is the monthly payment amount needed to amortize a loan of $1000 and this loan is for $150,000, the required monthly payment is

$$\frac{\$150000}{\$1000} \times \$7.75299 = \$1162.95.$$

14. Subtract the down payment from the purchase price of the house to find the principal.

$$\$218000 - (.20 \times \$218000) = \$174{,}400$$

In Table 4 find the 8.5% row and read across to the column for 30 years to find entry $7.68913. Since this is the monthly payment amount needed to amortize a loan of $1000, and this loan is for $174,400, the required monthly payment is

$$\frac{\$174400}{\$1000} \times \$7.68913 = \$1340.98.$$

The monthly tax and insurance payment is

$$\frac{\$1500 + \$750}{12} = \$187.50.$$

The total monthly payment, including taxes and insurance, is

$$\$1340.98 + \$187.50 = \$1528.48.$$

15. The "two points" charged by the lender mean 2%. This additional cost is the percentage taken of the amount borrowed, the principal from Exercise 14.

$$.02 \times \$174400 = \$3488$$

16. Writing exercise

17. If the index started at 7.85%, adding the margin of 2.25% gives

$$7.85\% + 2.25\% = 10.1\%.$$

Because the periodic rate cap is 2%, the interest rate during the second year cannot exceed

$$10.1\% + 2.00\% = 12.1\%.$$

18. 112,300 shares

19. First find how much she paid for the stock by multiplying the price per share by 1000.

$$\$12.75 \times 1000 = \$12,750$$

Calculate the amount of money she received from both dividends.

$$\$1.38 \times 1000 + \$1.02 \times 1000 = \$2400$$

Subtract this amount from the total price she paid.

$$\$12750 - \$2400 = \$10350$$

Finally subtract this amount from the price for which she sold the stock.

$$(\$10.36 \times 1000) - \$10350 = \$10$$

20. Compare her amount of return to how much she originally paid for the stock.

$$\frac{\$10}{\$12750} = .00078$$

This is .078%. This percentage is not the annual rate of return, because the time period was for more than one year.

15.1 EXERCISES

1. By counting, there are 7 vertices and 7 edges.
2. By counting, there are 4 vertices and 4 edges.
3. There are 10 vertices and 9 edges.
4. There are 6 vertices and 9 edges.
5. There are 6 vertices and 9 edges.
6. There are 5 vertices and 8 edges.
7. Two vertices have degree 3, three have degree 2, and two have degree 1. The sum of the degrees is

   $$3 + 3 + 2 + 2 + 2 + 1 + 1 = 14.$$

   This is twice the number of edges, which is 7.

8. One vertex has degree 3. Two vertices have degree 2. One vertex has degree 1. The sum of the degrees is

   $$3 + 2 + 2 + 1 = 8.$$

   This is twice the number of edges, which is 4.

9. Six vertices have degree 1. Four vertices have degree 3. The sum of the degrees is

   $$1 + 1 + 1 + 1 + 1 + 1 + 3 + 3 + 3 + 3 = 18.$$

   This is twice the number of edges, which is 9.

10. All of the vertices have degree 3. Since there are 6 vertices, there are

    $$6 \times 3 = 18$$

    vertices. This is twice the number of edges, which is 9.

11. No, the two graphs are not isomorphic. There is one more vertex in the (b) graph.

12. No, the two graphs are not isomorphic. The degree of each vertex in (a) is 2, and this is not true in graph (b).

13. Yes, the graphs are isomorphic.

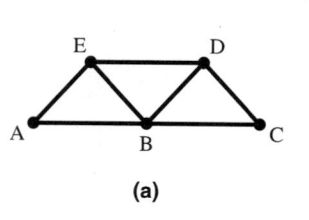 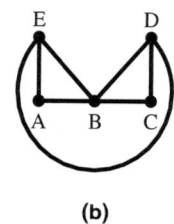

(a)   (b)

14. Yes, the graphs are isomorphic.

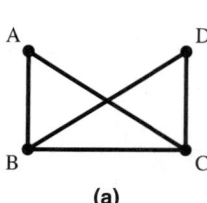 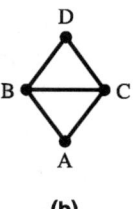

(a)   (b)

15. No, the two graphs are not isomorphic. Two vertices in graph (b) have degree 1, while only one vertex in graph (a) has a degree of 1.

16. Yes, the graphs are isomorphic.

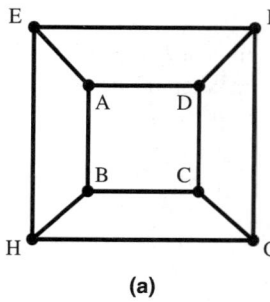 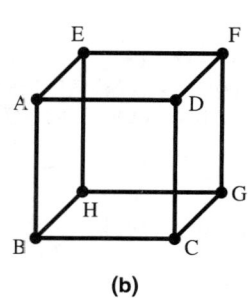

(a)   (b)

17. The graph is connected with only 1 component.
18. The graph is disconnected with 2 components.
19. The graph is disconnected with 3 components.
20. The graph is connected with only 1 component.
21. The graph is disconnected with 2 components.
22. The graph is disconnected with 4 components.
23. Since the sum of the degrees is given by

    $$4 \times 3 = 12,$$

    there are

    $$\frac{1}{2} \times 12 = 6 \text{ edges.}$$

24. Since the sum of the degrees is given by

    $$8 \times 4 = 32,$$

    there are

    $$\frac{1}{2} \times 32 = 16 \text{ edges.}$$

25. Since the sum of the degrees is given by
$$1+1+1+2+3 = 8,$$
there are
$$\frac{1}{2} \times 8 = 4 \text{ edges.}$$

26. Since the sum of the degrees is given by
$$1+1+2+2+2+3+5+6 = 22,$$
there are
$$\frac{1}{2} \times 22 = 11 \text{ edges.}$$

27. (a) Yes, A→B→C is a walk.

    (b) No, B→A→D is not a walk, since there is no edge from A to D.

    (c) No, E→F→A→E is not a walk, since there is no edge from A to E.

    (d) Yes, B→D→F→B→D is a walk.

    (e) Yes, D→E is a walk.

    (f) Yes, C→B→C→B is a walk.

28. (a) Yes, B→D→E→F is a path.

    (b) Yes, D→F→B→D is a path.

    (c) No, B→D→F→B→D is not a path, since edge B to D is used twice.

    (d) No, D→E→F→G→F→D is not a path, since edge G to F is used twice.

    (e) Yes, B→C→D→B→A is a path.

    (f) No, A→B→E→F→A is not a path, since there is no edge from B to E.

29. (a) No, A→B→C→D→E→F is not a circuit, since the path does not return to the starting vertex.

    (b) Yes, A→B→D→E→F→A is a circuit.

    (c) No, C→F→E→D→C is not a circuit, since there is no edge from C to F.

    (d) No, G→F→D→E→F is not a circuit, since the path does not return to the starting vertex.

    (e) No, F→D→F→E→D→F is not a circuit, since the edge from F to D is used more than once.

30. (a) No, F→G→J→H→F is not a walk, since there is no edge from J to H.

    (b) Yes, D→F is a walk.

    (c) Yes, B→A→D→F→H is a walk.

    (d) Yes, B→A→D→E→D→F→H is a walk.

    (e) Yes, I→G→J is a walk.

    (f) No, I→G→J→I is not a walk, since there is no edge from J to I.

31. (a) No, A→B→C is not a path, since there is no edge from B to C.

    (b) No, J→G→I→G→F is not a path, since the edge from I to G is used more than once.

    (c) No, D→E→I→G→F is not a path, since it is not a walk (there is no edge from E to I).

    (d) Yes, C→A is a path.

    (e) Yes, C→A→D→E is a path.

    (f) No, C→A→D→E→D→A→B is not a path, since the edges A to D and D to E are used more than once.

32. A→B→C→D→E is not a walk (since there is no edge from D to E) and hence, it is not a path (to be a path it must be a walk) nor is it a circuit (to be a circuit, it must be a path).

33. A→B→C is a walk and also a path (no edges are repeated) but not a circuit (since the path does not return to the starting vertex).

34. A→B→C→D→A is a walk, a path, and a circuit.

35. A→B→A→C→D→A is a walk, not a path (since the edge A to B is used twice), and hence, not a circuit.

36. A→B→C→A→D→C→E→A is a walk, a path, and a circuit.

37. C→A→B→C→D→A→E is a walk and a path but not a circuit (since it doesn't end at C).

38. Yes, this is a complete graph, since there is exactly one edge going from each vertex to each other vertex in the graph.

39. No, this is not a complete graph, since there is no edge going from A to C (nor B to D).

40. Yes, this is a complete graph, since there is exactly one edge going from each vertex to each other vertex in the graph.

41. No, this is not a complete graph, since there is no edge going from A to F, for example.

42. No, this is not a complete graph, since there is no edge going from A to D.

43. Yes, this is a complete graph, since there is exactly one edge going from each vertex to each other vertex in the graph.

44.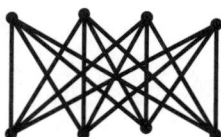

Counting the edges, there are 16 games to be played in this competition.

45.

Counting the edges, there are 6 games to be played in this competition.

46.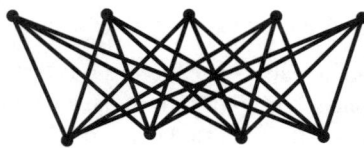

There are 20 edges in this graph.

47. One can draw a graph and count the edges or one may reason that each of the members of one team must shake the hand of each of the 6 members of the second team. Thus, there are $6 \times 6 = 36$ handshakes in total.

48. Each of the 7 people represents a vertex. Vertex 1(first person) has degree 4 since he/she shakes hands with 4. Vertices 2–5 have degree 2. Vertices 6 and 7 have degree 3. Thus, the sum of the degrees is

$$4 + 2 + 2 + 2 + 2 + 3 + 3 = 18.$$

Since there are one half as many edges as the sum of the degrees, we arrive at 9 edges or handshakes.

49. Each of the 8 people represents a vertex. Two of these people had 4 conversations. Thus, the degree for each of these vertices is 4. Another person had 3 conversations. Thus, the degree for this vertex is 3. Four people had 2 conversations each. Therefore, each of these 4 vertices have degree 2. One person had 1 conversation, so the degree of this vertex is 1. Thus, the sum of the degrees is

$$4 + 4 + 3 + 2 + 2 + 2 + 2 + 1 = 20.$$

Since there are one half as many edges as the sum of the degrees, we arrive at 10 edges, or telephone conversations.

50.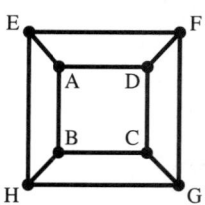

One of several circuits could be A→B→C→D→A. This corresponds to tracing round the edges of a single face.

51.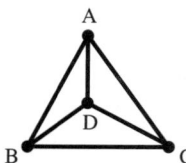

One of several circuits could be A→B→D→A. This corresponds to tracing round the edges of a single face.

52. (a)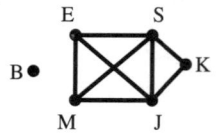

(b) This graph is disconnected with two components.

(c)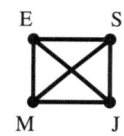

(d) One of several possible sets of graphs is as follows.

 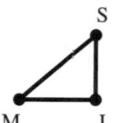 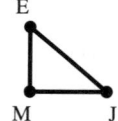

53. The sum of degrees is twice the number of edges, so it must be even. But if there were an odd number of vertices with odd degree, the sum of degrees would be odd.

54. (a) The following represents one set of such graphs.

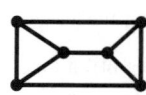 

(b) The first graph has circuits of length three. Since the second does not, these graphs are not isomorphic.

55.

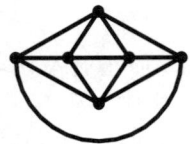

The degree of each vertex is 4, which shows that each face has common boundaries with 4 other faces.

56. (a)

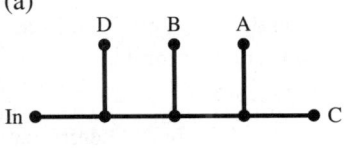

(b)

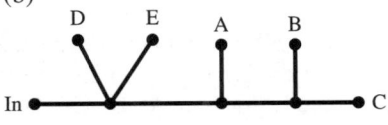

(c)

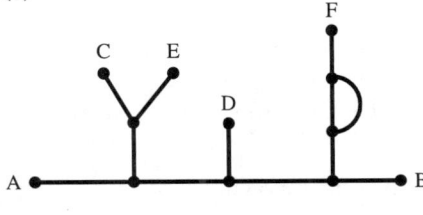

(d)

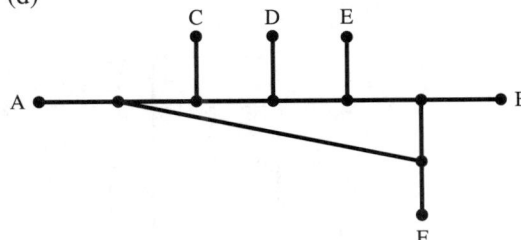

57. Writing exercise
58. Writing exercise
59. Writing exercise

15.2 EXERCISES

1. (a) No, the sequence of vertices, A→B→C→D→A→B→C→D→A, does not represent an Euler circuit, since it is not a path. Remember that a path can traverse an edge only once.

   (b) Yes, the sequence of vertices, C→B→A→D→C, is an Euler circuit, since it is a circuit and each edge is traversed only once.

   (c) No, the sequence, A→C→D→B→A, is not an Euler circuit, since it is not a path.

   (d) No, the sequence, A→B→C→D, is not an Euler circuit, since it is not a circuit.

2. (a) No, the sequence of vertices, A→B→C→D→E→F→A, does not represent an Euler circuit, since it doesn't use all of the edges (e.g. B→F).

   (b) No, the sequence, F→B→D→B, is not an Euler circuit since it is not a circuit (nor a path).

   (c) Yes, the sequence of vertices, A→B→C→D→E→F→B→D→F→A, is an Euler circuit, since it is a circuit and each edge is traversed only once.

   (d) Yes, the sequence of vertices, A→B→F→D→B→C→D→E→F→A, is an Euler circuit, since it is a circuit and each edge is traversed only once.

3. (a) No, the sequence of vertices, A→B→C→D→E→F→A, does not represent an Euler circuit, since it doesn't use all of the edges (e.g. B→F).

   (b) Yes, the sequence of vertices, A→B→C→D→E→G→C→E→F→G→B→F→A, is an Euler circuit since it is a circuit, and each edge is traversed only once.

   (c) No, the sequence of vertices, A→B→C→D→E→C→G→E→F→G→E→F→A, is not an Euler circuit, since it is not a path (the edge E to F is traversed twice).

   (d) Yes, the sequence of vertices, A→B→G→E→D→C→G→F→B→C→E→F→A, is an Euler circuit, since it is a circuit and each edge is traversed only once.

4. No, the graph will not have an Euler circuit, since some vertices (e.g. A) have odd degree.

5. Yes, the graph will have an Euler circuit, since all vertices have even degree.

6. No, the graph will not have an Euler circuit, since some vertices (e.g. A) have odd degree.

7. No, the graph will not have an Euler circuit, since some vertices (e.g. G) have odd degree.

8. Yes, the graph will have an Euler circuit, since all vertices have even degree.

9. If we assume that a vertex occurs at each intersection of curves and check the degree of each such vertex, we see that all vertices have an even degree. Thus, by Euler's theorem, all edges are traversed exactly one time —that is, we can find an Euler circuit. The answer is yes since this is equivalent to tracing the pattern without lifting your pencil nor going over any line more than one time.

10. If we treat each intersection of curves as a vertex, we see that several vertices are of odd degree (e.g. the center intersection) and hence, an Euler circuit cannot be found. Therefore, the answer is no, we will not be able to trace the pattern without lifting the pencil nor going over any curve more than one time.

11. Since all vertices have even degree, the graph has an Euler circuit. Because one must pass through vertex I more than once to complete any circuit, no circuit will visit each vertex exactly once.

12. All vertices have odd degree. Therefore, no Euler circuit exists. The sequence, A→B→C→D→E→F→A, visits each vertex exactly once.

13. Since all vertices have even degree, the graph has an Euler circuit. The sequence, A→B→H→C→G→D→F→E→A, visits each vertex exactly once.

14. Some vertices (e.g. B) have odd degree. Therefore, no Euler circuit exists. No circuit exists that will pass through each vertex exactly once.

15. Some vertices (e.g. B) have odd degree. Therefore, no Euler circuit exists. No circuit exists that will pass through each vertex exactly once.

Exercises 16–19 correspond to deciding if an Euler circuit exists when we assume that each intersection of line segments forms a vertex.

16. The corner (and other) vertices are of odd degree. Hence, there is no Euler circuit (or continuous path to apply the grout).

17. The upper left (and lower right) corner is of odd degree. Hence, there is no Euler circuit (or continuous path to apply the grout).

18. Since there are vertices of odd degree, there is no Euler circuit (or continuous path to apply the grout).

19. Since all vertices are of even degree, there exists an Euler circuit, and thus, a continuous path to apply the grout without retracing.

20. The cut edges are CD and FG, since a break along either edge will disconnect the graph.

21. There are no cut edges, since no break along an edge will disconnect the graph.

22. All edges are cut edges, since a break along any edge will disconnect the graph.

23. The student has a choice of B→E or B→D. Choosing B→F, for example, is not allowed, since it is a cut edge.

24. The student has a choice of B→D or B→F. Choosing B→A, for example, is not allowed, since it is a cut edge.

25. The student has a choice of B→C or B→H. Choosing B→G, for example, is not allowed since it is a cut edge.

There are many different correct answers for the following exercises.

26. Beginning at A, one Euler circuit that results is A→I→H→G→F→E→K→J→H→E→D→C→B→D→I→B→A.

27. Beginning at A, one Euler circuit that results is A→C→B→F→E→D→C→F→D→A.

28. Beginning at A, one Euler circuit that results is A→B→C→D→E→F→D→G→F→C→G→H→J→B→H→A.

29. Since each vertex is of even degree, there is an Euler circuit. Beginning at A, one such circuit is A→G→H→J→I→L→J→K→I→H→F→G→E→F→D→E→C→D→B→C→A→E→B→A.

30. Since each vertex is of even degree, there is an Euler circuit. Beginning at E, one such circuit is E→F→G→H→I→G→E→H→F→I→E.

31. Some vertices (e.g. C) have odd degree. Therefore, no Euler circuit exists.

32. Such a route exists. Since all vertices are of even degree, an Euler circuit exists for the graph. Using Fleury's algorithm, one such route is given by A→B→C→D→E→F→G→H→I→G→E→I→J→D→A→J→C→A.

33. Such a route exists. Since all vertices are of even degree, an Euler circuit exists for the graph. Using Fleury's algorithm, one such route is given by A→D→B→C→A→H→D→E→B→H→G→E→F→H→J→L→C→M→A→K→M→L→K→J→A.

34. Draw a map such as the following.

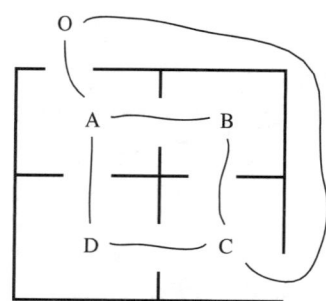

Since several vertices have odd degree (e.g. A), an Euler circuit does not exist. Thus, it will not be possible to walk through each door exactly once and end up back outside.

35. Draw a map such as the following.

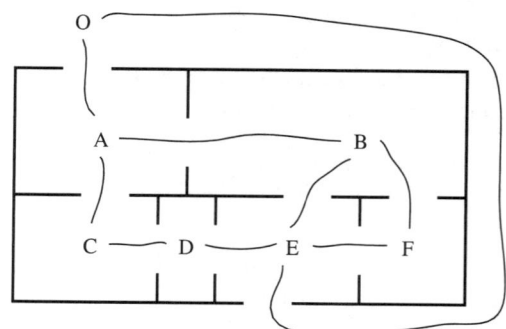

Since several vertices have odd degree (e.g. A), an Euler circuit does not exist. Thus, it will not be possible to walk through each door exactly once and end up back outside.

36. Draw a map such as the following.

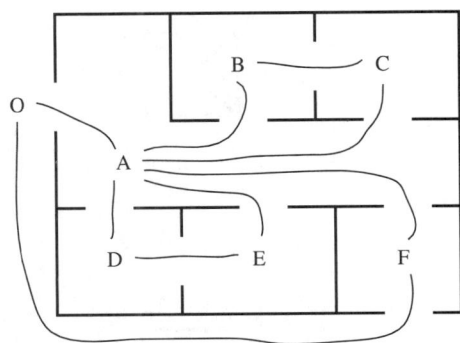

Since all vertices have even degree, an Euler circuit does exist. Thus, it will be possible to walk through each door exactly once and end up back outside.

37. There is no Euler path, since there are more than two vertices with odd degree.

38. Yes, there is an Euler path, since there are exactly two vertices with odd degree. The path must begin or end at H or I, the two vertices with odd degree.

39. Yes, there is an Euler path, since there are exactly two vertices with odd degree. The path must begin or end at B or G, the two vertices with odd degree.

40. No, it is not possible, since there are no vertices with odd degree.

Exercises 41–43 are solved by thinking of the corresponding graphs for Exercises 34–36. Remember that the rooms represent vertices and the doors are edges. To answer each question in the affirmative, an Euler path must exist.

41. Yes, it is possible because there are exactly two rooms with an odd number of doors (odd degree).

42. Yes, it is possible because there are exactly two rooms with an odd number of doors (odd degree).

43. No, there is no Euler path, due to the fact that all rooms have an even number of doors (or degree).

44. (a) The answer is no. Consider each land mass as a vertex and the bridges and tunnels as edges. There are several vertices with an odd degree (e.g. Randall's Island).

    (b) The answer is yes. There are just two vertices with an odd degree, Randall's Island and Manhattan. Thus, the drive must begin or end on one of these islands.

45. One possible circuit is A→B→D→C→A. There are 4 edges in any such circuit.

46. One possible circuit is A→B→C→D→E→C→F→B→E→F→A. There are 10 edges in any such circuit.

47. One possible circuit is A→B→C→D→E→F→G→H→C→A→H→I→A. There are 12 edges in any such circuit.

48. Only the octahedron, a regular solid with 8 faces, has such a tracing. The octahedron is the only regular solid with an even number of edges (12) meeting each vertex.

49. Only those *complete graphs* for which the number of vertices is an odd number greater than or equal to 3 have an Euler circuit. In a complete graph with n vertices, the degree of each vertex is $n - 1$. And $n - 1$ is even if, and only if, n is odd.

50. Writing exercise

51. Writing exercise

52. Writing exercise

15.3 EXERCISES

1. (a) No, A→E→C→D→E→B→A, is not a Hamilton circuit, since it visits vertex E twice.

    (b) Yes, A→E→C→D→B→A, is a Hamilton circuit, since it visits all vertices (except the first) only once.

    (c) No, D→B→E→A→B, is not a Hamilton circuit, since it does not visit C.

    (d) No, E→D→C→B→E, is not a Hamilton circuit, since it does not visit A.

2. (a) No, A→B→C→D→E→C→A→E→F→A, is not a Hamilton circuit, since it visits several vertices (e.g. C) twice.

    (b) No, A→C→D→E→F→A is not a Hamilton circuit, since it does not visit B.

    (c) No, F→A→C→E→F is not a Hamilton circuit, since it does not visit B or D.

    (d) No, C→D→E→F→A→B, is not a Hamilton circuit, since it does not return to its starting point.

3. (a) The path, A→B→C→D→E→A, is not a circuit (since BC is not an edge) and, hence, is not an Euler circuit nor a Hamilton circuit.

    (b) The path, B→E→C→D→A→B, is a circuit, is an Euler circuit (travels each edge only once), and is a Hamilton circuit (travels through each vertex only once).

    (c) The path, E→B→A→D→A→D→C→E, is not a circuit, is not an Euler circuit (travels edge AD twice), and is not a Hamilton circuit (travels through several vertices more than once).

4. (a) The path, A→B→C→D→E→F→G→A, is a circuit (travels any edge only once and ends up at starting vertex), is not an Euler circuit (since it does not travel across every edge), and is not a Hamilton circuit (since it does not travel through every vertex).

   (b) The path, B→I→G→F→E→D→H→F→I→D→C→B→G→A→B, is a circuit (travels any edge only once and ends up at starting vertex), is an Euler circuit (travels each edge only once), and is not a Hamilton circuit (since it travels through several vertices more than once).

   (c) The string, A→B→C→D→E→F→G→H→I→A, is not a circuit (since GH is not an edge) and, hence, is not an Euler circuit nor a Hamilton circuit.

5. This graph has a Hamilton circuit. One example is A→B→D→E→F→C→A.

6. Since this graph is disconnected, it is impossible to show a Hamilton circuit by moving through all vertices.

7. This graph has a Hamilton circuit. One example is G→H→J→I→G.

8. This graph has a Hamilton circuit. One example is A→B→F→C→E→D→A.

9. This graph has a Hamilton circuit. One example is X→T→U→W→V→X.

10. No Hamilton circuit can be shown.

11.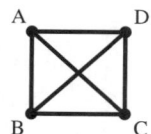

    A→B→C→D→A is a Hamilton circuit. There is no Euler circuit, since at least one of the vertices has odd degree. (In fact, all have odd degree.)

12.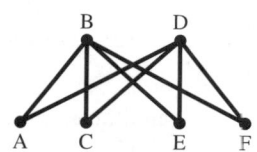

    This graph has no Hamilton circuit. A→B→C→D→E→B→F→D→A is an Euler circuit for the graph.

13.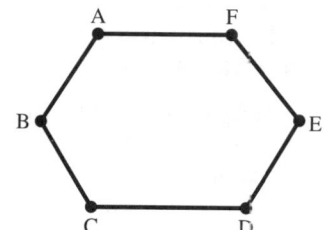

    A→B→C→D→E→F→A is both a Hamilton and an Euler circuit.

14. (a) True

    (b) True

    (c) False

    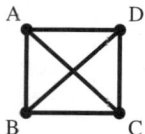

    A→B→C→D→A is a Hamilton circuit that does not use each edge.

    (d) True

    (e) False

    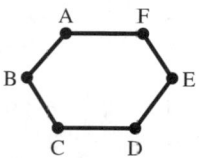

    A→B→C→D→E→F→A is both a Hamilton and an Euler circuit.

    (f) False

    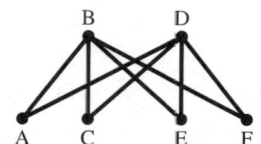

    A→B→C→D→E→B→F→D→A is an Euler circuit, that visits vertex B and D more than once.

15. This exercise could be solved by a Hamilton circuit, since each vertex represents a bandstand.

16. This exercise could be solved by a Hamilton circuit, since each vertex represents a team member.

17. This exercise could be solved by an Euler circuit, since each path corresponds to an edge.

18. This exercise could be solved by an Euler circuit, since an edge represents each border crossing.

19. This exercise could be solved by a Hamilton circuit, since each vertex represents a country.

20. This exercise could be solved by a Hamilton circuit, since each vertex represents a position for each measurement.

21. $4! = 4 \cdot 3 \cdot 2 \cdot 1 = 24$.

22. $6! = 6 \cdot 5 \cdot 4 \cdot 3 \cdot 2 \cdot 1 = 720$.

23. $9! = 9 \cdot 8 \cdot 7 \cdot 6 \cdot 5 \cdot 4 \cdot 3 \cdot 2 \cdot 1 = 362,880$.

24. $14! = 14 \cdot 13 \cdot 12 \cdot 11 \cdot 10 \cdot 9!$
    $= 240,240 \cdot 362,880$
    $\approx 8.72 \times 10^{10}$

25. There are
$$(10-1)! = 9!$$
Hamilton circuits.

26. There are
$$(15-1)! = 14!$$
Hamilton circuits.

27. There are
$$(18-1)! = 17!$$
Hamilton circuits.

28. There are
$$(60-1)! = 59!$$
Hamilton circuits.

29. Since this is a complete graph with 4 vertices, there are
$$(4-1)! = 3! = 3 \cdot 2 \cdot 1 = 6$$
Hamilton circuits. Choosing P as a beginning vertex, they are:

P→Q→R→S→P; P→Q→S→R→P; P→R→Q→S→P;
P→R→S→Q→P; P→S→Q→R→P; P→S→R→Q→P.

Note: A tree diagram might be helpful here. P would represent the first node; Q, R and S, the second node; the third node would represent each of the remaining two vertices.

30. Hamilton circuits starting with E→F→G would include:

E→F→G →H→I→E and E→F→G →I→H→E.

31. Hamilton circuits starting with E→H→I would include:

E→H→I→F→G→E and E→H→I →G→F→E.

32. Hamilton circuits starting with E→I→H would include:

E→I→H→F→G→E and E→I→H →G→F→E.

33. Hamilton circuits starting with E→F would include:

E→F→G →H→I→E; E→F→G →I→H→E;
E→F→H→G→I→E; E→F →H→I→G→E;
E→F→I→G→H→E; E→F→I→H→G→E.

34. Hamilton circuits starting with E→I would include:

E→I→F→G→H→E; E→I→F→H→G→E;
E→I→G→F→H→E; E→I →G→H→F→E;
E→I→H→F→G→E; E→I→H→G→F→E.

35. Hamilton circuits starting with E→G would include:

E→G→F→H→I→E; E→G→F→I→H→E;
E→G→H→F→I→E; E→G →H→I→F→E;
E→G→I→F→H→E; E→G→I→H→F→E.

36. Hamilton circuits starting with E would include:

E→F→G→H→I→E; E→G→F→H→I→E;
E→F→G→I→H→E; E→G→F→I→H→E;
E→F→H→G→I→E; E→G→H→F→I→E;
E→F→H→I→G→E; E→G→H→I→F→E;
E→F→I→G→H→E; E→G→I→F→H→E;
E→F→I→H→G→E; E→G→I→H→F→E;
E→H→F→G→I→E; E→I→F→G→H→E;
E→H→F→I→G→E; E→I→F→H→G→E;
E→H→G→F→I→E; E→I→G→F→H→E;
E→H→G→I→F→E; E→I→G→H→F→E;
E→H→I→F→G→E; E→I→H→F→G→E;
E→H→I→G→F→E; E→I→H→G→F→E.

37. Hamilton circuits starting with E would include:

A→B→C→D→E→A; A→C→B→D→E→A;
A→B→C→E→D→A; A→C→B→E→D→A;
A→B→D→C→E→A; A→C→D→B→E→A;
A→B→D→E→C→A; A→C→D→E→B→A;
A→B→E→C→D→A; A→C→E→B→D→A;
A→B→E→D→C→A; A→C→E→D→B→A;
A→D→B→C→E→A; A→E→B→C→D→A;
A→D→B→E→C→A; A→E→B→D→C→A;
A→D→C→B→E→A; A→E→C→B→D→A;
A→D→C→E→B→A; A→E→C→D→B→A;
A→D→E→B→C→A; A→E→D→B→C→A;
A→D→E→C→B→A; A→E→D→C→B→A.

38. Using Brute Force Algorithm:

Circuit:	Total weight of circuit:
1. F→I→H→G→F	$5 + 4 + 7 + 6 = 22$
2. F→G→I→H→F	$6 + 3 + 4 + 2 = 15$
3. F→I→G→H→F	$5 + 3 + 7 + 2 = 17$
4. F→H→I→G→F (opposite of 2)	15
5. F→G→H→I→F (opposite of 1)	22
6. F→H→G→I→F (opposite of 3)	17

Thus, the Minimum Hamilton circuit is F→G→I→H→F and the weight is 15.

39. Using the Brute Force Algorithm:

Circuit:	Total weight of circuit:
1. P→Q→R→S→P	$550 + 640 + 500 + 510 = 2200$
2. P→Q→S→R→P	$550 + 790 + 500 + 600 = 2440$
3. P→S→Q→R→P	$510 + 790 + 640 + 600 = 2540$
4. P→R→S→Q→P (opposite of 2)	2440
5. P→S→R→Q→P (opposite of 1)	2200
6. P→R→Q→S→P (opposite of 3)	2540

Thus, the Minimum Hamilton circuit is P→Q→R→S→P and the weight is 2200.

516 CHAPTER 15 GRAPH THEORY

40. Circuit: Total weight of circuit:
    1. A→B→C→D→E→A $4+7+3+7+6=27$
    2. A→B→C→E→D→A $4+7+2+7+5=25$
    3. A→B→D→E→C→A $4+6+7+2+1=20$
    4. A→B→D→C→E→A $4+6+3+2+6=21$
    5. A→B→E→D→C→A $4+8+7+3+1=23$
    6. A→B→E→C→D→A $4+8+2+3+5=22$
    7. A→C→B→D→E→A $1+7+6+7+6=27$
    8. A→C→B→E→D→A $1+7+8+7+5=28$
    9. A→C→D→B→E→A $1+3+6+8+6=24$
    10. A→D→C→B→E→A $5+3+7+8+6=29$
    11. A→D→B→C→E→A $5+6+7+2+6=26$
    12. A→D→B→E→C→A $5+6+8+2+1=22$
    13. – 24. (opposites of above)

    Thus, the Minimum Hamilton circuit is
    A→B→D→E→C→A and the weight is 20.

41. Observe that for a complete graph with 5 vertices we will have $(5-1) = 4! = 24$ Hamilton circuits.

    Circuit: Total weight of circuit:
    1. C→D→E→F→G→C $12+10+17+15+10=64$
    2. C→D→E→G→F→C $12+10+13+15+15=65$
    3. C→F→D→E→G→C $15+21+10+13+10=69$
    4. C→D→F→G→E→C $12+21+17+13+10=73$
    5. C→E→D→F→G→C $15+10+21+15+10=71$
    6. C→D→G→E→F→C $12+13+13+17+15=70$
    7. C→E→F→D→G→C $15+17+21+13+10=76$
    8. C→E→D→G→F→C $15+10+13+15+15=68$
    9. C→F→E→D→G→C $15+17+10+13+10=65$
    10. C→F→G→D→E→C $15+15+13+10+15=68$
    11. C→D→F→E→G→C $12+21+17+13+10=73$
    12. C→D→F→G→E→C $12+21+15+13+10=71$
    13. – 24. (opposites of above)

    Thus, the Minimum Hamilton circuit is
    C→D→E→F→G→C and the weight is 64.

42. (a) Using the nearest neighbor algorithm, choose the first edge with the minimum weight, F$\xrightarrow{2}$H. Keep track of weight by noting its value over the arrow. Choose the second edge, H$\xrightarrow{4}$I, which has the minimum weight. Choose the third edge, I$\xrightarrow{3}$G, which has the minimum weight. Finally, choose G$\xrightarrow{6}$F. Note that this is your only choice for the last remaining edge. The resulting approximate minimum Hamilton circuit is, therefore,

    F→H→I→G→F.

    The circuit has a total weight of

    $2+4+3+6=15.$

    (b) Using the nearest neighbor algorithm, choose the first edge with the minimum weight, G$\xrightarrow{3}$I. Choose the second edge with minimum weight, I$\xrightarrow{4}$H. Choose the third edge with minimum weight, H$\xrightarrow{2}$F. Finally, choose the remaining edge F $\xrightarrow{6}$ G. The resulting approximate minimum Hamilton circuit is, therefore, G→I→H→F→G.

    The circuit has a total weight of

    $3+4+2+6=15.$

    (c) Using the nearest neighbor algorithm, choose the first edge with the minimum weight, H$\xrightarrow{2}$F. Choose the second edge with minimum weight, F$\xrightarrow{5}$I. Choose the third edge with minimum weight, I$\xrightarrow{3}$G. Finally, choose the remaining edge G$\xrightarrow{7}$H. Thus, the resulting approximate minimum Hamilton circuit is

    H→F→I→G→H.

    The circuit has a total weight of

    $2+5+3+7=17.$

    (d) Using the nearest neighbor algorithm, choose the first edge with the minimum weight, I$\xrightarrow{3}$G. Choose the second edge with minimum weight, G$\xrightarrow{6}$F. Choose the third edge with minimum weight, F$\xrightarrow{2}$H. Finally, choose the remaining edge, H$\xrightarrow{4}$I. The resulting approximate minimum Hamilton circuit is, therefore,

    I→G→F→H→I.

    The circuit has a total weight of

    $3+6+2+4=15.$

43. (a) Using the nearest neighbor algorithm, choose the first edge with the minimum weight, A$\xrightarrow{1}$C. Keep track of weight by noting its value over the arrow. Choose the second edge, C$\xrightarrow{2}$E, which has the minimum weight. Choose the third edge, E$\xrightarrow{7}$D, which has the minimum weight. Choose the fourth edge, D$\xrightarrow{6}$B, which has the minimum weight. Finally, choose the remaining edge, B$\xrightarrow{4}$A. Note that this is your only choice for the last remaining edge. The resulting approximate minimum Hamilton circuit is, therefore,

    A→C→E→D→B→A.

    The circuit has a total weight of

    $1+2+7+6+4=20.$

    (b) Using the nearest neighbor algorithm, choose the edge which has the minimum weight, C$\xrightarrow{1}$A. Keep track of weight by noting its value over the arrow. Choose the second edge with corresponding minimum weight, A$\xrightarrow{4}$B. Choose the third edge, which has the minimum weight, B$\xrightarrow{6}$D. Choose the fourth edge, which has the minimum weight, D$\xrightarrow{7}$E. Finally, choose the remaining edge, E$\xrightarrow{2}$C. The resulting approximate minimum Hamilton circuit is, therefore,

    C→A→B→D→E→C.

    The circuit has a total weight of

    $1+4+6+7+2=20.$

(c) Using the nearest neighbor algorithm, choose the edge which has the minimum weight, D$\xrightarrow{3}$C. Choose the second edge with corresponding minimum weight, C$\xrightarrow{1}$A. Choose the third edge, which has the minimum weight, A$\xrightarrow{4}$B. Choose the fourth edge, which has the minimum weight, B$\xrightarrow{8}$E. Finally, choose the remaining edge, E$\xrightarrow{7}$D. The resulting approximate minimum Hamilton circuit is, therefore,

$$D \to C \to A \to B \to E \to D.$$

The circuit has a total weight of

$$3 + 1 + 4 + 8 + 7 = 23.$$

(d) Using the nearest neighbor algorithm, choose the edge which has the minimum weight, E$\xrightarrow{2}$C. Choose the second edge with corresponding minimum weight, C$\xrightarrow{1}$A. Choose the third edge, which has the minimum weight, A$\xrightarrow{4}$B. Choose the fourth edge, which has the minimum weight, B$\xrightarrow{6}$D. Finally, choose the remaining edge, D$\xrightarrow{7}$E. The resulting approximate minimum Hamilton circuit is, therefore,

$$E \to C \to A \to B \to D \to E.$$

The circuit has a total weight of

$$2 + 1 + 4 + 6 + 7 = 20.$$

44. (a) Using the nearest neighbor algorithm, choose the first edge with the minimum weight, A$\xrightarrow{1.6}$E. Choose the second edge, E$\xrightarrow{1.95}$D, which has the minimum weight. Choose the third edge, D$\xrightarrow{1.8}$C, which has the minimum weight. Choose the fourth edge, C$\xrightarrow{2.3}$F, which has the minimum weight. Choose the remaining edge, F$\xrightarrow{1.7}$B. Finally, to the last vertex, A, we must choose B$\xrightarrow{2.5}$A The resulting approximate minimum Hamilton circuit is, therefore,

$$A \to E \to D \to C \to F \to B \to A.$$

The circuit has a total weight of

$$1.6 + 1.95 + 1.8 + 2.3 + 1.7 + 2.5 = 11.85.$$

(b) Using the nearest neighbor algorithm, choose the first edge with the minimum weight, B$\xrightarrow{1.7}$F. Choose the second edge with corresponding minimum weight, F$\xrightarrow{2.3}$C. Choose the third edge, which has the minimum weight, C$\xrightarrow{1.8}$D. Choose the fourth edge, which has the minimum weight, D$\xrightarrow{1.95}$E. Choose the fifth edge, E$\xrightarrow{1.6}$A. Finally, choose the remaining edge, A$\xrightarrow{2.5}$B. The resulting approximate minimum Hamilton circuit is, therefore,

$$B \to F \to C \to D \to E \to A \to B.$$

The circuit has a total weight of

$$1.7 + 2.3 + 1.8 + 1.95 + 1.6 + 2.5 = 11.85.$$

(c) Using the nearest neighbor algorithm, choose the first edge with the minimum weight, C$\xrightarrow{1.8}$D. Choose the second edge with corresponding minimum weight, D$\xrightarrow{1.95}$E. Choose the third edge, which has the minimum weight, E$\xrightarrow{1.6}$A. Choose the fourth edge, which has minimum weight, A$\xrightarrow{2.5}$B. Choose the fifth edge, with minimum weight, B$\xrightarrow{1.7}$F. Finally, choose the remaining edge, F$\xrightarrow{2.3}$C. The resulting approximate minimum Hamilton circuit is, therefore,

$$C \to D \to E \to A \to B \to F \to C.$$

The circuit has a total weight of

$$1.8 + 1.95 + 1.6 + 2.5 + 1.7 + 2.3 = 11.85.$$

(d) Using the nearest neighbor algorithm, choose the first edge with the minimum weight, D$\xrightarrow{1.8}$C. Choose the second edge with corresponding minimum weight, C$\xrightarrow{1.9}$A. Choose the third edge, which has the minimum weight, A$\xrightarrow{1.6}$E. Choose the fourth edge, which has minimum weight, E$\xrightarrow{2.7}$B. Choose the fifth edge, with minimum weight, B$\xrightarrow{1.7}$F. Finally, choose the remaining edge, F$\xrightarrow{3.2}$D. The resulting approximate minimum Hamilton circuit is, therefore,

$$D \to C \to A \to E \to B \to F \to D.$$

The circuit has a total weight of

$$1.8 + 1.9 + 1.6 + 2.7 + 1.7 + 3.2 = 12.9.$$

(e) Using the nearest neighbor algorithm, choose the first edge with the minimum weight, E$\xrightarrow{1.6}$A. Choose the second edge with corresponding minimum weight, A$\xrightarrow{1.9}$C. Choose the third edge, which has the minimum weight, C$\xrightarrow{1.8}$D. Choose the fourth edge, which has minimum weight, D$\xrightarrow{3}$B. Choose the fifth edge, with minimum weight, B$\xrightarrow{1.7}$F. Finally, choose the remaining edge, F$\xrightarrow{4.1}$E. The resulting approximate minimum Hamilton circuit is, therefore,

$$E \to A \to C \to D \to B \to F \to E.$$

The circuit has a total weight of

$$1.6 + 1.9 + 1.8 + 3 + 1.7 + 4.1 = 14.1.$$

(f) Using the nearest neighbor algorithm, choose the first edge with the minimum weight, F$\xrightarrow{1.7}$B. Choose the second edge with corresponding minimum weight, B$\xrightarrow{2.4}$C. Choose the third edge, which has the minimum weight, C$\xrightarrow{1.8}$D. Choose the fourth edge, which has minimum weight, D$\xrightarrow{1.95}$E. Choose the fifth edge, with minimum weight, E$\xrightarrow{1.6}$A. Finally, choose the remaining edge, A$\xrightarrow{2.8}$F. The resulting approximate minimum Hamilton circuit is, therefore,

$$F \to B \to C \to D \to E \to A \to F.$$

The circuit has a total weight of

$$1.7 + 2.4 + 1.8 + 1.95 + 1.6 + 2.8 = 12.25.$$

45. (a) Beginning with vertex A, choose the edge with the minimum weight, A$\xrightarrow{5}$C. Choose the second edge, C$\xrightarrow{7}$D. Choose the third edge, D$\xrightarrow{10}$E. Choose the fourth edge, E$\xrightarrow{50}$B. Finally, to the last vertex, A, we must choose B$\xrightarrow{15}$A. The resulting approximate minimum Hamilton circuit is

$$A \to C \to D \to E \to B \to A.$$

The circuit has a total weight of

$$5 + 7 + 10 + 50 + 15 = 87.$$

Beginning with vertex B, choose the edge with the minimum weight, B$\xrightarrow{11}$C. Choose the second edge, C$\xrightarrow{5}$A. Choose the third edge, A$\xrightarrow{9}$E. Choose the fourth edge, E$\xrightarrow{10}$D. Finally, to the last vertex, B, we must choose D$\xrightarrow{60}$B. The resulting approximate minimum Hamilton circuit is

$$B \to C \to A \to E \to D \to B.$$

The circuit has a total weight of

$$11 + 5 + 9 + 10 + 60 = 95.$$

Beginning with vertex C, choose the edge with the minimum weight, C$\xrightarrow{5}$A. Choose the second edge, A$\xrightarrow{9}$E. Choose the third edge, E$\xrightarrow{10}$D. Choose the fourth edge, D$\xrightarrow{60}$B. Finally, to the last vertex, C, we must choose B$\xrightarrow{11}$C. The resulting approximate minimum Hamilton circuit is

$$C \to A \to E \to D \to B \to C.$$

The circuit has a total weight of

$$5 + 9 + 10 + 60 + 11 = 95.$$

Beginning with vertex D, choose the edge with the minimum weight, D$\xrightarrow{7}$C. Choose the second edge, C$\xrightarrow{5}$A. Choose the third edge, A$\xrightarrow{9}$E. Choose the fourth edge, E$\xrightarrow{50}$B. Finally, to the last vertex, D, we must choose B$\xrightarrow{60}$D. The resulting approximate minimum Hamilton circuit is

$$D \to C \to A \to E \to B \to D.$$

The circuit has a total weight of

$$7 + 5 + 9 + 50 + 60 = 131.$$

Beginning with vertex E, choose the edge with the minimum weight, E$\xrightarrow{8}$C. Choose the second edge, C$\xrightarrow{5}$A. Choose the third edge, A$\xrightarrow{14}$D. Choose the fourth edge, D$\xrightarrow{60}$B. Finally, to the last vertex, D, we must choose B$\xrightarrow{50}$E. The resulting approximate minimum Hamilton circuit is

$$E \to C \to A \to D \to B \to E.$$

The circuit has a total weight of

$$8 + 5 + 14 + 60 + 50 = 137.$$

(b) The best solution is the circuit

$$A \to C \to D \to E \to B \to A.$$

since its total weight, 87, is the smallest.

(c) One example would be

$$A \to B \to C \to D \to E \to A$$

with a total weight of

$$15 + 11 + 7 + 10 + 9 = 52.$$

46. (a)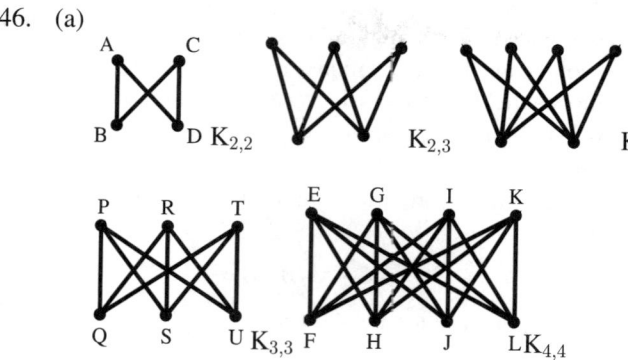

(b) The graphs, $K_{2,3}$ and $K_{2,4}$ have no Hamilton circuit. For $K_{2,2}$, A$\to$B$\to$C$\to$D$\to$A is a Hamilton circuit. For $K_{3,3}$, P$\to$Q$\to$R$\to$S$\to$T$\to$U$\to$P is a Hamilton circuit. For $K_{4,4}$, E$\to$F$\to$G$\to$H$\to$I$\to$J$\to$K$\to$L$\to$E is a Hamilton circuit.

(c) A conjecture: For $K_{m,n}$ to have a Hamilton circuit, $m = n$. The Hamilton circuit must alternate between the two groups of vertices until the circuit is complete. This can be done without visiting any vertex more than once only if $m = n$.

(d) For $K_{m,n}$ to have an Euler circuit, both m and n must be even numbers (not necessarily equal!). The degree of each vertex in the graph is m or n, so by Euler's theorem of section 2, we have an Euler circuit if and only if m and n are both even.

47. All Hamilton circuits include:

A→B→C→D→E→F→A; A→B→C→F→E→D→A;
A→B→E→D→C→F→A; A→B→E→F→C→D→A;
A→D→E→F→C→B→A; A→D→E→B→C→F→A;
A→D→C→B→E→F→A; A→D→C→F→E→B→A;
A→F→E→B→C→D→A; A→F→E→D→C→B→A;
A→F→C→D→E→B→A; A→F→C→B→E→D→A.

A tree diagram, such as below, can be a help in creating all of the Hamilton circuits.

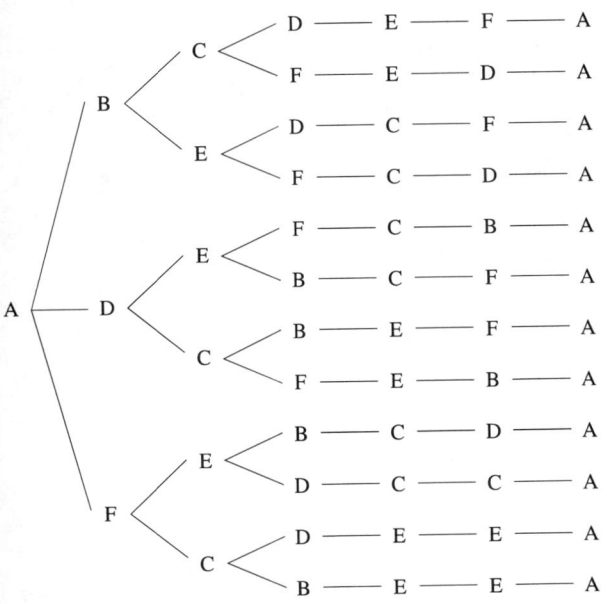

48. All Hamilton circuits include:

    A→B→C→D→E→A;   A→B→C→E→D→A;
    A→B→D→C→E→A;   A→B→D→E→C→A;
    A→C→B→D→E→A;   A→C→E→D→B→A;
    A→D→B→C→E→A;   A→D→E→C→B→A;
    A→E→C→B→D→A;   A→E→C→D→B→A;
    A→E→D→B→C→A;   A→E→D→C→B→A.

49. All Hamilton circuits include:

    A→B→C→D→E→F→A;
    A→B→C→E→D→F→A;
    A→B→C→E→F→D→A;
    A→B→C→F→E→D→A;
    A→D→E→F→C→B→A;
    A→D→F→E→C→B→A;
    A→F→D→E→C→B→A;
    A→F→E→D→C→B→A.

50. All Hamilton circuits include:

    A→B→C→G→F→H→D→E→A;
    A→E→D→H→F→G→C→B→A.

51. (a) For graph (1), $n/2: 6/2 = 3$. Condition not satisfied. For example, vertex A has degree 2
    For graph (2), $n/2: 6/2 = 3$. Condition is satisfied, since all vertices are 3 or larger in degree.
    For graph (3), $n/2: 5/2 = 2.5$. Condition not satisfied. For example, vertex G has degree 2.
    For graph (4), $n/2: 5/2 = 2.5$. Condition is satisfied, since all vertices are 3 or larger in degree.
    For graph (5), $n/2: 7/2 = 3.5$. Condition not satisfied. For example, vertex B has degree 3.

    (b) By Dirac's theorem, graphs (2) and (4) are predicted to have Hamilton circuits.

    (c) We can not be sure that a graph which doesn't satisfy Dirac's theorem will not have a Hamilton circuit. For example, graph (1), which doesn't satisfy Dirac's theorem, has a Hamilton circuit

    A→B→C→D→E→F→A.

    (d) Dirac's theorem is not true for $n < 3$ since such a graph will not have any Hamilton circuit at all.

    (e) The degree of each vertex in a complete graph with n vertices is $(n - 1)$. If $n \geq 3$, then $(n - 1) \geq n/2$. Thus, we can conclude that the graph has a Hamilton circuit.

Note that there are different possible answers for questions 52–54.

52. A→B→C→D→E→N→M→L→K→J→I→H→G→R→S→T→U→Q→P→F→A is a Hamilton circuit.

53. A→F→G→R→S→T→U→Q→P→N→M→L→K→J→I→H→B→C→D→E→A is a Hamilton circuit.

54. A→B→H→G→F→P→N→M→L→K→T→U→Q→R→S→I→J→C→D→E→A is a Hamilton circuit.

55. Writing exercise

56. Writing exercise

57. Writing exercise

15.4 EXERCISES

1. The graph is a tree, since it is connected and has no circuits.

2. The graph is a not a tree, since it has a circuit.

3. The graph is a not a tree, since it is not connected.

4. The graph is a not a tree, since it has a circuit.

5. The graph is a tree, since it is connected and has no circuits.

6. Such a graph is a not a tree, since it has circuits.

7. Such a graph is a not a tree, since it has a circuit. Remember that Euler's theorem tells us that all connected graphs with all vertices of even degree will have an Euler circuit.

8. An added edge such as AE will form a tree.

9. It is not possible to form a tree by adding an extra edge, since the graph has a circuit.

10. By adding edges RS and RT, a tree is formed.

11. Yes, such a graph is a tree, since it is connected and has no circuits.

12. Yes, such a graph is a tree. Note that an edge is established only on first visit to a site!

520 CHAPTER 15 GRAPH THEORY

13. No, such a graph would not necessarily be a tree since circuits may be formed.

14. The statement "Every graph with no circuits is a tree" is false, since disconnected graphs such as the following

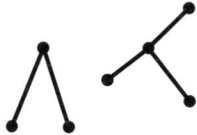

are not trees.

15. The statement "Every connected graph in which each edge is a cut edge is a tree" is true. Since all edges are cut edges, there can be no circuits.

16. The statement "Every graph in which there is a path between each pair of vertices is a tree" is false. For example,

is not a tree.

17. The statement "Every graph in which each edge is a cut edge is a tree" is false. For example, disconnected graphs as the following

are not trees.

There are many different correct answers for 18–20.

18.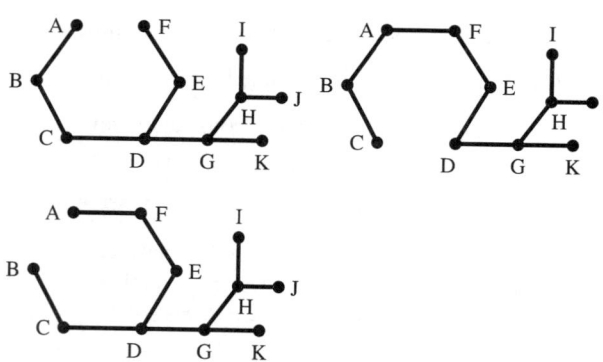

19.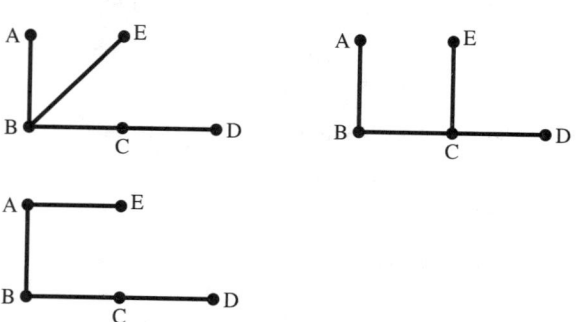

20.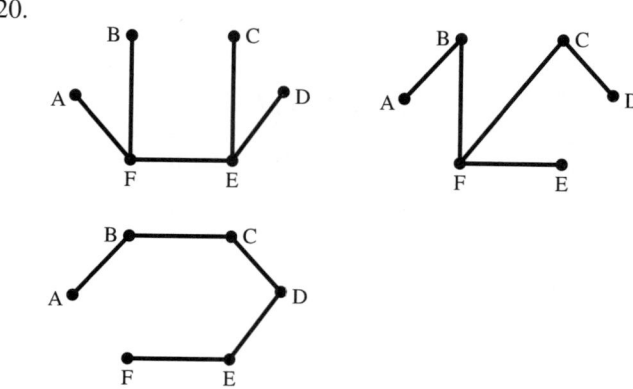

21.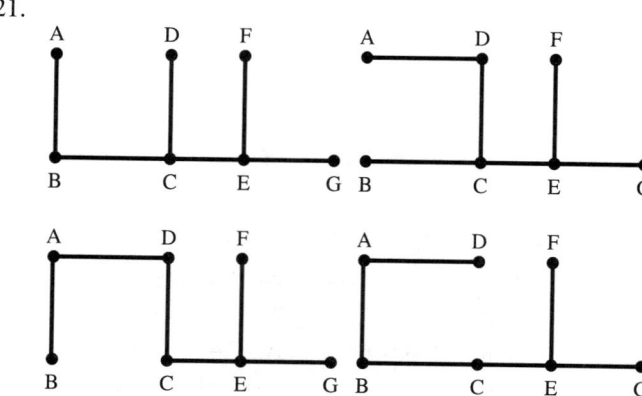

22.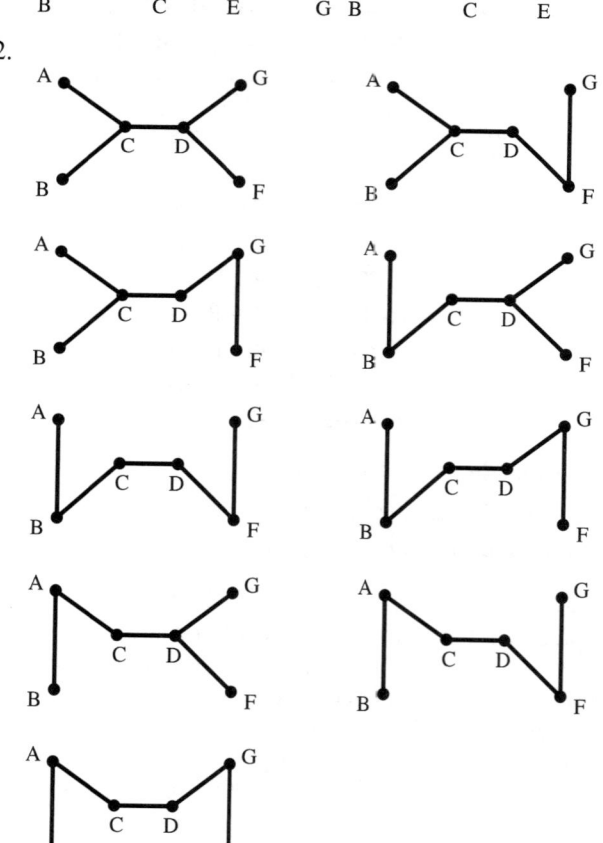

23.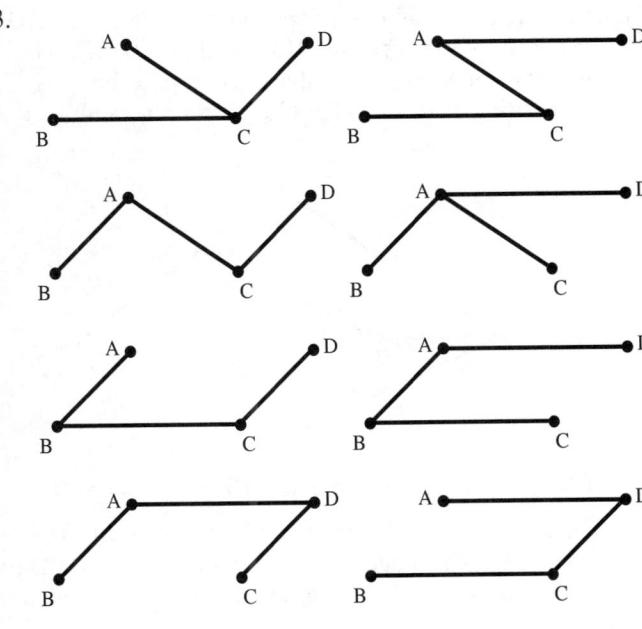

24. There are two circuits in this graph. We can drop any one of 4 edges from the first circuit and any one of 3 edges from the second circuit to form a spanning tree. Thus, there would be a total of $4 \times 3 = 12$ different spanning trees.

25. There are two circuits in this graph. We can drop any one of 4 edges from the first circuit and any one of 5 edges from the second circuit to form a spanning tree. Thus, there would be a total of $4 \times 5 = 20$ different spanning trees.

26. There are three circuits in this graph. We can drop any one of 5 edges from the first circuit, any one of 6 edges from the second circuit, and any one of 3 edges from the third circuit to form a spanning tree. Thus, there would be a total of $5 \times 6 \times 3 = 90$ different spanning trees.

27. If a connected graph has circuits, none of which have common edges, the number of spanning trees for the graph is the product of the number of edges in each circuit.

28. (a) All spanning trees for graphs (i)–(iii) are as follows.

    (i)

    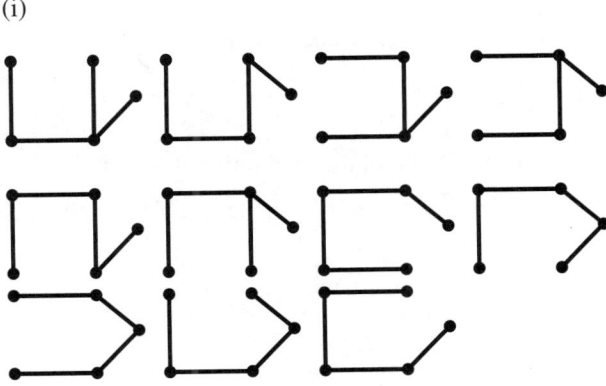

    (ii)

    (iii)

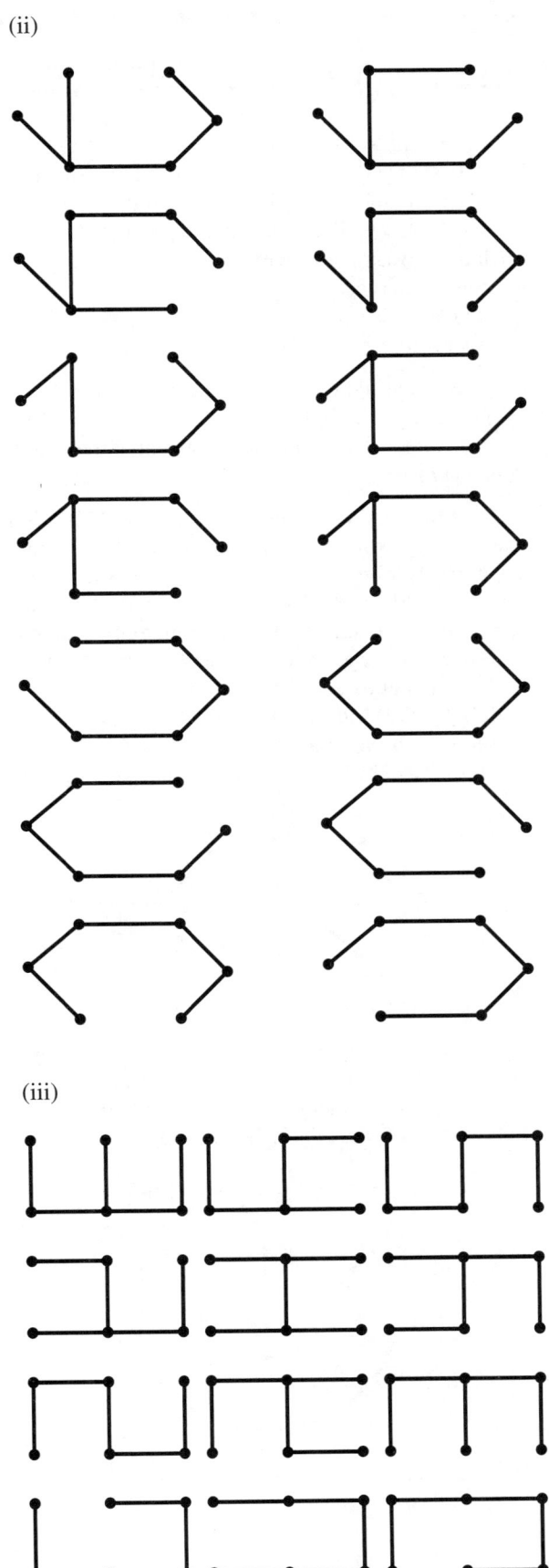

(b) The number of spanning trees for:

(i) is $6 \cdot 3 - 1 = 17$. Note that we have to subtract the number of common edges from the product, since the number of trees represented by the product alone includes the same tree, with common edge(s), disconnected twice.

(ii) is $5 \cdot 6 - 1 = 29$.

(iii) is $6 \cdot 8 - 1 = 47$.

(c) If a connected graph has two circuits having a common edge and if the circuits have m and n edges respectively, then the number of spanning trees for the graph is $mn - 1$.

29. Choose the edge with minimum weight GF (2). Choose the next edge with minimum weight, FC (3). The next edge with minimum weight is FD (8). The next edge of choice would be DE (10). Note here that, although DC represents a remaining edge with minimum weight (9), if it were chosen, we would then have a circuit, FDC, which is not allowed. Choose next edge with minimum weight FB (12) followed by BA (16). Note BG (15) is not allowed because the circuit, BGF, is then formed. We now have the following minimum spanning tree:

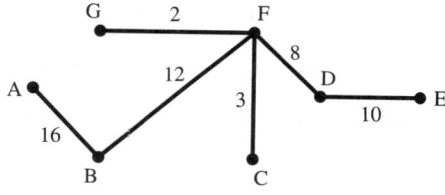

The total weight is $2 + 3 + 8 + 10 + 12 + 16 = 51$.

30. Choose initial minimum edge FC (10), followed by FD (11), then AC (12), and BD (14). The last remaining minimum edge to choose, without completing a circuit, is ED (27). Thus, the minimum spanning tree is:

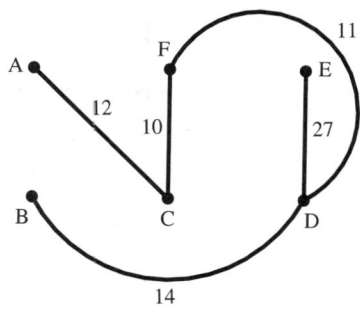

The total weight is $10 + 11 + 12 + 14 + 27 = 74$.

31. Choose initial minimum edge AD (6), followed by BG (8), then EF (9), then DC (10), and then AB (13). The last remaining minimum edge to choose, without completing a circuit, is AE (20). Thus, the minimum spanning tree is:

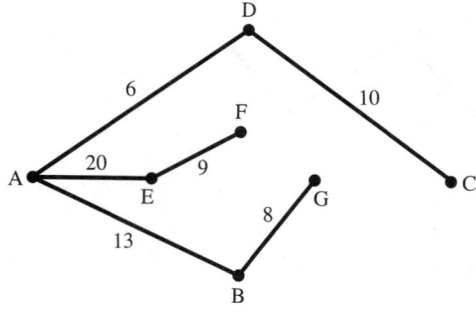

The total weight is $6 + 8 + 9 + 10 + 13 + 20 = 66$.

32. Choose initial minimum edge AC (8) or CE (8). Then choose the other edge whose weight is 8. Choose CD (9) followed by AB (10). Thus, the minimum spanning tree is:

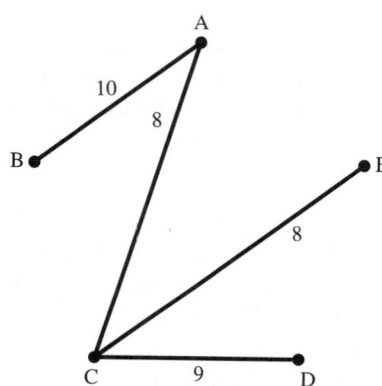

The total weight is $8 + 8 + 9 + 10 = 35$.

33. Choose the initial minimum edge, BF (23), followed by either AE (25) or EF (25). Then choose the other edge with weight 25. Finally, choose AD (32) followed by DC (35).

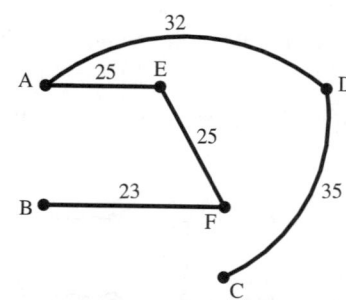

The total length of the minimum pathway is

$23 + 25 + 25 + 32 + 35 = 140$ ft.

34. Choose initial minimum edge AE (1.6), followed by BF (1.7), then CD (1.8), AC (1.9) and CF (2.3).
    Thus, the minimum spanning tree is:

    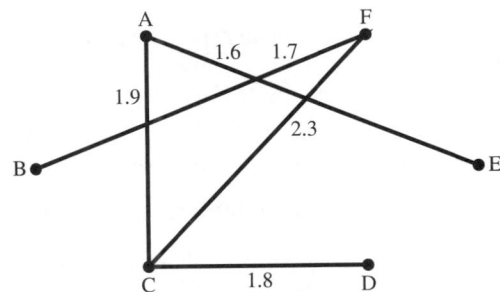

    The total cost of the minimum pathway is

    $1.6 + 1.7 + 1.8 + 1.9 + 2.3 = 9.3$ million dollars.

35. A tree with 34 vertices will have one less, or 33 edges.

36. A tree with 40 edges will have one more, or 41 vertices.

37. A spanning tree for a complete graph with 63 vertices will have one less, or 62 edges.

38. To have a spanning tree for a graph with 27 vertices we must have 26 edges. Thus, we will have to remove 17 ($43 - 17 = 26$) edges.

39. Different spanning trees must have the same number of edges, since the number of vertices in the tree is the number of vertices in the original graph, and the number of edges has to be one less than the number of vertices.

40. Consider a tree with 9 vertices.

    (a) There will be $9 - 1 = 8$ edges.

    (b) The sum of the degrees for all vertices will be twice the number of edges, or $2 \times 8 = 16$.

    (c) The smallest number of vertices of degree one on this graph will be 2 (the beginning vertex and the ending vertex).

    (d) The largest number of vertices of degree one in this graph is $9 - 1 = 8$. All may be degree one except for a central vertex with edges to the other 8 vertices.

    (e) For the general case of a tree with n vertices: There are $n - 1$ edges; $2(n - 1)$ is the sum of the degrees for the vertices; 2 is the smallest number of vertices of degree one; $n - 1$ is the largest possible number of vertices of degree one.

41. Consider a tree with 10 vertices.
    (a) There will be $10 - 1 = 9$ edges.

    (b) The sum of the degrees for all vertices will be twice the number of edges, or $2 \times 9 = 18$.

    (c) The smallest number of vertices of degree four on this graph will be 0. For example, a tree may be drawn with the first and last vertex of degree 1, and the remaining 8 vertices of degree 2.

    (d) The largest number of vertices of degree four in this graph is 2. Let us use the strategy of trial and error. If we use 4 vertices with degree 4, this will contribute 16 to the total degree and would leave $18 - 16 = 2$ as the degree sum of the remaining 6 vertices. But this means that some of those vertices would have no edges joined to them, so that our graph would not be connected and would not be a tree. Similarly, using 3 vertices with degree 4 would leave $18 - 12 = 6$ as the sum of (the remaining vertex) degrees. Thus, of the last 7 vertices, at least one would have no degree (or connecting edge). This would not be a problem if we choose two vertices with degree 4. The following tree, for example, would satisfy our conditions.

42. Consider a tree with 17 vertices.
    (a) There will be $17 - 1 = 16$ edges.

    (b) The sum of the degrees for all vertices will be twice the number of edges, or $2 \times 16 = 32$.

    (c) The largest number of vertices of degree five in this graph is, by trial and error, 3. If, for example, we use 4 vertices with degree 5, this will contribute 20 to the total degree and would leave $32 - 20 = 12$ as the degree sum of the remaining 13 vertices. But this means that at least one vertex would have no edges joined to it, so that our graph would not be connected and, hence, not a tree. This would not be a problem if we choose three vertices with degree 5. The following tree, for example, would satisfy our conditions.

    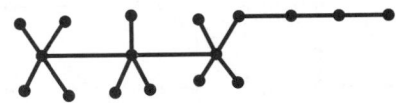

43. Treating each of the 23 employee's computers as a vertex and the network as a tree, there will be one less edge than the number of vertices. Hence, there will be 22 cables that will have to be run between the computers.

44. Treating each of the 12 vegetable and flower beds along with her front door as a vertex in a tree graph, there would be one less edge than the resulting 13 vertices. Thus, it will take 12 edges to complete the tree, and she will have to build a minimum of 12 paths.

45. It is possible to draw in the same number of vertices as edges. The graph must be a tree, since we still have one less edge than vertex.

46. It is not possible to draw in more vertices than edges. If we retain connectedness of the graph, each vertex we draw in has to have at least one edge associated with it.

47. It is possible to draw in more edges than vertices. But the graph cannot be a tree, since we would now have at least as many edges as vertices.

48. It is possible to draw in the same number of vertices as edges and end up with a disconnected graph. Just draw a graph such as the following next to the original tree.

49. Using Cayley's theorem with $n = 3$, "A complete graph with n vertices has n^{n-2} spanning trees," we arrive at $3^{3-2} = 3^1 = 3$ spanning trees. A complete graph is as follows.

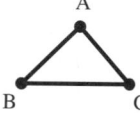

The three spanning trees are as follows.

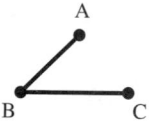 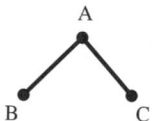 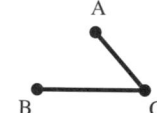

50. Using $n = 4$, Cayley's theorem suggests that there will be $4^{4-2} = 4^2 = 16$. A complete graph is as follows.

The sixteen spanning trees are as follows.

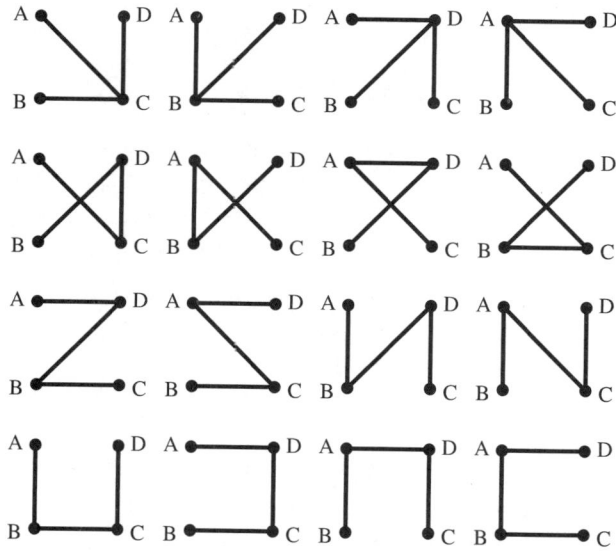

51. For a graph with 5 vertices, we will have $5^{5-2} = 5^3 = 125$ spanning trees.

52. There are just 2 non-isomorphic trees with 4 vertices. They are:

53. There are just 3 non-isomorphic trees with 5 vertices. They are:

54. There are 6 non-isomorphic trees with 6 vertices. They are:

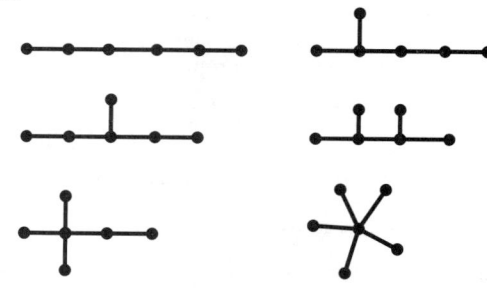

55. There are 11 non-isomorphic trees with 7 vertices. They are:

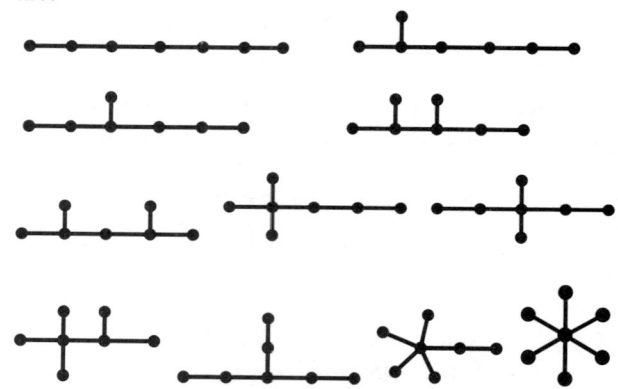

56. (a) The number of components (connected graphs) for any tree is 1.

(b) The graph must have at least one edge, since the tree is connected and has more than one vertex.

(c) If we remove one edge from a tree, the resulting graph will have two components. Remember that each edge is a cut edge.

(d) Removing a second edge from a tree will create one more, or three components.

(e) Removing 3 edges will create 4 components. Removing 4 edges will create 5 components.

(f) By the above pattern, if we remove n edges from the tree, we will be left with $n + 1$ components.

(g) The components left will be the vertices of the original tree.

(h) Thus, there must be $n+1$ vertices to begin with, if the tree has n edges to start with.

57. Writing exercise

58. Writing exercise

CHAPTER 15 TEST

1. By counting, there are 7 vertices.

2. The sum of the degrees of the vertices are:
$$4+2+2+2+6+2+2=20.$$

3. By counting, there are 10 edges.

4. (a) No, B→A→C→E→B→A is not a path, since edge AB is used twice.

    (b) Yes, A→B→E→A is a path.

    (c) No, A→C→D→E is not a path, since there is no edge from C to D.

5. (a) Yes, A→B→E→D→A is a circuit.

    (b) No, A→B→C→D→E→F→G→A is not a circuit, since, for example, there is no edge from B to C.

    (c) Yes, A→B→E→F→G→E→D→A→E→C→A is a circuit.

6. A graph with 2 components, for example, is:

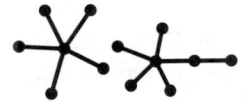

7. The sum of degrees of the vertices is:
$$4+4+4+2+2+2+2+2+2+2=26.$$
Thus, there are $26 \div 2 = 13$ edges.

8. The graphs are isomorphic.

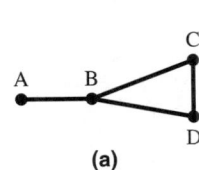 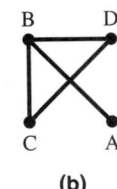

9. The graph is:

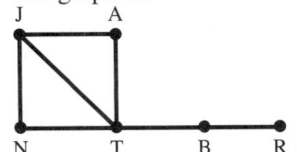

The graph is connected. Tina knows the largest number of other guests.

10. Let each of the 8 contestants represent a vertex of a complete graph (graph with each vertex connected by an edge to all other vertices). The degree of each vertex is 7. Thus, the sum of the degrees is $8 \cdot 7 = 56$. Since the sum of the degrees is twice the number of edges, there must by $56 \div 2 = 28$ edges. Since each edge represents a game to be played, there are 28 games in the competition.

11. Yes, the graph is a complete graph, since there is an edge from each vertex to each of the remaining 6 vertices.

12. (a) No, A→B→E→D→A is not an Euler circuit, since it does not use all of the edges.

    (b) No, A→B→C→D→E→F→G→A is not a circuit because, for example, there is no edge from B to C.

    (c) Yes, A→B→E→F→G→E→D→A→E→C→A is an Euler circuit.

13. No, the graph will not have an Euler circuit, since some of the vertices have odd degree.

14. Yes, the graph will have an Euler circuit, since all vertices have even degree.

15. No, since two of the rooms have an odd number of doors. Note that we are considering each room to be a vertex and asking the question "Can an Euler circuit be formed?"

16. A resulting Euler circuit is F→B→E→D→B→C→D→K→B→A→H→G→F→A→G→J→F. Note, BF is the only cut edge after F→B, thus you may choose any vertex in the right subgraph after B.

17. (a) No, A→B→E→D→A, is not a Hamilton circuit since it does not visit all vertices.

    (b) No, A→B→C→D→E→F→G→A is not a circuit because, for example, there is no edge from B to C.

    (c) No, A→B→E→F→G→E→D→A→E→C→A is not an Hamilton circuit since it visits some vertices twice before returning to starting vertex.

18. F→G→H→I→E→F; F→G→H→E→I→F;
    F→G→I→H→E→F; F→G→I→E→H→F;
    F→G→E→H→I→F; F→G→E→I→H→F.
    There are 6 such Hamilton circuits.

19. Using the Brute force algorithm and P as the starting vertex, we get the following circuits. Use a tree diagram as an aid.

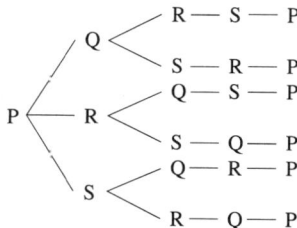

Circuit: Total weight of circuit:
1. P→Q→R→S→P $7+8+7+7=29$
2. P→Q→S→R→P $7+9+7+4=27$
3. P→R→Q→S→P $4+8+9+7=28$
4. P→R→S→Q→P (opposite of 2)   27
5. P→S→Q→R→P (opposite of 3)   28
6. P→S→R→Q→P (opposite of 1).   29.

Thus, P→Q→S→R→P is the minimum Hamilton circuit with a weight of 27.

20. Using the nearest neighbor algorithm, choose the first edge with the minimum weight, A$\overset{1.6}{\to}$E. Keep track of weight by noting its value over the arrow. Choose the second edge, E$\overset{1.95}{\to}$D, which has the minimum weight. Choose the third edge, D$\overset{1.8}{\to}$C, which has the minimum weight. Choose C$\overset{2.3}{\to}$F. Choose F$\overset{1.7}{\to}$B. Finally, choose B$\overset{2.5}{\to}$A. Note that this is your only choice for the last remaining edge. The resulting approximate minimum Hamilton circuit is, therefore,

A→E→D→C→F→B→A.

The circuit has a total weight of

$1.6 + 1.95 + 1.8 + 2.3 + 1.7 + 2.5 = 11.85$.

21. For a complete graph with 25 vertices, there will be $(25-1)! = 24!$ Hamilton circuits.

22. This problem calls for a Hamilton circuit, since the band wants to visit each city (vertex) only once.

23. Any three of the following, for example, would satisfy the stated conditions:

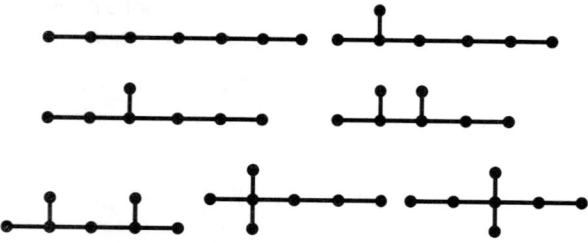

24. The statement "Every tree has a Hamilton circuit" is false, since in many trees, you will have to visit the same vertex more than once.

25. The statement "In a tree each edge is a cut edge" is true.

26. The statement "Every tree is connected" is true, since one can always move from each vertex of the graph along edges to every other vertex.

27. The following represent the different spanning trees for the accompanying graph.

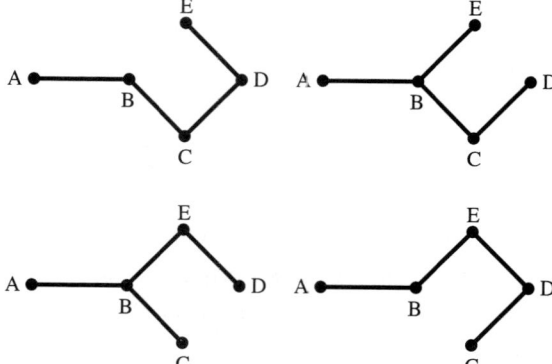

There are 4 spanning trees.

28. Using Kruskal's algorithm, choose the edge with minimum weight, B$\overset{3}{\to}$E. Choose the next edge with minimum weight, E$\overset{5}{\to}$D. Continuing in this fashion, we choose C$\overset{7}{\to}$D followed by D$\overset{9}{\to}$A.

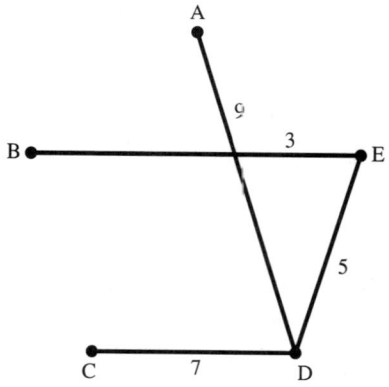

The weight is $3 + 5 + 7 + 9 = 24$.

29. The number of edges in a tree is one less than the number of vertices. Thus, there are $50 - 1 = 49$ edges.

Many of the exercises in this chapter are done most efficiently by using a spreadsheet.

16.1 EXERCISES

1. (a) Because there are four breeds, the number of ways that a staff member can complete her ballot is
$$4! = 4 \cdot 3 \cdot 2 \cdot 1 = 24.$$

   (b) The voter profile is

Votes	Ranking	Voters
3	$b > c > a > d$	1, 2, 9
2	$a > d > c > b$	3, 11
1	$c > d > b > a$	4
1	$d > c > b > a$	5
1	$d > a > b > c$	6
2	$c > a > d > b$	7, 8
1	$a > c > d > b$	10
2	$a > b > c > d$	12, 13

   (c) By the plurality method, the breed with the most votes wins. The Australian shepherd wins.

Breed	1st place votes
a	5
b	3
c	3
d	2

2. (a) Because there are four breeds, the number of ways that a staff member can complete her ballot is
$$4! = 4 \cdot 3 \cdot 2 \cdot 1 = 24.$$

   (b) The voter profile is

Votes	Ranking	Voter
3	$a > b > c > d$	1, 10, 13
1	$b > c > a > d$	2
2	$d > b > c > a$	3, 9
1	$c > b > a > d$	4
2	$a > d > c > b$	5, 6
1	$b > c > d > a$	7
1	$c > b > d > a$	8
2	$a > b > d > c$	11, 12

   (c) By the plurality method, the breed with the most votes wins. The Australian shepherd wins.

Breed	1st-place votes
a	7
b	2
c	2
d	2

3. For $n = 5$, the number of rankings is
$$5! = 5 \cdot 4 \cdot 3 \cdot 2 \cdot 1 = 120.$$
   For $n = 7$, the number of rankings is
$$7! = 7 \cdot 6 \cdot 5 \cdot 4 \cdot 3 \cdot 2 \cdot 1 = 5040.$$

4. For $n = 6$, the number of rankings is
$$6! = 6 \cdot 5 \cdot 4 \cdot 3 \cdot 2 \cdot 1 = 720.$$
   For $n = 8$, the number of rankings is
$$8! = 8 \cdot 7 \cdot 6 \cdot 5 \cdot 4 \cdot 3 \cdot 2 \cdot 1 = 40,320.$$

5. Writing exercise

6. The number of pairwise comparisons needed to learn the outcome of an election involving $n = 5$ candidates is the number of combinations of 5, taken 2 at a time. Mathematically this means
$$_5C_2 = \frac{5!}{2!(5-2)!} = \frac{5 \cdot 4 \cdot 3 \cdot 2 \cdot 1}{2 \cdot 1 \cdot 3 \cdot 2 \cdot 1} = 10.$$

   For $n = 7$ candidates, find the combinations of 7, taken 2 at a time.
$$_7C_2 = \frac{7!}{2!(7-2)!} = \frac{7 \cdot 6 \cdot 5 \cdot 4 \cdot 3 \cdot 2 \cdot 1}{2 \cdot 1 \cdot 5 \cdot 4 \cdot 3 \cdot 2 \cdot 1} = 21.$$

7. The number of pairwise comparisons needed to learn the outcome of an election involving $n = 6$ candidates is the number of combinations of 6, taken 2 at a time. Mathematically this means
$$_6C_2 = \frac{6!}{2!(6-2)!} = \frac{6 \cdot 5 \cdot 4 \cdot 3 \cdot 2 \cdot 1}{2 \cdot 1 \cdot 4 \cdot 3 \cdot 2 \cdot 1} = 15.$$

   For $n = 8$ candidates, find the combinations of 8, taken 2 at a time.
$$_8C_2 = \frac{8!}{2!(8-2)!} = \frac{8 \cdot 7 \cdot 6 \cdot 5 \cdot 4 \cdot 3 \cdot 2 \cdot 1}{2 \cdot 1 \cdot 6 \cdot 5 \cdot 4 \cdot 3 \cdot 2 \cdot 1} = 28.$$

8. Writing exercise

9. Writing exercise

10. Use the voter profile to answer the questions.

Votes	Ranking	Row
4	$a > b > c$	1
2	$b > c > a$	2
4	$b > a > c$	3
3	$c > a > b$	4

    (a) Using the plurality method to determine the chairperson, the winner is candidate b with 6 first-place votes in rows 2 and 3. Notice from the table that candidate a received 4 votes in row 1, and candidate c received 3 votes in row 4.

(b) First, the number of comparisons is

$$_3C_2 = \frac{3!}{2!(3-2)!} = \frac{3\cdot 2\cdot 1}{2\cdot 1\cdot 1} = 3.$$

Make a table to compute the votes in comparisons.

Comparison	Votes	Rows
$a > b$	$4 + 3 = 7$	1, 4
$b > a$	$2 + 4 = 6$	2, 3
$a > c$	$4 + 4 = 8$	1, 3
$c > a$	$2 + 3 = 5$	2, 4
$b > c$	$4 + 2 + 4 = 10$	1, 2, 3
$c > b$	$3 = 3$	4

Make another table to award pairwise points.

Pairs	Votes	Pairwise Points
$a:b$	7:6	a, 1
$a:c$	8:5	a, 1
$b:c$	10:3	b, 1

Using the pairwise comparison method, the winner is candidate a with 2 pairwise points. Candidate b received only 1 point, and candidate c received 0 points.

(c) Use the information from part (a) to make a table to examine the points in the Borda method.

		Points		
Votes	Ranking	2	1	0
4	$a > b > c$	a	b	c
2	$b > c > a$	b	c	a
4	$b > a > c$	b	a	c
3	$c > a > b$	c	a	b

Examine the last three columns and compute the weighted sum for each candidate.

Candidate	Borda Points
a	$4\cdot 2 + 2\cdot 0 + 4\cdot 1 + 3\cdot 1 = 15$
b	$4\cdot 1 + 2\cdot 2 + 4\cdot 2 + 3\cdot 0 = 16$
c	$4\cdot 0 + 2\cdot 1 + 4\cdot 0 + 3\cdot 2 = 8$

The winner is candidate b with 16 Borda points.

(d) There are 13 voters, so a candidate must receive a majority of the votes or 7 votes to win by the Hare method. No candidate received a majority; candidates a, b, and c received 4, 6, and 3 votes respectively. (See part a.) By this method, candidate c is eliminated, and another vote is taken. Now compare only candidates a and b.

Comparison	Votes	Rows
$a > b$	$4 + 3 = 7$	1, 4
$b > a$	$2 + 4 = 6$	2, 3

Candidate a now receives a majority and wins.

(e) Using the predetermined order c, b, a, begin by comparing c and b in the original voting.

Comparison	Votes	Row
$b > c$	$4 + 2 + 4 = 10$	1, 2, 3
$c > b$	$3 = 3$	4

Candidate b wins this competition, 10 to 3. Now compare b to a.

Comparison	Votes	Row
$b > a$	$2 + 4 = 6$	2, 3
$a > b$	$4 + 3 = 7$	1, 4

Candidate a wins the next competition, 7 to 6, and also wins the entire competition in the sequential pairwise comparison method.

11. Use the voter profile to answer the questions.

Votes	Ranking	Row
3	$a > c > b$	1
4	$c > b > a$	2
2	$b > a > c$	3
4	$b > c > a$	4

(a) Using the plurality method to determine the chairperson, the winner is candidate b with 6 first-place votes in rows 3 and 4. Notice from the table that candidate a received 3 votes in row 1, and candidate c received 4 votes in row 2.

(b) First, the number of comparisons is

$$_3C_2 = \frac{3!}{2!(3-2)!} = \frac{3\cdot 2\cdot 1}{2\cdot 1\cdot 1} = 3.$$

Make a table to compute the votes in comparisons.

Comparison	Votes	Rows
$a > b$	$3 = 3$	1
$b > a$	$4 + 2 + 4 = 10$	2, 3, 4
$a > c$	$3 + 2 = 5$	1, 3
$c > a$	$4 + 4 = 8$	2, 4
$b > c$	$2 + 4 = 6$	3, 4
$c > b$	$3 + 4 = 7$	1, 2

Make another table to award pairwise points.

Pairs	Votes	Pairwise Points
$a:b$	3: 10	b, 1
$a:c$	5: 8	c, 1
$b:c$	6: 7	c, 1

Using the pairwise comparison method, the winner is candidate c with 2 pairwise points. Candidate b received only 1 point, and candidate a received 0 points.

(c) Use the information from part (a) to make a table to examine the points in the Borda method.

		Points		
Votes	Ranking	2	1	0
3	$a>c>b$	a	c	b
4	$c>b>a$	c	b	a
2	$b>a>c$	b	a	c
4	$b>c>a$	b	c	a

Examine the last three columns and compute the weighted sum for each candidate.

Candidate	Borda Points
a	$3 \cdot 2 + 4 \cdot 0 + 2 \cdot 1 + 4 \cdot 0 = 8$
b	$3 \cdot 0 + 4 \cdot 1 + 2 \cdot 2 + 4 \cdot 2 = 16$
c	$3 \cdot 1 + 4 \cdot 2 + 2 \cdot 0 + 4 \cdot 1 = 15$

The winner is candidate b with 16 Borda points.

(d) There are 13 voters, so a candidate must receive a majority of the votes or 7 votes to win by the Hare method. No candidate received a majority; candidates a, b, and c received 3, 6, and 4 votes respectively. (See part a.) By this method, candidate a is eliminated, and another vote is taken. Now compare only candidates b and c.

Comparison	Votes	Rows
$b>c$	$2+4=6$	3, 4
$c>b$	$3+4=7$	1, 2

Candidate c now receives a majority and wins.

(e) Using the predetermined order c, b, a, begin by comparing c and b in the original voting.

Comparison	Votes	Row
$b>c$	$2+4=6$	3, 4
$c>b$	$3+4=7$	1, 2

Candidate c wins this competition, 7 to 6. Now compare c to a.

Comparison	Votes	Row
$c>a$	$4+4=8$	2, 4
$a>c$	$3+2=5$	1, 3

Candidate c wins this competition, 8 to 5, and also wins the entire competition in the sequential pairwise comparison method.

12. Use the voter profile to answer the questions.

Votes	Ranking	Row
2	$a>c>b$	1
3	$c>b>a$	2
4	$b>a>c$	3
4	$b>c>a$	4

(a) Using the plurality method to determine the logo, the winner is logo b with 8 first-place votes in rows 3 and 4. Notice from the table that logo a received 2 votes in row 1, and logo c received 3 votes in row 2.

(b) First, the number of comparisons is

$$_3C_2 = \frac{3!}{2!(3-2)!} = \frac{3 \cdot 2 \cdot 1}{2 \cdot 1 \cdot 1} = 3.$$

Make a table to compute the votes in comparisons.

Comparison	Votes	Rows
$a>b$	$2=2$	1
$b>a$	$3+4+4=11$	2, 3, 4
$a>c$	$2+4=6$	1, 3
$c>a$	$3+4=7$	2, 4
$b>c$	$4+4=8$	3, 4
$c>b$	$2+3=5$	1, 2

Make another table to award pairwise points.

Pairs	Votes	Pairwise Points
$a:b$	2: 11	b, 1
$a:c$	6: 7	c, 1
$b:c$	8: 5	b, 1

Using the pairwise comparison method, the winner is logo b with 2 pairwise points. Logo c received only 1 point, and logo a received 0 points.

(c) Use the information from part (a) to make a table to examine the points in the Borda method.

		Points		
Votes	Ranking	2	1	0
2	$a>c>b$	a	c	b
3	$c>b>a$	c	b	a
4	$b>a>c$	b	a	c
4	$b>c>a$	b	c	a

Examine the last three columns and compute the weighted sum for each logo.

Logo	Borda Points
a	$2 \cdot 2 + 3 \cdot 0 + 4 \cdot 1 + 4 \cdot 0 = 8$
b	$2 \cdot 0 + 3 \cdot 1 + 4 \cdot 2 + 4 \cdot 2 = 19$
c	$2 \cdot 1 + 3 \cdot 2 + 4 \cdot 0 + 4 \cdot 1 = 12$

The winner is logo b with 19 Borda points.

(d) There are 13 voters, so a logo must receive a majority of the votes or at least 7 votes to win by the Hare method. Refer to part (a) to see that logo b received 8 votes to win by the Hare method.

(e) Using the predetermined order a, c, b, begin by comparing a and c in the original voting.

Comparison	Votes	Row
$a>c$	$2+4=6$	1, 3
$c>a$	$3+4=7$	2, 4

Logo c wins this competition, 7 to 6. Now compare c to b.

Comparison	Votes	Row
$c > b$	$2 + 3 = 5$	1, 2
$b > c$	$4 + 4 = 8$	3, 4

Logo b wins this competition, 8 to 5, and also wins the entire competition in the sequential pairwise comparison method.

13. Use the voter profile to answer the questions.

Votes	Ranking	Row
6	$a > b > c$	1
1	$b > c > a$	2
3	$b > a > c$	3
3	$c > a > b$	4

(a) Using the plurality method to determine the logo, the winner is logo a with 6 first-place votes in row 1. Notice from the table that logo b received a total of 4 votes in rows 2 and 3, and logo c received 3 votes in row 4.

(b) First, the number of comparisons is

$$_3C_2 = \frac{3!}{2!(3-2)!} = \frac{3 \cdot 2 \cdot 1}{2 \cdot 1 \cdot 1} = 3.$$

Make a table to compute the votes in comparisons.

Comparison	Votes	Rows
$a > b$	$6 + 3 = 9$	1, 4
$b > a$	$1 + 3 = 4$	2, 3
$a > c$	$6 + 3 = 9$	1, 3
$c > a$	$1 + 3 = 4$	2, 4
$b > c$	$6 + 1 + 3 = 10$	1, 2, 3
$c > b$	$3 = 3$	4

Make another table to award pairwise points.

Pairs	Votes	Pairwise Points
$a : b$	9 : 4	a, 1
$a : c$	9 : 4	a, 1
$b : c$	10 : 3	b, 1

Using the pairwise comparison method, the winner is logo a with 2 pairwise points. Logo b received only 1 point, and logo c received 0 points.

(c) Use the information from part (a) to make a table to examine the points in the Borda method.

Votes	Ranking	Points 2	1	0
6	$a > b > c$	a	b	c
1	$b > c > a$	b	c	a
3	$b > a > c$	b	a	c
3	$c > a > b$	c	a	b

Examine the last three columns and compute the weighted sum for each logo.

Logo	Borda Points
a	$6 \cdot 2 + 1 \cdot 0 + 3 \cdot 1 + 3 \cdot 1 = 18$
b	$6 \cdot 1 + 1 \cdot 2 + 3 \cdot 2 + 3 \cdot 0 = 14$
c	$6 \cdot 0 + 1 \cdot 1 + 3 \cdot 0 + 3 \cdot 2 = 7$

The winner is logo a with 18 Borda points.

(d) There are 13 voters, so a logo must receive a majority of the votes or at least 7 votes to win by the Hare method. No logo received a majority. Logo a received 6 votes; logo b received 4 votes; logo c received 3 votes. Eliminate logo c and compare the remaining logos.

Votes	Ranking	Row
6	$a > b$	1
1	$b > a$	2
3	$b > a$	3
3	$a > b$	4

Now logo a receives $6 + 3 = 9$ votes; logo b receives $1 + 3 = 4$ votes. Logo a now receives a majority and wins.

(e) Using the predetermined order b, c, a, begin by comparing b and c in the original voting.

Comparison	Votes	Row
$b > c$	$6 + 1 + 3 = 10$	1, 2, 3
$c > b$	$3 = 3$	4

Logo b wins this competition, 10 to 3. Now compare b to a.

Comparison	Votes	Row
$a > b$	$6 + 3 = 9$	1, 4
$b > a$	$1 + 3 = 4$	2, 3

Logo a wins this competition, 9 to 4, and also wins the entire competition in the sequential pairwise comparison method.

14. Use the voter profile to answer the questions.

Votes	Ranking	Row
3	$e > h > g > j$	1
6	$h > e > g > j$	2
5	$j > g > e > h$	3
4	$g > e > h > j$	4
3	$j > e > h > g$	5

(a) Using the plurality method to determine the highest priority issue, the winner is issue j with a total of 8 first-place votes in rows 3 and 5. Notice from the table that the remaining issues received the following number of first-place votes: e, 3 votes; h, 6 votes; g, 4 votes.

(b) First, the number of comparisons is

$$_4C_2 = \frac{4!}{2!(4-2)!} = \frac{4 \cdot 3 \cdot 2 \cdot 1}{2 \cdot 1 \cdot 2 \cdot 1} = 6.$$

Make a table to compute the votes in comparisons.

Comparison	Votes	Rows
$e > h$	$3 + 5 + 4 + 3 = 15$	1, 3, 4, 5
$h > e$	$6 = 6$	2
$e > g$	$3 + 6 + 3 = 12$	1, 2, 5
$g > e$	$5 + 4 = 9$	3, 4
$e > j$	$3 + 6 + 4 = 13$	1, 2, 4
$j > e$	$5 + 3 = 8$	3, 5
$h > g$	$3 + 6 + 3 = 12$	1, 2, 5
$g > h$	$5 + 4 = 9$	3, 4
$h > j$	$3 + 6 + 4 = 13$	1, 2, 4
$j > h$	$5 + 3 = 8$	3, 5
$j > g$	$5 + 3 = 8$	3, 5
$g > j$	$3 + 6 + 4 = 13$	1, 2, 4

Make another table to award pairwise points.

Pairs	Votes	Pairwise Points
$e : h$	$15 : 6$	$e, 1$
$e : g$	$12 : 9$	$e, 1$
$e : j$	$13 : 8$	$e, 1$
$h : g$	$12 : 9$	$h, 1$
$h : j$	$13 : 8$	$h, 1$
$j : g$	$8 : 13$	$g, 1$

Using the pairwise comparison method, the winner is issue e with 3 pairwise points. Issue h received 2 points; issue g received 1 point; issue j received no points.

(c) Use the information from part (a) to make a table to examine the points in the Borda method.

Votes	Ranking	Points 3	2	1	0
3	$e > h > g > j$	e	h	g	j
6	$h > e > g > j$	h	e	g	j
5	$j > g > e > h$	j	g	e	h
4	$g > e > h > j$	g	e	h	j
3	$j > e > h > g$	j	e	h	g

Examine the last four columns, and compute the weighted sum for each issue.

Issue	Borda Points
e	$3 \cdot 3 + 6 \cdot 2 + 5 \cdot 1 + 4 \cdot 2 + 3 \cdot 2 = 40$
h	$3 \cdot 2 + 6 \cdot 3 + 5 \cdot 0 + 4 \cdot 1 + 3 \cdot 1 = 31$
g	$3 \cdot 1 + 6 \cdot 1 + 5 \cdot 2 + 4 \cdot 3 + 3 \cdot 0 = 31$
j	$3 \cdot 0 + 6 \cdot 0 + 5 \cdot 3 + 4 \cdot 0 + 3 \cdot 3 = 24$

The winner is issue e with 40 Borda points.

(d) There are 21 voters, so an issue must receive a majority of the votes or at least 11 votes to win by the Hare method. No issue received a majority. See part (a) that issue e received 3 first-place votes; issue h received 6 votes; issue g received 4 votes; and issue j received 8 votes. Eliminate issue e because it has the least number of votes, and compare only h, g, and j.

Votes	Ranking	Row
3	$h > g > j$	1
6	$h > g > j$	2
5	$j > g > h$	3
4	$g > h > j$	4
3	$j > h > g$	5

Issue h receives $3 + 6 = 9$ votes.
Issue g receives 4 votes.
Issue j receives $5 + 3 = 8$ votes.

Again, no issue has received a majority. Therefore, eliminate g with the least number of votes, and compare only h and j.

Votes	Ranking	Row
3	$h > j$	1
6	$h > j$	2
5	$j > h$	3
4	$h > j$	4
3	$j > h$	5

Issue h receives $3 + 6 + 4 = 13$ votes.
Issue j receives $5 + 3 = 8$ votes.
Finally, issue h is the winner.

(e) Using the predetermined order h, j, e, g, begin by comparing h and j in the original voting.

Comparison	Votes	Row
$h > j$	$3 + 6 + 4 = 13$	1, 2, 4
$j > h$	$5 + 3 = 8$	3, 5

Issue h wins this competition, 13 to 8. Now compare h to e.

Comparison	Votes	Row
$h > e$	$6 = 6$	2
$e > h$	$3 + 5 + 4 + 3 = 15$	1, 3, 4, 5

Issue e wins this competition, 15 to 6. Now compare e and g.

Comparison	Votes	Row
$e > g$	$3 + 6 + 3 = 12$	1, 2, 5
$g > e$	$5 + 4 = 9$	3, 4

Finally, issue e wins this competition, 12 to 9, and also wins the entire competition in the sequential pairwise comparison method.

15. Use the voter profile to answer the questions.

Votes	Ranking	Row
6	$h > j > g > e$	1
5	$e > g > j > h$	2
4	$g > j > h > e$	3
3	$j > h > g > e$	4
3	$e > j > h > g$	5

(a) Using the plurality method to determine the highest priority issue, the winner is issue e with a total of 8 first-place votes in rows 2 and 5. Notice from the table that the remaining issues received the following number of first-place votes: h, 6 votes; g, 4 votes; j, 3 votes.

(b) First, the number of comparisons is

$$_4C_2 = \frac{4!}{2!(4-2)!} = \frac{4 \cdot 3 \cdot 2 \cdot 1}{2 \cdot 1 \cdot 2 \cdot 1} = 6.$$

Make a table to compute the votes in comparisons.

Comparison	Votes	Rows
$e > h$	$5 + 3 = 8$	2, 5
$h > e$	$6 + 4 + 3 = 13$	1, 3, 4
$e > g$	$5 + 3 = 8$	2, 5
$g > e$	$6 + 4 + 3 = 13$	1, 3, 4
$e > j$	$5 + 3 = 8$	2, 5
$j > e$	$6 + 4 + 3 = 13$	1, 3, 4
$h > g$	$6 + 3 + 3 = 12$	1, 4, 5
$g > h$	$5 + 4 = 9$	2, 3
$h > j$	$6 = 6$	1
$j > h$	$5 + 4 + 3 + 3 = 15$	2, 3, 4, 5
$j > g$	$6 + 3 + 3 = 12$	1, 4, 5
$g > j$	$5 + 4 = 9$	2, 3

Make another table to award pairwise points.

Pairs	Votes	Pairwise Points
$e : h$	8 : 13	h, 1
$e : g$	8 : 13	g, 1
$e : j$	8 : 13	j, 1
$h : g$	12 : 9	h, 1
$h : j$	6 : 15	j, 1
$j : g$	12 : 9	j, 1

Using the pairwise comparison method, the winner is issue j with 3 pairwise points. Issue h received 2 points; issue g received 1 point; issue e received no points.

(c) Use the information from part (a) to make a table to examine the points in the Borda method.

Votes	Ranking	Points 3	2	1	0
6	$h > j > g > e$	h	j	g	e
5	$e > g > j > h$	e	g	j	h
4	$g > j > h > e$	g	j	h	e
3	$j > h > g > e$	j	h	g	e
3	$e > j > h > g$	e	j	h	g

Examine the last four columns, and compute the weighted sum for each issue.

Issue	Borda Points
h	$6 \cdot 3 + 5 \cdot 0 + 4 \cdot 1 + 3 \cdot 2 + 3 \cdot 1 = 31$
j	$6 \cdot 2 + 5 \cdot 1 + 4 \cdot 2 + 3 \cdot 3 + 3 \cdot 2 = 40$
g	$6 \cdot 1 + 5 \cdot 2 + 4 \cdot 3 + 3 \cdot 1 + 3 \cdot 0 = 31$
e	$6 \cdot 0 + 5 \cdot 3 + 4 \cdot 0 + 3 \cdot 0 + 3 \cdot 3 = 24$

The winner is issue j with 40 Borda points

(d) There are 21 voters, so an issue must receive a majority of the votes or at least 11 votes to win by the Hare method. No issue received a majority. See part (a) that issue e received 8 first-place votes; issue h received 6 votes; issue g received 4 votes; and issue j received 3 votes. Eliminate issue j because it has the least number of votes, and compare only h, g, and e.

Votes	Ranking	Row
6	$h > g > e$	1
5	$e > g > h$	2
4	$g > h > e$	3
3	$h > g > e$	4
3	$e > h > g$	5

Issue h receives $6 + 3 = 9$ votes.
Issue g receives 4 votes.
Issue e receives $5 + 3 = 8$ votes.
No issue has received a majority, so another vote is taken with issue g eliminated.

Votes	Ranking	Row
6	$h > e$	1
5	$e > h$	2
4	$h > e$	3
3	$h > e$	4
3	$e > h$	5

Issue h receives $6 + 4 + 3 = 13$ votes.
Issue e receives $5 + 3 = 8$ votes.
Now issue h has a majority and wins.

(e) Using the predetermined order h, j, e, g, begin by comparing h and j in the original voting.

Comparison	Votes	Row
$h > j$	$6 = 6$	1
$j > h$	$5 + 4 + 3 + 3 = 15$	2, 3, 4, 5

Issue j wins this competition, 15 to 6. Now compare j to e.

Comparison	Votes	Row
$j > e$	$6 + 4 + 3 = 13$	1, 3, 4
$e > j$	$5 + 3 = 8$	2, 5

Issue j wins this competition, 13 to 8. Now compare j and g.

Comparison	Votes	Row
$j > g$	$6 + 3 + 3 = 12$	1, 4, 5
$g > j$	$5 + 4 = 9$	2, 3

Finally, issue j wins this competition, 12 to 9, and also wins the entire competition in the sequential pairwise comparison method.

16. The voter profile from Exercise 2 (b) is

Votes	Ranking	Voter	Row
3	$a > b > c > d$	1, 10, 13	1
1	$b > c > a > d$	2	2
2	$d > b > c > a$	3, 9	3
1	$c > b > a > d$	4	4
2	$a > d > c > b$	5, 6	5
1	$b > c > d > a$	7	6
1	$c > b > d > a$	8	7
2	$a > b > d > c$	11, 12	8

(a) Using the pairwise comparison method, first make a table to compute the votes in comparison.

Compare	Votes	Rows
$a > b$	$3 + 2 + 2 = 7$	1, 5, 8
$b > a$	$1 + 2 + 1 + 1 + 1 = 6$	2, 3, 4, 6, 7
$a > c$	$3 + 2 + 2 = 7$	1, 5, 8
$c > a$	$1 + 2 + 1 + 1 + 1 = 6$	2, 3, 4, 6, 7
$a > d$	$3 + 1 + 1 + 2 + 2 = 9$	1, 2, 4, 5, 8
$d > a$	$2 + 1 + 1 = 4$	3, 6, 7
$b > c$	$3 + 1 + 2 + 1 + 2 = 9$	1, 2, 3, 6, 8
$c > b$	$1 + 2 + 1 = 4$	4, 5, 7
$b > d$	$3 + 1 + 1 + 1 + 1 + 2 = 9$	1, 2, 4, 6, 7, 8
$d > b$	$2 + 2 = 4$	3, 5
$c > d$	$3 + 1 + 1 + 1 + 1 = 7$	1, 2, 4, 6, 7
$d > c$	$2 + 2 + 2 = 6$	3, 5, 8

Make another table to award pairwise points.

Pairs	Votes	Pairwise Points
$a : b$	7 : 6	a, 1
$a : c$	7 : 6	a, 1
$a : d$	9 : 4	a, 1
$b : c$	9 : 4	b, 1
$b : d$	9 : 4	b, 1
$c : d$	7 : 6	c, 1

Using the pairwise comparison method, the winner is a with 3 pairwise points. Breed b received 2 points; breed c received 1 point; breed d received no points.

(b) Use the information from part (a) to make a table to examine the points in the Borda method.

			Points			
Votes	Ranking	3	2	1	0
3	$a > b > c > d$	a	b	c	d
1	$b > c > a > d$	b	c	a	d
2	$d > b > c > a$	d	b	c	a
1	$c > b > a > d$	c	b	a	d
2	$a > d > c > b$	a	d	c	b
1	$b > c > d > a$	b	c	d	a
1	$c > b > d > a$	c	b	d	a
2	$a > b > d > c$	a	b	d	c

Examine the last four columns and compute the weighted sum for each dog breed.

Breed	Borda Points
a	$3 \cdot 3 + 1 \cdot 1 + 2 \cdot 0 + 1 \cdot 1$ $+ 2 \cdot 3 + 1 \cdot 0 + 1 \cdot 0 + 2 \cdot 3 = 23$
b	$3 \cdot 2 + 1 \cdot 3 + 2 \cdot 2 + 1 \cdot 2$ $+ 2 \cdot 0 + 1 \cdot 3 + 1 \cdot 2 + 2 \cdot 2 = 24$
c	$3 \cdot 1 + 1 \cdot 2 + 2 \cdot 1 + 1 \cdot 3$ $+ 2 \cdot 1 + 1 \cdot 2 + 1 \cdot 3 + 2 \cdot 0 = 17$
d	$3 \cdot 0 + 1 \cdot 0 + 2 \cdot 3 + 1 \cdot 0$ $+ 2 \cdot 2 + 1 \cdot 1 + 1 \cdot 1 + 2 \cdot 1 = 14$

The winner is breed b with 24 Borda points.

(c) Calculate how many votes each breed obtained from the thirteen first-place votes from the voter profile.

Votes	Ranking	Voter
3	$a > b > c > d$	1, 10, 13
1	$b > c > a > d$	2
2	$d > b > c > a$	3, 9
1	$c > b > a > d$	4
2	$a > d > c > b$	5, 6
1	$b > c > d > a$	7
1	$c > b > d > a$	8
2	$a > b > d > c$	11, 12

Breed	1st-place votes
a	7
b	2
c	2
d	2

Using the Hare method, breed a has a majority of the votes and wins.

17. The voter profile from Exercise 1 (b) is

Votes	Ranking	Voter	Row
3	$b > c > a > d$	1, 2, 9	1
2	$a > d > c > b$	3, 11	2
1	$c > d > b > a$	4	3
1	$d > c > b > a$	5	4
1	$d > a > b > c$	6	5
2	$c > a > d > b$	7, 8	6
1	$a > c > d > b$	10	7
2	$a > b > c > d$	12, 13	8

(a) Using the pairwise comparison method, first make a table to compute the votes in comparison.

Compare	Votes	Rows
$a > b$	$2+1+2+1+2 = 8$	2, 5, 6, 7, 8
$b > a$	$3+1+1 = 5$	1, 3, 4
$a > c$	$2+1+1+2 = 6$	2, 5, 7, 8
$c > a$	$3+1+1+2 = 7$	1, 3, 4, 6
$a > d$	$3+2+2+1+2 = 10$	1, 2, 6, 7, 8
$d > a$	$1+1+1 = 3$	3, 4, 5
$b > c$	$3+1+2 = 6$	1, 5, 8
$c > b$	$2+1+1+2+1 = 7$	2, 3, 4, 6, 7
$b > d$	$3+2 = 5$	1, 8
$d > b$	$2+1+1+1+2+1 = 8$	2, 3, 4, 5, 6, 7
$c > d$	$3+1+2+1+2 = 9$	1, 3, 6, 7, 8
$d > c$	$2+1+1 = 4$	2, 4, 5

Make another table to award pairwise points.

Pairs	Votes	Pairwise Points
$a : b$	8 : 5	a, 1
$a : c$	6 : 7	c, 1
$a : d$	10 : 3	a, 1
$b : c$	6 : 7	c, 1
$b : d$	5 : 8	d, 1
$c : d$	9 : 4	c, 1

Using the pairwise comparison method, the winner is c with 3 pairwise points. Breed a received 2 points; breed d received 1 point; breed b received no points.

(b) Use the information from part (a) to make a table to examine the points in the Borda method.

Votes	Ranking	Points 3	2	1	0
3	$b > c > a > d$	b	c	a	d
2	$a > d > c > b$	a	d	c	b
1	$c > d > b > a$	c	d	b	a
1	$d > c > b > a$	d	c	b	a
1	$d > a > b > c$	d	a	b	c
2	$c > a > d > b$	c	a	d	b
1	$a > c > d > b$	a	c	d	b
2	$a > b > c > d$	a	b	c	d

Examine the last four columns and compute the weighted sum for each dog breed.

Breed	Borda Points
a	$3 \cdot 1 + 2 \cdot 3 + 1 \cdot 0 + 1 \cdot 0 + 1 \cdot 2 + 2 \cdot 2 + 1 \cdot 3 + 2 \cdot 3 = 24$
b	$3 \cdot 3 + 2 \cdot 0 + 1 \cdot 1 + 1 \cdot 1 + 1 \cdot 1 + 2 \cdot 0 + 1 \cdot 0 + 2 \cdot 2 = 16$
c	$3 \cdot 2 + 2 \cdot 1 + 1 \cdot 3 + 1 \cdot 2 + 1 \cdot 0 + 2 \cdot 3 + 1 \cdot 2 + 2 \cdot 1 = 23$
d	$3 \cdot 0 + 2 \cdot 2 + 1 \cdot 2 + 1 \cdot 3 + 1 \cdot 3 + 2 \cdot 1 + 1 \cdot 1 + 2 \cdot 0 = 15$

The winner is breed a with 24 Borda points.

(c) Calculate how many points each breed obtained from the thirteen first-place votes from the voter profile.

Votes	Ranking	Voter	Row
3	$b > c > a > d$	1, 2, 9	1
2	$a > d > c > b$	3, 11	2
1	$c > d > b > a$	4	3
1	$d > c > b > a$	5	4
1	$d > a > b > c$	6	5
2	$c > a > d > b$	7, 8	6
1	$a > c > d > b$	10	7
2	$a > b > c > d$	12, 13	8

Breed	1st-place votes
a	5
b	3
c	3
d	2

None of the dog breeds received a majority, 7, of the votes. Therefore, eliminate breed d, the breed that received the least number of votes. Compare the other three.

Votes	Ranking	Voter	Row
3	$b > c > a$	1, 2, 9	1
2	$a > c > b$	3, 11	2
1	$c > b > a$	4	3
1	$c > b > a$	5	4
1	$a > b > c$	6	5
2	$c > a > b$	7, 8	6
1	$a > c > b$	10	7
2	$a > b > c$	12, 13	8

Breed a: $2 + 1 + 1 + 2 = 6$ votes.
Breed b: 3 votes.
Breed c: $1 + 1 + 2 = 4$ votes.
None of the breeds received a majority; therefore, eliminate b and compare first-place points again for a and c.

Votes	Ranking	Voter	Row
3	$c > a$	1, 2, 9	1
2	$a > c$	3, 11	2
1	$c > a$	4	3
1	$c > a$	5	4
1	$a > c$	6	5
2	$c > a$	7, 8	6
1	$a > c$	10	7
2	$a > c$	12, 13	8

Breed a: $2 + 1 + 1 + 2 = 6$
Breed c: $3 + 1 + 1 + 2 = 7$
Breed c now has a majority and wins.

18. The voter profile is

Votes	Ranking	Row
18	$t > k > h > b > c$	1
12	$c > h > k > b > t$	2
10	$b > c > h > k > t$	3
9	$k > b > h > c > t$	4
4	$h > c > k > b > t$	5
2	$h > b > k > c > t$	6

(a) If the plurality method is used, activity t is selected with 18 first-place votes.

(b) Using the pairwise comparison method, first make a table to compute the votes in comparison.

Compare	Votes	Rows
$t > k$	$18 = 18$	1
$k > t$	$12 + 10 + 9 + 4 + 2 = 37$	2, 3, 4, 5, 6
$t > h$	$18 = 18$	1
$h > t$	$12 + 10 + 9 + 4 + 2 = 37$	2, 3, 4, 5, 6
$t > b$	$18 = 18$	1
$b > t$	$12 + 10 + 9 + 4 + 2 = 37$	2, 3, 4, 5, 6
$t > c$	$18 = 18$	1
$c > t$	$12 + 10 + 9 + 4 + 2 = 37$	2, 3, 4, 5, 6
$k > h$	$18 + 9 = 27$	1, 4
$h > k$	$12 + 10 + 4 + 2 = 28$	2, 3, 5, 6
$k > b$	$18 + 12 + 9 + 4 = 43$	1, 2, 4, 5
$b > k$	$10 + 2 = 12$	3, 6
$k > c$	$18 + 9 + 2 = 29$	1, 4, 6
$c > k$	$12 + 10 + 4 = 26$	2, 3, 5
$h > b$	$18 + 12 + 4 + 2 = 36$	1, 2, 5, 6
$b > h$	$10 + 9 = 19$	3, 4
$h > c$	$18 + 9 + 4 + 2 = 33$	1, 4, 5, 6
$c > h$	$12 + 10 = 22$	2, 3
$b > c$	$18 + 10 + 9 + 2 = 39$	1, 3, 4, 6
$c > b$	$12 + 4 = 16$	2, 5

Make another table to award pairwise points.

Pairs	Votes	Pairwise Points
$t : k$	$18 : 37$	$k, 1$
$t : h$	$18 : 37$	$h, 1$
$t : b$	$18 : 37$	$b, 1$
$t : c$	$18 : 37$	$c, 1$
$k : h$	$27 : 28$	$h, 1$
$k : b$	$43 : 12$	$k, 1$
$k : c$	$29 : 26$	$k, 1$
$h : b$	$36 : 19$	$h, 1$
$h : c$	$33 : 22$	$h, 1$
$b : c$	$39 : 16$	$b, 1$

Using the pairwise comparison method, the winner is h with 4 pairwise points. Activity k received 3 points; activity b received 2 points; activity c received 1 point; activity t received no points.

(c) Use the information from part (a) to make a table to examine the points in the Borda method.

Votes	Ranking	Points				
		4	3	2	1	0
18	$t > k > h > b > c$	t	k	h	b	c
12	$c > h > k > b > t$	c	h	k	b	t
10	$b > c > h > k > t$	b	c	h	k	t
9	$k > b > h > c > t$	k	b	h	c	t
4	$h > c > k > b > t$	h	c	k	b	t
2	$h > b > k > c > t$	h	b	k	c	t

Examine the last five columns and compute the weighted sum for each activity.

Breed	Borda Points
t	$18 \cdot 4 + 12 \cdot 0 + 10 \cdot 0 + 9 \cdot 0$ $+ 4 \cdot 0 + 2 \cdot 0 = 72$
k	$18 \cdot 3 + 12 \cdot 2 + 10 \cdot 1 + 9 \cdot 4$ $+ 4 \cdot 2 + 2 \cdot 2 = 136$
h	$18 \cdot 2 + 12 \cdot 3 + 10 \cdot 2 + 9 \cdot 2$ $+ 4 \cdot 4 + 2 \cdot 4 = 134$
b	$18 \cdot 1 + 12 \cdot 1 + 10 \cdot 4 + 9 \cdot 3$ $+ 4 \cdot 1 + 2 \cdot 3 = 107$
c	$18 \cdot 0 + 12 \cdot 4 + 10 \cdot 3 + 9 \cdot 1$ $+ 4 \cdot 3 + 2 \cdot 1 = 101$

The winner is k with 136 Borda points.

(d) Calculate how many points each activity obtained from the 55 first-place votes from the voter profile.

Votes	Ranking	Row
18	$t > k > h > b > c$	1
12	$c > h > k > b > t$	2
10	$b > c > h > k > t$	3
9	$k > b > h > c > t$	4
4	$h > c > k > b > t$	5
2	$h > b > k > c > t$	6

Activity	1st-place votes
t	18
c	12
b	10
k	9
h	6

None of the activities received a majority, 28, of the votes. Therefore, eliminate activity h, the activity that received the least number of votes. Compare the first-place votes of the others.

Votes	Ranking	Row
18	$t > k > b > c$	1
12	$c > k > b > t$	2
10	$b > c > k > t$	3
9	$k > b > c > t$	4
4	$c > k > b > t$	5
2	$b > k > c > t$	6

Activity t: 18 votes.
Activity c: $12 + 4 = 16$ votes.
Activity b: $10 + 2 = 12$ votes.
Activity k: 9 votes.
There is no activity that has received a majority; therefore, eliminate k with only 9 votes, and compare the first-place votes of the others.

Votes	Ranking	Row
18	$t > b > c$	1
12	$c > b > t$	2
10	$b > c > t$	3
9	$b > c > t$	4
4	$c > b > t$	5
2	$b > c > t$	6

Activity t: 18 votes.
Activity c: $12 + 4 = 16$ votes.
Activity b: $10 + 9 + 2 = 21$ votes.
Again there is not a majority; therefore, eliminate c and compare the first-place votes of t and b.

Votes	Ranking	Row
18	$t > b$	1
12	$b > t$	2
10	$b > t$	3
9	$b > t$	4
4	$b > t$	5
2	$b > t$	6

Activity t: 18 votes.
Activity b: $12 + 10 + 9 + 4 + 2 = 37$ votes.
Finally b wins with 37 votes.

(e) Using the predetermined order b, t, c, k, h, begin by comparing b and t in the original voting.

Comparison	Votes	Row
$b > t$	$12 + 10 + 9 + 4 + 2 = 37$	2, 3, 4, 5, 6
$t > b$	$18 = 18$	1

Activity b wins this competition, 37 to 18. Now compare b to c.

Comparison	Votes	Row
$b > c$	$18 + 10 + 9 + 2 = 39$	1, 3, 4, 6
$c > b$	$12 + 4 = 16$	2, 5

Activity b wins this competition, 39 to 16. Now compare b and k.

Comparison	Votes	Row
$b > k$	$10 + 2 = 12$	3, 6
$k > b$	$18 + 12 + 9 + 4 = 43$	1, 2, 4, 5

Activity k wins this competition, 43 to 12. Now compare k and h.

Comparison	Votes	Row
$k > h$	$18 + 9 = 27$	1, 4
$h > k$	$12 + 10 + 4 + 2 = 28$	2, 3, 5, 6

Finally, activity h wins this competition, 28 to 27, and also wins the entire competition in the sequential pairwise comparison method.

19. The voter profile is

Votes	Ranking	Row
18	$t > b > h > k > c$	1
12	$c > h > b > k > t$	2
10	$k > c > h > b > t$	3
9	$b > k > h > c > t$	4
4	$h > c > b > k > t$	5
2	$h > k > b > c > t$	6

(a) If the plurality method is used, activity t is selected with 18 first-place votes.

(b) Using the pairwise comparison method, first make a table to compute the votes in comparison.

Compare	Votes	Rows
$t > b$	$18 = 18$	1
$b > t$	$12 + 10 + 9 + 4 + 2 = 37$	2, 3, 4, 5, 6
$t > h$	$18 = 18$	1
$h > t$	$12 + 10 + 9 + 4 + 2 = 37$	2, 3, 4, 5, 6
$t > k$	$18 = 18$	1
$k > t$	$12 + 10 + 9 + 4 - 2 = 37$	2, 3, 4, 5, 6
$t > c$	$18 = 18$	1
$c > t$	$12 + 10 + 9 + 4 - 2 = 37$	2, 3, 4, 5, 6
$b > h$	$18 + 9 = 27$	1, 4
$h > b$	$12 + 10 + 4 + 2 = 28$	2, 3, 5, 6
$b > k$	$18 + 12 + 9 + 4 = 43$	1, 2, 4, 5
$k > b$	$10 + 2 = 12$	3, 6
$b > c$	$18 + 9 + 2 = 29$	1, 4, 6
$c > b$	$12 + 10 + 4 = 26$	2, 3, 5
$h > k$	$18 + 12 + 4 + 2 = 36$	1, 2, 5, 6
$k > h$	$10 + 9 = 19$	3, 4
$h > c$	$18 + 9 + 4 + 2 = 33$	1, 4, 5, 6
$c > h$	$12 + 10 = 22$	2, 3
$k > c$	$18 + 10 + 9 + 2 = 39$	1, 3, 4, 6
$c > k$	$12 + 4 = 16$	2, 5

Make another table to award pairwise points.

Pairs	Votes	Pairwise Points
$t : b$	$18 : 37$	$b, 1$
$t : h$	$18 : 37$	$h, 1$
$t : k$	$18 : 37$	$k, 1$
$t : c$	$18 : 37$	$c, 1$
$b : h$	$27 : 28$	$h, 1$
$b : k$	$43 : 12$	$b, 1$
$b : c$	$29 : 26$	$b, 1$
$h : k$	$36 : 19$	$h, 1$
$h : c$	$33 : 22$	$h, 1$
$k : c$	$39 : 16$	$k, 1$

Using the pairwise comparison method, the winner is h with 4 pairwise points. Activity b received 3 points; activity k received 2 points; activity c received 1 point; activity t received no points.

(c) Use the information from part (a) to make a table to examine the points in the Borda method,

		Points				
Votes	Ranking	4	3	2	1	0
18	$t > b > h > k > c$	t	b	h	k	c
12	$c > h > b > k > t$	c	h	b	k	t
10	$k > c > h > b > t$	k	c	h	b	t
9	$b > k > h > c > t$	b	k	h	c	t
4	$h > c > b > k > t$	h	c	b	k	t
2	$h > k > b > c > t$	h	k	b	c	t

Examine the last five columns and compute the weighted sum for each activity.

Breed	Borda Points
t	$18 \cdot 4 + 12 \cdot 0 + 10 \cdot 0 + 9 \cdot 0 + 4 \cdot 0 + 2 \cdot 0 = 72$
b	$18 \cdot 3 + 12 \cdot 2 + 10 \cdot 1 + 9 \cdot 4 + 4 \cdot 2 + 2 \cdot 2 = 136$
h	$18 \cdot 2 + 12 \cdot 3 + 10 \cdot 2 + 9 \cdot 2 + 4 \cdot 4 + 2 \cdot 4 = 134$
k	$18 \cdot 1 + 12 \cdot 1 + 10 \cdot 4 + 9 \cdot 3 + 4 \cdot 1 + 2 \cdot 3 = 107$
c	$18 \cdot 0 + 12 \cdot 4 + 10 \cdot 3 + 9 \cdot 1 + 4 \cdot 3 + 2 \cdot 1 = 101$

The winner is b with 136 Borda points.

(d) Calculate how many votes each activity obtained from the 55 first-place votes from the voter profile.

Votes	Ranking	Row
18	$t > b > h > k > c$	1
12	$c > h > b > k > t$	2
10	$k > c > h > b > t$	3
9	$b > k > h > c > t$	4
4	$h > c > b > k > t$	5
2	$h > k > b > c > t$	6

Activity	1st-place votes
t	18
c	12
k	10
b	9
h	6

None of the activities received a majority, 28, of the votes. Therefore, eliminate activity h, the activity that received the least number of votes. Compare the first-place votes of the others.

Votes	Ranking	Row
18	$t > b > k > c$	1
12	$c > b > k > t$	2
10	$k > c > b > t$	3
9	$b > k > c > t$	4
4	$c > b > k > t$	5
2	$k > b > c > t$	6

Activity t: 18 *votes.*
Activity c: $12 + 4 = 16$ *votes.*
Activity k: $10 + 2 = 12$ *votes.*
Activity b: 9 *votes.*
There is no activity that has received a majority; therefore, eliminate b with only 9 votes, and compare the first-place votes of the others.

Votes	Ranking	Row
18	$t > k > c$	1
12	$c > k > t$	2
10	$k > c > t$	3
9	$k > c > t$	4
4	$c > k > t$	5
2	$k > c > t$	6

Activity t: 18 *votes.*
Activity c: $12 + 4 = 16$ *votes.*
Activity k: $10 + 9 + 2 = 21$ *votes.*
Again, there is not a majority; therefore, eliminate c and compare the first-place votes of t and k.

Votes	Ranking	Row
18	$t > k$	1
12	$k > t$	2
10	$k > t$	3
9	$k > t$	4
4	$k > t$	5
2	$k > t$	6

Activity t: 18 *votes.*
Activity k: $12 + 10 + 9 + 4 + 2 = 37$ *votes.*
Finally, k wins with 37 votes.

(e) Using the predetermined order b, t, c, k, h, begin by comparing b and t in the original voting.

Comparison	Votes	Row
$b > t$	$12 + 10 + 9 + 4 + 2 = 37$	2, 3, 4, 5, 6
$t > b$	$18 = 18$	1

Activity b wins this competition, 37 to 18. Now compare b to c.

Comparison	Votes	Row
$b > c$	$18 + 9 + 2 = 29$	1, 4, 6
$c > b$	$12 + 10 + 4 = 26$	2, 3, 5

Activity b wins this competition, 29 to 26. Now compare b and k.

Comparison	Votes	Row
$b > k$	$18 + 12 + 9 + 4 = 43$	1, 2, 4, 5
$k > b$	$10 + 2 = 12$	3, 6

Activity b wins this competition, 43 to 12. Now compare b and h.

Comparison	Votes	Row
$b > h$	$18 + 9 = 27$	1, 4
$h > b$	$12 + 10 + 4 + 2 = 28$	2, 3, 5, 6

Finally, activity h wins this competition, 28 to 27, and also wins the entire competition in the sequential pairwise comparison method.

20. The voter profile is

Votes	Ranking	Row
18	$t > k > h > b > c$	1
12	$c > h > k > b > t$	2
10	$b > c > h > k > t$	3
9	$k > b > h > c > t$	4
4	$h > c > k > b > t$	5
2	$h > b > k > c > t$	6

The activities with the most votes are t and c. Eliminate the remaining candidates to examine a second runoff election between these two candidates.

Votes	Ranking	Row
18	$t > c$	1
12	$c > t$	2
10	$c > t$	3
9	$c > t$	4
4	$c > t$	5
2	$c > t$	6

Activity t has 18 votes.
Activity c: $12 + 10 + 9 + 4 + 2 = 37$.
Activity c wins.

21. The voter profile is

Votes	Ranking	Row
18	$t > b > h > k > c$	1
12	$c > h > b > k > t$	2
10	$k > c > h > b > t$	3
9	$b > k > h > c > t$	4
4	$h > c > b > k > t$	5
2	$h > k > b > c > t$	6

The activities with the most votes are t and c. Eliminate the remaining candidates to examine a second runoff election between these two candidates.

Votes	Ranking	Row
18	$t > c$	1
12	$c > t$	2
10	$c > t$	3
9	$c > t$	4
4	$c > t$	5
2	$c > t$	6

Activity t has 18 votes.
Activity c: $12 + 10 + 9 + 4 + 2 = 37$.
Activity c wins.

22. The voter profile is:

Votes	Ranking	Row
3	$e > h > g > j$	1
6	$h > e > g > j$	2
5	$j > g > e > h$	3
4	$g > e > h > j$	4
3	$j > e > h > g$	5

The issues with the most votes are h and j. Eliminate the remaining candidates to examine a second runoff election between these two candidates.

Votes	Ranking	Row
3	$h > j$	1
6	$h > j$	2
5	$j > h$	3
4	$h > j$	4
3	$j > h$	5

Issue h: $3 + 6 + 4 = 13$
Issue j: $5 + 3 = 8$.
Issue h wins.

23. The voter profile is:

Votes	Ranking	Row
6	$h > j > g > e$	1
5	$e > g > j > h$	2
4	$g > j > h > e$	3
3	$j > h > g > e$	4
3	$e > j > h > g$	5

The issues with the most votes are h and e. Eliminate the remaining candidates to examine a second runoff election between these two issue.

Votes	Ranking	Row
6	$h > e$	1
5	$e > h$	2
4	$h > e$	3
3	$h > e$	4
3	$e > h$	5

Issue h: $6 + 4 + 3 = 13$
Issue e: $5 + 3 = 8$.
Issue h wins.

16.1 THE POSSIBILITIES OF VOTING

24. The voter profile is

Votes	Ranking	Row
18	$t > k > h > b > c$	1
12	$c > h > k > b > t$	2
10	$b > c > h > k > t$	3
9	$k > b > h > c > t$	4
4	$h > c > k > b > t$	5
2	$h > b > k > c > t$	6

The activities that rank second and third in first-place votes are c and b, respectively. Here is a runoff election between these two activities:

Votes	Ranking	Row
18	$b > c$	1
12	$c > b$	2
10	$b > c$	3
9	$b > c$	4
4	$c > b$	5
2	$b > c$	6

Activity b: $18 + 10 + 9 + 2 = 39$
Activity c: $12 + 4 = 16$
Now b faces the first-place activity, t:

Votes	Ranking	Row
18	$t > b$	1
12	$b > t$	2
10	$b > t$	3
9	$b > t$	4
4	$b > t$	5
2	$b > t$	6

Activity t has 18 votes.
Activity b: $12 + 10 + 9 + 4 + 2 = 37$
Activity b wins with 37 votes.

25. The voter profile is

Votes	Ranking	Row
18	$t > b > h > k > c$	1
12	$c > h > b > k > t$	2
10	$k > c > h > b > t$	3
9	$b > k > h > c > t$	4
4	$h > c > b > k > t$	5
2	$h > k > b > c > t$	6

The activities that rank second and third in first-place votes are c and k, respectively. Here is a runoff election between these two activities:

Votes	Ranking	Row
18	$k > c$	1
12	$c > k$	2
10	$k > c$	3
9	$k > c$	4
4	$c > k$	5
2	$k > c$	6

Activity k: $18 + 10 + 9 + 2 = 39$
Activity c: $12 + 4 = 16$
Now k faces the first-place candidate, t:

Votes	Ranking	Row
18	$t > k$	1
12	$k > t$	2
10	$k > t$	3
9	$k > t$	4
4	$k > t$	5
2	$k > t$	6

Activity t has 18 votes.
Candidate k: $12 + 10 + 9 + 4 + 2 = 37$
Activity k wins with 37 votes.

26. The voter profile is as follows.

Votes	Ranking	Row
2	$a > c > b$	1
3	$c > b > a$	2
4	$b > a > c$	3
4	$b > c > a$	4

The logos that rank second and third in first-place votes are a and c, respectively. Here is a run off election between these two logos.

Votes	Ranking	Row
2	$a > c$	1
3	$c > a$	2
4	$a > c$	3
4	$c > a$	4

Logo a: $2 + 4 = 6$
Logo c: $3 + 4 = 7$
Now c faces the first-place logo, b:

Votes	Ranking	Row
2	$c > b$	1
3	$c > b$	2
4	$b > c$	3
4	$b > c$	4

Logo c: $2 + 3 = 5$
Logo b: $4 + 4 = 8$
Logo b wins.

27. The voter profile is

Votes	Ranking	Row
6	$a > b > c$	1
1	$b > c > a$	2
3	$b > a > c$	3
3	$c > a > b$	4

The logos that rank second and third in first-place votes are b and c, respectively. Here is a runoff election between these two logos.

Votes	Ranking	Row
6	$b > c$	1
1	$b > c$	2
3	$b > c$	3
3	$c > b$	4

Logo b: $6 + 1 + 3 = 10$
Logo c: 3
Now b faces the first-place logo, a:

Votes	Ranking	Row
6	$a > b$	1
1	$b > a$	2
3	$b > a$	3
3	$a > b$	4

Logo a: $6 + 3 = 9$
Logo b: $1 + 3 = 4$
Logo a wins.

28. (a) *Count the votes in the table shown in the text to see that Joan receives 8 votes; Lori receives 7; Mary receives 6; and Alison receives 5. Joan wins by the approval method.*

    (b) *Joan and Lori win with 8 and 7 votes, respectively.*

29. (a) *Count the votes in the table shown in the text to see that Joan receives 8 votes; Lori receives 7; Mary receives 10; and Alison receives 9. Mary wins by the approval method.*

    (b) *Mary and Alison win with 10 and 9 votes, respectively.*

30. (a) *For $n = 7$ candidates, find the combinations of 7, taken 2 at a time:*

    $$_7C_2 = \frac{7!}{2!(7-2)!} = \frac{7 \cdot 6 \cdot 5 \cdot 4 \cdot 3 \cdot 2 \cdot 1}{2 \cdot 1 \cdot 5 \cdot 4 \cdot 3 \cdot 2 \cdot 1} = 21.$$

    The sum of the number of comparisons listed in the table in the text is 20. Only one comparison remains, so g wins one point.

    (b) *Examine the table in the text to see that b wins with 6 pairwise points.*

31. (a) *For $n = 7$ candidates, find the combinations of 7, taken 2 at a time:*

    $$_7C_2 = \frac{7!}{2!(7-2)!} = \frac{7 \cdot 6 \cdot 5 \cdot 4 \cdot 3 \cdot 2 \cdot 1}{2 \cdot 1 \cdot 5 \cdot 4 \cdot 3 \cdot 2 \cdot 1} = 21.$$

    The sum of the number of comparisons listed in the table in the text is:

    $$3 + 5 + 7 + 1 + 2 + 1 = 19.$$

    Only two comparisons remain, so f wins two points.

    (b) *Examine the table in the text to see that c wins with 7 pairwise points.*

32. (a) *For $n = 8$ candidates, find the combinations of 8, taken 2 at a time:*

    $$_8C_2 = \frac{8!}{2!(8-2)!} = \frac{8 \cdot 7 \cdot 6 \cdot 5 \cdot 4 \cdot 3 \cdot 2 \cdot 1}{2 \cdot 1 \cdot 6 \cdot 5 \cdot 4 \cdot 3 \cdot 2 \cdot 1} = 28.$$

    The sum of the number of comparisons listed in the table in the text is:

    $$4 + 5 + 2 + 3 + 1 + 2 + 3 = 20.$$

    $28 - 20 = 8$ *comparisons remain, so d wins eight points.*

    (b) *Candidate d wins with 8 pairwise points*

33. (a) *For $n = 8$ candidates, find the combinations of 8, taken 2 at a time:*

    $$_8C_2 = \frac{8!}{2!(8-2)!} = \frac{8 \cdot 7 \cdot 6 \cdot 5 \cdot 4 \cdot 3 \cdot 2 \cdot 1}{2 \cdot 1 \cdot 6 \cdot 5 \cdot 4 \cdot 3 \cdot 2 \cdot 1} = 28.$$

    The sum of the number of comparisons listed in the table in the text is:

    $$2 + 6 + 3 + 4 + 2 + 2 + 2 = 21.$$

    $28 - 21 = 7$ *comparisons remain, so e wins seven points.*

    (b) *Candidate e wins with 7 pairwise points.*

34. *Notice that in the Borda method, the sum of all the points must equal the product of the number of voters and the number of possible points for each voter's selection. For example, in this exercise there are 15 voters choosing among 3 candidates. Each voter has a total of 3 points to assign: 2 points for first-place and 1 point for second place. Therefore, the total number of points is*

    $$15 \cdot 3 = 45.$$

    If Candidate a receives 17 points and Candidate b receives 14 points, then Candidate c must receive

    $$45 - (17 + 14) = 14 \text{ points.}$$

    Candidate a wins the Borda election.

35. *Notice that in the Borda method, the sum of all the points must equal the product of the number of voters and the number of possible points for each voter's selection. For example, in this exercise there are 15 voters choosing among 3 candidates. Each voter has a total of 3 points to assign: 2 points for first place and 1 point for second place. Therefore, the total number of points is*

    $$15 \cdot 3 = 45.$$

    If Candidate a receives 15 points and Candidate b receives 14 points, then Candidate c must receive

    $$45 - (15 + 14) = 16 \text{ points.}$$

    Candidate c wins the Borda election.

16.1 THE POSSIBILITIES OF VOTING 541

36. Notice that in the Borda method, the sum of all the points must equal the product of the number of voters and the number of possible points for each voter's selection. For example, in this exercise there are 20 voters choosing among 5 candidates. Each voter has a total of 10 points to assign: 4 points for first place, 3 points for second place, 2 points for third place, and 1 point for fourth place. Therefore, the total number of points is

$$20 \cdot 10 = 200.$$

Candidate a must receive the difference between 200 and the sum of the points of the other candidates.

$$200 - (50 + 25 + 40 + 55) = 30 \text{ points}.$$

Candidate e wins the Borda election with 55 points.

37. Notice that in the Borda method, the sum of all the points must equal the product of the number of voters and the number of possible points for each voter's selection. For example, in this exercise there are 20 voters choosing among 5 candidates. Each voter has a total of 10 points to assign: 4 points for first place, 3 points for second place, 2 points for third place, and 1 point for fourth place. Therefore, the total number of points is

$$20 \cdot 10 = 200.$$

Candidate c must receive the difference between 200 and the sum of the points of the other candidates.

$$200 - (35 + 40 + 40 + 30) = 55 \text{ points}.$$

Candidate c wins the Borda election with 55 points.

38. The voter profile is:

Votes	Ranking	Row
3	$e > h > g > j$	1
6	$h > e > g > j$	2
5	$j > g > e > h$	3
4	$g > e > h > j$	4
3	$j > e > h > g$	5

Using the Coombs method, eliminate the candidate with the most last place votes, which is j, with 13 last place votes. Compare the remaining issues.

Votes	Ranking	Row
3	$e > h > g$	1
6	$h > e > g$	2
5	$g > e > h$	3
4	$g > e > h$	4
3	$e > h > g$	5

There is no majority, so again compare totals of last place votes.
Issue g: $3 + 6 + 3 = 12$
Issue h: $5 + 4 = 9$
Eliminate g because it has the most last-place votes.
Now h and e remain.

Votes	Ranking	Row
3	$e > h$	1
6	$h > e$	2
5	$e > h$	3
4	$e > h$	4
3	$e > h$	5

Issue e has a majority, so it wins.

39. The voter profile is:

Votes	Ranking	Row
6	$h > j > g > e$	1
5	$e > g > j > h$	2
4	$g > j > h > e$	3
3	$j > h > g > e$	4
3	$e > j > h > g$	5

Using the Coombs method, eliminate the issue with the most last place votes, which is e, with 13 last place votes. Compare the remaining issues.

Votes	Ranking	Row
6	$h > j > g$	1
5	$g > j > h$	2
4	$g > j > h$	3
3	$j > h > g$	4
3	$j > h > g$	5

There is no majority, so again compare totals of last-place votes.
Issue g: $6 + 3 + 3 = 12$
Issue h: $5 + 4 = 9$
Eliminate g because it has the most last-place votes.
Now h and j remain.

Votes	Ranking	Row
6	$h > j$	1
5	$j > h$	2
4	$j > h$	3
3	$j > h$	4
3	$j > h$	5

Issue j has a majority, so it wins.

40. Make a table to compute the votes in comparisons.

Comparison	Pairwise Win	Rows
$n > a_x$	n	1
$a_y > n$	a_y	2
$n > e$	n	3
$a_x > a_y$	a_x	4
$e > a_x$	e	5
$a_y > e$	a_y	6

If the only bills in the competition are the new law and the existing law, the table in the text indicates $n > e$. Therefore, the new law, n, wins.

41. Using the predetermined order, n, a_x, e, examine the table to see that:

$$n \text{ beats } a_x$$
$$n \text{ beats } e$$

The new bill, n, wins.

42. Using the predetermined order, n, a_y, e, examine the table to see that:

$$a_y \text{ beats } n$$
$$a_y \text{ beats } e$$

The amended bill, a_y, wins.

43. Using the predetermined order, n, a_x, a_y, e, examine the table to see that:

$$n \text{ beats } a_x$$
$$a_y \text{ beats } n$$
$$a_y \text{ beats } e$$

The amended bill, a_y, wins.

44. Using the predetermined order, n, a_y, a_x, e, examine the table to see that:

$$a_y \text{ beats } n$$
$$a_x \text{ beats } a_y$$
$$e \text{ beats } a_x$$

The existing bill, e, wins.

45. Here is one possible arrangement.

Votes	Ranking	Row
2	$a > b > d > c$	1
4	$b > c > d > a$	2
5	$c > d > a > b$	3
7	$d > a > b > c$	4
3	$a > d > c > b$	5

Plurality method
Candidate a receives $2 + 3 = 5$ votes.
Candidate b receives 4 votes.
Candidate c receives 5 votes.
Candidate d receives 7 votes.

Pairwise Comparison method
First make a table to compute the votes in comparison.

Compare	Votes	Rows
$a > b$	$2 + 5 + 7 + 3 = 17$	1, 3, 4, 5
$b > a$	$4 = 4$	2
$a > c$	$2 + 7 + 3 = 12$	1, 4, 5
$c > a$	$4 + 5 = 9$	2, 3
$a > d$	$2 + 3 = 5$	1, 5
$d > a$	$4 + 5 + 7 = 16$	2, 3, 4
$b > c$	$2 + 4 + 7 = 13$	1, 2, 4
$c > b$	$5 + 3 = 8$	3, 5
$b > d$	$2 + 4 = 6$	1, 2
$d > b$	$5 + 7 + 3 = 15$	3, 4, 5
$c > d$	$4 + 5 = 9$	2, 3
$d > c$	$2 + 7 + 3 = 12$	1, 4, 5

Make another table to award pairwise points.

Pairs	Votes	Pairwise Points
$a : b$	$17 : 4$	$a, 1$
$a : c$	$12 : 9$	$a, 1$
$a : d$	$5 : 16$	$d, 1$
$b : c$	$13 : 8$	$b, 1$
$b : d$	$6 : 15$	$d, 1$
$c : d$	$9 : 12$	$d, 1$

Using the pairwise comparison method, the winner is d with 3 pairwise points. Candidate a received 2 points; candidate b received 1 point; candidate c received no points.

Borda method
Make a table to examine the points in the Borda method.

		Points			
Votes	Ranking	3	2	1	0
2	$a > b > d > c$	a	b	d	c
4	$b > c > d > a$	b	c	d	a
5	$c > d > a > b$	c	d	a	b
7	$d > a > b > c$	d	a	b	c
3	$a > d > c > b$	a	d	c	b

Examine the last four columns, and compute the weighted sum for each candidate.

Issue	Borda Points
a	$2 \cdot 3 + 4 \cdot 0 + 5 \cdot 1 + 7 \cdot 2 + 3 \cdot 3 = 34$
b	$2 \cdot 2 + 4 \cdot 3 + 5 \cdot 0 + 7 \cdot 1 + 3 \cdot 0 = 23$
c	$2 \cdot 0 + 4 \cdot 2 + 5 \cdot 3 + 7 \cdot 0 + 3 \cdot 1 = 26$
d	$2 \cdot 1 + 4 \cdot 1 + 5 \cdot 2 + 7 \cdot 3 + 3 \cdot 2 = 43$

The winner is d with 43 Borda points.

Hare method
There are 21 voters, so a candidate must receive a majority of the votes or at least 11 votes to win by the Hare method. Examine the original ranking to see that no candidate received a majority. Eliminate b because it has the least number of votes, and compare only a, c, and d.

Votes	Ranking	Row
2	$a > d > c$	1
4	$c > d > a$	2
5	$c > d > a$	3
7	$d > a > c$	4
3	$a > d > c$	5

Candidate a receives $2 + 3 = 5$ votes.
Candidate c receives $4 + 5 = 9$ votes.
Candidate d receives 7 votes.
Again no candidate has received a majority, so another vote is taken with a eliminated.

16.1 THE POSSIBILITIES OF VOTING

Votes	Ranking	Row
2	$d > c$	1
4	$c > d$	2
5	$c > d$	3
7	$d > c$	4
3	$d > c$	5

Candidate c receives $4 + 5 = 9$ votes.
Candidate d receives $2 + 7 + 3 = 12$ votes.
Finally, d has a majority and wins.

46. Here is one possible arrangement.

Votes	Ranking	Row
2	$d > b > a > c$	1
4	$b > c > a > d$	2
5	$c > a > d > b$	3
7	$a > d > b > c$	4
3	$d > a > c > b$	5

Plurality method
Candidate a receives 7 votes.
Candidate b receives 4 votes.
Candidate c receives 5 votes.
Candidate d receives $2 + 3 = 5$ votes.

Pairwise Comparison method
First make a table to compute the votes in comparison.

Compare	Votes	Rows
$a > b$	$5 + 7 + 3 = 15$	3, 4, 5
$b > a$	$2 + 4 = 6$	1, 2
$a > c$	$2 + 7 + 3 = 12$	1, 4, 5
$c > a$	$4 + 5 = 9$	2, 3
$a > d$	$4 + 5 + 7 = 16$	2, 3, 4
$d > a$	$2 + 3 = 5$	1, 5
$b > c$	$2 + 4 + 7 = 13$	1, 2, 4
$c > b$	$5 + 3 = 8$	3, 5
$b > d$	$4 = 4$	2
$d > b$	$2 + 5 + 7 + 3 = 17$	1, 3, 4, 5
$c > d$	$4 + 5 = 9$	2, 3
$d > c$	$2 + 7 + 3 = 12$	1, 4, 5

Make another table to award pairwise points.

Pairs	Votes	Pairwise Points
$a : b$	15 : 6	a, 1
$a : c$	12 : 9	a, 1
$a : d$	16 : 5	a, 1
$b : c$	13 : 8	b, 1
$b : d$	4 : 17	d, 1
$c : d$	9 : 12	d, 1

Using the pairwise comparison method, the winner is a with 3 pairwise points. Candidate d received 2 points; candidate b received 1 point; candidate c received no points.

Borda method
Make a table to examine the points in the Borda method.

Votes	Ranking	Points 3	2	1	0
2	$d > b > a > c$	d	b	a	c
4	$b > c > a > d$	b	c	a	d
5	$c > a > d > b$	c	a	d	b
7	$a > d > b > c$	a	d	b	c
3	$d > a > c > b$	d	a	c	b

Examine the last four columns, and compute the weighted sum for each candidate.

Issue	Borda Points
a	$2 \cdot 1 + 4 \cdot 1 + 5 \cdot 2 + 7 \cdot 3 + 3 \cdot 2 = 43$
b	$2 \cdot 2 + 4 \cdot 3 + 5 \cdot 0 + 7 \cdot 1 + 3 \cdot 0 = 23$
c	$2 \cdot 0 + 4 \cdot 2 + 5 \cdot 3 + 7 \cdot 0 + 3 \cdot 1 = 26$
d	$2 \cdot 3 + 4 \cdot 0 + 5 \cdot 1 + 7 \cdot 2 + 3 \cdot 3 = 34$

The winner is a with 43 Borda points.

Hare method
There are 21 voters so a candidate must receive a majority of the votes or at least 11 votes to win by the Hare method. Examine the original ranking to see that no candidate received a majority. Eliminate b because it has the least number of votes, and compare only a, c, and d.

Votes	Ranking	Row
2	$d > a > c$	1
4	$c > a > d$	2
5	$c > a > d$	3
7	$a > d > c$	4
3	$d > a > c$	5

Candidate a receives 7 votes.
Candidate c receives $4 + 5 = 9$ votes.
Candidate d receives $2 + 3 = 5$ votes.
Again no candidate has received a majority, so another vote is taken with d eliminated.

Votes	Ranking	Row
2	$a > c$	1
4	$c > a$	2
5	$c > a$	3
7	$a > c$	4
3	$a > c$	5

Candidate c receives $4 + 5 = 9$ votes.
Candidate a receives $2 + 7 + 3 = 12$ votes.
Finally, a has a majority and wins.

47. Writing exercise

48. Writing exercise

49. Writing exercise

50. Writing exercise

51. Writing exercise

16.2 EXERCISES

1. (a) Read the table in the text to see that a has the majority of first-place votes. Because there are 11 voters, 6 votes constitutes a majority.

   (b) Examine the following table to calculate the Borda points.

		Points		
Votes	Ranking	2	1	0
6	$a>b>c$	a	b	c
3	$b>c>a$	b	c	a
2	$c>b>a$	c	b	a

   Alternative a: $6 \cdot 2 + 3 \cdot 0 + 2 \cdot 0 = 12$
   Alternative b: $6 \cdot 1 + 3 \cdot 2 + 2 \cdot 1 = 14$
   Alternative c: $6 \cdot 0 + 3 \cdot 1 + 2 \cdot 2 = 7$
   Alternative b wins with 14 Borda points.

   (c) The Borda method violates the majority criterion because it fails to select the majority candidate.

2. (a) Read the table in the text to see that b has the majority of first-place votes. Because there are 11 voters, 6 votes constitutes a majority.

   (b) Examine the following table to calculate the Borda points.

		Points		
Votes	Ranking	2	1	0
6	$b>a>c$	b	a	c
3	$a>c>b$	a	c	b
2	$c>a>b$	c	a	b

   Alternative a: $6 \cdot 1 + 3 \cdot 2 + 2 \cdot 1 = 14$
   Alternative b: $6 \cdot 2 + 3 \cdot 0 + 2 \cdot 0 = 12$
   Alternative c: $6 \cdot 0 + 3 \cdot 1 + 2 \cdot 2 = 7$
   Alternative a wins with 14 Borda points.

   (c) The Borda method violates the majority criterion because it fails to select the majority candidate.

3. (a) Read the table in the text to see that a has the majority of first-place votes. Because there are 36 voters, at least 19 votes constitutes a majority. Alternative a received 20 votes.

   (b) Use the information from the text to make a table to examine the points in the Borda method.

		Points			
Votes	Ranking	3	2	1	0
20	$a>b>c>d$	a	b	c	d
6	$b>c>d>a$	b	c	d	a
5	$c>b>d>a$	c	b	d	a
5	$d>b>a>c$	d	b	a	c

   Examine the last four columns, and compute the weighted sum for each alternative.

Alternative	Borda Points
a	$20 \cdot 3 + 6 \cdot 0 + 5 \cdot 0 + 5 \cdot 1 = 65$
b	$20 \cdot 2 + 6 \cdot 3 + 5 \cdot 2 + 5 \cdot 2 = 78$
c	$20 \cdot 1 + 6 \cdot 2 + 5 \cdot 3 + 5 \cdot 0 = 47$
d	$20 \cdot 0 + 6 \cdot 1 + 5 \cdot 1 + 5 \cdot 3 = 26$

   The winner is alternative b with 78 Borda points.

   (c) The Borda method violates the majority criterion because it fails to select the majority candidate.

4. (a) Read the table in the text to see that d has the majority of first-place votes. Because there are 36 voters, at least 19 votes constitutes a majority. Alternative d received 19 votes.

   (b) Use the information from the text to make a table to examine the points in the Borda method.

		Points			
Votes	Ranking	3	2	1	0
4	$b>c>a>d$	b	c	a	d
9	$c>a>d>b$	c	a	d	b
4	$a>c>b>d$	a	c	b	d
19	$d>c>b>a$	d	c	b	a

   Examine the last four columns, and compute the weighted sum for each alternative.

Alternative	Borda Points
a	$4 \cdot 1 + 9 \cdot 2 + 4 \cdot 3 + 19 \cdot 0 = 34$
b	$4 \cdot 3 + 9 \cdot 0 + 4 \cdot 1 + 19 \cdot 1 = 35$
c	$4 \cdot 2 + 9 \cdot 3 + 4 \cdot 2 + 19 \cdot 2 = 81$
d	$4 \cdot 0 + 9 \cdot 1 + 4 \cdot 0 + 19 \cdot 3 = 66$

   The winner is alternative c with 81 Borda points.

   (c) The Borda method violates the majority criterion because it fails to select the majority candidate.

5. (a) Read the table in the text to see that a has the majority of first-place votes. Because there are 30 voters, at least 16 votes constitutes a majority. Alternative a received 16 votes.

   (b) Use the information from the text to make a table to examine the points in the Borda method.

		Points				
Votes	Ranking	4	3	2	1	0
16	$a>b>c>d>e$	a	b	c	d	e
3	$b>c>d>e>a$	b	c	d	e	a
5	$c>d>b>e>a$	c	d	b	e	a
3	$d>b>c>a>e$	d	b	c	a	e
3	$e>c>d>a>b$	e	c	d	a	b

   Examine the last five columns, and compute the weighted sum for each alternative.

Alternative	Borda Points
a	$16 \cdot 4 + 3 \cdot 0 + 5 \cdot 0 + 3 \cdot 1 + 3 \cdot 1 = 70$
b	$16 \cdot 3 + 3 \cdot 4 + 5 \cdot 2 + 3 \cdot 3 + 3 \cdot 0 = 79$
c	$16 \cdot 2 + 3 \cdot 3 + 5 \cdot 4 + 3 \cdot 2 + 3 \cdot 3 = 76$
d	$16 \cdot 1 + 3 \cdot 2 + 5 \cdot 3 + 3 \cdot 4 + 3 \cdot 2 = 55$
e	$16 \cdot 0 + 3 \cdot 1 + 5 \cdot 1 + 3 \cdot 0 + 3 \cdot 4 = 20$

The winner is alternative b with 79 Borda points.

(c) The Borda method violates the majority criterion because it fails to select the majority candidate.

6. (a) Read the table in the text to see that b has the majority of first-place votes. Because there are 30 voters, at least 16 votes constitutes a majority. Alternative b received 16 votes.

(b) Use the information from the text to make a table to examine the points in the Borda method.

		Points				
Votes	Ranking	4	3	2	1	0
16	$b > e > c > d > a$	b	e	c	d	a
3	$a > c > b > d > e$	a	c	b	d	e
6	$c > a > e > b > d$	c	a	e	b	d
2	$d > a > c > e > b$	d	a	c	e	b
3	$e > c > a > d > b$	e	c	a	d	b

Examine the last five columns, and compute the weighted sum for each alternative.

Alternative	Borda Points
a	$16 \cdot 0 + 3 \cdot 4 + 6 \cdot 3 + 2 \cdot 3 + 3 \cdot 2 = 42$
b	$16 \cdot 4 + 3 \cdot 2 + 6 \cdot 1 + 2 \cdot 0 + 3 \cdot 0 = 76$
c	$16 \cdot 2 + 3 \cdot 3 + 6 \cdot 4 + 2 \cdot 2 + 3 \cdot 3 = 78$
d	$16 \cdot 1 + 3 \cdot 1 + 6 \cdot 0 + 2 \cdot 4 + 3 \cdot 1 = 30$
e	$16 \cdot 3 + 3 \cdot 0 + 6 \cdot 2 + 2 \cdot 1 + 3 \cdot 4 = 74$

The winner is alternative c with 78 Borda points.

(c) The Borda method violates the majority criterion because it fails to select the majority candidate.

7. (a) Make a table to compute the votes in comparisons.

Comparison	Votes	Rows
$a > b$	$4 + 3 = 7$	1, 4
$b > a$	$2 + 4 = 6$	2, 3
$a > c$	$4 + 4 = 8$	1, 3
$c > a$	$2 + 3 = 5$	2, 4
$b > c$	$4 + 2 + 4 = 10$	1, 2, 3
$c > b$	$3 = 3$	4

Make another table to award pairwise points.

Pairs	Votes	Pairwise Points
$a : b$	$7 : 6$	a, 1
$a : c$	$8 : 5$	a, 1
$b : c$	$10 : 3$	b, 1

Using the pairwise comparison method, the winner is candidate a with 2 pairwise points and is the Condorcet candidate.

(b) Examine the table in the text to see that a receives 4 first-place votes; c receives 3 first-place votes; b receives $2 + 4 = 6$ first-place votes to win by the plurality method.

(c) There are 13 votes, which makes 7 votes a majority. No candidate wins a majority on this vote. Candidate c has the fewest votes, so compare only a and b.

Comparison	Votes	Rows
$a > b$	$4 + 3 = 7$	1, 4
$b > a$	$2 + 4 = 6$	2, 3

Candidate a now has a majority and is selected by the Hare method.

(d) Use the information from the text to make a table to examine the points in the Borda method.

		Points		
Votes	Ranking	2	1	0
4	$a > b > c$	a	b	c
2	$b > c > a$	b	c	a
4	$b > a > c$	b	a	c
3	$c > a > b$	c	a	b

Examine the last three columns, and compute the weighted sum for each alternative.

Alternative	Borda Points
a	$4 \cdot 2 + 2 \cdot 0 + 4 \cdot 1 + 3 \cdot 1 = 15$
b	$4 \cdot 1 + 2 \cdot 2 + 4 \cdot 2 + 3 \cdot 0 = 16$
c	$4 \cdot 0 + 2 \cdot 1 + 4 \cdot 0 + 3 \cdot 2 = 8$

Candidate b wins with 16 Borda points.

(e) The plurality and the Borda methods violate the Condorcet criterion because neither method selects candidate a, the Condorcet candidate. The Hare method does not violate the criterion because the Condorcet candidate wins.

8. (a) Make a table to compute the votes in comparisons.

Comparison	Votes	Rows
$a > b$	$3 = 3$	1
$b > a$	$4 + 2 + 4 = 10$	2, 3, 4
$a > c$	$3 + 2 = 5$	1, 3
$c > a$	$4 + 4 = 8$	2, 4
$b > c$	$2 + 4 = 6$	3, 4
$c > b$	$3 + 4 = 7$	1, 2

Make another table to award pairwise points.

Pairs	Votes	Pairwise Points
$a : b$	$3 : 10$	b, 1
$a : c$	$5 : 8$	c, 1
$b : c$	$6 : 7$	c, 1

Using the pairwise comparison method, the winner is candidate c with 2 pairwise points and is the Condorcet candidate.

(b) Examine the table in the text to see that a receives 3 first-place votes; c receives 4 first-place votes; b receives $2 + 4 = 6$ first-place votes to win by the plurality method.

(c) There are 13 votes, which makes 7 votes a majority. No candidate wins a majority on this vote. Candidate a has the fewest votes, so compare only b and c.

Comparison	Votes	Rows
$b > c$	$2 + 4 = 6$	3, 4
$c > b$	$3 + 4 = 7$	1, 2

Candidate c now has a majority and is selected by the Hare method.

(d) Use the information from the text to make a table to examine the points in the Borda method.

		Points		
Votes	Ranking	2	1	0
3	$a > c > b$	a	c	b
4	$c > b > a$	c	b	a
2	$b > a > c$	b	a	c
4	$b > c > a$	b	c	a

Examine the last three columns, and compute the weighted sum for each candidate.

Alternative	Borda Points
a	$3 \cdot 2 + 4 \cdot 0 + 2 \cdot 1 + 4 \cdot 0 = 8$
b	$3 \cdot 0 + 4 \cdot 1 + 2 \cdot 2 + 4 \cdot 2 = 16$
c	$3 \cdot 1 + 4 \cdot 2 + 2 \cdot 0 + 4 \cdot 1 = 15$

Candidate b wins with 16 Borda points.

(e) The plurality and the Borda methods violate the Condorcet criterion because neither method selects candidate c, the Condorcet candidate. The Hare method does not violate the criterion because the Condorcet candidate wins.

9. (a) Make a table to compute the votes in comparisons.

Comparison	Votes	Rows
$e > h$	$3 + 5 + 4 + 3 = 15$	1, 3, 4, 5
$h > e$	$6 = 6$	2
$e > g$	$3 + 6 + 3 = 12$	1, 2, 5
$g > e$	$5 + 4 = 9$	3, 4
$e > j$	$3 + 6 + 4 = 13$	1, 2, 4
$j > e$	$5 + 3 = 8$	3, 5
$h > g$	$3 + 6 + 3 = 12$	1, 2, 5
$g > h$	$5 + 4 = 9$	3, 4
$h > j$	$3 + 6 + 4 = 13$	1, 2, 4
$j > h$	$5 + 3 = 8$	3, 5
$j > g$	$5 + 3 = 8$	3, 5
$g > j$	$3 + 6 + 4 = 13$	1, 2, 4

Make another table to award pairwise points.

Pairs	Votes	Pairwise Points
$e : h$	$15 : 6$	e, 1
$e : g$	$12 : 9$	e, 1
$e : j$	$13 : 8$	e, 1
$h : g$	$12 : 9$	h, 1
$h : j$	$13 : 8$	h, 1
$j : g$	$8 : 13$	g, 1

Using the pairwise comparison method, the winner is candidate e with 3 pairwise points and is the Condorcet candidate.

(b) Using the plurality method to determine the highest priority issue, the winner is issue j with a total of 8 first-place votes in rows 3 and 5.

(c) There are 21 votes, which makes 11 votes a majority. No candidate wins a majority on this vote. Eliminate issue e because it has the least number of votes, and compare only h, g, and j.

Votes	Ranking	Row
3	$h > g > j$	1
6	$h > g > j$	2
5	$j > g > h$	3
4	$g > h > j$	4
3	$j > h > g$	5

Issue h receives $3 + 6 = 9$ votes.
Issue g receives 4 votes.
Issue j receives $5 + 3 = 8$ votes.
Again no issue has received a majority. Therefore, eliminate g with the least number of votes, and compare only h and j.

Votes	Ranking	Row
3	$h > j$	1
6	$h > j$	2
5	$j > h$	3
4	$h > j$	4
3	$j > h$	5

Issue h receives $3 + 6 + 4 = 13$ votes.
Issue j receives $5 + 3 = 8$ votes.
Finally, issue h is the winner.

(d) Make a table to examine the points in the Borda method.

		Points			
Votes	Ranking	3	2	1	0
3	$e > h > g > j$	e	h	g	j
6	$h > e > g > j$	h	e	g	j
5	$j > g > e > h$	j	g	e	h
4	$g > e > h > j$	g	e	h	j
3	$j > e > h > g$	j	e	h	g

Examine the last four columns and compute the weighted sum for each issue.

Issue	Borda Points
e	$3 \cdot 3 + 6 \cdot 2 + 5 \cdot 1 + 4 \cdot 2 + 3 \cdot 2 = 40$
h	$3 \cdot 2 + 6 \cdot 3 + 5 \cdot 0 + 4 \cdot 1 + 3 \cdot 1 = 31$
g	$3 \cdot 1 + 6 \cdot 1 + 5 \cdot 2 + 4 \cdot 3 + 3 \cdot 0 = 31$
j	$3 \cdot 0 + 6 \cdot 0 + 5 \cdot 3 + 4 \cdot 0 + 3 \cdot 3 = 24$

The winner is issue e with 40 Borda points.

(e) The plurality and Hare methods violate the Condorcet criterion because issue e, the Condorcet candidate, does not win. The Borda method does not violate the criterion.

10. (a) *Make a table to compute the votes in comparisons.*

Comparison	Votes	Rows
$e > h$	$5 + 3 = 8$	2, 5
$h > e$	$6 + 4 + 3 = 13$	1, 3, 4
$e > g$	$5 + 3 = 8$	2, 5
$g > e$	$6 + 4 + 3 = 13$	1, 3, 4
$e > j$	$5 + 3 = 8$	2, 5
$j > e$	$6 + 4 + 3 = 13$	1, 3, 4
$h > g$	$6 + 3 + 3 = 12$	1, 4, 5
$g > h$	$5 + 4 = 9$	2, 3
$h > j$	$6 = 6$	1
$j > h$	$5 + 4 + 3 + 3 = 15$	2, 3, 4, 5
$j > g$	$6 + 3 + 3 = 12$	1, 4, 5
$g > j$	$5 + 4 = 9$	2, 3

Make another table to award pairwise points.

Pairs	Votes	Pairwise Points
e: h	8: 13	h, 1
e: g	8: 13	g, 1
e: j	8: 13	j, 1
h: g	12:9	h, 1
h: j	6:15	j, 1
j: g	12:9	j, 1

Using the pairwise comparison method, the winner is issue j with 3 pairwise points. This is the Condorcet candidate.

(b) Using the plurality method to determine the highest priority issue, the winner is issue e with a total of 8 first-place votes in rows 2 and 5.

(c) There are 21 voters, so an issue must receive a majority of the votes or at least 11 votes to win by the Hare method. No issue received a majority. See part (a) that issue e received 8 first-place votes; issue h received

6 votes; issue g received 4 votes; and issue j received 3 votes. *Eliminate issue j because it has the least number of votes, and compare only h, g, and e.*

Votes	Ranking	Row
6	$h > g > e$	1
5	$e > g > h$	2
4	$g > h > e$	3
3	$h > g > e$	4
3	$e > h > g$	5

Issue h receives $6 + 3 = 9$ votes.
Issue g receives 4 votes.
Issue e receives $5 + 3 = 8$ votes.
No issue has received a majority, so another vote is taken with issue g eliminated.

Votes	Ranking	Row
6	$h > e$	1
5	$e > h$	2
4	$h > e$	3
3	$h > e$	4
3	$e > h$	5

Issue h receives $6 + 4 + 3 = 13$ votes.
Issue e receives $5 + 3 = 8$ votes.
Now issue h has a majority and wins.

(d) Make a table to examine the points in the Borda method.

Votes	Ranking	Points 3	2	1	0
6	$h > j > g > e$	h	j	g	e
5	$e > g > j > h$	e	g	j	h
4	$g > j > h > e$	g	j	h	e
3	$j > h > g > e$	j	h	g	e
3	$e > j > h > g$	e	j	h	g

Examine the last four columns, and compute the weighted sum for each issue.

Issue	Borda Points
h	$6 \cdot 3 + 5 \cdot 0 + 4 \cdot 1 + 3 \cdot 2 + 3 \cdot 1 = 31$
j	$6 \cdot 2 + 5 \cdot 1 + 4 \cdot 2 + 3 \cdot 3 + 3 \cdot 2 = 40$
g	$6 \cdot 1 + 5 \cdot 2 + 4 \cdot 3 + 3 \cdot 1 + 3 \cdot 0 = 31$
e	$6 \cdot 0 + 5 \cdot 3 + 4 \cdot 0 + 3 \cdot 0 + 3 \cdot 3 = 24$

The winner is issue j with 40 Borda points.

(e) The plurality and Hare methods violate the Condorcet criterion because neither method selects issue j, the Condorcet candidate. The Borda method does not violate the criterion.

11. (a) Using the pairwise comparison method, first make a table to compute the votes in comparison.

Compare	Votes	Rows
$t > k$	$18 = 18$	1
$k > t$	$12 + 10 + 9 + 4 + 2 = 37$	2, 3, 4, 5, 6
$t > h$	$18 = 18$	1
$h > t$	$12 + 10 + 9 + 4 + 2 = 37$	2, 3, 4, 5, 6
$t > b$	$18 = 18$	1
$b > t$	$12 + 10 + 9 + 4 + 2 = 37$	2, 3, 4, 5, 6
$t > c$	$18 = 18$	1
$c > t$	$12 + 10 + 9 + 4 + 2 = 37$	2, 3, 4, 5, 6
$k > h$	$18 + 9 = 27$	1, 4
$h > k$	$12 + 10 + 4 + 2 = 28$	2, 3, 5, 6
$k > b$	$18 + 12 + 9 + 4 = 43$	1, 2, 4, 5
$b > k$	$10 + 2 = 12$	3, 6
$k > c$	$18 + 9 + 2 = 29$	1, 4, 6
$c > k$	$12 + 10 + 4 = 26$	2, 3, 5
$h > b$	$18 + 12 + 4 + 2 = 36$	1, 2, 5, 6
$b > h$	$10 + 9 = 19$	3, 4
$h > c$	$18 + 9 + 4 + 2 = 33$	1, 4, 5, 6
$c > h$	$12 + 10 = 22$	2, 3
$b > c$	$18 + 10 + 9 + 2 = 39$	1, 3, 4, 6
$c > b$	$12 + 4 = 16$	2, 5

Make another table to award pairwise points.

Pairs	Votes	Pairwise Points
$t : k$	$18 : 37$	$k, 1$
$t : h$	$18 : 37$	$h, 1$
$t : b$	$18 : 37$	$b, 1$
$t : c$	$18 : 37$	$c, 1$
$k : h$	$27 : 28$	$h, 1$
$k : b$	$43 : 12$	$k, 1$
$k : c$	$29 : 26$	$k, 1$
$h : b$	$36 : 19$	$h, 1$
$h : c$	$33 : 22$	$h, 1$
$b : c$	$39 : 16$	$b, 1$

Using the pairwise comparison method, the winner is h with 4 pairwise points. This is the Condorcet candidate.

(b) If the plurality method is used, activity t is selected with 18 first-place votes.

(c) Calculate how many points each activity obtained from the 55 first-place votes from the voter profile.

Votes	Ranking	Row
18	$t > k > h > b > c$	1
12	$c > h > k > b > t$	2
10	$b > c > h > k > t$	3
9	$k > b > h > c > t$	4
4	$h > c > k > b > t$	5
2	$h > b > k > c > t$	6

Activity	1st-place votes
t	18
c	12
b	10
k	9
h	6

None of the activities received a majority, 28, of the votes. Therefore, eliminate activity h, the activity that received the least number of votes. Compare the first-place votes of the others.

Votes	Ranking	Row
18	$t > k > b > c$	1
12	$c > k > b > t$	2
10	$b > c > k > t$	3
9	$k > b > c > t$	4
4	$c > k > b > t$	5
2	$b > k > c > t$	6

Activity t: 18 votes.
Activity c: $12 + 4 = 16$ votes.
Activity b: $10 + 2 = 12$ votes.
Activity k: 9 votes.
There is no activity that has received a majority; therefore, eliminate k with only 9 votes, and compare the first-place votes of the others.

Votes	Ranking	Row
18	$t > b > c$	1
12	$c > b > t$	2
10	$b > c > t$	3
9	$b > c > t$	4
4	$c > b > t$	5
2	$b > c > t$	6

Activity t: 18 votes.
Activity c: $12 + 4 = 16$ votes.
Activity b: $10 + 9 + 2 = 21$ votes.
Again there is not a majority; therefore, eliminate c and compare the first-place votes of t and b.

Votes	Ranking	Row
18	$t > b$	1
12	$b > t$	2
10	$b > t$	3
9	$b > t$	4
4	$b > t$	5
2	$b > t$	6

Activity t: 18 votes.
Activity b: $12 + 10 + 9 + 4 + 2 = 37$ votes.
Finally, b wins with 37 votes in the Hare method.

(d) *Make a table to examine the points in the Borda method.*

Votes	Ranking	Points				
		4	3	2	1	0
18	$t>k>h>b>c$	t	k	h	b	c
12	$c>h>k>b>t$	c	h	k	b	t
10	$b>c>h>k>t$	b	c	h	k	t
9	$k>b>h>c>t$	k	b	h	c	t
4	$h>c>k>b>t$	h	c	k	b	t
2	$h>b>k>c>t$	h	b	k	c	t

Examine the last five columns and compute the weighted sum for each activity.

Activity	Borda Points
t	$18 \cdot 4 + 12 \cdot 0 + 10 \cdot 0 + 9 \cdot 0 + 4 \cdot 0 + 2 \cdot 0 = 72$
k	$18 \cdot 3 + 12 \cdot 2 + 10 \cdot 1 + 9 \cdot 4 + 4 \cdot 2 + 2 \cdot 2 = 136$
h	$18 \cdot 2 + 12 \cdot 3 + 10 \cdot 2 + 9 \cdot 2 + 4 \cdot 4 + 2 \cdot 4 = 134$
b	$18 \cdot 1 + 12 \cdot 1 + 10 \cdot 4 + 9 \cdot 3 + 4 \cdot 1 + 2 \cdot 3 = 107$
c	$18 \cdot 0 + 12 \cdot 4 + 10 \cdot 3 + 9 \cdot 1 + 4 \cdot 3 + 2 \cdot 1 = 101$

The winner is k with 136 Borda points.

(e) All three methods violate the Condorcet criterion because none of them select the Condorcet candidate.

12. *(a) Using the pairwise comparison method, first make a table to compute the votes in comparison.*

Compare	Votes	Rows
$t>b$	$18 = 18$	1
$b>t$	$12+10+9+4+2 = 37$	2, 3, 4, 5, 6
$t>h$	$18 = 18$	1
$h>t$	$12+10+9+4+2 = 37$	2, 3, 4, 5, 6
$t>k$	$18 = 18$	1
$k>t$	$12+10+9+4+2 = 37$	2, 3, 4, 5, 6
$t>c$	$18 = 18$	1
$c>t$	$12+10+9+4+2 = 37$	2, 3, 4, 5, 6
$b>h$	$18+9 = 27$	1, 4
$h>b$	$12+10+4+2 = 28$	2, 3, 5, 6
$b>k$	$18+12+9+4 = 43$	1, 2, 4, 5
$k>b$	$10+2 = 12$	3, 6
$b>c$	$18+9+2 = 29$	1, 4, 6
$c>b$	$12+10+4 = 26$	2, 3, 5
$h>k$	$18+12+4+2 = 36$	1, 2, 5, 6
$k>h$	$10+9 = 19$	3, 4
$h>c$	$18+9+4+2 = 33$	1, 4, 5, 6
$c>h$	$12+10 = 22$	2, 3
$k>c$	$18+10+9+2 = 39$	1, 3, 4, 6
$c>k$	$12+4 = 16$	2, 5

Make another table to award pairwise points.

Pairs	Votes	Pairwise Points
$t:b$	18 : 37	b, 1
$t:h$	18 : 37	h, 1
$t:k$	18 : 37	k, 1
$t:c$	18 : 37	c, 1
$b:h$	27 : 28	h, 1
$b:k$	43 : 12	b, 1
$b:c$	29 : 26	b, 1
$h:k$	36 : 19	h, 1
$h:c$	33 : 22	h, 1
$k:c$	39 : 16	k, 1

Using the pairwise comparison method, the winner is h with 4 pairwise points. This is the Condorcet candidate.

(b) If the plurality method is used, activity t is selected with 18 first-place votes.

(c) Calculate how many votes each activity obtained from the 55 first-place votes from the voter profile.

Votes	Ranking	Row
18	$t>b>h>k>c$	1
12	$c>h>b>k>t$	2
10	$k>c>h>b>t$	3
9	$b>k>h>c>t$	4
4	$h>c>b>k>t$	5
2	$h>k>b>c>t$	6

Activity	1st-place votes
t	18
c	12
k	10
b	9
h	6

None of the activities received a majority, 28, of the votes. Therefore, eliminate activity h, the activity that received the least number of votes. Compare the first-place votes of the others.

Votes	Ranking	Row
18	$t>b>k>c$	1
12	$c>b>k>t$	2
10	$k>c>b>t$	3
9	$b>k>c>t$	4
4	$c>b>k>t$	5
2	$k>b>c>t$	6

Activity t: 18 votes.
Activity c: $12 + 4 = 16$ votes.
Activity k: $10 + 2 = 12$ votes.
Activity b: 9 votes.
There is no activity that has received a majority; therefore, eliminate b with only 9 votes, and compare the first-place votes of the others.

Votes	Ranking	Row
18	$t > k > c$	1
12	$c > k > t$	2
10	$k > c > t$	3
9	$k > c > t$	4
4	$c > k > t$	5
2	$k > c > t$	6

Activity t: 18 votes.
Activity c: $12 + 4 = 16$ votes.
Activity k: $10 + 9 + 2 = 21$ votes.
Again there is not a majority; therefore, eliminate c and compare the first-place votes of t and k.

Votes	Ranking	Row
18	$t > k$	1
12	$k > t$	2
10	$k > t$	3
9	$k > t$	4
4	$k > t$	5
2	$k > t$	6

Activity t: 18 votes.
Activity k: $12 + 10 + 9 + 4 + 2 = 37$ votes.
Finally, k wins with 37 votes in the Hare method.

(d) Make a table to examine the points in the Borda method.

		Points			
Votes	Ranking	4	3	2	1	0
18	$t > b > h > k > c$	t	b	h	k	c
12	$c > h > b > k > t$	c	h	b	k	t
10	$k > c > h > b > t$	k	c	h	b	t
9	$b > k > h > c > t$	b	k	h	c	t
4	$h > c > b > k > t$	h	c	b	k	t
2	$h > k > b > c > t$	h	k	b	c	t

Examine the last five columns, and compute the weighted sum for each activity.

Activity	Borda Points
t	$18 \cdot 4 + 12 \cdot 0 + 10 \cdot 0 + 9 \cdot 0 + 4 \cdot 0 + 2 \cdot 0 = 72$
b	$18 \cdot 3 + 12 \cdot 2 + 10 \cdot 1 + 9 \cdot 4 + 4 \cdot 2 + 2 \cdot 2 = 136$
h	$18 \cdot 2 + 12 \cdot 3 + 10 \cdot 2 + 9 \cdot 2 + 4 \cdot 4 + 2 \cdot 4 = 134$
k	$18 \cdot 1 + 12 \cdot 1 + 10 \cdot 4 + 9 \cdot 3 + 4 \cdot 1 + 2 \cdot 3 = 107$
c	$18 \cdot 0 + 12 \cdot 4 + 10 \cdot 3 + 9 \cdot 1 + 4 \cdot 3 + 2 \cdot 1 = 101$

The winner is b with 136 Borda points.

(e) All three methods violate the Condorcet criterion because none of them select the Condorcet candidate.

13. (a) Using the pairwise comparison method, first make a table to compute the votes in comparison.

Compare	Votes	Rows
$m > c$	$5 + 3 = 8$	1, 3
$c > m$	$4 + 2 = 6$	2, 4
$m > s$	$5 + 2 = 7$	1, 4
$s > m$	$4 + 3 = 7$	2, 3
$m > b$	$5 + 2 = 7$	1, 4
$b > m$	$4 + 3 = 7$	2, 3
$c > s$	$5 + 2 = 7$	1, 4
$s > c$	$4 + 3 = 7$	2, 3
$c > b$	$5 + 2 = 7$	1, 4
$b > c$	$4 + 3 = 7$	2, 3
$s > b$	$5 + 2 = 7$	1, 4
$b > s$	$4 + 3 = 7$	2, 3

Make another table to award pairwise points.

Pairs	Votes	Pairwise Points
$m:c$	$8:6$	$m, 1$
$m:s$	$7:7$	$m, \frac{1}{2}; s, \frac{1}{2}$
$m:b$	$7:7$	$m, \frac{1}{2}; b, \frac{1}{2}$
$c:s$	$7:7$	$c, \frac{1}{2}; s, \frac{1}{2}$
$c:b$	$7:7$	$c, \frac{1}{2}; b, \frac{1}{2}$
$s:b$	$7:7$	$s, \frac{1}{2}; b, \frac{1}{2}$

Using the pairwise comparison method, here are the total points each city has received.

$$m: \quad 1 + \frac{1}{2} + \frac{1}{2} = 2$$
$$s: \quad \frac{1}{2} + \frac{1}{2} + \frac{1}{2} = 1\frac{1}{2}$$
$$b: \quad \frac{1}{2} + \frac{1}{2} + \frac{1}{2} = 1\frac{1}{2}$$
$$c: \quad \frac{1}{2} + \frac{1}{2} = 1$$

Montreal, m, is selected.

(b) Make a table to compute the votes in comparison.

Compare	Votes	Rows
$m > c$	$5 + 3 + 2 = 10$	1, 3, 4
$c > m$	$4 = 4$	2
$m > s$	$5 + 2 = 7$	1, 4
$s > m$	$4 + 3 = 7$	2, 3
$m > b$	$5 + 2 = 7$	1, 4
$b > m$	$4 + 3 = 7$	2, 3
$c > s$	$5 + 2 = 7$	1, 4
$s > c$	$4 + 3 = 7$	2, 3
$c > b$	$5 = 5$	1
$b > c$	$4 + 3 + 2 = 9$	2, 3, 4
$s > b$	$5 = 5$	1
$b > s$	$4 + 3 + 2 = 9$	2, 3, 4

Make another table to award pairwise points.

Pairs	Votes	Pairwise Points
m: c	10: 4	m, 1
m: s	7: 7	m, $\frac{1}{2}$; s, $\frac{1}{2}$
m: b	7: 7	m, $\frac{1}{2}$; b, $\frac{1}{2}$
c: s	7: 7	c, $\frac{1}{2}$; s, $\frac{1}{2}$
c: b	5: 9	b, 1
s: b	5: 9	b, 1

Using the pairwise comparison method, here are the total points each city has received for this vote.

$$m:\ 1 + \frac{1}{2} + \frac{1}{2} = 2$$
$$s:\ \frac{1}{2} + \frac{1}{2} = 1$$
$$b:\ \frac{1}{2} + 1 + 1 = 2\frac{1}{2}$$
$$c:\ \frac{1}{2}$$

Boston, b, is selected.

(c) The Monotonicity criterion is violated because city m does not win the second election.

14. (a) Using the pairwise comparison method, first make a table to compute the votes in comparison.

Compare	Votes	Rows
c > m	5 + 3 = 8	1, 3
m > c	4 + 2 = 6	2, 4
c > b	5 + 2 = 7	1, 4
b > c	4 + 3 = 7	2, 3
c > s	5 + 2 = 7	1, 4
s > c	4 + 3 = 7	2, 3
m > b	5 + 2 = 7	1, 4
b > m	4 + 3 = 7	2, 3
m > s	5 + 2 = 7	1, 4
s > m	4 + 3 = 7	2, 3
b > s	5 + 2 = 7	1, 4
s > b	4 + 3 = 7	2, 3

Make another table to award pairwise points.

Pairs	Votes	Pairwise Points
c: m	8: 6	c, 1
c: b	7: 7	c, $\frac{1}{2}$; b, $\frac{1}{2}$
c: s	7: 7	c, $\frac{1}{2}$; s, $\frac{1}{2}$
m: b	7: 7	m, $\frac{1}{2}$; b, $\frac{1}{2}$
m: s	7: 7	m, $\frac{1}{2}$; s, $\frac{1}{2}$
b: s	7: 7	b, $\frac{1}{2}$; s, $\frac{1}{2}$

Using the pairwise comparison method, here are the total points each city has received.

$$c:\ 1 + \frac{1}{2} + \frac{1}{2} = 2$$
$$b:\ \frac{1}{2} + \frac{1}{2} + \frac{1}{2} = 1\frac{1}{2}$$
$$s:\ \frac{1}{2} + \frac{1}{2} + \frac{1}{2} = 1\frac{1}{2}$$
$$m:\ \frac{1}{2} + \frac{1}{2} = 1$$

Chicago, c, is selected.

(b) Make a table to compute the votes in comparison.

Compare	Votes	Rows
c > m	5 + 3 + 2 = 10	1, 3, 4
m > c	4 = 4	2
c > b	5 + 2 = 7	1, 4
b > c	4 + 3 = 7	2, 3
c > s	5 + 2 = 7	1, 4
s > c	4 + 3 = 7	2, 3
m > b	5 + 2 = 7	1, 4
b > m	4 + 3 = 7	2, 3
m > s	5 = 5	1
s > m	4 + 3 + 2 = 9	2, 3, 4
b > s	5 = 5	1
s > b	4 + 3 + 2 = 9	2, 3, 4

Make another table to award pairwise points.

Pairs	Votes	Pairwise Points
c : m	10 : 4	c, 1
c : b	7 : 7	c, $\frac{1}{2}$; b, $\frac{1}{2}$
c : s	7 : 7	c, $\frac{1}{2}$; s, $\frac{1}{2}$
m : b	7 : 7	m, $\frac{1}{2}$; b, $\frac{1}{2}$
m : s	5 : 9	s, 1
b : s	5 : 9	s, 1

Using the pairwise comparison method, here are the total points each city has received for this vote.

$$c:\ 1 + \frac{1}{2} + \frac{1}{2} = 2$$
$$b:\ \frac{1}{2} + \frac{1}{2} = 1$$
$$s:\ \frac{1}{2} + 1 + 1 = 2\frac{1}{2}$$
$$m:\ \frac{1}{2} = \frac{1}{2}$$

San Francisco, s, is selected.

(c) The Monotonicity criterion is violated because city c does not win the second election.

15. (a) Make a table to examine the points in the Borda method.

Votes	Ranking	Points			
		3	2	1	0
7	$s > b > c > d$	s	b	c	d
5	$b > d > c > s$	b	d	c	s
3	$d > s > c > b$	d	s	c	b
4	$c > s > d > b$	c	s	d	b

Examine the last four columns, and compute the weighted sum for each city.

City	Borda Points
s	$7 \cdot 3 + 5 \cdot 0 + 3 \cdot 2 + 4 \cdot 2 = 35$
b	$7 \cdot 2 + 5 \cdot 3 + 3 \cdot 0 + 4 \cdot 0 = 29$
c	$7 \cdot 1 + 5 \cdot 1 + 3 \cdot 1 + 4 \cdot 3 = 27$
d	$7 \cdot 0 + 5 \cdot 2 + 3 \cdot 3 + 4 \cdot 1 = 23$

The winner is s with 35 Borda points in the preliminary non-binding decision.

(b) Make a table to examine the points in the Borda method.

Votes	Ranking	Points			
		3	2	1	0
7	$s > b > c > d$	s	b	c	d
5	$b > d > c > s$	b	d	c	s
3	$s > b > d > c$	s	b	d	c
4	$s > b > c > d$	s	b	c	d

Examine the last four columns, and compute the weighted sum for each city.

City	Borda Points
s	$7 \cdot 3 + 5 \cdot 0 + 3 \cdot 3 + 4 \cdot 3 = 42$
b	$7 \cdot 2 + 5 \cdot 3 + 3 \cdot 2 + 4 \cdot 2 = 43$
c	$7 \cdot 1 + 5 \cdot 1 + 3 \cdot 0 + 4 \cdot 1 = 16$
d	$7 \cdot 0 + 5 \cdot 2 + 3 \cdot 1 + 4 \cdot 0 = 13$

The winner is b with 43 Borda points.

(c) The Monotonicity criterion is violated because city s does not win the second election.

16. (a) Make a table to examine the points in the Borda method.

Votes	Ranking	Points			
		3	2	1	0
7	$c > d > s > b$	c	d	s	b
5	$d > b > s > c$	d	b	s	c
3	$b > c > s > d$	b	c	s	d
4	$s > c > b > d$	s	c	b	d

Examine the last four columns, and compute the weighted sum for each city.

City	Borda Points
c	$7 \cdot 3 + 5 \cdot 0 + 3 \cdot 2 + 4 \cdot 2 = 35$
d	$7 \cdot 2 + 5 \cdot 3 + 3 \cdot 0 + 4 \cdot 0 = 29$
s	$7 \cdot 1 + 5 \cdot 1 + 3 \cdot 1 + 4 \cdot 3 = 27$
b	$7 \cdot 0 + 5 \cdot 2 + 3 \cdot 3 + 4 \cdot 1 = 23$

The winner is c with 35 Borda points in the preliminary non-binding decision.

(b) Make a table to examine the points in the Borda method.

Votes	Ranking	Points			
		3	2	1	0
7	$c > d > s > b$	c	d	s	b
5	$d > b > s > c$	d	b	s	c
3	$c > d > b > s$	c	d	b	s
4	$c > d > s > b$	c	d	s	b

Examine the last four columns, and compute the weighted sum for each city.

City	Borda Points
c	$7 \cdot 3 + 5 \cdot 0 + 3 \cdot 3 + 4 \cdot 3 = 42$
d	$7 \cdot 2 + 5 \cdot 3 + 3 \cdot 2 + 4 \cdot 2 = 43$
s	$7 \cdot 1 + 5 \cdot 1 + 3 \cdot 0 + 4 \cdot 1 = 16$
b	$7 \cdot 0 + 5 \cdot 2 + 3 \cdot 1 + 4 \cdot 0 = 13$

The winner is d with 43 Borda points.

(c) The monotonicity criterion is violated because candidate c does not win the second election.

17. (a) Calculate how many votes each city obtained from the 17 first-place votes from the voter profile.

City	1st-place votes
a	6
c	$4 + 2 = 6$
b	5
d	0

None of the cities received a majority, 9, of the votes. Therefore, eliminate city d, the city that received the least number of votes. Compare the first-place votes of the others.

Votes	Ranking	Row
6	$a > c > b$	1
5	$b > a > c$	2
4	$c > b > a$	3
2	$c > a > b$	4

a: 6 = 6
b: 5 = 5
c: 4 + 2 = 6

Again, no city has received a majority; therefore, eliminate b with only 5 points and compare the first-place votes of a and c.

16.2 THE IMPOSSIBILITIES OF VOTING

Votes	Ranking	Row
6	$a > c$	1
5	$a > c$	2
4	$c > a$	3
2	$c > a$	4

a: $6 + 5 = 11$
c: $4 + 2 = 6$

Finally, a wins with 11 votes in the Hare method.

(b) Calculate how many votes each city obtained from the 17 first-place votes from the voter profile.

City	1st-place votes
a	$6 + 2 = 8$
b	5
c	4
d	0

None of the cities received a majority, 9, of the votes. Therefore, eliminate city d, the city that received the least number of votes. Compare the first-place votes of the others.

Votes	Ranking	Row
6	$a > c > b$	1
5	$b > a > c$	2
4	$c > b > a$	3
2	$a > c > b$	4

a: $6 + 2 = 8$
b: $5 = 5$
c: $4 = 4$

Again, no city has received a majority; therefore, eliminate c with only 4 votes and compare the first-place votes of the a and b.

Votes	Ranking	Row
6	$a > b$	1
5	$b > a$	2
4	$b > a$	3
2	$a > b$	4

a: $6 + 2 = 8$
b: $5 + 4 = 9$

Finally, b wins with 9 votes in the Hare method.

(c) Yes, the monotonicity criterion is violated because candidate a does not win the second election.

18. (a) Calculate how many votes each city obtained from the 17 first-place votes from the voter profile.

City	1st-place votes
a	5
b	6
c	$4 + 2 = 6$
d	0

None of the cities received a majority, 9, of the votes. Therefore, eliminate city d, the city that received the least number of votes. Compare the first-place votes of the others.

Votes	Ranking	Row
6	$b > c > a$	1
5	$a > b > c$	2
4	$c > a > b$	3
2	$c > b > a$	4

a: $5 = 5$
b: $6 = 6$
c: $4 + 2 = 6$

Again, no city has received a majority; therefore, eliminate a with only 5 votes and compare the first-place votes of b and c.

Votes	Ranking	Row
6	$b > c$	1
5	$b > c$	2
4	$c > b$	3
2	$c > b$	4

b: $6 + 5 = 11$
c: $4 + 2 = 6$

Finally, b wins with 11 points in the Hare method.

(b) Calculate how many votes each city obtained from the 17 first-place votes from the voter profile.

City	1st-place votes
b	$6 + 2 = 8$
a	5
c	4
d	0

None of the cities received a majority, 9, of the votes. Therefore, eliminate city d, the city that received the least number of votes. Compare the first-place votes of the others.

Votes	Ranking	Row
6	$b > c > a$	1
5	$a > b > c$	2
4	$c > a > b$	3
2	$b > c > a$	4

a: $5 = 5$
b: $6 + 2 = 8$
c: $4 = 4$

Again, no city has received a majority; therefore, eliminate c with only 4 votes and compare the first-place votes of the a and b.

Votes	Ranking	Row
6	b > a	1
5	a > b	2
4	a > b	3
2	b > a	4

$a: 5 + 4 = 9$
$b: 6 + 2 = 8$

Finally, a wins with 9 votes in the Hare method.

(c) Yes, the monotonicity criterion is violated because candidate b does not win the second election.

19. (a) Calculate how many votes each candidate obtained from the 13 first-place votes from the voter profile.

City	1st-place votes
a	6
b	5
c	2

Candidate a receives the most votes, 6, and wins by the plurality method.

(b) Compare the rankings if candidate c drops out.

Votes	Ranking	Row
6	a > b	1
5	b > a	2
2	b > a	3

Now candidate a has 6 first-place votes, and candidate b has 7 first-place votes; b is the winner.

(c) Yes, the Independence of Irrelevant Alternatives criterion is violated because candidate a does not win the second election.

20. (a) Calculate how many votes each candidate obtained from the 17 first-place votes from the voter profile.

City	1st-place votes
a	6
b	8
c	3

Candidate b receives the most votes, 8, and wins by the plurality method.

(b) Compare the rankings if candidate a drops out.

Votes	Ranking	Row
6	c > b	1
8	b > c	2
3	c > b	3

Now candidate b has 8 first-place votes, and candidate c has 9 first-place votes; c is the winner.

(c) Yes, the Independence of Irrelevant Alternatives criterion is violated because candidate b does not win the second election.

21. (a) Calculate how many votes each candidate obtained from the 175 first-place votes from the voter profile.

City	1st-place votes
a	75
b	30
c	50
d	20

Candidate a receives the most votes, 75, and wins by the plurality method.

(b) Compare the rankings if candidate b drops out.

Votes	Ranking	Row
75	a > c > d	1
50	c > a > d	2
30	c > d > a	3
20	d > c > a	4

$a \quad 75 = 75.$
$c \quad 50 + 30 = 80$
$d \quad 20 = 20$

Now c wins with 80 votes.

(c) Yes, the Independence of Irrelevant Alternatives criterion is violated because candidate a does not win the second election.

22. (a) Calculate how many votes each candidate obtained from the 95 first-place votes from the voter profile.

City	1st-place votes
a	40
b	15
c	30
d	10

Candidate a receives the most votes, 40, and wins by the plurality method.

(b) Compare the rankings if candidate b drops out.

Votes	Ranking	Row
40	a > c > d	1
30	c > a > d	2
15	c > d > a	3
10	d > c > a	4

$a \quad 40 = 40.$
$c \quad 30 + 15 = 45$
$d \quad 10 = 10$

Now c wins with 45 votes.

(c) Yes, the Independence of Irrelevant Alternatives criterion is violated because candidate a does not win the second election.

23. (a) Using the pairwise comparison method, first make a table to compute the votes in comparison.

Compare	Votes	Rows
$e > j$	$5+2+2=9$	1, 3, 4
$j > e$	$2=2$	2
$e > h$	$5+2+2=9$	1, 2, 4
$h > e$	$2=2$	3
$e > g$	$5+2+2=9$	1, 2, 3
$g > e$	$2=2$	4
$e > m$	$5=5$	1
$m > e$	$2+2+2=6$	2, 3, 4
$j > h$	$5+2+2=9$	1, 2, 4
$h > j$	$2=2$	3
$j > g$	$5+2=7$	1, 2
$g > j$	$2+2=4$	3, 4
$j > m$	$5=5$	1
$m > j$	$2+2+2=6$	2, 3, 4
$h > g$	$5+2+2=9$	1, 2, 3
$g > h$	$2=2$	4
$h > m$	$5+2=7$	1, 3
$m > h$	$2+2=4$	2, 4
$g > m$	$5+2=7$	1, 4
$m > g$	$2+2=4$	2, 3

Make another table to award pairwise points.

Pairs	Votes	Pairwise Points
$e : j$	9 : 2	e, 1
$e : h$	9 : 2	e, 1
$e : g$	9 : 2	e, 1
$e : m$	5 : 6	m, 1
$j : h$	9 : 2	j, 1
$j : g$	7 : 4	j, 1
$j : m$	5 : 6	m, 1
$h : g$	9 : 2	h, 1
$h : m$	7 : 4	h, 1
$g : m$	7 : 4	g, 1

Using the pairwise comparison method, here are the total points each issue has received.

$$e: \quad 1+1+1=3$$
$$m: \quad 1+1=2$$
$$j: \quad 1+1=2$$
$$h: \quad 1+1=2$$
$$g: \quad 1=1$$

The education issue, e, was voted most important with 3 pairwise points.

(b) Now omitting h and g, make a table to compute the votes in comparison.

Compare	Votes	Rows
$e > j$	$5+2+2=9$	1, 3, 4
$j > e$	$2=2$	2
$e > m$	$5=5$	1
$m > e$	$2+2+2=6$	2, 3, 4
$j > m$	$5=5$	1
$m > j$	$2+2+2=6$	2, 3, 4

Make another table to award pairwise points.

Pairs	Votes	Pairwise Points
$e : j$	9 : 2	e, 1
$e : m$	5 : 6	m, 1
$j : m$	5 : 6	m, 1

Using the pairwise comparison method, m has two pairwise points and e has only one pairwise point, so the military spending issue, m wins.

(c) Yes, the Independence of Irrelevant Alternatives criterion is violated because candidate e fails to receive the highest priority in the second comparison.

24. (a) See part (a) in Exercise 23.

(b) Now omitting h, j, and g, make a table to compute the votes in comparison.

Compare	Votes	Rows
$e > m$	$5=5$	1
$m > e$	$2+2+2=6$	2, 3, 4

Issue m wins with one pairwise vote.

(c) Yes the Independence of Irrelevant Alternatives criterion is violated because candidate e fails to receive the highest priority in the second comparison.

25. In Example 4 percussionist x wins the most comparisons in the pairwise selection. If w had dropped out instead of y, here are the comparisons

Pairs	Pairwise Points
$v : x$	v, 1
$y : v$	y, 1
$v : z$	v, 1
$x : y$	x, 1
$x : z$	x, 1
$z : y$	z, 1

Using the pairwise comparison method, here are the total points each percussionist has received.

$$v: \quad 1+1=2$$
$$x: \quad 1+1=2$$
$$y: \quad 1=1$$
$$z: \quad 1=1$$

No, the Independence of Irrelevant Alternatives criterion is not violated; this second pairwise vote has resulted in a tie between v and x.

26. In Example 4 percussionist x wins the most comparisons in the pairwise selection. If z had dropped out instead of y, here are the comparisons

Pairs	Pairwise Points
v: w	v, 1/2; w, 1/2
v: x	v, 1
y: v	y, 1
x: w	x, 1
y: w	y, 1
x: y	x, 1

Using the pairwise comparison method, here are the total points each percussionist has received.

$v: \quad \frac{1}{2} + 1 = 1\frac{1}{2}$

$w: \quad \frac{1}{2} = \frac{1}{2}$

$x: \quad 1 + 1 = 2$

$y: \quad 1 + 1 = 2$

No, the Independence of Irrelevant Alternatives criterion is not violated; this second pairwise vote has resulted in a tie between x and y.

27. (a) Make a table to examine the points in the Borda method.

Votes	Ranking	Points		
		2	1	0
13	$c > b > a$	c	b	a
8	$b > a > c$	b	a	c
4	$b > c > a$	b	c	a

Examine the last three columns, and compute the weighted sum for each candidate.

Candidate	Borda Points
a	$13 \cdot 0 + 8 \cdot 1 + 4 \cdot 0 = 8$
b	$13 \cdot 1 + 8 \cdot 2 + 4 \cdot 2 = 37$
c	$13 \cdot 2 + 8 \cdot 0 + 4 \cdot 1 = 30$

The winner is candidate b with 37 Borda points.

(b) If a drops out:

Votes	Ranking	Points	
		1	0
13	$c > b$	c	b
8	$b > c$	b	c
4	$b > c$	b	c

Examine the last two columns, and compute the weighted sum for each candidate.

Candidate	Borda Points
b	$13 \cdot 0 + 8 \cdot 1 + 4 \cdot 1 = 12$
c	$13 \cdot 1 + 8 \cdot 0 + 4 \cdot 0 = 13$

Candidate c wins this election with 13 Borda points.

(c) Yes, according to the Independence of Irrelevant Alternatives criterion, candidate b should win the second election; therefore, the Borda method violates the criterion.

28. (a) Make a table to examine the points in the Borda method.

Votes	Ranking	Points		
		2	1	0
9	$b > a > c$	b	a	c
6	$a > c > b$	a	c	b
2	$a > b > c$	a	b	c

Examine the last three columns, and compute the weighted sum for each candidate.

Candidate	Borda Points
a	$9 \cdot 1 + 6 \cdot 2 + 2 \cdot 2 = 25$
b	$9 \cdot 2 + 6 \cdot 0 + 2 \cdot 1 = 20$
c	$9 \cdot 0 + 6 \cdot 1 + 2 \cdot 0 = 6$

The winner is candidate a with 25 Borda points.

(b) If c drops out:

Votes	Ranking	Points	
		1	0
9	$b > a$	b	a
6	$a > b$	a	b
2	$a > b$	a	b

Examine the last two columns, and compute the weighted sum for each candidate.

Candidate	Borda Points
a	$9 \cdot 0 + 6 \cdot 1 + 2 \cdot 1 = 8$
b	$9 \cdot 1 + 6 \cdot 0 + 2 \cdot 0 = 9$

Candidate b wins this election with 9 Borda points.

(c) Yes, according to the Independence of Irrelevant Alternatives criterion, candidate a should win the second election; therefore, the Borda method violates the criterion.

29. (a) Calculate how many votes each candidate obtained from the 34 first-place votes from the voter profile.

City	1st-place votes
a	12
b	10
c	$8 + 4 = 12$

None of the candidates received a majority, 18, of the votes. Therefore, eliminate candidate b, the candidate that received the least number of votes. Compare the first-place votes of a and c..

Votes	Ranking	Row
12	$a > c$	1
10	$a > c$	2
8	$c > a$	3
4	$c > a$	4

$$a: \quad 12 + 10 = 22$$
$$c: \quad 8 + 4 = 12$$

Now a wins with 22 votes.

(b) If candidate c drops out, the rankings are:

Votes	Ranking	Row
12	$a > b$	1
10	$b > a$	2
8	$b > a$	3
4	$a > b$	4

$$a: \quad 12 + 4 = 16$$
$$b: \quad 10 + 8 = 18$$

Now b wins with 18 votes.

(c) Yes, according to the Independence of Irrelevant Alternatives criterion, candidate a should win the second election; therefore, the Hare method violates the criterion.

30. (a) Calculate how many votes each candidate obtained from the 54 first-place votes from the voter profile.

City	1st-place votes
a	17
b	15
c	$13 + 9 = 22$

None of the candidates received a majority, 28, of the votes. Therefore, eliminate candidate b, the candidate that received the least number of votes. Compare the first-place votes of a and c.

Votes	Ranking	Row
17	$a > c$	1
15	$a > c$	2
13	$c > a$	3
9	$c > a$	4

$$a: \quad 17 + 15 = 32$$
$$c: \quad 13 + 9 = 22$$

Now a wins with 32 votes.

(b) If candidate c drops out, the rankings are:

Votes	Ranking	Row
17	$a > b$	1
15	$b > a$	2
13	$b > a$	3
9	$a > b$	4

$$a: \quad 17 + 9 = 26$$
$$b: \quad 15 + 13 = 28$$

Now b wins with 28 votes.

(c) Yes, according to the Independence of Irrelevant Alternatives criterion, candidate a should win the second election; therefore, the Hare method violates the criterion.

31. (a) This is the original non-binding vote.

Voters	Ranking
6	$a > c > b > d$
5	$b > d > a > c$
4	$c > d > b > a$
2	$c > a > b > d$

If Chicago withdraws from the vote, compare the first-place votes of the others.

Votes	Ranking	Row
6	$a > b > d$	1
5	$b > d > a$	2
4	$d > b > a$	3
2	$a > b > d$	4

$$a: \quad 6 + 2 = 8$$
$$b: \quad 5 = 5$$
$$d: \quad 4 = 4$$

No city has received a majority; therefore, eliminate d with only 4 votes and compare the first-place votes of a and b.

Votes	Ranking	Row
6	$a > b$	1
5	$b > a$	2
4	$b > a$	3
2	$a > b$	4

$$a: \quad 6 + 2 = 8$$
$$b: \quad 5 + 4 = 9$$

Finally, b wins with 9 votes in the Hare method.

(b) Yes, according to the Independence of Irrelevant Alternatives criterion, candidate a should win the second election; therefore, the Hare method violates the criterion.

32. (a) This is the original non-binding vote.

Voters	Ranking
6	$b > c > a > d$
5	$a > d > b > c$
4	$c > d > a > b$
2	$c > b > a > d$

If Chicago withdraws from the vote, compare the first-place votes of the others.

Votes	Ranking	Row
6	$b > a > d$	1
5	$a > d > b$	2
4	$d > a > b$	3
2	$b > a > d$	4

$$a: \quad 5 = 5$$
$$b: \quad 6 + 2 = 8$$
$$d: \quad 4 = 4$$

No city has received a majority; therefore, eliminate d with only 4 votes, and compare the first-place votes of a and b.

Votes	Ranking	Row
6	$b > a$	1
5	$a > b$	2
4	$a > b$	3
2	$b > a$	4

b: $6 + 2 = 8$
a: $5 + 4 = 9$

Finally, a wins with 9 points in the Hare method.

(b) Yes, according to the Independence of Irrelevant Alternatives criterion, candidate b should win the second election; therefore, the Hare method violates the criterion.

33. Writing exercise

34. Original winner x gained a pairwise point by beating Percussionist y, so the withdrawal of y hurts x. Percussionist v was beaten originally by y, so the withdrawal of y does not alter the number of pairwise points for v.

35. One possibility is

Votes	Ranking
10	$a > b > c > d > e > f$
9	$b > f > e > c > d > a$

Candidate a wins the majority of the 19 votes. If a pairwise comparison is made, candidate a beats each of the others in turn to earn 5 points, from the 10 votes in the top row. From the information in the second row, notice that b beats f, e, c, and d and will receive 4 points in the pairwise comparison.

36. One possibility is

Votes	Ranking	Points 3	2	1	0
21	$d > b > c > a$	d	b	c	a
8	$b > c > a > d$	b	c	a	d
6	$c > b > d > a$	c	b	d	a
5	$a > c > b > d$	a	c	b	d

Examine the last four columns, and compute the weighted sum for each candidate.

Candidate	Borda Points
d	$21 \cdot 3 + 8 \cdot 0 + 6 \cdot 1 + 5 \cdot 0 = 69$
b	$21 \cdot 2 + 8 \cdot 3 + 6 \cdot 2 + 5 \cdot 1 = 83$
c	$21 \cdot 1 + 8 \cdot 2 + 6 \cdot 3 + 5 \cdot 2 = 65$
a	$21 \cdot 0 + 8 \cdot 1 + 6 \cdot 0 + 5 \cdot 3 = 23$

Although candidate d has the majority of the votes with a total of 21, the winner of the Borda election is b with 83 Borda points.

37. One possibility is

Votes	Ranking	Points 3	2	1	0
21	$a > b > c > d$	a	b	c	d
5	$b > c > a > d$	b	c	a	d
9	$c > b > d > a$	c	b	d	a
5	$d > c > b > a$	d	c	b	a

Examine the last four columns, and compute the weighted sum for each candidate.

Candidate	Borda Points
a	$21 \cdot 3 + 5 \cdot 1 + 9 \cdot 0 + 5 \cdot 0 = 68$
b	$21 \cdot 2 + 5 \cdot 3 + 9 \cdot 2 + 5 \cdot 1 = 80$
c	$21 \cdot 1 + 5 \cdot 2 + 9 \cdot 3 + 5 \cdot 2 = 68$
d	$21 \cdot 0 + 5 \cdot 0 + 9 \cdot 1 + 5 \cdot 3 = 24$

Although candidate a has the majority of the votes with a total of 21, the winner of the Borda election is b with 80 Borda points.

38. One possible profile is

Votes	Ranking
3	$a > b > c > d$
1	$b > c > a > d$
2	$d > b > c > a$
1	$c > b > a > d$
2	$a > d > c > b$
1	$b > c > d > a$
1	$c > b > d > a$
2	$a > b > d > c$

Candidate a is the Condorcet candidate because a is preferred over each of the other candidates.

Borda method
Here is the table from Exercise 15 of Section 16.1 to examine the points in the Borda method.

Votes	Ranking	Points 3	2	1	0
3	$a > b > c > d$	a	b	c	d
1	$b > c > a > d$	b	c	a	d
2	$d > b > c > a$	d	b	c	a
1	$c > b > a > d$	c	b	a	d
2	$a > d > c > b$	a	d	c	b
1	$b > c > d > a$	b	c	d	a
1	$c > b > d > a$	c	b	d	a
2	$a > b > d > c$	a	b	d	c

Examine the last four columns, and compute the weighted sum for each.

Candidate	Borda Points
a	$3 \cdot 3 + 1 \cdot 1 + 2 \cdot 0 + 1 \cdot 1$ $+ 2 \cdot 3 + 1 \cdot 0 + 1 \cdot 0 + 2 \cdot 3 = 23$
b	$3 \cdot 2 + 1 \cdot 3 + 2 \cdot 2 + 1 \cdot 2$ $+ 2 \cdot 0 + 1 \cdot 3 + 1 \cdot 2 + 2 \cdot 2 = 24$
c	$3 \cdot 1 + 1 \cdot 2 + 2 \cdot 1 + 1 \cdot 3$ $+ 2 \cdot 1 + 1 \cdot 2 + 1 \cdot 3 + 2 \cdot 0 = 17$
d	$3 \cdot 0 + 1 \cdot 0 + 2 \cdot 3 + 1 \cdot 0$ $+ 2 \cdot 2 + 1 \cdot 1 + 1 \cdot 1 + 2 \cdot 1 = 14$

The winner is b with 24 Borda points. The Condorcet candidate, a, has not won.

Hare method
Here is the table of 1st-place votes from Exercise 16 in 16.1.

Breed	1st-place votes
a	7
b	2
c	2
d	2

Using the Hare method, breed a has a majority of the votes and wins.

Plurality method
From the table above, it can be seen that a has the most first-place votes to win.

39. One possible profile is from Exercise 17 in 16.1.

Votes	Ranking	Voter	Row
3	$b > c > a > d$	1, 2, 9	1
2	$a > d > c > b$	3, 11	2
1	$c > d > b > a$	4	3
1	$d > c > b > a$	5	4
1	$d > a > b > c$	6	5
2	$c > a > d > b$	7, 8	6
1	$a > c > d > b$	10	7
2	$a > b > c > d$	12, 13	8

Candidate c is the Condorcet candidate because c is preferred over each of the other candidates (see Exercise 17).

Borda method
Here is the table from Exercise 17 to examine the points in the Borda method.

Votes	Ranking	Points 3	2	1	0
3	$b > c > a > d$	b	c	a	d
2	$a > d > c > b$	a	d	c	b
1	$c > d > b > a$	c	d	b	a
1	$d > c > b > a$	d	c	b	a
1	$d > a > b > c$	d	a	b	c
2	$c > a > d > b$	c	a	d	b
1	$a > c > d > b$	a	c	d	b
2	$a > b > c > d$	a	b	c	d

Examine the last four columns and compute the weighted sum for each.

Candidate	Borda Points
a	$3 \cdot 1 + 2 \cdot 3 + 1 \cdot 0 + 1 \cdot 0$ $+ 1 \cdot 2 + 2 \cdot 2 + 1 \cdot 3 + 2 \cdot 3 = 24$
b	$3 \cdot 3 + 2 \cdot 0 + 1 \cdot 1 + 1 \cdot 1$ $+ 1 \cdot 1 + 2 \cdot 0 + 1 \cdot 0 + 2 \cdot 2 = 16$
c	$3 \cdot 2 + 2 \cdot 1 + 1 \cdot 3 + 1 \cdot 2$ $+ 1 \cdot 0 + 2 \cdot 3 + 1 \cdot 2 + 2 \cdot 1 = 23$
d	$3 \cdot 0 + 2 \cdot 2 + 1 \cdot 2 + 1 \cdot 3$ $+ 1 \cdot 3 + 2 \cdot 1 + 1 \cdot 1 + 2 \cdot 0 = 15$

The winner is Candidate a with 24 Borda points. The Condorcet candidate has not won.

Hare method
Here is the information from Exercise 17.
Calculate how many votes each candidate obtained from the thirteen first-place votes from the voter profile.

Votes	Ranking	Voter	Row
3	$b > c > a > d$	1, 2, 9	1
2	$a > d > c > b$	3, 11	2
1	$c > d > b > a$	4	3
1	$d > c > b > a$	5	4
1	$d > a > b > c$	6	5
2	$c > a > d > b$	7, 8	6
1	$a > c > d > b$	10	7
2	$a > b > c > d$	12, 13	8

Candidate	1st-place votes
a	5
b	3
c	3
d	2

None of the candidates received a majority, 7, of the votes. Therefore, eliminate d, the candidate that received the least number of votes. Compare the other three.

Votes	Ranking	Voter	Row
3	$b > c > a$	1, 2, 9	1
2	$a > c > b$	3, 11	2
1	$c > b > a$	4	3
1	$c > b > a$	5	4
1	$a > b > c$	6	5
2	$c > a > b$	7, 8	6
1	$a > c > b$	10	7
2	$a > b > c$	12, 13	8

Candidate a: $2 + 1 + 1 + 2 = 6$ votes.
Candidate b: 3 votes.
Candidate c: $1 + 1 + 2 = 4$ votes.
None has received a majority; therefore, eliminate b and compare first-place points again for a and c.

560 CHAPTER 16 VOTING AND APPORTIONMENT

Votes	Ranking	Voter	Row
3	$c > a$	1, 2, 9	1
2	$a > c$	3, 11	2
1	$c > a$	4	3
1	$c > a$	5	4
1	$a > c$	6	5
2	$c > a$	7, 8	6
1	$a > c$	10	7
2	$a > c$	12, 13	8

Candidate a: $2 + 1 + 1 + 2 = 6$
Candidate c: $3 + 1 + 1 + 2 = 7$
Candidate c now has a majority and wins.

Plurality method
See from the Hare method above that a has the most first-place votes to win.

40. (a) Using the pairwise comparison method, first make a table to compute the votes in comparison.

Compare	Votes	Rows
$a > x$	$6 + 4 = 10$	1, 3
$x > a$	$5 + 3 = 8$	2, 4
$a > y$	$6 + 3 = 9$	1, 4
$y > a$	$5 + 4 = 9$	2, 3
$a > z$	$6 + 3 = 9$	1, 4
$z > a$	$5 + 4 = 9$	2, 3
$x > y$	$6 + 3 = 9$	1, 4
$y > x$	$5 + 4 = 9$	2, 3
$x > z$	$6 + 3 = 9$	1, 4
$z > x$	$5 + 4 = 9$	2, 3
$y > z$	$6 + 3 = 9$	1, 4
$z > y$	$5 + 4 = 9$	2, 3

Make another table to award pairwise points.

Pairs	Votes	Pairwise Points
$a: x$	$10 : 8$	$a, 1$
$a: y$	$9 : 9$	$a, \frac{1}{2}; y, \frac{1}{2}$
$a: z$	$9 : 9$	$a, \frac{1}{2}; z, \frac{1}{2}$
$x: y$	$9 : 9$	$x, \frac{1}{2}; y, \frac{1}{2}$
$x: z$	$9 : 9$	$x, \frac{1}{2}; z, \frac{1}{2}$
$y: z$	$9 : 9$	$y, \frac{1}{2}; z, \frac{1}{2}$

Using the pairwise comparison method, here are the total points each has received.

$$a: \quad 1 + \frac{1}{2} + \frac{1}{2} = 2$$
$$x: \quad \frac{1}{2} + \frac{1}{2} = 1$$
$$y: \quad \frac{1}{2} + \frac{1}{2} + \frac{1}{2} = 1\frac{1}{2}$$
$$z: \quad \frac{1}{2} + \frac{1}{2} + \frac{1}{2} = 1\frac{1}{2}$$

Candidate a wins with 2 pairwise points.

(b) A possible new ranking is

Voters	Ranking
6	$a > x > y > z$
5	$z > y > x > a$
4	$z > y > a > x$
3	$a > z > x > y$

Now make a table to compute the votes in comparison.

Compare	Votes	Rows
$a > x$	$6 + 4 + 3 = 13$	1, 3, 4
$x > a$	$5 = 5$	2
$a > y$	$6 + 3 = 9$	1, 4
$y > a$	$5 + 4 = 9$	2, 3
$a > z$	$6 + 3 = 9$	1, 4
$z > a$	$5 + 4 = 9$	2, 3
$x > y$	$6 + 3 = 9$	1, 4
$y > x$	$5 + 4 = 9$	2, 3
$x > z$	$6 = 6$	1
$z > x$	$5 + 4 + 3 = 12$	2, 3, 4
$y > z$	$6 = 6$	1
$z > y$	$5 + 4 + 3 = 12$	2, 3, 4

Make another table to award pairwise points.

Pairs	Votes	Pairwise Points
$a: x$	$13 : 5$	$a, 1$
$a: y$	$9 : 9$	$a, \frac{1}{2}; y, \frac{1}{2}$
$a: z$	$9 : 9$	$a, \frac{1}{2}; z, \frac{1}{2}$
$x: y$	$9 : 9$	$x, \frac{1}{2}; y, \frac{1}{2}$
$x: z$	$6 : 12$	$z, 1$
$y: z$	$6 : 12$	$z, 1$

Using the pairwise comparison method, here are the total points each has received.

$$a: \quad 1 + \frac{1}{2} + \frac{1}{2} = 2$$
$$x: \quad \frac{1}{2} = \frac{1}{2}$$
$$y: \quad \frac{1}{2} + \frac{1}{2} = 1$$
$$z: \quad \frac{1}{2} + 1 + 1 = 2\frac{1}{2}$$

Candidate z wins with $2\frac{1}{2}$ pairwise points.

41. (a) Use the information from the text to make a table to examine the points in the Borda method.

		Points			
Votes	Ranking	3	2	1	0
5	$m > b > s > c$	m	b	s	c
4	$b > s > c > m$	b	s	c	m
3	$s > m > c > b$	s	m	c	b
2	$c > m > s > b$	c	m	s	b

Examine the last three columns, and compute the weighted sum for each candidate.

Candidate	Borda Points
m	$5 \cdot 3 + 4 \cdot 0 + 3 \cdot 2 + 2 \cdot 2 = 25$
b	$5 \cdot 2 + 4 \cdot 3 + 3 \cdot 0 + 2 \cdot 0 = 22$
s	$5 \cdot 1 + 4 \cdot 2 + 3 \cdot 3 + 2 \cdot 1 = 24$
c	$5 \cdot 0 + 4 \cdot 1 + 3 \cdot 1 + 2 \cdot 3 = 13$

Candidate m wins with 25 Borda points.

(b) *Make a table to examine the points in the Borda method with the bottom two rows changed as follows:*

		Points			
Votes	Ranking	3	2	1	0
5	$m > b > s > c$	m	b	s	c
4	$b > s > c > m$	b	s	c	m
3	$m > b > s > c$	m	b	s	c
2	$m > b > c > s$	m	b	c	s

Examine the last three columns and compute the weighted sum for each candidate.

Candidate	Borda Points
m	$5 \cdot 3 + 4 \cdot 0 + 3 \cdot 3 + 2 \cdot 3 = 30$
b	$5 \cdot 2 + 4 \cdot 3 + 3 \cdot 2 + 2 \cdot 2 = 32$
s	$5 \cdot 1 + 4 \cdot 2 + 3 \cdot 1 + 2 \cdot 0 = 16$
c	$5 \cdot 0 + 4 \cdot 1 + 3 \cdot 0 + 2 \cdot 1 = 6$

Candidate b wins with 32 Borda points.

42. *Here is the original voter profile from Exercise 29.*

Votes	Ranking
12	$a > c > b$
10	$b > a > c$
8	$c > b > a$
4	$c > a > b$

In Exercise 35 it is shown that candidate a wins by the Hare method. If the 4 voters in the bottom row switch their ranking to $a > c > b$, the new voter profile will be:

Votes	Ranking
16	$a > c > b$
10	$b > a > c$
8	$c > b > a$

There are 34 voters, so a candidate must receive a majority of the votes or at least 18 votes to win by the Hare method. Since none of the candidates has received a majority, candidate c drops out with the fewest votes. Now the ranking is:

Votes	Ranking
16	$a > b$
10	$b > a$
8	$b > a$

Notice that candidate b now has a majority of votes with $10 + 8 = 18$. The monotonicity criterion has been violated.

43. *Here is one possibility. If candidate c is deleted, the voter profile is:*

Votes	Ranking
15	$a > b > d$
8	$b > a > d$
9	$b > a > d$
6	$d > b > a$

Now candidate b wins with $8 + 9 = 17$ votes. Because candidate a wins in the original profile, the plurality method violates the Independence of Irrelevant Alternatives criterion in the second election.

44. *Using the pairwise comparison method, first make a table to compute the votes in comparison.*

Compare	Votes	Rows
$a > c$	$15 + 6 + 6 = 27$	1, 3, 4
$c > a$	$6 = 6$	2
$a > b$	$15 + 6 + 6 = 27$	1, 2, 4
$b > a$	$6 = 6$	3
$a > d$	$15 + 6 + 6 = 27$	1, 2, 3
$d > a$	$6 = 6$	4
$a > e$	$15 = 15$	1
$e > a$	$6 + 6 + 6 = 18$	2, 3, 4
$b > c$	$6 = 6$	3
$c > b$	$15 + 6 + 6 = 27$	1, 2, 4
$b > d$	$15 + 6 + 6 = 27$	1, 2, 3
$d > b$	$6 = 6$	4
$b > e$	$15 + 6 = 21$	1, 3
$e > b$	$6 + 6 = 12$	2, 4
$c > d$	$15 + 6 = 21$	1, 2
$d > c$	$6 + 6 = 12$	3, 4
$c > e$	$15 = 15$	1
$e > c$	$6 + 6 + 6 = 18$	2, 3, 4
$d > e$	$15 + 6 = 21$	1, 4
$e > d$	$6 + 6 = 12$	2, 3

Make another table to award pairwise points.

Pairs	Votes	Pairwise Points
$a : b$	$27 : 6$	$a, 1$
$a : c$	$27 : 6$	$a, 1$
$a : d$	$27 : 6$	$a, 1$
$a : e$	$15 : 18$	$e, 1$
$b : c$	$6 : 27$	$c, 1$
$b : d$	$27 : 6$	$b, 1$
$b : e$	$21 : 12$	$b, 1$
$c : d$	$21 : 12$	$c, 1$
$c : e$	$15 : 18$	$e, 1$
$d : e$	$21 : 12$	$d, 1$

Using the pairwise comparison method, here are the total points each issue has received.

$$a:\quad 1+1+1=3$$
$$b:\quad 1+1=2$$
$$c:\quad 1+1=2$$
$$d:\quad 1=1$$
$$e:\quad 1+1=2$$

Candidate a wins with 3 pairwise points.

If both b and d are deleted, here is the new ranking:

Votes	Ranking
15	$a > c > e$
6	$e > c > a$
6	$e > a > c$
6	$e > a > c$

Now make a table to compute the votes in comparison.

Compare	Votes	Rows
$a > c$	$15 + 6 + 6 = 27$	1, 3, 4
$c > a$	$6 = 6$	2
$a > e$	$15 = 15$	1
$e > a$	$6 + 6 + 6 = 18$	2, 3, 4
$c > e$	$15 = 15$	1
$e > c$	$6 + 6 + 6 = 18$	2, 3, 4

Make another table to award pairwise points.

Pairs	Votes	Pairwise Points
$a:c$	$27:6$	$a, 1$
$a:e$	$15:18$	$e, 1$
$c:e$	$15:18$	$e, 1$

Using the pairwise comparison method, here are the total points each has received.

$$a:\quad 1=1$$
$$c:\quad 0=0$$
$$e:\quad 1+1=2$$

In this second election candidate e wins with 2 pairwise points, which violates the Independence of Irrelevant Alternatives criterion.

45. Here is one possible profile:

Votes	Ranking	Points 2	1	0
21	$g > j > e$	g	j	e
12	$j > e > g$	j	e	g
8	$j > g > e$	j	g	e

Examine the last three columns, and compute the weighted sum for each candidate.

Candidate	Borda Points
g	$21 \cdot 2 + 12 \cdot 0 + 8 \cdot 1 = 50$
j	$21 \cdot 1 + 12 \cdot 2 + 8 \cdot 2 = 61$
e	$21 \cdot 0 + 12 \cdot 1 + 8 \cdot 0 = 12$

The winner is candidate j with 61 Borda points.

For the second election, candidate e drops out.

Votes	Ranking	Points 1	0
21	$g > j$	g	j
12	$j > g$	j	g
8	$j > g$	j	g

Examine the last two columns, and compute the weighted sum for each candidate.

Candidate	Borda Points
g	$21 \cdot 1 + 12 \cdot 0 + 8 \cdot 0 = 21$
j	$21 \cdot 0 + 12 \cdot 1 + 8 \cdot 1 = 20$

Candidate g wins with 21 Borda points, which violates the Independence of Irrelevant Alternatives criterion.

46. Writing exercise
47. Writing exercise
48. Writing exercise

16.3 EXERCISES

1. (a)

State Park	Acres
a	1429
b	8639
c	7608
d	6660
e	5157
Totals	29,493

The standard divisor, d, is found by dividing the total number of acres by the number of trees.

$$d = \frac{29493}{239} \approx 123.4017$$

(b) Using the Hamilton method to apportion the trees, set up a table as seen below. Remember that the standard quota, Q, is found by dividing the number of acres of land by the standard divisor, d, from part (a).

Park	Acres	mQ	Rounded Q
a	1429	$\frac{1429}{123.4017} \approx 11.580$	11
b	8639	$\frac{8639}{123.4017} \approx 70.007$	70
c	7608	$\frac{7608}{123.4017} \approx 61.652$	61
d	6660	$\frac{6660}{123.4017} \approx 53.970$	53
e	5157	$\frac{5157}{123.4017} \approx 41.790$	41
Totals	29,493		236

There are 3 trees remaining to be apportioned to those parks that have the largest fractional parts of the standard quota: c, d, and e. Here are the final numbers.

Park	Rounded Q	Trees Apportioned
a	11	11
b	70	70
c	61	61 + 1 = 62
d	53	53 + 1 = 54
e	41	41 + 1 = 42
Totals	236	239

(c) Using the Jefferson method to apportion the trees, first calculate the standard divisor, md, by dividing the total number of acres by the number of trees as in part (a).

$$d = \frac{29493}{239} \approx 123.4017$$

Set up a table as seen below by using the table in part (a). Remember that the standard quota, Q, is found by dividing the number of acres of land by the standard divisor, d, from part (a). If d is decreased to 123, the modified quotas add up to approximately 239.8, which is too high. An md of 122 works.

Park	Acres	mQ	Trees Apportioned
a	1429	$\frac{1429}{122} \approx 11.713$	11
b	8639	$\frac{8639}{122} \approx 70.811$	70
c	7608	$\frac{7608}{122} \approx 62.361$	62
d	6660	$\frac{6660}{122} \approx 54.590$	54
e	5157	$\frac{5157}{122} \approx 42.270$	42
Totals	29,493		239

(d)

Standard Quota	Rounded Traditionally
11.580	12
70.007	70
61.652	62
53.970	54
41.790	42
	240

This sum is greater than the number of trees to be apportioned.

(e) The value of md for the Webster method should be greater than the standard divisor, d, because larger divisors create smaller quotas with a smaller total sum.

(f) Using the Webster method, again build a table using a modified divisor of 124.

Park	Acres	mQ	Trees Apportioned
a	1429	$\frac{1429}{124} \approx 11.524$	12
b	8639	$\frac{8639}{124} \approx 69.669$	70
c	7608	$\frac{7608}{124} \approx 61.355$	61
d	6660	$\frac{6660}{124} \approx 53.710$	54
e	5157	$\frac{5157}{124} \approx 41.589$	42
Totals	29,493		239

(g) The apportionment for the Hamilton and Jefferson methods are the same; the apportionment for the Webster method is different.

2. (a) Using the Jefferson method to apportion the computers, first calculate the standard divisor, d, by dividing the total enrollment by the number of computers.

$$d = \frac{2263}{109} \approx 20.76147.$$

Set up a table as seen below. Remember that the standard quota, mQ, is found by dividing the enrollment of each school by a modified divisor, md. If d is decreased to 20.25, the modified quotas add up appropriately.

School	Enrollment	mQ	Computers Apportioned
Applegate	335	$\frac{335}{20.25} \approx 16.543$	16
Bayshore	456	$\frac{456}{20.25} \approx 22.519$	22
Claypool	298	$\frac{298}{20.25} \approx 14.716$	14
Delmar	567	$\frac{567}{20.25} = 28$	28
Edgewater	607	$\frac{607}{20.25} \approx 29.975$	29
Totals	2263		109

(b) The Jefferson and Webster methods have the same apportionment; the Hamilton method differs.

(c) Claypool received an additional computer by the Hamilton method.

3. (a) The total enrollment is

$$56 + 35 + 78 + 100 = 269.$$

The standard divisor is

$$d = \frac{269}{11} = 24.\overline{45}.$$

(b) Using the Hamilton method, we have the following.

Course	Enrollment	Q	Rounded Q
Fiction	56	$\frac{56}{24.45455} \approx 2.290$	2
Poetry	35	$\frac{35}{24.45455} \approx 1.431$	1
Short Story	78	$\frac{78}{24.45455} \approx 3.190$	3
Multicultural	100	$\frac{100}{24.45455} \approx 4.089$	4
Totals	269		10

There is 1 section remaining to be apportioned to the course that has the largest fractional parts of the standard quota, Poetry. Here are the final numbers.

Course	Rounded Q	Sections Apportioned
Fiction	2	2
Poetry	1	1 + 1 = 2
Short Story	3	3
Multicultural	4	4
Totals	10	11

(c) Using the Jefferson method to apportion, the standard divisor must be modified. Set up a table as seen below. An md of 20 works.

Course	Enrollment	mQ	Sections Apportioned
Fiction	56	$\frac{56}{20} = 2.8$	2
Poetry	35	$\frac{35}{20} = 1.75$	1
Short Story	78	$\frac{78}{20} = 3.9$	3
Multicultural	100	$\frac{100}{20} = 5$	5
Totals	269		11

(d)

Standard Quota	Rounded Traditionally
2.290	2
1.431	1
3.190	3
4.089	4
	10

This sum is less the number of sections to be apportioned.

(e) The value of md for the Webster method should be less than the standard divisor, d, because smaller divisors create larger quotas with a larger total sum

(f) Use the Webster method with an md of 23.

Course	Enrollment	Q	mQ
Fiction	56	$\frac{56}{23} \approx 2.435$	2
Poetry	35	$\frac{35}{23} \approx 1.522$	2
Short Story	78	$\frac{78}{23} \approx 3.391$	3
Multicultural	100	$\frac{100}{23} \approx 4.348$	4
Totals	269		11

(g) The Hamilton and Webster apportionments are the same; the Jefferson apportionment differs.

(h) A Poetry student would hope that either the Hamilton or Webster method would be used, because both apportion 2 sections of the class rather than 1. This would create a smaller class size.

(i) A Multicultural student would hope that the Jefferson method be used, because this apportions 5 sections rather than 4.

4. (a) In Example 3, the standard divisor is

$$d = \frac{2013}{30} = 67.1$$

Use the Webster method with an md of 66.

Resort	Number of Rooms	mQ	Sailboats Apportioned
Anna	315	$\frac{315}{66} \approx 4.773$	5
Bob	234	$\frac{234}{66} \approx 3.545$	4
Cathy	420	$\frac{420}{66} \approx 6.364$	6
David	330	$\frac{330}{66} = 5$	5
Ellen	289	$\frac{289}{66} \approx 4.379$	4
Floyd	395	$\frac{395}{66} \approx 5.985$	6
Totals	2013		30

(b) The Hamilton and Webster apportionments are the same; the Jefferson is different.

5. (a) Use the Hamilton method to apportion the 131 seats. The standard divisor is found by dividing the entire population by the number of seats

$$d = \frac{47841}{131} \approx 365.1985$$

State	Population	Q	Seats Apportioned
Abo	5672	$\frac{5672}{365.1985} \approx 15.531$	15
Boa	8008	$\frac{8008}{365.1985} \approx 21.928$	21
Cio	2400	$\frac{2400}{365.1985} \approx 6.572$	6
Dao	6789	$\frac{6789}{365.1985} \approx 18.590$	18
Effo	4972	$\frac{4972}{365.1985} \approx 13.615$	13
Foti	20,000	$\frac{20,000}{365.1985} \approx 54.765$	54
Totals	47,841		127

There are $131 - 127 = 4$ seats remaining to be apportioned to those states that have the largest fractional parts of the standard quota: Boa, Foti, Effo, and Dao. Here are the final numbers.

State	Rounded Q	Seats Apportioned
Abo	15	15
Boa	21	$21 + 1 = 22$
Cio	6	6
Dao	18	$18 + 1 = 19$
Effo	13	$13 + 1 = 14$
Foti	54	$54 + 1 = 55$
Totals	127	131

(b) Use the Jefferson method with an md of 356 (found by trial and error).

State	Population	Modified Quota mQ	Rounded-down mQ Seats Apportioned
Abo	5672	$\frac{5672}{357} \approx 15.933$	15
Boa	8008	$\frac{8008}{357} \approx 22.494$	22
Cio	2400	$\frac{2400}{357} \approx 6.742$	6
Dao	6789	$\frac{6789}{357} \approx 19.070$	19
Effo	4972	$\frac{4972}{357} \approx 13.966$	13
Foti	20,000	$\frac{20,000}{357} \approx 56.180$	56
Totals	47,841		131

(c) The Hamilton, Jefferson, and Webster apportionments are all different.

6. Use the Webster method with an md of 366.9724.

State	Population	mQ	Rounded Q
Abo	5672	$\frac{5672}{366.9724} \approx 15.456$	15
Boa	8008	$\frac{8008}{366.9724} \approx 21.822$	22
Cio	2400	$\frac{2400}{366.9724} \approx 6.540$	7
Dao	6789	$\frac{6789}{366.9724} \approx 18.500$	19
Effo	4972	$\frac{4972}{366.9724} \approx 13.549$	14
Foti	20,000	$\frac{20,000}{366.9724} \approx 54.500$	55
Totals	47,841		**132**

Notice that the total of the last column is 132 seats.

7. Use the Webster method with an md of 366.9730.

State	Population	mQ	Rounded Q
Abo	5672	$\frac{5672}{366.9730} \approx 15.456178$	15
Boa	8008	$\frac{8008}{366.9730} \approx 21.821769$	22
Cio	2400	$\frac{2400}{366.9730} \approx 6.539991$	7
Dao	6789	$\frac{6789}{366.9730} \approx 18.4999999$	18
Effo	4972	$\frac{4972}{366.9730} \approx 13.548681$	14
Foti	20,000	$\frac{20,000}{366.9730} \approx 54.499922$	54
Totals	47,841		**130**

Notice that the total of the last column is 130 seats.

8. Use the Webster method for Example 2 with an md of 34,400.

State	Population	mQ	Rounded Q
VA	630560	$\frac{630560}{34400} \approx 18.330$	18
MA	475327	$\frac{475327}{34400} \approx 13.818$	14
PA	432879	$\frac{432879}{34400} \approx 12.584$	13
NC	353523	$\frac{353523}{34400} \approx 10.277$	10
NY	331589	$\frac{331589}{34400} \approx 9.639$	10
MD	278514	$\frac{278514}{34400} \approx 8.096$	8
CT	236841	$\frac{236841}{34400} \approx 6.885$	7
SC	206236	$\frac{206236}{34400} \approx 5.995$	6
NJ	179570	$\frac{179570}{34400} \approx 5.220$	5
NH	141822	$\frac{141822}{34400} \approx 4.123$	4
VT	85533	$\frac{85533}{34400} \approx 2.486$	2
GA	70835	$\frac{70835}{34400} \approx 2.059$	2
KY	68705	$\frac{68705}{34400} \approx 1.997$	2
RI	68446	$\frac{68446}{34400} \approx 1.990$	2
DE	55540	$\frac{55540}{34400} \approx 1.615$	2
Totals	3615920		105

9. (a) The total number of beds is
$$137 + 237 + 337 + 455 + 555 = 1721.$$
The standard divisor for the apportionment of nurses is
$$d = \frac{1721}{40} = 43.025.$$

(b) Use the Hamilton method to apportion the nurses.

Hosp.	No. of Beds	Q	Rounded Q
A	137	$\frac{137}{43.025} \approx 3.184$	3
B	237	$\frac{237}{43.025} \approx 5.508$	5
C	337	$\frac{337}{43.025} \approx 7.833$	7
D	455	$\frac{455}{43.025} \approx 10.575$	10
E	555	$\frac{555}{43.025} \approx 12.899$	12
Totals	1721		37

There are 3 more nurses to apportion to those hospitals with the greatest fractional part remaining in the Q column. The final apportionment is

Hospital	Q	mQ
A	$\frac{137}{43.025} \approx 3.184$	3
B	$\frac{237}{43.025} \approx 5.508$	5
C	$\frac{337}{43.025} \approx 7.833$	7 + 1 = 8
D	$\frac{455}{43.025} \approx 10.575$	10 + 1 = 11
E	$\frac{555}{43.025} \approx 12.899$	12 + 1 = 13
Totals		40

(c) Use the Jefferson method with an md of 40.

Hospital	No. of Beds	mQ	Rounded mQ
A	137	$\frac{137}{40} = 3.425$	3
B	237	$\frac{237}{40} = 5.925$	5
C	337	$\frac{337}{40} = 8.425$	8
D	455	$\frac{455}{40} = 11.375$	11
E	555	$\frac{555}{40} = 13.875$	13
Totals	1721		40

(d)

Standard Quota	Rounded Traditionally
3.184	3
5.508	6
7.833	8
10.575	11
12.899	13
Total	41

The traditionally rounded values add to 41, which is greater than the number of nurses to be apportioned.

(e) The value of md for the Webster method should be greater than the standard divisor, $d = 43.025$, because larger divisors create smaller modified quotas with a smaller total sum.

(f) Use the Webster method with an md of 43.1.

Hospital	No. of Beds	mQ	Rounded mQ
A	137	$\frac{137}{43.1} \approx 3.179$	3
B	237	$\frac{237}{43.1} \approx 5.499$	5
C	337	$\frac{337}{43.1} \approx 7.819$	8
D	455	$\frac{455}{43.1} \approx 10.557$	11
E	555	$\frac{555}{43.1} \approx 12.877$	13
Totals	1721		40

(g) All three apportionments are the same.

10. (a) From Exercise 1, the total number of acres in the parks is 29,493. The standard divisor is

$$d = \frac{29493}{439} \approx 67.18223.$$

(b) Use the Hamilton method to apportion the trees.

Park	Acres	Q	Rounded Q
a	1429	$\frac{1429}{67.18223} \approx 21.271$	21
b	8639	$\frac{8639}{67.18223} \approx 128.591$	128
c	7608	$\frac{7608}{67.18223} \approx 113.244$	113
d	6660	$\frac{6660}{67.18223} \approx 99.133$	99
e	5157	$\frac{5157}{67.18223} \approx 76.761$	76
Totals	29493		437

There are 2 more trees to apportion to those parks with the greatest fractional part remaining in the Q column. The final apportionment is as follows.

Park	Q	Rounded Q
a	21.271	21
b	128.591	128 + 1 = 129
c	113.244	113
d	99.133	99
e	76.761	76 + 1 = 77
Totals		439

(c) Use the Jefferson method with an md of 66.8.

Park	Acres	mQ	Trees Apportioned
a	1429	$\frac{1429}{66.8} \approx 21.392$	21
b	8639	$\frac{8639}{66.8} \approx 129.326$	129
c	7608	$\frac{7608}{66.8} \approx 113.892$	113
d	6660	$\frac{6660}{66.8} \approx 99.701$	99
e	5157	$\frac{5157}{66.8} \approx 77.201$	77
Totals	29493		439

(d)

Standard Quota	Rounded Traditionally
21.271	21
128.591	129
113.244	113
99.133	99
76.761	77
Total	439

The sum is equal to the number of trees to be apportioned.

(e) The value of md for the Webster method should equal the value of $d = 67.18223$, because the traditionally rounded values of Q already sum to exactly the number of trees to be apportioned.

(f) Using the Webster method with an md of 67.18223, the table will be exactly like the first table in part (b); however, the values of Q should be rounded traditionally as in part (d).

(g) All three apportionments are the same. The Governor was politically wise in making sure that all three apportionments under consideration would produce the same result.

11. Here is one possible population profile.

State	a	b	c	d	e	Total
Population	50	230	280	320	120	1000

<u>Hamilton</u> method

The standard divisor is

$$d = \frac{1000}{100} = 10.$$

State	Pop.	Q
a	50	$\frac{50}{10} = 5$
b	230	$\frac{230}{10} = 23$
c	280	$\frac{280}{10} = 28$
d	320	$\frac{320}{10} = 32$
e	120	$\frac{120}{10} = 12$
Totals	1000	100

It is unnecessary to find a modified divisor for the Jefferson and Webster methods, because the value of d divides into each population evenly. No rounding is necessary. That is, the modified divisor is 10 for both methods.

12. Here is one possible enrollment profile.

Course	Bio	Chem	Physics	Math	Total
Enrollment	35	40	28	95	198

<u>Hamilton</u> method

The standard divisor is

$$d = \frac{198}{9} = 22.$$

Course	Enr.	Q	Rounded Q
Bio	35	$\frac{35}{22} \approx 1.591$	1
Chem	40	$\frac{40}{22} \approx 1.818$	1
Physics	28	$\frac{28}{22} \approx 1.273$	1
Math	95	$\frac{95}{22} \approx 4.318$	4
Totals	198		7

The remaining two teaching assistants will be assigned to Chemistry and Biology, the courses with the largest fractional portion of their quotient. The final apportionment is as follows.

Course	Enr.	Q	Number of TA's
Bio	35	1.591	1 + 1 = 2
Chem	40	1.818	1 + 1 = 2
Physics	28	1.273	1
Math	95	4.318	4
Totals	198		9

<u>Webster</u> method

Using the standard divisor as the modified divisor, round each value of Q according to normal rules of rounding.

Course	Enr.	Q	Rounded Q
Bio	35	$\frac{35}{22} \approx 1.591$	2
Chem	40	$\frac{40}{22} \approx 1.818$	2
Physics	28	$\frac{28}{22} \approx 1.273$	1
Math	95	$\frac{95}{22} \approx 4.318$	4
Totals	198		9

<u>Jefferson</u> method

Using a modified divisor of 18, here is the apportionment.

Course	Enr.	mQ	T.A. Apportioned
Bio	35	$\frac{35}{18} \approx 1.944$	1
Chem	40	$\frac{40}{18} \approx 2.222$	2
Physics	28	$\frac{28}{18} \approx 1.556$	1
Math	95	$\frac{95}{18} \approx 5.278$	5
Totals	198		9

13. Here is one possible ridership profile.

Bus Route	a	b	c	d	e	Total
No. of Riders	131	140	303	178	197	949

<u>Hamilton</u> method

The standard divisor is

$$d = \frac{949}{16} = 59.3125.$$

Bus Route	Riders	Q	Rounded Q
a	131	$\frac{131}{59.3125} \approx 2.209$	2
b	140	$\frac{140}{59.3125} \approx 2.360$	2
c	303	$\frac{303}{59.3125} \approx 5.109$	5
d	178	$\frac{178}{59.3125} \approx 3.001$	3
e	197	$\frac{197}{59.3125} \approx 3.321$	3
Totals	949		15

The remaining bus will be assigned to route b because it has the largest fractional portion in its quotient. The final apportionment is as follows.

Bus Route	Riders	Q	Rounded Q
a	131	2.209	2
b	140	2.360	2 + 1 = 3
c	303	5.109	5
d	178	3.001	3
e	197	3.321	3
Totals	949		16

<u>Jefferson</u> method

Using a modified divisor of 50, here is the apportionment.

Bus Route	Riders	mQ	Bus Apportioned
a	131	$\frac{131}{50} = 2.62$	2
b	140	$\frac{140}{50} = 2.8$	2
c	303	$\frac{303}{50} = 6.06$	6
d	178	$\frac{178}{50} = 3.56$	3
e	197	$\frac{197}{50} = 3.94$	3
Totals	949		16

<u>Webster</u> method

Using a modified divisor of 56.2, round each value of Q according to normal rules of rounding.

Bus Route	Riders	mQ	Bus Apportioned
a	131	$\frac{131}{56.2} \approx 2.331$	2
b	140	$\frac{140}{56.2} \approx 2.491$	2
c	303	$\frac{303}{56.2} \approx 5.391$	5
d	178	$\frac{178}{56.2} \approx 3.167$	3
e	197	$\frac{197}{56.2} \approx 3.505$	4
Totals	949		16

14. The modified divisor for the Adams method is always found by slowly increasing the value of the standard divisor, because the modified quotas are always rounded up to the nearest integer. Rounding up increases the sum. A larger divisor produces modified quotas that are small enough so that they sum to the number of seats to be apportioned even when rounded up.

15. (a) Using the Adams method with an md of 29

Course	Enrollment	Q	mQ
Fiction	56	$\frac{56}{29} \approx 1.931$	2
Poetry	35	$\frac{35}{29} \approx 1.207$	2
Short Story	78	$\frac{78}{29} \approx 2.690$	3
Multicultural	100	$\frac{100}{29} \approx 3.448$	4
Totals	269		11

(b) The Adams apportionment is the same as the Hamilton and Webster apportionments. It is different from the Jefferson apportionment.

16. (a) Use the Adams method with an md of 46.

Hosp.	No. of Beds	mQ	Nurses Apportioned
A	137	$\frac{137}{46} \approx 2.978$	3
B	237	$\frac{237}{46} \approx 5.152$	6
C	337	$\frac{337}{46} \approx 7.236$	8
D	455	$\frac{455}{46} \approx 9.891$	10
E	555	$\frac{555}{46} \approx 12.065$	13
Totals	1721		40

(b) The Hamilton, Jefferson, and Webster apportionments all agree; the Adams apportionment is different.

17. (a) Use the Adams method with an md of 377.3.

State	Population	mQ	Seats Apportioned
Abo	5672	$\frac{5672}{377.3} \approx 15.033$	16
Boa	8008	$\frac{8008}{377.3} \approx 21.224$	22
Cio	2400	$\frac{2400}{377.3} \approx 6.361$	7
Dao	6789	$\frac{6789}{377.3} \approx 17.994$	18
Effo	4972	$\frac{4972}{377.3} \approx 13.178$	14
Foti	20,000	$\frac{20,000}{377.3} \approx 53.008$	54
Totals	47,841		131

(b) Each method produces a different apportionment.

18. The cutoff point for rounding the modified quota of 5.470 up to 6 is calculated by finding the geometric mean of 5 (the integer part of 5.470) and 6.

$$\sqrt{5 \cdot 6} = \sqrt{30} \approx 5.477.$$

19. The cutoff point for rounding the modified quota of 56.498 up to 57 is calculated by finding the geometric mean of 56 (the integer part of 56.498) and 57.

$$\sqrt{56 \cdot 57} = \sqrt{3192} \approx 56.498.$$

20. The cutoff point for rounding the modified quota of 11.71 up to 12 is calculated by finding the geometric mean of 11 (the integer part of 11.71) and 12.

$$\sqrt{11 \cdot 12} = \sqrt{132} \approx 11.489.$$

21. The cutoff point for rounding the modified quota of 32.497 up to 33 is calculated by finding the geometric mean of 32 (the integer part of 32.497) and 33.

$$\sqrt{32 \cdot 33} = \sqrt{1056} \approx 32.496.$$

22. If the sum of the traditionally rounded Q values is less than the number of objects being apportioned, the modified divisor is found by slowly decreasing the value of d, because a smaller divisor produces larger modified quotas with a larger sum.

23. If the sum of the traditionally rounded Q values is greater than the number of objects being apportioned, then the modified divisor is found by slowly increasing the value of d. A larger divisor produces smaller modified quotas with a smaller sum.

24. If the sum is equal to the number of objects to be apportioned, then no modification is needed; the standard divisor, d, is used.

25. (a) Use the Hill-Huntington method with an md of 24.

Course	Enrollment	mQ	Seats Apportioned
Fiction	56	$\frac{56}{24} = 2.\overline{3}$	2
Poetry	35	$\frac{35}{24} \approx 1.458$	2
Short Story	78	$\frac{78}{24} = 3.25$	3
Multicultural	100	$\frac{100}{24} = 4.1\overline{6}$	4
Totals	269		11

    (b) The Hill-Huntington apportionment is the same as the Hamilton, Webster, and Adams apportionments. It is different from the Jefferson apportionment.

26. (a) Use the Hill-Huntington method with an md of 43.3.

Hosp.	No. of Beds	mQ	Nurses Apportioned
A	137	$\frac{137}{43.3} \approx 3.164$	3
B	237	$\frac{237}{43.3} \approx 5.473$	5
C	337	$\frac{337}{43.3} \approx 7.783$	8
D	455	$\frac{455}{43.3} \approx 10.508$	11
E	555	$\frac{555}{43.3} \approx 12.818$	13
Totals	1721		40

    (b) The Hill-Huntington apportionment is the same as the Hamilton, Webster, and Jefferson apportionments. It is different from the Adams apportionment.

27. (a) Using the Hill-Huntington method with an md of 367, each value of mQ is found by dividing the population figure by 367. See the explanation in the text for computation of the geometric mean.

State	Population	mQ	Geo. Mean	Seats Apportioned
Abo	5672	15.4550	15.4919	15
Boa	8008	21.8202	21.4942	22
Cio	2400	6.5395	6.4807	7
Dao	6789	18.4986	18.4932	19
Effo	4972	13.5477	13.4907	14
Foti	20,000	54.4959	54.4977	54
Totals	47,841			131

    (b) The Hill-Huntington apportionment is the same as the Webster apportionment. The other three are all different.

28. Writing exercise

29. Writing exercise

30. Writing exercise

31. Writing exercise

16.4 EXERCISES

1. The sum of the populations is
   $$17179 + 7500 + 49400 + 5824 = 79{,}903.$$

   The standard divisor is
   $$d = \frac{79903}{132} = 605.326.$$

   Using the standard quota, we have the following.

State	Population	Q	Rounded Q
a	17179	$\frac{17179}{605.326} \approx 28.380$	28
b	7500	$\frac{7500}{605.326} \approx 12.390$	12
c	49400	$\frac{49400}{605.326} \approx 81.609$	**81**
d	5824	$\frac{5824}{605.326} \approx 9.621$	9
Totals	79,903		130

   Use the Jefferson method with $md = 595$.

State	Population	mQ	Rounded mQ
a	17179	$\frac{17179}{595} \approx 28.872$	28
b	7500	$\frac{7500}{595} \approx 12.605$	12
c	49400	$\frac{49400}{595} \approx 83.025$	**83**
d	5824	$\frac{5824}{595} \approx 9.788$	9
Totals	79,903		132

The Jefferson method violates the Quota Rule because state c receives receives two more seats than its apportionment from the standard quota.

2. The sum of the populations is
$$67000 + 35000 + 15000 + 9900 = 126900.$$

The standard divisor is
$$d = \frac{126900}{200} = 634.5.$$

Use the standard quota.

State	Population	Q	Rounded Q
a	67000	$\frac{67000}{634.5} \approx 105.595$	**105**
b	35000	$\frac{35000}{634.5} \approx 55.162$	55
c	15000	$\frac{15000}{634.5} \approx 23.641$	23
d	9900	$\frac{9900}{634.5} \approx 15.603$	15
Totals	126900		198

Use the Jefferson method with $md = 626$.

State	Population	mQ	Rounded mQ
a	67000	$\frac{67000}{626} \approx 107.029$	**107**
b	35000	$\frac{35000}{626} \approx 55.911$	55
c	15000	$\frac{15000}{626} \approx 23.962$	23
d	9900	$\frac{9900}{626} \approx 15.815$	15
Totals	79,903		200

The Jefferson method violates the Quota Rule because state a receives two more seats than its apportionment from the standard quota.

3. From Exercise 5 in Section 16.3, the standard divisor is found by dividing the entire population by the number of seats:
$$d = \frac{47841}{131} \approx 365.1985.$$

State	Population	Q	Rounded Q
Abo	5672	$\frac{5672}{365.1985} \approx 15.531$	15
Boa	8008	$\frac{8008}{365.1985} \approx 21.928$	21
Cio	2400	$\frac{2400}{365.1985} \approx 6.572$	6
Dao	6789	$\frac{6789}{365.1985} \approx 18.590$	18
Effo	4972	$\frac{4972}{365.1985} \approx 13.615$	13
Foti	20000	$\frac{20,000}{365.1985} \approx 54.765$	**54**
Totals	47841		127

Use the Jefferson method with $md = 356$.

State	Population	mQ	Rounded mQ
Abo	5672	$\frac{5672}{356} \approx 15.933$	15
Boa	8008	$\frac{8008}{356} \approx 22.494$	22
Cio	2400	$\frac{2400}{356} \approx 6.742$	6
Dao	6789	$\frac{6789}{356} \approx 19.070$	19
Effo	4972	$\frac{4972}{356} \approx 13.966$	13
Foti	20,000	$\frac{20,000}{356} \approx 56.180$	**56**
Totals	47,841		131

The Jefferson method violates the Quota Rule because Foti receives two more seats than its apportionment from the standard quota.

4. The sum of the populations is
$$4589 + 1515 + 2013 + 1118 + 1111 = 10,346.$$

The standard divisor is
$$d = \frac{10346}{200} = 51.73.$$

Use the standard quota

State	Population	Q	Rounded Q
a	4589	$\frac{4589}{51.73} \approx 88.711$	**88**
b	1515	$\frac{1515}{51.73} \approx 29.289$	29
c	2013	$\frac{2013}{51.73} \approx 38.914$	38
d	1118	$\frac{1118}{51.73} \approx 21.612$	21
e	1111	$\frac{1111}{51.73} \approx 21.477$	21
Totals	10346		197

Use the Jefferson method with $md = 50.9$.

State	Population	mQ	Rounded mQ
a	4589	$\frac{4589}{50.9} \approx 90.157$	**90**
b	1515	$\frac{1515}{50.9} \approx 29.764$	29
c	2013	$\frac{2013}{50.9} \approx 39.548$	39
d	1118	$\frac{1118}{50.9} \approx 21.965$	21
e	1111	$\frac{1111}{50.9} \approx 21.827$	21
Totals	10346		200

The Jefferson method violates the Quota Rule because state a receives two more seats than its apportionment from the standard quota.

5. The sum of the populations is

$$2567 + 1500 + 8045 + 950 + 1099 = 14161.$$

The standard divisor is

$$d = \frac{14161}{290} \approx 48.8310.$$

Use the standard quota

State	Population	mQ	Rounded mQ
a	2567	$\frac{2567}{48.8310} \approx 52.569$	52
b	1500	$\frac{1500}{48.8310} \approx 30.718$	30
c	8045	$\frac{8045}{48.8310} \approx 164.752$	**164**
d	950	$\frac{950}{48.8310} \approx 19.455$	19
e	1099	$\frac{1099}{48.8310} \approx 22.506$	22
Totals	14,161		287

Use the Jefferson method with $md = 48.4$.

State	Population	mQ	Rounded mQ
a	2567	$\frac{2567}{48.4} \approx 53.037$	53
b	1500	$\frac{1500}{48.4} \approx 30.992$	30
c	8045	$\frac{8045}{48.4} \approx 166.219$	**166**
d	950	$\frac{950}{48.4} \approx 19.628$	19
e	1099	$\frac{1099}{48.4} \approx 22.707$	22
Totals	14,161		290

The Jefferson method violates the Quota Rule because state c receives two more seats than its apportionment from the standard quota.

6. The sum of the populations is

$$1720 + 3363 + 6960 + 24223 + 8800 = 45,066.$$

The standard divisor is

$$d = \frac{45066}{150} = 300.44.$$

Use the standard quota

State	Population	Q	Rounded Q
a	1720	$\frac{1720}{300.44} \approx 5.725$	5
b	3363	$\frac{3363}{300.44} \approx 11.194$	11
c	6960	$\frac{6960}{300.44} \approx 23.166$	23
d	24223	$\frac{24223}{300.44} \approx 80.625$	**80**
e	8800	$\frac{8800}{300.44} \approx 29.290$	29
Totals	45066		148

Use the Jefferson method with $md = 295$.

State	Population	mQ	Rounded mQ
a	1720	$\frac{1720}{295} \approx 5.831$	5
b	3363	$\frac{3363}{295} = 11.4$	11
c	6960	$\frac{6960}{295} \approx 23.593$	23
d	24223	$\frac{24223}{295} \approx 82.112$	**82**
e	8800	$\frac{8800}{295} \approx 29.831$	29
Totals	45066		150

The Jefferson method violates the Quota Rule because state d receives two more seats than its apportionment from the standard quota.

7. First calculate the apportionment with $n = 204$. The total population is

$$3462 + 7470 + 4265 + 5300 = 20,497.$$

The standard divisor for 204 seats is

$$d = \frac{20497}{204} \approx 100.4755.$$

Use the Hamilton method.

State	Population	Q	Rounded Q
a	3462	$\frac{3462}{100.4755} \approx 34.456$	34
b	7470	$\frac{7470}{100.4755} \approx 74.346$	74
c	4265	$\frac{4265}{100.4755} \approx 42.448$	42
d	5300	$\frac{5300}{100.4755} \approx 52.749$	52
Totals	20497		202

The two remaining seats will be apportioned to State d and State a, because the fractional parts of their quotas are the largest. The final apportionment is as follows.

State	Population	Q	Number of seats
a	3462	34.456	$34 + 1 = \mathbf{35}$
b	7470	74.346	74
c	4265	42.448	42
d	5300	52.749	$52 + 1 = 53$
Totals	20497		204

Now increase the number of seats to 205. The standard divisor is

$$d = \frac{20497}{205} \approx 99.9854.$$

Use the Hamilton method with the new d.

16.4 THE IMPOSSIBILITIES OF APPORTIONMENT

State	Population	Q	Rounded Q
a	3462	$\frac{3462}{99.98537} \approx 34.625$	**34**
b	7470	$\frac{7470}{99.98537} \approx 74.711$	74
c	4265	$\frac{4265}{99.98537} \approx 42.656$	42
d	5300	$\frac{5300}{99.98537} \approx 53.008$	53
Totals	20497		203

The two remaining seats will be apportioned to State b and State c, because the fractional parts of their quotas are the largest. The final apportionment is as follows.

State	Q	Number of seats
a	34.623	**34**
b	74.711	74 + 1 = 75
c	42.656	42 + 1 = 43
d	53.008	53
Totals		205

State a is a victim of the Alabama Paradox, because it has lost a seat despite the fact that the overall number of seats has increased.

8. First calculate the apportionment with $n = 71$. The total population is

$$1050 + 2040 + 3060 + 4050 = 10,200.$$

The standard divisor for 71 seats is

$$d = \frac{10200}{71} \approx 143.66197.$$

Use the Hamilton method.

State	Population	Q	Rounded Q
a	1050	$\frac{1050}{143.66197} \approx 7.309$	7
b	2040	$\frac{2040}{143.66197} \approx 14.200$	14
c	3060	$\frac{3060}{143.66197} \approx 21.300$	21
d	4050	$\frac{4050}{143.66197} \approx 28.191$	28
Totals	10200		70

The remaining seat will be apportioned to State a, because the fractional parts of its quotas is the largest. The final apportionment is as follows.

State	Population	Q	Number of seats
a	1050	7.309	7 + 1 = **8**
b	2040	14.200	14
c	3060	21.300	21
d	4050	28.191	28
Totals	10200		71

Now increase the number of seats to 72. The standard divisor is

$$d = \frac{10200}{72} \approx 141.66667.$$

Use the Hamilton method with the new d.

State	Population	Q	Rounded Q
a	1050	$\frac{1050}{141.66667} \approx 7.412$	**7**
b	2040	$\frac{2040}{141.66667} \approx 14.400$	14
c	3060	$\frac{3060}{141.66667} \approx 21.600$	21
d	4050	$\frac{4050}{141.66667} \approx 28.588$	28
Totals	10200		70

The two remaining seats will be apportioned to State c and State d, because the fractional parts of their quotas are the largest. The final apportionment is as follows.

State	Q	Number of seats
a	7.412	**7**
b	14.400	14
c	21.600	21 + 1 = 22
d	28.588	28 + 1 = 29
Totals		72

State a is a victim of the Alabama Paradox, because it has lost a seat despite the fact that the overall number of seats has increased.

9. First calculate the apportionment with $n = 126$. The total population is

$$263 + 808 + 931 + 781 + 676 = 3459.$$

The standard divisor for 126 seats is

$$d = \frac{3459}{126} \approx 27.4524.$$

Use the Hamilton method.

State	Population	Q	Rounded Q
a	263	$\frac{263}{27.4524} \approx 9.580$	9
b	808	$\frac{808}{27.4524} \approx 29.433$	29
c	931	$\frac{931}{27.4524} \approx 33.913$	33
d	781	$\frac{781}{27.4524} \approx 28.449$	28
e	676	$\frac{676}{27.4524} \approx 24.624$	24
Totals	3459		123

The three remaining seats will be apportioned to States c, e, and a, because the fractional parts of their quotas are the largest. The final apportionment is

State	Q	Number of seats
a	9.580	9 + 1 = **10**
b	29.433	29
c	33.913	33 + 1 = 34
d	28.449	28
e	24.624	24 + 1 = 25
Totals		126

Now increase the number of seats to 127. The standard divisor is

$$d = \frac{3459}{127} \approx 27.2362.$$

Use the Hamilton method with the new d.

State	Population	Q	Rounded Q
a	263	$\frac{263}{27.2362} \approx 9.656$	9
b	808	$\frac{808}{27.2362} \approx 29.666$	29
c	931	$\frac{931}{27.2362} \approx 34.182$	34
d	781	$\frac{781}{27.2362} \approx 28.675$	28
e	676	$\frac{676}{27.2362} \approx 24.820$	24
Totals	3459		124

The three remaining seats will be apportioned to States e, d and b, because the fractional parts of their quotas are the largest. The final apportionment is

State	Q	Number of seats
a	9.656	9
b	29.666	29 + 1 = 30
c	34.182	34
d	28.675	28 + 1 = 29
e	24.820	24 + 1 = 25
Totals		127

State a is a victim of the Alabama Paradox, because it has lost a seat despite the fact that the overall number of seats has increased.

10. First calculate the apportionment with $n = 45$. The total population is

    $$309 + 289 + 333 + 615 + 465 = 2011.$$

    The standard divisor for 45 seats is

    $$d = \frac{2011}{45} \approx 44.68889.$$

    Using the Hamilton method

State	Population	Q	Rounded Q
a	309	$\frac{309}{44.68889} \approx 6.914$	6
b	289	$\frac{289}{44.68889} \approx 6.467$	6
c	333	$\frac{333}{44.68889} \approx 7.452$	7
d	615	$\frac{615}{44.68889} \approx 13.762$	13
e	465	$\frac{465}{44.68889} \approx 10.405$	10
Totals	2011		42

The three remaining seats will be apportioned to States a, d, and b, because the fractional parts of their quotas are the largest. The final apportionment is as follows.

State	Q	Number of seats
a	6.914	6 + 1 = 7
b	6.467	6 + 1 = **7**
c	7.452	7
d	13.762	13 + 1 = 14
e	10.405	10
Totals		45

Now increase the number of seats to 46. The standard divisor is

$$d = \frac{2011}{46} \approx 43.71739.$$

Use the Hamilton method with the new d.

State	Population	Q	Rounded Q
a	309	$\frac{309}{43.71739} \approx 7.068$	7
b	289	$\frac{289}{43.71739} \approx 6.611$	**6**
c	333	$\frac{333}{43.71739} \approx 7.617$	7
d	615	$\frac{615}{43.71739} \approx 14.068$	14
e	465	$\frac{465}{43.71739} \approx 10.636$	10
Totals	3459		44

The two remaining seats will be apportioned to States e and c, because the fractional parts of their quotas are the largest. The final apportionment is as follows.

State	Q	Number of seats
a	7.068	7
b	6.611	**6**
c	7.617	7 + 1 = 8
d	14.068	14
e	10.636	10 + 1 = 11
Totals		46

State b is a victim of the Alabama Paradox, because it has lost a seat despite the fact that the overall number of seats has increased.

11. First calculate the apportionment with $n = 149$. The total population is

$$5552 + 8260 + 5968 + 6256 + 5150 = 31186.$$

The standard divisor for 149 seats is

$$d = \frac{31186}{149} \approx 209.302.$$

Use the Hamilton method.

State	Population	Q	Rounded Q
a	5552	$\frac{5552}{209.302} \approx 26.526$	26
b	8260	$\frac{8260}{209.302} \approx 39.465$	39
c	5968	$\frac{5968}{209.302} \approx 28.514$	28
d	6256	$\frac{6256}{209.302} \approx 29.890$	29
e	5150	$\frac{5150}{209.302} \approx 24.606$	24
Totals	31,186		146

The three remaining seats will be apportioned to States d, e, and a, because the fractional parts of their quotas are the largest. The final apportionment is as follows.

State	Q	Number of seats
a	26.526	$26 + 1 = 27$
b	39.465	39
c	28.514	28
d	29.890	$29 + 1 = 30$
e	24.606	$24 + 1 = 25$
Totals		149

Now increase the number of seats to 150. The standard divisor is

$$d = \frac{31186}{150} \approx 207.9067.$$

Use the Hamilton method with the new d.

State	Population	Q	Rounded Q
a	5552	$\frac{5552}{207.9067} \approx 26.704$	26
b	8260	$\frac{8260}{207.9067} \approx 39.729$	39
c	5968	$\frac{5968}{207.9067} \approx 28.705$	28
d	6256	$\frac{6256}{207.9067} \approx 30.090$	30
e	5150	$\frac{5150}{207.9067} \approx 24.771$	24
Totals	31186		147

The three remaining seats will be apportioned to States e, b, and c, because the fractional parts of their quotas are the largest. The final apportionment is

State	Q	Number of seats
a	26.704	**26**
b	39.729	$39 + 1 = 40$
c	28.705	$28 + 1 = 29$
d	30.090	30
e	24.771	$24 + 1 = 25$
Totals		150

State a is a victim of the Alabama Paradox, because it has lost a seat despite the fact that the overall number of seats has increased.

12. First calculate the apportionment with $n = 160$. The total population is

$$5613 + 6958 + 7434 + 5085 = 25,090.$$

The standard divisor for 160 seats is

$$d = \frac{25090}{160} = 156.8125.$$

Use the Hamilton method.

State	Population	Q	Rounded Q
a	5613	$\frac{5613}{156.8125} \approx 35.794$	35
b	6958	$\frac{6958}{156.8125} \approx 44.371$	44
c	7434	$\frac{7434}{156.8125} \approx 47.407$	47
d	5085	$\frac{5085}{156.8125} \approx 32.427$	32
Totals	25090		158

The two remaining seats will be apportioned to States a and d, because the fractional parts of their quotas are the largest. The final apportionment is as follows.

State	Q	Number of seats
a	35.794	$35 + 1 = 36$
b	44.371	44
c	47.407	47
d	32.427	$32 + 1 = \mathbf{33}$
Totals		160

Now increase the number of seats to 161. The standard divisor is

$$d = \frac{25090}{161} \approx 155.8385.$$

Use the Hamilton method with the new d.

State	Population	Q	Rounded Q
a	5613	$\frac{5613}{155.8385} \approx 36.018$	36
b	6958	$\frac{6958}{155.8385} \approx 44.649$	44
c	7434	$\frac{7434}{155.8385} \approx 47.703$	47
d	5085	$\frac{5085}{155.8385} \approx 32.630$	32
Totals	25090		159

The two remaining seats will be apportioned to States c and b, because the fractional parts of their quotas are the largest. The final apportionment is as follows.

State	Q	Number of seats
a	36.018	36
b	44.649	44 + 1 = 45
c	47.703	47 + 1 = 48
d	32.630	**32**
Totals		161

State d is a victim of the Alabama Paradox, because it has lost a seat despite the fact that the overall number of seats has increased.

13. First calculate the apportionment for the initial populations. The total population is

$$55 + 125 + 190 = 370.$$

The standard divisor for 11 seats is

$$d = \frac{370}{11} = 33.6364.$$

Use the Hamilton method.

State	Population	Q	Rounded Q
a	55	$\frac{55}{33.6364} \approx 1.635$	1
b	125	$\frac{125}{33.6364} \approx 3.716$	3
c	190	$\frac{190}{33.6364} \approx 5.649$	5
Totals	370		9

The two remaining seats will be apportioned to States b and c, because the fractional parts of their quotas are the largest. The final apportionment is as follows.

State	Q	Number of seats
a	1.635	1
b	3.716	3 + 1 = 4
c	5.649	5 + 1 = 6
Totals		11

Now apportion the seats for the revised populations.

$$61 + 148 + 215 = 424$$

The new standard divisor is

$$d = \frac{424}{11} \approx 38.5455.$$

Use the Hamilton method with the new d.

State	Population	Q	Rounded Q
a	61	$\frac{61}{38.5455} \approx 1.583$	1
b	148	$\frac{148}{38.5455} \approx 3.840$	3
c	215	$\frac{215}{38.5455} \approx 5.578$	5
Totals	424		9

The two remaining seats will be apportioned to States a and b, because the fractional parts of their quotas are the largest. The final apportionment is as follows.

State	Q	Number of seats
a	1.583	1 + 1 = 2
b	3.840	3 + 1 = 4
c	5.578	5
Totals		11

Here is a final summary.

State	Old Pop	New Pop	% Inc.	Old No. of Seats	New No. of Seats
a	55	61	**10.91**	1	**2**
b	125	148	18.40	4	4
c	190	215	**13.16**	6	**5**
Totals	370	424		11	11

Notice from the table that there was a greater percent increase in growth for State c than for State a. Yet, State a gained a seat, and State c lost a seat. This is an example of the Population Paradox.

14. The calculations for the initial number of seats for the states have been done in Exercise 13.

State	Q	Number of seats
a	1.635	1
b	3.716	3 + 1 = 4
c	5.649	5 + 1 = 6
Totals		11

Now apportion the seats for the revised populations.

$$62 + 150 + 218 = 430$$

The new standard divisor is

$$d = \frac{430}{11} \approx 39.0909.$$

Use the Hamilton method with the new d.

16.4 THE IMPOSSIBILITIES OF APPORTIONMENT 577

State	Population	Q	Rounded Q
a	62	$\frac{62}{39.0909} \approx 1.586$	1
b	150	$\frac{150}{39.0909} \approx 3.837$	3
c	218	$\frac{218}{39.0909} \approx 5.577$	5
Totals	430		9

The two remaining seats will be apportioned to States a and b, because the fractional parts of their quotas are the largest. The final apportionment is as follows.

State	Q	Number of seats
a	1.586	1 + 1 = 2
b	3.837	3 + 1 = 4
c	5.577	5
Totals		11

Here is a final summary.

State	Old Pop	New Pop	% Inc.	Old No. of Seats	New No. of Seats
a	55	62	12.73	1	2
b	125	150	20.00	4	4
c	190	218	14.74	6	5
Totals	370	430		11	11

Notice from the table that there was a greater percent increase in growth for State c than for State a. Yet State a gained a seat, and State c lost a seat. This is an example of the Population Paradox.

15. First calculate the apportionment for the initial populations. The total population is

$$930 + 738 + 415 = 2083.$$

The standard divisor for 13 seats is

$$d = \frac{2083}{13} = 160.2308.$$

Use the Hamilton method.

State	Population	Q	Rounded Q
a	930	$\frac{930}{160.2308} \approx 5.804$	5
b	738	$\frac{738}{160.2308} \approx 4.606$	4
c	415	$\frac{415}{160.2308} \approx 2.590$	2
Totals	2083		11

The two remaining seats will be apportioned to States a and b, because the fractional parts of their quotas are the largest. The final apportionment is as follows.

State	Q	Number of seats
a	5.804	5 + 1 = 6
b	4.606	4 + 1 = 5
c	2.590	2
Totals		13

Now apportion the seats for the revised populations.

$$975 + 750 + 421 = 2146.$$

The new standard divisor is

$$d = \frac{2146}{13} \approx 165.0769.$$

Use the Hamilton method with the new values.

State	Population	Q	Rounded Q
a	975	$\frac{975}{165.0769} \approx 5.906$	5
b	750	$\frac{750}{165.0769} \approx 4.543$	4
c	421	$\frac{421}{165.0769} \approx 2.550$	2
Totals	2146		11

The two remaining seats will be apportioned to States a and c, because the fractional parts of their quotas are the largest. The final apportionment is as follows

State	Q	Number of seats
a	5.906	5 + 1 = 6
b	4.543	4
c	2.550	2 + 1 = 3
Totals		13

Here is a final summary.

State	Old Pop	New Pop	% Inc.	Old No. of Seats	New No. of Seats
a	930	975	4.84	6	6
b	738	750	1.63	5	4
c	415	421	1.45	2	3
Totals	2083	2146		13	13

Notice from the table that there was a greater percent increase in growth for State b than for State c. Yet, State c gained a seat, and State b lost a seat. This is an example of the Population Paradox.

16. The calculations for the initial number of seats for the states have been done in Exercise 15.

State	Q	Number of seats
a	5.804	5 + 1 = 6
b	4.606	4 + 1 = 5
c	2.590	2
Totals		13

Now apportion the seats for the revised populations.

$$975 + 752 + 422 = 2149.$$

The new standard divisor is

$$d = \frac{2149}{13} \approx 165.3077.$$

Use the Hamilton method with the new values.

State	Population	Q	Rounded Q
a	975	$\frac{975}{165.3077} \approx 5.898$	5
b	752	$\frac{752}{165.3077} \approx 4.549$	4
c	422	$\frac{422}{165.3077} \approx 2.553$	2
Totals	2149		11

The two remaining seats will be apportioned to States a and c, because the fractional parts of their quotas are the largest. The final apportionment is as follows.

State	Q	Number of seats
a	5.898	5 + 1 = 6
b	4.549	4
c	2.553	2 + 1 = 3
Totals		13

Here is a final summary.

State	Old Pop	New Pop	% Inc.	Old No. of Seats	New No. of Seats
a	930	975	4.84	6	6
b	738	752	**1.90**	5	4
c	415	422	**1.69**	2	3
Totals	2083	2150		13	13

Notice from the table that there was a greater percent increase in growth for State b than for State c. Yet, State c gained a seat, and State b lost a seat. This is an example of the Population Paradox

17. First calculate the apportionment for the initial populations. The total population is

$$89 + 125 + 225 = 439.$$

The standard divisor for 13 seats is

$$d = \frac{439}{13} \approx 33.7692.$$

Use the Hamilton method.

State	Population	Q	Rounded Q
a	89	$\frac{89}{33.7692} \approx 2.636$	2
b	125	$\frac{125}{33.7692} \approx 3.702$	3
c	225	$\frac{225}{33.7692} \approx 6.663$	6
Totals	439		11

The two remaining seats will be apportioned to States b and c, because the fractional parts of their quotas are the largest. The final apportionment is as follows.

State	Q	Number of seats
a	2.636	2
b	3.702	3 + 1 = 4
c	6.663	6 + 1 = 7
Totals		13

Now apportion the seats for the revised populations.

$$97 + 145 + 247 = 489.$$

The new standard divisor is

$$d = \frac{489}{13} \approx 37.6154.$$

Use the Hamilton method with the new values.

State	Population	Q	Rounded Q
a	97	$\frac{97}{37.6154} \approx 2.579$	2
b	145	$\frac{145}{37.6154} \approx 3.855$	3
c	247	$\frac{247}{37.6154} \approx 6.566$	6
Totals	489		11

The two remaining seats will be apportioned to States a and b, because the fractional parts of their quotas are the largest. The final apportionment is as follows.

State	Q	Number of seats
a	2.579	2 + 1 = 3
b	3.855	3 + 1 = 4
c	6.566	6
Totals		13

Here is a final summary.

State	Old Pop	New Pop	% Inc.	Old No. of Seats	New No. of Seats
a	89	97	**8.99**	2	3
b	125	145	16.00	4	4
c	225	247	**9.78**	7	6
Totals	439	489		13	13

Notice from the table that there was a greater percent increase in growth for State c than for State a. Yet, State a gained a seat, and State c lost a seat. This is an example of the Population Paradox.

18. The calculations for the initial number of seats for the states have been done in Exercise 17.

State	Q	Number of seats
a	2.636	2
b	3.702	3 + 1 = 4
c	6.663	6 + 1 = 7
Totals		13

Now apportion the seats for the revised populations.

$$98 + 145 + 249 = 492.$$

The new standard divisor is

$$d = \frac{492}{13} \approx 37.8462.$$

Use the Hamilton method with the new values.

State	Population	Q	Rounded Q
a	98	$\frac{98}{37.8462} \approx 2.589$	2
b	145	$\frac{145}{37.8462} \approx 3.831$	3
c	249	$\frac{249}{37.8462} \approx 6.579$	6
Totals	492		11

The two remaining seats will be apportioned to States a and b, because the fractional parts of their quotas are the largest. The final apportionment is as follows.

State	Q	Number of seats
a	2.589	2 + 1 = 3
b	3.831	3 + 1 = 4
c	6.579	6
Totals		13

Here is a final summary.

State	Old Pop	New Pop	% Inc.	Old No. of Seats	New No. of Seats
a	89	98	**10.11**	2	3
b	125	145	16.00	4	4
c	225	249	**10.67**	7	6
Totals	439	492		13	13

Notice from the table that there was a greater percent increase in growth for State c than for State a. Yet, State a gained a seat, and State c lost a seat. This is an example of the Population Paradox.

19. *First calculate the apportionment for the initial populations. The total population is*

$$134 + 52 = 186.$$

The standard divisor for 16 seats is

$$d = \frac{186}{16} = 11.625.$$

Use the Hamilton method.

State	Population	Q	Rounded Q
a	134	$\frac{134}{11.625} \approx 11.527$	11
b	52	$\frac{52}{11.625} \approx 4.473$	4
Totals	186		15

The one remaining seat will be apportioned to State a because the fractional parts of its quota is larger. The final apportionment is as follows.

State	Q	Number of seats
a	11.527	11 + 1 = 12
b	4.473	4
Totals		16

Now apportion the seats to include the new state. The standard quota of the new state is

$$Q = \frac{38}{11.625} \approx 3.269.$$

Rounded down to 3, add 3 new seats to the original 16 to obtain 19 seats to be apportioned. Now the new population is

$$134 + 52 + 38 = 224.$$

The new standard divisor is

$$d = \frac{224}{19} \approx 11.7895.$$

State	Population	Q	Rounded Q
a	134	$\frac{134}{11.7895} \approx 11.366$	11
b	52	$\frac{52}{11.7895} \approx 4.411$	4
c	38	$\frac{38}{11.7895} \approx 3.223$	3
Totals	224		18

The one remaining seat will be apportioned to State b because the fractional part of its quota is the largest. The final apportionment is as follows.

State	Q	Number of seats
a	11.366	11
b	4.411	4 + 1 = 5
c	3.223	3
Totals		19

The New State Paradox has occurred because the addition of the new state has caused a shift in the apportionment of the original states. States a and b originally had 12 and 4 seats, respectively; now they have 11 and 5, respectively.

20. *In Exercise 19, the apportionment is calculated for the original states.*

State	Q	Number of seats
a	11.527	11 + 1 = 12
b	4.473	4
Totals		16

Now apportion the seats to include the new state. The standard quota of the new state is

$$Q = \frac{39}{11.625} \approx 3.3548.$$

Rounded down to 3, add 3 new seats to the original 16 to obtain 19 seats to be apportioned. Now the new population is

$$134 + 52 + 39 = 225.$$

The new standard divisor is

$$d = \frac{225}{19} \approx 11.8421.$$

State	Population	Q	Rounded Q
a	134	$\frac{134}{11.8421} \approx 11.316$	11
b	52	$\frac{52}{11.8421} \approx 4.391$	4
c	39	$\frac{38}{11.8421} \approx 3.209$	3
Totals	225		18

The one remaining seat will be apportioned to State b because the fractional part of its quota is the largest. The final apportionment is as follows.

State	Q	Number of seats
a	11.316	11
b	4.391	4 + 1 = 5
c	3.209	3
Totals		19

The New State Paradox has occurred because the addition of the new state has caused a shift in the apportionment of the original states. States a and b originally had 12 and 4 seats, respectively; now they have 11 and 5, respectively.

21. First calculate the apportionment for the initial populations. The total population is

$$3184 + 8475 = 11{,}659.$$

The standard divisor for 75 seats is

$$d = \frac{11659}{75} \approx 155.4533.$$

Use the Hamilton method.

State	Population	Q	Rounded Q
a	3184	$\frac{3184}{155.4533} \approx 20.482$	20
b	8475	$\frac{8475}{155.4533} \approx 54.518$	54
Totals	11,659		74

The one remaining seat will be apportioned to State b because the fractional parts of its quota is larger. The final apportionment is as follows.

State	Q	Number of seats
a	20.482	20
b	54.518	54 + 1 = 55
Totals		75

Now apportion the seats to include the new state. The standard quota of the new state is

$$Q = \frac{330}{155.4533} \approx 2.123.$$

Rounded down to 2, add 2 new seats to the original 75 to obtain 77 seats to be apportioned. Now the new population is

$$3184 + 8475 + 330 = 11{,}989.$$

The new standard divisor is

$$d = \frac{11989}{77} \approx 155.7013.$$

State	Population	Q	Rounded Q
a	3184	$\frac{3184}{155.7013} \approx 20.449$	20
b	8475	$\frac{8475}{155.7013} \approx 54.431$	54
c	330	$\frac{330}{155.7013} \approx 2.119$	2
Totals	11,989		76

The one remaining seat will be apportioned to State a because the fractional part of its quota is the largest. The final apportionment is as follows.

State	Q	Number of seats
a	20.449	20 + 1 = 21
b	54.431	54
c	2.119	2
Totals		77

The New State Paradox has occurred because the addition of the new state has caused a shift in the apportionment of the original states. States a and b originally had 20 and 55 seats, respectively; now they have 21 and 54, respectively.

22. In Exercise 20, the apportionment is calculated for the original states.

State	Q	Number of seats
a	20.482	20
b	54.518	54 + 1 = 55
Totals		75

Now apportion the seats to include the new state. The standard quota of the new state is

$$Q = \frac{350}{155.4533} \approx 2.2515.$$

Rounded down to 2, add 2 new seats to the original 75 to obtain 77 seats to be apportioned. Now the new population is

$$3184 + 8475 + 350 = 12{,}009.$$

The new standard divisor is

$$d = \frac{12009}{77} \approx 155.9610.$$

State	Population	Q	Rounded Q
a	3184	$\frac{3184}{155.9610} \approx 20.415$	20
b	8475	$\frac{8475}{155.9610} \approx 54.341$	54
c	350	$\frac{350}{155.9610} \approx 2.244$	2
Totals	12,009		76

The one remaining seat will be apportioned to State a because the fractional part of its quota is the largest. The final apportionment is as follows.

State	Q	Number of seats
a	20.415	20 + 1 = 21
b	54.341	54
c	2.244	2
Totals		77

The New State Paradox has occurred because the addition of the new state has caused a shift in the apportionment of the original states. States a and b originally had 20 and 55 seats, respectively; now they have 21 and 54, respectively.

23. First calculate the apportionment for the initial populations. The total population is

$$7500 + 9560 = 17{,}060.$$

The standard divisor for 83 seats is

$$d = \frac{17060}{83} \approx 205.5422.$$

Use the Hamilton method.

State	Population	Q	Rounded Q
a	7500	$\frac{7500}{205.5422} \approx 36.489$	36
b	9560	$\frac{9560}{205.5422} \approx 46.511$	46
Totals	17,060		82

The one remaining seat will be apportioned to State b because the fractional parts of its quota is larger.

The final apportionment is as follows.

State	Q	Number of seats
a	36.489	36
b	46.511	46 + 1 = 47
Totals		83

Now apportion the seats to include the new state. The standard quota of the new state is

$$Q = \frac{1500}{205.5422} \approx 7.298.$$

Rounded down to 7, add 7 new seats to the original 83 to obtain 90 seats to be apportioned. Now the new population is

$$7500 + 9560 + 1500 = 18{,}560.$$

The new standard divisor is

$$d = \frac{18560}{90} \approx 206.2222.$$

State	Population	Q	Rounded Q
a	7500	$\frac{7500}{206.2222} \approx 36.369$	36
b	9560	$\frac{9560}{206.2222} \approx 46.358$	46
c	1500	$\frac{1500}{206.2222} \approx 7.274$	7
Totals	18,560		89

The one remaining seat will be apportioned to State a because the fractional part of its quota is the largest. The final apportionment is as follows.

State	Q	Number of seats
a	36.369	36 + 1 = 37
b	46.358	46
c	7.274	7
Totals		90

The New State Paradox has occurred because the addition of the new state has caused a shift in the apportionment of the original states. States a and b originally had 36 and 47 seats, respectively; now they have 37 and 46, respectively.

24. In Exercise 23, the apportionment is calculated for the original states.

State	Q	Number of seats
a	36.489	36
b	46.511	46 + 1 = 47
Totals		83

Now apportion the seats to include the new state. The standard quota of the new state is

$$Q = \frac{1510}{205.5422} \approx 7.346.$$

Rounded down to 7, add 7 new seats to the original 83 to obtain 90 seats to be apportioned. Now the new population is

$$7500 + 9560 + 1510 = 18,570.$$

The new standard divisor is

$$d = \frac{18570}{90} \approx 206.3333.$$

State	Population	Q	Rounded Q
a	7500	$\frac{7500}{206.3333} \approx 36.349$	36
b	9560	$\frac{9560}{206.3333} \approx 46.333$	46
c	1510	$\frac{1510}{206.3333} \approx 7.318$	7
Totals	18,570		89

The one remaining seat will be apportioned to State a because the fractional part of its quota is the largest. The final apportionment is as follows.

State	Q	Number of seats
a	36.349	36 + 1 = 37
b	46.333	46
c	7.318	7
Totals		90

The New State Paradox has occurred because the addition of the new state has caused a shift in the apportionment of the original states. States a and b originally had 36 and 47 seats, respectively; now they have 37 and 46, respectively.

25. Use the Jefferson method to apportion the teaching assistants for the following enrollment profile.

Class	Enrollment
a	225
b	45
c	35
d	30
Total	335

First, calculate the standard divisor, d, by dividing the total enrollment by the number of teaching assistants, 25.

$$d = \frac{335}{25} = 13.4.$$

Calculate the standard quota, Q, for each class by dividing the enrollment by the standard divisor, d. If d is used, the modified quotas total to only 23. If d is slowly decreased to 12.4, the modified quotas add up to 25.

Class	Enrollment	Standard Q	Rounded Q	New Q	mQ
a	225	$\frac{225}{13.4} \approx 16.8$	16	$\frac{225}{12.4} \approx 18.1$	18
b	45	$\frac{45}{13.4} \approx 3.4$	3	$\frac{45}{12.4} \approx 3.6$	3
c	35	$\frac{35}{13.4} \approx 2.6$	2	$\frac{35}{12.4} \approx 2.8$	2
d	30	$\frac{30}{13.4} \approx 2.2$	2	$\frac{30}{12.4} \approx 2.4$	2
Totals	335		23		25

For class a, rounding 16.8 up to 17 indicates that this class could indeed receive one more teaching assistant without violating the quota rule. However, class a receives two additional teaching assistants by the Jefferson method.

26. Use the Jefferson method to apportion the council seats for the following population profile.

County	Population
a	56000
b	15000
c	7080
d	9500
Total	87,580

First, calculate the standard divisor, d, by dividing the total population by the number of council seats, 50.

$$d = \frac{87580}{50} = 1751.6.$$

Calculate the standard quota, Q, for each county by dividing the population by the standard divisor, d. If d is used, the modified quotas total to only 48. If d is slowly decreased to 1690, the modified quotas add up to 50.

Class	Pop	Standard Q	Rounded down Q	New Q	mQ
a	56000	$\frac{56000}{1751.6} \approx 31.97$	31	$\frac{56000}{1690} \approx 33.1$	33
b	15000	$\frac{15000}{1751.6} \approx 8.56$	8	$\frac{15000}{1690} \approx 8.9$	8
c	7080	$\frac{7080}{1751.6} \approx 4.04$	4	$\frac{7080}{1690} \approx 4.2$	4
d	9500	$\frac{9500}{1751.6} \approx 5.42$	5	$\frac{9500}{1690} \approx 5.6$	5
Totals	87,580		48		50

For class a, rounding 31.97 up to 32 indicates that this class could indeed receive one more council seat without violating the quota rule. However, class a receives two additional seats by the Jefferson method.

27. Here is the profile from Exercise 5, using 200 seats to be apportioned. The standard divisor is

$$d = \frac{14161}{200} = 70.805.$$

State	Population	Q	Rounded Q
a	2567	$\frac{2567}{70.805} \approx 36.255$	36
b	1500	$\frac{1500}{70.805} \approx 21.185$	21
c	8045	$\frac{8045}{70.805} \approx 113.622$	**113**
d	950	$\frac{950}{70.805} \approx 13.417$	13
e	1099	$\frac{1099}{70.805} \approx 15.522$	15
Totals	14,161		198

Use the Jefferson method with $md = 69.5$.

State	Population	Q	Rounded Q
a	2567	$\frac{2567}{69.5} \approx 36.94$	36
b	1500	$\frac{1500}{69.5} \approx 21.58$	21
c	8045	$\frac{8045}{69.5} \approx 115.76$	**115**
d	950	$\frac{950}{69.5} \approx 13.67$	13
e	1099	$\frac{1099}{69.5} \approx 15.81$	15
Totals	14,161		200

The Jefferson method violates the Quota Rule because state c receives two more seats than its apportionment from the standard quota.

28. Here is the profile from Exercise 6, using 100 seats to be apportioned. The standard divisor is

$$d = \frac{45066}{100} = 450.66.$$

Use the standard quota.

State	Population	Q	Rounded Q
a	1720	$\frac{1720}{450.66} \approx 3.817$	3
b	3363	$\frac{3363}{450.66} \approx 7.462$	7
c	6960	$\frac{6960}{450.66} \approx 15.444$	15
d	24223	$\frac{24223}{450.66} \approx 53.750$	**53**
e	8800	$\frac{8800}{450.66} \approx 19.527$	19
Totals	45066		97

Use the Jefferson method with $md = 440$.

State	Population	Q	Rounded Q
a	1720	$\frac{1720}{440} \approx 3.909$	3
b	3363	$\frac{3363}{440} \approx 7.643$	7
c	6960	$\frac{6960}{440} \approx 15.818$	15
d	24223	$\frac{24223}{440} \approx 55.052$	**55**
e	8800	$\frac{8800}{440} = 20$	20
Totals	45066		100

The Jefferson method violates the Quota Rule because state d receives two more seats than its apportionment from the standard quota

29. First calculate the apportionment with $n = 49$. The total population is

$$465 + 552 + 385 + 251 = 1653.$$

The standard divisor for 49 seats is

$$d = \frac{1653}{49} \approx 33.7347.$$

Use the Hamilton method.

State	Population	Q	Rounded Q
a	465	$\frac{465}{33.7347} \approx 13.784$	13
b	552	$\frac{552}{33.7347} \approx 16.363$	16
c	385	$\frac{385}{33.7347} \approx 11.413$	11
d	251	$\frac{251}{33.7347} \approx 7.440$	7
Totals	1653		47

The two remaining seats will be apportioned to State a and State d, because the fractional parts of their quotas are the largest. The final apportionment is as follows.

State	Population	Q	Number of seats
a	465	13.784	13 + 1 = 14
b	552	16.363	16
c	385	11.413	11
d	251	7.440	7 + 1 = **8**
Totals	1653		49

Now increase the number of seats to 50. The standard divisor is

$$d = \frac{1653}{50} = 33.06.$$

Use the Hamilton method with the new d.

State	Population	Q	Rounded Q
a	465	$\frac{465}{33.06} \approx 14.065$	14
b	552	$\frac{552}{33.06} \approx 16.697$	16
c	385	$\frac{385}{33.06} \approx 11.645$	11
d	251	$\frac{251}{33.06} \approx 7.592$	**7**
Totals	1653		48

The two remaining seats will be apportioned to State b and State c, because the fractional parts of their quotas are the largest. The final apportionment is as follows.

State	Q	Number of seats
a	14.065	14
b	16.697	16 + 1 = 17
c	11.645	11 + 1 = 12
d	7.592	7
Totals		50

State d is a victim of the Alabama Paradox, because it has lost a seat despite the fact that the overall number of seats has increased.

30. First calculate the apportionment with $n = 17$. The total population is

$$777 + 418 + 664 + 280 = 2139.$$

The standard divisor for 49 seats is

$$d = \frac{2139}{17} \approx 125.8235.$$

Use the Hamilton method.

State	Population	Q	Rounded Q
a	777	$\frac{777}{125.8235} \approx 6.175$	6
b	418	$\frac{418}{125.8235} \approx 3.322$	3
c	664	$\frac{664}{125.8235} \approx 5.277$	5
d	280	$\frac{280}{125.8235} \approx 2.225$	2
Totals	2139		16

The one remaining seat will be apportioned to State b because the fractional part of its quota is the largest. The final apportionment is as follows.

State	Population	Q	Number of seats
a	777	6.175	6
b	418	3.322	3 + 1 = **4**
c	664	5.277	5
d	280	2.225	2
Totals	2139		17

Now increase the number of seats to 18. The standard divisor is

$$d = \frac{2139}{18} \approx 118.8333.$$

Use the Hamilton method with the new d.

State	Population	Q	Rounded Q
a	777	$\frac{777}{118.8333} \approx 6.539$	6
b	418	$\frac{418}{118.8333} \approx 3.518$	3
c	664	$\frac{664}{118.8333} \approx 5.588$	5
d	280	$\frac{280}{118.8333} \approx 2.356$	2
Totals	2139		16

The two remaining seats will be apportioned to State c and State a, because the fractional parts of their quotas are the largest. The final apportionment is as follows.

State	Q	Number of seats
a	6.539	6 + 1 = 7
b	3.518	**3**
c	5.588	5 + 1 = 6
d	2.356	2
Totals		18

State b is a victim of the Alabama Paradox, because it has lost a seat despite the fact that the overall number of seats has increased.

31. From Exercise 13, calculate the apportionment for the preliminary enrollments. The total enrollment is

$$55 + 125 + 190 = 370.$$

The standard divisor for 11 teaching assistants is

$$d = \frac{370}{11} = 33.6364.$$

Use the Hamilton method.

Course	Enrollment	Q	Rounded Q
a	55	$\frac{55}{33.6364} \approx 1.635$	1
b	125	$\frac{125}{33.6364} \approx 3.716$	3
c	190	$\frac{190}{33.6364} \approx 5.649$	5
Totals	370		9

The two remaining teaching assistants will be apportioned to Courses b and c, because the fractional parts of their quotas are the largest. The final apportionment is as follows.

16.4 THE IMPOSSIBILITIES OF APPORTIONMENT

Course	Q	Number of TA's
a	1.635	1
b	3.716	3 + 1 = 4
c	5.649	5 + 1 = 6
Totals		11

Now try the following actual enrollments.

$$61 + 145 + 213 = 419.$$

The new standard divisor is

$$d = \frac{419}{11} \approx 38.0909.$$

Use the Hamilton method with the new d.

Course	Enrollment	Q	Rounded Q
a	61	$\frac{61}{38.0909} \approx 1.601$	1
b	145	$\frac{145}{38.0909} \approx 3.807$	3
c	213	$\frac{213}{38.0909} \approx 5.592$	5
Totals	419		9

The two remaining teaching assistants will be apportioned to Courses a and b, because the fractional parts of their quotas are the largest. The final apportionment is as follows.

Course	Q	Number of TA's
a	1.601	1 + 1 = 2
b	3.807	3 + 1 = 4
c	5.592	5
Totals		11

Here is a final summary.

Course	Pre. Enr.	Actual Enr.	% Inc.	Old No. of TA's	New No. of TA's
a	55	61	10.9	1	2
b	125	145	16.0	4	4
c	190	213	12.1	6	5
Totals	370	419		11	11

Notice from the table that there was a greater percent increase in growth for Course c than for Course a. Yet, Course a gained a teaching assistant, and Course c lost one. This is an example of the Population Paradox.

32. First calculate the apportionment for the preliminary enrollments. The total enrollment is

$$89 + 125 + 225 = 439.$$

The standard divisor for 13 teaching assistants is

$$d = \frac{439}{13} = 33.7692.$$

Use the Hamilton method.

Course	Enrollment	Q	Rounded Q
a	89	$\frac{89}{33.7692} \approx 2.636$	2
b	125	$\frac{125}{33.7692} \approx 3.702$	3
c	225	$\frac{225}{33.7692} \approx 6.663$	6
Totals	439		11

The two remaining teaching assistants will be apportioned to Courses b and c, because the fractional parts of their quotas are the largest. The final apportionment is as follows.

Course	Q	Number of TA's
a	2.636	2
b	3.702	3 + 1 = 4
c	6.663	6 + 1 = 7
Totals		13

Now try the following actual enrollments.

$$96 + 145 + 245 = 486.$$

The new standard divisor is

$$d = \frac{486}{13} \approx 37.3846.$$

Use the Hamilton method with the new d.

Course	Enrollment	Q	Rounded Q
a	96	$\frac{96}{37.3846} \approx 2.568$	2
b	145	$\frac{145}{37.3846} \approx 3.879$	3
c	245	$\frac{245}{37.3846} \approx 6.554$	6
Totals	486		11

The two remaining teaching assistants will be apportioned to Courses a and b, because the fractional parts of their quotas are the largest. The final apportionment is as follows.

Course	Q	Number of TA's
a	2.568	2 + 1 = 3
b	3.879	3 + 1 = 4
c	6.554	6
Totals		13

Here is a final summary.

Course	Pre. Enr	Actual Enr	% Inc.	Old No. of TA's.	New No. of TA's.
a	89	96	7.87	2	3
b	125	145	16.0	4	4
c	225	245	8.89	7	6
Totals	439	486		13	13

Notice from the table that there was a greater percent increase in growth for Course c than for Course a. Yet, Course a gained a teaching assistant, and Course c lost one. This is an example of the Population Paradox.

33. *Here is the apportionment from Example 4, with a population of 531 for the second subdivision. The total population would be*

$$8500 + 1671 + 531 = 10,702.$$

The standard divisor would be

$$d = \frac{10702}{105} \approx 101.9238.$$

Community	Pop.	Standard Quota	Rounded down Q
Original	8500	$\frac{8500}{101.9238} \approx 83.396$	83
1st Annexed	1671	$\frac{1671}{101.9238} \approx 16.395$	16
2nd Annexed	531	$\frac{531}{101.9238} \approx 5.210$	5
Totals	10,702		104

The one remaining seat would be apportioned to the original community because the fractional parts of its quota is the largest. The final apportionment is as follows.

Community	Q	Number of seats
a	83.396	83 + 1 = 84
b	16.395	16
c	5.210	5
Totals		105

Now increasing the population of the second annexed subdivision to 532, here are the calculations. The total population would be

$$8500 + 1671 + 532 = 10,703.$$

The standard divisor would be

$$d = \frac{10703}{105} \approx 101.9333.$$

Community	Pop.	Standard Quota	Rounded down Q
Original	8500	$\frac{8500}{101.9333} \approx 83.388$	83
1st Annexed	1671	$\frac{1671}{101.9333} \approx 16.393$	16
2nd Annexed	532	$\frac{532}{101.9333} \approx 5.219$	5
Totals	10,703		104

Now the one remaining seat would be apportioned to the 1st annexed community because the fractional parts of its quota is the largest. The final apportionment is as follows.

Community	Q	Number of seats
a	83.388	83
b	16.396	16 + 1 = 17
c	5.219	5
Totals		105

The New State Paradox occurs when the population of the 2nd annexed subdivision increases to 532.

34. *Here is the apportionment from Example 4, with a population of 548 for the second subdivision. The total population would be*

$$8500 + 1671 + 548 = 10,719.$$

The standard divisor would be

$$d = \frac{10719}{105} \approx 102.0857.$$

Community	Pop.	Standard Quota	Rounded down Q
Original	8500	$\frac{8500}{102.0857} \approx 83.263$	83
1st Annexed	1671	$\frac{1671}{102.0857} \approx 16.369$	16
2nd Annexed	548	$\frac{548}{102.0857} \approx 5.368$	5
Totals	10,719		104

The one remaining seat would be apportioned to the 1st annexed community because the fractional parts of its quota is the largest. The final apportionment is as follows.

Community	Q	Number of seats
a	83.263	83
b	16.369	16 + 1 = 17
c	5.368	5
Totals		105

Now increasing the population of the second annexed subdivision to 549, here are the calculations. The total population would be

$$8500 + 1671 + 549 = 10,720.$$

The standard divisor would be

$$d = \frac{10720}{105} \approx 102.0952.$$

Community	Pop.	Standard Quota	Rounded down Q
Original	8500	$\frac{8500}{102.0952} \approx 83.256$	83
1st Annexed	1671	$\frac{1671}{102.0952} \approx 16.367$	16
2nd Annexed	549	$\frac{549}{102.0952} \approx 5.377$	5
Totals	10,720		104

Now the one remaining seat would be apportioned to the 2nd annexed community because the fractional parts of its quota is the largest. The final apportionment is as follows.

Community	Q	Number of seats
a	83.256	83
b	16.367	16
c	5.377	5 + 1 = 6
Totals		105

35. The sum of the populations is

$$1720 + 3363 + 6960 + 24223 + 8800 = 45066.$$

The standard divisor is

$$d = \frac{45066}{220} \approx 204.8455.$$

State	Population	mQ	Rounded mQ
a	1720	$\frac{1720}{204.8455} \approx 8.397$	8
b	3363	$\frac{3363}{204.8455} \approx 16.417$	16
c	6960	$\frac{6960}{204.8455} \approx 33.977$	33
d	24223	$\frac{24223}{204.8455} \approx 118.250$	**118**
e	8800	$\frac{8800}{204.8455} \approx 42.959$	42
Totals	45066		217

Use the Adams method with a modified divisor of 208.

State	Population	mQ	Rounded mQ
a	1720	$\frac{1720}{208} \approx 8.269$	9
b	3363	$\frac{3363}{208} \approx 16.168$	17
c	6960	$\frac{6960}{208} \approx 33.462$	34
d	24223	$\frac{24223}{208} \approx 116.457$	**117**
e	8800	$\frac{8800}{208} \approx 42.308$	43
Totals	45066		220

This is a violation of the Quota Rule because State d receives only 117 seats, although it should receive at least 118.

36. The sum of the populations is

$$1720 + 3363 + 6960 + 24223 + 8800 = 45066.$$

The standard divisor is

$$d = \frac{45066}{219} \approx 205.7808.$$

State	Population	Q	Rounded Q
a	1720	$\frac{1720}{205.7808} \approx 8.358$	8
b	3363	$\frac{3363}{205.7808} \approx 16.343$	16
c	6960	$\frac{6960}{205.7808} \approx 33.822$	33
d	24223	$\frac{24223}{205.7808} \approx 117.713$	**117**
e	8800	$\frac{8800}{205.7808} \approx 42.764$	42
Totals	45066		216

Use the Adams method with a modified divisor of 209.

State	Population	Q	Rounded Q
a	1720	$\frac{1720}{209} \approx 8.230$	9
b	3363	$\frac{3363}{209} \approx 16.091$	17
c	6960	$\frac{6960}{209} \approx 33.301$	34
d	24223	$\frac{24223}{209} \approx 115.900$	**116**
e	8800	$\frac{8800}{209} \approx 42.105$	43
Totals	45066		219

This is a violation of the Quota Rule because State d receives only 116 seats, although it should receive at least 117.

37. Writing exercise
38. Writing exercise
39. Writing exercise
40. Writing exercise
41. Writing exercise
42. Writing exercise

CHAPTER 16 TEST

1. The voter profile is as follows.

Votes	Ranking
5	$a > b > d > c$
6	$b > c > a > d$
5	$c > d > b > a$
7	$d > a > b > c$
4	$c > a > d > b$

By the plurality method the destination with the most votes wins. Cancun receives $5 + 4 = 9$ first-place votes to win.

2. Use the ranking information from Exercise 1 to make a table to examine the points in the Borda method.

Votes	Ranking	Points 3	2	1	0
5	$a>b>d>c$	a	b	d	c
6	$b>c>a>d$	b	c	a	d
5	$c>d>b>a$	c	d	b	a
7	$d>a>b>c$	d	a	b	c
4	$c>a>d>b$	c	a	d	b

Examine the last four columns, and compute the weighted sum for each candidate.

Candidate	Borda Points
a	$5 \cdot 3 + 6 \cdot 1 + 5 \cdot 0 + 7 \cdot 2 + 4 \cdot 2 = 43$
b	$5 \cdot 2 + 6 \cdot 3 + 5 \cdot 1 + 7 \cdot 1 + 4 \cdot 0 = 40$
c	$5 \cdot 0 + 6 \cdot 2 + 5 \cdot 3 + 7 \cdot 0 + 4 \cdot 3 = 39$
d	$5 \cdot 1 + 6 \cdot 0 + 5 \cdot 2 + 7 \cdot 3 + 4 \cdot 1 = 40$

The winner is Aruba with 43 Borda points.

3. There are 27 sorority sisters voting, so a destination must receive a majority of the votes or at least 14 votes to win by the Hare method. No destination received a majority. See from Exercise 1 that Aruba received 5 first-place votes; the Bahamas received 6 votes; Cancun received 9 votes; and the Dominican Republic received 7 votes. Eliminate Aruba because it has the least number of votes, and compare only b, c, and d.

Votes	Ranking	Row
5	$b>d>c$	1
6	$b>c>d$	2
5	$c>d>b$	3
7	$d>b>c$	4
4	$c>d>b$	5

The Bahamas received $5 + 6 = 11$ votes.
Cancun receives $5 + 4 = 9$ votes.
The Dominican Republic receives 7 votes.
Again, no destination has received a majority. Therefore, eliminate d with the least number of votes, and compare only b and c.

Votes	Ranking	Row
5	$b>c$	1
6	$b>c$	2
5	$c>b$	3
7	$b>c$	4
4	$c>b$	5

The Bahamas receives $5 + 6 + 7 = 18$ votes.
Cancun receives $5 + 4 = 9$ votes.
Finally, the Bahamas is the winner.

4. Using the predetermined order a, c, b, d begin by comparing a and c in the original voting.

Comparison	Votes	Row
$a>c$	$5 + 7 = 12$	1, 4
$c>a$	$6 + 5 + 4 = 15$	2, 3, 5

Cancun wins this competition, 15 to 12. Now compare c to b.

Comparison	Votes	Row
$c>b$	$5 + 4 = 9$	3, 5
$b>c$	$5 + 6 + 7 = 18$	1, 2, 4

The Bahamas wins this competition, 18 to 9. Now compare b and d.

Comparison	Votes	Row
$b>d$	$5 + 6 = 11$	1, 2
$d>b$	$5 + 7 + 4 = 16$	3, 4, 5

Finally, Dominican Republic wins this competition, 16 to 11, and also wins the entire competition in the sequential pairwise comparison method.

5. One possible answer: You would want to select the method that selects your first choice. Since you are a dictator.

6. Using the approval voting method with each sister voting for her first- and second-choice, here are the number of votes for each destination.

Aruba $5 + 7 + 4 = 16$,
Bahamas $5 + 6 = 11$,
Cancun $6 + 5 + 4 = 15$,
Dominican Republic $5 + 7 = 12$;

Aruba wins.

7. Using the approval voting method with the first 5 sisters giving approval to all destinations except Cancun, the approvals look like the following.

Number of Voters	Ranking
5	a, b, d
6	b, c
5	c, d
7	d, a
4	c, a

Here are the number of votes for each destination.

Aruba $5 + 7 + 4 = 16$,
Bahamas $5 + 6 = 11$,
Cancun $6 + 5 + 4 = 15$,
Dominican Republic $5 + 5 + 7 = 17$;

the Dominican Republic wins.

8. Using the pairwise comparison method make a table to compute the votes in comparisons.

Comparison	Votes	Rows
$a > b$	$5 + 7 + 4 = 16$	1, 4, 5
$b > a$	$6 + 5 = 11$	2, 3
$a > c$	$5 + 7 = 12$	1, 4
$c > a$	$6 + 5 + 4 = 15$	2, 3, 5
$a > d$	$5 + 6 + 4 = 15$	1, 2, 5
$d > a$	$5 + 7 = 12$	3, 4
$b > c$	$5 + 6 + 7 = 18$	1, 2, 4
$c > b$	$5 + 4 = 9$	3, 5
$b > d$	$5 + 6 = 11$	1, 2
$d > b$	$5 + 7 + 4 = 16$	3, 4, 5
$c > d$	$6 + 5 + 4 = 15$	2, 3, 5
$d > c$	$5 + 7 = 12$	1, 4

Make another table to award pairwise points.

Pairs	Votes	Pairwise Points
$a : b$	16 : 11	a, 1
$a : c$	12 : 15	c, 1
$a : d$	15 : 12	a, 1
$b : c$	18 : 9	b, 1
$b : d$	11 : 16	d, 1
$c : d$	15 : 12	c, 1

Using the pairwise comparison method, here are the final point totals.

$$a: \quad 1 + 1 = 2$$
$$b: \qquad\;\; 1 = 1$$
$$c: \quad 1 + 1 = 2$$
$$d: \qquad\;\; 1 = 1$$

There is a tie between Aruba and Cancun with 2 points each.

9. One possibility: Using the predetermined order a, b, d, c, begin by comparing a and b in the original voting.

Comparison	Votes	Row
$a > b$	$5 + 7 + 4 = 16$	1, 4, 5
$b > a$	$6 + 5 = 11$	2, 3

Aruba wins this competition, 16 to 11. Now compare a to c.

Comparison	Votes	Row
$a > c$	$5 + 7 = 12$	1, 4
$c > a$	$6 + 5 + 4 = 15$	2, 3, 5

Cancun wins this competition, 15 to 12. Now compare c and d.

Comparison	Votes	Row
$c > d$	$6 + 5 + 4 = 15$	2, 3, 5
$d > c$	$5 + 7 = 12$	1, 4

Finally, Cancun wins this competition, 15 to 12, and also wins the entire competition in the sequential pairwise comparison method.

10. Writing exercise
11. Writing exercise
12. Writing exercise
13. Writing exercise
14. Use the ranking information from the text to make a table to examine the points in the Borda method.

Votes	Ranking	Points 2	1	0
16	$a > b > c$	a	b	c
8	$b > c > a$	b	c	a
7	$c > b > a$	c	b	a

Examine the last three columns, and compute the weighted sum for each candidate.

Candidate	Borda Points
a	$16 \cdot 2 + 8 \cdot 0 + 7 \cdot 0 = 32$
b	$16 \cdot 1 + 8 \cdot 2 + 7 \cdot 1 = 39$
c	$16 \cdot 0 + 8 \cdot 1 + 7 \cdot 2 = 22$

Although a has the majority of first-place votes in the ranking, candidate b wins by the Borda method. This violates the majority criterion.

15. Make a table to compute the votes in comparisons.

Comparison	Votes	Rows
$a > b$	$5 = 5$	1
$b > a$	$6 + 4 + 6 = 16$	2, 3, 4
$a > c$	$5 + 4 = 9$	1, 3
$c > a$	$6 + 6 = 12$	2, 4
$b > c$	$4 + 6 = 10$	3, 4
$c > b$	$5 + 6 = 11$	1, 2

Make another table to award pairwise points.

Pairs	Votes	Pairwise Points
$a : b$	5 : 16	b, 1
$a : c$	9 : 12	c, 1
$b : c$	10 : 11	c, 1

Using the pairwise comparison method, the winner is c with 2 pairwise points. Alternative b received only 1 point, and a received 0 points.

16. Alternative c is the Condorcet candidate.

<u>Plurality</u> method

$$a: \quad 5 = 5$$
$$b: \quad 4 + 6 = 10$$
$$c: \quad 6 = 6$$

Alternative b is selected, which violates the Condorcet criterion.

Borda method
Use the ranking information from the text to make a table to examine the points in the Borda method.

Votes	Ranking	Points		
		2	1	0
5	$a > c > b$	a	c	b
6	$c > b > a$	c	b	a
4	$b > a > c$	b	a	c
6	$b > c > a$	b	c	a

Examine the last three columns, and compute the weighted sum for each candidate.

Candidate	Borda Points
a	$5 \cdot 2 + 6 \cdot 0 + 4 \cdot 1 + 6 \cdot 0 = 14$
b	$5 \cdot 0 + 6 \cdot 1 + 4 \cdot 2 + 6 \cdot 2 = 26$
c	$5 \cdot 1 + 6 \cdot 2 + 4 \cdot 0 + 6 \cdot 1 = 23$

Alternative b wins by the Borda method. This violates the Condorcet criterion.

Hare method
None of the candidates has a majority of first-place votes, which is at least 11. Eliminate a because it has the least number of votes at 5, and compare only b and c.

Votes	Ranking
5	$c > b$
6	$c > b$
4	$b > c$
6	$b > c$

Now, alternative c receives $5 + 6 = 11$ votes and wins. This does not violate the Condorcet criterion.

17. Make a table to compute the votes in comparisons.

Comparison	Votes	Rows
$c > m$	$8 + 6 = 14$	1, 3
$m > c$	$7 + 5 = 12$	2, 4
$c > b$	$8 + 5 = 13$	1, 4
$b > c$	$7 + 6 = 13$	2, 3
$c > s$	$8 + 5 = 13$	1, 4
$s > c$	$7 + 6 = 13$	2, 3
$m > b$	$8 + 5 = 13$	1, 4
$b > m$	$7 + 6 = 13$	2, 3
$m > s$	$8 + 5 = 13$	1, 4
$s > m$	$7 + 6 = 13$	2, 3
$b > s$	$8 + 5 = 13$	1, 4
$s > b$	$7 + 6 = 13$	2, 3

Make another table to award pairwise points.

Pairs	Votes	Pairwise Points
$c : m$	$14 : 12$	$c, 1$
$c : b$	$13 : 13$	$c, \frac{1}{2}; b, \frac{1}{2}$
$c : s$	$13 : 13$	$c, \frac{1}{2}; s, \frac{1}{2}$
$m : b$	$13 : 13$	$m, \frac{1}{2}; b, \frac{1}{2}$
$m : s$	$13 : 13$	$m, \frac{1}{2}; s, \frac{1}{2}$
$b : s$	$13 : 13$	$b, \frac{1}{2}; s, \frac{1}{2}$

Using the pairwise comparison method, here are the final point totals.

$$c: \quad 1 + \frac{1}{2} + \frac{1}{2} = 2$$

$$b: \quad \frac{1}{2} + \frac{1}{2} + \frac{1}{2} = 1\frac{1}{2}$$

$$m: \quad \frac{1}{2} + \frac{1}{2} = 1$$

$$s: \quad \frac{1}{2} + \frac{1}{2} + \frac{1}{2} = 1\frac{1}{2}$$

Using the pairwise comparison method, the winner is c with 2 pairwise points.

Now, change the ranking of the last 5 voters as shown in the text and make a table to compute the votes in comparisons.

Comparison	Votes	Rows
$c > m$	$8 + 6 + 5 = 19$	1, 3, 4
$m > c$	$7 = 7$	2
$c > b$	$8 + 5 = 13$	1, 4
$b > c$	$7 + 6 = 13$	2, 3
$c > s$	$8 + 5 = 13$	1, 4
$s > c$	$7 + 6 = 13$	2, 3
$m > b$	$8 + 5 = 13$	1, 4
$b > m$	$7 + 6 = 13$	2, 3
$m > s$	$8 = 8$	1
$s > m$	$7 + 6 + 5 = 18$	2, 3, 4
$b > s$	$8 = 8$	1
$s > b$	$7 + 6 + 5 = 18$	2, 3, 4

Make another table to award pairwise points.

Pairs	Votes	Pairwise Points
$c{:}m$	$19{:}7$	$c, 1$
$c{:}b$	$13{:}13$	$c, \frac{1}{2}; b, \frac{1}{2}$
$c{:}s$	$13{:}13$	$c, \frac{1}{2}; s, \frac{1}{2}$
$m{:}b$	$13{:}13$	$m, \frac{1}{2}; b, \frac{1}{2}$
$m{:}s$	$8{:}18$	$s, 1$
$b{:}s$	$8{:}18$	$s, 1$

Using the pairwise comparison method, here are the final point totals.

$$c: \quad 1 + \frac{1}{2} + \frac{1}{2} = 2$$
$$b: \quad \phantom{1 + {}}\frac{1}{2} + \frac{1}{2} = 1$$
$$m: \quad \phantom{1 + \frac{1}{2} + {}}\frac{1}{2} = \frac{1}{2}$$
$$s: \quad \frac{1}{2} + 1 + 1 = 2\frac{1}{2}$$

Using the pairwise comparison method, the winner is s with $2\frac{1}{2}$ pairwise points. Although c was moved to the top of the ranking when the 5 voters changed their votes, c did not win. The pairwise comparison method has violated the monotonicity criterion.

18. Count the number of first-place votes to see that choice a has 9; choice b has 8, choice c has 12, and choice d has 0. None of the candidates has a majority of first-place votes, which is at least 15. The elimination of d does not change the number of first-place votes of a, b, and c. Therefore, eliminate b and compare only a and c.

Votes	Ranking
9	$a > c$
8	$a > c$
7	$c > a$
5	$c > a$

Now a receives $9 + 8 = 17$ votes and wins.

Now change the ranking of the last 5 voters as shown in the text and apply the Hare method again. Count the number of first-place votes to see that choice a now has 14, choice b still has 8, and choice c now has 7. Because none of the candidates has a majority, eliminate c and compare only a and b.

Votes	Ranking
9	$a > b$
8	$b > a$
7	$b > a$
5	$a > b$

This time b received $8 + 7 = 15$ votes and wins. Although a was moved to the top of the ranking when the 5 voters changed their votes, a did not win. The Hare method has violated the monotonicity criterion.

19. Examine the voter profile in the text to see that candidate a receives the most votes at 10 and is the winner by the plurality method. If alternative c is dropped from the selection process the rankings are as follows.

Votes	Ranking
10	$a > b$
7	$b > a$
5	$b > a$

Now alternative b has 12 votes and wins. Although losing alternative c is dropped, candidate a does not win the second election. This is a violation of the Independence of Irrelevant Alternatives criterion.

20. Use the ranking information from the text to make a table to examine the points in the Borda method.

Votes	Ranking	Points 3	2	1	0
7	$a > b > c > d$	a	b	c	d
5	$b > c > d > a$	b	c	d	a
4	$d > c > a > b$	d	c	a	b

Examine the last four columns, and compute the weighted sum for each candidate.

Candidate	Borda Points
a	$7 \cdot 3 + 5 \cdot 0 + 4 \cdot 1 = 25$
b	$7 \cdot 2 + 5 \cdot 3 + 4 \cdot 0 = 29$
c	$7 \cdot 1 + 5 \cdot 2 + 4 \cdot 2 = 25$
d	$7 \cdot 0 + 5 \cdot 1 + 4 \cdot 3 = 17$

Alternative b wins by the Borda method.

Now repeat the Borda method after eliminating alternative d. Make a table to examine the points in the Borda method.

Votes	Ranking	Points 2	1	0
7	$a > b > c$	a	b	c
5	$b > c > a$	b	c	a
4	$c > a > b$	c	a	b

Examine the last three columns, and compute the weighted sum for each candidate.

Candidate	Borda Points
a	$7 \cdot 2 + 5 \cdot 0 + 4 \cdot 1 = 18$
b	$7 \cdot 1 + 5 \cdot 2 + 4 \cdot 0 = 17$
c	$7 \cdot 0 + 5 \cdot 1 + 4 \cdot 2 = 13$

This time alternative a wins by the Borda method. This is a violation of the Independence of Irrelevant Alternatives criterion.

21. Count the number of first-place votes to see that choice a has 7; choice b has 5, choice d has 4, and choice c has 0. None of the candidates has a majority of first-place votes, which is at least 9. The elimination of c does not change the number of first-place votes of a, b, and d. Therefore, eliminate d and compare only a and b.

Votes	Ranking
7	$a > b$
5	$b > a$
4	$a > b$

Now a receives $7 + 4 = 11$ votes and wins.

Now eliminate candidate b, and apply the Hare method again.

Here is a new voter profile.

Votes	Ranking
7	$a > c > d$
5	$c > d > a$
4	$d > c > a$

Count the number of first-place votes to see that choice a now has 7, choice c has 5, and choice d now has 4.

Because none of the candidates has a majority, eliminate d and compare only a and c.

Votes	Ranking
7	$a > c$
5	$c > a$
4	$c > a$

This time c received $5 + 4 = 9$ votes and wins. Although b was dropped from the ranking, the originally preferred candidate a did not win. This is a violation of the Independence of Irrelevant Alternative criterion.

22. Writing exercise

23. Using the Hamilton method, the total population is
$$1429 + 8639 + 7608 + 6660 + 1671 = 26007.$$
The standard divisor is
$$d = \frac{26007}{195} \approx 133.3692.$$

Ward	Population	Q	Rounded Q
1st	1429	$\frac{1429}{133.369} \approx 10.715$	10
2nd	8639	$\frac{8639}{133.369} \approx 64.775$	64
3rd	7608	$\frac{7608}{133.369} \approx 57.045$	57
4th	6660	$\frac{6660}{133.369} \approx 49.937$	49
5th	1671	$\frac{1671}{133.369} \approx 12.529$	12
Totals	26007		192

The three remaining seats will be apportioned to the 1st, 2nd and 4th wards, because the fractional parts of their quotas are the largest. The final apportionment is as follows.

Ward	Q	Number of seats
1st	10.715	$10 + 1 = 11$
2nd	64.775	$64 + 1 = 65$
3rd	57.045	57
4th	49.937	$49 + 1 = 50$
5th	12.529	12
Totals		195

24. Use the Jefferson method with $md = 131$.

Ward	Population	mQ	Rounded mQ
1st	1429	$\frac{1429}{131} \approx 10.908$	10
2nd	8639	$\frac{8639}{131} \approx 65.947$	65
3rd	7608	$\frac{7608}{131} \approx 58.076$	58
4th	6660	$\frac{6660}{131} \approx 50.840$	50
5th	1671	$\frac{1671}{131} \approx 12.756$	12
Totals	26007		195

25. Use the Webster method with $md = 133.7$.

Ward	Population	Q	Rounded Q
1st	1429	$\frac{1429}{133.7} \approx 10.688$	11
2nd	8639	$\frac{8639}{133.7} \approx 64.615$	65
3rd	7608	$\frac{7608}{133.7} \approx 56.904$	57
4th	6660	$\frac{6660}{133.7} \approx 49.813$	50
5th	1671	$\frac{1671}{133.7} \approx 12.498$	12
Totals	26007		195

26. Writing exercise
27. Writing exercise
28. Writing exercise
29. Writing exercise

30. The sum of the populations is
$$2354 + 4500 + 5598 + 23000 = 35452.$$
The standard divisor is
$$d = \frac{35452}{100} = 354.52.$$

Using the standard quota, we have the following.

State	Population	Q	Rounded Q
a	2354	$\frac{2354}{354.52} \approx 6.640$	6
b	4500	$\frac{4500}{354.52} \approx 12.693$	12
c	5598	$\frac{5598}{354.52} \approx 15.790$	15
d	23000	$\frac{23000}{354.52} \approx 64.876$	**64**
Totals	35,452		97

Use the Jefferson method with $md = 347$.

State	Population	mQ	Rounded mQ
a	2354	$\frac{2354}{347} \approx 6.784$	6
b	4500	$\frac{4500}{347} \approx 12.968$	12
c	5598	$\frac{5598}{347} \approx 16.133$	16
d	23000	$\frac{23000}{347} \approx 66.282$	**66**
Totals	35,452		100

The Jefferson method violates the Quota Rule because state d receives two more seats than its apportionment from the standard quota.

31. First, calculate the apportionment with $n = 126$. The total population is
$$263 + 809 + 931 + 781 + 676 = 3460.$$
The standard divisor for 126 seats is
$$d = \frac{3460}{126} \approx 27.4603.$$
Use the Hamilton method.

State	Population	Q	Rounded Q
a	263	$\frac{263}{27.4603} \approx 9.577$	9
b	809	$\frac{809}{27.4603} \approx 29.461$	29
c	931	$\frac{931}{27.4603} \approx 33.903$	33
d	781	$\frac{781}{27.4603} \approx 28.441$	28
e	676	$\frac{676}{27.4603} \approx 24.617$	24
Totals	3460		123

The three remaining seats will be apportioned to States c, e, and a, because the fractional parts of their quotas are the largest. The final apportionment is as follows.

State	Q	Number of seats
a	9.577	$9 + 1 = $ **10**
b	29.461	29
c	33.903	$33 + 1 = 34$
d	28.441	28
e	24.617	$24 + 1 = 25$
Totals		126

Now increase the number of seats to 127. The standard divisor is
$$d = \frac{3460}{127} \approx 27.2441.$$
Use the Hamilton method with the new d.

State	Population	Q	Rounded Q
a	263	$\frac{263}{27.2441} \approx 9.653$	9
b	809	$\frac{809}{27.2441} \approx 29.695$	29
c	931	$\frac{931}{27.2441} \approx 34.173$	34
d	781	$\frac{781}{27.2441} \approx 28.667$	28
e	676	$\frac{676}{27.2441} \approx 24.813$	24
Totals	3460		124

The three remaining seats will be apportioned to States e, b, and d, because the fractional parts of their quotas are the largest. The final apportionment is as follows.

State	Q	Number of seats
a	9.653	**9**
b	29.695	$29 + 1 = 30$
c	34.173	34
d	28.667	$28 + 1 = 29$
e	24.813	$24 + 1 = 25$
Totals		127

State a is a victim of the Alabama Paradox, because it has lost a seat despite the fact that the overall number of seats has increased.

32. First, calculate the apportionment for the initial populations. The total population is
$$55 + 125 + 190 = 370.$$
The standard divisor for 11 seats is
$$d = \frac{370}{11} = 33.6364.$$
Use the Hamilton method.

State	Population	Q	Rounded Q
a	55	$\frac{55}{33.6364} \approx 1.635$	1
b	125	$\frac{125}{33.6364} \approx 3.716$	3
c	190	$\frac{190}{33.6364} \approx 5.649$	5
Totals	370		9

The two remaining seats will be apportioned to States b and c, because the fractional parts of their quotas are the largest. The final apportionment is as follows.

State	Q	Number of seats
a	1.635	1
b	3.716	$3 + 1 = 4$
c	5.649	$5 + 1 = 6$
Totals		11

Now apportion the seats for the revised populations.
$$63 + 150 + 220 = 433.$$
The new standard divisor is
$$d = \frac{433}{11} \approx 39.3636.$$

Use the Hamilton method with the new values.

State	Population	Q	Rounded Q
a	63	$\frac{63}{39.3636} \approx 1.600$	1
b	150	$\frac{150}{39.3636} \approx 3.811$	3
c	220	$\frac{220}{39.3636} \approx 5.589$	5
Totals	433		9

The two remaining seats will be apportioned to States a and b, because the fractional parts of their quotas are the largest. The final apportionment is as follows.

State	Q	Number of seats
a	1.600	$1 + 1 = 2$
b	3.811	$3 + 1 = 4$
c	5.589	5
Totals		11

Here is a final summary.

State	Old Pop	New Pop	% Inc.	Old No. of Seats	New No. of Seats
a	55	63	**14.55**	1	2
b	125	150	20.00	4	4
c	190	220	**15.79**	6	5
Totals				11	11

Notice from the table that there was a greater percent increase in growth for State c than for State a. Yet, State a gained a seat, and State c lost a seat. This is an example of the Population Paradox.

33. First, calculate the apportionment for the initial populations. The total population is
$$49 + 160 = 209.$$
The standard divisor for 100 seats is
$$d = \frac{209}{100} = 2.09.$$
Use the Hamilton method.

State	Population	Q	Rounded Q
a	49	$\frac{49}{2.09} \approx 23.445$	23
b	160	$\frac{160}{2.09} \approx 76.555$	76
Totals	209		99

The one remaining seat will be apportioned to State b because the fractional parts of its quota is larger. The final apportionment is as follows.

State	Q	Number of seats
a	23.445	23
b	76.555	$76 + 1 = 77$
Totals		100

Now apportion the seats to include the new state. The standard quota of the new state is:
$$Q = \frac{32}{2.09} \approx 15.311.$$

Rounded down to 15, add 15 new seats to the original 100 to obtain 115 seats to be apportioned. Now the new population is
$$49 + 160 + 32 = 241.$$
The new standard divisor is
$$d = \frac{241}{115} \approx 2.0957.$$

State	Population	Q	Rounded Q
a	49	$\frac{49}{2.0957} \approx 23.381$	23
b	160	$\frac{160}{2.0957} \approx 76.347$	76
c	32	$\frac{32}{2.0957} \approx 15.269$	15
Totals	241		114

The one remaining seat will be apportioned to State a because the fractional part of its quota is the largest. The final apportionment is as follows.

State	Q	Number of seats
a	23.381	$23 + 1 = 24$
b	76.347	76
c	15.269	15
Totals		115

The New State Paradox has occurred because the addition of the new state has caused a shift in the apportionment of the original states. States a and b originally had 23 and 77 seats, respectively; now they have 24 and 76, respectively.

34. *Writing exercise*

THE METRIC SYSTEM

1. $\dfrac{8 \text{ m}}{1} \cdot \dfrac{1000 \text{ mm}}{1 \text{ m}} = 8000 \text{ mm}$

2. $\dfrac{14.76 \text{ m}}{1} \cdot \dfrac{100 \text{ cm}}{1 \text{ m}} = 1476 \text{ cm}$

3. $\dfrac{8500 \text{ cm}}{1} \cdot \dfrac{1 \text{ m}}{100 \text{ cm}} = 85 \text{ m}$

4. $\dfrac{250 \text{ mm}}{1} \cdot \dfrac{1 \text{ m}}{1000 \text{ mm}} = .25 \text{ m}$

5. $\dfrac{68.9 \text{ cm}}{1} \cdot \dfrac{10 \text{ mm}}{1 \text{ cm}} = 689 \text{ mm}$

6. $\dfrac{3.25 \text{ cm}}{1} \cdot \dfrac{10 \text{ mm}}{1 \text{ cm}} = 32.5 \text{ mm}$

7. $\dfrac{59.8 \text{ mm}}{1} \cdot \dfrac{1 \text{ cm}}{10 \text{ mm}} = 5.98 \text{ cm}$

8. $\dfrac{3.542 \text{ mm}}{1} \cdot \dfrac{1 \text{ cm}}{10 \text{ mm}} = .3542 \text{ cm}$

9. $\dfrac{5.3 \text{ km}}{1} \cdot \dfrac{1000 \text{ m}}{1 \text{ km}} = 5300 \text{ m}$

10. $\dfrac{9.24 \text{ km}}{1} \cdot \dfrac{1000 \text{ m}}{1 \text{ km}} = 9240 \text{ m}$

11. $\dfrac{27,500 \text{ m}}{1} \cdot \dfrac{1 \text{ km}}{1000 \text{ m}} = 27.5 \text{ km}$

12. $\dfrac{14,592 \text{ m}}{1} \cdot \dfrac{1 \text{ km}}{1000 \text{ m}} = 14.592 \text{ km}$

13. 2.54 cm; 25.4 mm

14. 3.3 cm; 33 mm

15. 5 cm; 50 mm

16. 2.54 cm; 25.4 mm

17. $\dfrac{6 \text{ L}}{1} \cdot \dfrac{10^2 \text{ cl}}{1 \text{ L}} = 6 \times 100 = 600 \text{ cl}$

18. $\dfrac{4.1 \text{ L}}{1} \cdot \dfrac{10^3 \text{ ml}}{1 \text{ L}} = 4.1 \times 1000 = 4100 \text{ ml}$

19. $\dfrac{8.7 \text{ L}}{1} \cdot \dfrac{10^3 \text{ ml}}{1 \text{ L}} = 8.7 \times 1000 = 8700 \text{ ml}$

20. $\dfrac{12.5 \text{ L}}{1} \cdot \dfrac{10^2 \text{ cl}}{1 \text{ L}} = 12.5 \times 100 = 1250 \text{ cl}$

21. $\dfrac{925 \text{ cl}}{1} \cdot \dfrac{1 \text{ L}}{10^2 \text{ cl}} = \dfrac{925}{100} = 9.25 \text{ L}$

22. $\dfrac{412 \text{ ml}}{1} \cdot \dfrac{1 \text{ L}}{10^3 \text{ ml}} = \dfrac{412}{1000} = .412 \text{ L}$

23. $\dfrac{8974 \text{ ml}}{1} \cdot \dfrac{1 \text{ L}}{10^3 \text{ ml}} = \dfrac{8974}{1000} = 8.974 \text{ L}$

24. $\dfrac{5639 \text{ cl}}{1} \cdot \dfrac{1 \text{ L}}{10^2 \text{ cl}} = \dfrac{5639}{100} = 56.39 \text{ L}$

25. $\dfrac{8000 \text{ g}}{1} \cdot \dfrac{1 \text{ kg}}{10^3 \text{ g}} = \dfrac{8000}{1000} = 8 \text{ kg}$

26. $\dfrac{25000 \text{ g}}{1} \cdot \dfrac{1 \text{ kg}}{10^3 \text{ g}} = \dfrac{25000}{1000} = 25 \text{ kg}$

27. $\dfrac{5.2 \text{ kg}}{1} \cdot \dfrac{10^3 \text{ g}}{1 \text{ kg}} = 5.2 \times 1000 = 5200 \text{ g}$

28. $\dfrac{12.42 \text{ kg}}{1} \cdot \dfrac{10^3 \text{ g}}{1 \text{ kg}} = 12.42 \times 1000 = 12,420 \text{ g}$

29. $\dfrac{4.2 \text{ g}}{1} \cdot \dfrac{10^3 \text{ mg}}{1 \text{ g}} = 4.2 \times 1000 = 4200 \text{ mg}$

30. $\dfrac{3.89 \text{ g}}{1} \cdot \dfrac{10^2 \text{ cg}}{1 \text{ g}} = 3.89 \times 100 = 389 \text{ cg}$

31. $\dfrac{598 \text{ mg}}{1} \cdot \dfrac{1 \text{ g}}{10^3 \text{ mg}} = \dfrac{598}{1000} = .598 \text{ g}$

32. $\dfrac{7634 \text{ cg}}{1} \cdot \dfrac{1 \text{ g}}{10^2 \text{ cg}} = \dfrac{7634}{100} = 76.34 \text{ g}$

33. $C = \dfrac{5}{9}(F - 32) = \dfrac{5}{9}(86 - 32) = \dfrac{5}{9}(54) = 30°$

34. $C = \dfrac{5}{9}(F - 32) = \dfrac{5}{9}(536 - 32) = \dfrac{5}{9}(504) = 280°$

35. $C = \dfrac{5}{9}(F - 32) = \dfrac{5}{9}(-114 - 32) = \dfrac{5}{9}(-146) = -81°$

36. $C = \dfrac{5}{9}(F - 32) = \dfrac{5}{9}(-40 - 32) = \dfrac{5}{9}(-72) = -40°$

37. $F = \dfrac{9}{5}C + 32 = \dfrac{9}{5} \cdot 10 + 32 = 18 + 32 = 50°$

38. $F = \dfrac{9}{5}C + 32 = \dfrac{9}{5} \cdot 25 + 32 = 45 + 32 = 77°$

39. $F = \dfrac{9}{5}C + 32 = \dfrac{9}{5} \cdot -40 + 32 = -72 + 32 = -40°$

40. $F = \dfrac{9}{5}C + 32 = \dfrac{9}{5} \cdot -15 + 32 = -27 + 32 = 5°$

41. $\dfrac{1 \text{ kg}}{1} \cdot \dfrac{1000 \text{ g}}{1 \text{ kg}} \cdot \dfrac{1 \text{ nickel}}{5 \text{ g}} = 200 \text{ nickels}$

42. $\dfrac{1 \text{ L}}{1} \cdot \dfrac{1000 \text{ ml}}{1 \text{ L}} \cdot \dfrac{3.5 \text{ g}}{1000 \text{ ml}} = 3.5 \text{ g}$

43. $\dfrac{1 \text{ L}}{1} \cdot \dfrac{1000 \text{ ml}}{1 \text{ L}} \cdot \dfrac{.0002 \text{ g}}{1 \text{ ml}} = .2 \text{ g}$

44. $\dfrac{1 \text{ ml}}{1} \cdot \dfrac{1 \text{ L}}{1000 \text{ ml}} \cdot \dfrac{1500 \text{ g}}{1 \text{ L}} = 1.5 \text{ g}$

45. $\dfrac{7 \text{ strips}}{1} \cdot \dfrac{67 \text{ cm}}{1 \text{ strip}} \cdot \dfrac{1 \text{ m}}{100 \text{ cm}} \cdot \dfrac{\$8.74}{1 \text{ m}} = \$40.99$

46. $\dfrac{15 \text{ pieces}}{1} \cdot \dfrac{384 \text{ mm}}{1 \text{ piece}} \cdot \dfrac{1 \text{ m}}{1000 \text{ mm}} \cdot \dfrac{\$54.20}{1 \text{ m}} = \$312.19$

47. $A = 128 \text{ cm} \cdot 174 \text{ cm} = 22,272 \text{ cm}^2$

$\dfrac{22,272 \text{ cm}^2}{1} \cdot \dfrac{(1 \text{ m})^2}{(100 \text{ cm})^2} \cdot \dfrac{\$174.20}{1 \text{ m}^2} = \$387.98$

48. $A = 9 \text{ cm} \cdot 14 \text{ cm} = 126 \text{ cm}^2$

$\dfrac{80 \text{ pieces}}{1} \cdot \dfrac{126 \text{ cm}^2}{1 \text{ piece}} \cdot \dfrac{(1 \text{ m})^2}{(100 \text{ cm})^2} \cdot \dfrac{\$63.79}{1 \text{ m}^2} = \$64.30$

49. $\dfrac{82 \text{ cm}}{1} \cdot \dfrac{1 \text{ m}}{100 \text{ cm}} = .82 \text{ m}$

$V = .82 \text{ m} \cdot 1.1 \text{ m} \cdot 1.2 \text{ m} = 1.0824 \text{ m}^3$

$\dfrac{1.0824 \text{ m}^3}{1} \cdot \dfrac{(100 \text{ cm})^3}{(1 \text{ m})^3} = 1,082,400 \text{ cm}^3$

50. $\dfrac{1.5 \text{ m}}{1} \cdot \dfrac{100 \text{ cm}}{1 \text{ m}} = 150 \text{ cm}$

$V = 150 \text{ cm} \cdot 74 \text{ cm} \cdot 97 \text{ cm} = 1,076,700 \text{ cm}^3$

$\dfrac{1,076,700 \text{ cm}^3}{1} \cdot \dfrac{(1 \text{ m})^3}{(100 \text{ cm})^3} = 1.0767 \text{ m}^3$

51. $\dfrac{160 \text{ L}}{1} \cdot \dfrac{1000 \text{ ml}}{1 \text{ L}} \cdot \dfrac{1 \text{ bottle}}{800 \text{ ml}} = 200 \text{ bottles}$

52. $\dfrac{80 \text{ people}}{1} \cdot \dfrac{400 \text{ ml}}{1 \text{ person}} \cdot \dfrac{1 \text{ L}}{1000 \text{ ml}} \cdot \dfrac{1 \text{ bottle}}{2 \text{ L}} = 16 \text{ bottles}$

53. $\dfrac{982 \text{ yd}}{1} \cdot \dfrac{.9144 \text{ m}}{1 \text{ yd}} = 897.9 \text{ m}$

54. $\dfrac{12.2 \text{ km}}{1} \cdot \dfrac{.6214 \text{ mi}}{1 \text{ km}} = 7.581 \text{ mi}$

55. $\dfrac{125 \text{ mi}}{1} \cdot \dfrac{1.609 \text{ km}}{1 \text{ mi}} = 201.1 \text{ km}$

56. $\dfrac{1000 \text{ mi}}{1} \cdot \dfrac{1.609 \text{ km}}{1 \text{ mi}} = 1609 \text{ km}$

57. $\dfrac{1816 \text{ g}}{1} \cdot \dfrac{.0022 \text{ lb}}{1 \text{ g}} = 3.995 \text{ lb}$

58. $\dfrac{1.42 \text{ lb}}{1} \cdot \dfrac{454 \text{ g}}{1 \text{ lb}} = 644.7 \text{ g}$

59. $\dfrac{47.2 \text{ lb}}{1} \cdot \dfrac{454 \text{ g}}{1 \text{ lb}} = 21,428.8 \text{ g}$

60. $\dfrac{7.68 \text{ kg}}{1} \cdot \dfrac{2.2 \text{ lb}}{1 \text{ kg}} = 16.90 \text{ lb}$

61. $\dfrac{28.6 \text{ L}}{1} \cdot \dfrac{1.0567 \text{ qt}}{1 \text{ L}} = 30.22 \text{ qt}$

62. $\dfrac{59.4 \text{ L}}{1} \cdot \dfrac{1.0567 \text{ qt}}{\text{L}} = 62.77 \text{ qt}$

63. $\dfrac{28.2 \text{ gal}}{1} \cdot \dfrac{3.785 \text{ L}}{1 \text{ gal}} = 106.7 \text{ L}$

64. $\dfrac{16 \text{ qt}}{1} \cdot \dfrac{.9464 \text{ L}}{1 \text{ qt}} = 15.14 \text{ L}$

65. Unreasonable; $\dfrac{2 \text{ kg}}{1} \cdot \dfrac{2.2 \text{ lb}}{1 \text{ kg}} = 4.4 \text{ lb}$

66. Unreasonable; $\dfrac{4 \text{ L}}{1} \cdot \dfrac{.2642 \text{ gal}}{1 \text{ L}} \approx 1.1 \text{ gal}$

67. Reasonable;

$\dfrac{25 \text{ ml}}{1} \cdot \dfrac{1 \text{ L}}{1000 \text{ ml}} \cdot \dfrac{1.0567 \text{ qt}}{1 \text{ L}} \cdot \dfrac{32 \text{ oz}}{1 \text{ qt}} \cdot \dfrac{2 \text{ T}}{1 \text{ oz}} \approx 1.7 \text{ T}$

68. Reasonable; $\dfrac{6 \text{ L}}{1} \cdot \dfrac{.2642 \text{ gal}}{1 \text{ L}} \approx 1.6 \text{ gal}$

69. Unreasonable; $\dfrac{.5 \text{ L}}{1} \cdot \dfrac{1.0567 \text{ qt}}{1 \text{ L}} \approx .5 \text{ qt}$

70. Unreasonable; $\dfrac{40 \text{ g}}{1} \cdot \dfrac{.0022 \text{ lb}}{1 \text{ g}} \cdot \dfrac{16 \text{ oz}}{1 \text{ lb}} \approx 1.4 \text{ oz}$

71. B; $\dfrac{3 \text{ m}}{1} \cdot \dfrac{3.2808 \text{ ft}}{1 \text{ m}} \approx 9.8 \text{ ft}$

72. C; $\dfrac{5 \text{ m}}{1} \cdot \dfrac{3.2808 \text{ ft}}{1 \text{ m}} \approx 16 \text{ ft}$

73. B; $\dfrac{5000 \text{ km}}{1} \cdot \dfrac{.6214 \text{ mi}}{1 \text{ km}} \approx 3100 \text{ mi}$

74. A; $\dfrac{3 \text{ cm}}{1} \cdot \dfrac{1 \text{ in}}{2.54 \text{ cm}} \approx 1.2 \text{ in}$

75. C; $\dfrac{193 \text{ mm}}{1} \cdot \dfrac{1 \text{ cm}}{10 \text{ mm}} \cdot \dfrac{1 \text{ in}}{2.54 \text{ cm}} \approx 7.6 \text{ in}$

76. A; $\dfrac{1 \text{ kg}}{1} \cdot \dfrac{2.20 \text{ lb}}{1 \text{ kg}} = 2.2 \text{ lb}$

77. A; $\dfrac{1300 \text{ kg}}{1} \cdot \dfrac{2.20 \text{ lb}}{1 \text{ kg}} \approx 2900 \text{ lb}$

78. B; $\dfrac{355 \text{ ml}}{1} \cdot \dfrac{1 \text{ L}}{1000 \text{ ml}} \cdot \dfrac{1.0567 \text{ qt}}{1 \text{ L}} \cdot \dfrac{32 \text{ oz}}{1 \text{ qt}} = 12.0 \text{ oz}$

79. A; $\dfrac{180 \text{ cm}}{1} \cdot \dfrac{1 \text{ m}}{100 \text{ cm}} \cdot \dfrac{3.2808 \text{ ft}}{1 \text{ m}} \approx 5.9 \text{ ft}$

80. C; $\dfrac{13,000 \text{ km}}{1} \cdot \dfrac{.6214 \text{ mi}}{1 \text{ km}} \approx 8100 \text{ mi}$

81. C; $\dfrac{800 \text{ m}}{1} \cdot \dfrac{1 \text{ km}}{1000 \text{ m}} \cdot \dfrac{.6214 \text{ mi}}{1 \text{ km}} \approx .5 \text{ mi}$

82. A; $\dfrac{1 \text{ L}}{1} \cdot \dfrac{1.0567 \text{ qt}}{1 \text{ L}} = 1.1 \text{ qt}$

83. A; $\dfrac{70 \text{ cm}}{1} \cdot \dfrac{1 \text{ m}}{100 \text{ cm}} \cdot \dfrac{39.37 \text{ in}}{1 \text{ m}} \approx 28 \text{ in}$

84. C; $\dfrac{70 \text{ kg}}{1} \cdot \dfrac{2.20 \text{ lb}}{1 \text{ kg}} = 154 \text{ lb}$

85. B; $\dfrac{50 \text{ cm}}{1} \cdot \dfrac{1 \text{ m}}{100 \text{ cm}} \cdot \dfrac{39.37 \text{ in}}{1 \text{ m}} \approx 20 \text{ in}$

86. A; $\dfrac{1 \text{ m}}{1} \cdot \dfrac{3.2808 \text{ ft}}{1 \text{ m}} = 3.3 \text{ ft}$

87. B; $\dfrac{9 \text{ mm}}{1} \cdot \dfrac{1 \text{ in}}{25.4 \text{ mm}} \approx .35 \text{ in}$

88. A; $\dfrac{300 \text{ mm}}{1} \cdot \dfrac{1 \text{ in}}{25.4 \text{ mm}} \approx 12 \text{ in}$

89. A; This is the freezing temperature of water in Celsius.

90. B; $F = \dfrac{9}{5} \cdot 40 + 32 = 72 + 32 = 104°F$

91. B; $F = \dfrac{9}{5} \cdot 60 + 32 = 108 + 32 = 140°F$

92. A; This is the boiling temperature of water in Celsius.

93. C; $F = \dfrac{9}{5} \cdot 10 + 32 = 18 + 32 = 50°F$

94. A; $F = \dfrac{9}{5} \cdot 30 + 32 = 54 + 32 = 86°F$

95. B; $F = \dfrac{9}{5} \cdot 170 + 32 = 306 + 32 \approx 340°F$

96. A; $F = \dfrac{9}{5} \cdot 35 + 32 = 63 + 32 = 95°F$